Lecture Notes in Control and Information Sciences 441

For further volumes:
http://www.springer.com/series/642

Selim S. Hacısalihzade

Biomedical Applications of Control Engineering

Springer

Selim S. Hacısalihzade
Bosporus University
Istanbul
Turkey
selim.hacisalihzade@boun.edu.tr

ISSN 0170-8643 ISSN 1610-7411 (electronic)
ISBN 978-3-642-37278-0 ISBN 978-3-642-37279-7 (eBook)
DOI 10.1007/978-3-642-37279-7
Springer Heidelberg New York Dordrecht London

Library of Congress Control Number: 2013933986

Printed on acid-free paper

Springer is part of Springer Science+Business Media (www.springer.com)

For

İlhan Kandan because she taught me to read and write,
Erol İnelmen because he taught me about science,
Vasfi Bingöl because he taught me to think independently,
Mohamed Mansour because he taught me control engineering,
Emel and Salih Hacısalihzade because they were my parents.

Preface

Countless scientists and engineers have contributed untold volumes of work applying control engineering techniques to biomedical problems since the pioneering efforts of Norbert Wiener who coined the term "cybernetics" more than half a century ago. Although that word has gradually lost its glamor in the decades that followed, ingenious researchers have not failed to come up with new fields of application. Therefore, even a superficial review of the field which claims to be exhaustive would have to cover several volumes. Hence, any textbook in this field has either to take one specific application area and explore it in depth or be an eclectic collection of several application areas. I have chosen to take the latter path even tough many examples of both are available in the press.

This book presents different control engineering applications in the medical field. It is intended for senior undergraduate or graduate students in both control engineering and biomedical engineering programs. For control engineering students, it shows the application of various techniques they have already learnt in theoretical lectures, in the biomedical arena. For biomedical engineering students, it shows solutions to various problems in their field using methods commonly used by control engineers. On the other hand, it should also appeal to medical students and practitioners who want to enhance their quantitative understanding of physiological processes and perhaps get a new vantage point from which they can attack their current problems.

There are no *sine qua non* prerequisites for reading, enjoying and learning from this book other than basic engineering mathematics and a rudimentary understanding of control engineering concepts.

The book is organized in seven Chapters and two Appendices. Every chapter includes three types of exercises at its conclusion which help the reader to better understand the topic. Type A exercises are mostly verbal and review the main points of the chapter; they help the reader to gauge her[1] understanding of the topic. Type B

[1] To avoid the clumsy constructs like "his/her", where appropriate, both male and female personal pronouns are used throughout the book alternatingly and they are interchangeable with no preference for or a prejudice against either gender.

exercises require some calculation and they are intended to deepen the understanding of the methods discussed in the chapter. Type C exercises involve open ended questions and their solutions typically require significant time, effort and creativity; they might be solved as term projects or graduation projects and indicate directions for thesis work.

Chapter 1, Introduction to Systems, begins by introducing some basic definitions and concepts of mathematical system theory. Mathematical description and analysis of systems like input-output relationships in time and frequency domains as well as in the state space are discussed. Some important system properties like controllability, observability and stability are introduced. Students familiar with basic control engineering should know everything in this Chapter and can skip it altogether. However, medical and biomedical engineering students not familiar with the contents of this Chapter should study it carefully.

Chapter 2, Modeling and Identification, gives an introduction to a very important topic which is often neglected in most engineering curricula, namely mathematical modeling of physical and physiological systems. It begins by discussing what a model is and gives various classification methods for different types of models. The actual modeling process is illustrated using the popular inverted pendulum and the stock prices. The determination of the parameters in the system once the mathematical structure of the model is fixed, the so called parameter identification, is handled together with the question of experiment design. Finally, the importance of integrating model validation in the modeling process is emphasized. A small philosophical excursion in this Chapter is intended to motivate the reader to increase his awareness of the difference between reality and models.

Chapter 3, The Human Operator, looks into the mathematical modeling of the human operator and manual control with emphasis on control of dynamic systems like aircraft by human beings. Different types of manual control systems and popular human operator models like the second order linear model, the describing function model and the optimal control model are discussed in detail. In this Chapter, the reader is also given a concise description of how and why an airplane flies.

Chapter 4, Drug Administration and Dosage Optimization, studies the question of drug administration with special emphasis on the optimization of the "when" and "how much" of the administered drug. This problem is analyzed using various tools and a number of different solutions are described in detail. Specifically, optimization of drug delivery based on both pharmacokinetic and pharmacodynamic data is handled. It is shown how the choice of the objective function leads to different mathematical problems like linear programming, Gaussian least squares, dynamic programming or Prony algorithms. In short, numerical optimization methods are discussed for the cases where analytical solutions are not possible.

Chapter 5, Parkinsons Disease, starts off by defining the disease, its etiology, pathophysiology, symptoms and treatment. Measuring Parkinsonian symptoms like tremor and bradykinesia objectively and quantitatively is explained in detail. Both a low tech approach and a computer based measurement technique assessing the tracking ability of the patients is introduced and evaluated. Finally, some of the techniques explained in Chapter 4 are applied to optimize drug therapy with respect to both pharmacokinetic and pharmacodynamic criteria.

Chapter 6, Diabetes and Control of Blood Glucose, begins by reviewing a global epidemic, diabetes, and methods to measure blood glucose levels applying invasive techniques, minimally-invasive techniques and non-invasive techniques. Various methods of insulin delivery are discussed and a broad spectrum of mathematical models of varying complexity ranging from linear models to nonlinear and comprehensive methods for patient dynamics are reviewed. Also, a choice of control algorithms for regulating the blood glucose levels in diabetes are explained. Those algorithms are grouped under the headings rule based control (making use of black box models to PID controllers) and model based control (making use of pole placement, adaptive control and optimal control methods). Finally, a number of common commercial insulin pumps and advisory devices are listed and a market leader product's functioning is described in detail.

Chapter 7, Controlling Depth of Anaesthesia, introduces the problem of anaesthesia. Various methods of measuring depth of anaesthesia are discussed before talking about the most common anaesthetic agents. A simple fuzzy controller which controls the blood pressure during surgery as a proxy for depth of anaesthesia is described and results from its use in real life conditions are discussed. A more advanced and complex multivariable controller which is used as a mean arterial pressure (MAP) controller, as an endtidal anaesthetic gas controller, as an inspired anaesthetic gas controller and as a cascade controller where a MAP or an endtidal anaesthetic gas controller defines an inspired anaesthetic gas set value that is controlled by an internal loop is described in detail. A 12th order model of the patient is developed and reduced to a third order model for practical purposes. Similarly comprehensive models of the breathing system and of the patient's reaction to surgical stimulation are developed. Subsequently, all these models are integrated to be used in the control loop. In the end an observer based state feedback controller with reference tracking is designed and implemented using optimal control theory.

The Appendices deal with topics which are often not covered in undergraduate courses. They are intended for the reader to get an idea about some advanced concepts used in the book with which she might not be familiar. The Appendices are neither meant to be mathematically rigorous nor as substitutes for corresponding courses. Appendix A is a refresher for optimal control. Calculus of variations is used to derive the Pontryagin - Bellman solutions and the optimality principle for the deterministic case. The extension of the solutions to the stochastic case is also briefly reviewed. Several simple examples are used for illustrative purposes. Appendix B covers fuzzy control which is explained using the example of speed control of a car.

As always, there is a long list of people to acknowledge and thank for their support in preparing this book. I want to begin by thanking Jürg Tödtli who initiated my interest in the field while we were at the Institute of Automatic Control at the ETH Zurich many years ago and who later became my boss at Landis & Gyr and a life long friend in the mean while. Special thanks are certainly due to late Günter Baumgartner and Carlo Albani at the Zurich University Hospital's Neurological Clinic for introducing me to interdisciplinary research through his coaching and mentoring in

Parkinson's disease. I also have to thank Heinz Lienhard, the head of corporate R&D at Landis & Gyr for encouraging this project and Manfred Morari at the Institute of Automatic Control at the ETH Zurich for suggesting that this book appears in the Lecture Notes in Control and Information Sciences, Springer Verlag.

I want to thank Alex Zbinden at the Institute of Critical Care and Anaesthesiology of the University of Bern's Inselspital for introducing me to the field of anaesthesiology and a very pleasant collaboration over years which bore many fruits. One of these fruits was a Ph.D. thesis by my former student Marco Derighetti on which the Chapter on the Control of Depth of Anaesthesia is modeled. I must thank Marco also for reading the manuscript and suggesting corrections and changes.

Special thanks go to my diabetician Giatgen Spinas, head of the Zurich University Hospital's Endocrinology and Diabetology Clinic who inspired the Chapter on diabetes and control of blood sugar and who also read a draft of the manuscript critically to suggest numerous improvements. In that context, I must also thank several employees and researchers of Roche Diabetes Care, primarily Yaprak Yay in Istanbul for connecting me to the right people in Burgdorf, namely Ulrich Haueter, Thomas Kleppisch, and Robert Hellwig who kindly explained the workings of their insulin pump and also suggested improvements on the manuscript's corresponding Section.

Many students at Bosporus University have contributed by critically reading parts of the manuscript but most notably Onur Cihan to whom I am much indebted for helping to write the Appendix on fuzzy control and for supplying a good deal of the figures; he also helped tirelessly with my never ending LaTeXproblems. Special thanks are due to my students Daniel Buffet and Roman Strotz as well as my colleague Mika Senning at ETH Zurich and Rui de Figueiredo at Rice University in Houston, Texas for their contributions on parts of Chapter 4 on dosage optimization and to my students at ETH Zurich Ruedi Meier and Jacques Nieuwland for their contributions on parts of Chapter 7 on fuzzy control of depth of anaesthesia.

I also want to thank Ewart R. Carson of the City University in London for many fruitful discussions and for supplying an original figure. Significant parts of the Chapters on modeling methodology and glucose control modeling are based on his seminal works.

And, of course, my dear wife Hande deserves special thanks for her encouragement and endless support during the preparation of this book.

Selim S. Hacısalihzade
Istanbul, 2013

Contents

1 Introduction to Systems 1
- 1.1 Introduction 1
- 1.2 Some Special Types of Systems 3
- 1.3 Mathematical Description and Analysis of Systems 4
  - 1.3.1 Input-Output Description of Systems in Time Domain 4
  - 1.3.2 Input-Output Description of Systems in Frequency Domain 7
  - 1.3.3 State Space Description of Systems 8
- 1.4 Some Important System Properties 14
- 1.5 Stability 15
  - 1.5.1 BIBO Stability 15
  - 1.5.2 Lyapunov Stability 17
- 1.6 Exercises 19

2 Modeling and Identification 23
- 2.1 Introduction 23
- 2.2 What Is a Model? 24
- 2.3 Modeling Process 31
  - 2.3.1 Inverted Pendulum 31
  - 2.3.2 Stock Prices 33
  - 2.3.3 Lessons Learned 36
- 2.4 Parameter Identification 37
- 2.5 Model Validation 38
- 2.6 Summary 39
- 2.7 Exercises 41

3 The Human Operator 45
- 3.1 Introduction 45
- 3.2 Types of Manual Control Systems 51
- 3.3 Human Operator Models 53
  - 3.3.1 Second Order Linear Model 53
  - 3.3.2 Describing Function Model 54

3.3.3 Optimal Control Model ... 54
3.4 Exercises ... 58

4 Drug Administration and Dosage Optimization ... 61
4.1 Introduction ... 61
4.2 Optimization of Drug Delivery Based on Pharmacokinetic Data ... 64
4.2.1 Modeling Pharmacokinetic Relationships ... 65
4.2.2 The Multiple Dosing Problem ... 66
4.2.3 Linear Programming Solutions of the Multiple Dosing Problem ... 68
4.2.4 Solution of the Multiple Dosing Problem with a Quadratic Objective Function ... 83
4.2.5 Semi-analytical Solution of the Multiple Dosing Problem ... 86
4.2.6 Optimization of Drug Administration By a Tauberian Approach ... 88
4.2.7 Remarks on Pharmacokinetic Optimization ... 92
4.3 Optimization of Drug Delivery Based on Pharmacodynamic Data ... 95
4.4 Exercises ... 97

5 Parkinson's Disease ... 99
5.1 Introduction ... 99
5.2 Etiology ... 99
5.3 Pathophysiology ... 101
5.4 Symptoms ... 103
5.5 Treatment ... 104
5.6 Measuring Parkinsonian Symptoms ... 106
5.6.1 Measuring Tremor ... 109
5.6.2 Measuring Bradykinesia ... 112
5.6.3 Low-Tech Approach ... 115
5.7 Dosage Optimization in Parkinson's Disease ... 116
5.7.1 Parkinson Patients Are Poor Controllers ... 117
5.7.2 Modeling the Effects of Levodopa on Parkinson Symptoms ... 121
5.8 Optimal Control of Levodopa Therapy ... 127
5.9 A Case Study ... 130
5.10 Summary ... 133
5.11 Exercises ... 134

6 Diabetes and Control of Blood Glucose ... 137
6.1 Introduction ... 137
6.2 Measuring Blood Glucose Levels ... 141
6.2.1 Invasive Techniques ... 141
6.2.2 Minimally-Invasive Techniques ... 142

6.2.3 Non-invasive Techniques 145
6.2.4 Clarke's Error Grid Analysis 146
6.3 Insulin Delivery 148
6.4 Patient Dynamics 149
6.4.1 Mathematical Models 152
6.5 The Controller 159
6.5.1 Rule Based Control 159
6.5.2 Model Based Control Algorithms 161
6.6 Commercial Insulin Pumps 164
6.6.1 Accu-Chek® 165
6.7 Exercises 172

7 Controlling Depth of Anaesthesia 175
7.1 Introduction 175
7.2 Measuring Depth of Anaesthesia 175
7.3 Anaesthetic Agents 176
7.4 The Basic Control Loop 177
7.5 Fuzzy Control of Anaesthesia 178
7.6 Multivariable Control of Anaesthesia 182
7.7 Implementation of the Multivariable Controller 184
7.8 The Patient Model 185
7.9 The Breathing System 194
7.10 Surgical Stimulation 204
7.11 Model Integration 210
7.12 Controller Design and Implementation 213
7.12.1 Observer Based State Feedback Controller with Reference Tracking 213
7.12.2 Artifact Handling 215
7.12.3 Implementation 216
7.13 Exercises 218

Appendix A Optimal Control 219
A.1 Introduction 219
A.2 Calculus of Variations 220
A.2.1 Subject Matter 220
A.2.2 Fixed Endpoint Problem 221
A.2.3 Variable Endpoint Problem 222
A.2.4 Variation Problem with Constraints 224
A.3 Optimal Dynamic Systems 224
A.3.1 Fixed Endpoint Problem 224
A.3.2 Variable Endpoint Problem 227
A.3.3 Generalized Objective Function 229
A.4 Pontryagin's Minimum Principle 230
A.5 Hamilton-Jacobi Method 233
A.6 Optimal Control of Linear Systems 237

A.6.1 Riccati Equation 237
A.6.2 Optimal Control When Not All States Are Measurable 241
A.6.3 Dynamic Programming 243

Appendix B Fuzzy Logic Control 247
B.1 Fuzzy Logic Architecture 247
B.1.1 Membership Functions 248
B.1.2 Fuzzification 250
B.1.3 Fuzzy Rule Base 251
B.1.4 Inference Engine 251
B.1.5 Defuzzification 252
B.2 Fuzzy Logic Toolbox of MATLAB 253
B.3 Application: Fuzzy Control of a System 254

References 259

Chapter 1
Introduction to Systems

I look at everything as a stochastic system with clearly defined inputs and outputs.
Life is just trying to find models mapping those inputs to outputs.
The rest is easy!

1.1 Introduction

The concepts of "system" and "signal" are very general and encompass a broad range of applications. Therefore, it is not easy to define them precisely without making compromises in terms of generality. However, by means of descriptions and by giving some examples, it is possible to create a common and acceptable understanding of these concepts[1].

We shall use the term **system** for a bounded and ordered collection of elements that fulfil a common purpose in unison. A system interacts with its surroundings through **signals** at its inputs and outputs. These signals can be in the form of energy, matter or information. A system converts its input signals to its output signals. Accordingly, a system can also be seen as a mapping which maps its input variables to its output variables.

This description of a system brings with itself the difficulty of defining the border where a system ends and its surrounding begins. Since this border is arbitrary and application dependent, it makes sense to define the border such that the interface between a system and its outputs are as clearly defined as possible. The inputs which cannot be influenced are conveniently labeled as disturbances.

A common method of depicting systems is by means of **block diagrams** in which systems are symbolized as rectangular blocks and signals as arrows entering (inputs) or leaving (outputs) a block (see Figure 1.1).

Example: Let us consider a simple electrical low pass without external load as a system with its primary voltage u_i as the system input, and its secondary voltage u_o

[1] This Chapter follows the general structure of [208].

S.S. Hacısalihzade: *Biomedical Applications of Control Engineering*, LNCIS 441, pp. 1–21.
DOI: 10.1007/978-3-642-37279-7_1

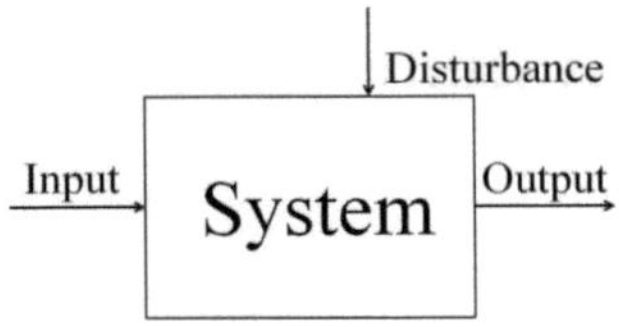

Fig. 1.1 The block diagram.

as the system output (see Figure 1.2). Calculation of the tension over each element using Ohm's law results in

$$i = C\frac{du_o}{dt} \tag{1.1}$$

$$u_i = u_o + RC\frac{du_o}{dt} \quad \Rightarrow \quad \frac{du_o}{dt} = \frac{1}{RC}(u_i - u_o) \tag{1.2}$$

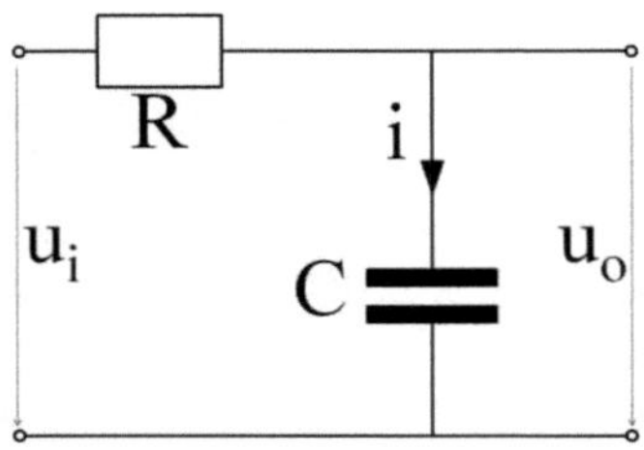

Fig. 1.2 Electrical low pass.

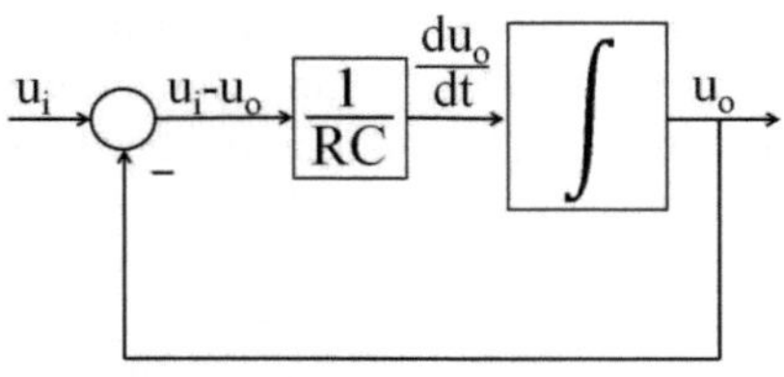

Fig. 1.3 A block diagram of the low pass in Figure 1.2.

The corresponding block diagram is shown in Figure 1.3.

1.2 Some Special Types of Systems

The system in the previous example is a so called single-input-single-output (SISO) system. There are, of course, also multi-input-multi-output (MIMO) systems (such systems are also called multivariable systems). Figure 1.4 shows such a system with p inputs $u_1, u_2, \ldots, u_p$ and q outputs $y_1, y_2, \ldots, y_q$.

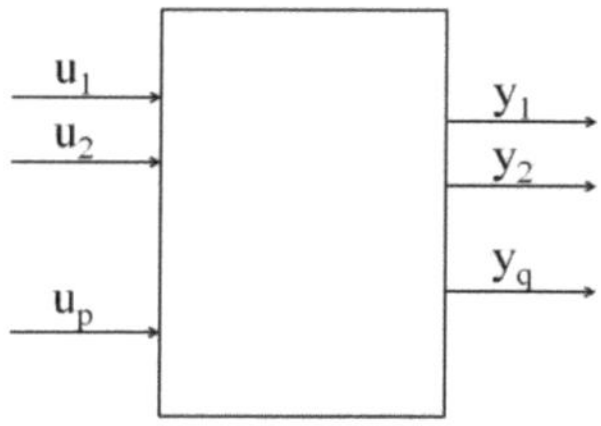

Fig. 1.4 A multivariable system with p inputs and q outputs.

For multivariable systems, it is convenient to aggregate all inputs $u_1, u_2, \ldots, u_p$ in a p-dimensional input vector $\mathbf{u}$ and all the outputs $y_1, y_2, \ldots, y_q$ in a q-dimensional output vector $\mathbf{y}$. The system is thus defined by the functional relationship

$$\mathbf{y}(t) = \mathbf{f}\big(\mathbf{u}(t), t\big) \tag{1.3}$$

In other words, any output of a multivariable system is a time function of a combination of its inputs. Some examples of systems, their inputs and outputs are shown in Table 1.1.

Table 1.1 Some systems with their inputs and outputs. The examples are chosen to demonstrate the broad spectrum of application areas of the system concept.

System	Input	Output
Motor	Electrical current	Shaft position
Motor	Electrical current	Angular velocity
Electronic amplifier	Input voltage	Output voltage
Pharmacokinetic compartment	Drug delivery	Blood concentration of the drug
Pharmacokinetic compartment	Drug delivery	Urine concentration of the drug
Man	Drug delivery	Blood pressure
Man	Drug delivery	Body weight
Bioculture	Temperature	Mass
Respiratory assist device	Electrical voltage	Inhaled air volume
Pupil	Light intensity	Pupil diameter
National economy	Interest rates	Inflation

We now define several important special cases of (1.3):

A system for which the superposition of the inputs results in the superposition of the outputs is called a **linear system**. Formally,

$$\mathbf{f}\big(\alpha_1\mathbf{u}_1(t)+\alpha_2\mathbf{u}_2(t)\big)=\alpha_1\mathbf{f}\big(\mathbf{u}_1(t)\big)+\alpha_2\mathbf{f}\big(\mathbf{u}_2(t)\big) \tag{1.4}$$

A system for which the choice of the initial time t_0 has no effect on the output is called a **time invariant system**. Formally,

$$\mathbf{y}(t-t_0)=\mathbf{f}\big(\mathbf{u}(t-t_0)\big) \qquad \forall t_0 \tag{1.5}$$

A system for which two identical input signals (starting with identical initial conditions) result in two identical output signals is called a **causal system**. Formally,

$$\mathbf{u}_1(t)=\mathbf{u}_2(t) \quad \Rightarrow \quad \mathbf{y}_1(t)=\mathbf{y}_2(t) \tag{1.6}$$

A system whose output depends only on the momentary value of the input is called a **static system**. Formally,

$$\mathbf{y}(t_1)=\mathbf{f}\big(\mathbf{u}(t_1)\big), \qquad -\infty<t_1<\infty \tag{1.7}$$

A system whose output depends on the previous as well as the momentary values of its input is called a **dynamic system**. Formally,

$$\mathbf{y}(t_1)=\mathbf{f}\big(\mathbf{u}(t)\big), \qquad t_0\le t\le t_1 \tag{1.8}$$

1.3 Mathematical Description and Analysis of Systems

1.3.1 Input-Output Description of Systems in Time Domain

In general, a system defined as in (1.3) cannot easily be handled analytically. Therefore, most of the analysis methods assume the system in question to be linear, causal and time invariant. A possible way of describing such a system is by means of its impulse response.

Let us first look at the concept of the impulse. The impulse in Figure 1.5 can be described as

$$\delta_\Delta(t,t_1)=\begin{cases}1/\Delta & t_1\le t\le t_1+\Delta\\ 0 & \text{else}\end{cases} \tag{1.9}$$

The shaded area is 1. We now let $\Delta\to 0$ while keeping the area as 1. This results in the so called Dirac impulse function[2]

$$\delta(t,t_1)=\lim_{\Delta\to 0}\delta_\Delta(t,t_1) \tag{1.10}$$

[2] Mathematicians prefer the term “distribution”, because this limit case is not strictly a function.

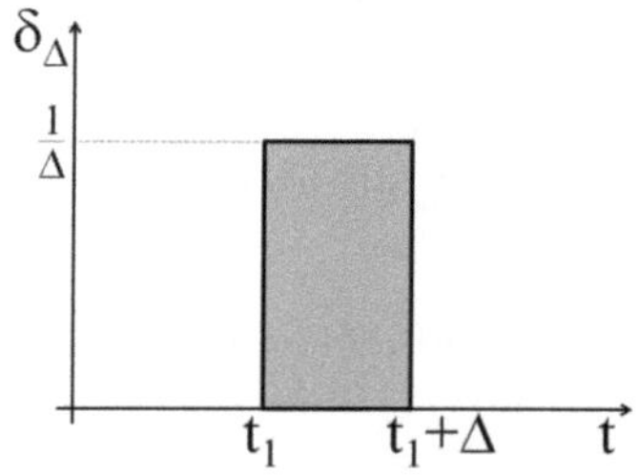

Fig. 1.5 An impulse function.

This definition implies

$$\int_{-\infty}^{\infty} \delta(t, t_1) dt = 1 \tag{1.11}$$

and

$$\int_{-\infty}^{\infty} u(t)\delta(t, t_1) dt = u(t_1) \tag{1.12}$$

Why is this relevant? Well, as shown in Figure 1.6, the input $u(t)$ of any SISO system can be approximated as a sum of impulses

$$u(t) = \sum_{i=0}^{\infty} u(t_i)\delta_\Delta(t, t_i)\Delta \tag{1.13}$$

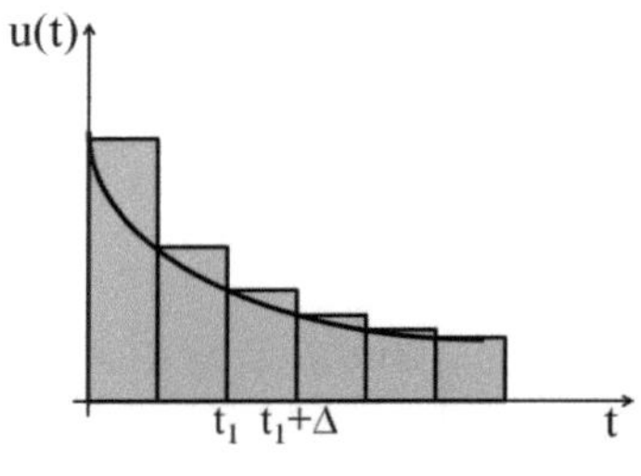

Fig. 1.6 Approximation of a system's input as a sum of impulses.

The output of a linear, causal and time invariant system with the input in (1.13) is

$$
\begin{aligned}
y(t_k) &= f\big(u(t),t_k\big) && (1.14)\\
&= f\big(u(0,\ldots,t_k),t_k\big) && (1.15)\\
&= f\Big(\sum_{i=0}^{k-1} u(t_i)\delta_\Delta(t,t_i)\Delta,t_k\Big) && (1.16)\\
&= \sum_{i=0}^{k-1} u(t_i) f\big(\delta_\Delta(t,t_i),t_k\big)\Delta && (1.17)\\
&= \sum_{i=0}^{k-1} u(t_i) f\big(\delta_\Delta(t,0),t_k-t_i\big)\Delta && (1.18)
\end{aligned}
$$

and by letting $\Delta \to 0$ we get

$$
y(t) = \int_0^t f\big(\delta(t,0),t-\tau\big)u(\tau)d\tau \tag{1.19}
$$

$f\big(\delta(t,0),t\big)$ is called the impulse response of the system and is often symbolized as $g(t)$. Thus, the output of a system with vanishing initial conditions is the convolution of the systems's impulse response with its input

$$
\begin{aligned}
y(t) &= g(t) \circ u(t) && (1.20)\\
&= \int_0^t g(t-\tau)u(\tau)d\tau && (1.21)
\end{aligned}
$$

It is important to note that knowing the response of a linear, causal and time invariant system to an impulse is sufficient to know what the system's output will be for an arbitrary input function since the latter can be decomposed to a sum of impulses. The output is, according to (1.4), computable as the sum of the single impulse responses.

For multivariable systems (1.20) simply becomes

$$
\mathbf{y}(t) = \int_0^t \mathbf{G}(t-\tau)\mathbf{u}(\tau)d\tau \tag{1.22}
$$

with

$$
\mathbf{G}(t) = \begin{pmatrix} g_{1,1} & g_{1,2} & \cdots & g_{1,p} \\ g_{2,1} & g_{2,2} & \cdots & g_{2,p} \\ \vdots & \vdots & \ddots & \vdots \\ g_{q,1} & g_{q,2} & \cdots & g_{q,p} \end{pmatrix}
$$

where $g_{i,j}$ denotes the response of the i-th output to an impulse at the j-th input.

1.3.2 Input-Output Description of Systems in Frequency Domain

Taking the Laplace transform of (1.20) results in the much simpler relationship between the input and the output of a system with vanishing initial conditions, namely

$$Y(s) = G(s)U(s) \tag{1.23}$$

where $G(s)$ is called the transfer function of the system. Similarly, the Laplace transform of (1.22) results in

$$\mathbf{Y}(s) = \mathbf{G}(s)\mathbf{U}(s) \tag{1.24}$$

where $\mathbf{G}(s)$ is called the transfer function matrix of the system which has the elements $G_{i,j}(s)$ which are the transfer functions between the i-th output and the j-th input.

The transfer function can be determined as the Laplace transform of the impulse response or as the quotient of the system output and the system input in the Laplace domain. The transfer functions of a linear, causal and time invariant system with lumped parameters is in the form of a proper rational function or in other words, $G(s)$ is a fraction with polynomials as its nominator and denominator with the order of the nominator being less than the order of the denominator.

The differential equation of such a system is

$$y^{(n)} + a_{n-1}y^{(n-1)} + \cdots + a_1\dot{y} = b_0 u + b_1\dot{u} + \cdots + b_{n-1}u^{(n-1)} \tag{1.25}$$

The Laplace transforms of $y^{(n)}$ and $u^{(n)}$ (the n-th derivative of $y(t)$ and $u(t)$) are

$$\mathscr{L}\{y^{(n)}\} = s^n Y(s) - s^{n-1}y(0) - s^{n-2}\dot{y}(0) - \cdots sy^{(n-2)}(0) - y^{(n-1)}(0) \tag{1.26}$$

$$\mathscr{L}\{u^{(n)}\} = s^n U(s) - s^{n-1}u(0) - s^{n-2}\dot{u}(0) - \cdots su^{(n-2)}(0) - u^{(n-1)}(0) \tag{1.27}$$

Substituting (1.26) in (1.25) and rearranging results in

$$Y(s) = G(s)U(s) + G_0(s) \tag{1.28}$$

with

$$G(s) = \frac{b_{n-1}s^{n-1} + \cdots + b_1 s + b_0}{s^n + a_{n-1}s^{n-1} + \cdots + a_1 s + a_0}$$

$$G_0(s) = \frac{\beta_{n-1}s^{n-1} + \cdots + \beta_1 s + \beta_0}{s^n + a_{n-1}s^{n-1} + \cdots + a_1 s + a_0}$$

where

$$\begin{aligned}
\beta_{n-1} &= y(0) \\
\beta_{n-2} &= \dot{y}(0) + a_{n-1}y(0) - b_{n-1}u(0) \\
\beta_{n-3} &= \ddot{y}(0) + a_{n-2}y(0) + a_{n-1}\dot{y}(0) - b_{n-2}u(0) - b_{n-1}\dot{u}(0) \\
&\vdots \\
\beta_1 &= y^{(n-2)}(0) + \sum_{i=2}^{n-1} a_i y^{(i-2)}(0) - \sum_{i=2}^{n-1} b_i u^{(i-2)}(0) \\
\beta_0 &= y^{(n-1)}(0) + \sum_{i=2}^{n-1} a_i y^{(i-1)}(0) - \sum_{i=2}^{n-1} b_i u^{(i-1)}(0)
\end{aligned} \tag{1.29}$$

Clearly, $G(s)$ corresponds to the transfer function in (1.23) and is independent of the initial conditions. $G_0(s)$ on the other hand depends on the initial conditions. Since the system is linear, both parts can be handled separately and their responses can ultimately be superimposed as Figure 1.7 shows.

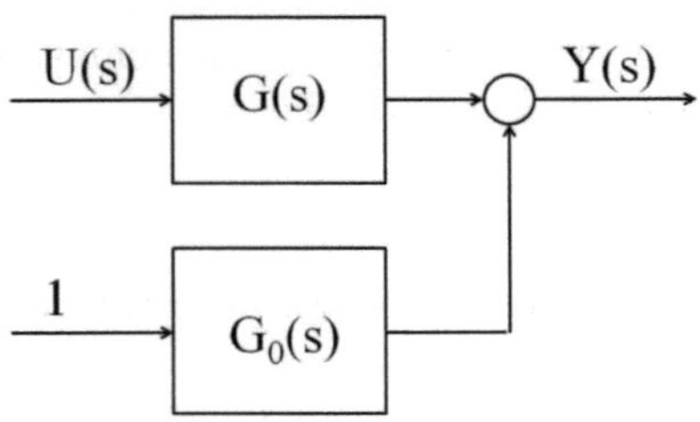

Fig. 1.7 A linear system's output is the result of the superimposition of a part dependent on the input and a part dependent on the initial conditions.

An interesting consequence of (1.23) is that systems and signals can be "multiplied" in the frequency domain. Figure 1.8 shows some basic combinations of systems and their equivalent transfer functions.

1.3.3 State Space Description of Systems

The description of a linear system with a differential equation as in (1.25) is not suitable for the solution of many problems like numerical integration *etc.* For some purposes it is more convenient to replace this single n-th order differential equation by n first order differential equations. A possible way of achieving this is by defining

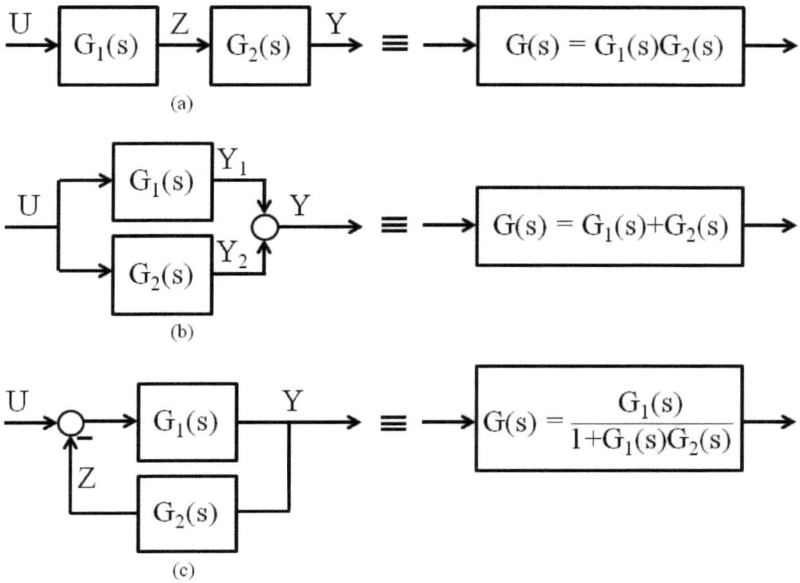

Fig. 1.8 Block diagrams and equivalent transfer functions of systems a) in series, b) in parallel and c) with feedback. A circle block builds the sum of its inputs. For instance, in b) $Y_1 + Y_2$, in c) $E = U - Z$.

$$
\begin{aligned}
y &= x_1 \\
\frac{dy}{dt} &= x_2 \\
&\vdots \\
\frac{d^{n-1}y}{dt^{n-1}} &= x_n
\end{aligned}
\tag{1.30}
$$

and by substituting $\dot{x}_i$ for dx_i/dt as

$$
\begin{aligned}
\dot{x}_1 &= x_2 \\
\dot{x}_2 &= x_3 \\
&\vdots \\
\dot{x}_n &= -a_0x_1 - a_1x_2 - \cdots - a_{n-1}x_n + b_0u + b_1\dot{u} + \cdots + b_{n-1}u^{(n)}
\end{aligned}
\tag{1.31}
$$

Or in matrix notation with **x** denoting the state vector

$$\dot{\mathbf{x}} = \mathbf{A}\mathbf{x} + \mathbf{b}_0 u + \mathbf{b}_1 \dot{u} + \cdots + \mathbf{b}_{n-1} u^{(n)}$$
$$y = \mathbf{c}^T \mathbf{x} + du$$

with

$$\mathbf{A} = \begin{pmatrix} 0 & 1 & 0 & \cdots & 0 \\ 0 & 0 & 1 & 0 & \cdots \\ 0 & 0 & 0 & 1 & \cdots \\ \vdots & \vdots & \vdots & \vdots & 1 \\ -a_0 & -a_1 & \cdots & -a_{n-2} & -a_{n-1} \end{pmatrix}$$

$$\mathbf{b}_0 = \mathbf{b}_1 = \cdots = \mathbf{b}_{n-1} = \begin{pmatrix} 0 \\ 0 \\ \vdots \\ 0 \\ 1 \end{pmatrix}$$

$$\mathbf{c}^T = (1 \ 0 \ 0 \ \cdots \ 0), \qquad d = 0$$

In most practical cases the derivatives of the input $u(t)$ do not appear explicitly, therefore $\mathbf{b}_i$ are zero except for $\mathbf{b}_0$ which is frequently abbreviated simply as $\mathbf{b}$[3]. Thus, the general state space description for a linear multivariable system is

$$\begin{aligned} \dot{\mathbf{x}} &= \mathbf{A}\mathbf{x} + \mathbf{B}\mathbf{u} \\ \mathbf{y} &= \mathbf{C}\mathbf{x} + \mathbf{D}\mathbf{u} \end{aligned} \tag{1.32}$$

(The direct coupling between the input and the output, **D**, can be left out in most cases without loss of generality.) For the names and the dimensions of the matrices **A**, **B**, **C** and **D** see Table 1.2. Figure 1.9 shows the block diagram of a linear system in state space description.

Table 1.2 Common names and dimensions of the matrices **A**, **B**, **C** and **D** in (1.32). n is the system order, p is the number of inputs and q is the number of outputs.

Symbol	Name	Dimension
A	System matrix	[n,n]
B	Input matrix	[n,p]
C	Output matrix	[q,n]
D	Direct input-output coupling matrix	[p,q]

[3] Lower case boldface letters indicate vectors. Upper case boldface letters indicate matrices. Lower case italics indicate scalars.

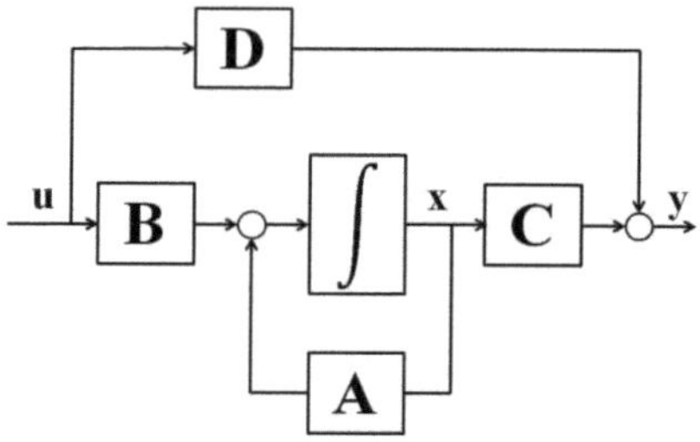

Fig. 1.9 The general block diagram of a system in the state space. **u** is the input vector, **y** is the output vector, and **x** is the state vector.

For a nonlinear time variant system (1.32) generalizes to

$$\begin{aligned}\dot{\mathbf{x}} &= \mathbf{f}\big(\mathbf{x}(t),\mathbf{u}(t),t\big)\\ \mathbf{y} &= \mathbf{g}\big(\mathbf{x}(t),\mathbf{u}(t),t\big)\end{aligned} \tag{1.33}$$

The concept of a system's state is very important and plays a central role in system theory. The number of states of a system is equal to the number of energy storage elements of the system. This number can be finite or infinite. Any system with lumped parameters which can be described by an ordinary differential equation of the order n can be equivalently described in the stet space with n equations. A system with distributed parameters which can only be described by a partial differential equation needs an infinite number of states to be described in the state space. For practical purposes, even a distributed parameter system is often approximated by a finite number of states.

The choice of the states for a given system is **not** unique. In other words, any set of system states $x_1, x_2, \ldots, x_n$ can be transformed to any other set of states $\hat{x}_1, \hat{x}_2, \ldots, \hat{x}_n$ by means of a nonsingular affine transformation

$$\hat{\mathbf{x}} = \mathbf{P}\mathbf{x} \qquad \det(\mathbf{P}) \neq 0$$

without changing the input-output behavior of the system. Of course, the matrices **A**, **B**, **C** and **D** must also be transformed accordingly.

If a system is presented in a state space description, the corresponding n-th order differential equation can be reconstructed by computing the transfer function matrix of the system according to

$$\mathbf{G}(s) = \mathbf{C}(s\mathbf{I} - \mathbf{A})^{-1}\mathbf{B} + \mathbf{D} \tag{1.34}$$

(**I** is the $[n,n]$ identity matrix) and subsequently by substituting s in (1.23) by the differential operator d/dt.

In order to solve the vector differential equation in (1.32) we have to introduce the matrix exponential:

$$e^{\mathbf{A}t} = \mathbf{I} + \mathbf{A}t + \mathbf{A}^2\frac{t^2}{2!} + \ldots + \mathbf{A}^n\frac{t^n}{n!} + \ldots \tag{1.35}$$

Most of the rules for handling scalar exponentials hold in the matrix case as well. Specifically,

$$e^{\mathbf{A}t_1}e^{\mathbf{A}t_2} = e^{\mathbf{A}(t_1+t_2)} \tag{1.36}$$

and

$$\frac{d}{dt}e^{\mathbf{A}t} = \mathbf{A}e^{\mathbf{A}t} = e^{\mathbf{A}t}\mathbf{A} \tag{1.37}$$

The matrix $e^{\mathbf{A}t}$ is called the transition matrix of the system and is often written as $\Phi(t)$. Different ways of computing it can be found, for instance in [230]. Some of the interesting properties of the transition matrix are shown in Table 1.3.

Table 1.3 Some properties of the transition matrix $\Phi(t)$.

$\frac{d\Phi}{dt} = A\Phi$
$\Phi(0) = \mathbf{I}$
$\Phi(t-p)\Phi(p-q) = \Phi(t-q)$
$\Phi(t-p)\Phi(p-t) = \Phi(t-t) = \mathbf{I}$

To solve

$$\dot{\mathbf{x}} = \mathbf{A}\mathbf{x} + \mathbf{b}u \tag{1.38}$$

we first find the solution of the homogenous differential equation

$$\frac{d}{dt}\mathbf{x}(t) = \mathbf{A}\mathbf{x}(t)$$

as

$$\mathbf{x}(t) = e^{\mathbf{A}t}\mathbf{v} \tag{1.39}$$

and use this solution as an *Ansatz* for the solution of the inhomogeneous equation. By substituting (1.39) in (1.38) we get with (1.37)

$$\begin{aligned}
\dot{\mathbf{x}}(t) &= e^{\mathbf{A}t}\dot{\mathbf{v}}(t) + \mathbf{A}e^{\mathbf{A}t}\mathbf{v}(t) = \mathbf{A}e^{\mathbf{A}t}\mathbf{v}(t) + \mathbf{b}u \\
\dot{\mathbf{v}}(t) &= e^{-\mathbf{A}t}\mathbf{b}u
\end{aligned}$$

and integrating $\dot{\mathbf{v}}(t)$

$$\mathbf{v}(t) = \mathbf{v}_0 + \int_{t_0}^{\tau} e^{-\mathbf{A}t}\mathbf{b}u(\tau)d\tau$$

Thus the general solution of (1.38) is

$$\mathbf{x}(t) = e^{\mathbf{A}t}\mathbf{v}_0 + \int_{t_0}^{t} e^{\mathbf{A}(t-\tau)}\mathbf{b}u(\tau)d\tau \tag{1.40}$$

and the particular solution for the initial conditions $\mathbf{x}(t = t_0) = \mathbf{x}_0$ is

$$\mathbf{x}(t) = e^{\mathbf{A}(t-t_0)}\mathbf{x}_0 + \int_{t_0}^{t} e^{\mathbf{A}(t-\tau)}\mathbf{b}u(\tau)d\tau \tag{1.41}$$

This solution is easily transferable to multivariable systems by substituting $\mathbf{b}$ by $\mathbf{B}$ and u by $\mathbf{u}$. Also, this result should not depend on the choice of t_0 for a time invariant system. Therefore, it can be simplified by taking $t_0 = 0$. Obviously, the solution has two terms, one depending on the initial conditions but not on the input and the other depending on the input but not on the initial conditions. This is another manifestation of the system's output being the sum of the system's response to the input and to the initial conditions as we have seen in (1.28) and in Figure 1.7.

The output of the system can now be calculated from (1.41) to be

$$y(t) = \underbrace{\mathbf{c}^T e^{\mathbf{A}t} x_0}_{(a)} + \underbrace{\int_{t_0}^{t} \mathbf{c}^T e^{\mathbf{A}(t-\tau)}\mathbf{b}u(\tau)d\tau}_{(b)} \tag{1.42}$$

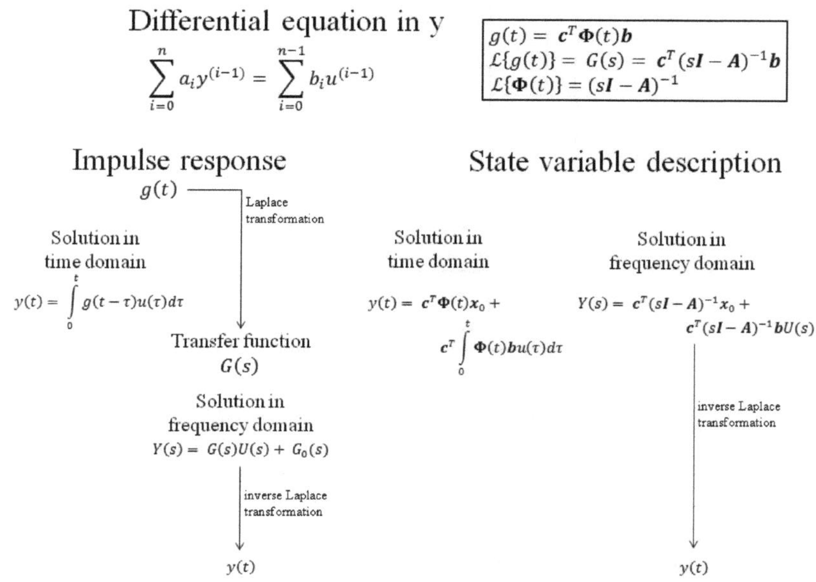

Fig. 1.10 Different ways of solving linear, causal and time invariant systems and relationships between different forms of describing them.

For vanishing initial conditions

$$y(t) = \int_{t_0}^{t} \mathbf{c}^T e^{\mathbf{A}(t-\tau)} \mathbf{b} u(\tau) d\tau \qquad (1.43)$$

As can be seen, this equation has the same form as (1.20)! This observation gives us a way of computing a system's impulse response from the state space description. Namely,

$$g(t) = \mathbf{c}^T e^{\mathbf{A}t} \mathbf{b} = \mathbf{c}^T \Phi(t) \mathbf{b}$$

To sum up this Section, the relationships between different forms of describing linear systems are shown in Figure 1.10.

1.4 Some Important System Properties

There are several qualitative properties of a system which decide on its applicability for a given problem. Those are *controllability*, *observability* and *stability*. These central concepts will be explained mainly for linear, causal and time invariant systems. Another important system property, namely *identifiability* will be handled in the next Chapter.

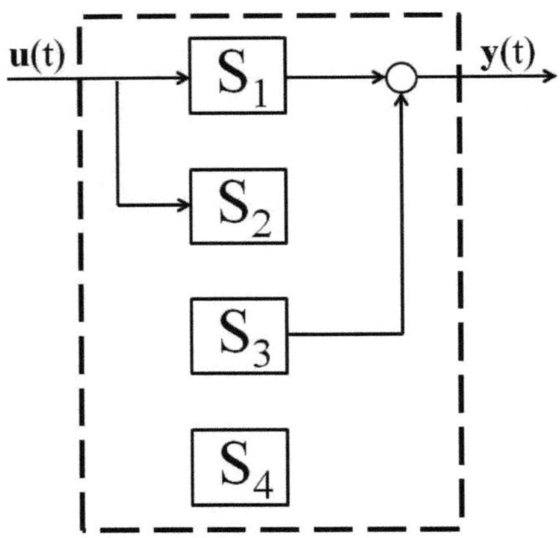

Fig. 1.11 Decomposition of a system in a controllable and observable part (S_1), a controllable but not observable part (S_2), a not controllable but observable part (S_3), and a not controllable and not observable part (S_4).

A system of order n is called **controllable** if a control variable $\mathbf{u}(t)$ exists such that the system can be brought from an arbitrary starting point in the state space (initial condition) $\mathbf{x}_0$ to an arbitrary end point $\mathbf{x}_e$ in finite time. This is the case if and only if the controllability matrix

$$\mathbf{Q}_c = \left(\mathbf{B}, \mathbf{AB}, \mathbf{A}^2\mathbf{B}, \ldots, \mathbf{A}^{n-1}\mathbf{B}\right)$$

has the rank n. For SISO systems this requirement is equivalent to the determinant of $\mathbf{Q}_c$ (which has the dimensions $[n,n]$ not being equal to zero. For MIMO systems $\mathbf{Q}_c$ has the dimensions $[n,np]$ (with p being the number of inputs) and the rank requirement is for the largest minor in $\mathbf{Q}_c$ to have a nonzero determinant. For a proof see [161].

A system is called **observable** if the initial conditions $\mathbf{x}_0$ can be determined unambiguously from the measurement of the output $\mathbf{y}(t)$ for a finite time with known input $\mathbf{u}(t)$. This is the case if and only if the observability matrix

$$\mathbf{Q}_o = \begin{pmatrix} \mathbf{C} \\ \mathbf{CA} \\ \mathbf{CA}^2 \\ \vdots \\ \mathbf{CA}^{n-1} \end{pmatrix}$$

has the rank n.

With these definitions of controllability and observability, any system can be divided into a) a controllable and observable part, b) a controllable but not observable part, c) a not controllable but observable part, and d) a not controllable and not observable part as schematized in Figure 1.11 [162].

1.5 Stability

1.5.1 BIBO Stability

A property which is perhaps more important than the system properties described in the previous Section is *stability*. Stability is one of the most studied issues in system theory. Consequently, it has several different definitions. Let us first consider the bounded-input-bounded-output (BIBO) stability.

A system is said to be BIBO stable if and only if its output is bounded for all bounded inputs. This statement is equivalent to

$$\int_0^\infty |g(t)|dt \le p < \infty \tag{1.44}$$

with $g(t)$ denoting the impulse response of the system and p a real number. However, in general, this does not necessarily imply that $|g(t)| < m$ or that $g(t \to \infty) = 0$, except for systems with proper rational transfer functions. For a multivariable system to be BIBO stable all of the impulse responses $g_{i,j}(t)$ must fulfil (1.44). For $t \to \infty$ a BIBO stable system's output approaches a periodic function with the same period as the input function. If, on the other hand, the input is constant, the output also approaches a constant but not necessarily the same value.

There is a simple way we can tell whether a system is BIBO stable: a SISO system with a proper rational transfer function $G(s)$ is BIBO stable if and only if all of the poles of $G(s)$ (zeros of the denominator polynomial) are in the open left half complex plane (to which the imaginary axis does not belong). Figure 1.12 shows the pole configuration of some BIBO stable or unstable systems. For a multivariable system with the transfer function matrix $\mathbf{G}(s)$ to be BIBO stable, all of its transfer functions $G_{i,j}$ must be BIBO stable.

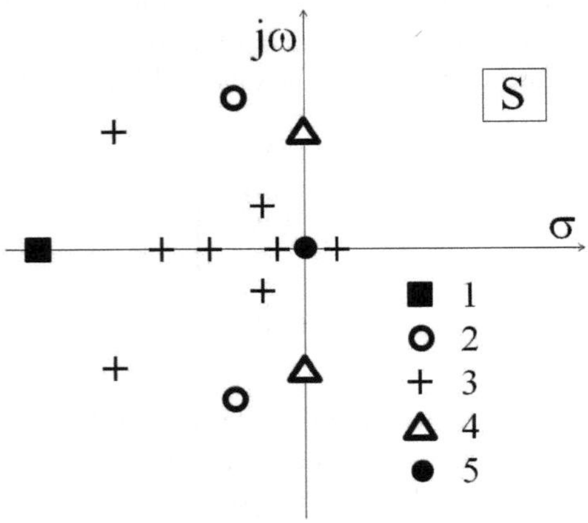

Fig. 1.12 Pole configurations of five systems of varying orders. System 1 and 2 are BIBO stable, the others are not.

To determine the exact location of the zeros of a polynomial is not always easy; for polynomials of the order 5 or higher it is, in general, even impossible. Nevertheless, it is possible to determine whether a polynomial of any given order has zeros in the right half plane or not. The most common way to solve this problem goes back more than a century to Routh and Hurwitz, who, independent of each other, developed methods to determine the number of zeros a polynomial has in the right half plane.

The polynomial under study is given as

$$P(s) = a_0 s^n + a_1 s^{n-1} + \cdots + a_{n-1} s + a_n \tag{1.45}$$

A necessary but not sufficient condition for all of the zeros of $P(s)$ to be in the open left half plane is that all coefficients of the polynomial in (1.45) must be positive (this condition is also sufficient for polynomials up to the second degree). In other words, if even a single coefficient is negative or missing this means that $P(s)$ has at least one zero in the right half plane. On the other hand, if all coefficients are positive, it still cannot be concluded that $P(s)$ has no zeros in the right half plane. So, how can be sure that it does not?

Let us decompose $P(s)$ in two polynomials

$$\begin{aligned} P_1(s) &= a_0 s^n + a_2 s^{n-2} + \cdots \\ P_2(s) &= a_1 s^{n-1} + a_3 s^{n-3} + \cdots \end{aligned}$$

The quotient of these polynomials can be expressed as a continuous fraction

$$\frac{P_1(s)}{P_2(s)} = \alpha_1 s + \cfrac{1}{\alpha_2 s + \ddots \cfrac{1}{\alpha_{n-1}s + \frac{1}{\alpha_n s}}} \tag{1.46}$$

It can be shown that the polynomial in (1.45) has all its roots in the open left half plane if and only if all α_i in (1.46) are positive. See Table 1.4 for an easy way of calculating α_i.

Table 1.4 Routh scheme for the calculation of the coefficients in the continuous fraction representation of the quotient of even and odd parts of a polynomial.

$\alpha_1 = \frac{a_0}{a_1}$	s^n	$a_0\ a_2\ a_4\ a_6\ \ldots$	
			$b_1 = \frac{a_1a_2 - a_0a_3}{a_1}$
$\alpha_2 = \frac{a_1}{b_1}$	s^{n-1}	$a_1\ a_3\ a_5\ \ldots$	
$\alpha_3 = \frac{b_1}{c_1}$	s^{n-2}	$b_1\ b_2\ b_3\ \ldots$	$b_2 = \frac{a_1a_4 - a_0a_5}{a_1}$
	s^{n-3}	$c_1\ c_2\ c_3\ \ldots$	
	$\vdots$		$c_1 = \frac{b_1a_3 - a_1b_2}{b_1}$
			$\vdots$
$\alpha_n = \frac{j_1}{k_1}$	s	$j_1\ 0$	
	1	$k_1\ 0$	

The polynomial $P(s)$ has all its roots in the left half plane if all the elements $a_1, b_1, c_1, \ldots, k_1$ in the Routh table are positive. Furthermore, the number of roots in the right half plane is equal to the change of signs in the first column of the Routh table.

1.5.2 Lyapunov Stability

So far we have studied the stability of a system given by its input-output description in the frequency domain. How can the results be transferred for a system described in its state space? We know that the solution of the set of equations in (1.32) as given in (1.42) consists of two terms, the first one depending on the initial conditions but not on the input and the second term depending on the input but not on the initial conditions. Therefore, the question of stability should be studied separately for both parts.

We cannot apply the BIBO stability on the first part since it is not dependent on the input. The study of the effect of initial conditions on the system response leads to a new concept: the stability of an equilibrium point. An equilibrium point $\mathbf{x}_e$ of a system $\dot{\mathbf{x}} = \mathbf{f}(\mathbf{x})$ is given by $\dot{\mathbf{x}} = \mathbf{f}(\mathbf{x}_e) = \mathbf{0}$. In other words, a system is at equilibrium if its state does not change.

We now define an equilibrium point $\mathbf{x}_e$ of a system $\dot{\mathbf{x}} = \mathbf{f}(\mathbf{x})$ to be Lyapunov stable if and only if for an arbitrarily small ε a $\delta(\varepsilon)$ exists such that

$$\|\mathbf{x}_e - \mathbf{x}_0\| < \delta \quad \Rightarrow \quad \|\mathbf{x}(t) - \mathbf{x}_e\| < \varepsilon \qquad \forall t \geq 0 \tag{1.47}$$

(see Figure 1.13). Further, the equilibrium point is said to be asymptotically stable if it is Lyapunov stable and for any $\gamma > 0$, $\mu > 0$ a time $T(\mu, \gamma)$ exists, such that

$$\|\mathbf{x}_e - \mathbf{x}_0\| < \gamma \quad \Rightarrow \quad \|\mathbf{x}(t) - \mathbf{x}_e\| < \mu \qquad \forall t \geq T \tag{1.48}$$

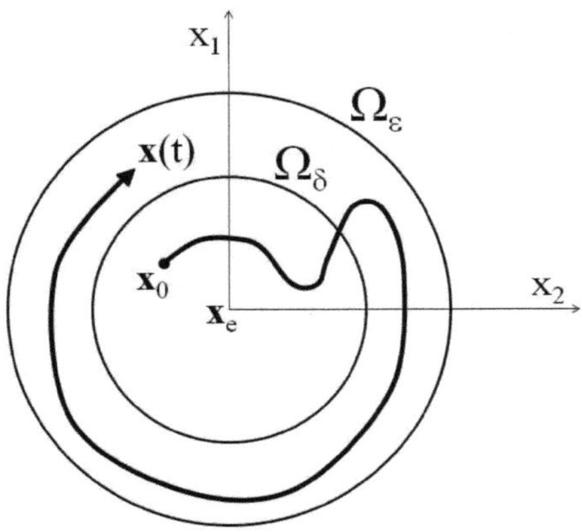

Fig. 1.13 An equilibrium point of a system is Lyapunov stable if and only if for any arbitrary spherical domain Ω_ε around it another spherical domain Ω_δ exists such that any system trajectory starting in Ω_δ never leaves Ω_ε as shown in the example of a second order system.

Note that those definitions are also valid for a non-linear system which can have none, one, several or an infinite number of equilibrium points. A linear system, on the other hand, has either an equilibrium point at the origin of the state space (if the system matrix $\mathbf{A}$ is regular) or an infinite number of equilibrium points (if the system matrix $\mathbf{A}$ is singular). Furthermore, the stability of an equilibrium point is a local property of a non-linear system. However, for a linear system, the stability (or instability) is a global property.

The equilibrium point $\mathbf{x}_e$ of a linear system $\dot{\mathbf{x}} = \mathbf{A}\mathbf{x}$, thus the linear system itself is asymptotically stable if and only if all the eigenvalues of $\mathbf{A}$ have negative real parts. (The Routh-Hurwitz scheme can be used to determine the position of the eigenvalues, in other words, the zeros of the characteristic polynomial of $\mathbf{A}$.)

It is important to note that each pole of a system's transfer function $G(s)$ is also an eigenvalue of the system's **A** matrix. However, the reversal of this statement is in general not true, because $G(s)$ describes the relationship the input and the output of only the observable and controllable part of the system. Therefore, asymptotic stability of a linear system implies BIBO stability, but a BIBO stable system is not necessarily Lyapunov stable as the following example demonstrates.

Example: The state and output equations of a second order system is given as

$$\begin{aligned} \dot{x}_1 &= x_1 \\ \dot{x}_2 &= -0.5x_2 + u \\ y &= x_2 \end{aligned}$$

The system matrix of this system is

$$\mathbf{A} = \begin{pmatrix} 1 & 0 \\ 0 & -0.5 \end{pmatrix}$$

which has the eigenvalues $\lambda_1 = 1$ and $\lambda_2 = -0.5$, thus is not Lyapunov stable. The transfer function can be computed with (1.34) to be

$$G(s) = \frac{1}{s+0.5}$$

which has its only pole in the left half plane, thus is BIBO stable. □

1.6 Exercises

A1: Explain the concepts of observability and controllability and give examples.

A2: What are the differences between BIBO stability and Lyapunov stability?

A3: In Figure 1.8 a) $G(s)$ is given as $G_1(s)G_2(s)$. Is $G(s) = G_2(s)G_1(s)$ also true? What about multivariable systems?

A4: Are the following systems static or dynamic, linear or linear, time variant or time invariant? Explain your answers. a) $t\dot{y} + y = u$, b) $y\dot{y} + y = u$, c) $\ddot{y} + y = u$, d) $y = mu + b$, e) $\ddot{y}\cos(t) + y\sin(2t) = u$, f) $y = \min(u_1, u_2)$, g) $y = \dot{u} + u^2$.

A5: Answer without any calculations. Is the electrical network in Figure 1.14 observable and controllable?

A6: Are the following systems which are characterized by their impulse responses $g(t)$ BIBO stable? a) e^{-t}, b) te^{-t}, c) 1, d) $e^{-t}\sin 3t$, e) $\sin \omega t$.

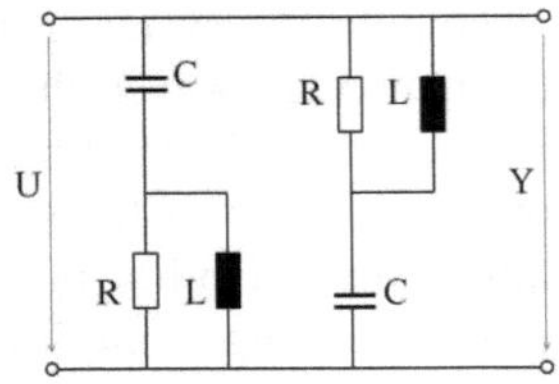

Fig. 1.14 Electrical network for Exercise A5.

B1: Find the equivalent transfer function matrix of the system in Figure 1.15. The transfer function matrices $\mathbf{G}_i(s)$ are given.

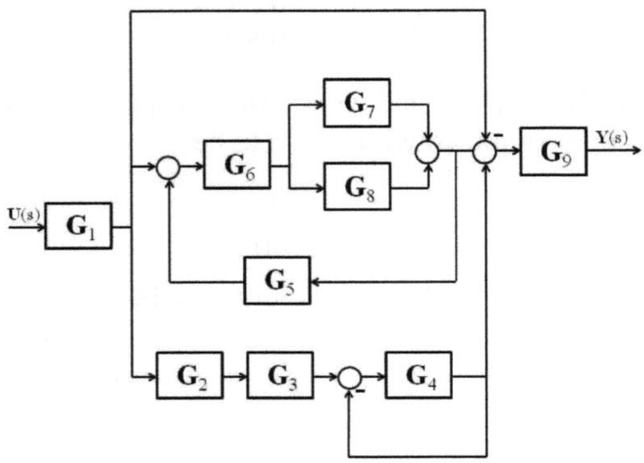

Fig. 1.15 What is the equivalent transfer function matrix $\mathbf{G}(s)$ which maps $\mathbf{U}(s)$ onto $\mathbf{Y}(s)$?

B2: Find a state space description for the system in Figure 1.16. Is the system a) BIBO stable, b) controllable, c) observable, d) Lyapunov stable?

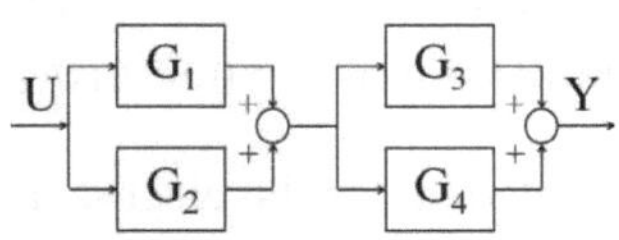

Fig. 1.16 System for the exercise B2. $G_1(s) = \frac{-2}{s+1}$, $G_2 = \frac{3}{s+2}$, $G_3 = \frac{0.5}{s-1}$, $G_4 = \frac{0.5}{s+3}$.

B3: For which values of K is the system in Figure 1.17 stable?

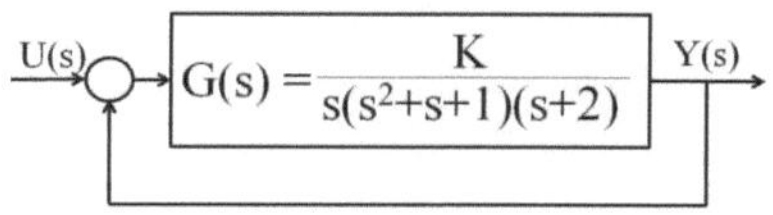

Fig. 1.17 Figure for the exercise B3.

B4: Find a state space description of the system in Figure 1.18.

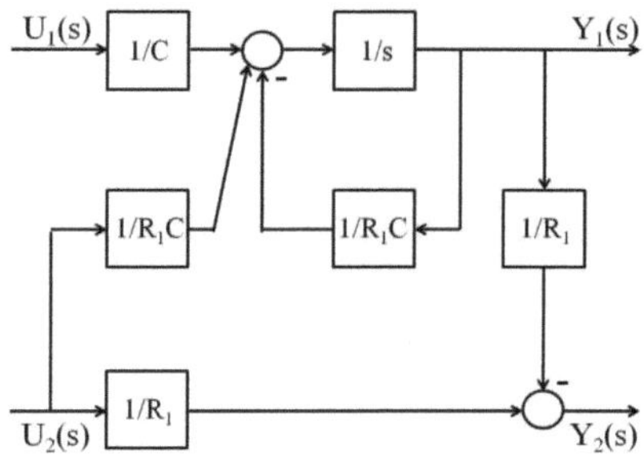

Fig. 1.18 Figure for the exercise B4.

C1: You will have some problem with the Routh-Hurwitz scheme finding out whether the following polynomials have any roots in the right half plane:

$$P(s) = s^3 + 2s^2 + s + 2$$
$$P(s) = s^5 + 2s^4 + 5s^3 + 10s^2 + 4s + 8$$

Extend the Routh-Hurwitz scheme to make it applicable for these cases as well (the solution is known).

C2: A system may be controllable but it might take an impractical control variable $u(t)$ to bring it from a given starting point in the state space to another given point. Find some measures of controllability for controllable systems and discuss their respective merits and drawbacks.

Chapter 2
Modeling and Identification

"That's another thing we've learned from your Nation," said Mein Herr, "map-making. But we've carried it much further than you. What do you consider the largest map that would be really useful?"
"About six inches to the mile."
"Only six inches!" exclaimed Mein Herr. "We very soon got to six yards to the mile. Then we tried a hundred yards to the mile. And then came the grandest idea of all! We actually made a map of the country, on the scale of a mile to the mile!"
"Have you used it much?" I enquired.
"It has never been spread out, yet," said Mein Herr: "the farmers objected: they said it would cover the whole country, and shut out the sunlight! So we now use the country itself, as its own map, and I assure you it does nearly as well.

— *Lewis Carroll, Sylvie and Bruno Concluded*

A designer knows he has achieved perfection not when there is nothing left to add, but when there is nothing left to take away.

— *Antoine de Saint-Exupery*

2.1 Introduction

Most control engineering exercises begin with the phrase "Given is the system with the transfer function ...". In engineering practice, alas, a mathematical description of the plant which is to be controlled is seldom available. The plant first needs to be modeled mathematically as a system. In spite of this fact, control engineering courses generally do not spend much time on modeling. Perhaps this is because modeling can be said to be more of an art than a science.

We will begin this Chapter by defining what we mean by the words model and modeling. We shall then encounter various types of models and study the process

S.S. Hacısalihzade: *Biomedical Applications of Control Engineering*, LNCIS 441, pp. 23–43.
DOI: 10.1007/978-3-642-37279-7_2

of modeling. Finally, we will look at the question of parameter identification and model validation.

2.2 What Is a Model?

Indeed, what is a model? Probably you will get as many different responses as the number of persons you pose this question. Therefore, it is not surprising that Merriam-Webster Dictionary offers 19 different definitions under that one entry. The fourth definition reads "a usually miniature representation of something" as in a "model helicopter". The ninth definition reads, rather unglamorously, "one who is employed to display clothes or other merchandize" as in Twiggy, Claudia Schiffer or Gisele Bündchen depending on which decade you came off age. Well, if your interest in models is limited to these definitions you can close this book right now...

Definitions 11 and 12 read "a description or analogy used to help visualize something (as an atom) that cannot be directly observed" and "a system of postulates, data and inferences presented as a mathematical description of an entity or state of affairs; also: a computer simulation based on such a system <climate models>". These definitions are closer to the sense of the word model we shall be using throughout this book.

These definitions might be considered too general to be of any practical use. Let us therefore think of a model as an approximative representation of reality. One obvious fact, forgotten surprisingly often, is that a model is an abstraction and that any model is, by necessity, an *approximation* of reality. Consequently, there is no one "true model", rather there are models which are better than others. But what do we mean when we say "better". This clearly depends on the context and the problem at hand.

Example: An illustrative example which can be found in many high school physics books is dropping an object from the edge of a table and calculating the time it will take for the object to hit the floor. Assuming no air friction, here you can write the well known Newtonian equation of motion which states that the object will move with a constant acceleration caused by the weight of the object, which again is caused by the Earth's gravitational attraction:

$$m\ddot{x}(t) = mg \tag{2.1}$$

where $\ddot{x}(t)$ denotes the second derivative of the distance x (acceleration) of the object from the edge of the table as a function of time after the drop; g is the Earth's acceleration constant. m is the mass of the object but it is not relevant, because it can be canceled away. Solving the differential equation (2.1) with the initial conditions $x(0) = 0$ and $\dot{x}(0) = 0$ results in

$$x(t) = \frac{1}{2}gt^2 \tag{2.2}$$

and solving (2.2) for the impact time t_i with the height of the table denoted as h results in the well known formula

$$t_i = \sqrt{\frac{2h}{g}} \tag{2.3}$$

Let us now consider the problem of dropping an object attached to a parachute from an aircraft. Here, you can no longer assume there is no air friction. The very reason of deploying a parachute is to make use of the air friction to slow down the impact velocity of the drop. Hence, to calculate the impact time with better accuracy, (2.1) needs to be modified to account for the air friction, a force which increases with the square of velocity:

$$\ddot{x}(t) = g - kS\dot{x}^2(t) \tag{2.4}$$

where $\dot{x}(t)$ denotes the first derivative of the distance x (velocity) of the object from the drop height, k is the viscous friction coefficient and S is the effective surface area of the parachute. Clearly it is no longer possible to have a simple analytical solution of this non-linear differential equation and one has to revert to numerical means to solve it. □

This example shows that the same physical phenomenon can be modeled in different ways with varying degrees of complexity. It is not always possible to know as in this case which model will give better results *a priori*. Therefore one often speaks of the "art" of modeling. An engineer's approach to modeling might be to begin with certain restrictive assumptions leading to a simple model. At further steps, these assumptions can be relaxed to modify the initial simple model and to account for further complications until the purpose of the model is reached.

There are many different types of models in science and philosophy. The reader might have heard of mental models, conceptual models, epistemological models, statistical models, scientific models, economic models or business models just to name a few. In the sequel we shall limit ourselves to scientific and mathematical models. Scientific modeling is the process of generating abstract, conceptual, graphical or mathematical models using a number of methods, techniques and theories. The general purpose of a scientific model is to represent empirical phenomena in a logical and objective way. We shall study the process of modeling in the next Section.

A mathematical model helps describe a system using mathematical tools. The use of mathematical models is certainly not limited to engineering or natural sciences applications. Mathematical models are increasingly being used in social sciences like economics, psychology and sociology[1].

[1] Which science fiction enthusiast is unaffected by Hari Seldon's "psychohistory", which combines history, sociology and statistics to make general predictions about the future behavior of large groups of populations – like the Galactic Empire with a quintillion citizens [17]?

Mathematical models can be classified under the following dichotomic headings:

Inductive versus **deductive** models: A deductive model is based on a physical theory. In the example above, we have used Newton's laws of movement to model the movement of a falling object. An inductive model, on the other hand, is based on empirical findings and their generalizations without forming a generally applicable law of nature. A well known example is the set of Lotka-Volterra equations used in modeling the dynamics of biological systems in which two species interact as predator and prey [202]. Here the change of rate of the number of preys (say, rabbits) is proportional to the number of preys who can find ample food and who can breed, well, like rabbits. This exponential growth is corrected by prey-predator encounters (say, foxes). Dually, the number of predators decreases in proportion to the number of predators either because they starve off or emigrate. This exponential decay is corrected by the encounters of predators with preys.

Deterministic versus **stochastic** models: In a deterministic model no randomness is involved in the development of future states of the modeled system. In other words, a deterministic model will always produce the same output from a given initial condition like in the example above. A stochastic model, on the other hand, includes randomness. The states of the modeled system do not have unique values but rather be described by probability distribution functions. A good example is the movement of small particles in a liquid – Brownian motion – resulting from their bombardment by a vast number of fast moving molecules [100].

Static versus **dynamic** models: In static models, as the name suggests, the effect of time is not considered. Dynamic models specifically account for time. Such models make use of difference or differential equations with time as a free variable. Looking at the input-output relationship of an amplifier far from saturation, a static model will simply consist of the amplification factor. A dynamic model, on the other hand, will use a time function to describe how the output changes dynamically, including transients and all, as a consequence of changes in the input [70].

Discrete versus **continuous** models: Some recent models of quantum gravity [286] not withstanding, in our current *Weltbild* time flows smoothly and continuously. Models building on that – often tacit – assumptions make use of differential equations, solutions of which are time continuous functions. However, the advent of the digital computer which works with a clock and spews out results at discrete time points made it necessary to work with discrete time models which are best described using difference equations. Whereas the classical speed control by means of a flyball governor is based upon a continuous time model of a steam engine [299], modern robot movement controllers employing digital processors use discrete time models [258].

Lumped parameter models versus **distributed parameter** models: Distributed parameter or distributed element models assume that the attributes of the modeled system are distributed continuously throughout the system. This is in contrast to lumped parameter or lumped element models, which assume that these values are lumped into discrete elements. One example of a distributed parameter model is the transmission line model which begins by looking at the electrical properties of an infinitesimal length of a transmission line and results in the telegrapher's equations, which are partial differential equations involving partial derivatives with respect to both space and time variables [229]. A lumped parameter model simplifies the description of the behavior of spatially distributed physical systems into a topology consisting of discrete entities that approximate the behavior of the distributed system. The mathematical tools required to analyze such models are ordinary differential equations involving derivatives with respect to the time variable alone. A simple example for a lumped parameter models is an electrical circuit consisting of a resistor and an inductor in series.

Linear models versus **non-linear** models: Although a rigorous definition might involve some sophisticated mathematical concepts as discussed in the previous Chapter, it can be said that a model is linear if the net response caused by two stimuli is the sum of the responses which would have been caused by each stimulus individually [161]. Most physical phenomena are inherently non-linear. However, they may legitimately be modeled linearly either in limited operational ranges (*e.g.*, if an amplifier working far from saturation amplifies 1 to 5, it amplifies 2 to 10) or linearized around an operating point (*e.g.*, a spring extended far away from its breaking point). While writing this Chapter, it was suggested by a student that dividing all models to linear and non-linear models is like dividing the world to bananas and non-bananas. Such a remark is clearly facetious, because there is a closed theory for linear systems and many tools and methods exist for dealing with them. On the other hand, non-linear systems are analyzed mostly with approximative numerical methods specific to the problem at hand [314].

It can be said that nature is inherently dynamic, stochastic, continuous, and non-linear with distributed parameters, hence difficult to model accurately. However, depending on the application area, many simplifying assumptions can be made to make the model more amenable to analytical techniques as indicated above.

To sum it up, a good model should provide some insight which goes beyond what is already known from direct investigation of the phenomenon being studied. The more observations it can explain and the more accurate predictions it can make the better is the model. Combine that with the quote "Everything should be made as simple as possible, but not simpler" attributed to Albert Einstein, you have all the elements of a good model. A corollary to this statement is that the level of detail incorporated in the model is determined by the intended purpose of the model itself.

The four distinctly different types of models depending on their purposes can be categorized as descriptive, interpretive, predictive or explanatory models. **Descriptive** models represent relationships between various variables using a concise mathematical language. Hooke's law stated as

$$F = -kx \tag{2.5}$$

is a well known example here with F representing the force which lengthens a spring by the distance x. One could describe this law in words or using diagrams, but the expression in (2.5) is a pithy description.

Suppose you want to model the purchasing power of money in an inflationary economy. You might plot the number of loaves of bread one can buy with a constant salary every month and see that this number drops with time and that the rate of the drop is constant as in Figure 2.1. You can use this exponential decay model to **interpret** the decay rate as a measure of the inflation.

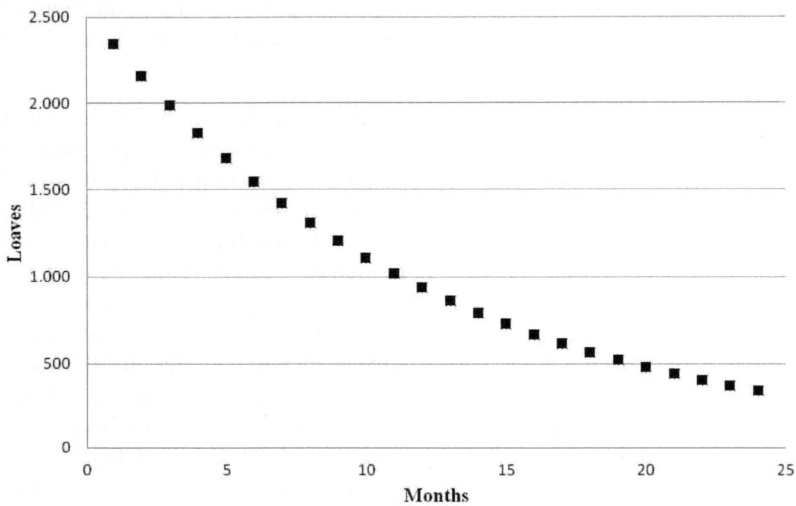

Fig. 2.1 Number of breads one can buy with a fixed salary in an inflationary economy.

We need a **predictive** model if we want to use the model to see how a system will react to a certain stimulus. A typical example is a linear electrical circuit with a resistor, an inductor and a capacitor. The input-output transfer function of such a circuit can be described using the Laplace transform as

$$G(s) = \frac{2}{s^2 + 3s + 8} \tag{2.6}$$

Using this model we can now simulate the response of the system to a unit step at its input which is shown in Figure 2.2.

A typical example of an **explanatory** model is given in [79]:

> ... if the parameters of the model correspond to explicit physiological processes or effects, then changes in observed behavior can be interpreted in terms of changing parameter values. This means that the model can be used to help provide a physiological explanation for observed dynamic effects. In this way a model can, for instance, be used to help understand how changes in physiological parameters can bring about changes in the uptake of substances, including drugs, by various organs of the body.

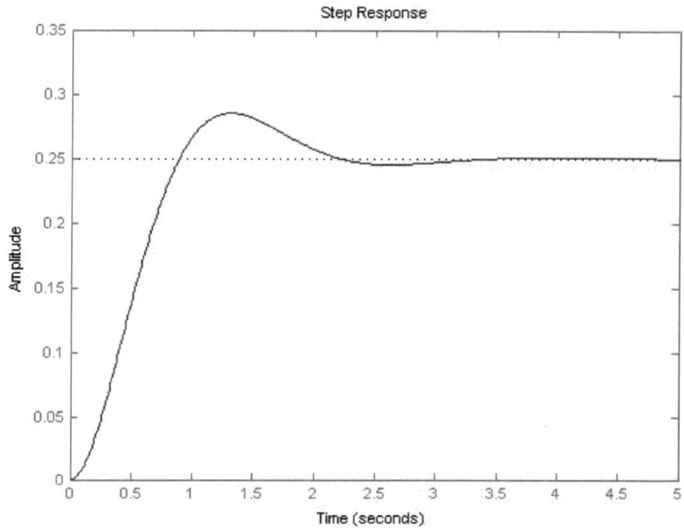

Fig. 2.2 Response of the system in (2.6) to a unit step at its input.

A mathematical model can be represented with a block diagram as shown in Figure 2.3.

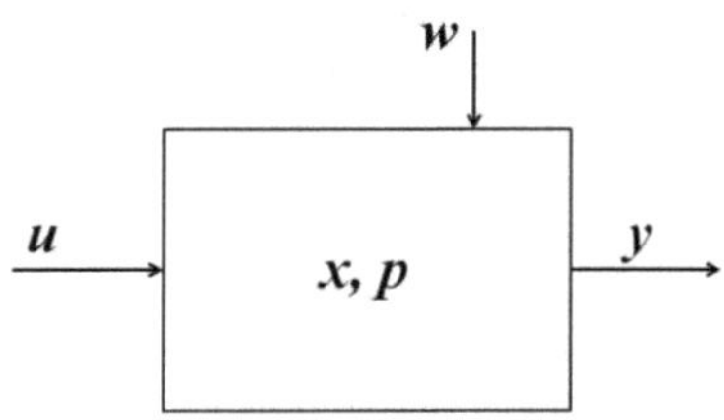

Fig. 2.3 Block diagram representation of a mathematical model has different types of variables: input variables which can be controlled $\mathbf{u}$, input variables which cannot be controlled (often random and called disturbances) $\mathbf{w}$, output variables $\mathbf{y}$, internal state variables $\mathbf{x}$, model parameters $\mathbf{p}$. Since there can be several of each type of variables, those are aggregated in vectors. The state space representation of such a system is given by the equations $\dot{\mathbf{x}}(t) = \mathbf{f}(\mathbf{x}, \mathbf{u}, \mathbf{w}, \mathbf{p}, t)$ and $\mathbf{y}(t) = \mathbf{g}(\mathbf{x}, \mathbf{u}, \mathbf{w}, \mathbf{p}, t))$.

A philosophical excursion

Let us consider the Newtonian model of gravity. Newton posits an invisible force able to act over distances which attracts bodies. This force is proportional to the masses of the bodies and inversely proportional to the square of the distance between the bodies. Note that Newton made no attempt at explaining the cause of that attraction. This model is immensely successful. One can even argue that, together with his three laws of motion, Newtonian mechanics constitutes one of the pillars on which the industrial revolution and the ensuing increase in the welfare of the human population during the last three centuries builds. Construction machines, the houses these machines build, the cars that travel among these houses, the aircraft that move people from one continent to another, the spacecraft that brought the Apollo astronauts to the moon, just to give a couple of examples, all are built using Newtonian mechanics. So, does Newtonian mechanics constitute a correct model of reality?

On the other hand, if you want to use the satellite based global positioning system (GPS) to determine your position on earth accurately, Newtonian mechanics will fail you. Einstein's general relativity theory no longer talks about invisible forces of attraction between bodies but posits the very fabric of space being curved by the presence of matter. A consequence of this theory is that "time flows faster" in the vicinity of matter. Concretely, this means that the clocks on the surface of the earth and on a satellite in earth orbit do not tick to the same rhythm. Since GPS makes use of accurate time intervals between signals received and sent by satellites, not taking this relativistic effect into account reduces its accuracy. So, does Einstein's mechanics constitute a correct model of reality?

In case of building an aircraft, you might argue that Newtonian mechanics is demonstrably correct since aircraft fly all the time. In case of building an accurate GPS, you might argue that general relativity is demonstrably correct since, using GPS, hikers or soldiers in unfamiliar terrain know exactly where they are. It follows that the question about which model is better or correct depends on the problem at hand. Clearly, one does not need to deal with the complexity of Einstein's field equations when a Newtonian model suffices to build a car.

The pesky question arises inevitably: yes, but how is it in reality? Do objects attract each other or is the space curved due to the presence of matter or is it another yet to be developed theory? The answer is most probably none of the above. Depending on the problem at hand it will be more appropriate to use one model of reality or other. The point to remember is that a model of reality is just that: a model. Like all models, that model also has its limitations. If you want to have a model which incorporates *all* aspects of reality that model will have to be the reality itself as Mein Herr pointed out at the beginning of this Chapter.

Then pops up the next question: which model is closer to reality? Even if there is an absolute reality out there, it is highly questionable that we will ever know how it is in reality. As our observations and intellectual capability improve we will most likely be able to develop models which will explain more phenomena, more of the observations and make more precise predictions. However, we will probably never know in the epistemological sense whether we are any closer to the reality. Besides, and here comes the provocative statement, science is not interested in finding the reality, it is content to find progressively better models of it; search for reality is outside the scope of science!

2.3 Modeling Process

Maybe it will relax the reader to hear that there are no exact rules in mathematical modeling and there are no "correct answers" as should be apparent by now. But how do we get a mathematical model of a process? Let us begin with a couple of examples.

2.3.1 Inverted Pendulum

Consider the inverted pendulum problem: there is a pole mounted with a ball bearing fulcrum on a cart as shown in Figure 2.4. The pole's vertical equilibrium point is unstable and it will fall down due to the smallest of disturbances. Therefore, the task is to move the cart left and right in such a way as to keep the pole stabilized at its vertical position, much like a jongleur on a country fair who keeps a sword balanced at the tip of his finger by moving his hand back and forth. How could one achieve this target?

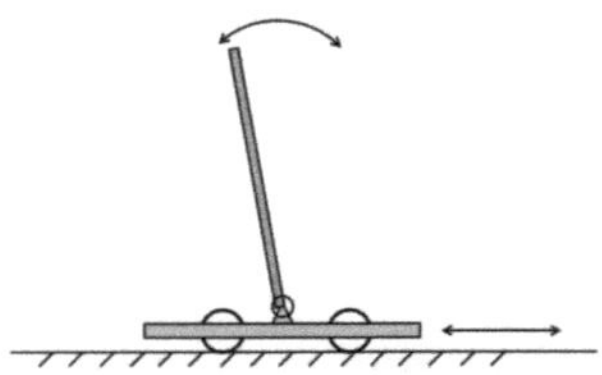

Fig. 2.4 How to move the cart left and right such that the pole on the cart stays vertical?

The engineers among the readers will immediately recognize this as a control problem and they will want to apply various techniques they have learned at control engineering classes. They will also remember that a mathematical model of the "plant" they should control is required. Therefore, they will begin to solve the problem by building a mathematical model. To build the model, they will firstly analyze the forces involved. A diagram as in Figure 2.5 is very helpful here. Summing the forces in the diagram of the cart in the horizontal direction gives

$$M\ddot{x} + k\dot{x} + N = F \tag{2.7}$$

with M as the mass of the cart and k as the friction coefficient. Summing the forces in the diagram of the pole in the horizontal direction gives

$$N = m\ddot{x} + ml\ddot{\theta}cos\theta - ml\dot{\theta}^2 sin\theta \tag{2.8}$$

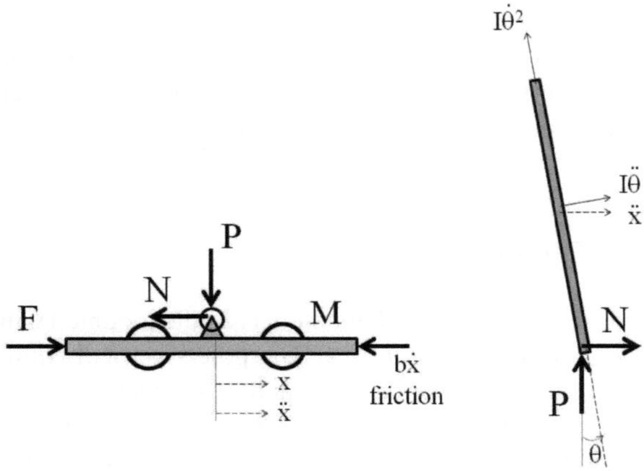

Fig. 2.5 Analyzing the forces in the cart-pole system.

with m as the mass of the pole, l as the length of the pole and θ as the angle of the pole from the vertical. Substituting (2.8) in (2.7) yields the first equation of motion for the system:

$$F = (M+m)\ddot{x} + k\dot{x} + ml\ddot{\theta}cos\theta - ml\dot{\theta}^2 sin\theta \tag{2.9}$$

We now sum up the forces acting on the longitudinal axis of the pole:

$$Psin\theta + Ncos\theta - mgsin\theta = ml\ddot{\theta} + m\ddot{x}cos\theta \tag{2.10}$$

We also need to sum the moments around the center of gravity of the pole to get rid of the terms N and P:

$$-Plsin\theta - Nlcos\theta = I\dot{\theta}^2 \tag{2.11}$$

with I as the moment of inertia of the pole. Combining (2.11) with (2.10) we get the second equation of motion as

$$(I+ml^2)\dot{\theta}^2 + mglsin\theta = -ml\ddot{x}cos\theta \tag{2.12}$$

At this stage we make the assumption that the pole does not deviate too much from the vertical and linearize about $\theta = \pi$. With $\theta = \pi + \phi$ and ϕ small we get $cos\theta \approx -1$, $sin\theta \approx -\phi$ and $(\frac{d\theta}{dt})^2 \approx 0$ we get the linearized equations of motion with u as the control input for the motion of the cart

$$(I+ml^2)\ddot{\phi} - mgl\phi = ml\ddot{x} \tag{2.13}$$

$$u = (M+m)\ddot{x} + k\dot{x} - ml\ddot{\phi} \tag{2.14}$$

We can stop now with these equations. We have modeled the system with a linear 4th order ordinary differential equation. The rest is a straight forward exercise in control engineering which can be solved using, for instance, MATLAB. □

2.3.2 *Stock Prices*

An interesting problem that arises in financial engineering is how to determine the relative weights of stocks in a diversified portfolio [210]. This requires estimates of future stock prices. Most of the time, these estimates are based on past prices of the stocks. In an illustrative example, imagine that the closing prices of a stock during the last 24 months is as shown in Table 2.1.

Table 2.1 Closing prices of a stock during 24 months.

Month	Closing price	Month	Closing price
1	32.12	13	39.85
2	32.12	14	35.21
3	32.45	15	37.11
4	32.48	16	38.00
5	30.37	17	39.25
6	33.61	18	44.52
7	31.03	19	43.01
8	31.56	20	49.05
9	35.75	21	50.07
10	32.39	22	55.22
11	33.19	23	54.82
12	33.29	24	56.46

In order to be able to estimate the closing prices of this stock for the next few months we can naively model the price as a linear function of time and find the equation of the line that best fits the data of the past 24 months. A simple linear regression analysis based on Gaussian least squares method results in

$$p(t) = 1.0596t + 25.635 \tag{2.15}$$

with p as the stock price in \$ and t as the time in months.

Plotting the equation (2.15) together with the past data is shown in Figure 2.6.

Equation (2.15) is a very rough model and the closeness of its fit to available data measured by Pearson product-moment correlation coefficient is $r^2 = 0.7935$ (the closer r^2 to 1, the better the fit [269]). When we use (2.15) to estimate the closing prices of this stock for months 25 – 28 (in other words, when we extrapolate) we get the values in Table 2.2.

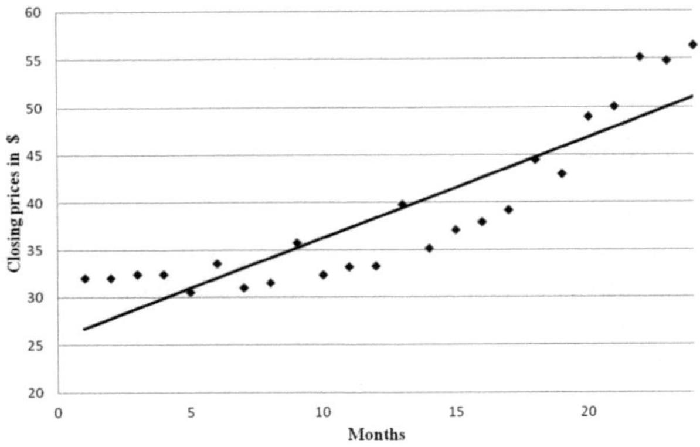

Fig. 2.6 Closing prices of a stock during 24 months and their regression line given by equation (2.15).

Table 2.2 Closing prices estimated using (2.15).

Month	Closing price
25	52.13
26	53.18
27	54.24
28	55.30

Observing the general tendency of the prices in Figure 2.6 to increase faster with time, we can hope to get a better fit of the data by using an exponential rather than a linear model. The regression analysis results in

$$p(t) = 27.593e^{0.0258t} \tag{2.16}$$

Plotting the equation (2.16) together with the past data is shown in Figure 2.7.

Just by looking at their respective graphs, equation (2.16) seems to be a better model than equation (2.15). Also, here $r^2 = 0.8226$. When we now use equation (2.16) to estimate the closing prices of this stock for months 25 – 28, we get the values in Table 2.3.

Can we improve on these results? We can try to model the stock prices with a polynomial to get a better fit. Indeed, for a sixth order polynomial we get the best curve fit with the equation

$$p(t) = -0.000008t^6 + 0.0006t^5 - 0.0149t^4 + 0.1829t^3 - 1.0506t^2 + 2.489t + 30.408 \tag{2.17}$$

The fit is impressive with $r^2 = 0.9666$. Since the fit is better, we expect to get better estimates for the future values of the stocks. We now use (2.17) to estimate the closing prices of this stock for months 25 – 28 and get the values in Table 2.4.

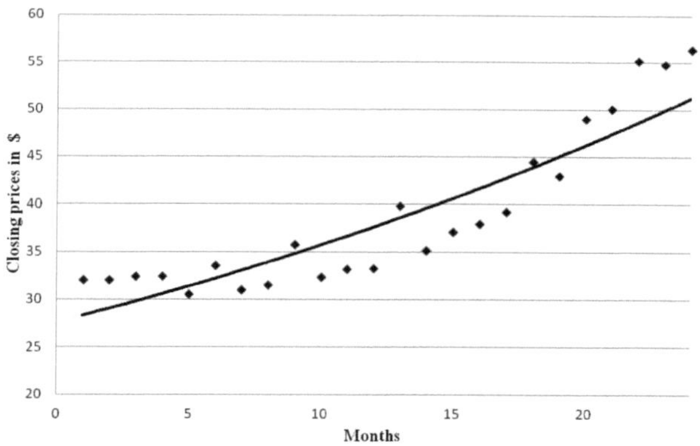

Fig. 2.7 Closing prices of a stock during 24 months and their regression line given by equation (2.16).

Table 2.3 Closing prices estimated using equation (2.16) are very similar to the result of the previous estimation.

Month	Closing price
25	52.59
26	53.97
27	55.38
28	56.82

Table 2.4 Closing prices estimated using equation (2.17) look very different from the previous estimations.

Month	Closing price
25	53.20
26	45.36
27	30.33
28	5.21
29	-33.41

Plotting the equation (2.17) together with the past data is shown in Figure 2.8.

Clearly, the model in equation (2.17) is not an appropriate model for forecasting future stock prices. Actually, one can see that there is always at least one $n-1$ or higher degree polynomial that goes through n points. This means that one can always find a polynomial model that fits the given data points perfectly ($r^2 = 1$ or zero residuals). However, such models are seldom useful for interpolating and hardly ever useful for extrapolating. Actually, the stock prices are successfully modeled as so called log-normal Brownian motions with impressive success [7]. □

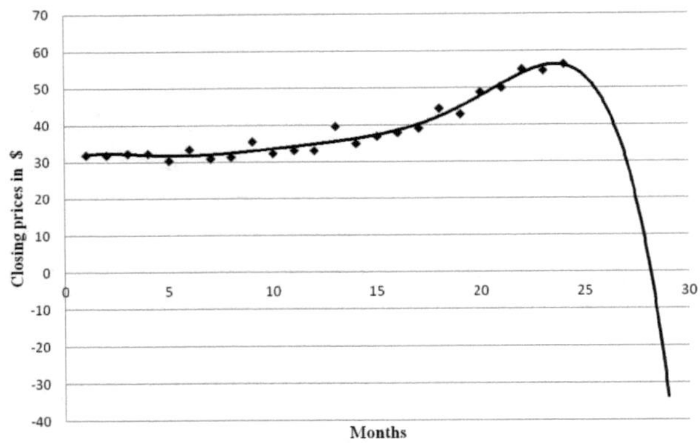

Fig. 2.8 The model given by equation (2.16) fits available data very well. However, this model results in forecasts with negative values for the stock and is therefore, completely useless for making predictions.

2.3.3 Lessons Learned

We have seen two modeling problems which we attacked with two very different approaches. In the first example, we have applied Newton's laws of motion to the problem, deriving the differential equations governing the physical system in a fairly straight forward manner. This approach can be generalized to problems concerning the modeling of physical systems. We can analyze the given system using known physical laws like continuity or conservation of energy and obtain equations pertaining to the behavior of the system. Whenever possible, this is the preferred approach to modeling, because it gives us insight into the workings of the system. This way we can "understand" what is actually going on and we can use the model to make fairly reliable predictions. The parameters have physical meanings like the mass of a body or the diameter of a pipe which can either be measured or estimated.

In the second example we have approached the relationship between the input of the system (time) and its output (closing price of the stock) as a black-box without any knowledge of its inner workings. This approach does not give any insights and is really an act of desperation. The modeling problem is quickly reduced to a curve fitting problem which can be solved mechanically. For instance, empirical evidence gained over centuries in the stock markets indicates the use of an exponential model [8]. Neither in this case nor in the case of polynomial models, the parameters of the model have any physical meaning. Also, as should be apparent by now, a good curve fit does not necessarily imply a good model which can be used to make predictions.

We shall see many examples of both types of modeling in the following Chapters.

While modeling, either way, one always has to keep in mind the underlying assumptions. In the first example we have linearized the equations of motion close to the vertical position of the pole. That means the model describes the behavior of the system for small deviations of the pole from the vertical in a fairly accurate way. On the other hand, as the pole moves away from its vertical position, the predictive power of the model deteriorates rapidly.

A common error in modeling and while using models is to forget the underlying assumptions and not be aware of the limitations or the applicability range of the model and try to use it to make predictions. One humorous case in point is the marketing model used by a fast growing toy company some years ago which predicted an increase of 1% in their market share with every new shop they opened. A facetious executive made ambitious growth plans to open 110 shops, because he was not informed (!) about the underlying assumptions and the limitations of the model...

2.4 Parameter Identification

When physical laws are used to model a system, the parameters of the model (*e.g.*, the length of the pole) are often readily measurable. If not directly measurable, they can often be calculated with sufficient precision (*e.g.*, the mass of the pole can be calculated by first measuring its dimensions, using these dimensions to calculate its volume and multiplying the volume with the density of the material used to produce the pole). However, one might still end up with parameters that are neither directly measurable nor calculable (*e.g.*, the friction coefficient). Such parameters must be determined experimentally. The process of determining those model parameters is called identification. Actually, while using black box models, identification is often the only way to determine the model parameters.

In order to be able to perform a reasonable parameter identification, a sufficiently large number of observations are necessary. Let us consider again the second example above. Imagine that the data available is limited to the first five data points. The result of a linear regression analysis on the available data is shown in Figure 2.9. A comparison with Figure 2.6 shows a radically different course of the regression line. Whereas the regression based on five data points predicts falling stock prices, a similar regression analysis using 24 data points predicts a rising course of the stock prices. This example demonstrates the necessity of using a sufficiently large number of data points for being able to perform parameter identification. In most practical cases the consequence of using a small number of data points is not as dramatic as in this example. Rather, assuming that the model structure is correct, increasing the number of data points will increase the accuracy of the estimates of the parameters. Most widely used identification methods make use of linear or non-linear least squares methods. For instance [86] gives a detailed analysis of this problem.

How can one identify the parameters of a model when there are no or insufficient data for identification? The solution lies in conducting experiments to obtain sufficient data. Those experiments generally consist of exciting the system with a sufficiently rich set of inputs and observing its output. For that strategy to be successful, the model must be identifiable. A model is said to be identifiable if it is theoretically possible to infer the true value of the models parameters after obtaining an infinite number of observations from it. In other words there must be a unique solution to

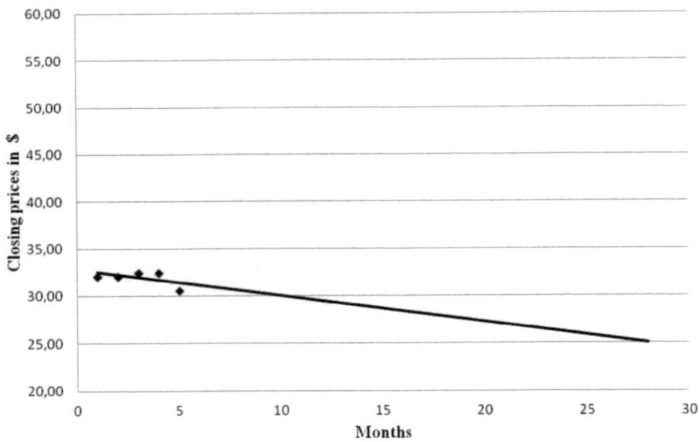

Fig. 2.9 Regression line based on the first five data points predicts falling stock prices.

the problem of finding the mapping how the input is mapped to the output of the model (see, for instance, [192] for a formal definition of identifiability).

If a system is modeled as a linear system, in an experiment to identify the parameters of the model, it is sufficient to excite the system with an impulse or a step function and to observe its output at different times [2]. Since measurements cost time and money, one cannot make an infinite number of measurements. Therefore, one has to agree on a number of measurements. The next question to be resolved is when to conduct those measurements. Clearly, it does not help to make meany measurements once the transient behavior is over. Last two points make sense intuitively. The general problem of experiment design is more complicated and treated in detail in *e.g.*, [18], [259] and [105].

2.5 Model Validation

Model validation is the process of checking whether a model is appropriate for the intended purpose. Remembering that a model is, by its very nature, an inexact approximation of reality, it becomes clear that it cannot replicate all the features of the actual system. Therefore, to validate a model means to verify that it can reproduce the features that are essential for the purpose for which the model was developed. One has to keep in mind that model validation is an integral part of the modeling process. In other words, one has to verify the model's appropriateness both during the modeling process and at the end of it. This is done primarily by simulating the system behavior by solving the model equations with the identified parameters. In most cases, the model consists of a set of differential equations which must be

[2] A unit impulse function contains all frequencies, because its Laplace transform is simply 1. Therefore, it is a sufficiently rich input for identification purposes.

integrated numerically. Many techniques and algorithms and even dedicated simulation languages are available [70], [71].

Another important requirement of a model is its parsimony. It is quite possible to develop models that describe the same phenomena appropriately but with varying degrees of complexity. Among such models, the one with the lowest order or smallest number of parameters, the most parsimonious one, is to be preferred. This principle is a direct application of Occam's razor [3].

A common problem in connection with modeling and particularly model validation is model reduction. In many applications, the available model of a system, although appropriate for most purposes of modeling, might be too complex for controller design purposes. It then makes sense to reduce the order of the model. There are several ways of achieving model reduction. Some are obvious, straight forward and simple. Others are more sophisticated and can be rather tedious to perform (*e.g.*, balanced model reduction [128]).

2.6 Summary

We began this Chapter by elucidating what we mean by a model. A model is always an approximation of reality. Therefore, there is no one "true model". Depending on the context and purpose, there are models which are better than others. Using the example of a falling object we have seen that depending on the context, it might make sense to neglect certain aspects of reality in the modeling process. After all, modeling is an abstraction of reality.

Although there are many different types of models, we shall focus on mathematical models in the coming Chapters. Mathematical models can be classified under several dichotomic headings: inductive versus deductive models, deterministic versus stochastic models, static versus dynamic models, discrete versus continuous models, lumped parameter models versus distributed parameter models and linear models versus non-linear models. Nature is inherently dynamic, stochastic, continuous, and non-linear with distributed parameters, hence difficult to model properly. Therefore, simplifying assumptions are necessary to make the model more amenable to available mathematical techniques. We have also seen that there are four different types of models depending on their purposes which can be categorized as descriptive, interpretive, predictive or explanatory models.

The modeling process is best summarized using Figure 2.10. Firstly, the model structure for the unknown system is determined analyzing the input-output data of the unknown system. Secondly, an experiment is designed to obtain further data which is then used to identify the unknown parameters in the system model. Finally, in the validation phase, the model's appropriateness is examined by comparing the available data with the results obtained from simulations of the system with the

---

[3] Occam's razor is the law of parsimony, economy or succinctness. It is a principle urging one to select from among competing hypotheses the one which makes the fewest assumptions. Although the principle was known earlier, Occam's razor is attributed to the 14th-century English Franciscan friar William of Ockham.

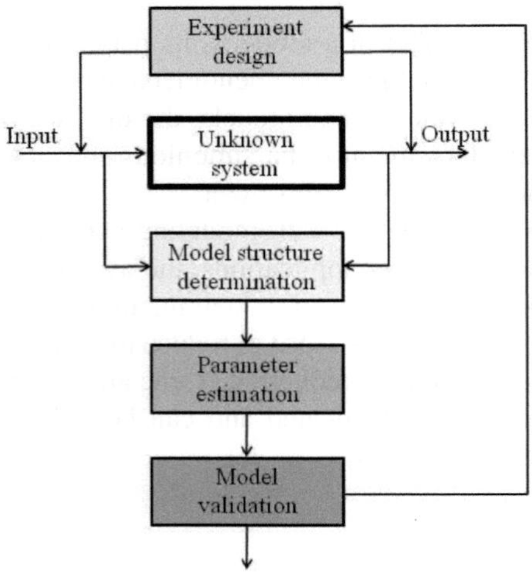

Fig. 2.10 Complete modeling process as adapted from [79]. Darker shades indicate more advanced phases of modeling.

parameters identified in the previous phase. Further experiments might be necessary to obtain additional data to improve parameter identification if the results are not satisfactory. Of course, if continued experiments do not yield a parameter set which results in a satisfactory model, it might be necessary to modify the model structure.

An important characteristic of physiological systems are their complexity. Any model of a physiological system must both take the complexity of the system into account and the availability of data for estimating the parameters of the model. A difficulty which arises in modeling physiological systems is that, due to ethical and practical considerations, we cannot make as many experiments with living organisms as we would like for modeling or parameter identification purposes. A simplified conceptual model of the human body is shown in Figure 2.11. This model barely begins to give an inkling of how complex the human body actually is. Although simplified, it reflects many important dynamic processes and control mechanisms present in the human body. The multi-input-multi-output (MIMO) model includes inputs like information entering the system through sensory organs as well as learned or inborn traits like moral constraints and survival instinct. It includes a large number of subsystems like the brain or the cardiovascular system which in turn are further divided into simpler subsystems. The interconnections and cross couplings among the many systems and subsystems further complicate the modeling efforts. One cannot simply ignore those interconnections, because, as in the example of billions and billions of connections among the brain cells, it is their very existence which causes the functionality of the brain [211].

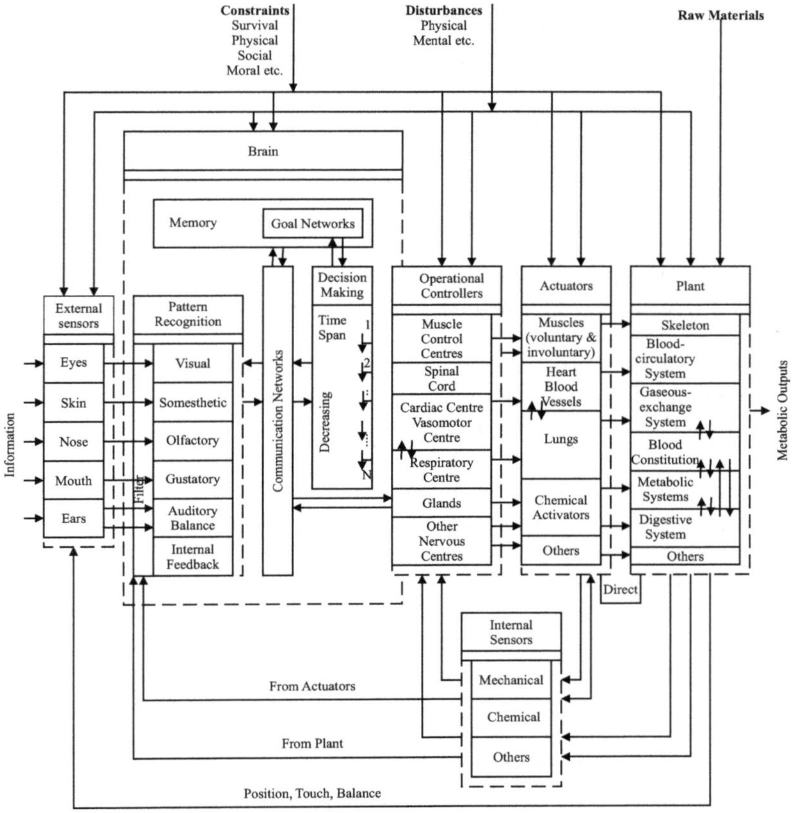

Fig. 2.11 A simplified conceptual model of the human body (printed with permission from [156]).

2.7 Exercises

A1: Explain in your own words what you understand when you hear the word "model".

A2: Explain how you would go about modeling an electro-mechanical system.

A3: Why is it difficult to develop a model to predict stock prices?

A4: How would you proceed to model a physiological system?

B: (*The problems **B1–B7** are based on the premise described here and should be solved successively.*) When viral particles of a certain virus enter the human body, they replicate rapidly. In about four hours, the number of viral particles has doubled. The immune system does not respond until there are about 1 million viral particles in the body. The first response of the immune system is fever. The rise in temperature lowers the rate at which the viral particles replicate to 160% every four hours, but the immune system can only eliminate these particular viral particles at the rate of about 50'000 viral particles per hour. Often people do not seek medical attention immediately as they think they have a common cold. If the number of viral particles, however, reaches 10^{12} the person dies.

B1: Model the initial phase of the illness for a person infected with 10'000 viral particles to determine how long it will take for the body's immune response to begin.

B2: Develop a model for the next phase of the illness, when the immune response has begun but no medications have yet been administered. Use the model to determine how long it will be before the patient dies if the infection is left untreated.

B3: An antiviral medication can be administered as soon as a person seeks medical attention. The medication does not affect the growth rate of the viruses but together with the immune response can eliminate 1.2 million viral particles per hour. If the person is to make a full recovery, explain why effective medication must be administered before the number of viral particles reaches 9 to 10 million.

B4: The antiviral medication is difficult for the body to adapt to, so it must initially be carefully introduced to the body over a four hour time period of continuous intravenous dosing. This means the same amount of medication is entering the body at any given time during the first four hours. At the same time, however, the kidneys eliminate about 2.5% of this medication per hour. The physician has calculated that the patient needs at least 90 micrograms of medication to begin and maintain the rate of elimination of 1.2 million viral particles. Create a mathematical model for this four hour period so that by the end of the four hour period the patient has 90 micrograms of medication in her body. Find the solution to your model analytically, or estimate its solution numerically.

B5: One the level of medication has reached 90 micrograms the patient is taken off the intravenous phase and is given and given injections every four hours. The kidneys will still be working to eliminate the medication, so the physician must calculate the additional dosage D accordingly. Dosage D should allow for maintenance of a minimum of 90 micrograms within the patient's bloodstream throughout the treatment regimen. What dosage, D, administered every four hours from the end of the first continual intravenous phase, would allow for the patient to maintain at least 90 micrograms of the medication in his system? Make sure you take into account the kidneys' rate of elimination. Explain carefully how you came to this number.

B6: Determine the last possible time from the onset of infection to start the regimen of medication. How long will it take to clear the viral particles from the patient's system? Show on a graph the entire treatment regimen from the time the treatment begins until the viral particles are eliminated.

B7: Analyze all your models discussing any assumptions you made, the strengths and weaknesses of the models and the reliability of your results.

C1: A meteor approaching the earth is detected at a great distance and velocity. You are given the task to determine the time if and when it will hit the earth. Clearly, you cannot say that it is in free fall. For one, the gravitational constant g changes with distance. The Sun, the other planets and even the Moon also exert gravitational forces on the meteor. Develop the differential equations that govern the movement of the meteor. Note explicitly the simplifying assumptions you are making. Drop these assumptions one by one. How do the non-linearities present in your model affect the outcome? How confident are you in your calculations?

C2: Historically, some concepts like submarines or space travel were first introduced in science fiction literature and became science fact with time. Others, like time travel or Startrek style "beaming" have not (yet?) made that transition. Consider Hari Seldon's "psychohistory". Do you think it might one day become a fact? Argue your case using concepts like chaos and stochastic models.

C3: Study your own circadian rhythm and develop a mathematical model for it. How would you proceed? How long are your measurements going to take? How do you propose to validate your model?

Chapter 3
The Human Operator

To invent an airplane is nothing. To build one is something. To fly is everything.

— Otto Lilienthal

The natural function of the wing is to soar upwards and carry that which is heavy up to the place where dwells the race of gods. More than any other thing that pertains to the body it partakes of the nature of the divine.

— Plato, Phaedrus

3.1 Introduction

In manual control systems as found, for instance, in vehicle control, process control, manipulator control and teleoperations, the human operator functions as a controller. The controller is often characterized by its transfer function $G_c(s)$ which determines the control variable $u(t)$ acting on the process that is to be controlled. The process is characterized by its transfer function $G(s)$ in the linear case. The aim is for the process or plant output $y(t)$ to mimic the reference input $r(t)$ as closely as possible in spite of the disturbance $v(t)$ acting upon the process (see Figure 3.1).

In some cases there might be several control variables, several reference inputs and several outputs that may be coupled with each other. A good example of such a multivariable system is an aircraft. If one reference input, $r_1(t)$ is a heading different from the current heading, $y_1(t)$ and another reference input $r_2(t)$ is the same as the current altitude $y_2(t)$, the pilot activates the ailerons, $u_1(t)$ to change the lift and drag characteristics of each wing, thus causing the aeroplane to bank and to turn in the direction of the bank. However, the bank reduces the effective surface area of the wings, thus reducing the total lift. Therefore, the pilot must also activate the elevators, $u_2(t)$ in order to keep the aircraft's altitude constant. It might even be necessary to increase engine power, $u_3(t)$ in steep turns to compensate for the loss of lift (see Box for definitions and a detailed description of flight dynamics).

S.S. Hacısalihzade: *Biomedical Applications of Control Engineering*, LNCIS 441, pp. 45–59.
DOI: 10.1007/978-3-642-37279-7_3

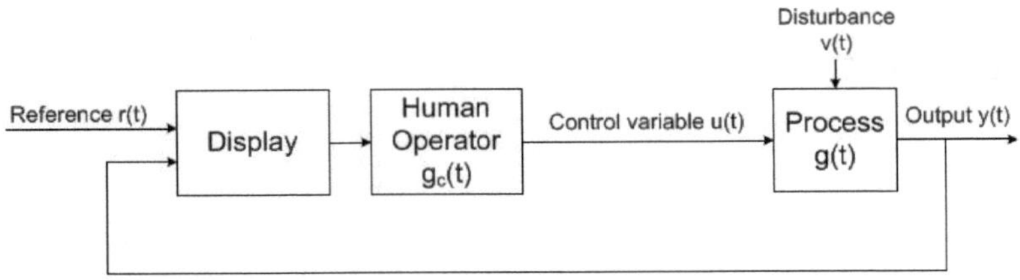

Fig. 3.1 The block diagram of a manual control system.

Some typical examples of process control situations involving human operators and their transfer functions are listed below [280].

Aiming a Garden Hose or Steering a Lawn Mower: When you aim the water flowing out of a garden hose at a point on the ground by moving the tip of the hose with your hand, assuming that you don't manipulate the tap to which the hose is attached, the water will keep hitting the earth at that point as long as you don't move your hand. Hence, there is a static relationship between your control variable (the position of your hand) and the process output (the point where the water hits the earth). Such a system can be adequately modeled with the transfer function $G(s) = K$.

Controlling an Aircraft's Pitch Angle by Its Elevator or a Car's Direction Control: When the pilot pulls the control wheel of the aircraft towards herself and keep it at a constant position, the nose of the aircraft will start to move up, gradually increasing the pitch angle (this is, of course, true only in a limited range). Hence, there is a dynamic relationship between the control variable (the position of the control wheel) and the process output (the pitch angle) where the control variable affects the rate of change of the process output. Similarly, when the driver of a car moves the driving wheel to a position other than the neutral position and keeps it there, the heading of the car gradually changes. In other words, there is again a dynamic relationship between the control variable (the position of the wheel) and the process output (the direction of the car) where the control variable affects the rate of change of the process output. In both cases the systems can be modeled with the transfer function $G(s) = K/s$.

Controlling an Aircraft's Heading by Its Ailerons: When the pilot turns the control wheel of the aircraft to a position left of the neutral position and keeps it there, this causes the aileron at the left wing to move up and the aileron at the right wing to move down. At this constellation of the ailerons, the left wing has an increased drag and the right wing an increased lift. As a consequence, the left wing keeps going down and the right wing keeps going up. In other words, the banking of the aeroplane to the left keeps increasing and it keeps turning to the left at an ever increasing rate. In other words, there is again a dynamic relationship between the control variable (the left-right position of the control wheel) and the process output (the heading of the aircraft) where the control variable affects the rate of change of the rate of change of the process output. In this case the system can be modeled with the transfer function $G(s) = K/s^2$. What does the pilot have to do if she wants to turn the aircraft to a new heading at a constant rate? She first turns the control wheel in the direction of the turn and once the aircraft reaches the desired rate of turn she briskly turns the control wheel at its neutral position and keeps it there. With the control wheel and consequently the ailerons at their neutral positions and the aircraft in a bank and gradually turning towards the desired heading, she waits until the desired new heading is reached. She then moves the control wheel opposite to the direction of the turn until the aircraft comes out of the bank and the aircraft is again flying straight. Finally, she finishes off the manoeuvre by moving the control wheel back to its neutral position.

Why does an aeroplane fly?

There are four primary forces that act on an aeroplane in flight: thrust, weight, drag and lift. It is the interplay between these four forces that result in an aeroplane's motion (see Figure 3.2 for the definition of these forces).

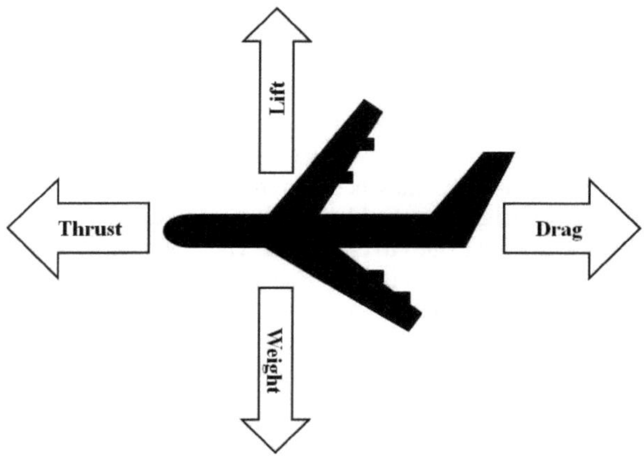

Fig. 3.2 The four forces acting on an aircraft.

The **weight** force is caused by the mass of the aircraft accelerating towards the center of the earth due to the gravitational pull of the earth.

In order for an aeroplane to fly, a force must be generated that is stronger than the weight force. That force is called lift. The **lift** force is generated by air flowing over an object. The direction of the lift force will always be perpendicular to the direction of the air that is flowing. As an aeroplane flies, air flows over its wings, from the front (or leading edge) to the back (or trailing edge). This generates a lift force perpendicular to the direction of the airflow.

How is lift generated? The airflow around a curved wing results in the air above the wing to flow faster than the air below the wing. According to the Bernoulli principle, the pressure the air generates on the wing where it flows faster is less than where it flows slower. This means that the pressure below the wing is higher than the pressure above the wing. When the pressure differential is sufficiently large, sufficient lift is generated to counteract the weight of the aircraft and the aircraft flies.

When an aeroplane is on the ground not moving, there is not enough air flowing around it to create lift. Another force is needed to get the aeroplane moving through the air, so that the airflow can do its job of creating lift. This force is called thrust. **Thrust** propels an object in a particular direction. A jet engine (or a propeller) generates thrust and, because it is attached to the aeroplane, its thrust will be applied to the aeroplane. So, as the engines thrust the aeroplane in the direction that they are pointed, air flows over and under the wings which creates the lift force. If enough lift is generated, the aeroplane will fly.

The fourth primary force is drag. **Drag** is the force that resists any object trying to move through a fluid. The drag on an aeroplane is the result of, among other things, the energy needed to move the air out of the way of the aeroplane. Any motion or movement by the aeroplane will always be resisted by a drag force. The direction of the drag force is opposite to the direction of flight. The thrust force is aligned to counter the drag force.

Flying an aircraft by a pilot is thus a balancing act among these four primary forces. If no net force acts upon the aircraft during flight it is said to be in straight and level flight. The manoeuvres executed by the aircraft in air are the result of the pilot's reconfiguring these primary forces through the controlled movement of the control surfaces and engine power. This is easier said than done and requires considerable skill attained through practice.

A three-dimensional coordinate system is used for all references to motion and direction of an aircraft. As Figure 3.3 shows, the center of the coordinate system is found at the centroid of the aircraft – roughly between the wings in the middle of the fuselage (the centroid is a fictive point around which all rotation takes place). The three axes intersect at the centroid at right angles to each other. The longitudinal axis runs the length of the aircraft from the nose through the tail. The lateral axis runs across the wings from tip to tip. The vertical axis runs from the ceiling of the aeroplane to its floor. These axes extend indefinitely from the center of the aircraft. Movement along or around one axis does not necessarily involve any movement on or around the other two.

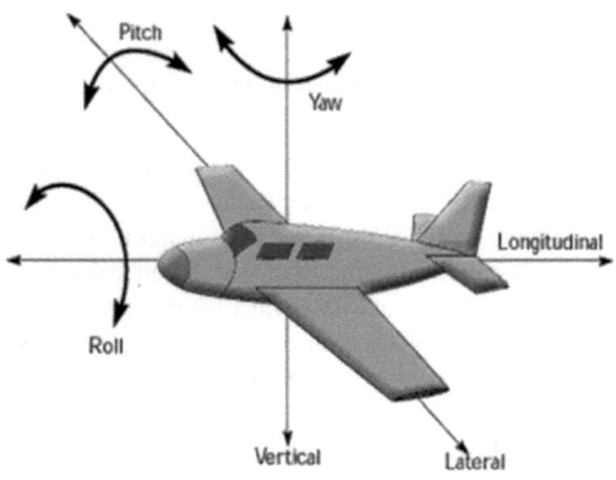

Fig. 3.3 Movements of an aircraft along its principal axes.

During flight, an aeroplane can have six motions along and around the three axes. Three of the movements are linear: front and back along the longitudinal axis; side to side along the lateral axis; up and down along the vertical axis. The other three movements are rotational: movement around the longitudinal axis, called **roll**; movement around the lateral axis, called **pitch**; movement around the vertical axis, called **yaw**.

An aeroplane has basically three types of control surfaces: ailerons, elevators and a rudder. Within the cockpit, two controls operate the control surfaces. The control wheel controls the ailerons and elevators. The rudder pedals control the rudder.

The **ailerons** are flap-like structures on the trailing edge of the wings – one on each side. When the pilot turns the control wheel to the right, the right aileron tilts up and the left aileron tilts down. This will cause the aircraft to roll to the right. When the pilot turns the control wheel to the left, the left aileron tilts up, the right aileron tilts down and the aeroplane rolls to the left. This happens because as the aileron tilts downward more lift is created and the wing rises. As it tilts upward, less lift will be created and the wing will descend. If the wing on one side of the aeroplane rises and the other descends, the aeroplane will roll towards the side of the decrease in lift (see Figure 3.4).

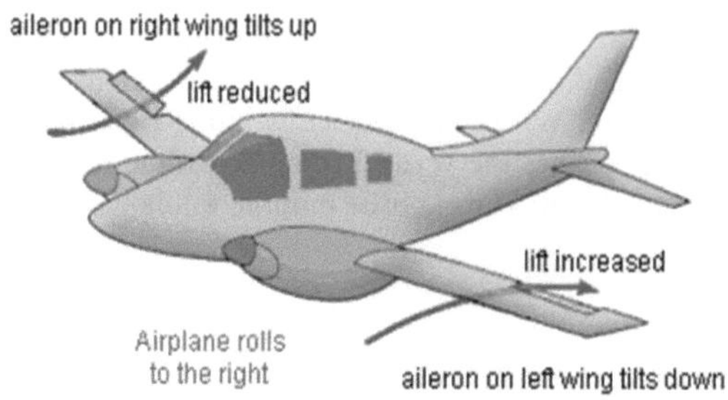

Fig. 3.4 Movements of the ailerons change the configuration of the wings and consequently their lifts. Different lifts on each wing cause the aircraft to roll.

The **elevators** are also flap-like structures that are mounted on each side of the horizontal stabilizer. When the pilot pushes the control wheel forward, the elevators tilt downward. This causes the tail of the aeroplane to rise and the fuselage to tilt down – this is called pitching down. As Figure 3.5 shows, when the pilot pulls the control wheel back, the elevators tilt upward, the tail goes down and the fuselage pitches nose-up. When the elevator tilts downward more lift is created (like the ailerons) and the tail rises. When the elevator tilts upward, less lift is created and the tail descends.

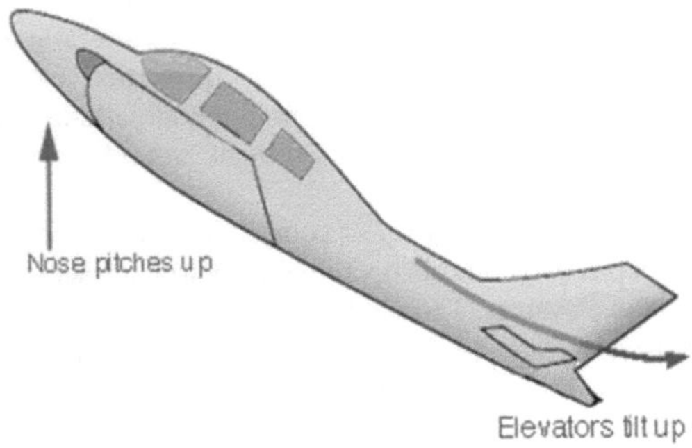

Fig. 3.5 Movements of the elevators change the configuration of the stabilizers and consequently their lifts. Elevators tilting up decrease their lift. This causes the tail to move downwards with respect to the fuselage and causes the nose to pitch up.

The **rudder** is located on the fin. The two rudder pedals are located at the pilot's feet. When the pilot pushes on the right rudder pedal, the rudder tilts to the right and the aeroplane yaws "nose-right." As Figure 3.6 shows, when the pilot pushes on the left rudder pedal, the rudder tilts to the left and the aeroplane yaws "nose-left." Again this is due to lift. However, the direction of this lift force is different than the lift force that causes the aeroplane to rise. When the rudder tilts to the right, more lift is created on the right, which"lifts" or pushes the vertical stabilizer to the left. This, in turn, causes the aeroplane to yaw nose-right. The opposite motion occurs when the rudder tilts to the left.

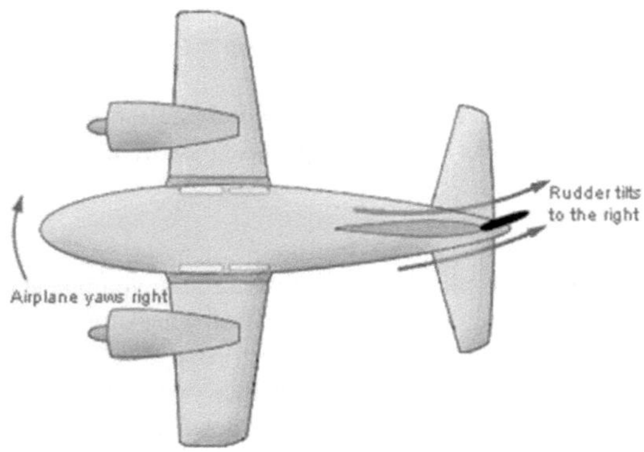

Fig. 3.6 Movements of the rudder changes the configuration of the fin and causes the aircraft to yaw.

The information in this Box is supplied by NASA Ames Research Center for educational purposes.

3.2 Types of Manual Control Systems

Basically, one can differentiate between four distinctly different classes of closed-loop manual control systems:

In a **compensatory** system (see Figure 3.7) the human operator receives the input $e(t)$ which is the error between the actual system response $y(t)$ and the reference input $r(t)$ without being privy to either of these two. This corresponds to the classical servomechanism. The task of the human operator is to keep the error as close as possible to zero without reference to the instantaneous values of the system response and the reference signal. An example of a compensatory tracking task is flying an aeroplane in straight and level flight with reference to the "artificial horizon" instrument alone, which shows the pilot the aeroplane's pitch and roll angles but not the heading or the altitude.

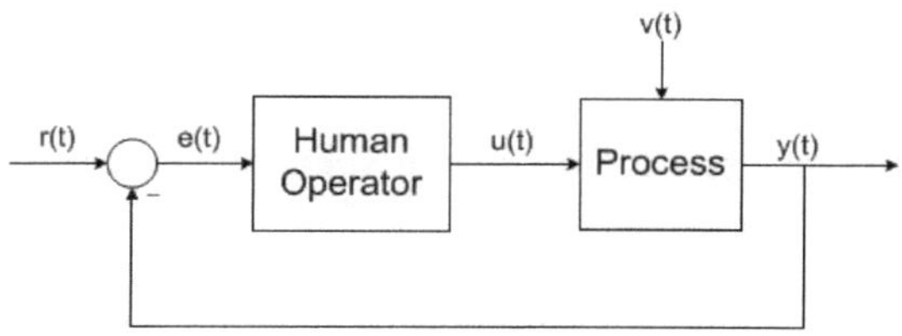

Fig. 3.7 Block diagram of a compensatory system (after [280]).

In a **pursuit** system (see Figure 3.8) the momentary values of both the reference input $r(t)$ and the system response $y(t)$ are available to the human operator independent of each other. The task of the operator is to make the system response follow the reference input as closely as possible. An example of a pursuit tracking task is driving a car and adjusting its speed to the maximum allowed speed as indicated by the road signs on a stretch of the road.

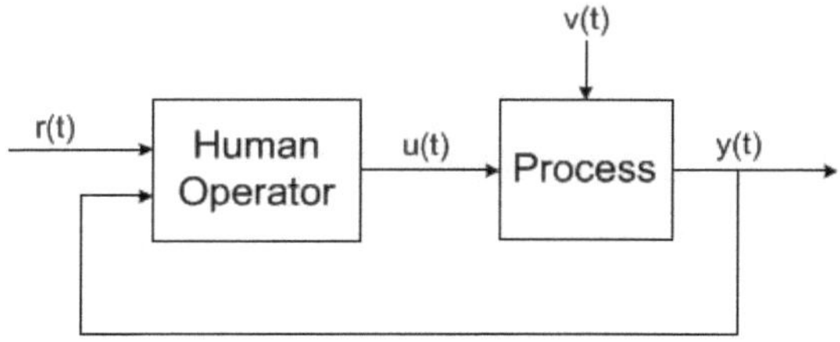

Fig. 3.8 Block diagram of a pursuit system (after [280]).

A **preview** system (see Figure 3.9) is very similar to a pursuit system with the difference that the human operator knows the future values of the reference input, up to a certain point in time, in addition to its instantaneous value. An example of a preview tracking task is driving a car and adjusting its speed to the maximum allowed speed as indicated by the road signs on a stretch of the road. The difference to pursuit tracking arises from additional signs which indicate that the speed limit is going to change at a specified distance.

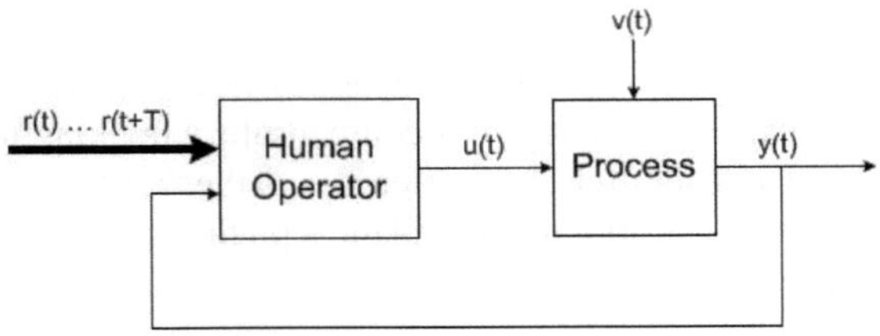

Fig. 3.9 Block diagram of a preview system (after [280]).

A **precognitive** system (see Figure 3.10) is similar to a preview system with the difference that the human operator has a certain but not complete foreknowledge of the reference input other than by visual and direct means. This knowledge can be, for instance, about the amplitude constraints of the reference input or contingency information. An example of a precognitive tracking task is driving a car and adjusting its speed to the maximum allowed speed as indicated by the road signs on a stretch of the road. The difference to pursuit and preview tracking arises from the knowledge that the driver is approaching a town and that the speed limits are usually lower in a town then they are on a freeway.

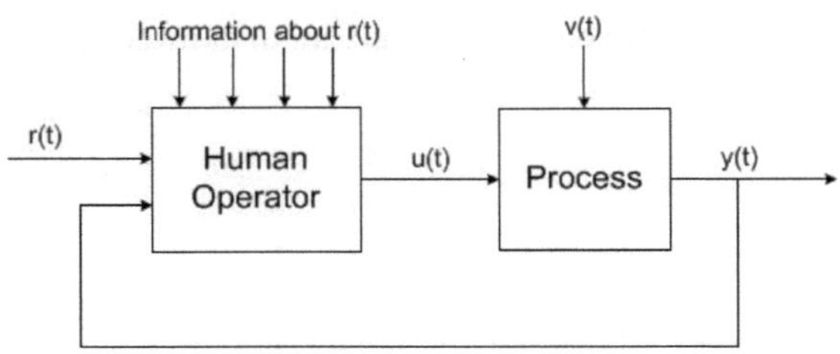

Fig. 3.10 Block diagram of a precognitive system (after [280]).

Since the manual control tasks are executed by the human operator through sensory feedback, two principal types of tracking tasks can be defined:

- The systems depicted in Figures 3.7 – 3.10 are all **closed-loop** systems where the behavior of the process which is to be controlled is fed back to the human operator visually and/or through somatosensory or proprioceptive feed-back.

- One speaks of an **open-loop** system if the process output is not fed back to the human operator. In many cases the absence of visual information about the behavior of the process is enough to interrupt this feedback.

3.3 Human Operator Models

Many models of the human operator depicted in Figure 3.1 have been developed by various researchers in the past with the bulk of the work concentrating during the space race between United States of America and the Soviet Union in the years 1955 – 1975. These models vary considerably in their complexity and according to their intended fields of application.

3.3.1 Second Order Linear Model

One of the most simple but useful descriptions used in representing the human operator is the linear second order time invariant model [219, 220]. The transfer function of this model which is well suited for modeling the compensatory or pursuit tracking behavior of the human operator can be expressed as

$$G(s) = \frac{K(1+T_H s)}{(1+T_1 s)(1+T_2 s)} e^{-sT_L} \tag{3.1}$$

where K is the loop gain adjustable by the operator over a wide range (up to four decades) depending on the units of display and control, T_L is a reaction time delay (0.1s – 0.3s), T_1 is the coefficient of first order neuromuscular lag (0.2s – 0.5s). T_H and T_2 are respectively the lead and lag coefficients which can be adjusted by the human operator to result in dynamic behavior such that the integral of $e^2(t)$ is more or less minimized over the task duration.

The step response of the system given by (3.1) is

$$g(t) = a\cdot K\cdot H(t-T_L) + b\cdot K\cdot H(t-T_L)\cdot e^{\left(-\frac{(t-T_L)}{T_2}\right)} + c\cdot K\cdot H(t-T_L)e^{\left(-\frac{(t-T_L)}{T_1}\right)} \tag{3.2}$$

with

$$a = 1 \qquad b = \frac{(T_H - T_1)}{(T_1+T_2)} - 1 \qquad c = \frac{(T_1 - T_H)}{(T_1+T_2)}$$

$H(.)$ is the Heaviside step function.

Figure 3.11 shows the step response of a human operator (averaged over 30 steps) following a reference signal consisting of unpredictable steps with the movements of his thumb and the simulation of the model in (3.1) with optimized parameters $K = 0.98$, $T_L = 0.36s$, $T_1 = 0.35s$, $T_2 = 0.93s$, and $T_H = 1.13s$ [134]. The reader is referred to Section 5.6 on measuring Parkinsonian symptoms for a more detailed description of the measurement set-up and protocol.

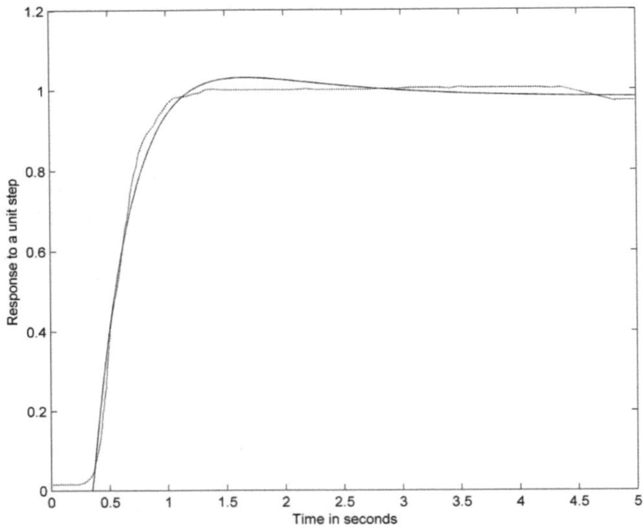

Fig. 3.11 Step response of a human operator. Continuous curve is the modeled and the dashed curve is the measured step response.

3.3.2 Describing Function Model

Another well known human operator model is the so called describing function model (see Figure 3.12). The human operator is modeled as a linear system with output noise. The output noise, called the remnant, can be interpreted as the subject's non-linear behavior and/or their limited observation capabilities and/or their limited motor control abilities and/or their non-stationary behavior mainly due to fatigue. The output of the linear part of the model is linearly correlated with its output but the remnant is uncorrelated with the input. Many researchers have used this model for different purposes [186, 232]. For a broad review of different variations of the describing function method see [221].

3.3.3 Optimal Control Model

A well trained, well motivated human operator acts in a near optimal way while controlling a process if the dynamic characteristics of the system are similar to those of the operator. Figure 3.13 shows the structure of a very successful model of the human operator by Kleinman and colleagues [175].

The human operator is not privy to all the states of the system $\mathbf{x}(t)$ which he has to control. Rather, the operator is shown a linear combination of the system states as the system output $\mathbf{y}(t)$ through a display. The operator perceives this information not as it is shown but distorted by an additive observation noise $\mathbf{v}_o(t)$. The operator cannot

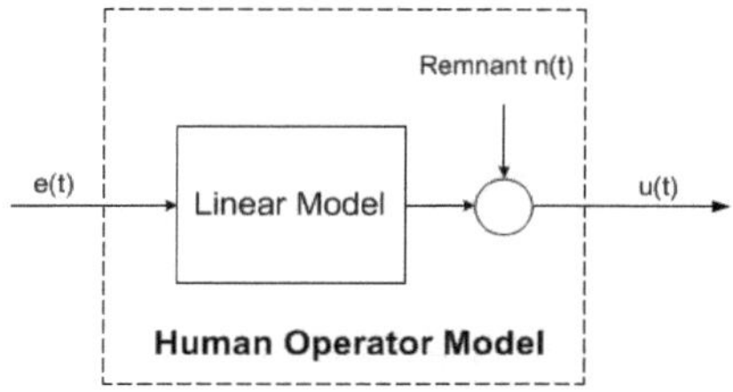

Fig. 3.12 The human describing function model.

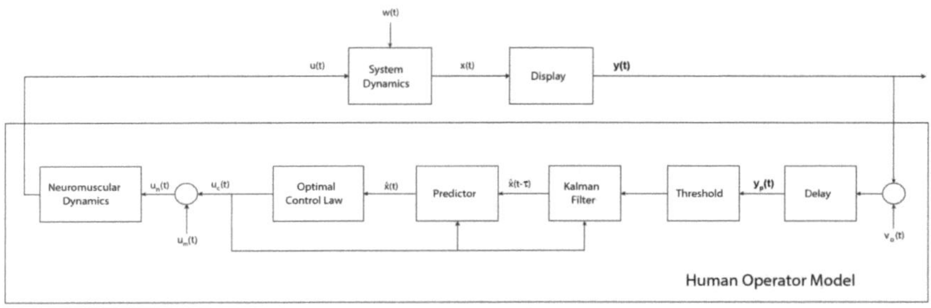

Fig. 3.13 The optimal control model of the human operator adapted from [175]. $u(t)$ is the actual control input, $\mathbf{x}(t)$ is the state of the system to be controlled, $\mathbf{y}(t)$ is the observed output of the system. $\mathbf{y_p}(t)$ is the perceived output. $\mathbf{w}(t)$, $u_m(t)$, $\mathbf{v_o}(t)$ are the system, motor and observation noises respectively. $\hat{\mathbf{x}}(t-\tau)$ is the estimated system state. $\hat{\mathbf{x}}(t)$ is the predicted system state. $u_c(t)$ is the computed control input. $u_n(t)$ is the control input for the neuromuscular dynamics. The threshold is not considered in the model described in the text.

act upon this perceived system output immediately. Rather his action is delayed by τ. Furthermore, the operator cannot act upon very small changes in this perceived system output. Rather he reacts only if the change is above a certain threshold. His processing algorithm can be modeled as a Kalman filter which first computes an estimate of the delayed system state. This estimate is then used to predict the current system state with a predictor. This optimal estimate of the predicted current state is finally used to compute an optimal control input as if the estimate were the actual system state. A motor noise is superimposed on this control input which constitutes the actual input for the neuromuscular (actuator) system with its own dynamic structure. What the system to be controlled receives as the control input is the actual output of the neuromuscular dynamics.

All this can be modeled adequately using a well known optimal control model. Modeling the human operator this way allows the identification of various parameters like gain factors in this model. This, in turn allows to optimize an electronic controller by tuning its parameters to mimic a well trained human operator.

It is assumed that the system to be controlled, perhaps a vehicle, can be adequately described by a linear time invariant equation in the state space by

$$\dot{\mathbf{x}} = \mathbf{A}\mathbf{x}(t) + \mathbf{b}u(t) + \mathbf{E}\mathbf{w}(t)$$

where $\mathbf{x}(t)$ is the n dimensional state vector, $u(t)$ is the human operator's actual control input to the system and $\mathbf{w}(t)$ represents random (zero-mean normalized Gaussian white-noise) external disturbances. The optimal control model of the human operator controlling a time varying system is slightly more complicated [176]. For a discrete time model see, for example, [244].

$$\mathbf{y}(t) = \mathbf{C}\mathbf{x}(t) + \mathbf{d}u(t)$$

denotes the system outputs which are displayed to the operator (not necessarily visually). The human operator can extract the rate of change of a displayed variable but not its higher derivatives.

The limitations of the human operator can be summed up in four elements: 1) the linear neuromuscular dynamics, 2) the remnant of the neuromuscular dynamics represented as motor noise, 3) the pure time delays inherent in the human operator's response which can be lumped in one time delay, and 4) the errors introduced by the operator in "reading" the display, represented as the observation noise $\mathbf{v}_o(t)$ where a single noise $v_{o_i}(t)$ is associated with each of the r displayed outputs $y_i(t)$. $v_{o_i}(t)$ are independent Gaussian white-noise processes with known autocovariances. Thus the human perceives a delayed and a noisy version of the system outputs

$$\mathbf{y}_p(t) = \mathbf{C}\mathbf{x}(t-\tau) + \mathbf{d}u(t-\tau) + \mathbf{v}_o(t-\tau)$$

$\mathbf{y}_p(t)$ is then processed by the human operator to result in a commanded control input $u_c(t)$ to which the motor noise $u_m(t)$ is added before it acts on the system through neuromuscular dynamics, such that $u_n(t) = u_c(t) + u_m(t)$. $u_m(t)$ is assumed to be a first order noise generated by

$$\dot{u}_m(t) + \gamma u_m(t) = \gamma v_m(t) \tag{3.3}$$

where $v_m(t)$ is a Gaussian white noise with known autocovariance.

Various experiments have confirmed that the control task is adequately reflected in the human operator's choice of a feedback control $u_c(t)$ which, in the steady state, minimizes the performance index

$$J(u) = E\left\{\sum_{i=1}^{r} q_i y_i^2 + g\dot{u}_c^2\right\}$$

conditioned on the perceived information $\mathbf{y}_p(t)$. E denotes the statistical expected value. The weighting factors $q_i \geq 0$ and $g > 0$ are selected by the human operator depending on her actual control objectives. Note that the performance index includes a cost on control rate. This makes sense when one considers that well trained human operators like pilots or drivers rarely make abrupt control movements.

Within the framework described above, the human operator's control characteristics are determined by the solution of an optimal linear regulator problem with time delay and observation noise as described in [174].

$u_c(t)$ that minimizes the performance index $J(u)$ conditioned on the perceived information $\mathbf{y}_p(t)$, is generated by the feedback law

$$\tau_N \dot{u}_c(t) + u_c(t) = -\mathbf{k}^* \hat{\mathbf{x}}(t) - k_m^* \hat{u}_m(t) \tag{3.4}$$

where $\hat{\mathbf{x}}(t)$ is the optimal estimate of the system state $\mathbf{x}(t)$ based on the observed data $\mathbf{y}_p(\sigma)$, $\sigma < t$. Similarly, $\hat{u}_m(t)$ is the optimal estimate of $u_m(t)$. Note that $\hat{u}_c(t) = u_c(t)$.

The time constant τ_N and the optimal gains in $\mathbf{k}^*$ are given by

$$\tau_N = 1/f_{n+1}$$

$$k_i^* = \tau_N f_i, \qquad i = 1, ..., n$$

where $\mathbf{f} = [f_1, f_2, ..., f_{n+1}]$ is obtained from $\mathbf{f} = \mathbf{b}_0^T \mathbf{K}_0 / g$ with $\mathbf{K}_0$ as the unique positive definite solution of the algebraic Riccati equation

$$\mathbf{A}_0^T \mathbf{K}_0 + \mathbf{K}_0 \mathbf{A}_0 + \mathbf{C}_0^T \mathbf{Q}\, \mathbf{C}_0 - \mathbf{K}_0 \mathbf{b}_0 \mathbf{b}_0^T \mathbf{K}_0 / g = 0$$

with $\mathbf{Q} = \text{diag}\ [q_1, q_2, ..., q_r]$, $\mathbf{b}_0 = \text{col}[0, 0, ..., 0, 1]$ and

$$\mathbf{A}_0 = \begin{bmatrix} \mathbf{A} & \mathbf{b} \\ \mathbf{0} & \mathbf{0} \end{bmatrix}$$

$$\mathbf{C}_0 = [\mathbf{C} \quad \mathbf{d}]$$

The bandwidths of $u_c(t)$ and $u_m(t)$ are approximately equal. Therefore, $\gamma = 1/\tau_N$. Furthermore, the term $k_m^* \hat{u}_m(t)$ is negligible relative to $\mathbf{k}^* \hat{\mathbf{x}}(t)$ because $u_m(t)$ is a wide band noise with the best estimate for any time instant $u_m(t) \approx 0$.

With these assumptions (3.3) and (3.4) yield

$$\tau_N \dot{u}(t) + u(t) = m(t) + v_m(t)$$

$$m(t) = -\mathbf{k}^* \hat{\mathbf{x}}(t)$$

Note that the Kalman filter and the predictor together generate optimal estimates of the system state $\hat{\mathbf{x}}(t)$ and that the optimal control variable $u(t)$ is computed as if $\hat{\mathbf{x}}(t)$ were the actual state. Derivation of the Kalman filter and the optimal predictor coefficients as well as extensive discussions of the so called certainty equivalence control and the separation principle which allows the problems of optimal estimation of the state feedback control to be handled independently are to be found in Appendix A.

The division of the human operator model into two parts is reasonable from a behavioral point of view too. Johannsen [160] divides the optimal control model of the human operator model in Figure 3.13 in three distinct parts. These parts have different psychological functions. The threshold and the delay constitute the perception and the attention allocation part. The Kalman filter and the predictor constitute the central information processing part. The optimal control law and the neuromuscular dynamics constitute the action part.

3.4 Exercises

A1: What are the differences between pursuit racking, compensatory tracking, preview tracking and precognitive tracking?

A2: What are the possible interpretations of the remnant in the describing function model of the human operator?

A3: Why should anyone prefer to use the optimal control models of the human operator which is fairly complicated compared with the simpler models described above?

A4: We have seen some relatively simple general purpose models of the human operator in this chapter. Sometimes more complicated and application specific models are also used (see [145] for example). What are the advantages and disadvantages of such models?

A5: A possible (though not practical) way of identifying the parameters of linear systems like an electrical circuit or a motor is to apply pure sine waves with different single frequencies and normalized amplitudes at their inputs one after the other, measure the amplitude of the output signals at the corresponding frequencies, build the ratio of the output amplitudes to the input amplitudes at given frequencies and perform a curve fit in the frequency domain. Why would this method not be appropriate for determining the parameters of a human operator model?

B1: Consider the configuration in Figure 3.12. We are interested in identifying the parameters of the impulse response of the linear part of the system, $g_c(t)$. A possible way of doing this is indicated in [280]: get the autocovariance of the input $e(t)$ with itself

$$\phi_{ee}(\theta) = \lim_{T\to\infty}\left[\frac{1}{2T}\int_{-T}^{T} e(t)e(t+\theta)dt\right]$$

And get the crosscovariance of the input $e(t)$ with the output $u(t)$

$$\phi_{eu}(\theta) = \lim_{T\to\infty}\left[\frac{1}{2T}\int_{-T}^{T} e(t)u(t+\theta)dt\right]$$

Since $u(t) = v(t) + n(t)$, the crosscovariance consists of two terms

$$\phi_{ee}(\theta) = \lim_{T\to\infty}\left[\frac{1}{2T}\int_{-T}^{T} e(t)v(t+\theta)dt\right] + \lim_{T\to\infty}\left[\frac{1}{2T}\int_{-T}^{T} e(t)n(t+\theta)dt\right]$$

The second term vanishes since, by definition, $e(t)$ and $u(t)$ are not correlated. Further, by replacing $v(t)$ by its convolution integral definition

$$v(t) = \int_{-\infty}^{\infty} g_c(t)e(t-\tau)d\tau$$

ϕ_{eu} can be written as

$$\begin{aligned}
\phi_{ee}(\theta) &= \lim_{T\to\infty}\left[\frac{1}{2T}\int_{-T}^{T} e(t)\int_{-\infty}^{\infty} g_c(t)e(t+\theta-\tau)d\tau dt\right] \\
&= \int_{-\infty}^{\infty} g_c(t)\lim_{T\to\infty}\left[\frac{1}{2T}\int_{-T}^{T} e(t)e(t+\theta-\tau)dt\right]d\tau \\
&= \int_{-\infty}^{\infty} g_c(t)\phi_{ee}(\theta-\tau)d\tau \qquad (3.5)
\end{aligned}$$

In an experiment, $e(t)$ and $u(t)$ can be recorded for a "long" sample and ϕ_{ee} and ϕ_{eu} can be computed. The problem is thus reduced to finding the set of $g_c(i\Delta)$ in a discrete approximation of (3.5) as

$$\hat{\phi}_{eu}(\theta) = \sum_{i=0}^{N} g_c(i\Delta)\hat{\phi}_{ee}(\theta - i\Delta)$$

(where Δ signifies a unit time delay), that results in the best approximation of $\phi_{eu}(\theta)$. What are the difficulties inherent in this approach? Can you think of better methods of identifying the parameters of a given human operator model?

C1: The threshold element in Figure 3.13 is not actually present in the optimal control behavior model of the human operator described in this chapter. One of the reasons for its absence is the difficulty in applying the optimal control theory on non-linear systems. Some of the other hard non-linearities inherent in the human behavior are saturation, signal limiting and hysteresis. Develop an optimal control model of the human operator which accounts for such non-linearities.

C2: An astonishing capability of the human operator is their ability to adapt to changing dynamic characteristics of the system to be controlled. This is not only true for slow changes as in changes of an aircraft's aerodynamic response as a function of its speed and altitude but also in very fast changes as in the steering response of a car with failing power steering. Conduct experiments with abruptly changing system parameters and develop adaptive control models of the human operator with special emphasis on understanding the adaptation algorithms used by humans.

C3: An alternative approach to modeling control systems is the application of models using fuzzy or rule based controllers. Study their usefulness and applicability in modeling the human operator's dynamic properties [164]. Study Appendix B for additional thoughts and references.

Chapter 4
Drug Administration and Dosage Optimization

The dose makes the poison.

— *Paracelsus*

Nearly all men die of their medicines, not of their diseases.

— *Molière*

Doctors are men who prescribe medicines of which they know little, to cure diseases of which they know less, in human beings of whom they know nothing.

— *Voltaire*

4.1 Introduction

There are several intrinsic feedback control mechanisms in most therapeutic situations that cover a wide range from the acute to the chronic (see Figure 4.1). First the physician prescribes a therapy to the patient. This therapy can include several drugs that can be administered orally, intravenously, *etc.* For instance, if certain drugs are to be given orally, the physician can include other instructions, like special dietary considerations, exercise, physiotherapy, and so on.

This therapy has measurable and immeasurable affects on the patient. The measurable factors can be, for instance, heart rate, blood pressure and blood or urine levels of certain substances. The immeasurable factors can be pain, skin color, mood, *etc.* The physician can adjust the therapy based on the measured information. This is the first feedback loop.

Often, there is a controller, which determines the therapy based on the measurable and the immeasurable attributes of the patient. This controller can, for instance, adjust the speed of a pump to deliver a drug with a certain rate based on the time-history of the variable that is to be controlled. The controller can also be an abstract

S.S. Hacısalihzade: *Biomedical Applications of Control Engineering*, LNCIS 441, pp. 61–97.
DOI: 10.1007/978-3-642-37279-7_4

rule like allowing the patient to take a second dose of an analgesic if the pain is not relieved after the initial dose. In this generalized sense, the controller can even be a nurse or a paramedic who can administer drugs according to a rigid scheme based on the assessment of the measurable and immeasurable factors. This is the second feedback loop.

In the general case, the controller is also privy to the therapy the patient receives. Since the controller may include a model of the patient (this model can vary from the most crude to the very sophisticated), it can compare the measured outcome of the therapy with the predicted one. This comparison can be used to modify the controller itself if it has adaptive features built into it, or the result of the comparison is communicated to the physician, who can then modify the controller. At any rate, the output of the controller is utilized by the physician to adjust the therapy. This constitutes the main feedback loop.

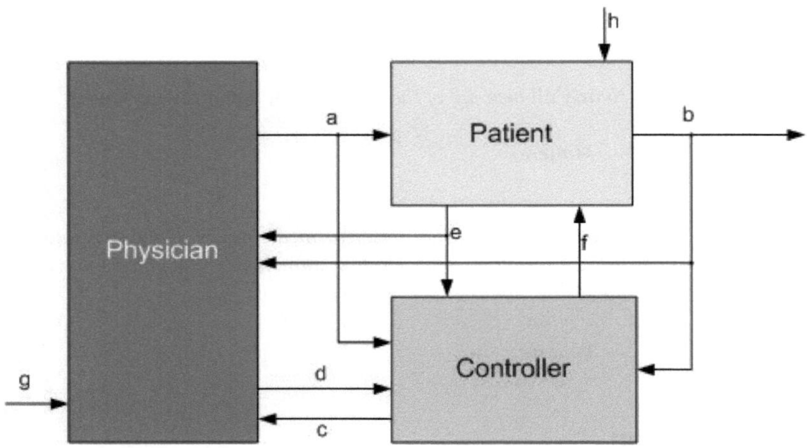

Fig. 4.1 General therapeutic situation: a) Therapy, b) Measurable effects of the therapy, c) Controllers communication to the physician, d) Physicians modification of the controller, e) Immeasurable effects of the therapy, f) Direct treatment of the patient by the controller, g) Physicians continuing education, h) Uncontrollable effects acting on the patient.

At times, certain aspects of this complex loop system may be missing. When a physician tells a patient on the phone to "take two aspirins (therapy) and call me back in the morning if your headache persists (feedback of the immeasurable factors)", he is basically hoping to have an "open-loop" case. On the other hand, when a controller regulates the mixture of various gases which the patient inhales (therapy) to maintain a prescribed level of anaesthesia based on the relaxation level of the patent's muscles (measurable factors), it is a simple feedback controller. There are applications, in which the controller includes a model of how the patient should react to anaesthesia. This model (especially its parameters) can constantly be modified according to a preset scheme based on the individual properties of the patient like his weight, age, actual anaesthetic situation and delivered drugs. In that case we are dealing with a (model-) adaptive controller. It is possible to have a further level of adaptation. This is the case where the anaesthetist switches from one adaptation scheme to another based on some information (like the pupil size) to which the controller has no direct access. Of course, it is possible for the physician to learn more about the therapeutic situation at hand from outside sources, which might influence the way to map all the available information to a therapy. Also, there are external variables (like modeling errors or most commonly failure to comply with the physician's prescription of drugs and diet), which might affect the patients's response to the therapy in an unpredictable fashion. Note that all the subsystems involved in the general therapeutic situation are in general time-varying and non-linear.

It was not until 1970's that the mathematical closeness of drug delivery and control engineering problems was discovered [56], [30], [285], [279]. Different applications with varying degrees of complexity and sophistication have since been in use. By far, the most common application of control engineering methods is in the field of open-loop drug delivery, in which mathematical models of the dose-effect relationship of the used drugs, with parameters identified from population statistics are utilized. This field employs mostly oral or rectal administrations, which are rather insensitive to inaccuracies in dosage or timings of the administration. However, the problem of subcutaneous delivery of insulin for the purpose of blood glucose level control in patients with diabetes is the single application that has received the greatest attention (see Chapter 6). It would be best, of course, if the insulin administration could be made a function of the actual blood glucose level (closed-loop control). However, miniaturized glucose sensors that can last a long time in the body are still not available. Also, closed-loop control is a tremendously difficult task to accomplish, because of the difficulties arising from large delays in the body's reaction to glucose intake and insulin administration. The best solutions today rely on individually programmable pocket size insulin pumps, which can deliver insulin according to a preset time-history.

Necessarily more complicated are closed-loop controllers, which regulate the amount of insulin to be administered based on the measured blood glucose levels (artificial pancreas). The use of such devices, however, is indicated in the case of stationary patients only. Inherently, even more complex are adaptive controllers used during surgery for anaesthesia. Much research has gone into different types of adaptation schemes like gain scheduling or model reference adaptation after it became clear that, due to special characteristics of the dose-effect relationship, simple proportional controllers can become unstable under certain circumstances. Another technique employed here is the use of Smith predictors, which is indicated by the presence of pure delays in the dose-effect relationships of the used anaesthetic agents (see Chapter 7).

The idea of intelligent controllers, using knowledge or rule based control laws relying on previous actions of experts have been applied with moderate success in fields like anaesthesia or fluid retention control. Such controllers are waiting to be applied to other fields of drug delivery. The very nature of drug delivery problems, with its difficulties in measurement or assessment of pharmacological effects and awkward simplifications in the modeling process as well as the difficulty in transforming the physicians' decisions to simple input-output relationships, makes this field an excellent candidate for further applications of intelligent control (see Appendix B).

Obviously, a pressing concern during the use of computer controlled drug delivery devices is safety. It can be argued that once industry standards have been set for safety features and related subjects, the use of automatically controlled drug delivery devices will increase dramatically in the coming years.

Some review papers which give a good feel of the problems and the accomplishments in this field are [165], [166], [305], [159], [201].

4.2 Optimization of Drug Delivery Based on Pharmacokinetic Data

Pharmacokinetics is the study of bodily absorption, distribution, metabolism and excretion of drugs. Historically, pharmacokinetic studies of drug concentration level problems have long been considered to be in the realm of established compartmental models alone (the first quantitative analysis with compartmental models was made by Widmark almost a century ago, who studied the pharmacokinetics of several narcotics [312]). A compartment is a hypothetical space bound by an unspecified membrane across which drugs are transferred. Formally, a variable $x(t)$ of a system is called a compartment if it is governed by the differential equation

$$\frac{dx}{dt} = -Kx + g(t)$$

with K constant and $g(t)$ denoting the addition of the drug to the compartment. Note that this definition of a "compartment" is mathematically very similar to the definition of a "state" as used in control engineering. Some books which cover pharmacokinetics with this approach are [122], [307], [246], [187] and less mathematically [261], [50], [309].

Compartmental analysis is a tool used in describing materials transfer in biological systems. A compartment is a store of material. Compartmental analysis studies the exchange of material among the compartments. The material exchange takes place either by physical transport from one location to another or by chemical reactions. The mathematical model is derived by writing down the mass balance equations for each compartment and relations describing the rate of material transfer between compartments. As an illustrative example consider the two compartments 1 and 2 shown in Figure 4.2.

It is more recent that several researchers introduced unconventional approaches, resulting in non-compartmental techniques which were applicable for the same class of problems [315], [148], [149]. In this section we will introduce a system theoretical approach with which problems concerning the blood level concentrations of administered drugs can be solved [132].

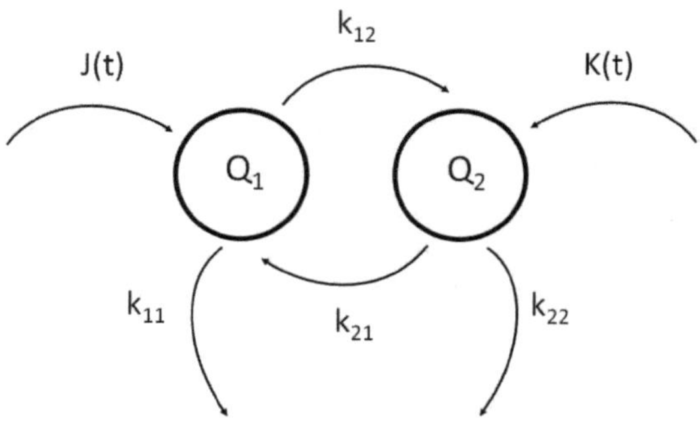

Fig. 4.2 Schematic representation of a compartmental model with two compartments.

4.2.1 Modeling Pharmacokinetic Relationships

The word "system" is one of those words which are understood in totally different ways by different people according to their professional backgrounds. For instance, Webster's Dictionary defines a system as

> ... a regularly interacting or interdependent group of items forming a unified whole.

Therefore, it is appropriate to remember the definition of this word as it is used in the context of this book. A definition given by Schaufelberger [195] which defines a system as

> ... a configuration of objects on which energy, information or matter acts through the inputs and which gives out energy, information or matter through its outputs

is adopted throughout this book.

Mathematically speaking, a system can be seen as a mapping which maps the input onto the output (for a rigorous mathematical definition of a system see, for example, [161]). This can be formalized as

$$y(t) = f[u(t)] \tag{4.1}$$

with $u(t)$ denoting the input, $y(t)$ the output and $f(.)$ the mapping which maps the function space $\mathscr{U}$ of the functions $u(t)$ into the function space $\mathscr{Y}$ of the functions $y(t)$. As we have seen in Chapter 1, a system may have several inputs and several outputs and the number of inputs and the number of outputs do not necessarily have to be the same. In this case, $\mathbf{u}(t)$ and $\mathbf{y}(t)$ are vectors of respective dimensions (this is called the multi-input-multi-output, the MIMO or the multivariable case). In the most general case, $\mathbf{f}(.)$ can be represented as a time-varying, non-linear vector.

Drug administration can now be considered as the input of a system and plasma concentration level as its output. The system itself is nothing but the pharmacokinetic relationship between the administered drug and its plasma concentration. The multivariable case corresponds to the case in which several drugs are administered simultaneously and several plasma concentration levels are of interest. An example may illustrate how these considerations can be formalized.

Example: There are cases in which a drug is administered in conjunction with another one (in many cases an inhibitor) which is meant to control the metallization rate of the first one. It is possible that the clinician is interested in the plasma concentration of, say, three metabolites as functions of the administered doses and their times of administration. In this case, the pharmacokinetic relationship can be modeled as a system with two inputs (the drug d and the inhibitor c) and three outputs (the metabolites m_1, m_2 and m_3). Formally,

$$\begin{aligned} m_1(t) &= f_1[d(t), c(t), t] \\ m_2(t) &= f_2[d(t), c(t), t] \\ m_3(t) &= f_3[d(t), c(t), t] \end{aligned} \tag{4.2}$$

In terms of (4.1) $\mathbf{y}(t)$ is a vector with the components m_1, m_2 and m_3 and $\mathbf{u}(t)$ is a vector with the components $c(t)$ and $d(t)$, and $f_1[.]$, $f_2[.]$ and $f_3[.]$ comprise the mapping with which the pharmacokinetic dependencies are expressed. This is a very general expression; normally a much simpler linear, time-invariant model would be used. □

4.2.2 *The Multiple Dosing Problem*

In many cases, clinicians are interested in attaining a desired drug level time-history of a substance in the patient's plasma in order to obtain some therapeutic effect. A convenient way of doing this is by means of multiple administrations of the given drug. The therapy is prescribed such that the actual time-history approximates the desired time history. As mentioned above, this problem has traditionally been solved by the use of compartmental models, which correspond to a particular state space representation of $f[.]$ in (4.1). However, to solve this particular problem, we will chose not a state space description but an input-output representation. Although some (mostly numerical) techniques are known to handle problems concerning nonlinear systems, we will assume here the linearity of the mapping $f[.]$.

The input-output relationship of a multivariable linear system with p inputs and q outputs is given by

$$\begin{pmatrix} y_1(t) \\ y_2(t) \\ \vdots \\ y_q(t) \end{pmatrix} = \begin{pmatrix} f_{11}(t) & f_{12}(t) & \cdots & f_{1p}(t) \\ f_{21}(t) & f_{22}(t) & \cdots & f_{2p}(t) \\ \vdots & \vdots & \ddots & \vdots \\ f_{q1}(t) & f_{q2}(t) & \cdots & f_{qp}(t) \end{pmatrix} \circ \begin{pmatrix} u_1(t) \\ u_2(t) \\ \vdots \\ u_p(t) \end{pmatrix}$$

or, more compactly

$$\mathbf{y} = \mathbf{F} \circ \mathbf{u} \tag{4.3}$$

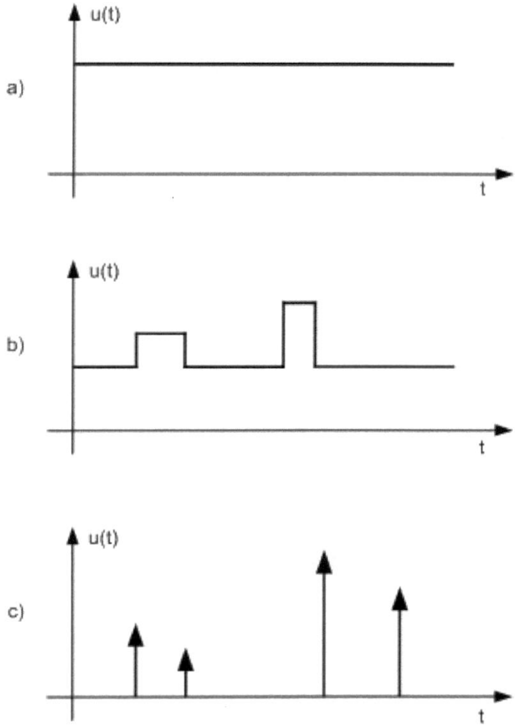

Fig. 4.3 Drug administration as a function of time. a) Constant rate infusion, b) Two bolus infusions superimposed on a base-line constant rate infusion, c) Four oral administrations.

where '∘' symbolizes the convolution operation. The k-th output can be written as

$$y_k(t) = \sum_{i=1}^{p} f_{ki}(t) \circ u_i(t) \tag{4.4}$$

Normally, drug administrations are represented as time functions with the area under the curve corresponding to the total administered dose (see Figure 4.3).

In case of an oral administration, it can be assumed that the administration itself takes practically no time at all. Therefore, the administration can be represented by a Dirac-pulse, the amplitude of which is equal to the dose. Therefore, a sequence of such oral administrations can be formalized as

$$u(t) = \sum_{i=1}^{n} d_i \delta(t - \tau_i) \tag{4.5}$$

where d_i denotes the dose and τ_i the timing of the i-th administration. $\delta(.)$ represents the Dirac-pulse and n is the number of administrations during the time interval of interest. In the multivariable case of (4.4) the j-th input can be expressed as

$$u_j(t) = \sum_{i=1}^{n_j} d_{ji}\delta(t - \tau_{ji}) \tag{4.6}$$

where d_{ji} denotes the dose and τ_{ji} the timing of the i-th administration of the j-th.drug. n_j is the number of doses of drug j during the time interval of interest. Substituting (4.6) in (4.4) results in

$$\begin{aligned} y_k(t) &= \sum_{j=1}^{p} \int_0^t u_j(\tau) f_{kj}(t-\tau)d\tau \\ &= \sum_{j=1}^{p} \int_0^t \sum_{i=1}^{n_j} d_{ji}\delta(\tau - \tau_{ji}) f_{kj}(t-\tau)d\tau \\ &= \sum_{j=1}^{p} \sum_{i=1}^{n_j} d_{ji} \int_0^t \delta(\tau - \tau_{ji}) f_{kj}(t-\tau)d\tau \\ &= \sum_{j=1}^{p} \sum_{i=1}^{n_j} d_{ji} f_{kj}(t - \tau_{ji}) \end{aligned} \tag{4.7}$$

For the single drug case (4.7) simplifies to

$$y(t) = \sum_{i=1}^{n} d_i f(t - \tau_i) \tag{4.8}$$

The matrix **F** in (4.3) is composed of elements which are responses to Dirac-pulses. In this case the element f_{kj} of **F** is the time-history of the substance k's blood level after the administration of a unit dose of drug j. **F** can be determined element by element through several blood level analyses after the administration of standard pills.

4.2.3 Linear Programming Solutions of the Multiple Dosing Problem

4.2.3.1 Minimization of the Total Dosage

Often, physicians are interested in attaining a desired profile of a drug's plasma concentration. Although this profile may be constant, in general it is a function of time. On the other hand, they are often also interested in keeping the total amount of the administered drug at a minimum. Obviously, the therapy with the minimal total dose would be not to administer any drugs at all! Therefore, it makes sense to state a constraint that should additionally be fulfilled, like asking for the absolute value of the discrepancy between the desired profile and the actual one not to exceed a given value. The question here is how to minimize the total daily dosage of the medication used while keeping the deviation of the instantaneous value of the drug's plasma concentration from its set trajectory small. In other words, what is the treatment which requires the least daily amount of the drug, for the deviation of the instantaneous value of the drug's plasma concentration from its set trajectory not to exceed a permissible error? The main idea of the solution is based on the fact that the timings cannot be arbitrarily finely prescribed (meaning that we have temporal discretization).

The system output in that case can be given as

$$y(t) = \sum_{i=1}^{n} d_i f(t - \tau_i) \qquad \forall t \in T$$

with $y(t)$ the instantaneous value of the drug's plasma concentration, $f(t)$ the pharmacokinetic pulse response. d_i is the dose of the i-th administration and τ_i is the corresponding timing. T is not a continuum but rather a set consisting of a finite number of points in the temporal continuum. We will call these points "time-points". There are $n = (t_f - t_0)m$ such time-points in T where t_f is the final time of interest and t_0 the initial time of interest. m is the number of discretization steps per hour. With quarter of an hour as temporal discretization, m is 4.

The condition that the deviation of the instantaneous value from its set-trajectory should not be too big can now be quantified as

$$|y(t) - y_{set}(t)| \leq e_{max} \qquad \forall t \in T$$

where e_{max} is the largest permissible such deviation and which has to be specified for each application.

With the introduction of the new function

$$e(t) = y(t) - y_{set}(t)$$

we can rewrite the previous equation as

$$|e(t)| \leq e_{max} \qquad \forall t \in T$$

or equivalently as

$$-e_{max} \leq e(t) \leq e_{max} \qquad \forall t \in T$$

so that we can pose the problem as a well known linear programming problem:

$$e(t) - \sum_{i=1}^{n} d_i f(t - \tau_i) = -y_{set}(t) \ , \quad t = \tfrac{1}{4}, \tfrac{1}{2}, \ldots, t_f$$

$$-e_{max} \leq e(t) \leq e_{max} \qquad , \quad t = \tfrac{1}{4}, \tfrac{1}{2}, \ldots, t_f$$

$$0 \leq d_i \leq d_{max} \qquad , \quad i = 1, 2, \ldots, n$$

$$\sum_{i=1}^{n} d_i \rightarrow Minimum$$

One severe drawback of this formulation is that it might result in too many d_i which are not zero, meaning that the patient has to take practically every quarter of an hour one very small dose (which does not exist anyway). There is one easy way out of this dilemma and it is to prescribe the timings of the administrations, number of which should reasonably be less than, say, $n_p = 10$ ($n_p \ll n$) per day. This has the effect that the linear programming problem can be rewritten to

$$
\begin{aligned}
& e(t) - \sum_{i=1}^{n_p} d_i f(t-\tau_i) = -y_{set}(t)\ , && t = \tfrac{1}{4}, \tfrac{1}{2}, \ldots, t_f \\
& -e_{max} \le e(t) \le e_{max} && ,\quad t = \tfrac{1}{4}, \tfrac{1}{2}, \ldots, t_f \\
& 0 \le d_i \le d_{max} && ,\quad i = 1, 2, \ldots, n_p \\
& \sum_{i=1}^{n_p} d_i \to Minimum
\end{aligned}
$$

In order to get the classical form of the linear programming problem we have to introduce a new variable

$$e^*(t) = e(t) + e_{max}$$

so that we have as our linear programming problem in the classical form

n-Equations:

$$e^*(t_j) - e_{max} - \sum_{i=1}^{n_p} d_i f(t_j - \tau_i) = -y_{set}(t_j) \quad j = 1, 2, \ldots, n$$

n-Inequations:

$$0 \le e^*(t_j) \le 2e_{max} \qquad j = 1, 2, \ldots, n$$

n_p-Inequations:

$$0 \le d_i \le d_{max} \qquad i = 1, 2, \ldots, n_p$$

Objective function

$$-\sum_{i=1}^{n_p} d_i \to Maximum \tag{4.9}$$

Those equations and inequalities written out explicitly look as follows:

$$\begin{array}{llll}
d_1 & & & \leq d_{max} \\
\quad d_2 & & & \leq d_{max} \\
\quad\quad \ddots & & & \vdots \\
\quad\quad\quad d_{n_p} & & & \leq d_{max} \\
e^*(t_1) & & & \leq 2e_{max} \\
\quad e^*(t_2) & & & \leq 2e_{max} \\
\quad\quad \ddots & & & \vdots \\
\quad\quad\quad e^*(t_n) & & & \leq 2e_{max} \\
-f(t_1-\tau_1)d_1 - f(t_1-\tau_2)d_2 - \ldots - f(t_1-\tau_{n_p})d_{n_p} + e^*(t_1) & & & = -y_{set}(t_1) + e_{max} \\
-f(t_2-\tau_1)d_1 - f(t_2-\tau_2)d_2 - \ldots - f(t_2-\tau_{n_p})d_{n_p} + e^*(t_2) & & & = -y_{set}(t_2) + e_{max} \\
\vdots & & & = \vdots \\
-f(t_n-\tau_1)d_1 - f(t_n-\tau_2)d_2 - \ldots - f(t_n-\tau_{n_p})d_{n_p} + e^*(t_n) & & & = -y_{set}(t_n) + e_{max}
\end{array}$$

This means that, with $e_i^* = e^*(t_i)$, we have

$$\mathbf{Ax} \leq \mathbf{b}$$
$$\mathbf{c}^T\mathbf{x} \rightarrow maximum$$

where

$$\mathbf{x} = \begin{pmatrix} d_1 \\ d_2 \\ \vdots \\ d_{n_p} \\ e_1^* \\ e_2^* \\ \vdots \\ e_n^* \end{pmatrix}$$

with

$$\mathbf{A} = \begin{pmatrix} & \mathbf{I}_{n_p} & & \mathbf{0}_{n,n_p} \\ & \mathbf{0}_{n_p,n} & & \mathbf{I}_n \\ -f_1^1 & -f_1^2 & \cdots & -f_1^{n_p} & \\ -f_2^1 & -f_2^2 & \cdots & -f_2^{n_p} & \mathbf{I}_n \\ & & \cdots & & \\ -f_n^1 & -f_n^2 & \cdots & -f_n^{n_p} & \end{pmatrix}, \quad \mathbf{b} = \begin{pmatrix} d_{max} \\ d_{max} \\ \vdots \\ 2e_{max} \\ 2e_{max} \\ \vdots \\ -y_{set}(t_1) + e_{max} \\ \vdots \\ -y_{set}(t_n) + e_{max} \end{pmatrix}$$

$$\mathbf{c}^T = [\underbrace{-1, ..., -1}_{n_p}, \underbrace{0, ..., 0}_{n}]$$

where $f_i^j = f(t_i - \tau_j)$, $\mathbf{I}_k$ is the $k \times k$ unit matrix, $\mathbf{0}_{a,b}$ is the $a \times b$ zero matrix.

Some comments on the linear programming problem (4.9): T is not a temporal continuum but rather a set consisting of a finite number of points in time. Since it is desirable to prescribe round numbers as the timings of the administrations, *e.g.*, *not* 8hrs 18min 32sec, a temporal discretization with a quarter of an hour long steps makes sense. This results in $n = 4(t_f - t_0)$ such points during the time frame of interest. (The temporal discretization steps must be chosen such that they are much smaller than the smallest time constant involved in the pharmacokinetics of the drug.) t_0 and t_f might denote the times when the patient wakes up, say 6am, and goes to sleep, say 9pm, respectively. Therefore, the set of equations and the first set of inequations stand for about sixty equations and sixty inequations. As to the second set of equations, it is obvious that the doses may not be negative and it also makes sense to limit the size of a single dose with d_{max}.

Example: Parkinson's disease is a progressive neurological disorder the symptoms of which can be tracked back to the lack of neurotransmitter dopamine in the basal ganglia of the brain. Therefore, it is fairly standard to treat this disorder symptomatically by the oral administration of a drug containing the precursor of dopamine, namely levodopa [226] (see Chapter 5 for a detailed treatment of this subject). It is also known that the motor activity level which changes during the day in healthy humans correlates with the blood level of levodopa [233]. Consequently, the clinicians try to attain a desired time-history of levodopa's plasma concentration which has "highs" when there is need for fine motor ability like shaving or eating (*e.g.*, early in the morning). The level may be allowed to sink during the night while the patient sleeps, hence is inactive.

A common commercial form of four parts of levodopa combined with one part of benserazide (to inhibit the fast decarboxylase of levodopa) is Madopar® [1]. It is available in multiples of a basic dose of 62.5mg. The maximum allowable single dose is 6 (=375mg/62.5mg). The blood levels of levodopa shown in Figure 4.4 were measured after administering 250mg Madopar® to the patient AB26 at $t = 0$ (pulse response).

Let us start with $e_{max} = 4000$ for the set trajectory shown in Figure 4.5. The application of (4.9) for this patient with $e_{max} = 4000$ results in the treatment in Table 4.1. The simulation of this treatment is also shown in Figure 4.5. Solving (4.9) for this patient with $e_{max} = 3500$ results in the treatment in Table 4.2 The simulation of this treatment is shown in Figure 4.6. Solving (4.9) for this patient with $e_{max} = 3000$ results in the treatment in Table 4.3 The simulation of this treatment is shown in Figure 4.7. We can see that the discrepancy between the actual and desired trajectories get smaller as the maximum allowed deviation is reduced from 4000 to

[1] Madopar® is a registered trademark of Hoffmann-LaRoche.

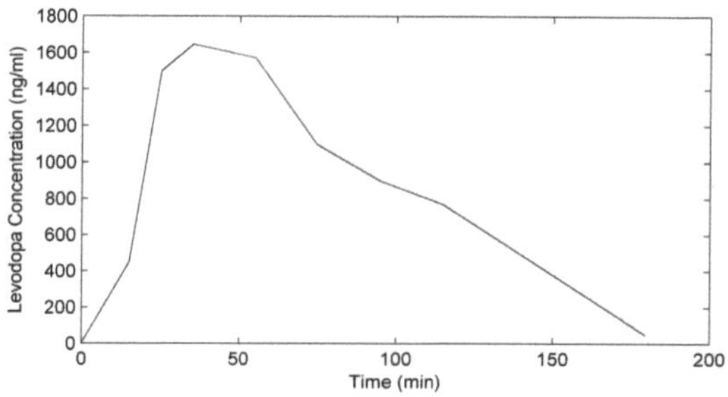

Fig. 4.4 Levodopa concentration in the blood of a patient (AB26) as a function of time after the administration of 62.5mg Madopar®.

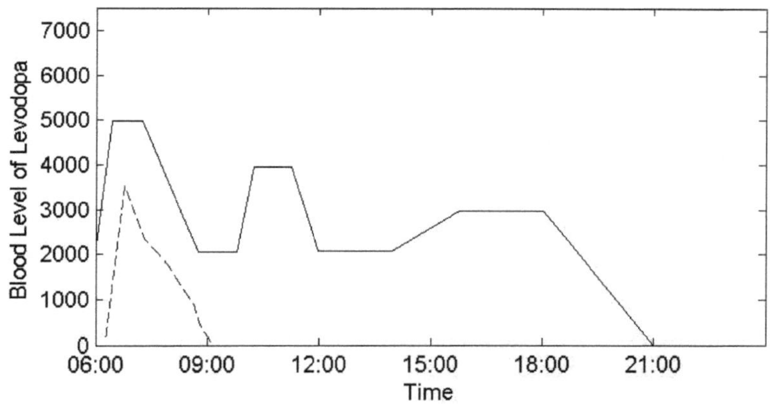

Fig. 4.5 Simulation of the therapy in Table 4.1 (optimal for e_{max} = 4000, with daily 135mg Madopar®). In this figure and the rest of the figures in this chapter solid lines represent the desired time-histories and the dashed lines represent the simulated time-histories.

Table 4.1 Optimal treatment for patient AB26 with $e_{max} = 4000$.

Timings:	06:15
Doses:	2.16

3000. The price we have to pay, however, is an increase of the total amount of the administered drug.

There are several questions the interested reader should be asking by now at the latest: firstly, how do we get the timings τ_i? For didactic purposes, the timings here were determined heuristically. For a detailed answer of this interesting question see Subsection 4.2.6. Then, the daily amounts of Madopar® is very low; isn't that great? Yes, that is indeed wonderful but the therapy effects we are getting are not that great

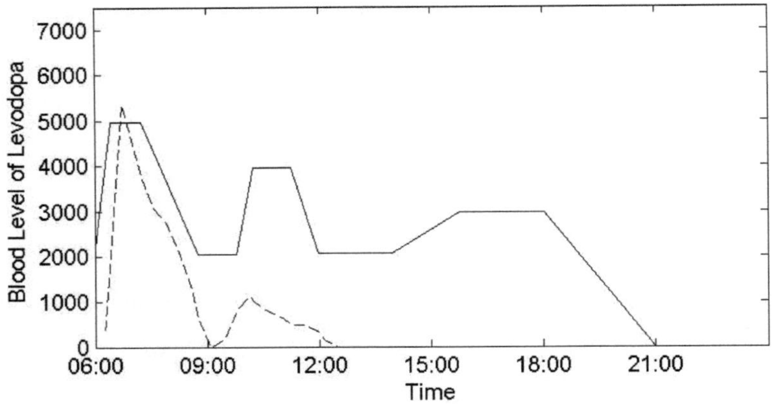

Fig. 4.6 Simulation of the therapy in Table 4.2 (optimal for e_{max} = 3500, with daily 245mg Madopar®).

Table 4.2 Optimal treatment for patient AB26 with $e_{max} = 3500$.

Timings:	06:15	09:30
Doses:	3.24	0.68

(actually rather useless)! Well, no problem, one might say, let's just decrease e_{max} further until we get a reasonable therapy.

Not so fast! What is the geometric interpretation of the equations and inequations in (4.9)? We are looking for the intersection of several hyperplanes (equations) in a convex region of the hyperspace spanned by the axes $e(t_0)$, $e(t_0+\Delta t)$, $e(t_0+2\Delta t)$, ..., $e(t_f-\Delta t)$, $e(t_f)$ and $d_1, d_2, ..., d_n$ (inequations). If we look at a case, for instance, in two dimensions with the set of conditions

$$
\begin{aligned}
a_1x_1 + b_1 &= 0 \\
a_2x_1 + b_2 &= 0 \\
a_3x_1 + b_3 &= 0 \\
x_1 &\geq 0 \\
x_2 &\geq 0
\end{aligned}
$$

it becomes clear that only certain values of a_i and b_i can fulfil these conditions. Figure 4.8 shows cases where these conditions are satisfied and not satisfied.

It is clear that we might not have a solution at all if we look in too small a region, *i.e.*, if we give e_{max} as too small. Here arises an interesting question: how small can we ask e_{max} to be and still get a feasible solution to (4.9)?

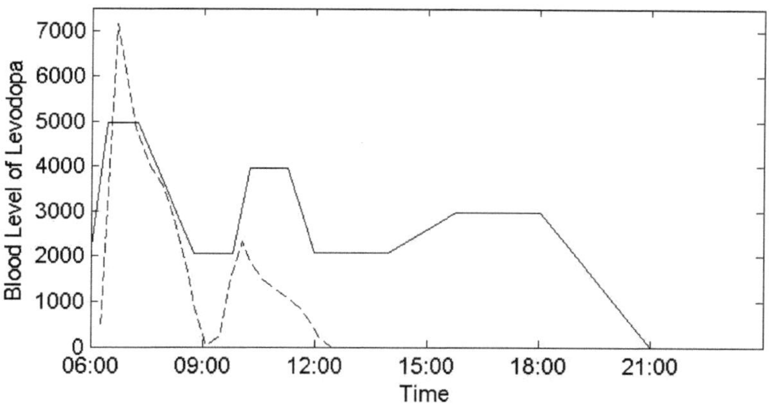

Fig. 4.7 Simulation of the therapy in Table 4.3 (optimal for e_{max} = 3000, with daily 335mg Madopar®).

Table 4.3 Optimal treatment for patient AB26 with $e_{max} = 3000$.

Timings:	06:15	09:30
Doses:	4.32	1.36

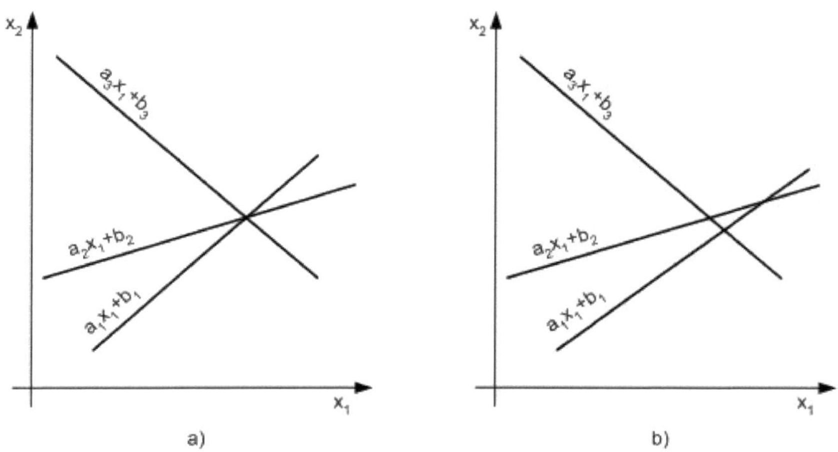

Fig. 4.8 a) The combination of a_i and b_i allow a solution of the simultaneous equations, b) The combination of a_i and b_i do not allow a solution of the simultaneous equations.

Obviously, in our case with roughly 70 equations and inequalities in (4.9), it cannot be expected that the conditions in the linear programming problem will always be satisfied. If we think of the problem in physical terms, it means that clearly there is a lower limit to the maximum admissible error below which no solution exists. Explicitly, with ten doses we cannot expect to come arbitrarily close to the set-value trajectory. Furthermore, there is a limit to how close we can come to it. □

Another, almost self-evident comment is about the limit in the opposite direction: how big can we ask the deviation to be? Obviously, we can ask it to be as big as we like. For any value less than the highest value in the desired set-value trajectory we will get a treatment with non-vanishing doses, but if we ask the highest permissible deviation to be anything bigger than that level, the treatment with the minimum total daily dosage will degenerate to a (non-)treatment with no doses.

4.2.3.2 How Small Can e_{max} Be?

Example: An interesting question which arises at this point is how small can we ask e_{max} to be and still expect to get a solution? For this, we will study another patient (RS34) and a different set trajectory shown in Figure 4.9. Solving the linear programming problem (4.9) for this patient with $e_{max} = 2000$ results in the treatment in Table 4.4. The simulation of this treatment is also shown in Figure 4.9.

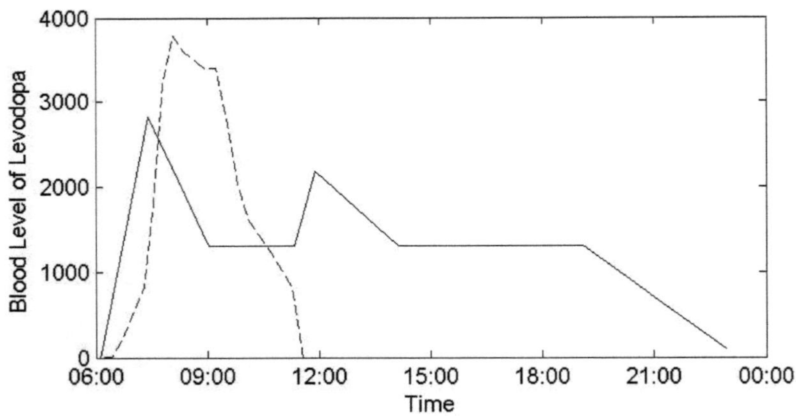

Fig. 4.9 Simulation of the therapy in Table 4.4 (optimal for e_{max} = 2000, with daily 376mg Madopar®).

Table 4.4 Optimal treatment for patient RS34 with $e_{max} = 2000$.

Timings:	06:15	06:45	07:00	08:15	08:30	10:00
Doses:	3.27	0.57	0.50	0.02	1.60	0.06

Since we are looking for the treatment which minimizes the maximum deviation, this is a typical minimax problem. We can again reduce the problem to that of a linear programming. The idea is basically the same as in the previous example but we have in addition one more variable, namely the maximum error and, of course, our objective function is modified such that we want to minimize the maximum error.

We again have to start with

n-Equations:

$$e^*(t_j) - \sum_{i=1}^{n} d_i f(t_j - \tau_i) = -y_{set}(t_j) \qquad j = 1,2,...,n$$

n-Inequations:

$$0 \leq e^*(t_j) \leq 2e_{max} \qquad j = 1,2,...,n$$

n_p-Inequations:

$$0 \leq d_i \leq d_{max} \qquad i = 1,2,...,n_p$$

Objective function

$$z = e_{max}$$

$$-z \rightarrow maximum \tag{4.10}$$

z is a new variable so that with the nomenclature above the equations and the inequalities can be written explicitly as follows:

$$\begin{array}{ll}
d_1 & \leq d_{max} \\
\quad d_2 & \leq d_{max} \\
\qquad \ddots & \vdots \\
\qquad\quad d_{n_p} & \leq d_{max} \\
-2z + e^*(t_1) & \leq 0 \\
-2z + \quad e^*(t_2) & \leq 0 \\
\qquad \ddots & \vdots \\
-2z + \qquad e^*(t_n) & \leq 0 \\
-z & \leq 0 \\
-f(t_1-\tau_1)d_1 - f(t_1-\tau_2)d_2 - ... - f(t_1-\tau_{n_p})d_{n_p} + e^*(t_1) - z & = -y_{set}(t_1) \\
-f(t_2-\tau_1)d_1 - f(t_2-\tau_2)d_2 - ... - f(t_2-\tau_{n_p})d_{n_p} + e^*(t_2) - z & = -y_{set}(t_2) \\
\vdots - z & = \vdots \\
-f(t_n-\tau_1)d_1 - f(t_n-\tau_2)d_2 - ... - f(t_n-\tau_{n_p})d_{n_p} + e^*(t_n) - z & = -y_{set}(t_n)
\end{array}$$

Formalized in the classical format, this again means

$$\mathbf{A}\mathbf{x} \leq \mathbf{b}$$
$$\mathbf{c}^T\mathbf{x} \rightarrow maximum$$

where

$$\mathbf{x} = \begin{pmatrix} d_1 \\ d_2 \\ \vdots \\ d_{n_p} \\ e_1^* \\ e_2^* \\ \vdots \\ e_n^* \\ z \end{pmatrix}$$

with

$$\mathbf{A} = \begin{pmatrix} & \mathbf{I}_{n_p} & & & \mathbf{0}_{n,n_p} & & 0 \\ & & & & & & \vdots \\ & & & & & & 0 \\ & & & & & & -2 \\ & \mathbf{0}_{n_p,n} & & & \mathbf{I}_n & & \vdots \\ & & & & & & -2 \\ 0 & & \cdots & 0 & 0 \cdots & 0 & -1 \\ -f_1^1 & -f_1^2 & \cdots & -f_1^{n_p} & & & -1 \\ -f_2^1 & -f_2^2 & \cdots & -f_2^{n_p} & \mathbf{I}_n & & \vdots \\ & & \cdots & & & & \vdots \\ -f_n^1 & -f_n^2 & \cdots & -f_n^{n_p} & & & -1 \end{pmatrix}, \quad \mathbf{b} = \begin{pmatrix} d_{max} \\ \vdots \\ d_{max} \\ 0 \\ \vdots \\ 0 \\ -y_{set}(t_1) \\ \vdots \\ -y_{set}(t_n) \end{pmatrix}$$

and

$$\mathbf{c}^T = [\underbrace{0,\ldots,0}_{n+n_p},-1]$$

Solving (4.10) for this patient results in the treatment with the smallest maximum error in Table 4.5. The simulation of this treatment is shown in Figure 4.10. □

All these simulations actually show us that optimizing the total dosage or the error does not result in desired therapeutic effects. Therefore, one should not try to minimize the total dosage but rather a measure of the discrepancy between the desired and the actual time-histories of levodopa's blood level. Formally,

$$\sum_{\forall t \in T} |y(t) - y_{set}(t)| \rightarrow Minimum \tag{4.11}$$

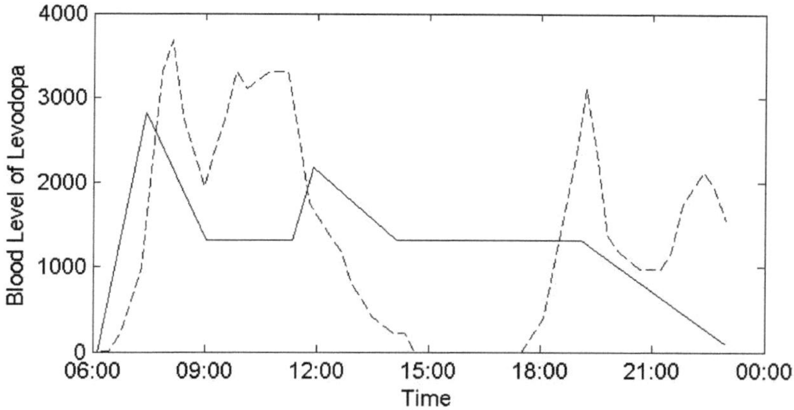

Fig. 4.10 Simulation of the therapy in Table 4.5 results in the smallest attainable $e_{max} = 1911$ for this patient and this set trajectory, with daily 895mg Madopar®).

Table 4.5 Optimal treatment for patient RS34 which results in the smallest maximum error for the desired time history of levodopa concentration in blood.

Timings:	06:15	08:00	08:15	08:30	08:45	09:15	09:30	17:15	20:15	20:30	20:45
Doses:	3.98	1.29	0.29	0.59	0.52	0.25	1.77	3.08	1.20	0.06	1.29

This needs a little trick to solve which is shown in the formulation of the linear programming problem (4.12).

n-Equations:

$$e(t) - \sum_{i=1}^{n} d_i f(t - \tau_i) = -y_{set}(t) \qquad \forall t \in T$$

n-Inequations:

$$-e_1(t) \le e(t) \le e_2 \qquad \forall t \in T$$

$$e_1(t) \ge 0 \qquad \forall t \in T \tag{4.12}$$

$$e_2(t) \ge 0 \qquad \forall t \in T$$

$$0 \le d_i \le d_{max} \qquad i = 1, 2, \ldots, n$$

Objective function

$$\sum_{\forall t \in T} [e_1(t) + e_2(t)] \rightarrow Minimum$$

Remember that in minimizing the total daily dose above, we wrote down the conditions for $e(t)$ never to be below $-e_{max}$ and above e_{max}. Here, to minimize the L_1 norm of the error we are forcing $e(t)$ to be within asymmetric bounds. In other words, minimizing the sum of two non-negative auxiliary variables e_1 and e_2 which constitute the limits of the discrepancy between the desired and the actual time histories of the drug plasma concentration is equivalent to minimizing the absolute value of the discrepancy summed over the period of interest. Note that there are some 240 inequality constraints in this linear programming problem.

Example: The application of the linear programming problem (4.12) again on the patient RS34 results in the therapy shown in Table 4.6. The simulation of this therapy is depicted in Figure 4.11 which shows an excellent agreement of the actual trajectory with the desired one.

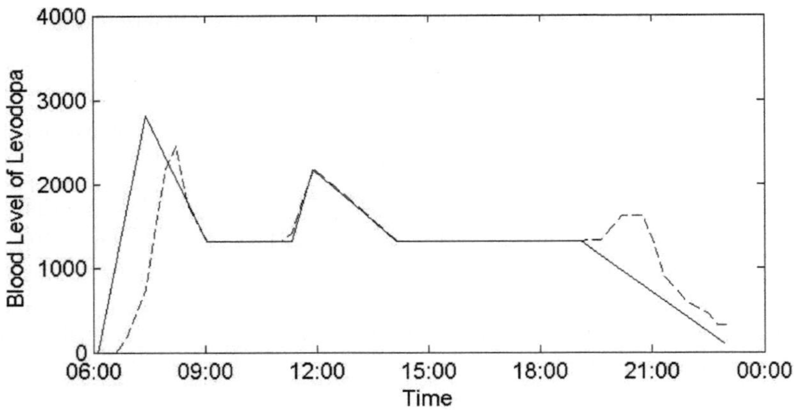

Fig. 4.11 Simulation of the therapy in Table 4.6 shows an excellent matching of the desired and the actual time-histories of the levodopa plasma levels. Is that realistic?

Table 4.6 The treatment for patient RS34 which minimizes the absolute value of the discrepancy between the desired and actual time-histories of levodopa concentration in blood. The patient has to take 48 individual doses per day!

Timings:	06:15	07:15	07:30	07:45	...	19:00	19:15	19:30	19:45
Doses:	2.70	0.06	0.09	0.16	...	0.06	0.06	0.05	0.06

4.2.3.3 Is That Realistic?

A glance at Table 4.6 shows that the patient has to take very small doses practically every quarter of an hour, a regimen which is clearly out of the question. Therefore, one should prescribe the timings first and perform the optimization with those given timings. This means that the index i in (4.12) should vary from 1 to m which is to be, say, not more than 10. Also, τ_1, τ_2, ..., τ_m should be specified.

The result of this modified optimization algorithm's application on patient RS34 (for $m = 7$) are to be seen in Table 4.7 and Figure 4.12 which shows a fair matching of the desired and actual profiles. □

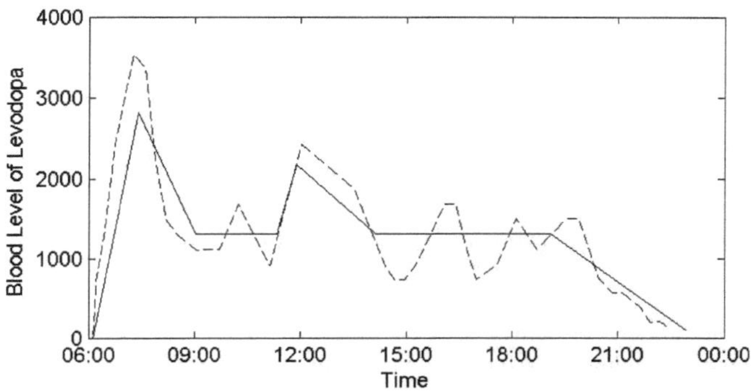

Fig. 4.12 Simulation of the therapy in Table 4.7 shows a fair matching of the desired and the actual time-histories of the levodopa plasma levels.

Table 4.7 Optimal therapy for patient RS34 with 7 individual doses per day.

Timings:	05:30	08:30	10:15	11:30	14:00	16:00	17:30
Doses:	3.92	1.32	2.22	0.94	1.73	1.13	1.30

4.2.3.4 What about the Total Dosage?

The last optimization does not consider the total dosage at all. However, a slight modification of the objective function (4.11) can account for it as can be seen in (4.13).

$$\alpha \sum_{i=1}^{m} d_i + \sum_{\forall t \in T} |y(t) - y_{set}(t)| \rightarrow Minimum \tag{4.13}$$

where α is an appropriate, non-negative, weighting factor for the total dosage (α must also have the right dimension to make the two additive terms in (4.13) to have the same dimension). This only results in a corresponding alteration in the linear programming problem (4.12).

Example: The application of the linear programming problem (4.12) modified by the objective function (4.13) on a new patient (MK33) and the set trajectory in Figure 4.13 with given timings and different choices of α result in the dosage schemes shown in Tables 4.8, 4.9 and 4.10. Figures 4.13, 4.14 and 4.15 show the simulations of these dosage schemes.

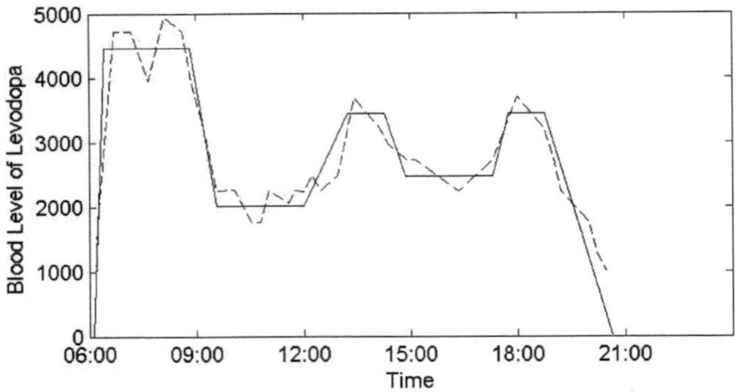

Fig. 4.13 Simulation of the therapy in Table 4.8 ($\alpha = 0$).

Table 4.8 Optimal therapy for patient MK33 with 9 individual doses per day at given times without taking the total administered dose into account ($\alpha = 0$).

Timings:	05:00	06:30	09:15	11:15	11:30	13:15	14:00	15:30	16:00
Doses:	4.90	3.30	2.19	2.42	1.42	1.14	1.26	1.50	2.56

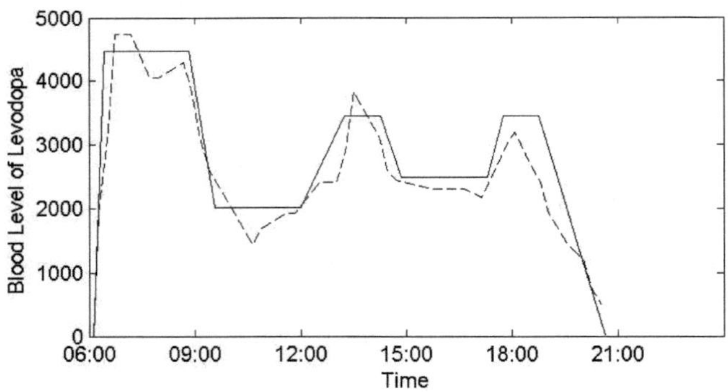

Fig. 4.14 Simulation of the therapy in Table 4.9 ($\alpha = 6000$).

Table 4.9 Optimal therapy for patient MK33 with 9 individual doses per day at given times with some consideration for the total administered dose ($\alpha = 6000$).

Timings:	05:00	06:30	09:15	11:15	11:30	13:15	14:00	15:30	16:00
Doses:	4.61	2.69	2.10	0.86	2.62	0.91	1.30	2.35	0.53

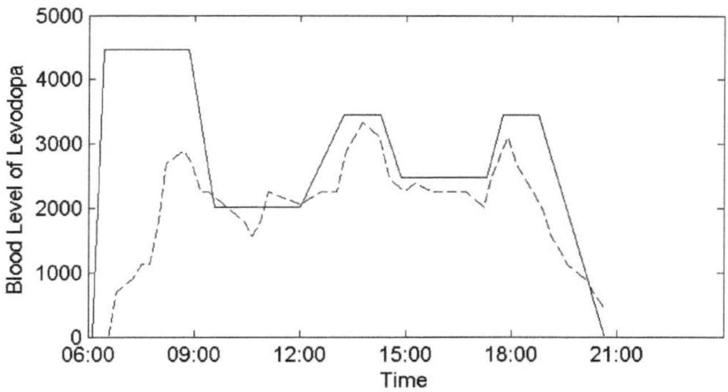

Fig. 4.15 Simulation of the therapy in Table 4.10 ($\alpha = 7000$) shows a poor agreement of the actual and desired time-histories as a result of missing doses at the beginning and the end.

Table 4.10 Optimal therapy for patient MK33 with 9 individual doses per day at given times with heavier consideration for the total administered dose ($\alpha = 7000$) results in the first and the last administration to vanish.

Timings:	05:00	06:30	09:15	11:15	11:30	13:15	14:00	15:30	16:00
Doses:	0.00	3.18	2.03	1.14	2.38	0.98	1.23	2.58	0.00

How is the value of α in the objective function (4.13) to be determined? The clinician will assign a different value to α according to his assessment of the side-effects. The more critical the side-effects are the less drugs should be prescribed, meaning a larger value for α. If α is assigned the value zero it means that the side-effects are not considered at all in the optimization. For instance, taking α as 6000 in our example reduces the daily dosage by 13% as can be seen in Table 4.9. If α is taken as 7000 it results in a 35% reduction of the daily dosage as can be seen in Table 4.10. However, as the simulations of these therapies show, attaining the desired therapeutic effect becomes more difficult as α grows. □

Before we move on, the reader should recognize that the fact that the resulting doses in this section have non-integer values makes the results hence the methods described herein rather unrealistic.

4.2.4 *Solution of the Multiple Dosing Problem with a Quadratic Objective Function*

In the previous Subsection we have studied various optimization problems based on the L_1 norm of the error between the actual and desired time-histories of a drug's blood concentration. We have seen that such problems can be reduced to a classical linear programming problem. What happens when we look at the L_2 norm of the error? In this Subsection we will try to minimize

$$Z = \int_{t_0}^{t_f} (y(t) - y_{set}(t))^2 dt \tag{4.14}$$

What we know is $f(t)$, $y_{set}(t)$ as well as t_0 and t_f.

The doses must realistically be integer multiples of a basic weight. Therefore, we can designate this integer value to each possible dose. Due to the potential immediate side-effects of very high doses, a single dose cannot be arbitrarily large. Thus, the possible set of doses can be given as

$$d_i \in \{0, 1, 2, ..., d_{max}\} \tag{4.15}$$

Similarly, the set of possible timings can be given as

$$\tau_i \in \{..., -30min, -15min, 0, 15min, 30min, 45min, ..., t_f\} \tag{4.16}$$

Note that the negative values for τ_i result from the consideration that sometimes the drug must be administered before t_0 in order for the plasma concentration to reach the desired value by t_0.

What we now will try to do is to find the parameters $d_1, d_2, ..., d_n$ and $\tau_1, \tau_2, ..., \tau_n$ such that

$$Z(d_1, d_2, ..., d_n, \tau_1, \tau_2, ..., \tau_n) \rightarrow Minimum \tag{4.17}$$

In other words, we look for a set of parameters in the $2n$-dimensional parameter space which minimizes the value of the objective function in (4.14).

One way to perform parameter optimization in a higher dimensional parameter space is the so called "cyclic coordinate search method". In our case this means that starting from a reasonable combination of doses and timings, all parameters are kept constant except for the first dose. The first dose is varied till a minimum value for Z is reached. Then, all the parameters are kept constant (including the first dose) and the first timing is varied till a minimum value for Z is reached. This process is continued till all of the doses and the timings are varied one by one, keeping the rest of the parameters constant. Then, starting again with the first dose, the whole procedure is repeated again till the value of Z stops decreasing, meaning that the therapy which results in the minimum value of Z is reached.

The computation of $y(t)$ needed to calculate Z can be done using equation (4.8). Note that neither $y_{set}(t)$ nor $f(t)$ have to be known analytically. For instance, if the desired time-history and the pulse-response are known at certain given points in time only, the values of these functions for the times in between those points can be computed through linear interpolation. Of course, spline interpolation or curve fitting, especially for the pulse-response, could also be used [257], [130].

So far, the objective function has been assumed to be unimodal, that is, having a single minimum. In general, this is obviously not the case. However, when we accept the unpleasant fact that the objective function is multimodal, meaning that

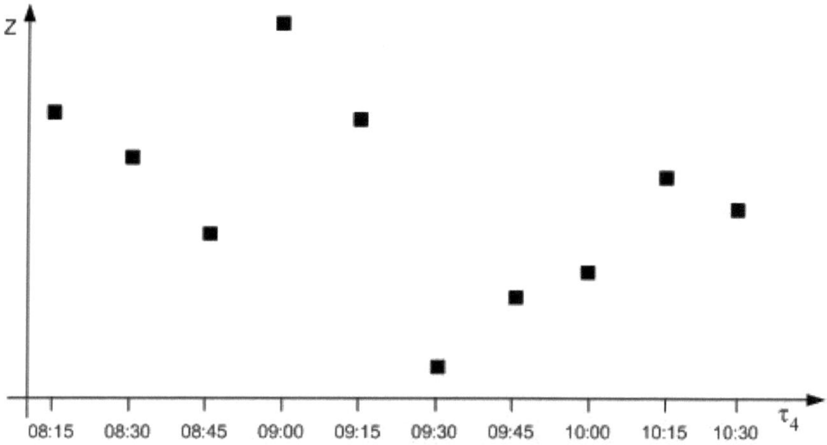

Fig. 4.16 The objective function may have several local minima (here at 08:45 and 09:30).

it has several local minima, the situation shown in Figure 4.16 might occur. This figure shows the plot of a hypothetical objective function versus, say, τ_4. Suppose we start with τ_4=9:00. We indeed have a case as just mentioned, whereas

$$Z(08:45) < Z(09:00) > Z(09:15)$$

and

$$Z(08:45) < Z(09:15)$$

The obvious choice of 08:45 would (since we do not know the values of $Z(09:30$ *etc.*) lead us to only a local minimum. There is no easy solution to this particular problem. Therefore the following trick – which sounds somewhat contradictory – can be used: assume the objective function to be unimodal and search the minimum as described above. However, since it is known that the minimum which is found is a local one depending on the starting values of the doses and the timings, change the initial guess several times and apply this algorithm to each new combination. Then, the optimal therapy should be taken as the set of doses and timings resulting in the smallest value of the objective function out of these results. Such an approach of varying the starting values randomly is a special case of the Monte Carlo method [190]. [275] gives an extensive survey of such methods. A more recent approach based on particle swarm optimization is given in [278].

Example: Table 4.11 shows the results of the repeated application of the cyclic coordinate search method with different initial guesses on a typical Parkinsonian patient. As can be seen, the optimal therapy reached in this manner is the one with the underlined value of the objective function in Table 4.11. However, it can be observed that most of the optimal therapies result in values of the objective function which differ from each other only within the magnitude of the accuracy of this method.

Table 4.11 Optimizing Z with the cyclic coordinate search method starting with different therapies results in different optimal therapies.

									Value of the objective function
Initial guess	Timings:	06:00	08:00	10:00	12:00	14:00	16:00	18:00	
	Doses:	3	1	1	1	1	1	1	618.14
Optimal therapy	Timings:	05:30	09:00	10:30	13:00	14:30	16:00	17:30	
	Doses:	4	2	2	1	1	1	1	332.35
Initial guess	Timings:	05:30	08:45	10:00	10:45	13:00	14:45	16:45	
	Doses:	4	1	1	2	1	1	2	354.21
Optimal therapy	Timings:	05:30	08:30	09:45	10:45	13:00	14:30	16:45	
	Doses:	4	1	1	2	1	1	2	338.31
Initial guess	Timings:	05:30	08:45	10:00	10:45	13:00	14:45	16:45	
	Doses:	3	1	1	1	1	1	1	457.75
Optimal therapy	Timings:	05:30	09:15	10:30	11:15	13:00	15:00	17:15	
	Doses:	4	2	1	1	1	2	1	375.96
Initial guess	Timings:	05:30	08:00	10:15	12:00	14:00	16:00	17:30	
	Doses:	3	1	3	1	2	1	1	352.67
Optimal therapy	Timings:	05:30	08:15	10:15	12:00	14:00	16:00	17:30	
	Doses:	4	1	3	1	2	1	1	335.44
Initial guess	Timings:	06:00	08:00	10:00	12:00	14:00	16:00	18:00	
	Doses:	3	1	3	1	2	1	1	486.45
Optimal therapy	Timings:	05:30	08:15	10:15	12:15	14:15	16:15	17:15	
	Doses:	4	1	3	1	2	1	1	335.77

This enables the physician to choose from different possible therapies according to other considerations not included in the optimization process. □

4.2.5 Semi-analytical Solution of the Multiple Dosing Problem

In this Subsection, we will assume that the number of the doses as well as the timings are prescribed and try to find the corresponding doses such that the performance index in (4.14) is as small as possible [131].

Since we assumed linearity in our system, we can say that the error $e(t) = y(t) - y_{set}(t)$ is a linear function of the doses $\mathbf{d}$:

$$e(t) = \mathbf{a}^T(t)\mathbf{d} + b(t) \tag{4.18}$$

Substituting (4.18) in the performance index (4.14) results in

$$Z = \int_{t_0}^{t_f} \left\{ [\mathbf{a}^T(t)\mathbf{d}]^2 + b^2(t) + 2\mathbf{a}^T(t)\mathbf{d}b(t) \right\} dt$$

Since Z is quadratic in $\mathbf{d}$, the condition

$$\frac{\partial Z}{\partial \mathbf{d}} = 0$$

is necessary and sufficient for the minimum, thus the **d** fulfilling this condition is the **d** for which the error is minimum. We can rewrite this condition through the substitution of the previous equation in it and through the exchange of the integration and the derivation as

$$\frac{\partial Z}{\partial \mathbf{d}} = \frac{\partial}{\partial \mathbf{d}} \int_{t_0}^{t_f} [\mathbf{a}^T(t)\mathbf{d}]^2 dt + \frac{\partial}{\partial \mathbf{d}} \int_{t_0}^{t_f} 2\mathbf{a}^T(t)\mathbf{d}b(t)dt = 0$$

The term with $b^2(t)$ vanishes since it is not dependent on **d**. Remembering the rule of differentiation with respect to a vector

$$\frac{\partial(\mathbf{a}^T\mathbf{x})}{\partial \mathbf{x}} = \mathbf{a}$$

this equation is equivalent to

$$\int_{t_0}^{t_f} [2\mathbf{a}(t)\mathbf{a}^T(t)\mathbf{d}]dt = -\int_{t_0}^{t_f} [2b(t)\mathbf{a}(t)]dt$$

We can now divide both sides by 2 and get **d** out of the integrand since it is not a function of time. This gives us

$$\int_{t_0}^{t_f} [\mathbf{a}(t)\mathbf{a}^T(t)]dt\mathbf{d} = -\int_{t_0}^{t_f} [b(t)\mathbf{a}(t)]dt$$

This is a linear equation in **d** of the form

$$\mathbf{F}\mathbf{d} = \mathbf{g} \tag{4.19}$$

Let's remember

$$e(t) = y_{set}(t) - \sum_{i=1}^{n_p} d_i f(t - \tau_i)$$

This equation, written without the summation sign is

$$e(t) = -f(t-\tau_1)d_1 - f(t-\tau_2)d_2 \cdots - f(t-\tau_{n_p})d_{n_p} + y_{set}(t)$$

When we look at the equation (4.18) which defines the error as a linear function of the doses we see that

$$\begin{aligned} a_i &= -f(t-\tau_i) \\ b(t) &= y_{set}(t) \end{aligned}$$

and this gives us the elements of **F** and **g** as

$$\begin{aligned} f_{ij} &= \int_{t_o}^{t_f} f(t-\tau_i)f(t-\tau_j)dt \\ g_i &= \int_{t_o}^{t_f} y_{set} f(t-\tau_i)dt \end{aligned}$$

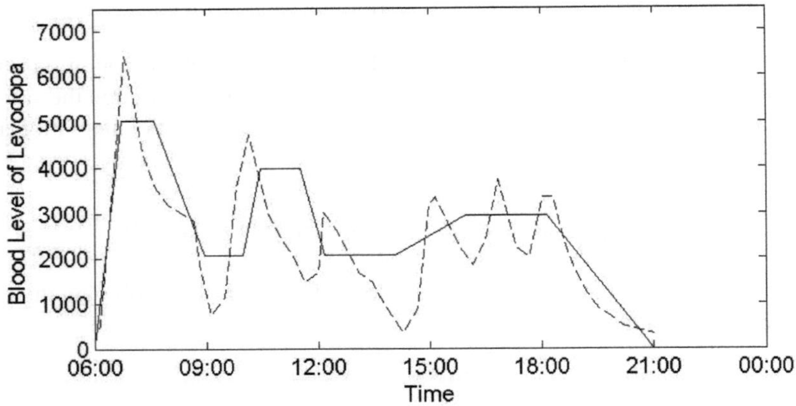

Fig. 4.17 Simulation of the therapy in Table 4.12.

Table 4.12 Quadratically optimal therapy for patient MB31 with 8 individual doses per day at given times calculated analytically.

Timings:	06:15	08:00	09:30	11:45	14:30	16:15	17:30	19:45
Doses:	3.89	0.76	2.64	1.70	2.19	1.66	1.38	0.28

Example: For patient MB31 and the desired trajectory shown in Figure 4.17 the timings are given by the physician for $n_p = 8$ administrations during a day as shown in Table 4.12. The optimal doses in Table 4.12 are calculated by solving equation (4.19). The simulation of the resulting therapy which minimizes the objective function (4.14) is also shown in Figure 4.17. □

4.2.6 Optimization of Drug Administration By a Tauberian Approach

The simplicity of the analytical dosage optimization lies in the fact that the dose-drug level relationship is linear. However, the relationship between the timings and the drug level is highly non-linear, thus rendering itself unsuitable for an analogous analytical solution. Nevertheless, another approach which is based on a Tauberian approximation of the drug level set-trajectory results in an acceptable solution.

The desired profile of the drug concentration can be expressed by means of a Tauberian approximation as a sum of weighted and time shifted pulse-responses [313]:

$$y_{set}(t) = \sum_{i=1}^{n} d_i f(t - \tau_i) \tag{4.20}$$

the Fourier transform of which is

$$Y_{set}(\omega) = \sum_{i=1}^{n} d_i e^{-j\tau_i\omega} F(\omega) \tag{4.21}$$

or rewriting this equation

$$H(\omega) = \frac{Y_{set}(\omega)}{F(\omega)} = \sum_{i=1}^{n} d_i e^{-j\tau_i\omega} \tag{4.22}$$

$H(\omega)$ is computable since $Y_{set}(\omega)$ and $F(\omega)$ can both be determined by Fourier transforming the given functions $y_{set}(t)$ and $f(t)$. Introducing the notation $H_k = H(\omega_k)$ where $\omega_k = \omega_0 + k\Delta\omega$ for $k \in \{0,1,2,...,K-1\}$ with $K > 2n$ results in

$$H_k = \sum_{i=1}^{n} d_i e^{-j\tau_i\omega_k} \tag{4.23}$$

Prony has shown that the solution of the equation system in (4.23) can be reduced to the problem of finding the zeros of one single function. It can be shown that the optimal timings can be computed with the following scheme [91], where

$$c_0 = 1, \quad \mathbf{c} = \begin{pmatrix} c_1 \\ c_2 \\ \vdots \\ c_n \end{pmatrix}, \quad \mathbf{H} = \begin{pmatrix} H_{n-1} & \dots & H_0 \\ \vdots & \vdots & \vdots \\ H_{k-2} & \dots & H_{k-n-1} \end{pmatrix}, \quad \mathbf{h} = \begin{pmatrix} H_n \\ \vdots \\ H_{k-1} \end{pmatrix} \tag{4.24}$$

1) Compute c with the equation

$$\mathbf{Hc} = -\mathbf{h} \tag{4.25}$$

The general solution of this equation is given by

$$\mathbf{c} = -(\mathbf{H}^*\mathbf{H})^{-1}\mathbf{H}^*\mathbf{h} \tag{4.26}$$

with $*$ denoting the conjugate complex transpose of a matrix.
2) Compute the n solutions of the equation

$$\sum_{i=0}^{n} c_i z^{n-i} = 0 \tag{4.27}$$

Due to (4.22), the roots of the polynomial in (4.27) have a simple structure containing the timings alone as

$$z_i = e^{(-j\tau_i\Delta\omega)} \quad \Rightarrow \quad |z_i| = 1 \tag{4.28}$$

3) Compute the timings from the roots in (4.28) with

$$\tau_i = \frac{1}{\Delta\omega}[\arctan(-\Im\{z_i\}/\Re\{z_i\}) + (\pi)] \tag{4.29}$$

This scheme has the following distinct advantages : a) If (4.20) holds, the optimal timings τ_i can be computed exactly, b) The equation system for the computation of the timings contains a single non-linear equation, namely (4.27), c) The method is lucid and results in a unique solution, d) The computational expense for the solution of the equation is small.

On the other hand this method brings with itself certain problems: as mentioned before, the desired profile, in general, cannot be attained exactly. Therefore (4.22) has to be modified as follows

$$H(\omega) = \frac{Y_{set}(\omega)}{F(\omega)} = \sum_{i=1}^{n} d_i e^{-j\tau_i\omega} + V(\omega) \tag{4.30}$$

where $V(\omega)$ is the Fourier transform of $v(t)$ which is assumed to be a zero mean Gaussian noise. This means, however, that the elements of the matrix **H** and the vector **h** in (4.24) are not exact, but noisy. Because of the noise imposed on the H_k's, a substantial property of the Prony's algorithm is lost, i.e., the absolute values of the roots z_i in (4.28) are different from unity. As a results of this and in order to achieve a simple solution for the computation of the timings, the absolute values of the roots z_i should be normalized to unity in the direction of the complex pointer as illustrated in Fig. 4.18.

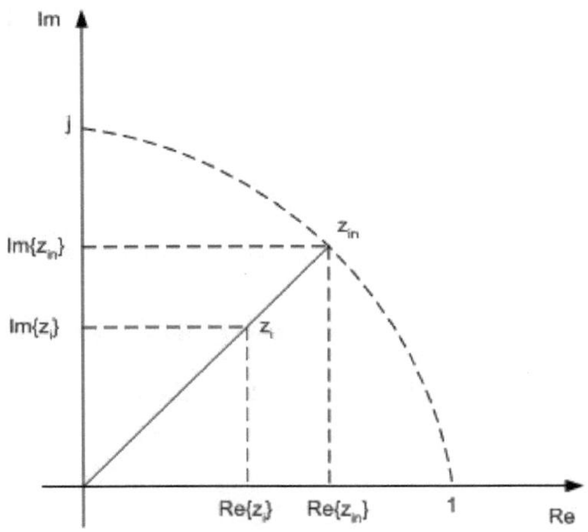

Fig. 4.18 The roots of (4.27) have to be normalized to unity to make use of the Prony algorithm.

However, the negative influence of the noisy roots can be substantially reduced by a time shift of the desired profile relative to the pulse response. As can be seen in Fig. 4.19 $y_{set}(t)$ has to be shifted such that the resulting timings are as large as possible. Of course this off-set has to be corrected later during the computation of the timings.

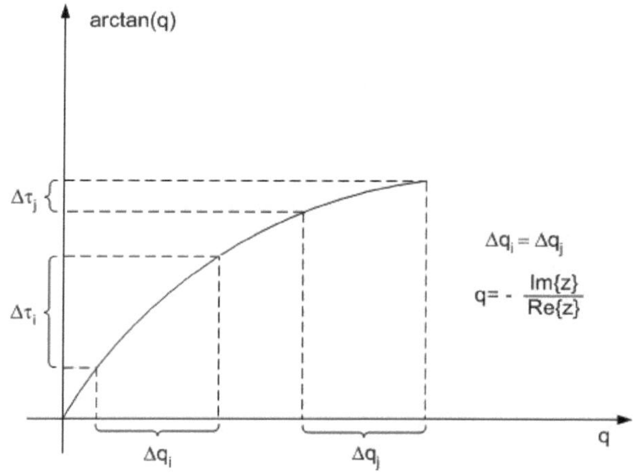

Fig. 4.19 Visualization of the fact that that the influence of the noise on the roots z_i is bigger for small τ_i.

Another possibility to reduce the effect of the noisy H_k's on the computation of the timings makes use of the fact that at least $2n$ H_k's are needed to solve (4.25).The minimum number of H_k's to be used in the linear regression can be estimated by means of the concept of the power of a signal. This can best be explained with the aid of a periodic desired profile $y_{set}(t)$ with the period T. The pulse response can also be thought to continue with the same period. The Fourier coefficients can now be used to compute the H_k's. With ω_T as the basic harmonic.

$$\omega_k = k\omega_T = k\frac{2\pi}{T}, \qquad k = 0, ..., K-1, \quad K > 2n \tag{4.31}$$

$$H(\omega_k) = \frac{Y_{set}(\omega_k)}{F(\omega_k)} = \frac{Y_{set}(k\omega_T)}{F(k\omega_T)} = \frac{Tc_{ky}}{Tc_{kf}} = H(k\omega_T) \tag{4.32}$$

where

c_{ky} =k-th Fourier coefficient of $y_{set}(t)$ and c_{kf} =k-th Fourier coefficient of $f(t)$.

The power of the signal $y_{set}(t)$ can now be estimated as

$$P_y = \frac{1}{T}\int_0^T y_{set}^2(t)dt = c_{0y}^2 + 2\sum_{k=0}^{\infty}|c_{ky}|^2 > 2\sum_{k=0}^{\infty}|c_{ky}|^2 \tag{4.33}$$

P_f can also be estimated in the same manner. If, now, a maximum permissible deviation of the power is given arbitrarily, the minimum number of H_k's for that deviation is fixed.

Yet another way of reducing the adverse effect of noisy H_k's on the optimization of the timings is to compute more than n necessary timings. Afterwards, by eliminating, for instance, the timings which are outside the period of interest or coalescing the timings which are very close to each other their number can be reduced to n.

Another problem which arises is that the timings which are computed in this manner have yet to be rounded, because the timings must have discrete values. However, it has been found out that this rounding error is negligible for discretization steps of a quarter hours.

Example: Application of the method explained above to patient MB31 was executed in the following way: first, the timings were computed for a set of given dosage sequence. This was accomplished by the algorithm described above. Then the doses were minimized for the resulting timings. This was done by dynamic programming [271]. The whole procedure was repeated till the objective function stopped getting smaller and a local minimum was achieved (obviously, this approach is equivalent to a cyclic coordinate search method where the doses and the timings build the "meta-axes"). This process was repeated starting with different initial guesses of the dosage sequence – the so called Monte Carlo method.

The fact that the doses must have discrete values was exploited in the application of dynamic programming for their optimization [133]. The reason why dynamic programming was not used to optimize the timings as well as the doses lies in the enormous computational expense it would require in this case – the so called "curse of dimensionality".

Obviously, a high number of administrations is rather uncomfortable for the patient. Therefore, the number of the administrations was also included in the optimization process, thus modifying the objective function to

$$Z = \int_{t_0}^{t_f} (y_{set}(t) - y(t))^2 dt + \beta n \tag{4.34}$$

β denotes the weighting factor for the number of administrations. The optimal therapy schemes for a patient for β different from zero and equal to zero result in eight, respectively thirteen administrations in a day, which is the periodicity of the desired profile.

The resulting therapy schemes are given in Table 4.13 and Table 4.14. The simulations of these sequences of administrations are shown in Figure 4.20 and 4.21 respectively. It can be seen in these figures and tables that a decrease in the value of β results, up to a certain point, in an increasing number of administrations in a day, which facilitates attaining the desired profile. □

4.2.7 Remarks on Pharmacokinetic Optimization

In this Section, we have seen the application of several optimization techniques to the problem of dosage optimization with respect to an objective function containing the pharmacokinetics of the used drug. It can be said that each of these techniques, varying from analytical optimization to dynamic programming, has its advantages and drawbacks. Not surprisingly, the more exact methods require longer computational time. It soon became apparent that the more sophisticated algorithms resulted in an overkill due to the errors in modeling. Experience resulting from extensive work with different algorithms and techniques shows that the cyclic

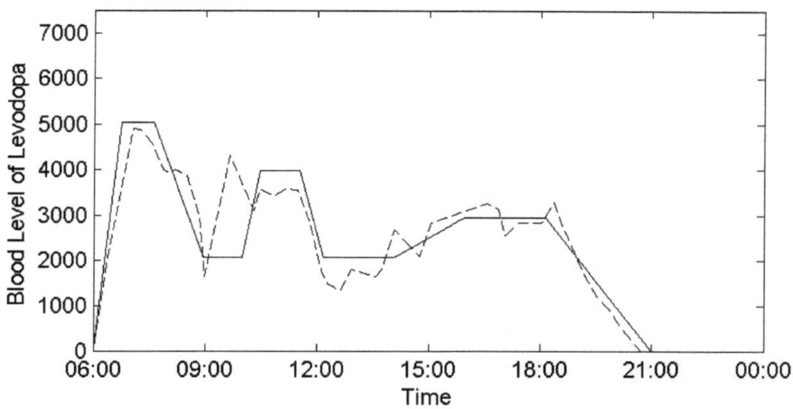

Fig. 4.20 Simulation of the therapy in Table 4.13.

Table 4.13 Optimal therapy for patient MB31 according to (4.34) with $\beta = 0$ results in thirteen doses a day.

Timings:	06:00	06:45	07:45	09:00	10:00	10:45	12:30	13:30	14:30	15:15	16:00	17:00	17:45
Doses:	3	1	1	1	2	1	1	1	1	1	1	1	1

coordinate search method is the most simple and time efficient method to optimize the discrete doses and timings.

It has to be emphasized at this point that the methods used in this Section use input-output relations of a system and not its state space description, which is used in the classical compartmental techniques. Furthermore, the input-output relation does not have to be known analytically. As a matter of fact, all the methods in this chapter relied on the description of this relation in table form with interpolations. The price one has to pay for this convenience is the loss of knowledge about the internal states of the system.

It should be pointed out that (4.8), i.e., the linearity of the pharmacokinetic relationship holds, must be tested thoroughly before the above explained algorithms can be applied. One way of doing this can be sketched as follows: the blood level of the relevant substance should be measured at different times after the administration of a unit pill. After waiting until the level sinks back to its original value, a multiple of the unit pill should be administered and the profile of the relevant substance's blood level recorded again. If both of the time histories are proportional and the proportionality factor is the multiplicity of the second pill, one should proceed with the next test where the blood level profile should be measured after temporally shifted multiple administrations of the drug to validate the assumption of linearity. Only if both of these tests result in satisfactory findings can the explained algorithms be

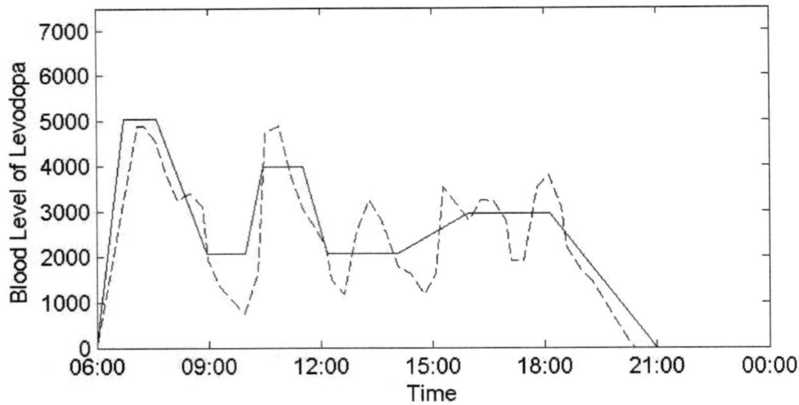

Fig. 4.21 Simulation of the therapy in Table 4.14.

Table 4.14 Optimal therapy for patient MB31 according to (4.34) with $\beta \neq 0$ results in eight doses a day.

Timings:	06:00	06:45	08:00	10:00	12:30	14:45	15:45	17:15
Doses:	3	1	1	3	2	2	1	2

used sensibly. The second test is especially important if the drug contains several chemicals which affect each others absorption or excretion rates.

The results of the above mentioned optimizations must be handled with a healthy dose of scepticism since they in no way account for disturbances like varying effects of different nourishment on the absorption of the drug. Other optimization strategies which are based on stochastic optimal control are more suitable for cases when those effects have to be taken into consideration (see, for instance, [37]).

In case the pharmacodynamic response can be modeled as a linear system, the therapy can be optimized the same way to achieve a desired time history of the relevant dynamic variable like the body temperature or the blood pressure rather than the plasma concentration of the medication. However, such a relation is seldom, if ever, linear. It is more likely that it can be approximated as linear in a limited zone or domain.

Therefore, it is very interesting to establish a relationship between the drug administration and its therapeutic response. In order to do this, the effect has to be made measurable. Consequently, the next Chapter deals with the problem of measuring the neurological functions and more specifically the effects of dopaminergic substitution in Parkinson's disease.

4.3 Optimization of Drug Delivery Based on Pharmacodynamic Data

As mentioned in the previous Section, a mathematical model which describes the mathematical relationship between the drug administration and the effects of the administered drug is needed to compute an optimal therapy. In order to determine the parameters of such a model the effects have to be or made measurable.

So far, the term *pharmacokinetics* have been used to describe the process controlling the drug concentration at any time after a single or multiple dose and the usual pharmacokinetic models describe drug concentration as a function of both dose and time [148]. However, most common pharmacodynamic models are essentially independent of time and they describe the time-independent relationship between the drug's plasma concentration and its effect as Figure 4.22 shows [149]. Various pharmacodynamic models have been proposed and used for different purposes. The most common ones will be briefly discussed here. (Actually the term pharmacodynamics is a misnomer since it usually contains no dynamics!)

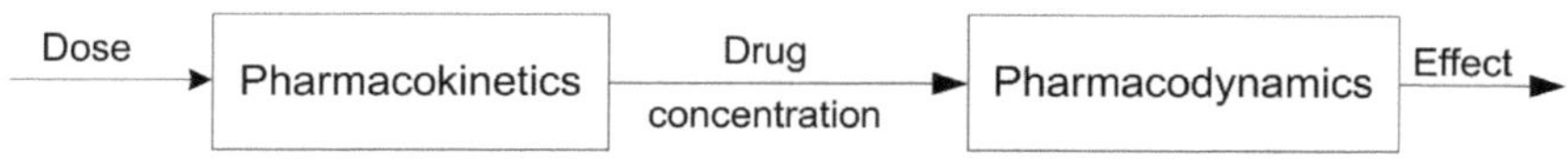

Fig. 4.22 Schematic representation of the dose-effect relationship using pharmacokinetic and pharmacodynamic models.

The **fixed effect** model can be seen as a two-point relay, stating whether the effect is present or not. This simple model can be used by stating a threshold for the effect and giving the concentration level of the used drug above which the effect is present. An example of such a model for the levodopa pharmacokinetics and pharmacodynamics in fluctuating parkinsonian patients was used in [237].

The $\mathbf{E_{max}}$ model is another simple model which, however, describes the drug effect over the whole range of concentrations. It is based on the hyperbolic relationship

$$E = \frac{E_{max}C}{EC_{50}+C} \tag{4.35}$$

where E is the effect, C the concentration, E_{max} the maximum effect due to the drug and EC_{50} is the concentration producing 50% of E_{max}. This model which has long been in use in enzyme kinetics and protein binding has a sound basis in the molecular level [16]. Two very important properties of this model is that it predicts no effect when no drug is present and that it predicts the maximum effect attainable by the drug. This model is so successful, because it incorporates the biological law of diminishing returns – which simply means that ever higher concentrations are needed to increase the effect by a given amount.

The **linear model** is described by

$$E = S \cdot C \tag{4.36}$$

where S is the slope of the line relating effect to concentration. It is applicable when the drug concentrations are much smaller than E_{50} in the E_{max} model.

The **sigmoid E$_{\mathbf{max}}$** model differs from the E_{max} model only in an additional parameter and it is described by

$$E = \frac{E_{max} C^n}{EC_{50}^n + C^n} \tag{4.37}$$

n does not have to have a physical meaning. It just influences the slope of the curve described by (4.35).

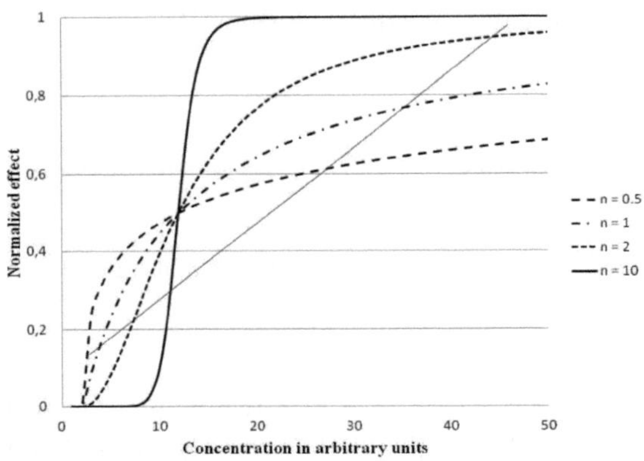

Fig. 4.23 Various pharmacodynamic models are depicted in this figure. Note that, except for the straight line corresponding to the linear model, all curves pass through the point (10, 0.5) which means $EC_{50} = 10$; in other words, the concentration level which produces 50% of the maximum effect is 10.

Figure 4.23 shows some of these models. The **linear model** is depicted by the straight line in this figure. It shows that as the concentration increases so does the effect proportional to the concentration. The curves correspond to various values of n in the **sigmoid** E_{max} model. $n = 1$ corresponds to the **E$_{\mathbf{max}}$** model. One can also see that as n increases, the **sigmoid E$_{\mathbf{max}}$** model approaches the **fixed effect** model with the threshold at 10 arbitrary units of concentration, meaning that there is no effect if the concentration is below 10. Above that level the full effect of the drug is present.

Another possibility of modeling the pharmacodynamics is by means of a **dynamic model** although this approach is not often used. Such a model can be described by

$$E = f(C) \tag{4.38}$$

where f is the functional which maps the concentration and its derivatives on the effect.

No matter which of these pharmacodynamic models is used, an objective function involving the therapeutic effects of the used drug has to be defined to perform any optimization. An elaborate form of (4.38) will be used in the next Chapter on Parkinson's disease.

4.4 Exercises

A1: Explain the concepts of open-loop control, closed-loop control, adaptive control and knowledge-based control in connection with therapeutic situations and give several examples of each.

A2: What is the main difference between optimizing a therapy with respect to pharmacokinetic and pharmacodynamic data? What are the advantages and drawbacks of these methods?

A3: What is the main difference between optimizing an objective function involving the L_1 and L_2 norms of the discrepancy between the desired and actual time-histories of a drug's plasma concentration?

B1: How could an objective function look which makes sure that a drug's plasma concentration a) *never* exceeds a toxic level y_{max} and b) is *preferably* over a minimum y_{min} to be effective and c) should have a better agreement with a desired profile y_{set} later during the day.

C1: Use integer dynamic programming to solve the problem of finding the optimal doses and their optimal administration times, beginning with the practical fact that these have to be integer multiples of a basic dose administered with intervals which are multiples of, say, one hour.

C2: An elegant method of computing the times and sizes of step inputs for a linear system which optimizes a quadratic objective function is given in [236]. Modify this theory for impulse inputs and apply it to a problem of the type described in Section 4.2.

C3: An "exact" approach to drug delivery problems was used in this Chapter. Develop the corresponding algorithms which use "fuzzy logic" to solve the same kind of problems (see also Chapter 7 and Appendix B).

Chapter 5
Parkinson's Disease

Involuntary tremulous motion, with lessened muscular power, in parts not in action and even when supported; with a propensity to bend the trunk forwards, and to pass from walking to a running pace: the senses and the intellect being uninjured.

— *James Parkinson, An Essay on the Shaking Palsy*

With Parkinson's, it's like you're in the middle of the street and you're stuck there in cement shoes and you know a bus is coming at you, but you don't know when. You think you can hear it rumbling, but you have a lot of time to think. And so you just don't live that moment of the bus hitting you until it happens. There's all kinds of room in that space.

— *Michael J. Fox*

5.1 Introduction

The parkinsonian syndrome, or parkinsonism, is a syndrome characterized by tremor, muscular rigidity and loss of postural reflexes [226]. The largest number of cases of parkinsonism falls in the category of Parkinson's disease (also known as primary parkinsonism, paralysis agitans or shaking palsy) which has no definable cause. Patients with the so called secondary parkinsonism (known also as symptomatic parkinsonism) have other symptoms besides the parkinsonian ones.

5.2 Etiology

The parkinsonian syndrome is a leading cause of neurological disability in adults over sixty years of age and it affects both sexes and all races. The overall prevalence rate of the parkinsonian syndrome has been estimated to be about 100 to 150 per 100,000 population with an incidence of 20 cases per 100,000 per year [226].

S.S. Hacısalihzade: *Biomedical Applications of Control Engineering*, LNCIS 441, pp. 99–135.
DOI: 10.1007/978-3-642-37279-7_5

The occurrence of the parkinsonian syndrome is believed to be correlated with the loss of neurons in the basal ganglia [34]. It is not yet understood whether the parkinsonian syndrome is a disorder of the extrapyramidal motor system alone or if all catecholamine neurons – noradrenergic as well as dopaminergic – are affected. The majority of the cases with parkinsonism are idiopathic, *i.e.*, the etiology of neuronal degeneration in these patients is unclear. These cases, *i.e.*, the patients with Parkinson's disease, are in most countries – inexplicably – more prevalent in the male in the ratio of 10:7 [92] in spite of exceptions. Also the average age of onset of Parkinson's disease is about seven years earlier in the male. Parkinson's disease is about four times more common in people with fair complexion compared with dark colored races [185]. A widely accepted view attributes this unevenness to the abundance of melanin, the loss of which is seen by some researchers as a cause of Parkinson's disease [214], in the pigmentation of dark skinned individuals.

Studies investigating the role of aging in the etiology of Parkinson's disease show that the dopamine levels in the corpus striatum of healthy adults diminish by 13% per decade. The appearance of parkinsonian symptoms coincides with this level falling 70% below the normal [266]. It has also been suggested that the onset of parkinsonian symptoms is due to the combined effects of normal aging and a lesion which has been acquired earlier (maybe even during infancy) and which resulted in a loss of more than 20% of the dopaminergic neurons [63].

Recent epidemiologic studies suggest an increasing prevalence of Parkinson's disease during the industrial age [217] and a correlation between incidence rates and environmental toxins has been widely established [270] [117], [26]. Unusual cases of drug abuse with synthesized designer drugs [88] led to the positive identification of MPTP as a parkinsonian neurotoxin [188], [64]. The metabolite of MPTP, MPP^+, is taken up by catecholaminergic neurons in the striatum and the substantia nigra and thus causes the cell death [155], [87], [287], [288]. Presently, this drug is employed to induce parkinsonism in animals to create a valid animal model for Parkinson's disease [322] which may help towards a better understanding of the etiology of Parkinson's disease and a therapy of patients suffering from it.

Many researchers from the late nineteenth century [194] on till more recently [282] believed in the significance of hereditary factors in the etiology of Parkinson's disease. Although many neurologic conditions are genetically determined, no such influence has been positively demonstrated in Parkinson's disease [306], [98] except for early-onset cases [25]. The erroneous belief in the role of hereditary factors may be due to the fact that the environmental factors affect all members of a family in a similar fashion.

The cases of the parkinsonian syndrome with a definable disease process are called secondary or symptomatic parkinsonism. Reasons for the cases with secondary parkinsonism include encephalitis lethargica [58], cerebral arteriosclerosis [184], [254], intoxication with chemicals like manganese [1], carbon monoxide, [281] and carbon disulfide [235], the use of some neuroleptic agents [249] and – in a smaller scale – brain tumors, trauma, syphilis or Behçet's syndrome [240].

5.3 Pathophysiology

Already researchers of the first half of last century had pinpointed the substantia nigra as the locality of pathogenesis [302], [139]. Ehringer and Hornykiewicz [99] discovered, that dopamine was depleted in the brains of patients with parkinsonism. Subsequent research confirmed that the death of pigmented neurons in the substantia nigra, locus ceruleus and brain stem nuclei led to a degeneration of the nigrostriatal dopaminergic pathway and that the corresponding dopamine deficiency accounts for the major symptoms of the parkinsonian syndrome [147], [14], [35].

Biochemical, pharmacological and histochemical studies have shown that the major chemicals involved in the electrochemical transmission of the neural impulses in the central nervous system include dopamine, noradrenaline, acetylcholine and serotonin which gives them their generic name neurotransmitters [47], [118]. The concentration of acetylcholine – the transmitter of the cholinergic system – is not diminished in the parkinsonian striata [226] and earlier studies showed a relationship between dopamine and extrapyramidal motor activity [69], [36]. These two facts combined with the proof of dopaminergic depletion in substantia nigra led to the hypothesis that the loss of dopamine leads to a predominance of acetylcholine, and that the symptoms of parkinsonism are caused by the disturbed balance between the dopaminergic and the cholinergic activity in the striatum [266], [226].

However, it is not yet certain that the loss of dopamine is the only defect in the brains of patients with parkinsonism. Indeed it has been suggested that noradrenaline and serotonin may also be involved. On the other hand, it is well established that restoration of striatal dopamine activity has an anti-parkinsonian effect and this constitutes the basis of the symptomatic therapy of patients with parkinsonian syndrome.

In the beginnings of the last century many researchers believed that coordinated movements could be explained in terms of sequential activation of reflexes. The concept of command interneurons and the notion of motor programs introduced in the 1970's, however, brought the attention away from reflex systems [183]. Different models of varying detail and complexity have been proposed to explain the various aspects of the pathophysiology of Parkinson's disease [253], [103], [213], [295], [31], [57], [247], [191], [296].

Clinical evidence suggests that the basal ganglia are responsible for the automatic execution of learned motor programs [214]. Figure 5.1 illustrates schematically the relationships of the basal ganglia to motor output structures.

Figure 5.2 shows the roles of several brain structures in movement. It has been suggested that the basal ganglia and the cerebellar hemispheres are involved with the association cortex in the programming of voluntary movements. When motor command from the basal ganglia, the cerebellum and the association cortex reaches the motoneurons, the pars intermedia updates the intended movement, based on the motor command and somatosensory information. Follow-up correction can be performed by motor cortex when cerebellar hemisphere and pars intermedia malfunction [10].

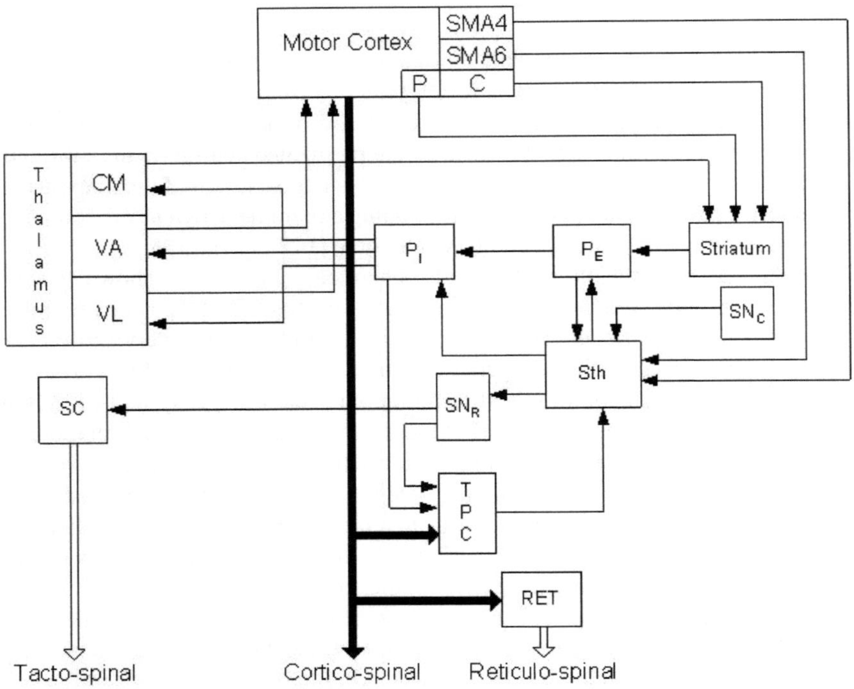

Fig. 5.1 Relationships of the basal ganglia to motor output structures. The arrows show pathways between structures. CM: centromedian nucleus of thalamus, P_I, P_E: globus pallidus internus, externus, Ret: reticular nuclei, SC: superior colliculus, SNc, SNr: substantia nigra compacta, reticulata, SMA: supplementary motor area, Sth: subthalamic nucleus, TPC: pedunculopontine nucleus, VA,VL: ventroanterior and ventrolateral nucleus of thalamus after [296].

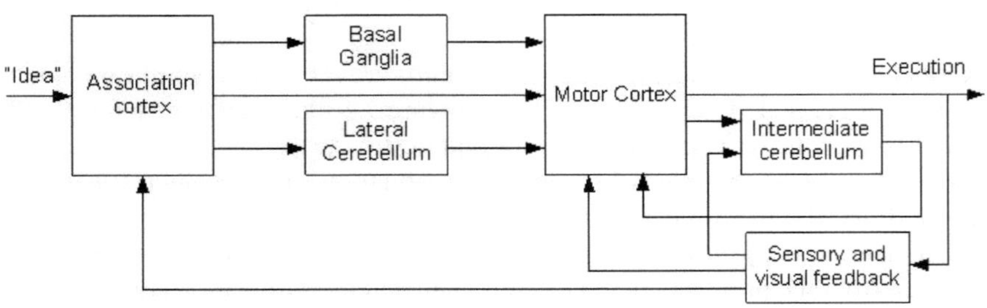

Fig. 5.2 Schematic diagram of the most important physiological relationships in movement initiation and control after [52].

Studies comparing the tracking performance of control subjects with patients suffering from parkinsonism [112], [113], [114] revealed striking differences. Pre-programmed movements in parkinsonians were slow and they showed a decreased repertoire of movements. Parkinson patients did not increase velocity of the movement proportional to the amplitude of the reference signal. Furthermore, parkinsonian movements were broken to several smaller movements when the amplitude became too big. When visual feed-back was interrupted, the parkinsonian movements went "off-track" – suggesting the somatosensory feedback to be insufficient. Finally, tracking performance in parkinsonians did not improve with exercise. It was concluded that the parkinsonians have a defect in their capacity to use kinesthetic feedback to predict the subsequent course of movement.

A long known rule of control engineering states that position control (as in moving a hand from one position to another) can be accomplished with a proportional-differential-integral controller [150]. The controller's differential part can be seen as a predictor in this context. Flower's research in conjunction with studies of cats and MPTP treated monkeys [294] [28] seems to confirm the mathematical model where the parkinsonian symptoms are attributed to a deficiency of the differential part of the controller responsible for the control of motor activity. This exhibits itself in the slowing down of the movements.

5.4 Symptoms

The classical triad of parkinsonian symptoms are **tremor**, **rigidity**, and **akinesia**, although Parkinson [243] did not mention rigidity as one of the cardinal signs of parkinsonian syndrome. Akinesia is characterized by the inability to transform potential energy to kinetic energy, thus rendering the patient immobile. Rigidity is caused by the simultaneous innervation of the agonist and antagonist muscle pair. Tremor is the uncontrolled shaking of the different parts of the body (Parkinson called the disease "shaking palsy"). In Parkinson's disease disturbances of posture, equilibrium and autonomic function are of comparable frequency and prominence [226].

Tremor is described as a pathognomonic symptom of Parkinson's disease [243] but it is by no means *sine qua non* although present together with the loss of fine motoric ability in some 80% of cases [276]. Parkinson tremor is a resting tremor which especially affects distal extremities with a frequency of about 5 Hz, an average velocity of 45 cm/s and a maximum acceleration of 16.8 m/s^2 [45] (the average velocity and maximum acceleration of intention tremor are about four respectively five times bigger). The tremor can virtually disappear during sleep or rest [15], whereas excitement amplifies it (this fact has been traced back to adrenaline production [82]). Parkinson tremor is in many cases initially unilateral and it becomes bilateral as the disease progresses [282].

Muscular rigidity is present almost in all cases of Parkinson's disease [45]. It can be defined as increasing resistance to passive movement. The resistance can be felt throughout the movement in contrast to "clasp knife" rigidity caused by pyramidal lesions. The parkinsonian rigor especially affects the proximity of a member. Rigor is amplified during active movements and during rest and excitement – unlike tremor – it virtually disappears [282].

Akinesia is probably the most complex symptom in Parkinson's disease. It can be described as a liability to sudden arrest of voluntary movements while carrying out purposive action [240]. Constant awareness of fatigue, difficulty in shifting from one motor contraction pattern to another and inability to complete tasks and hence never beginning others have also been reported as manifestations of akinesia [277]. In the progress of Parkinson's disease akinesia is usually preceded by bradykinesia (slowing down of the movements) and hypokinesia (deterioration of the ability to execute voluntary movements) [282]. Probably, akinesia is unrelated to rigidity, because it has been reported to be present without clinical hypertonus.

The physiology behind posture and equilibrium disturbances in Parkinson's disease are controversial. Some researchers see these disturbances as a consequence of rigor [240], whereas others hold a lesion of the nucleus caudatus responsible [38]. Some additional symptoms which can be observed in patients with Parkinson's disease are kinesia and akinesia paradoxa, clog-wheel phenomenon, "on-off" phenomenon and yo-yoing [61] as well as micrography. Although Parkinson himself reported the intellect to be unaffected, some researchers believe that dementia is an intrinsic characteristic of Parkinson's disease [226].

5.5 Treatment

No cure for Parkinson's disease has so far been established, mainly because – as previous Sections indicate – its etiology and pathogenesis are not yet fully understood. Therefore, any treatment can only be symptomatic or palliative. The accepted methods of treatment nowadays include drug treatment, surgical treatment, physiotherapy and psychotherapy whereas the last two are meant to alleviate secondary symptoms like deformities caused by rigor and emotional disturbances.

Surgical treatment of Parkinson's disease includes section of various tracts in the spinal cord [260], ablation of parts of caudate nucleus [228], resection of the premotor cortex [177], destruction of the globus pallidus [234], occlusion of the anterior choroidal artery [83] and stereotaxic surgery [181]. For a detailed review of surgical treatment of Parkinson's disease (especially stereotaxic operations and instruments) see [282].

Although surgery does not relieve akinesia, it can dramatically alleviate tremor and rigidity although these symptoms tend to reappear in the months following the operation [240]. Because of the risks of an intracranial operation (hemiplegia, speech defects, mental confusion and emotional disturbances [268]) surgery is advisable only in very carefully selected cases.

First effective drug treatment of Parkinson's disease, which made use of preparations from solanaceous plants resulting in symptomatic relief, dates back more than a century [72]. Synthetic anticholinergic drugs with anti-parkinson effect were introduced after the second World War [284]. Dopaminergic substitution through intravenous [42] and oral [24] administration of its precursor levodopa constituted a major breakthrough in the treatment of Parkinson's disease. Addition of benserazide to inhibit the decarboxylase of levodopa [40] made it possible to diminish the latter's side-effects by means of increasing the anti-parkinson effect of much smaller doses (in some countries, where benserazide is not commercially available due to its presumed carcinogenic effect, carbidopa is used for the same purpose [252]). Combination of levodopa with levotryptophane helped to decrease the psychologic side-effects of dopaminergic therapy [43]. Hyperkinetic side effects of dopaminergic therapy may be reduced by combining levodopa with taurine and litoralone [41].

Levodopa

Levodopa, often shortened as L-DOPA, is formally known as L-3,4-dihydroxyphenylalanine. Its systematic (IUPAC) name is (S)-2-amino-3-(3,4-dihydroxyphenyl) propanoic acid. Levodopa is a chemical that is produced naturally by some plants and animals. Humans produce it via biosynthesis from the amino acid L-tyrosine. L-DOPA is a precursor of catecholamine class neurotransmitters like dopamine, norepinephrine, and epinephrine. L-DOPA can also be manufactured industrially and is sold under many trade names like Atamet®, Madopar®, Parcopa®, Prolopa®, Sinemet®, Stalevo®. As a drug it is used in the clinical treatment of Parkinson's disease and dopamine-responsive dystonia. The reason why levodopa rather than dopamine is used in dopaminergic substitution therapy is that it can cross the blood-brain barrier whereas dopamine cannot. The chemical structure of levodopa is shown in Figure 5.3.

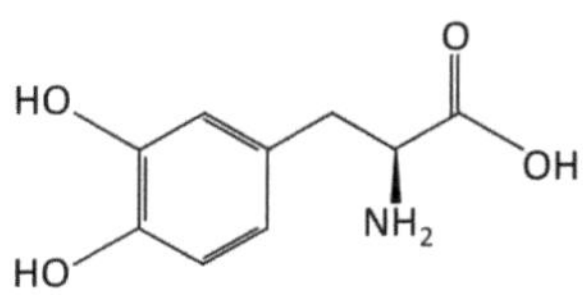

Fig. 5.3 Structure of levodopa.

Parallel to the development of a dopaminergic therapy, research concerning the monoamine oxidase (MAO) inhibitors resulted in alternative drugs. Studies of combined levodopa and MAO inhibitor therapy showed a remarkable decrease in on-off periods [44]. Research with dopamine antagonists which increase the serotonin concentration in the brain showed significant relief of akinesia, rigor and tremor, however, with far more side-effects than levodopa therapy [62], [216].

The dosage in levodopa therapy of Parkinson's disease is still controversial. Some researchers tend to prescribe huge amounts [85], whereas others give so little as to be ineffective [267]. It has been reported that levodopa therapy may induce nausea, vomiting, anorexia, heightened nervous tension, anxiety, delusions, hallucinations, hypomania, depression, cardiac dysrthymia and hypotension [226]. One of the most marked side effect of long-term levodopa therapy is the development of adventitious involuntary movements. Moreover, there is evidence to believe that the higher doses tend to increase the speed with which the therapeutic effect of levodopa diminishes over the years [226].

Therefore, it makes sense to prescribe levodopa in such a way that it is always just barely effective, thus keeping the dosage relatively low. The merits of individual dosage prescription has been demonstrated in several studies [3]. Because of the large variability in response to therapy among the patients, it is necessary to prescribe the dosage matched individually to the specific needs of the patient and this requires a process which might take many months.

A possible therapy for Parkinson's disease is neural transplants. Various clinical studies were performed towards the end of last century, mostly in Mexico and Scandinavia [200], [205]. Dopaminergic neurons derived from the human embryonic brain cells were grafted into the striatum of patients with Parkinson's disease. These studies provided proof-of-principle that therapeutic benefits can be achieved. Subsequent studies, however, showed variable and mostly negative results [46], [116], [238]. They also revealed that some patients develop involuntary movements, so called dyskinesias, as side effects [137]. Thus, while nigral transplants clearly work well in some selected cases, the technique still needs refinement before it can successfully be performed in a large series of patients. Future possibilities are offered by stem cells as potential sources of dopamine neurons that can be used for transplantation in Parkinson's disease [53].

However, probably the most effective therapies for the motor features of advanced Parkinson's disease to date are deep brain stimulation (a surgical treatment involving the implantation of a medical device called a brain pacemaker, which sends electrical impulses to specific parts of the brain) [167] and apomorphine or Duodopa® infusion therapies [196].

5.6 Measuring Parkinsonian Symptoms

Measuring symptoms as described above is not as easy as measuring a voltage with a voltmeter or a weight by a balance! Traditionally, physicians "measure" Parkinson symptoms when they clinically examine a patient. The physician has to establish the momentary extent of different symptoms from the "ability to stand up from a chair without armrests" to "building of dandruff on the face" and rate them to belong to one of the predefined classes like "moderate", "severe", and so on. An example of such scales is the Columbia Scale (see Table 5.1).

The hardware of the device basically consists of a screen on which a reference signal is displayed. This is a circular target with a centimeter radius. A potentiometer transduces the angular position of the end phalanx of the thumb, which is also displayed on the screen as a circular dot with 2mm radius. The subject is instructed to move the dot by moving his thumb in such a way as to follow the target which moves up and down on the screen. It is made sure that the tracking is performed by the movement of the end phalanx of the thumb alone by strapping the arm on an ergonomic armrest and placing the hand on an individually formed plaster of Paris cast, leaving only the end phalanx free to move. Why the thumb? There are two main reasons for that choice: 1) Hand or arm movements need a coordinated co-activation of various muscle groups; the end phalanx of the thumb needs the activation of extensor pollicis longus and flexor pollicis longus muscle pair alone. This way we measure a movement with a single degree of freedom as opposed to multi-degree of freedom movements which can be accomplished by and indeterminate combination of muscle activations. 2) The rotational range of the thumb is about double that of the index finger resulting in higher measurement resolution.

This method of using scales obviously relies on the necessarily subjective evaluation of the physician. What is really needed is a reliable and objective measurement method. The device shown in Figure 5.4 was developed exactly for that purpose.

Table 5.1 Some questions chosen from the Columbia University Rating Scale for parkinsonian patients.

Facial expression	
Normal	0
Minimal hypomimia, could be normal "poker face"	1
Slight but definitely abnormal diminution of facial expression	2
Moderate hypomimia	3
Masked or fixed faces with severe loss of facial expression	4
Foot tapping	
Normal	0
Slightly slow	1
Slow	2
Markedly slow	3
Unable	4
Bradykinesia of hands including handwriting	
No involvement	0
Detectable slowing of the supination-pronation rate evidenced by beginning difficulty in handling tools, buttoning clothes, and with handwriting	1
Moderate slowing of supination-pronation rate, one or both sides, evidenced by moderate impairment of the hand function	2
Severe slowing of supination-pronation rate; unable to write or button clothes; marked difficulty in handling utensils	3
Rigidity	
Non-detectable	0
Detectable rigidity in neck and shoulders; activation phenomenon is present; one or both arms show mild, negative, resting rigidity	1
Moderate rigidity in neck and shoulders; resting rigidity is positive when patient not on medication	2
Severe rigidity in neck and shoulders; resting rigidity cannot be reversed by medication	3

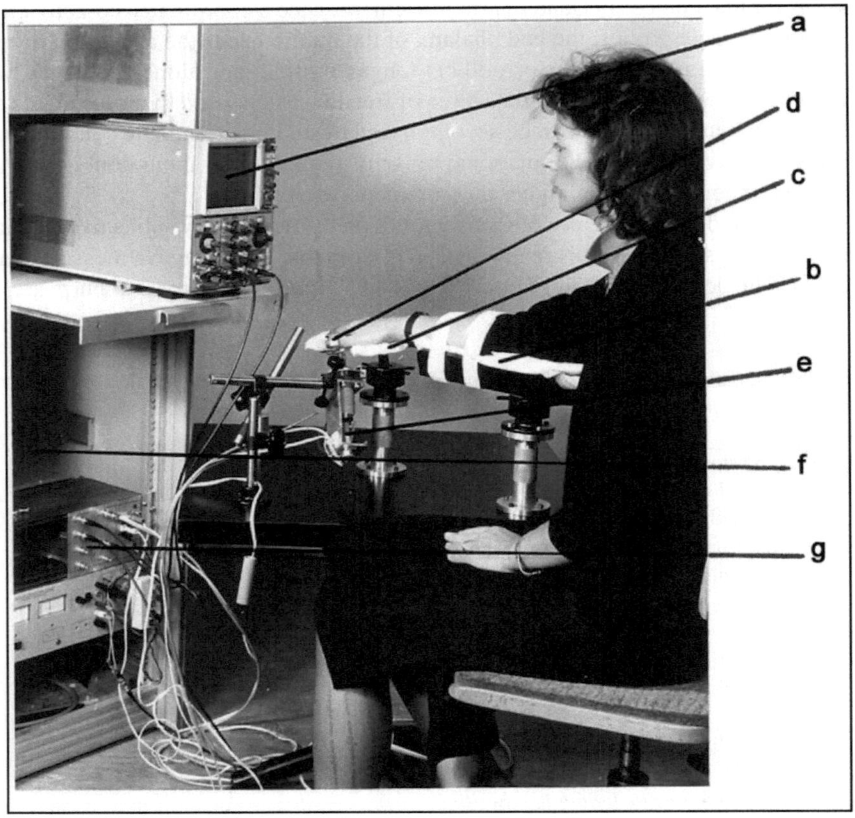

Fig. 5.4 The measurement set-up consists of a screen (a), an armrest (b), a hand platform (c), a thumbscrew (d), a potentiometer (e), a computer (f) and the A/D – D/A converters (g).

Figure 5.5 depicts the block diagram of the measurement set-up. The computer delivers a pre-computed reference signal to be displayed on the screen and records the actual position of the thumb 100 times a second. The reference signal is a pseudo random binary sequence of 6cm high steps viewed from 50cm distance. The tracking data is recorded to be analyzed later off-line. For certain tests it is necessary to apply a counter torque on the shaft which is moved by the thumb to see how well the patient can cope with disturbances (not used here).

Before any measurements with motorically impaired patients were done, a series of tests were performed to gather comparative data about the tracking performance of healthy control subjects.

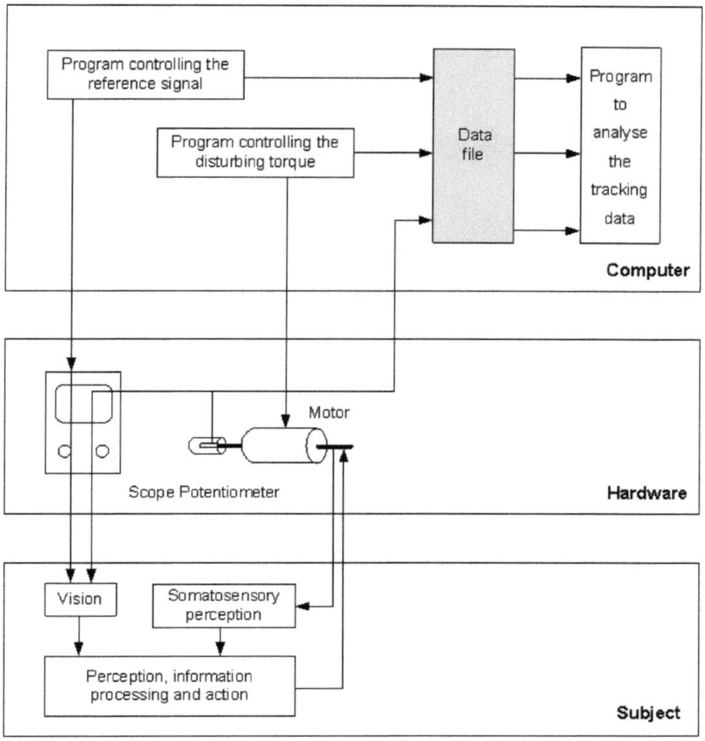

Fig. 5.5 A schematic diagram depicting the measurement set-up. The motor shown in the drawing is not used for the measurements described in the text.

5.6.1 Measuring Tremor

Tracking data of a healthy subject measured by this device is shown in Figure 5.6. The dashed line is the reference signal and the continuous line is the tracking data of the subject. One can see a fairly good tracking performance.

Tracking data of a Parkinson patient with tremor is shown in Figure 5.7. One can actually see that the patient has no difficulty following the up or down movements while following the steps. However, she cannot maintain a steady constant position. Her thumb moves rapidly, about a constant position. This is tremor. This is why another name for Parkinson's disease is "shaking palsy".

Tremor manifests itself as relatively high frequency components in the frequency spectrum of the tracking response. We can look at the power spectral densities (PSD) of the tracking response to quantify tremor.

With the tracking response $r(t)$ and $R(f)$, the Fourier transform of $r(t)$, $P(f)$ is the power spectral density of $r(t)$ calculated as $P(f) = R(f)R^*(f)$. Also when we compare the PSD of a control subject and a patient with tremor the differences are

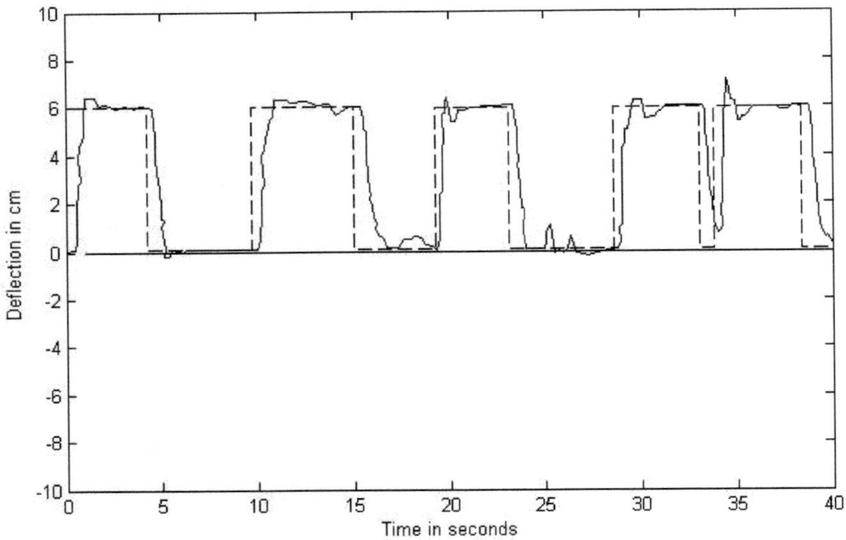

Fig. 5.6 Tracking performances of a healthy subject.

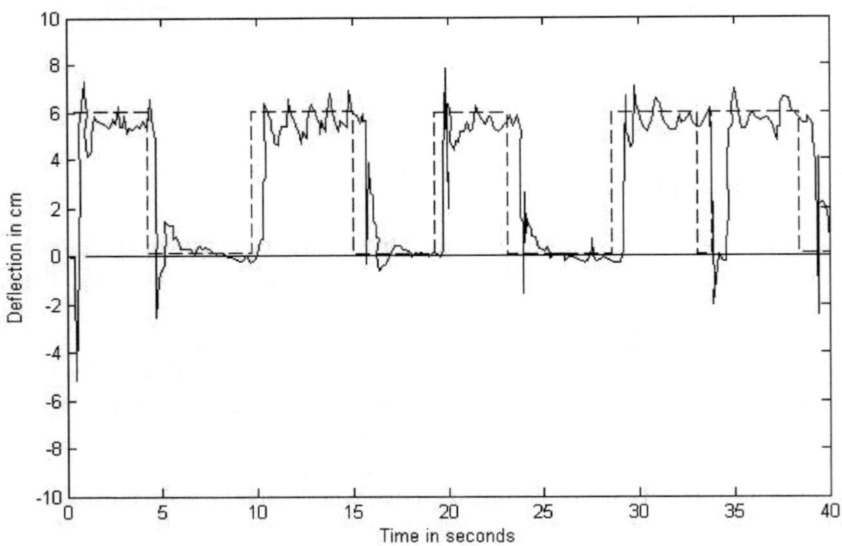

Fig. 5.7 Tracking performances of a parkinsonian patient with tremor as the main symptom.

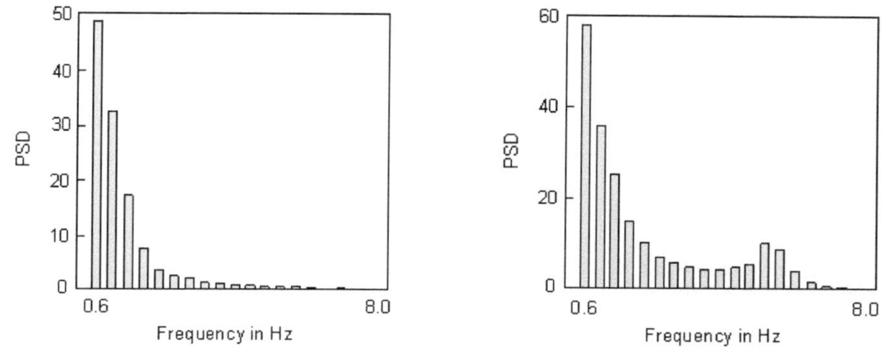

Fig. 5.8 PSD of a control subject (a) and a Parkinson patient with tremor (b).

obvious. One can see the higher frequency components present in the PSD of the tremor patient in Figure 5.8.

When we analyze the PSD we can see that some main characteristics that can be extracted from the PSD are 1) The total energy E_T , 2) The energy in the 4 – 6 Hz band E_B, and 3) The power of the largest frequency component in the 4 – 6 Hz band P_{max}. This band is where most of parkinsonian tremor is located. The energy is computed as the integral of the PSD in the corresponding frequency band.

$$E = \int_{f_d}^{f_u} P(f)df$$

After normalizing them with their control values, we can now blend all those three characteristics in a single scalar and call it the tremor index I_t. For the patient whose tracking performance we have seen above, the tremor index I_t is hence computed the following way (Figure 5.9 shows the results of this analysis for a typical tremor patient):

$$\begin{aligned} I_t &= \frac{1}{3}\left(\frac{E_T^p}{E_T^c} + \frac{E_B^p}{E_B^c} + \frac{P_{max}^p}{P_{max}^c}\right) \\ &= \frac{1}{3}\left(\frac{211}{122} + \frac{36.1}{6.1} + \frac{10.7}{0.8}\right) \\ &= 7.01 \end{aligned}$$

The superscript p stands for parkinsonian patient and the superscript c stands for control group average values. Obviously, the higher the tremor index the more impaired is the patient.

Theory states that for the Fourier transform to work properly, the signal has to be infinitely long. So we use a simple engineering trick, dividing the signal $r(t)$ in sub-signals $r_i(t)$, calculating the PSD of each sub-signal $P_i(f)$ and finally taking the weighted average of these PSDs.

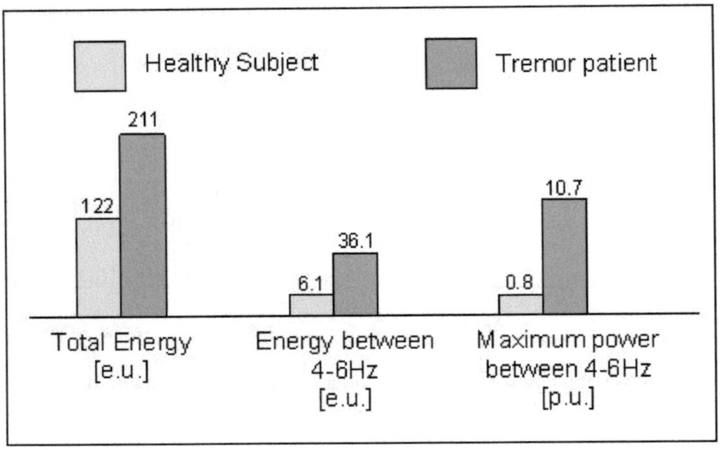

Fig. 5.9 Comparison of data extracted from the power spectrum densities of a healthy subject and a parkinsonian patient with tremor as the main symptom.

$$P(f) = \frac{1}{n}\sum_{i=1}^{n} \eta_i P_i$$

5.6.2 *Measuring Bradykinesia*

Another leading symptom of Parkinson's disease is bradykinesia, the slowing down of the motor functions. The tracking response of a patient presenting bradykinesia in the right hand in a session identical to the one in Figure 5.7 is depicted in Figure 5.10. A remarkable slowing down of the movements, thus the disability to perform fast movements becomes visible.

Let us now look at Figure 5.11 which depicts hypothetical tracking data as response to a step of the reference signal. The characteristics which can be extracted from such a step response are the reaction time, the movement time and the error area. In order for the recorded tracking performances to be processed meaningfully, they are first filtered with a tapped delay filter (FIR) of 100th order and 1Hz cut-off frequency.

The reaction time is defined as the time interval between the appearance of the stimulus and the detection of a reaction to the stimulus. It comprises the time it takes for the stimulus to be perceived, processed, a reaction to be decided upon and a certain amount of minimal action to be executed (see Chapter 3, especially Subsection 3.3.3). Since the reaction time includes a certain time in which a movement has to occur, its length gives an idea about the severity of motor impediment, among others, in parkinsonian patients. Reaction time is computed as the time it takes for the response to pass a threshold of 10% of the stimulus step and not to fall below this threshold for another half a second. Of course, one cannot expect the response to be

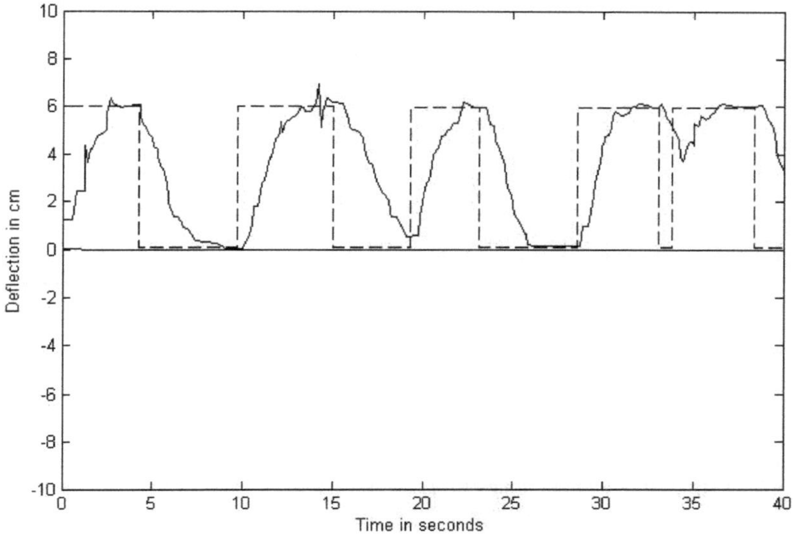

Fig. 5.10 The tracking performance of a parkinsonian patient with severe bradykinesia compared with the healthy subject in Figure 5.6 shows a marked slowing down of the movements.

exactly zero when the stimulus appears. Therefore, an increase of 10% of the step size from the momentary position is to be detected.

The movement time is the time between the end of the reaction time and the time the response passes a threshold of 90% of the stimulus step. But since the initial position of the response and the overshoot can distort the length of the movement time it has yet to be normalized, for instance, according to

$$T'_M = T_M(H+h)(H+a)/H^2$$

This characteristic of the response gives valuable information about the ability of the subject to execute fast movements. For instance, parkinsonian patients with rigor and bradykinesia have invariably very long movement times.

The error area is another characteristic which gives information about the subject's ability to track the stimulus signal. It is defined according to

$$A = \int_{t_0}^{t_f} |s(t) - r(t)| dt$$

where t_0 is the beginning and t_f is the end of the stimulus step. $s(t)$ and $r(t)$ denote the stimulus and the response time functions respectively.

The comparison of the reaction times, the movement times and the error areas as described above between the bradykinetic patient and the average of his age group results in the chart shown in Figure 5.12. Analogous to tremor index I_t, we can now

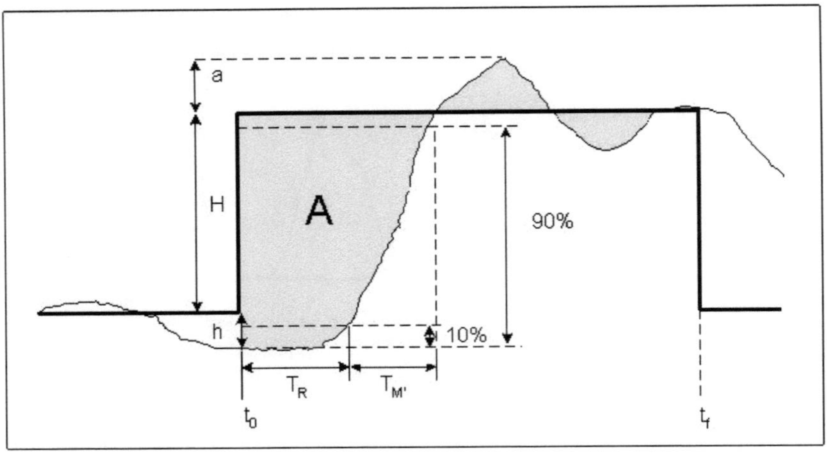

Fig. 5.11 Hypothetical tracking response to a step reference signal. H is the height of the step which starts at t_0 and continues till t_f. h and a are the values of the response at t_0 and the overshoot respectively. T_I denotes the reaction time, and T_M the movement time. A is the error area.

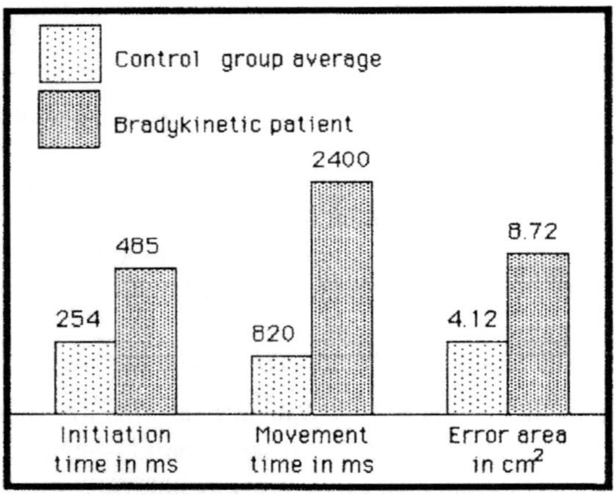

Fig. 5.12 The comparison of some characteristics extracted from the tracking performance of the bradykinetic patient in Figure 5.10 and the corresponding age control group average.

define a bradykinesia index I_b. Here we combine the reaction (or initiation) time, the movement time and the error area after normalizing them with the corresponding control group averages.

In the case of the patient, whose response we have seen in Figure 5.10, the reaction time is 485 ms as opposed to 254 ms of the control group average. The movement time is 2400ms as opposed to 820 ms of the control group average and the error area is 8.72 cm^2 as opposed to 4.12 cm^2 of the control group average. Thus, I_b computes to 2.32:

$$\begin{aligned} I_b &= \frac{1}{3}\left(\frac{T_R^p}{T_R^c} + \frac{T_M^p}{T_M^c} + \frac{A^p}{A^c}\right) \\ &= \frac{1}{3}\left(\frac{485}{254} + \frac{2400}{820} + \frac{8.72}{4.12}\right) \\ &= 2.32 \end{aligned}$$

Clearly, the higher the bradykinesia index the more impaired are the movements of the patient.

An important drawback of these methods described above is their reliance on the tracking device and the very limited number of movements which can be analyzed. Other attempts at analyzing tracking data generated by well defined movements with several degrees of freedom have also been made using more complex tracking devices (see Figure 5.13).

5.6.3 Low-Tech Approach

A totally different, low-tech approach to measuring parkinsonian symptoms is, however, also possible [135]. A possible index of parkinsonian symptoms is measured as follows:

1. Measure the time needed for 10 taps with regularly alternating fingers (thumb, index, middle, ring, little),
2. Measure the time needed for writing in connected hand writing with a soft (HB1) pencil on three lines underneath each other three words which are the same in most languages (Istanbul, Paris, Berlin),
3. Measure the time needed for walking 5 meters,
4. Measure the number of paces for walking 5 meters,
5. Measure the time needed for transferring 10 white beans (one by one) from one plate to another placed 20cm apart,
6. Measure the number of dyskinetic movements during any 30 seconds,
7. The index results as the sum of all the numbers (without dimensions) obtained in the steps 1 to 6,
8. Since most movements slow down with age, controls from different decades have indices ranging from 40 to 60,
9. Finally, the comparable Parkinson index thus measured is calculated by dividing the figure obtained in step 7 by the average index of the control group with the corresponding age.

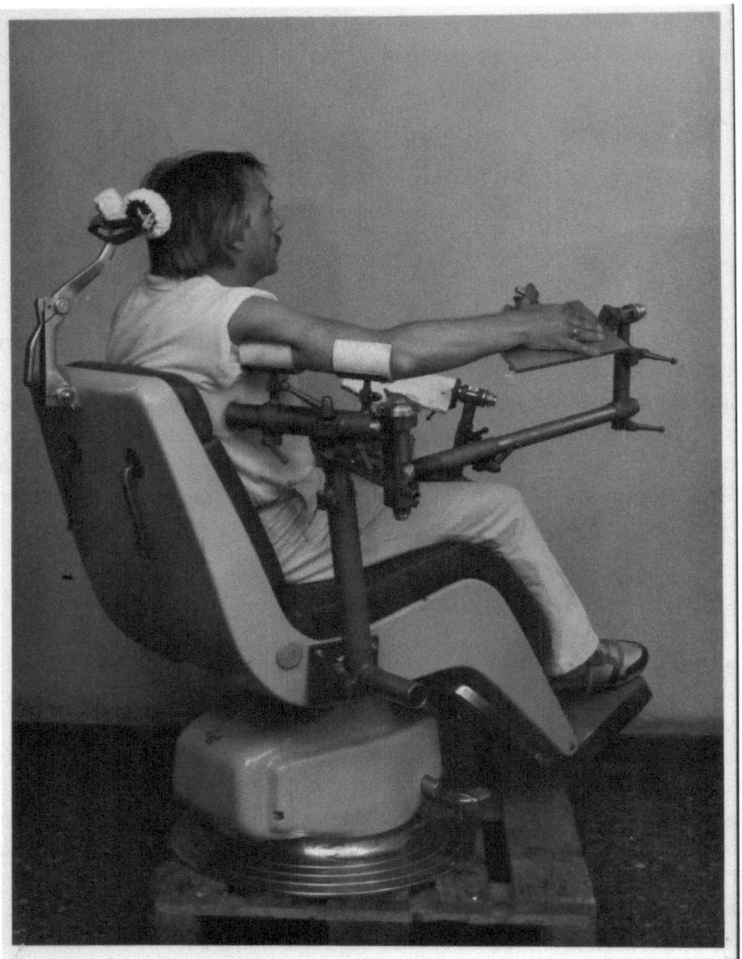

Fig. 5.13 Measurements with this device, which captured movements of the elbow combined with the movements of the end phalanx of the thumb proved to be too difficult to be of practical value.

5.7 Dosage Optimization in Parkinson's Disease

In Chapter 4 we have seen a number of methods to calculate optimal dosage regimens with respect to pharmacokinetic data. We have derived and analyzed methods to calculate the timings and dosages of a daily drug therapy such that the blood level concentrations of the administered drug follow a prescribed trajectory as closely as possible. As concrete examples we have used levodopa therapy in Parkinson's disease. In this Chapter we will go a step further and determine the optimal timings and dosages of levodopa therapy such that the desired therapeutic effect can be attained throughout the day according to the individual needs of the patient and based on the individual pharmacokinetic and pharmacodynamic response of the patients.

5.7.1 Parkinson Patients Are Poor Controllers

In order to optimize the therapy with respect to pharmacodynamics, we need to model the dose effect relationship of levodopa. But before that, let's have a quick look at the tracking behavior of Parkinson patients.

Many studies, especially by USAF and NASA analyzing the tracking data by means of mathematical models have shown that in most case simple linear second order models are sufficient to duplicate the tracking behavior of humans (see Chapter 3). Here, such a model will be used to study parkinsonian deficits as deviations of the system parameters from their normal values. The experiments discussed here are based on the measurements in Section 5.6.

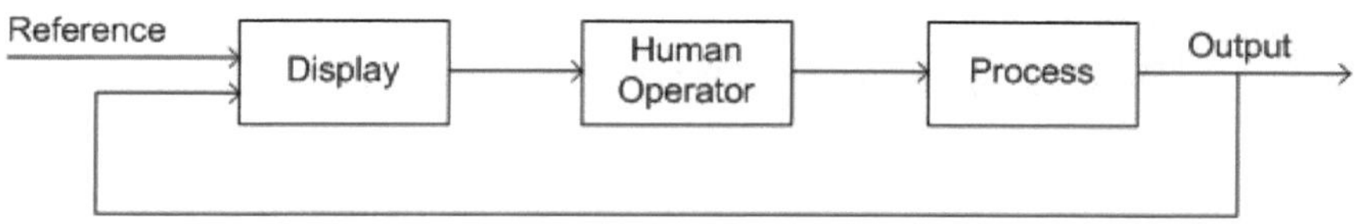

Fig. 5.14 Human operator acts as a controller in a tracking task.

As Figure 5.14 indicates, the role of the human operator in tracking can be considered to be that of a controller. As we have seen in detail in Chapter 3, studies on the dynamic behavior of the human operator resulted in a broad spectrum of mathematical models ranging from the very simple to the extremely complex.

The model used here is the input-output model of a linear, time-invariant system (input: the reference signal on the display, output: the position of the thumb) with the transfer function

$$G(s) = \frac{K(1+T_H s)}{(1+T_1 s)(1+T_2 s)} e^{-T_L s}$$

where K denotes the gain, T_L the residual transmission time lag, T_H the high pass filter time constant, T_1 and T_2 the low pass time constants. s is the complex frequency.

The step response of the system given by this transfer function is

$$g(t) = KAH(t-T_L) + KBH(t-T_L)e^{-\frac{t-T_L}{T_2}} + KCH(t-T_L)e^{-\frac{t-T_L}{T_1}}$$

with

$$A = 1, \quad B = \frac{(T_H - T_1)}{(T_1 + T_2)} - 1, \quad C = \frac{(T_1 - T_H)}{(T_1 + T_2)}$$

$H(.)$ is the Heaviside step function.

The problem of finding the values of the parameters K, T_L, T_H, T_1 and T_2 of the model such that the step response of the model approximates the measured tracking step response as closely as possible can be shown to be equivalent to

$$\min_{K,T_L,T_H,T_1,T_2} \int_0^{\infty} (g(t) - m(t))^2 dt$$

Where $g(t)$ denotes the simulated model step response and $m(t)$ the measured step response. This minimization (identification of the parameters) can be done, for instance, with a modified, finite difference, Levenberg-Marquardt algorithm [212].

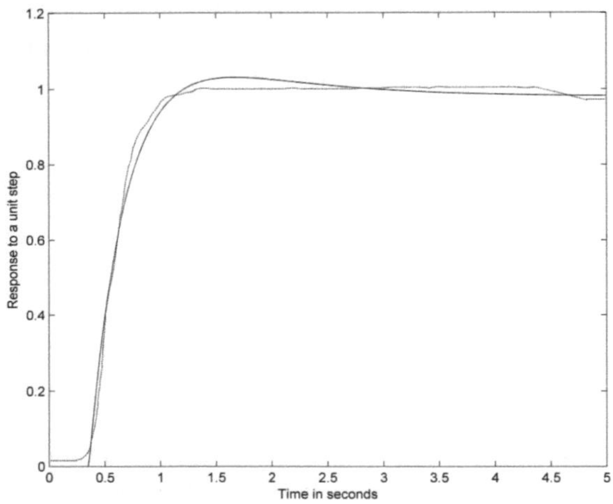

Fig. 5.15 Averaged step response of a healthy subject. Continuous curve is the modeled and the dashed one is the measured step response.

Experiments with a control group of 26 healthy subjects (10 female, 16 male), with an average age of 47.9 (s.d. 13.0, range 25 to 63) as well as with a parkinsonian population of 11 subjects (4 female, 7 male) with an average age of 50.8 (s.d.11.7, range 34 to 72) presenting bradykinesia as the major symptom have been conducted. In each case at least 30 step responses were averaged before performing the identification. Figure5.15 and Figure 5.16 show typical averaged step responses of a healthy and a bradykinetic subject. Continuous curves are measured step responses. Dashed curves correspond to the simulation of the model with identified parameters. The identified tracking parameters in the two cases are shown in the Table 5.2.

Figure 5.18 shows the comparison of the identified parameters of the control group with the parkinsonian population. The parameters are normalized by the mean values of the control group. It can be seen that the two low pass time constants are identical. This holds for parkinsonians as well as the control group. Further

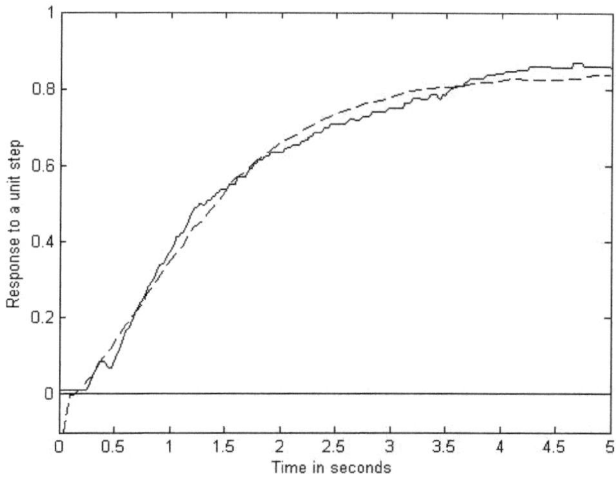

Fig. 5.16 Averaged step response of a bradykinetic patient without medication. Continuous curve is the measured and the dashed one is the modeled step response.

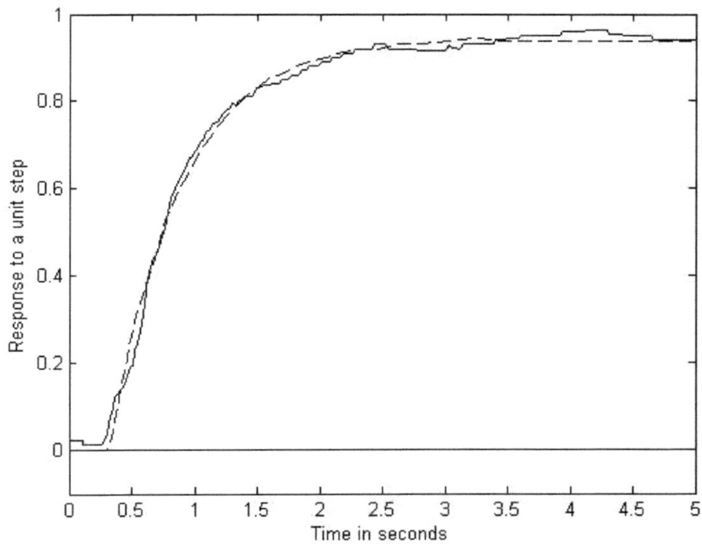

Fig. 5.17 Averaged step response of the bradykinetic patient in Figure 5.16 with adequate medication. Comparison of the two figures shows the remarkable therapeutic effect.

Table 5.2 Identified tracking parameters of the healthy group and the parkinsonian population.

	Control group					Parkinsonians				
	$K[-]$	$T_L[s]$	$T_1[s]$	$T_2[s]$	$T_H[s]$	$K[-]$	$T_L[s]$	$T_1[s]$	$T_2[s]$	$T_H[s]$
$\bar{x}$	0.98	0.26	0.43	0.43	156.6	0.93	0.41	0.54	0.54	143.7
σ	0.01	0.07	0.11	0.11	78.9	0.03	0.09	0.12	0.12	79.0
x_{max}	1.00	0.41	0.50	0.50	344.7	0.97	0.49	0.88	0.88	267.3
x_{min}	0.96	0.13	0.22	0.22	14.5	0.86	0.16	0.37	0.37	8.9

analysis of the parameters shows that the transmission time lag as well as the time constants of the low pass filters are significantly longer in the parkinsonian patients, whereas no significant differences in the gain and the high pass filter time constants are detectable.

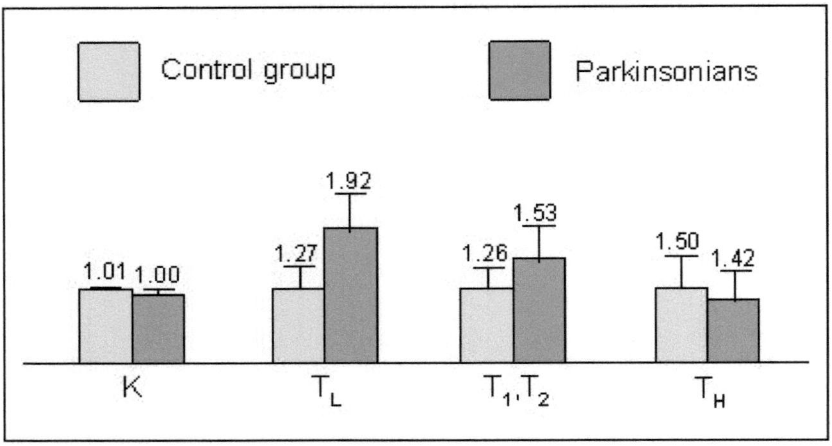

Fig. 5.18 Comparison of the identified parameters of the healthy group with the parkinsonian population. For the purpose of clarity, the parameters are normalized by the mean values of the parameters of the healthy group.

Finally, Figure 5.17 shows the step response of the bradykinetic patient shown above after adequate medication. Comparing Figure 5.16 with Figure 5.17 shows the remarkable improvement in the motor abilities with medication.

5.7.2 Modeling the Effects of Levodopa on Parkinson Symptoms

As mentioned earlier, a mathematical model which describes the dynamic relationship between the drug administration and the effects of the drug is needed to compute an optimal therapy. In order to determine the parameters of such a model the effects have to be made measurable. The device and the method described in Section 5.6 were developed exactly for that purpose.

So far, the term pharmacokinetics has been used to describe the process controlling drug concentration at any time after a single or multiple dose. Traditionally pharmacokinetic models describe drug concentrations as a function of both dose and time [148]. However, as discussed in detail at the end of Chapter 4, most pharmacodynamic models are essentially independent of time and they describe the time-independent relationship between concentration and effect as Figure 5.19 shows [149].

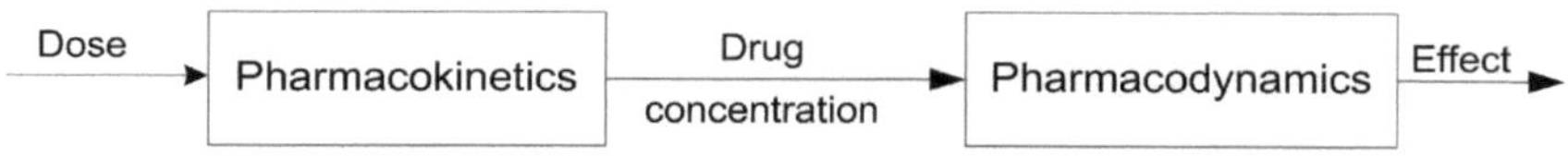

Fig. 5.19 Schematic representation of the dose-effect relationship using pharmacokinetic and pharmacodynamic models as already seen in Chapter 2.

5.7.2.1 Pharmacokinetics of Levodopa

Producers of Madopar® have found out that the pharmacokinetics of the drug can be adequately modeled using a two-compartment model. The transfer function of second order kinetics with drug administration as input and blood concentration of levodopa as output is given by

$$G(s) = \frac{K}{(1+sT_1)(1+sT_2)} \tag{5.1}$$

where K denotes the amplification factor and has the unit ng/ml/mg, T_1 and T_2 are the time constants in hours.

In a study [3] the blood level of levodopa was measured several times after the administration of a 125mg Madopar® capsule in order to identify the parameters K, T_1 and T_2 in this model. The results of the parameter identification and the simulation of the model with the identified parameters is depicted in Figure 5.20.

Although there is a fair correspondence between the simulated concentration profile and the measurements, a slight modification of the model in (5.1) to include a time lag between the drug administration and an increase in the levodopa concentration ameliorates the correspondence between the modeled and measured concentrations. This can be formulated as

$$G(s) = \frac{Ke^{-sT}}{(1+sT_1)(1+sT_2)} \tag{5.2}$$

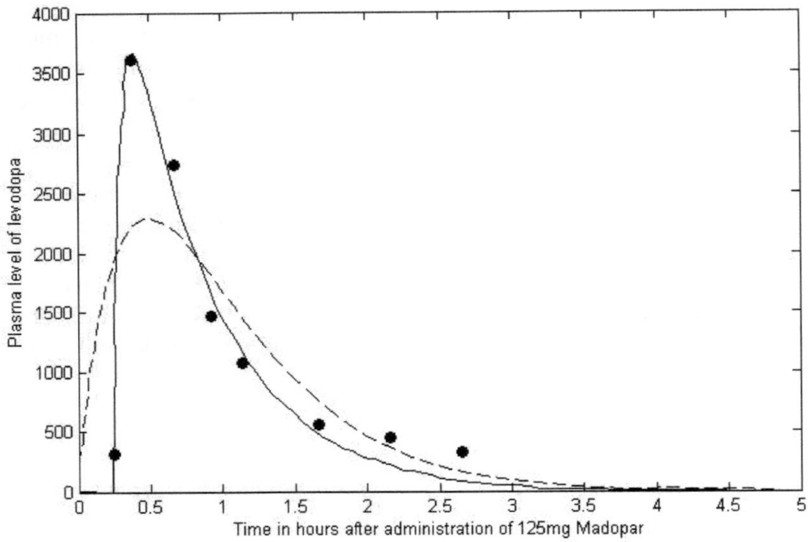

Fig. 5.20 Comparison of the two-compartment pharmacokinetic model of levodopa with $K = 1572$, $T_1 = 0.503$ and $T_2 = 0.503$ (dashed line) with the second order model with a time lag with $K = 1418$, $T = 0.246$, $T_1 = .0547$ and $T_2 = 0.607$ (solid line). Black points indicate actual measurements.

with T denoting the time lag needed for the levodopa to start to be absorbed in the blood. The improvement of the correspondence is due to the sounder physical justification of the augmented model. The results of the parameter identification and the simulation of this extended model with the identified parameters is also depicted in Figure 5.20. A comparison of the two curves convinces that the second model describes the pharmacokinetics of levodopa much better.

5.7.2.2 Pharmacodynamics of Levodopa

We have seen a number of different pharmacodynamic models at the end of Chapter 4. In the sequel we shall be using a really dynamic model for the pharmacodynamics of levodopa. Such a model can abstractly be described by

$$E = f(C) \tag{5.3}$$

where f is the functional which maps the concentration and its derivatives on the effect.

It was found out during the pretrials described above that the attributes which are most representative of the symptoms of Parkinson's disease and which are best measurable are error area, reaction time and movement time as defined in Section 5.6.

In a measurement session with the patient ASG45 (suffering from Parkinson's disease since a decade, Hoehn & Yahr rating 2), a tracking task consisting of random step sequences during a minute was repeated every three to fifteen minutes for a duration of more than three hours after the administration of 125mg Madopar®. After a qualitative study of the evaluations of these measurements, it was decided to model the dose effect relationship as a single input (dose) and multi output (reaction time, movement time, and error area) system, whereas the pharmacodynamic part of each output was modeled as a first order linear system with saturation as shown in Figure 5.21.

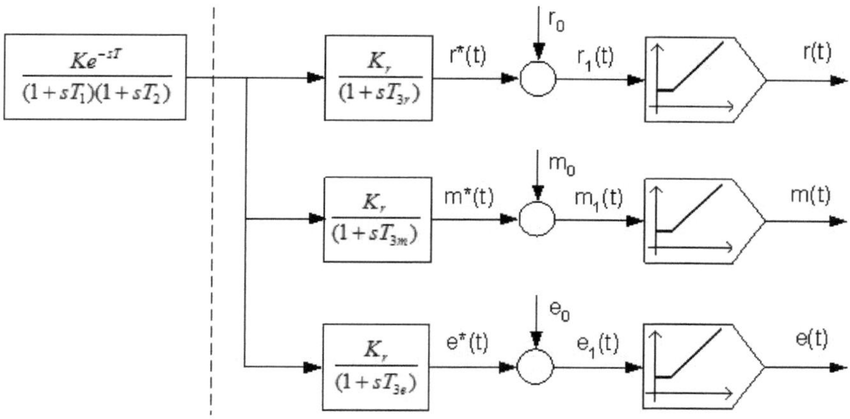

Fig. 5.21 Block diagram of the effect of levodopa on the reaction time, the movement time and the error area. The pharmacokinetic part has the parameters which were already identified in Paragraph 5.7.2.1.

The linear and purely dynamic part of the relationships between the levodopa concentration C and the reaction time r, the movement time m and the error area e can be formalized as

$$\begin{aligned} \dot{r}^*(t) &= -r^*(t)T_r + K_rC/T_r \\ \dot{m}^*(t) &= -m^*(t)T_m + K_mC/T_m \\ \dot{e}^*(t) &= -e^*(t)T_e + K_eC/T_e \end{aligned} \tag{5.4}$$

where K_i denote the pharmacodynamic amplifications and T_i denote the pharmacodynamic time constants of the corresponding effects. Note that K_i must have negative values, because an increase of the concentration results in a decrease of the measured attributes.

The non-linear and static part of the pharmacodynamics can be written as

$$r(t) = \begin{cases} r^*(t) + r_0 & , r_1(t) > r_{min} \\ r_{min} & , r_1(t) \le r_{min} \end{cases}$$
$$m(t) = \begin{cases} m^*(t) + m_0 & , m_1(t) > m_{min} \\ m_{min} & , m_1(t) \le m_{min} \end{cases} \quad (5.5)$$
$$e(t) = \begin{cases} e^*(t) + e_0 & , e_1(t) > e_{min} \\ e_{min} & , e_1(t) \le e_{min} \end{cases}$$

where the index 0 denotes the values of the attributes without medication; the index *min* denotes the minimum values of the corresponding attribute which can be achieved through medication.

The identification of the parameters in equations (5.4) and (5.5) using the measurement session described above results in the values of the parameters shown in Table 5.3.

Table 5.3 Identified parameters of the pharmacodynamic relationship between levodopa concentration in plasma and the reaction time, the movement time and the error area.

	T_3	K	Limit values
Reaction time	$T_{3r} = 1.439$	$K_r = 0.2048$	$r_{min} = 122.3$ $r_0 = 238.0$
Movement time	$T_{3m} = 1.660$	$K_m = 0.9318$	$m_{min} = 772.0$ $m_0 = 1300.0$
Error area	$T_{3e} = 1.604$	$K_e = 0.0024$	$e_{min} = 2.9$ $e_0 = 4.8$

The simulation of the system described by the equations in (5.4) and (5.5) with the parameter values shown in Table 5.3 are shown in Figures 5.22, 5.23 and 5.24. Each black dot in Figures 5.22 to 5.27 denotes the average of an attribute (reaction time, movement time, error area) measured during a tracking task consisting of random step sequences. The tracking tasks all had a duration of one minute.

A thorough verification of the model derived above would have to include numerous tests with other doses as well. However, this would mean further hardship for the patient who voluntarily agreed to participate in the measurements described in this chapter. Therefore, it was decided to perform just one more similar test session with a substantially higher dose of Madopar®. The results of a measurement session conducted after the administration of a 375 mg capsule under identical conditions as above are depicted in Figures 5.25, 5.26, and 5.27. Note that the effect of the triple dosage continues about two and a half times as long, which agrees well with the simulations.

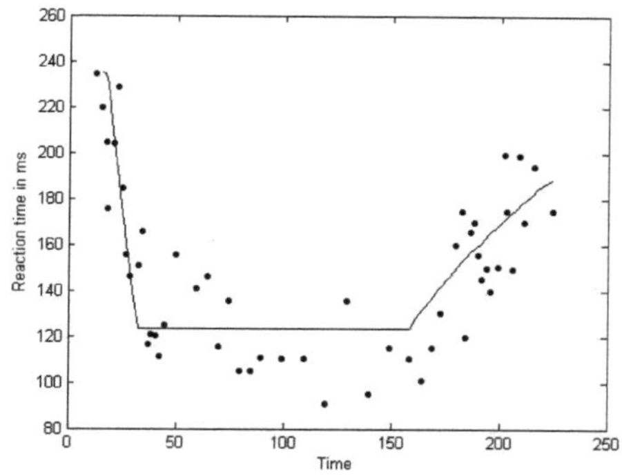

Fig. 5.22 The effect of 125mg Madopar® on the reaction time as a function of time after administration. Black points indicate measured values. The continuous curve is the result of parameter identification.

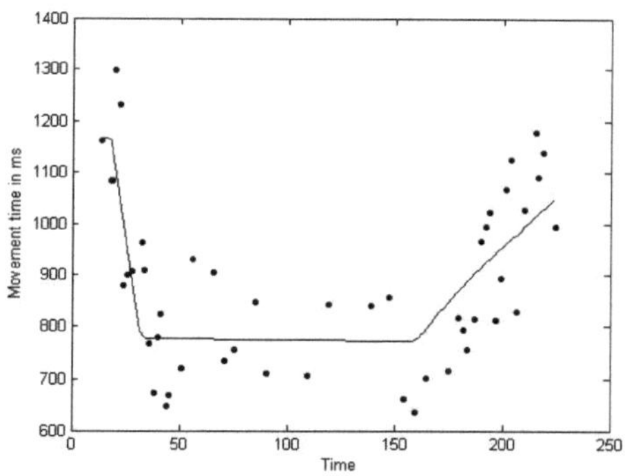

Fig. 5.23 The effect of 125mg Madopar® on the movement time as a function of time after administration. Black points indicate measured values. The continuous curve is the result of parameter identification.

Before we go any further, limitations of this model have to be discussed. First of all, the model derived in this Subsection is based upon a single dose response. No tests were conducted to test how the model performs in predicting the therapeutic effect after several bolus administrations at different times. However, the measurement sessions with 125mg and 375mg Madopar® seem to verify the effects of a higher single dose. Furthermore, it is assumed that the measurable effects are a function of levodopa alone, whereas other metabolites of the drug (like carbidopa or 3-methyl-o-dopa) might play a therapeutic role as well.

It must also be stressed that Madopar® is composed of 80% levodopa and 20% benserazide, a decarboxylase inhibitor. Although no assays with benserazide were made, it is known to have a much slower metabolization rate than levodopa.

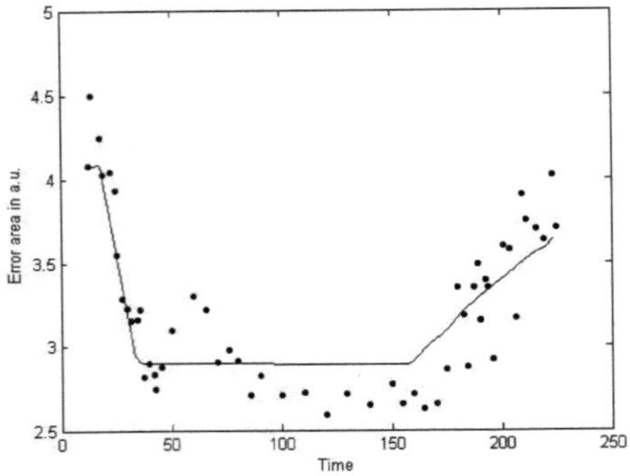

Fig. 5.24 The effect of 125mg Madopar® on the error area as a function of time after administration. Black points indicate measured values. The continuous curve is the result of parameter identification.

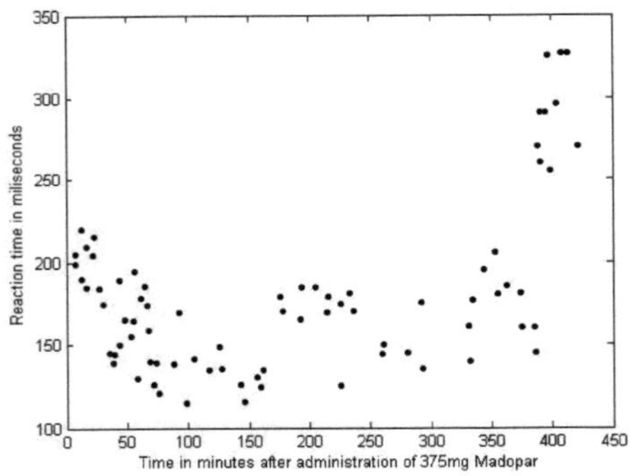

Fig. 5.25 The effect of 375mg Madopar® on the reaction time.

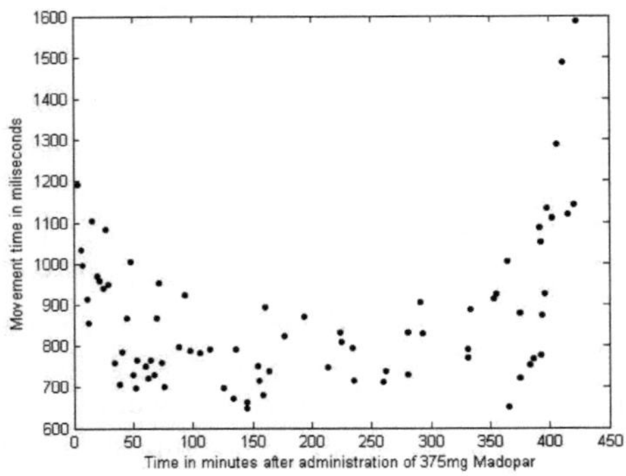

Fig. 5.26 The effect of 375mg Madopar® on the movement time.

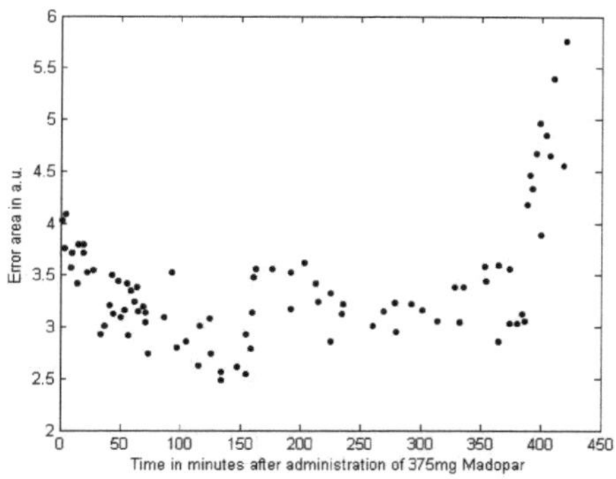

Fig. 5.27 The effect of 375mg Madopar® on the error area.

Therefore, subsequent doses of Madopar® can be reasonably expected to be metabolized in a slower speed than the first one. Moreover, the measurements on which the model is based were performed after a day of "drug holiday", whereas in daily use it can be expected that benserazide concentration in the blood is in a so called "steady state".

5.8 Optimal Control of Levodopa Therapy

Certain theoretical considerations have to be made before the model developed in the previous chapter describing the dose-effect relationship of levodopa and parkinsonian symptoms can be used to compute an optimal dosage regimen [136].

First of all an adequate objective function has to be defined. In our case, it makes sense to try to minimize the reaction time, the movement time and the error area over the whole period of optimization. On the other hand the total amount of levodopa which has to be administered should also be minimized. The parameters of the input function with which the objective function can be minimized are the doses d_i and the timings τ_i of each drug administration. These objectives can be formalized as

$$\min_u \int_{t_0}^{t_f} (\alpha_1 r(t) + \alpha_2 m(t) + \alpha_3 e(t))dt + \alpha_4 \sum_{i=1}^{n} d_i, \quad \alpha_i \geq 0 \tag{5.6}$$

with

$$u(t) = \sum_{i=1}^{n} d_i \delta(t - \tau_i) \tag{5.7}$$

where d_i is the size of the i-th administration with τ_i being the corresponding timing. t_0 and t_f denote the beginning and the end of the time window of interest. n is the number of administrations during this time window.

It was assumed that the dose-effect relationship which was modeled in Section 5.7 is deterministic. However, a look at Figures 5.22 to 5.27 shows that there is a fairly large discrepancy between the model and the measurements. Taking that

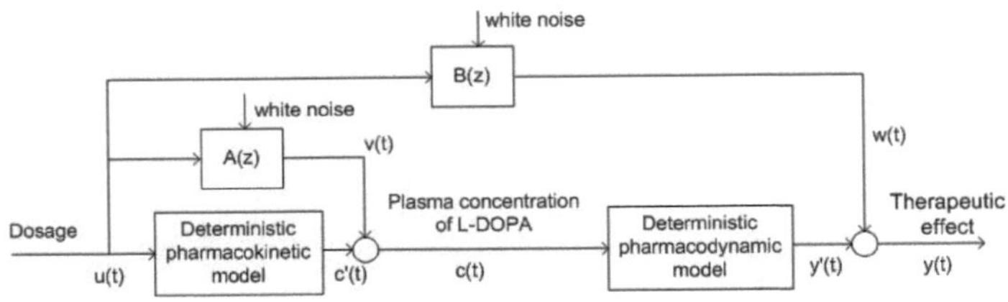

Fig. 5.28 The dose-effect relationship modeled as a stochastic system. It is assumed that the discrepancies between the measured and the modeled blood concentration of levodopa and the therapeutic effect can be modeled in both cases as additive white noise which are produced by Gaussian processes driven by the input of the system.

discrepancy into consideration, the dose-effect relationship should be drawn correctly as in Figure 5.28 where $\mathbf{y}(t)$ is composed of the measured attributes $r(t)$, $m(t)$ and $e(t)$.

Notice that the modeling errors in the case of pharmacokinetics v and in the case of pharmacodynamics $\mathbf{w}$ are modeled as autoregressive functions of the input $u(t)$ (dosage regimen) as

$$\begin{aligned} v &= Au \\ &= (a_0 + a_1 z + a_2 z^2 + \cdots + a_m z^m)u \\ w_i &= B_i u \\ &= (b_{i0} + b_{i1} z + b_{i2} z^2 + \cdots + b_{in_i} z^{n_i})u \end{aligned} \tag{5.8}$$

with z as the shift operator and with a_j and b_{ij} as random variables with normal distributions. This means that the dose-effect relationship is inherently stochastic, which further complicates the minimization of the objective function Z as described in (5.6).

Let us now analyze the random functions $v(t)$ and $\mathbf{w}(t)$ before we go any further. They can be computed as the difference of the modeled and the measured values of the blood concentration of levodopa and the attributes $r(t)$, $m(t)$ and $e(t)$ respectively as

$$\begin{aligned} v(t) &= c(t) - c'(t) \\ w_i(t) &= y_i(t) - y_i'(t) \end{aligned} \tag{5.9}$$

Figure 5.29 shows the plot of $w_r(t)$ as an example. A correlation analysis of these random functions and the Dirac-pulse at the input of the system shows that the correlations between $u(t)$ and $w_i(t)$ are negligible. On the other hand, there is a strong correlation between $u(t)$ and $v(t)$. Further analysis shows that the cause of this correlation is the poor agreement of the last two measurements with the model in Figure 5.20, which hardly warrants the use of an autoregressive relationship between the dosage regimen and the blood concentration of levodopa [51], [13]. This means that the random functions $v(t)$ and $w_i(t)$ can be modeled as noise produced by random processes which are **not** driven by $u(t)$ (see Figure 5.30).

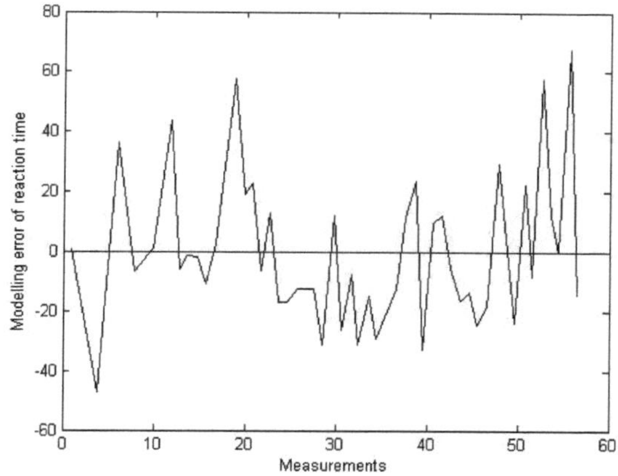

Fig. 5.29 The difference between the modeled and the measured course of the reaction time after the administration of 125mg Madopar®. This difference can be modeled as a random function with a normal distribution with the mean value $\mu = .5077$ and the standard deviation $\sigma^2 = 24.10$.

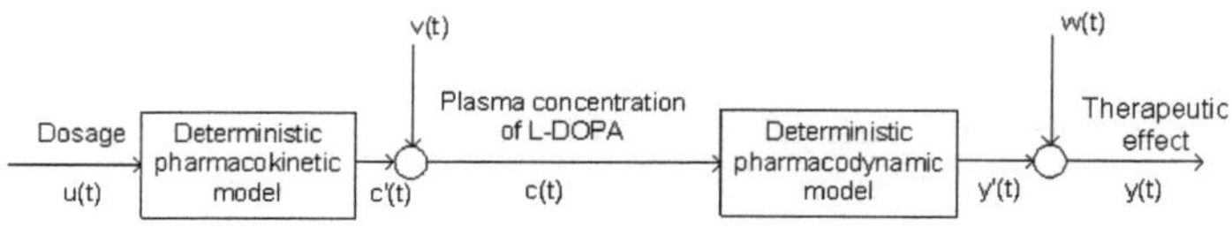

Fig. 5.30 The dose-effect relationship modeled as a stochastic system. It is assumed that the discrepancies between the measured and the modeled blood concentration of levodopa and the therapeutic effect are in each case caused by an additive white noise which is not correlated with the input of the system.

This observation has far reaching consequences. Let us have a closer look at the equation (5.6). If the dose-effect model shown in Figure 5.28 were valid, the objective function which is separable in a deterministic part Z_d and a stochastic part Z_s

$$Z = Z(u, f, v, \mathbf{w}) = Z_d(u, f) + Z_s(u, f, v, \mathbf{w}) \tag{5.10}$$

(with f describing the deterministic part of the combined pharmacokinetics and pharmacodynamics) would be a random functional and we would have to minimize its expected value. Note that both Z_d and Z_s would be a function of u.

However, since the measurements have shown that we can use the model depicted in Figure 5.30, the objective function Z can now be written as

$$Z = Z(u, f, v, \mathbf{w}) = Z_d(u, f) + Z_s(v, \mathbf{w}) \tag{5.11}$$

with a stochastic part which is **not** a function of u. Therefore,

$$\begin{aligned}\min_u Z &= \min_u \left(Z_d(u,f) + Z_s(v,\mathbf{w})\right)\\ &= \min_u \left(Z_d(u,f)\right)\end{aligned} \tag{5.12}$$

which is a purely deterministic optimization problem! The solution of this problem was discussed in detail in Chapter 4.

This surprising result is, of course, possible only if the objective function is linear or the non-linearity in the objective function does not produce products of input and noise. Furthermore, it has to be noted that this result does not assume the linearity of f or any other restricting conditions on the noise.

5.9 A Case Study

In this Section, we will see the results of the application of the treatment optimization which was developed in Section 5.8 on a patient. Concluding remarks concerning the applicability of the methodology described in this Chapter as well as its limitations will be discussed. Finally, further research directions for the refinement of levodopa therapy in Parkinson's disease will be briefly indicated.

ASG45, the patient in the Paragraph 5.7.2.2, had been suffering from Parkinson's disease for about a decade. He had been treated over the years with various drugs and he was prescribed Madopar® in the last couple of years.

The optimization of the dosage scheme with the procedure explained in the previous Section by applying the cyclic coordinate search method which was studied in Chapter 4 (with the weighting factors α_i in equation (5.6) chosen such that the therapeutic effect and the total amount of the drug are weighted about the same in the objective function) results in the dosing regimen shown in Table 5.4. The simulation of this dosage scheme results in the measurable attributes depicted in Figures 5.31, 5.32 and 5.33.

Table 5.4 A simple dosage scheme with which the drug-effect model of Madopar® is verified.

Timings:	07:30	11:00	14:00
Doses:	125 mg	125 mg	125 mg

It is, of course, of great interest to see how the modeled and the actual therapeutic effects of this regimen compare with each other in daily life. To use the measurement device introduced earlier would be utterly impractical, because this would mean to perform frequent (every couple of minutes) measurements with the patient during a whole day (measuring once every hour or so would not yield the desired verification, because of the variance of every single measurement).

Therefore, it was decided to use a method based on rating scales to evaluate the real therapeutic effect of the dosage scheme in Table 5.4. The patient was given a set of prepared charts on which he was asked to note his subjective evaluation of his motor impediment in hourly intervals for a period of four weeks while pursuing

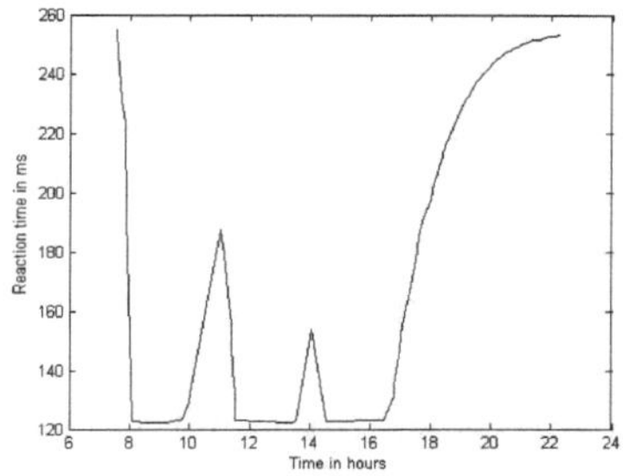

Fig. 5.31 The simulated course of the reaction time resulting from the administration of the dosage scheme in Table 5.4.

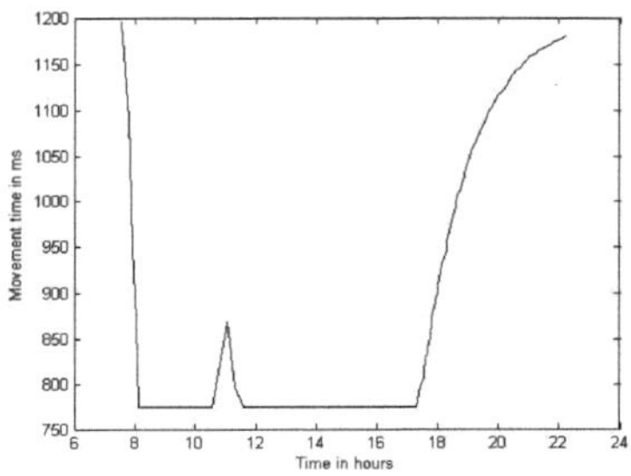

Fig. 5.32 The simulated course of the movement time resulting from the administration of the dosage scheme in Table 5.4.

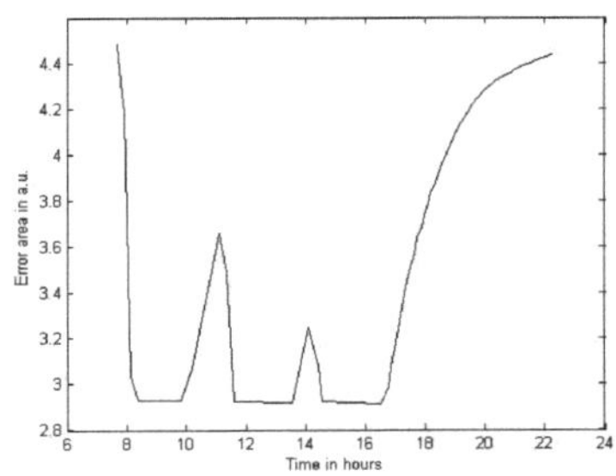

Fig. 5.33 The simulated course of the error area resulting from the administration of the dosage scheme in Table 5.4.

his daily life without any modifications. To facilitate his evaluation, he was asked to rate his momentary feeling from "better" (1) to "worse" (4) (see Figure 5.34).

It has to be pointed out that this method cannot be reliably used in most cases. However, the mental capabilities of the patient ASG45 rendered this method suitable

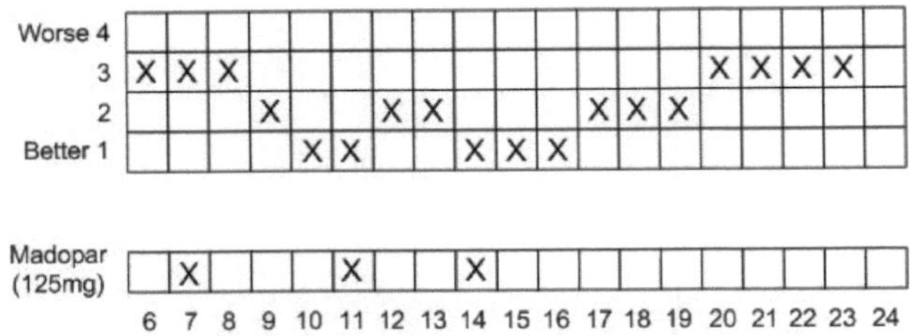

Fig. 5.34 The rating scale with which the patient ASG45 rated his movement index. The small crosses are his subjective hourly assessments of his ability to move on a typical day. (1–4: better–worse).

for the purpose of comparing the actual therapeutic effect and the effect which the model predicts.

The results are compared in Figure 5.35, where the subjective rating of the patient averaged over four weeks is shown on the same plot as the disability index. (The disability index is composed of the weighted sum of the measured attributes as in equation (5.6)). It can be seen that the model correctly predicts a rapid amelioration after the first dose at 7am in the morning and a diminishing therapeutic effect after 5pm. There is also a short period of lessened therapeutic effect just before noon which is also predicted by the model.

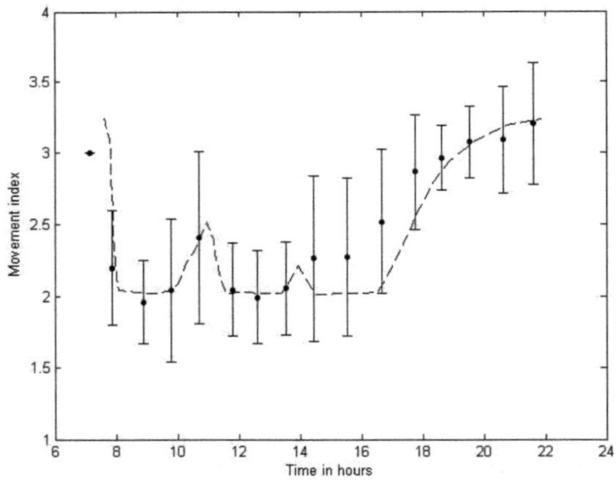

Fig. 5.35 Comparison of the modeled and (continuous curve) the subjectively assessed (dots with s.d. bars) movement index shows a remarkable congruence.

It has to be stressed at this point that these results are obtained from a single patient. However, it can be said that the results of a study with a single patient are very encouraging indeed.

5.10 Summary

The problem of dosage determination in the levodopa therapy of Parkinson's disease was formalized as a classical parameter optimization problem. The parameters were the doses and the timings of the drug administration. The drug used in this case was Madopar® (80% levodopa, 20% benserazide, a decarboxylase inhibitor). In Chapter 4 the relation between the drug administration and levodopa concentration in the blood was modeled as a linear system which was described by its input-output relationship. This model was used in the study of various optimization techniques and algorithms applied to the problem of finding the doses and the timings of drug administration, such that a predefined time history of levodopa concentration in the plasma was approximated as closely as possible. It was found that a cyclic coordinate search method is best suited for the case of discrete doses and timings (the only case which is clinically relevant).

However, it soon became apparent that the problem of achieving a desired concentration profile is only of relative interest, because the assumption that the levodopa level in the plasma is directly correlated with the antiparkinson effect of the drug does not generally hold. Therefore, it made more sense to optimize the treatment with respect to its therapeutic effect. However, the therapeutic effect of levodopa is not readily measurable in an objective manner. Therefore, a simple computer based tracking device was designed to render some parkinsonian symptoms measurable. The tracking was performed by moving the end phalanx of the thumb and the tracking data was recorded by the process computer which also controlled the reference signal. The development of some sophisticated filtering and signal processing algorithms were necessary to analyze the tracking data of the parkinsonian patients. These algorithms allowed the quantitative and objective measurement of tremor and bradykinesia in the same measurement sessions. Rigorous tests with a control group and a population of parkinsonians showed the measurements to be relevant in clinical use.

Furthermore, the device was used to model the pharmacodynamic relationship between the drug administration and its therapeutic effect. It resulted in a non-linear third order model with delay. This model was than used to optimize the timings and the doses of drug administration, such that maximum therapeutic effect is achieved. This maximization, of course, took the amount of the administered drug into consideration as well. The application of this method on a patient showed the model and the method to be pertinent in the clinical sense.

Although only thumb movements were analyzed – mainly because of their simplicity – maybe more complicated motor activity should also be studied and modeled to get a better clinical picture of the therapeutic effect of the used drug. In this context, the study of normalized movements as well as speech and gait analysis, can be named. Methodologies in these fields are so advanced that various instruments and software products are commercially available from companies like Lukotronic, Xsens, Vicon Motion systems just to name a few [125], [245].

As mentioned in Section 5.6, many researchers have tried to find ways of measuring the neurological functions objectively and quantitatively. Numerous batteries of tests to measure the neurological functions exist already. Basically, these batteries fall into two groups: a) general measurement of the neurological function and b) measurement of the neurological deficits in specific diseases like multiple sclerosis or Parkinson's disease. Most of these batteries use simulated daily activities like the time it takes to button a garment or to walk a given distance. However, all these batteries (which were mostly developed in university hospitals) are rather cumbersome to perform and the equipment needed can seldom be afforded by individual physicians.

It is, of course, possible to develop further programs to measure symptoms of other diseases as well. As mentioned above, the PC can also act as an excellent core for other measurements. Another possibility given by the use of PC's is the storage and classification of the measurement results which can facilitate studies concerning a population of patients. It has to be pointed out that the role of the PC in measuring symptoms is not limited to the therapy optimization. A very important use in this context is the assessment of the therapeutic effects of various drugs.

Further refinements of the mathematical drug-effect model in our case can include the consideration of the pharmacokinetics of various metabolites and chemicals relevant to the therapeutic effect of the drug. Moreover, the model can certainly be enriched in a manner which allows it to include the various side effects of the drug.

It must also be pointed out that the methodology described in this Chapter can be used very effectively in the education of neurologists. They can see the possible therapeutic as well as undesired side effects of a treatment they might have in mind by merely simulating it. Repeated use of this method can render itself eventually unnecessary if the physician becomes proficient enough to guess the optimal therapy before hand.

Last but not least, it also has to be stressed that although the drug used was Madopar®, the methodology described in this Chapter is applicable to any present and future drug used in the symptomatic treatment of Parkinson's disease.

5.11 Exercises

A1: What are the major symptoms of Parkinson's disease?

A2: What are the possible ways of measuring these symptoms?

A3: Which parts of the brain and which neurotransmitters are involved in Parkinson's disease?

A4: How is Parkinson's disease treated?

B1: Explain the workings of the pharmacodynamic model in Figure 5.21 and equations (5.4) and (5.5) qualitatively.

B2: The three effects of medication measured with the device shown in Figure 5.4, the reaction time, the movement time and the error area are highly correlated. Nevertheless, all of them are used in tandem for therapy optimization. Comment on that.

B3: Explain why therapy optimization can be treated as a deterministic problem in spite of the stochastic nature of the situation? Does it make sense to modify the objective function such that a deterministic approach fails?

B4: It might be very useful to modify the objective function such that it includes the dynamic side effects in a more direct way. It is also certainly possible to penalize each dose separately and not necessarily linearly. Improve the objective function accordingly. How would you go about modeling the side effects as functions of medication?

C1: One practical problem which may arise from complicated dosage regimens which require the patient to administer different doses at irregular times can be solved with a simple device. This device is a "smart pill box", which can be programmed to ring at preset times and to display the pills which the patient has to take. Develop such a "smart pill box".

C2: The optimization process may be further developed to include factors which are not easily quantifiable but clinically relevant. Develop an expert system which suggests levodopa dosage regimens for Parkinson patients with the help of artificial intelligence methodologies.

C3: Extend the system developed in C3 to deal with a number of treatments for different clinical conditions.

C4: The technology described in Section 5.6 for measuring Parkinsonian symptoms relies on a combination of PC with display. Develop a new measurement method based on tablet computers for measurements and analysis.

Chapter 6
Diabetes and Control of Blood Glucose

Let food be your medicine, and medicine your food.

— *Hippocrates*

The only way to keep your health is to eat what you don't want, drink what you don't like, and do what you'd rather not.

— *Mark Twain*

6.1 Introduction

In healthy persons, blood glucose concentration normally climbs up right after meals but it drops back to normal levels of 3.5 – 5.5 mmol/L in a couple of hours[1]. When healthy people do not eat, blood glucose concentration is normally within this range [129]. Hyperglycaemia is the term used for an elevated glucose concentration in the blood. Diabetic patients have impaired regulatory responses. Therefore, blood glucose remains high after the meals and even when the patients do not eat. Also, otherwise healthy people under surgical stress may show diabetic responses because of elevated levels of catecholamine, a hormone which has the opposite effect of insulin. This is the main reason why surgical operations on diabetic patients are riskier.

Blood glucose concentration is regulated mostly by two hormones: insulin and glucagon[2]. These hormones are both secreted by the islets of Langerhans in the pancreas.

[1] In some countries the unit mg/dL is used instead of mmol/L. The linear conversion factor is 18. For instance 11.1 mmol/L corresponds to 200 mg/dL.

[2] This Chapter follows the structure of [74].

S.S. Hacısalihzade: *Biomedical Applications of Control Engineering*, LNCIS 441, pp. 137–173.
DOI: 10.1007/978-3-642-37279-7_6

Insulin

Insulin is a peptide hormone, produced by the pancreas. Insulin is responsible for regulating carbohydrate and fat metabolism in the body. It helps cells to take up glucose from the blood. The way insulin works is by stoping the use of fat as an energy source by inhibiting the release of glucagon. In healthy humans insulin is provided within the body in a constant proportion to remove excess glucose from the blood. When blood glucose levels fall below a threshold, the body begins to use stored sugar as an energy source. Insulin also has several other anabolic effects throughout the body.

Insulin is stored in the body as a hexamer, while the active form is the monomer. A computer-generated image of six insulin molecules are shown in Figure 6.1.

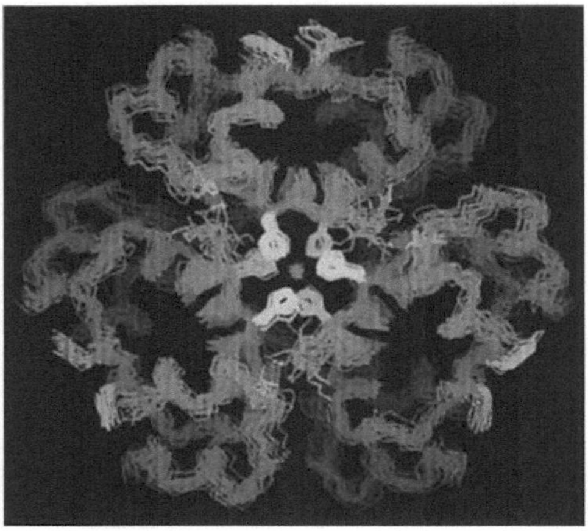

Fig. 6.1 Six insulin molecules are assembled in a hexamer, highlighting the threefold symmetry, the zinc ions holding it together, and the histidine residues involved in zinc binding [73].

The discovery of insulin is fairly recent. In 1923, two Canadian physicians from Toronto, Frederick G. Banting and John Macleod, were awarded the Nobel prize in physiology and medicine for their discovery and isolation of the insulin molecule [48]. In 1958 the same prize went to the British biologist, Frederick Sanger, for determining the primary structure of this peptide hormone composed of 51 amino acids. 1n 1969 a British chemist, Dorothy Hodgkin, was able to determine the tertiary structure (spatial conformation) of insulin using X-ray diffraction techniques. She had already received the Nobel prize in Chemistry for her development of X-ray crystallography and for having applied her techniques to deciphering the structure of vitamin B_{12}.

Insulin is anabolic (synthetic phase of metabolism) and causes rapid uptake and use of glucose by the tissues. Insulin also causes the storage of excess glucose as glycogen in the liver and muscles. If there is too much glucose so that it cannot all be stored as glycogen, it is converted to fatty acids and stored in adipose tissue, what we call fat. Glucagon, on the other hand is catabolic and helps to break down glucose and fatty acids. This couple of hormones, insulin and (its less famous sister) glucagon, have reciprocal affects and they are normally secreted to keep the blood glucose concentration within the normal range. When the glucose level is too high, insulin is secreted. This lowers the blood glucose concentration. If the blood glucose concentration is too low, glucagon is secreted and this increases the blood glucose concentration towards normal levels. Insulin and glucagon action is shown schematically in the block diagram in Figure 6.2.

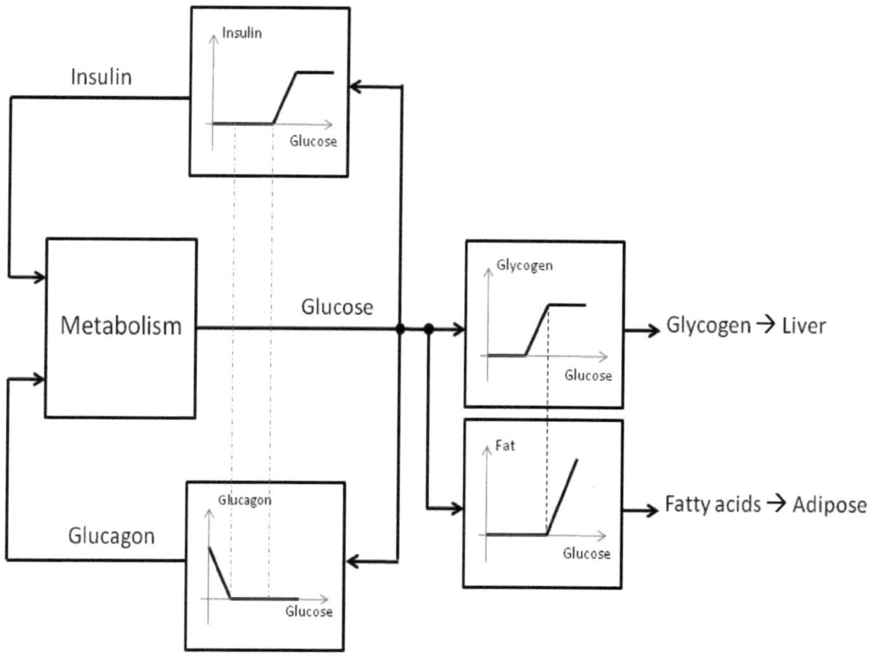

Fig. 6.2 A schematic depiction of the insulin and glucagon action.

What happens if the blood glucose concentration is too high? Why is this such a big problem? A high glucose concentration exerts osmotic pressure in the extracellular fluid and can cause cellular dehydration. When cells lose their water they shrivel and die. The risk of cardiovascular disease doubles with diabetes. Further complications arising from diabetes are bladder control problems for women and erectile dysfunction for men as well as retinopathy for both sexes, not to mention wounds that do not heal and resulting, in the extreme case, in the amputation of limbs, especially feet and legs.

There are two types of diabetes: Type 1 diabetes results from the body's failure to produce insulin (this was also known as insulin dependent diabetes mellitus IDDM or juvenile diabetes until recent years). Type 2 diabetes is caused by the cells not being able to use insulin properly. In both cases the pancreas is defective and produces no insulin in type 1 and not sufficient insulin in type 2. World Health Organization

(WHO) gives the number of diabetes patients worldwide as 346 million [311]. In 2004, an estimated 3.4 million people died from consequences of high blood sugar. More than 80% of diabetes deaths occur in low- and middle-income countries. WHO projects that diabetes deaths will double between 2005 and 2030. Healthy diet, regular physical activity, maintaining a normal body weight and avoiding tobacco use can prevent or delay the onset of type 2 diabetes.

Diabetes has probably been around for as long as man has lived. There is an Egyptian papyrus dating back to 1500 BC which alludes to frequent urination. Symptoms of diabetes are recorded as far back as 400 BC; Susruta wrote in an ancient Hindi document that patients were characterized by "honeyed urine" [180]. One of the best early descriptions is by Aretaeus of Cappadocia, a physician who lived during the first or second century AD in Alexandria, during the reign of Emperor Nero or Vespasian. He wrote around 150 AD [180]

> "Diabetes is a remarkable disorder, and not one very common to man. It consists of a moist and cold wasting of the flesh and limbs into urine, from a cause similar to that of dropsy; the secretion passes in the usual way, by the kidneys and the bladder. The patients never cease making water, but the discharge is as incessant as a sluice let off. This disease is chronic in character, and is slowly engendered, though the patient does not survive long when it is completely established for the marasmus produced is rapid and death is speedy."

Diabetes in old Greek meant siphon, because diabetic patients drink copiously, causing them to urinate, thus siphon. Mellitus comes from Latin. It means honey, because it was noted that the urine of diabetic patients contained a lot of sugar (probably flies preferred the urine of diabetics over healthy individuals). Many languages also use the expression "sugar disease" for diabetes mellitus.

Various clinical studies have shown that keeping the blood glucose concentration within the comfort zone of 3 – 6 mmol/L significantly reduces the complications caused by diabetes. So, how can that be accomplished? The classical answer to that question is the self monitoring of blood glucose levels and the patient adjusting the amount of exogenous insulin that she administers to herself. How does this work? The general therapeutic situation was described in Chapter 4. Looking at the problem of keeping the blood glucose levels in a given range from that point of view, one distinguishes between the closed-loop control and the quasi closed-loop control.

Most common therapy in early stages of diabetes consists of the physician seeing the patient with a frequency of 1 – 6 months and adjusting the medication according to how the patient is reacting to the treatment. A diet with limited caloric intake (especially low on carbohydrates) together with moderate but frequent physical exercise are part of the treatment. Also, the patient is instructed to measure her blood glucose several times a week when she gets up, with the hope that high readings will encourage the patient to remember her prescribed diet and exercise regimens. Glycated hemoglobin (often abbreviated as HbA1c) is also measured, albeit much less frequently, to identify the average plasma glucose concentration over prolonged periods of time (1 – 3 months). Whereas early morning measurements (usually using finger prick methods) give information about preprandial glucose levels, HbA1c levels hold clues to postprandial glucose levels as well. This type of treatment can be seen as quasi closed-loop control as defined in Chapter 4, because the treatment is adjusted only a couple of times per year.

However, the infrequent sampling cannot take care of peaks in blood glucose levels after meals or troughs caused by fasting which might be dangerous if the blood glucose levels drop too much. This is called hypoglycaemia and may be deadly if not taken care of immediately. Clearly, one cannot keep doing this the whole time. Therefore, one solution is continuous measuring of blood glucose and adjusting the infusion rate of a pump which continuously administers insulin to the patient. But this can be done only if the patient is hospitalized and connected to intravenous drip. However, portable pumps with integrated sensors and controllers also exist since a couple of decades, albeit with moderate success. Although there are various in-between solutions as well, the basic loop always has the structure shown in Figure 6.3. We will study each block in the loop separately in some detail in the coming sections.

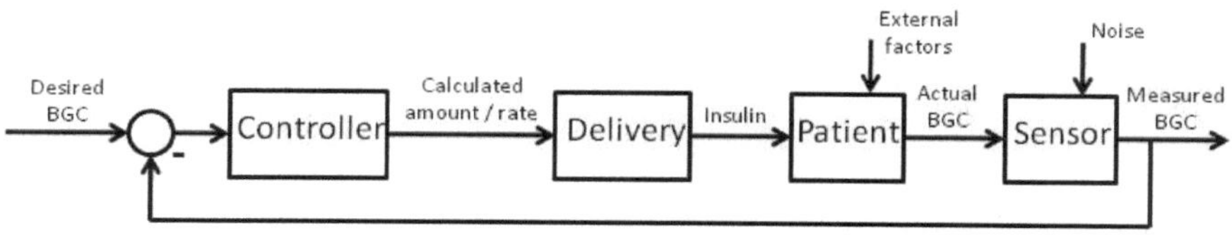

Fig. 6.3 The controller determines how much insulin should be infused based on the difference between the desired and the measured blood glucose concentrations (BGC). A pump infuses the amount of insulin determined by the controller. The patients dynamics and the external factors acting upon the patient (*e.g.*, what she eats or her exertion level) determine her BGC. Finally, this BGC is measured by an imperfect sensor to result in the measured BGC.

6.2 Measuring Blood Glucose Levels

Different ways of measuring the blood glucose concentration can be divided in three categories: invasive, minimally-invasive, and non-invasive techniques as shown in Figure 6.4.

6.2.1 Invasive Techniques

Invasive techniques use direct venous access to blood and ex-vivo determination of blood glucose levels by a glucose sensor. Clearly, this can be applied only for hospitalized patients [5]. Various techniques are used for determining the glucose level of the blood taken from the patient like amperometric, pH or color measurement methods as well as potentiometric and calorimetric detection techniques [107]. 20 years ago the time needed to determine the glucose level was about 10 minutes; nowadays it takes less than a minute. This way of measuring blood glucose is very accurate, because whole blood is analyzed for its glucose content. On the other hand, it continuously withdraws blood from the patient and is associated with an increased risk

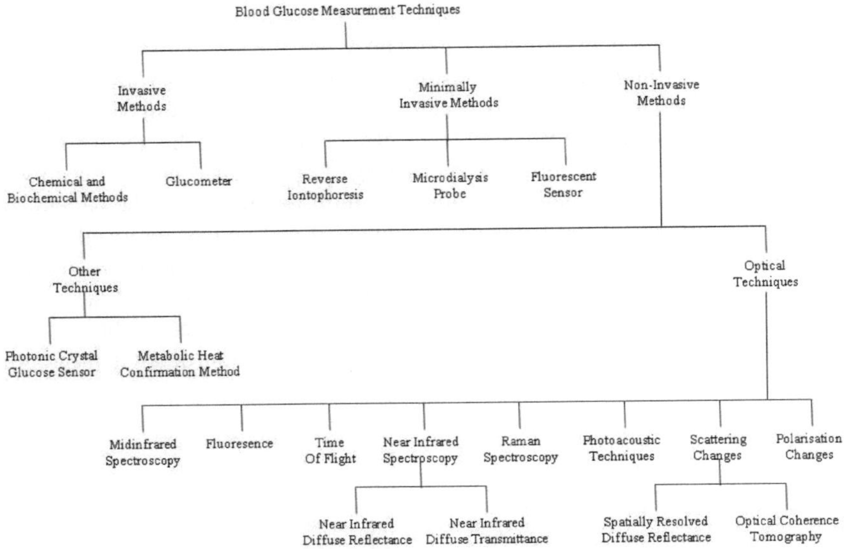

Fig. 6.4 Different techniques of measuring blood glucose concentrations after [251].

of infection and thrombosis. Also, this kind of measurement cannot be performed by the patient himself and needs proper supervision, thus limiting its applicability to hospitalised patients.

6.2.2 *Minimally-Invasive Techniques*

Minimally-invasive techniques are more convenient, because the measurement is done outside the circulation. That way the risk of pain, infection and thrombosis is significantly reduced. Minimally-invasive techniques measure the blood glucose level indirectly using the glucose concentration in the interstitial fluid surrounding the cells under the skin. It is based on the permissible assumption that the subcutaneous glucose concentration and plasma (whole blood) glucose concentration are strongly correlated [263]. Adipose tissue is especially well suited for that purpose, because fewer pain receptors exist in those areas. The measurement of subcutaneous glucose can be done by the implantation of a glucose sensor directly in the subcutaneous tissue or by means of fluid extraction from the subcutaneous space by microdialysis. See [124] for an overview of minimally-invasive techniques.

6.2.2.1 Implanted Sensors

[300] provides a tutorial introduction to biosensors for engineers and research and development workers involved in sensor development and application. It reviews

the fundamental principles before considering the practical aspects of design and fabrication for each of the major classes of biosensor. In this book amperometry is defined as

> "... the determination of the intensity of the current crossing an electrochemical cell under an imposed potential. This intensity is a function of the concentration of the electrochemically active species in the sample."

Glucose oxydase dehydrogenates glucose ($C_6H_{12}O_6$), transforming it to glucono-lactone ($C_6H_{10}O_6$). This reaction creates hydrogen peroxide (H_2O_2). By applying a small voltage (*e.g.*, -700 mV) with respect to the working platinum electrode, hydrogen peroxide is electrolyzed into O_2 and $2H^+$ which frees electrons . The flow of these electrons can be measured as an electrical current which is directly proportional to the concentration of glucose.

Such implantable amperometric sensors are commercially available since a couple of years, for instance, with the trade name MiniMed CGMS, the sensor is designed for continuous blood glucose monitoring in diabetic patients for periods of up to 3 days. The device takes a subcutaneous glucose reading every 10 seconds and stores an averaged reading every 5 minutes .

Fluorescence detection is a totally different principle for measuring blood glucose. The method relies on measuring a change in light emission when glucose binds to or reacts with certain molecules. Various aspects of this technique are explained in detail, for instance, in [120].

The major advantages of implantable sensors are that they are small in size, portable, and do not require extraction of fluids for operation. Their major drawback is their relatively short life span. The most common problem here is membrane biofouling (adhesion of biological matter like proteins on the sensor surface). Biofouling results in drifting or diminishing of the sensor signal. It starts upon contact of the sensor with the body. Another major problem with implantable sensors is immune cells' reaction to them, causing inflammation. In case of amperometric sensors, the body's saline environment corrodes the implanted metal electrodes. Of course, measurement artifacts due the patients movement and noise (most notably created by antioxidants like vitamin C) are also facts of life to be reckoned with.

6.2.2.2 Finger Prick Methods

> "Regular diabetic testing is essential for proper management of diabetes. Depending on the severity of the diabetes, a diabetic may be instructed to test their blood sugar levels as often as every two hours. Diabetic testing equipment is available from most medical supply stores and pharmacies. There are many options available in diabetic testing equipment. The oldest technique still in use involves blood glucose testing strips that are compared against a color chart like pH testing paper. There are also many glucose monitors available that will electronically read glucose levels off their own type of testing strips. All of these diabetic testing techniques require collecting blood for the test, most often by pricking the fingertip with a special lancet, and squeezing a drop of blood out of the finger. The finger is then pressed against the diabetic testing strip,

allowing measurement of blood glucose either with an electronic blood glucose meter or testing strip color chart". [3]

There are many such commercially available systems, requiring minuscule amounts of blood ($\leq 10\mu L$) to perform the blood glucose measurement and they are regulated by the norm DIN EN ISO 15197. However, their accuracies show a remarkable variance [115], [182]. Classified by working principle, these devices are either electrochemical glucose oxidase meters (*e.g.*, One Touch Ultra® and Precision X-tra®) or glucose dehydrogenase meters (*e.g.*, Ascensia® Contour, Accu-Chek Complete®, and Abbott Freestyle®). As an example, technical and engineering details of such a system can be found in [19].

6.2.2.3 Fluid Extraction

Microdialysis is a method that allows blood or interstitial fluid to be sampled and analyzed by a sensor outside the body. Fibres of dialysis membrane are implanted subcutaneously and perfused at a slow flow rate with isotonic fluid as shown schematically in Figure 6.5.

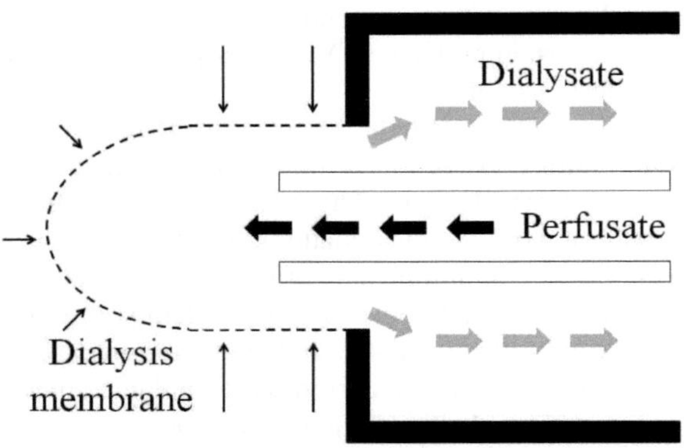

Fig. 6.5 The dialysate is pumped to a flow chamber that incorporates a glucose sensor. The inner cannula of the dialysis probe is perfused with an isotonic fluid (perfusate), which is transported to the tip of the dialysis probe. The probe is placed under the skin and glucose diffuses through the dialysis membrane down the concentration gradient due to the difference of concentrations between the perfusate and the interstitial fluid. (Adapted from [227].)

The major advantage of using microdialysis for measuring blood glucose lies in its preventing many other substances present in the interstitial fluid from entering the measuring cell and creating additional measurement noise. On the other hand, the catheters used for microdialysis are fragile and difficult to handle. Also, biofouling

[3] Taken from the popular website
`http://www.informationaboutdiabetes.com`

is still an issue (not at the glucose sensor but at the exchange membrane of the dialysis probe). Another potential hazard relates to the viral particles in the dialysate: they need to be disposed of properly due to the potential for viral transmission such as hepatitis and HIV.

6.2.3 Non-invasive Techniques

Non-invasive techniques are clearly the favorite way of blood glucose measurement, because they cause no pain or discomfort to the patient. Such measurements require a method of examining the glucose content in the tissue without puncturing the skin. Optical spectroscopy is a common such technique. Glucose concentration can be measured by optical means, because glucose exhibits properties such as light absorption, reflection, polarization, and other responses to radiation. Practically any method that allows the optical properties of glucose to be determined can be used to measure glucose concentration non-invasively. Most optical measurement techniques known to analytical chemistry can also be employed to measure blood glucose levels. However, most of these techniques are still experimental and not yet part of routine clinical practice. A non-invasive glucose meter approved by the U.S. FDA is the GlucoWatch G2 Biographer. It was designed to be worn on the wrist and to use electric fields to draw out body fluid for testing. The device was never meant to replace conventional blood glucose monitoring. An important limitation was that the GlucoWatch was not able to cope with perspiration at the measurement site. Due to limitations like this, the product is no longer on the market. It can, however, be reasonably expected that within the next decade or so, finger pricking may be replaced with non-invasive continuous glucose sensors for many diabetes patients [169], [12].

Explaining various experimental measurement techniques is beyond the scope of this book. Therefore, several references are given in Table 6.1 for the reader who wants to get in depth knowledge of the corresponding method.

Table 6.1 References for various experimental blood glucose measurement techniques.

Technique	References
Infrared absorptiometry	[140], [138], [222], [171], [142], [283], [170]
Mid-infrared emission spectrometry	[55], [304] [206]
Light scattering	[54], [141], [179], [256]
Optical coherence tomography	[274], [102], [189], [121], [173]
Raman spectroscopy	[119], [32], [101], [178]
Polarimetry	[84], [218], [203]
Photoacoustic	[293], [75], [204], [65]
Light diffraction	[215], [97], [318]
Dielectric (impedance) spectroscopy	[59], [60], [297]

6.2.4 *Clarke's Error Grid Analysis*

An important issue while self-monitoring the blood glucose is the accuracy of the employed sensor. In other words, the relationship between the measured value indicated by the sensor and the actual glucose concentration. Clarke and his colleagues have devised an analysis method to assess the accuracy of a particular glucometer [76]. This method evaluates the accuracy of a particular sensor system over the entire range of blood glucose values. This tool was designed to take into account the value of the glucometer's glucose measurement, the actual blood glucose value and the clinical significance of these two values. The x-axis is taken as the actual blood glucose and the y-axis as the value generated by the monitoring system. The diagonal represents the perfect agreement between the two. Data points above the diagonal overestimate and data points below the diagonal underestimate the actual values. It is assumed that a) the target blood glucose range for the patients is between 3.9 – 10 mmol/L, b) the patients will attempt to correct blood glucose readings that are outside this target range, c) the patients will not try to correct the readings that fall within the range, d) corrective treatment by the patient is considered to be wrong if the treatment causes the blood glucose reading to fall outside the target range, and finally, e) failure to treat blood glucose values below 3.9 mmol/L and above 13.3 mmol/L is inappropriate.

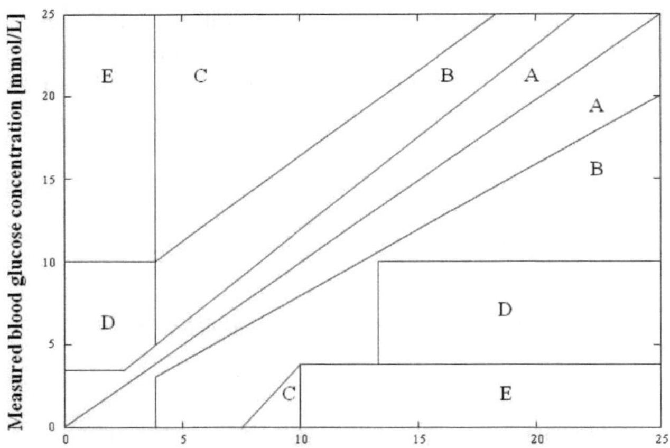

Fig. 6.6 Clarke Error Grid Analysis [76].

As can be seen in Figure 6.6, the grid can be divided in five regions with different characteristics:

Zone A: Glucose levels deviate either less than 20% from reference levels or are in the hypoglycaemic range if the reference is also below 3.8 mmol/L. In other words, measurements falling within this zone are clinically accurate in the sense that they would lead to clinically correct treatment decisions.

Zone B: Glucose levels deviate more than 20% from reference levels but this would lead to benign or no treatment based on the assumptions above.

Zone C: Glucose levels deviate more than 20% from reference levels and this would lead to overcorrecting acceptable glucose levels and such treatments might cause the actual blood glucose to fall outside the desired range.

Zone D: Actual blood glucose levels are outside the target range but sensor values are within the target range. This would cause a dangerous failure to detect and treat actually too high or too low blood glucose levels.

Zone E: Reference blood glucose levels and the measured values are opposite to each other. Therefore, the treatment decisions are opposite to those which are required. This is the "erroneous treatment zone".

In summary, values in Zones A and B are clinically acceptable while values in Zones C, D and E are potentially dangerous and are clinically wrong [74][4].

Figure 6.7 shows the analysis of a sensor using Clarke's error grid analysis method.

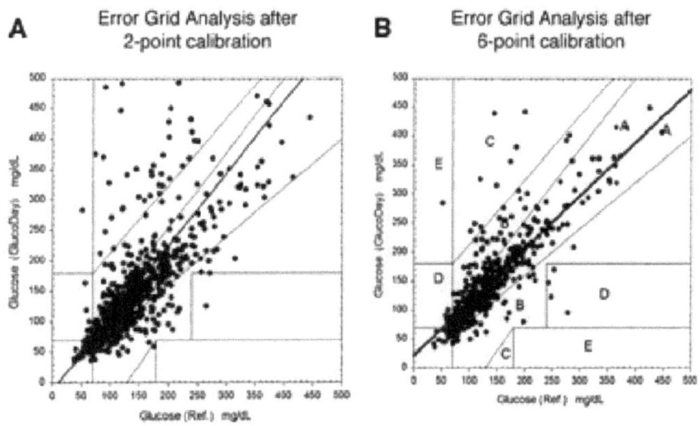

Fig. 6.7 Using Clarke Error Grid Analysis compares the accuracy of a sensor with 2 point and 6 point calibrations (from [90]).

[4] A newer version of Clarke's grid can be found, for instance, in [242]. The major improvements over the classical error grid are a symmetrical grid border (around the 1:1 axis) and no grid borders that jump *e.g.*, from zone A to C without any B in-between.

6.3 Insulin Delivery

Insulin delivery from the pancreas is a continuously regulated process which responds to changes in glucose concentration within seconds. In a healthy person, insulin is secreted by the pancreas directly into the venous circulation. It undergoes partial (40% – 80%) extraction by the liver. After this clearance, it is diluted into the systemic insulin pool and is distributed within the plasma as free insulin. The free insulin is then diffused into the extravascular compartment. It finally reaches insulin receptors on the surface of the target cells. The interaction of insulin with its receptor enables fuel to enter cells and activate enzymes for the storage or direct metabolization of those fuels in the cell. The fuels are mostly carbohydrates or amino acids. Excess insulin is degraded mainly by the liver and cleared from blood by kidneys. Insulin is normally cleared rapidly from the blood. Therefore, circulating insulin levels and insulin's ultimate delivery to the target tissues depends on the entry of insulin to the circulation.

In healthy humans insulin enters the circulation mainly from pancreatic release. If the endogenous production does not work properly (like in diabetes), exogenous insulin must be administered. This can be done intraperitoneally, subcutaneously or intravenously. Also possible routes are intramuscular, oral, nasal, or transdermal.

Intraperitoneal delivery is the most obvious way to deliver exogenous insulin only at first sight, though. In a healthy individual, insulin is first delivered to the circulation by the pancreas. Therefore, delivery of the insulin to the liver should produce normal metabolization of glucose. Also, this reduces peripheral insulin levels and the risk of arteriosclerosis. Furthermore, intraperitoneal delivery results in the metabolization of carbohydrates and particularly lipid in a more similar way to the normal physiological state than resulting from subcutaneous injection of insulin. On the other hand, intraperitoneal delivery requires an insulin pump to be surgically implanted in the peritoneum. Clearly, this is not the most practical solution. Firstly the pump has to be implanted surgically, which is associated with an increased risk of infection, bleeding, *etc.*. A mechanical failure of the pump poses additional risks to the patient, because he has to go through another surgery for its removal. Nevertheless, research continues towards the implantable artificial pancreas [157], [123], [264].

Subcutaneous delivery is the preferred route by patients to receive their daily insulin doses. This can be either by direct injection or via an insulin pump. Subcutaneously injected insulin forms a depot at the injection site. The compound then transforms to an absorbable state through dissociation and dissolution and diffuses from the depot into the circulation [74]. Similar to endogenous insulin, the injected insulin is distributed within the plasma as unbounded insulin, diffusing further into the extravascular compartment to reach the target cells.

Although very popular, subcutaneous delivery also has several shortcomings. The absorption rate is unpredictable and might change due to the exact location and depth of injection, exercise and vascularity of tissue. This might cause large swings in blood glucose levels. The insulin peaks can be adjusted to match food absorption by giving insulin before meals. However, if the patient does not eat after the injection, this might cause hypoglycaemia. It has to be remembered that there is a lag time of more than 10 minutes between the injection of insulin and its appearance in the plasma.

Intravenous delivery solves the problem of time lag but requires venous access which is difficult without medical supervision. Also, this increases the infection risk. Intravenous insulin may be given as either repeated boluses or continuous infusion. The bolus ensures insulin absorption and permits rapid adjustment of insulin to suit the patients needs. However, the effect of each bolus of short acting insulin dissipates within an hour or so. Therefore, it is necessary to measure blood glucose frequently and to deliver insulin boluses as needed. Continuous intravenous insulin delivery, on the other hand, allows less frequent blood glucose measurements (albeit, at the cost of increased risk of hypoglycaemia) and permits fine tuning of blood glucose levels.

Oral [172], nasal [317], or transdermal [96] delivery are more modern techniques, all in early stages of development and each technique with their advantages and drawbacks.

This completes the treatment of the block labeled "Delivery" in Figure 6.3. In the next Section we will study the block labeled "Patient".

6.4 Patient Dynamics

We have seen that in healthy humans, the pancreas keeps the blood glucose concentration in a tight range by secreting the hormones insulin and glucagon. In diabetic patients pancreas cannot fulfil this task. Therefore, an artificial external controller is required to fulfill that task. Control engineers know that the design of a controller must begin with understanding the dynamic relationship between the inputs and the outputs of the system that is to be controlled (often called "Plant"). In this case the "Plant" is the patient's glucose dynamics. In other words, we have to understand how the nature solves the control problem.

The bulk of the pancreas is an exocrine gland secreting pancreatic fluid into the duodenum after a meal. Scattered through the pancreas are 1 – 2 million clusters of cells called islets of Langerhans. The islets are endocrine tissue containing four types of cells: alpha, beta, gamma (also called pp cells) and delta. The beta cells constitute about 60% of the cells and secrete insulin. The alpha cells constitute about 25% of the total and secrete glucagon. The delta cells, constitute about 10% of the total, and secrete somatostatin (somatostatin has a variety of functions, but taken together, they work to reduce the rate at which food is absorbed from the contents of the intestines). The gamma cells, constitute the rest of the total and secrete pancreatic polypeptide (which, basically reduces appetite). Insulin and glucagon, as we have seen above, play the most important roles in the glucose regulatory system which is not an isolated system. It is coupled with many other metabolic pathways in the body.

In a healthy person, blood glucose concentration is controlled to move between 3.5 mmol/L and 5.6 mmol/L in the fasting state. During fasting, insulin secretion is at a low basal level and glucagon is released to allow the liver to mobilize glucose from its glycogen stores and to synthesize glucose from amino acids. When insulin levels are low, the uptake of glucose by muscles is minimized and free fatty acids are released by the muscles. When fasting continues more than 12 to 18 hours, these free fatty acids become the main energy carrier used by all tissues of the body except for the brain, whose glucose is supplied by amino acids [74].

Glucose is an essential nutrient for the brain. The processes described above allow the fasting blood glucose concentration to stay more or less constant so that the brain, which has no energy stores, has a sufficient fuel for its normal activity. When a person prepares to eat a meal, insulin secretion occurs in two phases: an anticipatory phase and a glucose sensitive phase. In the first phase, the sight of the food, its smell and the first bite of a meal cause the brain to send signals to the pancreas. These signals cause the pancreas to release insulin into the hepatic circulation. Once the insulin is in the hepatic circulation, the liver stops breaking down glycogen into glucose. As the food enters the stomach, the release of insulin is further facilitated by gastrointestinal hormones which increase the sensitivity of the islet cells to glucose. As nutrients are absorbed into the circulation, the glucose sensitive phase begins. After the absorption of all the carbohydrates, the feedback system for control of blood glucose returns the glucose concentration back to the control level, usually within a couple of hours.

This biphasic response is also observed in the plasma compartment during glucose clamp studies, where a step function of hyperglycaemia (sugar loading) is introduced intravenously in a normal individual. The first phase of the insulin response consists of a rapid rise in the insulin level during the first 10 minutes with a peak response at 4 minutes. The plasma insulin level then falls and reaches a local minimum at 10 minutes. After that, the plasma insulin level increases gradually and persists for the duration of the stimulus as shown in Figure 6.8.

Diabetic patients have a greatly reduced number of insulin producing beta cells. Therefore, the first phase of this biphasic response disappears. Without this anticipatory phase, the liver does not get a message to stop breaking down glycogen into glucose. This, in turn, results in continued hepatic glucose production. Consequently, when there is no insulin release coupled with further glucose being absorbed from the meal hyperglycaemia results. The major difference between type 1 and type 2 diabetes mellitus is that type 1 patients absolutely require exogenous insulin because they have no beta cells. Therefore, type 1 diabetes was also known as insulin dependent diabetes mellitus (IDDM).

It is very difficult to mimic the body's reaction to imminent food intake (which causes the first phase insulin release) by means of an external closed-loop controller, because this anticipatory phase in a healthy person is initiated by neuro- and gastrointestinal signals and those signals are not easily measurable. In open-loop systems this signal can be given by the patient by injecting insulin boluses manually

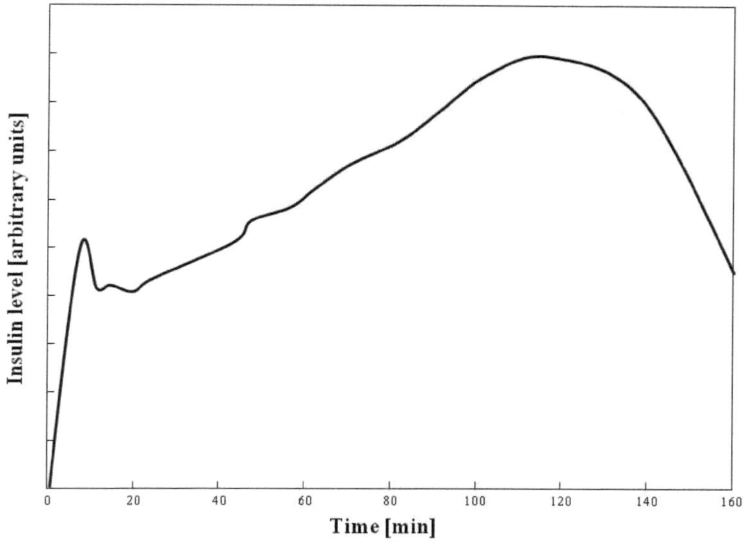

Fig. 6.8 Schematic biphasic insulin response. (Adapted from [227].)

before a meal. A solution to emulate the first phase insulin release using a closed-loop system would be to use sensors with response times in the order of seconds and to immediately deliver insulin when blood glucose levels start rising. This method was used in early artificial pancreas experiments. For the second phase insulin release, the matching of the insulin close to the blood sugar intake depends on the knowledge how much glucose is ingested and the responsiveness of the insulin receptors on target cells to enable glucose to enter and be utilized by the cells [74].

But why is it so important to keep the blood glucose level low? There are several reasons for that: 1) Elevated glucose concentration levels exert an osmotic pressure in the extracellular fluid and cause dehydration of the cells. 2) High levels cause loss of glucose through urination. This depletes the body further of fluids and electrolytes. 3) Constantly elevated levels of blood glucose concentration are "toxic" for cells and tissues and may cause functional damages in the cells and affect the wound healing process negatively [308]. 4) Therapy that keeps the blood glucose levels below 11 mmol/L improves long term outcome of diabetic patients with acute myocardial infarction [303]. 5) Keeping the pre-prandial blood glucose levels below 6.7 mmol/L delays the onset of diabetic microvascular complications like visual impairment, erectile dysfunction, kidney disease and nerve damage [298].

So why not be on the safe side and keep the blood glucose level very low? Mainly because low blood glucose levels carry the risk of hypoglycaemic coma. The levels should not drop below a threshold, because glucose is the only nutrient that can be used for energy by the brain or the retina.

6.4.1 Mathematical Models

So far we have looked at a qualitative description of how the body regulates the blood glucose concentration. Many researchers in the past have developed a multitude of mathematical models to study the problem of blood glucose control mathematically with varying degrees of complexity and applicability. [74] groups these models under the headings linear models, non-linear models and comprehensive models. We shall study a selected few of these models in some detail.

6.4.1.1 Linear Models

Linear models can be used if the dynamics of the metabolic system can be modeled appropriately as a linear system. As we have seen in Chapter 1, a system for which the superposition of the inputs results in the superposition of the outputs is called a linear system. Such systems can be described by a set of linear differential equations in the state space

$$\dot{\mathbf{x}}(t) = \mathbf{A}\mathbf{x}(t) + \mathbf{B}\mathbf{u}(t) \tag{6.1}$$

$$\mathbf{y}(t) = \mathbf{C}\mathbf{x}(t) + \mathbf{D}\mathbf{u}(t) \tag{6.2}$$

where $\mathbf{x}(t)$ is the state vector, $\mathbf{u}(t)$ is the combination of controllable inputs and uncontrollable disturbances and $\mathbf{y}(t)$ represents the outputs.

In Chapter 4 we had seen that compartmental analysis is a practical and simple method of modeling material exchange. For a system consisting of two compartments depicted as in Figure 4.2 the mass balance equations can hence be written as

$$\frac{dQ_1}{dt} = -k_{11}Q_1 - k_{12}Q_1 + k_{21}Q_2 + J(t) \tag{6.3}$$

$$\frac{dQ_2}{dt} = k_{12}Q_1 - k_{21}Q_2 - k_{22}Q_2 + K(t) \tag{6.4}$$

where Q_1 and Q_2 indicate the quantity of materials in compartment 1 and 2 respectively. k_{ij} indicates the rates at which the materials in compartment i is transferred to compartment j. $J(t)$ and $K(t)$ are flows of material from exogenous sources. These equations can be simplified as

$$\frac{dQ_1}{dt} = -m_1Q_1 + k_{21}Q_2 + J(t) \tag{6.5}$$

$$\frac{dQ_2}{dt} = k_{12}Q_1 - m_3Q_2 + K(t) \tag{6.6}$$

where $m_1 = (k_{11} + k_{12})$ and $m_3 = (k_{21} + k_{22})$. This is the most common form of linearized models of glucose-insulin, where glucose and insulin are the primary compartments.

Ackerman came up with this model in early 1960s which was to become very popular. Using Ackerman's nomenclature [2], where G is the glucose concentration, G_0 is the fasting glucose concentration, H is the blood hormone concentration (including insulin), H_0 is the fasting blood hormone concentration, the variables g and h are defined as $g = G - G_0$ and $h = H - H_0$, equations (6.5) and (6.6) become

$$\frac{dg}{dt} = -m_1 g - m_2 h + J \tag{6.7}$$

$$\frac{dh}{dt} = m_4 g - m_3 h + K \tag{6.8}$$

m_1 is the rate constant for the removal of glucose above the fasting level due to its own excess above the fasting level (glucose effectiveness). m_2 is the rate constant for the removal of glucose above the fasting level due to blood hormone concentration above the fasting level. m_3 is the rate constant for the removal of the hormone above the fasting level due to its own excess above the fasting level. m_4 is the rate constant for the removal of hormone above the fasting level due to the blood glucose concentration above the fasting level. Typical values for these rate constants can be found in [320].

Bolie's model is very similar but non-linear [49]:

$$\frac{dH}{dt} = \frac{1}{V}\dot{U} - \frac{F_1(H)}{V} + \frac{F_2(G)}{V} \tag{6.9}$$

$$\frac{dG}{dt} = \frac{1}{V}\dot{P} - \frac{F_3(G,H)}{V} - \frac{F_4(G,H)}{V} \tag{6.10}$$

with

G : extracellular glucose concentration,
H : extracellular insulin concentration,
V : extracellular fluid volume,
$\dot{U}$: rate of insulin injection,
$\dot{P}$: rate of glucose injection,
$F_1(H)$: rate of insulin destruction,
$F_2(G)$: rate of insulin production,
$F_3(G,H)$: rate of liver accumulation of glucose,
$F_4(G,H)$: rate of tissue utilization of glucose.

For small variations of the insulin and glucose concentrations, (6.9) and (6.10) can be linearized around their nominal values G_0 and H_0 using a Taylor series expansion:

$$\frac{dH}{dt} = u - \left[\frac{1}{V}\left(\frac{\partial F_1}{\partial H}\right)_{H_0}\right]\cdot h + \left[\frac{1}{V}\left(\frac{\partial F_2}{\partial G}\right)_{G_0}\right]\cdot g \tag{6.11}$$

$$\frac{dG}{dt} = p - \left[\frac{1}{V}\left(\frac{\partial F_3}{\partial H} + \frac{\partial F_4}{\partial H}\right)_{H_0,G_0}\right]\cdot h - \left[\frac{1}{V}\left(\frac{\partial F_3}{\partial G} + \frac{\partial F_4}{\partial G}\right)_{H_0,G_0}\right]\cdot g \tag{6.12}$$

where $u = \frac{\dot{U}}{V}$ and $p = \frac{\dot{P}}{V}$. Comparing these equations with (6.7) and (6.8) shows the similarity of Ackerman's model and the linearized form of Bolie's model.

6.4.1.2 Non-linear Models

Linear models are results of oversimplification and often contain both unacceptable levels of modeling error and significant mismatch between the process and the model. Therefore, many non-linear models were developed over the decades which can emulate various aspects of the actual process with varying degrees of accuracy.

This Paragraph is adapted from [68] and explains the seven distinct models of glucose uptake proposed therein. The first three assume that glucose utilization is insulin independent. In the rest of the models, explicit relationships between insulin, glucose and glucose fluxes are assumed.

Model 1: This model assumes that glucose is produced constantly and that it distributes in a single compartment. It is further assumed that the rate of glucose utilization is a linear function of the plasma glucose concentration. The first order linear equation of this model is thus

$$\frac{dG(t)}{dt} = -k_1 + p_L \tag{6.13}$$

with the initial value of glucose concentration being that in steady state ($G(0) = G_{ss}$). k_1 is the glucose disappearance. p_L is the glucose production of the liver.

Model 2: In this model, the previous model is modified for the glucose utilization to be saturable and dependent on plasma glucose concentration:

$$\frac{dG(t)}{dt} = \frac{-V_m/V}{K_m + G(t)} + p_L \tag{6.14}$$

again with the initial value of glucose concentration being that in steady state ($G(0) = G_{ss}$). V is the glucose volume. V_m and k_m are Michaelis-Menten parameters (see Box).

Michaelis-Menten Kinetics
Michaelis - Menten kinetics is one of the simplest and most commonly employed model for enzyme kinetics. It is named after German biochemist Leonor Michaelis and Canadian physician Maud Menten. The model is an equation describing the rate of enzymatic reactions, by relating reaction rate v to the concentration of a substrate S $[S]$ simply as $v = \frac{v_{max}[S]}{K_m+[S]}$. v_{max} represents the maximum rate achieved by the system, at saturating substrate concentrations. The Michaelis constant K_m is the substrate concentration at which the reaction rate is half of v_{max}. Biochemical reactions involving a single substrate are often modeled with Michaelis - Menten kinetics [193]. The attentive reader might recognize the similarity with a pharmacodynamic model introduced in the final section of Chapter 4.

Model 3: This model assumes that glucose distribution is represented by two compartments. Disappearance from either compartment depends linearly on the concentration of glucose in that compartment. Again, glucose production is assumed to be constant. The corresponding equations are

$$\frac{dG(t)}{dt} = -(k_1 + k_3)G(t) + k_1k_2X(t) + p_L \tag{6.15}$$

$$\frac{dX(t)}{dt} = G(t) - (k_2 + k_4)X(t) \tag{6.16}$$

where $X(t) = G_2(t)/k_1$, $G(0) = G_{ss}$, $X(0) = 0$. G_2 is the glucose concentration in a compartment different from plasma. k_i represent either fractional turnover rates or control actions.

Model 4: In this model, which derives from Bolie's and Ackerman's models, glucose disappearance depends linearly on the concentrations of both glucose and insulin. Glucose distribution is constant and distributes in a single compartment as in Model 1.

$$\frac{dG(t)}{dt} = -k_1G(t) - k_2I(t) + p_L \tag{6.17}$$

where I is the plasma insulin concentration.

Model 5: This model assumes that glucose uptake depends directly on the insulin concentration, not in plasma but in the second compartment.

$$\frac{dG(t)}{dt} = -k_1G(t) - k_2k_4X(t) + p_L \tag{6.18}$$

$$\frac{dX(t)}{dt} = -k_3X(t) + i(t) \tag{6.19}$$

where $i = I - I_{ss}$ is the deviation from basal value of actual plasma insulin concentration, $i' = I' - I_{ss}$ is the deviation of insulin concentration I' in a compartment other than plasma, $X = i'/k_2$.

Models 1 – 5 assume the glucose production p_L to be constant. However, we know that insulin inhibits glucose production. The following two models take that inhibition into account.

Model 6: In this model glucose production and hepatic glucose uptake are lumped together as net hepatic glucose balance. The rate of change of glucose is the difference between the net hepatic glucose balance B and the disappearance of glucose into peripheral tissues only U_P. We know that [33] hepatic glucose balance varies according to

$$B(t) = B_0 - (k_5 + k_6 i'(t))G(t) \tag{6.20}$$

where B_0 is the value expected when plasma glucose concentration is extrapolated to zero. It is further assumed that the insulin acts from a remote compartment as in Model 5.

Similarly, the glucose utilization is given by

$$U_P(t) = (k_1 + k_4 i'(t))G(t) \tag{6.21}$$

where remote insulin is seen to increase glucose disappearance through a process at molecular level.

$$\frac{dG(t)}{dt} = (-[k_1 + k_5] - X(t))\,G(t) + B_0 \tag{6.22}$$

$$\frac{dX(t)}{dt} = -k_3 X(t) + k_2[k_4 + k_6]i(t) \tag{6.23}$$

again with the initial conditions $G(0) = G_{ss}$ and $X(0) = 0$.

Model 7: In this model hepatic and peripheral glucose utilization are lumped together and absolute glucose production is represented explicitly.

$$\frac{dG(t)}{dt} = (-k_1 - k_2 k_4 X(t))G(t) + \frac{p_L/k_5}{1 + [k_2/k_5]X(t)} \tag{6.24}$$

$$\frac{dX(t)}{dt} = -k_3 X(t) + i(t) \tag{6.25}$$

again with the same initial conditions as in Model 6. All the initial conditions in those models are basal values of glucose and insulin concentrations. The intravenous bolus injection of glucose or insulin can be modeled as an impulse function (see Chapter 4 for details of how an why this is done). It is important to note that all the parameters in Models 1 – 6 are uniquely identifiable whereas Model 7 is not (two solutions). The issue of identifiability is especially important in modeling biological systems. [79] deals with this issue in detail.

First and second phase insulin secretion modeled by [127] and [198] are described in detail in [68].

6.4.1.3 Comprehensive Models

As an example of comprehensive models, a model for glucose regulation, first introduced by Cobelli and coworkers in [78], will be described in detail here. The model describes how glucose concentration is controlled by insulin and glucagon secretion. The structure of the model is shown in Figure 6.9.

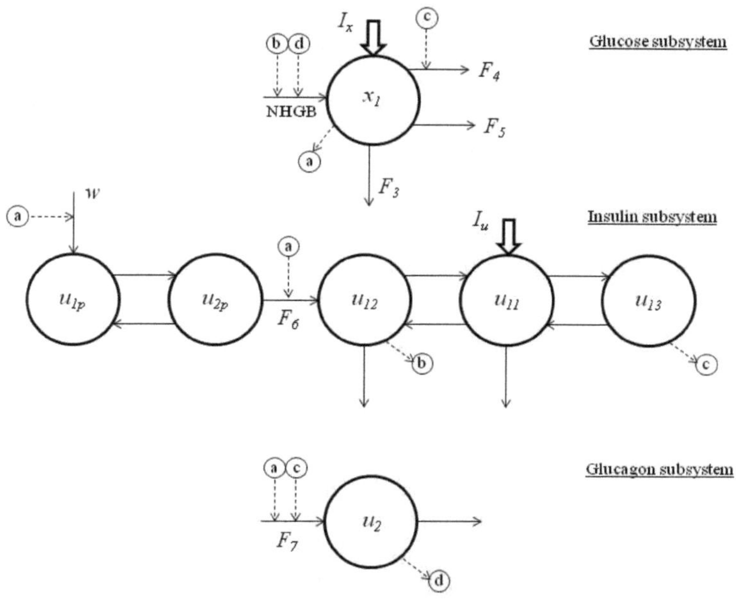

Fig. 6.9 Control system model of the glucose regulatory system with the subsystems a) glucose, b) insulin and c) Glucagon. Continuous lines represent material flow, dashed lines represent control signals. See text for the explanation of variables. (Adapted from [77].)

Glucose subsystem is modeled as a single compartment system with plasma and extracellular fluids as the compartment (x_1). Insulin subsystem is modeled as a five compartment system with stored pancreatic insulin (u_{1p}), promptly releasable pancreatic insulin (u_{2p}), insulin in plasma (u_{11}), insulin in the liver (u_{12}) and interstitial fluid insulin (u_{13}) as the compartmental variables. Glucagon subsystem is also modeled as a single compartment system with u_2 as glucagon in the plasma and interstitial fluids. Further variables used in the mathematical description have the following meaning:

W and F_i: non-linear functions,
I_x: test input of glucose,
I_u: test input of insulin,
k_{02}: a function of x_1
m_{ij}, h_{ij}, k_{ij}: parameters.

Model equations are

$$\dot{x}_1(t) = NHGB(x_1, u_{12}, u_2) - F_3(x_1) - F_4(x_1, u_{13}) - F_5(x_1) + I_x(t), \qquad x_1(0) = x_{10} \tag{6.26}$$

$$\dot{u}_{1p}(t) = -k_{21}u_{1p} + k_{12}u_{2p} + w(x_1), \qquad u_{1p}(0) = u_{1p0} \tag{6.27}$$

$$\dot{u}_{2p}(t) = k_{21}u_{1p} - [k_{12} + k_{02}(x_1)]u_{2p}, \qquad u_{2p}(0) = u_{2p0} \tag{6.28}$$

$$\dot{u}_{11}(t) = -(m_{01} + m_{21} + m_{31})u_{11} + m_{12}u_{12} + m_{13}u_{13} + I_u(t), \qquad u_{11}(0) = u_{110} \tag{6.29}$$

$$\dot{u}_{12}(t) = -(m_{02} + m_{12})u_{12} + m_{21}u_{11} + k_{02}(x_1)u_{2p}, \qquad u_{12}(0) = u_{120} \tag{6.30}$$

$$\dot{u}_{13}(t) = -m_{13}u_{13} + m_{31}u_{11}, \qquad u_{13}(0) = u_{130} \tag{6.31}$$

$$\dot{u}_2(t) = -h_{02}u_2 + F_7(x_1, u_{13}), \qquad u_2(0) = u_{20} \tag{6.32}$$

(6.26) represents the mass balance for the glucose subsystem. $NHGB$ is the net hepatic glucose balance given by $NHGB = F_1(x_1, u_{12}, u_2) - F_2(x_1, u_{12})$. F_1 is the liver glucose production and F_2 is the liver glucose uptake. F_3 is renal excretion. F_4 is the peripheral insulin-dependent glucose utilization. F_5 is the peripheral insulin-independent glucose utilization. $I_x(t)$ is the rate of exogenous intravenous glucose injection. (6.27) and (6.28) represent the beta cells of the pancreas. $w(x_1)$ is the insulin synthesis controlled by blood glucose concentration. $k_{02}(x_1)u_{2p} = F_6(u_{2p}, x_1)$ is the insulin secretion rate. (6.29), (6.30) and (6.31) describe insulin kinetics. u_{11} is the insulin quantity in plasma. u_{12} is the insulin quantity in liver. u_{13} is the insulin quantity in interstitial fluids. $I_u(t)$ is the insulin test input. (6.32) describes the glucagon dynamics where F_7 is the endogenous release of glucagon dependent on blood glucose and interstitial fluid insulin. The testing and the verification of this model (in spite of difficulties inherent in the model structure) with various measurements are described in detail in [68].

Comprehensive models with additional compartments which include the gut, the kidneys, the heart, the lungs and the brain have also been developed [289], [291].

However, the added complexity and the increase in the number of equations and the parameters involved opens up new issues in their identifiability, thus reducing the applicability of such comprehensive models in the solution of the related control problems.

6.5 The Controller

Basically there are two very different approaches to controlling the blood glucose levels. The first approach does not require an explicit mathematical model of the patient or the relationship between the glucose, insulin and glucagon concentrations. Those are called model-less control methods. The second approach makes use of models of varying detail level. A review of various controllers is given in [80].

6.5.1 Rule Based Control

The simplest controllers that do not require mathematical models of the relationships among the substances pertaining to the control of blood glucose levels like glucose concentrations or mechanisms for insulin secretion are the so called rule based controllers. Those controllers are based on previous experience and perform surprisingly well.

One such method makes use of lookup tables to determine the amount of insulin that has to be administered to the patient. Those tables are based on experience only. Most common controllers are (almost) open-loop controllers. Blood glucose levels are measured intermittently (as infrequently as 1 – 3 hours) and insulin delivery is determined according to a lookup table by hospital staff. The tables map the blood glucose level to the insulin delivery rate [310]. Variations of this method include smooth transitions from one delivery rate to the other.

Another method uses more sophisticated knowledge based controllers which might include a two-loop feedback strategy with the inner-loop providing the amount of both rapid (Lispro) and slow (NPH) insulin types that the patient has to program in a three-shots daily basis. The combined preparation is then injected to the patient through a subcutaneous route. Meanwhile, the outer-loop adjusts the maximum amounts of insulin provided to the patient. Thus, the amount of insulin required to maintain a normal glucose level is optimized. Both controllers are designed by gathering information from the treating physician [66].

More sophisticated methods using neural networks, fuzzy control and their combinations have also been reported [301], [153], [126], [89].

6.5.1.1 Black Box Models

This method is based on obtaining the relationship between the inputs and the outputs by selecting a suitable mathematical curve and determining its parameters by curve fitting techniques. As we have seen in the stock price prediction example in Chapter 2, the equation behind the curve does not necessarily have a physical meaning and can be seen as the input-output model of a black box (a black box is an object which can be viewed solely in terms of its input, output and transfer characteristics without any knowledge of its internal workings). One of the first applications of this method on the control of blood glucose concentrations was introduced by Albisser [6] which later became the basis for the successful product Biostator®. In a brief to a journal in 1974 Albisser was both exuberant and realistic in his assessment of the "Artificial Pancreas" as he called it:

> The artificial pancreas brings us one step nearer to the diabetics' dream of eliminating injections and hypoglycaemia. However, years of development will certainly be needed for its miniaturization and implantation, and even then it seems unlikely to have universal application [4].

Albisser used the following algorithm for his artificial pancreas:

$$R_g = \frac{1}{2}M_g(1 - \tanh(S_g[G - B_g]) \tag{6.33}$$

$$R_i = \frac{1}{2}M_i(1 - \tanh(S_i[G_p - B_i]) \tag{6.34}$$

$$G_p = G + K_1(e^{(\frac{A}{K_2})} - 1) \tag{6.35}$$

where the hyperbolic tangent function was used to model a sigmoid curve. R_g and R_i denote the glucose and insulin infusion rates, respectively. G and G_p are the current and the projected blood glucose concentrations. M_g is the maximum glucose infusion rate, M_i is the maximum insulin infusion rate. B_g and B_i are the blood glucose concentration levels of glucose and insulin at which half of the maximum infusion rate of glucose and insulin rate is chosen to occur, respectively. A is the rate of change of blood glucose concentration (averaged over the last four measurements). K_1 and K_2 are parameters to adjust the magnitude of the difference factor and the sensitivity to changes in A, respectively. S_g and S_i are slope parameters.

Those equations were then gradually modified to address various observations made during their use and later algorithms dropped the infusion of exogenous glucose altogether.

6.5.1.2 Good Old PID-Controller

Not surprisingly, control engineers' favorite tool, the Proportional, Integral, Derivative Controller (PID), was also used to maintain the blood glucose concentration of diabetic patients by many researchers. Figure 6.10 depicts such a controller schematically.

The controller compares the output of the system to be controlled ("the plant") with its desired output and builds their difference as the error signal e. The controller, based on this error signal, determines the control signal of the plant. In case of a PID-controller, the error is a) multiplied by a proportionality factor K_P, b) integrated and multiplied by an integral factor K_I, c) differentiated and multiplied by a

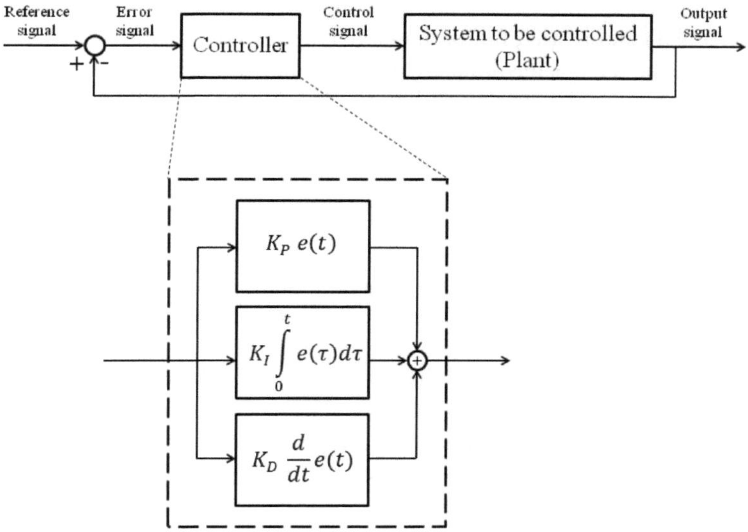

Fig. 6.10 Schematic diagram of a PID-Controller.

derivative factor K_D. Finally the signals gained by the parallel steps a), b) and c) are summed to result in the control signal which is used as the input for the plant. For a digital controller, the algorithm to calculate the control input $u(t)$ can be formalized as

$$u(t) = K_P e(t) + K_I \left(\sum_{i=1}^{n} e_i(t)\Delta t \right) + K_D \left(\frac{e(t) - e(t-\Delta t)}{\Delta t} \right) \tag{6.36}$$

Δt denotes the sampling time and n the integration time.

PID-Controllers are easy to design and their parameters can be fine tuned to account for the mismatch between the assumed and the actual models of the plant. The design process generally requires a model for the plant but even without a model, satisfactory results can be achieved in many cases. A major drawback of PID-Controllers is their sensitivity to time lags inherent in the plant. This might lead to instability of the closed-loop system which manifests itself by ever growing oscillations of the output. Needless to say, in any biomedical application area, such system behavior is simply not acceptable.

6.5.2 *Model Based Control Algorithms*

Black box models use a simple method of modeling the input-output relationship of a plant (such models do not even have to be quantitative). However, predictive power of such models is rather limited, because they do not give an insight on what is actually happening in the black box. Therefore, more successful controllers have been developed making use of a formal model of the plant.

In this case the plant is the human glucose-insulin interaction. If (and it turns out to be a big if) we can develop a good mathematical model for this interaction, we should be able to design controllers with excellent characteristics simply by solving the corresponding mathematical problem. If we can build a good mathematical model, apart from offering insight into the actual processes, it also allows us to simulate various situations and test the controller without any risk to the patient. As we have seen in Chapter 2, whether a mathematical model is appropriate depends on the purpose of the model. We have also seen that some of the parameters in the model may be inherently unidentifiable. Such structures limit the usefulness of the model.

For linear models, solutions of the mathematical problems resulting from the controller design problem can be obtained analytically. For non-linear models, the models are either linearized to be able to use known methods for linear models or numerical methods are used to solve the pertaining mathematical problems.

6.5.2.1 Pole Placement

If a linear system is controllable, it is known that the poles of the closed system can be placed anywhere in the complex plane by means of state feedback, to achieve the desired dynamical response [161]. Let the controllable system be defined through its state equations as

$$\dot{\mathbf{x}}(t) = \mathbf{A}\mathbf{x}(t) + \mathbf{b}u(t) \tag{6.37}$$

with $\mathbf{x}(t)$ being the state vector and $u(t)$ as the scalar control input. If we feed the complete states of the system back as the input ($u(t) = -\mathbf{k}^T\mathbf{x}(t)$) we get

$$\dot{\mathbf{x}}(t) = (\mathbf{A} - \mathbf{b}\mathbf{k}^T)\mathbf{x}(t) \tag{6.38}$$

If the vector $\mathbf{k}$ is chosen properly, the matrix $(\mathbf{A} - \mathbf{b}\mathbf{k}^T)$ can be made asymptotically stable. Actually, the design process first determines the positions of the poles of $(\mathbf{A} - \mathbf{b}\mathbf{k}^T)$ and with $\mathbf{A}$ and $\mathbf{b}$ known, goes on to determine the feedback weighting factors in $\mathbf{k}$.

Caveat: choosing the poles so that the system states tend to zero very quickly can result in feedback gains (elements of $\mathbf{k}$) which are too large. In our case, this might mean impractically high rates of insulin infusion.

6.5.2.2 Optimal Control

The optimal control problem aims to find a feedback controller resulting in the control signal $u(t)$ such that in a linear system as in the previous subsection, a quadratic performance criterion

$$J(u) = \int_0^\infty [\mathbf{x}(t) - \mathbf{x}_b]^2 dt \tag{6.39}$$

is minimized. This is equivalent to finding the control signals which make the system states as close as possible to the desired states (see Appendix A).

Applying optimal control theory in glucose control historically (and sensibly) uses a quadratic performance criterion of the form

$$J(u) = \int_0^{\infty} x_1^2(t) + Ru^2(t)dt \tag{6.40}$$

where $x_1(t)$ is the glucose level above a basal value (*i.e.*, $x(t) - x_b(t)$), $R > 0$ is a scalar weighting factor included to adjust the sensitivity of the control to be achieved (large values of R penalizes the use of large values of $u(t)$. Note that this way both the glucose level and insulin usage are included in the performance index to be minimized. Optimal control theory gives us the solution of this minimization problem as $u(t) = Kx_1(t)$ where $K = -\frac{1}{R}bP$ where P is a positive scalar. The generalization of this solution for multivariable systems makes use of the concept of positive definite matrices **P** and **R** for the weighting of the states and the control variables in the objective function.

Caveat: the solution of the optimal control problem this way might result in negative values for $u(t)$ sometimes. This would mean negative values for administered insulin, which is obviously not practicable. One solution is to administer glucose where negative insulin values result. However, this is not really a good solution, because insulin and glucose concentration obey different dynamical equations. Another way out of this dilemma is to modify (6.40) with the constraint $u(t) \geq 0, \forall t$. This, on the other hand, changes the mathematical problem in a way that an analytical solution is no longer possible.

Many researchers have applied optimal control theory to the solution of blood glucose concentration control problem. Some examples can be found in [152], [109],[239], [108], [292].

6.5.2.3 Adaptive Control

In adaptive control of glucose level, a glucose-insulin model is used as internal model and the parameters of the model are estimated using algorithms like recursive least squares [86]. Then, these estimated parameters are plugged into the algorithm which consists of familiar mathematical techniques like optimal control. The algorithm, finally, determines the insulin infusion rate to be applied by the pump to the patient. Figure 6.11 depicts such a controller schematically.

Other than being an accurate method of optimizing the therapy for that specific patient, this set-up also can deal with changing parameter values of the patient. Experience with adaptive controllers for the control of blood glucose in diabetics can be found, for example, in [199], [151], [67], [106], [241].

Caveat: care has to be given that the changing of the parameters by the patient model does not lead to unstable behavior of the controller!

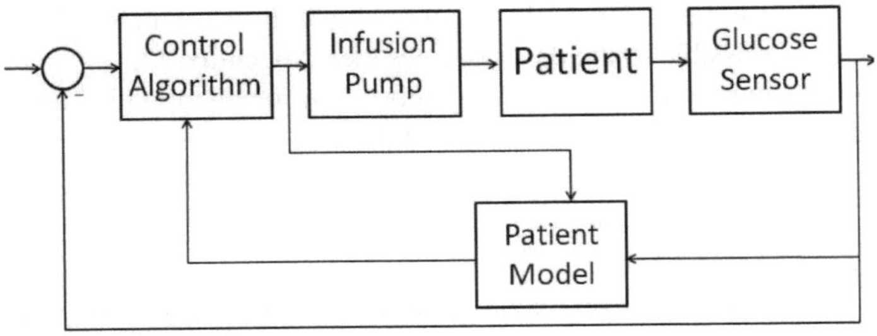

Fig. 6.11 Schematic diagram of a model adaptive controller.

6.6 Commercial Insulin Pumps

Several companies have developed, certified and marketed a number of commercial insulin pumps over the years. Some leading producers are Medtronic, Animas, Roche (Disetronic), Sooil. However, the dream of a portable artificial pancreas has proven to be elusive in spite of concentrated academic and commercial research and development during the last half century.

There are two major engineering issues which have proven to be hard nuts to crack. The first one is the continuous sensing of the blood glucose concentration levels. As we have seen in Section 6.2 there are many methods for measuring the blood glucose levels. What is needed for a continuous closed-loop control is a sensor which is either placed on the skin or implanted inside the body of the patient. The non-invasive techniques are still experimental and invasive sensors do not work reliably for extended periods of time. The measurements are corrupted due to drift and biofouling.

The second major problem which makes the realization of an artificial pancreas so difficult is a control engineering problem. The closed-loop control of the complex insulin-glucose system inherently includes significant time delays in the order of tens of minutes to an hour. Any driver knows the difficulty in steering a car if the car does not react instantaneously to the movements of the steering wheel. The driver's natural reaction in such a case is to oversteer. At the next instant the driver oversteers even more and so on. This is a typical manifestation of instability. Analogously, the controller decides to administer a high bolus of insulin to reduce the glucose level resulting from the intake of carbohydrates but this does not immediately result in a reduction of the glucose level thus causing the controller to administer even more insulin. This process might quickly lead to very high amounts of insulin administration which in turn might cause potentially deadly hypoglycaemia. Therefore, smart algorithms are required which go far beyond "natural reactions".

Furthermore, several studies have shown that use of insulin pumps do not always improve glycemic control [93], [146], [154], [262], [144]. This is probably because not all patients profit in the same way from using an insulin pump in terms of HbA1c. Even if there may be patients showing only a small (or even no) improvement in HbA1c with insulin pump therapy versus injection therapy, benefits are the larger for patients having a higher HbA1c [250], [158]. With these caveats, in this Section we shall discuss Accu-Chek®, a leading product by Roche in some detail.

6.6.1 *Accu-Chek®*

Roche's Accu-Chek® Combo insulin pump system circumvents the two major problems described above with a smart concept which uses two separate, soap bar sized devices. Accu-Chek® Aviva/Performa Combo (Figure 6.12) is basically a finger prick blood glucose meter built around standard processors that run a proprietary firmware and powered by three 1.5 AAA batteries [20], [21], [22]. It also serves as remote control for the insulin pump and diabetes management device Accu-Chek® Spirit Combo (Figure 6.13), which is a programmable insulin pump. It also uses standard processors that run a proprietary firmware and is powered by a 1.5 AA battery [290]. These two devices which we shall call, for simplicity's sake, Aviva Combo and Spirit Combo, communicate with each other using Bluetooth technology. The patient measures his blood glucose level several times a day. Aviva Combo receives all data about recent insulin delivery from the pump. This info on recent insulin deliveries is then considered for the calculation of the appropriate insulin dosage which are then transmitted to the insulin pump. Based on its personalized settings, dietary and physical activity information fed into it and these measurements, the insulin pump delivers the computed amount of insulin to the patient. The delivery is through a very thin cannula (0.36mm) inserted into fatty tissue subcutaneously as a part of an infusion set attached to the patient's belly with an adhesive tape.

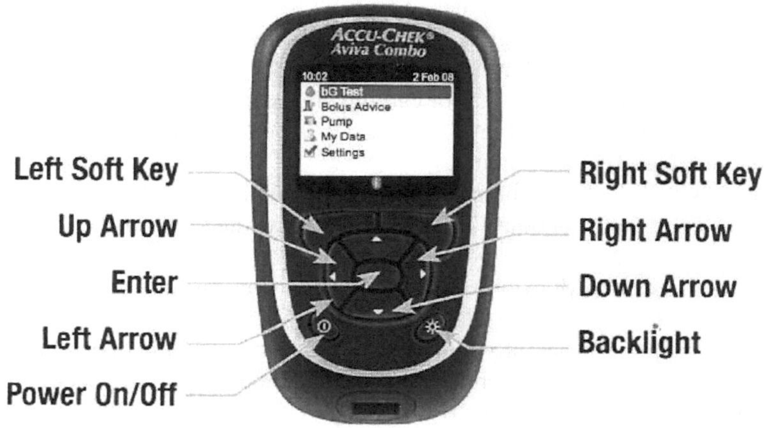

Fig. 6.12 The dimensions of Accu-Chek® Aviva Combo are 94 x 55 x 25 mm.

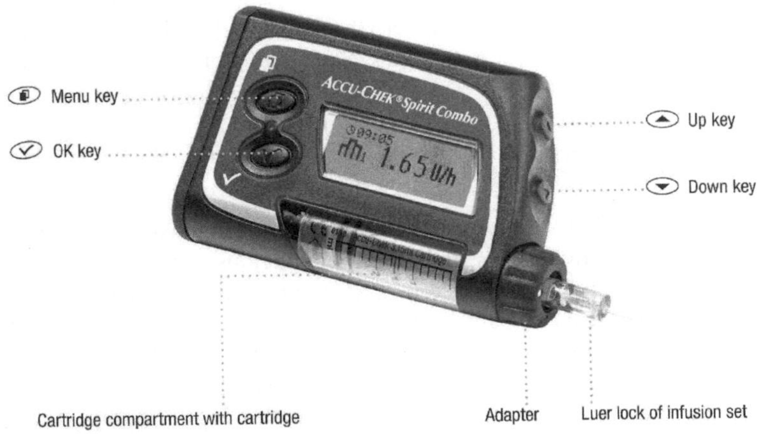

Fig. 6.13 The dimensions of Accu-Chek® Spirit Combo are 83 x 56 x 21 mm.

6.6.1.1 Accu-Chek® Aviva/Performa Combo

The patient initializes and individualizes the diabetes management systems prior to their first use. He begins by selecting the language (more than 15 languages including English are currently available) and setting the time (the importance of this will become apparent below). The individual warning limits for hypoglycaemia and hyperglycaemia are entered (typically 3 mmol/L and 12 mmol/L respectively). The day is divided in up to 8 time blocks which are defined by their beginning times. Those time blocks may, for instance, correspond to sleeping, morning, afternoon, evening and night with typical values of 0h – 6h, 6h – 11h, 11h – 17h, 17h – 22h and 22h – 0h. The patient then goes on to define the target range for each time block (*e.g.*, 4 mmol/L – 8 mmol/L for the sleeping time block), the carb ratio (the amount of insulin needed to cover one unit of carbohydrates, *e.g.*, 1U for 10g) and the insulin sensitivity (the amount of insulin needed to lower the blood glucose level by a given amount, *e.g.*, 1U for 2 mmol/L).

Bolus Advice calculates a bolus for the patient that is adapted to the time of the day and his changing situations. This function is activated only if the patient sets up Bolus Advice on his meter. For this function, first the effects of health events are individualized. **Health events** correspond to percentages that allow the patient to adjust the calculated insulin amount to allow for individual needs and health statuses, for example, when physical activity is planned (subtract percentages) or when the patient is ill (add percentages). Typically, moderate exercise could reduce the calculated insulin amount by 10% and vigorous exercise by 20%. For illness or stress an increase of 10% to 25% could be programmed. Finally, advice option settings are individualized. Those are **Meal Rise**, **Snack Size**, **Acting Time** and **Offset Time**.

During or after meals, an increase in blood glucose levels is considered normal within a certain range even though a meal bolus has been delivered. Meal Rise is the maximum increase in blood glucose level that is to be tolerated without an additional correction bolus. Its value could be around 6 mmol/L.

Snack Size is the amount of carbohydrates that is not to be counted as a regular meal with the expected Meal Rise. An increase in the patient's blood glucose level is not tolerated since the Meal Rise factor is not triggered for bolus advice calculations. This could be around 10g.

The length of time from the start of the Meal Rise or the delivery of the correction bolus until the patient's blood glucose level returns to the target level is called Acting Time. Its length can be adjusted for the patient's individual needs between 1.5 hours and 8 hours.

Offset Time takes into account the expected delay for the blood glucose level to actually fall due to the acting insulin in the body. This could be around 1 hour.

The device can also be set up to remind the patient to retest his blood glucose level after a high blood glucose test, after a low blood glucose test or after a meal. The after meal blood glucose test reminder is set up with a carb value for the snack size and occurs after the snack size is exceeded. Any of these reminders can be turned on or off individually, as needed.

After conducting the set ups mentioned above, the patient is now ready to use the Accu-Chek® Aviva Combo. Of course, the Bluetooth connection with his Accu-Chek® Spirit Combo has first to be paired.

All the commands and advisories for bolus insulin deliveries are generated according to the algorithms described in [143]. Both devices are controlled by high performance processors and supervisor processors with failsafe mechanisms.

6.6.1.2 Accu-Chek® Spirit Combo

Just like the Accu-Chek® Aviva Combo the patient initializes and individualizes the Accu-Chek® Spirit Combo – which is, in essence, a programmable insulin pump – prior to its first use. It must first be programmed with the personal basal rate profile of the patient.

The pump typically delivers insulin held in a 3.15 mL (315U) cartridge every 3 minutes in 20 equal doses every hour, 24 hours a day. This flow of insulin, measured in international units per hour (U/h), is called the basal rate profile and is calculated to meet the patient's basic insulin needs. A basal rate profile consists of up to 24 different hourly basal rates. Each hourly rate may be changed independently. The total of all 24 basal rates in one basal rate profile is called the daily basal rate total. As an example, see Figure 6.14 for 18U/24 hours[5]. All the other commands and advisories for bolus insulin deliveries generated by the Accu-Chek® Aviva Combo are superimposed on this basal rate profile.

[5] Note the similarity of this profile with the required motor ability profile for Parkinsonian patients in the previous Chapter.

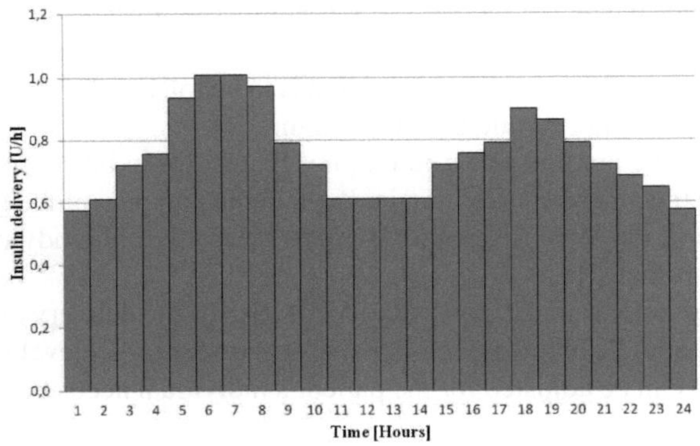

Fig. 6.14 An example of a basal rate profile for 18U/24 hours after [23]. Note the higher infusion rates before and after breakfast and dinner times.

6.6.1.3 Accu-Chek® Aviva Expert

Another product by Roche in the same family is Accu-Chek® Aviva Expert [104] which is used for management of MDI (Multiple Daily Injection) therapy with an insulin pen or syringes. Usually, patients inject long lasting insulin 1 to 2 times a day and short acting insulin according to demand for covering food intake or correcting their blood glucose values back to normal. The device provides an algorithm to calculate short-acting insulin dosages (boluses). One could describe this device as an insulin pump without the pump. In other words, this device, which is very similar to Accu-Chek® Aviva/Performa Combo generates advisories for bolus administrations based on the individual characteristics and needs of the patient and the actual blood glucose measurements obtained by finger pricks.

This product is well suited for patients who use a long lasting insulin analogue and multiple daily short acting injections of insulin. The users of this device claim to have their blood glucose levels under better control and their fear of hypoglycaemia to be reduced. Also, clinical studies have shown that using this device reduces HbA1c and significantly increases patient satisfaction with the therapy [273].

6.6.1.4 The Algorithm

We shall use excerpts from the original patent [143] in order to understand the actual algorithm used by Accu-Chek® Aviva/Performa Combo. Let us begin by defining some concepts and variables used by the algorithm. Figure 6.15 is an illustrative plot of blood glucose vs. time representing two examples of insulin bolus administrations relative to a current and next adjacent time interval.

A current time interval begins at a "begin current interval" time, T_{BCI} and ends at a subsequent "end current interval" time, T_{ECI}. The target blood glucose *(3)* defined during the current interval is designated BGT_{CI} and the insulin sensitivity defined during the current interval is called IS_{CI}. The next adjacent time interval begins at

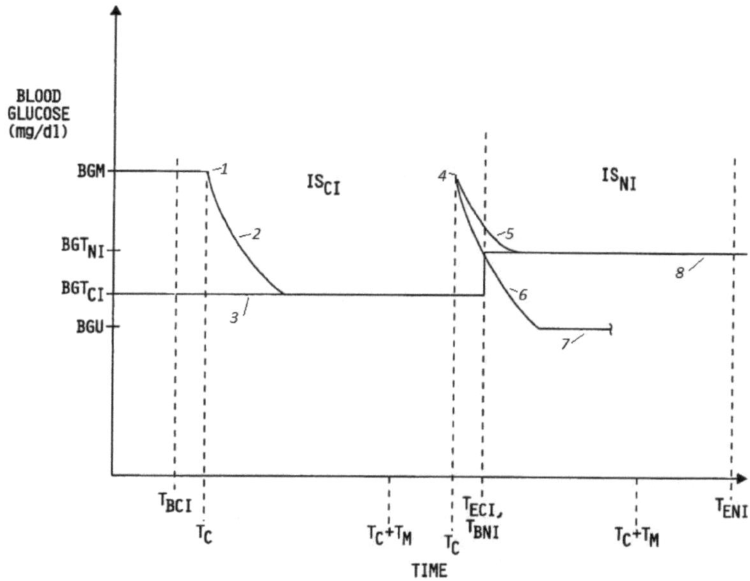

Fig. 6.15 Illustration of two adjacent time intervals used for the definition of variables used in the algorithm used by Accu-Chek® Aviva/Performa Combo (after [143]).

"begin next interval" time, T_{BNI}, which coincides with the "end current interval" time T_{ECI} and ends at a subsequent "end next interval" time, T_{ENI}. The target blood glucose (*8*) defined during the next adjacent interval is BGT_{NI} and the insulin sensitivity defined during the next adjacent interval is called IS_{NI}.

If the current measurement of the glucose level exceeds the target blood glucose level for the current interval, BGT_{CI}, conventional bolus recommendation systems are typically operable to compute a recommended correction insulin bolus, *CB*, according to the equation

$$CB = (BGM - BGT_{CI})/IS_{CI} \tag{6.41}$$

where *BGM* is the blood glucose level measured at the current time, T_C. The patient then administers the recommended correction insulin bolus at or near the time T_C and the administered insulin bolus functions in a known manner to lower the glucose level over a time period T_M, the duration of the glucose lowering action of an administered insulin bolus. In Figure 6.15, the foregoing scenario is illustrated in the current interval where a blood glucose value (*1*) is measured at the current time, T_C. With a correction bolus calculated with equation (6.41) and administered at or near T_C, the blood glucose decreases over time (*2*) in this example to the target glucose value (*3*), which corresponds to the target blood glucose value for the current interval, BGT_{CI}, and remains at that level for the duration T_M.

As illustrated in this example, the conventional correction bolus equation (6.41) works well as long as the correction bolus is administered sufficiently early in the current time interval so that the duration of the glucose lowering action of the administered insulin bolus, T_M, is confined to the current interval. However using the equation (6.41) as a basis for computing and administering a correction bolus at a current time, T_C, that occurs later in the current time interval, so that the duration of the glucose lowering action of the administered bolus, T_M, spans the current and the next adjacent time intervals may yield undesirable results. For example, consider the case where a blood glucose value (*4*) is measured at the now current time T_C near the end, T_{ECI}, of the current time interval (corresponding to the beginning time, T_{BNI}, of the next adjacent time interval). With a correction bolus CB computed using equation (6.41) and administered at or near T_C, the blood glucose decreases over time (*6*) into the next adjacent time interval and to a glucose level (*7*) that may be significantly below the target glucose value (*8*) of the next adjacent time interval.

One effective technique for improving the accuracy of the correction bolus determination in cases where the time duration, T_M, of the glucose lowering action of the administered insulin bolus spans the current and the next time interval is to take into account the time-dependent nature of the target glucose level and of the insulin sensitivity. For example, if $h(\tau)$ is the relative amount of insulin activity used up for a bolus of insulin administered at $\tau = 0$, at $\tau \geq T_M$, $h(\tau) = 1$. If $IS(T)$ is the insulin sensitivity as a function of time and assuming linearity, then a bolus of insulin administered at an arbitrary time $T = T_A$ has the following impulse response

$$H(T, T_A) = \int_0^{T-T_A} \dot{h}(\tau) \cdot IS(T_A + \tau) d\tau \tag{6.42}$$

where $\dot{h}(\tau)$ is the time derivative of $h(\tau)$. If $D(T)$ is the time-dependent insulin bolus infusion rate function, the glucose drop at a given time relative to $BG_{-\infty} = BG(T \rightarrow -\infty)$ is

$$\Delta BG(T) = \int_{-\infty}^{T} D(T_A) \int_0^{T-T_A} \dot{h}(\tau) \cdot IS(T_A + \tau) d\tau dT_A \tag{6.43}$$

To calculate at the current time $T = T_C$ the blood glucose drop ΔBG_{pb} that will happen in the future caused by the already administered past insulin boli, the following equation is used

$$\Delta BG_{pb}(T_C) = \int_{T_C-T_M}^{T_C} D(T_A) \int_{T_C-T_A}^{T_C-T_A+T_M} \dot{h}(\tau) \cdot IS(T_A + \tau) d\tau dT_A \tag{6.44}$$

To calculate the correction bolus amount, CB, to be given soon after $T = T_C$ with a bolus infusion duration that is negligible compared to T_M, the following equation is used

$$CB = \frac{\Delta BG_n - \Delta BG_{pb}(T_C)}{\int_0^{T_M} \dot{h}(\tau) \cdot IS(T_C + \tau) d\tau} \tag{6.45}$$

The needed blood glucose drop, ΔBG_n, must be calculated based on the target blood glucose value at $T = T_C + T_M$, or BGT_{NI}. Combining the results of (6.41), (6.42) and (6.45) and assuming that no insulin was administered during the interval $[T_C - T_M, T_C]$, i.e., $BG_{pb} = 0$, the equation for calculating a correction bolus amount, *CB*, to be administered at or near the current time, $T = T_C$, and when the time duration, T_M, spans the current time interval *and* the next time interval which begins at time $T = T_{BNI}$, is thus given by

$$CB = \frac{BGM - BGT_{NI}}{IS_{CI} \cdot \int_0^{T_{BNI}-T_C} \dot{h}(\tau) + IS_{NI} \cdot \int_{T_{BNI}-T_C}^{T_M} \dot{h} d\tau} \tag{6.46}$$

Solving the integrals in (6.46) yields

$$CB = \frac{BGM - BGT_{NI}}{IS_{CI} \cdot [h(T_{BNI} - T_C) - h(0)] + IS_{NI} \cdot [h(T_M) - h(T_{BNI} - T_C)]} \tag{6.47}$$

where $[h(T_{BNI} - T_C) - h(0)]$ corresponds to the fraction, P_{CI}, of the insulin action of the bolus given at or near $T = T_C$ that will be spent or used to lower the measured blood glucose value, *BGM*, during the current time interval, and $[h(T_M) - h(T_{BNI} - T_C)]$ corresponds to the fraction, P_{NI}, of the insulin action of the bolus given at or near $T = T_C$ that will be spent or used to lower the measured blood glucose value, *BGM*, during the next adjacent time interval. Substituting the insulin action fractions into the equation (6.47) yields the equation

$$CB = \frac{BGM - BGT_{NI}}{IS_{CI} \cdot P_{CI} + IS_{NI} \cdot P_{NI}} \tag{6.48}$$

Let us look at a numerical example to get a better understanding of the effect of equation (6.48) on the blood glucose measurement value (*4*) in reference to Figure 6.15. Assume that the measured blood glucose value, *BGM*, corresponding to point (*4*), is 200 mg/dl, the target blood glucose value during the current interval BGT_{CI}, is 120 mg/dl, the insulin sensitivity during the current interval, IS_{CI}, is 30 mg/dl/IU, the target blood glucose value during the next adjacent time interval, BGT_{NI}, is 150 mg/dl, and the insulin sensitivity during the next interval, IS_{NI}, is 40 mg/dl/IU. Also assume that at $T = T_{BNI}$ 40% of the insulin action by the bolus given at or near $T = T_C$ (corresponding to point (*4*)) will have been spent to lower the glucose level during the current time interval. This leaves $100\% - 40\% = 60\%$ of the insulin action of the bolus given at or near $T = T_C$ to be spent lowering the glucose level during the next adjacent time interval. According to equation (6.48) then, the correction bolus, *CB*, that should be administered at or near $T = T_C$ is $CB = (200mg/dl - 150mg/dl)/[(30mg/dl/IU \cdot 0.4) + (40mg/dl/IU \cdot 0.6)] = 1.389IU$ or approximately $1.4IU$. With an insulin sensitivity during the next adjacent time interval of $40mg/dl$, administering $1.4IU$ of insulin at or near $T = T_C$ results in a decrease (*5*) in blood glucose of $[(30mg/dl) \cdot (1.4IU) \cdot 0.4] + [(40mg/dl) \cdot (1.4IU) \cdot 0.6] = 50.4mg/dl$ corresponding to a decrease in blood glucose from $200mg/dl$ at point (*4*) to a final glucose level at the next adjacent time interval equal to the target blood glucose level, BGT_{NI}, (*8*).

Let us now contemplate that the time duration, T_M, of the glucose lowering action of the insulin bolus to be administered at or near T_C may span one or more of the next adjacent time intervals. This may occur as a result of any combination of a sufficiently long time duration, T_M, sufficiently short durations of one or more of the current time interval and/or any series of next adjacent time intervals, the timing of the current time, T_C, relative to the current time interval, and the like. In any case, equation (6.48) may be modified to account for the time duration, T_M, spanning any number, j of next adjacent time intervals, wherein j may be any positive integer. One form of such an equation, based on equation (6.48), is the following

$$CB = \frac{BGM - BGT_{NI+j-1}}{\sum_{n=-1}^{j-1} (IS_{NI+n} \cdot P_{NI+n})} \tag{6.49}$$

where $IS_{NI+n} = IS_{CI}$ and $P_{NI+n} = P_{CI}$ when $n = -1$. According to equation (6.49), the blood glucose target value, BGT, used is that of the last, *i.e.*, j-th, one of the next adjacent time intervals, and the denominator of the equation (6.49) represents the sum of IS and P products for all of the time intervals that T_M spans, including the current time interval and all j of the next adjacent time intervals.

6.7 Exercises

A1: Explain the blood glucose level control in the body. What are the control signals? What are the disturbances?

A2: Explain what diabetes is and why it is a significant disease both for personal and public health.

A3: Compare and contrast different ways of measuring blood glucose levels.

A4: Compare and contrast different ways of controlling exogenous blood glucose levels.

B1: Explain Albisser's control algorithm for controlling the blood glucose levels in diabetic patients in your own words using equations and defining all the variables. Why does this algorithm include a glucose infusion rate? What sigmoid functions could you use other than the hyperbolic tangent function? Why do you think Albisser chose that function?

B2: Explain Bolie's linearized model of glucose insulin interaction. What is the physical meaning of the variables in this model?

B3: Consider the system

$$\dot{\mathbf{x}} = \begin{pmatrix} 1 & 1 \\ 0 & 2 \end{pmatrix} \mathbf{x} + \begin{pmatrix} 1 \\ 0 \end{pmatrix} u$$

$$y = (0 \quad 1)\mathbf{x}$$

Is the system stable? Find the state feedback gains K_1 and K_2 such that the closed-loop system has its poles at $\lambda_1 = -1$ and $\lambda_2 = -2$. Why are you encountering a problem?

B4: A sensor indicates the blood glucose level as 3 mmol/L when it is actually 12 mmol/L. Using the Clarke's error grid analysis, please explain in which zone this sensor falls and how you rate the quality of the sensor from excellent (10) to useless (0).

C1: What are the problems in controlling systems with pure time delays? Conduct a theoretical study of the problem and develop a tool to help you design controllers for systems with pure time delays.

C2: Design and build a fuzzy controller using commercially available components like sensors and processors. What has been done to that end already? What design steps do you need to solve this engineering problem? What practical and ethical problems do you expect to encounter while building and testing your controller?

Chapter 7
Controlling Depth of Anaesthesia

To be in love is merely to be in a perpetual state of anesthesia – to mistake an ordinary young man for a Greek god or an ordinary young woman for a goddess.

— Henry Louis Mencken

Alcohol is the anesthesia by which we endure the operation of life.

— George Bernard Shaw

7.1 Introduction

Anaesthesia comes from the Greek word *anaisthesia* which means insensibility. Today it is used to mean loss of sensation with or without loss of consciousness. More specifically, it describes the condition of having sensation, including the feeling of pain, being blocked. Most of us are familiar with local anaesthesia applied through an injection before dental treatment like a filling or a root canal treatment. General anaesthesia which is a pharmacologically induced state, allows patients to undergo surgery without the pain they would otherwise experience. Oliver Wendell Holmes first used the term anesthesia in 1846 to describe drug-induced insensibility to sensation (particularly pain) following the demonstration of inhaled ether rendering a patient unresponsive during a surgical procedure [248].

7.2 Measuring Depth of Anaesthesia

The depth of anaesthesia is a functional state of the central nervous system resulting from stimulation by surgical stimuli and inhibitions of the cerebral activity through anaesthetic agents [325]. It depends on the actuation degree of **hypnosis**, the loss of consciousness, **amnesia**, the inhibition of awareness and remembering of pain, **analgesia**, the inhibition of the perception of pain and **muscle relaxation**.

S.S. Hacısalihzade: *Biomedical Applications of Control Engineering*, LNCIS 441, pp. 175–218.
DOI: 10.1007/978-3-642-37279-7_7

Consequently, measuring the depth of anaesthesia is not a trivial task. However, in order to be controlled, it must somehow be measured either directly or through proxies.

Most commonly used observations for measuring the depth of anaesthesia are

Mean Arterial Pressure (MAP): Changes of MAP can entail valuable clues concerning the depth of anaesthesia. A surgical stimulation like an incision can increase MAP. Often, the deeper the cut the more vigorous the increase. However, MAP is also affected by other intraoperative changes like blood loss or dehydration. Increased concentration of some anaesthetic agents in the blood might reduce MAP. Nevertheless, this is a commonly used measure for the depth of anaesthesia, because MAP is very easy to observe and because it is always continuously measured during surgery to follow the patient's general condition.

Heart Rate (HR): Changes of HR can also give indications concerning the depth of anaesthesia. HR might increase due to perceived pain. But it might also be increased by increased concentration of some volatile anaesthetic agents or with dehydration. Since HR is always monitored during surgery, this is a very commonly used measure for depth of anaesthesia.

Electroencephalogram (EEG): EEG is the measurement of cortical electrical activity and it reflects the changes of cerebral perfusion and exchange activities. Adequate processing of EEG can be used as a measure of depth of anaesthesia. This method is less commonly used, because it is cumbersome and electrical artifacts in the operating room environment can corrupt the EEG.

Evoked Potentials (EP): Sensory, auditory or acoustic stimulation of the patient evokes certain electrical responses in the brain. Those electrical signals called EP can be extracted from the EEG. This method is also less commonly used, because of the difficulties in isolating the stimuli to evoke the EP and because of the same reasons that lead to difficulties in measuring EEG.

7.3 Anaesthetic Agents

Prior to the introduction of ether by Clarke and Long as an anaesthetic agent in 1842 [81] surgery patients were held still by physical means and they often felt the full pain of the surgery. Today, a wide spectrum of drugs are used in anesthetic practice. Anesthetics can be categorized as general anesthetics that cause a reversible loss of consciousness and local anesthetics that cause reversible local loss of pain perception without causing unconsciousness.

Local anesthetics are ester or amide based chemicals that prevent the transmission of nerve impulses by binding to fast sodium channels. The most commonly used local anaesthetics are procaine (well known by its commercial name novocaine, a favorite of dental patients with a low threshold of pain), tetracaine (also known as amethocaine which is mainly used topically in ophthalmology), lidocaine, prilocaine and many more with the suffix "-caine" in their names.

The biological mechanisms of the action of general anaesthetics are still not well understood. They are administered either as gases (inhalational anaesthetics) or as injections (intravenous anaesthetics). Most commonly both of these forms are combined, with an injection to induce anaesthesia and a gas used to maintain it. Most common injection anaesthetics are propofol, etomidate, ketamine, benzodiazepine (better known as benzo and made famous by their commercial names Valium or Diazem) and barbiturates.

Inhalational anaesthetic substances are usually delivered using an anaesthesia machine which allows to compose a mixture of oxygen, anaesthetics, ambient air and delivers it to the patient. Desflurane and sevoflurane which are gradually replacing isoflurane in more affluent countries are the most widely used volatile anaesthetics. They are often administered in combination with nitrous oxide.

The standard measure of potency for inhalatory anaesthetics is the Minimum Alveolar Concentration (MAC). It is defined as the minimum alveolar concentration of an anaesthetic that prevents movement in response to skin incision in 50% of the patients. Hence, the potency of an anaesthetic increases with decreasing MAC. Table 7.1 gives a comparison of the potency of some anaesthetics.

Table 7.1 Potency of some anaesthetics indicated by their Minimum Alveolar Concentrations (MAC). The lower the MAC the more potent is the agent. MAC usually drops with age. The figures given here are for patients in the age bracket 30 – 50.

Anaesthetic	**MAC**
Nitrous oxide	104 vol %
Desflurane	6 vol %
Sevoflurane	1.8 vol %
Isoflurane	1.15 vol %
Halothane	0.75 vol %

7.4 The Basic Control Loop

Let us now look at the situation in the operating theater in the context of the first Section of Chapter 4 and specifically Figure 4.1. The anaesthetic situation is schematized in Figure 7.1. The anaesthetist determines the administered anaesthetic agents and the settings of the breathing and monitoring system based on measurable factors like the patient's exhaled gas parameters, heart rate and blood pressure as well as unmeasurable factors like skin color together with the actual breathing parameters. The surgical actions of the surgeon like incisions act as disturbances which need to be compensated by proper actions taken by the anaesthetist.

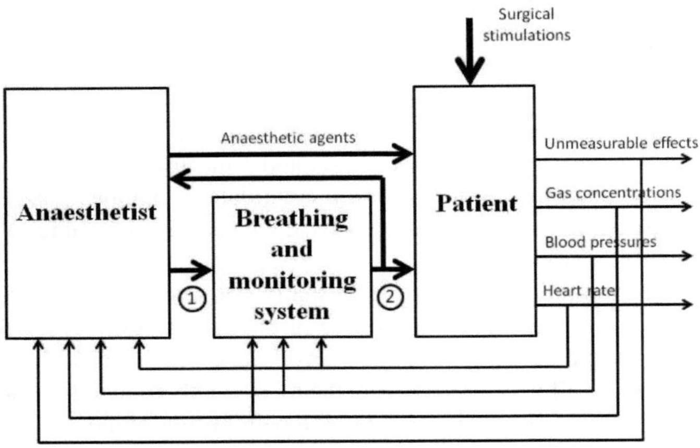

Fig. 7.1 General anaesthetic situation with the anaesthetist controlling depth of anaesthesia (measured through heart rate and blood pressure of the patient) in spite of disturbances caused by actions of the surgeon. 1 and 2 are the set and actual breathing parameters like inhaled air volume per minute.

7.5 Fuzzy Control of Anaesthesia

In a first attempt at implementing a controller which controls MAP as the proxy for depth of anaesthesia, the situation was analyzed as shown in Figure 7.3. The anaesthetist adjusts the settings of the vaporizer such that the correct mixture of oxygen, nitrous oxide and isoflurane are administered to the patient (here oxygen and nitrous oxide flow are kept constant and only the isoflurane concentration, F_{inf}, is varied by the anaesthetist). The mixture is inhaled by the patient and then exhaled. The exhaled mixture, now rich in carbon dioxide is filtered through an absorber which removes the carbon dioxide and mixes the remaining gas with the gas from the vaporizer. The reason for this closed circuit is to keep the isoflurane in the circuit both for environmental and economic reasons. The concentration of isoflurane in the gas mixture that the patient inhales is F_{ins}. The concentration of isoflurane in the gas mixture that the patient exhales is F_{end}. The anaesthetist observing these gas concentrations and the MAP adjusts a dial on the vaporizer to keep the MAP close to the desired figure in spite of the incisions made by the surgeon [324]. Such a vaporizer is shown in Figure 7.2.

Letting a fuzzy controller do the task of the anaesthetist results in the corresponding classical simplified block diagram shown in Figure 7.4.

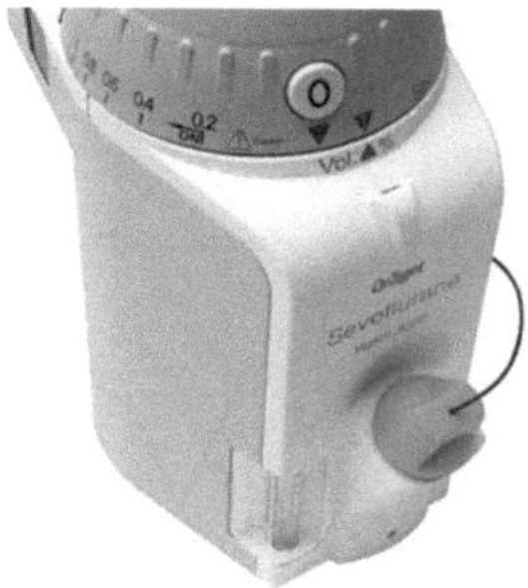

Fig. 7.2 A vaporizer used in anaesthesia with sevoflurane. The dial which the anaesthetist adjusts to choose the concentration of the anaesthetic agent inhaled by the patient is visible on top.

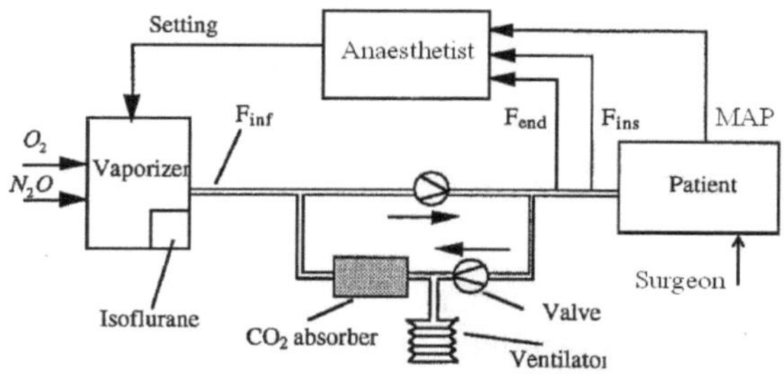

Fig. 7.3 Flow and signal diagram of the configuration used during anaesthesia with the anaesthetist acting as the controller. F_{inf} is the inflowing isoflurane concentration, F_{ins} is the inspired isoflurane concentration and F_{end} is the endtidal isoflurane concentration.

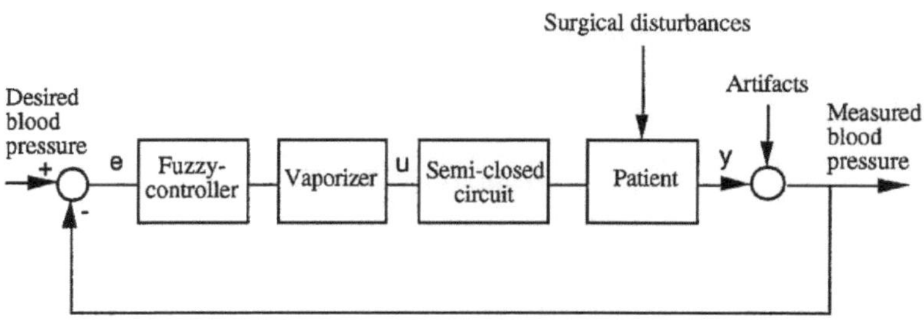

Fig. 7.4 The block diagram of the control loop for blood pressure control during anaesthesia.

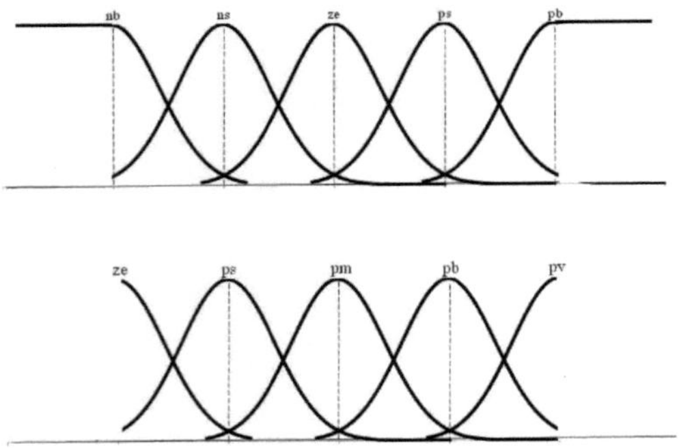

Fig. 7.5 The membership functions for the error μ_e, the integral of the error μ_{ie} (upper curves) and the control variable μ_u (lower curves). *nb*: negative big, *ns*: negative small, *ze*: zero, *ps*: positive small, *pm*: positive medium, *pb*: positive big, *pv*: positive very big. The center points for the membership functions are listed in Table 7.2. Note that negative values for the isoflurane concentrations are not possible. Also, concentrations above a maximum of 4% are not allowed.

Table 7.2 The center points for the fuzzy membership functions for the error μ_e, the integral of the error μ_{ie} and the control variable μ_u.

INPUTS			OUTPUT	
	e [mmHg]	**ie** [mmHg s]		**u** [%]
nb	-10	-160	**ze**	0
ns	-5	-90	**ps**	1
ze	0	0	**pm**	2
ps	5	90	**pb**	3
pb	10	160	**pv**	4

Table 7.3 Linguistic rules for determining the control variable *u* based on the input variables *e* and *ie*. For instance Rule 7 says "If the error is around 0 (ze) and if the integral of the error is around -90 (ns), then set the inflow concentration of isoflurane about 3% (pb).

	INPUTS		OUTPUT
Rule	**e**	**ie**	**u**
1	ns		pb
2	ps		ps
3	nb		pv
4	pb		ze
5	ze	ze	pm
6	ze	ps	ps
7	ze	ns	pb
8		nb	pv
9		pb	pb

Various simulations have shown that best results in terms of control and noise rejection characteristics are achieved with bell shaped fuzzy membership functions for the error and the integral of the error as well as the control variable shown in Figure 7.5 and fuzzy control rules shown in Table 7.3 (for a refresher on fuzzy logic and fuzzy control see Appendix B).

The variation of MAP and the inflow isoflurane concentration as adjusted by the fuzzy controller described above during the first 2000 seconds of an abdominal operation as depicted in Figure 7.6 shows that the controller can adjust the blood

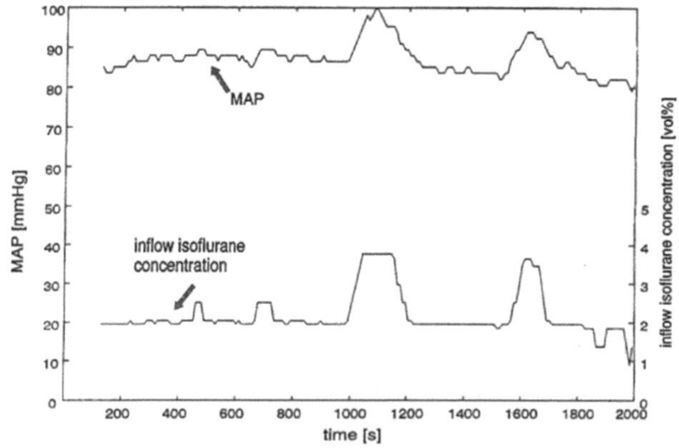

Fig. 7.6 Fuzzy controller in action. The upper curve is the MAP (left hand scale) and the lower curve is the inflow concentration of isoflurane (right hand scale) during an abdominal operation. The set value of MAP is 85 mmHg. The disturbances at 1000 seconds and 1500 seconds are due to skin incisions.

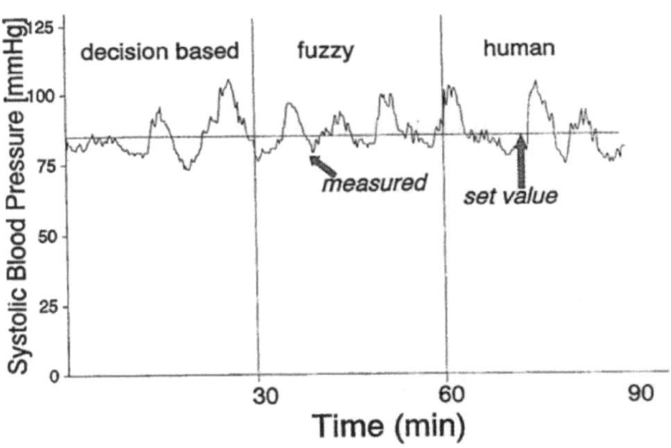

Fig. 7.7 The control of blood pressure during surgery by a decision based controller (not described here), by a fuzzy controller and by a human controller. The set value of the systolic blood pressure is 85 mmHg.

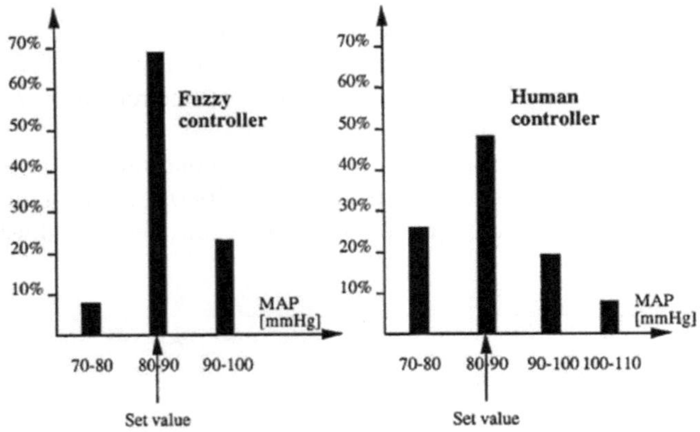

Fig. 7.8 The fuzzy controller manages to keep the blood pressure at the set value of 85±5 mmHg 70% of the time whereas the human controller achieves that 50% of the time. Also, the fuzzy controller always keeps the blood pressure within ±15 mmHg of the set value whereas the human controller achieves that 92% of the time.

pressure adequately [223]. Looking at how the performance of the fuzzy controller fares compared to an experienced anaesthetist during an operation in Figure 7.7 shows that the anaesthetist tends to overshoot or undershoot more often than the fuzzy controller. The comparison of the frequency distributions of the blood pressure of a typical patient while the fuzzy controller was active and while the anaesthetist controlled the blood pressure shows that the fuzzy controller was measurably more accurate in keeping the blood pressure within a small range of its set value of 85 mmHg as depicted in Figure 7.8.

To summarize, a proportional-integral fuzzy controller which controls the MAP during anaesthesia with isoflurane was designed and implemented on a personal computer. The controller was tested in 11 surgical cases (all abdominal surgery on patients in the 20 – 55 years age group). The anaesthetist always supervised the controller vigilantly. He could intervene or override it at any time but he never had to. The quality of control achieved by the fuzzy controller proved to be better than manual control. Therefore, it can be concluded that such a controller can be routinely used during anaesthesia with an agent like isoflurane. However, the controller has to be enhanced to account for the use of more than one anaesthetic agent and variable gas flow which necessitates multivariable control.

7.6 Multivariable Control of Anaesthesia

The primary aim of an anaesthesia control system is the automatization of some routine tasks of the anaesthetist in normal situations. Furthermore, to avoid dangerous situations, the anaesthetist should have a clear overview of the current state of all automated processes at all times. Since there is only a limited amount of measurements available for the anaesthesia system, it is very important for the anaesthetist to monitor the available information. Also, while designing an anaesthesia control system, it has to be kept in mind that the safety of the patient is paramount. Therefore, to increase safety, the controller must permanently be supervised by an independent controller at a higher hierarchical level. Such a controller is called a supervisor and it performs consistency and plausibility checks on the inputs and outputs of the controller, it decides which controller configuration should be activated and it interacts closely with the anaesthetist. Figure 7.9 illustrates this hierarchical structure[1].

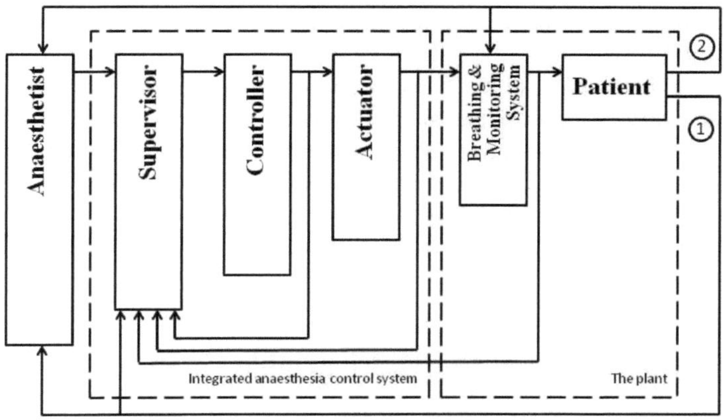

Fig. 7.9 Block diagram of an integrated anaesthesia control system. The anaesthetist has access to the measurable (1) and unmeasurable (2) aspects of the patient's state and determines the necessary anaesthetic procedure which she inputs to the integrated anaesthesia control system. The supervisor passes this information to the controller. The controller sets the inflow of anaesthetic agents to the patient through the actuator. The supervisor has access to all available measurable data and continuously performs plausibility and consistency checks to assure the safety of the patient. The anaesthetist monitors the situation and can intervene, disable or override the control system at any time.

Some of the major difficulties in designing an integrated anaesthesia control systems can be listed as follows:

- Not all mechanisms of how various anaesthetic agents act on the human body are completely understood.
- It is not easy to extract and understand the knowledge based on the education and experience of the anaesthetists. Many decisions are taken intuitively by the anaesthetists who evaluate various measurable and unmeasurable information about the state of the patient unconsciously.

[1] This Section generally follows the structure of [95] and is based on the results presented therein.

- Assuming that the reaction of the patients to anaesthesia can be modeled as dynamic systems, the interpatient variability of the parameters of such systems is large and the individual parameters of a patient are not easy to identify during surgery. Furthermore, some of these parameters can change during the time span of the surgery.
- The actual controller design is rather challenging due to non-linearities and inconsistencies in the reactions of the patient.
- For ethical and practical reasons, testing of the designed controller must be kept at a minimum. Furthermore, one cannot perform arbitrary tests during surgery and is limited to the requirements of the surgical operation.
- The implementation of the supervisor alone requires the solution of many several complex problems, which often are strongly correlated. These include "intelligent alarming" which involves alarming of all possible critical situations with as few false alarms as possible; "man machine interface" which offers the anaesthetist an overview of the actual situation and the ability to change any settings at all times efficiently in an intuitive manner; "artifact handling" which involves the detection and reporting of all system disturbances due to artifacts (those are very common in the 'noisy' operating room environment), also any reactions of the anaesthesia control system that might jeopardize the safety of the patient must be avoided.

7.7 Implementation of the Multivariable Controller

The controller implemented in [95] adjusts 1) the isoflurane concentration in the fresh gas flow, 2) the oxygen flow and 3) the nitrous oxide fresh gas flow as schematized in Figure 7.10. The controller was designed as a cascade controller where the controller of the external loop determines the set points of the controller of the internal loop. This structure improves safety in case of a failure and it allows to decouple the dynamics of the patient and the breathing system [224], [94]. It also includes the control of the endtidal carbon dioxide concentration. The anaesthesia gas controller can have different configurations:

1. As a **MAP controller** it calculates the concentration of the anaesthetic agent in the fresh gas flow directly from the actual set and measured values of MAP and the measured inspired and endtidal concentrations of the anaesthetic agent.
2. As an **endtidal anaesthetic gas controller** it calculates the anaesthetic gas concentration in the fresh gas flow directly from the actual endtidal anaesthetic gas concentration and the set and measured values of the inspired concentrations of the anaesthetic agent.
3. As an **inspired anaesthetic gas controller** it calculates the anaesthetic gas concentration in the fresh gas flow directly from the actual inspired anaesthetic gas concentration and its set value, and the measured endtidal concentrations of the anaesthetic agent.
4. As a **cascade controller** where a MAP or an endtidal anaesthetic gas controller defines an inspired anaesthetic gas set value that is controlled by an internal loop.

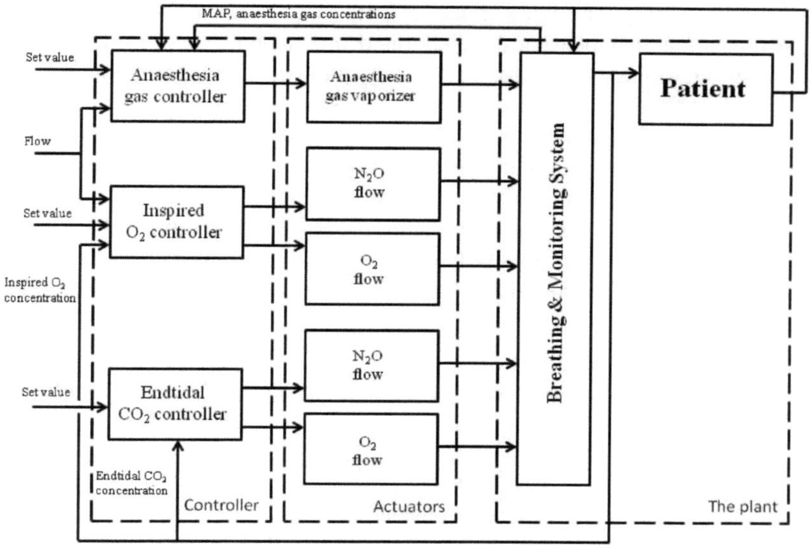

Fig. 7.10 Detailed block diagram of the anaesthesia controller depicts the controller, the actuators and "the plant". The different configurations of the controller as MAP controller, as endtidal anaesthetic gas controller and as inspired anaesthetic gas controller are indicated. In the configuration as cascade controller the set values are determined by an external loop.

7.8 The Patient Model

In order to be able to design the proper control algorithms for the anaesthesia controller we need to have a proper model of the plant which we want to control. We can split the plant in three main elements: the patient model which includes the pharmacokinetic and pharmacodynamic effects of the anaesthetic agents, the (extracorporeal) breathing system, and the hemodynamic effects of surgical stimulations.

The patient model describes the uptake and distribution (pharmacokinetics) and the effects of concentration levels of various gases in the human body (pharmacodynamics). The gases are inhaled by the lungs of the patient. The inspiratory gas concentration in the lung constitutes the main input of the model. We will begin by briefly describing the main physiological mechanisms involving different gases before presenting a compartment model for the pharmacokinetics and pharmacodynamics of anaesthetic gases. We shall develop these models which are non-linear and involve higher order differential equations. In the end we will linearize and reduce the model orders so that the models become suitable for control purposes.

Gases are transported from blood to tissue and from blood to the alveolar gas part by blood flow, by airflow and by active diffusion. Blood flow distributes the substances and exchange them with the tissue. In the breathing circuit (including the gas part of the lungs), mass transport is historically modeled using compartmental models [265]. The mass transport path is divided into subsystems called compartments (see Chapter 4). As elaborated in Chapter 2, a reasonable physiological interpretation generally requires a large number of compartments. However, generally, identifiability suffers as the model order increases.

Under reasonable assumptions, the uptake and distribution model in [328] describes the pharmacokinetics of the anaesthetic agent using 12 compartments, where each compartment represents organs or groups of organs with similar pharmacokinetic, pharmacodynamic and hemodynamic properties (see Table 7.4 for a list of compartments). The compartments 1 to 9 are called 'normal' compartments. Two compartments model the arterial and venous blood pool and one compartment models the lungs.

Table 7.4 List of different compartments and the corresponding indices in the compartmental equations.

Heart muscle	1
Brain gray matter	2
Brain white matter	3
Well perfused organs	4
Poorly perfused organs	5
Stomach, intestines	6
Skeletal muscle	7
Fat	8
Skin shunt	9
Lungs	L
Arterial system	A
Venous system	V

The normal compartments have the same structure. They consist of a tissue and a blood part. Both the tissue and the blood part have the same anaesthetic partial pressure p_i but the anaesthetic has different solubilities in tissue and blood. Anaesthetics enter the compartment with the arterial blood and partial pressure p_A and they leave the compartment with the venous blood and partial pressure p_i. Therefore, the partial pressure of the anaesthetic in the compartment can be described as

$$\dot{p}_i(t) = K_i \lambda_b q_i(t) [p_A(t) - p_i(t)] \tag{7.1}$$

$$K_i = \frac{1}{\lambda_b V_{i,b} + \lambda_i V_{i,t}} \tag{7.2}$$

λ_b and λ_i are the solubility coefficients for blood and tissue respectively and $V_{i,b}$ and $V_{i,t}$ are the blood volumes in equilibrium with the tissue of the compartment i and the tissue volume of the compartment i respectively. q_i are the blood flows through the compartment i.

The partial pressure of the anaesthetic in the lungs compartment depends, in addition to the blood and tissue volumes, on the so called functional residual capacity V_a which is the volume of gas remaining in the lung after normal expiration. This is due to the gas exchange between blood and gas in the lungs. However, not all of the blood flowing through the lungs is subject to the gas exchange. Therefore, the venous blood flow is split in a first part which represents the effective blood flow subject to the gas exchange and a second part which goes directly from the veins to the arteries (the lung shunt, ls, which is typically around 3%). This leads to the equation

$$\dot{p}_L(t) = K_L\lambda_b q_L(t)[p_V(t) - p_L(t)] + q_{Air}(t)[p_{Air}(t) - p_V(t)] \tag{7.3}$$

$$K_L = \frac{1}{\lambda_b V_{L,b} + \lambda_L V_{L,t} + V_A} \tag{7.4}$$

$$q_L(t) = CO(t)(1 - ls) \tag{7.5}$$

where q_{Air} denotes the minute volume and p_{Air} denotes the anaesthetic gas concentration (partial pressure) of the inhaled air. CO is the cardiac output.

The flows entering and leaving the arterial and venous compartments comes from or goes to different compartments. Therefore, they obey the equations

$$\dot{p}_A(t) = K_A\lambda_b CO(t)[p_V(t)ls + p_L(t)(1 - ls) - p_A(t)] \tag{7.6}$$

$$K_A = \frac{1}{\lambda_b V_{A,b} + \lambda_A V_{A,t}} \tag{7.7}$$

$$\dot{p}_V(t) = K_V\lambda_b\left[\sum_{i=1}^{9} q_i(t)p_i(t) - CO(t)p_V(t)\right] \tag{7.8}$$

$$K_V = \frac{1}{\lambda_b V_{V,b} + \lambda_V V_{V,t}} \tag{7.9}$$

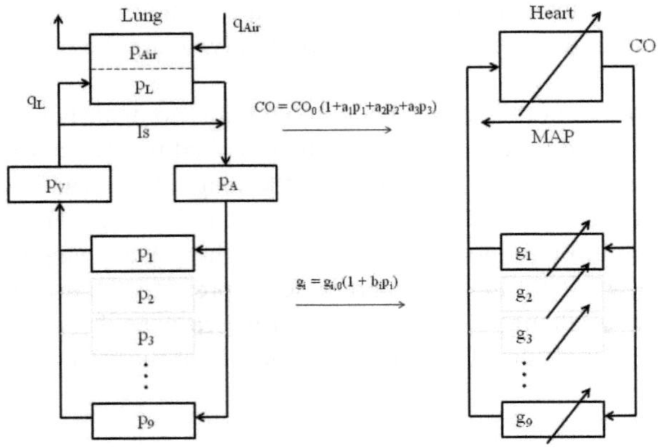

Fig. 7.11 Compartmental model of pharmacokinetic and pharmacodynamic of anaesthetic gas after [94].

This compartmental model is schematized in Figure 7.11. The right part of this figure represents the circulation model which describes the blood flow. The heart produces an average blood outflow called the cardiac output. The total CO is distributed to the various 'normal' compartments. Each one of these 9 compartments exhibits a certain conductivity g_i. Neglecting all dynamic effects, in analogy to Ohm's law, MAP is then given by

$$MAP = \frac{CO}{\sum_{i=1}^{9} g_i} \tag{7.10}$$

Effects of the anaesthetic agents on the conductivity of each compartment and the cardiac output can be modeled as linear combinations of partial pressure of the anaesthetic in the compartment as the following equations show.

$$g_i = g_{i,0}(1 + b_i p_i) \tag{7.11}$$

$$CO = CO_0(1 + a_1 p_1 + a_2 p_2 + a_A p_A) \tag{7.12}$$

The resulting non-linear model is of the form

$$\begin{aligned} \dot{\mathbf{p}}(t) &= \mathbf{f}(\mathbf{p}(t), u(t)) \\ MAP(t) &= h(\mathbf{p}(t), u(t)) \end{aligned} \tag{7.13}$$

with the state vector $\mathbf{p}(t)$ describing the partial pressure of the anaesthetic gas in each compartment. The control input $u(t)$ is the concentration of anaesthetic gas in the inspired air. MAP is the output of the system. Substituting all the previous equations in the last set finally gives us the 12th order system equations with Q_i as the blood flow going through compartment i as

$$Q_i = MAPg_i \tag{7.14}$$

$$\dot{p}_i = k_i Q_i (p_A - p_i) \tag{7.15}$$

$$\dot{p}_L = k_L\{\lambda_b(1-ls)CO(p_V - p_L) + q_{Air}(p_{Air} - p_L)\} \tag{7.16}$$

$$\dot{p}_A = k_A CO\{p_V ls + p_L(1-ls) - p_A\} \tag{7.17}$$

$$\dot{p}_V = k_V CO \left[\frac{\sum_{i=1}^{9} g_i p_i}{\sum_{j=1}^{9} g_j} - p_V \right] \tag{7.18}$$

$$MAP = \frac{CO}{\sum_{j=1}^{9} g_j} \tag{7.19}$$

Some interesting properties of these state equations become apparent when we write them in the matrix form with $q_L = \lambda_b CO(1-ls)$ as

$$\dot{\mathbf{p}} = \mathbf{A}(\mathbf{p})\mathbf{p} + \mathbf{B}\mathbf{u} \tag{7.20}$$

with

$$\mathbf{A} = \left[\begin{array}{ccc|ccc} -k_1Q_1 & & 0 & 0 & k_1Q_1 & 0 \\ & \ddots & & \vdots & \vdots & \vdots \\ 0 & & -k_9Q_9 & 0 & k_9Q_9 & 0 \\ \hline 0 & \cdots & 0 & -k_L(q_L + q_{Air}) & 0 & k_L q_L \\ 0 & \cdots & 0 & -k_A CO(1-ls) & -k_A CO & k_A COls \\ k_V Q_1 & \cdots & k_V Q_9 & 0 & 0 & -k_V CO \end{array}\right] \tag{7.21}$$

$$\mathbf{B} = \left[\begin{array}{cc} 0 & 0 \\ \vdots & \vdots \\ 0 & 0 \\ \hline \frac{1}{2}k_L q_{Air} & \frac{1}{2}k_L p_{Air} \\ 0 & 0 \\ 0 & 0 \end{array}\right] \tag{7.22}$$

$\mathbf{A}(\mathbf{p})$ is the state matrix of a non-linear compartmental system with state dependent transfer coefficients. The elements of $\mathbf{A}(\mathbf{p})$, $a_{ij}(\mathbf{p})$, describe the flow from compartment j to compartment i. Assuming no metabolic activity in all compartments, the only mass transfer to the outside is the transfer through the lung. Due to the connection of the lung compartment, the system is not autonomous (**B** matrix consists of zeros except for the non-zero elements in the 10th (lung) row). The volume of each compartment is approximatively constant

$$a_{jj}(\mathbf{p}) = \sum_{\substack{i=1 \\ i \neq j}}^{n} a_{ij}(\mathbf{p}) \qquad \forall j \neq j_L \tag{7.23}$$

where j_L is the index of the lung compartment. In the lung compartment the term $-k_L q_{Air}$ is added to equation (7.23). Equation (7.20) conserves the main compartment structure of the pharmacokinetics. Thee non-linearities due to the pharmacodynamics with the circulatory system are expressed in Q_i and CO.

At this stage it looks as if this model requires the identification of many parameters. However, some parameters like blood volumes or solubility can be found in the literature. Hence, actually only a few have parameters have to be identified. How this can be done is explained in detail in [95] and numerical values are given.

The nice thing about this model is that it can generally be used for all major anaesthesia gasses. Work has been also reported about use of this model for combination of gases. Specifically, it was implemented for mixing N_2O with isoflurane. The model was first extended with the pharmacokinetic equations for N_2O and the pharmacodynamic equations were modified to include the additive effect of N_2O as

$$g_i = g_{i,0}(1 + b_i p_i + b_{i,N_2O} p_{i,N_2O}) \tag{7.24}$$

$$CO = CO_0(1 + a_1 p_1 + a_2 p_2 + a_A p_A + a_4 p_{1,N_2O} + a_5 p_{2,N_2O} + a_6 p_{A,N_2O}) \tag{7.25}$$

However, the observed effect is not quite additive. The transport of anaesthesia gas from the lung to the blood is influenced by concentration changes of gases that are used in high concentrations [326]. In this case N_20 can act as carrier for the other anaesthesia gases, thus reducing the equilibrium time. This effect was compensated by increasing the effective air flow q_{Air} in the differential equation of the lung compartment by an affine factor depending on the gas concentration mixture in the lung and in the veins as

$$\hat{q}_{Air} = q_{Air}[1 + K_{q_{Air}}(p_{Air}, p_{Air,N_2O}, p_{V,N_2O})] \tag{7.26}$$

In [9] $K_{q_{Air}}$ was set to 0.235 independent of p_{V,N_2O} for the usual operating point with $p_{Air,N_2O} = 70[vol\%]$ and $p_{Air,O_2} = 30[vol\%]$.

Most available controller design methods work best with linear models. Therefore, it is common to linearize a non-linear model around an operating point which are usually steady state conditions. In the state space representation of a system this simply means taking the derivative of the state vector and equating it to the zero vector:

$$\dot{\mathbf{x}} = \mathbf{0} \tag{7.27}$$

The resulting steady state and the corresponding input are indicated with bars:

$$\mathbf{f}(\bar{\mathbf{x}}, \bar{\mathbf{u}}) = \mathbf{0} \tag{7.28}$$

For the model described above, there are three types of steady state conditions:

$$\bar{\mathbf{p}}_1 = \mathbf{p} | \begin{cases} CO = 0 \\ p_{Air} = p_L \end{cases} \tag{7.29}$$

$$\bar{\mathbf{p}}_2 = \mathbf{p} | \; p_i = p_A = p_V = p_L = p_{Air} \quad \forall i \tag{7.30}$$

$$\bar{\mathbf{p}}_3 = \mathbf{p} | \begin{cases} p_i = -\frac{1}{b_i} \quad \forall i \\ p_A = p_V = p_L = p_{Air} \end{cases} \tag{7.31}$$

The steady state condition $\bar{\mathbf{p}}_1$ implies zero cardiac output and can be safely ignored. The steady state condition $\bar{\mathbf{p}}_3$ corresponds to all normal compartments having zero conductivity. In this case, no organ is perfused any longer. This means no change of the concentration is happening and we can also ignore this (very!) steady state. Therefore, we will concentrate on the steady state condition $\bar{\mathbf{p}}_2$ for the linearization of the model and call it the reference point.

Developing the system equations (7.21) and (7.22) in a Taylor series around the reference point and ignoring the higher order terms gives us

$$\dot{\mathbf{p}}(t) \approx \mathbf{f}(\bar{\mathbf{p}}, \bar{\mathbf{u}}) + \left.\frac{\partial \mathbf{f}}{\partial \mathbf{p}}\right|_{\bar{\mathbf{p}}, \bar{\mathbf{u}}} \delta \mathbf{p}(t) + \left.\frac{\partial \mathbf{f}}{\partial \mathbf{u}}\right|_{\bar{\mathbf{p}}, \bar{\mathbf{u}}} \delta \mathbf{u}(t) \tag{7.32}$$

$$MAP(t) \approx h(\bar{\mathbf{p}}, \bar{\mathbf{u}}) + \left.\frac{\partial h}{\partial \mathbf{p}}\right|_{\bar{\mathbf{p}}, \bar{\mathbf{u}}} \delta \mathbf{p}(t) + \left.\frac{\partial h}{\partial \mathbf{u}}\right|_{\bar{\mathbf{p}}, \bar{\mathbf{u}}} \delta \mathbf{u}(t) \tag{7.33}$$

where $\delta \dot{\mathbf{p}}(t) = \mathbf{p}(t) - \bar{\mathbf{p}}$ and $\delta MAP(t) = MAP(t) - \overline{MAP}$ represent the difference of the actual value to the quasi steady state value at the reference point (the bars on the variables indicate constant values calculated at the reference point). Rewriting the system equations with the new system matrices yields the model linearized around $\bar{\mathbf{p}}_2$ as

$$\delta \dot{\mathbf{p}}(t) = \bar{\mathbf{A}} \delta \mathbf{p}(t) + \bar{\mathbf{B}} \delta \mathbf{u}(t) \tag{7.34}$$

$$\delta MAP(t) = \bar{\mathbf{C}} \delta \mathbf{p}(t) + \bar{\mathbf{D}} \delta \mathbf{u}(t) \tag{7.35}$$

with

$$\bar{\mathbf{A}} = \left[\begin{array}{ccc|ccc} -k_1\bar{Q}_1 & & 0 & 0 & k_1\bar{Q}_1 & 0 \\ & \ddots & & \vdots & \vdots & \vdots \\ 0 & & -k_9\bar{Q}_9 & 0 & k_9\bar{Q}_9 & 0 \\ \hline 0 & \cdots & 0 & -k_L\bar{q}_L & 0 & k_L\bar{q}_L \\ 0 & \cdots & 0 & -k_A\overline{CO}(1-ls) & -k_A\overline{CO} & k_A\overline{CO}ls \\ k_V\bar{Q}_1 & \cdots & k_V\bar{Q}_9 & 0 & 0 & -k_V\overline{CO} \end{array}\right] \tag{7.36}$$

$$\bar{\mathbf{B}} = \left[\begin{array}{cc} 0 & 0 \\ \vdots & \vdots \\ 0 & 0 \\ \hline \frac{1}{2}k_L\bar{q}_{Air} & \frac{1}{2}k_L\bar{p}_{Air} \\ 0 & 0 \\ 0 & 0 \end{array}\right] \tag{7.37}$$

$$\bar{\mathbf{C}} = [\bar{m}_1 - \bar{n}_1 \;\; \bar{m}_2 - \bar{n}_2 \;\; -\bar{n}_3 \;\; \cdots \;\; -\bar{n}_9 \;\; 0 \;\; \bar{m}_{11} \;\; 0] \tag{7.38}$$

$$\bar{\mathbf{D}} = \mathbf{0} \tag{7.39}$$

with

$$\bar{m}_i = \frac{CO_0 a_i}{\sum_{j=1}^{9} \bar{g}_j} \qquad \bar{n}_i = -\frac{CO g_{i,0}\bar{b}_i}{\left[\sum_{j=1}^{9} \bar{g}_j\right]^2} \qquad \bar{q}_L = \lambda_b\overline{CO}(1-ls) \tag{7.40}$$

At this stage one can use *e.g.*, Allgöwer's method [11] to check how much the behavior of the linearized model deviates from the non-linear model. How various parameters in the model can be estimated and how the quality of the linearization can be checked with this method is shown in detail in [95]. However, we still have to deal with a 12th order system here for which controller design is rather difficult. Therefore, the next step is to reduce the model order.

The objective of model reduction here is to obtain a reduced order single-input-multiple-output model with MAP, P_i and P_e as the outputs and use that model to design a model based controller. One has to remember that not all mathematical models derived by the application of physical laws are suitable for closed-loop control (see Chapters 1 and 2). The model has to be in its minimal realization [163]. This means that all states of the model must be controllable and observable. The linearized model above meets those conditions. However, the level of contribution of the states to the input/output behavior differs significantly. One can thus neglect the weakly observable and controllable states for control purposes.

The so called controllability and the observability gramians defined as

$$\mathbf{G}_c = \int_0^\infty e^{\mathbf{A}\tau}\mathbf{B}\mathbf{B}^T e^{\mathbf{A}^T\tau} d\tau \tag{7.41}$$

$$\mathbf{G}_o = \int_0^\infty e^{\mathbf{A}\tau}\mathbf{C}^T\mathbf{C} e^{\mathbf{A}^T\tau} d\tau \tag{7.42}$$

allow to define some measure of controllability or observability, but it is not possible to give an accurate measure of how the system states influence the input-output behavior. One speaks of a balanced realization of the system if $\hat{\mathbf{G}}_\mathbf{c} = \hat{\mathbf{G}}_\mathbf{o} = \Sigma$ [231]. This realization is obtained by a linear transformation $\mathbf{T}$ of the state space such that $\hat{\mathbf{x}} = \mathbf{T}\mathbf{x}$. For a stable system, the balanced realization is

$$\left(\begin{array}{c|c} \hat{\mathbf{A}} & \hat{\mathbf{B}} \\ \hline \hat{\mathbf{C}} & \hat{\mathbf{D}} \end{array}\right) = \left(\begin{array}{c|c} \mathbf{T}\mathbf{A}\mathbf{T}^{-1} & \mathbf{T}\mathbf{B} \\ \hline \mathbf{C}\mathbf{T}^{-1} & \mathbf{D} \end{array}\right) \tag{7.43}$$

when T is chosen such that $\hat{\mathbf{G}}_\mathbf{c} = \hat{\mathbf{G}}_\mathbf{o} = \Sigma$ with

$$\Sigma = \begin{bmatrix} \sigma_1 & & & \\ & \sigma_2 & & 0 \\ 0 & & \ddots & \\ & & & \sigma_n \end{bmatrix} \tag{7.44}$$

where $\sigma_1 \geq \sigma_2 \geq \cdots \geq \sigma_n$ are called the Hankel singular values [327]. This singular value decomposition of the gramians allows the identification of the states which are weakly controllable and observable. It can be shown that the truncation of these states result in a reduced stable system [231].

Applying this method to the linearized 12th order model of the combined pharmacokinetic and pharmacodynamic model of anaesthesia gases results in Hankel singular values which can be clustered in two groups in Table 7.5. A measure of the truncation error is given in [327]. With a truncated system order of 5 an error of less than 1% is achieved. Even with a truncated system order of 3 an error of less than 3% can be achieved.

Table 7.5 The Hankel singular values of the linearized model of the combined pharmacokinetic and pharmacodynamic model of anaesthesia gases.

σ_1	1.03×10^1
σ_2	2.65×10^0
σ_3	4.73×10^{-1}
σ_4	3.81×10^{-1}
σ_5	2.40×10^{-2}

σ_6	6.26×10^{-3}
σ_7	1.81×10^{-3}
σ_8	5.49×10^{-5}
σ_9	8.61×10^{-6}
σ_{10}	1.16×10^{-6}
σ_{11}	2.06×10^{-7}
σ_{12}	3.32×10^{-8}

7.9 The Breathing System

The second element in the control loop in Figure 7.4 is the breathing system. It is basically composed of actuators and mechanical devices ensuring artificial breathing and the lungs. Figure 7.12 shows the basic structure of the breathing system (Cicero® by Drägerwerk) used in this study.

The structure of this subsystem changes through the opening and closing of the valves during the breathing cycle. During the inspiratory phase the gas flows through the inspiratory pipe path into the lungs. Both V_2 (excess valve) and V_1 (fresh gas path valve) are closed. After the inspiratory pause, V_2 is opened. The expiratory phase can be characterized by five different valve configurations. However, not all of these configurations occur during every breathing cycle. During the very beginning of the expiratory phase gas from the lungs flows into the manual ventilation bag (0.2 seconds). After that, the piston of the pump begins to move to fill the pump cylinder. This backward movement of the piston is completed within 1.2 seconds. If V_2 is opened while the pump piston is moving, the configuration is changed. V_2 is opened at least during 0.1 seconds and opens as soon as the pressure at pump output increases during 0.25 seconds. At the end V_2 is closed and V_1 is opened as the piston finishes its movement.

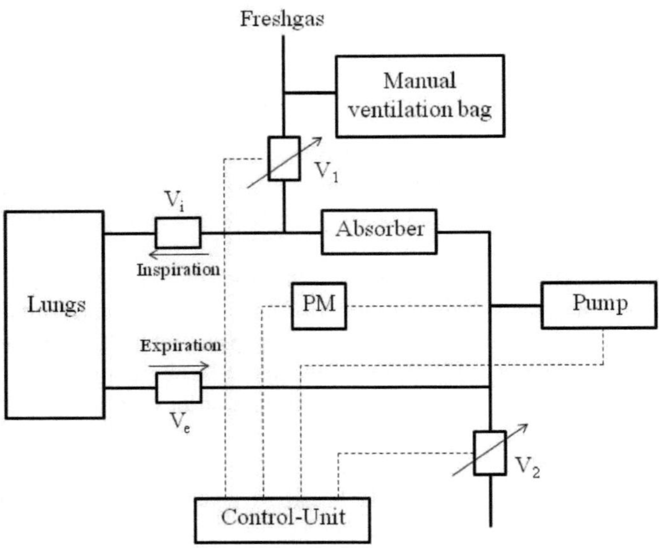

Fig. 7.12 Schematic diagram of the breathing system. *PM* is the pressure measurement, V_i and V_e are directional valves. V_1 and V_2 are valves controlled by the control-unit.

Table 7.6 shows the different configurations during the respiratory cycle. There are two possible sequences of configuration switching. Sequence I: V_1 opens, pump piston moves, movement stops, V_2 opens, both V_1 and V_2 are open, V_2 closes, V_1 is open. Sequence II: V_1 opens, pump piston moves, V_2 opens while the piston is

Table 7.6 Possible configurations of the valves in the breathing system.

Breathing phase	V_1	V_2	V_i	V_e	Pump flow
Inspiration	closed	closed	open	closed	F_i
Expiration I	open	closed	closed	open	0
Expiration II	open	closed	closed	open	F_e
Expiration III	open	open	closed	open	F_e
Expiration IV	open	open	closed	open	0
Expiration V	open	closed	closed	open	0

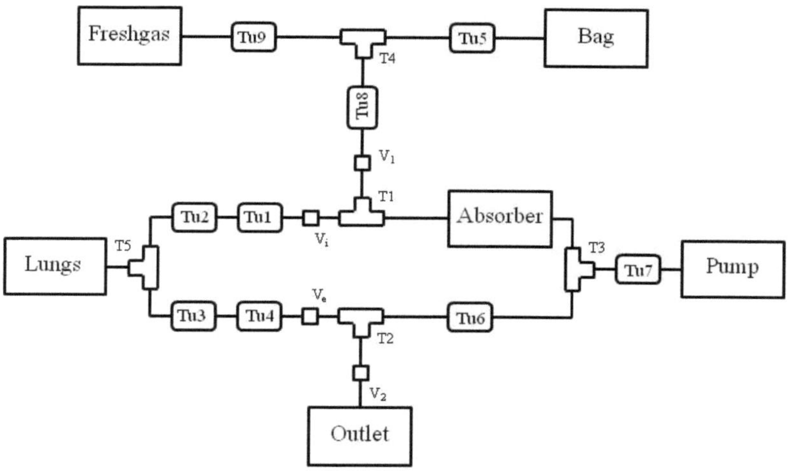

Fig. 7.13 Block diagram of the Cicero breathing system. Tu_1 - Tu_9 are the nine tubes. T_1 to T_4 are the four $T-$junctions. *Bag* is the manual ventilation bag. *Freshgas* is the fresh gas connection. *Lungs* are the lungs and the connection with the patient. V_j are the different valves.

moving, movement stops, V_2 opens, both V_1 and V_2 are open, V_2 closes, V_1 is open. A more detailed block diagram of Figure 7.12 is given in Figure 7.13.

To develop a mathematical model of the breathing system we simply apply fundamental physical laws beginning with the ideal gas equation

$$P\,V = N\,R\,T \tag{7.45}$$

where N (the mole fraction of gas) is the ratio of the gas mass (m) to the mole mass (m^*). P is the pressure, V the volume, R the ideal gas constant ($8.314J/K/mol$) and T the absolute temperature. Dalton's law says that the total pressure exerted by the mixture of non-reactive gases is equal to the sum of the partial pressures of individual gases:

$$P = \sum_i p_i \tag{7.46}$$

$$N = \sum_i N_i \tag{7.47}$$

This leads to

$$P = \sum_i \frac{m_i}{m_i^*} \frac{RT}{V} \tag{7.48}$$

Assuming that each segment is an ideal compartment, what is the amount of incoming and outgoing gas to and from each compartment? Each section type has its own dynamic behavior even if all sections are based on the same gas laws. A general compartment with n connections has the structure shown in Figure 7.14.

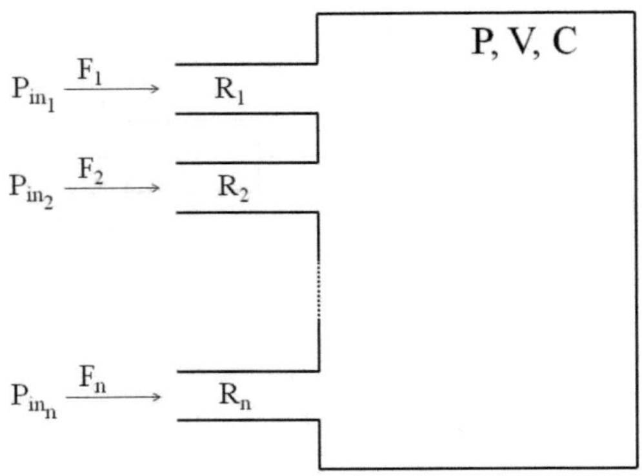

Fig. 7.14 General compartment for the modeling of the respiratory cycle. P, V, C are the pressure, volume and compliance of the general compartment. P_{in_i} are the pressures, F_i are the flows and R_i are the resistances at the contact point i.

Gas exchange among elements is possible only through these connection points and only if a pressure difference exists. Exchange through diffusion is neglected. Gas in the compartment is assumed to mix instantly and uniformly. The relation between flows and pressure is governed by

$$F_i = \frac{P_{in_i} - P}{R_i} \tag{7.49}$$

where F_i is the gas flow from the i-th neighbor element, P_{in_i} is the pressure of this element and R_i is the resistance of the i-th connection. If there are k gases in a gas mixture, the flow of any single gas is proportional to the gas composition of the source compartment

$$F_{ij} = \begin{cases} F_i \frac{P_j}{P} & \text{if} \quad F_i < 0 \\ \\ F_i \frac{P_{in_{ij}}}{P_{in_i}} & \text{if} \quad F_i \geq 0 \end{cases} \tag{7.50}$$

for $i = 1,...,n,\ j = 1,...,k$. The continuity equation (7.51) is simply the sum of all incoming flows multiplied by their densities.

$$\dot{m} = \sum_{i=1}^{n} \rho_i F_i \tag{7.51}$$

The flow density corresponds approximately to the mean value of source and target densities.

$$\rho_i = \sum_{j=1}^{k} \frac{m_j^*}{RT} P_{\rho_{ij}} \tag{7.52}$$

$$\approx \frac{\bar{m}^*}{RT} P_{\rho_i} = \bar{\rho}_i \tag{7.53}$$

$$P_{\rho_i} = \frac{P_{in_i} + P}{2} \approx P_\rho \tag{7.54}$$

This leads to the general description of the compartments

$$\dot{P}V + P\dot{V} = P_\rho \sum_{i=1}^{n} \frac{P_{in_i} - P}{R_i} \tag{7.55}$$

With (7.53) the equivalent equation of partial pressures is

$$\dot{P}_j V + P_j \dot{V} = P_\rho \sum_{i=1}^{n} F_{ij} \tag{7.56}$$

It follows, with $c_j = P_j/P$ as the ratio of partial to total gas pressure, $\dot{P}_j = \dot{c}_j P + c_j \dot{P}$. Using this result, (7.56) can be written as

$$\dot{c}_j \frac{P}{P\rho} = \frac{1}{V} \sum_{i=1}^{n} F_{ij} - c_j \left[\frac{\dot{P}}{P_\rho} + \frac{P}{P_\rho} \frac{\dot{V}}{V} \right] \tag{7.57}$$

With $P \approx P_\rho$ and using (7.49) and (7.55) we get

$$\dot{c}_j = \frac{1}{V} \sum_{i=1}^{n} (F_{ij} - F_i c_j) \tag{7.58}$$

We now know everything we need to be able to model the breathing system. We must proceed as follows: 1) Calculate F_i with (7.49), 2) Calculate F_{ij} with (7.50), 3)

Integrate (7.55) to find P and V, 4) Integrate (7.58) to find c_j. All blocks in Figure 7.13 are governed by the equations (7.55), (7.56) and (7.58) (of course, except for Fresh gas and Outlet), only their number of connections and how their volumes are defined are different as shown in Table 7.7.

Table 7.7 Different block units in the breathing system. V_0 is the volume of the chamber at atmospheric conditions. At the reference point $V = V_0 + C\Delta P$ with $\Delta P = P_{out} - P$. C denotes the so called compliance which, in general, is a function of P. We linearize C around its reference point and consider it a constant.

Block unit	Connections	Volume
Tube	2	$V_0 + CP(t)$
T-junction	3	V_0
Bag	1	$V(P)$
Pump	1	$V(t)$
Absorber	3	$V_0 + CP(t)$
Lungs	>4	$V_0 + CP(t)$

A **tube** is the connection piece between different compartments. The length of the tubes vary from 7.5cm to 150cm. They have some time delays caused by the actual transport of the gas and very small compliances. We will consider only segments of 7.5cm. Solving equation (7.55) for a tube results in the dynamic model of this block unit

$$\dot{P} = \frac{P_\rho}{V_0 + C(2P - P_{atm})} \left[\frac{P_{in_1} - P}{R_1} + \frac{P_{in_2} - P}{R_2} \right] \approx K_{tube} \left[\frac{P_{in_1} - P}{R_1} + \frac{P_{in_2} - P}{R_2} \right] \tag{7.59}$$

with $K_{tube} = P/(V_0 + CP)$. A common value for K_{tube} in the literature is 1.315×10^5.

A T**-junction** has negligible compliance and no time delay. It can be seen as a very fast tube with three ends. Hence the dynamic model can be given as

$$\dot{P} = K_T \left[\frac{P_{in_1} - P}{R_1} + \frac{P_{in_2} - P}{R_2} + \frac{P_{in_3} - P}{R_3} \right] \tag{7.60}$$

with $K_T = P_\rho / V_0$.

The **manual ventilation bag** has a variable volume which is changed by the manual action of the anaesthetist. The volume as a function of the pressure can be modeled as a polynomial. Using the chain rule and (7.55), the dynamic model of the bag can be given as

$$\dot{P} = \frac{P_\rho}{V + \frac{dV}{dP} P} \frac{P_{in_1} - P}{R_1} \tag{7.61}$$

The **pump** has a time varying volume with a single connection to the inspiratory cycle. Both $V(t)$ and $\dot{V}(t)$ are therefore known. During the inspiratory phase the volume of the chamber decreases linearly with time and during the second expiratory phase the volume increases linearly with time. Again using (7.55), the dynamic model of the pump can be given as

$$\dot{P} = \frac{P_\rho}{V} \frac{P_{in_1} - P}{R_1} - P\frac{\dot{V}}{V} \tag{7.62}$$

The **absorber** eliminates CO_2 from the circuit with a given rate. Other than that it behaves exactly like a tube element. Therefore, the dynamic model of the absorber is essentially the same as the dynamic model of the tube corrected by this additive term which represents an additional outlet. With P_{CO_2} as the partial pressure of CO_2 in the absorber, K_{abs} as the absorber constant and K_A derived from (7.59), the dynamic model of the absorber can be written as

$$\dot{P} = K_A \left(\frac{P_{in_1} - P}{R_1} + \frac{P_{in_2} - P}{R_2} \right) - K_{abs} P_{CO_2} \tag{7.63}$$

The **lungs** constitute the final element in the circuit and it cannot be modeled as simply as the other elements. Much research has gone into the dynamics of respiration [110]. The upper respiratory tract which is composed of the nose, the pharynx and the larynx is bridged by the intubation and can be seen as a tube with low compliance. The lower respiratory tract begins with the trachea. It then forks into the two main bronchi which lead to the lungs. There are 23 further layers of subdivisions before the alveoli are reached. The first 16 are just conducting airways without gas exchange. The first 11 layers are of cartilage tissue and have low compliance. The 12th to the 16th divisions are called bronchioles and have a higher compliance. The final 6 divisions are the respiratory bronchioles with some gas exchange and the alveolar ducts with the main respiratory zone composed by the alveoli. Therefore, it is reasonable to model the lung with 4 compartments as schematized in Figure 7.15, each with increasing compliances.

The dynamic model of each compartment is derived by applying the general tube model (7.59):

$$\dot{P}_{l_1} = K_{t_1} \left(F_{in_1} - \frac{P_{l_1} - P_{l_2}}{R_{l_1 l_2}} \right) \tag{7.64}$$

$$\dot{P}_{l_2} = K_{t_2} \left(\frac{P_{l_1} - P_{l_2}}{R_{l_1 l_2}} - \frac{P_{l_2} - P_{l_3}}{R_{l_2 l_3}} \right) \tag{7.65}$$

$$\dot{P}_{l_3} = K_{t_3} \left(\frac{P_{l_2} - P_{l_3}}{R_{l_2 l_3}} - \frac{P_{l_3} - P_{l_4}}{R_{l_3 l_4}} \right) \tag{7.66}$$

$$\dot{P}_{l_4} = K_{t_4} \left(\frac{P_{l_3} - P_{l_4}}{R_{l_3 l_4}} - F_{Out_4} \right) \tag{7.67}$$

$$F_{Out_4} = F_{O_2} + F_{N_2O} + F_{CO_2} + F_{agas} + F_{others} \tag{7.68}$$

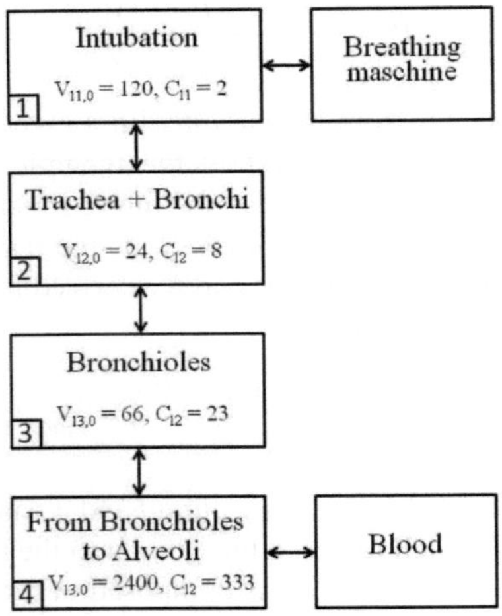

Fig. 7.15 The model of the lung with 4 compartments with increasing compliance. $V_{li,0}$ is the functional residual capacity in compartment i expressed in $[ml]$ and C_{li} is the compliance of the i-th compartment expressed in $[ml/mbar]$.

$$K_{t_i} = \frac{P}{V_0 + CP} \tag{7.69}$$

where V_0 is the volume of the considered part of the lung, C is the compliance under anaesthesia, P is the atmospheric pressure. All the K_{t_i} are assumed to be constant. $R_{l_il_j}$ are the flow resistance factors. F_{in_1} is the total flow amount into the lung and F_{Out_4} the flow balance of the gas exchange between the lung and the blood.

With all this knowledge we can finally simulate the rather complex breathing system depicted in Figure 7.12. However, the great differences among the time constants involved (the stiffness of the system) might lead to problems during the numerical integration ([71] gives clues about how to deal with these kind of problems).

The model described here is very useful to describe the most important flow and concentration changes phenomena in the whole breathing circuit. However, it is too complex for the implementation of model based controllers. The valves' very function of switching cannot be linearized. It would theoretically be possible to model the whole system as a piecewise linear system but then issues would pop up about when to switch between various models. Actually, the input-output relation between the fresh gas concentrations and the inspiratory and endtidal concentrations are the only information needed by the controller. This can be approximated by two blocks representing the breathing machine and the lung respectively (see Figure 7.16).

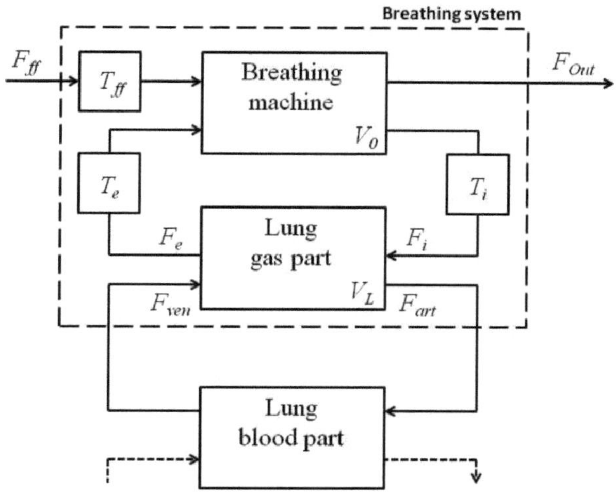

Fig. 7.16 Mass flow diagram of a model structure of the breathing and lung system. T_j are time delays approximated by second order elements. Both the breathing machine and the gas part of the lung represent first order systems.

The mass flow into the first block is composed of the anaesthesia gas flowing in the fresh gas ($F_i = P_f F_{ff}$) and the left over anaesthesia gas in the expired gas ($F_e = P_e AMV$). AMV stands for assisted mechanical ventilation. The inspired anaesthesia gas flow ($F_i = P_i AMV$) together with some unused anaesthesia gas flow out from the first block (F_{Out}). The second block represents the gas part of the lung. Inspiratory and expiratory anaesthesia flows describe the breathing movement, while the gas exchange between the blood part and the gas part of the lung is described through F_{ven} for the gas flow from the venous compartment and F_{art} for the flow to the arterial compartment. The transportation time from the lung to the breathing machine (T_e) and from the breathing machine to the lung (T_i) and for the fresh gas flow to the breathing machine (T_{ff}) are modeled as second order approximations of (infinite order) pure time delays e^{-sT} as

$$G(s) = \frac{1}{1 + Ts + \frac{T^2}{4} s^2} \tag{7.70}$$

Both the breathing machine and the gas part of the lung in Figure 7.16 are compartments. In the former case, the relations between the input and output flows of the block and the input and output of the ideally stirred tank depend on the parameters of the breathing machine set by the anaesthetist. Typical values of these parameters are listed in Table 7.8.

Table 7.8 Typical settings of the breathing machine.

Parameter	Description	Range	Unit
FF	Fresh gas flow	0.5 – 20	liter/min
f_r	Breathing frequency	6 – 60	1/min
V_T	Tidal volume	0.05 – 1.4	liter
T_i/T_e	Ratio of durations of inspiratory to expiratory phases	1:3 – 2:1	-
T_{ip}/T_i	Ratio of durations of inspiratory movement to pause	0 – 60	%
P_e	Endtidal pressure	0 – 20	mbar

Depending on these parameters $(1-K_\beta)$ part of the fresh gas flow goes directly to the outlet as shown in Figure 7.17. K_β is calculated as

$$K_\beta = \min\left(AMV, \max(0, \frac{K_{T_iT_e}}{1+K_{T_iT_e}}\frac{FF}{f_R} - V_T)\right) \tag{7.71}$$

Another $(1-K_\gamma)$ part flows directly into the lung. Therefore, the amount of anaesthesia gas coming from the fresh gas is

$$F_{fresh} = F_{ff}K_\beta K_\gamma \tag{7.72}$$

The stirred tank is ventilated by a total flow amount of AMV. Therefore, the reused gas flow is

$$F_{reused} = \begin{cases} AMV - F_{ff}K_\beta & \text{if} \quad AMV > F_{ff}K_\beta \\ 0 & \text{if} \quad AMV \leq F_{ff}K_\beta \end{cases} \tag{7.73}$$

$$K_\alpha = \frac{F_{reused}}{AMV} \tag{7.74}$$

The parameters in Table 7.9 are set to describe the model.

Table 7.9 The variables of the breathing machine.

Variable	Description
K_α	Reused gas amount from AMV
K_β	Used gas amount from fresh gas
K_γ	Part of gas amount described by K_β that is mixed in the tank
K_s	Part of gas amount that flows directly into the endtidal path
V_0	Volume of the virtual gas tank representing the breathing machine
V_L	Gas volume of the lung taking active part in the gas exchanges

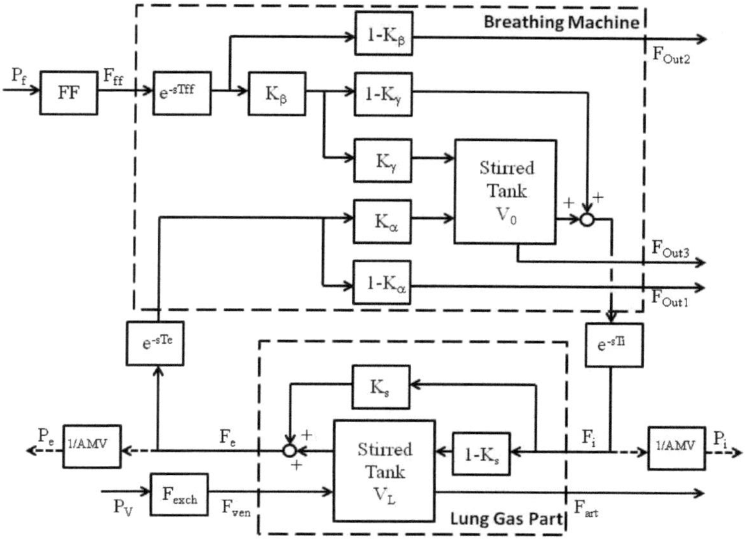

Fig. 7.17 Signal flow diagram of the breathing system.

The three parameters K_γ, K_s and V_0 are very sensitive to the actual operating condition of the system. A first approximation is given by

$$K_\gamma = \begin{cases} 0 & FF < 1 \\ \frac{FF-1}{3} & 1 \le FF < 4 \\ 1 & FF \ge 4 \end{cases} \tag{7.75}$$

$$K_s = 0.3 \tag{7.76}$$

$$V_0 = \begin{cases} 2 - 0.4FF & FF < 4 \\ 0.4 & FF \ge 4 \end{cases} \tag{7.77}$$

The lung is modeled as a first order system and approximated as a stirred tank with two input flows and two output flows generated by the breathing system and by the gas exchange between blood and the alveolar gas mixture (F_{exch}). For V_L we assume the functional residual capacity. K_s is used to model the gas shunt not involved in the mixing. The pressure to volume relation of the manual ventilation bag is modeled as a polynomial $V(P) = \sum_{i=0}^{5} a_i P^i$. The values of the models parameters are shown in Table 7.10.

Table 7.10 Overview of the parameter values of the breathing system elements used in [95].

Element	**R** $\left[\frac{g}{cm^4 s}\right] \approx \left[\frac{mbar}{litre/s}\right]$	$\mathbf{V_0}$ $[ml]$	**C** $\left[\frac{cm^4 s^2}{g}\right] \approx \left[\frac{litre}{mbar}\right]$	Other
Tube (1 – 4)	0.1	750	5×10^{-4}	
Tube (5 – 8)	0.1	75	5×10^{-5}	
Ventilation bag	0.1	-	-	$a_0 = 126000$ $a_1 = -289000$ $a_2 = 212000$ $a_3 = -49000$ $a_4 = 1000$ $a_5 = 130$
Pump	0.1	-	-	
Absorber	0.1	1500	0.001	$K_{abs} = 0.95$
Lung	$R_{l_0 l_1} = 1.00$ $R_{l_1 l_2} = 0.33$ $R_{l_2 l_3} = 0.30$ $R_{l_3 l_4} = 0.80$	$V_{01} = 43$ $V_{12} = 8$ $V_{23} = 22$ $V_{34} = 800$	$C_1 = 0.002$ $C_2 = 0.008$ $C_3 = 0.230$ $C_4 = 0.333$	

7.10 Surgical Stimulation

One of the main objectives during anaesthesia is to keep mean arterial blood pressure within certain bounds. For the control of MAP it is necessary to understand what influences it. We know that anaesthesia generally lowers the MAP and can be used as a control input. But what about the other types of inputs which cannot be controlled? Such inputs are known as disturbances and they can only be compensated indirectly through the control inputs. In case of inhalational anaesthesia, the main disturbance is the surgical stimulation and the control variable is the anaesthesia fresh gas flow as we have already seen in Section 6.5 on fuzzy control of depth of anaesthesia. How MAP can be affected is shown as a linear system in Figure 7.18. Both $G_g(s)$ and $G_d(s)$ act additively on the base level mean arterial pressure MAP_0 and are only slightly influenced by their respective operating conditions [323]. There are other disturbance sources such as infusion drugs, blood loss, dehydration, *etc.* in addition to these two factors. Each disturbance has a different effect on the MAP. For instance, infusion drugs can have a strong influence on the effect of surgical stimulations. In the sequel we will limit ourselves to these two main influence factors. Exact measurements of surgical stimuli are not possible. Therefore, we begin with a qualitative analysis by assuming excitation by intubation, a well defined pulse stimulus, and compare MAP reactions to the model.

Surgical stimulations under anaesthesia cause similar hemodynamic reactions to painful stimulations in the awake state. The autonomous nervous system passes a stimulus from where the stimulation occurs to the spinal nerves. The processed signal is then transmitted to the brain, followed by a stimulation of the sympathetic nervous system which can be considered as an emergency mechanism usually

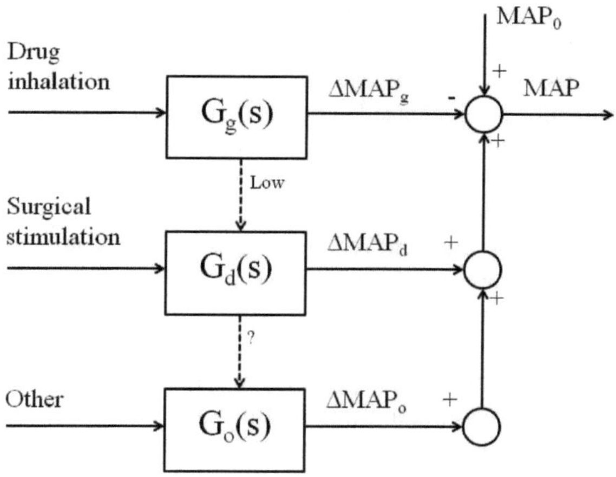

Fig. 7.18 Main influence factors on mean arterial pressure (MAP). The dashed lines represent the influence of the actual operating condition on the dynamic of the respective transfer function.

activated under conditions of potential danger like stress, infection, bleeding or pain. Such stimulation prepares the body for the threatening situation. It leads to physiological reactions like heart rate and blood pressure increase and contraction of skin vessels to decrease blood loss in case of injury. This means, the body is being prepared for fight or flight. The sympathetic system evokes both a neuronal and humoral reaction as illustrated in Figure 7.19 that causes specific reactions of effector cells [81]. Mostly changes in peripheral resistance and cardiac output (*CO*) are relevant for modeling the hemodynamic reaction. Therefore we shall focus on the neurotransmitters epinephrine (adrenaline) and norepinephrine (noradrenaline) [316].

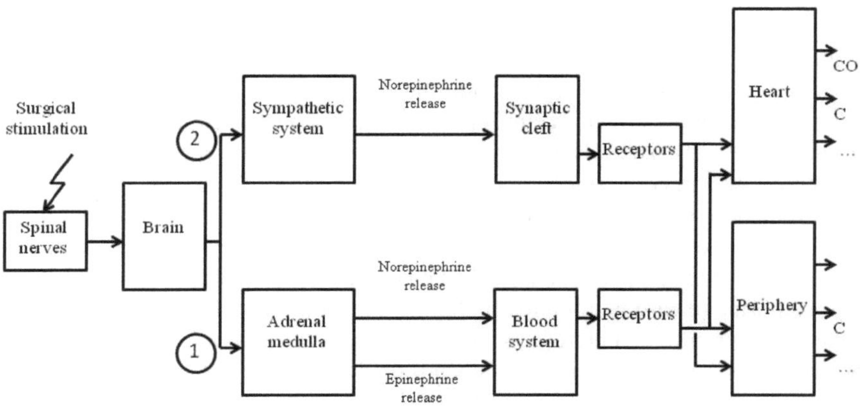

Fig. 7.19 Qualitative description of the physiology of how surgical stimulation influences the MAP. 1) is the humoral reaction, 2) is the neuronal reaction. At the output of the periphery and the heart *CO* is the cardiac output and *C* is the conductivity.

The neuronal reaction is very fast. The stimulation propagates through the sympathetic system to the sympathetic nerve endings where norepinephrine is released into the space between the nerve endings and the effector cells called the synaptic cleft. How much norepinephrine is released depends on the epinephrine concentration detected by the adrenergic receptors at the presynaptic nerve endings. Norepinephrine in the sympathetic cleft decreases rapidly as soon as the neuronal reaction is terminated. This is because of very fast and almost complete absorption of epinephrine by the nerve endings. The humoral reaction, on the other hand, is slower than the neuronal reaction and consists of an increase of epinephrine and norepinephrine release from the adrenal medulla. These neurotransmitters are discharged into the blood stream and exert their effects when they are carried to the effector cells.

As mentioned above, for the humoral reaction epinephrine and norepinephrine produced in the adrenal medulla are released into the circulatory system. It makes sense to model these pharmacokinetics with the same compartments as the ones used to model the distribution of anesthetic drugs. Therefore, the model needs to be extended by another 24 (*sigh!*) compartments. Unlike volatile anesthetic agents, epinephrine and norepinephrine are rapidly inactivated through specialized enzymes which can be modeled by a Michaelis-Menten equation. This means that we can use (7.1) but need to correct it with an additional clearance term A_n. With n_i denoting the norepinephrine concentration in the i-th compartment (index A corresponds to the arterial compartment) and with V_n and k_n constants governing the Michaelis-Menten dynamics, the model is described by the following equations. Similar equations hold for the epinephrine concentration e.

$$\dot{n}_i(t) = \frac{q_i(t)}{V_{i,b}}[n_A(t) - n_i(t)] - A_n \tag{7.78}$$

$$A_n = \frac{V_n n_i}{k_n + n_i} \tag{7.79}$$

As illustrated in Figure 7.19 in the neuronal path, the stimulation signal is transmitted by the release of norepinephrine into the synaptic cleft. The number of receptors stimulated at the effector cells is proportional to the amount of norepinephrine present in the cleft. The release of norepinephrine, P_c, depends on the presynaptic stimulation of adrenergic receptors, β_2, reached by the epinephrine in the blood. The stimulation of these receptors is proportional to the epinephrine concentration. There are control mechanisms which prevent too high norepinephrine concentrations in the synaptic cleft (norepinephrine is reabsorbed by the synapse itself and norepinephrine release is inhibited by high concentration in the cleft). The change of the concentration in a cleft, c, is the difference between the release rate and the absorption rate A_C. The release rate depends on the intensity of the surgical stimulation, u_s, which can be modeled as an affine function of the epinephrine concentration in the blood of the compartment e_i. The norepinephrine absorption is assumed to be limited by a Michaelis-Menten dynamic.

$$\dot{c} = P_c - A_c \tag{7.80}$$

$$P_c = u_s[1 + K(e_i)] \tag{7.81}$$

$$A_c = \frac{cV_c}{c + k_c} \tag{7.82}$$

The two effects of sympathetic stimulation which concern us here are changes in the cardiac output and peripheral conductivity. Assuming that CO is only affected by the humoral and neuronal reaction in the heart, it makes sense to extend the pharmacodynamic equations (7.11) and (7.12) with epinephrine and norepinephrine dependent terms as follows

$$g_i = g_{i,0}(1 + b_i p_i + \gamma_i e_i + \delta_i n_i + \varepsilon_i c_i) \tag{7.83}$$

$$CO = CO_0(1 + \alpha_1 p_1 + \alpha_2 p_2 + \alpha_3 p_A + \alpha_4 e_1 + \alpha_5 n_1 + \alpha_6 c_1) \tag{7.84}$$

The model described here makes physical sense. However, it clearly cannot be used as a model for disturbance rejection during control, because it is of too high order with too many unknown parameters. Furthermore, it is non-linear and coupled with the pharmacokinetic and pharmacodynamic model of anaesthesia gases. Therefore, the model has to be reduced significantly. A solution with three compartments as developed in [95] is given below.

The three compartments are 1) The mean humoral epinephrine blood concentration e, 2) The mean neuronal norepinephrine concentration in the synaptic cleft c, and 3) The mean humoral norepinephrine blood concentration n. The epinephrine compartment is described with one input representing the release from the adrenal medulla and one output given by the clearance C_e due to the rapid inactivation through enzymes. Because there is a basic epinephrine concentration in the blood, it is assumed that the Michaelis-Menten equation above can be linearized around a mean reference point. Hence,

$$\dot{e} = -C_e e + P_e \tag{7.85}$$

The epinephrine production P_e is taken as proportional to the intensity of the actual surgical stimulation:

$$P_e = K_{P_e} u_s \tag{7.86}$$

A similar structure can be assumed for the humoral epinephrine compartment. In addition, there is a significant epinephrine exchange with the surgical cleft (terms with K_{cn} and K_{nc}) due to diffusion phenomena called the spillover effect:

$$\dot{n} = -C_n n + K_{cn} c - K_{nc} n + P_n \tag{7.87}$$

$$P_n = K_{P_n} u_s \tag{7.88}$$

The only non-linearity in this model describes the release mechanism of norepinephrine in the cleft:

$$P_c = u_s[1 + \Delta P_c] \tag{7.89}$$

$$\Delta P_c = K_\beta e - K_\alpha f(c) \tag{7.90}$$

$$f(c) = \begin{cases} 0 & 0 \leq c \leq c_{min} \\ c - c_{min} & c > c_{min} \end{cases} \tag{7.91}$$

The release is amplified by K_β by the epinephrine concentration in the blood (through β_2 receptors) and inhibited affine to the cleft concentration itself. The inhibition acts only above a minimal concentration c_{min}. The compartment dynamic of epinephrine concentration in the cleft is therefore characterized by the release P_c and absorption A_c of epinephrine from the presynapsis in the cleft and by the spillover effect. As in the other compartment, the Michaelis-Menten kinetic is linearized around a mean reference point.

$$\dot{c} = -A_c - K_{cn}c + K_{nc}n + P_c \tag{7.92}$$

This equation is non-linear due to the release mechanism in (7.89). However, the non-linearity is composed of two regions. Therefore, the non-linear equation can be split into two continuous non-linear differential equations.

$$\dot{c} = \begin{cases} \begin{bmatrix} 0 & K_{nc} & -(K_{cn}+K_\alpha) \end{bmatrix} \begin{bmatrix} e \\ n \\ c \end{bmatrix} + K_\beta u_s e & 0 \leq c \leq c_{min} \\ \begin{bmatrix} 0 & K_{nc} & -(K_{cn}+K_\alpha) \end{bmatrix} \begin{bmatrix} e \\ n \\ c \end{bmatrix} + K_\beta u_s e - K_\alpha u_s (c - c_{min}) & c > c_{min} \end{cases} \tag{7.93}$$

Linearizing around the reference points $\bar{e}, \bar{n}, \bar{c}$ yields

$$\delta\dot{c} = \begin{cases} \begin{bmatrix} K_\beta \bar{u}_s & K_{nc} & -(K_{cn}+K_\alpha) \end{bmatrix} \begin{bmatrix} \delta e \\ \delta n \\ \delta c \end{bmatrix} & 0 \leq c \leq c_{min} \\ \begin{bmatrix} K_\beta \bar{u}_s & K_{nc} & -(K_{cn}+K_\alpha+K_\alpha \bar{u}_s) \end{bmatrix} \begin{bmatrix} \delta e \\ \delta n \\ \delta c \end{bmatrix} & c > c_{min} \end{cases} \tag{7.94}$$

with

$$e(t) = \bar{e} + \delta e(t) \qquad n(t) = \bar{n} + \delta n(t) \qquad c(t) = \bar{c} + \delta c(t) \tag{7.95}$$

Simplifying the hemodynamic equations with the assumption of same concentration values in all compartments

$$e_i = e \qquad c_i = c \qquad n_i = n \qquad \forall i \tag{7.96}$$

Further simplification can be achieved by decoupling the effect of humoral and neuronal reaction to MAP from the effect coming from the anaesthetic gas. This means the non-linear output equation

$$MAP = \frac{CO_0(1+\alpha_1 p_1 + \alpha_2 p_2 + \alpha_3 p_A + \alpha_4 e + \alpha_5 n + \alpha_6 c)}{\sum_{i=1}^{9} g_{i,0}(1+b_i p_i + \gamma_i e_i + \delta_i n_i + \varepsilon_i c_i)} \tag{7.97}$$

should be split as

$$MAP = \overline{MAP}(\bar{p},\bar{e},\bar{n},\bar{c}) + \delta MAP_g + \delta MAP_d(\delta e, \delta n, \delta c) \tag{7.98}$$

where MAP is the MAP value at the reference point $\bar{p},\bar{e},\bar{n}$ and $\bar{c}$, δMAP_g is the hemodynamic effect of anaesthetic gas concentration and δMAP_d is the hemodynamic effect of surgical stimulation. We define the changes in the cardiac output and conductivities at the reference levels as

$$\Delta\overline{CO} = (\alpha_1 + \alpha_2 + \alpha_3)\bar{p} \qquad \Delta\bar{g}_i = b_i\bar{p} \tag{7.99}$$

such that

$$\overline{MAP} = \frac{CO_0(1+\Delta\overline{CO} + \alpha_4\bar{e} + \alpha_5\bar{n} + \alpha_6\bar{c})}{\sum_{i=1}^{9} g_{i,0}(1+\Delta\bar{g}_i + \gamma_i\bar{e} + \delta_i\bar{n} + \varepsilon_i\bar{c})} \tag{7.100}$$

We can now linearize (7.97) as

$$\delta MAP_d = \mathbf{c}_d^T \mathbf{x}_d \tag{7.101}$$

$$= \frac{1}{\overline{G}^2} \begin{bmatrix} \overline{G}CO_0\alpha_4 - \overline{CO}\sum_{i=1}^{9} g_{i,0}\gamma_i \\ \overline{G}CO_0\alpha_5 - \overline{CO}\sum_{i=1}^{9} g_{i,0}\delta_i \\ \overline{G}CO_0\alpha_6 - \overline{CO}\sum_{i=1}^{9} g_{i,0}\varepsilon_i \end{bmatrix}^T \begin{bmatrix} \delta e \\ \delta n \\ \delta c \end{bmatrix} \tag{7.102}$$

$$\overline{CO} = CO_0(1 + \Delta\overline{CO} + \alpha_4\bar{e} + \alpha_5\bar{n} + \alpha_6\bar{c}) \tag{7.103}$$

$$\overline{G} = \sum_{i=1}^{9} g_{i,0}(1 + \Delta\bar{g}_i + \gamma_i\bar{e} + \delta_i\bar{n} + \varepsilon_i\bar{c}) \tag{7.104}$$

The variables with bars are steady state values of the linearization reference point. This results in the classical state space representation of the resulting linear MAP disturbance model.

$$\dot{\mathbf{x}} = \mathbf{A}_d\mathbf{x}_d + \mathbf{B}_d u_s \tag{7.105}$$

$$\delta MAP_d = \mathbf{c}_d^T\mathbf{x}_d \tag{7.106}$$

$$\mathbf{A}_d = \begin{bmatrix} -C_e & 0 & 0 \\ 0 & -K_{nc} - C_n & K_{cn} \\ K_\beta\bar{u}_s & K_{nc} & -(K_{cn} + K_\alpha(1 + \kappa\bar{u}_s)) \end{bmatrix} \tag{7.107}$$

$$\mathbf{B}_d = \begin{bmatrix} K_{P_e} \\ K_{P_n} \\ A_c - \kappa K_\alpha(c - c_{min}) \end{bmatrix} \tag{7.108}$$

where

$$\kappa = \begin{cases} 0 & 0 \le c \le c_{min} \\ 1 & c > c_{min} \end{cases} \tag{7.109}$$

7.11 Model Integration

We have finally arrived at the point where we can have a complete model of "the plant" in Figure 7.10. Figure 7.20 shows the connections between the model of the breathing system as described in Section 7.9, the model of the pharmacokinetics and pharmacodynamics of the delivery of anaesthesia gases in Section 7.8 and the model of hemodynamic reactions to surgical stimulations in Section 7.10. We define the linear model realizations as follows:

The matrix quadruplet $\{\mathbf{A}_p, \mathbf{B}_p, \mathbf{C}_p, \mathbf{D}_p\}$ represents the simplified 8th order model (two main compartments and three second order time delay approximations) of the breathing system. The model has two inputs, the anaesthesia gas concentration in the fresh gas flow P_{ff} and in the venous body compartment p_V, and two outputs, the inspired P_i and expired P_e anaesthesia gas concentrations. For the working points described in Section 7.9 we obtain

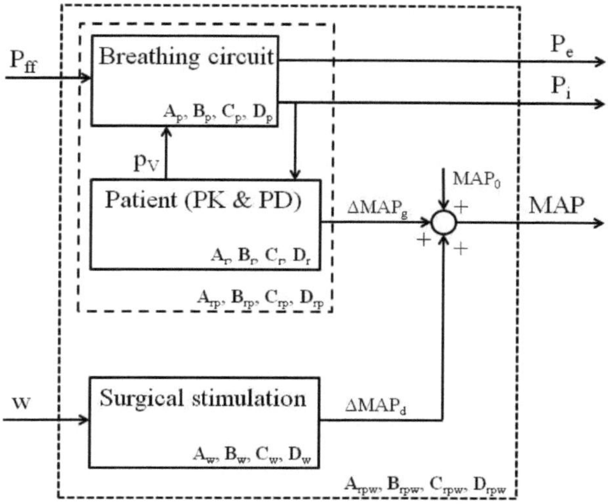

Fig. 7.20 The sub-models developed in the last three Sections make up the integrated model of "the plant" which can be used for purposes of controller design. The linear system description of each block is given by the quadruplet $\{\mathbf{A_x}, \mathbf{B_x}, \mathbf{C_x}, \mathbf{D_x}\}$ matrices. P_i and P_e are the inspired and the endtidal anaesthesia gas concentrations respectively. p_V is the venous anaesthesia gas partial pressure.

$$\mathbf{A}_p = \begin{bmatrix} -32 & -256 & 0 & 0 & 0 & 0.8611 & 0 & 2.222 \\ 1 & 0 & 0 & 0 & 0 & 0 & 0 & 0 \\ 0 & 496.2 & -4.708 & 0 & 0 & 0 & 0 & 0 \\ 0 & 76.80 & 0.7000 & -32 & -256 & 0 & 0 & 0 \\ 0 & 0 & 0 & 1 & 0 & 0 & 0 & 0 \\ 0 & 0 & 0 & 0 & 992 & -3.875 & 0 & 0 \\ 0 & 0 & 0 & 0 & 0 & 0 & -8 & -16 \\ 0 & 0 & 0 & 0 & 0 & 0 & 1 & 0 \end{bmatrix} \tag{7.110}$$

$$\mathbf{B}_p = \begin{bmatrix} 0 & 0 \\ 0 & 0 \\ 0 & 2.7692 \\ 0 & 0 \\ 0 & 0 \\ 0 & 0 \\ 1 & 0 \\ 0 & 0 \end{bmatrix} \tag{7.111}$$

$$\mathbf{C}_p = \begin{bmatrix} 0 & 256 & 0 & 0 & 0 & 0 & 0 & 0 \\ 0 & 76.80 & 0.7 & 0 & 0 & 0 & 0 & 0 \end{bmatrix} \tag{7.112}$$

$$\mathbf{D}_p = \begin{bmatrix} 0 & 0 \\ 0 & 0 \end{bmatrix} \tag{7.113}$$

The matrix quadruplet $\{\mathbf{A}_r, \mathbf{B}_r, \mathbf{C}_r, \mathbf{D}_r\}$ represents the reduced linear model of the pharmacokinetics and pharmacodynamics of the delivery of anaesthesia gases. The model order depends on the reduction degree. The input of the model is the inspiratory anaesthesia concentration and the outputs are the effect on MAP, ΔMAP_g, and the venous isoflurane partial pressure p_V, which are required to define the anaesthesia mass flow through the lung. The coefficients for a reduced system of fifth order are given in [95]. The main parameter values using isoflurane as anaesthesia gas are $W_{Kg} = 70\text{kg}, FF = 1\text{liter/min}, f_R = 6/\text{min}$ and $V_T = 1.2\text{liters}$. The resulting system matrices are

$$\mathbf{A}_r = \begin{bmatrix} -0.0973 & -0.0694 & 0.2338 & -0.0106 & -0.0184 \\ -0.0694 & -0.0661 & 0.3816 & -0.0140 & -0.0246 \\ -0.2338 & -0.3818 & -2.0969 & 0.4611 & 0.6585 \\ -0.0121 & -0.0164 & -0.4777 & -0.0128 & -0.0234 \\ -0.0204 & -0.0280 & -0.6835 & -0.0236 & -0.0438 \end{bmatrix} \tag{7.114}$$

$$\mathbf{B}_r = \begin{bmatrix} 1.1868 \\ 0.5648 \\ 1.3418 \\ 0.0757 \\ 0.1276 \end{bmatrix} \tag{7.115}$$

$$\mathbf{C}_r = \begin{bmatrix} -1.1855 & -0.5645 & 1.3391 & -0.0653 & -0.1124 \\ 0.0565 & 0.0171 & -0.0857 & 0.0383 & 0.0602 \end{bmatrix} \tag{7.116}$$

$$\mathbf{D}_r = \begin{bmatrix} 0 \\ 0 \end{bmatrix} \tag{7.117}$$

The matrix quadruplet $\{\mathbf{A}_w, \mathbf{B}_w, \mathbf{C}_w, \mathbf{D}_w\}$ represents the model of the hemodynamic reactions to surgical stimulations. The model has one input, the surgical disturbance signal w and one output, the effect on the MAP, ΔMAP_d. Following benchmark values are given in [95]:

$$\mathbf{A}_w = \begin{bmatrix} -1.2714 & 0 & 0 \\ 0.4114 & -3 & 0.5 \\ 0 & 0.5 & -0.5571 \end{bmatrix} \tag{7.118}$$

$$\mathbf{B}_w = \begin{bmatrix} 24.4898 \\ 87.9241 \\ 15.3061 \end{bmatrix} \tag{7.119}$$

$$\mathbf{C}_w = \begin{bmatrix} 2.9035 & 0.1936 & 0.1563 \end{bmatrix} \tag{7.120}$$

$$\mathbf{D}_w = \begin{bmatrix} 0 \end{bmatrix} \tag{7.121}$$

The matrix quadruplet $\{\mathbf{A}_{rp}, \mathbf{B}_{rp}, \mathbf{C}_{rp}, \mathbf{D}_{rp}\}$ represents the conjunction of the breathing system and the model of the patient as shown in Figure 7.20.

The matrix quadruplet $\{\mathbf{A}_{rpw}, \mathbf{B}_{rpw}, \mathbf{C}_{rpw}, \mathbf{D}_{rpw}\}$ represents the conjunction of all three sub-models in Figure 7.20.

7.12 Controller Design and Implementation

7.12.1 Observer Based State Feedback Controller with Reference Tracking

The derivation of optimal controllers where unmeasurable states are also estimated optimally with an observer is explained in Appendix A. Let us enhance the optimal controller structure to include a filter at its input as shown in Figure 7.21.

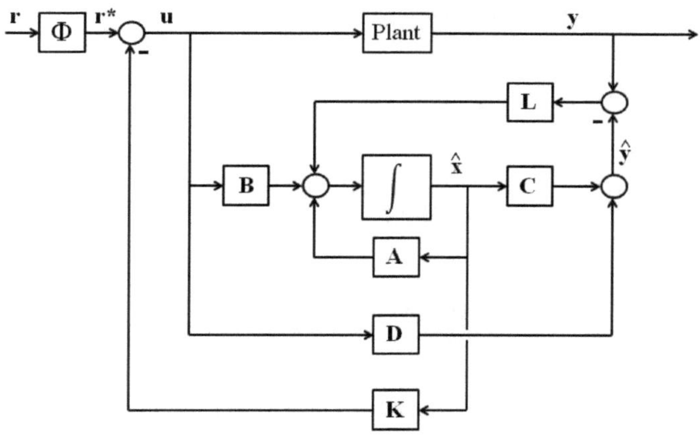

Fig. 7.21 Structure of an observer based state feedback controller with reference tracking as derived in the Appendix A. **K** is the controller gain matrix, **r** is the reference signal, $\mathbf{r}^*$ is the filtered reference signal, **u** is the control input, **y** is the measured output, $\hat{\mathbf{x}}$ is the estimated state and $\hat{\mathbf{y}}$ is the estimated output.

With $\hat{\mathbf{x}}$ as the estimated state vector and with $\hat{\mathbf{y}}$ as the estimated output, the observer equations are

$$\dot{\hat{\mathbf{x}}} = (\mathbf{A} - \mathbf{LC})\hat{\mathbf{x}} + (\mathbf{B} - \mathbf{LD})\mathbf{u} + \mathbf{Ly} \tag{7.122}$$

$$\hat{\mathbf{y}} = \mathbf{C}\hat{\mathbf{x}} + \mathbf{DU} \tag{7.123}$$

It is also explained in the Appendix that the control signal of a state feedback controller is the weighted sum of the actual system state:

$$\mathbf{u} = \mathbf{r}^* - \mathbf{K}\hat{\mathbf{x}} \tag{7.124}$$

where the controller gain matrix K is calculated to optimize the objective function

$$J = \int_0^T (\mathbf{x}^T \mathbf{Q} \mathbf{x} + \mathbf{u}^T \mathbf{R} \mathbf{u}) dt \tag{7.125}$$

The controller has to follow a reference state which, in general, is different from the zero steady state. To this end, an external compensation $\boldsymbol{\Phi}$ of the reference signal is necessary so that the input-output gain is equal to 1. This means

$$\boldsymbol{\Phi} = \left(\lim_{s \to 0} \mathbf{G}_{tot}(s) \right) \tag{7.126}$$

$$\lim_{s \to 0} \mathbf{G}_{tot}(s) = -\mathbf{C}_{tot} \mathbf{A}_{tot}^{-1} \mathbf{B}_{tot} + \mathbf{D}_{tot} \tag{7.127}$$

$\mathbf{G}_{tot}(s)$ is the closed-loop transfer function of the system with the observer and the state feedback loop and is calculated according to

$$\begin{pmatrix} \dot{\mathbf{x}} \\ \dot{\hat{\mathbf{x}}} \end{pmatrix} = \underbrace{\begin{pmatrix} \mathbf{A} & -\mathbf{BK} \\ \mathbf{LC} & (\mathbf{A} - \mathbf{LC} - \mathbf{BK}) \end{pmatrix}}_{\mathbf{A}_{tot}} \begin{pmatrix} \mathbf{x} \\ \hat{\mathbf{x}} \end{pmatrix} + \underbrace{\begin{pmatrix} \mathbf{B} \\ \mathbf{B} \end{pmatrix}}_{\mathbf{B}_{tot}} \mathbf{r}^* \tag{7.128}$$

$$\mathbf{y} = \underbrace{(\mathbf{C} \quad -\mathbf{DK})}_{\mathbf{C}_{tot}} \begin{pmatrix} \mathbf{x} \\ \hat{\mathbf{x}} \end{pmatrix} + \underbrace{\mathbf{D}}_{\mathbf{D}_{tot}} \mathbf{r}^* \tag{7.129}$$

$$\mathbf{G}_{tot} = \mathbf{C}_{tot} (s\mathbf{I} - \mathbf{A}_{tot})^{-1} \mathbf{B}_{tot} + \mathbf{D}_{tot} \tag{7.130}$$

We want the output $\mathbf{y}$ to track a desired trajectory $\mathbf{y}_{set}$. A standard solution making use of optimal control theory yet again is known [27]. It begins by defining a new objective function involving the tracking error $\mathbf{y} - \mathbf{y}_{set}$: [2]

$$J = \int_0^\infty \left(\bar{\mathbf{y}}^T \mathbf{Q}_1 \bar{\mathbf{y}} + (\mathbf{y} - \mathbf{y}_{set})^T \mathbf{Q}_2 (\mathbf{y} - \mathbf{y}_{set}) + \mathbf{u}^T \mathbf{R} \mathbf{u} \right) dt \tag{7.131}$$

where $\mathbf{Q}_1$ and $\mathbf{Q}_2$ are nonnegative definite symmetric weighting matrices. The objective function is derived from

$$J = \int_0^\infty \left((\mathbf{x} - \mathbf{x}_{set})^T \mathbf{Q} (\mathbf{x} - \mathbf{x}_{set}) + \mathbf{u}^T \mathbf{R} \mathbf{u} \right) dt \tag{7.132}$$

with

$$\mathbf{L} = \mathbf{C}^T (\mathbf{C}\mathbf{C}^T)^{-1} \tag{7.133}$$

$$\bar{\mathbf{C}} = \mathbf{I} - \mathbf{LC} \tag{7.134}$$

$$\bar{\mathbf{y}} = \bar{\mathbf{C}} \mathbf{x} \tag{7.135}$$

[2] Note the similarity to the problem of finding the timings and the doses of administered drug to achieve a desired trajectory of the drug concentration in blood as discussed in Chapter 4.

$$\mathbf{Q} = \bar{\mathbf{C}}^T \mathbf{Q}_1 \bar{\mathbf{C}} + \bar{\mathbf{C}}^T \mathbf{Q}_2 \bar{\mathbf{C}} \tag{7.136}$$

$$\mathbf{x}_{set} = \mathbf{L}\mathbf{y}_{set} \tag{7.137}$$

A state feedback controller as described above is able to compensate for unknown disturbance impulses. However, it fails to perform properly in case of constant disturbances. To deal with constant disturbances, the controller must have integral action. His can be achieved by adding an integral part on the input. Furthermore, because of natural limitations on the output (saturation), an anti reset windup mechanism in the loop is also necessary [319].

7.12.2 Artifact Handling

In the operating room environment, measurement errors due to invalid working conditions of the measurement device like disconnections or wash outs occur very frequently. When detected, such errors are classified as artifacts. It is not always easy to detect artifacts automatically. Nevertheless, detection of artifacts by the supervisor system in Figure 7.9 is of paramount importance in controlling the depth of anesthesia automatically.

The complexity of the problem is illustrated in [95] using the example of artifact handling for *MAP*. There are three types of artifacts that have to be detected during surgery so that the measurement values which are classified as such can be ignored: the extraction of a blood sample, the wash out process of the sensor, and a switch to measurements of the central venous catheter pressure. All these artifacts are caused by the anaesthetist but not registered by electronic equipment. Therefore, these artifacts must be identified by analyzing the measured signals.

The artifact handler has to make sure that the measured *MAP* is a valid *MAP*. The following rules were used for this purpose:

- *MAP* cannot be larger than a maximum value, MAP_{max}
- *MAP* cannot be smaller than a minimum value, MAP_{min}
- ΔMAP defined as the absolute value of the difference between the current value and the last valid value MAP_o should not be larger than a maximum value ΔMAP_{max}
- MAP_o is updated only if
  - a new value has been accepted
  - ΔMAP is outside the permitted range, but the last value was outside the range $[MAP_{min}\ \ MAP_{max})$ and the actual value is in this range
  - if more than Counter_{max} consecutive values were rejected (This additional rule was inserted in order to avoid blocking. This might lead to incorrect control behavior and must be detected by the human supervisor.)

The artifact detection parameters were set as

$$\Delta MAP_{max} = 5\ mmHg$$
$$\text{Counter}_{max} = 8$$

$$MAP_{min} = 50\ mmHg$$

$$MAP_{max} = 150\ mmHg$$

Artifact detection and suppression worked fairly well in routine situations as can be seen in Figure 7.22. However, the algorithm filters also some actual effects of surgical disturbances as can be seen in the bottom left graphic in Figure 7.22 at 381 min.

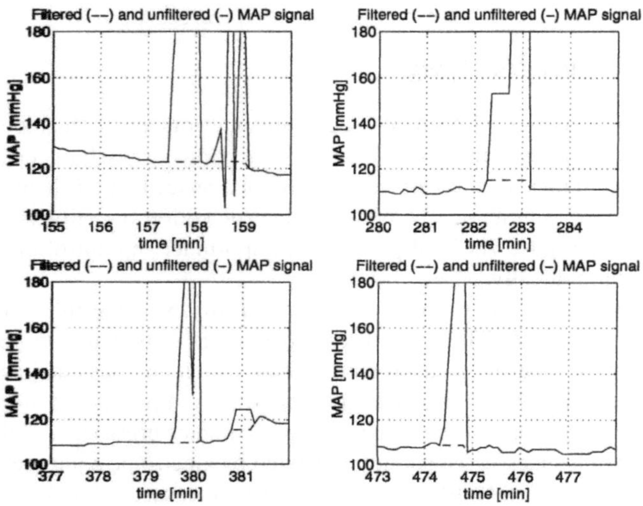

Fig. 7.22 The application of the artifact detection and correction algorithm on four cases of MAP measurement during surgery.

7.12.3 Implementation

The controller described in the previous Subsection has been implemented and used in several clinical tests (see Figure 7.23. The implemented structure was composed of

- a MATLAB simulation environment
- a monitoring system (host, with OBERON[3] as operating system)
- a real time system (target, with XOBERON[4] as operating system)

[3] Oberon is the name of a programming language in the Pascal/Modula tradition. The Oberon project was launched in 1985 by Niklaus Wirth and Jürg Gutknecht at ETH Zurich. Although the project was originally targeted towards in-house hardware, the language and system have now been ported to many computer platforms.

[4] XOberon (a.k.a. Denia) is a real-time operating system, based on Oberon.

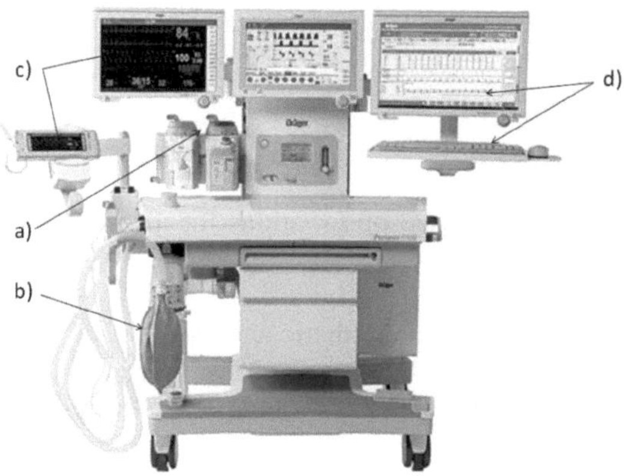

Fig. 7.23 An integrated controller for controlling the depth of anaesthesia looks like the one shown here as an example: Perseus® A500 by Dräger. Clearly visible are a) the two vaporizers for two different inhalational anaesthetic agents, b) the automatic/manual ventilation bag, c) the displays for vital signals of the patient, and d) the computer input and display for operating the controller.

The MATLAB environment was used for simulations and definition of the controller's parameters. The host provides the man machine interface, the communication with the real time system and serves as backup utility. The feedback loop is implemented on the real time system (target) with all its components (input/output devices, filters, supervisor and controllers). The control parameters depend on the patient's weight and respiratory parameters. The code generation of a parameter set is done through the following semiautomatic procedure:

- In MATLAB the controller parameter obtained from the design process are automatically transformed in Oberon source file.
- Transfer of this source file to host.
- Compilation of the generated source code on the host
- Transfer of the compiled code from host to target
- The controller program can be started

The implemented controller was used in a number of separate surgical operations for the control of inspired anaesthesia concentration, for the control of the endtidal anaesthesia concentration and for the control of *MAP*. The results are discussed in detail in [95] which concludes with the remarks

> "The first clinical results reflects the behavior observed in simulation. State feedback control can therefore be applied to control in anaesthesia. Comparing to fuzzy control, the tuning effort was reduced considerably, while a big modeling effort was necessary. Nevertheless, this additional knowledge can be systematically used to improve systems performance: *e.g.*, expanding the patients group considered or changing the anaesthetic

gas type. The implementation of reference tracking, the handling of saturations on the controlled variables and on the output variables could be implemented by introduction of additional functions."

7.13 Exercises

A1: What are the commonly used observations for measuring the depth of anaesthesia?

A2: What are the most common anaesthetic agents?

A3: How does one measure the potency of an anaesthetic agent?

A4: Explain the hierarchical structure and the working principles of an automatic controller for the depth of anaesthesia. In your opinion, what is the most important issue such a controller must take in to account?

B1: The reduction of the patient model in Section 7.8 makes use of controllability and the observability gramians and Hankel singular values. Find another method of reducing system order and apply it on the patient model.

B2: When an automatic controller takes over a task from a human being, a very important source for errors is introduced: artifacts. A human operator quickly recognizes such measurement errors and ignores them. An automatic artifact handling algorithm in the loop for anesthesia control is described in Subsection 7.12.2. Find additional tests you could do to recognize artifacts while giving special attention to "false alarms" (valid measurements which are classified as artifacts).

C1: The patient model described in Section 7.8 is quite elaborate. Develop a simplified patient model to be used in multivariable control of depth of anaesthesia.

C2: Develop a mathematical model for the physiology of how surgical stimulation influences the MAP as schematized in Figure 7.19. What difficulties do you foresee and how do you propose to master them?

Appendix A
Optimal Control

Optimal Control is a very important and broad branch of control engineering. Clearly, no appendix can possibly claim to cover all of this field. Therefore, this Appendix which is based on [209] is intended solely as a refresher. For rigorous definitions, derivations and proofs, the reader is referred to standard textbooks used in courses on Optimal Control. Some examples are [27], [111], [272], [197].

A.1 Introduction

Most control engineering problems are underdetermined. Therefore, it makes sense to demand additional conditions for the solution of such problems. Consider linear systems in the state space representation

$$\dot{\mathbf{x}} = \mathbf{A}\mathbf{x} + \mathbf{B}\mathbf{u} \qquad \mathbf{x}(0) = \mathbf{x}_0 \tag{A.1}$$

with $\mathbf{x}(t)$ as the system state, $\mathbf{u}(t)$ as the input and $\mathbf{A}$ and $\mathbf{B}$ as the system matrix and the input matrix respectively. A reasonable (and very common) approach is to define a cost functional and solve the control problem subject to the minimization of the cost functional (also called the performance index). It makes sense to include the system states' trajectories, inputs and final states in the cost functionals. Probably the most popular cost functional is the quadratic form

$$Z = \frac{1}{2}\mathbf{x}^T(t_f)\mathbf{S}\mathbf{x}(t_f) + \frac{1}{2}\int_{t_0}^{t_f} \mathbf{x}^T(t)\mathbf{Q}\mathbf{x}(t) + \mathbf{u}^T(t)\mathbf{R}\mathbf{u}(t)dt \tag{A.2}$$

where t_0 and t_f are the initial and final times of interest respectively. $\mathbf{Q}$ is a positive semidefinite matrix and $\mathbf{R}$ and $\mathbf{S}$ are positive definite matrices. Those weighting matrices are often chosen as diagonal matrices.

The first integrand $\mathbf{x}^T(t)\mathbf{Q}\mathbf{x}(t) \geq 0$ because $\mathbf{Q}$ is positive semidefinite. It represents the cost incurred at time t for the state trajectories which deviate from 0. The second integrand $\mathbf{u}^T(t)\mathbf{R}\mathbf{u}(t) > 0$ for $\mathbf{u}(t) \neq \mathbf{0}$ because $\mathbf{R}$ is positive definite. It represents the cost at time t incurred by the control effort to get $\mathbf{x}(t)$ to 0. The total cost is the cumulative cost incurred during the time interval of interest. The choice of the weighting matrices **Q**, **R** and **S** affects the tradeoff between the requirements of controlling the state and the effort (energy) to do so as well as how closely we reach a desired final state.

Some special cases are of particular interest.

$$Z = \int_{t_0}^{t_f} dt = t_f - t_0 \tag{A.3}$$

For a given t_0 optimizing Z results in the shortest time.

$$Z = \int_{t_0}^{t_f} \mathbf{u}^T(t)\mathbf{R}\mathbf{u}(t)dt \tag{A.4}$$

For given t_0 and t_f as well as **R** this corresponds to the minimal energy solution.

$$Z = \mathbf{x}^T(t_f)\mathbf{S}\mathbf{x}(t_f) \tag{A.5}$$

For given t_f and **S** this cost functional corresponds to attaining the final state $\mathbf{x}(t_f)$ no matter how much the states deviate from it and how much energy it requires.

A.2 Calculus of Variations

A.2.1 Subject Matter

Let us begin with an illustrative example. Assume that the functional V represents the length of a curve $x(t)$.

$$V(x) = \int_a^b \sqrt{1 + \dot{x}(t)^2} dt \tag{A.6}$$

An interesting problem is to find $x(t)$ that minimizes V, whereas, in general

$$V(x) = \int_a^b L[x(t), \dot{x}(t), t] dt \tag{A.7}$$

Definition: A functional V has a relative minimum at x_0 if a neighborhood Ω of x_0 exists such that $V(x) \geq V(x_0)$ $\forall x \in \Omega$ for small values of $x - x_0$ and $\dot{x} - \dot{x}_0$.
Theorem: If the variation of the functional V exists and has a relative minimum at x_0, then $\delta V = 0$ at x_0.

$$\delta V = \left.\frac{\partial V}{\partial x}\right|_{x_0} \delta x \quad \text{is the first variation of } V \tag{A.8}$$

where $\delta x = x_1 - x_0$, $x_1 \in \Omega$. The total variation $\Delta V = V(x_1) - V(x_0) = V(x_0 + \delta x) - V(x_0)$ is nonnegative, because $V(x_1) \geq V(x_0)$.

The Taylor series expansion of the total variation around x_0 is

$$\Delta V = \left.\frac{\partial V}{\partial x}\right|_{x_0} \delta x + \frac{1}{2!} \cdot \left.\frac{\partial^2 V}{\partial x^2}\right|_{x_0} (\delta x)^2 + \frac{1}{3!} \cdot \left.\frac{\partial^3 V}{\partial x^3}\right|_{x_0} (\delta x)^3 + \ldots \tag{A.9}$$

$$= \delta V + \frac{1}{2!}\delta^2 V + \frac{1}{3!}\delta^3 V + \ldots \tag{A.10}$$

where

$$\delta V = \left.\frac{\partial V}{\partial x}\right|_{x_0} \delta x \quad \text{is the first variation of } V$$

$$\delta^2 V = \left.\frac{\partial^2 V}{\partial x^2}\right|_{x_0} (\delta x)^2 \quad \text{is the second variation of } V$$

$$\delta^3 V = \ldots$$

A.2.2 Fixed Endpoint Problem

Given

$$V(x) = \int_a^b L[x(t), \dot{x}(t), t]dt$$
$$x(a) = A$$
$$x(b) = B$$

where L has continuous first and second derivatives. V has a minimum at x. $x_1 = x + \delta x$ and $\delta x(a) = \delta x(b) = 0$ (fixed endpoint). Defining ΔV as the difference between the functional values with x_1 and x as

$$\Delta V = \int_a^b L[x + \delta x, \dot{x} + \delta \dot{x}, t]dt - \int_a^b L[x, \dot{x}, t]dt \tag{A.11}$$

and first writing down its Taylor series expansion, then writing down its first variation with the boundary conditions $\delta x(a) = \delta x(b) = 0$ and finally equating it to 0 to obtain the minimum value for any variation in x results in the so called Euler equation

$$\frac{\partial L(x, \dot{x}, t)}{\partial x} - \frac{d}{dt}\frac{\partial L(x, \dot{x}, t)}{\partial \dot{x}} = 0 \tag{A.12}$$

Theorem: For a continuous and differentiable function x and the functional $V(x) = \int_a^b L(x, \dot{x}, t)dt$ which is minimized by x where L has continuous first and second partial derivatives and the beginning and end points are fixed, then x satisfies the Euler equation.

Getting back to the example at the beginning of this Section, where $x(a) = A$, $x(b) = B$ and $V(x) = \int_a^b \sqrt{1+\dot{x}(t)^2}dt$ the Euler equation becomes (because L is not a function of x the first term vanishes)

$$\frac{d}{dt}\frac{\partial L(x,\dot{x},t)}{\partial \dot{x}} = 0$$

Integrating this equation with respect to time yields

$$\frac{\partial L(x,\dot{x},t)}{\partial \dot{x}} = constant$$

$$\frac{\partial L(x,\dot{x},t)}{\partial \dot{x}} = \frac{\dot{x}}{\sqrt{1+\dot{x}^2}} = constant$$

Therefore, $\dot{x} = K_1$ hence $x(t) = K_1 t + K_2$. As expected, the answer is a straight line and the boundary conditions can be used to calculate its slope and y-intercept K_1 and K_2.

For multivariable systems with the state vector $\mathbf{x}$ comprising n continuous and differentiable components, the Euler equation becomes n equations

$$\frac{\partial L(\mathbf{x},\dot{\mathbf{x}},t)}{\partial x_i} - \frac{d}{dt}\frac{\partial L(\mathbf{x},\dot{\mathbf{x}},t)}{\partial \dot{x}_i} = 0 \qquad i = 1,\ldots,n \tag{A.13}$$

A.2.3 Variable Endpoint Problem

Variable endpoint problems in calculus of variations occur when e.g., we are looking for the shortest distance from a given point to a given curve. Consider the situation in Figure A.1.

Again with $V(x) = \int_a^b L[x(t),\dot{x}(t),t]dt$ the Euler equation satisfies

$$\Delta V = V(x+h) - V(x) \tag{A.14}$$

$$= \int_{t_0}^{t_1+\delta t_1} L[x+h,\dot{x}+\dot{h},t]dt - \int_{t_0}^{t_1} L[x,\dot{x},t]dt \tag{A.15}$$

Using the mean value theorem, Taylor series expansion of ΔV and the Euler equation we get the generalized boundary condition

$$\left.\frac{\partial L(x,\dot{x},t)}{\partial \dot{x}}\right|_{t_1} \delta x_1 + \left.\left[L(x,\dot{x},t) - \dot{x}\frac{\partial L(x,\dot{x},t)}{\partial \dot{x}}\right]\right|_{t_1} \delta t_1 = 0 \tag{A.16}$$

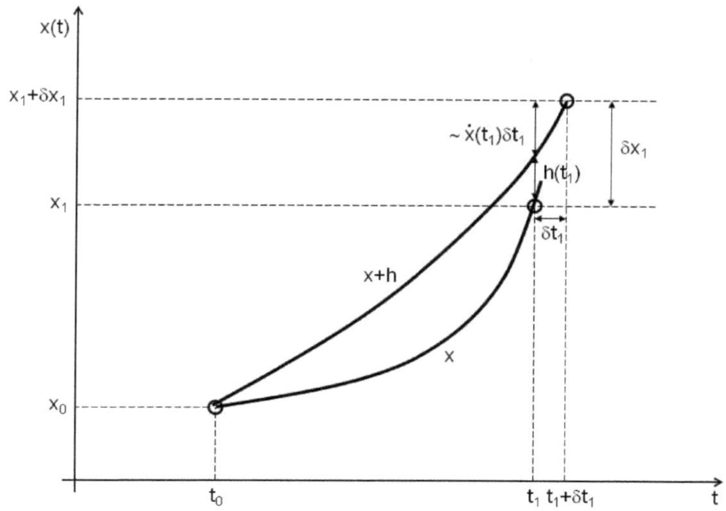

Fig. A.1 The variable endpoint problem. x_0 is the fixed starting point. The endpoint is not known. x represents the unknown function which optimizes the given functional.

Several special cases of the generalized boundary condition are of particular interest.

When δt_1 and δx_1 are independent (A.16) becomes

$$\left.\frac{\partial L(x,\dot{x},t)}{\partial \dot{x}}\right|_{t_1} = 0 \tag{A.17}$$

$$\left.\left[L(x,\dot{x},t) - \dot{x}\frac{\partial L(x,\dot{x},t)}{\partial \dot{x}}\right]\right|_{t_1} = 0 \tag{A.18}$$

If t_1 is given (*i.e.*, $\delta t_1 = 0$) (A.16) becomes

$$\left.\frac{\partial L(x,\dot{x},t)}{\partial \dot{x}}\right|_{t_1} = 0 \tag{A.19}$$

If x_1 is given (*i.e.*, $\delta x_1 = 0$) (A.16) becomes

$$\left.\left[L(x,\dot{x},t) - \dot{x}\frac{\partial L(x,\dot{x},t)}{\partial \dot{x}}\right]\right|_{t_1} = 0 \tag{A.20}$$

If the endpoint lies on a curve $y(t)$ (*i.e.*, $x(t_1) = y(t_1)$ and $\delta x_1 \approx \dot{y}(t_1)\delta t_1$) (A.16) becomes the transversality condition

$$\left[L(x,\dot{x},t) + (\dot{y} - \dot{x})\frac{\partial L(x,\dot{x},t)}{\partial \dot{x}}\right]\Bigg|_{t_1} = 0 \tag{A.21}$$

An illustrative example is looking for the minimal length curve between the origin (0, 0) and the line $y(t) = 2 - t$. We know from Euler equation that the curve has the form $x(t) = K_1 t + K_2$. Since $x(0) = 0, K_2 = 0$. Transversality condition (A.21) can be written as

$$\left[\sqrt{1+\dot{x}^2} + (\dot{y} - \dot{x})\frac{\dot{x}}{\sqrt{1+\dot{x}^2}}\right]\Bigg|_{t_1} = 0$$

and because $\dot{x}(t_1) = K_1$ and $\dot{y}(t_1) = -1$

$$\sqrt{1+K_1^2} + (-1 - K_1)\frac{K_1}{\sqrt{1+K_1^2}} = 0 \qquad \Rightarrow \qquad K_1 = 1$$

As expected, the answer is the line $x(t) = t$ which is perpendicular to the line $y(t) = 2 - t$.

A.2.4 Variation Problem with Constraints

Consider the problem

$$\min_x V(x) = \int_{t_0}^{t_f} L[x(t), \dot{x}(t), t]dt \tag{A.22}$$

subject to the constraint

$$g(x, \dot{x}, t) = 0 \tag{A.23}$$

The solution can be obtained by using Lagrange's method and defining an auxiliary function $F = L + \lambda g$. We then apply Euler's method to F as described above. Note that λ is not a constant but a function.

The generalization and solution of (A.22) for m constraints $g_1 = 0, g_2 = 0, \ldots g_m = 0$ is straight forward. We first define an auxiliary function $F = L + \lambda_1 g_1 + \lambda_2 g_2 + \ldots + \lambda_m g_m$ with the Lagrange multipliers $\lambda_1, \lambda_2, \ldots, \lambda_m$ (again remembering that λ_i are functions) and apply Euler's method.

A.3 Optimal Dynamic Systems

A.3.1 Fixed Endpoint Problem

Consider the system given by its state space description as $\dot{\mathbf{x}} = \mathbf{f}(\mathbf{x}, \mathbf{u}, t)$ with $\mathbf{x}(t)$ as its state vector and $\mathbf{u}(t)$ its input vector. Also given is an objective function $Z = \int_{t_0}^{t_1} L(\mathbf{x}, \mathbf{u}, t)$ (with L positive definite) as well as the initial and final states $\mathbf{x}(t_0)$ and $\mathbf{x}(t_1)$ respectively. The interesting question is which $\mathbf{u}(t)$ minimizes Z?

Applying what we have learned of calculus of variations so far we can proceed as follows to answer that question. The constraints are given by the system description

$$\mathbf{g}(\mathbf{x},\dot{\mathbf{x}},\mathbf{u},\dot{\mathbf{u}},t) = \mathbf{f}(\mathbf{x},\mathbf{u},t) - \dot{\mathbf{x}} = \mathbf{0} \tag{A.24}$$

Let us now define a new objective function

$$\Gamma = \int_{t_0}^{t_1} \Lambda(\mathbf{x},\dot{\mathbf{x}},\mathbf{u},\dot{\mathbf{u}},t)dt \tag{A.25}$$

with

$$\Lambda(\mathbf{x},\dot{\mathbf{x}},\mathbf{u},\dot{\mathbf{u}},t) = L(\mathbf{x},\mathbf{u},t) + \sum_{j=1}^{n} \lambda_j \left[f_j(\mathbf{x},\mathbf{u},t) - \dot{x}_j\right] \tag{A.26}$$

The resulting Euler equations are

$$\frac{\partial}{\partial x_i}\Lambda - \frac{d}{dt}\frac{\partial}{\partial \dot{x}_i}\Lambda = 0 \qquad i = 1,2,\ldots,n \tag{A.27}$$

$$\frac{\partial}{\partial u_k}\Lambda - \frac{d}{dt}\frac{\partial}{\partial \dot{u}_k}\Lambda = 0 \qquad k = 1,2,\ldots,r \tag{A.28}$$

The first part of the Euler equations can be written more explicitly as

$$\frac{\partial}{\partial x_i}\left\{L(\mathbf{x},\mathbf{u},t) + \sum_{j=1}^{n} \lambda_j \left[f_j(\mathbf{x},\mathbf{u},t) - \dot{x}_j\right]\right\} - \frac{d}{dt}\frac{\partial}{\partial \dot{x}_i}\left\{L(\mathbf{x},\mathbf{u},t) + \sum_{j=1}^{n} \lambda_j \left[f_j(\mathbf{x},\mathbf{u},t) - \dot{x}_j\right]\right\} = 0 \tag{A.29}$$

But $\dot{x}_j$ is not a function of x_i. Also $L(\mathbf{x},\mathbf{u},t)$ and $f_j(\mathbf{x},\mathbf{u},t)$ are not not functions of $\dot{x}_i$. Therefore, (A.29) becomes

$$\dot{\lambda}_i = -\frac{\partial}{\partial x_i}\left[L(\mathbf{x},\mathbf{u},t) + \sum_{j=1}^{n} \lambda_j f_j(\mathbf{x},\mathbf{u},t)\right] \tag{A.30}$$

The second part of the Euler equation can be written more explicitly as

$$\frac{\partial}{\partial u_k}\left\{L(\mathbf{x},\mathbf{u},t) + \sum_{j=1}^{n} \lambda_j \left[f_j(\mathbf{x},\mathbf{u},t) - \dot{x}_j\right]\right\} - \frac{d}{dt}\frac{\partial}{\partial \dot{u}_k}\left\{L(\mathbf{x},\mathbf{u},t) + \sum_{j=1}^{n} \lambda_j \left[f_j(\mathbf{x},\mathbf{u},t) - \dot{x}_j\right]\right\} = 0 \tag{A.31}$$

But the second large braces in (A.31) is not a function of $\dot{\mathbf{u}}$. Therefore, (A.31) becomes

$$\frac{\partial}{\partial u_k}\left[L(\mathbf{x},\mathbf{u},t)+\sum_{j=1}^{n}\lambda_j f_j(\mathbf{x},\mathbf{u},t)\right]=0 \tag{A.32}$$

The large brackets in (A.32) is called a Hamiltonian and can be written as

$$H(\mathbf{x},\mathbf{u},\lambda,t)=L(\mathbf{x},\mathbf{u},t)+\lambda^T\mathbf{f}(\mathbf{x},\mathbf{u},t) \tag{A.33}$$

Thus, the final result of the interesting question posed at the beginning of this section can be written as

$$\dot{\lambda}=-\frac{\partial H}{\partial \mathbf{x}} \tag{A.34}$$

$$\mathbf{0}=\frac{\partial H}{\partial \mathbf{u}} \tag{A.35}$$

$$\dot{\mathbf{x}}=\frac{\partial H}{\partial \lambda} \tag{A.36}$$

The system of n equations in (A.34) describes an adjunct system. The system of n equations in (A.36) are the actual system description and constitute the constraints in the mathematical problem. Particular solutions can be found to these $2n$ equations making use of the n initial conditions and n final conditions of the system's states. The equation (A.35) delivers the optimal $\mathbf{u}(t)$.

To solve the problem of finding the input of a system which optimizes a given objective function it is enough to follow the recipe below:

1: Build the Hamiltonian $H(\mathbf{x},\mathbf{u},\lambda,t)=L(\mathbf{x},\mathbf{u},t)+\lambda^T\mathbf{f}(\mathbf{x},\mathbf{u},t)$.

2: Solve $\frac{\partial H}{\partial \mathbf{u}}=\mathbf{0}$ to find $\mathbf{u}^*=\mathbf{u}^*(\mathbf{x},\lambda,t)$.

3: Insert $\mathbf{u}^*$ to get $H^*(\mathbf{x},\lambda,t)=H(\mathbf{x},\mathbf{u}^*,\lambda,t)$

4: Solve the $2n$ equations $\dot{\mathbf{x}}=\frac{\partial H^*}{\partial \lambda}$ and $\dot{\lambda}=-\frac{\partial H^*}{\partial \mathbf{x}}$ using the initial and final conditions $\mathbf{x}(t_0)$ and $\mathbf{x}(t_1)$.

5: Insert $\mathbf{x}(t)$ and $\lambda(t)$ from the previous step to get $\mathbf{u}^*=\mathbf{u}^*(\mathbf{x},\lambda,t)=\mathbf{u}^*(t)$.

An example might help to illustrate the procedure. Given the system

$$\dot{x}=-x+u$$

and the objective function

$$Z=\int_0^1(x^2+u^2)dt$$

with the initial and final conditions $x(0)=1$ and $x(1)=0$. We begin by building the Hamiltonian

$$H(x,u,\lambda,t) = x^2 + u^2 + \lambda(-x+u)$$

We equate the derivative of the Hamiltonian to zero to find u^*

$$\frac{\partial H}{\partial u} = 2u + \lambda = 0 \qquad \Rightarrow \qquad u^* = -\lambda/2$$

We build the Hamiltonian with this u^* as

$$H^* = x^2 + \lambda^2/4 - \lambda x - \lambda^2/2 = x^2 - \lambda^2/4 - \lambda x$$

Hence the system equation

$$\frac{\partial H^*}{\partial \lambda} = \dot{x} = -x - \lambda/2$$

and the equation of the adjunct system

$$-\frac{\partial H^*}{\partial x} = \dot{\lambda} = \lambda - 2x$$

Solving the last two equations yields

$$\begin{aligned} x(t) &= K_1 e^{-\sqrt{2}t} + K_2 e^{\sqrt{2}t} \\ \lambda(t) &= -2K_1(1-\sqrt{2})e^{-\sqrt{2}t} - 2K_2(1+\sqrt{2})e^{\sqrt{2}t} \end{aligned}$$

With $x(0) = 1$ and $x(1) = 0$ we get $1 = K_1 + K_2$ and $0 = K_1 e^{-\sqrt{2}} + K_2 e^{\sqrt{2}}$. Solving these for K_1 and K_2 results in

$$\begin{aligned} K_1 &= \frac{1}{1-e^{-2\sqrt{2}}} \\ K_2 &= \frac{1}{1-e^{2\sqrt{2}}} \end{aligned}$$

Finally, we get the optimal control variable

$$u^* = \frac{1-\sqrt{2}}{1-e^{-2\sqrt{2}}} e^{-\sqrt{2}t} + \frac{1+\sqrt{2}}{1-e^{2\sqrt{2}}} e^{\sqrt{2}t}$$

A.3.2 Variable Endpoint Problem

For the variable endpoint problem the generalized boundary condition with constraints becomes

$$\left.\frac{\partial \Lambda(\mathbf{x},\dot{\mathbf{x}},\mathbf{u},\dot{\mathbf{u}},t)}{\partial \dot{\mathbf{x}}}\right|_{t_1} \delta \mathbf{x}_1 + \left.\left[\Lambda(\mathbf{x},\dot{\mathbf{x}},\mathbf{u},\dot{\mathbf{u}},t) - \dot{\mathbf{x}}\frac{\partial \Lambda(\mathbf{x},\dot{\mathbf{x}},\mathbf{u},\dot{\mathbf{u}},t)}{\partial \dot{\mathbf{x}}}\right]\right|_{t_1} \delta t_1 = 0 \quad \text{(A.37)}$$

with

$$\Lambda = L(\mathbf{x},\mathbf{u},t) + \lambda^T[\mathbf{f}(\mathbf{x},\mathbf{u},t) - \dot{\mathbf{x}}]$$

Since neither $L(.)$ nor $\mathbf{f}(.)$ are functions of $\dot{\mathbf{x}}$ (A.37) simplifies to

$$-\lambda^T\Big|_{t_1} \delta\mathbf{x}_1 + \left\{L(\mathbf{x},\mathbf{u},t) + \lambda^T[\mathbf{f}(\mathbf{x},\mathbf{u},t) - \dot{\mathbf{x}}] + \lambda^T\dot{\mathbf{x}}\right\}\Big|_{t_1} \delta t_1 = 0 \tag{A.38}$$

or with the Hamiltonian

$$H^* = L(\mathbf{x},\mathbf{u},t) + \lambda^T[\mathbf{f}(\mathbf{x},\mathbf{u},t) - \dot{\mathbf{x}}] + \lambda^T\dot{\mathbf{x}}$$

even simpler to

$$-\lambda^T\Big|_{t_1} \delta\mathbf{x}_1 \; + \; H^*\Big|_{t_1} \delta t_1 = 0 \tag{A.39}$$

To solve the problem of finding the input of a system which optimizes a given objective function where the final state is not given it is enough to follow the modified recipe below:

1: Build the Hamiltonian $H(\mathbf{x},\mathbf{u},\lambda,t) = L(\mathbf{x},\mathbf{u},t) + \lambda^T\mathbf{f}(\mathbf{x},\mathbf{u},t)$.

2: Solve $\frac{\partial H(\mathbf{x},\mathbf{u},\lambda,t)}{\partial \mathbf{u}} = \mathbf{0}$ to find $\mathbf{u}^* = \mathbf{u}^*(\mathbf{x},\lambda,t)$.

3: Insert $\mathbf{u}^*$ to get $H^*(\mathbf{x},\lambda,t) = H(\mathbf{x},\mathbf{u}^*,\lambda,t)$

4: Solve the $2n$ equations $\dot{\mathbf{x}} = \frac{\partial H^*(\mathbf{x},\lambda,t)}{\partial \lambda}$ and $\dot{\lambda} = -\frac{\partial H^*(\mathbf{x},\lambda,t)}{\partial \mathbf{x}}$ using the initial conditions $\mathbf{x}(t_0)$ and the generalized boundary conditions

$$-\lambda^T\delta\mathbf{x}\Big|_{t_1} \; + \; H^*(\mathbf{x},\lambda,t)\delta t\Big|_{t_1} = 0 \tag{A.40}$$

5: Insert $\mathbf{x}(t)$ and $\lambda(t)$ from the previous step to get $\mathbf{u}^* = \mathbf{u}^*(\mathbf{x},\lambda,t) = \mathbf{u}^*(t)$.

Following special cases are of particular interest:

If $\delta\mathbf{x}(t_1)$ and δt_1 are independent

$$\lambda^T\Big|_{t_1} = \mathbf{0} \tag{A.41}$$

$$H^*(\mathbf{x},\lambda,t)\Big|_{t_1} = 0 \tag{A.42}$$

If t_1 is given

$$\lambda^T\Big|_{t_1} = \mathbf{0}$$

If $\mathbf{x}(t_1)$ is given

$$H^*(\mathbf{x},\lambda,t)\Big|_{t_1} = 0$$

If the endpoint lies on a curve (so called rendez-vous problem), then $\delta\mathbf{x} = \mathbf{y}(t_1)dt$

$$\lambda^T\dot{\mathbf{y}}\Big|_{t_1} + H^*(\mathbf{x},\lambda,t)\Big|_{t_1} = 0 \tag{A.43}$$

$$x(t_1) = y(t_1) \tag{A.44}$$

An example might again help to illustrate the procedure. Given the system

$$\dot{x} = -2\sqrt{2}x + u$$

$$Z = \int_0^1 (x^2 + u^2)dt$$

$x(0) = 2$ and $x(1)$ free

$$H = x^2 + u^2 + \lambda(-2\sqrt{2}x + u)$$

$$\frac{\partial H}{\partial u} = 2u + \lambda = 0 \qquad \Rightarrow \qquad u^* = -\lambda/2$$

$$\begin{aligned} \dot{x} &= -2\sqrt{2}x - \lambda/2 \\ \dot{\lambda} &= -2x + 2\sqrt{2}\lambda \end{aligned}$$

$$H^* = x^2 + \lambda^2/4 - 2\sqrt{2}\lambda x - \lambda^2/2$$

With the initial condition $x(0) = 2$ and (because $t_1 = 1$ is given) with $\lambda(1) = 0$

$$\lambda(t) = 0.687(e^{-3t} - e^{-6}e^{3t})$$

$$u^*(t) = -\lambda(t)/2 = -0.343(e^{-3t} - e^{-6}e^{3t})$$

A.3.3 Generalized Objective Function

Let us now consider a new objective function where the endpoint (the final state) is variable and is penalized

$$Z = S(\mathbf{x}(t_1),t_1) + \int_{t_0}^{t_1} L(\mathbf{x},\mathbf{u},t)dt \tag{A.45}$$

Our recipe for the solution changes only slightly:

1: Build the Hamiltonian $H(\mathbf{x},\mathbf{u},\lambda,t) = L(\mathbf{x},\mathbf{u},t) + \lambda^T\mathbf{f}(\mathbf{x},\mathbf{u},t)$.

2: Solve $\frac{\partial H(\mathbf{x},\mathbf{u},\lambda,t)}{\partial \mathbf{u}} = \mathbf{0}$ to find $\mathbf{u}^* = \mathbf{u}^*(\mathbf{x},\lambda,t)$.

3: Insert $\mathbf{u}^*$ to get $H^*(\mathbf{x},\lambda,t) = H(\mathbf{x},\mathbf{u}^*,\lambda,t)$

4: Solve the $2n$ equations $\dot{\mathbf{x}} = \frac{\partial H^*(\mathbf{x},\lambda,t)}{\partial \lambda}$ and $\dot{\lambda} = -\frac{\partial H^*(\mathbf{x},\lambda,t)}{\partial \mathbf{x}}$ using the initial conditions $\mathbf{x}(t_0)$ and the generalized boundary conditions

$$\left(\frac{\partial S}{\partial \mathbf{x}} - \lambda\right)^T \delta\mathbf{x}\Big|_{t_1} + \left(H^*(\mathbf{x},\lambda,t) + \frac{\partial S}{\partial t}\right)\delta t\Big|_{t_1} = 0 \tag{A.46}$$

5: Insert $\mathbf{x}(t)$ and $\lambda(t)$ from the previous step to get $\mathbf{u}^* = \mathbf{u}^*(\mathbf{x},\lambda,t) = \mathbf{u}^*(t)$.

A.4 Pontryagin's Minimum Principle

In the previous Section we have seen that finding the optimal $\mathbf{u}(t)$ that minimizes a given penalty function is achieved firstly by building a Hamiltonian and equating the partial derivative of the Hamiltonian with respect to $\mathbf{u}(t)$ to zero. What happens if the resulting control variable gets too big? One way of dealing with this problem is penalizing $\mathbf{u}(t)$ in the objective function more. A more general solution for the case that $\mathbf{u}$ belongs to a set U a was given by Lev Pontryagin [255] who has shown that the solution can be reformulated as

$$\min_{\mathbf{u}\in U} H \to \mathbf{u}^* \tag{A.47}$$

Again an example shall illustrate the procedure.

Given is the system

$$\dot{x}_1 = x_2$$

$$\dot{x}_2 = u$$

where the control variable is limited to $|u| \leq 1$. Further, the initial and final states are given as $x_1(0) = x_{10},\ x_2(0) = x_{20},\ x_1(T) = 0,\ x_2(T) = 0$. Our objective is to bring the system in the shortest possible time to the given endpoint (0, 0). In other words, the objective function to minimize is

$$Z = \int_0^T 1dt$$

We again begin by building the Hamiltonian

$$H = 1 + \lambda_1 x_2 + \lambda_2 u$$

and find the $u^*(t)$ which minimizes the Hamiltonian

$$\min_{u \in U} H \rightarrow u^* = -\text{signum}(\lambda_2) = \pm 1$$

Therefore

$$H^* = 1 + \lambda_1 x_2 - \lambda_2 \text{signum}(\lambda_2)$$

$$\dot{x}_1 = \frac{\partial H^*}{\partial \lambda_1} = x_2$$

$$\dot{x}_2 = \frac{\partial H^*}{\partial \lambda_2} = -\text{signum}(\lambda_2)$$

$$\dot{\lambda}_1 = -\frac{\partial H^*}{\partial x_1} = 0$$

$$\dot{\lambda}_2 = -\frac{\partial H^*}{\partial x_2} = -\lambda_1$$

With the given initial and final conditions the generalized boundary conditions become (because the endpoint is given)

$$H^*\big|_T = 0 \quad \Rightarrow \quad |\lambda_2(T)| = 1$$

$$\lambda_1 = K_1 \qquad \lambda_2 = -K_1 t + K_2$$

u^* clearly has the value either $+1$ or -1 as Figure A.2 shows.

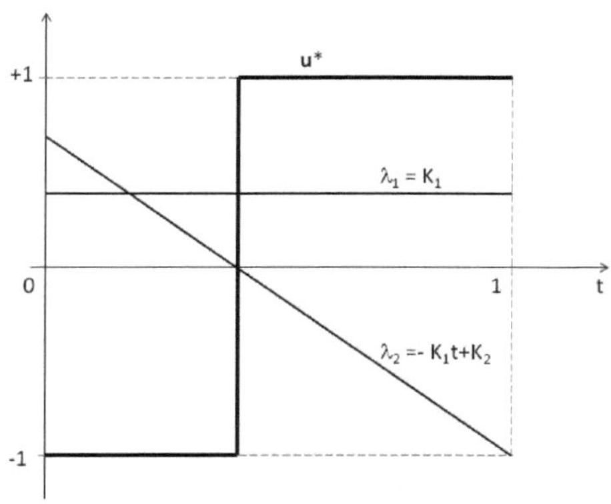

Fig. A.2 Optimal $u(t)$ is shown together with the adjunct variables.

Let us now look at the phase portrait of the system as shown in Figure A.3. When $u = +1$

$$\frac{dx_2}{dx_1} = \frac{1}{x_2}$$

$$\Rightarrow \frac{1}{2}x_2^2 = x_1 + C$$

and when $u = -1$

$$\frac{dx_2}{dx_1} = -\frac{1}{x_2}$$

$$\Rightarrow \frac{1}{2}x_2^2 = -x_1 + C$$

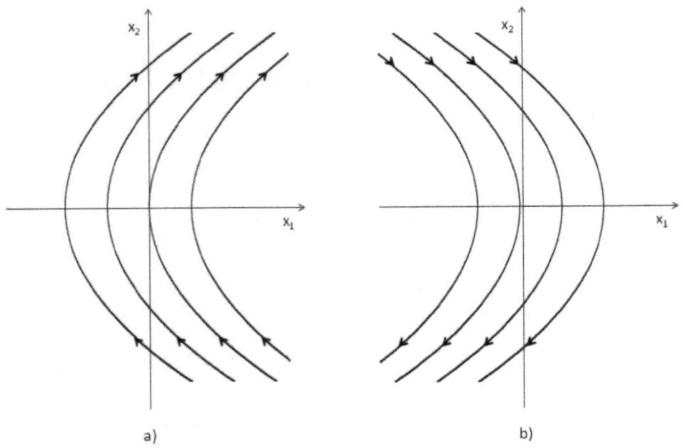

Fig. A.3 Phase portrait of the system a) when $u = +1$ and b) when $u = -1$. The system should reach the origin in the shortest possible time.

Figure A.4 illustrates how the optimal controller works. If the initial condition is below the switching curve the controller works "full gas" with $u = +1$ until the system reaches the so called switching curve. At that point the controller switches to "full brake" with $u = -1$ until the origin is reached. If the initial condition is above the switching curve the controller works "full brake" with $u = -1$ until the system reaches the so called switching curve. At that point the controller switches to "full gas" with $u = +1$ until the origin is reached. Such a controller is called a bang-bang controller.

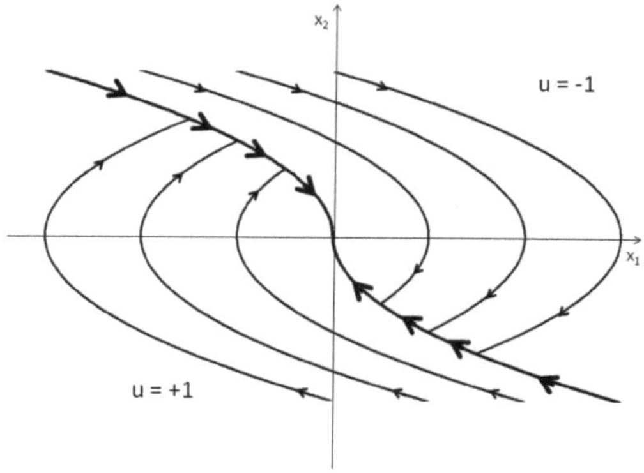

Fig. A.4 Phase portrait of the optimal bang-bang controller with the switching curve.

A.5 Hamilton-Jacobi Method

In the previous Section we have seen how to find the control variable $\mathbf{u}^*(t)$ which optimizes a given objective function (open-loop control). Consider now the problem of finding the control variable $\mathbf{u}^*(\mathbf{x}(t),t)$ as a function of the state of the system which optimizes a given objective function (closed-loop control) as illustrated in Figure A.5.

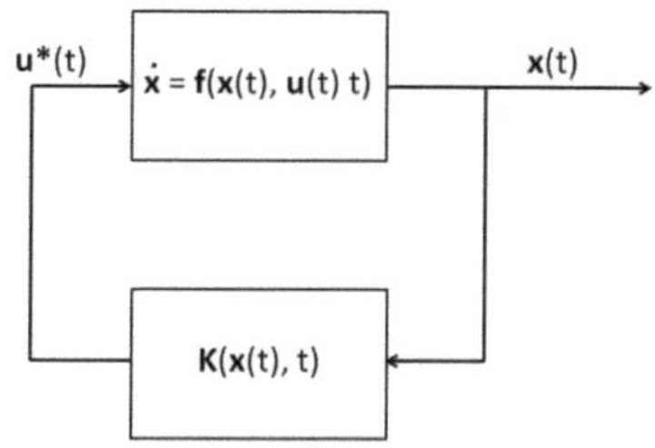

Fig. A.5 Structure of the optimal closed-loop controller.

Let us again begin by defining a scalar function

$$V(\mathbf{x},t) = \int_t^{t_1} L[\mathbf{x}(\tau),\mathbf{u}^*(\mathbf{x},\tau),\tau]d\tau \tag{A.48}$$

as the minimal value of the performance index for a starting state $\mathbf{x}$ and the time t. In other words, $V(\mathbf{x},t)$ is the objective function for the optimal trajectory which starts at $\mathbf{x}(t)$ and ends at $\mathbf{x}(t_1)$. Assuming that $V(\mathbf{x},t)$ has continuous second derivatives, it follows

$$\dot{V}(\mathbf{x},t) = \frac{\partial V(\mathbf{x},t)}{\partial \mathbf{x}} \cdot \dot{\mathbf{x}} + \frac{\partial V(\mathbf{x},t)}{\partial t} \tag{A.49}$$

or, with $\nabla\mathbf{V}(\mathbf{x},t)$ as the gradient vector of $V(\mathbf{x},t)$

$$\dot{V}(\mathbf{x},t) = \nabla\mathbf{V}^T(\mathbf{x},t) \cdot \dot{\mathbf{x}} + \frac{\partial V(\mathbf{x},t)}{\partial t} \tag{A.50}$$

Since the system is optimal along the trajectory

$$\dot{\mathbf{x}} = \mathbf{f}(\mathbf{x},\mathbf{u}^*(\mathbf{x},t),t) \tag{A.51}$$

(A.50) becomes

$$\dot{V}(\mathbf{x},t) = \nabla\mathbf{V}^T(\mathbf{x},t) \cdot \mathbf{f}(\mathbf{x},\mathbf{u}^*(\mathbf{x},t),t) + \frac{\partial V(\mathbf{x},t)}{\partial t} \tag{A.52}$$

From the definition of $V(\mathbf{x},t)$ follows

$$\dot{V}(\mathbf{x},t) = -L(\mathbf{x},\mathbf{u}^*(\mathbf{x},t),t) \tag{A.53}$$

Finally, from (A.52) and (A.53) follows

$$\nabla\mathbf{V}^T(\mathbf{x},t) \cdot \mathbf{f}(\mathbf{x},\mathbf{u}^*(\mathbf{x},t),t) + \frac{\partial V(\mathbf{x},t)}{\partial t} + L(\mathbf{x},\mathbf{u}^*(\mathbf{x},t),t) = 0 \tag{A.54}$$

The partial derivative of (A.54) gives

$$\frac{\partial}{\partial \mathbf{x}}(\nabla\mathbf{V}^T \cdot \mathbf{f}) + \frac{\partial^2 V}{\partial \mathbf{x} \partial t} + \frac{\partial L}{\partial \mathbf{x}} = 0 \tag{A.55}$$

rewriting we get

$$\nabla\mathbf{V}^T \cdot \frac{\partial \mathbf{f}}{\partial \mathbf{x}} + \mathbf{f}^T \cdot \frac{\partial \nabla\mathbf{V}}{\partial \mathbf{x}} + \frac{\partial^2 V}{\partial \mathbf{x} \partial t} + \frac{\partial L}{\partial \mathbf{x}} = 0 \tag{A.56}$$

Finally with

$$\frac{d}{dt}\nabla\mathbf{V} = \mathbf{f}^T \cdot \frac{\partial \nabla\mathbf{V}}{\partial \mathbf{x}} + \frac{\partial^2 V}{\partial \mathbf{x} \partial t} \tag{A.57}$$

we get

$$\frac{d}{dt}\nabla\mathbf{V} = -\left(\nabla\mathbf{V}^T \cdot \frac{\partial \mathbf{f}}{\partial \mathbf{x}} + \frac{\partial L}{\partial \mathbf{x}}\right) \tag{A.58}$$

Building the Pontryagin function according to the Pontryagin minimum principle as

$$H(\mathbf{x},\lambda,t) = H(\mathbf{x},\mathbf{u}^*(\mathbf{x},t),\lambda,t) = \lambda^{\mathbf{T}} \cdot \mathbf{f}(\mathbf{x},\mathbf{u}^*(\mathbf{x},t)) + L(\mathbf{x},\mathbf{u}^*(\mathbf{x},t)) \tag{A.59}$$

and the Euler equations for the adjunct system as

$$\dot{\lambda} = -\frac{\partial H(\mathbf{x},\lambda,t)}{\partial \mathbf{x}} \quad \text{or} \tag{A.60}$$

$$\frac{d}{dt}\lambda = -\left(\lambda^T \cdot \frac{\partial \mathbf{f}}{\partial \mathbf{x}} + \frac{\partial L}{\partial \mathbf{x}}\right) \tag{A.61}$$

Looking at equations (A.58) and (A.60) we can see that $\nabla \mathbf{V} = \lambda$. Since we get $\mathbf{u}^*$ according to the Pontryagin principle from

$$\min_{\mathbf{u}\in U} H(\mathbf{x},\mathbf{u},\nabla \mathbf{V},t)$$

it follows that

$$\begin{aligned} \mathbf{u}^* &= \mathbf{u}^*(\mathbf{x},\nabla \mathbf{V},t) \quad \text{and} \\ H^*(\mathbf{x},\nabla \mathbf{V},t) &= \nabla \mathbf{V}^T \cdot \mathbf{f}(\mathbf{x},\mathbf{u}^*(\mathbf{x},\nabla \mathbf{V},t),t) + L(\mathbf{x},\mathbf{u}^*(\mathbf{x},\nabla \mathbf{V},t),t) \end{aligned}$$

Comparing this equation with equation (A.54) finally results in the Hamilton-Jacobi equation:

$$H^*(\mathbf{x},\nabla \mathbf{V}(\mathbf{x},t),t) + \frac{\partial V(\mathbf{x},t)}{\partial t} = 0 \tag{A.62}$$

Hence, the optimal closed-loop control problem is reduced to the solution of this non-linear first order differential equation for $V(\mathbf{x},t)$ with the boundary conditions $V(\mathbf{x}_1,t_1) = 0$ for a fixed final state problem and $V(\mathbf{x},t_1) = 0$ for a variable final state problem.

Let us again write down the recipe for the solution of finding the closed-loop optimal control problem.

1: For the dynamic system given in its state representation $\dot{\mathbf{x}} = \mathbf{f}(\mathbf{x},\mathbf{u},t)$ and the performance index in the form $\Lambda = \int_{t_0}^{t_1} L(\mathbf{x},\mathbf{u},t)dt$ build the Pontryagin function

$$H(\mathbf{x},\mathbf{u},\nabla \mathbf{V},t) = \nabla \mathbf{V}^T \cdot \mathbf{f}(\mathbf{x},\mathbf{u},t) + L(\mathbf{x},\mathbf{u},t)$$

2: Minimize H with respect to permissible control variables

$$\min_{\mathbf{u}\in U} H(\mathbf{x},\mathbf{u},\nabla \mathbf{V},t) \qquad \rightarrow \qquad \mathbf{u}^* = \mathbf{u}^*(\mathbf{x},\nabla \mathbf{V},t)$$

3: Build the optimal Pontryagin function with that optimal $\mathbf{u}^*$

$$H^*(\mathbf{x},\nabla \mathbf{V},t) = H(\mathbf{x},\mathbf{u}^*(\mathbf{x},\nabla \mathbf{V},t),\nabla \mathbf{V},t)$$

4: Solve the Hamilton-Jacobi equation

$$H^*(\mathbf{x}, \nabla\mathbf{V}, t) + \frac{\partial V}{\partial t} = 0$$

with the boundary conditions $V(\mathbf{x}_1, t_1) = 0$ for a fixed final state problem and $V(\mathbf{x}, t_1) = 0$ for a variable final state problem resulting in $V(\mathbf{x}, t)$ and $\nabla V(\mathbf{x}, t)$.

5: Substitute the $V(\mathbf{x}, t)$ and $\nabla V(\mathbf{x}, t)$ from the previous step in $\mathbf{u}^*$ to get

$$\mathbf{u}^*(\mathbf{x}, t) = \mathbf{u}^*(\mathbf{x}, \nabla V, t)$$

An example should help to illustrate the procedure. A controller is sought for the system $\dot{x} = -2x + u$, which optimizes the performance index $\Lambda = \int_0^\infty (x^2 + u^2)dt$. The control variable u is unbounded.

We begin by building the Pontryagin function

$$H = \nabla V \cdot (-2x + u) + x^2 + u^2$$

To optimize H with respect to u we can simply equate the derivative of H to zero because u is unbounded.

$$\frac{\partial H}{\partial u} = 0 \qquad \Rightarrow \qquad \nabla V + 2u = 0 \qquad \Rightarrow \qquad u^* = -\nabla V/2$$

We insert this optimal u^* in the Pontryagin function

$$H^* = -2\nabla Vx - \frac{(\nabla V)^2}{2} + x^2 + \frac{(\nabla V)^2}{4}$$
$$H^* = -2\nabla Vx - \frac{(\nabla V)^2}{4} + x^2$$

Using the optimal Pontryagin function we build the Hamilton-Jacobi equation

$$-2\nabla Vx - \frac{(\nabla V)^2}{4} + x^2 + \frac{\partial V}{\partial t} = 0$$

We can solve this equation with the *Ansatz* $V = \alpha x^2$ and consequently $\nabla V = 2\alpha x$

$$-4\alpha x^2 - \alpha^2 x^2 + x^2 = x^2(1 - 4\alpha - \alpha^2) = 0$$

α must be positive so that V is always positive. Hence, $\alpha = \sqrt{5} - 2$ and $\nabla V = 2(\sqrt{5} - 2)x$. This gives us the optimal controller as

$$u^*(x, t) = -(\sqrt{5} - 2)x$$

The block diagram of this optimally controlled system is shown in Figure A.6.

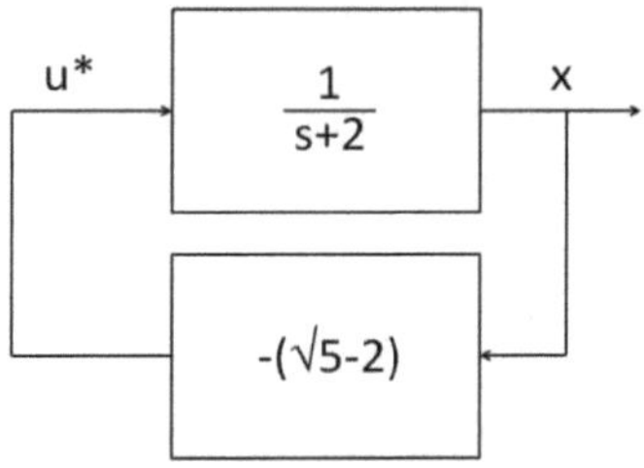

Fig. A.6 Optimal closed-loop controller.

Some remarks are in order here.

1) The optimal control variable $\mathbf{u}^*(\mathbf{x},t)$ as a function of the state vector $\mathbf{x}(t)$ is optimal for *all* initial conditions $\mathbf{x}(0) = \mathbf{x}_0$ and represents the solution of the optimal closed-loop control problem.
2) The control variable can be bounded or unbounded.
3) We do *not* have to solve a two point boundary condition problem. Rather, a partial differential equation has to be solved. The Hamilton-Jacobi method is especially well suited for linear or continuously linearized systems with a quadratic performance index. In that case, the partial differential equation can be solved analytically and the solution is given by the matrix Riccati equation as we shall see in the next Section.
4) The Hamilton-Jacobi equation can also be derived using the optimality principle [29] instead of using calculus of variations as we have done.
5) To achieve an optimal control *all* states have to be fed back.

A.6 Optimal Control of Linear Systems

A.6.1 Riccati Equation

In the previous Section we have found a general solution for the optimal feedback control problem. In this Section we will study a special case which is widely used. We are looking for the optimal closed-loop control variable $\mathbf{u}^*(\mathbf{x},t)$ for the linear system given by its state space equation

$$\dot{\mathbf{x}} = \mathbf{A}\mathbf{x} + \mathbf{B}\mathbf{u} \tag{A.63}$$

which optimizes the performance index

$$\Lambda = \int_0^T (\mathbf{x}^T\mathbf{Q}\mathbf{x} + \mathbf{u}^T\mathbf{P}\mathbf{u})dt \tag{A.64}$$

Let us remember several points before applying the Hamilton-Jacobi method: there are n states and r inputs. The final state $\mathbf{x}(T)$ is not fixed. $\mathbf{P}$ is a symmetrical, positive definite matrix. $\mathbf{Q}$ is a symmetrical, positive semidefinite matrix. The control variable $\mathbf{u}$ is unbounded but by choosing a large matrix $\mathbf{P}$ it can be made arbitrarily small.

1: The Pontryagin function for the given system and the objective function is

$$H(\mathbf{x},\mathbf{u},\nabla\mathbf{V},t) = \nabla\mathbf{V}^T \cdot (\mathbf{A}\mathbf{x}+\mathbf{B}\mathbf{u}) + \mathbf{x}^T\mathbf{Q}\mathbf{x} + \mathbf{u}^T\mathbf{P}\mathbf{u} \tag{A.65}$$

2: Let us minimize the Pontryagin function by equating its derivative with respect to $\mathbf{u}$ to zero. We can do that because $\mathbf{u}$ is unbounded.

$$\mathbf{B}^T \cdot \nabla\mathbf{V} + 2\cdot\mathbf{P}\cdot\mathbf{u} = \mathbf{0} \qquad \Rightarrow \qquad \mathbf{u}^*(\mathbf{x},t) = -\frac{1}{2}\mathbf{P}^{-1}\cdot\mathbf{B}^T\cdot\nabla\mathbf{V}(\mathbf{x},t) \tag{A.66}$$

$\mathbf{P}^{-1}$ exists always, because $\mathbf{P}$ is positive definite.
3: The optimal Pontryagin function is therefore

$$\begin{aligned} H^*(\mathbf{x},\nabla\mathbf{V},t) &= \nabla\mathbf{V}^T\cdot\mathbf{A}\mathbf{x} - \frac{1}{2}\nabla\mathbf{V}^T\cdot\mathbf{B}\cdot\mathbf{P}^{-1}\cdot\mathbf{B}^T\cdot\nabla\mathbf{V} + \mathbf{x}^T\mathbf{Q}\mathbf{x} \\ &\qquad + \frac{1}{4}\nabla\mathbf{V}^T\cdot\mathbf{B}\cdot\mathbf{P}^{-1}\cdot\mathbf{P}\cdot\mathbf{P}^{-1}\cdot\mathbf{B}^T\cdot\nabla\mathbf{V} \\ &= \nabla\mathbf{V}^T\cdot\mathbf{A}\mathbf{x} - \frac{1}{4}\nabla\mathbf{V}^T\cdot\mathbf{B}\cdot\mathbf{P}^{-1}\cdot\mathbf{B}^T\cdot\nabla\mathbf{V} + \mathbf{x}^T\mathbf{Q}\mathbf{x} \end{aligned} \tag{A.67}$$

4: To solve the Hamilton-Jacobi equation

$$\nabla\mathbf{V}^T\cdot\mathbf{A}\cdot\mathbf{x} - \frac{1}{4}\nabla\mathbf{V}^T\cdot\mathbf{B}\cdot\mathbf{P}^{-1}\cdot\mathbf{B}^T\cdot\nabla\mathbf{V} + \mathbf{x}^T\mathbf{Q}\mathbf{x} + \frac{\partial V}{\partial t} = 0 \tag{A.68}$$

we make use of the *Ansatz* $V(\mathbf{x},t) = \mathbf{x}^T\mathbf{R}(t)\mathbf{x}$ which is a quadratic form and always nonnegative. The integrand in the performance index

$$V(\mathbf{x},t) = \int_t^T (\mathbf{x}^T\mathbf{Q}\mathbf{x} + \mathbf{u}^T\mathbf{P}\mathbf{u})d\tau \tag{A.69}$$

is a positive definite function. Hence the matrix $\mathbf{R}(t)$ is a positive definite symmetrical matrix for $t < T$ and for $t = T$ $\mathbf{R}(T) = \mathbf{0}$. Substituting the partial derivatives of the *Ansatz*

$$\nabla\mathbf{V} = 2\mathbf{R}(t)\mathbf{x} \qquad \text{and} \qquad \frac{\partial V}{\partial t} = \mathbf{x}^T\dot{\mathbf{R}}(t)\mathbf{x} \tag{A.70}$$

in the Hamilton-Jacobi equation yields

$$2\mathbf{x}^T\mathbf{R}(t)\mathbf{A}\mathbf{x} - \mathbf{x}^T\mathbf{R}(t)\mathbf{B}\mathbf{P}^{-1}\mathbf{B}^T\mathbf{R}(t)\mathbf{x} + \mathbf{x}^T\mathbf{Q}\mathbf{x} + \mathbf{x}^T\dot{\mathbf{R}}(t)\mathbf{x} = 0 \tag{A.71}$$

or rewriting (A.71) as a quadratic form

$$\mathbf{x}^T\left[2\mathbf{R}(t)\mathbf{A} - \mathbf{R}(t)\mathbf{B}\mathbf{P}^{-1}\mathbf{B}^T\mathbf{R}(t) + \mathbf{Q} + \dot{\mathbf{R}}(t)\right]\mathbf{x} = 0 \tag{A.72}$$

For a quadratic form, we know that

$$\mathbf{x}^T\mathbf{M}\mathbf{x} = \mathbf{x}^T\mathbf{M}_s\mathbf{x} \tag{A.73}$$

where $\mathbf{M}_s = (\frac{\mathbf{M}+\mathbf{M}^T}{2})$ is the symmetrical part of $\mathbf{M}$. Therefore, the Hamilton-Jacobi equation is fulfilled for all states $\mathbf{x}$ if only the symmetrical part of the matrix in the large brackets in (A.72) is zero. All terms of that bracket is symmetrical except for the first one. The symmetrical part of the first term can be calculated as

$$\text{sym}(2\mathbf{R}(t)\mathbf{A}) = 2 \cdot \frac{\mathbf{R}(t)\mathbf{A} + \mathbf{A}^T\mathbf{R}^T(t)}{2} = \mathbf{R}(t)\mathbf{A} + \mathbf{A}^T\mathbf{R}^T(t) \tag{A.74}$$

Since $\mathbf{R}(t)$ is symmetrical, the matrix Riccati equation follows from (A.72)

$$\dot{\mathbf{R}}(t) + \mathbf{Q} - \mathbf{R}(t)\mathbf{B}\mathbf{P}^{-1}\mathbf{B}^T\mathbf{R}(t) + \mathbf{R}(t)\mathbf{A} + \mathbf{A}^T\mathbf{R}(t) = 0 \tag{A.75}$$

with the boundary condition $\mathbf{R}(T) = \mathbf{0}$.

This first order ordinary non-linear differential equation can be solved for $\mathbf{R}(t)$ numerically, whereas the integration must be performed backwards in time, starting with the final state.

5: The optimal closed-loop control variable is hence

$$\mathbf{u}^*(\mathbf{x},t) = -\mathbf{P}^{-1}\mathbf{B}^T\mathbf{R}(t)\mathbf{x} \tag{A.76}$$

This equation is of the form

$$\mathbf{u}^*(\mathbf{x},t) = -\mathbf{K}^T(t)\mathbf{x} \tag{A.77}$$

with $\mathbf{K}(t) = \mathbf{R}(t)\mathbf{B}\mathbf{P}^{-1}$. The elements of the matrix $\mathbf{K}(t)$ are the variable feedback gains of the optimal system. Equation (A.77) gives the structure of the optimal controller as shown in Figure A.7. The components of the control vector $\mathbf{u}$ in an optimally closed-loop controlled system are linear combinations of the state vector $\mathbf{x}$.

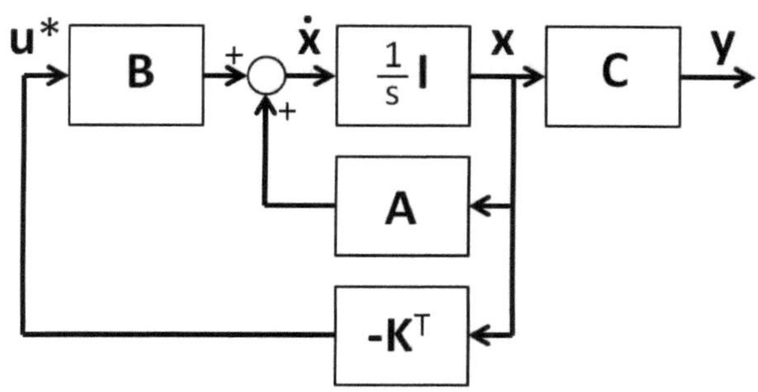

Fig. A.7 Optimal closed-loop controller requires all of the system states to be fed back and not just the output.

This is an important insight, because for the controller to be optimal, *all* states have to be fed back and not only the output as is the case in classical control theory.

Let us again use an example to demonstrate the solution of the problem. We are looking for the optimal controller which minimizes

$$\Lambda = \int_0^T (x^2 + u^2)dt$$

for the system

$$\dot{x} = -x + u$$

In this problem, all the matrices are reduced to scalars such that $\mathbf{A} = a = -1$, $\mathbf{B} = b = 1$, $\mathbf{Q} = q = 1$, $\mathbf{P} = p = 1$. Consequently, the matrix $\mathbf{R}(t)$ is reduced to a scalar time function $r(t)$. The matrix Riccati equation becomes a first order scalar non-linear differential equation.

$$\dot{r}(t) + 1 - r^2(t) - 2r(t) = 0$$

with the boundary condition $r(T) = 0$. Actually, in this special case, there is an analytical solution of the Riccati equation which is given by

$$r(t) = \frac{e^{-\sqrt{2}(t-T)} - e^{\sqrt{2}(t-T)}}{(\sqrt{2}+1)e^{-\sqrt{2}(t-T)} + (\sqrt{2}-1)e^{\sqrt{2}(t-T)}}$$

Using this $r(t)$ the optimal controller is

$$u^*(x,t) = -r(t)x$$

An interesting special case of the matrix Riccati equation (A.75) is obtained when the final time in the performance index (A.69) $T = \infty$. In that case the scalar function $V(\mathbf{x},t)$ must be independent of t. In other words

$$V(\mathbf{x},t_1) = V(\mathbf{x},t_2) \qquad \forall\, t_1 \text{ and } t_2 \tag{A.78}$$

Consequently, the Riccati matrix $\mathbf{R}(t)$ becomes a constant matrix $\mathbf{R}_0$ and the feedback gains become constant and much easier to implement. Also, the Riccati (differential) equation becomes an algebraic equation, because $\mathbf{R}_0$ is constant, hence $\dot{\mathbf{R}}_0 = 0$. In other words (A.75) becomes

$$\mathbf{A}^T\mathbf{R}_0 + \mathbf{R}_0\mathbf{A} - \mathbf{R}_0\mathbf{B}\mathbf{P}^{-1}\mathbf{B}^T\mathbf{R}_0 + \mathbf{Q} = \mathbf{0} \tag{A.79}$$

which is a system of $n(n+1)/2$ non-linear algebraic equations. There are, in general, several solutions for $\mathbf{R}_0$. However, only one of the solutions is a positive define matrix $\mathbf{R}_0$. This solution is actually the same one as one would get by solving (A.75) and letting $T \to \infty$.

Let us now revisit the previous example, this time taking the limit of the integral in the performance index as $T = \infty$. The scalar function $r(t)$ becomes a constant r_0

and the optimal state feedback control becomes $u^*(x,t) = -r_0 x$. We can compute r_0 by calculating $r(t \to \infty)$.

$$\begin{aligned} r_0 &= \lim_{T\to\infty} r(t) \\ &= \lim_{T\to\infty} \frac{e^{\sqrt{2}T} - e^{-\sqrt{2}T}}{(\sqrt{2}+1)e^{\sqrt{2}T} + (\sqrt{2}-1)e^{-\sqrt{2}T}} \\ &= \lim_{T\to\infty} \frac{1 - e^{-2\sqrt{2}T}}{(\sqrt{2}+1) + (\sqrt{2}-1)e^{-2\sqrt{2}T}} \\ &= \frac{1}{\sqrt{2}+1} \\ &= \sqrt{2}-1 \end{aligned}$$

We get the same result by solving the algebraic Riccati equation.

$$-2r_0 - r_0^2 + 1 = 0 \Rightarrow r_{0_{1,2}} = -1 \pm \sqrt{2}$$

Taking the positive solution as $r_0 = \sqrt{2} - 1$ gives the same result.

A.6.2 Optimal Control When Not All States Are Measurable

As pointed out in the previous Subsection, optimal closed-loop control requires *all* of the states of the plant to be fed back with appropriate gains. However, in practical applications not all states are always measurable. How do we proceed in such cases? The solution to this problem lies in estimating the unmeasurable states with an observer and using those estimates as if they were true measurements of the states as illustrated in Figure A.8. However, a necessary condition for this to be possible, is that the plant has to be observable. In other words, it must be possible to determine $\mathbf{x}$ from given $\mathbf{u}$ and measured $\mathbf{y}$. For a linear system this condition is met if the observability matrix $\mathbf{Q}_o$

$$\mathbf{Q}_o = \begin{bmatrix} \mathbf{C}^T \\ \mathbf{C}^T\mathbf{A} \\ \mathbf{C}^T\mathbf{A}^2 \\ \vdots \\ \mathbf{C}^T\mathbf{A}^{n-1} \end{bmatrix} \tag{A.80}$$

has the rank n.

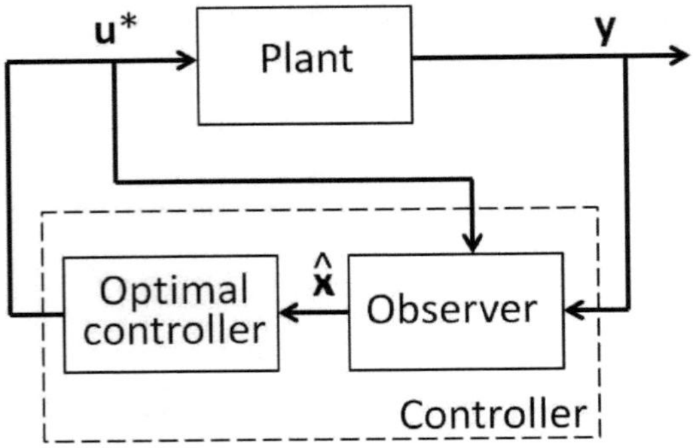

Fig. A.8 When not all system states of the plant are available they must be estimated and used as true states for optimal closed-loop control. The observer is essentially a mathematical model of the plant. It makes use of the plant's input vector $\mathbf{u}^*$ and its output vector $\mathbf{y}$ to compute the estimated system states $\hat{\mathbf{x}}$. However, the initial values of the unmeasurable states are not known. This constitutes an additional problem.

In cases where the plant can be modeled as a linear system, its matrices $\mathbf{A}, \mathbf{B}, \mathbf{C}$ are known. However, the initial conditions of the plant and the model are not necessarily the same, because the initial conditions of the plant states that cannot be measured are not known. Again, using a quadratic performance index for the error between the measured and modeled system outputs leads to a Riccati equation for the calculation of the observer gains if one assumes the structure shown in Figure A.9.

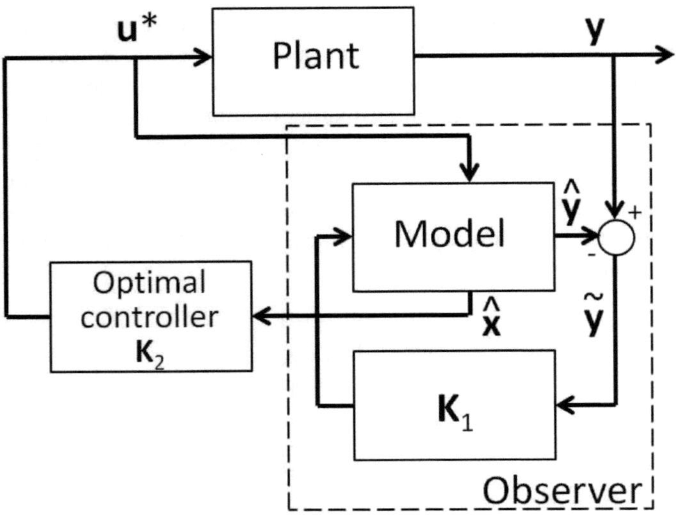

Fig. A.9 The observer which is not privy to all initial states of the plant computes an estimated output vector $\hat{\mathbf{y}}$ which is compared with the measured system output $\mathbf{y}$. The difference $\tilde{\mathbf{y}} = \mathbf{y} - \hat{\mathbf{y}}$ is then fed back to the model after being amplified by $\mathbf{K}_1$. The eigenvalues of this inner observer loop must be far in the left half plane, so that the estimation error $\tilde{\mathbf{y}} \to \mathbf{0}$ quickly.

Hence, there are two optimization problems: 1) Optimize the observer where the observer gains $\mathbf{K}_1$ are computed as the solution of a Riccati equation, 2) Optimize the controller gains $\mathbf{K}_2$ also through the solution of a Riccati equation. Separation principle says that these two problems can be solved independent of each other.

A.6.3 Dynamic Programming

There are many applications where a system given by its state space description

$$\dot{\mathbf{x}} = \mathbf{f}(\mathbf{x}, \mathbf{u}) \tag{A.81}$$

with $\mathbf{u}$ being the decision vector and $\mathbf{x}$ the state vector, is controlled by a controller which generates piecewise constant decision values as shown in Figure A.10. Correspondingly, $\mathbf{x}_1$ is the initial state, $\mathbf{u}_1$ is the first decision, $\mathbf{x}_2 = \psi(\mathbf{x}_1, \mathbf{u}_1)$ is the state after the first decision, $\mathbf{u}_2$ is the second decision, $\mathbf{x}_3 = \psi(\mathbf{x}_2, \mathbf{u}_2)$ is the state after the second decision, ..., $\mathbf{x}_N = \psi(\mathbf{x}_{N-1}, \mathbf{u}_{N-1})$ is the final state in a process with N steps. We are now looking for the decision values which minimize the objective function

$$Z = Z(\mathbf{x}_1, \mathbf{x}_2, \ldots, \mathbf{x}_N, \mathbf{u}_1, \mathbf{u}_2, \ldots, \mathbf{u}_N) \tag{A.82}$$

However, assuming $\mathbf{x}_1$ is known and because all the following states $\mathbf{x}_i, i > 1$ depend on $\mathbf{x}_1$, (A.82) becomes

$$Z = Z(\mathbf{u}_1, \mathbf{u}_2, \ldots, \mathbf{u}_N) \tag{A.83}$$

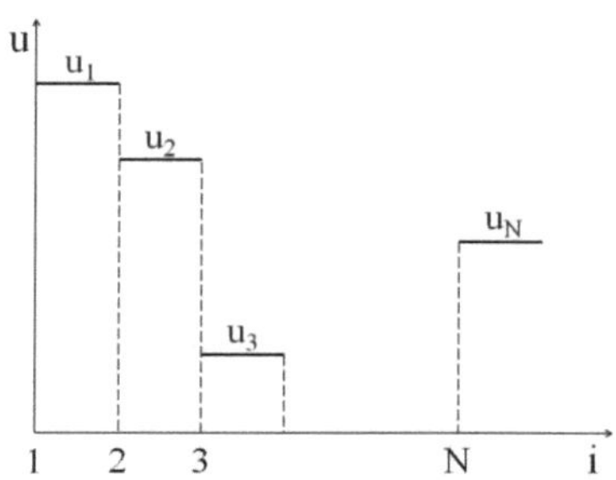

Fig. A.10 Piecewise constant decision values.

Let us further assume that in a process with N steps, the contribution of the remaining $N-k$ steps to the objective function after k decisions depend only on the state of the system after the k-th decision and the decisions after the k-th decision (this is known as the Markov property). With that assumption we can now derive a central result in Optimal Control Theory, namely the celebrated optimality principle:

$$Z = L(\mathbf{x}_1, \mathbf{u}_1) + L(\mathbf{x}_2, \mathbf{u}_2) + \ldots + L(\mathbf{x}_N, \mathbf{u}_N)$$

$$\mathbf{x}_i = \psi(\mathbf{x}_{i-1}, \mathbf{u}_{i-1}) \qquad i = 2, 3, \ldots, N$$

$$\begin{aligned}
Z_1(\mathbf{x}_1) &= \min_{\mathbf{u}_1} L(\mathbf{x}_1, \mathbf{u}_1) \\
Z_2(\mathbf{x}_1) &= \min_{\mathbf{u}_1, \mathbf{u}_2} [L(\mathbf{x}_1, \mathbf{u}_1) + L(\mathbf{x}_2, \mathbf{u}_2)] \\
&\vdots \\
Z_N(\mathbf{x}_1) &= \min_{\mathbf{u}_i} [L(\mathbf{x}_1, \mathbf{u}_1) + L(\mathbf{x}_2, \mathbf{u}_2) + \ldots + L(\mathbf{x}_N, \mathbf{u}_N)] \\
&= \min_{\mathbf{u}_1} \min_{\mathbf{u}_2} \ldots \min_{\mathbf{u}_N} [L(\mathbf{x}_1, \mathbf{u}_1) + L(\mathbf{x}_2, \mathbf{u}_2) + \ldots + L(\mathbf{x}_N, \mathbf{u}_N)] \\
&= \min_{\mathbf{u}_1} [\min_{\mathbf{u}_2} \min_{\mathbf{u}_3} \ldots \min_{\mathbf{u}_N} \{L(\mathbf{x}_2, \mathbf{u}_2) + \ldots + L(\mathbf{x}_N, \mathbf{u}_N)\}] \\
&= \min_{\mathbf{u}_1} [L(\mathbf{x}_1, \mathbf{u}_1) + Z_{N-1}(\mathbf{x}_2)] \\
&= \min_{\mathbf{u}_1} [L(\mathbf{x}_1, \mathbf{u}_1) + Z_{N-1}(\psi(\mathbf{x}_1, \mathbf{u}_1))] \qquad N = 2, 3, \ldots
\end{aligned}$$

This means that an optimal strategy has the property that, independent of the initial state and the first decision, the remaining control decisions build an optimal strategy starting with the state resulting from the first control decision. In other words, each control decision of an optimal strategy is itself optimal.

The main advantages of dynamic programming are

- Absolute minimum can be found.
- No two point boundary condition problem has to be solved.
- Limitations of the control variable simplify the solution of the problem, because the condition $|u| \leq m$ limits the number of possible control decisions at each step.
- The functions $f(.)$ and $L(.)$ do not need to be differentiable (actually these functions can even be given as tables).
- Optimality principle excludes a large number of unnecessary strategies automatically.

The major disadvantage of dynamic programming lies in the number of calculations that are necessary for higher order systems (the so called "curse of dimensioanality"), because the number of operations increase exponentially with the number of variables. Let us again consider an example to see how dynamic programming actually works.

$$\dot{x} = u \qquad x(0) = x_0, \quad x(t_1) = 0$$

$$Z = \int_0^{t_1} (x^2 + 5u^2)dt \rightarrow \text{Minimum}$$

We can find the continuous solution using the maximum principle:

$$H = x^2 + 5u^2 + \lambda u$$

$$\frac{\partial H}{\partial u} = 0 \qquad 10u + \lambda = 0 \quad \Rightarrow \quad u^* = -\frac{\lambda}{10}$$

$$\dot{\lambda} = -\frac{\partial H}{\partial x} = -2x$$

$$\dot{x} = -\frac{\lambda}{10}$$

$$\ddot{x} = \frac{x}{5} \qquad \omega = \sqrt{\frac{1}{5}}$$

$$x(t) = x_0 \frac{\sinh(\omega(t_1 - t))}{\sinh(t_1)}$$

$$\lambda(t) = 10\omega x_0 \frac{\cosh(\omega(t_1 - t))}{\sinh(t_1)}$$

$$u^* = -\omega x_0 \frac{\cosh(\omega(t_1 - t))}{\sinh(t_1)}$$

For the discrete (approximative) solution with $N = 5$ steps and $t_1 = N\Delta$

$$-6 \le u \le 0 \qquad u \text{ integer}$$

$$0 \le x \le 15 \qquad x \text{ integer}$$

The system equation becomes, with $\Delta = 1$

$$x(k+1) = x(k) + u(k)$$

and the objective function at the beginning of each step is approximated by

$$Z = \sum_{k=1}^{N} [x^2(k) + 5u^2(k)]$$

Table A.1 shows the complete calculation with dynamic programming. Figure A.11 shows the resulting optimal control signal.

Table A.1 The first column holds all the possible initial values. The second column holds the values for the objective function $Z1 = x_1^2 + 5u_1^2$. The third column holds the optimal values for u_1. The fourth column holds the optimal value of the objective function Z_2 and so on.

x	Z_1	u_1	Z_2	u_2	Z_3	u_3	Z_4	u_4	Z_5	u_5
0	0	0	0	0	0	0	0	0	0	0
1	1	0	2	0	3	0	4	0	5	0
2	4	0	8	0	11	-1	12	-1	13	-1
3	9	0	18	0/-1	22	-1	25	-1	26	-1
4	16	0	30	-1	39	-1	43	-1	46	-1
5	25	0	46	-1	60	-1	67	-2	70	-2
6	36	0	66	-1	86	-2	95	-2	99	-2
7	49	0	90	-1	115	-2	129	-2	136	-2
8	64	0	118	-1	150	-2	169	-3	176	-3
9	81	0	150	-1/-2	191	-2	212	-3	224	-3
10	100	0	184	-2	235	-3	260	-3	274	-3
11	121	0	222	-2	274	-3	316	-3/-4	330	-4
12	144	0	264	-2	339	-3	374	-4	393	-4
13	169	0	310	-2	398	-3	440	-4	461	-4
14	196	0	360	-2	460	-4	511	-4	533	-5
15	225	0	414	-2/-3	527	-4	585	-5	610	-5

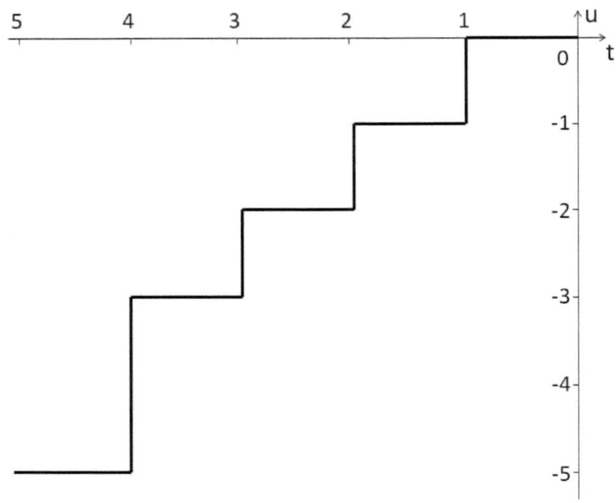

Fig. A.11 Optimal $u(t)$ using the dynamic programming in Table A.1 is calculated as follows: beginning with the last row, we can see that the optimal control value for u_5 is -5. This brings us to row 10 which corresponds to the state $x = 10$. There we see that the optimal control value for u_4 is -3. This brings us to row 7. There we see that the optimal control value for $u_3$is -2. This brings us to row 5. There we see that the optimal control value for u_2 is -1. This brings us to row 4. There we see that the optimal control value for u_1 is 0.

Appendix B
Fuzzy Logic Control

The Fuzzy Logic concept was introduced by Lotfi Zadeh as a way of processing data in 1965 [321]. However, it was not used in control engineering until the computational capability increased in the 1970s. Fuzzy control was then introduced by Ebrahim Mamdani in 1975 [207]. Advances in microcontroller technology allow numerous calculations in a very limited time. This makes fuzzy logic very suitable for control applications. Use of fuzzy logic in control engineering mimics how people take their decisions in a very fast manner. A fuzzy logic system can be thought of as a non-linear mapping of an input data array into an output vector.

Conventional logic deals with exact classifications, such as being black or white; up, down, left or right etc. But if something contains both black and white, it is neither of them or rather both of them to varying extents. A gray object can be defined in terms of black and white when black and white are defined in a fuzzy manner.

Fuzzy systems are especially suitable for two types of applications: (i) Applications where a mathematical model of the plant to be controlled does not exist and cannot be easily derived (complex systems), (ii) Applications where fast but not exact solutions are needed. Therefore, although linear systems can be controlled in a fuzzy way, it is not meaningful to do so. Once an exact linear model is obtained, classical controllers with better performances can be implemented easily. As an example, consider a car. We know that if we press the gas pedal the car accelerates. This limited information may be very beneficial when we want to control its speed. In the next Section, the structure of a fuzzy logic controller and its components are explained and illustrated with examples. In Section B.2, Fuzzy Logic Toolbox of MATLAB environment is introduced. With the information given in these two Sections, a fuzzy logic controller is then designed and simulated using the Fuzzy Logic Toolbox of MATLAB in Section B.3.

B.1 Fuzzy Logic Architecture

Block diagram of a classical Fuzzy Logic System is depicted in Figure B.1.

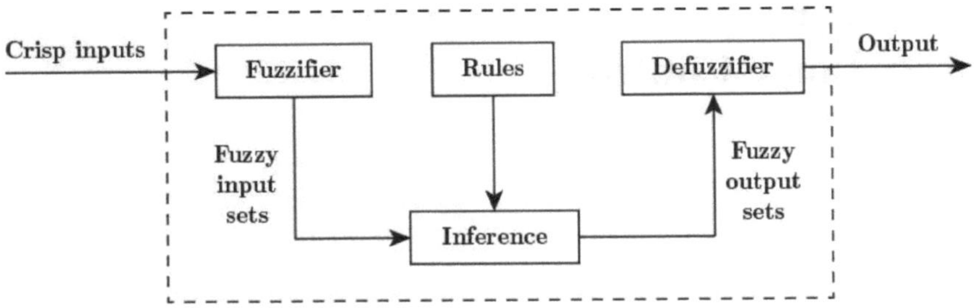

Fig. B.1 Block diagram of a classical Fuzzy Logic System.

Fuzzy Logic Architecture is based on of 5 main concepts: membership functions, fuzzification process, fuzzy rule base, inference engine and the defuzzification process.

B.1.1 Membership Functions

Physical variables are described by linguistic values and expressions in fuzzy logic. As an example, 'very fast' and 'big' can be designated as linguistic terms of the physical values 'speed' and 'size'. An input variable is referred to as a linguistic value when it is described in linguistic terms. Table B.1 shows some engineering terms and their common contextual usages.

Table B.1 Some engineering terms and their contextual usages [225].

Term	Contextual Usage
Error	large, medium, small, almost zero
Frequency	low, high, ultra-high
Resolution	low,high
Noise	very big, big, small
Kurtosis	big, medium, small

Non-fuzzy input values are mapped to fuzzy linguistic terms using membership functions (and vice versa). For ease of understanding, consider the speed control problem of a car, where we want to determine the amount of gas to be given depending on the current speed and the slope of the road. Slope of the road in degrees (crisp value) can be expressed in linguistic terms: positive big slope, positive small slope, zero slope, negative small slope or negative big slope. Similarly, one can define the speed of the car as being fast, normal and slow.

Consider the membership function 'fast' defined as in Figure B.2. According to this membership function, if the speed of the car is 75km/h, we can say speed of the car is '0.5 fast'. Here 0.5 is called 'the truth value' for the membership function.

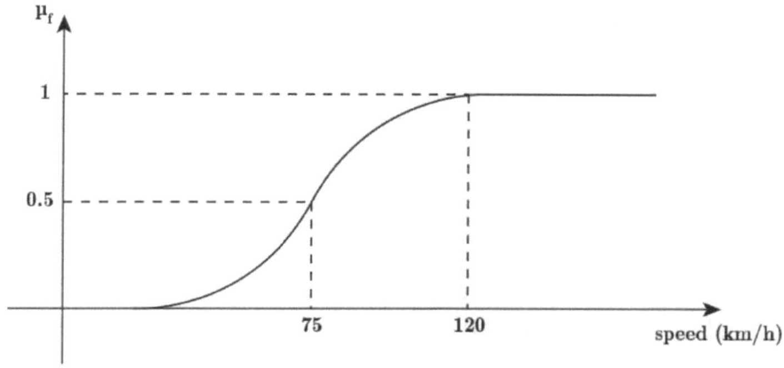

Fig. B.2 The fuzzy membership function for speed of the car: 'fast'.

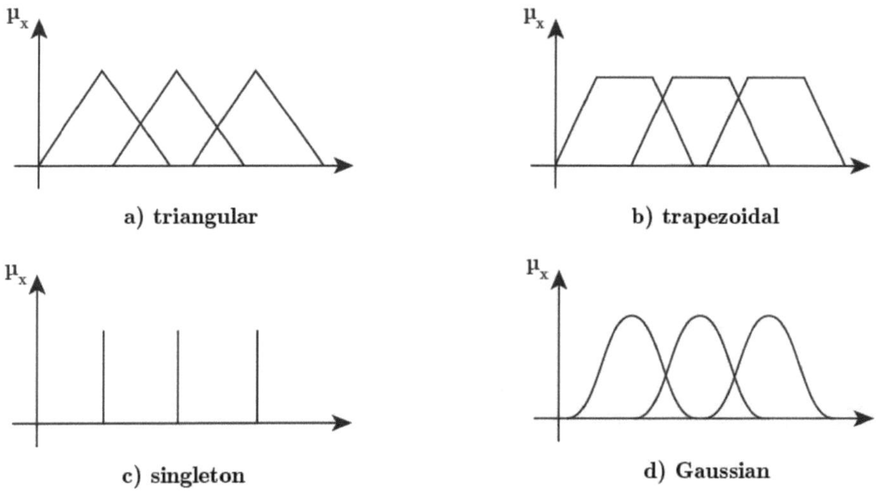

Fig. B.3 Different forms of membership functions.

A truth value is always between 0 and 1. If the speed of the car is greater than 120km/h, we say the speed of the car is 1.0 fast.

There are different forms of membership functions such as triangular, trapezoidal, singleton, Gaussian or piecewise linear just to name the most commonly used ones (Figure B.3).

For a long time period, membership functions were arbitrarily chosen by users. Therefore, two different users might have had quite different membership functions for the same problem. However, there are now optimization procedures that helps the user to choose membership functions systematically. The number of membership functions is chosen by the user as well. A higher number provides higher resolution while a lower number provides less computational complexity. The user decides the number based on her experience in Fuzzy Logic.

B.1.2 Fuzzification

The process of expressing crisp values (*e.g.*, speed) in terms of linguistic variables (*e.g.*, slow, normal and fast) is called 'fuzzification'. Often there are several membership functions for each input and they define the input in linguistic terms with truth values ranging from 0 to 1. For instance, if we define membership functions 'slow', 'normal' and 'fast' for the speed of the car as in Figure B.4, 70km/h can be expressed as 0.1 slow, 1.0 normal and 0.1 fast and 95km/h as 0 slow, 0.25 normal and 0.75 fast. This provides a better explanation, since most people call such speeds as just 'normal'. When the membership functions overlap (as in this example), the inputs are not turned into absolute linguistic terms which makes the Fuzzy Logic Systems more robust.

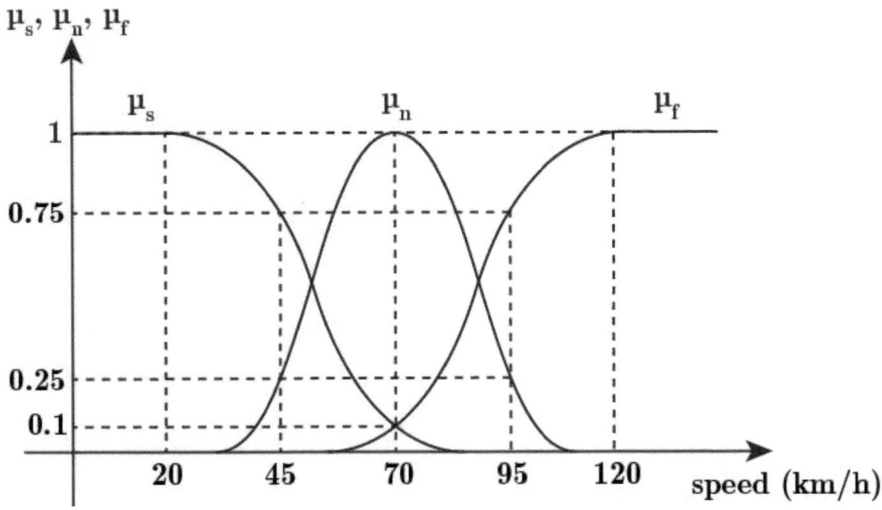

Fig. B.4 Three membership functions to express the speed of a car.

The membership functions denoted by μ_s, μ_n and μ_f classify the speed as slow, normal and fast respectively. Here are some examples to express the speed of a car in terms of fuzzy membership functions in Figure B.4:

- 20 km/h is 1.0 slow.
- 45 km/h is 0.75 slow and 0.25 normal.
- 70 km/h is 0.1 slow, 1.0 normal and 0.1 fast.
- 95 km/h is 0.25 normal and 0.75 fast.
- 120 km/h is 1.0 fast.

An important note about truth values is that they do not have to sum up to one.

B.1.3 Fuzzy Rule Base

The relationship between the inputs and the outputs are defined in the fuzzy rule base which contains empirical knowledge about the operation of a given process. It is a collection of IF–THEN statements where the IF part can be a combination of inputs connected by 'AND'/'OR' operators. In the car speed control problem, assuming that we want to keep the speed of the car 'normal', the amount of gas to be given can be expressed linguistically as following:

- IF current speed is high AND inclination is downhill, THEN break.
- IF current speed is normal AND inclination is zero, THEN give a little gas just enough to compensate for friction.
- IF current speed is normal AND inclination is uphill, THEN give gas.
- IF current speed is low AND inclination is uphill, THEN give lots of gas.
- ...

Here, the required action to be taken has four membership functions (break, give a little gas, give gas and give lots of gas).

B.1.4 Inference Engine

Inference engine uses the fuzzy rule–base to produce fuzzy outputs. There are three subfunctions in the inference engine, namely, aggregation, activation and accumulation.

Aggregation: The 'IF' part of a fuzzy rule may consist of more than one subcondition. In this case, for each fuzzy rule, the degree of accomplishment of each condition is determined by an aggregation algorithm. Different algorithms can be used for the same operator and aggregation algorithm should be chosen properly according to the problem. Some commonly used algorithms are listed in Table B.2.

Table B.2 List of conjunctions, corresponding set operations and operators.

Conjunction	Set operation	Operators
AND	$A \cap B$ (Intersection)	$\mu_{A\wedge B}(x) = min\{\mu_A(x);\mu_B(x)\}$ $\mu_{A\wedge B}(x) = \mu_A(x)\cdot\mu_B(x)$ $\mu_{A\wedge B}(x) = max\{0;\mu_A(x)+\mu_B(x)-1\}$
OR	$A \cup B$ (Union)	$\mu_{A\vee B}(x) = max\{\mu_A(x);\mu_B(x)\}$ $\mu_{A\vee B}(x) = \mu_A(x)+\mu_B(x)-\mu_A(x)\cdot\mu_B(x)$ $\mu_{A\vee B}(x) = min\{1;\mu_A(x)+\mu_B(x)\}$
NOT	$\bar{A}$ (Negation)	$\mu_{\bar{A}}(x) = 1-\mu_A(x)$

Activation: On the basis of the degree of accomplishment of the condition determined in aggregation, the next step is to determine to degree of membership of the conclusion. MIN (the minimum of input values) or PROD (the product of input values) are widely used as activation algorithms.

Accumulation: The last subfunction of the inference engine is used to combine the rules in order to obtain an overall result. Table B.3 shows the most commonly used accumulation methods and their formulas.

Table B.3 Commonly used accumulation methods.

Operation	Formula
Maximum	$Max\{\mu_A(x), \mu_B(x)\}$
Bounded sum	$Min\{1, \mu_A(x) + \mu_B(x)\}$

B.1.5 *Defuzzification*

The output of the inference engine is a fuzzy variable. However, a control element cannot process the fuzzy information directly, hence the input of the system we want to control (i.e., the control signal) must be a crisp value. This crisp value should provide a good representation of the fuzzy control information. The process used to determine a crisp control value from a fuzzy control value is called "defuzzification". There are many defuzzification methods such as COG (center of gravity), LMM (Left Most Maximum), RMM (Right Most Maximum), etc.

Center of Gravity Method: The crisp output calculated by this method is given by the equation

$$U = \frac{\int_{min}^{max} u\mu(u)du}{\int_{min}^{max} \mu(u)du} \tag{B.1}$$

where u is the output variable and μ is the membership function after accumulation [39].

Left Most Maximum Method: Let $\mu(u') = sup(\mu(u))$. Then the output calculated by this method is $U = \min(u')$.

Right Most Maximum Method: The output formula is $U = \max(u')$.

Different methods may lead to different crisp values and therefore different performances. Selecting the best defuzzification method depends on the application and experience of the control engineer.

In the speed control problem, Figure B.5 shows the application of center of gravity method for defuzzification. The output of the fuzzy logic controller is the amount of gas that is required to be applied.

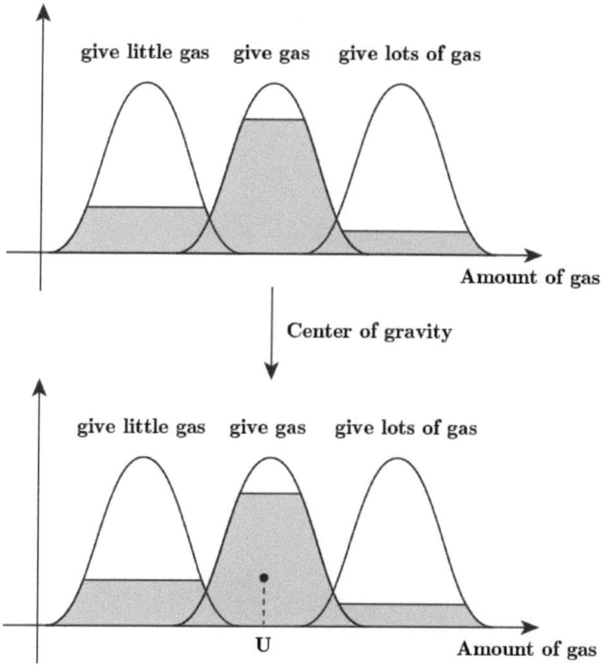

Fig. B.5 Center of gravity method of defuzzification.

B.2 Fuzzy Logic Toolbox of MATLAB

Fuzzy logic control needs repeated realtime calculations. Therefore, computer programs provide convenience. In this section Fuzzy Logic Toolbox of MATLAB will be briefly introduced.

Fuzzy Inference System (FIS) Editor can be initialized from the command window of MATLAB by entering the command "fuzzy". The default fuzzy model has a single input and a single output. The number of inputs and outputs can be changed through the "Edit" menu. The default fuzzy model is mamdani which can be changed to sugeno through the "File" menu. The methods for aggregation and defuzzification can be chosen in this window as well.

The membership functions can be defined by double clicking an input or an output in the Membership Function Editor. There are quite a few function types from which one can select a membership function. The fuzzy rules are entered in the Rule Editor window (Figure B.6) by double clicking the white box in the FIS Editor window.

The relationship between the inputs and the outputs are entered in this window. The relationships are given as a collection of IF–THEN statements. Not only new rules can be entered in this window, but also existing rules can be modified or deleted.

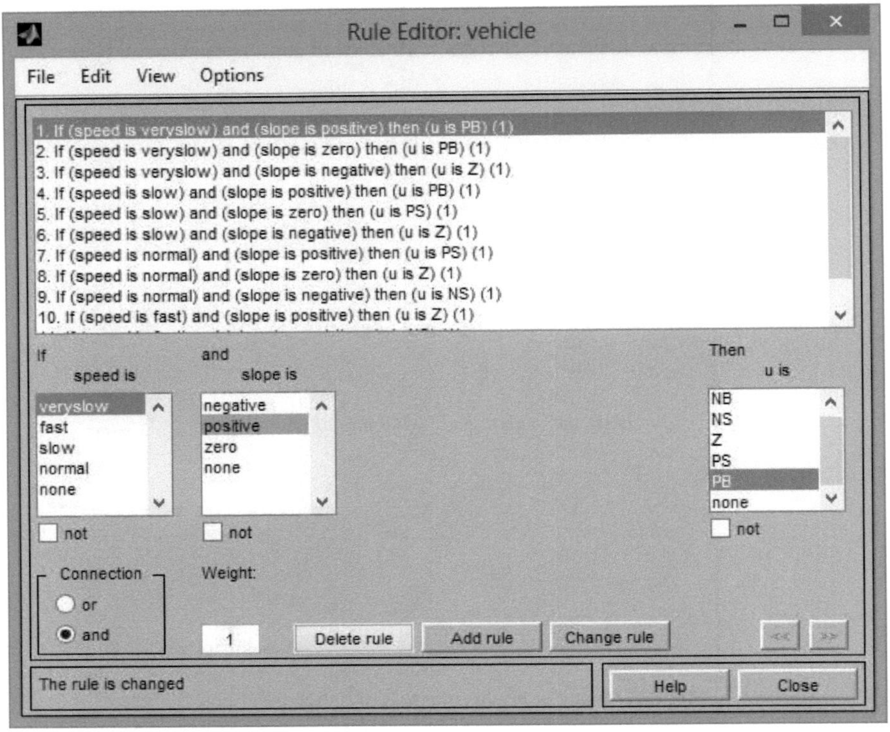

Fig. B.6 Rule Editor of Fuzzy Logic Toolbox.

After the model is constructed properly, it must be exported to a '.fis' file or to the MATLAB workspace so as to be used in Simulink. This can be done from the File > Export menu.

B.3 Application: Fuzzy Control of a System

In this section, we are going to design a fuzzy logic controller for the speed control problem of a vehicle. Control of linear systems is relatively simple and can be solved in numerous ways, i.e., using the optimal control techniques discussed in Appendix A. In cases where the system is too complex to be modeled as a linear system, the use of intelligent control techniques are indicated. The relationship between the gas/break amount and the speed of the car is strictly non-linear and a fuzzy logic controller may work fine for such a problem.

A fuzzy logic controller can be designed in 5 steps:

1. Construct the Simulink Model of the Feedback Control System

The speed of the vehicle may change as the slope of the road and the gas/break amount changes. Hence the vehicle model should have two inputs: the gas/break amount and the slope of the road. Before designing a controller, we have to define the inputs and the output of the fuzzy logic controller. Here, the fuzzy logic controller will decide on the gas/break amount by looking at the current speed and the slope of the road. Therefore it has to have two inputs and a single output.

The feedback control model can be constructed as shown in Figure B.7.

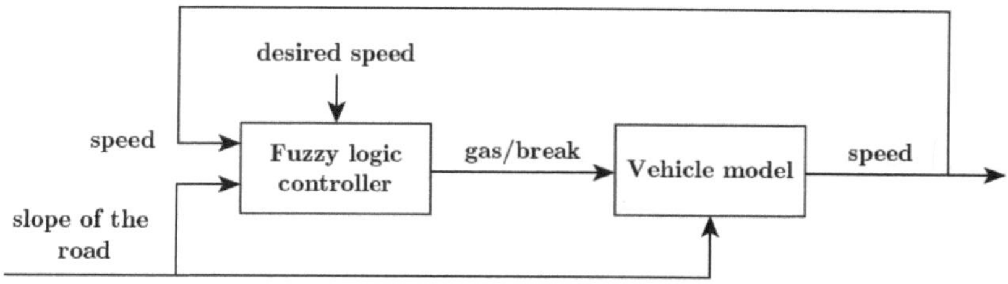

Fig. B.7 Fuzzy logic control of a system.

2. Define the Fuzzy Controller Structure

First of all, we have to open the FIS Editor by typing "fuzzy" in the command window of MATLAB. By default, the fuzzy controller has a single input and a single output. Therefore we must add the second input using the "Edit" menu. Let the names of these inputs be 'speed' and 'slope', respectively. Since we need a single output for the controller we do not have to add another output. The FIS Editor should look like Figure B.8.

3. Define the Fuzzy Membership Functions

Fuzzy sets representing the meaning of linguistic values are then to be defined for the inputs and the output. The choice of membership functions is an important task in the fuzzy controller design process. The shapes and the ranges of the membership functions are crucial and different for every application. The shapes, ranges and the number of membership functions are to be determined by the designer. In this problem we can choose four membership functions for 'speed' (very slow, slow, normal and fast), and three membership functions for 'slope' (negative, zero and positive) as depicted in Figures B.9 and B.10.

The output is the amount of gas/break and we introduce five membership functions (negative big, negative small, zero, positive small and positive big). Here, positive and negative correspond to gas and break, respectively. Figure B.11 illustrates our choice of membership functions.

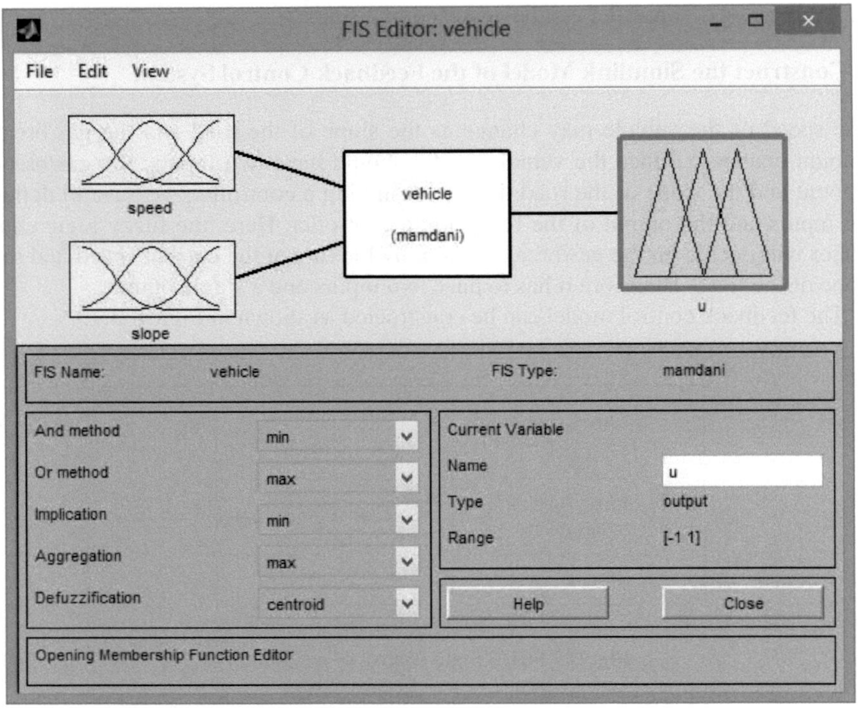

Fig. B.8 Fuzzy controller with two inputs and single output.

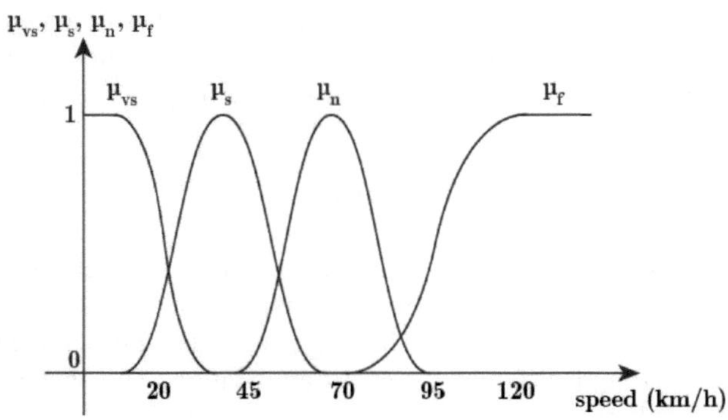

Fig. B.9 Membership functions for the 'speed' input.

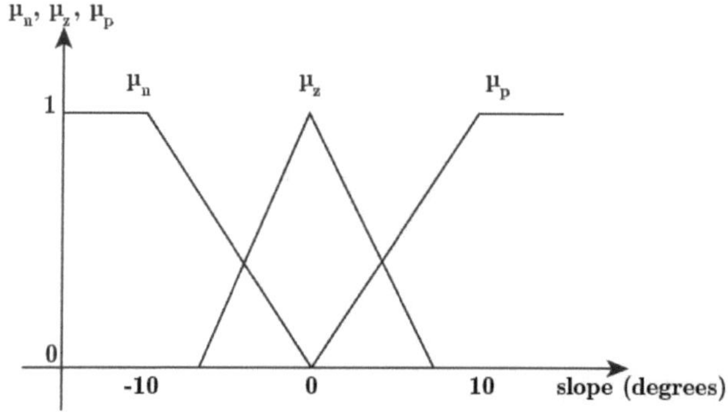

Fig. B.10 Membership functions for the 'slope' input.

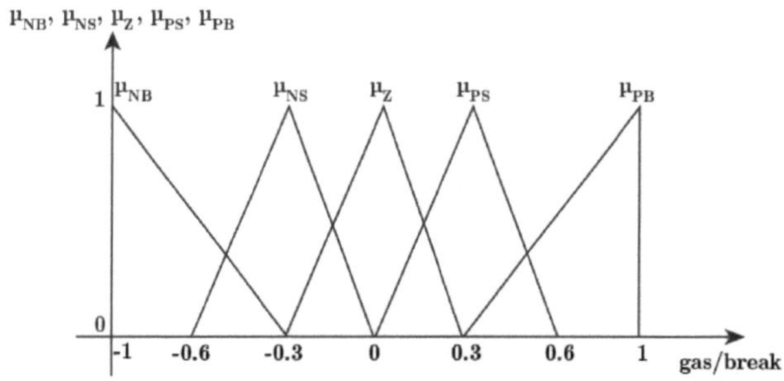

Fig. B.11 Membership functions for the output.

4. Define the Fuzzy Rules

The relationship between the inputs and the output must be stated properly. Since we have one input with three and one input with four membership functions, we have to define 4 x 3 = 12 rules for their combinations. The rules can be determined as given in Table B.4 and are to be entered in the Rule editor of Fuzzy Logic Toolbox of MATLAB as shown in Figure B.6.

5. Simulate

When the all the fuzzy rules are defined using the Rule editor, the fuzzy controller design is complete. It can now be exported to a file or to the workspace of MATLAB. The next step is associating the controller with the Simulink model. By double clicking the Fuzzy Logic Controller block in the Simulink model, we can enter the name of the fuzzy logic controller we designed (enter the name of the variable exported to the workspace). The next step is to generate a random looking slope of the road. Simulation can be started from the menu Simulation > Start. When the vehicle

Table B.4 Rule table for the fuzzy output.

		slope of the road		
		negative	zero	positive
speed	very slow	PB	PB	Z
	slow	PB	PS	Z
	normal	PS	Z	NS
	fast	Z	NS	NB

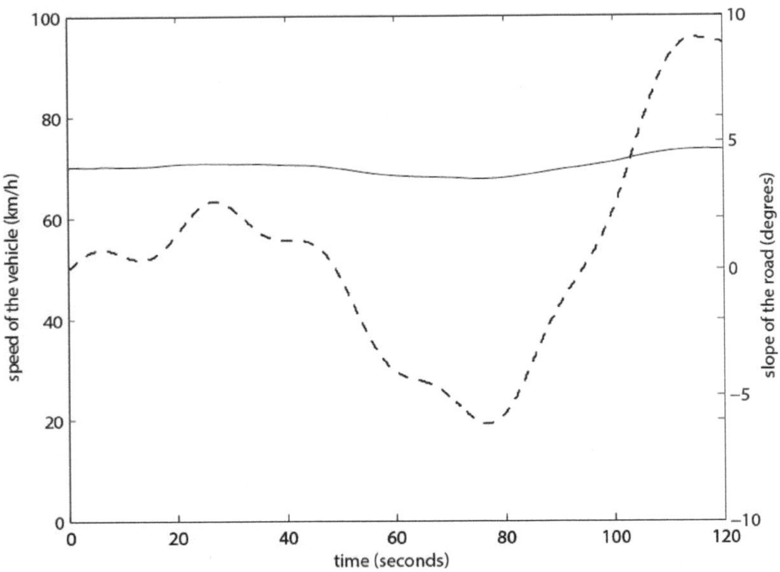

Fig. B.12 Speed of the vehicle (continuous line, LHS) and the slope of the road (dashed line, RHS).

is controlled with the fuzzy logic controller, its speed and the slope of the road are depicted together in Figure B.12. We can see that for the set speed of 75km/h and the initial speed of 75km/h, the controller kicks in immediately and maintains the set speed fairly well in spite of the changing slope of the road.

References

[1] Abdelnaby, S., Hassanelin, M.: Neuropsychiatry Manifestations of Chronic Manganese Poisoning. J. Neurol. Neurosurg. Psychiat. 28 (1965)

[2] Ackerman, E., Rosevear, J., McGuckin, W.: A Mathematical Model of Glucose-tolerance Test. Physics in Medicine and Biology 9 (1964)

[3] Albani, C., Asper, R., Hacısalihzade, S.S., Baumgartner, G.: Individual Levodopa Therapy in Parkinson's Disease. In: Proc. 8th Int. Symp. on Parkinson's Disease. Raven Press, New York (1986)

[4] Albisser, A.M.: An Artificial Pancreas? British Medical Journal 4 (1974)

[5] Albisser, A.M., Leibel, B.S., Ewart, T.G., Davidovac, Z., Botz, C.K., Zingg, W.: Clinical Control of Diabetes by the Artificial Pancreas. Diabetes 23 (1974)

[6] Albisser, A.M., Leibel, B.S.: Diabetes Control and the Artificial Pancreas. Journal of Medical Systems 1 (1977)

[7] Alexander, C.: Quantitative Methods in Finance. John Wiley and Sons, Chichester (2010)

[8] Alexander, C.: Market Models. John Wiley and Sons, Chichester (2011)

[9] Alexandre, N.: Programme de Simulation de la Pharmacocinétique et de la Pharmacodynamique des Anesthésiques par Inhalation. MD thesis, University of Bern (1995)

[10] Allen, G.I., Tsukahara, N.: Cerebocerebellar Communication Systems. Physiol. Rev. 54 (1974)

[11] Allgöwer, F.: Näherungsweise Ein-/Ausgangs-Linearisierung nichthnearer Systeme. PhD thesis, Institut für Systemdynamik und Regelungstechnik der Universität Stuttgart (1996)

[12] Ferrante do Amaral, C.E., Wolf, B.: Current Development in Non-invasive Glucose Monitoring. Medical Engineering & Physics 30 (2008)

[13] Anderson, O.D.: Time Series Analysis and Forecasting: The Box-Jenkins Approach. Butterworths, London (1977)

[14] Andrews, J.M., Terry, R.D., Spataro, J.: Striatonigral Degeneration: Clinical-Pathological Correlations Response to Stereotaxic Surgery. Arch. Neurol. 23 (1970)

[15] April, R.S.: Observations on Parkinson Tremor in All-Night Sleep. Neurology 16 (1966)

[16] Ariens, E.J., Simonis, A.M.: A Molecular Basis for Drug Action. Journal of Pharmacy and Pharmacology 27 (1964)

[17] Asimov, I.: Foundation. Gnome Press Publishers, New York (1951)

[18] Atkinson, A.C., Donev, A.N., Tobias, R.: Optimum Experimental Designs, With SAS. Oxford University Press (2007)

[19] Evaluation Report: ACCU-CHEK® Aviva System. Roche Diagnostics (2007)

[20] ACCU-CHEK® Aviva Combo, Blood Glucose Meter, Getting Started Guide, Roche Diagnostics, Mannheim (2009)

[21] ACCU-CHEK® Aviva Combo, Blood Glucose Meter, Standard Owner's Booklet, Roche Diagnostics, Mannheim (2009)

[22] ACCU-CHEK® Aviva Combo, Blood Glucose Meter, Advanced Owner's Booklet, Roche Diagnostics, Mannheim (2009)

[23] Bachran, R., Beyer, P., Klinkert, C., Heidtmann, B., Rosenbauer, J., Holl, R.W.: Basal Rates and Circadian Profiles in Continuous Subcutaneous Insulin Infusion (CSII) Differ for Preschool Children, Prepubertal Children, Adolescents and Young Adults. Pediatric Diabetes 13 (2012)

[24] Barbeau, A., Murphy, C.F., Sourkes, T.L.: Excretion of Dopamine in Diseases of Basal Ganglia. Science 133 (1961)

[25] Barbeau, A., Pourcher, E.: New Data on the Genetics of Parkinson's Disease. Can. J. Neurol. Sci. 9 (1982)

[26] Barbeau, A., Roy, M.: Genetic Susceptibility, Environmental Factors and Parkinson's Disease. In: Proc. 8th Int. Symp. on Parkinson's Disease, Raven Press, New York (1986)

[27] Anderson, B.D.O., Moore, J.B.: Linear Quadratic Methods. Dover Books on Engineering (1990)

[28] Bedingham, W., Tatton, W.G.: Kinesthetic "Encoding" by Motor Cortical Neurons in the Awake Cat. Soc. Nerosci. Abstr. 9 (1983)

[29] Bellman, R.E.: Dynamic Programming. Princeton University Press, Princeton (1957)

[30] Bellman, R.: Topics in Pharmacokinetics III: Repeated Dosage and Impulse Control. Mathematical Biosciences 12 (1971)

[31] Berardelli, A., Sabra, A.F., Hallet, M.: Physiological Mechanisms of Rigidity in Parkinson's Disease. J. Neurol. Neurosurg. and Psychiat. 46 (1983)

[32] Berger, A.J., Itzkan, I., Feld, M.S.: Feasibility of Measuring Blood Glucose Concentration by Near-Infrared Raman Spectroscopy. Spectrochimica Acta Part A: Molecular and Biomolecular Spectroscopy 53 (1997)

[33] Bergman, R.N., Ider, Y.Z., Bowden, C.R., Cobelli, C.: Quantitative Estimation of Insulin Sensitivity. American Journal of Physiology 236 (1979)

[34] Bernheimer, H., Birkmayer, W., Hornykiewicz, O.: Zur Biochemie des Parkinson-Syndroms des Menschen. Wien. Klin. Wschr. 75 (1963)

[35] Bernheimer, H., Birkmayer, W., Hornykiewicz, O., Jellinger, K., Seitelberger, F.: Brain Dopamine and the Syndromes of Parkinson and Huntington: Clinical, Morphological and Neurochemical Correlations. J. Neurol. Sci. 20 (1973)

[36] Bertler, A., Rosengren, E.: Occurrence and Distribution of Cathecolamines in Brain. Acta Physiol. Scand. 47 (1959)

[37] Bertsekeas, D.P., Shreve, S.E.: Stochastic Optimal Control: The Discrete-Time Case. Athena Scientific (2007)

[38] Billenkamp, K.: Untersuchungen zur Psychologie des Parkinsonisms. Arch. Psychiat. Nervenkr. 198 (1959)

[39] Bingchuan, Y., Herbert, J.: Fuzzy CARA – A Fuzzy-Based Context Reasoning System For Pervasive Healthcare. Procedia Computer Science 10 (2012)

[40] Birkmayer, W.: Experimentelle Ergebnisse über die Kombinationsbehandlung des Parkinson-Syndroms mit L-DOPA und einem Decarboxylasehemmer. Wien. Klin. Wschr. 81 (1969)

[41] Birkmayer, W.: Personal communication (1984)

[42] Birkmayer, W., Hornykiewicz, O.: Der L-Dioxyphenylalanin-Effekt bei der Parkinson-Akinesie. Wien. Klin. Wschr. 73 (1961)
[43] Birkmayer, W., Neumayer, E.: Die Behandlung der DOPA-Psychosen mit L-Tryptophan. Nervenarzt 43 (1972)
[44] Birkmayer, W., Riederer, P., Youdim, M.B.H., Linauer, W.: Potentiation of Anti-Akinetic Effect after L-DOPA Treatment by an Inhibitor of MAO-B, Deprenil. J. Neural. Transm. 36 (1975)
[45] Birkmayer, W., Riederer, P.: Die Parkinson-Krankheit. Springer, Wien-New York (1980)
[46] Bjorklund, A., Dunnett, S.B., Brundin, P., Stoessl, A.J., Freed, C.R., Breeze, R.E.: Neural Transplantation for the Treatment of Parkinson's Disease. Lancet Neurology 2(7) (2003)
[47] Blaschko, H., Muscholl, E.: Cathecolamines (Handb. Exp. Pharm. XXXIII). Springer, Heidelberg (1972)
[48] Bliss, M.: The Discovery of Insulin. Blackwell (1991)
[49] Bolie, V.W.: Coefficients of Normal Blood Glucose Regulation. Journal of Applied Physiology 16 (1961)
[50] Bourne, D.W.A., Triggs, E.J., Eadie, M.J.: Pharmacokinetics for the Non-Mathematical. MTP Press, Boston (1986)
[51] Box, G.E.P., Jenkins, G.M.: Time Series Analysis Forecasting and Control. Holden-Day, San Fransisco (1970)
[52] Brooks, V.B.: Control of Intruded Limb Movements by the Lateral and Intermediate Cerebellum. In: Asanuma, H., Wilson, V.J. (eds.) Integration in the Nervous System, Tokyo, Igaku-Shoin (1979)
[53] Brundin, P., Barker, R.A., Parmar, M.: Neural Grafting in Parkinson's Disease: Problems and Possibilities. Progress in Brain Research 184 (2010)
[54] Bruulsema, J.T., Hayward, J.E., Farrell, T.J., Patterson, M.S., Heinemann, L., Berger, M., Koschinsky, T., Sandahl-Christiansen, J., Orskov, H., Essenpreis, M., Schmelzeisen-Redeker, G., Böcker, D.: Correlation Between Blood Glucose Concentration in Diabetics and Noninvasively Measured Tissue Optical Scattering Coefficient. Optics Letters 22 (1997)
[55] Budinova, G., Salva, J., Volka, K.: Application of Molecular Spectroscopy in the Mid-Infrared Region to the Determination of Glucose and Cholesterol in Whole Blood and in Blood Serum. Applied Spectroscopy 51 (1997)
[56] Buell, J., Jellife, R., Kalaba, R., Sridhar, R.: Modern Control Theory and Optimal Drug Regimens. Mathematical Biosciences 5 (1969)
[57] Burns, S.R., Chiueh, C.C., Markey, S.P., Ebert, M.H., Jacobowitz, D.M., Kopin, L.: A Primate Model of Parkinsonism: Selective Destruction of Dopaminergic Neurons in the Pars Compacta of the Substantia Nigra by MPTP. Proc. Nat. Acad. Sci. USA 80 (1983)
[58] Buzzard, E.F., Greenfield, J.G.: Lethargic Encephalitis, its Sequelae and Morbid Anatomy. Brain 42 (1919)
[59] Caduff, A., Hirt, E., Feldman, Y., Ali, Z., Heinemann, L.: First Human Experiments with a Novel Non-invasive, Non-optical Continuous Glucose Monitoring System. Biosensors and Bioelectronics 19 (2003)
[60] Caduff, A., Dewarrat, F., Talary, M., Stalder, G., Heinemann, L., Yu, F.: Non-invasive Glucose Monitoring in Patients with Diabetes: A Novel System Based on Impedance Spectroscopy. Biosensors and Bioelectronics 22 (2006)
[61] Calne, D.B.: Dopaminergic Agonists in Parkinsonism. In: Birkmayer, W., Hornykiewicz, O. (eds.) Advances in Parkinsonism. Editiones Roche, Basel (1976)

[62] Calne, D.B., Teychenne, P.F., Leigh, P.N., Bamji, A.N., Greenacre, J.K.: Treatment of Parkinsonism with Bromocriptine. Lancet II (1974)
[63] Calne, D.B., Langston, J.W.: Etiology of Parkinson's Disease. Lancet 2 (1983)
[64] Calne, D.B., Langston, J.W., Martin, W.R.W., Stoessl, A.J., Ruth, T.J., Adam, M.J., Pate, B.D., Schulzer, M.: Positron Emission Tomography After MPTP: Observations Relating to the Cause of Parkinson's Disease. Nature 317 (1985)
[65] Camou, S., Ueno, Y., Tamechika, E.: Towards Non-invasive and Continuous Blood Sugar Sensor: Detection of Aqueous Glucose Based on CW-photoacoustic Protocol. In: Proceedings of IEEE Sensors 2010, Daegu, South Korea (2010)
[66] Campos-Delgado, D.U., Hernandez-Ordonez, M., Femat, R., Gordillo-Moscoso, A.: Fuzzy-Based Controller for Glucose Regulation in Type-1 Diabetic Patients by Subcutaneous Route. IEEE Transactions on Biomedical Engineering, BME-53 (2006)
[67] Candas, B., Radziuk, J.: An Adaptive Plasma Glucose Controller Based on a Nonlinear Insulin/Glucose Model. IEEE Transactions on Biomedical Engineering BME-41 (1994)
[68] Carson, E.R., Cobelli, C., Finkelstem, L.: The Mathematical Modeling of Metabolic and Endocrine Systems. John Wiley and Sons, New York (1983)
[69] Carlsson, A., Lindquist, M., Magnusson, T., Waldeck, B.: On the Presence of 3-hydroxytyramine in Brain. Science 137 (1958)
[70] Cellier, F.E., Greifeneder, J.: Continuous System Modeling. Springer (1991)
[71] Cellier, F.E., Kofman, E.: Continuous System Simulation. Springer (2006)
[72] Charcot, J.M.: Leçons sur les Maladies du Système Nerveux Faites a la Salpêtrière. Recueillies et publiees par Bourneville A. Delahaye et Lecrosnier, Paris (1892)
[73] Chang, X., Jørgensen, A.M., Bardrum, P., Led, J.J.: Solution Structures of the R_6 Human Insulin Hexamer. Biochemistry 36 (1997)
[74] Chee, F., Fernando, T.: Closed-Loop Control of Blood Glucose. Springer, Heidelberg (2007)
[75] Christison, G.B., MacKenzie, H.A.: Laser Photoacoustic Determination of Physiological Glucose Concentrations in Human Whole Blood. Medical and Biological Engineering and Computing 31 (1993)
[76] Clarke, W.L., Cox, D., Gonder-Frederick, L.A., Carter, W., Pohl, S.L.: Evaluating Clinical Accuracy of Systems for Self-Monitoring of Blood Glucose. Diabetes Care 10 (1987)
[77] Cobelli, C., Federspil, G., Pacini, G., Salvan, A., Scandellari, C.: An Integrated Mathematical Model of the Dynamics of Blood Glucose and its Hormonal Control. Mathematical Bioscience 58 (1982)
[78] Cobelli, C., Mari, A.: Validation of Mathematical Models of Complex Endocrine-Metabolic Systems. A Case Study on a Model of Glucose Regulation. Medical and Biological Engineering and Computing 21 (1983)
[79] Cobelli, C., Carson, E.R.: Introduction to Modeling in Physiology and Medicine. Elsevier, London (2008)
[80] Cobelli, C., Renard, E., Kovatchev, B.: Artificial Pancreas: Past, Present, Future. Diabetes 60 (2011)
[81] Collins, V.: Physiologic and Pharmacologic Bases of Anesthesia. Williams & Wilkins, Philadelphia (1996)
[82] Constas, C.: The Effects of Adrenaline, Noradrenaline and Isoprenaline on Parkinsonian Tremor. J. Neurol. Neurosurg. Psychiat. 25 (1962)
[83] Cooper, I.S., Riklan, N., Davidson, N.W., Gorek, E.A.: Chemopallidectomy and Chemothalemectomy for Parkinsonism. Geriatrics 13 (1958)

[84] Coté, G.L., Fox, M.D., Northrop, R.B.: Noninvasive Optical Polarimetric Glucose Sensing Using a True Phase Measurement Technique. IEEE Transactions on Biomedical Engineering, BME 39 (1992)
[85] Cotzias, G.C., van Woert, M.H., Schiffer, L.M.: Aromatic Amino Acids and Modification of Parkinsonism. New England Journal of Medicine 276 (1967)
[86] Crassidis, J.L., Junkins, J.L.: Optimal Estimation of Dynamic Systems. CRC Press, Boca Raton (2004)
[87] D'Amoto, R.J., Lipman, Z.P., Snyder, S.H.: Selectivity of the Parkinsonian Neurotoxin MPTP: High Affinity Binding of the Active Metabolite MPP* to Neuromelanin. Science 233 (1986)
[88] Davis, G.C., Williams, A.C., Markey, S.P., Ebert, M.H., Caine, E.D., Reichert, C.M., Kopin, L.J.: Chronic Parkinsonism Secondary to Intravenous Injection of Mepredine Analogues. Psychiatry Res. 1 (1979)
[89] Dazzi, D., Taddei, F., Gavarini, A., Uggeri, E., Negro, R., Pezzarossa, A.: The Control of Blood Glucose in the Critical Diabetic Patient – A Neuro-Fuzzy Method. Journal of Diabetes and Its Complications 15 (2001)
[90] De Block, C., Manuel-y-Keenoy, B., Van Gaal, L., Rogiers, P.: Intensive Insulin Therapy in the Intensive Care Unit: Assessment by Continuous Glucose Monitoring. Diabetes Care 29 (2006)
[91] DeFigueiredo, R.J.P., Hu, C.L.: Waveform Feature Extraction Based on Tauberian Approximation. IEEE Transactions on Pattern Analysis and Machine Intelligence PAMI-4 (1982)
[92] DeJong, R.N.: Parkinson's Disease: Statistics. J. Neurosurg. 24(suppl.) (1966)
[93] Deiss, D., Bolinder, J., Riveline, J.P., Battelino, T., Bosi, E., Tubiana-Rufi, N.: Improved Glycemic Control in Poorly Controlled Patients with Type 1 Diabetes Using Real-Time Continuous Glucose Monitoring. Diabetes Care 29 (2006)
[94] Derighetti, M.P.: Multivariable Fuzzy-Regelung in der Anästhesie. Post Graduate Thesis, Automatic Control Laboratory, Swiss Federal Institute of Technology (ETH), Zurich (1993)
[95] Derighetti, M.P.: Feedback Control in Anaesthesia. Doctoral Dissertation 12898 at the Swiss Federal Institute of Technology (ETH), Zurich (1999)
[96] Dixit, N., Bali, V., Baboota, S., Ahuja, A., Ali, J.: Iontophoresis – An Approach for Controlled Drug Delivery: A Review. Current Drug Delivery 4 (2007)
[97] Domschke, A., March, W.F., Kabilan, S., Lowe, C.: Initial Clinical Testing of a Holographic Non-Invasive Contact Lens Glucose Sensor. Diabetes Technology & Therapeutics 8 (2006)
[98] Duvoisin, R.: Is Parkinson's Disease Acquired or Inherited? Can. J. Neurol. Sci. 11 (1984)
[99] Ehringer, H., Hornykiewicz, O.: Verteilung von Noradrenalin und Dopamin im Gehirn des Menschen und ihr Verhalten bei Erkrankungen des extrapyramidal Systems. Wien. Klin. Wschr. 72 (1960)
[100] Einstein, A.: Über die von der molekularkinetischen Theorie der Wärme geforderte Bewegung von in ruhenden Flüssigkeiten suspendierten Teilchen. Annalen der Physik 17 (1905)
[101] Enejder, A.M.K., Scecina, T.G., Hunter, M., Shih, W.-C., Feld, M.S., Oh, J., Sasic, S., Horowitz, G.L.: Raman Spectroscopy for Noninvasive Glucose Measurements. Journal of Biomedical Optics 10 (2005)
[102] Esenaliev, R.O., Larin, K.V., Larina, I.V., Motamedi, M.: Noninvasive Monitoring of Glucose Concentration with Optical Coherence Tomography. Optical Letters 26 (2001)

[103] Evarts, E.V., Teravainen, H., Beuchert, D.E., Calne, D.B.: Pathophysiology of Motor Performance in Parkinson's Disease. In: Fuxe, K., Calne, D.B. (eds.) Dopaminergic Ergot Derivatives and Motor Functions, Pergamon Press, London (1979)

[104] ACCU-CHEK® Aviva Expert, Blutzuckermessgerät, Standardgebrauchsanweisung, Roche Diagnostics, Mannheim (2009)

[105] Fedorov, V.V., Hackl, P.: Model-Oriented Design of Experiments. Springer, New York (1997)

[106] Fischer, U., Schenk, W., Salzsieder, E., Albrecht, G., Abel, P., Freyse, E.J.: Does Physiological Blood Glucose Control Require an Adaptive Strategy? IEEE Transactions on Biomedical Engineering BME-34 (1987)

[107] Fischer, U.: Continuous in vivo Monitoring in Diabetes: The Subcutaneous Glucose Concentration. Acta Anaesthesiologica Scandinavica 39(suppl. s104) (March 1995)

[108] Fisher, M.E., Teo, K.L.: Optimal Insulin Infusion Resulting from a Mathematical Model of Blood Glucose Dynamics. IEEE Transactions on Biomedical Engineering BME-36 (1989)

[109] Fisher, M.E.: A Semiclosed-Loop Algorithm for the Control of Blood Glucose Levels in Diabetics. IEEE Transactions on Biomedical Engineering BME-38 (1991)

[110] Fishman, A.P., Macklem, P.T., Mead, J., Geiger, S.R. (eds.): Handbook of Physiology: The Respiratory System. American Physiological Society, Bethesda (1986)

[111] Fleming, W.M., Rishel, R.W.: Deterministic and Stochastic Otimal Control. Springer, New York (1975)

[112] Flowers, K.: Ballistic and Corrective Movements in an Aiming Task: Intention Tremor and Parkinsonian Movement Disorders Compared. Neurology 25 (1975)

[113] Flowers, K.: Visual 'Closed-loop' and 'Open-loop' Characteristics of Voluntary Movement in Patients with Parkinsonism and Intention Tremor. Brain 99 (1976)

[114] Flowers, K.: Some Frequency Response Characteristics of Parkinsonism in Pursuit Tracking. Brain 101 (1978)

[115] Freckmann, G., Baumstark, A., Jendrike, N., Zschornack, E., Kocher, S., Tshiananga, J., Heister, F., Haug, C.: System Accuracy Evaluation of 27 Blood Glucose Monitoring Systems According to DIN EN ISO 15197. Diabetes Technology & Therapeutics 12 (2010)

[116] Freed, C.R., Greene, P.E., Breeze, R.E., Tsai, W.Y., DuMouchel, W., Kao, R.: Transplantation of Embryonic Dopamine Neurons for Severe Parkinson's Disease. New England Journal of Medicine 344(10) (2001)

[117] Gajdusek, D.C., Salazar, A.M.: Amyotrophic Lateral Sclerosis and Parkinsonian Syndromes in High Incidence among the Auyu and Jakai People of West New Guinea. Neurology 32 (1982)

[118] Ganong, W.F.: Review of Medical Physiology. Lange Medical Publications, Los Altos (1977)

[119] Gardiner, D.J., Graves, P.R. (eds.): Practical Raman Spectroscopy. Springer, New York (1989)

[120] Geddes, C.D., Lakowicz, J.R.: Glucose Sensing. Springer, New York (2006)

[121] Ghosn, M.G., Tuchin, V.V., Larin, K.V.: Depth-resolved Monitoring of Glucose Diffusion in Tissues by Using Optical Coherence Tomography 31 (2006)

[122] Gibaldi, M., Perrier, D.: Pharmacokinetics. Marcel Dekker, New York (1982)

[123] Gin, H., Renard, E., Melki, V., Boivin, S., Schaepelynck-Bélicar, P., Guerci, B., Selam, J.L., Brun, J.M., Riveline, J.P., Estour, B., Catargi, B., The EVADIAC Study Group: Combined Improvements in Implantable Pump Technology and Insulin Stability Allow Safe and Effective Long Term Intraperitoneal Insulin Delivery in Type I Diabetic Patients: The EVADIAC Experience. Diabetes & Metabolism 29 (2003)

[124] Ginsberg, B.H.: An Overview of Minimally Invasive Technologies. Clinical Chemistry 38 (1992)

[125] Gorton, G., Hebert, D., Goode, B.: Assessment of the Kinematic Variability Between Twelve Shriners Motion Analysis Laboratories. Gait Posture 13 (2001)

[126] Grant, P.: A New Approach to Diabetic Control: Fuzzy Logic and Insulin Pump Technology. Medical Engineering & Physics 29 (2007)

[127] Grodsky, G.M.: A Threshold Distribution Hypothesis for Packet Storage of Insulin and its Mathematical Modelling. Journal of Clinical Invetigation 51 (1972)

[128] Gugercin, S., Anthoulas, A.C.: A Survey of Model Reduction by Balanced Truncation and Some New Results. International Journal of Control 77 (2004)

[129] Guyton, A.C., Hall, J.E.: Textbook of Medical Physiology. Saunders (1996)

[130] Hacısalihzade, S.S.: A Simple Algorithm for the Solution of the Multiple Dosing Problem. Computer Programs in Biomedicine 20 (1985)

[131] Hacısalihzade, S.S.: Optimal Determination of Dosage in Parkinson's Disease. Kybernetes 14 (1985)

[132] Hacısalihzade, S.S., Albani, C., Tödtli, J., Mansour, M.: A System Theoretical Approach to Drug Concentration Level Problems. Automedica 6 (1986)

[133] Hacısalihzade, S.S., Senning, M.F., Strotz, R., DeFigueiredo, R.J.P.: Optimization of Drug Administration by a Tauberian Approach. IEEE Transactions in Biomedical Engineering BME-34 (1987)

[134] Hacısalihzade, S.S., Mansour, M., Albani, C.: A Second Order Model of the Tracking Behaviour of Parkinsonian Patients. Biomedical Measurement, Informatics and Control 2 (1988)

[135] Hacısalihzade, S.S.: Counting Beans Improve Levodopa Therapy in Parkinson's Disease. In: Proceedings of the Annual International Conference of the IEEE Engineering in Medicine and Biology Society, New Orleans (1988)

[136] Hacısalihzade, S.S., Albani, C., Mansour, M.: Optimization of Symptomatic Therapy in Parkinson's Disease. IEEE Transactions in Biomedical Engineering BME-36 (1989)

[137] Hagell, P., Piccini, P., Bjorklund, A., Brundin, P., Rehncrona, S., Widner, H.: Dyskinesias following neural transplantation in Parkinson's disease. Nature Neuroscience 5(7) (2002)

[138] Hall, J.W., Pollard, A.: Near-Infrared Spectrophotometry: A New Dimension in Clinical Chemistry. Clinical Chemistry 38 (1992)

[139] Hassler, R.: Zur Pathologie der Paralysis agitans und des postenzephalitischen Parkinsonismus. J. Psychol. Neurol. 48 (1938)

[140] Heinemann, L., Schmelzeisen-Redeker, G.: Non-invasive Continuous Glucose Monitoring in Type I Diabetic Patients with Optical Glucose Sensors. Diabetologia 41 (1998)

[141] Heinemann, L., Krämer, U., Klötzer, H.-M., Hein, M., Volz, D., Hermann, M., Heise, T., Rave, K.: Noninvasive Glucose Measurement by Monitoring of Scattering Coefficient During Oral Glucose Tolerance Tests. Diabetes Technology & Therapeutics 2 (2000)

[142] Heise, H.M.: Non-invasive Monitoring of Metabolites Using Near Infrared Spectroscopy: State of the Art. Hormone and Metabolic Reserach 28 (1996)

[143] Hellwig, R., Weinert, S.: System and Method for Determining Insulin Bolus Quantities. United States Patent US7,869,851 B2 (2011)

[144] Hermanides, J., DeVries, J.H.: Sense and Nonsense in Sensors. Diabetologia 53 (2010)

[145] Hess, R.A., Modjtahedzadeh, A.: A Control Theoretical Model of Driver Steering Behaviour. IEEE Control Systems Magazine 10(5) (1990)

[146] Hirsch, I.B., Abelseth, J., Bode, B.W., Fischer, J.S., Kaufman, F.R., Mastrototaro, J.: Sensor-Augmented Insulin Pump Therapy: Results of the First Randomized Treat-to-Target Study. Diabetes Technology & Therapeutics 10 (2008)
[147] Hoehn, M.M., Yahr, M.D.: Parkinsonism: Onset, Progression and Mortality. Neurology 17 (1967)
[148] Holford, N.H.G., Sheiner, L.B.: Understanding the Dose-Effect Relationship: Clinical Application of Pharmacokinetic-Pharmacodynamic Models. Clinical Pharmacokinetics 6 (1981)
[149] Holford, N.H.G., Sheiner, L.B.: Pharmacokinetic and Pharmacodynamic Modelling in Vivo. CRC Critical Reviews in Bioengineering 7 (1981)
[150] Hostetter, G.H., Savant, C.J., Stefani, R.T.: Design of Feedback Control Systems. Holt-Saunders, Tokyo (1982)
[151] Hovorka, R.: Management of Diabetes Using Adaptive Control. International Journal of Adaptive Control and Signal Processing 19 (2005)
[152] Ibbini, M.S., Masadeh, M.A., Bani Amer, M.M.: A Semiclosed-Loop Optimal Control System for Blood Glucose Level in Diabetics. Journal of Medical Engineering & Technology 28 (2004)
[153] Ibbini, M.S., Masadeh, M.A.: A Fuzzy Logic Based Closed-Loop Control System for Blood Glucose Level Regulation in Diabetics. Journal of Medical Engineering & Technology 29 (2005)
[154] The Juvenile Diabetes Research Foundation Continuous Glucose Monitoring Study Group "Continuous Glucose Monitoring and Intensive Treatment of Type 1 Diabetes". New England Journal of Medicine 359 (2008)
[155] Javitch, J.A., D'Amoto, R.J., Strittmatter, S.M., Snyder, S.H.: Parkinsonism-inducing Neurotoxin, N-methyl-4-phenyl-1,2,3,6-tetrahydropyridine: Uptake of the Metabolite N-methyl-4-phenylpyridine by Dopamine Neurons Explains Selective Toxicity. Proc. Natl. Acad. Sci. USA 82 (1985)
[156] Janes, F.R., Carson, E.R.: Modelling Biological Systems. IEE Electronics and Power 17 (1971)
[157] Jaremko, J., Rorstad, O.: Advances Toward the Implantable Artificial Pancreas for Treatment of Diabetes. Diabetes Care 21 (1998)
[158] Jeitler, K., Horvath, K., Berghold, A., Gratzer, T.W., Neeser, K., Pieber, T.R., Siebenhofer, A.: Continuous Subcutaneous Insulin Infusion versus Multiple Daily Insulin Injections in Patients with Diabetes Mellitus: Systematic Review and Meta-Analysis. Diabetologia 51 (2008)
[159] Jellife, R.W., Schumitzky, A.: Modelling, Adaptive Control and Optimal Drug Delivery. Medical Progress Through Technology 16 (1990)
[160] Johannsen, G.: Man Machine Systems – Introduction and Background. In: IFAC/IFIP/IFORS Conf. on Analysis, Design and Evaluation of Man-Machine Systems (1982)
[161] Kailath, T.: Linear Systems. Prentice Hall, Englewood Cliffs (1982)
[162] Kalman, R.E.: Mathematical Description of Linear Dynamical Systems. Journal of SIAM Control 1 (1963)
[163] Kalman, R.: Irreducible Realizations and the Degree of Rational Matrix. Journal of the Society for Industrial and Applied Mathematics (SIAM) 13 (1965)
[164] Karwowski, W., Mital, A.: Applications of Fuzzy Set Theory in Human Factors. Elsevier, Amsterdam (1986)
[165] Katona, P.G.: Automated Control of Physiological Variables and Clinical Therapy. Critical Reviews in Bioengineering 8 (1982)

[166] Katona, P.G.: Closed-Loop Control of Physiological Variables. In: Proceedings of the IFAC Symposium on Modelling and Control of Biomedical Systems, Venice, Italy (April 1988)
[167] Kleiner-Fisman, G., Herzog, J., Fisman, D.N., Tamma, F., Lyons, K.E., Pahwa, R., Lang, A.E., Deuschl, G.: Subthalamic Nucleus Deep Brain Stimulation: Summary and Meta-Analysis of Outcomes. Movement Disorders 21(suppl. 14) (2006)
[168] Khalil, O.S.: Spectroscopic and Clinical Aspects of Noninvasive Glucose Measurements. Clinical Chemistry 45 (1999)
[169] Khalil, O.S.: Non-invasive Glucose Measurement Technologies: An Update from 1999 to the Dawn of the New Millennium. Diabetes Technology & Therapeutics 6 (2004)
[170] Kim, Y.-J., Hahn, S., Yoon, G.: Determination of Glucose in Whole Blood Samples by Mid-Infrared Spectroscopy. Applied Optics 42 (2003)
[171] Kim, Y.-J., Yoon, G.: Prediction of Glucose in Whole Blood by Near-Infrared Spectroscopy: Influence of Wavelength Region, Preprocessing, and Hemoglobin Concentration. Journal of Biomedical Optics 11 (2006)
[172] Kinesh, V.P., Neelam, D.P., Punit, B.P., Bhavesh, S.B., Pragna, K.S.: Novel Approaches for Oral Delivery of Insulin and Current Status of Oral Insulin Products. International Journal of Pharmaceutical Sciences and Nanotechnology 3 (2010)
[173] Kinnunen, M., Myllylä, R., Jokela, T., Vainio, S.: In Vitro Studies Toward Noninvasive Glucose Monitoring with Optical Coherence Tomography. Applied Optics 45 (2006)
[174] Kleinman, D.L.: Optimal Control of Linear Systems with Time Delay and Observation Noise. IEEE Transactions on Automatic Control AC-14 (1969)
[175] Kleinman, D.L., Baron, S., Levison, W.H.: A Control Theoretic Approach to Manned-Vehicle Systems Analysis. IEEE Transactions on Automatic Control AC-16 (1971)
[176] Kleinman, D.L., Perkins, T.: Modelling the Human in a Time-Varying Anti-Aircraft Tracking Loop. IEEE Transactions on Automatic Control AC-19 (1974)
[177] Klemme, R.M.: Oculogyric Crises: A Therapeutic Approach. Amer. J. Opth. 24 (1941)
[178] Kong, C.-R., Barman, I., Dingari, N.C., Kang, J.W., Galindo, L., Dasari, R.R., Feld, M.S.: A Novel Non-imaging Optics Based Raman Spectroscopy Device for Transdermal Blood Analyte Measurement. American Institute of Physics Advances 1 (2011)
[179] Koschinsky, T., Heinemann, L.: Sensors for Glucose Monitoring: Technical and Clinical Aspects. Diabetes/Metabolism Research and Reviews 17 (2001)
[180] Krall, L.P., Levine, R., Barnett, D.: The History of Diabetes. In: Kahn, C.R., Weir, G.C. (eds.) Joslin's Diabetes Mellitus. Lea & Febiger, Philadelphia (1994)
[181] Krayenbühl, H., Yaşargil, M.G.: Bilateral Operations on the Thalamus and Pallidum for Parkinsonism. J. Neurol. Neurosurg. Psychiat. 23 (1960)
[182] Kristensen, G.B.B., Monsen, G., Skeie, S., Sandberg, S.: Standardized Evaluation of Nine Instruments for Self-Monitoring of Blood Glucose. Diabetes Technology & Therapeutics 10 (2008)
[183] Kupfermann, L., Weiss, K.R.: The Command Neuron Concept. Behav. Brain Sci. 1 (1978)
[184] Kurland, L.T.: Epidemiology: Incidence, Geographic Distribution, and Genetic Considerations. In: Fields, W.S. (ed.) Pathogenesis and Treatment of Parkinsonism. C.C. Thomas, Springfield (1958)
[185] Kurland, L.T., Darrell, R.W.: Epidemiologic and Genetic Characteristics of Parkinsonism: A Review. Int. J. Neurol. 2 (1962)
[186] Kvalseth, T.O.: Describing Function Modelling of Pursuit Tracking. In: IEEE Proceedings of the International Conference on Cybernetics and Society (1981)
[187] Labaune, J.-P.: Handbook of Pharmacokinetics: Toxicity Assessment of Chemicals. Ellis Herwood, Chichester (1989)

[188] Langston, J.W., Forno, L.S., Rebert, C.S., Irwin, I.: Selective Nigral Toxicity After Systemic Administration of l-methyl-4-phenyl-1,2,3,6-tetrahydropyridine (MPTP) in the Squirrel Monkey. Brain Res. 292 (1983)
[189] Larin, K.V., Eledrisi, M.S., Motamedi, M., Esenaliev, R.O.: Noninvasive Blood Glucose Monitoring With Optical Coherence Tomography. Diabetes Care 25 (2002)
[190] Law, A.M., Kelton, W.D.: Simulation, Modeling and Analysis. McGraw-Hill, New York (1982)
[191] Lee, R.G.: Physiology of the Basal Ganglia: An Overview. Can. J. Neurol. Sci. 11 (1984)
[192] Lehmann, E.L., Casella, G.: Theory of Point Estimation. Springer, New York (1998)
[193] Lehninger, A.L., Nelson, D.L., Cox, M.M.: Lehninger Principles of Biochemistry. W.H. Freeman, New York (2005)
[194] Leroux, P.D.: Contribution a L'etude des Causes de la Paralisysie Agitante. These Med., Paris (1880)
[195] Leuthold, P., Schaufelberger, W.: System- und Nachrichtentechnik. AMIV Verlag, Zurich (1975)
[196] Lewis, S.J., Caldwell, M.A., Barker, R.A.: Modern Therapeutic Approaches in Parkinson's Disease. Expert Reviews in Molecular Medicine 5(10) (2003)
[197] Liberzon, D.: Calculus of Variations and Optimal Control Theory: A Concise Introduction. Princeton University Press, Princeton (2012)
[198] Ličko, V.: Threshold Secretory Mechanism: A Model of Derivative Element in Biological Control. Bulletin of Mathematical Biology 35 (1973)
[199] Lin, J., Lee, D., Chase, J.G., Shaw, G.M., Le Compte, A., Lotz, T., Wong, J., Lonergan, T., Hann, C.E.: Stochastic Modelling of Insulin Sensitivity and Adaptive Glycemic Control for Critical Care. Computer Methods and Programs in Biomedicine 89 (2008)
[200] Lindvall, O., Rehncrona, S., Brundin, P., Gustavii, B., Astedt, B., Widner, H.: Human Fetal Dopamine Neurons Grafted into the Striatum in two Patients with Severe Parkinson's Disease. A Detailed Account of Methodology and a 6-Month Follow-up. Archives of Neurology 46(6) (1989)
[201] Linkens, D.A., Hacısalihzade, S.S.: Computer Control Systems in Pharmacological Drug Administration: A Survey. Journal of Medical Engineering and Technology 14 (1990)
[202] Lotka, A.J.: Elements of Mathematical Biology. Dover Press, New York (1956)
[203] Lo, Y.-L., Yu, T.-C.: A Polarimetric Glucose Sensor Using a Liquid-Crystal Polarization Modulator Driven by a Sinusoidal Signal. Optics Communications 259 (2006)
[204] MacKenzie, H.A., Ashton, H.S., Spiers, S., Shen, Y., Freeborn, S.S., Hannigan, J., Lindberg, J., Rae, P.: Advances in Photoacoustic Noninvasive Glucose Testing. Clinical Chemistry 45 (1999)
[205] Madrazo, I., Leon, V., Torres, C., Aguilera, M.C., Varela, G., Alvarez, F.: Transplantation of Fetal Substantia Nigra and Adrenal Medulla to the Caudate Nucleus in two Patients with Parkinson's Disease. New England Journal of Medicine 318(1) (1988)
[206] Malchoff, C.D., Shoukri, K., Landau, J.I., Buchert, J.M.: A Novel Non-invasive Blood Glucose Monitoring. Diabetes Care 25 (2002)
[207] Mamdani, E.H., Assilian, S.: An Experiment in Linguistic Synthesis with a Fuzzy Logic Controller. International Journal of Man-Machine Studies 7 (1975)
[208] Mansour, M.: Lineare Regelsysteme. Lecture Notes, Swiss Federal Institute of Technology (1977)
[209] Mansour, M.: Optimierung dynamischer Systeme. Lecture Notes, Swiss Federal Institute of Technology (1979)
[210] Markowitz, H.: Portfolio Selection. Journal of Finance 7 (1952)

[211] Markram, H.: The Human Brain Project. Scientific American 306 (2012)
[212] Marquardt, D.: An Algorithm for Least-Squares Estimation of Nonlinear Parameters. SIAM Journal on Applied Mathematics 11 (1963)
[213] Marsden, C.D., Merton, P.A., Morton, H.B.: Servo Action in the Human Thumb. J. Physiol. 257 (1976)
[214] Marsden, C.D.: The Mysterious Motor Function of the Basal Ganglia. Neurology 32 (1982)
[215] Marshall, A.J., Young, D.S., Blyth, J., Kabilan, S., Lowe, C.R.: Metabolite-Sensitive Holographic Biosensors. Analytical Chemistry 76 (2004)
[216] Marttila, R.J., Rinne, U.K., Sonninen, V.: Dopaminergic Agonist Effects on Brain Monoamine Metabolism in Parkinsonism. In: Birkmayer, W., Hornykiewicz, O. (eds.) Advances in Parkinsonism. Editiones Roche, Basel (1976)
[217] Marttila, R.J., Rinne, U.K., Sonninen, V.: Dopaminergic Agonist Effects on Brain Monoamine Metabolism in Parkinsonism. In: Birkmayer, W., Hornykiewicz, O. (eds.) Advances in Parkinsonism. Editiones Roche, Basel (1976)
[218] McNichols, R.J., Coté, G.L.: Optical Glucose Sensing in Biological Fluids: An Overview. Journal of Biomedical Optics 5 (2000)
[219] McRuer, D.T., Krendel, E.S.: Dynamic Response of Human Operators, USAF Wright Air Development Center. Technical Report 56-524, Ohio (1957)
[220] McRuer, D.T., Krendel, E.S.: The Human Operator as a Servo System Element. Journal of the Franklin Institute 267 (1959)
[221] McRuer, D.T., Jex, H.R.: A Review of Quasilinear Pilot Models. IEEE Transactions on Human Factors in Electronics HFE-8 (1967)
[222] Mendelson, Y., Clermont, A.C., Peura, R.A., Lin, B.C.: Blood Glucose Measurement by Multiple Attenuated Total Reflection and Infrared Absorption Spectroscopy. IEEE Transactions on Biomedical Engineering BME-37 (1990)
[223] Meier, R., Nieuwland, J., Hacısalihzade, S., Steck, D., Zbinden, A.: Fuzzy Control of Blood Pressure During Anesthesia with Isoflurane. In: Proceedings IEEE Conference on Fuzzy Systems (1992)
[224] Meier, R., Nieuwland, J.: Multivariable Fuzzy-Regelung der Anästhesietiefe. Diplomarbeit, Automatic Control Laboratory, Swiss Federal Institute of Technology (ETH), Zurich, IfA 8696 (1992)
[225] Mendel, J.M.: Fuzzy Logic Systems for Engineering: A Tutorial. Proc. of the IEEE 83 (March 1995)
[226] Merritt, H.H.: A Textbook of Neurology. Lea & Febiger, Philadelphia (1979)
[227] Meyerhoff, C., Bischoff, E., Sternberg, F., Zier, H., Pfeiffer, E.F.: On line Continuous Monitoring of Subcutaneous Tissue Glucose in Men by Combining Portable Glucosensor with Microdialysis. Diabetologia 35 (1992)
[228] Meyers, H.R.: Surgical Procedure for Post-Encephalitic Tremor with Notes on the Physiology of Premotor Fibres. Arch. Neurol. Psychiat. 44 (1940)
[229] Miano, G., Maffuci, A.: Transmission Lines and Lumped Circuits. Academic Press, London (2001)
[230] Moler, C.B., van Loan, C.F.: Nineteen Dubious Ways to Compute the Exponential of a Matrix, Twenty-Five Years Later. SIAM Review 45 (2003)
[231] Moore, B.C.: Principal Component Analysis in Linear Systems: Controllability, Observability and Model Reduction. IEEE Transactions on Automatic Control 26 (1981)
[232] Müller, U., Henning, K., Rake, H.: Estimation of Stability Factors of the Human Operator in Closed Loop Control Tasks. In: Willumeit, H.P. (ed.) Human Decision Making and Manual Control. Nort Holland, Amsterdam (1986)

[233] Munter, M.D., Tyce, G.M.: L-DOPA Therapy in Parkinson's Disease: Plasma Concentration, Therapeutic Response and Side-Effects. Mayo Clinic Proceedings 46 (1971)

[234] Narabayashi, H., Okuma, T., Shikuba, S.: Procaine Oil Block of the Globus Pallidus. Arch. Neurol. Psychiat. 75 (1956)

[235] Negro, M.F.: Le Phénomène de la Roue Dentée. Encephale 23 (1928)

[236] Nour Eldin, H.A.: Optimierung linearer Regelsysteme mit quadratischer Zielfunktion. Springer, New York (1971)

[237] Nutt, J.G., Woodward, W.R.: Levodopa Pharmacokinetics and Pharmacodynamics in Fluctuating Parkinsonian Patients. Neurology 36 (1986)

[238] Olanow, C.W., Goetz, C.G., Kordower, J.H., Stoessl, A.J., Sossi, V., Brin, M.F.: A Double-Blind Controlled Trial of Bilateral Fetal Nigral Transplantation in Parkinson's Disease. Annals of Neurology 54(3) (2003)

[239] Ollerton, R.L.: Application of Optimal Control Theory to Diabetes Mellitus. International Journal of Control 50 (1989)

[240] Onuaguluchi, G.: Parkinsonism. Butterworths, London (1964)

[241] Pacini, G., Bergman, R.N.: PACBERG: An Adaptive Program for Controlling the Blood Sugar. Computer Programs in Biomedicine 16 (1983)

[242] Parkes, J.L., Slatin, S.L., Pardo, S., Ginsberg, B.H.: A New Consensus Error Grid to Evaluate the Clinical Significance of Inaccuracies in the Measurement of Blood Glucose. Diabetes Care 23 (2000)

[243] Parkinson, J.: Essay on the Shaking Palsy. Sherwood, London (1817)

[244] Paskin, H.M.: A Discrete Stochastic Optimal Control Model of the Human Operator in a Closed-Loop Tracking Task. Doctoral Dissertation, USAF Institute of Technology Air University (1970)

[245] Paulis, W.D., Horemans, H.L., Brouwer, B.S., Stam, H.J.: Excellent Test-Retest and Inter-Rater Reliability for Tardieu Scale Measurements with Inertial Sensors in Elbow Flexors of Stroke Patients. Gait Posture 33 (2011)

[246] Pecile, A., Rescigno, A. (eds.): Pharmacokinetics: Mathematical and Statistical Approaches to Metabolism and Distribution of Chemicals and Drugs. Plenum Press, New York (1988)

[247] Penney, J.B., Young, A.B.: Speculations on the Functional Anatomy of Basal Ganglia Disorders. Ann. Rev. Neurosci. 6 (1983)

[248] Perkins, W.: How does anesthesia work? Scientific American 292 (2005)

[249] Perrottet, E.: Tables Psycholeptique. Medicine et Hygiene, Geneve (1965)

[250] Pickup, J.C.: Management of Diabetes Mellitus: Is the Pump Mightier than the Pen? Nature Reviews Endocrinology 8 (2012)

[251] Poddar, R., Andrews, J.T., Shukla, P., Sen, P.: Non-Invasive Glucose Monitoring Techniques: A Review and Current Trends. Medeley (2008), `http://arxiv.org/0810.5755`

[252] Podiwinsky, H., Mentasti, M., Riederer, P., Birkmayer, W.: Zur Behandlung des Parkinson-Syndroms mit Kombinationspraparaten von L-DOPA mit Decarboxylase-Hemmern (Carbidopa, Benserazide). Wien. Klin. Wschr. 91 (1979)

[253] Poirier, L.J.: The Development of Animal Models for Studies in Parkinson's Disease. In: McDowell, F.H., Markham, C.H., Davis, F.A. (eds.) Recent Advances in Parkinson's Disease, Philadelphia (1971)

[254] Pollock, M., Hornabrook, R.W.: The Prevalence, Natural History and Dementia of Parkinson's Disease. Brain 89 (1966)

[255] Pontryagin, L.S., Boltyanskii, V.G., Gamkrelidze, R.V., Mishchenko, E.F.: The Mathematical Theory of Optimal Processes. Interscience Publishers, New York (1962)

[256] Popp, J., Tuchin, V.V., Chiou, A., Heinemann, S.H. (eds.): Handbook of Biophotonics. Wiley, Weinheim (2012)

[257] Powell, M.J.D.: Approximation Theory and Methods. Cambridge University Press, Cambridge (1981)

[258] Predko, M.: Programming Robot Controllers. McGraw-Hill, New York (2003)

[259] Pukelsheim, F.: Optimal Design of Experiments. SIAM, Philadelphia (2006)

[260] Putnam, T.J.: Results of Treatment of Athetosis by Section of Extrapyramidal Tracts in the Spinal Cord. Arch. Neurol. Psychiat. 38 (1938)

[261] Raaflaub, J.: Pharmacokinetics: A Guide for Beginners. Editiones Roche, Basel (1985)

[262] Raccah, D., Sulmont, V., Reznik, Y., Guerci, B., Renard, E., Hanaire, H.: Incremental Value of Continuous Glucose Monitoring when Starting Pump Therapy in Patients with Poorly Controlled Type 1 Diabetes: The Real Trend study. Diabetes Care 32 (2009)

[263] Rebrin, K., Steil, G.M., van Antwerp, W.P., Mastrototaro, J.J.: Subcutaneous Glucose Predicts Plasma Glucose Independent of Insulin: Implications for Continuous Monitoring. American Journal of Physiology Endocrinology and Metabolism 277 (1999)

[264] Renard, E., Place, J., Cantwell, M., Chevassus, H., Palerm, C.C.: Closed-Loop Insulin Delivery Using a Subcutaneous Glucose Sensor and Intraperitoneal Insulin Delivery: Feasibility Study Testing a New Model for the Artificial Pancreas. Diabetes Care 33 (2010)

[265] Rideout, V.C.: Mathematical Computer Modeling of Physiological Systems. Prentice Hall, Englewood Cliffs (1991)

[266] Riederer, P., Birkmayer, W., Seemann, D., Wuketich, S.: Brain Noradrenaline and 3-methoxy-4-hydroxyphenylglycol in Parkinson's Syndrome. 3. Neural Transm. 41 (1977)

[267] Riederer, P.: Personal communication (1984)

[268] Riklan, M., Diller, L.: Follow Up Studies in Chemopallidectomy for Paralysis Agitans. 3. Amer. Med. Ass. 167 (1958)

[269] Rodgers, J.L., Nicewander, A.: Thirteen Ways to Look at the Correlation Coefficient. The American Statistician 42 (1988)

[270] Rosati, G., Granieri, E., Pinna, L., Aiello, L., Tola, R., DeBastiani, P., Pirisi, A., Devoto, M.C.: The Risk of Parkinson's Disease in Mediterranean People. Neurology 30 (1980)

[271] Ross, S.M.: Introduction to Stochastic Dynamic Programming. Academic Press, San Diego (1983)

[272] Sage, A.P., White, C.C.: Optimum Systems Control. Prentice Hall, Englewood Cliffs (1968)

[273] Schmidt, S., Meldgaard, M., Serifovski, N., Storm, C., Christensen, T.M., Gade-Rasmussen, B., Norgaard, K.: Use of an Automated Bolus Calculator in MDI-Treated Type 1 Diabetes: The BolusCal Study, a Randomized Controlled Pilot Study. Diabetes Care 35 (2012)

[274] Schmitt, J.M.: Optical Coherence Tomography (OCT): A Review. IEEE Journal of Selected Topics in Quantum Electronics 5 (1999)

[275] Schoen, F.: Stochastic Techniques for Global Optimization: A Survey of Recent Advances. Journal of Global Optimization 1 (1991)

[276] Schwab, R.S.: Symptomatology and Medical Treatment of Parkinson's Disease. Int. J. Neurol. 2 (1961)

[277] Schwab, R.S., England, A.L., Peterson, E.: Akinesia in Parkinson's Disease. Neurology 9 (1959)

[278] Schutte, J.F., Groenwold, A.A.: A Study of Global Optimization Using Particle Swarms. Journal of Global Optimization 31 (2005)

[279] Sheiner, L.B., Rosenberg, B., Melmon, K.L.: Modelling of Individual Pharmacokinetics for Computer-Aided Drug Dosage. Computers and Biomedical Research 5 (1972)

[280] Sheridan, T.B., Ferrel, W.R.: Man-Machine Systems. MIT Press, Cambridge (1974)

[281] Shillito, F.H., Drinker, C.R., Shaughnessy, T.J.: The Problem of Nervous and Mental Sequelae in Carbon Monoxyde Poisoning. J.A.M.A. 106 (1936)

[282] Siegfried, J.: Die Parkinsonische Krankheit und ihre Behandlung. Springer, Wien-New York (1968)

[283] Siesler, H.W., Ozaki, Y., Kawata, S., Heise, H.M.: Near-Infrared Spectroscopy. Wiley, Weinheim (2002)

[284] Sigwald, J., Bovet, D., Dumont, G.: Le Traitement de la Maladie de Parkinson par le Chlorhydrate de Diethylaminoethyl-N-thiodiphenylalanine. Rev. Neurol. 78 (1946)

[285] Smolen, V.F., Turrie, B.D., Weigand, W.A.: Drug Input Optimization: Bioavailability-effected Time-Optimal Control of Multiple, Simultaneous Pharmacological Effects and their Interrelationships. Journal of Pharmaceutical Sciences 61 (1972)

[286] Smolin, L.: Three Roads to Quantum Gravity. Basic Books, New York (2001)

[287] Snyder, S.H., D'Amoto, R.: MPTP: A Neurotoxin Relevant to the Pathophysiology of Parkinson's Disease. Neurology 36 (1986)

[288] Snyder, S.H., D'Amoto, R., Nye, J., Javitch, J.: Selective Uptake of MPP+ by Dopamine Neurons is Required for MPTP Toxicity: Studies in Brain Synaptosomes and PC-12 Cells. In: Markey, S.P., Castagnoli, N., Kopin, I.J., Trevor, A.T. (eds.) MPTP: a Neurotoxin Producing a Parkinsonian Syndrome. Academic Press, New York (1986)

[289] Sorensen, J.T., Colton, C.K., Hillman, R.S., Soeldner, J.S.: Use of a Physiologic Pharmacokinetic Model of Glucose Homeostasis for Assessment of Performance Requirements for Improved Insulin Therapies. Diabetes Care 5 (1982)

[290] ACCU-CHEK® Spirit Combo, Insulin Pump, User Guide, Roche Diabetes Care, Burgdorf (2010)

[291] Steil, G.M., Clark, B., Kanderian, S., Rebrin, K.: Modelling Insulin Action for Development of Close-loop Artificial Pancreas. Diabetes Technology & Therapeutics 7 (2005)

[292] Swan, G.W.: An Optimal Control Model of Diabetes Mellitus. Bulletin of Mathematical Biology 44 (1982)

[293] Tam, A.C.: Applications of Photoacoustic Sensing Techniques. Reviews of Modern Physics 58 (1986)

[294] Tatton, W.G., Bawa, P., Bruce, I.C.: Altered Motor Cortical Activity in Motor Cortical Rigidity. In: Poirier, L.J., Sourkes, T.L., Bedard, P.J. (eds.) "The Extrapyramidal System", Adv. in Neurol. 10 (1979)

[295] Tatton, W.G., Bruce, I.C.: A Schema for the Interactions between Motor Programs and Sensory Input. Can. J. Physiol. Pharmacol. 59 (1981)

[296] Tatton, W.G., Eastman, M.J., Bedingham, W., Verrier, M.C., Bruce, I.C.: Defective Utilization of Sensory Input as the Basis for Bradykinesia, Rigidity and Decreased Movement Repertoire in Parkinson's Disease: A Hypothesis. Can. J. Neurol. Sci. 11 (1984)

[297] Teson, S.A., Gifford, R., Fei, J., Reynolds, J.S.: Electrochemical Impedance Spectroscopy Enabled Continuous Glucose Monitoring Sensor Systems. United States Patent Application 20110040163 (2011)

[298] The Diabetes Control and Complications Research Group "The Effect of Intensive Treatment of Diabetes on the Development and Progression of Long Term Complications in Insulin Dependent Diabetes Mellitus". New England Journal of Medicine 329 (1993)
[299] Thompson, D.E.: Design Analysis: Mathematical Modeling of Nonlinear Systems. Cambridge University Press, Cambridge (1999)
[300] Tran, M.C.: Biosensors. Chapman and Hall, London (1993)
[301] Tresp, V., Briegel, T., Moody, J.: Neural-Network Models for the Blood Glucose Metabolism of a Diabetic. IEEE Transactions on Neural Networks 10 (1999)
[302] Tretiakof, F.C.: Contribution a L'étude de L'anatomie Pathologique du Locus Niger de Sommering avec quelques Déductions Relatives à la Pathogenic des Trobles du Tonus Musculaire et de la Maladie de Parkinson. These Med. (293) (1919)
[303] Van den Berghe, G., Wouters, P., Weekers, F., Verwaest, C., Bruyninckx, F., Schetz, M., Vlasselaers, D., Ferdinande, P., Lauwers, P., Bouillon, R.: Intensive Insulin Therapy in Critically Ill Patients. New England Journal of Medicine 345 (2001)
[304] Vonach, R., Buschmann, J., Falkowski, R., Schindler, R., Lendl, B., Keller, R.: Application of Mid-Infrared Transmission Spectrometry to the Direct Determination of Glucose in Whole Blood. Applied Spectroscopy 52 (1998)
[305] Vozeh, S., Steimer, J.-L.: Feedback Control Methods for Drug Dosage Optimization. Clinical Pharmacokinetics 10 (1985)
[306] Ward, C., Duvoisin, R.C., Ince, S.E., Nutt, J.D., Eldridge, R., Calne, D.B.: Parkinson's Disease in 65 pairs of Twins and in a set of Quadruplets. Neurology 33 (1983)
[307] Wartak, J.: Clinical Pharmacokinetics. Praeger, New York (1983)
[308] Watts, N.B., Gebhart, S.S.P., Clark, R.V., Phillips, L.S.: Postoperative Management of Diabetes Mellitus. Diabetes Care 10 (1987)
[309] Welling, P.G.: Pharmacokinetics: Processes and Mathematics. American Chemical Society, Washington D.C. (1986)
[310] White, N., Skor, D., Santiago, J.V.: Practical Closed-Loop Insulin Delivery: A System for the Maintenance of Overnight Euglycemia and the Calculation of Basal Insulin Requirements in Insulin-Dependent Diabetics. Annals of Internal Medicine 97 (1982)
[311] World Health Organisation, Fact Sheet No. 312 (August 2011)
[312] Widmark, E.M.P.: Studies in the Concentration of Indifferent Narcotics in Blood and Tissues. Acta Medica Scandinavia 52 (1919)
[313] Wiener, N.: The Fourier Integral and Certain of its Applications. Cambridge University Press, Cambridge (1933)
[314] Wiggins, S.: Introduction to Applied Nonlinear Dynamical Systems and Chaos. Springer, New York (2003)
[315] Wilkinson, G.R.: Pharmacokinetics of Drug Deposition: Hemodynamic Considerations. Annual Review of Pharmacology 15 (1975)
[316] Wood, M., Wood, A.: Drugs and Anesthesia, Pharmacology for Anesthesiologists. Williams & Wilkins, Philadelphia (1998)
[317] Wong, T.W.: Chitosan and Its Use in Design of Insulin Delivery System. Recent Patents on Drug Delivery & Formulation 3 (2009)
[318] Worsley, G.J., Tourniaire, G.A., Medlock, K.E.S., Sartain, F.K., Harmer, H.E., Thatcher, M., Horgan, A.M., Pritcharda, J.: Continuous Blood Glucose Monitoring with a Thin-Film Optical Sensor. Clinical Chemistry 53 (2007)
[319] Wu, Q.: Stabilitätsanalyse von Regelsystemen mit Begrenzungen. Ph.D. Thesis No. 7930, Institut für Automatik, ETH Zurich (1986)

[320] Yipintsoi, T., Gatewood, L.C., Ackerman, E., Spivak, P.L., Molnar, G.D., Rosevear, J.W., Service, F.J.: Mathematical Analysis of Blood Glucose and Plasma Insulin Responses to Insulin Infusion in Healthy and Diabetic Subjects. Computers in Biology and Medicine 3 (1973)
[321] Zadeh, L.: Fuzzy Sets. Information and Control 8 (1965)
[322] Zamir, N., Skofitsch, G., Bannon, M.J., Helke, C.J., Kopin, I.J., Jacobowitz, D.M.: Primate Model of Parkinson's Disease: Alterations in Multiple Opioid Systems in the Basal Ganglia. Brain Res. 322 (1984)
[323] Zbinden, A., Petersen-Felix, S., Thomson, D.: Anesthetic Depth Defined Using Multiple Noxious Stimuli During Isoflurane/Oxygen Anesthesia: Hemodynamic Responses. Anesthesiology 80 (1994)
[324] Zbinden, A., Feigenwinter, P., Petersen-Felix, S., Hacısalihzade, S.: Arterial Pressure Control with Isoflurane Using Fuzzy Logic. British Journal of Anaesthesia 74 (1995)
[325] Zbinden, A.M.: Inhalationsanästhetika: Aufnahme und Verteilung: Allgemeine Grundlagen. Wissenschaftliche Verlagsabteilung Deutsche Abbott GmbH, Wiesbaden (1984)
[326] Zbinden, A.: Pharmacokinetics of Inhaled anaesthetics. Baillière's Clinical Anesthesiology 5 (1991)
[327] Zhou, K., Doyle, J.C., Glover, K.: Robust and Optimal Control. Prentice Hall, Englewood Cliffs (1996)
[328] Zwart, A.N., Smith, N.T., Beneken, J.E.W.: Multiple Model Approach to Uptake and Distribution of Halothane: The Use of an Analog Computer. Computers and Biomedical Research 5 (1972)

More information about this series at https://link.springer.com/bookseries/558

Augusto Sampaio · Marielle Stoelinga
Editors

Formal Methods

27th International Symposium, FM 2026
Tokyo, Japan, May 18–22, 2026
Proceedings, Part II

Springer

Editors
Augusto Sampaio
Universidade Federal de Pernambuco
Recife, Brazil

Marielle Stoelinga
University of Twente
Enschede, The Netherlands

ISSN 0302-9743 ISSN 1611-3349 (electronic)
Lecture Notes in Computer Science
ISBN 978-3-032-26219-6 ISBN 978-3-032-26220-2 (eBook)
https://doi.org/10.1007/978-3-032-26220-2

This Springer imprint is published by the registered company Springer Nature Switzerland AG
The registered company address is: Gewerbestrasse 11, 6330 Cham, Switzerland

Preface

We proudly present two volumes with the papers accepted for publication and presented at the 27th International Symposium on Formal Methods (FM 2026), in Tokyo, Japan, from May 18–22, 2026.

FM 2026 is the 27th edition of the symposium series organized by Formal Methods Europe (FME), an independent association that promotes the effective use of formal methods in system development. The FM symposia serve as a forum for researchers and practitioners from academia, industry, and government to present new results, report on experience, and explore emerging directions. FM 2026 continues this role, supporting interaction across different parts of the community.

Tracks. The current two LNCS volumes contain the papers presented in four different tracks at FM 2026:

- *The Main track,* co-chaired by us, Augusto Sampaio (Universidade Federal de Pernambuco, Brazil) and Marielle Stoelinga (University of Twente & Radboud University, The Netherlands),
- *The Tests and Proofs (TAP) track,* co-chaired by Achim Brucker (University of Exeter, UK) and Marie-Christine Jakobs (Ludwig-Maximilians-Universität München, Germany),
- *Industry Day,* co-chaired by Jeroen Keiren (Eindhoven University of Technology, The Netherlands) and Naoto Sato (Hitachi, Japan), and
- *The Tutorials track,* co-chaired by Kazuhiro Ogata (Japan Advanced Institute of Science and Technology, Japan) and Neeraj Kumar Singh (University of Toulouse, France).

In addition, the FM program features several tracks published at other venues:

- *Journal First Track,* chaired by Anne Haxthausen (Technical University of Denmark),
- Doctoral Symposium, co-chaired by Toshiaki Aoki (JAIST, Japan) and Cyrille Artho (KTH Royal Institute of Technology, Sweden),
- International Conference on Rigorous State-Based Methods (ABZ, co-chaired by Elvinia Riccobene, Università degli Studi di Milano, Italy) and Yamine Ait-Ameur (University of Toulouse, France).
- *Workshops:* AIPV, FMTea, Overture, and Rodin.

Keynotes. FM 2026 featured four keynotes by internationally renowned researchers,

- Cristian Cadar (Imperial College London, UK)
- Daniel Kroening (Amazon Web Services, AWS, USA)
- Ichiro Hasuo (National Institute of Informatics, Japan)
- Ruzica Piskac (Yale University, USA).

Submission numbers. FM 2026 again received a record number of submissions, namely, **239** submissions for the Main and TAP tracks. Each submission underwent review by at least three reviewers, followed by a discussion within the program committee. The Main and TAP tracks accepted **49 papers** (43 papers in the Main track and 6 papers in the TAP track), resulting in an acceptance rate of **21%.** Furthermore, 16 tutorials were submitted, of which 8 were accepted for publication. The Industry Day received 17 submissions and accepted 9.

Artifact evaluations. The authors of all accepted regular papers were strongly encouraged to submit an artifact: additional material, such as software, datasets, log files, machine-checkable proofs, etc., that substantiates the claims made in the paper. Ideally, artifacts allow for the full reproduction of all results in the corresponding paper by providing details on all relevant steps, inputs, configurations, and parameters used. For tools, an artifact typically includes the source code or a binary of the tool and corresponding documentation on how to use, reuse, and possibly extend it.

Artifacts were evaluated by an Artifact Evaluation Committee (AEC), chaired by Étienne André (Nantes Université, France) and Masaki Waga (Kyoto University, Japan). After a short quick-check phase, at least two and, for most artifacts, three AEC members reviewed each artifact in terms of consistency with and reproducibility of results presented in the paper, completeness, documentation, and ease of (re)use, and availability in an online repository with a DOI.

The authors of the accepted tutorial papers were also invited to the artifact evaluation. Although the quick-check phase was skipped due to timing constraints, the tutorial artifacts were evaluated under the same criteria as the regular papers' artifacts by two AEC members.

Based on these reviews, and in strict adherence to the EAPLS artifact badging guidelines, artifacts were awarded a badge in the following categories:

- **Available.** Artifacts that are publicly archived permanently with a DOI that are in some way "*relevant to*" *and* "*add value beyond the text in the article*" were awarded the available badge.
- **Functional.** Artifacts that are documented (containing at least an inventory and "*sufficient description to allow artifacts to be exercised*"), consistent (i.e., "relevant to the associated paper and significantly contribute to the generation of its main results"), complete ("as far as possible") and exercisable, received the functional badge.
- **Reusable.** Functional and available artifacts that are "*very carefully documented and well-structured to the extent that reuse and repurposing is facilitated*" received the reusable badge instead.

Of the **33 submitted artifacts**, 2 received the available badge, 21 were functional, and 10 were awarded the (functional and) reusable badge.

Thank you! We thank all who contributed to FM 2026. We are especially grateful to the FME board for their support and prompt responses to all our questions. We are grateful to the general chair Fuyuki Ishikawa (National Institute of Informatics, Japan) and his team for the smooth organization. We thank the program committees and all reviewers for their careful work, and the organizers of the various tracks and co-located events. Finally, a big thanks to the authors of all submissions.

March 2026

Augusto Sampaio
Marielle Stoelinga

Organization

Program Committees

Main Track and TAP Track

Augusto Sampaio (Co-chair Main Track)	Federal University of Pernambuco, Brazil
Marielle Stoelinga (Co-chair Main Track)	University of Twente, Netherlands
Achim D. Brucker (Co-chair TAP)	University of Exeter, UK
Marie-Christine Jakobs (Co-chair TAP)	LMU Munich, Germany
Erika Abraham	RWTH Aachen University, Germany
Wolfgang Ahrendt	Chalmers University of Technology, Sweden
Bernhard Aichernig	Institute for Formal Models and Verification, Johannes Kepler University Linz, Austria
Alessandro Aldini	University of Urbino "Carlo Bo", Italy
Oana Andrei	University of Glasgow, UK
Pedro Antonino	The Blockhouse Technology Limited, UK
Toshiaki Aoki	JAIST, Japan
Alessandro Armando	University of Genova & Fondazione Bruno Kessler, Italy
Christel Baier	TU Dresden, Germany
Clark Barrett	Stanford University, USA
Ezio Bartocci	TU Wien, Austria
Gustavo Betarte	InCo, Universidad de la República, Uruguay, Uruguay
Mohamed Bettaz	Czech Technical University in Prague, Czechia
Dirk Beyer	LMU Munich, Germany
Roderick Bloem	Graz University of Technology, Austria
Patricia Bouyer	CNRS, France
Gustavo Carvalho	Universidade Federal de Pernambuco, Brazil
Darion Cassel	Amazon Web Services, USA
Valentina Castiglioni	Eindhoven University of Technology, Netherlands
Ana Cavalcanti	University of York, UK
Milan Ceska	Brno University of Technology, Czechia
Yu-Fang Chen	Academia Sinica, Taiwan

Pedro R. D'Argenio	Universidad Nacional de Córdoba, CONICET, Argentina
Nancy Day	University of Waterloo, Canada
Frank de Boer	Centrum Wiskunde & Informatica, Netherlands
Jin Song Dong	National University of Singapore, Singapore
Brijesh Dongol	University of Surrey, UK
Catherine Dubois	ENSIIE-Samovar, France
João F. Ferreira	University of Porto, Portugal
Marco Faella	Università di Napoli "Federico II", Italy
Marie Farrell	The University of Manchester, UK
Natasha Fernandes	Macquarie University, Australia
Carla Ferreira	Universidade NOVA de Lisboa, Portugal
Wan Fokkink	Vrije Universiteit Amsterdam, Netherlands
Jose Fragoso Santos	INESC-ID/IST, Portugal
Carlo A. Furia	USI - Università della Svizzera italiana, Switzerland
Mario Gleirscher	University of Bremen, Germany
Jan Friso Groote	Eindhoven University of Technology, Netherlands
Ernst Moritz Hahn	University of Twente, Netherlands
Ichiro Hasuo	National Institute of Informatics, Japan
Paula Herber	University of Münster, Germany
Falk Howar	TU Clausthal/IPSSE, Germany
Marieke Huisman	University of Twente, Netherlands
Reiner Hähnle	TU Darmstadt, Germany
Atsushi Igarashi	Kyoto University, Japan
Einar Broch Johnsen	University of Oslo, Norway
Sun Jun	Singapore Management University, Singapore
Jeroen Keiren	Eindhoven University of Technology, Netherlands
Alexander Knapp	Universität Augsburg, Germany
Tsutomu Kobayashi	Japan Aerospace Exploration Agency, Japan
Nikolai Kosmatov	Thales Research & Technology, France
Laura Kovács	TU Wien, Austria
Juneyoung Lee	Amazon Web Services, USA
Zhiming Liu	Southwest University, China
Nuno Macedo	University of Porto & INESC TEC, Portugal
Diego Marmsoler	University of Exeter, UK
Mieke Massink	CNR-ISTI, Pisa, Italy
Anastasia Mavridou	KBR/NASA Ames Research Center, USA
Annabelle McIver	Macquarie University, Australia
Hernan Melgratti	Universidad de Buenos Aires, Argentina
Alexandra Mendes	University of Porto, Portugal
Carroll Morgan	The University of New South Wales, Australia

Alexandre Mota	Universidade Federal de Pernambuco, Brazil
Mohammad Reza Mousavi	King's College London, UK
David Naumann	Stevens Institute of Technology, USA
Daniel Neider	TU Dortmund University, Germany
Gethin Norman	University of Glasgow, UK
Kazuhiro Ogata	JAIST, Japan
Jose Oliveira	University of Minho, Portugal
Luigia Petre	Åbo Akademi University, Finland
Laure Petrucci	Université Paris 13, France
Jean-Francois Raskin	Université libre de Bruxelles, Belgium
Matteo Rossi	Politecnico di Milano, Italy
Marco Roveri	University of Trento, Italy
Pierluigi San Pietro	Politecnico di Milano, Italy
Gerardo Schneider	Chalmers University of Technology and University of Gothenburg, Sweden
Daniel Schoepe	Amazon, USA
Cristina Seceleanu	Mälardalen University, Sweden
Emil Sekerinski	McMaster University, Canada
Adenilso Simao	ICMC/USP, Brazil
Graeme Smith	The University of Queensland, Australia
Ana Sokolova	University of Salzburg, Austria
Silvia Lizeth Tapia Tarifa	University of Oslo, Norway
Maurice ter Beek	CNR, Italy
Jan Tretmans	TNO - Embedded Systems Innovation, Netherlands
Mattias Ulbrich	Karlsruhe Institute of Technology, Germany
Frits Vaandrager	Radboud University, Netherlands
Jaco van de Pol	Aarhus University, Denmark
Matthias Volk	Eindhoven University of Technology, Netherlands
Heike Wehrheim	University of Oldenburg, Germany
Kirsten Winter	The University of Queensland, Australia
Burkhart Wolff	University Paris-Saclay, France
Jim Woodcock	University of York, UK
Naijun Zhan	Institute of Software, Chinese Academy of Sciences, China
Min Zhang	East China Normal University, China
Zhenya Zhang	Kyushu University, Japan

Tutorials

Kazuhiro Ogata (Co-chair)	JAIST, Japan
Neeraj Kumar Singh (Co-chair)	INPT-ENSEEIHT/IRIT, University of Toulouse, France
Erika Abraham	RWTH Aachen University, Germany
Toshiaki Aoki	JAIST, Japan
Nikolaj Bjørner	Microsoft, USA
Sylvie Boldo	Inria and Université Paris-Saclay, France
Michael Butler	University of Southampton, UK
Ana Cavalcanti	University of York, UK
Yunja Choi	Kyungpook National University, South Korea
Alcino Cunha	University of Minho, Portugal
Deepak D'Souza	Indian Institute of Science, India
Thi Thu Ha Doan	Japan Advanced Institute of Science and Technology, Japan
Brijesh Dongol	University of Surrey, UK
Catherine Dubois	ENSIIE-Samovar, France
Marie Farrell	The University of Manchester, UK
Leo Freitas	Newcastle University, UK
Kim Guldstrand Larsen	Aalborg University, Denmark
Raju Halder	IIT, Patna, India
Marieke Huisman	University of Twente, Netherlands
Fuyuki Ishikawa	National Institute of Informatics, Japan
Sun Jun	Singapore Management University, Singapore
Regine Laleau	Paris Est Creteil University, France
Thierry Lecomte	CLEARSY, France
Michael Leuschel	University of Düsseldorf, Germany
Stephan Merz	Inria Nancy, France
Canh Minh Do	Japan Advanced Institute of Science and Technology, Japan
Antoine Miné	LIP6, UPMC, France
Rosemary Monahan	Maynooth University, Ireland
Cesar Munoz	NASA, USA
Fumiko Nagoya	Nihon University, Japan
Masaki Nakamura	Toyama Prefectural University, Japan
Anh Nguyen	Japan Advanced Institute of Science and Technology, Japan
S. Ramesh	General Motors R&D, India
Elvinia Riccobene	University of Milan, Italy
Graeme Smith	The University of Queensland, Australia
Tsubasa Takagi	Japan Advanced Institute of Science and Technology, Japan

Carolyn Talcott	SRI International, USA
Maurice ter Beek	CNR, Italy
Tatsuhiro Tsuchiya	The University of Osaka, Japan
Laurent Voisin	Systerel, France
Tim Willemse	Eindhoven University of Technology, Netherlands

Artifact Evaluation

Étienne André (Co-chair)	Université de Nantes, France
Masaki Waga (Co-chair)	Kyoto University, Japan
Augustin Albert	LIX, France
Ricardo Almeida	University of Glasgow, UK
Thom Badings	University of Oxford, UK
Lydia Bakiri	Université Sorbonne Paris Nord, France
Hugo Bazille	LRE, France
Laura Bussi	University of Amsterdam, Netherlands
Shaowei Cai	Institute of Software, Chinese Academy of Sciences, China
Juan Diego Campo	Universidad de la República, Uruguay
Agnishom Chattopadhyay	Imiron, Japan
Po-Chun Chien	LMU Munich, Germany
Susmoy Das	University of Glasgow, India
Sarah Dépernet	Inria Nancy, Université de Lorraine, Loria, France
Yanni Dong	University of Twente, Netherlands
Daniel Drodt	TU Darmstadt, Germany
Paul Eichler	CISPA - Helmholtz Center for Information Security, Germany
Mohamed Elwakil	United States Coast Guard Academy, USA
Eduardo Fernandes	University of Southern Denmark, Denmark
Chao Gao	Kyoto University, Japan
Gloria Gori	University of Florence, Italy
Matthias Güdemann	UAS Munich, Germany
Amit Gurung	UPES, Dehradun, India
Niklas Heidler	Technical University Darmstadt, Germany
Hsi-Ming Ho	University of Sussex, UK
Chiao Hsieh	Kyoto University, Japan
Noa Izsak	CISPA Helmholtz Center for Information Security, Germany
Philipp Kern	KIT, Germany
Elvis Gerardin Konjoh Selabi	Gran Sasso Science Institute & University of Camerino, Italy
Marian Lingsch-Rosenfeld	LMU Munich, Germany

Tianhai Liu	ITI KIT, Germany
Wei Liu	Concordia University, Canada
Deyun Lyu	National Institute of Informatics, Japan
Dara MacConville	Maynooth University, Ireland
Misato Nakabayashi	NTT Social Informatics Laboratories, Japan
Tiago Oliveira	University of Minho, Portugal
Rémi Parrot	École Centrale de Nantes, LS2N, France
Samuel Pastva	Masaryk University, Slovakia
Reuven Peleg	Hebrew University of Jerusalem Israel
Bram Pellen	Radboud University, Netherlands
Kittiphon Phalakarn	National Institute of Informatics, Japan
Tephilla Prince	Indian Institute of Technology, Dharwad, India
Lisa Pühl	RWTH Aachen University, Germany
Robert Rubbens	University of Twente, Netherlands
Sota Sato	Kyoto University, Japan
Rohith Kumar Shanmuganathan	Fraunhofer IEM, Germany
Fabien Siron	Université Paris-Saclay, CEA, List, France
Soaibuzzaman	Bauhaus University Weimar, Germany
Reza Soltani	University of Twente, Netherlands
Gaëtan Staquet	École Centrale Nantes, France
Amy Stell	University of Exeter, UK
Han Su	National Institute of Informatics, Japan
Takashi Suwa	Graduate School of Informatics, Kyoto University, and National Institute of Informatics, Japan
Landon Taylor	Utah State University, USA
Benjamin von Berg	IST, Graz University of Technology, Austria
Masaki Waga	Kyoto University, Japan
Ryo Wakizaka	Kyoto University, Japan
Yilong Yang	Beihang University, China
Shenghao Yuan	Nanjing University of Aeronautics and Astronautics (NUAA), China
Guanqin Zhang	University of New South Wales, Australia
Lei Zhang	University of Maryland, Baltimore County, USA
Zicen Zhao	Kyushu University, Japan
Kevin Zhou	National Institute of Informatics, Japan

Industry Day Track

Jeroen Keiren (Co-chair)	Eindhoven University of Technology, Netherlands
Naoto Sato (Co-chair)	Hitachi, Ltd., Japan
Klaus Havelund	Jet Propulsion Laboratory, California Institute of Technology, USA

Peter Höfner	Australian National University, Australia
Joseph Kiniry	Galois, Inc., USA
Nikolai Kosmatov	Thales Research & Technology, France
Cong Liu	Collins Aerospace, USA
Cristina Seceleanu	Mälardalen University, Sweden

Doctoral Symposium

Toshiaki Aoki (Co-chair)	Japan Advanced Institute of Science and Technology, Japan
Cyrille Artho (Co-chair)	KTH Royal Institute of Technology, Sweden
Kyungmin Bae	Pohang University of Science and Technology (POSTECH), South Korea
Alcino Cunha	University of Minho, Portugal
Guillaume Dupont	Institut de Recherche en Informatique de Toulouse, France
Dimitra Giannakopoulou	AWS, USA
Daisuke Ishii	Japan Advanced Institute of Science and Technology, Japan
Yi Li	Nanyang Technological University, Singapore
Shaoying Liu	East China Normal University, China
Peter Ölveczky	University of Oslo, Norway
Elvinia Riccobene	University of Milan, Italy
Kristin Yvonne Rozier	Iowa State University, USA
Gerardo Schneider	Chalmers University of Technology and University of Gothenburg, Sweden
Martina Seidl	Johannes Kepler University Linz, Austria
Masaki Waga	Kyoto University, Japan

Journal First Track

Anne Haxthausen (Chair)	Technical University of Denmark, Denmark
Jeannette Wing	Columbia University, USA
Cliff Jones	Newcastle University, UK
Michael Butler	University of Southampton, UK

FME Board

Ana Cavalcanti	University of York, UK
Maurice ter Beek	CNR, Italy
Nico Plat	Thanos, Netherlands
Lars-Henrik Eriksson	Uppsala University, Sweden

Einar Broch Johnsen	University of Oslo, Norway

Organization Committee

General Chair

Fuyuki Ishikawa	National Institute of Informatics, Japan

Financial Chair

Paolo Arcaini	National Institute of Informatics, Japan

Publicity Chair

Tsutomu Kobayashi	Japan Aerospace Exploration Agency (JAXA), Japan

Web Chair

Deyun Lyu	National Institute of Informatics, Japan

Workshop Chairs

Elena Troubitsyna	KTH, Sweden
Tomoyuki Yokogawa	Okayama Prefectural University, Japan

Additional Reviewers

Aadithya Srinivasan Anand
Alex Ozdemir
Alexander Stekelenburg
Alexander Weigl
Alexandre Moine
Alvaro Miyazawa
Álvaro Silva
Álvaro Tasistro
Amir Goharshady
Andrea Esposito
Andrea Manini
Andreas Bremer
Anirban Majumdar
Anna Stramaglia
Armando Tacchella
Arnaud Grivet Sébert
B. Srivathsan
Benjamin Lucien Kaminski
Benjamin von Berg
Benoît Delahaye
Bharti Chimdyalwar
Bohua Zhan

Boutheina Bannour
Bram Pellen
Calvin Chau
Carlos Tavares
Caroline Lemke
Ciaran Dunne
Clemens Eisenhofer
Daisuke Ishii
Daniel Baier
Daniel Daniel
Daniel Drodt
Daniel Luick
David Lidell
David Pape
Davide Basile
Dejan Nickovic
Delphine Longuet
Diego Latella
Dirk Pattinson
Dominik Haneberg
Dominik Mascherbauer
Dominik Schmidt
Dorde Žikelić
Duong Dinh Tran
Eduard Kamburjan
Emilio Tuosto
Fabian Vu
Fabio Mogavero
Fang Yan
Fang-Yi Lo
Federico Olmedo
Felix Gigler
Felix Wallner
Filip Macák
Filipe Arruda
Florian Sextl
Francesca Randone
Francesco Girardello
François Bobot
Franziska Alber
Gabriele Costa
Gaëtan Staquet
George Granberry
Gerhard Schellhorn
Giorgio Delzanno
Giovanni Bacci
Harshit Jitendra Motwani
Hengjun Zhao
Henrik Wachowitz
Hubert Garavel
Jacques-Henri Jourdan
Jakob Piribauer
Jie An
João Pascoal Faria
Johanna Schwartzentruber
Johannes Schoisswohl
Jonas Becker-Kupczok
Jorge Sousa Pinto
José Proença
Juan García-Garland
Julius Adelt
Juraj Síč
Kais Klai
Kevin van de Glind
Lara Bargmann
Laurent Fuchs
Leandro Buss Becker
Leopoldo Teixeira
Livia Lestingi
Lorenz Winkler
Luca Di Stefano
Luca Marzari
Luca Verderame
Ludovico Fusco
Luigi D. C. Soares
Luigi Sauro
Lukas Grätz
Lukas Holik
Lukas Panneke
Malte Lochau
Marco Scaletta
Marek Jankola
María Belén Rodríguez
Maria João Frade
Marian Lingsch-Rosenfeld
Mario Carneiro
Mark van Wijk
Martim Afonso
Martín Pedemonte
Martin Tappler

Massimo Benerecetti
Matthias Kettl
Michele Chiari
Mihaela-Larisa Clement
Mingyue Zhang
Moamin Abughazala
Mohammad Kakooei
Nazareno Aguirre
Nestan Tsiskaridze
Nicholas Smallbone
Nicola Thoben
Nicolas Berthier
Nuno Rodrigues
Ondřej Lengál
Pauline Blohm
Pedro Carrott
Peixin Wang
Peter Riviere
Philip Tasche
Philipp Kern
Philipp Wendler
Philippe Heim
Po-Chun Chien
Prince Mathew
Renato Neves
Richard Bubel
Richard Stewing
Robert J. Colvin
Robert Mensing
Robert Rubbens
Robin Coutelier
Rocio Aldeco-Perez
Rohit Kumar Shanmuganathan
Roman Andriushchenko
S. Akshay
Saina Sunny
Samuel Teuber
Sascha Klüppelholz
Sebastián Betancourt
Shuling Wang
Simon Foster
Simon Jantsch
Simone Tini
Stéphane Demri
Takashi Tomita
Teofil Sidoruk
Terru Stübinger
Thom Badings
Thomas Hader
Thomas Lemberger
Tianhai Liu
Till Schallau
Tobias Nießen
Valentin Promies
Vincenzo Ciancia
Wanwei Liu
Wasim Essbai
Wei-Lun Tsai
Wolfram Pfeifer
Xiong Xu
Yicheng Qian
Zhenbang Chen
Zhixuan Yang
Ziggy Attala

Contents

Main Track (2)

Pono 2.0: A Versatile SMT-Based Model Checker for Safety and Liveness (Long Tool Paper) 3
Áron Ricardo Perez-Lopez, Po-Chun Chien, Florian Lonsing, Samantha Archer, Ahmed Irfan, and Clark Barrett

Asynchronous Team Automata 27
Davide Basile, Maurice H. ter Beek, and José Proença

Maximum Realizability for LTL Modulo Theories 50
Andoni Rodríguez Herrero and César Sánchez

Exact Moment Estimation of Stochastic Differential Dynamics 71
Shenghua Feng, Jie An, Naijun Zhan, and Fanjiang Xu

A Framework for the Interoperable Specification and Verification of Encapsulated Data Structures 90
Wolfram Pfeifer, Werner Dietl, and Mattias Ulbrich

Can LLM Aid in Solving Constraints with Inductive Definitions? 111
Weizhi Feng, Shidong Shen, Jiaxiang Liu, Taolue Chen, Fu Song, and Zhilin Wu

Fast SMT-Based Fault Tolerance Verification for Wide Area Networks 133
Ning Kang, Peng Zhang, Hao Li, and Jianyuan Zhang

Pacing Types for Asynchronous Stream Equations 154
Florian Kohn, Arthur Correnson, Jan Baumeister, and Bernd Finkbeiner

Performance Heuristics for GR(1) Unrealizable Core Computation 174
Shachaf Cohen and Shahar Maoz

A General Framework for Robust Quantitative Semantics of Signal Temporal Logic 193
Jiawei Chen, José Luiz Vargas de Mendonça, Konstantinos Mamouras, and Jean-Baptiste Jeannin

Quantitative Monitoring of Signal First-Order Logic 214
Marek Chalupa, Thomas A. Henzinger, N. Ege Saraç, and Emily Yu

Complexity of Consistency Testing for the Release-Acquire Semantics 234
R. Govind, S. Krishna, Sanchari Sil, and B. Srivathsan

TAP Track

Towards Formally Verified Smart Contracts Compilation 255
Elvira Albert, Samir Genaim, and Enrique Martin-Martin

Selective Concolic Testing 275
Guofeng Zhang, Zhenbang Chen, Ziqi Shuai, Jun Sun, Weijiang Hong, Yufeng Zhang, Ji Wang, and Yang Liu

EUF-based Solving Dyck-Reachability with Applications to Static Analysis ... 296
Yide Du, Zhenbang Chen, Kunlin Liu, Guofeng Zhang, Xudong Wang, Ke Ma, Wei Dong, and Ji Wang

From Execution to Necessity: Proof-Based Metrics for Code Coverage (Short Paper) 317
Karl Mattsson and Peter Backeman

Active Automata Learning with Noisy Data: From Big to Small Data 326
Felix Wallner, Bernhard K. Aichernig, Benjamin von Berg, and Maximilian Rindler

Correct-by-Construction Dynamic Reachability: A Galois-Connected Approach to Bidirected Dyck Languages 347
Xiaofei Zhao

Tutorial Track

Digital Twins: a Briefing for Formalists 369
John S. Fitzgerald, Cláudio Gomes, Peter Gorm Larsen, Mikkel Schmidt Andersen, Santiago Gil, and Morten Haahr Kristensen

Reasoning over Relaxed Shared Memory Models: A Tutorial 394
Brijesh Dongol

Requirements Elicitation, Formalization, and Analysis with FRET: A Tutorial 420
Anastasia Mavridou, Andreas Katis, Mari A. Aoki, and Marie Farrell

Mixed Flow-Sensitive Static Analysis: Engineering Modularity 446
Helmut Seidl, Vesal Vojdani, Julian Erhard, and Michael Schwarz

Formally Explaining Neural Network Classification 471
Tomáš Kolárik, Grigory Fedyukovich, Faezeh Labbaf, Fabrizio Leopardi, Natasha Sharygina, and Michael Wand

The SLEEC Framework for Normative Requirements Engineering 497
Pedro Ribeiro, Radu Calinescu, Ana Cavalcanti, Marsha Chechik, Sinem Getir Yaman, Lina Marsso, Isobel Standen, and Beverley Townsend

Probabilistic Model Checking Taken by Storm: A Tutorial on the Probabilistic Model Checker Storm 524
Matthias Volk, Linus Heck, Sebastian Junges, Joost-Pieter Katoen, and Tim Quatmann

Distributed Runtime Verification in Proximity-Based Networks: A Tutorial on the Aggregate Programming Approach 550
Giorgio Audrito, Ferruccio Damiani, Giordano Scarso, Volker Stolz, and Gianluca Torta

Industry Day Track

Verifiably UX Compliant and User-Intent Layout Generation (Extended Abstract) 579
Joana Coutinho, Alexandre Lemos, and Pedro Resende

ConFixer: Robustness Semantics Based Configuration Bug Fixing for Automated Driving Systems 585
Xiaodong Zhang, Songyang Yan, and Zijiang Yang

Pragmatic Uses of AI in Formal Methods–Based Railway Projects: Early Lessons and Perspectives (Extended Abstract) 605
Thierry Lecomte, Dalay Almeida, and Norman Maury

Formal Verification for Security Certification: From a First Success to Sustainable Industrial Usage 610
Adel Djoudi and Nikolai Kosmatov

Shift-Left Requirements Verification: Integrating LLMs and Formal Methods for Automotive Systems 631
Zipong Lim, Bozhi Wu, Yon Shin Teo, Shang-wei Lin, and Yi Li

Formal Verification of Functional Correctness for the OpenHarmony LiteOS-M Kernel 651
Tianqi Zhao, Qinxiang Cao, Shenghua Feng, Minghui Zhou, Naijun Zhan, Yongzhi Cao, Junfeng Zhao, Haiyan Zhao, Hao Wang, and Zhenjiang Hu

Automated LTL Specification Generation from Industrial Aerospace Requirements 670
Zhi Ma, Xiao Liang, Cheng Wen, Rui Chen, Bin Gu, Shengchao Qin, Cong Tian, and Mengfei Yang

SeaCoral: A Collaborative Test Generation Toolset for Industrial Orchestration of Testing Tools 691
Nicolas Berthier, Steven de Oliveira, Nikolai Kosmatov, and Delphine Longuet

Efficient Multi-level Mine Dewatering Using UPPAAL STRATEGO 711
Muhammad Naeem, Cristina Seceleanu, Alf J. Isaksson, and Tiberiu J. Seceleanu

Correction to: Asynchronous Team Automata C1
Davide Basile, Maurice H. ter Beek, and José Proença

Author Index 731

Main Track (2)

Pono 2.0: A Versatile SMT-Based Model Checker for Safety and Liveness (Long Tool Paper)

Áron Ricardo Perez-Lopez[1(✉)], Po-Chun Chien[1,2], Florian Lonsing[3], Samantha Archer[1], Ahmed Irfan[4], and Clark Barrett[1]

[1] Stanford University, Stanford, USA
{arpl,barrett}@cs.stanford.edu, samanthaarcher@stanford.edu
[2] LMU Munich, Munich, Germany
po-chun.chien@sosy.ifi.lmu.de
[3] Linz, Austria
fml@florianlonsing.com
[4] SRI International, Menlo Park, USA
ahmed.irfan@sri.com

Abstract. We introduce an updated version of the Pono model checker. Pono is a versatile SMT-based model checker that integrates multiple verification algorithms and interfaces with a wide range of SMT solvers through a solver-agnostic back end. It emphasizes usability, offering support for commonly used input formats and providing C++ and Python APIs for programmatic access. The new version 2.0 introduces several important new features, including support for liveness properties, new interpolation-based safety-checking engines, a new VMT-LIB front end, and a number of usability and performance enhancements. An evaluation of the new version demonstrates significant improvements in performance over its previous version and comparable performance to other state-of-the-art model checkers. These results highlight Pono 2.0's effectiveness as a general-purpose and easily extensible verification platform.

Keywords: Model checking · Safety · Liveness · Craig interpolation · IC3 · SMT · CEGAR · Verification witness · Btor2 · VMT-LIB

1 Introduction

Model checking is a widely-used technique for automatically verifying hardware and software systems. Early algorithms for symbolic verification of finite-state systems used Binary Decision Diagrams [26]. Gradually, methods based on Boolean satisfiability (SAT) have become more prominent due to their robustness and scalability [12,22,44,62]. More recently, solvers for satisfiability modulo theories (SMT) have been employed, enabling the representation of finite-state and even infinite-state transition systems with richer structural information. For

A. Sampaio and M. Stoelinga (Eds.): FM 2026, LNCS 16557, pp. 3–26, 2026.
https://doi.org/10.1007/978-3-032-26220-2_1

example, a hardware register can be modeled as a single bit-vector rather than as individual bits, lifting reasoning from the bit level to the word level.

PONO is an open-source SMT-based model checker [58].[1] Its name, "Pono," is Hawaiian for *proper*, *correct*, or *goodness*, reflecting our goal of providing a reliable tool for verifying the correctness of systems. The tool is designed to be flexible and extensible, enabling both experimentation and competitive performance with state-of-the-art model checkers. PONO can model theory-rich transition systems, including bit-vectors, integers, reals, and arrays, and supports popular input formats such as BTOR2 [68] and SMV [28,32,61] and, as of version 2.0, VMT-LIB [38], with support for MOXI [70] forthcoming. It is solver-agnostic and interfaces with multiple SMT solvers via SMT-SWITCH [59].[2] By default, PONO uses BITWUZLA [65] for bit-vector and array solving, MATHSAT5 [36] for interpolation queries, and CVC5 [4] for everything else.

PONO supports checking safety and, as of version 2.0, also liveness properties. It boasts a versatile set of algorithms based on bounded model checking [12], k-induction [71], interpolation [62,76,77], IC3/PDR [22,35,49], liveness-to-safety reduction [11,39], and counterexample-guided abstraction refinement [40,52,57].

Since the initial release in 2021 [58], we have made numerous improvements that benefit both the user experience and performance.

1.1 Use Cases

PONO is designed to support three primary use cases: push-button verification, expert verification, and model checking research, goals described in the PONO 1.0 paper [58]. With the release of PONO 2.0, we can now point to many concrete examples and success stories in these three categories.

Push-Button Verification. Several recent verification frameworks, including μARCHIFI [74], MOXI-MC-FLOW [53], DREAMMINER [79], and BTOR2-SELECT [56], as well as industrial tool suites like TABBYCAD[3] integrate PONO as a back-end model checker for fully automated verification tasks. These systems directly invoke PONO's binary executable on model-checking problems without user intervention, demonstrating its robustness and ease of use. Notably, PONO achieved **first place** in the word-level array track of the Hardware Model Checking Competition (HWMCC) 2025 [17], further confirming its competitiveness in push-button verification.

Expert Verification. For advanced users, PONO provides flexible C++ and Python APIs for manipulating and instrumenting transition systems and properties. This enables fine-grained control over the verification process, as demonstrated by works on automated system configuration [75] and translation validation [63] for agile hardware design workflows.

[1] https://github.com/stanford-centaur/pono
[2] https://github.com/stanford-centaur/smt-switch
[3] https://www.yosyshq.com/tabby-cad-datasheet

Model Checking Research. PONO's modular and solver-agnostic architecture makes it a convenient platform for evaluating new verification algorithms and solver configurations. For instance, it has been used to explore different interpolation configurations in BITWUZLA [66]. In the evaluation part (Sect. 6) of this paper, PONO serves as a unified framework for exercising multiple model-checking algorithms. Its extensible design continues to make it an effective foundation for both research and tool development.

1.2 Contributions

To the best of our knowledge, PONO 2.0 is the first tool to support verification of liveness properties for systems represented using the BTOR2 format. We also introduce new interpolation-based model-checking engines [76, 77] as well as improvements to existing engines, enhancing both flexibility and performance. Usability has also been improved, with a new VMT-LIB [38] front end and a more robust witness-generation component. PONO 2.0 integrates seamlessly with state-of-the-art SMT solvers such as BITWUZLA [65] and CVC5 [4], allowing users to exploit advanced solver capabilities. Finally, we present a comprehensive evaluation on a diverse benchmark set, demonstrating PONO's effectiveness across a variety of verification tasks.

The rest of the paper is organized as follows. In Sect. 2, we provide a more formal description of the model checking problem addressed by PONO. Section 3 provides an overview of PONO's architecture, and Sect. 4 goes into more detail about the features that were added or improved between versions 1.0 and 2.0. We compare PONO with other state-of-the-art model checkers in Sect. 6 and conclude the paper in Sect. 7.

2 Background

Every SMT variable has a *sort* derived from a background theory, e.g., `Int`—an integer, or `(_ BitVec 32)`—a 32-bit vector. These variables can be combined into Boolean-valued *atoms* using predefined theory-specific functions, e.g., the atom `(= (bvmul a b) #b100)` states that the result of performing bit-vector multiplication on `a` and `b` equals 4. Finally, atoms can be connected with Boolean operators like "and" ($\wedge$) and "implies" ($\implies$) to create *formulas.*

The input of PONO is a *transition system* $\langle X, I, T\rangle$ consisting of: (1) X, a set of state variables (2) $I(X)$, the set of initial states, expressed as a formula over the state variables, and (3) $T(X, X')$, the transition relation, expressed as a formula over the state variables and their primed copies, which represent the current and next state, respectively. PONO allows the designation of some variables as *input* variables, which do not appear in the initial state condition or the next-state set X' of the transition relation. This makes them effectively free variables whose value can be chosen arbitrarily at every step.

If the domain of every variable in a transition is finite, e.g., a bit-vector, as defined by the theory of fixed-size bit-vectors, the transition system is finite. If

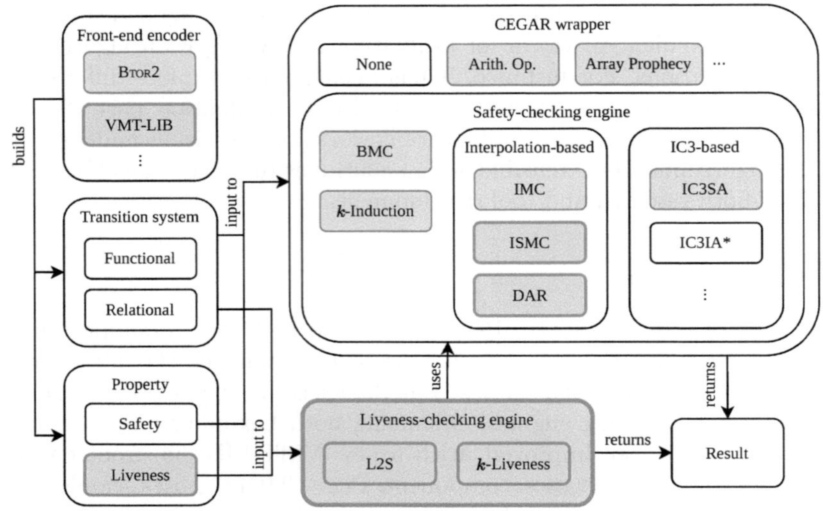

Fig. 1. System diagram of PONO (new and enhanced components in 2.0 are highlighted in green and blue, respectively; *: IC3IA is also interpolation-based)

there are some variables with infinite domain, e.g., integers from the theory of Linear Integer Arithmetic, the transition system will have an infinite number of states. This distinction is important in several model checking algorithms.

The goal of a model-checking procedure is to determine whether a certain temporal property holds for a given transition system. We restrict ourselves to properties expressible in Linear Temporal Logic (LTL), which covers many useful properties. LTL properties can be divided into two classes. *Safety* properties, also called invariant properties, assert that every reachable state of the system satisfies a predicate $P(X)$. In contrast, *liveness* properties [2] assert that some condition holds *eventually* along any path (rather than always).

The BTOR2 input format [68] is a word-level extension of the AIGER format [18] and represents safety properties with specially designated `bad` formulas, which are negations of safety properties. In other words, a `bad` property is something that should *never* hold in any reachable state. Liveness properties are expressed using lists of `justice` formulas. As with the `bad` keyword, the `justice` keyword indicates the set of formulas that should *not* hold infinitely often. More precisely, the liveness property does not hold if there exists a path starting from an initial state along which each of the justice formulas holds infinitely often. This is equivalent to saying that the property does hold if there is some path along which eventually at least one of the justice formulas becomes false forever. The space of reachable states can be restricted with `constraint` formulas—only states for which all constraint formulas are true are considered reachable.

```
IMPORT smt_switch, pono
solver = smt_switch.create_bitwuzla_solver(False)
solver.set_opt("incremental", "true")
ts = pono.FunctionalTransitionSystem(solver)
# build the transition system here ...
prop = pono.SafetyProperty(solver, solver.make_term(...))
engine = pono.KInduction(prop, ts, solver)
result = engine.check_until(10)  # unroll at most 10 steps
```

Fig. 2. Example Python API usage

3 Architecture

Figure 1 provides a high-level overview of the system and its core components. The architecture of Pono is described in detail in its previous tool paper [58]. Here, we revisit the major components and highlight the changes in version 2.0. Pono can be either executed directly as a command-line tool or linked into another application as a library. With Pono 2.0, the command-line interface (CLI) can be used without manual compilation, by running the OCI container built from the main branch.[4] This section describes the C++ API that Pono provides for other applications, but all of this functionality is also available via the CLI.

Pono builds on Smt-Switch [59] for constructing SMT formulas and solving queries. Accordingly, most operations described below use Smt-Switch `Terms` or their collections (`TermVec`, `UnorderedTermSet`, etc.) for input arguments and return values.

The core data structure in Pono is the `TransitionSystem`. As the name implies, this represents a transition system, and all other principal operations are carried out on this class. `TransitionSystems` can be constructed by manually calling `add_statevar()` and `add_inputvar()` or by using a `Frontend`. The original version of Pono provided `Btor2Frontend` and `SMVFrontend` for constructing `TransitionSystems` from Btor2 [68] and SMV [28,32,61] files, respectively. Pono 2.0 further adds support for the VMT-LIB [38] input format.

The sets of state and input variables in a `TransitionSystem` can be obtained using the `statevars()` and `inputvars()` methods, respectively, while `init()` returns the initial state condition. In functional systems, `assign_next()` can be used to set the values for each variable's next state and `state_updates()` can be used to retrieve them. In a relational style, `trans()` and `set_trans()` can be used to query and set the transition relation, respectively.

Pono supports all background theories provided by the underlying SMT solvers via Smt-Switch. Thus, when using Pono's APIs, users can construct transition systems over arbitrary solver-supported sorts. However, certain engines only support a subset of these theories, and when loading models through formats such as Btor2 or VMT-LIB, the available theories are restricted by the expressiveness of the respective input format.

[4] https://ghcr.io/stanford-centaur/pono:latest

Table 1. Summary of new and improved features in Pono 2.0

Feature	New in version 2.0	Description
Front ends	VMT-LIB, BTOR2 `justice`	Sect. 4.1
BMC engine	bound start, bound step	Sect. 4.2
k-Induction engine	bound step	
IMC engine	backward interpolation, frontier-set simplification	
ISMC engine	entirely new	
DAR engine	entirely new	
IC3SA engine	array support	
L2S translation	entirely new	Sect. 4.3
k-liveness translation	entirely new	
Arith. op. abstraction	abs. with free variables, support for more engines	Sect. 4.4
CEG prophecy	finite-index arrays	
Witness generation	support for all engines	Sect. 4.5
SMT backends	configurable options	Sect. 4.7

The workhorses of PONO are the `Provers`. These are the classes that implement the various supported model checking algorithms. The inputs to a `Prover` are a `TransitionSystem` and a `Property`. The `Property` is the temporal property that the `Prover` is asked to prove or disprove. In PONO 1.x, this is always a safety property encoded by an SMT-SWITCH `Term` representing a formula. PONO 2.0 additionally supports liveness properties and introduces two distinct classes, `SafetyProperty` and `LivenessProperty`. Since many model-checking algorithms are specialized to either safety or liveness properties, PONO 2.0 has `SafetyProver` and `LivenessProver` subclasses that take (and attempt to prove) the corresponding type of `Property`. New and updated `Provers` in PONO 2.0 are described in detail in Sect. 4.

There are also Python bindings for accessing most of the C++ API. An example using the Python bindings is illustrated in Fig. 2. A BITWUZLA solver instance `solver` is created using SMT-SWITCH. Then, a transition system `ts` and a safety property `prop` are constructed and given to a k-induction engine. Finally, the `check_until()` method is invoked to perform model checking.

4 New Features in Pono 2.0

In this section, we highlight the new features and improvements in Pono 2.0, which are summarized in Table 1.

4.1 Front-End Parsers

In Pono 1.0, the supported front ends are Btor2, SMV, and CoreIR formats, but only for safety properties. Version 2.0 adds support for liveness properties in Btor2 (using the keyword `justice`) and extends safety verification to the VMT-LIB format. VMT-LIB is based on the SMT-LIB [5] format, and supports reasoning over a richer set of theories than does Btor2.

4.2 Safety Model Checking

Pono provides a diverse set of model-checking engines for verifying safety properties, including approaches based on unrolling, IC3/PDR [22,44], and interpolation. In version 2.0, two new interpolation-based engines have been added, interpolation sequence-based model checking and dual-approximated reachability, and many of the other engines have been improved.

Bounded Model Checking (BMC) and k -Induction. BMC [12] is a common verification technique for checking safety properties. It works by unrolling the transition system $T(X, X')$ from the initial state $I(X)$ and encoding the paths that reach bad states $\neg P(X)$ as SAT/SMT queries. A satisfiable result indicates a reachable bad state (a counterexample), while an unsatisfiable result means no bad state is reachable within the unrolled depth. k-induction [71] extends BMC by adding an inductive step which attempts to prove that no bad state is reachable beyond the unrolling depth, provided the property holds for the previous k steps. If the inductive step succeeds, the property holds for all reachable states, but if not, the result is inconclusive, as the property may still hold but just not be k-inductive.

Both BMC and k-induction were supported in Pono 1.0. In version 2.0, we introduce additional options to control the unrolling process. Users can specify the step size for each unrolling, and BMC can optionally increase the unrolling bound exponentially. For BMC, there is now also an option to set the starting point for the unrolling. These enhancements have proven useful for bug finding. Notably, these engines helped Pono find the highest number of satisfiable instances in the word-level array track at HWMCC '25 [17].

Interpolation-Based Model Checking (IMC). IMC [62] is a well-known algorithm for safety verification that combines bounded model checking [12] with Craig interpolation [41]. The algorithm incrementally unrolls the transition system and checks for reachability of a bad state. When the BMC check is unsatisfiable, it iteratively extracts interpolants to overapproximate the set of reachable states. The algorithm continues until a counterexample is found (i.e., the BMC check is satisfiable) or the disjunction of computed interpolants forms a fixed point (i.e., an inductive invariant).

IMC was already available in Pono 1.0 [58]. In Pono 2.0, we extend it with two new features. First, we introduce backward derivation of interpolants. That is, instead of calling `get_interpolant(A, B)`, where `A` and `B` are the two partitions of an unsatisfiable BMC formula [62], we compute `not(get_interpolant( B, A))`. This allows the SMT solver to keep the formula `B`, which is usually more complicated than `A` (as `B` involves several copies of the unrolled transition relation T) and remains identical across multiple interpolation queries, on the solver stack, facilitating incremental solving. Prior work [8] also shows that this approach often produces "better" interpolants, making IMC converge to a fixed point faster. Second, we apply frontier-set simplification, an optimization suggested in the original IMC paper [62]. Rather than taking the disjunction of previous interpolants, we retain only the latest one for the next interpolation query, which reduces solving effort while preserving the correctness.

Interpolation Sequence-Based Model Checking (ISMC). A new addition in Pono 2.0 is the ISMC engine [76]. Similar to IMC, ISMC is also based on unrolling of the transition system and Craig interpolation, but differs in how the interpolants are computed and used to construct a fixed point. Instead of deriving a single interpolant at each step, which represents a one-step overapproximation, ISMC derives a length n *inductive sequence* of interpolants, with each interpolant overapproximating reachable states at steps 1 to n. At each iteration, ISMC unrolls the system one step further and checks for reachability of bad states. If the BMC check is unsatisfiable, a new interpolation sequence is computed and conjoined with previously derived ones to form an evolving reachability sequence (a concept similar to *frames* in IC3/PDR [22,44]). The algorithm continues until either a counterexample is found or the reachability sequence forms a fixed point. Unlike IMC, ISMC preserves previously computed interpolants when the unrolling bound is increased, allowing information to be reused across iterations. Furthermore, ISMC issues only a single interpolation query per unrolling and immediately continues with the next unrolling, whereas IMC may pose multiple queries. This often enables ISMC to detect counterexamples more quickly in practice, although convergence to a fixed point may take longer [7,76].

Dual Approximated Reachability (DAR). Another new engine in Pono 2.0 is DAR [77], which is also based on interpolation. Compared to ISMC, which maintains only a forward reachability sequence (a sequence of overapproximations of states reachable from the initial state), DAR additionally maintains a backward reachability sequence (a sequence of overapproximations of states from which a bad state is reachable). Moreover, DAR interpolates in both forward and backward directions to simultaneously refine these sequences.

DAR primarily poses local interpolation queries that pair one forward and one backward element with a single step of the transition relation. Interpolants derived from these local queries extend and refine both the forward and backward sequences. When the information from local queries is insufficient for further refinement, DAR resorts to global queries that use multiple unrollings to derive stronger interpolants. The algorithm proceeds until either a counterexample is found or one of the two reachability sequences forms a fixed point. Although DAR typically requires more interpolation queries than IMC or ISMC, these queries tend to be smaller and more tractable for the SMT solver. Previous studies [7,77] have shown that DAR complements IMC and ISMC by solving problem instances that are hard for the other two algorithms.

IC3 with Syntax-Guided Abstraction (IC3SA). IC3SA [49] is a word-level IC3 variant originally implemented in AVR [50]. Instead of reasoning at the bit level, it abstracts the state space using equivalence classes between the terms in the syntax of the transition system and the property. This abstraction ensures that the abstract state space has a size that is independent of the registers' bit widths, while at the same time capturing equality relations among the syntactic terms of the model. It can be further combined with uninterpreted functions for data abstraction. In PONO 1.0, the IC3SA implementation supported only bit-vectors, which is the version described in the original paper. Version 2.0 extends it to handle arrays (with bit-vector indices and values).

4.3 Liveness Model Checking

PONO 2.0 introduces support for verifying liveness properties in finite-state transition systems. Two approaches are implemented, both of which reduce liveness verification to safety verification and leverage the existing safety-checking infrastructure in PONO.

Liveness to Safety (L2S). L2S [11] is a standard reduction technique that transforms liveness checking into safety checking by detecting *revisited states* to identify lassos. It introduces a nondeterministic *save* oracle, modeled by a primary input variable, and a copy of the state variables to record the saved state. When the save variable is triggered, the current state is stored in the copied variables as the saved state. If the system later revisits the same state, a lasso is detected, indicating a counterexample to the liveness property. If no such revisit is possible, the property is proven. Our implementation supports multiple justice signals and can find both counterexamples and proofs. In PONO, L2S is implemented as a preprocessing step that transforms the transition system in place, allowing any safety-checking engine to be used directly on the resulting system. This design also allows other components, such as k-liveness, to reuse the transformation procedure seamlessly.

k-Liveness. The k-liveness algorithm [39] reduces liveness checking to safety checking by instrumenting the transition system with a counter that tracks how many times a justice signal has been triggered. Our implementation currently supports only a single justice signal. If no execution exists that triggers the signal k times, the algorithm reports that the property holds. Otherwise, it increments k and repeats the process. While k-liveness, in its original form, can only determine when a property holds, PONO extends it with additional procedures to also detect cases when the property is violated. Specifically, when a trace is found that triggers the justice signal k times, Pono checks for the existence of a lasso (i.e., a revisited state) in two ways: (1) by analyzing the trace found in the previous step, and (2) by performing BMC in lock-step on the L2S-transformed system, as suggested in the original publication [39]. The first approach is incomplete but more lightweight, as it requires no additional solving. In practice, k-liveness has been shown to be complementary to L2S, particularly for proving when a liveness property holds. In Pono, users can configure the increment step for k as well as the counter encoding (binary or one-hot).

4.4 Counterexample-Guided Abstraction Refinement (CEGAR)

Pono provides CEGAR [40] wrappers on top of its safety-checking engines. In general, CEGAR works by constructing an abstract system that overapproximates the transition relation and then checking safety using this abstraction. If the property is proven (i.e., the bad state is unreachable), it holds for the concrete system as well. Otherwise, the obtained counterexample is analyzed: if it is spurious, the abstraction is refined, and the process repeats.

Arithmetic Operation Abstraction. A common approach to speeding up hardware model checking is to abstract complex arithmetic operations such as multiplication, division, or remainder using uninterpreted functions (UFs) [52]. During refinement, Pono reconstructs a BMC query guided by the abstract counterexample and adds concretization lemmas for each UF as assumptions. If the BMC query is unsatisfiable, an unsat core is extracted, and the operations involved in the core are concretized.

Although CEGAR with UF abstraction was already implemented in the first version of Pono [58], it was only integrated with the IC3SA and IC3IA engines. In Pono 2.0, we extend CEGAR integration to additional engines, including BMC, k-induction, IMC, and the newly-implemented ISMC and DAR. In addition to UF-based abstraction, Pono 2.0 also supports abstracting arithmetic operations using free variables, providing a coarser abstraction.

To be integrated into this CEGAR framework, a reachability engine must support two capabilities: exporting counterexamples (via `Prover::witness()`) for refinement and resetting its internal state (via `Prover::reset_env()`) to start a new CEGAR iteration. This design enables a largely black-box integration of engines. In the general case, the solver is reinitialized between iterations, favoring modularity over performance. For selected engines such as IC3IA and IC3SA, Pono provides a tighter integration that reuses solver state across refinements, improving efficiency while requiring additional implementation effort.

Counterexample-Guided Prophecy (CEGP). CEGP is a CEGAR-based approach for systems with arrays [57]. Array variables are abstracted using uninterpreted sorts, and array operations such as `select` and `store` are replaced with uninterpreted functions (UFs). This abstraction enables the use of SMT solvers that do not natively support the theory of arrays or do not have the functionality to compute interpolants. During refinement, it is not always possible to eliminate spurious counterexamples using only quantifier-free lemmas over existing variables. To avoid the need for potentially expensive quantified lemmas, the approach introduces auxiliary *history* and *prophecy* variables [1]. These variables rule out spurious traces at the given bound and have been shown to help prove properties that would otherwise require universally quantified invariants [57].

Building on the CEGP implementation in PONO 1.0 [58], which only supported arrays with infinite-domain indices (e.g., mathematical integers), the current version extends support to finite-domain indices such as bit-vectors, which are prevalent in hardware model checking. Although our current implementation of CEGP in PONO 2.0 is incomplete for finite-domain indices due to its reliance on the refiner based on the array-property fragment [23], it remains practically useful. In particular, at HWMCC '25 [17], CEGP contributed to PONO solving the most uniquely-proved instances in the word-level array track. Additionally, PONO 2.0 also allows CEGP to be combined with arithmetic-operation abstraction, further enhancing its applicability to complex hardware-verification tasks.

4.5 Witness Generation

PONO can export violation witnesses for safety properties in BTOR2 and VCD formats, and for liveness properties in BTOR2 format. The witnesses can also be retrieved programmatically via the `Prover::witness()` method of the API. However, in the original version, this functionality was limited to unrolling-based engines such as BMC, k-induction, and IMC. In PONO 2.0, witness generation for violated properties has been extended to all engines, including IC3-based and CEGAR-based engines, as well as the newly integrated ISMC and DAR. Extensive testing has been conducted to improve the robustness of witness generation. Notably, PONO achieved a 100% witness confirmation rate at HWMCC '25 [17].

PONO does not yet support exporting correctness proofs as *witness circuits* for safety properties [46]. However, advanced users can retrieve the inductive invariant via the `Prover::invar()` API call. When the `check-invar` option is enabled, PONO can also internally validate the inductiveness of the computed invariant to ensure its correctness.

4.6 Parallel Portfolio

Parallel portfolios are a common approach for combining the strengths of multiple engines to improve overall performance. In PONO 2.0, we provide improved, separate scripts to run portfolios on BTOR2 tasks with and without arrays, supporting safety properties and the exporting of violation witnesses. The portfolios were constructed with a data-driven approach: we first benchmarked a representative set of engines on HWMCC benchmark tasks from 2024 [16] and earlier, then applied an iterative greedy selection to choose engines that maximize the number of solved tasks when combined with the existing selections. Expert knowledge was used to further refine the portfolios to avoid overfitting. The portfolios are designed to be run on machines with at least 16 cores but can be easily adapted to different hardware and use cases. PONO's parallel portfolio was essential to its strong performance in the word-level array track in HWMCC '25 [17].

4.7 SMT Solver Backends

PONO relies on SMT-SWITCH [59] to interface with SMT solvers in a uniform and solver-agnostic manner. For PONO 2.0, both SMT-SWITCH and the solver-integration layer have been extended to improve performance and configurability.

Updates in Smt-Switch. PONO 2.0 builds on the latest version of SMT-SWITCH, which has received several major updates since its initial release [59]. First, the CVC4 [6] backend has been upgraded to CVC5 [4], providing better performance, modernized APIs, and continued solver maintenance. Second, BITWUZLA [65] has been fully integrated with an updated API and extended to support interpolation, replacing BOOLECTOR [68] (which is no longer actively maintained) as the default SMT solver in PONO 2.0. Finally, incremental interpolation support has been added for MATHSAT5 [36], allowing formulas to be reused across interpolation queries for improved efficiency. (This is also supported by BITWUZLA.) Together, these developments enable PONO to leverage the latest progress in SMT technology.

Configurability. To complement the backend updates, PONO 2.0 introduces several configurability enhancements. Users can now specify a dedicated interpolating solver that differs from the main SMT solver, providing additional flexibility for experimentation and tuning. Advanced users can also pass custom solver options (the options that can be set via `set-option` in SMT-LIB) through PONO, allowing fine-grained control over backend configurations. This feature has been particularly useful in evaluating the impact of different interpolation heuristics and parameters in BITWUZLA [66].

5 Related Work

This section reviews several model-checking tools related to PONO.

AVR [50] is a word-level model checker for safety properties expressed in the BTOR2 and VMT-LIB formats. Its main engine is IC3SA [49], for which it is the original implementation. It also has basic implementations of BMC and k-induction, as well as a number of arithmetic and data abstraction techniques. AVR interfaces with BOOLECTOR [68], MATHSAT5 [36], YICES2 [43], and Z3 [64] as SMT solver backends. In comparison, PONO additionally supports CVC5 [4] and BITWUZLA [65]. PONO also includes several additional verification engines (e.g., interpolation-based), offers a Python API, provides greater configurability, and supports liveness checking.

NUXMV [28] is a model-checking tool suite developed at Fondazione Bruno Kessler along with the MATHSAT5 solver. It is the successor to NUSMV [32], which in turn evolved from the original SMV solver. Accordingly, its native input language is SMV, though it also supports VMT-LIB and Btor2. nuXmv offers a diverse set of verification engines, including BMC, k-induction, IMC, IC3, and IC3IA, and supports both liveness and safety properties. However, it is closed-source and exclusively relies on MathSAT5 as its SMT backend, thereby limiting its extensibility.

Kind2 [29] is an SMT-based model checker that uses the Lustre input format [27]. It implements BMC, k-induction, IC3, and IC3IA and supports a wide variety of SMT solvers (Bitwuzla, cvc5, MathSAT5, OpenSMT [25], SMT-Interpol [31], Yices2, Yices, and Z3). Kind2 calls the solver binaries directly, passing the queries in SMT-LIB format. Although this provides similar flexibility to that provided by Smt-Switch in Pono, as evidenced by the number of supported solvers, it can incur a significant overhead and restricts solver usage to interactions via the command line.

rIC3 [72] and ABC [24] are two very successful bit-level model checkers for AIGER [18], which represents finite-state systems over the QF_BV theory. Btor2 extends AIGER to the word level and adds support for arrays, allowing higher-level system descriptions. While rIC3 also accepts Btor2 as an input, it does not support arrays and preprocesses the model by bit-blasting it to a purely propositional representation. rIC3 primarily handles safety properties and has a preliminary implementation of liveness analysis[5], whereas ABC supports both safety and liveness verification.

CoSA [60], an early prototype for what eventually became Pono, is an SMT-based model checker implemented in Python and built on PySMT [47]. It takes CoreIR [42] as input and supports only safety verification. Similarly, MoXIchecker [3] and PyVMT [48] are two other model-checking frameworks built on PySMT, which take MoXI [70] and VMT models as input, respectively. As of the latest release (0.9.6), PySMT supports a set of SMT solvers similar to that supported by Smt-Switch, except that it does not yet support Bitwuzla.

Constrained Horn Clauses (CHCs) offer an alternative way of encoding model checking problems. Instead of explicitly unrolling or symbolically exploring the transition system, CHC-based approaches encode verification tasks as sets of logical implications whose satisfiability corresponds to system correctness, i.e., the property holds. Spacer [54] and FreqHorn [45] are CHC solvers built on top of Z3, while the more recent Golem [20] employs OpenSMT as its back end.

6 Evaluation

Pono is intended to be used as an automated verification tool in hardware and software design flows. Therefore, we aim to make its performance competitive with other state-of-the-art tools. Pono 2.0 finished in first place in the bit-vectors with arrays track of HWMCC 2025 and in fifth place in the bit-vector-only track (behind two bit-level tools [55,72], a portfolio tool whose portfolio includes Pono itself [56], and AVR [50]). Here we provide a more comprehensive

[5] https://github.com/gipsyh/rIC3/pull/53

experimental evaluation of Pono's performance on various model checking tasks in both finite (bit-vectors) and infinite (integers and reals) domains.

Evaluated Tools. We compared Pono 2.0 (commit 60dd6b3)[6] against its previous version 1.0, along with other state-of-the-art model checkers, including nuXmv (2.1.0), and the HWMCC '25 versions of rIC3 (commit 5253520) and AVR (with fixes to enable it to run in our environment; commit 03bc7f6).

All compared tools are top contenders in recent editions of HWMCC. AVR was the overall winner at HWMCC '20. rIC3 won the bit-vectors-only tracks at HWMCC '24 and '25. nuXmv consistently ranked second or third in various tracks in HWMCC '20 and '24. Pono 1.0 was the overall winner at HWMCC '19 (under its previous name CoSA2) and the third-place finisher at HWMCC '20.

In the following, we present the portfolio results based on the virtual best solver computed from 16 configurations per tool, approximating the HWMCC setup. For Pono 2.0, we used modified HWMCC '25 portfolios; for Pono 1.0, an expanded HWMCC '20 portfolio (the original one has only 6 engines). We employed rIC3's 16-engine portfolio for HWMCC '25. AVR's HWMCC '25 portfolio contains 22 engines; we selected the top-performing 16. nuXmv's competition portfolio is not publicly available, so we evaluated all combinations of configuration options in its user manual [21] and selected the best-performing 16.

Pono 2.0 in this evaluation uses Bitwuzla [65,66] for bit-vector and array solving and interpolation,[7] cvc5 [4] for arithmetic logic solving, and MathSAT5 [36] for arithmetic logic interpolation. Pono 1.0 uses Boolector [67] for bit-vector and array solving and MathSAT5 for interpolation. rIC3 uses SAT solvers GipSAT [73], CaDiCaL [13], and Kissat [14] for bit-vector solving. AVR uses both Boolector [68] and Yices2 [43] for bit-vector and array solving and only Yices2 for arithmetic logic solving. nuXmv uses MathSAT5 for solving and interpolation for all logics. The evaluated portfolios are provided in the accompanying reproduction package [69].

Experimental Setup. All experiments were conducted on machines running Ubuntu 24.04 (64-bit), each equipped with an AMD Ryzen 9 7950X 16-core processor and 128 GiB of RAM. Each benchmark process corresponded to a run of a single model-checking engine, and was limited to one CPU core, 20 min of CPU time, and 8 GiB of memory. BenchExec [9] was employed to ensure reliable resource limits and measurement.

---

[6] While the experimental evaluation was conducted using a specific pre-release development commit of Pono, all features and engines described in this paper are fully integrated and available in the official 2.0 release.

[7] Bitwuzla has full interpolation support for bit-vectors and limited support for interpolation in other theories including uninterpreted functions and arrays; see their paper for details. MathSAT5 has similar limitations.

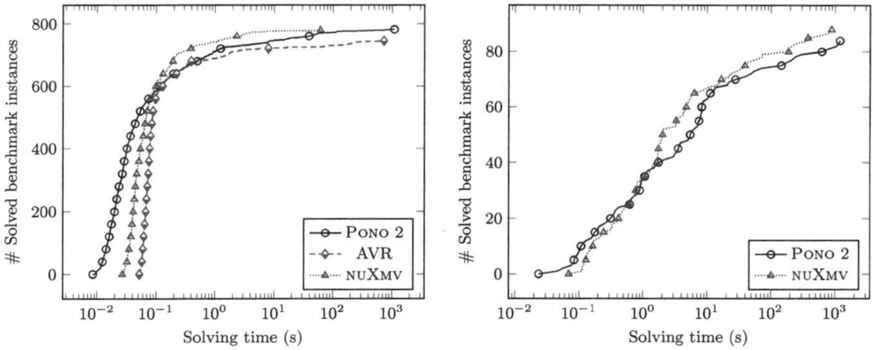

(a) 796 QF_LIA tasks (b) 99 QF_LRA tasks

Fig. 3. Number of VMT-LIB safety tasks solved by each tool

6.1 Safety Checking

We evaluated Pono's performance on a range of safety benchmarks covering integer and real arithmetic, fixed-size bit-vectors, and bit-vector arrays. For each category, we compared Pono 2.0 with state-of-the-art model checkers that support the corresponding input formats and theories, measuring both the number of instances solved and run times. The results highlight the improvements introduced in Pono 2.0 over its predecessor, as well as its competitive performance against other tools across diverse benchmark families.

Integer Arithmetic. We collected 796 verification tasks from the VMT-LIB benchmark set [29, 37] using quantifier-free linear integer arithmetic (QF_LIA). Pono 2.0 was compared with AVR and nuXmv, both of which support the VMT-LIB format natively. (Pono 1.0 was excluded as it does not support VMT-LIB.) As shown in Fig. 3a, Pono solved 783 instances, outperforming AVR (746) and slightly surpassing nuXmv within the time limit. All three tools solved the majority of the tasks in 10 s. Notably, Pono was able to solve two instances uniquely and was the fastest on 551 tasks.

Real Arithmetic. Pono 2.0 was compared with nuXmv on 99 VMT-LIB tasks featuring quantifier-free linear real arithmetic (QF_LRA) [33, 37]. We exclude AVR from the comparison as it does not support QF_LRA. The results shown in Fig. 3b indicate that Pono's performance (85 solved) is comparable to nuXmv (89 solved). In particular, Pono completed 37 tasks faster than nuXmv and was able to solve one instance that nuXmv could not.

Fixed-Size Bit-Vectors. We compared Pono 2.0 with the tools that natively support Btor2 (rIC3, AVR, and Pono 1.0) on two benchmark sets using quantifier-free bit-vectors (QF_BV). These comprise all 839 available word-level Btor2 tasks from previous HWMCCs [15, 16, 19], as well as the SV-COMP '25 [10] ReachSafety tasks translated to Btor2 by CPV [30, 51]. Since

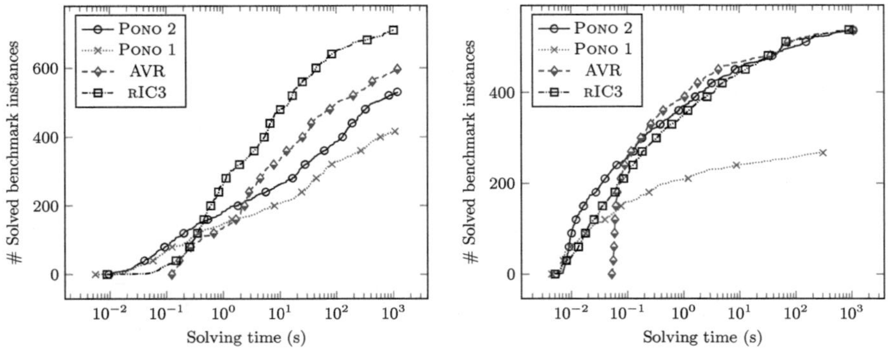

(a) 839 HWMCC tasks (b) 698 SV-COMP ReachSafety tasks

Fig. 4. Number QF_BV safety tasks solved by each tool

the HWMCC and SV-COMP sets overlap, and the latter contains several thousand instances, we sampled up to 16 benchmarks per family, 698 in total, from the SV-COMP set after excluding those already present in the HWMCC set.

Figures 4a and 4b summarize the results on the HWMCC and SV-COMP benchmark set, respectively. On the HWMCC benchmark set, rIC3 (709 solved) and AVR (597 solved) both outperformed Pono 2.0 (534 solved). Despite this, Pono 2.0 was the fastest on 212 of the tasks and uniquely solved 51 of them. On the SV-COMP benchmark set, Pono 2.0 solved the same number of instances as rIC3 (537) and one more than AVR. It was the fastest on 157 (more than any other tool) and uniquely solved six instances. Across both benchmark sets, Pono 2.0 consistently performed better than 1.0, which solved only 417 HWMCC and 268 SV-COMP instances.

Several new features in Pono 2.0 are crucial for this improved performance, including arbitrary starting points and exponentially increasing step size in BMC, backward interpolation in IMC (Sect. 4.2), and arithmetic abstraction using free variables (Sect. 4.4) .

Bit-Vector Arrays. Following a similar process as for QF_BV, we collected benchmark tasks using quantifier-free arrays with bit-vector indices and values (QF_ABV) from previous HWMCCs (863) and the SV-COMP '25 ReachSafety track, removed duplicate instances, and sampled up to 16 benchmarks per family from the SV-COMP set (yielding 502). Pono 2.0 was compared with AVR and Pono 1.0, the only other tools that natively support QF_ABV in Btor2. On the HWMCC set (Fig. 5a, Pono 2.0 solved 677 tasks (fastest on 171; uniquely solved 13), outperforming Pono 1.0 (587) and closely matching AVR (690). On the SV-COMP set (Fig. 5b), Pono 2.0 solved 132 instances (fastest on 92; uniquely solved 8), surpassing both Pono 1.0 (54) and AVR (129).

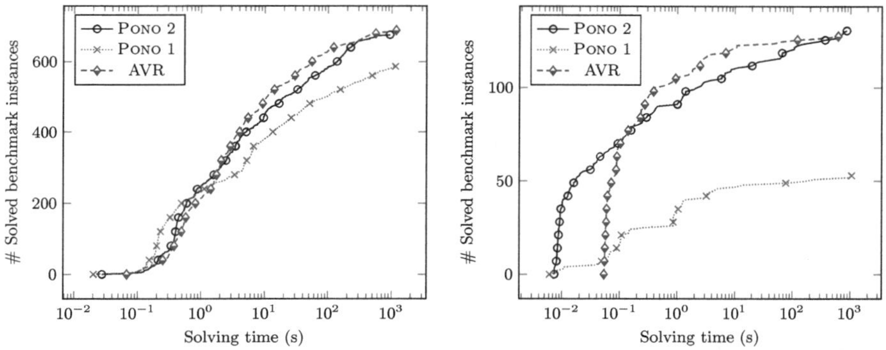

(a) 863 HWMCC tasks (b) 502 SV-COMP ReachSafety tasks

Fig. 5. Number of QF_ABV safety tasks solved by each tool

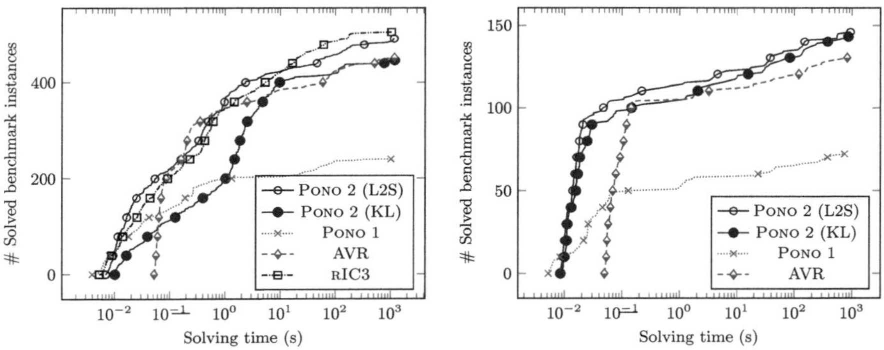

(a) 839 QF_BV tasks (b) 220 QF_ABV tasks

Fig. 6. Number of translated SV-COMP '25 Termination tasks solved by each tool

6.2 Liveness Checking

PONO's implementations of L2S and k-liveness (KL), both utilizing the same underlying safety-checking engines as in the previous experiments, were evaluated alongside PONO 1.0, AVR, and RIC3. The benchmark tasks were collected from the SV-COMP '25 Termination track and translated from C to BTOR2 by CPV [30]. Since the other tools do not support liveness (the keyword `justice` in BTOR2) directly, they were run on tasks preprocessed with L2S by CPV .

Fixed-Size Bit-Vectors. Figure 6a compares PONO 2.0 with AVR, RIC3, and PONO 1.0 on 839 QF_BV liveness tasks. PONO 2.0 (running L2S) outperformed AVR and its predecessor, and is second only to RIC3. PONO's L2S also solved 48 unique tasks and was fastest on 488. Among the two liveness algorithms, L2S performed noticeably better than k-liveness (494 vs. 447 solved), although one task was solved exclusively by k-liveness, and it was the fastest for 7.

Bit-Vector Arrays. As shown in Fig. 6b, PONO 2.0 achieved better performance than either AVR or PONO 1.0 on liveness benchmark tasks using QF_ABV. L2S and k-liveness exhibited similar performance, solving 147 and 144 instances, respectively. L2S solved 3 uniquely and 114 fastest, while k-liveness had 33 fastest but no unique solves.

7 Conclusion and Future Work

We present PONO 2.0, a major evolution of our SMT-based model checker. It introduces support for liveness, new and improved safety engines, a new input format, and usability and performance enhancements. Thanks to its modular and extensible design, PONO 2.0 is both a dependable tool for push-button verification and a foundation for exploring novel verification techniques. In the future, we plan to continue expanding PONO's capabilities, including support for the MoXI [70] language, exporting correctness proofs as witness circuits [46], and integrating new liveness-checking algorithms such as rlive [34, 78].

Acknowledgements. We would like to thank Makai Mann, the original author of PONO [58]; Luiza Pires Ribeiro, who helped identify issues in PONO's witness-generation component; and Anna Eaton, who initiated work on MoXI support for Pono. We also thank all other contributors listed in the public commit log. We are especially grateful to Aina Niemetz and Mathias Preiner for their work on making BITWUZLA [65] a powerful SMT solver for PONO, including its support for producing interpolants for bit-vectors [66]. We thank Yoni Zohar for his work on SMT-SWITCH [59] and for contributions to several solvers used by PONO. Finally, we thank the developers of CVC5 [4] and MATHSAT5 [36], which are also used by PONO.

Funding Information. Po-Chun Chien is supported by the German Research Foundation (DFG) under grants 378803395 (ConVeY) and 536040111 (Bridge), as well as a fellowship from the German Academic Exchange Service (DAAD). This work was also supported by the Stanford Center for Portable Accelerated Learning (Portal) and the Stanford Center for Automated Reasoning.

Data Availability Statement. A reproduction package for this article is available on Zenodo [69]. The source code for both PONO and SMT-SWITCH is hosted on GitHub under the permissive BSD 3-Clause License.

References

1. Abadi, M., Lamport, L.: The existence of refinement mappings. Theor. Comput. Sci. **82**(2), 253–284 (1991). https://doi.org/10.1016/0304-3975(91)90224-P
2. Alpern, B., Schneider, F.B.: Defining liveness. Inf. Process. Lett. **21**(4), 181–185 (1985). https://doi.org/10.1016/0020-0190(85)90056-0
3. Ates, S., Beyer, D., Chien, P.C., Lee, N.Z.: MoXIchecker: An extensible model checker for MoXI. In: Protzenko, J., Raad, A. (eds.) Proc. VSTTE. LNCS, vol. 15525, pp. 1–14. Springer (2024). https://doi.org/10.1007/978-3-031-86695-1_1

4. Barbosa, H., et al.: cvc5: a versatile and industrial-strength SMT solver. In: Fisman, D., Rosu, G. (eds.) Proc. TACAS (1). LNCS, vol. 13243, pp. 415–442. Springer (2022). https://doi.org/10.1007/978-3-030-99524-9_24
5. Barrett, C., Fontaine, P., Tinelli, C.: The satisfiability modulo theories library (SMT-LIB) (2016). https://smt-lib.org
6. Barrett, C.W., et al.: CVC4. In: Gopalakrishnan, G., Qadeer, S. (eds.) Proc. CAV. LNCS, vol. 6806, pp. 171–177. Springer (2011). https://doi.org/10.1007/978-3-642-22110-1_14
7. Beyer, D., Chien, P.C., Jankola, M., Lee, N.Z.: A transferability study of interpolation-based hardware model checking for software verification. Proc. ACM Softw. Eng. **1**(FSE), 2028–2050 (2024). https://doi.org/10.1145/3660797
8. Beyer, D., Lee, N.Z., Wendler, P.: Interpolation and SAT-based model checking revisited: adoption to software verification. J. Autom. Reason. **69**(1), 5 (2025). https://doi.org/10.1007/S10817-024-09702-9
9. Beyer, D., Löwe, S., Wendler, P.: Reliable benchmarking: requirements and solutions. Int. J. Softw. Tools Technol. Transf. **21**(1), 1–29 (2019). https://doi.org/10.1007/S10009-017-0469-Y
10. Beyer, D., Strejcek, J.: Improvements in software verification and witness validation: SV-COMP 2025. In: Gurfinkel, A., Heule, M. (eds.) Proc. TACAS (3). LNCS, vol. 15698, pp. 151–186. Springer (2025). https://doi.org/10.1007/978-3-031-90660-2_9
11. Biere, A., Artho, C., Schuppan, V.: Liveness checking as safety checking. Electr. Notes Theor. Comput. Sci. **66**(2), 160–177 (2002). https://doi.org/10.1016/S1571-0661(04)80410-9
12. Biere, A., Cimatti, A., Clarke, E.M., Strichman, O., Zhu, Y.: Bounded model checking. Adv. Comput. **58**, 117–148 (2003). https://doi.org/10.1016/S0065-2458(03)58003-2
13. Biere, A., Faller, T., Fazekas, K., Fleury, M., Froleyks, N., Pollitt, F.: CaDiCaL 2.0. In: Gurfinkel, A., Ganesh, V. (eds.) Proc. CAV (1). LNCS, vol. 14681, pp. 133–152. Springer (2024). https://doi.org/10.1007/978-3-031-65627-9_7
14. Biere, A., Fazekas, K., Fleury, M., Heisinger, M.: CaDiCaL, Kissat, Paracooba, Plingeling and Treengeling entering the SAT competition 2020. In: Balyo, T., Froleyks, N., Heule, M., Iser, M., Järvisalo, M., Suda, M. (eds.) Proc. SAT Competition – Solver and Benchmark Descriptions. Department of Computer Science Report Series B, vol. B-2020-1, pp. 51–53. University of Helsinki (2020). https://tuhat.helsinki.fi/ws/files/142452772/sc2020_proceedings.pdf.#page=50
15. Biere, A., Froleyks, N., Preiner, M.: Hardware model checking competition 2020. https://hwmcc.github.io/2020/. Accessed 15 Oct 2025
16. Biere, A., Froleyks, N., Preiner, M.: Hardware model checking competition 2024. In: Narodytska, N., Rümmer, P. (eds.) Proc. FMCAD, p. 7. TU Wien Academic Press (2024). https://doi.org/10.34727/2024/ISBN.978-3-85448-065-5_6
17. Biere, A., Froleyks, N., Preiner, M.: Hardware model checking competition 2025. In: Irfan, A., Kaufmann, D. (eds.) Proc. FMCAD. p. 7. TU Wien Academic Press (2025). https://doi.org/10.34727/2025/isbn.978-3-85448-084-6_6
18. Biere, A., Heljanko, K., Wieringa, S.: AIGER 1.9 and beyond. Tech. Rep. 11/2, Institute for Formal Models and Verification, Johannes Kepler University (2011). https://doi.org/10.35011/fmvtr.2011-2
19. Biere, A., Preiner, M.: Hardware model checking competition 2019. https://fmv.jku.at/hwmcc19/, Accessed 15 Oct 2025

20. Blicha, M., Britikov, K., Sharygina, N.: The Golem Horn solver. In: Enea, C., Lal, A. (eds.) Proc. CAV (2). LNCS, vol. 13965, pp. 209–223. Springer (2023). https://doi.org/10.1007/978-3-031-37703-7_10
21. Bozzano, M., et al.: nuXmv 2.1.0 user manual. Tech. rep., Fondazione Bruno Kessler (2024). https://nuxmv.fbk.eu/downloads/nuxmv-user-manual.pdf
22. Bradley, A.R.: SAT-based model checking without unrolling. In: Jhala, R., Schmidt, D. (eds.) Proc. VMCAI. LNCS, vol. 6538, pp. 70–87. Springer (2011). https://doi.org/10.1007/978-3-642-18275-4_7
23. Bradley, A.R., Manna, Z.: The calculus of computation - decision procedures with applications to verification. Springer (2007). https://doi.org/10.1007/978-3-540-74113-8
24. Brayton, R.K., Mishchenko, A.: ABC: An academic industrial-strength verification tool. In: Touili, T., Cook, B., Jackson, P.B. (eds.) Proc. CAV. LNCS, vol. 6174, pp. 24–40. Springer (2010). https://doi.org/10.1007/978-3-642-14295-6_5
25. Bruttomesso, R., Pek, E., Sharygina, N., Tsitovich, A.: The OpenSMT Solver. In: Esparza, J., Majumdar, R. (eds.) Proc. TACAS. LNCS, vol. 6015, pp. 150–153. Springer (2010). https://doi.org/10.1007/978-3-642-12002-2_12
26. Burch, J.R., Clarke, E.M., McMillan, K.L., Dill, D.L., Hwang, L.J.: Symbolic model checking: 10^{20} states and beyond. Inf. Comput. **98**(2), 142–170 (1992). https://doi.org/10.1016/0890-5401(92)90017-A
27. Caspi, P., Pilaud, D., Halbwachs, N., Plaice, J.: LUSTRE: a declarative language for programming synchronous systems. In: Proc. POPL, pp. 178–188. ACM (1987). https://doi.org/10.1145/41625.41641
28. Cavada, R., et al.: The nuXmv Symbolic Model Checker. In: Biere, A., Bloem, R. (eds.) Proc. CAV. LNCS, vol. 8559, pp. 334–342. Springer (2014). https://doi.org/10.1007/978-3-319-08867-9_22
29. Champion, A., Mebsout, A., Sticksel, C., Tinelli, C.: The Kind 2 Model Checker. In: Chaudhuri, S., Farzan, A. (eds.) Proc. CAV (2). LNCS, vol. 9780, pp. 510–517. Springer (2016). https://doi.org/10.1007/978-3-319-41540-6_29
30. Chien, P.C., Lee, N.Z.: CPV: A circuit-based program verifier. In: Finkbeiner, B., Kovács, L. (eds.) Proc. TACAS (3). LNCS, vol. 14572, pp. 365–370. Springer (2024). https://doi.org/10.1007/978-3-031-57256-2_22
31. Christ, J., Hoenicke, J., Nutz, A.: SMTInterpol: An interpolating SMT solver. In: Donaldson, A.F., Parker, D. (eds.) Proc. SPIN. LNCS, vol. 7385, pp. 248–254. Springer (2012). https://doi.org/10.1007/978-3-642-31759-0_19
32. Cimatti, A., et al.: NuSMV 2: An Opensource tool for symbolic model checking. In: Brinksma, E., Larsen, K.G. (eds.) Proc. CAV. LNCS, vol. 2404, pp. 359–364. Springer (2002). https://doi.org/10.1007/3-540-45657-0_29
33. Cimatti, A., Griggio, A.: Software Model Checking via IC3. In: Madhusudan, P., Seshia, S.A. (eds.) Proc. CAV. LNCS, vol. 7358, pp. 277–293. Springer (2012). https://doi.org/10.1007/978-3-642-31424-7_23
34. Cimatti, A., Griggio, A., Johannsen, C., Rozier, K.Y., Tonetta, S.: Infinite-state liveness checking with rlive. In: Piskac, R., Rakamaric, Z. (eds.) Proc. CAV (1). LNCS, vol. 15931, pp. 215–236. Springer (2025). https://doi.org/10.1007/978-3-031-98668-0_11
35. Cimatti, A., Griggio, A., Mover, S., Tonetta, S.: IC3 modulo theories via implicit predicate abstraction. In: Ábrahám, E., Havelund, K. (eds.) Proc. TACAS. LNCS, vol. 8413, pp. 46–61. Springer (2014). https://doi.org/10.1007/978-3-642-54862-8_4

36. Cimatti, A., Griggio, A., Schaafsma, B.J., Sebastiani, R.: The MathSAT5 SMT solver. In: Piterman, N., Smolka, S.A. (eds.) Proc. TACAS. LNCS, vol. 7795, pp. 93–107. Springer, Heidelberg (2013). https://doi.org/10.1007/978-3-642-36742-7_7
37. Cimatti, A., Griggio, A., Tonetta, S.: Verification Modulo Theories (language, benchmarks and tools) (2015). https://vmt-lib.fbk.eu/
38. Cimatti, A., Griggio, A., Tonetta, S.: The VMT-LIB language and tools. In: Déharbe, D., Hyvärinen, A.E.J. (eds.) Proc. SMT. CEUR Workshop Proceedings, vol. 3185, pp. 80–89. CEUR-WS.org (2022). https://ceur-ws.org/Vol-3185/extended9547.pdf
39. Claessen, K., Sörensson, N.: A liveness checking algorithm that counts. In: Cabodi, G., Singh, S. (eds.) Proc. FMCAD, pp. 52–59. IEEE (2012). https://ieeexplore.ieee.org/document/6462555/
40. Clarke, E.M., Grumberg, O., Jha, S., Lu, Y., Veith, H.: Counterexample-guided abstraction refinement. In: Emerson, E.A., Sistla, A.P. (eds.) Proc. CAV. LNCS, vol. 1855, pp. 154–169. Springer (2000). https://doi.org/10.1007/10722167_15
41. Craig, W.: Linear reasoning. A new form of the Herbrand-Gentzen theorem. J. Symb. Log. **22**(3), 250–268 (1957). https://doi.org/10.2307/2963593
42. Daly, R.: CoreIR: A simple LLVM-style hardware compiler. https://github.com/rdaly525/coreir, Accessed 15 Oct 2025
43. Dutertre, B.: Yices 2.2. In: Biere, A., Bloem, R. (eds.) Proc. CAV. LNCS, vol. 8559, pp. 737–744. Springer (2014). https://doi.org/10.1007/978-3-319-08867-9_49
44. Eén, N., Mishchenko, A., Brayton, R.K.: Efficient implementation of property directed reachability. In: Bjesse, P., Slobodová, A. (eds.) Proc. FMCAD, pp. 125–134. IEEE (2011). https://ieeexplore.ieee.org/document/6148886
45. Fedyukovich, G., Prabhu, S., Madhukar, K., Gupta, A.: Quantified invariants via syntax-guided synthesis. In: Dillig, I., Tasiran, S. (eds.) Proc. CAV (1). LNCS, vol. 11561, pp. 259–277. Springer (2019). https://doi.org/10.1007/978-3-030-25540-4_14
46. Froleyks, N., Yu, E., Preiner, M., Biere, A., Heljanko, K.: Introducing certificates to the hardware model checking competition. In: Piskac, R., Rakamaric, Z. (eds.) Proc. CAV (1). LNCS, vol. 15931, pp. 281–295. Springer (2025). https://doi.org/10.1007/978-3-031-98668-0_14
47. Gario, M., Micheli, A.: PySMT: A solver-agnostic library for fast prototyping of SMT-based algorithms. In: Proc. SMT (2015). https://andrea.micheli.website/papers/pysmt.pdf
48. Girardi, A., Bombardelli, A., Griggio, A., Tomassi, J.: PyVMT: A Python library to interact with transition systems. https://github.com/pyvmt/pyvmt, accessed: 2025-10-15
49. Goel, A., Sakallah, K.A.: Model checking of Verilog RTL using IC3 with syntax-guided abstraction. In: Badger, J.M., Rozier, K.Y. (eds.) Proc. NFM. LNCS, vol. 11460, pp. 166–185. Springer (2019). https://doi.org/10.1007/978-3-030-20652-9_11
50. Goel, A., Sakallah, K.: AVR: Abstractly Verifying Reachability. In: Proc. TACAS (1). LNCS, vol. 7795, pp. 413–422. Springer (2020). https://doi.org/10.1007/978-3-030-45190-5_23
51. Griggio, A., Jonás, M.: Kratos2: An SMT-based model checker for imperative programs. In: Enea, C., Lal, A. (eds.) Proc. CAV (3). LNCS, vol. 13966, pp. 423–436. Springer (2023). https://doi.org/10.1007/978-3-031-37709-9_20
52. Ho, Y.S., Chauhan, P., Roy, P., Mishchenko, A., Brayton, R.K.: Efficient uninterpreted function abstraction and refinement for word-level model checking. In:

Piskac, R., Talupur, M. (eds.) Proc. FMCAD, pp. 65–72. IEEE (2016). https://doi.org/10.1109/FMCAD.2016.7886662

53. Johannsen, C., et al.: The MoXI model exchange tool suite. In: Gurfinkel, A., Ganesh, V. (eds.) Proc. CAV (1). LNCS, vol. 14681, pp. 203–218. Springer (2024). https://doi.org/10.1007/978-3-031-65627-9_10
54. Komuravelli, A., Gurfinkel, A., Chaki, S.: SMT-Based Model Checking for Recursive Programs. In: Biere, A., Bloem, R. (eds.) Proc. CAV. LNCS, vol. 8559, pp. 17–34. Springer, Cham (2014). https://doi.org/10.1007/978-3-319-08867-9_2
55. Li, J., Zhu, S., Zhang, Y., Pu, G., Vardi, M.Y.: Safety model checking with complementary approximations. In: Parameswaran, S. (ed.) Proc. ICCAD. pp. 95–100. IEEE (2017). https://doi.org/10.1109/ICCAD.2017.8203765
56. Lu, Z., Chien, P.C., Lee, N.Z., Gurfinkel, A., Ganesh, V.: Btor2-Select: Machine learning based algorithm selection for hardware model checking. In: Piskac, R., Rakamaric, Z. (eds.) Proc. CAV (1). LNCS, vol. 15931, pp. 296–311. Springer (2025). https://doi.org/10.1007/978-3-031-98668-0_15
57. Mann, M., Irfan, A., Griggio, A., Padon, O., Barrett, C.W.: Counterexample-guided prophecy for model checking modulo the theory of arrays. Log. Methods Comput. Sci. **18**(3) (2022). https://doi.org/10.46298/LMCS-18(3:26)2022
58. Mann, M., et al.: Pono: A flexible and extensible SMT-based model checker. In: Silva, A., Leino, K.R.M. (eds.) Proc. CAV (2). LNCS, vol. 12760, pp. 461–474. Springer, Cham (2021). https://doi.org/10.1007/978-3-030-81688-9_22
59. Mann, M., et al.: Smt-Switch: A solver-agnostic C++ API for SMT solving. In: Li, C.M., Manyà, F. (eds.) Proc. SAT. LNCS, vol. 12831, pp. 377–386. Springer (2021). https://doi.org/10.1007/978-3-030-80223-3_26
60. Mattarei, C., Mann, M., Barrett, C.W., Daly, R.G., Huff, D., Hanrahan, P.: CoSA: Integrated verification for agile hardware design. In: Bjørner, N.S., Gurfinkel, A. (eds.) Proc. FMCAD, pp. 1–5. IEEE (2018). https://doi.org/10.23919/FMCAD.2018.8603014
61. McMillan, K.L.: Symbolic model checking: An approach to the state explosion problem. Ph.D. thesis, Carnegie Mellon University, Pittsburgh, PA, USA (1992). https://apps.dtic.mil/sti/tr/pdf/ADA250924.pdf
62. McMillan, K.L.: Interpolation and SAT-based model checking. In: Jr., W.A.H., Somenzi, F. (eds.) Proc. CAV. LNCS, vol. 2725, pp. 1–13. Springer (2003). https://doi.org/10.1007/978-3-540-45069-6_1
63. Melchert, J., Terrill, C., Perez-Lopez, Á.R., Barrett, C., Raina, P.: Automated translation validation of a compiler for statically scheduled accelerators. In: Irfan, A., Kaufmann, D. (eds.) Proc. FMCAD, pp. 198–208. TU Wien Academic Press (2025). https://doi.org/10.34727/2025/isbn.978-3-85448-084-6_26
64. de Moura, L.M., Bjørner, N.S.: Z3: an efficient SMT solver. In: Ramakrishnan, C.R., Rehof, J. (eds.) Proc. TACAS. LNCS, vol. 4963, pp. 337–340. Springer (2008). https://doi.org/10.1007/978-3-540-78800-3_24
65. Niemetz, A., Preiner, M.: Bitwuzla. In: Enea, C., Lal, A. (eds.) Proc. CAV (2). LNCS, vol. 13965, pp. 3–17. Springer (2023). https://doi.org/10.1007/978-3-031-37703-7_1
66. Niemetz, A., Preiner, M.: Bit-precise interpolation in Bitwuzla. In: Junges, S., Katz, G. (eds.) Proc. TACAS. LNCS, vol. 16505, pp. 66–88. Springer (2026). https://doi.org/10.1007/978-3-032-22752-2_4
67. Niemetz, A., Preiner, M., Biere, A.: Boolector 2.0. J. Satisf. Boolean Model. Comput. **9**(1), 53–58 (2014). https://doi.org/10.3233/SAT190101

68. Niemetz, A., Preiner, M., Wolf, C., Biere, A.: Btor2, BtorMC and Boolector 3.0. In: Chockler, H., Weissenbacher, G. (eds.) Proc. CAV (1). LNCS, vol. 10981, pp. 587–595. Springer (2018). https://doi.org/10.1007/978-3-319-96145-3_32
69. Perez-Lopez, Á.R., Chien, P.C., Lonsing, F., Archer, S., Irfan, A., Barrett, C.: FM 2026 artifact for Pono 2.0: A versatile SMT-based model checker for safety and liveness. Zenodo (2026). https://doi.org/10.5281/zenodo.18680797
70. Rozier, K.Y., et al.: MoXI: An intermediate language for symbolic model checking. In: Neele, T., Wijs, A. (eds.) Proc. SPIN. LNCS, vol. 14624, pp. 26–46. Springer (2024). https://doi.org/10.1007/978-3-031-66149-5_2
71. Sheeran, M., Singh, S., Stålmarck, G.: Checking safety properties using induction and a SAT-solver. In: Hunt, W.A., Jr., Johnson, S.D. (eds.) Proc. FMCAD. LNCS, vol. 1954, pp. 108–125. Springer (2000). https://doi.org/10.1007/3-540-40922-X_8
72. Su, Y., Yang, Q., Ci, Y., Bu, T., Huang, Z.: The rIC3 hardware model checker. In: Piskac, R., Rakamaric, Z. (eds.) Proc. CAV (1). LNCS, vol. 15931, pp. 185–199. Springer (2025). https://doi.org/10.1007/978-3-031-98668-0_9
73. Su, Y., Yang, Q., Ci, Y., Li, Y., Bu, T., Huang, Z.: Deeply optimizing the SAT solver for the IC3 algorithm. In: Piskac, R., Rakamaric, Z. (eds.) Proc. CAV (1). LNCS, vol. 15931, pp. 237–257. Springer (2025).https://doi.org/10.1007/978-3-031-98668-0_12
74. Tollec, S., Asavoae, M., Couroussé, D., Heydemann, K., Jan, M.: μArchiFI: formal modeling and verification strategies for microarchitectural fault injections. In: Nadel, A., Rozier, K.Y. (eds.) Proc. FMCAD, pp. 101–109. IEEE (2023). https://doi.org/10.34727/2023/ISBN.978-3-85448-060-0_18
75. Tsiskaridze, N., et al.: Automating system configuration. In: Proc. FMCAD, pp. 102–111. TU Wien Academic Press (2021). https://doi.org/10.34727/2021/ISBN.978-3-85448-046-4_19
76. Vizel, Y., Grumberg, O.: Interpolation-sequence based model checking. In: Proc. FMCAD, pp. 1–8. IEEE (2009). https://doi.org/10.1109/FMCAD.2009.5351148
77. Vizel, Y., Grumberg, O., Shoham, S.: Intertwined forward-backward reachability analysis using interpolants. In: Piterman, N., Smolka, S.A. (eds.) Proc. TACAS. LNCS, vol. 7795, pp. 308–323. Springer (2013). https://doi.org/10.1007/978-3-642-36742-7_22
78. Xia, Y., Cimatti, A., Griggio, A., Li, J.: Avoiding the shoals - A new approach to liveness checking. In: Gurfinkel, A., Ganesh, V. (eds.) Proc. CAV (1). LNCS, vol. 14681, pp. 234–254. Springer (2024). https://doi.org/10.1007/978-3-031-65627-9_12
79. Yan, Z., Zhang, H.: Word-level augmentation of formal proof by learning from simulation traces. In: Xiong, J., Wille, R. (eds.) Proc. ICCAD, pp. 115:1–115:9. ACM (2024). https://doi.org/10.1145/3676536.3676686

Asynchronous Team Automata

Davide Basile[1], Maurice H. ter Beek[1], and José Proença[2(✉)]

[1] Formal Methods and Tools lab, CNR–ISTI, Pisa, Italy
{davide.basile,maurice.terbeek}@isti.cnr.it
[2] CISTER & INESC TEC & University of Porto, Porto, Portugal
jose.proenca@fc.up.pt

Abstract. Team automata were introduced as a flexible extension of I/O automata to model collaborative behaviour in component-based and distributed systems. Their distinctive features include multi-party communication and a liberal synchronisation mechanism: components may jointly execute shared actions according to synchronisation policies that specify which subsets of components participate as senders or receivers. While this makes team automata well suited for modelling coordination, existing communication is synchronous and therefore insufficient for capturing certain behavioural aspects (e.g., due to message reordering) of modern networks and distributed systems, in which communication is typically asynchronous and message delays are unpredictable.

In this paper, we introduce asynchronous team automata (ATeams), which extend team automata with buffers to model asynchronous communication, in addition to conventional synchronous interaction. ATeams support individual interactions involving multiple senders and receivers, unlike well-known asynchronous models such as communicating finite-state machines and multi-party session types. We formalise the syntax and operational semantics of ATeams, study well-formedness and well-behavedness conditions, and present the prototypical A-Team tool that supports specification, animation and automated checks. This proposes ATeams as a unifying semantic foundation for modelling and analysis of heterogeneous synchronous–asynchronous multi-party interactions.

1 Introduction

Team automata have originally been proposed for modelling components of groupware systems and their interconnections [38]. They were inspired by Input/Output (I/O) automata [53] and in particular inherit their distinction between internal (private, i.e., not observable by other automata) actions and external (i.e., input and output) actions used for communication with the environment (i.e., other automata). The underlying philosophy was that automata cooperate and collaborate by jointly executing (synchronising) transitions with the same action label (but possibly of different nature, i.e., input or output) as agreed upfront. Team automata were formally defined in terms of component automata that synchronise on certain executions of (common) actions according to specific

---

The original version of the chapter has been revised. A correction to this chapter can be found at https://doi.org/10.1007/978-3-032-26220-2_36

A. Sampaio and M. Stoelinga (Eds.): FM 2026, LNCS 16557, pp. 27–49, 2026.
https://doi.org/10.1007/978-3-032-26220-2_2

synchronisation policies [19]. Technically, team automata are an extension of I/O automata [24], since a number of the restrictions of I/O automata (in particular input-enabledness and disjointness of output actions) were dropped for more flexible modelling of several kinds of interactions in groupware systems. Unlike I/O automata, team automata impose hardly any restrictions on the role of actions in components and their composition is not limited to the synchronous product. Composing a team automaton requires defining its transitions by providing the actions and synchronisations that can take place from the combined states of the components. Each team automaton is thus a composite automaton, defined over component automata, the transitions of which are exactly those combinations of component transitions that represent a synchronisation on a common action by all the components that share that action (possible involving multiple sending and receiving actions). The effect of a synchronously executed action on the state of the composed automaton is described in terms of the local state changes of the automata that take part in the synchronisation. The distinguishing feature of team automata is this very loose nature of synchronisation determined by synchronisation policies that define how many component automata can participate in the synchronised execution of a shared external action, either as a sender (i.e., via an output action) or as a receiver (i.e., via an input action). This flexibility makes team automata capable of capturing in a precise manner a variety of notions related to coordination in distributed systems (of systems).

Motivation. Most distributed systems and real networks (e.g., the Internet) are asynchronous, with unpredictable message delays. Modelling this explicitly thus allows the verification of subtle properties (e.g., deadlocks, causal consistency) that depend on message reordering and delay. Synchronous models, on the other hand, can silently hide certain behaviour (e.g., long message delays, message reordering) that does occur in practice and may lead to bugs in actual deployment. This has been acknowledged in actor-based and object-based programming models and languages such as Erlang, Rebeca, ABS, and Akka [1,4,41,44,46,47,52].

Contribution. We investigate *asynchronous* team automata (ATeams), in which component automata use either FIFO channels or multisets to act as buffers for asynchronous communication (like in asynchronous multi-party session types [43] or communicating finite state machines (CFSMs) [29,34]), in addition to synchronous communication based on simultaneous execution of shared actions (as common in related models like contract automata [11,12] or choreography automata [7,8]). However, CFSMs and related asynchronous models are based on peer-to-peer communication. Therefore, as sketched in Fig. 1, the challenge is how to generalise their approach by taking into account multi-party communication specified by synchronisation types. Moreover, multi-party session types and other choreography models typically enrich local actions with the names of the communication partners (e.g., a binary interaction $n \rightarrow m : \mathsf{a}$ leads to a local output action $n\,m!\mathsf{a}$ for n and a local input action $n\,m?\mathsf{a}$ for m). In team automata and other component-based development approaches supporting interface-based design [2,19,51,53], on the other hand, transitions describing local behaviour are often labelled just by message names accompanied by information on whether the message is output or input of a component (e.g., a binary interaction $n \rightarrow m : \mathsf{a}$ leads to a local output action $\mathsf{a}!$ for n and a local input action $\mathsf{a}?$ for m). In [22], we coined the former "rich local actions" and the latter "poor local actions". From a software designer's point of view, the poor localisation style better supports the principle of loose coupling, whereas the rich style better avoids undesired synchronisations.

To address the challenge outlined above and sketched in Fig. 1, we formalise the syntax and operational semantics of ATeams, provide well-formedness and well-behavedness conditions that guarantee desirable properties regarding the structure and behaviour of ATeams, and present the prototypical A-Team tool that supports the specification, animation and automated checks of ATeams.

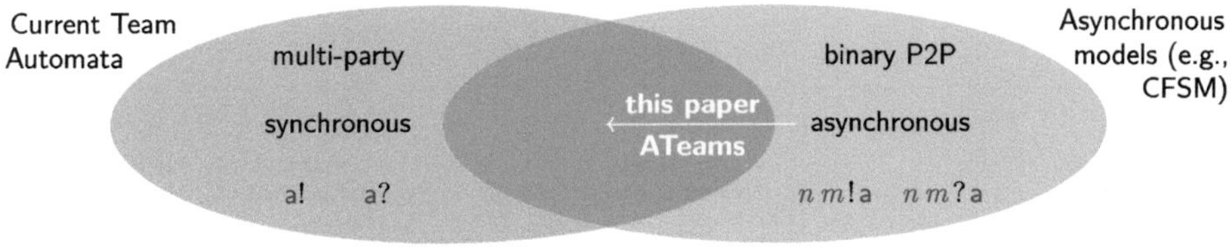

Fig. 1. This paper's challenge: equip team automata with asynchronous communication

Outline. We start by motivating ATeams via elementary illustrative examples (in Sect. 2.1), followed by defining its syntax (in Sect. 2.2) and semantics (in Sect. 2.3). We then present desirable properties of ATeams with respect to its structure (well-formedness, in Sect. 2.4) and its behaviour (well-behavedness, in Sect. 2.5). and describe the A-Team tool to animate and analyse systems in ATeams, which supports fast experimentation (in Sect. 3). We discuss related work in Sect. 4 and conclude in Sect. 5.

2 Asynchronous Team Automata

This section introduces *asynchronous team automata* (ATeams), starting with a motivating example before presenting the syntax, operational semantics, and some desirable properties. We refer to a concrete instance (automaton) of ATeams as a *system*, consisting of a set of communicating concurrent components (automata) enriched with synchronisation types restricting their communication.

We have developed in Scala and JavaScript a prototypical companion tool, called the A-Team, to analyse ATeams. This open-source tool can be used online or downloaded at https://fm-dcc.github.io/a-team and is described in Sect. 3. All examples in this paper are predefined in the tool, like the synchronous Race, including a hyperlink to open the tool with the specific example.

2.1 Motivating Race Example

We illustrate ATeams using different variations of the Race example, specified in Fig. 2 and borrowed from recent work on Team Automata [18,23]. The leftmost version is the synchronous version of Team Automata, in which a controller `Ctr` simultaneously asks two runners `R` to `start` a race, who then reply one at a time once they `finish`. The initial **`acts`** block declares that the action `start` can be used to communicate between one sender and two receivers (`1->2`), and uses `synchronous` communication, while the `finish` uses 1-to-1 communication. The second **`proc`** block defines a set of recursive processes using a variation of CCS [54], and the third **`init`** block specifies the agents being executed.

```
// Synchronous Race
acts
 start:  1->2, sync;
 finish: 1->1, sync;
proc
 Ctr = start!.finish?.finish?.Ctr
 R = start?.finish!.R
init
  Ctr || R || R
```

```
// Async. Race (sender)
acts
 start:  1->2, fifo@snd;
 finish: 2->1, fifo@snd;
proc
 Ctr = start!.finish?r1,r2.Ctr
 R = start?c.finish!.R
init
  c:Ctr || r1:R || r2:R
```

```
// Async. Race (global)
acts
 start: 1->2, fifo@global;
 finish:2->1, fifo@global;
proc
 Ctr = start!.finish?.Ctr
 R = start?.run!.finish!.R
init
 Ctr || R || R
```

Fig. 2. Variations of the Race example in ATeams

The other Race examples use FIFO buffers to communicate asynchronously. The middle Race example specifies that the `start` actions must be buffered at the sender side (`fifo@snd`). Consequently, the runners must specify who is the sender, writing `start`? c to state that the buffer can be found in agent c. Similarly for `finish`. The rightmost Race example instead uses a single global FIFO queue, shared by all messages, hence no references to senders or receivers are needed by the components. This freedom to intertwine different interaction patterns, and even different buffer types, facilitate experimenting with the impact of modifying the communication channels. One can specify, e.g., the notion of mailbox in Rebeca [1], and the notions of multicast and broadcast in computer networks.

The semantics of these three Race systems are not the same. The synchronous semantics can more compactly describe, in a single transition, the execution of actions from multiple agents, while the asynchronous semantics requires more actions for each sending and receiving. More importantly, these versions can have undesirable behaviour. For instance, the middle Race example can deadlock if the same runner reads the `start` message before the other runner. The rightmost Race example does not deadlock, but allows a runner to start twice before the other runner starts, which may not be the intention. All such undesired behaviour can be detected using our A-Team implementation online (cf. Section 3).

2.2 Syntax

We consider a global set of agent names $n \in \mathcal{N}$, action names $\mathsf{a} \in \mathcal{A}$, process names $K \in \mathcal{K}$. We also write $P \in \mathcal{P}$ to range over process definitions and $st \in \mathcal{S}$ to range over synchronisation types, defined in Fig. 3.

$$
\begin{aligned}
P &:= K \mid \mathbf{0} \mid \alpha.P \mid P + P && \text{(process)}\\
\alpha &:= \mathsf{a}! \mid \mathsf{a}? \mid \mathsf{a}!\overline{n} \mid \mathsf{a}?\overline{n} && \text{(action)}\\
st &:= (\mathsf{sync}, \mathit{intrv}, \mathit{intrv}) \mid (\mathsf{async}, \mathit{intrv}, \mathit{intrv}, \mathit{buft}, \mathit{loct}) && \text{(synchronisation type)}\\
\mathit{intrv} &:= (i, i) \mid (i, \infty) && \text{(interval)}\\
\mathit{buft} &:= \mathsf{fifo} \mid \mathsf{bag} \mid \ldots && \text{(buffer type)}\\
\mathit{loct} &:= \mathsf{@snd} \mid \mathsf{@rcv} \mid \mathsf{@global} \mid \mathsf{@snd\text{-}rcv} && \text{(location type)}
\end{aligned}
$$

Fig. 3. Syntax of processes and synchronisation types, where $i \in \mathbb{N}$ and $\overline{n}$ denotes a non-empty sequence of agent names in $\mathcal{N}$

A *process* P is a variation of a CCS process [54] where (1) send and receive actions can explicitly state who are the target processes, and (2) each interaction can involve multiple senders and receivers (as in team automata). Internal actions, which are actions not used in the interactions [18], are not considered here for simplicity. A *synchronisation type* describes the *synchronisation mode* of a given send/receive action, i.e., whether communication is (1) synchronous (requiring interaction between all senders and receivers to occur atomically), or (2) asynchronous (using a buffer for interaction between senders and receivers). In the asynchronous case, actions are sent without blocking and are later consumed by their corresponding receivers. Depending on the *location type*, a new buffer is instantiated for each sender (@snd), for each receiver (@rcv), globally (@global), or for each pair of sender and receiver (@snd-rcv).

Finally, a *system* is defined as a triple $sys = \langle \gamma; \sigma; \rho \rangle$, in which $\gamma : \mathcal{K} \to \mathcal{P}$ is a function mapping process names to their (recursive) definitions, $\sigma : \mathcal{A} \to \mathcal{S}$ maps action names to their synchronisation types, and $\rho : \mathcal{N} \to \mathcal{P}$ maps agent names to their process definitions.

2.3 Operational Semantics

We define the semantics of a system using a small-step (operational) semantics over a configuration, which intuitively captures the current state of the processes and of the buffers of the asynchronous channels. We consider global sets of *locations* $\mathcal{L} = (\mathcal{N} \cup \{\bot\}) \times (\mathcal{N} \cup \{\bot\})$ and *buffers* $\mathcal{B}$, writing ℓ and B to range over $\mathcal{L}$ and $\mathcal{B}$, respectively.

Intuitively, each location consists of an optional sender $snd \in \mathcal{N}$ and an optional receiver $rcv \in \mathcal{N}$. A location type denotes whether the sender, the receiver, or both are specified. For example, we write $\ell = (n, m)$ to denote the location where n is the sender and m is the receiver (i.e., ℓ has location type @snd-rcv). In this case, both the sender and the receiver know each other. In locations of type @snd (e.g., $\ell = (n, \bot)$) the receiver knows who deposited the message in the buffer, i.e., the sender n, but the sender does not know who will consume this message, i.e., it could be any receiver. This is analogous for locations of type @rcv (e.g., $\ell = (\bot, m)$). In locations of type @global, where $\ell = (\bot, \bot)$, the sender is unknown to the receiver and the receiver is unknown to the sender.

Each buffer B is an algebraic structure such that $B' = B \oplus \mathsf{a}$ adds an element a to the buffer, and $B' = B \ominus \mathsf{a}$ consumes a from the buffer. Our semantics considers that $\oplus$ always succeeds[1] (non-blocking additions) and $\ominus$ only succeeds if the given value is available. Our prototypical implementation (cf. Section 3) currently supports unbounded FIFO queues (describing order-preserving communication) and multisets (describing unsorted message passing, called bags).

Finally, a *configuration* in our semantics is a pair $\langle \rho, \beta \rangle$, with ρ as before and $\beta : \mathcal{L} \to \mathcal{B}$ maps locations to buffers. A configuration evolves by the transition $\langle \rho, \beta \rangle \xrightarrow{\lambda}_{\gamma,\sigma} \langle \rho', \beta' \rangle$, where λ is a global label for ATeams that can either be an (asynchronous) action α or a (synchronous) interaction $\overline{n} \to \overline{m} : \mathsf{a}$, and γ, σ provide the context with a fixed set of definitions of processes and synchronisation types. The operational rules are presented in Fig. 4 (for synchronous communication) and Figs. 5 and 6 (for asynchronous communication), using the notation introduced below. Given a system $\langle \gamma, \sigma, \rho \rangle$, the initial configuration is $\langle \rho, \beta_0 \rangle$, where β_0 initialises each buffer to an empty buffer, based on the location types and on the set of existing agent names, and the context is given by γ and σ.

$$\text{CONST}\frac{P \xrightarrow{\alpha} P'}{K \xrightarrow{\alpha} P'} K{:}P \in \gamma \qquad \text{PREFIX}\frac{}{\alpha.P \xrightarrow{\alpha} P} \qquad \text{SUM-L}\frac{P \xrightarrow{\alpha} P'}{P+Q \xrightarrow{\alpha} P'} \qquad \text{SUM-R}\frac{Q \xrightarrow{\alpha} Q'}{P+Q \xrightarrow{\alpha} Q'}$$

$$\text{SYNC}\frac{\begin{array}{c}\mathsf{a}{:}(\mathsf{sync}, \mathit{from}, \mathit{to}) \in \sigma \qquad\qquad \#\mathit{snds} \in \mathit{from} \qquad\qquad \#\mathit{rcvs} \in \mathit{to} \\ \mathit{snds} \subseteq \{n{:}s' \mid n{:}s \in \rho, s \xrightarrow{\mathsf{a}!} s'\} \cup \{n{:}s' \mid n{:}s \in \rho, s \xrightarrow{\mathsf{a}!\overline{m}} s', \overline{m} \subseteq \mathsf{dom}(\mathit{rcvs})\} \\ \mathit{rcvs} \subseteq \{n{:}r' \mid n{:}r \in \rho, r \xrightarrow{\mathsf{a}?} r'\} \cup \{n{:}r' \mid n{:}r \in \rho, r \xrightarrow{\mathsf{a}?\overline{m}} r', \overline{m} \subseteq \mathsf{dom}(\mathit{snds})\}\end{array}}{\langle \rho, \beta \rangle \xrightarrow{\mathsf{dom}(\mathit{snds}) \to \mathsf{dom}(\mathit{rcvs}) : a} \langle \rho \uplus \mathit{snds} \uplus \mathit{rcvs}, \beta \rangle}$$

Fig. 4. Operational rules for processes and synchronous communication

Notation. Given a function M, we write $M \uplus M'$ to denote the function M updated with the function M', i.e., $(M \uplus M')(k) = v$ iff $M'(k) = v$ or $M(k) = v$

[1] Exceptionally, $\oplus$ fails when trying to add an action associated with a buffer type that differs from the existing buffer. Well-formedness conditions (in Sect. 2.4) guarantee that this will never happen.

$$\text{SEND@RCV}\ \frac{\mathsf{a}{:}(\mathsf{async}, \_, \mathit{to}, \_, \mathsf{@rcv}) \in \sigma \quad n{:}s \in \rho \quad s \xrightarrow{\mathsf{a}!\overline{m}} s' \quad \#\overline{m} \in \mathit{to} \quad \text{all} \oplus \text{succeed} \quad \overline{m} \subseteq \gamma}{\langle \rho, \beta \rangle \xrightarrow{\mathsf{a}!\overline{m}} \langle \rho \uplus \{n{:}s'\}, \beta \uplus \{(\bot, m){:}(\beta(\bot, m) \oplus \mathsf{a}) \mid m \in \overline{m}\}\rangle}$$

$$\text{SEND@SND}\ \frac{\mathsf{a}{:}(\mathsf{async}, \_, (t,t), \_, \mathsf{@snd}) \in \sigma \quad n{:}s \in \rho \quad s \xrightarrow{\mathsf{a}!} s' \quad \text{all} \oplus \text{succeed}}{\langle \rho, \beta \rangle \xrightarrow{\mathsf{a}!} \langle \rho \uplus \{n{:}s'\}, \beta \uplus \{(n, \bot){:}(\beta(n, \bot) \oplus \underbrace{\mathsf{a} \oplus \cdots \oplus \mathsf{a}}_{t \text{ times}})\}\rangle}$$

$$\text{SEND@GLOB}\ \frac{\mathsf{a}{:}(\mathsf{async}, \_, (t,t), \_, \mathsf{@global}) \in \sigma \quad n{:}s \in \rho \quad s \xrightarrow{\mathsf{a}!} s' \quad \text{all} \oplus \text{succeed}}{\langle \rho, \beta \rangle \xrightarrow{\mathsf{a}!} \langle \rho \uplus \{n{:}s'\}, \beta \uplus \{(\bot, \bot){:}(\beta(\bot, \bot) \oplus \underbrace{\mathsf{a} \oplus \cdots \oplus \mathsf{a}}_{t \text{ times}})\}\rangle}$$

$$\text{SEND@SND-RCV}\ \frac{\mathsf{a}{:}(\mathsf{async}, \_, \mathit{to}, \_, \mathsf{@snd\text{-}rcv}) \in \sigma \quad n{:}s \in \rho \quad s \xrightarrow{\mathsf{a}!\overline{m}} s' \quad \#\overline{m} \in \mathit{to} \quad \text{all} \oplus \text{succeed} \quad \overline{m} \subseteq \gamma}{\langle \rho, \beta \rangle \xrightarrow{\mathsf{a}!\overline{m}} \langle \rho \uplus \{n{:}s'\}, \beta \uplus \{(n, m){:}(\beta(n, m) \oplus \mathsf{a}) \mid m \in \overline{m}\}\rangle}$$

Fig. 5. Operational rules for asynchronous communication (sending)

whenever $M'(k)$ is undefined. Moreover, we write $k{:}v \in M$ whenever $M(k) = v$ and we write $\mathsf{dom}(M)$ to denote the domain of M. Finally, we write $\langle \rho, \beta \rangle \xrightarrow{\lambda} \langle \rho', \beta' \rangle$ instead of $\langle \rho, \beta \rangle \xrightarrow{\lambda}_{\gamma,\sigma} \langle \rho', \beta' \rangle$, omitting the context for simplicity.

Example. Our Race example with fifo@snd communication is formalised by the system $\langle \gamma, \sigma, \rho \rangle$, with $\gamma = \{\mathsf{start}{:}(\mathsf{async}, (1,1), (2,2), \mathsf{fifo}, \mathsf{@snd}), \mathsf{finish}{:}(\mathsf{async}, (2,2), (1,1), \mathsf{fifo}, \mathsf{@snd})\}$, $\sigma = \{\mathtt{Ctr}{:}\mathsf{start!}.\mathsf{finish?}\mathit{r1}, \mathit{r2}.\mathtt{Ctr}, \mathsf{R}{:}\mathsf{start?}.\mathsf{finish!}\mathit{c}.\mathsf{R}\}$, and $\rho = \{c{:}\mathtt{Ctr}, \mathit{r1}{:}\mathsf{R}, \mathit{r2}{:}\mathsf{R}\}$. The initial configuration is formalised by the pair $\langle \rho, \beta_0 \rangle$, with $\beta_0 = \{ (\bot, n){:}\mathsf{EmptyFifo} \mid n \in \{c, \mathit{r1}, \mathit{r2}\} \}$. Initially, only rule SEND@SND can be applied, evolving by $\langle \rho, \beta_0 \rangle \xrightarrow{\mathsf{c{:}start!}} \langle \rho', \beta' \rangle$, where ρ' maps c to $\mathsf{finish?}\mathit{r1}, \mathit{r2}.\mathtt{Ctr}$ and β' maps $(c, \bot)$ to a FIFO buffer with two occurrences of start.

Synchronous Rules. Rules CONST, PREFIX, SUM-L, and SUM-R in Fig. 4 are standard in CCS [54]. The empty process **0** is a special process that does not evolve. Rule SYNC models the synchronisation over an action named a sent by the senders in $\mathsf{dom}(\mathit{snds})$ and to the receivers in $\mathsf{dom}(\mathit{rcvs})$. The synchronisation type of a, given by $\sigma(\mathsf{a}) = (\mathsf{sync}, \mathit{from}, \mathit{to})$, captures that a is synchronous and the number of senders (receivers, resp.) must be in the interval *from* (*to*, resp.). Synchronously sending to or receiving from multiple agents does not guarantee the maximal number of participants in a single action;[2] instead, any number permitted by the synchronisation type and the available participants is allowed.

[2] This differs from other approaches in the literature such as, e.g., broadcast communication in UPPAAL [25].

$$\text{RECV@SND}\ \frac{\mathsf{a}{:}(\mathsf{async}, \mathit{fr}, \_, \_, \texttt{@snd}) \in \sigma \quad n{:}s \in \rho \quad s \xrightarrow{\mathsf{a}?\overline{m}} s' \quad \#\overline{m} \in \mathit{fr} \quad \text{all} \ominus \text{succeed} \quad \overline{m} \subseteq \gamma}{\langle \rho, \beta \rangle \xrightarrow{\mathsf{a}?\overline{m}} \langle \rho \uplus \{n{:}s'\}, \beta \uplus \{(m, \bot){:}(\beta(m, \bot) \ominus \mathsf{a}) \mid m \in \overline{m}\}\rangle}$$

$$\text{RECV@RCV}\ \frac{\mathsf{a}{:}(\mathsf{async}, (f, f), \_, \_, \texttt{@rcv}) \in \sigma \quad n{:}s \in \rho \quad s \xrightarrow{\mathsf{a}?} s' \quad \text{all} \ominus \text{succeed}}{\langle \rho, \beta \rangle \xrightarrow{\mathsf{a}?} \langle \rho \uplus \{n{:}s'\}, \beta \uplus \{(\bot, n){:}(\beta(\bot, n) \ominus \underbrace{\mathsf{a} \ominus \cdots \ominus \mathsf{a}}_{f \text{ times}})\}\rangle}$$

$$\text{RECV@GLOB}\ \frac{\mathsf{a}{:}(\mathsf{async}, (f, f), \_, \_, \texttt{@global}) \in \sigma \quad n{:}s \in \rho \quad s \xrightarrow{\mathsf{a}?} s' \quad \text{all} \ominus \text{succeed}}{\langle \rho, \beta \rangle \xrightarrow{\mathsf{a}?} \langle \rho \uplus \{n{:}s'\}, \beta\{(\bot, \bot){:}(\beta(\bot, \bot) \ominus \underbrace{\mathsf{a} \ominus \cdots \ominus \mathsf{a}}_{f \text{ times}})\}\rangle}$$

$$\text{RECV@SND-RCV}\ \frac{\mathsf{a}{:}(\mathsf{async}, \mathit{fr}, \_, \_, \texttt{@snd-rcv}) \in \sigma \quad n{:}s \in \rho \quad s \xrightarrow{\mathsf{a}?\overline{m}} s' \quad \#\overline{m} \in \mathit{fr} \quad \text{all} \ominus \text{succeed} \quad \overline{m} \subseteq \gamma}{\langle \rho, \beta \rangle \xrightarrow{\mathsf{a}?\overline{m}} \langle \rho \uplus \{n{:}s'\}, \beta \uplus \{(m, n){:}(\beta(m, n) \ominus \mathsf{a}) \mid m \in \overline{m}\}\rangle}$$

Fig. 6. Operational rules for asynchronous communication (receiving)

The sets of senders *snds* (receivers *rcvs*, resp.) are obtained by updating the state of the corresponding process to the target state of the sending (receiving, resp.) transitions. Additionally, the send or receive transition may identify a set of recipients $\overline{m}$. In this case, it is required that these recipients be involved in performing the synchronisation (e.g., $\overline{m} \subseteq \mathsf{dom}(\mathit{snds})$).

Asynchronous Rules. The asynchronous rules for sending are presented in Fig. 5, and those for receiving are presented in Fig. 6. These are dual to each other and are guided by a few guiding principles, explained below, describing desirable properties of successful executions.

Principle 1: *Asynchronous sending messages should not block.*

This is mainly a consequence from requiring buffers to only block when being consumed, and to always succeed when adding new elements. In rule SEND@RCV, a sender n sends a to a set of receivers $\overline{m}$ with cardinality within the interval *to* defined by the synchronisation type $\sigma(a)$. Since the location type is @rcv, for each receiver $m \in \overline{m}$, the message is added to its buffer (i.e., $\beta(\bot, m) \oplus \mathsf{a}$). Rule SEND@SND-RCV is similar, but the message is added to the buffer $\beta(n, m)$, identifying both sender and receiver. Rule RCV@SND is dual to SEND@RCV. In this case, instead of the receivers, the interval of senders *fr* is considered, and for each sender m the buffer $\beta(m, \bot)$ is updated by removing a message. Similarly, rule RCV@SND-RCV is the dual of SEND@SND-RCV. In this case, messages are removed from the buffer $\beta(n, m)$. This behaviour follows our next principle.

Principle 2: *The sender or receiver of an asynchronous action must not be explicitly mentioned when this is disregarded by the buffer location.*

Rule SEND@SND deposits the message in the buffer $\beta(n, \bot)$ of the sender, and does not need to know the receiver. However, our semantics requires the number of receivers to be known, leading to the next principle.

Principle 3: *The number of asynchronous messages sent or received must be deterministic.*

In other words, the number of senders (receivers, resp.) must be known by explicitly specifying the senders (receivers, resp.), or by using a singleton interval in the corresponding synchronisation type. Hence, no set of receivers is identified in rule SEND@SND and it is required that the interval of receivers in synchronisation type $\sigma(a)$ is a singleton (t, t). The buffer $\beta(n, \bot)$ is updated by adding t copies of the message a. Rule SEND@GLOB is similar, but global buffer $\beta(\bot, \bot)$, which does not identify any sender or receiver, is updated. The receiving-side dual of SEND@SND is RECV@RCV. Similarly, RCV@GLOB is dual to SEND@GLOB. In both cases, the interval of senders is a singleton (f, f), and f copies of the message are removed from the buffer of the receiver n in the first case (i.e., $\beta(\bot, n)$), and from the global buffer $\beta(\bot, \bot)$ in the second case. Note that requiring a singleton interval is only necessary to achieve a deterministic number of asynchronous messages. This constraint can be lifted at the cost of introducing non-determinism in the number of possible messages exchanged in the above cases, thereby increasing the state space.

2.4 Well-Formedness

The operational semantics presented in Figs. 4 to 6 perform some well-formedness checks at runtime. These include type checking the buffers (i.e., operations $\oplus$ and $\ominus$ must succeed, cf. Sect. 2.3) and enforcing Principles 2 and 3, i.e., they constrain the synchronisation types of actions to singleton intervals *from* or *to*. Indeed, problems that arise from violating the expected synchronisation types, or by using unknown names of actions or agents, can lead to unexpected scenarios in which no rule can be applied (triggering runtime errors in the A-Team tool, cf. Section 3). To avoid these problems, we formalise below a set of well-formedness conditions that can statically detect them, all of them implemented in A-Team.

Given a system $\langle\gamma; \sigma; \rho\rangle$, a process P occurring in γ or in ρ is said to be well-formed if the 9 conditions in Fig. 7 and the 10 conditions in Fig. 8 hold. We first discuss Fig. 7. Conditions (1), (8), and (9) are straightforward. Condition (2) checks that an agent never tries to send a message to more senders than specified in its synchronisation type; Condition (3) is the dual for the receiving side. Conditions (4) to (7) avoid violating Principle 2, while Conditions (6) and (7) also avoid violating Principle 3 (cf. Section 2.3).

Figure 8 displays the conditions for checking the well-formedness of buffer types. The A-Team tool statically checks these conditions↗. They ensure that the buffer types declared by the actions are compliant with their usage within processes: in particular, a specific buffer at a given location should have a unique type. This in turn ensures that, in the semantic rules in Sect. 2.3, operation

$$
\begin{array}{lll}
P = \alpha.P' & \Rightarrow P' \text{ is well-formed} & (1)\\
P = \mathsf{a}!\overline{n}.P' & \Rightarrow \mathsf{a}{:}(\_,\_, \mathit{to}, \ldots) \in \sigma \wedge \#\overline{n} \in \mathit{to} \wedge \overline{n} \subseteq \mathsf{dom}(\rho) & (2)\\
P = \mathsf{a}?\overline{n}.P' & \Rightarrow \mathsf{a}{:}(\_, \mathit{from}, \ldots) \in \sigma \wedge \#\overline{n} \in \mathit{from} \wedge \overline{n} \subseteq \mathsf{dom}(\rho) & (3)\\
P = \mathsf{a}!\overline{n}.P' & \Rightarrow (\mathsf{a}{:}(\mathsf{sync}, \ldots) \in \sigma \vee \mathsf{a}{:}(\mathsf{async}, \_,\_,\_, @\mathsf{rcv}) \in \sigma & \\
 & \quad \vee\, \mathsf{a}{:}(\mathsf{async}, \_,\_,\_, @\mathsf{snd\text{-}rcv}) \in \sigma) \wedge \overline{n} \subseteq \mathsf{dom}(\rho) & (4)\\
P = \mathsf{a}?\overline{n}.P' & \Rightarrow (\mathsf{a}{:}(\mathsf{sync}, \ldots) \in \sigma \vee \mathsf{a}{:}(\mathsf{async}, \_,\_,\_, @\mathsf{snd}) \in \sigma & \\
 & \quad \vee\, \mathsf{a}{:}(\mathsf{async}, \_,\_,\_, @\mathsf{snd\text{-}rcv}) \in \sigma) \wedge \overline{n} \subseteq \mathsf{dom}(\rho) & (5)\\
P = \mathsf{a}!.P' & \Rightarrow \mathsf{a}{:}(\mathsf{sync}, \ldots) \in \sigma \vee \mathsf{a}{:}(\mathsf{async}, \_, (t,t), \_, @\mathsf{snd}) \in \sigma & \\
 & \quad \vee\, \mathsf{a}{:}(\mathsf{async}, \_, (t,t), \_, @\mathsf{glob}) \in \sigma & (6)\\
P = \mathsf{a}?.P' & \Rightarrow \mathsf{a}{:}(\mathsf{sync}, \ldots) \in \sigma \vee \mathsf{a}{:}(\mathsf{async}, (f,f)\_, \_, @\mathsf{rcv}) \in \sigma & \\
 & \quad \vee\, \mathsf{a}{:}(\mathsf{async}, (f,f)\_, \_, @\mathsf{glob}) \in \sigma & (7)\\
P = P_1 + P_2 & \Rightarrow P_1 \text{ and } P_2 \text{ are well-formed} & (8)\\
P = K & \Rightarrow K{:}\_ \in \gamma & (9)
\end{array}
$$

Fig. 7. Conditions for well-formedness

$\oplus$ is always enabled (guaranteeing Principle 1), whilst operation $\ominus$ is never disabled by mismatching buffer types. Here, $act(P)$ denotes the actions occurring in process P, structurally defined as $act(\alpha.P) = \{\alpha\} \cup act(P)$, $act(P_1 + P_2) = act(P_1) \cup act(P_2)$, $act(\mathbf{0}) = \emptyset$, and $act(P) = act(K)$ when $P = K$.

Condition (1) ensures that all asynchronous actions written into (or read from) the global buffer must declare the same (global) buffer type. Conditions (2) to (4) deal with (asynchronous) actions with the same location type snd but different buffer types. Condition (2) guarantees that every process n does not place in its buffer $\beta(n, \perp)$ actions declaring different buffer type, whilst Condition (3) checks that whenever a process n writes into its buffer $\beta(n, \perp)$ an action with a specific buffer type, no other process may read from $\beta(n, \perp)$ an action with a different buffer type. Condition (4) ensures that all actions read from a sender buffer $\beta(s, \perp)$ by processes declare the same buffer type.

Conditions (5) to (7) are used for (asynchronous) actions with the same location type rcv but different buffer types. Condition (5) guarantees that all actions inserted into a receiver buffer $\beta(\perp, r)$ by processes declare the same buffer type. Similarly, Condition (6) requires that all actions read from buffer $\beta(\perp, r)$ by process r must have a declared buffer type consistent with that of the actions previously written into $\beta(\perp, r)$, whilst Condition (7) guarantees that every process n does not read from its buffer $\beta(\perp, n)$ actions declaring different buffer types.

Finally, Conditions (8) to (10) are used for (asynchronous) actions with the same location type snd-rcv but different buffer types. Condition (8) stipulates that a process n cannot place in a buffer $\beta(n, r)$ shared with a receiver r actions declaring different buffer types. Likewise, Condition (9) demands that all actions read from buffer $\beta(n, r)$ by process r must have a declared buffer type consistent with that of the actions written into $\beta(n, r)$ by process n, whilst Condition (10) guarantees that a process n cannot read from a buffer $\beta(r, n)$ shared with a sender r actions declaring different buffer types.

$$\forall \mathsf{a}{:}(\mathsf{async},\_,\_,\mathsf{ba},@\mathsf{glob}),\mathsf{b}{:}(\mathsf{async},\_,\_,\mathsf{bb},@\mathsf{glob}) \in \sigma \Rightarrow \mathsf{ba} = \mathsf{bb} \tag{1}$$

$$\forall n{:}P \in \gamma, \forall \mathsf{a}{:}(\mathsf{async},\_,\_,\mathsf{ba},@\mathsf{snd}),\mathsf{b}{:}(\mathsf{async},\_,\_,\mathsf{bb},@\mathsf{snd}) \in \sigma, \mathsf{ba} \neq \mathsf{bb}.\; \mathsf{a}! \in act(P) \Rightarrow \mathsf{b}! \notin act(P) \tag{2}$$

$$\forall n{:}P, r{:}Q \in \gamma, \forall \mathsf{a}{:}(\mathsf{async},\_,\_,\mathsf{ba},@\mathsf{snd}),\mathsf{b}{:}(\mathsf{async},\_,\_,\mathsf{bb},@\mathsf{snd}) \in \sigma, \mathsf{ba} \neq \mathsf{bb}.\; \mathsf{a}! \in act(P) \land \mathsf{b}?\overline{m} \in act(Q) \Rightarrow n \notin \overline{m} \tag{3}$$

$$\forall n{:}P, r{:}Q \in \gamma, \forall \mathsf{a}{:}(\mathsf{async},\_,\_,\mathsf{ba},@\mathsf{snd}),\mathsf{b}{:}(\mathsf{async},\_,\_,\mathsf{bb},@\mathsf{snd}) \in \sigma, \mathsf{ba} \neq \mathsf{bb}.\; \mathsf{a}?\overline{m_1} \in act(P) \land \mathsf{b}?\overline{m_2} \in act(Q) \Rightarrow \overline{m_1} \cap \overline{m_2} = \emptyset \tag{4}$$

$$\forall n{:}P, r{:}Q \in \gamma, \forall \mathsf{a}{:}(\mathsf{async},\_,\_,\mathsf{ba},@\mathsf{rcv}),\mathsf{b}{:}(\mathsf{async},\_,\_,\mathsf{bb},@\mathsf{rcv}) \in \sigma, \mathsf{ba} \neq \mathsf{bb}.\; \mathsf{a}!\overline{m_1} \in act(P) \land \mathsf{b}!\overline{m_2} \in act(Q) \Rightarrow \overline{m_1} \cap \overline{m_2} = \emptyset \tag{5}$$

$$\forall n{:}P, r{:}Q \in \gamma, \forall \mathsf{a}{:}(\mathsf{async},\_,\_,\mathsf{ba},@\mathsf{rcv}),\mathsf{b}{:}(\mathsf{async},\_,\_,\mathsf{bb},@\mathsf{rcv}) \in \sigma, \mathsf{ba} \neq \mathsf{bb}.\; \mathsf{a}!\overline{m_1} \in act(P) \land \mathsf{b}? \in act(Q) \Rightarrow r \notin \overline{m} \tag{6}$$

$$\forall n{:}P \in \gamma, \forall \mathsf{a}{:}(\mathsf{async},\_,\_,\mathsf{ba},@\mathsf{rcv}),\mathsf{b}{:}(\mathsf{async},\_,\_,\mathsf{bb},@\mathsf{rcv}) \in \sigma, \mathsf{ba} \neq \mathsf{bb}.\; \mathsf{a}? \in act(P) \Rightarrow \mathsf{b}? \notin act(P) \tag{7}$$

$$\forall n{:}P \in \gamma, \forall \mathsf{a}{:}(\mathsf{async},\_,\_,\mathsf{ba},@\mathsf{snd\text{-}rcv}),\mathsf{b}{:}(\mathsf{async},\_,\_,\mathsf{bb},@\mathsf{snd\text{-}rcv}) \in \sigma, \mathsf{ba} \neq \mathsf{bb}.\; \mathsf{a}!\overline{m_1}, \mathsf{b}!\overline{m_2} \in act(P) \Rightarrow \overline{m_1} \cap \overline{m_2} = \emptyset \tag{8}$$

$$\forall n{:}P, r{:}Q \in \gamma, \forall \mathsf{a}{:}(\mathsf{async},\_,\_,\mathsf{ba},@\mathsf{snd\text{-}rcv}),\mathsf{b}{:}(\mathsf{async},\_,\_,\mathsf{bb},@\mathsf{snd\text{-}rcv}) \in \sigma, \mathsf{ba} \neq \mathsf{bb}.\; \mathsf{a}!\overline{m_1} \in act(P) \land \mathsf{b}?\overline{m_2} \in act(Q) \Rightarrow n \notin \overline{m_2} \tag{9}$$

$$\forall n{:}P \in \gamma, \forall \mathsf{a}{:}(\mathsf{async},\_,\_,\mathsf{ba},@\mathsf{snd\text{-}rcv}),\mathsf{b}{:}(\mathsf{async},\_,\_,\mathsf{bb},@\mathsf{snd\text{-}rcv}) \in \sigma, \mathsf{ba} \neq \mathsf{bb}.\; \mathsf{a}?\overline{m_1}, \mathsf{b}?\overline{m_2} \in act(P) \Rightarrow \overline{m_1} \cap \overline{m_2} = \emptyset \tag{10}$$

Fig. 8. Conditions for well-formedness of buffer types

Note that some process may violate these conditions even if at runtime it will never raise buffer type errors. For instance, the process a! + b! such that both a and b have location type @glob but different buffer types, violates Condition (1). Indeed, the global buffer has an inconsistent type declaration by actions a and b. However, at runtime this error will never occur.

2.5 Well-Behavedness

This section presents conditions that guarantee well-behavedness, to complement the previous well-formedness conditions. While well-formedness conditions can be checked syntactically, well-behavedness conditions require traversing all reachable states. These conditions are organised below into five subsections and are inspired on existing asynchronous models like CFSMs [29], i.e., with only binary peer-to-peer asynchronous communication and with rich local actions. We include several literature references to motivate and justify these conditions, anticipating the related work in Sect. 4.

Compliance and compatibility of components have been recognised as fundamental properties for component-based and distributed systems for guaranteeing safe (successful) communication [2,9,10,14–16,21,28,31–33,36]. We extensively investigated requirements for safe communication in the context of (synchronous) team automata [15,20,21]. We defined a generic procedure to derive requirements for *receptiveness* and *responsiveness* that guarantee absence of message loss (output actions of one component not accepted as input by some other component) and indefinite waiting (for input to be received in the form of an appropriate output action provided by another component), respectively. A team automaton is defined to be *compliant* with a given set of communication requirements (in the form of receptiveness and responsiveness) if in each of its reachable states, the desired communications can occur immediately.[3] We implemented both notions in the A-Team tool, which is described in Sect. 3. However, in asynchronous communications the messages spend time in buffers, hence we focus on those in our quest for well-behavedness conditions for ATeams.

Receptiveness and Responsiveness. According to guiding Principle 1, the sending action in an asynchronous communication in ATeams is never blocked, meaning that receptiveness properties (ensuring that a message is eventually sent) are always satisfied. Receiving actions, however, block until they are available in the buffers, giving room to responsiveness problems (when none of the receiving options is available at a given point). For CFSMs, in [29], it was studied how to avoid *unspecified reception* configurations, in which there is a receiving state unable to receive a message from its buffers because in each channel from which it could consume there is a message which it cannot receive in that state.

The A-Team tool detects both (synchronous) receptiveness and (synchronous and asynchronous) responsiveness problems. For example, in our synchronous Race example↗, the tool produces the warning that *"1 can receive start, but the system cannot (yet) send it: {0: finish?.Ctr, 1: R, 2: finish!.R}."*, where *1* is the name given to the first runner. The asynchronous versions raise similar warnings, e.g., the version with global buffers gives six responsiveness warnings, including that *"r1 can receive start, but the system cannot (yet) send it"*. No receptiveness warnings are detected in these examples. The weaker versions of these communication properties[5] are not implemented, i.e., to detect that a sending or receiving action can eventually be performed after the rest of the system evolves, which would require a more complex analysis (cf. [18]). In our experience, most asynchronous examples are not (strongly) responsive, i.e., it is likely that a process has to wait until a given receiving action can be performed, suggesting that ensuring this property may not be useful in most scenarios. Hence we expect the weaker form to be more useful.

Absence of Orphan Messages. For multi-party session types, it was studied in [35] how to avoid *orphan message configurations* (with each component in a

[3] A team automaton is defined to be *weakly compliant* if the communication can eventually occur, possibly after some other actions have been performed (cf., e.g., [23]).

final state, but still at least one non-empty buffer). For I/O transition systems, in [42] a notion of *asynchronous compatibility* was defined, which guarantees that in any reachable configuration, if there is a channel (buffer) with a message on its top position, then some component can consume (receive) it. However, as for CFSMs, in both cases only binary peer-to-peer communication is considered.

The A-Team tool detects scenarios with orphan messages when there are non-empty buffers with messages sent to agents that successfully terminated. We added an example of a coffee machine that starts with an extra coin and can terminate after receiving coffee. In this example, a coffee can be left in the buffers, leading to the warning *"Agent u terminated and has incoming messages"*.

Absence of Infinite Buffers. The semantics of asynchronous communication in ATeams uses unbounded buffers, assuming that adding new messages never fails. Consequently, it is possible to model a scenario in which a buffer keeps on growing indefinitely. In previous work, we addressed this problem in the context of asynchronous choreographies, using a pomset structure for the semantics [37] and borrowing the notion of *dependent guardedness* to restrict recursion from Rensink and Wehrheim [58]. The actor-based Rebeca model [1] also includes analyses to detect if the buffers storing messages between agents are bounded.

Inspired by these approaches, the A-Team tool can detect simple scenarios with infinite buffers, namely when a sender sends a message asynchronously but its state remains unchanged. This can happen, e.g., in a variation of our asynchronous race with an unbounded buffer that can grow indefinitely, whereas the controller can ask the runners to rest without waiting for an acknowledgement. This leads to the warning: *"can send rest!r1,r2 forever: c: Ctr, r1: R, r2: R"*. Our tool can also detect a similar loop in a larger case study by Pal et al. in the context of Italian Healthcare [56]. This healthcare example specifies the local behaviour of each agent, using 1-to-1 interactions, used by Pal et al. to push the limits of their approach to reason about realisability based on pomsets. The authors compare the performance using both a synchronous and an asynchronous semantics, without comparing behavioural differences, and unfolding the loops a small number of times. In our example, these interactions are specified as being synchronous, but changing them instead to being asynchronous triggers a warning that the sending of proc!gDH can be executed indefinitely.

Absence of Deadlocks. The relationship between *deadlock-freedom*, used in many different disguises in the literature, and communication properties for team automata is subtle, mainly because the distinction between input and output actions is not relevant for the more traditional notions of deadlock and because typically *global* and *local* deadlocks are distinguished [5]. In case of some specific assumptions on the synchronisation types (cf. [21]), weak versions of receptiveness and responsiveness can be used to guarantee (global) deadlock-freedom, in the sense of BIP [40], choreography automata [7], or *stuck-freeness* in multi-party session types [59]. However, global deadlock-freedom does not imply receptiveness, as we discussed in previous work [23]. For CFSMs, finally, it was recently studied in [6] how to guarantee *lock-freedom* (for any reachable configuration, there is no component in a receiving state that never receives a message in all possible transition sequences from that configuration).

The A-Team tool detects some global deadlocks by searching for states where no action can be performed, but some individual agent can still evolve; e.g., the version of the Race example that uses fifo@snd has a deadlock, resulting in the message *"Deadlock found: c: finish?r1,r2.Ctr, r1: R, r2: R, r2->_=>[finish,finish]"* (and a symmetric one in which *r1* rather than *r2* is started twice in a row).

3 Tool Support

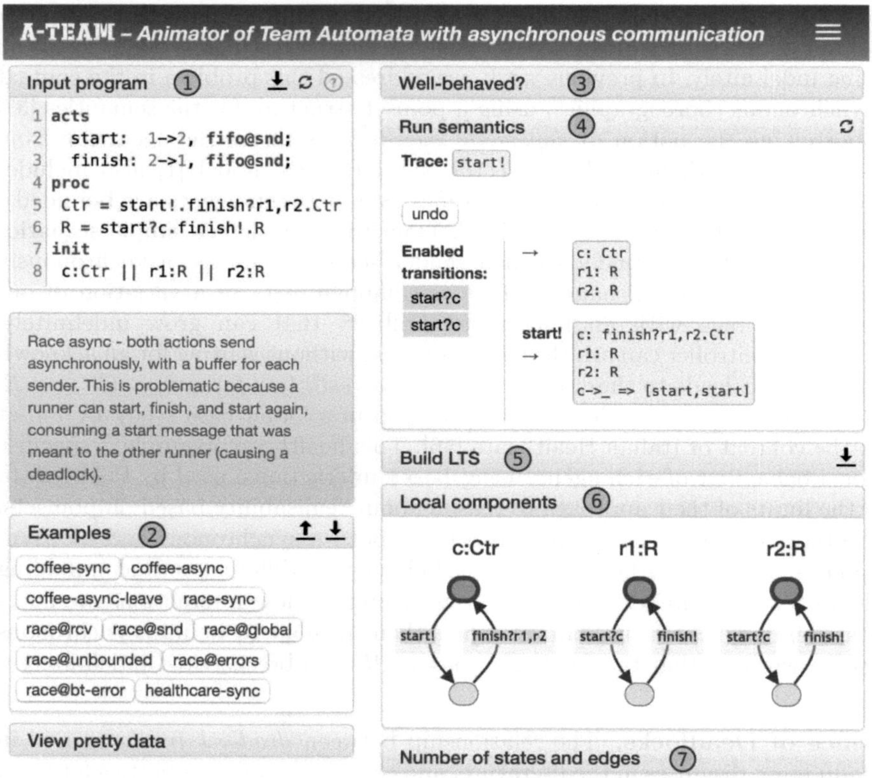

Fig. 9. Screenshot of the prototypical A-Team online tool

We implemented a prototypical tool, called the A-Team, to specify ATeams, to animate its semantics, and to analyse well-formedness and well-behavedness conditions. A-Team is open-source and can be used online by loading the static website https://fm-dcc.github.io/a-team. Figure 9 illustrates its main interface, organised into several widgets: ① is a text editor to specify a system, using a dedicated domain-specific language; ② loads examples of ATeams, including the ones presented in this paper; ③ lists all the well-behavedness errors found (up to a bounded number of transitions); ④ allows the guided step-by-step execution of the operational rules from Figs. 4 to 5; ⑤ unfolds and draws all the operational steps in a graph (until a fixed bound); ⑥ draws the local automata from the `init` block; and ⑦ counts the number of edges and states from widget ⑤ (useful for larger systems). A hidden widget verifies all well-formedness conditions (cf. Section 2.4) and throws warnings when these do not hold. Some of these widgets are collapsed (e.g., ③, ⑤, and ⑦) and the others are expanded (e.g., ① and ②); clicking the header alternates between these states. Only expanded widgets are evaluated, and expanding a collapsed widget triggers its (re-)evaluation.

A-Team is written in Scala and uses the CAOS libraries [57] to generate an interactive website without any server, whereas all functionality is compiled to JavaScript using `Scala.js`. A-Team is meant to provide quick insights on the expressivity of our semantics, experiment with different communication types, and have a small code-base easy to adapt and maintain. For example, the full semantics is encoded in file https://github.com/FM-DCC/a-team/blob/main/src/main/scala/ateams/backend/Semantics.scala. Here the function `nextState(st:St)` receives a current `st` with the action and process declarations, and returns a set of possible steps, each consisting of an action and its associated new state (including the updated buffers). The behavioural analyses are currently not meant to be particularly efficient or scalable, but to showcase how to traverse the state space to search for desirable properties, and to provide early insights on some of these.

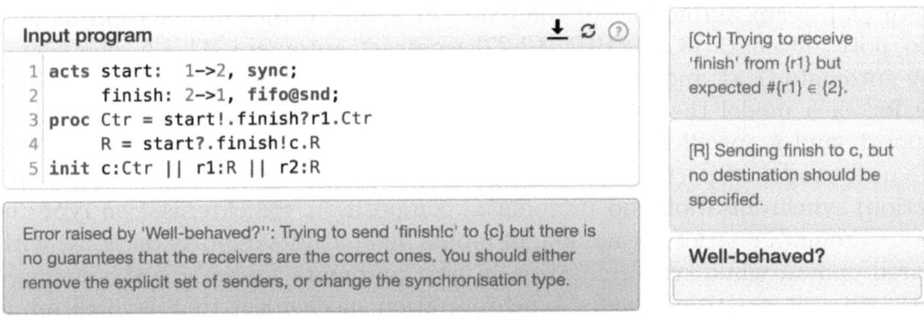

Fig. 10. A-Team providing feedback regarding well-formedness and well-behavedness

Example with Errors. A screenshot from A-Team with an asynchronous variation of the synchronous Race example with well-formedness and well-behavedness problems is depicted in Fig. 10. In this variation, start is synchronous and finish uses a FIFO buffer; however, finish is used incorrectly in both process definitions, as described in the warning messages on the top-right: *Ctr* tries to receive from a single runner (instead of two) and *R* tries to send to an explicit agent, which should be omitted. The well-behavedness widget does not succeed because the tool raises an error (cf. bottom left in Fig. 10) while traversing the state space, due to a well-formedness violation (cf. top right in Fig. 10). Errors from any of the widgets are displayed below the input program, which can be seen at the bottom left in Fig. 10.

4 Related Work

In this section, we discuss and provide pointers to related work on synchronous automata-based or coordination models and on asynchronous communication models, and we briefly recall related work on properties that guarantee safe communication in component-based and distributed systems.

In the introduction, we related team automata with I/O automata [53]. Further closely related I/O automata-based models include I/O systems [49], interface automata [2], reactive transition systems [31], interacting state machines [55], and component-interaction (CI) automata [30]. All these models distinguish input, output, and internal actions. However, I/O automata are *input-enabled* (in each state of the automaton, every input action of the automaton is enabled). Most of these models define composition as the synchronous product of automata. However, I/O automata only allow the compositon of I/O automata whose output actions are disjoint. Moreover, interface automata restrict product states to compatible states. Finally, CI automata share the distinguishing feature of multi-party composition à la team automata, but communication is restricted to binary synchronisation between a pair of input and output actions.

In [23], team automata are compared in detail to the coordination models Reo/port automata [3,50], BIP [13,27], contract automata [11,12], choreography automata [7,8], and (synchronous) multi-party session types [59,60], including for each model the definition of composition, an overview of existing tool support, and a specification of the Race example in the specific model. Reo's semantic model of port automata abstracts from data constraints to focus on (action) synchronisation and (automata) composition. Synchronisation types in team automata restrict the number of inputs and outputs of ports (actions) with shared names, while synchronisation in port automata forces how this is done. BIP (without priorities) has a synchronisation mechanism that ignores input and output but, similar to synchronisation types in team automata, it has a formalisation parameterised on the number of components. Ports (actions) can have multiple instances and bounds on the number of allowed components they can synchronise with and on the number of interactions they can be involved in. Both contract and choreography automata support only binary synchronisations in which each interaction has a single sender (the component that offers) and a single receiver (the component that requests). Synchronous multi-party session types, finally, support a more limited set of communication patterns than team automata. They typically support only binary synchronisation, yet variants exist that support multiple receivers [26] or multiple senders [45].

In the introduction, we mentioned actor-based and object-based programming models and languages like Erlang [4], Rebeca [1,44], Creol/ABS [46,47], and Akka [41], as well as asynchronous communication models like asynchronous multi-party session types [43] and CFSMs [29], as part of our inspiration for addressing asynchronous communication in team automata in this paper. Erlang/Akka, Rebeca, and Creol/ABS are actor-based models with no shared memory that instead model systems as actors (components) communicating by asynchronous message-passing, making them particularly suitable for specifying concurrent and distributed systems. The Basic Actor-based Language (Babel) [39] is a core actor-based language with a semantics with variation points for communication and coordination that can be instantiated to obtain the concrete semantics of actor-based languages like Rebeca, ABS, and Erlang/Akka. In particular, the operations of sending and receiving messages are distinguished by variation points based on when a message can be added to the buffer (i.e., based on the buffer's type and capacity) and when it can be consumed from the buffer (e.g., FIFO, EDF (Earliest Deadline First), based on priority or pattern matching, or nondeterministically). In the future, we could consider adding some of these policies to our buffer types. CFSMs and multi-party session types, on the other hand, are meant for specifying and reasoning (through formal verification and type checking) about safe communication in message-passing concurrent and distributed systems rather than actor-based models. However, communication in CFSMs and most related asynchronous models is limited to peer-to-peer communication. This is part of the challenge of addressing asynchronous communication in team automata, as sketched in Fig. 1.

In Sect. 2.5, we discussed in detail related studies on guaranteeing safe communication in terms of properties like receptiveness, responsiveness, absence of orphan messages, infinite buffers, and deadlocks in particular in models such as CFSMs, BIP, choreography automata, and multi-party session types.

5 Conclusion and Future Work

Building on the extensive body of work on team automata [23], including work presented at FM [17,18], we introduced ATeams, extending team automata with asynchronous communication. Unlike established asynchronous models such as CFSMs, ATeams naturally support multi-party communication with multiple senders and multiple receivers. We formalised their syntax and operational semantics, identified conditions for well-formed and well-behaved specifications, and provided tool support for modelling, animation, and automated checks for well-formedness and well-behavedness. Together, these contributions position ATeams as a unifying semantic foundation for the modelling and analysis of heterogeneous synchronous–asynchronous multi-party systems.

Future Work. In particular the well-behavedness conditions for ATeams need to be fine-tuned, expanded and implemented in A-Team to offer further analyses guaranteeing a rich set of communication properties, such as deadlock detection (considering the decidability issues mentioned below). These analyses could be further improved by developing back-ends that rely on efficient tools, such as mCRL2 for team automata [18]. The proposed heterogeneous communication channels with multi-party interactions could also be exploited when describing systems at a *global* rather than local level. These specifications could be analysed to reason about realisability, similar to our previous work on realising team automata from global specifications [22], or about multi-party session types for multilingual programming [48].

We conjecture that CFSMs can be encoded into ATeams, by only allowing actions whose synchronisation type has buffer type FIFO and location type SND-RCV. It would then follow that, in general, deadlock detection is undecidable. As future work, we plan to investigate whether decidability may be recovered under suitable restrictions, such as when FIFO ordering is ruled out (i.e., only unordered buffers are permitted), or when buffers are finite (although checking for buffer finiteness may itself be undecidable).

Acknowledgments. The first and second authors were partially funded by MUR PRIN 2020TL3X8X project T-LADIES (Typeful Language Adaptation for Dynamic, Interacting and Evolving Systems) and by CNR project "Formal Methods in Software Engineering 2.0", CUP B53C24000720005. The third author was funded by National Funds through the FCT (Portuguese Foundation for Science and Technology) within the project POISE, with reference 2024.15783.PEX (https://doi.org/10.54499/2024.15783.PEX).

The authors would like to thank Rolf Hennicker for inspiring discussions on adding asynchronous communication to team automata, and the anonymous reviewers for their useful comments and suggestions.

Disclosure of Interests. The authors have no competing interests to declare that are relevant to the content of this article.

References

1. Aceto, L., Cimini, M., Ingólfsdóttir, A., Reynisson, A.H., Sigurdarson, S.H., Sirjani, M.: Modelling and simulation of asynchronous real-time systems using timed Rebeca. In: Mousavi, M.R., Ravara, A. (eds.) Proceedings of the 10th International Workshop on the Foundations of Coordination Languages and Software Architectures (FOCLASA 2011). EPTCS, vol. 58, pp. 1–19 (2011). https://doi.org/10.4204/EPTCS.58.1
2. de Alfaro, L., Henzinger, T.A.: Interface automata. In: Proceedings of the 8th European Software Engineering Conference held jointly with 9th ACM SIGSOFT International Symposium on Foundations of Software Engineering (ESEC/FSE 2001), pp. 109–120. ACM (2001). https://doi.org/10.1145/503209.503226
3. Arbab, F.: Reo: a channel-based coordination model for component composition. Math. Struct. Comput. Sci. **14**(3), 329–366 (2004). https://doi.org/10.1017/S0960129504004153

4. Armstrong, J., Virding, R., Williams, M.: Concurrent Programming in Erlang. Prentice Hall, 2 edn. (1996)
5. Attie, P.C., Bensalem, S., Bozga, M., Jaber, M., Sifakis, J., Zaraket, F.A.: Global and local deadlock freedom in BIP. ACM Trans. Softw. Eng. Methodol. **26**(3), 9:1–9:48 (2018). https://doi.org/10.1145/3152910
6. Barbanera, F., Hennicker, R.: Safe composition of systems of communicating finite state machines. In: Aubert, C., Di Giusto, C., Fowler, S., Ka I Pun, V. (eds.) Proceedings of the 17th Interaction and Concurrency Experience (ICE 2024). EPTCS, vol. 414, pp. 39–57 (2024). https://doi.org/10.4204/EPTCS.414.3
7. Barbanera, F., Lanese, I., Tuosto, E.: Choreography automata. In: Bliudze, S., Bocchi, L. (eds.) COORDINATION 2020. LNCS, vol. 12134, pp. 86–106. Springer, Cham (2020). https://doi.org/10.1007/978-3-030-50029-0_6
8. Barbanera, F., Lanese, I., Tuosto, E.: A Theory of Formal Choreographic Languages. Log. Meth. Comp. Sci. **19**(3), 9:1–9:36 (2023). https://doi.org/10.46298/LMCS-19(3:9)2023
9. Bartoletti, M., Cimoli, T., Zunino, R.: Compliance in behavioural contracts: a brief survey. In: Bodei, C., Ferrari, G.-L., Priami, C. (eds.) Programming Languages with Applications to Biology and Security. LNCS, vol. 9465, pp. 103–121. Springer, Cham (2015). https://doi.org/10.1007/978-3-319-25527-9_9
10. Basile, D., ter Beek, M.H., Degano, P., Legay, A., Ferrari, G.L., Gnesi, S., Di Giandomenico, F.: Controller synthesis of service contracts with variability. Sci. Comput. Program. **187** (2020). https://doi.org/10.1016/j.scico.2019.102344
11. Basile, D., Degano, P., Ferrari, G.L.: Automata for specifying and orchestrating service contracts. Log. Meth. Comp. Sci. **12**(4:6), 1–51 (2016). https://doi.org/10.2168/LMCS-12(4:6)2016
12. Basile, D., Degano, P., Ferrari, G., Tuosto, E.: From orchestration to choreography through contract automata. In: Lanese, I., Lluch Lafuente, A., Sokolova, A., Torres Vieira, H. (eds.) Proceedings of the 7th Interaction and Concurrency Experience (ICE 2014). EPTCS, vol. 166, pp. 67–85 (2014). https://doi.org/10.4204/EPTCS.166.8
13. Basu, A., Bozga, M., Sifakis, J.: Modeling heterogeneous real-time components in BIP. In: Proceedings of the 4th IEEE International Conference on Software Engineering and Formal Methods (SEFM 2006), pp. 3–12. IEEE (2006). https://doi.org/10.1109/SEFM.2006.27
14. Bauer, S.S., Mayer, P., Schroeder, A., Hennicker, R.: On weak modal compatibility, refinement, and the MIO workbench. In: Esparza, J., Majumdar, R. (eds.) TACAS 2010. LNCS, vol. 6015, pp. 175–189. Springer, Heidelberg (2010). https://doi.org/10.1007/978-3-642-12002-2_15
15. ter Beek, M.H., Carmona, J., Hennicker, R., Kleijn, J.: Communication requirements for team automata. In: Jacquet, J.-M., Massink, M. (eds.) COORDINATION 2017. LNCS, vol. 10319, pp. 256–277. Springer, Cham (2017). https://doi.org/10.1007/978-3-319-59746-1_14
16. ter Beek, M.H., Carmona, J., Kleijn, J.: Conditions for compatibility of components: The case of masters and slaves. In: Margaria, T., Steffen, B. (eds.) ISoLA 2016. LNCS, vol. 9952, pp. 784–805. Springer, Cham (2016). https://doi.org/10.1007/978-3-319-47166-2_55
17. ter Beek, M.H., Cledou, G., Hennicker, R., Proença, J.: Featured team automata. In: Huisman, M., Păsăreanu, C., Zhan, N. (eds.) FM 2021. LNCS, vol. 13047, pp. 483–502. Springer, Cham (2021). https://doi.org/10.1007/978-3-030-90870-6_26

18. ter Beek, M.H., Cledou, G., Hennicker, R., Proença, J.: Can we communicate? Using dynamic logic to verify team automata. In: Chechik, M., Katoen, J.P., Leucker, M. (eds.) FM 2023. LNCS, vol. 14000, pp. 122–141. Springer (2023). https://doi.org/10.1007/978-3-031-27481-7_9
19. ter Beek, M.H., Ellis, C.A., Kleijn, J., Rozenberg, G.: Synchronizations in team automata for groupware systems. Comput. Sup. Coop. Work **12**(1), 21–69 (2003). https://doi.org/10.1023/A:1022407907596
20. ter Beek, M.H., Hennicker, R., Kleijn, J.: Compositionality of safe communication in systems of team automata. In: Pun, V.K.I., Stolz, V., Simao, A. (eds.) ICTAC 2020. LNCS, vol. 12545, pp. 200–220. Springer, Cham (2020). https://doi.org/10.1007/978-3-030-64276-1_11
21. ter Beek, M.H., Hennicker, R., Kleijn, J.: Team automata@work: on safe communication. In: Bliudze, S., Bocchi, L. (eds.) COORDINATION 2020. LNCS, vol. 12134, pp. 77–85. Springer, Cham (2020). https://doi.org/10.1007/978-3-030-50029-0_5
22. ter Beek, M.H., Hennicker, R., Proença, J.: Realisability of global models of interaction. In: Ábrahám, E., Dubslaff, C., Tapia Tarifa, S.L. (eds.) ICTAC 2023. LNCS, vol. 14446, pp. 236–255. Springer (2023). https://doi.org/10.1007/978-3-031-47963-2_15
23. ter Beek, M.H., Hennicker, R., Proença, J.: Team automata: overview and roadmap. In: Castellani, I., Tiezzi, F. (eds.) COORDINATION 2024. LNCS, vol. 14676, pp. 161–198. Springer (2024). https://doi.org/10.1007/978-3-031-62697-5_10
24. ter Beek, M.H., Kleijn, J.: Modularity for teams of I/O automata. Inf. Process. Lett. **95**(5), 487–495 (2005). https://doi.org/10.1016/j.ipl.2005.05.012
25. Behrmann, G., David, A., Larsen, K.G.: A tutorial on UPPAAL. In: Bernardo, M., Corradini, F. (eds.) SFM-RT 2004. LNCS, vol. 3185, pp. 200–236. Springer, Heidelberg (2004). https://doi.org/10.1007/978-3-540-30080-9_7
26. Bejleri, A., Yoshida, N.: Synchronous multiparty session types. Electr. Notes Theor. Comput. Sci. **241**, 3–33 (2008). https://doi.org/10.1016/j.entcs.2009.06.002
27. Bliudze, S., Sifakis, J.: The algebra of connectors: structuring interaction in BIP. IEEE Trans. Comput. **57**(10), 1315–1330 (2008). https://doi.org/10.1109/TC.2008.26
28. Bordeaux, L., Salaün, G., Berardi, D., Mecella, M.: When are two web services compatible? In: Shan, M.-C., Dayal, U., Hsu, M. (eds.) TES 2004. LNCS, vol. 3324, pp. 15–28. Springer, Heidelberg (2005). https://doi.org/10.1007/978-3-540-31811-8_2
29. Brand, D., Zafiropulo, P.: On communicating finite-state machines. J. ACM **30**(2), 323–342 (1983). https://doi.org/10.1145/322374.322380
30. Brim, L., Černá, I., Vařeková, P., Zimmerová, B.: Component-interaction automata as a verification-oriented component-based system specification. ACM Softw. Eng. Notes **31**(2), (2006). https://doi.org/10.1145/1118537.1123063
31. Carmona, J., Cortadella, J.: Input/output compatibility of reactive systems. In: Aagaard, M.D., O'Leary, J.W. (eds.) FMCAD 2002. LNCS, vol. 2517, pp. 360–377. Springer, Heidelberg (2002). https://doi.org/10.1007/3-540-36126-X_22
32. Carmona, J., Kleijn, J.: Compatibility in a multi-component environment. Theor. Comput. Sci. **484**, 1–15 (2013). https://doi.org/10.1016/j.tcs.2013.03.006
33. Castagna, G., Gesbert, N., Padovani, L.: A theory of contracts for web services. ACM Trans. Program. Lang. Syst. **31**(5), 19:1–19:61 (2009). https://doi.org/10.1145/1538917.1538920

34. Clemente, L., Herbreteau, F., Sutre, G.: Decidable topologies for communicating automata with FIFO and bag channels. In: Baldan, P., Gorla, D. (eds.) CONCUR 2014. LNCS, vol. 8704, pp. 281–296. Springer, Heidelberg (2014). https://doi.org/10.1007/978-3-662-44584-6_20
35. Deniélou, P.-M., Yoshida, N.: Multiparty session types meet communicating automata. In: Seidl, H. (ed.) ESOP 2012. LNCS, vol. 7211, pp. 194–213. Springer, Heidelberg (2012). https://doi.org/10.1007/978-3-642-28869-2_10
36. Durán, F., Ouederni, M., Salaün, G.: A generic framework for n-protocol compatibility checking. Sci. Comput. Program. **77**(7–8), 870–886 (2012). https://doi.org/10.1016/j.scico.2011.03.009
37. Edixhoven, L., Jongmans, S.S., Proença, J., Castellani, I.: Branching pomsets: design, expressiveness and applications to choreographies. J. Log. Algebr. Methods Program **136**, (2024). https://doi.org/10.1016/j.jlamp.2023.100919
38. Ellis, C.S.: Team automata for groupware systems. In: Proceedings of the International ACM SIGGROUP Conference on Supporting Group Work: The Integration Challenge (GROUP 1997), pp. 415–424. ACM (1997). https://doi.org/10.1145/266838.267363
39. Ghassemi, F., Sirjani, M., Khamespanah, E., Mirani, M., Hojjat, H.: Transparent actor model. In: Proceedings of the 11th International Conference on Formal Methods in Software Engineering (FormaliSE 2023). pp. 97–107. IEEE (2023). https://doi.org/10.1109/FORMALISE58978.2023.00018
40. Gössler, G., Sifakis, J.: Composition for component-based modeling. Sci. Comput. Program. **55**, 161–183 (2005). https://doi.org/10.1016/j.scico.2004.05.014
41. Haller, P.: On the integration of the actor model in mainstream technologies: the Scala perspective. In: Agha, G.A., Bordini, R.H., Marron, A., Ricci, A. (eds.) Proceedings of the 2nd edition on Programming systems, languages and applications based on actors, agents, and decentralized control abstractions (AGERE! 2012). pp. 1–6. ACM (2012). https://doi.org/10.1145/2414639.2414641
42. Hennicker, R., Bidoit, M.: Compatibility properties of synchronously and asynchronously communicating components. Log. Meth. Comp. Sci. **14**(1), 1–31 (2018). https://doi.org/10.23638/LMCS-14(1:1)2018
43. Honda, K., Yoshida, N., Carbone, M.: Multiparty asynchronous session types. In: Proceedings of the 35th ACM SIGPLAN-SIGACT Symposium on Principles of Programming Languages (POPL 2008), pp. 273–284. ACM (2008). https://doi.org/10.1145/1328438.1328472
44. Jahandideh, I., Ghassemi, F., Sirjani, M.: An actor-based framework for asynchronous event-based cyber-physical systems. Softw. Syst. Model. **20**(3), 641–665 (2021). https://doi.org/10.1007/S10270-021-00877-Y
45. Ji, Z., Wang, S., Xu, X.: Session types with multiple senders single receiver. In: Hermanns, H., Sun, J., Bu, L. (eds.) SETTA 2023. LNCS, vol. 14464, pp. 112–131. Springer (2023). https://doi.org/10.1007/978-981-99-8664-4_7
46. Johnsen, E.B., Hähnle, R., Schäfer, J., Schlatte, R., Steffen, M.: ABS: a core language for abstract behavioral specification. In: Aichernig, B.K., de Boer, F.S., Bonsangue, M.M. (eds.) FMCO 2010. LNCS, vol. 6957, pp. 142–164. Springer, Heidelberg (2011). https://doi.org/10.1007/978-3-642-25271-6_8

47. Johnsen, E.B., Owe, O.: An asynchronous communication model for distributed concurrent objects. Softw. Syst. Model. **6**(1), 39–58 (2007). https://doi.org/10.1007/S10270-006-0011-2
48. Jongmans, S., Proença, J.: ST4MP: A blueprint of multiparty session typing for multilingual programming. In: Margaria, T., Steffen, B. (eds.) ISoLA 2022. LNCS, vol. 13701, pp. 460–478. Springer (2022). https://doi.org/10.1007/978-3-031-19849-6_26
49. Jonsson, B.: Compositional Verification of Distributed Systems. Ph.D. thesis, Uppsala University (1987)
50. Koehler, C., Clarke, D.: Decomposing port automata. In: Shin, S.Y., Ossowski, S. (eds.) Proceedings of the 24th ACM Symposium on Applied Computing (SAC 2009), pp. 1369–1373. ACM (2009). https://doi.org/10.1145/1529282.1529587
51. Larsen, K.G., Nyman, U., Wąsowski, A.: Modal I/O automata for interface and product line theories. In: De Nicola, R. (ed.) ESOP 2007. LNCS, vol. 4421, pp. 64–79. Springer, Heidelberg (2007). https://doi.org/10.1007/978-3-540-71316-6_6
52. Lynch, N.A.: Distributed Algorithms. Morgan Kaufmann (1996)
53. Lynch, N.A., Tuttle, M.R.: An Introduction to Input/Output Automata. CWI Q. **2**(3), 219–246 (1989), https://ir.cwi.nl/pub/18164, reprinted in ACM Form. Asp. Comput. (2026). https://doi.org/10.1145/3788687
54. Milner, R.: Communication and Concurrency. Prentice Hall (1989)
55. von Oheimb, D.: Interacting state machines: a stateful approach to proving security. In: Abdallah, A.E., Ryan, P., Schneider, S. (eds.) FASec 2002. LNCS, vol. 2629, pp. 15–32. Springer, Heidelberg (2003). https://doi.org/10.1007/978-3-540-40981-6_4
56. Pal, S., Guanciale, R., Lanese, I., Tuosto, E., Clo, M.: Pomsets for process management: A healthcare case study. In: Liu, Z., Saoud, A., Wehrheim, H. (eds.) ICTAC 2025. LNCS, vol. 16237, pp. 378–395. Springer (2025). https://doi.org/10.1007/978-3-032-11176-0_22
57. Proença, J., Edixhoven, L.: Caos: a reusable Scala web animator of operational semantics. In: Jongmans, S.S., Lopes, A. (eds.) COORDINATION 2023. LNCS, vol. 13908, pp. 163–171. Springer (2023). https://doi.org/10.1007/978-3-031-35361-1_9
58. Rensink, A., Wehrheim, H.: Process algebra with action dependencies. Acta Inform. **38**(3), 155–234 (2001). https://doi.org/10.1007/s002360100070
59. Scalas, A., Yoshida, N.: Less is more: multiparty session types revisited. Proc. ACM Program. Lang. **3**, 30:1–30:29 (2019). https://doi.org/10.1145/3290343
60. Severi, P., Dezani-Ciancaglini, M.: Observational equivalence for multiparty sessions. Fundam. Inform. **170**(1–3), 267–305 (2019). https://doi.org/10.3233/FI-2019-1863

Maximum Realizability for LTL Modulo Theories

Andoni Rodríguez Herrero[1,2] and César Sánchez[1(✉)]

[1] IMDEA Software Institute, Madrid, Spain
cesar.sanchez@imdea.org
[2] Universidad Politécnica de Madrid, Madrid, Spain

Abstract. The synthesis of systems from formal specifications is a fundamental problem in symbolic AI and formal methods where the goal is to automatically construct implementations that meet desired requirements. In practice, specifications often include both hard constraints (critical requirements) and soft constraints (desirable properties). However, when specifications are unrealizable (due to conflicts between requirements) traditional synthesis methods fail to provide meaningful implementations or guidance. This problem has been addressed with *maximum realizability*, a framework for synthesizing systems that satisfy hard constraints while maximizing the satisfaction of soft constraints. However, the literature only solves this technique for classic discrete Linear Temporal Logic (LTL), whereas its extension to richer LTL modulo theories ($\text{LTL}_{\mathcal{T}}$) remains unexplored. In this paper, we bridge this gap and we propose two approaches: (1) a method based on exhaustively traversing a set of abstractions and (2) an alternative method that incrementally refines abstractions during synthesis. Additionally, (3) we introduce lattice-based optimization techniques to further improve scalability by pruning uninteresting combinations of soft constraints. Our methods are evaluated on benchmarks from synthesis competitions and practical case studies, demonstrating their scalability and effectiveness.

1 Introduction

The synthesis of systems from formal specifications is a cornerstone of automated system design [28,31], where the goal is to automatically construct a correct implementation from high-level requirements. LTL [32] has long been a preferred formalism for specifying temporal properties of systems due to its expressiveness and compatibility with synthesis tools. However, as specifications become increasingly complex and intertwined with numeric or environmental dynamics, the challenge of *realizability*—determining whether a system that satisfies the specification can be synthesized—has grown significantly. In particular, *unrealizability* is common in modern LTL synthesis problems due to the inherent

This work was funded in part by the DECO Project (PID2022-138072OB-I00) funded by MCIN/AEI/10.13039/501100011033 and by ESF+.

A. Sampaio and M. Stoelinga (Eds.): FM 2026, LNCS 16557, pp. 50–70, 2026.
https://doi.org/10.1007/978-3-032-26220-2_3

complexity of environments and dependencies that designers might overlook or underestimate.

Traditional synthesis approaches often treat specifications in a monolithic manner: either the whole specification is realizable, and a system is synthesized, or it is unrealizable, and the designer is informed of the failure. However, this dichotomous perspective is insufficient in many practical scenarios, because, as specifications grow in complexity, the likelihood of conflicts among requirements increases. Factors such as numeric intricacies, complex dependencies, and dynamic environments make it nearly impossible to account for every possible scenario during the specification phase, which is more notable in $\mathrm{LTL}_{\mathcal{T}}$ [19,20,36,37], which extends LTL with predicates in first-order theories $\mathcal{T}$.

In order to bridge this gap, specifications have been sometimes composed of both *critical requirements* (hard constraints)—that must always be gurateed—and *desirable but non-essential properties* (soft constraints). For example, in the design of safety-critical systems, safety (hard constraint) is non-negotiable, but optimizing performance or minimizing energy consumption might be treated as secondary, yet valuable, goals.

This gap is precisely where *maximum realizability* [16] becomes crucial. Maximum realizability aims to synthesize systems that satisfy the hard constraints while optimizing the satisfaction of soft constraints *to the greatest extent possible*. This approach not only provides a best-effort implementation but also offers insights into the trade-offs between conflicting requirements, effectively guiding system designers in refining their specifications. In other words, maximum realizability not only strives to construct the best possible implementation but also provides valuable feedback to system designers. By identifying which soft constraints were partially or completely ignored, designers can better understand the reasons behind unrealizability and refine their specifications accordingly. However, this approach has not gained widespread popularity, primarily because of its computational complexity and the community's focus on alternative methods, such as repair [24], bounded synthesis [42], and probabilistic synthesis [33].

However, maximum realizability becomes particularly necessary in the context of $\mathrm{LTL}_{\mathcal{T}}$, due to the richer expressiveness of its specifications. Unlike LTL—which operates on purely Boolean conditions—$\mathrm{LTL}_{\mathcal{T}}$ integrates numeric constraints and theories, allowing designers to express dependencies and dynamics involving real-world quantities, such as time, resources, or physical measurements. This expressiveness, while powerful, introduces new challenges: the interaction between hard and soft constraints often becomes more intricate, and the environment's power to create unrealizable scenarios is significantly amplified. This makes it difficult to anticipate all potential conflicts during the specification phase. Therefore, maximum realizability has not been studied for symbolic (potentially infinite) scenarios, like those specified in $\mathrm{LTL}_{\mathcal{T}}$.

In this paper, we bridge this gap. First, we provide a rigorous definition of the maximum realizability problem for $\mathrm{LTL}_{\mathcal{T}}$ given a concrete quantitative semantics comparing strategies based on their satisfaction of soft constraints. In addition:

1. We propose two algorithms for solving maximum realizability: the first method constructs abstractions for all possible combinations of hard and soft constraints, while the second one builds upon the first one, but adds incremental abstraction refinement during synthesis.
2. To address the scalability issue inherent in these algorithms, we introduce a lattice-based method that exploits the structure of *mutations* (combinations of hard and soft constraints). These methods uses SMT-based techniques to optimize the search.
3. We validate our methods through experiments on recent benchmarks for infinite-state synthesis, demonstrating their feasibility for real-world applications.

This paper paves the way for future maximum realizability techniques that introduce complex criteria to be maximized for the optimal implementation.

Related Work. The main inspiration of this work is [16,17] in which they introduce and solve the problem of maximum realizability for propositional LTL specifications. However, their approach is based in bounded synthesis (whereas no bounded synthesis technique is known for $\text{LTL}_{\mathcal{T}}$ yet). The comparison of maximal realizability with similar approaches is also studied in [16], so we refer the reader to that paper for positioning questions.

In the broader landscape of reactive synthesis, several approaches have been developed to handle unrealizable specifications. For instance, [11,12] decompose specifications into components that can be realized separately. Another direction is parameterized synthesis [3,23], which allows tuning parameters in the specification to make it realizable while preserving the original intent. In general, many recent works have been proposed for reactive synthesis beyond Booleans [7,21,22,25,27,35,38,41]. However, they do not solve **maximum** realizability.

On the contrary, *Best-effort synthesis* is a synthesis paradigm that, like maximum realizability, addresses the case when a specification is unrealizable by classical means [4,5]. Rather than requiring full satisfaction of a temporal objective under all environment behaviors, best-effort synthesis seeks maximal strategies: those that succeed against a largest feasible set of environment reactions and are not strictly dominated by any other strategy. Compared to maximum realizability, best-effort synthesis differs in both goal and formalism. Maximum realizability assumes an explicit partition into hard and soft constraints and defines a preference ordering over strategies based on the degree to which soft requirements are satisfied, yielding a quantitative optimization problem over soft satisfaction levels. Best-effort synthesis, in contrast, operates without such a partition and instead induces a dominance order on strategies based on how many environment behaviors they can handle with respect to accomplishing the overall goal. In maximum realizability, the specification itself determines the optimization objective; in best-effort synthesis, the optimization objective arises from the comparison of strategy performance across environment responses.

Last, our *guess-check-learn* idea for the efficient algorithms is inspired by [37], which shows a method to enhance search performance, but for a different task (finding so-called *reactions* to compute abstractions and decide realizability), which in turn also resembles other similar directions [8,9].

2 Preliminaries

Classic Reactive Synthesis. The syntax of LTL formulas over a set of atomic propositions AP is defined as:

$$\varphi ::= \top \mid p \mid \neg\varphi \mid \varphi \vee \varphi \mid \bigcirc\varphi \mid \varphi\,\mathcal{U}\,\varphi$$

where $p \in AP$, $\bigcirc$ is the *next* operator, and $\mathcal{U}$ is the *until* operator. We derive the commonly used operators like $\Diamond\varphi \equiv \top\mathcal{U}\varphi$ and $\Box\varphi \equiv \neg\Diamond\neg\varphi$. The semantics are defined over infinite traces $\sigma = \sigma_0\sigma_1 \ldots \in (2^{AP})^\omega$:

$$\begin{array}{lll}
(\sigma, i) \models \top & & \text{always holds} \\
(\sigma, i) \models p & \text{iff } p \in \sigma_i & \\
(\sigma, i) \models \neg\varphi & \text{iff } (\sigma, i) \not\models \varphi & \\
(\sigma, i) \models \varphi \vee \psi & \text{iff } (\sigma, i) \models \varphi \text{ or } (\sigma, i) \models \psi & \\
(\sigma, i) \models \bigcirc\varphi & \text{iff } (\sigma, i+1) \models \varphi & \\
(\sigma, i) \models \varphi\,\mathcal{U}\,\psi & \text{iff for some } j \geq i \text{ s.t. } (\sigma, j) \models \psi \text{ and} & \\
 & \quad\quad \text{for all } i \leq k < j, (\sigma, k) \models \varphi &
\end{array}$$

Given an LTL formula φ with AP partitioned into inputs I (environment) and outputs O (system), the synthesis problem asks whether there exists a *strategy* $\rho : (2^I)^+ \to 2^O$ such that all traces generated by ρ satisfy φ. Formally, given $\sigma \in (2^I)^\omega$, ρ induces an infinite trace $\sigma^\rho = (i_0 \cup o_0)(i_1 \cup o_1) \ldots$ where $o_k = \rho(i_0 \ldots i_{k-1})$. A strategy ρ is winning whenever $\sigma^\rho \models \varphi$ for all $i_0 i_1 \ldots \in (2^I)^\omega$, in which case we write $\rho \models \varphi$.

Reactive Synthesis Modulo Theories. LTL$_\mathcal{T}$ extends LTL with first-order theory atoms:

$$\varphi ::= \top \mid l \mid \neg\varphi \mid \varphi \vee \varphi \mid \bigcirc\varphi \mid \varphi\,\mathcal{U}\,\varphi$$

where l is a theory literal (e.g., $x > y$). The semantics extend LTL with: $(\sigma, i) \models l$ iff $\sigma_i \models_\mathcal{T}$ 1 Also, the *fetch* operator $\lhd x$ retrieves x's previous value; i.e., a trace in LTL$_\mathcal{T}$ has to hold that $\sigma_{i-1}(x) = \sigma_i(\lhd x)$.

The synthesis problem for LTL$_\mathcal{T}$ is undecidable in general, but there are some methods mentioned in the related work. In this paper, we use CounterExample-Guided REactive Synthesis (CEGRES) [36] as a building block for Algorithm 2. In a nutshell, the CEGRES approach handles works as follows:

1. Given a formula φ^+ in a expressive logic LTL$^+$, abstract φ^+ into φ^- in the less expressive logic LTL$^-$ (e.g., from LTL$_\mathcal{T}$ to LTL).
2. Synthetize ρ for the abstracted φ^- formula (e.g., using Strix [29] if LTL$^-$ is propositional LTL).

3. Validate ρ against the semantics of LTL$^+$ (e.g., no contradiction in $\mathcal{T}$).
4. When validation fails, add *reactive tautologies* to φ^+, which are like strengthening invariants, to ensure the same spurious ρ is pruned. An example tautology is $(x > 10) \rightarrow \bigcirc(\lhd x > 4)$.
5. Iterating until convergence.

This yields a sound procedure that terminates for decidable fragments ($\mathcal{T}$ with bounded domains or fetch-free formulae) and is guaranteed to always progress (while adding tautologies) otherwise.

2.1 Maximum Realizability for LTL

Often, specifications consist of multiple requirements that may not be realizable together. In such cases, in addition to notifying the system designer about the unrealizability, we aim for the synthesis procedure to create an implementation that satisfies the specification to the greatest extent possible. This approach is especially valuable when some requirements represent desirable but non-essential properties of the system.

We use a quantitive semantics to formally capture what "to the greatest extent possible" means. For simplicity and continuity, our quantitative interpretation is based on the semantics of LTL formulae of the form $\Box\varphi$ (see [16]), but other weighted variants can be designed without altering the methods that will be presented.

Let $\varphi_1, \ldots, \varphi_n$ be LTL specifications, where each φ_i is an LTL formula. In order to formalize the maximal satisfaction of $\varphi_1 \wedge \ldots \wedge \varphi_n$, we first give a quantitative semantics of formulas of the form φ.

Definition 1. *Given an LTL formula φ and a strategy ρ, we define* the value *$val(\rho, \varphi)$* of *φ* in *ρ as:*

$$val(\rho,\varphi) = \begin{cases} (1,1,1,1) & \text{if } \rho \models \Box\varphi, \\ (0,1,1,1) & \text{else if } \rho \models \Diamond\Box\varphi \text{ but } \rho \not\models \Box\varphi, \\ (0,0,1,1) & \text{else if } \rho \models \Box\Diamond\varphi \text{ but } \rho \not\models \Diamond\Box\varphi, \\ (0,0,0,1) & \text{else if } \rho \models \Diamond\varphi \text{ but } \rho \not\models \Box\Diamond\varphi, \\ (0,0,0,0) & \text{otherwise.} \end{cases}$$

Now, we extend Definition 1 in order to compare strategies ρ with respect to their satisfaction of a set L of specifications.

Definition 2. *Given a set $L = \{\varphi_0, \varphi_1, ..., \varphi_n\}$ of LTL formulae and a strategy ρ, we define* the value *$val(\rho, L)$* of *φ* in *ρ as:*

$$val(\rho, L) = \Big(\sum_{i=0}^{n} v_{i,1}, \sum_{i=0}^{n} v_{i,2}, \sum_{i=0}^{n} v_{i,3}, \sum_{i=0}^{n} v_{i,4}\Big),$$

where $val(\rho, \varphi_i) = (v_{i,1}, v_{i,2}, v_{i,3}, v_{i,4})$ for $i \in \{1, \ldots, n\}$.

Given $val(\rho, L) = (w_1, w_2, w_3, w_4)$ and $val(\rho', L) = (w'_1, w'_2, w'_3, w'_4)$ we say that $val(\rho, L) \geq val(\rho', L)$ whenever $w_i \geq w'_i$ for $i = 1, 2, 3, 4$. Note that Definition 2 intuitively gives higher values to strategies in which a fewer number of soft specifications have low values. Also, note that the specification does not express any preference for each element in L.

We are now ready to define our problem formally:

Definition 3. *Given an LTL formula φ (the hard constraint) and formulae $L = \{\varphi_0, \ldots, \varphi_n\}$ (the soft constraints), where each φ_i is an LTL formula, the maximum realizability problem asks to determine if there exists a strategy ρ such that $\rho \models \varphi$, and if the answer is positive, to synthesize ρ such that $\rho \models \varphi$, and such that for every strategy ρ' with $\rho' \models \varphi$ it holds that $val(\rho, L) \geq val(\rho', L)$.*

Note that the classical realizability problem corresponds to the maximum realizability where $L = \{\emptyset\}$. Also, note that the outcome of this method is not unique, and that will be different depending on the criteria to maximize (e.g., inverse lexicographic).

3 Maximum Realizability for $\text{LTL}_\mathcal{T}$

3.1 Problem Overview

In classic realizability modulo theories, the method receives a $\text{LTL}_\mathcal{T}$ formula $\varphi^\mathcal{T}$ and checks whether it is realizable or not. Intuitively, maximum realizability extends classic realizability because it receives a hard formula $\varphi_h^\mathcal{T}$ and a list of soft $\text{LTL}_\mathcal{T}$ formulae $L = \{\varphi_{s_0}^\mathcal{T}, \varphi_{s_1}^\mathcal{T}, ..., \varphi_{s_n}^\mathcal{T}\}$, determines the strategy ρ that realizes $\varphi_h^\mathcal{T}$ and maximally realizes L. Note that the outcome of this method will be different depending on the criteria to maximize. In this paper we solve the following problem.

Given an $\text{LTL}_\mathcal{T}$ formula $\varphi^\mathcal{T}$ and formulae $L = \{\varphi_{s_0}^\mathcal{T}, \varphi_{s_1}^\mathcal{T}, ..., \varphi_{s_n}^\mathcal{T}\}$, where each $\varphi_i^\mathcal{T}$ is an $\text{LTL}_\mathcal{T}$ formula, the *maximum realizability problem* $(\varphi^\mathcal{T}, L)$ consists of determining whether there exists a strategy ρ such that $\rho \models \varphi^\mathcal{T}$, and if the answer is positive, to synthesize a (symbolic) strategy ρ that (1) $\rho \models \varphi^\mathcal{T}$, and (2) for strategy ρ' with $\rho' \models \varphi^\mathcal{T}$ it holds that $val(\rho, L) \geq val(\rho', L)$.

Given a maximum realizability problem $(\varphi^\mathcal{T}, L)$ we call $\varphi^\mathcal{T}$ the *hard constraint* and L the *soft constraints*.

Example 1. Consider pair $(\varphi_h^\mathcal{T}, L = \{\varphi_{s_0}^\mathcal{T}, \varphi_{s_1}^\mathcal{T}\})$ where:

$$\begin{aligned}
\varphi_h^\mathcal{T} &= \Box[(x > 10 \wedge y < 0) \rightarrow \bigcirc(x > y)] \\
\varphi_{s_0}^\mathcal{T} &= y < 0 \\
\varphi_{s_1}^\mathcal{T} &= \bigcirc(y > x)
\end{aligned}$$

It is easy to see that $\varphi = \Box\varphi_h^{\mathcal{T}} \wedge \Box\varphi_{s_0}^{\mathcal{T}} \wedge \Box\varphi_{s_1}^{\mathcal{T}}$ is unrealizable, because whenever the environment plays a value of x s.t. $(x > 10)$ and the system plays a value of y with $y < 0$, then it cannot happen that $\bigcirc(y > x)$ and $\bigcirc(x > y)$ hold at the same time. However, $\varphi_h^{\mathcal{T}}$ is realizable by itself, but at the cost ignoring soft formulae in L. In this example, however, some of the soft constraints can be guaranteed along $\varphi_h^{\mathcal{T}}$. For example, $\Box(y < 0)$ can be guaranteed along $\varphi_h^{\mathcal{T}}$, and boteh $\Box\Diamond(y < 0)$ and $\Box\Diamond\bigcirc(y > x)$ can be guaranteed together with $\varphi_h^{\mathcal{T}}$.

In this paper, we propose a method to construct an *optimal* controller of Example 1 solving maximum realizability. In order to do so, we first introduce the following definition.

Definition 4 (Mutation). *Given a hard constraint $\varphi_h^{\mathcal{T}}$ and soft constraints $L = \{\varphi_{s_0}^{\mathcal{T}}, \varphi_{s_1}^{\mathcal{T}}, ..., \varphi_{s_n}^{\mathcal{T}}\}$, a mutation m is a tuple $\langle G, P, R, F, N\rangle$ that is a partition of L (i.e. $G \cup P \cup R \cup F \cup N = L$ and the sets G, P, R, F, N are pairwise disjoint).*

The characteristic formula of m is

$$f(m) = \varphi_h^{\mathcal{T}} \wedge \psi_0 \wedge \psi_1 \wedge \cdots \wedge \psi_n, \text{ where}$$

$$\psi_i = \begin{cases} \Box\varphi_i & \text{if } \varphi_{s_i}^{\mathcal{T}} \in G \\ \Diamond\Box\varphi_i & \text{if } \varphi_{s_i}^{\mathcal{T}} \in P \\ \Box\Diamond\varphi_i & \text{if } \varphi_{s_i}^{\mathcal{T}} \in R \\ \Diamond\varphi_i & \text{if } \varphi_{s_i}^{\mathcal{T}} \in F \\ \mathit{true} & \text{if } \varphi_{s_i}^{\mathcal{T}} \in N \end{cases}$$

We use $\mathcal{M}$ for the set of all mutations.

The names G, P, R, F, N come from *Globally*, *Persistence*, *Recurrence*, *Finally* and *None*, respectively, because, for example, a formula $\varphi_{s_i}^{\mathcal{T}} \in R$ then it will be treated as $\Box\Diamond\varphi_{s_i}^{\mathcal{T}}$. Given a maximum realizability problem $(\varphi_h^{\mathcal{T}}, L)$ there exist $5^{|L|}$ mutations. Now, we propose a method (Algorithm 1) that exhaustively traverses the set of $5^{|L|}$ mutations, checking realizability for each of them. Later we will present alternatives that enhance performance reusing theory facts learned previously (Algorithm 2). We will abuse notation and say "check the realizability of a mutation" m to mean the realizability of $f(m)$. Finally, we can define the value of a mutation as follows.

Definition 5. *Given mutation $m : (G, P, R, F, N)$, the value of a mutation is $val(m) = (|G|, |P|, |R|, |F|)$.*

Every strategy ρ satisfies some mutations m and violates others. Hence there is some mutation with maximal value, the mutation that corresponds to the soft constraint with the highest of G, P, R, F and N that ρ satisfies. It is easy to see that if two strategies ρ and ρ' are such that $val(\rho, L) \geq val(\rho', L)$ then the corresponding mutations m and m' are such that $val(m) \geq val(m')$. Therefore, finding a strategy with optimal value in the original formulation of maximal realizability (see Sect. 2) corresponds to finding the strategy for a mutation that provides the maximum value.

3.2 General Method

We now present a general method for solving the maximum realizability problem for LTL$_\mathcal{T}$. Given an LTL$_\mathcal{T}$ formula φ, we use a black-box procedure, that we call *synth*(φ), which decides the realizability of φ and produces strategy ρ. Informally, Algorithm 1 works as follows: (1) it receives the hard specification $\varphi_h^\mathcal{T}$ and the list of soft specifications L; (2) it considers the set $\mathcal{M}$ of all mutations; (3) it initializes *best* with value 0 and then checks for each $m \in \mathcal{M}$, whether *val*(m) $>$ *best*; in that case it uses *synth*($f(m)$) to determine if the system wins (i.e., $f(m)$ is realizable) and then stores the ρ and the better *val*. At the end of the loop, *best*$_\rho$ is the best possible strategy with value *best*.

Algorithm 1: Exhaustive exploration

```
FUNCTION Exhaustive()
Input: φ_h^T, L = {φ_s0^T, ..., φ_sn^T}
best, best_ρ ← 0, ∅
for m ∈ M do
    if val(m) > best then
        (is_real, ρ) ← synth(f(m))
        if is_real then
            best, best_ρ ← val(m), ρ
return best, best_ρ
```

Theorem 1. *Algorithm 1 solves the maximum realizability problem returning the maximal strategy ρ if there is one (provided all realizability queries terminate).*

The complexity of the problem depends on the considered temporal logic fragment as it performs $|\mathcal{M}|$ queries to a the synthsis procedure (which for LTL is 2-EXPTIME in the general case).

Example 2. Consider again Example 1. There are $5^{|L|} = 25$ mutations of the problem:

$$m_0 = \varphi_h^\mathcal{T} \wedge \Box\varphi_{s_0}^\mathcal{T} \wedge \Box\varphi_{s_1}^\mathcal{T}$$
$$m_1 = \varphi_h^\mathcal{T} \wedge \Box\varphi_{s_0}^\mathcal{T} \wedge \Diamond\Box\varphi_{s_1}^\mathcal{T}$$
$$\dots$$
$$m_{24} = \varphi_h^\mathcal{T} \wedge \mathit{true} \wedge \mathit{true}$$

Algorithm 1 will consider all the 25 mutations, each of them invoking an LTL$_\mathcal{T}$ synthesis problem. In this particular case, Algorithm 1 can effectively solve all the instances and find that the mutation m_8 with formula $f(m_8) = \Box\varphi_h^\mathcal{T} \wedge \Box\Diamond\varphi_{s_0}^\mathcal{T} \wedge \Diamond\varphi_{s_1}^\mathcal{T}$ is the mutation with a controller ρ of maximal value *val*(m_8) $=$ *val*(ρ, L).

3.3 Pre-processing Via Lazy Abstractions

Algorithm 1 quickly becomes intractable due to the agnostic relationships between different loop iterations. In particular, if the synthesis process of a mutation m needs to discover that a tautology like $(x > 10) \rightarrow \bigcirc(\lhd x > 4)$ is crucial to decide realizability of $f(m)$, then the synthesis problem for another mutation m' can use this tautology, which typically results in an improved performance.

We propose now a refined method that leverages tautology propagation. The key concept for this method is a function *getLast*, which returns the tautologies discovered during a synthesis procedure *solve* (where the word *last* simply refers to the fact that there was a last conclusive abstraction[1] in the synthesis process).

Algorithm 2: Reusing tautologies

FUNCTION *ReuseTauto*()
Input: $\varphi_h^{\mathcal{T}}$, $L = \{\varphi_{s_0}^{\mathcal{T}}, \varphi_{s_1}^{\mathcal{T}}, ..., \varphi_{s_n}^{\mathcal{T}}\}$
$\varphi_h^{\mathcal{T}'} \leftarrow \varphi_h^{\mathcal{T}} \wedge getLast(solve(\varphi_h^{\mathcal{T}}))$
for $\varphi_{s_i}^{\mathcal{T}} \in L$ **do**
  $\varphi_{s_i}^{\mathcal{T}'} \leftarrow \varphi_{s_i}^{\mathcal{T}} \wedge getLast(solve(\varphi_{s_i}^{\mathcal{T}}))$
best, $best_\rho \leftarrow$
  Algorithm $1(\varphi_h^{\mathcal{T}'}, \{\varphi_{s_0}^{\mathcal{T}'}, \ldots, \varphi_{s_n}^{\mathcal{T}'}\})$
return *best*, $best_\rho$

Algorithm 2 includes a preprocessing phase in which each hard and soft specification is enriched with tautologies using an invocation to *getLast* with the formula without the additional temporality as input. This typically computes the tautologies that are necessary for later computing realizability of the more complicated mutation formulas. After the preprocessing, Algorithm 2 invokes Algorithm 1, and the following holds:

Theorem 2. *Algorithm 2 solves the maximum realizability problem returning the maximal strategy ρ if there is one (provided all realizability queries terminate).*

Example 3. Recall Example 1 again. Algorithm 1 requires checking all $5^{|L|} = 25$ mutations, and in each call the $\text{LTL}_{\mathcal{T}}$ realizability solver needs to rediscover the theory tautologies (e.g., $(y < 0) \rightarrow (y < 10)$) again, through expensive synthesis calls. Instead, Algorithm 2 first preprocesses each soft constraint individually. For instance, for $\varphi_{s_0}^{\mathcal{T}}$, the solver discovers and stores the following fact `tauto` : $(y < 0) \rightarrow (y < 10)$. Then, these precomputed tautologies accelerate subsequent mutation checks within Algorithm 1. For instance, evaluating mutation $\varphi_h^{\mathcal{T}'} \wedge \Box\Diamond\varphi_{s_0}^{\mathcal{T}'} \wedge \Diamond\varphi_{s_1}^{\mathcal{T}'}$ now reuses learned constraints like `tauto`, which is included in $\varphi_{s_0}^{\mathcal{T}'}$ rather than recomputed.

The preprocessing phase admits flexibility, since other preliminary checks may generate more useful tautologies than Algorithm 2 or be faster in obtaining useful facts. Section 5 shows empirically that Algorithm 2 substantially improves the performance of Algorithm 1.

4 Efficient Lattice Based Algorithms

The methods presented in Sect. 3 exhaustively traverse the set of mutations, one at a time, checking whether each mutation is realizable. Therefore, the body of

[1] Again, we refer the reader to [36] for further technical details on $\text{LTL}_{\mathcal{T}}$ synthesis, Boolean abstractions and tautology generation.

the loop in the previous algorithms is visited $5^{|L|}$ times, so the running time of these algorithms quickly becomes unfeasible with the number of soft constraints. We will now improve both Algorithms 1 and 2 with the following observation.

Proposition 1. *Given an* $\text{LTL}_\mathcal{T}$ *formula* φ*, the following are valid formulas:*

$$\Box\varphi \rightarrow \Diamond\Box\varphi \qquad \Diamond\Box\varphi \rightarrow \Box\Diamond\varphi \qquad \Box\Diamond\varphi \rightarrow \Diamond\varphi \qquad \Diamond\varphi \rightarrow \mathit{true}$$

We now use these temporal validities to improve the previous algorithms, by exploiting that the realizability outcome of a mutation reveals information about the realizability of other mutations, which allows us to prune the search space.

4.1 Mutations and Lattices

Our improved methods prune the set of interesting mutations to evaluate. To illustrate the potential pruning choices, consider again specification in Example 1.

Example 4. The specification in Example 1 is $(\varphi_h^\mathcal{T}, \{\varphi_{s_0}^\mathcal{T}, \varphi_{s_1}^\mathcal{T}\})$ where $\varphi_{s_0}^\mathcal{T} : x < 10$ and $\varphi_{s_1}^\mathcal{T} : \bigcirc(y > x)$. Consider all the 5^2 mutations. Let m_0 be the mutation $G = \{\varphi_{s_0}^\mathcal{T}, \varphi_{s_1}^\mathcal{T}\}$ and $P = R = F = N = \emptyset$ (whose formula is $\varphi_h^\mathcal{T} \wedge \Box\varphi_{s_0}^\mathcal{T} \wedge \Box\varphi_{s_1}^\mathcal{T}$) and m_{24} be the mutation $G = P = R = F = \emptyset$ and $N = \{\varphi_{s_0}^\mathcal{T}, \varphi_{s_1}^\mathcal{T}\}$ (whose formula is $\varphi_h^\mathcal{T} \wedge \mathit{true} \wedge \mathit{true}$). Clearly, if m_0 is realizable, so are all other mutations. Analogously, if m_{24} is unrealizable so are all other mutations.

Also, there are incomparable mutations (in the sense that the realizability or unrealizability of one does not imply the realizability or unrealizability of the other) like the mutations m_1 and m_{15} with formulas:

$$f(m_1) : \Box\varphi^\mathcal{T} \wedge \ \Box\varphi_{s_1}^\mathcal{T} \wedge \Box\Diamond\varphi_{s_2}^\mathcal{T}$$
$$f(m_{15}) : \Box\varphi^\mathcal{T} \wedge \ \Diamond\varphi_{s_1}^\mathcal{T} \wedge \Box\varphi_{s_1}^\mathcal{T}$$

Example 4 reveals that mutations satisfy an order relation that is captured by the sets G, P, R, F and N.

We will use the following notation: given a mutation (G, P, R, F, N) we use $\widehat{P}$ (read "P or more") for $G \cup P$, $\widehat{R}$ for $G \cup P \cup R$, $\widehat{F}$ for $G \cup P \cup R \cup F$ and $\widehat{N}$ for $G \cup P \cup R \cup F \cup N$. We can now define a relation $\preceq$ that captures whether a mutation is more easily realized than another mutation.

Definition 6. *Given mutations* $m : (G, P, R, F, N)$ *and* $m' : (G', P', R', F', N')$, $m \preceq m'$ *holds if and only if* $G \subseteq G'$, $\widehat{P} \subseteq \widehat{P}'$, $\widehat{R} \subseteq \widehat{R}'$ *and* $\widehat{F} \subseteq \widehat{F}'$.

It holds that $\preceq$ is a partial order, and that for every two elements m and m' there is a greatest lower bound, namely

$$m \sqcap m' = (G'', P'', F'', F'', N'') \text{ where } \begin{cases} G'' = G \cap G' \\ P'' = (\widehat{P} \cap P') \cup (P \cap \widehat{P}') \\ R'' = (\widehat{R} \cap R') \cup (R \cap \widehat{R}') \\ F'' = (\widehat{F} \cap F') \cup (F \cap \widehat{F}') \\ N'' = (\widehat{N} \cap N') \cup (N \cap \widehat{N}') \end{cases}$$

and minimum upper bound

$$m \sqcup m' = (G'', P'', F'', F'', N'') \text{ where } \begin{cases} G'' = G \cup G' \\ P'' = (\widehat{P} \cup \widehat{P}') \cap (P \cup P') \\ R'' = (\widehat{R} \cup \widehat{R}') \cap (R \cup R') \\ F'' = (\widehat{F} \cup \widehat{F}') \cap (F \cap F') \\ N'' = (\widehat{N} \cup \widehat{N}') \cap (N \cup N') \end{cases}$$

Both operations $\sqcap$ and $\sqcup$ are associative, commutative and idempotent. Note that $m \preceq m'$ if and only if $m \sqcap m' = m$.
Then, the following proposition follows:

Proposition 2. *$(M, \sqcap, \sqcup)$ is a lattice.*

Also, the following holds directly from Proposition 1 and the fact that adding conjuncts makes an $\text{LTL}_{\mathcal{T}}$ formula "less satisfiable":

Theorem 3. *Let m, m' be two mutations with $m' \preceq m$. If m is realizable then m' is realizable. If m' is unrealizable then m is unrealizable.*

4.2 A Guess, Check and Learn Algorithm

We now present a faster algorithm that replaces the main loop of Algorithm 1 by pruning mutations that need not be explored captured by $\preceq$. We capture the set of interesting mutations that remain to be explored with the use an SMT formula S, which characterizes the search space with variables $M : \{m_1, ..., m_{|L|}\}$ (one m_i for each temporal soft constraint φ_{s_i}) of type $[5] = \{0, 1, 2, 3, 4\}$. For example, $m_i = 1$ means $\varphi_{s_i}^{\mathcal{T}} \in P$. A valuation of M, that is $v : M \to [5]$, represents a mutation $m = (G, P, R, F, N)$. We will use $f(v)$ for the characteristic function $f(m)$ of the mutation that corresponds to v.

Models of S correspond to interesting valuations of M, that is, interesting mutations to explore. We will use the following formulas, that given a concrete valuation v create the symbolic formula for the variables M that captures the set of mutations above v and below v:

$$above(v) = \bigwedge_{i \in L} v(m_i) \preceq m_i \qquad\qquad below(v) = \bigwedge_{i \in L} m_i \preceq v(m_i)$$

Algorithm 3: *Guess-Check-Learn* with SMT

1 FUNCTION *GuessCheckLearn*()
2 **Input:** $\varphi_h^{\mathcal{T}}, L = \{\varphi_{s_0}^{\mathcal{T}}, \dots, \varphi_{s_n}^{\mathcal{T}}\}$
3 $best, best_\rho \leftarrow 0, \emptyset$
4 $S \leftarrow true$
5 **while** ***satisfiable***(S) **do**
6     $v \leftarrow model(S)$
7     $m \leftarrow mutation(v)$
8     $\rho \leftarrow synth(f(m))$
9   **if** $env(\rho)$ **then**
10         $S \leftarrow S \wedge \neg above(v)$
11   **else**
12   **if** $val(m) > best$ **then**
13             $best, best_\rho \leftarrow val(m), \rho$
14         $S \leftarrow S \wedge \neg below(v)$
15 **return** $best, best_\rho$

Theorem 3 ensures that whenever we find a valuation v whose mutation m is unrealizable, exploring any m' such that $m \preceq m'$ is unnecessary because m' is guaranteed to be unrealizable. Hence, we will add $\neg above(v)$ as a new constraint to S to remove from S all valuations of M that correspond to the mutations $m' \preceq m$. Analogously, if we find a valuation v whose mutation m is realizable, exploring any m' such that $m' \preceq m$ is unnecessary because $val(m) \leq val(m')$. In this case, the constraint to add to S is $\neg below(v)$.

This method appears in Algorithm 3, and we call it a *Guess-Check-Learn* algorithm following the nomenclature of [34] (in which a similar method was used to solve regular realizability modulo theories). Concretely, the search is initialized with $S = true$ (ln. 4) and then, iteratively, an SMT query is used to find a satisfying assignment v for S (line 5), which corresponds to a mutation m whose realizability is interesting to be explored. The loop stops when S is unsatisfiable (ln. 5). Then, the remaining building blocks are as follows:

- *Guess*: Obtain a satisfying assignment v for S (ln. 6) to get an unexplored interesting mutation (ln. 7).
- *Check*: Check the realizability the mutation (ln. 8).
- *Learn*: Prune S according to what can be learned, as follows:
  - If the mutation results into an unrealizable specification, then the algorithm prunes all mutations above (ln. 10).
  - If the resulting specification is realizable, ln. 14 prunes all mutations below (in this case the value of the mutation the value, if maximum, is checked in ln. 13).

Theorem 4. *Algorithm 3 terminates and returns the best mutation with respect to val (assuming that all realizability queries terminate).*

Note that termination of the exploration of the search space is guaranteed because each execution of the loop removes at least one mutation and the number of mutations is finite.

In addition, we acknowledge that a SAT formula could have been used to represent S instead of an SMT formula, but we considered that a symbolic representation is easier to follow and unifies tooling (i.e., we can use the same SMT solver to get mutation candidates for S and solve the underlying realizability queries). Moreover, a symbolic encoding is a more promising approach for S whenever the complexity of the search space for mutations grows.

4.3 Quasi-Mutations

It is often the case that a given an unrealizable mutation m a small sub-set of the sub-formulas within $f(m)$ are already unrealizable. Hence, every mutation that contains this sub-set is doomed to be unrealizable, which allows a much more aggressive pruning. We now introduce some preliminary definitions to formalize this intuition.

Definition 7 (quasi-mutations). *Given a mutation* $m : \langle G, P, R, F, N\rangle$, *another mutation* $m' : \langle G', P', R', F', N'\rangle$ *is a quasi-mutation of* m *if* $G' \subseteq G$, $P' \subseteq P$, $R' \subseteq R$ *and* $F' \subseteq F$. *We use* Q_m *for the set of quasi-mutations of* m.

Informally, a quasi-mutation of m removes from m some soft formula from G, P, R and F and make it true. We say that such formula is masked.

Example 5. Consider a case with soft specifications $\varphi^{\mathcal{T}}_{s_0}, \varphi^{\mathcal{T}}_{s_1}, ..., \varphi^{\mathcal{T}}_{s_6}$ and mutation $m : (\{\varphi^{\mathcal{T}}_{s_2}\}, \{\varphi^{\mathcal{T}}_{s_1}\}, \{\varphi^{\mathcal{T}}_{s_0}\}, \{\varphi^{\mathcal{T}}_{s_3}, \varphi^{\mathcal{T}}_{s_4}, \varphi^{\mathcal{T}}_{s_5}, \varphi^{\mathcal{T}}_{s_6}\}, \{\})$. Its quasi-mutation $m' : (\{\varphi^{\mathcal{T}}_{s_2}\}, \{\varphi^{\mathcal{T}}_{s_1}\}, \{\varphi^{\mathcal{T}}_{s_0}\}, \{\varphi^{\mathcal{T}}_{s_3}, \varphi^{\mathcal{T}}_{s_4}, \varphi^{\mathcal{T}}_{s_5}\}, \{\varphi^{\mathcal{T}}_{s_6}\})$ that masks $\varphi^{\mathcal{T}}_{s_6}$ has formula $f(m')$:

$$\Box\varphi^{\mathcal{T}}_{s_2} \wedge \Diamond\Box\varphi^{\mathcal{T}}_{s_1} \wedge \Box\Diamond\varphi^{\mathcal{T}}_{s_0} \wedge \Diamond\varphi^{\mathcal{T}}_{s_3} \wedge \Diamond\varphi^{\mathcal{T}}_{s_4} \wedge \Diamond\varphi^{\mathcal{T}}_{s_5}$$

Note that, in Example 5, nothing is stated in m' about $\varphi^{\mathcal{T}}_{s_6}$, precisely meaning that it does not belong to neither G, P, R or F.

We now define the quasi-mutation order $\preceq_Q$:

Definition 8. *We write* $m' \preceq_Q m$ *if* m' *is a quasi-mutation of* m.

Note that if $m' \preceq_Q m$ then $m' \preceq m$. Again, $\preceq_Q$ is a partial order with a lower greatest lower bound $\sqcap_Q$ defined as (for $m : \langle G, P, R, F, N\rangle$ and $n : \langle G', P', R', F', N'\rangle$):

$$m \sqcap_Q n = \langle G'', P'', R'', F'', N''\rangle \text{ where } \begin{cases} G'' = G \cap G' \\ P'' = P \cap P' \\ R'' = R \cap R' \\ F'' = F \cap F' \\ N'' = L \setminus (G' \cup P' \cup R' \cup F') \end{cases}$$

For every m, $(Q_m, \sqcap_Q)$ forms a lower semi-lattice. The following holds directly from the previous definitions (and the fact that adding conjuncts makes a first-order formula "less satisfiable").

Theorem 5. *Let m and m' be such that $m \preceq_Q m'$. If m is unrealizable then m' is unrealizable. If m' is realizable then m is realizable.*

4.4 A Guess, Check and Learn Algorithm with Quasi-Mutations

The intuition of this algorithm is to generalize unrealizable cases and prune more cases from the exploration. We now show how to use quasi-mutations to complement Algorithm 4 in this direction.

Example 6. Consider again Example 5, with m : $(\{\varphi^{\mathcal{T}}_{s_2}\}, \{\varphi^{\mathcal{T}}_{s_1}\}, \{\varphi^{\mathcal{T}}_{s_0}\}, \{\varphi^{\mathcal{T}}_{s_3}, \varphi^{\mathcal{T}}_{s_4}, \varphi^{\mathcal{T}}_{s_5}, \varphi^{\mathcal{T}}_{s_6}\}, \{\})$ and its quasi-mutation $m :' (\{\varphi^{\mathcal{T}}_{s_2}\}, \{\varphi^{\mathcal{T}}_{s_1}\}, \{\varphi^{\mathcal{T}}_{s_0}\}, \{\varphi^{\mathcal{T}}_{s_3}, \varphi^{\mathcal{T}}_{s_4}, \varphi^{\mathcal{T}}_{s_5}\}, \{\varphi^{\mathcal{T}}_{s_6}\})$ that masks $\varphi^{\mathcal{T}}_{s_6}$. Then, if m' is unrealizable, any mutation n that coincides with m' where $\varphi^{\mathcal{T}}_{s_6}$ is not masked and coincides with m in all other soft formulas is unrealizable.

We now refine Algorithm 3 with an additional *generalization* step for unrealizable instances (see Algorithm 4). For example, in Example 6, if m is found unrealizable we will try to generalize it to m' because if m' is also unrealizable, m' will prune many more mutations than m (note that this idea resembles unsatisfiability cores [6]). Also, crucially, m' is typically faster to check realizability because $f(m')$ has less conjuncts than $f(m)$.

In order to search for a minimal quasi-mutation of a mutation that is found unrealizable, we will use a SAT formula S' to encode the search space of the quasi-mutations below m, where each variable z_i for $\{z_1, \ldots, z_{|L|}\}$ represents whether the soft formula $\varphi^{\mathcal{T}}_{s_i}$ is masked or not. An assignment of S' is a map $z : \{z_1, \ldots, z_{|L|}\} \to \mathbb{B}$.

Function *generalize* performs a search for a promising collection of unrealizable quasi-mutations under an unrealizable mutation m. This function explores semi-lattice of quasi-mutations that are below m. We detail it now.

Algorithm 4: Search with generalization

```
FUNCTION SearchGen()
Input: φ_h^T, L = {φ_{s_0}^T, ..., φ_{s_n}^T}
best, best_ρ ← 0, ∅
S ← true
while satisfiable(S) do
    v ← model(S)
    m ← mutation(v)
    ρ ← synth(f(m))
    if env(ρ) then
        S ← S ∧ ¬generalize(m)
    else
        if val(m) > best then
            best, best_ρ ← val(m), ρ
        S ← S ∧ ¬below(v)
return best, best_ρ
________________________________
FUNCTION generalize(m)
Z ← true
prune ← false
while satisfiable(Z) do
    z ← model(Z)
    m' ← mask(m, z)
    ρ ← synth(m')
    if env(ρ) then
        Z ← Z ∧ ¬ ⋀_{z_i:true} z_i
        prune ←
            prune ∨ ⋀_{z_i:true} m_i ⪯ x_i
    else
        Z ← Z ∧ ¬z ∧ ¬ ⋀_{z_i:false} z_i
return prune
```

Given m, `FUNCTION` generalize(m) proceeds analogously to Algorithm 3: (1) if m' is realizable this means that all the quasi-mutations with more masking are uninteresting to explore (because they will necessarily also be realizable); whereas (2) if m' is unrealizable then all the quasi-mutations with less masking are also necessarily unrealizable.

Most importantly, when unrealizable, this core is added to the information to be returned to the main loop of Algorithm 4. Concretely, given a model z of Z, $\neg \bigwedge_{z_i:false} z_i$ is added to Z whenever m' is realizable and $\neg \bigwedge_{z_i:true} z_i$ when unrealizable. In the latter case, $\bigwedge_{u_i:\text{true}} m_i \preceq x_i$ is also added to *prune*, to cut the mutation state space in the outer function (the one that operates over S).

In addition, it is relevant to note that, as in Algorithm 3, this procedure finishes when Z is unsatisfiable. However, this time the loop can be exited before Z becomes unsatisfiable without affecting soundness, but this comes at the price of less pruning. In other words, we can equip *generalize* with further criteria to stop earlier and, in general, there is a trade-off between (1) using more exploratory versions of *generalize*, which find lower quasi-mutations that allows us to prune more, and (2) the additional overhead that these versions convey, because calculation of realizability queries is costly and there might be many (naturally, depending on the size of the lattice).

Theorem 6. *Algorithm 4 terminates and returns the best mutation with respect to val (assuming that all realizability queries terminate).*

5 Empirical Evaluation

Experimental Setting. In order to showcase the applicability of our method, we performed an empirical evaluation. We evaluated our proposed algorithms on a set of benchmarks drawn from the recent synthesis contributions for infinite-state synthesis [7,21,36], with the addition of soft constraints ensuring that our experiments covered the spectrum of complexity encountered in practical synthesis tasks. We developed a prototype implementation `MaxSynTheos` in Python3, using Z3 [15] as the SMT solver and Strix [29] as the synthesis solver. The experiments were conducted on a MacBook Air with M1 chip and 16 GB of RAM.

Results. The results of our experiments are reported in Table 1, where each element in the left-hand side (Misc, Elevator...) corresponds to a set of benchmarks that is solvable without soft constraints. Concretely, we make three variants for each benchmark set, with 2, 5 and 10 soft constraints respectively. Then, for each set of benchmarks, we measured the average time and average percentage of realizability queries performed (out of the 5^n maximum) for Algorithm 1 (A1), Algorithm 2 (A2) and Algorithm 4 (A4).

Table 1. Summary of main results comparing the performance of Algorithm 1 (A1), Algorithm 2 (A2), and Algorithm 4 (A4) across different benchmarks with 2, 5, and 10 soft constraints. A dash (-) indicates that the algorithm could not resolve the benchmark within reasonable time or resource limits.

	Time (minutes)									Queries (% of max)								
	2			5			10			2			5			10		
	A1	A2	A4	A1	A2	A4	A1	A2	A4	A1	A2	A4	A1	A2	A4	A1	A2	A4
Misc.	22	16	9	320	95	28	-	-	110	100	100	48	100	100	22	100	100	0.6
Elevator	28	20	11	390	110	32	-	-	135	100	100	50	100	100	24	100	100	0.7
Basics	19	14	7	265	80	22	-	-	95	100	100	45	100	100	19	100	100	0.5
Thermo	33	24	13	465	130	38	-	-	160	100	100	53	100	100	26	100	100	0.9
Robot	37	26	15	540	145	42	-	-	185	100	100	56	100	100	29	100	100	1.0
Heli	41	29	17	620	170	48	-	-	205	100	100	58	100	100	31	100	100	1.1
Heli +	46	33	19	700	190	55	-	-	230	100	100	61	100	100	33	100	100	1.2
Express	39	27	14	575	155	44	-	-	195	100	100	55	100	100	27	100	100	0.8
LTLt	43	31	18	645	180	51	-	-	215	100	100	59	100	100	32	100	100	1.0
LTLt +	52	38	22	765	210	60	-	-	255	100	100	63	100	100	36	100	100	1.3
Lift	34	25	13	495	135	39	-	-	170	100	100	54	100	100	25	100	100	0.8
Train C	49	35	20	735	205	58	-	-	245	100	100	62	100	100	35	100	100	1.2
C and	45	32	18	675	185	53	-	-	225	100	100	60	100	100	33	100	100	1.1
Usb C	38	27	15	560	150	43	-	-	190	100	100	57	100	100	30	100	100	0.9
Stages C	47	34	19	705	195	56	-	-	235	100	100	61	100	100	34	100	100	1.2

Our results indicate that the algorithms exhibit substantial and varied performance differences, particularly as the number of soft constraints increases. A1 consistently shows the highest runtimes and query percentages due to its exhaustive exploration of all mutations. Notably, for 10 soft constraints, A1 failed to resolve any benchmark within time or resource limits, and even with 5 soft constraints, it requires significantly longer times, often hundreds of seconds, underscoring its scalability issues. A2 reduces time compared to A1 for 2 and 5 soft constraints by 50–75%, but still explores the full set of mutations in terms of query percentages, and like A1, it fails to resolve any benchmark with 10 soft constraints within reasonable limits. The most impressive gains are observed in A4, which achieves reductions of 99% in time and queries compared to the theoretical maximum for A1 and A2, and remains the only viable approach for 10 soft constraints, solving problems in a fraction of the time and queries that A1 and A2 cannot handle.

6 Conclusion

In this paper, we introduced the problem of maximum synthesis and realizability for $\text{LTL}_{\mathcal{T}}$. This problem is interesting because many $\text{LTL}_{\mathcal{T}}$ specifications are unrealizable due to non-critical components, even though some realizable combinations can be achieved in order to not disregard the complete specification, finding them is difficult because of numeric intricacies. Thus, we proposed different algorithmic solutions, (1) an exhaustive method (2) a preprocessing method that leverages symbolic properties and (3) a lattice-based algorithm that makes use of SMT solvers. We also compared them empirically and showed superiority of the latter approach.

A natural future work is studying the bounded synthesis version of the problem, which was the original formulation of the problem in [16]. This would require methods based on MaxSMT [1,30], because the maximum realizability problem is closely related to Partial MaxSAT [18]. In a similar line, in this paper we only considered the 5^n maximum realizability case, but some generalizations are easy (e.g., weighted versions). Moreover, LTL modulo theories allows *val* to represent even richer properties. More ambitiously, we can consider the problem of this paper to be a fragment of the most general problem of parameterized synthesis under different families. In parallel, we would like to get more information from unrealizable specifications via algorithms that provide partial implementations along with clear feedback on unmet soft constraints.

Last, there is a relevant application of our techniques to safe navigation systems via shielding [2,10,14,33] for infinite-state systems [13,39,40]. In particular, if users write $\text{LTL}_{\mathcal{T}}$ specifications that turn out to be unrealizable, can then redefine them by splitting them into safety (avoid action that potentially leads to collision) and liveness properties (avoid action that potentially leads to revisiting a region that is banned for K steps). Previous papers like [26] show methods which in similar specifications are either realizable or unrealizable, but there are no pragmatic approaches towards best-effort solutions (for instance, using terminology of this paper, our maximum realizability method can find that region r_i can be avoided persistently, together with r_j being only avoided recurrently).

References

1. Albert, E., Genaim, S., Martin-Martin, E., Merayo, A., Rubio, A.: Lower-bound synthesis using loop specialization and max-SMT. In: Silva, A., Leino, K.R.M. (eds.) CAV 2021. LNCS, vol. 12760, pp. 863–886. Springer, Cham (2021). https://doi.org/10.1007/978-3-030-81688-9_40
2. Alshiekh, M., Bloem, R., Ehlers, R., Könighofer, B., Niekum, S., Topcu, U.: Safe reinforcement learning via shielding. In: Proceedings of the 32nd AAAI Conf. on Artificial Intelligence (AAAI'18), pp. 2669–2678 (2018)
3. Alur, R., Moarref, S., Topcu, U.: Compositional synthesis with parametric reactive controllers. In: Proceedings of the 19th International Conference on Hybrid Systems: Computation and Control (HSCC'16), pp. 215–224. ACM (2016)
4. Aminof, B., De Giacomo, G., Lomuscio, A., Murano, A., Rubin, S.: Synthesizing strategies under expected and exceptional environment behaviors. In: Proceedings of the 29th International Joint Conference on Artificial Intelligence, (IJCAI 2020). ijcai.org (2020)
5. Aminof, B., De Giacomo, G., Rubin, S.: Best-effort synthesis: Doing your best is not harder than giving up. In: Proceedings of the Thirtieth International Joint Conference on Artificial Intelligence, (IJCAI 2021), pp. 1766–1772. ijcai.org (2021)
6. Asín, R., Nieuwenhuis, R., Oliveras, A., Rodríguez-Carbonell, E.: Efficient generation of unsatisfiability proofs and cores in SAT. In: Cervesato, I., Veith, H., Voronkov, A. (eds.) LPAR 2008. LNCS (LNAI), vol. 5330, pp. 16–30. Springer, Heidelberg (2008). https://doi.org/10.1007/978-3-540-89439-1_2
7. Azzopardi, S., Di Stefano, L., Piterman, N., Schneider, G.: Full LTL synthesis over infinite-state arenas. In: Proceedings of the 37th International Conference on Computer Aided Verification (CAV'25), LNCS (2025)
8. Bendík, J., Meel, K.S.: Approximate counting of minimal unsatisfiable subsets. In: Lahiri, S.K., Wang, C. (eds.) CAV 2020. LNCS, vol. 12224, pp. 439–462. Springer, Cham (2020). https://doi.org/10.1007/978-3-030-53288-8_21
9. Bendík, J., Meel, K.S.: Counting maximal satisfiable subsets. In: Proceedings of the Thirty-Fifth AAAI Conference on Artificial Intelligence (AAAI 2021), pp. 3651–3660. AAAI Press (2021)
10. Bloem, R., Könighofer, B., Könighofer, R., Wang, C.: Shield Synthesis: In: Baier, C., Tinelli, C. (eds.) TACAS 2015. LNCS, vol. 9035, pp. 533–548. Springer, Heidelberg (2015). https://doi.org/10.1007/978-3-662-46681-0_51
11. Bloem, R., Chatterjee, K., Jacobs, S., Könighofer, R.: Assume-guarantee synthesis for concurrent reactive programs with partial information. In: Baier, C., Tinelli, C. (eds.) TACAS 2015. LNCS, vol. 9035, pp. 517–532. Springer, Heidelberg (2015). https://doi.org/10.1007/978-3-662-46681-0_50
12. Brizzio, M., Gorostiaga, F., Sánchez, C., Degiovanni, R.: Mode-based reactive synthesis. In: Proceedings of the 17th International Symposium of NASA Formal Methods (NFM'25), vol. 15682 , LNCS Science, pp. 116–137. Springer (2025)
13. Corsi, D., Amir, G., Rodríguez, A., Katz, G., Sánchez, C., Fox, R.: Verification-guided shielding for deep reinforcement learning. RLJ **4**, 1759–1780 (2024)

14. Corsi, D., Mallik, K., Rodríguez, A., Sánchez, C.: Efficient dynamic shielding for parametric safety specifications. In: Proceedings of the 23rd International Symposium in Automated Technology for Verification and Analysis (ATVA 2025), vol. 16145, LNCS, pp. 157–179. Springer (2025)
15. de Moura, L., Bjørner, N.: Z3: an efficient SMT solver. In: Ramakrishnan, C.R., Rehof, J. (eds.) TACAS 2008. LNCS, vol. 4963, pp. 337–340. Springer, Heidelberg (2008). https://doi.org/10.1007/978-3-540-78800-3_24
16. Dimitrova, R., Ghasemi, M., Topcu, U.: Maximum realizability for linear temporal logic specifications. In: Lahiri, S.K., Wang, C. (eds.) ATVA 2018. LNCS, vol. 11138, pp. 458–475. Springer, Cham (2018). https://doi.org/10.1007/978-3-030-01090-4_27
17. Dimitrova, R., Ghasemi, M., Topcu, U.: Reactive synthesis with maximum realizability of linear temporal logic specifications. Acta Info. **57**(1–2), 107–135 (2020)
18. Fu, Z., Malik, S.: On solving the partial MAX-SAT problem. In: Biere, A., Gomes, C.P. (eds.) SAT 2006. LNCS, vol. 4121, pp. 252–265. Springer, Heidelberg (2006). https://doi.org/10.1007/11814948_25
19. Geatti, L., Gianola, A., Gigante, N.: Linear temporal logic modulo theories over finite traces. In: Proceedings of the 31st International Joint Conference on Artificial Intelligence, (IJCAI'22), pp. 2641–2647. ijcai.org (2022)
20. Geatti, L., Gianola, A., Gigante, N., Winkler, S.: Decidable fragments of LTLF modulo theories (extended version). CoRR, abs/2307.16840 (2023)
21. Heim, P., Dimitrova, R.: Issy: a comprehensive tool for specification and synthesis of infinite-state reactive systems. In: Proceedings of the 37th International Conference on Computer Aided Verification (CAV'25), LNCS (2025)
22. Heim, P., Dimitrova, R.: Translation of temporal logic for efficient infinite-state reactive synthesis. Proc. ACM Program. Lang., 9(POPL) (2025)
23. Jacobs, S., Bloem, R.: Parameterized synthesis. In: Flanagan, C., König, B. (eds.) TACAS 2012. LNCS, vol. 7214, pp. 362–376. Springer, Heidelberg (2012). https://doi.org/10.1007/978-3-642-28756-5_25
24. Jobstmann, B., Griesmayer, A., Bloem, R.: Program repair as a game. In: Etessami, K., Rajamani, S.K. (eds.) CAV 2005. LNCS, vol. 3576, pp. 226–238. Springer, Heidelberg (2005). https://doi.org/10.1007/11513988_23
25. Katis, A., et al.: Validity-guided synthesis of reactive systems from assume-guarantee contracts. In: Beyer, D., Huisman, M. (eds.) TACAS 2018. LNCS, vol. 10806, pp. 176–193. Springer, Cham (2018). https://doi.org/10.1007/978-3-319-89963-3_10
26. Kim, K., et al.: Realizable continuous-space shields for safe reinforcement learning. In: Proceedings of the 7th Annual Learning for Dynamics & Control Conference (L4DC'25), PMLR (2025)
27. Maderbacher, B., Bloem, R.: Reactive synthesis modulo theories using abstraction refinement. In: Proceedings of the 22nd Conference on Formal Methods in Computer-Aided Design, (FMCAD 2022), pp. 315–324. IEEE (2022)
28. Manna, Z., Pnueli, A.: Temporal Verification of Reactive Systems: Safety. Springer-Verlag (1995)
29. Meyer, P.J., Sickert, S., Luttenberger, M.: Strix: explicit reactive synthesis strikes back! In: Chockler, H., Weissenbacher, G. (eds.) CAV 2018. LNCS, vol. 10981, pp. 578–586. Springer, Cham (2018). https://doi.org/10.1007/978-3-319-96145-3_31
30. Pinckney, D., Cassano, F., Guha, A., Bell, J., Culpo, M., Gamblin, T.: Flexible and optimal dependency management via max-SMT. In: Proceedings of the 45th IEEE/ACM International Conference on Software Engineering (ICSE 2023), Melbourne, Australia, May 14-20, 2023, pp. 1418–1429. IEEE (2023)

31. Piterman, N., Pnueli, A., Sa'ar, Y.: Synthesis of reactive(1) designs. In: Proceedings of the 7th International Conference on Verification, Model Checking, and Abstract Interpretation (VMCAI'06), vol. 3855 of LNCS, pp. 364–380. Springer (2006)
32. Pnueli, A.: The temporal logic of programs. In: Proceedings of 18th Annual Symposium on Foundations of Computer Science (FOCS'77), pp. 46–67. IEEE CS Press (1977)
33. Pranger, S., Könighofer, B., Posch, L., Bloem, R.: TEMPEST - synthesis tool for reactive systems and shields in probabilistic environments. In: Hou, Z., Ganesh, V. (eds.) ATVA 2021. LNCS, vol. 12971, pp. 222–228. Springer, Cham (2021). https://doi.org/10.1007/978-3-030-88885-5_15
34. Rodriguez, A., Sánchez, C.: Realizability modulo theories. J. Log. Algeb. Meth. Prog., 100971 (2024
35. Rodríguez, A., Gorostiaga, F., Sánchez, C.: Predictable and performant reactive synthesis modulo theories via functional synthesis. In: Proceedings of the 22nd International Symposium in Automated Technology for Verification and Analysis, (ATVA 2024) Part II, vol. 15055 of LNCS, pp. 28–50. Springer (2024)
36. Rodríguez, A., Gorostiaga, F., Sánchez, C.: Counter example guided reactive synthesis for LTL modulo theories. In: Proceedings of the 37th International Conference on Computer Aided Verification (CAV'25), LNCS (2025)
37. Rodriguez, A., Sánchez, C.: Boolean abstractions for realizabilty modulo theories. In: Proceedings of the 35th International Conference on Computer Aided Verification (CAV'23), vol. 13966, LNCS. Springer, Cham (2023)
38. Rodriguez, A., Sánchez, C.: Adaptive reactive synthesis for LTL and LTLF modulo theories. In: Proceedings of the 38th AAAI Conference on Artificial Intelligence (AAAI 2024) (2024)
39. Rodríguez, A., Shaik, I., Corsi, D., Fox, R., Sánchez, C.: Explanations for unrealizability of infinite-state safety shields. In: Proceedings of the 22nd International Conference on Principles of Knowledge Representation and Reasoning (KR 2025) (2025)
40. Rodríguez, A., Amir, G., Corsi, D., Sánchez, C., Katz, G.: Shield synthesis for LTL modulo theories. In: Proceedings of the 39th AAAI Conference on Artificial Intelligence (AAAI'25), pp. 15134–15142. AAAI Press (2025)
41. Samuel, S., D'Souza, D., Komondoor, R.: GenSys: a scalable fixed-point engine for maximal controller synthesis over infinite state spaces. In: 29th ACM Joint European Software Engineering Conference and Symposium on the Foundations of Software Engineering (ESEC/FSE '21), pp. 1585–1589. ACM (2021)
42. Schewe, S., Finkbeiner, B.: Bounded synthesis. In: Namjoshi, K.S., Yoneda, T., Higashino, T., Okamura, Y. (eds.) ATVA 2007. LNCS, vol. 4762, pp. 474–488. Springer, Heidelberg (2007). https://doi.org/10.1007/978-3-540-75596-8_33

Exact Moment Estimation of Stochastic Differential Dynamics

Shenghua Feng[1,2(✉)], Jie An[1,2], Naijun Zhan[3,4], and Fanjiang Xu[1,2]

[1] National Key Laboratory of Space Integrated Information System, Institute of Software Chinese Academy of Sciences, Beijing, China
{anjie,fanjiang}@iscas.ac.cn

[2] University of Chinese Academy of Sciences, Beijing, China
fengshenghua@iscas.ac.cn

[3] School of Computer Science, Peking University, Beijing, China
njzhan@pku.edu.cn

[4] Zhongguancun Laboratory, Beijing, China

Abstract. Moment estimation for stochastic differential equations (SDEs) is fundamental to the formal reasoning and verification of stochastic dynamical systems, yet remains challenging and is rarely available in closed form. In this paper, we study time-homogeneous SDEs with polynomial drift and diffusion, and investigate when their moments can be computed exactly. We formalize the notion of moment-solvable SDEs and propose a generic symbolic procedure that, for a given monomial, attempts to construct a finite-dimensional linear ordinary differential equation (ODE) system governing its moment, thereby enabling exact computation. We introduce a syntactic class of pro-solvable SDEs, characterized by a block-triangular structure, and prove that all polynomial moments of any pro-solvable SDE admit such finite ODE representations. This class strictly generalizes linear SDEs and includes many nonlinear models. Experimental results demonstrate the effectiveness of our approach.

Keywords: Stochastic Dynamical Systems · Stochastic Differential Equations (SDEs) · Moment Estimation · Pro-solvable SDEs

1 Introduction

Stochastic differential equations (SDEs) are foundational mathematical models for describing the evolution of stochastic systems across a wide range of fields, from physics and biology [24] to finance [4] and engineering [13]. Moments, defined as expectations of monomial functions of the system variables, provide critical insights into the behavior and stability of the underlying stochastic systems. Consequently, moment estimation is a central problem in various applications, including formal verification of stochastic dynamical systems [3,8,17], sensitivity analysis [2,11], and the derivation of bounds for safety-critical models [21,26].

A. Sampaio and M. Stoelinga (Eds.): FM 2026, LNCS 16557, pp. 71–89, 2026.
https://doi.org/10.1007/978-3-032-26220-2_4

Despite their importance, exact moment estimation, i.e., obtaining closed-form expressions for moments of SDEs, remains a formidable challenge. Since SDEs, even linear ones, are often difficult to solve in closed form, their moments are correspondingly hard to compute directly. To the best of our knowledge, existing approaches based on the synthesis of supermartingales [30] can provide bounds on moments [25], but these methods are often conservative and rely heavily on the choice of template functions for the supermartingale, which may require significant manual effort. As illustrated in our case study (cf. Sect. 5.1), such approaches may fail to yield tight estimates for the moments of interest. Currently, there is no general framework for the exact computation of moments in general SDEs.

In this paper, we focus on time-homogeneous SDEs with polynomial drift and diffusion terms, and seek to systematically characterize when the moments can be computed exactly. Motivated by advances in the verification of probabilistic programs [3], we introduce the notion of *moment-solvable* SDEs, for which every moment admits an explicit, closed-form solution. Central to our approach is a symbolic procedure that, given any monomial, attempts to construct a finite-dimensional linear ODE system governing the moment dynamics. This procedure iteratively expands the set of coupled moments by applying the infinitesimal generator of the SDE, and halts if the expansion closes after finitely many steps.

Our main theoretical contribution is the identification of a syntactic class of SDEs, termed *pro-solvable SDEs*, which are characterized by a block-triangular structure in their coefficients. We prove that for all pro-solvable SDEs, the symbolic closure procedure always terminates, ensuring that every moment can be computed by solving a finite-dimensional linear ODE system. Notably, this class strictly generalizes the linear SDEs, encompassing a broad array of nonlinear models encountered in practice.

We demonstrate the practical effectiveness of our approach through experiments on a diverse suite of SDE benchmarks. Our method efficiently computes exact moments for many linear and nonlinear systems of interest, including higher-order cases, illustrating its scalability and applicability to the formal analysis of stochastic systems.

In summary, our main contributions are as follows:

- After the problem formulation (Sect. 2), we introduce the concept of moment-solvable SDEs and establish a general procedure for constructing finite-dimensional ODE systems governing the moment dynamics. (Sect. 3)
- We introduce and characterize the class of pro-solvable SDEs, proving that they are moment-solvable and that all their moments can be computed exactly. Furthermore, we provide a complexity analysis of our method. (Sect. 4)
- We implement our method and conduct experiments to demonstrate the broad applicability and effectiveness of our approach. (Sect. 5)

Related Work. The closest related works are those on *prob-solvable loops* and related classes of probabilistic programs in discrete time [3,21], where loop moments satisfy solvable linear recurrences and can be computed exactly. Our setting differs in that we consider continuous-time SDEs and build on the infinitesimal generator. A second line of work provides structural exact-moment results for stochastic reaction networks [19] and jump Markov processes [5,29], where specific network topologies (i.e., feedforward structures) yield closed finite moment equations. These results, however, are tailored to particular classes of jump processes [10,28] and do not offer a uniform procedure for general polynomial SDEs. Finally, existing works employ martingales [6,12,25] and semidefinite relaxations [9,18] to approximate moments and provide bounds used for the verification and analysis of SDEs and stochastic hybrid systems.

2 Problem Formulation

Let $\mathbb{N}$, $\mathbb{Z}$, and $\mathbb{R}$ denote the sets of natural numbers, integers, and real numbers, respectively. Vectors are denoted in bold; for $\boldsymbol{x} \in \mathbb{R}^n$, x_i refers to its i-th component. For $\boldsymbol{x} = (x_1, \ldots, x_n)^\top$, let $\alpha \triangleq (\alpha_1, \ldots, \alpha_n) \in \mathbb{N}^n$ be a multi-index, with $|\alpha| \triangleq \sum_{i=1}^n \alpha_i$. We use the notation $\boldsymbol{x}^\alpha \triangleq x_1^{\alpha_1} \cdots x_n^{\alpha_n}$ for presenting the corresponding monomial.

Probability and Moments. Let $(\Omega, \mathcal{F}, P)$ be a probability space, where Ω is the sample space, $\mathcal{F} \subseteq 2^\Omega$ is a σ-algebra, and $P\colon \mathcal{F} \to [0,1]$ is a probability measure. A *random variable* X defined on $(\Omega, \mathcal{F}, P)$ is an $\mathcal{F}$-measurable function $X\colon \Omega \to \mathbb{R}^n$; its expectation (w.r.t. P) is denoted by $\mathbb{E}[X]$. Given a multi-index $\alpha = (\alpha_1, \ldots, \alpha_n) \in \mathbb{N}^n$, the α-moment of X is $\mathbb{E}[X^\alpha] \triangleq \mathbb{E}[(X^{(1)})^{\alpha_1} \cdots (X^{(n)})^{\alpha_n}]$, where $X^{(i)}$ denotes the i-th component of X. A (continuous-time) *stochastic process* is a collection of random variables $\{X_t\}_{t\in T}$, where unless otherwise noted, the index set T is the half-line $[0, \infty)$.

Stochastic Differential Equations (SDEs). We consider a class of stochastic dynamical systems governed by time-homogeneous stochastic differential equations (SDEs) of the form

$$dX_t = b(X_t)\,dt + \sigma(X_t)\,dW_t, \quad t \geq 0, \tag{1}$$

where $\{X_t\}$ is an n-dimensional continuous-time stochastic process, $\{W_t\}$ is an m-dimensional Wiener process (standard Brownian motion), $b\colon \mathbb{R}^n \to \mathbb{R}^n$ is a vector-valued polynomial drift coefficient modeling the deterministic part of the dynamics, and $\sigma\colon \mathbb{R}^n \to \mathbb{R}^{n\times m}$ is a matrix-valued polynomial diffusion coefficient encoding the system's coupling to Gaussian white noise dW_t.

Under standard regularity and growth conditions [22, Chap. 5.2], given an initial state (random variable) X_0, the SDE (1) admits a unique solution $X_t(\omega) = X(t,\omega)\colon [0,\infty) \times \Omega \to \mathbb{R}^n$ that satisfies the stochastic integral equation

$$X_t = X_0 + \int_0^t b(X_s)\,ds + \int_0^t \sigma(X_s)\,dW_s.$$

The solution process $\{X_t\}$ of (1) is also called an *(Itô) diffusion process*, and may be denoted X_t^{0,X_0} (or simply $X_t^{X_0}$) to indicate the initial condition X_0 at time $t = 0$. In the special case where $\sigma \equiv 0$, the SDE reduces to an ordinary differential equation (ODE), recovering the classical deterministic setting.

The *exact moment estimation* (EME) problem of SDEs studied in this paper reads as follows:

EME Problem. Let $\{X_t\}_{t\geq 0}$ denote the solution of SDE (1). Our objective is to compute the α-moment of the random variable X_t, that is,

$$m_\alpha(t) \triangleq \mathbb{E}[X_t^\alpha] = \mathbb{E}\big[\big(X_t^{(1)}\big)^{\alpha_1} \cdots \big(X_t^{(n)}\big)^{\alpha_n}\big]$$

for any given multi-index $\alpha = (\alpha_1, \ldots, \alpha_n) \in \mathbb{N}^n$ and any $t \geq 0$.

3 Reduction of EME Problem to ODE Solving

In this section, we formalize the concept of *moment-solvable* SDEs, which precisely delineates those systems for which the EME problem admits an explicit solution. We then present a generic symbolic procedure that, given a target monomial, systematically attempts to construct a finite-dimensional linear ODE system governing the evolution of its moment, thus enabling exact computation of the desired expectation.

Definition 1 (Moment-solvable SDE). *Given a multi-index $\alpha \in \mathbb{N}^n$, we say that the SDE* (1) *is* moment-solvable for α *if there exists an explicit function $h_\alpha \colon [0, \infty) \to \mathbb{R}$ such that*

$$m_\alpha(t) = \mathbb{E}[X_t^\alpha] = h_\alpha(t) \quad \textit{for all } t \geq 0.$$

If this property holds for all multi-indices α, then the SDE is moment-solvable.

Clearly, solving the EME problem for a multi-index α is equivalent to establishing that the SDE is moment-solvable for α. In practice, obtaining explicit formulas for moments is highly nontrivial. First, closed-form solutions for nonlinear SDEs are generally unavailable. Second, even when a stochastic representation of the solution is known, the evaluation of $\mathbb{E}[X_t^\alpha]$ typically involves high-dimensional integrals that do not admit simple analytical expressions.

To circumvent these challenges, rather than attempting to compute $\mathbb{E}[X_t^\alpha]$ directly, we instead consider the time evolution of moments. By deriving differential equations for moments using the infinitesimal generator, we seek to construct a closed, finite-dimensional linear ODE system for the evolution of a suitable collection of moments. This approach is grounded in Dynkin's formula, which serves as the stochastic analogue of the Newton–Leibniz rule and connects the dynamics of the process to the evolution of expected values.

Theorem 1 (Dynkin's formula [22]). *Let $\{X_t\}_{t\geq 0}$ be the solution of* (1). *If $f \in C^2(\mathbb{R}^n)$ has compact support, then for all $t \geq 0$,*

$$\frac{\mathrm{d}}{\mathrm{d}t}\mathbb{E}[f(X_t)] = \mathbb{E}[(\mathcal{A}f)(X_t)],$$

Algorithm 1. Construction of a finite moment system for a given multi-index α

Require: Drift b, diffusion σ, generator $\mathcal{A}$ of (1), initial multi-index α
Ensure: Multi-index set S which entails a closed linear ODE of moments if terminates.

$S \leftarrow \{\alpha\}$ ▷ *set of multi-indices (monomials)*
$\mathcal{P} \leftarrow \{\alpha\}$ ▷ *set of unprocessed multi-indices*
while $\mathcal{P} \neq \emptyset$ **do**
  Select and remove some $\beta \in \mathcal{P}$
  Compute the generator action on the monomial x^β: $\mathcal{A}\boldsymbol{x}^\beta = \sum_\gamma a_{\beta\gamma}\,\boldsymbol{x}^\gamma + c_\beta$
  for all γ such that $a_{\beta\gamma} \neq 0$ **do**
    if $\gamma \notin S$ **then**
      $S \leftarrow S \cup \{\gamma\}$
      $\mathcal{P} \leftarrow \mathcal{P} \cup \{\gamma\}$
Let $S = \{\alpha, \beta^1, \ldots, \beta^k\}$ be the final set of indices.
Form the moment vector $m(t) \leftarrow \left(\mathbb{E}[X_t^\alpha], \mathbb{E}[X_t^{\beta_1}], \ldots, \mathbb{E}[X_t^{\beta_k}]\right)^\top$.
Use Theorem 1 to obtain the closed linear ODE system $\dfrac{\mathrm{d}}{\mathrm{d}t}m(t) = A\,m(t) + c$.

where $\mathcal{A}$ is the infinitesimal generator of (1) *given by*

$$\mathcal{A}f(\boldsymbol{x}) = \sum_{i=1}^{n} b_i(\boldsymbol{x})\frac{\partial f}{\partial x_i}(\boldsymbol{x}) + \frac{1}{2}\sum_{i,j=1}^{n} (\sigma\sigma^\top)_{ij}(\boldsymbol{x})\,\frac{\partial^2 f}{\partial x_i \partial x_j}(\boldsymbol{x}).$$

Remark 1. The compact-support assumption in Theorem 1 is primarily a technical requirement. By a standard localization argument, Dynkin's formula can be extended to many unbounded functions under mild integrability/growth conditions. Specifically, the formula applies to polynomials whenever the corresponding moments exist; that is, if the relevant moments are finite, Dynkin's formula remains valid for monomials. A detailed formal derivation is available in the full version. Notably, the assumption of moment existence is appropriate here, as our primary objective is the computation of these moments. Such existence can be verified using a ranking supermartingale or a Lyapunov function, which are standard approaches in the literature (see, e.g., [16, Chap. 1, Sect. 1.4] and [20, Chap. 11]).

Since both the drift vector $b(\boldsymbol{x})$ and the diffusion matrix $\sigma(\boldsymbol{x})$ are polynomial, applying Dynkin's formula to the test function $f(\boldsymbol{x}) = \boldsymbol{x}^\alpha$ yields

$$\frac{d}{dt}m_\alpha(t) = \mathbb{E}[\mathcal{A}(\mathbf{x}^\alpha)|\mathbf{x} = X_t] \qquad \text{[Suppose } \mathcal{A}(\boldsymbol{x}^\alpha) = \sum_\gamma a_{\alpha\gamma}\boldsymbol{x}^\gamma + c_\alpha\text{]}$$

$$= \mathbb{E}\left[\sum_\gamma a_{\alpha\gamma}X_t^\gamma\right] + c_\alpha = \sum_\gamma a_{\alpha\gamma}\mathbb{E}[X_t^\gamma] + c_\alpha = \sum_\gamma a_{\alpha\gamma}m_\gamma(t) + c_\alpha \qquad (2)$$

This procedure reveals that the dynamics of any single moment are coupled to the dynamics of other moments. We can systematically uncover the full set of coupled moments by starting with our initial moment m_α and recursively applying the generator $\mathcal{A}$ to any new monomials that appear on the right-hand side of Eq. (2).

We formalize this process by constructing a set of monomials that is closed under the action of $\mathcal{A}$, as described in Algorithm 1 (the *moment closure algorithm*). The key steps are summarized as follows:

1. **Initialization.** Initialize the set of monomials $S := \{\boldsymbol{x}^{\alpha}\}$.
2. **Closure construction.** While there exists a monomial $\boldsymbol{x}^{\beta} \in S$ that has not yet been processed, compute $\mathcal{A}(\boldsymbol{x}^{\beta})$ and expand it as a linear combination of monomials (line 4–5 in Algorithm 1):
$$\mathcal{A}\boldsymbol{x}^{\beta} = \sum_{\gamma} a_{\beta\gamma}\, \boldsymbol{x}^{\gamma} + c_{\beta}.$$
For each new monomial $\boldsymbol{x}^{\gamma}$ appearing on the right-hand side that is not already in S, add $\boldsymbol{x}^{\gamma}$ to S. Mark $\boldsymbol{x}^{\beta}$ as processed. (line 6–9 in Algorithm 1)
3. **Exact moment calculation.** If this procedure terminates after finitely many steps, we obtain a finite multi-index set $S = \{\alpha, \beta_1, \ldots, \beta_k\}$. In this case, defining the moment vector $m(t) \triangleq \left(\mathbb{E}[X_t^{\alpha}], \mathbb{E}[X_t^{\beta_1}], \ldots, \mathbb{E}[X_t^{\beta_k}]\right)^{\top}$, Dynkin's formula yields a closed linear ODE system
$$\frac{\mathrm{d}}{\mathrm{d}t} m(t) = A\, m(t) \,+\, c\,, \tag{3}$$
where the matrix A and the vector c collects the coefficients from the generator expansions.

If Algorithm 1 terminates, the resulting finite-dimensional ODE system yields explicit expressions for all moments in the set S, and in particular for the target moment $\mathbb{E}[X_t^{\alpha}]$. Consequently, whenever the closure procedure terminates, the SDE is moment-solvable for α. We formalize it as the following theorem.

Theorem 2 (Explicit moment computation). *If Algorithm 1 terminates, then SDE* (1) *is moment-solvable for α. In particular, the moment vector $m(t)$ can be explicitly expressed as*
$$\left(\mathbb{E}[X_t^{\alpha}], \mathbb{E}[X_t^{\beta_1}], \ldots, \mathbb{E}[X_t^{\beta_k}]\right)^{\top} \;=\; m(t) \;=\; e^{At} m(0) + \int_0^t e^{A(t-s)}\, c \;\mathrm{d}s. \tag{4}$$
where $m(0) = \left(\mathbb{E}[X_0^{\alpha}],\, \mathbb{E}[X_0^{\beta_1}],\, \ldots,\, \mathbb{E}[X_0^{\beta_k}]\right)^{\top}$ is determined by the initial distribution, and the target moment $\mathbb{E}[X_t^{\alpha}]$ is given by the first component of $m(t)$.

Proof. Since Algorithm 1 terminates, we obtain a linear ODE system Eq. (3) for $m(t)$, whose solution is given explicitly by Eq. (4). This completes the proof. □

Remark 2. The matrix exponential and integral appearing on the right-hand side of (4) can typically be evaluated in closed form using standard symbolic computation tools (e.g., MATHEMATICA), making the explicit computation of $m(t)$ readily achievable in practice. We permit such closed-form representations to contain implicit symbolic expressions, as the evaluation of the matrix exponential generally entails computing eigenvalues, which may include implicit algebraic quantities.

Depending on the specific moment m_α and the structure of the drift and diffusion coefficients, Algorithm 1 may either terminate or diverge. The following example illustrates a nonlinear dynamics for which the closure procedure terminates, resulting in a finite-dimensional moment system.

Example 1 (Chemical process in an Ornstein–Uhlenbeck environment [14]*).* Consider a chemical system, where X_t models a fluctuating environment and Y_t denotes the concentration of a chemical species influenced by the environment:

$$\begin{cases} \mathrm{d}X_t = -X_t \, \mathrm{d}t + \mathrm{d}W_t^{(1)}, \\ \mathrm{d}Y_t = \left(-2Y_t + X_t + X_t^2\right) \mathrm{d}t + X_t \, \mathrm{d}W_t^{(2)}, \end{cases} \tag{5}$$

Suppose the initial state is $(X_0, Y_0) = (0, 0)$. We seek to compute the second moment of Y_t, i.e., $m_{(0,2)} = \mathbb{E}[Y_t^2]$. Applying Algorithm 1, we obtain a closed 8-dimensional linear ODE system for the collection of moments

$$m_{(0,2)}, \quad m_{(2,1)}, \quad m_{(2,0)}, \quad m_{(1,1)}, \quad m_{(4,0)}, \quad m_{(3,0)}, \quad m_{(0,1)}, \quad m_{(1,0)}$$

where $m_{(i,j)} \triangleq \mathbb{E}[X_t^i Y_t^j]$. Solving this system, we obtain an explicit formula for the second moment:

$$\mathbb{E}\left[Y_t^2\right] = \frac{1}{3} + \frac{2}{3}e^{-3t} + \left(-\frac{t}{4} - \frac{11}{8}\right) e^{-2t} + \left(\frac{3}{4}t^2 + t + \frac{3}{8}\right) e^{-4t}. \qquad \triangleleft$$

The following example illustrates a case in which Algorithm 1 does *not* terminate.

Example 2 (Double-well potential [7]*).* Consider the bistable Langevin system describing the dynamics of a particle in a double-well potential,

$$\mathrm{d}X_t = \left(aX_t - X_t^3\right) dt + \sigma \, dW_t,$$

where $a, \sigma \in \mathbb{R}$ are constants. Applying the generator $\mathcal{A}$ to the monomial x^n for $n \geq 2$, we obtain $\mathcal{A}x^n = -nx^{n+2} + anx^n + \frac{\sigma^2}{2}n(n-1)x^{n-2}$. Therefore, starting from x^n, each application of the generator introduces a new, higher-degree monomial x^{n+2}, and recursively, all monomials of the form x^{n+2k} for $k \geq 0$ are generated. As a result, the set S continues to expand indefinitely, and Algorithm 1 does not terminate. □

Combining Theorem 2 with Definition 1, we obtain the following characterization:

Theorem 3. *Given an SDE* (1)*, if Algorithm 1 terminates for every monomial* $\boldsymbol{x}^\alpha$*, then the SDE* (1) *is moment-solvable; that is, explicit closed-form expressions can be computed for all moments.*

4 A Class of SDEs with Moment-Solvable Property

In this section, we identify a class of SDEs, termed *pro-solvable SDEs*, for which Algorithm 1 terminates for any monomial $\boldsymbol{x}^{\alpha}$, thereby ensuring the moment-solvable property by Theorem 3. The class of pro-solvable SDEs contains both linear SDEs and certain nonlinear SDEs whose variables exhibit a triangular dependence structure. We present the definition of pro-solvable SDEs in Sect. 4.1 and prove Algorithm 1 terminates for any pro-solvable SDE and any multi-index α in Sect. 4.2.

4.1 Pro-Solvable SDEs

We begin by formalizing the concept of *ordered partition*, which serves as a foundation for introducing the notion of pro-solvable SDEs.

Definition 2 (Ordered partition). *We say that non-empty blocks* $G_1, G_2, \ldots, G_r$ *form an* ordered partition *of* $\{1, 2, \ldots, n\}$ *if*

$$\{1, 2, \ldots, n\} = G_1 \cup \cdots \cup G_r, \qquad G_i \cap G_j = \emptyset \quad \textit{for any } i \neq j,$$

and the blocks $G_1, G_2, \ldots, G_r$ *are equipped with the natural order* $G_1 \prec \cdots \prec G_r$. *For each* $p \in \{1, \ldots, r\}$, *let* $\boldsymbol{x}^{(p)} \triangleq (x_i)_{i \in G_p}$ *denote the collection of variables whose indices belong to block* G_p, *and* $\boldsymbol{x}^{(<p)} \triangleq (x_i)_{i \in G_1 \cup \cdots \cup G_{p-1}}$ *be the collection of all variables whose indices belong to the preceding blocks of* G_p.

Definition 3 (Pro-solvable SDEs). *An SDE* (1) *is* pro-solvable *if there exists an ordered partition* $G_1, G_2, \ldots, G_r$ *of* $\{1, 2, \ldots, n\}$, *such that its drift* $b = (b_1, \ldots, b_n)$ *and diffusion matrix* $\sigma = (\sigma_{ik})_{1 \leq i \leq n,\, 1 \leq k \leq m}$ *satisfy the* block-triangular affine *structure: for every block* G_p, *and for all* $i \in G_p$, $k = 1, \ldots, m$,

$$b_i(\boldsymbol{x}) = \sum_{j \in G_p} \Lambda_{ij} x_j + B_i\big(\boldsymbol{x}^{(<p)}\big), \qquad \Lambda_{ij} \in \mathbb{R}, \;\; B_i \in \mathbb{R}[\boldsymbol{x}^{(<p)}], \tag{6}$$

$$\sigma_{ik}(\boldsymbol{x}) = \sum_{j \in G_p} A_{ikj} x_j + P_{ik}\big(\boldsymbol{x}^{(<p)}\big), \qquad A_{ikj} \in \mathbb{R}, \;\; P_{ik} \in \mathbb{R}[\boldsymbol{x}^{(<p)}]. \tag{7}$$

In particular, each b_i *and* σ_{ik} *is affine-linear in the variables of its own block* $\boldsymbol{x}^{(p)}$, *and any nonlinearity depends only on variables from earlier blocks* $\boldsymbol{x}^{(<p)}$.

The block-triangular affine structure in pro-solvable SDEs imposes a natural hierarchy among the variables: within each block G_p, the drift and diffusion coefficients are affine-linear functions of the variables in that block, while any nonlinear dependence is restricted to variables in preceding blocks $\boldsymbol{x}^{(<p)}$. It ensures that the moment dynamics associated with variables in higher-indexed blocks only depend on moments of lower-indexed blocks, and never vice versa.

Example 3. Reconsider the SDE in Example 1, given in Eq. (5). Suppose its drift and diffusion coefficients are

$$b(x_1, x_2) = \begin{pmatrix} -x_1 \\ -2x_2 + x_1 + x_1^2 \end{pmatrix}, \qquad \sigma(x_1, x_2) = \begin{pmatrix} 1 & 0 \\ 0 & x_1 \end{pmatrix}.$$

It is straightforward to verify that SDE (5) is pro-solvable under the ordered partition $G_1 = \{1\}$, $G_2 = \{2\}$ with $r = 2$. □

The class of pro-solvable SDEs subsumes several important subclasses. In particular, when the ordered partition is taken as (i) $r = 1$ with $G_1 = \{1, 2, \ldots, n\}$, the pro-solvable SDEs specialize to the well-known class of linear SDEs, in which both the drift and diffusion coefficients are affine functions of all variables. On the other hand, when the partition is (ii) $r = n$ with $G_i = \{i\}$ for $1 \leq i \leq n$, the pro-solvable condition reduces to the strictly triangular case, where each variable may depend nonlinearly only on those variables with strictly smaller indices.

Remark 3. A trivial approach of checking pro-solvability of SDEs by enumerating all possible ordered partitions and then checking block triangular affine structure, is exponential in system dimension. A more efficient polynomial-time alternative, similar to the solvability checking of the recurrence relation in [1, Sect. 4], by constructing a dependency graph of variables, an SDE is pro-solvable iff no strongly connected component contains a nonlinear edge.

4.2 Pro-solvable SDEs are Moment-Solvable

In this subsection, we establish that pro-solvable SDEs guarantee termination of the iterative moment closure procedure described in Algorithm 1 (Lines 3–9). That is, for any monomial $\boldsymbol{x}^\alpha$, Algorithm 1 generates only finitely many new moments, thereby ensuring the moment-solvable property.

Clearly, the termination of Algorithm 1 hinges on whether infinitely many new monomials are added to the set $\mathcal{P}$. The intuitive strategy is to construct a ranking function over monomials such that, whenever a new monomial $\boldsymbol{x}^\gamma$ arises from the generator action $\mathcal{A}\boldsymbol{x}^\beta$, its rank is no greater than that of $\boldsymbol{x}^\beta$.

To formalize this idea, we examine in detail how the operator $\mathcal{A}$ acts on pro-solvable SDEs. For such systems, the generator $\mathcal{A}$ expands as

$$\begin{aligned} \mathcal{A}f &= \sum_{i=1}^{n} b_i(\boldsymbol{x})\partial_{x_i} f + \frac{1}{2}\sum_{i,j=1}^{n}\sum_{k=1}^{m} \sigma_{ik}(\boldsymbol{x})\sigma_{jk}(\boldsymbol{x})\, \partial_{x_i}\partial_{x_j} f \\ &= \sum_{i=1}^{n}\left(\sum_{l\in G_{I(i)}} A_{il}x_l + B_i\left(\boldsymbol{x}^{(<I(i))}\right)\right)\partial_{x_i} f \qquad (8) \\ &+\frac{1}{2}\sum_{i,j=1}^{n}\sum_{k=1}^{m}\left(\sum_{l\in G_{I(i)}} A_{ikl}x_l + P_{ik}\left(\boldsymbol{x}^{(<I(i))}\right)\right)\left(\sum_{l\in G_{I(j)}} A_{jkl}x_l + P_{jk}\left(\boldsymbol{x}^{(<I(j))}\right)\right)\partial_{x_i}\partial_{x_j} f \end{aligned}$$

where $I(i)$ denotes the unique block index for which i belongs to $G_{I(i)}$, and ∂_{x_i} is the partial differential operator that maps a function f to $\partial f/\partial x_i$. Thus, the operator $\mathcal{A}$ is a linear combination of primitive terms of the form

$$\boldsymbol{x}^{\gamma}\,\partial_{x_i}, \qquad \boldsymbol{x}^{\gamma}\,\partial_{x_i}^2, \qquad \boldsymbol{x}^{\gamma}\,\partial_{x_i}\partial_{x_j}, \tag{9}$$

where the monomial $\boldsymbol{x}^{\gamma}$ arises from the polynomial coefficients of b_i and σ_{ij}. Depending on how $\boldsymbol{x}^{\gamma}$ is produced, we distinguish two types of primitive terms.

Definition 4 (Classification of primitive terms in $\mathcal{A}$). *Suppose SDE* (1) *is pro-solvable with ordered partition $G_1, G_2, \ldots, G_r$, then the primitive terms in operator $\mathcal{A}$ are classified into*

(i) Linear-produced terms*: if monomial $\boldsymbol{x}^{\gamma}$ in primitive term comes entirely from the affine linear parts $\sum_{l\in G_p} A_{il}x_l$, $\sum_{l\in G_p} A_{ikl}x_l$ or $\sum_{l\in G_q} A_{jkl}x_l$ with no factor from any B_i, P_{ik} or P_{jk}, where $I(i) = p, I(j) = q$. A monomial in linear-produced terms is not necessarily linear, as we allow the product of two affine linear parts.*

(ii) Polynomial-produced terms*: if monomial $\boldsymbol{x}^{\gamma}$ in primitive term contains at least one factor from some $B_i(\boldsymbol{x}^{(<I(i))})$, $P_{ik}(\boldsymbol{x}^{(<I(i))})$, or $P_{jk}(\boldsymbol{x}^{(<I(j))})$.*

For a polynomial-produced primitive term, we define its *source block index* as follows. If the monomial $\boldsymbol{x}^{\gamma}$ contains a factor from B_i or P_{ik} for some $i \in G_p$, then p is regarded as a candidate source block index. In cases where multiple candidate source block indices exist (e.g., cross terms involving both P_{ik} and P_{jk}), we select the source block index as the larger one (e.g. $\max\{p, q\}$ if $i \in G_p$ and $j \in G_q$ in the cross term case). Consequently, every polynomial-produced primitive term is associated with a unique source block index $p \in \{1, \ldots, r\}$.

Example 4. Continuing Example 3, the generator $\mathcal{A}$ corresponding to SDE (5) simplifies to

$$\mathcal{A} \;=\; -\,x_1\,\partial_{x_1} + \left(-2x_2 + x_1 + x_1^2\right)\partial_{x_2} + \frac{1}{2}\,\partial_{x_1}^2 + \frac{1}{2}x_1^2\,\partial_{x_2}^2. \tag{10}$$

Following the classification in Definition 4, the linear-produced terms $\{x_1\,\partial_{x_1}, x_2\,\partial_{x_2}\}$ are obtained, and the polynomial-produced terms are $\{x_1\,\partial_{x_2}, x_1^2\,\partial_{x_2}, \partial_{x_1}^2, x_1^2\partial_{x_2}^2\}$. Moreover, the corresponding source block indices for the polynomial-produced terms $x_1\,\partial_{x_2}$, $x_1^2\,\partial_{x_2}$, $\partial_{x_1}^2$, and $x_1^2\,\partial_{x_2}^2$ are 2, 2, 1, and 2 respectively. ◁

We proceed to analyze how the exponents change when applying $\mathcal{A}$ to a monomial $\boldsymbol{x}^{\beta}$. To facilitate this analysis, we introduce the notion of the *block exponent sum* and *block exponent difference.*

Definition 5. *Given the notations above, let β be a multi-index with $\boldsymbol{x}^{\beta} = \prod_{i=1}^{n} x_i^{\beta_i}$. For each block G_p, the block exponent sum is defined as*

$$s_p(\beta) \;\triangleq\; \sum_{i\in G_p} \beta_i, \qquad \text{for } p = 1, \ldots, r.$$

Moreover, for any two multi-indices β and β', the block exponent difference is defined by $\Delta s_p(\beta, \beta') \triangleq s_p(\beta') - s_p(\beta)$ for $1 \leq p \leq r$.

Now, given a monomial $\boldsymbol{x}^\beta$, let $\boldsymbol{x}^{\beta'}$ be any monomial that appears in the expansion of $\mathcal{A}\boldsymbol{x}^\beta$. According to Eq. (8), there must exist a unique primitive term T of the form $\boldsymbol{x}^\gamma\,\partial_{x_i}$, $\boldsymbol{x}^\gamma\,\partial^2_{x_i}$, or $\boldsymbol{x}^\gamma\,\partial_{x_i}\partial_{x_j}$ such that $T\boldsymbol{x}^\beta = c\,\boldsymbol{x}^{\beta'}$ for some constant c. The following result shows that the change in exponents is bounded, and this bound is independent of the specific monomials $\boldsymbol{x}^\beta$ and $\boldsymbol{x}^{\beta'}$.

Lemma 1 (Bound on exponent change). *Given a monomial $\boldsymbol{x}^\beta$, let $\boldsymbol{x}^{\beta'}$ be any monomial that appears in the expansion of $\mathcal{A}\boldsymbol{x}^\beta$. Suppose $\boldsymbol{x}^{\beta'}$ is produced by a primitive term T (i.e., T is of the form given in Eq. (9), and $T\boldsymbol{x}^\beta = c\,\boldsymbol{x}^{\beta'}$ for some constant c). Then the following properties hold:*

- *If T is a linear-produced term, then $\Delta s_q(\beta, \beta') \leq 0$ for all $q = 1, \ldots, r$.*
- *If T is a polynomial-produced term with source block index p, then*
  (1) $\Delta s_p(\beta, \beta') \leq -1$;
  (2) $\Delta s_q(\beta, \beta') = 0$ for all $q > p$;
  (3) For each $q < p$, there exists a constant $C_{p,q} \in \mathbb{N}$ (independent of β and β', depending only on q and on the degrees of B_i and P_{ik} for all i in $G_1 \cup \cdots \cup G_p$) such that $0 \leq \Delta s_q(\beta, \beta') \leq C_{p,q}$.

Proof. The proof proceeds by a case-by-case analysis of the primitive term T.

Case 1: T is a linear-produced term. Suppose T is a linear-produced term, then T can only take the form in either $x_l\,\partial_{x_i}$ for some $i, l \in G_p$ or $x_l x_{l'}\,\partial_{x_i}\partial_{x_j}$ for some $i, l \in G_p$ and $j, l' \in G_{p'}$. A direct calculation shows that $\Delta s_q \leq 0$ for all $q = 1, \ldots, r$. Indeed, each derivative ∂_{x_i} reduces the exponent of some x_i in its block by 1, and the linear coefficient can reintroduce at most as many variables in that block as there are derivatives; hence, the total exponent in each block never increases under linear-produced terms.

Case 2: T is a polynomial-produced term. Suppose T is a polynomial-produced term with source block index p. By Eq. (8) and the definition of the source block index, T always contains at least one derivative with respect to some x_i in block G_p, and T can only take the form in the following three cases:

* $\boldsymbol{x}^\gamma\partial_{x_i}$ with $\boldsymbol{x}^\gamma$ a monomial in $B_i(\boldsymbol{x}^{(<p)})$;
* $\boldsymbol{x}^\gamma\partial^2_{x_i}$ with $\boldsymbol{x}^\gamma$ a monomial in $2x_l P_{ik}(\boldsymbol{x}^{(<p)})$ or $\left(P_{ik}(\boldsymbol{x}^{(<p)})\right)^2$ for some k and some $l \in G_p$;
* $\boldsymbol{x}^\gamma\partial_{x_i}\partial_{x_j}$ with $\boldsymbol{x}^\gamma$ a monomial in $P_{ik}(\boldsymbol{x}^{(<p)})P_{jk}(\boldsymbol{x}^{(<u)})$ for some k, and for some $j \neq i$ with $j \in G_u$ and $u \leq p$.

In either case, we can directly check that:

(1) $\Delta s_p \leq -1$, i.e. the total exponent in the source block G_p strictly decreases;
(2) $\Delta s_q = 0$ for all $q > p$, i.e. no later block is affected;
(3) For each $q < p$, there exists a constant $C_{p,q} \in \mathbb{N}$ (*depending only on* q and on the degrees of B_i and P_{ik} for all i in $G_1 \cup \cdots \cup G_p$) such that $0 \leq \Delta s_q \leq C_{p,q}$.

Intuitively, the above result implies each application of a polynomial-produced term from block p differentiates at least once in some variable of G_p, thus reducing s_p by one or two, while the coefficient can reintroduce at most one variable from G_p and some bounded amount of variables from earlier blocks $G_1, \ldots, G_{p-1}$. Combining both cases, we obtain the desired result. □

At the beginning of this subsection, we mention that our strategy is to construct a ranking function over monomials to prove the termination of Algorithm 1. Now, we introduce the following weighted block degree serving as a ranking function, such that its value over a newly added monomial does not increase.

Definition 6. *Given the notations above, the* weighted block degree *for a monomial is defined as*

$$\deg_W(\boldsymbol{x}^\beta) \triangleq \sum_{p=1}^{r} W_p\, s_p(\beta),$$

where the block weights $W_1, \ldots, W_r$ are chosen as follows: set $W_1 \triangleq 1$, and for each $p = 2, \ldots, r$, choose W_p inductively by

$$W_p > \sum_{q<p} C_{p,q}\, W_q.$$

The following lemma shows that the weighted block degree is indeed nonincreasing. Intuitively, by Lemma 1, no new higher-degree monomials are created within the same block. Any additional complexity can only arise from dependencies on variables in earlier blocks, whose closure is handled inductively. This behavior is precisely captured by the choice of block weights in Definition 6.

Lemma 2 (Non-increase of weighted block degree). *Given a monomial $\boldsymbol{x}^\beta$, let $\boldsymbol{x}^{\beta'}$ be any monomial that appears in the expansion of $\mathcal{A}\boldsymbol{x}^\beta$, then*

$$\deg_W(\boldsymbol{x}^{\beta'}) - \deg_W(\boldsymbol{x}^\beta) \;\leq\; 0\,.$$

Proof Given the notations above, suppose T is a polynomial-produced term with source block p, then by Lemma 1

$$\deg_W(\boldsymbol{x}^{\beta'}) - \deg_W(\boldsymbol{x}^\beta) = \sum_{q=1}^{r} W_q\, \Delta s_q \;\leq\; W_p \cdot (-1) + \sum_{q<p} W_q C_{p,q} \;<\; 0.$$

Therefore, any genuinely new monomial produced by a polynomial-produced term strictly *decreases* the weighted block degree. On the other hand, if T is a linear-produced term, we have, by Lemma 1

$$\deg_W(\boldsymbol{x}^{\beta'}) - \deg_W(\boldsymbol{x}^\beta) = \sum_{q=1}^{r} W_q\, \Delta s_q \;\leq\; 0,$$

Hence, the weighted degree is *nonincreasing* under linear-produced terms. This completes the proof. □

Consider the directed graph whose vertices are monomials and with an edge $\boldsymbol{x}^\beta \to \boldsymbol{x}^{\beta'}$ whenever $\boldsymbol{x}^{\beta'}$ appears in $\mathcal{A}\boldsymbol{x}^\beta$. Starting from $\boldsymbol{x}^\alpha$, any monomial that appears in some $\mathcal{A}^m(x^\alpha)$ is reachable via a directed path. Along any path of distinct vertices $\boldsymbol{x}^{\beta^{(0)}} = \boldsymbol{x}^\alpha \to \boldsymbol{x}^{\beta^{(1)}} \to \boldsymbol{x}^{\beta^{(2)}} \to \cdots$, the sequence $\deg_W(\boldsymbol{x}^{\beta^{(k)}})$ is non-increasing by Lemma 2. Thus, for every reachable monomial $\boldsymbol{x}^\beta$ we have

$$\deg_W(\boldsymbol{x}^\beta) \;\le\; \deg_W(\boldsymbol{x}^\alpha).$$

Since all weights W_p are positive, it implies that each block sum $s_p(\beta)$ is bounded:

$$0 \le s_p(\beta) \le \frac{\deg_W(\boldsymbol{x}^\alpha)}{W_p}, \qquad p = 1, \dots, r.$$

In particular, each individual exponent is bounded:

$$0 \le \beta_i \le s_p(\beta) \le \frac{\deg_W(\boldsymbol{x}^\alpha)}{W_p} \quad \text{for all } i \in G_p,\ p = 1, \dots, r.$$

Therefore, the set of all multi-indices $\beta \in \mathbb{N}^n$ with $\deg_W(\boldsymbol{x}^\beta) \le \deg_W(\boldsymbol{x}^\alpha)$ is finite, hence only finitely many monomials are reachable from $\boldsymbol{x}^\alpha$. Based on the above analysis, we present the main result for pro-solvable SDEs as follows.

Theorem 4. *If SDE* (1) *is pro-solvable, then Algorithm 1 terminates after finitely many iterations for any monomial* $\boldsymbol{x}^\alpha$, *and SDE* (1) *is moment-solvable.*

Proof. By Lemma 2, only finitely many monomials can appear when iteratively executing lines 3–9 of Algorithm 1, starting from any $\boldsymbol{x}^\alpha$. It implies that Algorithm 1 terminates for any multi-index α, hence SDE (1) is moment-solvable by Theorem 3. □

Complexity Analysis. The computational complexity of our method for exact moment estimation arises primarily from two sources: constructing the moment closure set S in Algorithm 1, and solving the resulting linear ODE system (3).

(1) Complexity of closure construction. For a fixed pro-solvable SDE and initial multi-index α, let S_α denote the finite set of monomials reachable from x^α by Algorithm 1. By Lemma 2 and the argument preceding Theorem 4, there exists a constant C_0 (depending only on the SDE) such that for every $x^\beta \in S_\alpha$,

$$|\beta| \;=\; \sum_{i=1}^{n} \beta_i \;\le\; \sum_{p=1}^{r} \frac{\deg_W(\boldsymbol{x}^\alpha)}{W_p} \;\le\; C_0|\alpha|\,.$$

Hence, S_α is contained within the set of all monomials of total degree at most $C_0|\alpha|$ in n variables, which gives the combinatorial bound

$$|S_\alpha| \le \binom{n + C_0|\alpha|}{C_0|\alpha|}.$$

In particular, for fixed dimension n, this yields $|S_\alpha| = O\big((|\alpha|)^n\big)$, while for fixed moment order $k = |\alpha|$, we have $|S_\alpha| = O\big(n^k\big)$. Since Algorithm 1 processes each element of S_α at most once, the time and memory complexity of closure construction (cf. lines 3–9 in Algorithm 1) is $O(|S_\alpha|)$. The hidden constants in the big-O notation depend only on the given SDE, in particular on the degrees and number of monomials in the polynomial drift and diffusion coefficients.

(2) Complexity of solving the ODE system. According to Eq. (4) in Theorem 2, solving the resulting ODE system reduces to computing the matrix exponential e^{At} where A is a $|S_\alpha|$ dimensional matrix. Using standard algebraic methods (e.g., Jordan or rational canonical form), this can be done in time polynomial in $|S_\alpha|$, with worst-case complexity $\mathcal{O}(|S_\alpha|^3)$. ◁

5 Experiments

To demonstrate the effectiveness and applicability of our EME framework, we implemented[1] Algorithm 1 in Python 3.13, leveraging standard symbolic and numerical linear algebra libraries. Given a polynomial SDE (1) and a target monomial x^α, our prototype automatically constructs the moment-closure set S_α, derives the corresponding linear ODE system, and computes the desired α-moment.

Benchmarks. We evaluated our method on a suite of SDE benchmarks, encompassing both linear and nonlinear examples in the literature as well as models with practical relevance. Specifically, we present in detail two cases: a consensus network with noise [23] and a nonlinear vehicle platoon system adapted from [15] to illustrate our method and demonstrate its usefulness for verification problems. All experiments were performed on a MacBook Pro with an Apple M4 processor, 16 GB of RAM, and running macOS Sequoia.

5.1 Case Studies

Consensus network with noise [23] This model describes an $n+1$ dimensional multi-agent consensus network (e.g., distributed sensors or robots) with noisy communication:

$$\mathrm{d}X_i(t) = \Big(a_i - \lambda X_i(t) + \kappa\big(X_{i+1}(t) - 2X_i(t) + X_{i-1}(t)\big)\Big)\,\mathrm{d}t + \sigma_i X_i(t)\,\mathrm{d}W_t^{(i)},$$

for $i = 1, \ldots, n+1$, with periodic boundary conditions $X_0(t) = X_n(t), X_{n+1}(t) = X_1(t)$. Since this dynamic is linear, it is pro-solvable by Definition 3. Consider the case $n = 2$ with parameter values $a_1 = 0$, $a_2 = 0$, $\lambda = 1$, $\kappa = 0.5$, $\sigma_1 = 1$, $\sigma_2 = 1$, and initial state $(1, 0)$. The verification objective is to ensure that, with probability at least $1 - e^{-t}$, the disagreement between the two agents, encoded by $|X_1(t) - X_2(t)|$, remains less than 0.1 for any $t \geq 10$, that is,

$$P\left(|X_1(t) - X_2(t)| \geq 0.1\right) \leq e^{-t} \quad \text{for } t \geq 10.$$

[1] Available at https://github.com/Shenghua-Feng/Exact_Moment_Estimation.

To this end, we compute $\mathbb{E}[(X_1(t) - X_2(t))^2]$. By explicitly calculating the moments $\mathbb{E}[X_1(t)^2]$, $\mathbb{E}[X_1(t)X_2(t)]$, and $\mathbb{E}[X_2^2(t)]$, we obtain

$$\mathbb{E}[(X_1(t) - X_2(t))^2] = \frac{(17 - 3\sqrt{17})e^{\frac{(\sqrt{17}-7)t}{2}} + (17 + 3\sqrt{17})e^{-\frac{(\sqrt{17}+7)t}{2}}}{34}.$$

Consequently, by applying Markov's inequality, we obtain, for any $t \gtrsim 10$,

$$P\left(|X_1(t) - X_2(t)| \geq 0.1\right) \leq \frac{\mathbb{E}[(X_1(t) - X_2(t))^2]}{0.1^2} \leq e^{-t}.$$

This verifies the goal. It is worth noting that, to the best of our knowledge, this verification problem cannot be solved directly by a martingale-based approach that seeks a polynomial $h(t, x_1, x_2)$ satisfying the supermartingale condition to upper bound $\mathbb{E}[(X_1(t) - X_2(t))^2]$. Specifically, if such a polynomial h existed, we have $P(|X_1(t) - X_2(t)| \geq 0.1) \leq 100\,\mathbb{E}[(X_1(t) - X_2(t))^2] \leq 100\,\mathbb{E}[h(t, X_1(t), X_2(t))] \leq 100\,h(0, x_1(0), x_2(0))$. Since there must exist some $T > 10$ such that $e^{-T} < 100\,h(0, x_1(0), x_2(0))$, the standard martingale-based method cannot certify the property holds for all $t \geq 10$. ◁

Vehicle Platoon [15]**.** We consider a nonlinear system adapted from [15] that involves two vehicles moving along a straight lane. For each vehicle, denote its position and velocity by $(p_1(t), v_1(t))$ and $(p_2(t), v_2(t))$, respectively. The first vehicle acts as the leader and follows a stochastic acceleration model:

$$\mathrm{d}p_1 = v_1\,\mathrm{d}t, \qquad \mathrm{d}v_1 = \left(-a_1 v_1 + u_1\right)\mathrm{d}t + \sigma_1\,\mathrm{d}W_t^{(1)},$$

where $a_1 > 0$ is a damping coefficient, u_1 is a control input (desired acceleration), and σ_1 scales the driving noise $W_t^{(1)}$. The second vehicle implements a nonlinear control law based on both its own velocity and the velocity of the leader:

$$\mathrm{d}p_2 = v_2\,\mathrm{d}t, \qquad \mathrm{d}v_2 = \left(-a_2 v_2 + (v_1 - 1)^2\right)\mathrm{d}t + \sigma_2\,\mathrm{d}W_t^{(2)},$$

where a_2 and σ_2 are parameters. It can be checked that this system is pro-solvable under the ordered partition $G_1 = \{v_1\}$, $G_2 = \{p_1, p_2, v_2\}$. Consider the parameter instantiation $a_1 = 1$, $u_1 = 1$, $\sigma_1 = 1$, $a_2 = 1$, and $\sigma_2 = 1$, with initial state $(p_1, v_1, p_2, v_2) = (1, 0, 0, 0)$. Suppose that the verification objective is to ensure that the expected distance between the vehicles, $\mathbb{E}[p_1(t) - p_2(t)]$, always remains between 0.5 and 1.5 for all $t \geq 0$.

To verify this, we compute the expected values $\mathbb{E}[p_1(t)]$ and $\mathbb{E}[p_2(t)]$, yielding

$$\frac{3}{4} \leq \mathbb{E}\left[p_1(t) - p_2(t)\right] = \frac{3}{4} + \frac{e^{-t}}{2} - \frac{e^{-2t}}{4} \leq 1,$$

which verifies the desired safety property. ◁

5.2 Evaluation of Effectiveness

Table 1 summarizes the experimental results of our method on a diverse suite of polynomial SDE benchmarks, which cover a range of system dimensions, polynomial degrees, and moment orders, illustrating the generality of our approach.

Efficiency and Scalability. For all pro-solvable SDEs, our method successfully constructs the finite closures and computes the exact moment for all tested cases. Closure construction times are consistently short, and the dimension $|S_\alpha|$ scales polynomially with the moment order and system size (cf. ou-env), consistent with our theoretical analysis. The subsequent ODE solving is also efficient for moderate dimensions, with larger $|S_\alpha|$ (e.g., high-order moments in gene) leading to higher computational cost primarily due to the complexity of matrix exponentiation. Note that developing more efficient symbolic solvers for linear ODEs is a complementary and orthogonal direction to our work; in our implementation, we simply rely on off-the-shelf symbolic packages for this step.

Comparison Across Models. Linear and low-dimensional systems (such as the consensus and oscillator) exhibit particularly fast closure and solution times. For nonlinear pro-solvable examples (e.g., ou-env and gene), the closure remains tractable even for moments of degree up to 10, validating the practical scalability of our framework. The benchmark coupled3d further shows that our method may still terminate for certain SDEs that *do not* satisfy the pro-solvable property; however, termination is not guaranteed in general.

Overall, the experimental results demonstrate that our approach is effective, broadly applicable to both linear and a wide class of nonlinear systems, and scales well in practice for pro-solvable SDEs.

Table 1. Experimental results for exact moment calculation.

	SDE System			Moment		Obtained Closure S_α			Solve ODE	
Benchmark	dim	deg	p-s	$\boldsymbol{x}^\alpha$	$\|\alpha\|$	succ	time	$\|S_\alpha\|$	succ	time
ou-env [14]	2	2	yes	$\mathbb{E}[x_2^2]$	2	✓	0.01s	8	✓	0.2s
				$\mathbb{E}[x_2^3]$	3	✓	0.02s	15	✓	0.6s
				$\mathbb{E}[x_2^4]$	4	✓	0.02s	24	✓	1.2s
				$\mathbb{E}[x_2^5]$	5	✓	0.04s	35	✓	2.8s
				$\mathbb{E}[x_2^{10}]$	10	✓	0.17s	120	✓	39.4s
gene [27]	5	3	yes	$\mathbb{E}[x_1x_5]$	2	✓	0.04s	23	✓	3.0s
				$\mathbb{E}[x_5^2]$	2	✓	0.14s	85	✓	79.6s
				$\mathbb{E}[x_1x_5^2]$	3	✓	0.17s	115	✓	164.1s
consensus [23]	2	1	yes	$\mathbb{E}[x_1x_2]$	2	✓	0.01s	3	✓	0.2s
vehicles [15]	4	2	yes	$\mathbb{E}[x_2^2]$	2	✓	0.01s	13	✓	0.5s
oscillator [12]	3	2	yes	$\mathbb{E}[x_2x_3^2]$	3	✓	0.01s	6	✓	2.3s
coupled3d	3	3	no	$\mathbb{E}[x_1^2x_2^2]$	4	✓	0.01s	3	✓	0.2s

dim: Dimension of the SDE system; **deg**: Maximum polynomial degree of drift/diffusion terms in the SDE; **p-s**: Whether the SDE is pro-solvable; $\boldsymbol{x}^\alpha$: Target moment to compute; $|\alpha|$: Degree of the target moment; **succ**: Whether a closed linear ODE system was successfully constructed and solved; **time**: Time required to obtain the closure S_α (i.e., construct or solve the linear ODE system); $|\mathbf{S}_\alpha|$: Dimension of the resulting linear ODE system.

6 Conclusion

We presented a general symbolic method for exact moment estimation of polynomial SDEs, and identified a broad class of pro-solvable systems in which all moments can be computed exactly via finite-dimensional linear ODEs. Both theoretical analysis and experimental results demonstrate its effectiveness and scalability for a wide range of linear and nonlinear models, paving the way for moment-based verification and analysis of stochastic dynamical systems.

Limitations. Despite these results, several limitations warrant further discussion. Our method is inherently conditional on the termination of the closure construction. Many polynomial SDEs induce an infinite moment hierarchy, in which case the exact finite-dimensional reduction is unavailable. Even when termination is guaranteed, scalability, particularly with respect to matrix exponentiation, may be limited by the size of the closed moment set. Finally, pro-solvability is sufficient but not necessary for termination.

Future Work. Firstly, when Algorithm 1 diverges, one could explore closure approximations by truncation, together with formal error bounds to preserve verification soundness. Secondly, since pro-solvability is not necessary, it would be of interest to characterize termination criteria beyond pro-solvability. Additionally, leveraging sparsity and block structure in the derived ODEs, along with more scalable matrix-exponential techniques, could substantially improve the scalability.

Acknowledgments. We thank the anonymous reviewers for their valuable comments and helpful suggestions. This work has been partially funded by the National Key R&D Program of China under grant No. 2022YFA1005101 and 2022YFA1005102, the Open Foundation of Key Laboratory of Cyberspace Security, Ministry of Education of China and Henan Key Laboratory of Network Cryptography under grant No. KLCS20240302, the National NSF of China under grant No. 62192732, W2511064, and 62502475, the CAS Project for Young Scientists in Basic Research, and the ISCAS Basic Research under Grant No. ISCAS-JCZD-202406.

Data Availability Statement. The artifact and data are available at https://doi.org/10.5281/zenodo.18630506.

References

1. Amrollahi, D., Bartocci, E., Kenison, G., Kovács, L., Moosbrugger, M., Stankovič, M.: (Un)Solvable loop analysis. Formal Methods Syst. Des. **65**(1), 163–194 (2025)
2. Armstrong, T.B., Kolesár, M.: Sensitivity analysis using approximate moment condition models. Quant. Econ. **12**(1), 77–108 (2021)
3. Bartocci, E., Kovács, L., Stankovič, M.: Automatic generation of moment-based invariants for prob-solvable loops. In: International Symposium on Automated Technology for Verification and Analysis, pp. 255–276. Springer (2019)
4. Black, F., Scholes, M.: The pricing of options and corporate liabilities. J. Polit. Econ. **81**(3), 637–654 (1973)

5. Borri, A., Carravetta, F., Palumbo, P.: Cubification of $\sigma\pi$-SDE and exact moment equations. Syst. Control Lett. **136**, 104602 (2020)
6. Feng, S., Chen, M., Xue, B., Sankaranarayanan, S., Zhan, N.: Unbounded-time safety verification of stochastic differential dynamics. In: International Conference on Computer Aided Verification, pp. 327–348. Springer (2020)
7. Gardiner, C.W., et al.: Handbook of Stochastic Methods, vol. 3. Springer, Berlin (2004)
8. Ghusinga, K.R., Lamperski, A., Singh, A.: Moment analysis of stochastic hybrid systems using semidefinite programming. Automatica **112**, 108634 (2020)
9. Ghusinga, K.R., Soltani, M., Lamperski, A., Dhople, S.V., Singh, A.: Approximate moment dynamics for polynomial and trigonometric stochastic systems. In: 2017 IEEE 56th Annual Conference on Decision and Control (CDC), pp. 1864–1869. IEEE (2017)
10. Guidoum, A.C., Boukhetala, K.: Exact higher-order moments for linear nonhomogeneous stochastic differential equation. Model. Assist. Stat. Appl. **18**(4), 321–329 (2023)
11. Gunawan, R., Cao, Y., Petzold, L., Doyle, F.J.: Sensitivity analysis of discrete stochastic systems. Biophys. J . **88**(4), 2530–2540 (2005)
12. Hafstein, S., Gudmundsson, S., Giesl, P., Scalas, E., et al.: Lyapunov function computation for autonomous linear stochastic differential equations using sum-of-squares programming. Discret. Continuous Dyn. Syst. Ser. B **23**(2), 939–956 (2018)
13. Hoogendoorn, S., Bovy, P.: Pedestrian route-choice and activity scheduling theory and models. Transp. Res. Part B Methodol. **38**(2), 169–190 (2004)
14. Kallianpur, G., Xiong, J.: Stochastic models of environmental pollution. Adv. Appl. Probab. **26**(2), 377–403 (1994)
15. Kavathekar, P., Chen, Y.: Vehicle platooning: a brief survey and categorization. In: International Design Engineering Technical Conferences and Computers and Information in Engineering Conference, vol. 54808, pp. 829–845 (2011)
16. Khasminskii, R.: Stochastic Stability of Differential Equations, vol. 66. Springer (2011)
17. Lamperski, A., Ghusinga, K.R., Singh, A.: Analysis and control of stochastic systems using semidefinite programming over moments. IEEE Trans. Autom. Control **64**(4), 1726–1731 (2018)
18. Lasserre, J.B.: The moment-SOS hierarchy. In: Proceedings of the International Congress of Mathematicians: Rio de Janeiro 2018, pp. 3773–3794. World Scientific (2018)
19. Lee, C.H., Kim, K.H., Kim, P.: A moment closure method for stochastic reaction networks. J. Chem. Phys. **130**(13) (2009)
20. Meyn, S.P., Tweedie, R.L.: Markov Chains and Stochastic Stability. Springer (2012)
21. Moosbrugger, M., Stankovič, M., Bartocci, E., Kovács, L.: This is the moment for probabilistic loops. Proc. ACM Program. Lang. **6**(OOPSLA2), 1497–1525 (2022)
22. Øksendal, B.: Stochastic Differential Equations: An Introduction with Applications. Springer (2013)
23. Olfati-Saber, R., Fax, J.A., Murray, R.M.: Consensus and cooperation in networked multi-agent systems. Proc. IEEE **95**(1), 215–233 (2007)
24. Panik, M.: Stochastic Differential Equations: An Introduction with Applications in Population Dynamics Modeling. Wiley (2017)

25. Prajna, S., Jadbabaie, A., Pappas, G.J.: A framework for worst-case and stochastic safety verification using barrier certificates. IEEE Trans. Autom. Control **52**(8), 1415–1428 (2007)
26. Scarciotti, G., Teel, A.R.: On moment matching for stochastic systems. IEEE Trans. Autom. Control **67**(2), 541–556 (2021)
27. Sinigh, A., Hespanha, J.P.: Stochastic analysis of gene regulatory networks using moment closure. In: 2007 American Control Conference, pp. 1299–1304. IEEE (2007)
28. Sontag, E.D.: Examples of computation of exact moment dynamics for chemical reaction networks. In: Emerging Applications of Control and Systems Theory: A Festschrift in Honor of Mathukumalli Vidyasagar, pp. 295–312. Springer (2018)
29. Sontag, E.D., Singh, A.: Exact moment dynamics for feedforward nonlinear chemical reaction networks. IEEE Life Sci. Lett. **1**(2), 26–29 (2015)
30. Williams, D.: Probability with Martingales. Cambridge University Press (1991)

A Framework for the Interoperable Specification and Verification of Encapsulated Data Structures

Wolfram Pfeifer[1(✉)], Werner Dietl[2], and Mattias Ulbrich[1]

[1] Karlsruhe Institute of Technology, Karlsruhe, Germany
{wolfram.pfeifer,mattias.ulbrich}@kit.edu
[2] University of Waterloo, Waterloo, Canada
wdietl@uwaterloo.ca

Abstract. Well-designed data structures are fundamental to the construction of robust programs, particularly when their correctness can be established using formal methods. Currently, various successful deductive verification techniques exist but necessitate distinct and non-interoperable specifications and verification methods for data structure implementations. This paper presents a technique enabling cross-paradigm interoperable specification and verification of encapsulated data structures. The novel approach builds on the shared mathematical foundations of algebraic data types (ADTs) across these diverse methodologies. The technique enables the coherent integration of components that have been verified using different deductive program verification approaches within a single project, provided that a regime of encapsulation is followed, which is slightly stricter than those originally imposed by the verification techniques.

We formally introduce and discuss the encapsulation principle using a simplified conceptual object-oriented language and subsequently instantiate it for the Java programming language. This integrates the approaches of KeY, VeriFast, and Universe Types, thereby enabling heterogeneous verification projects in Java that encompass Dynamic Frames, Separation Logic, and Ownership Types.

We demonstrate the applicability of our approach by conducting a cooperative verification of a client with three different data structures, each of which is verified with one of the aforementioned techniques.

Keywords: Deductive Program Verification · Collaborative Verification · Interoperability of Formal Methods · Algebraic Data Types (ADTs)

1 Introduction

Significant progress in deductive program verification has been made over recent decades [24], both theoretically and in verification tools capable of analysing

A. Sampaio and M. Stoelinga (Eds.): FM 2026, LNCS 16557, pp. 90–110, 2026.
https://doi.org/10.1007/978-3-032-26220-2_5

complex data structures and algorithms. Notable case studies include the verification of the TimSort implementation in Java libraries [23], the competitively efficient sorting routine IPSO [8], and the parallel nested depth-first search [43].

Nonetheless, prevailing approaches and tools tend to evolve in isolation, building on expressive and powerful formalisation and modelling concepts that lack compatibility with other verification techniques. This incompatibility exists not only at the language level but also conceptually; for instance, there is no clear method to integrate a program verified with a Separation Logic tool [45] with a client verified using Dynamic Frames [34].

In the domain of fully automatic software verification tools employing advanced static analysis techniques like property-driven reachability (PDR), k-induction, or bounded model checking, considerable progress has been made towards cooperative verification [3,9,11,25,46]. Standard exchange formats enable the sharing of artifacts [13], like witnesses and path invariants between different verification engines, thereby enhancing collaborative efforts. Witnesses have become integral to the SV-COMP verification competition [12].

Intensive cooperation between diverse tools is currently not available in heavyweight deductive program verification of sophisticated properties. A key reason is the required level of abstraction. Automatic verification operates close to the actual program state, whereas the functional verification of sophisticated properties requires modularisation, decomposition, and abstraction to subdivide tasks into more manageable units. The main issue here is the need to model memory, particularly the heap, and to reason about independent, encapsulated data units occupying disjoint memory areas. Current deductive techniques mandate that the entire program be verified with the same memory model within a single tool. Modern deductive tools rely on diverse, incompatible memory management philosophies, notably Dynamic Frames (e.g. KeY [1,7], Dafny) and Separation Logic (e.g. VeriFast [30,47], VerCors [14]). Furthermore, subtle differences in language design hinder the direct exchange of VeriFast SL specifications with those of VerCors.

As an example, consider the Java class `Client` in Lst. 1. It uses three data structures: a mutable `int` cell, a tree set of `int`s, and a linked list of `Cell` references. While Java verification tools such as KeY, VeriFast, or VerCors could be used to verify this project, they require verification of the entire project within a single tool. This is unfortunate, as the `Cell` library may have been previously verified with KeY, the linked list with VeriFast, and the tree set with an Ownership Types approach. To our knowledge, no unifying method exists that integrates verification results across such diverse memory modelling paradigms.

In this paper, we overcome this limitation and propose an interoperable specification and verification technique that enhances compatibility among various memory management ecosystems. It enforces stricter notions of *encapsulation* to ensure data structures are "well-behaved" for heterogeneous verification projects. We define encapsulation within this context and give sufficient criteria and methods to demonstrate it across the major memory management methodologies: *Separation Logic*, *Dynamic Frames*, and *Ownership Types*. Our approach facili-

```
class Client {                                     // Client: proven with DF + KeY
  void m() {
    Cell c0 = new Cell(); c0.set(5);               // Cell:   proven with DF + KeY
    Cell c1 = new Cell(); c1.set(9)
    LinkedCellList l = new LinkedCellList(); // List:   proven with VeriFast
    l.add(c0); l.add(c1);
    IntTreeSet s = new IntTreeSet();               // Set:    proven with UET + KeY
    s.add(5); s.add(9);
    Cell last1 = l.getLast();    int r1 = last1.value();
    assert s.contains(r1);       assert r1 == 9;
  }
}
```

Listing 1: Example of a client class which uses three different data structures. In particular, references to objects of type Cell are passed to LinkedCellList.

tates the distribution of tasks according to the strengths of different techniques and enables the transfer of proven specifications for library code between tools.

We formally introduce the encapsulation requirement in Sect. 3, applying it subsequently to Java. We refine the encapsulation concept for three relevant Java verification tools: Separation Logic in VeriFast (Sect. 4.1), Dynamic Frames in KeY (Sect. 4.2), and an Ownership Type System (Sect. 4.3). For KeY and VeriFast, we demonstrate that strict encapsulation can be guaranteed by adhering to certain specification patterns. Additionally, for the type system, we achieved interoperability by introducing a new type annotation, `payload` for objects that should not be part of the data structure, but are only passed around via opaque references. Each refinement includes a corresponding theorem stating that adhering to the specification schema guarantees encapsulation.

The approach has been prototypically implemented using Contract-LIB [21] as the exchange language for functional specifications (see Sect. 5). The example from Lst. 1 can be verified cooperatively across three tools. Figure 1 illustrates the heterogeneity of the project.[1] The classes in the hatched boxes are automatically derived from language-independent contract specifications and used in the three verification frameworks. Section 5 elaborates more on this.

An extended version of this paper [44] is available, containing more details about the programming language model as well as proofs for the lemmas and theorems.

2 Related Work

As mentioned above, an active research area focuses on the interoperability of automatic verification tools. But also in the domain of deductive software verification, the call for collaboration has a long-standing tradition, shown by numerous community challenges [27–29,38] over the decades. However, examples of tool collaborations or case studies using heterogeneous approaches remain scarce. The

[1] The full code of the example can be found in the artifact accompanying this paper (DOI: 10.5281/zenodo.18607514).

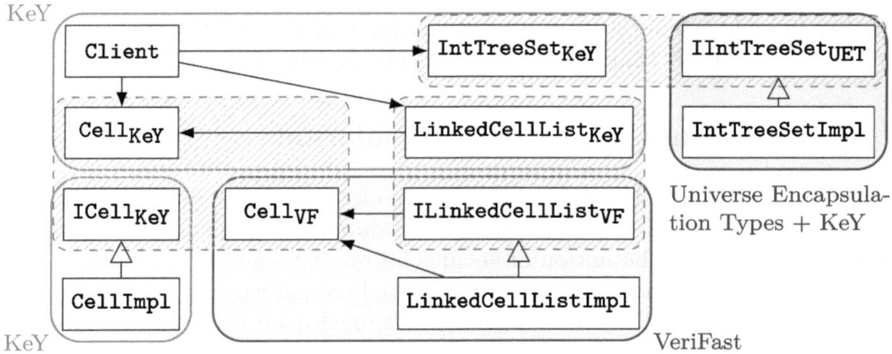

Fig. 1: Overview of the example, with the client from Lst. 1 in the top-left corner. Everything in the hatched red boxes is automatically generated from the corresponding Contract-LIB specifications. The subscripts indicate for which tool the specifications were generated. Filled arrows indicate usage in the code, open arrows denote implementation of an interface. The implementations are written manually and include the program code as well as additional specifications such as coupling invariants/predicates.

Karlsruhe Java Verification Suite [35] enables collaboration through a subset of JML [16,37] assertions but, unlike our work, does not address encapsulated objects and relies on a direct memory model. Armborst et al. [2] propose a method to connect fully automatic and deductive software verification for C programs, with a focus on enabling tools from both verification types to collaborate by exchanging invariants, rather than integrating verification solutions across different program components as we aim to do.

Jakobs [31] presents a technique for constructing correctness witnesses from various partial analyses, aimed at integrating model checkers and static analysers that maintain explicit records of visited and checked states through abstract reachability graphs. Bao et al. [5] unify Separation Logic and Region Logic [4] into a single formal reasoning framework, showcasing reconciliation of different principles at the logical level. In contrast, our approach leverages algebraic data types (ADTs) as a common semantic foundation, facilitating interoperability of encapsulated components verified across paradigms without altering the underlying tools.

Unconstrained aliasing and mutations are causes of errors and complicate program reasoning. Object ownership and limitations on the effects of computations are well-studied proposed solutions [15]. Static and dynamic object ownership systems, in particular Universe Types, have been designed for program verification [20].

Intermediate verification languages (IVLs) such as Why3 [22], Boogie [39], Viper [41], and SV-LIB [10] decouple reasoning engines from source code languages, akin to intermediate representations in compilers. While IVLs effectively enable a shared backend for various verification engines, they only permit cooperation between tools that use compatible translations into the IVL logic.

3 Formal Foundations of Encapsulation

This section formally introduces the principle of data structure encapsulation in a conceptually reduced object-oriented programming language, focusing on the concepts essential for the notion of encapsulation. The definitions include deliberate simplifications and abstractions compared to real-world programming languages. In Sect. 4, we illustrate how insights gained from this abstract language can be transferred onto a concrete language—we chose Java—and be integrated with existing state-of-the-art verification frameworks.

3.1 Model of an Object-Oriented Programming Language

There are five mutually disjoint infinite sets of names: $\mathcal{F}$ for field names, $\mathcal{L}$ for local variable names, $\mathcal{P}$ for procedure names, $\mathcal{Q}$ for query names, and $\mathcal{C}$ for class names. In addition, $\mathcal{O}$ denotes the infinite set of object references, which is disjoint from the name sets. The set of *values* contains at least the integers, booleans, and object references: $\mathrm{Val} = \mathbb{Z} \cup \mathbb{B} \cup \mathcal{O} \cup \ldots$. This leaves room for values of other types when instantiating the framework with concrete languages.

We need a memory model that is general enough to entail those of existing verification methodologies in an object-oriented context, such as Dynamic Frames and Separation Logic. Therefore, we define a *location* (o, f) as a pair of an object reference $o \in \mathcal{O}$ and a field name $f \in \mathcal{F}$. $\mathrm{Loc} = \mathcal{O} \times \mathcal{F}$ is the set of all locations. These locations are then used as indices of memories: A *memory* $M : \mathrm{Loc} \to \mathrm{Val}$ is a mapping from locations to values. Mem denotes the set of all memories. Note that M is a total function: We assume that there is always *some* value defined for each location. With this general memory model, we can define objects and their operations. We assume that there are two clearly distinguished types of operations. A *procedure* $p : \mathrm{Mem} \times \mathrm{Val} \to \mathrm{Mem}$ is a function that takes a memory and a value and returns a memory, but not a value. A *query* $q : \mathrm{Mem} \times \mathrm{Val} \to \mathrm{Val}$ is a function that takes a memory and a value and returns a value, but not a memory. This is a simplification, but mixed operations could always be split up into a sequence of procedures and queries. In addition, for keeping the presentation simple, operations only have a single value argument. Again, this is not a conceptual limitation, since it could be easily overcome by using Godelization or introducing tuples into the values.

Definition 1 (Object). *An object is a tuple* $o = (\mathbb{F}_o, \mathbb{P}_o, \mathbb{Q}_o)$ *of a finite set of field names* $\mathbb{F}_o \subseteq \mathcal{F}$ *and finite partial maps from procedure names to procedures* $\mathbb{P}_o : \mathcal{P} \nrightarrow (\mathrm{Mem} \times \mathrm{Val} \to \mathrm{Mem})$ *and query names to queries* $\mathbb{Q}_o : \mathcal{Q} \nrightarrow (\mathrm{Mem} \times \mathrm{Val} \to \mathrm{Val})$*. The set of all objects is* Obj.

Note that this notion of an object is more general than usual in programming languages. For example, a `struct` in C together with functions working only on it could also be an object according to our definition. So far, we do not have a possibility to track created objects, or create new ones. Inspired by virtual function tables in OO languages, we introduce *vTables* and constructors for that. A *vTable* $T\colon \mathcal{O} \nrightarrow \mathrm{Obj}$ is a finite partial map from object references to objects. The set of all vTables is $\mathcal{T}$. Note that we do not show how classes and object creation are modelled, but just assume that there is a way to create objects, and the objects created match the source code (for instance, they only have the operations defined in their class). More details can be found in the extended version [44]. In addition to normal objects, there are instances of data structures, which we model as a special case of objects.

Definition 2 (Data Structure Instance (DSI)). *A* data structure instance *(DSI) is a special object* $d = (\emptyset, \mathbb{P}_d, \mathbb{Q}_d)$ *without (public) fields.*

This notion serves as the basic building block for defining encapsulation. Intuitively, a DSI is an entity supposed to be encapsulated, and can thus be abstracted more than a normal object. Having no public fields is necessary for encapsulation; however, it is not sufficient, and we still need to prove that. This brings us to the definition of a footprint, which intuitively is the set of locations that (semantically) "belong" to a DSI in memory. We split the footprint into a read and a write part, and determine them via queries and procedures.

Definition 3 (Footprints). *The read and write footprint* $\mathrm{rfp}(d, M)$, $\mathrm{wfp}(d, M) \subseteq \mathrm{Loc}$ *for a DSI* $d = (\emptyset, \mathbb{P}_d, \mathbb{Q}_d)$ *in a memory* $M \in \mathrm{Mem}$ *are defined as:*

$$\begin{aligned}
\mathrm{rfp}(d, M) := \big\{l \in \mathrm{Loc} \mid \exists n \in \mathbb{N}, \bar{p} \in \mathbb{P}_d^n, q \in \mathbb{Q}_d, v \in \mathrm{Val}, a \in \mathrm{Val}, \bar{a} \in \mathrm{Val}^n \\
q(p_1(p_2(...(p_n(M \qquad\quad , a_n), ...), a_2), a_1), a) \\
\neq q(p_1(p_2(...(p_n(M[l \mapsto v], a_n), ...), a_2), a_1), a)\big\}
\end{aligned}$$

$$\mathrm{wfp}(d, M) := \big\{l \in \mathrm{Loc} \mid \exists p \in \mathbb{P}_d, a \in \mathrm{Val}\,.\; p(M, a)(l) \neq M(l)\big\}$$

$$\mathrm{fp}(d, M) := \mathrm{rfp}(d, M) \cup \mathrm{wfp}(d, M)$$

The notation $M[l \mapsto v]$ *denotes a function update, that is* $M[l \mapsto v](l) = v$ *and* $M[l \mapsto v](x) = M(x)$ *for all* $x \neq l$.

A location is in the rfp of a DSI in a memory M, iff changing the location can make a difference in a later observed state of the DSI. Note that it is not sufficient to look at only a single procedure at a time, since the state change could be hidden and only have an effect on a query result much later. Therefore, we need to consider chains of procedure applications. A location is in the wfp of a DSI, iff it can be changed by any of its procedures.

We refrain from giving formal definitions for programs and their semantics for space reasons; these can be found in the extended version [44]. The intuition is, however, that it is sufficient to look at sequences of atomic statements (inspired by computation sequences in literature [26, Sec. 5.3]). In particular, our program model contains no control structures. In addition, the statements do not contain method calls on non-DSI objects. Our programs can be thought of as having all statements outside of a DSI "inlined". For DSIs, on the other hand, we stop this "inlining" when we first reach one of their methods from the outside. This gives us a clear separation between the outside world of the rest of the program, and the DSIs that are supposed to be encapsulated (which will be proven later) and can thus be abstracted further. It suffices to consider only these restricted linear programs in our model, because later we always quantify over all terminating programs (for fixed code of the DSI under examination).

For understanding the definition of encapsulation, however, we need a notion of program state: A *program state* $s = (V, M, T) \in S$ is a triple of a partial function $V : \mathcal{L} \rightharpoonup \text{Val}$ of local variable names to values, a memory $M : Loc \to \text{Val}$, and a vTable $T : \mathcal{O} \to \text{Obj}$. We define a program $P = (s_0, \sigma)$ as a pair of a starting state s_0 and a sequence σ of atomic statements, and denote the set of all states reached during the execution of P by $R(P)$.

3.2 Reachable Locations and Encapsulation

Since we conceptually "inline" object procedures and queries, we define their variant of a footprint as the locations that are reachable via chained field accesses.

Definition 4 (Reachable Locations, Relevant Locations). *We define the set of locations reachable from an object o in a memory M as the smallest fixpoint of the following recursive function (*fst *selects the first element of a tuple):*

$$\mathit{reachLocs}(o, M) := \begin{cases} \bigcup_{f \in \text{fst}(o)} \{(o, f)\} \cup \mathit{reachLocs}(M(o,f), M) & (o \in \mathcal{O}) \\ \emptyset & (o \notin \mathcal{O}) \end{cases}$$

The relevant locations *of an object o in a memory M are defined as follows:*

$$\mathit{relLocs}(o, M) = \begin{cases} \text{fp}(o, M) & \text{if } o \text{ is a DSI} \\ \mathit{reachLocs}(o, M) & \text{if } o \text{ is not a DSI} \end{cases}$$

Note that for convenience, we define the function `reachLocs` also for values that are not object references (for instance integers), which makes the definition significantly shorter and easier to read.

Intuitively, *reachLocs* of an object o contains all the fields that are either fields of the object directly, or recursively reachable after following one of the references stored in one of the fields. The definition of *relLocs* avoids the need of more case distinctions later.

Definition 5 (Encapsulation). *A DSI d is called* encapsulated *if for all programs in all possible states of the program the footprint of d is disjoint from the relevant locations of objects referenced in local variables:*

$$\forall \text{ Program } (s_0, \sigma).\ \forall \text{ State } (V, M, T) \in R(s_0, \sigma).\ \forall \text{ Object } o.$$
$$o \in dom(T) \wedge o \neq d \wedge (\exists v \in dom(V).\ V(v) = o) \to \text{fp}(d, M) \cap \mathit{relLocs}(o, M) = \emptyset$$

Our central definition intuitively states that external references cannot access the footprint of an encapsulated DSI, nor can any external location be reached from within the DSI. Its procedures and queries serve as an interface, allowing the state of the DSI to be queried and modified solely through this interface. Note that it is sufficient to consider references stored in local variables, since (reachable) references from the heap could be obtained and immediately stored in those. Since in the definition there is a quantifier ranging over all programs, this also includes the program where this heap read and assignment to a local variable happens. Thus, the condition also indirectly covers any objects that are referenced in fields on the heap.

4 Encapsulation in Java

The formalizations in the previous section are agnostic to the concrete programming language, and only require some basic object-oriented features to be present. We instantiate the framework now for Java, and demonstrate how it ties in with different verification methodologies. The rather abstract DSI from the previous section is more concrete here: It is, in the simplest form, a Java object of a data structure class. We support a subset of the Java language. In the spirit of soundiness [40], we do not support Java features that involve dynamic change to the classes and code available, in particular reflection, `invokedynamic`, and code generation at runtime. Furthermore, we do not support methods declared as `native` and the Java Native Interface (JNI). However, these are restrictions commonly made in static program analysis. More importantly, we do not support subclassing—therefore, all classes are final—and generics at the moment, although we are confident that our methodology can be expanded to them. We require that all fields of data structure classes are private, which fits together with their abstraction (DSI) not having any public fields. For the sake of presentation, we ignore other visibility modifiers (`protected`, package private). In addition, we require that there are no unchecked downcasts in the program that could lead to runtime exceptions. Finally, for readability, we assume that procedures have only a single parameter, and constructors and queries do not have any. This is only for presentation, not a conceptual restriction.

In this section, we present schemata for the three verification and framing methodologies Separation Logic, Dynamic Frames, and Ownership Types. For each of them, we provide a theorem stating that if the data structure implementation adheres to the schema, all objects of the data structure are encapsulated according to Def. 5. For space reasons, we do not show proofs for the theorems and lemmas in this section, but instead refer to the extended version [44].

4.1 Separation Logic (VeriFast)

Separation Logic [42] (SL) is a deductive verification disclipline with a rather strict memory regime: Only those locations whose values have been explicitly captured in the preconditions are allowed to be read from or written to. This paradigm enforces strict memory separation, as the same locations must never be part of the memory footprint of more than one DSI at a time. SL hence enforces a strict notion of encapsulation—if the entire code base adheres to SL proofs. We will elaborate what this entails for approach-spanning verification attempts and describe a pattern that must be obeyed to guarantee encapsulation if not the entire code base adheres to SL. The challenge here is to remain open to "plugging in" data structures verified with other techniques.

We focus on the SL variant with permissions supported by VeriFast [30], which does not support quantifiers and magic wand in formulas. Data structure abstraction in SL is usually done using *abstraction predicates* which take a reference to the data implementation and the abstract value, capture the memory locations of the footprint of a DSI, and hold if the value computed from the data in the data structure is the same as the abstract argument value.

In interoperable contexts where some program components do not conform to SL, a significant challenge arises: an object may reference other objects in different data structures that should not be externally accessed. If the program were verified in one SL framework, access clauses for the relevant memory locations would be required, but these clauses cannot be generated. Without SL rigor, arbitrary Java code can access fields and traverse references, bypassing the abstract predicates defined by SL. The solution is to identify abstract predicates that include *all* reachable memory locations, thereby preventing access to prohibited memory.

Definition 6 (Deep Abstraction Predicates). *An abstraction predicate p for a class C is called* deep *if it refers to all fields of C and asserts a deep abstraction predicate for every field of reference type.*

Deep abstraction predicates represent heap chunks maximal wrt. reachability.

We introduce an abstract predicate `payload(x)` that captures the locations reachable from the Java object `x` via field dereferences. The predicate does not have a definition and is therefore intentionally underspecified: it exposes no concrete points-to facts and yields no functional guarantees about `x`. Its purpose is framing: with separating conjunction, `payload`$(x) * R$ asserts that R must be disjoint from the footprint of x. Because `payload` does not have a definition, clients cannot create initial instances. We therefore provide an introduction axiom `payloadCreate`, permitting `payload(x)` to be asserted without preconditions. This grants the `payload` token (for disjointness/framing), not any concrete field permissions. It reflects the ability of a client to pass arbitrary references into data structures as payloads. For soundness, we need to make sure that it is only used in clients, not in the data structure itself, which can be done with a simple syntactic check.

The encapsulation schema for a data structure class `C` is as follows (a template is shown in Lst. 2):

1) There is a deep abstraction predicate for `C` (`absC` in the example) with a parameter of type `C` and a second one of an immutable type (usually an ADT). It connects the heap-related type `C` to its heap-independent abstraction.

```
* /*@ predicate absC(C c, <adt> a) = ...; @*/
  class C {
    C()
      //@ requires true;
*     //@ ensures absC(this, φ1());
    { ... }

    <ret_type> query()          // obtains payload ref. from data structure
      //@ requires absC(this, ?x);
*     //@ ensures absC(this, φ2(x)) &*& payload(result);
    { ... }

    void procedure(<param_type> p) // stores payload ref. in data structure
      //@ requires absC(this, ?x) &*& payload(p);
*     //@ ensures absC(this, φ3(x, p));
    { ... }
  }
```

Listing 2: Schematic example for encapsulation with VeriFast Separation Logic. The ϕ_i are placeholders for expressions of the ADT type, possibly depending on the given parameters as well as fields of this. The $*$ indicates lines dealing with functional properties covered in Sect. 5.

2) Each constructor establishes well-formedness with a postcondition `absC (this, φ1())`, where ϕ_1 is an expression of the ADT type.
3) Each method requires and establishes the well-formedness of the data structure with a precondition `absC(this, ?x)` (the operator "?" binds `x` in the contract) and a postcondition `absC(this, φi(x))`, where ϕ_i is an expression of ADT type possibly depending on the input parameters and fields.
4) Each procedure receiving a parameter `p` adds a precondition `payload(p)`.
5) Each query returning a payload adds a postcondition `payload(result)`.

An observation is that an abstraction predicate of a DSI carries a set of permissions to memory locations, which is a superset of its footprint.

Definition 7 (Permissible Locations). *For an SL formula ϕ, we define the set $\texttt{pLocs}(\phi, d, M)$ as:*

$$\texttt{pLocs}(\phi, d, M) := \begin{cases} \emptyset & \phi \in \{\textbf{\textit{false}}, \textbf{\textit{true}}, \textbf{\textit{emp}}\} \\ \{(d, f)\} & \phi = f \mapsto y \\ \bigcup_\psi \texttt{pLocs}(\psi, d, M) & \textit{otherwise (with } \psi \textit{ subformula of } \phi) \end{cases}$$

*Here, **emp** denotes the SL formula that states that the heap is empty, and $f \mapsto y$ ("points-to") expresses that the heap is a single cell with address f and value y.*

We assume that there is exactly one abstraction predicate α_d defined for each DSI d and use $\texttt{pLocs}(d, M)$ as a shorthand for $\texttt{pLocs}(\alpha_d, d, M)$.

Lemma 1. *The set $\texttt{pLocs}(d, M)$ of a DSI d is a superset of its footprint* $\mathrm{fp}(d, M)$*:* $\forall$ DSI $d.\ \forall\ M \in \mathrm{Mem}\,.\ \texttt{pLocs}(d, M) \supseteq \mathrm{fp}(d, M)$

Theorem 1 (Encapsulation with the VeriFast Separation Logic Schema). *A verified DSI d that adheres to the above schema is encapsulated. That is, the permissible locations of the DSIs are disjoint from the relevant locations of all objects referenced in local variables:*

$$\forall \text{ Program } (s_0, \sigma).\ \forall \text{ State } (V, M, T) \in R(s_0, \sigma).\ \forall \text{ Object } o.$$
$$o \in dom(T) \wedge o{\neq}d \wedge (\exists v \in dom(V).\ V(v){=}o) \rightarrow \texttt{pLocs}(d, M) \cap \texttt{relLocs}(o, M) = \emptyset$$

The theorem is proven by induction on the program length (for details see [44]). With Lemma 1, this gives us that the DSI is encapsulated.

4.2 Dynamic Frames (KeY)

Dynamic Frames [34,48] provide a flexible methodology for specifying and reasoning about shared structures. However, this flexibility entails significant specification overhead. Additionally, proofs can be challenging due to the need for extensive reasoning about potentially overlapping memory locations.

We present the conditions for encapsulation with Dynamic Frames in the syntax of the Java Modeling Language (JML) [16,37], more specifically in the JML dialect of the Java verifier KeY [1,7]. We present two variants of the schema for encapsulation. First, we show the original pattern known from literature [33, 48], where only Dynamic Frames are used for all classes. Second, we present a generalized version that is suitable for our goal of combining multiple different specification and reasoning methodologies.

Pure Dynamic Frames With only Dynamic Frames, an encapsulated data structure has to adhere to the following schema (a template class is shown in Lst. 3):

1) There is a virtual (ghost) field of type locations set (`\locset`), which at least contains all fields of `this` (`this.*`). It is defined manually by the user to contain all heap locations which "belong" to the data structure.
2) The footprint locations are freshly created by each constructor (`ensures \fresh (fp)`), and no existing locations are written (`assignable \nothing`).
3) Procedures and constructors write only to locations in `fp` (`assignable fp`) or add freshly created objects to it (`ensures \new_elems_fresh(fp)`)[2]. This ensures that references of objects passed as parameters are not captured.
4) Query results depend only on the footprint (`accessible fp`).
5) No query can leak pointers into the data structure (`ensures \disjoint(fp, \result.fp)`).
6) Nothing is leaked or captured via any method parameter `p` of reference type (`ensures \disjoint (fp, p.fp)`). For proving, we usually need to know that this also holds in the pre-state (`requires \disjoint (fp, p.fp)`).

[2] This is known as the *swinging pivot condition* in literature [32].

```
   class C {
     //@ ghost \locset fp;
     //@ invariant fp == this.* ∪ ...
*    //@ ghost <adt> absVal;

*    //@ ensures \fresh(fp) && φ1(absVal);
     //@ assignable \nothing;
     C() { ... }

     //@ ensures \disjoint(fp, \result.fp);          // ⟵ \reachLocs(\result)
*    //@ ensures φ2(\result, absVal);
     //@ assignable \strictly_nothing;
     //@ accessible fp;
     <ret_type> query() { ... }

     //@ requires \disjoint(fp, p.fp);                // ⟵ \reachLocs(p)
     //@ ensures  \disjoint(fp, p.fp);                // ⟵ \reachLocs(p)
*    //@ ensures \new_elems_fresh(fp) && φ3(absVal, \old(absVal));
     //@ assignable fp;
     void procedure(<param_type> p) { ... }
   }
```

Listing 3: Schematic example for encapsulation with Dynamics Frames. When Dynamic Frames shall be used in cooperation with other techniques, the underlined parts can be replaced as indicated with the expressions in the comments to obtain the hybrid schema. The $*$ indicates lines dealing with functional properties (expressed by the placeholder formulas ϕ_i) covered in Sect. 5.

Hybrid approach for Dynamic Frames combined with other techniques To enable collaboration with other verification techniques, we cannot assume that each class has a clearly defined footprint field. Therefore, we require alternative methods to express the disjointness conditions (points 5/6 above). We introduce the function `\reachLocs`(o) on JML level, which has the semantics of *reachLocs* as defined in Definition 4 applied to the current heap state. The location set returned by `\reachLocs` is an overapproximation of the actual footprint, representing the "worst possible" footprint of the data structure while adhering to the programming language's capabilities, including visibility modifiers. With the `\reachLocs` function, we can now formulate the conditions for a data structure to be encapsulated (a template is shown as an alternative in Lst. 3):

1-4) These conditions are the same as in the pure Dynamic Frames technique.
5) No query can leak pointers into the data structure (`ensures \disjoint (fp, \reachLocs (\result ))`).
6) Nothing is leaked or captured via any method parameter `p` of reference type (`ensures \disjoint (fp, \reachLocs (p))`). For proving, we usually need to know that this also holds in the pre-state (`requires \disjoint (fp, \reachLocs (p))`).

Lemma 2. *For a DSI d adhering to the constraints above, the footprint field holds a superset of* $\mathrm{fp}(d, M)$*:* $\forall$ DSI d. $\forall \mathrm{Mem}\, M.\ M(d, fp) \supseteq \mathrm{fp}(d, M)$

Theorem 2 (Encapsulation with the Hybrid Dynamic Frames Schema). *If a DSI d specified with the hybrid Dynamic Frames schema is verified successfully in KeY, it is encapsulated. That is, the locations stored in the virtual footprint field* fp *of the DSI are disjoint from the relevant locations of all objects referenced in local variables:*

$$\forall \text{ Program } (s_0, \sigma).\ \forall \text{ State } (V, M, T) \in R(s_0, \sigma).\ \forall \text{ Object } o.$$
$$o{\in}dom(T) \wedge o{\neq}d \wedge (\exists v \in dom(V).\ V(v) = o) \rightarrow M(d, fp){\cap}\texttt{relLocs}(o, M){=}\emptyset$$

To give an intuition, this means that all the outside objects cannot have a reference into the data structure.

The proofs of Lemma 2 and Theorem 2 are very similar to the SL case, and can be found in the extended version [44]. Combined, we obtain the guarantee that the DSI is encapsulated if it adheres to either variant of the DF schema.

4.3 Ownership/Universe Types

In contrast to Separation Logic and Dynamic Frames, which combine reasoning about memory with reasoning about functional properties, Ownership Types only give guarantees about memory. Therefore, it is necessary to complement them with an additional technique for functional reasoning. One advantage of this is that the memory checks are done by the type system independently, and can then be assumed in the functional verifier, instead of being mixed-in with the rest of the verification. In our case, we use a type checker for a variant of Universe Types[3], implemented using the Checker Framework [17], in collaboration with KeY with simplified proof obligations only for the functional part[4].

Universe Encapsulation Types (UET). We define the new type and effect system *Universe Encapsulation Types (UET)*, which builds upon the existing Universe Types [18,19], but replaces their `any` types with `payload` types to obtain stronger guarantees. The type system adds a second dimension to the standard Java reference types. A type in UET is a pair (U, T), where T is the standard Java type and $U \in \{\texttt{rep}, \texttt{peer}, \texttt{payload}\}$ is an ownership modifier. The subtype relation is defined element-wise: $(U_s, T_s) \leq (U, T)$ iff $U_s \leq_{\text{UET}} U$ and $T_s \leq_{\text{Java}} T$. Here, $\leq_{\text{Java}}$ is the standard Java subtype relation, and $\leq_{\text{UET}}$ is the relation shown in Fig. 2.

Universe Types classify objects into contexts, allowing an object to read or modify everything in its context (directly or via methods). Contexts are indicated by ownership modifiers. The modifier `rep` denotes that the referenced object is (directly) owned by `this`, while `peer` indicates that the object shares the same owner as `this`. We replace the UT modifier `any` with `payload`, enhancing guarantees from the type system by prohibiting method calls or de-referencing of them (only equality comparison is allowed), in contrast to calls to `pure` methods

[3] https://github.com/WolframPfeifer/universe/tree/pfeifer/encapsulation.
[4] https://github.com/KeYProject/key/tree/pfeifer/universeEncapsulation.

```
  @DSI class C {
*   //@ ghost <adt> absVal;
    @Rep <field_type> f;

*   //@ ensures ϕ1(absVal);
    @RepOnly C() { ... }

*   //@ ensures ϕ2(result, absVal);
    @RepOnly @Payload <ret_type> query() { ... }

*   //@ ensures ϕ3(absVal, \old(absVal, p));
    @RepOnly void procedure(@Payload <param_type> p) { ... }
  }
```

Listing 4: Schematic example for encapsulation with UET. The functional part is expressed in JML and can be proven with KeY. The $*$ indicates lines with functional properties (see Sect. 5), and the ϕ_i stand for formulas expressing these properties.

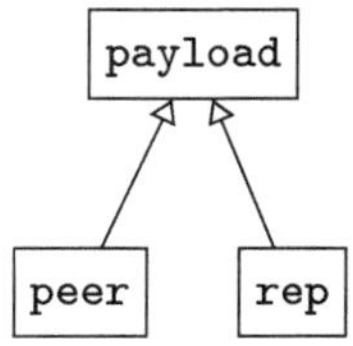

Fig. 2: Type hierarchy of UET.

in UT. They are treated as opaque, not depending on the state of underlying objects, allowing for storage and passing around. Similar to other type and effect systems, such as Universe Types, there are annotations in addition to the type modifiers. In UET, these are `DSI`, a marker for classes subject to encapsulation checking, and `RepOnly`, which denotes a method that operates solely on the `rep` footprint of the DSI. This is ensured by allowing assignments only to fields of `this`, and method calls where the receiver is `this` or a `rep` reference.

Encapsulation with UET. We can now use this type system for proving encapsulation. However, well-typedness alone is not sufficient. Instead, as for Separation Logic and Dynamic Frames, the DSI needs to comply to a specific schema. A DSI o (marked as such by the user with the corresponding annotation) of a class C is encapsulated if it adheres to the following schema, and it is UET well-typed (a template class following the schema is found in Lst. 4):

1) All methods and constructors in class C are `repOnly`.
2) All fields of reference type in C are `rep` or `payload`.
3) All parameters (of reference type) of methods are `payload`.
4) All (reference) return types of methods are `payload`.

Note that the internal classes of the DSI (for instance a class `Node` that is used internally in an implementation of `List`) do not have these restrictions. There, it is for example also allowed to have `peer` references and methods that operate on peers. Only the top-level "interface" class has to comply to the schema, internally, the full expressive power of the original Universe Types can be used.

Definition 8 (Rep Footprint). *The* rep footprint *repFP*(d, M) *of a DSI d is the set of locations that are directly or transitively owned by d. It is the smallest* *fixpoint of the following function, where x is used as a shortcut for $M(d, f)$:*

$$\textit{repFP}(d, M) := \bigcup_{f \in \text{fst}(d)} (d, f) \cup \bigcup_{\substack{f \in \text{fst}(d), \\ f \text{ is rep}}} \Big(\bigcup_{g \in \text{fst}(x)} (x, g) \cup \bigcup_{\substack{g \in \text{fst}(x), \\ g \text{ is not payload}}} \textit{repFP}(M(x, g), M) \Big)$$

Lemma 3. *For a DSI d adhering to the constraints above, the rep footprint of d is a superset of* fp(d, M)*:* $\forall$ DSI d. $\forall$ Mem M. $\textit{repFP}(d, M) \supseteq \text{fp}(d, M)$

Theorem 3 (Encapsulation with Universe Encapsulation Types). *If a DSI d is annotated with UET, and checked with the UET checker, it is encapsulated. That is, the rep footprint of the DSI is disjoint from the relevant locations of all objects referenced in local variables:*

$$\forall \text{ Program } (s_0, \sigma).\ \forall \text{State } (V, M, T) \in R(s_0, \sigma).\ \forall \text{ Object } o.$$
$$o \in dom(T) \wedge o \neq d \wedge (\exists v \in dom(V).\ V(v) = o) \rightarrow \textit{repFP}(d, M) \cap \texttt{relLocs}(o, M) = \emptyset$$

As with SL and DF, for the proofs we refer to the extended version [44], and obtain encapsulation of the DSI with Lemma 3.

5 Cooperation Between Verification Techniques

Up to here, we have demonstrated the verification of encapsulated data structures within a single verification framework. However, the strict encapsulation notion aims to verify heterogeneous client programs with parts verified using different verification tools. Achieving cooperative deductive verification also requires the exchange of functional specifications beyond framing across different tools. The programming language in Sect. 4 is consistently Java, such that code can be readily shared between tools. However, sharing functional specifications is challenging due to significant conceptual differences between the specification languages of KeY, VeriFast, and Universe Types.

The presented approach mainly focuses on integrating various seemingly incompatible memory management approaches, such that we only briefly outline how functional properties can be shared across verification paradigms.

Despite diverse specification mechanisms, many formalisms share a common foundational concept: *Algebraic Data Types* (ADTs). Rooted in the mathematical notion of inductive structures, ADTs are recognised across many verification methods, sharing identical semantics (namely that of free generated algebras). As encapsulated DSIs possess an independent state unaffected by external operations, ADTs are an ideal means to represent abstract values in encapsulated data types.

Contract-LIB [21] is a programming-language-agnostic interface specification language allowing the specification of data abstractions and operations via contracts. This makes it suitable for data exchange between various verification approaches, as it enables the declaration of a data abstraction for the data structure along with contracts that express the effects of operations in terms

```
; Abstractions
(declare-abstractions
    ((example.Cell 0))
    (((Cell (absVal Int)))))

(declare-abstractions
    ((example.LinkedCellList 0))
    (((LinkedCellList (absVal (Seq (Ref Cell)))))))

; Contracts
(define-contract example.LinkedCellList.init       ; constructor
    ((result (out LinkedCellList)))
    ((true
     (= (absVal result) seq.empty))))

(define-contract example.LinkedCellList.add        ; procedure
    ((this (inout LinkedCellList))
     (v (in (Ref Cell))))
    ((true
     (= (absVal this) (seq.++ (old (absVal this)) (seq.unit v))))))

(define-contract example.LinkedCellList.getLast    ; query
    ((this (inout LinkedCellList))
     (result (out (Ref Cell))))
    (((not (= (absVal this) seq.empty))
     (= result (seq.nth (absVal this) (- (seq.len (absVal this)) 1))))))
```

Listing 5: Specification of LinkedCellList in Contract-LIB.

of these abstractions. Algebraic and co-algebraic data types, either as built-in types of SMT-LIB [6] or defined using **declare-datatypes**, are supported as types for abstract data. The exchange language Contract-LIB and the specification schemata in Lsts. 2, 3 and 4 allow us to verify well-encapsulated data structures across different verification paradigms, using the following workflow:

1) Data structures and their operations are specified via ADTs in Contract-LIB.
2) For each data structure, one verification approach from Sect. 4 is selected.
3) The Contract-LIB specifications are used to generate a *verification interface* containing the translated specifications in the respective target language.
4) The manually written implementations are then verified against the interfaces.
5) The Contract-LIB specifications can also be used to produce *verification stubs*, verification-only classes bearing the same specifications that can be used on the client side to verify against them.

If we come back to the initial heterogeneous code example from Lst. 1, the overview in Fig. 1 shows how the Contract-LIB specification e.g. for the class `LinkedCellList` (as shown in Lst. 5), induces the verification interface $\texttt{ILinkedCellList}_{\texttt{VF}}$ providing the specification for the implementation in the class `LinkedCellListImpl`. Separately from verification of the data structure in VeriFast, the class is used in the context of the client, hence the verification stub $\texttt{LinkedCellList}_{\texttt{KeY}}$ is produced for client verification in KeY.

We implemented automation support for Contract-LIB in a framework named Contract-Chameleon[5], which encompasses a parser and a translator to/from the specification languages used by KeY (JML with Dynamic Frames) and VeriFast (permission-based Separation Logic). Contract-Chameleon also generates stub Java files within which the actual implementation and auxiliary specifications (such as coupling and loop invariants) can be integrated.

As established by Theorems 1, 2 and 3, verification entails encapsulation. Thus, it is sound to translate specifications from one language to another and employ the appropriate one for the client's verification proof. The functional guarantees are transported via abstraction predicates (SL) and ghost fields (DF and UET), as well as through the method contracts that describe how the abstract state evolves with each operation. In the schematic examples, this can be found in the lines marked with $*$ in Lsts. 2 to 4, where the expressions/formulas $\phi_1, \ldots, \phi_4$ describe the effects of the operations in terms of the abstract value. With the guarantee of encapsulation, it allows us to exchange the functional specifications soundly between the different verification methodologies.

6 Future Work and Conclusion

The encapsulation patterns discussed in this paper are restrictive. Some data structures may *behave* encapsulated, while sharing memory internally, for example, copy-on-write lists. We plan to extend our approach towards *observational encapsulation* to support such cases. Other structures are only partially encapsulated, such as a tree-set that uses hash codes for logarithmic-time operations, while remaining functionally indistinguishable from an abstract set. Techniques similar to declassification in information-flow verification may address this. Future work also includes support for subclassing and parametric polymorphism to make `payload` annotations of UET more useful. Combining UET with deductive verification [36] could further improve the expressiveness and usability of the system. We plan to extend current prototype tools, for example, to import VeriFast specifications into Contract-LIB, to support other Separation Logic variants (most notably the one supported by VerCors), and with a dedicated proof management tool.

To sum up, in this paper, we formalized encapsulation, and provided sufficient conditions in three common verification methodologies: Separation Logic (for the tool VeriFast), Dynamic Frames (KeY), and an Ownership Type system. We have proven that they are sufficient for each of the methodologies. In addition, we described how functional guarantees can be exchanged via Contract-LIB. We demonstrated the applicability of our framework by conducting a cooperative verification effort of a small example with three data structures, each verified with one of the methodologies.

Acknowledgments. We thank the anonymous reviewers for their valuable feedback on this paper. This work was supported by the DFG projects BE 2334/9-1, UL 433/3-1 and the KIT International Excellence Fellowships Program with funds granted to

[5] https://github.com/Contract-LIB/contract-chameleon.

the University of Excellence concept of Karlsruhe Institute of Technology. We also acknowledge the support of the Natural Sciences and Engineering Research Council of Canada (NSERC) Discovery Grants program, RGPIN-2020-05502, and an Early Researcher Award from the Government of Ontario.

Data Availability Statement. The full code of the example can be found in the artifact accompanying this paper (DOI: 10.5281/zenodo.18607514).

Disclosure of Interests. The authors have no competing interests to declare.

References

1. Ahrendt, W., Beckert, B., Bubel, R., Hähnle, R., Schmitt, P.H., Ulbrich, M. (eds.): Deductive Software Verification – The KeY Book, LNCS, vol. 10001. Springer International Publishing, Cham (2016). https://doi.org/10.1007/978-3-319-49812-6
2. Armborst, L., Beyer, D., Huisman, M., Lingsch-Rosenfeld, M.: AutoSV-Annotator: Integrating deductive and automatic software verification. In: 30th International Conference on Formal Methods for Industrial Critical Systems. Lecture Notes in Computer Science, vol. 16040, pp. 59–77. Springer, Berlin, Heidelberg (2025). https://doi.org/10.1007/978-3-032-00942-5_4
3. Ayaziová, P., Beyer, D., Lingsch-Rosenfeld, M., Spiessl, M., Strejček, J.: Software verification witnesses 2.0. In: Model Checking Software. pp. 184–203. Springer Nature Switzerland, Cham (2025). https://doi.org/10.1007/978-3-031-66149-5_11
4. Banerjee, A., Naumann, D.A.: Local reasoning for global invariants, Part II: dynamic boundaries. J. ACM **60**(3), 19:1–19:73 (2013). https://doi.org/10.1145/2485981
5. Bao, Y., Leavens, G.T., Ernst, G.: Unifying separation logic and region logic to allow interoperability. Formal Asp. Comput. **30**(3-4), 381–441 (2018). https://doi.org/10.1007/s00165-018-0455-5
6. Barrett, C., Fontaine, P., Tinelli, C.: The SMT-LIB standard: Version 2.7. Tech. rep., Department of Computer Science, The University of Iowa (2025)
7. Beckert, B., et al.: The Java verification tool KeY: a tutorial. In: 26th International Symposium on Formal Methods, FM 2024, pp. 597–623. Springer, Berlin, Heidelberg (2024). https://doi.org/10.1007/978-3-031-71177-0_32
8. Beckert, B., Sanders, P., Ulbrich, M., Wiesler, J., Witt, S.: Formally verifying an efficient sorter. In: Tools and Algorithms for the Construction and Analysis of Systems: 30th International Conference, TACAS 2024, pp. 268–287. Springer, Berlin, Heidelberg (2024). https://doi.org/10.1007/978-3-031-57246-3_15
9. Beyer, D., Dangl, M., Lemberger, T., Tautschnig, M.: Tests from witnesses. In: Tests and Proofs, pp. 3–23. Springer International Publishing, Cham (2018). https://doi.org/10.1007/978-3-319-92994-1_1
10. Beyer, D., Ernst, G., Jonáš, M., Lingsch-Rosenfeld, M.: SV-LIB 1.0: a standard exchange format for software-verification tasks (2025)
11. Beyer, D., Kanav, S., Richter, C.: Construction of verifier combinations based on off-the-shelf verifiers. In: Fundamental Approaches to Software Engineering, pp. 49–70. Springer International Publishing, Cham (2022). https://doi.org/10.1007/978-3-030-99429-7_3

12. Beyer, D., Strejček, J.: Improvements in software verification and witness validation: SV-COMP 2025. In: Tools and Algorithms for the Construction and Analysis of Systems: 31st International Conference, TACAS 2025, pp. 151–186. Springer, Berlin, Heidelberg (2025). https://doi.org/10.1007/978-3-031-90660-2_9
13. Beyer, D., Wehrheim, H.: Verification artifacts in cooperative verification: Survey and unifying component framework. In: Leveraging Applications of Formal Methods, Verification and Validation: Verification Principles. pp. 143–167. Springer International Publishing, Cham (2020). https://doi.org/10.1007/978-3-030-61362-4_8
14. Blom, S., Huisman, M.: The VerCors tool for verification of concurrent programs. In: FM 2014: Formal Methods, pp. 127–131. Springer International Publishing, Cham (2014). https://doi.org/10.1007/978-3-319-06410-9_9
15. Clarke, D., Noble, J., Wrigstad, T. (eds.): Aliasing in Object-Oriented Programming. Types, Analysis and Verification, Lecture Notes in Computer Science, vol. 7850. Springer Berlin Heidelberg (2013). https://doi.org/10.1007/978-3-642-36946-9
16. Cok, D.R., Leavens, G.T., Ulbrich, M.: JML Reference Manual. Second edition Edn. (2025)
17. Dietl, W., Dietzel, S., Ernst, M.D., Muşlu, K., Schiller, T.W.: Building and using pluggable type-checkers. In: International Conference on Software Engineering (ICSE), pp. 681–690. Association for Computing Machinery, New York, NY, USA (2011). https://doi.org/10.1145/1985793.1985889
18. Dietl, W., Drossopoulou, S., Müller, P.: Generic universe types. In: ECOOP 2007 – Object-Oriented Programming, pp. 28–53. Lecture Notes in Computer Science, Springer, Berlin, Heidelberg (2007). https://doi.org/10.1007/978-3-540-73589-2_3
19. Dietl, W., Ernst, M.D., Müller, P.: Tunable static inference for generic universe types. In: ECOOP 2011 – Object-Oriented Programming, pp. 333–357. Springer, Berlin, Heidelberg (2011). https://doi.org/10.1007/978-3-642-22655-7_16
20. Dietl, W., Müller, P.: Object ownership in program verification. In: Aliasing in Object-Oriented Programming. Types, Analysis and Verification, pp. 289–318. Lecture Notes in Computer Science, Springer, Berlin, Heidelberg (2013). https://doi.org/10.1007/978-3-642-36946-9_11
21. Ernst, G., Pfeifer, W., Ulbrich, M.: Contract-LIB: a proposal for a common interchange format for software system specification. In: Leveraging Applications of Formal Methods, Verification and Validation. Specification and Verification - 12th International Symposium, ISoLA 2024. Lecture Notes in Computer Science, vol. 15221, pp. 79–105. Springer Nature Switzerland, Cham (2025). https://doi.org/10.1007/978-3-031-75380-0_6
22. Filliâtre, J.C., Paskevich, A.: Why3 — where programs meet provers. In: Programming Languages and Systems, pp. 125–128. Springer, Berlin, Heidelberg (2013). https://doi.org/10.1007/978-3-642-37036-6_8
23. de Gouw, S., de Boer, F.S., Bubel, R., Hähnle, R., Rot, J., Steinhöfel, D.: Verifying OpenJDK's sort method for generic collections. J. Autom. Reason. **62**(1), 93–126 (2017). https://doi.org/10.1007/s10817-017-9426-4
24. Hähnle, R., Huisman, M.: Deductive software verification: from pen-and-paper proofs to industrial tools. In: Computing and Software Science: State of the Art and Perspectives, pp. 345–373. Springer International Publishing, Cham (2019). https://doi.org/10.1007/978-3-319-91908-9_18
25. Haltermann, J., Jakobs, M.C., Richter, C., Wehrheim, H.: Parallel program analysis on path ranges. Sci. Comput. Program. **238**, 103154 (2024). https://doi.org/10.1016/j.scico.2024.103154

26. Harel, D., Kozen, D., Tiuryn, J.: Parallel program analysis on path ranges. Dynamic Logic. The MIT Press (2000). https://doi.org/10.7551/mitpress/2516.001.0001
27. Hoare, C., Misra, J., Leavens, G.T., Shankar, N.: The verified software initiative: a manifesto. ACM Comput. Surv. **41**(4), 22:1–22:8 (2009). https://doi.org/10.1145/1592434.1592439
28. Hoare, T.: The verifying compiler: a grand challenge for computing research. J. ACM **50**(1), 63–69 (2003). https://doi.org/10.1145/602382.602403
29. Huisman, M., Monti, R., Ulbrich, M., Weigl, A.: The VerifyThis collaborative long term challenge. In: Deductive Software Verification: Future Perspectives: Reflections on the Occasion of 20 Years of KeY, Lecture Notes in Computer Science, vol. 12345, chap. 10, pp. 246–260. Springer, Berlin, Heidelberg (2020). https://doi.org/10.1007/978-3-030-64354-6_10
30. Jacobs, B., Smans, J., Philippaerts, P., Vogels, F., Penninckx, W., Piessens, F.: VeriFast: a powerful, sound, predictable, fast verifier for C and Java. In: NASA Formal Methods, vol. 6617, pp. 41–55. Springer, Berlin, Heidelberg (2011). https://doi.org/10.1007/978-3-642-20398-5_4
31. Jakobs, M.C.: PARTPW: from partial analysis results to a proof witness. In: Software Engineering and Formal Methods, pp. 120–135. Springer International Publishing, Cham (2017). https://doi.org/10.1007/978-3-319-66197-1_8
32. Kassios, I.T.: The dynamic frames theory. Formal Aspects Comput. **23**(3), 267–288 (2011). https://doi.org/10.1007/s00165-010-0152-5
33. Kassios, I.T.: Dynamic Frames and Automated Verification
34. Kassios, I.T.: Dynamic frames: support for framing, dependencies and sharing without restrictions. In: FM 2006: Formal Methods. pp. 268–283. Springer, Berlin, Heidelberg (2006). https://doi.org/10.1007/11813040_19
35. Klamroth, J., Lanzinger, F., Pfeifer, W., Ulbrich, M.: The Karlsruhe Java verification suite. In: The Logic of Software. A Tasting Menu of Formal Methods: Essays Dedicated to Reiner Hähnle on the Occasion of His 60th Birthday, pp. 290–312. Springer International Publishing, Cham (2022). https://doi.org/10.1007/978-3-031-08166-8_14
36. Lanzinger, F., Weigl, A., Ulbrich, M., Dietl, W.: Scalability and precision by combining expressive type systems and deductive verification. Proc. ACM Prog. Lang. **5**(OOPSLA) (2021). https://doi.org/10.1145/3485520
37. Leavens, G.T., et al.: JML Reference Manual (2013)
38. Leino, K.R.M., Moskal, M.: VACID-0: verification of ample correctness of invariants of data-structures, edition 0. In: Proceedings of Tools and Experiments Workshop at VSTTE (2010)
39. Leino, K.R.M., Rümmer, P.: A polymorphic intermediate verification language: design and logical encoding. In: Tools and Algorithms for the Construction and Analysis of Systems, pp. 312–327. Springer, Berlin, Heidelberg (2010). https://doi.org/10.1007/978-3-642-12002-2_26
40. Livshits, B., et al.: In defense of soundiness: a manifesto. Commun. ACM **58**(2), 44–46 (2015). https://doi.org/10.1145/2644805
41. Müller, P., Schwerhoff, M., Summers, A.J.: Viper: a verification infrastructure for permission-based reasoning. In: Jobstmann, B., Leino, K.R.M. (eds.) VMCAI 2016. LNCS, vol. 9583, pp. 41–62. Springer, Heidelberg (2016). https://doi.org/10.1007/978-3-662-49122-5_2
42. O'Hearn, P.: Separation logic. Commun. ACM **62**(2), 86–95 (2019). https://doi.org/10.1145/3211968

43. Oortwijn, W., Huisman, M., Sebastiaan J. C. Joosten, Jaco van de Pol: Automated verification of parallel nested DFS. In: Tools and Algorithms for the Construction and Analysis of Systems, pp. 247–265. Springer International Publishing, Cham (2020). https://doi.org/10.1007/978-3-030-45190-5_14
44. Pfeifer, W., Dietl, W., Ulbrich, M.: A framework for the interoperable specification and verification of encapsulated data structures. Tech. rep., Karlsruhe Institute of Technology (KIT) (2026). https://doi.org/10.5445/IR/1000191071
45. Reynolds, J.: Separation logic: a logic for shared mutable data structures. In: Proceedings 17th Annual IEEE Symposium on Logic in Computer Science, pp. 55–74 (2002). https://doi.org/10.1109/LICS.2002.1029817
46. Richter, C., Chalupa, M., Jakobs, M.C., Wehrheim, H.: Cooperative software verification via dynamic program splitting. In: 47th International Conference on Software Engineering (ICSE), pp. 2087–2099 (2025). https://doi.org/10.1109/ICSE55347.2025.00092
47. Smans, J., Jacobs, B., Piessens, F.: VeriFast for Java: a tutorial. In: Aliasing in Object-Oriented Programming. Types, Analysis and Verification, pp. 407–442. Springer, Berlin, Heidelberg (2013). https://doi.org/10.1007/978-3-642-36946-9_14
48. Weiß, B.: Deductive verification of object-oriented software: dynamic frames, dynamic logic and predicate abstraction (2011). https://doi.org/10.5445/KSP/1000021694

Can LLM Aid in Solving Constraints with Inductive Definitions?

Weizhi Feng[1,2], Shidong Shen[1,2], Jiaxiang Liu[1,2], Taolue Chen[3], Fu Song[1,2,4], and Zhilin Wu[1,2(✉)]

[1] Key Laboratory of System Software (Chinese Academy of Sciences), Institute of Software, Chinese Academy of Sciences, Beijing, China
{fengwz,shensd,liujx,songfu,wuzl}@ios.ac.cn
[2] University of Chinese Academy of Sciences, Beijing, China
[3] Birkbeck, University of London, London, UK
t.chen@bbk.ac.uk
[4] Nanjing Institute of Software Technology, Nanjing, China

Abstract. Solving constraints involving inductive (aka recursive) definitions is challenging. State-of-the-art SMT/CHC solvers and first-order logic provers provide only limited support for solving such constraints, especially when they involve, e.g., abstract data types. In this work, we leverage structured prompts to elicit Large Language Models (LLMs) to generate auxiliary lemmas that are necessary for reasoning about these inductive definitions. We further propose a neuro-symbolic approach, which synergistically integrates LLMs with constraint solvers: the LLM iteratively generates conjectures, while the solver checks their validity and usefulness for proving the goal. We evaluate our approach on a diverse benchmark suite comprising constraints originating from algebraic data types and recurrence relations. The experimental results show that our approach can improve the state-of-the-art SMT and CHC solvers, solving considerably more (around 25%) proof tasks involving inductive definitions, demonstrating its efficacy.

1 Introduction

Inductive definitions are prevalent in program verification. Typically, they exhibit in two forms, i.e., algebraic data types (ADTs) and recursively defined functions (RDFs). An ADT defines a type whose elements are constructed by a finite number of applications of a given set of rules. As an example, Listing 1.1 defines natural numbers (`Nat`) as an ADT, and two operations on natural numbers as RDFs. `Nat` is defined as either zero (`zero`) or the successor of another `Nat`, given by `succ(n)`. Both RDFs `plus` and `mult` are specified through assertions with universal quantifiers, serving as axioms.

- The addition operation (`plus`) (lines 2–5) is recursively defined by two cases: (i) adding `zero` to any natural number `y` yields `y`, and (ii) adding the successor of any natural number `x` to `y` yields the successor of `plus(x,y)`.

A. Sampaio and M. Stoelinga (Eds.): FM 2026, LNCS 16557, pp. 111–132, 2026.
https://doi.org/10.1007/978-3-032-26220-2_6

```
1  datatypes Nat := zero | succ(n: Nat)
2  fun plus(Nat, Nat): Nat {
3    ∀y∈Nat.plus(zero,y) = y
4    ∀x∈Nat,y∈Nat.plus(succ(x),y) = succ(plus(x,y))
5  }
6  fun mult(Nat, Nat): Nat {
7    ∀y∈Nat.mult(zero,y) = zero
8    ∀x∈Nat,y∈Nat.mult(succ(x),y) = plus(mult(x,y),y)
9  }
10 property: ∀x∈Nat,y∈Nat.mult(x,y) = mult(y,x)
```

Listing 1.1. A pseudocode program: the running example

- The multiplication operation (`mult`) (lines 6–9) is recursively defined by two cases as well: (i) multiplying `zero` by any natural number `y` results in `zero`, and (ii) multiplying the successor of any natural number `x` by `y` is the addition of `mult(x,y)` and `y`.

In this example, the proof goal is the commutativity of the multiplication operation `mult`, formulated as `property` at line 10 in Listing 1.1.

Such ADTs are ubiquitous in functional programming languages (e.g., Gallina) for interactive theorem provers (ITPs, e.g., Rocq [3]), and are increasingly being supported by other languages such as TypeScript and Rust.

Inductive Reasoning. Program verification requires reasoning about properties of inductive definitions, including ADTs and RDFs. The standard reasoning process typically consists of two steps: first proving the base case(s) and then proving the inductive case(s) with inductive hypotheses as premises. For example, in Listing 1.1, one may apply induction on the natural number `x`, leading to the base case where `x` is `zero` and an inductive case where `x` is `succ(x')` in which the property is assumed to hold for `x'`. These two cases yield two verification conditions (VCs):

$$\begin{aligned} &\forall\, y \in \texttt{Nat}.\ \texttt{mult}(\texttt{zero}, y) = \texttt{mult}(y, \texttt{zero}),\\ &\forall\, x', y \in \texttt{Nat}.\ \texttt{mult}(x', y) = \texttt{mult}(y, x') \rightarrow \\ &\qquad\qquad\qquad\qquad \texttt{mult}(\texttt{succ}(x'), y) = \texttt{mult}(y, \texttt{succ}(x')). \end{aligned} \tag{1}$$

Different techniques can be applied for inductive reasoning, ranging from manual to fully automated. In general, ITPs such as Rocq [3], Isabelle [34] and Lean [32], and semi-automated verifiers (aka deductive verifiers) such as Dafny [28], Stainless [19] and Verus [27], require user intervention.

In contrast, automated constraint solving, including SMT solving and first-order logic theorem proving, is designed to operate without user intervention. Typically, induction schemas are incorporated into SMT solvers [40] or into superposition-based first-order logic theorem provers [8,13,16–18,20,38], enabling automatic generation of base and inductive cases as verification conditions with inductive hypotheses as premises. However, many proof goals with inductive definitions cannot be solved by solely applying the inductive hypotheses and axioms. Consequently, additional *auxiliary lemmas* are required to complete the proof. Namely, to prove $A \to P$ where A and P are the premise and the conclusion, respectively, we may introduce a set of auxiliary lemmas $\{L_i\}_{i \in I}$ such that $A \to L_i$ holds for each $i \in I$, and $A \wedge \bigwedge_{i \in I} L_i \to P$ holds.

Lemma Generation Methods and Limitations. This paper tackles *automated generation of auxiliary lemmas* in constraint solving. The existing methods can roughly be classified into three main categories. First, the *theory exploration* method, implemented in solvers such as cvc5 [2], iteratively enumerates conjectures starting from small terms and applies heuristic checks to quickly filter out invalid ones before verifying their validity and usefulness for assisting in proving the goal [7,40,42]. Second, the *generalization* method, implemented in solvers such as Vampire [25], identifies common sub-terms in the proof goal and replaces them with new variables to find auxiliary lemmas that are simpler and can generalize the goal [16]. Third, *Constrained Horn Clause (CHC)*-based methods [22] treats inductive definitions as transition systems and solves the verification problem by synthesizing inductive invariants. (Cf. Sect. 5 for a brief review of the related work.)

Despite their effectiveness in certain scenarios, these methods have significant limitations. Theory exploration is effective for generating simple lemmas but often struggles with discovering more complicated lemmas required for inductive proofs. Indeed, the representative state-of-the-art SMT solver, cvc5, fails to prove the `property` given in Listing 1.1 (cf. the full version of the paper for details). Generalization methods have limited expressiveness and cannot handle more general problems. CHC-based methods have limited capability to handle inductive definitions, particularly RDFs. Overall, these traditional logic-based methods mostly rely on fixed mechanisms and heuristic strategies, which often suffer from limited scalability or expressiveness, restricting the usability and applicability of automated constraint solving in program verification.

LLM-Aided Solving. LLMs have demonstrated remarkable capabilities in code generation [15,45,46,52], program specification inference [31,49], loop-invariant generation [5,36,50], and interactive theorem proving [39,48]. Recent works have also explored LLM-assisted lemma/conjecture generation in deductive verification (e.g., Dafny [41] and Verus [51]), hardware model checking [35] and mathematical reasoning [1,6,47]. Overall, these methods largely follow a generate-then-verify paradigm, but they target either program-level proof construction or hardware/mathematical formula reasoning, rather than fully automated inductive reasoning and first-order logic constraint solving.

Inspired by these advances of LLMs in generative AI and formal reasoning, we introduce LLMs into automated inductive reasoning, in particular, to overcome the limitations of pure logic-based methods in lemma generation. However, there are two main technical challenges:

Challenge 1. While LLMs often exhibit emergent abilities across diverse problem types, they may not be very effective in specific downstream tasks. The primary challenge is to effectively unleash their capability for inductive reasoning, e.g., to guide them to recognize what constitutes *good* auxiliary lemmas therein.

Challenge 2. Unlike traditional logic-based methods, the outputs of LLMs exhibit randomness, and can even hallucinate. The second challenge is how to quickly eliminate invalid outputs and ensure their validity and usefulness for assisting in proving the target goal.

Our Contributions. We propose a neuro-symbolic approach that synergistically integrates LLMs and constraint solving for automated lemma generation in inductive reasoning. Our approach consists of three stages: *query*, *filter* and *validate*. Given an instance of the inductive reasoning problem that comprises inductive definitions and a proof goal, our approach automatically generates auxiliary lemmas and proves the goal. The three stages are designed to address the above two technical challenges. To address Challenge 1, we design two prompt strategies in the query stage: (i) the first strategy imitates human inductive reasoning through equational term rewriting, enabling goal-oriented lemma discovery, and (ii) the second strategy abstracts the proof goal into simpler forms and generates bridging lemmas between simplified forms and the original goal. Moreover, it incorporates heuristics, such as suggesting commonly used axioms, attempting to generate strengthened propositions that imply the original goal, and applying equivalent transformations to terms in the proof goal. The filter and validate stages together address Challenge 2, where the filter stage quickly eliminates incorrect and useless outputs using backend solvers with short timeouts, resulting in candidate lemmas, while the validate stage enforces validity and usefulness of candidate lemmas through a two-step process: checking whether the candidate lemmas can assist in proving the original goal, and then recursively verifying each candidate lemma if they suffice to prove the goal.

We evaluate our approach on 706 instances of the inductive reasoning problem collected from diverse benchmarks commonly used in the inductive reasoning literature [22,40,44]. We compare our method against state-of-the-art solvers (including cvc5 and Vampire). The experimental results show that our approach achieves (at least 25%) higher success rate than these solvers. Additionally, the ablation studies show that both the filtering and prompt designs contribute substantially to the performance improvement. Further experiments also suggest that our approach is robust across different LLMs, sampling temperatures and backend solvers, which demonstrates the generality of our approach.

The full version of the paper, as well as the tool, benchmarks and experimental data, is available at https://github.com/fengwz17/LLM4Ind.

2 Preliminary

2.1 Satisfiability Modulo Theories

Satisfiability Modulo Theory (SMT) refers to deciding whether a given logical formula is satisfiable, i.e., whether there exists an assignment of its variables over background theories, under which the formula holds.

Formally, an SMT formula is defined over a given signature Σ that consists of a set of sorts (types) $\mathcal{S}$, function symbols $\mathcal{F}$ and predicate symbols $\mathcal{P}$. A model $\mathcal{M}$ of a given theory $\mathcal{T}$ (e.g., Linear integer arithmetic (LIA), Bit-vector (BV), Uninterpreted functions (UFs), Algebraic data types (ADTs)) provides an interpretation of sorts, function symbols and predicate symbols in Σ. A set of clauses $F = \{C_1, \cdots, C_n\}$ represents the conjunction $C_1 \wedge \cdots \wedge C_n$, where each clause C_i may be universally quantified.

SMT solvers, such as Z3 [33], cvc5 [2] or Yices [12], combine SAT-solving techniques with specialized theory solvers. The SAT engine handles the propositional structure of the formula, while theory solvers reason about constraints in their respective domains. Through this integration, SMT solvers have become powerful tools for, among others, program verification and synthesis, model checking, and test-case generation. Besides, some first-order logic provers, such as Vampire [25], can also accept a fragment of SMTLIB2 inputs [37]. For simplicity, we refer to both SMT solvers and such first-order logic provers as *solvers* in this paper when no ambiguity arises.

2.2 Inductive Definitions

Algebraic Data Types. The theory of algebraic data types (ADTs) is defined over a signature Σ. For example, natural numbers and lists can be defined as:

```
datatypes Nat := zero | succ(n: Nat),
datatypes List := nil | cons(head: Nat, tail: List).
```

Σ includes sorts for representing ADTs (e.g., `Nat`, `List`), function symbols for constructors (e.g., `zero`, `succ` for `Nat`) and selectors (e.g., `head`, `tail` for `List`), and predicate symbols for testers.

Recursively Defined Functions. A recursively defined function (RDF) is a function whose definition refers to itself on smaller or simpler inputs. Formally, such a function is specified by a set of base cases, which provide values for the simplest inputs, together with recursive rules that reduce larger or more complex inputs to instances of the same function on smaller arguments. This structure ensures that every call to the function eventually reaches a base case, guaranteeing well-definedness and termination. RDFs are fundamental in mathematics and computer science, as they naturally capture inductive structures such as sequences, trees and lists.

2.3 Lemma Generation

Let A denote a set of axioms encoding RDFs over ADTs, and P denote a target property to be proved. A usually consists of universally quantified equations, while P is a universally quantified formula, e.g., $\forall \vec{x} \in \tau.\ t(\vec{x})$, where $\vec{x} = \{x_1, \cdots, x_n\}$, τ is an ADT sort (e.g., `Nat`), and t is a term over $\vec{x}$. The goal is to establish the validity of $A \rightarrow P$. To this end, we aim to generate a set of lemmas $\mathcal{L} = \{L_i\}_{i \in I}$ such that the following conditions hold:

$$A \wedge \bigwedge_{i \in I} L_i \rightarrow P, \quad \text{and} \quad \forall i \in I.\ A \rightarrow L_i.$$

In practice, to prove $A \rightarrow P$, SMT-based approaches negate the property P, and check whether the formula, given by the set of clauses $F = \{A, \neg P\}$, is unsatisfiable. Then the verification task is checking the satisfiability of F. The *lemma generation task* is to find a set of lemmas $\mathcal{L} = \{L_i\}_{i \in I}$ such that: (i) each lemma L_i is entailed by A, i.e., $\{A, \neg L_i\}$ is unsatisfiable, and (ii) the set of clauses $\{A, \neg P\} \cup \mathcal{L}$ is unsatisfiable.

SMT solving commonly copes with universally quantified formulas such as $\forall \vec{x} \in \tau.t(\vec{x})$ through instantiation-based techniques, and handles the negation of universally quantified formulas (effectively existentially quantified formulas since $\neg\forall \vec{x} \in \tau.t(\vec{x})$ is equivalent to $\exists \vec{x} \in \tau.\neg t(\vec{x})$) via skolemization. cvc5 implements an inductive strengthening technique in its skolemization module to incorporate the induction schema into SMT solving and a theory exploration method to automatically generate auxiliary lemmas. There are other solvers such as Vampire [18] and Racer [22] which support reasoning about inductive definitions and are reported to outperform cvc5 in some categories of benchmarks. However, in general, cvc5 represents the state-of-the-art in inductive reasoning, which is used as the backend solver.

Hereafter, we adopt the following terminology for the LLM-aided lemma generation task. Outputs generated by LLMs are referred to as *conjectures* if they are in the correct format. Conjectures $\{L_i\}_{i \in I}$ are said *useful* if they can assist in proving P when used as premises, namely, $A \wedge \bigwedge_{i \in I} L_i \rightarrow P$ can be proved. A conjecture L is referred to as a *lemma* if it can be proved under the axiom A. Useful conjectures $\{L_i\}_{i \in I}$ are referred to as *auxiliary lemmas* when all of them can be proved under the axiom A, i.e., $A \rightarrow L_i$ can be proved for all $i \in I$. To establish the validity of $A \rightarrow P$, our main goal is to elicit LLMs to generate auxiliary lemmas.

3 Our Neuro-Symbolic Approach

3.1 Motivating Example

We first illustrate the challenges of naively querying LLMs for lemma generation without carefully designed prompts. Consider the task of verifying the property in Listing 1.1 using the following naive prompt:

```
You are an expert in constraint solving, inductive
    reasoning, and functional program verification.
Please generate auxiliary lemmas in SMTLIB2 format to
    add to the following file, which can help the solver
     verify the property.

  { Input SMTLIB2 file }
```

Listing 1.2. A naive prompt for lemma generation

Due to randomness and possible hallucination, LLM-generated conjectures may not be auxiliary lemmas. Based on our experiment, these conjectures can be categorized into the following three types:

- **Incorrect Conjectures.** When the outputs of LLMs are complete and syntactically correct, forming conjectures, they may be semantically incorrect. Namely, an LLM-generated conjecture L contradicts the axiom A, making the premise $A \wedge L$ false. In this case, while $A \wedge L \rightarrow P$ is vacuously true, the conjecture L does not hold under the axiom A and is useless. For instance, LLMs may generate the following conjecture for the motivating example:

$$\forall\, x \in \texttt{Nat}.\ \texttt{plus}(x, \texttt{zero}) = \texttt{zero}$$

  which contradicts the axiom that $\texttt{plus}(x, \texttt{zero}) = x$.
- **Correct but Useless Conjectures.** Even when an LLM-generated conjecture L is correct under the axiom A, i.e., $A \rightarrow L$ holds, it may still be useless for proving the property P. For instance, an LLM generates a conjecture L that is identical to the property P or a generic property that is completely irrelevant to the property P. For instance, LLMs may generate the following conjectures for the motivating example:

$$\forall\, x, y, z \in \texttt{Nat}.\ \texttt{plus}(\texttt{plus}(x, y), z) = \texttt{plus}(x, \texttt{plus}(y, z)),$$
$$\forall\, x, y \in \texttt{Nat}.\ \texttt{plus}(x, y) = \texttt{plus}(y, x), \qquad \forall\, x \in \texttt{Nat}.\ \texttt{mult}(x, \texttt{zero}) = \texttt{zero}).$$

  These conjectures, while valid, provide no assistance in proving P.
- **Useful but Remaining to be Proved Conjectures.** While LLMs can generate useful conjectures for proving the property P, these conjectures themselves should be proved from the axiom A, to qualify as auxiliary lemmas. In practice, proving useful conjectures themselves is non-trivial, e.g., some cannot be directly proved via SMT solving, leaving them as new proof obligations. For instance, for the motivating example, LLMs may generate the following useful conjecture:

$$\forall\, x, y \in \texttt{Nat}.\ \texttt{mult}(x, \texttt{succ}(y)) = \texttt{plus}(\texttt{mult}(x, y), x).$$

  This conjecture cannot be directly proved via SMT solving using cvc5, and thus requires further verification with additional auxiliary lemmas.

This example demonstrates that state-of-the-art LLMs have the potential to understand inductive reasoning and generate useful conjectures beyond existing fixed mechanisms and heuristic strategies. (Note that this example cannot be proved by the representative state-of-the-art solver cvc5; cf. the full version of the paper.) However, LLMs still lack clear reasoning strategies and may generate conjectures which are incorrect, useless, or themselves non-trivial to prove. Effectively leveraging LLMs requires carefully designed prompts to guide lemma generation and a systematic workflow to validate and integrate these conjectures, which motivates our approach.

3.2 Workflow

To bring the best of LLM-aided solving and traditional logic-based solving, we propose a neuro-symbolic approach, which synergistically integrates LLMs and SMT solvers to iteratively generate conjectures, check their validity and usefulness for assisting in proving the target property. After preprocessing the input file in SMTLIB2 format, we invoke the `ProveRun` function as the main workflow to prove the property.

Preprocessing. Given an instance of the inductive reasoning problem as an SMTLIB2 file, we parse the input file and organize it as three parts: datatype definitions, recursive function definitions, and the verification target, each of which is labeled with a comment (e.g., `; datatype definitions`, `; function definitions`, and `; proof goal`) to help LLMs' understanding and reasoning.

Algorithm 1: Main Workflow

```
max_depth ← 3, max_iter_number ← 3; // In default
// Input: labeled SMTLIB2 file P and current recursion depth d;
// Output: True if proved, False otherwise;
Function ProveRun(P, d):
    if initialCheck(P) then
        return True;
    if d ≥ max_depth then return False ;
    foreach prompt ∈ prompt_pool do
        for i ← 1 to max_iter_number do
            r, subgoals ← Prove(P, prompt);
            if r = True then
                all_success ← True;
                parallel for subgoal ∈ subgoals do
                    if ¬ProveRun(subgoal, d + 1) then
                        all_success ← False;
                        break;
                if all_success then
                    return True;
    return False;
```

The `ProveRun` Function. The main workflow is implemented in the `ProveRun` function in Algorithm 1. It takes a labeled SMTLIB2 file P and the current recursion depth d (initialized to 0) as input, attempts to prove P (for clarity, the property specified in P is also referred to as P). The recursive execution of `ProveRun` forms a proof tree whose depth and width are bounded by the configurable parameters max_depth and $max\_iter\_number \times |\texttt{prompt\_pool}|$, where max_depth bounds the depth of recursive calls, max_iter_number bounds the number of queries to LLMs for each prompt, and `prompt_pool` comprises the designated prompts.

In detail, `ProveRun` first checks whether P can be *directly* proved via SMT solving, without LLM-generated conjectures. If it cannot be proved and the recursion has not yet reached the depth bound *max_depth*, `ProveRun` attempts to prove P using designated prompts from the prompt pool through three key stages *query*, *filter* and *validate*.

Each prompt is queried up to *max_iter_number* times (three times in this work, considering the cost and effectiveness), at lines 6–16. In this work, we design two effective prompt strategies (cf. Section 3.3) to form the prompt pool, while other new prompt strategies can be easily extended into our workflow.

For each query of a prompt `prompt`, `ProveRun` invokes the function `Prove` (line 8), which attempts to prove P assisted by conjectures generated by LLMs using `prompt`. If P is proved by `Prove` using `prompt`, with LLM-generated conjectures $\{L_i\}_{i \in I}$, i.e., $A \wedge \bigwedge_{i \in I} L_i \rightarrow P$ is proved, then these conjectures $\{L_i\}_{i \in I}$ are regarded as sub-goals and are validated by recursively invoking `ProveRun` in parallel. We can complete the proof of P if all the sub-goals can be proved, i.e., $A \rightarrow L_i$ for $i \in I$ are proved.

The `Prove` Function. The `Prove` function (Algorithm 2) first queries the specified LLM using `prompt`, which will produce a set of conjectures C. The LLM-generated conjectures are iteratively checked by invoking `isFiltered` (line 4), which is designed to quickly identify incorrect and useless conjectures. If none of these conjectures can be filtered out, `Prove` invokes the `Verify` function (line 6) which checks whether these conjectures suffice to prove P (line 6), i.e., checking the unsatisfiability of $\{A, \neg P\} \cup C$. If this is the case (line 7), these conjectures are returned, which will be validated as sub-goals by recursively invoking `ProveRun`. Otherwise, `Prove` returns `False`.

Algorithm 2: The Prove Function

```
   // Input: labeled SMTLIB2 file P, current prompt prompt;
   // Output: (True,C) if P is proved with useful conjectures C;
              (False, ∅) if conjectures cannot assist in proving P;
1  Function Prove(P, prompt):
2      C ← LLMQuery(P, prompt);
3      foreach c ∈ C do
4          if isFiltered(c) then
5              return (False, ∅);
6      if Verify(P, C) then
7          return (True, C);
8      return (False, ∅);
```

The `isFiltered` Function. The filtering stage, implemented via `isFiltered`, is designed to quickly identify incorrect and useless conjectures. An LLM-generated conjecture c is filtered out by `isFiltered`, i.e., `isFiltered`(c) returns `True`, if (i) c has syntactic errors checked by the parser of an SMT solver (e.g., cvc5), or (ii) c is identical to the proof goal P compared by syntactic equivalence checking, or (iii) c is inconsistent with the axiom A, i.e., $A \wedge c$ is unsatisfiable, via SMT solving. We check the inconsistency of A and c instead of the invalidity of $A \rightarrow c$ (equivalently, the satisfiability of $A \wedge \neg c$) because: $A \wedge c$ usually contains only the universal quantifier $\forall$ while $A \wedge \neg c$ contains both the universal quantifier $\forall$ and existential quantifier $\exists$, thus $A \wedge c$ is easier to be checked; moreover the unsatisfiability of $A \wedge c$ entails the satisfiability of $A \wedge \neg c$, thus c cannot be a lemma under the axiom A.

The `Verify` Function. The validation stage implemented via `Verify` is used to check whether the LLM-generated conjectures C are *useful* for proving the goal. It is done by checking the unsatisfiability of $\{A, \neg P\} \cup C$, which is equivalent to checking $A \wedge \bigwedge_{c \in C} c \rightarrow P$. If the unsatisfiability of $\{A, \neg P\} \cup C$ can be proved by SMT solving, `Verify` returns `True`, otherwise `False`.

3.3 Prompt Strategies for Lemma Generation

To effectively elicit LLMs to generate useful lemmas, we design two prompt strategies, each of which consists of three parts: `Task Description` (explaining the SMTLIB2 format and the task), `Chain of Thoughts` (providing the reasoning strategy), and `Output Format` (specifying how to format the output).

Strategy 1: Equational Reasoning. This strategy guides LLMs to perform step-by-step, human-like equational reasoning. Concretely, for a proof goal of the form $\forall x \forall \vec{y}.\ f(x, \vec{y}) = g(x, \vec{y})$, where f is inductively defined over the variable x, the prompt instructs the LLM to: (i) identify the inductive definition of f from the labeled SMTLIB2 file; (ii) determine whether the base case requires auxiliary lemmas and, if it can be proved by the solver's built-in simple induction, proceed directly to the inductive case; (iii) for the inductive case $\forall x' \forall \vec{y}.\ f(t(x'), \vec{y}) = g(t(x'), \vec{y})$, where x' is the variable used in the inductive hypothesis (e.g., $\forall x' \forall y.\ \mathtt{mult}(\mathtt{succ}(x'), y) = \mathtt{mult}(y, \mathtt{succ}(x'))$), transform the left-hand side $f(t(x'), \vec{y})$ step-by-step using known premises (axioms and the inductive hypothesis); and (iv) when a step cannot be derived directly from known premises, generate it as a conjecture.

Example 1. To prove $\forall x, y \in \mathtt{Nat}.\ \mathtt{mult}(x, y) = \mathtt{mult}(y, x)$ (i.e., the property in Listing 1.1), using Strategy 1, an LLM may reason as:

```
consider the inductive case:
        mult(succ(x'), y)) =:| axiom of mult |:
      plus(mult(x', y), y) =:| inductive hypothesis |:
      plus(mult(y, x'), y) =:| unknown conjecture |:
        mult(y, succ(x'))
```

The last highlighted step cannot be directly derived from known premises, so the following conjecture is generated by the LLM:

$$\forall\ x, y \in \mathtt{Nat}.\ \mathtt{plus}(\mathtt{mult}(y, x), y) = \mathtt{mult}(y, \mathtt{succ}(x)).$$

Though this conjecture suffices to prove the property, it cannot be directly proved via SMT solving (e.g., cvc5 with 1200 s time limit). Thus, it will be regarded as a sub-goal on which Strategy 1 is applied again, finally resulting in the following useful conjecture:

$$\forall\ x, y \in \texttt{Nat}.\texttt{plus}(\texttt{plus}(\texttt{mult}(y, x), x), \texttt{succ}(y)) = \texttt{plus}(\texttt{plus}(\texttt{mult}(y, x), y), \texttt{succ}(x)). \tag{2}$$

Strategy 2: Term Rewriting and Generalization. Unlike Strategy 1, this strategy does not instruct LLMs to think step-by-step, but provides common ideas in lemma generation, encouraging them to generate lemmas more flexibly and simplify the proof goal P. Concretely, the prompt instructs the LLM to: (i) generate basic axioms; (ii) strengthen the conclusion, i.e., find a stronger lemma that is easier to prove; (iii) identify a common term that appears on both sides of the proof goal P and replace it with a fresh variable, leading to a simplified proof goal P'; (iv) if no common terms exist, attempt to rewrite terms in P using function definitions to have a common term; and (v) if necessary (e.g., P' is not enough to prove P), generate bridging lemmas between P' and P as conjectures such as $P' \rightarrow L$ and $L \rightarrow P$.

Example 2. Suppose the LLM fails to generate useful lemmas for proving the sub-goal in Eq. (2), via Strategy 1. Then, Strategy 2 is used to guide the LLM, which identifies the common term $\texttt{mult}(y, x)$, rewrites the sub-goal, and generates the following simplified sub-goal:

$$\forall\ t, x, y \in \texttt{Nat}.\ \texttt{plus}(\texttt{plus}(t, x), \texttt{succ}(y)) = \texttt{plus}(\texttt{plus}(t, y), \texttt{succ}(x)).$$

This simplified sub-goal can be directly proved via SMT solving, thus we can complete the overall proof of the property. A further example of Strategy 2 is given in the full version of the paper.

4 Evaluation

We implement our approach in a tool, named LLM4Ind. We evaluate LLM4Ind to answer the following research questions (**RQs**):

RQ1. How effective is our approach at solving constraints involving inductive definitions compared with state-of-the-art solvers?

RQ2. What is the contribution of our prompt and filtering designs to the overall performance?

RQ3. Is our approach robust over different LLM models and sampling temperatures that control creativity and diversity of the output?

Benchmarks. We collected four inductive reasoning benchmarks from prior work: StandardDT, StandardDTLIA, AutoproofBM, and IndBen, which are commonly used in literature [10,22,26,40,44].

- StandardDT [40] comprises 241 proof tasks in the ADT theory which was collected from test suites of IsaPlanner [11], CLAM [21] and HipSpec [7], and verification conditions in functional program verification in Leon [4]. It is often regarded as a standard benchmark for automated inductive reasoning.

- StandardDTLIA [10] contains 168 proof tasks which were derived from StandardDT by converting the `Nat` type and all the related operations to `Int` in the LIA theory, covering both ADT and LIA theories.
- AutoProofBM [44] comprises 141 proof tasks, which use more complex ADT and LIA definitions and are relatively more challenging to solve.
- IndBen [16] provides 3,396 proof tasks among which we select 156 representative tasks, one task per group, as the tasks in the same group define the same definitions and property (differing only in syntax). The 156 proof tasks cover common ADT theories such as `List`, `Nat` and `Tree`.

Overall, we collected 706 proof tasks as our subject.

Experimental Setup. To answer **RQ1**, we compare LLM4IND with cvc5, Vampire and Racer in terms of the number of solved proof tasks.

- cvc5 [2] is one of the leading SMT solvers. As its performance may vary with options and versions [26,44], we run cvc5 (with three option configurations) and cvc4 (with the default inductive reasoning setting) in parallel (cf. the full version of the paper for detailed options), which serves as the baseline for cvc5 and acts as the backend SMT solver in our workflow (for clarity, it is referred to directly as cvc5).
- Vampire [18,25] is a leading automated first-order logic prover based on superposition calculus. It supports a fragment of the SMTLIB2 inputs. We run Vampire under the portfolio mode, which runs multiple proving strategies in parallel.
- Racer [22] extends the CHC solver Spacer [24] to better support ADTs and RDFs. As it only supports inputs in CHCs (not general SMTLIB2) form, we only compare with it using its provided benchmarks.

To answer **RQ2**, we conduct an ablation study. To answer **RQ3**, we evaluate LLM4IND using four different LLMs: DeepSeek-v3.2 (DeepSeek), Qwen3-235B-instruct (Qwen), Gemini 2.5 flash (Gemini) and GPT-5 and vary the sampling temperature. Note that we use Qwen as the default LLM unless stated explicitly otherwise, as a trade-off between the cost and performance.[1]

All experiments were conducted on a machine with an Intel(R) Xeon(R) Platinum 8368Q CPU (76 cores, 2.60 GHz), 1024 GiB RAM and Ubuntu 22.04.5 LTS (1024 GiB RAM is the memory for evaluation instead of consumption). The solving of each proof task is limited to 1200 s (wall time) unless stated explicitly otherwise. For LLM4IND, we use both prompt strategies described in Sect. 3.3, each invoked for up to 3 iterations as in Algorithm 1 (line 6 and line 7). The backend solver is given 60 s for initial checking (Algorithm 1, line 3) and for each invocation of the `Verify` function to verify candidate conjectures (Algorithm 2, line 6). The filtering checking (Algorithm 2, line 4) is given 1 s per conjecture. At most 4 threads are used to verify sub-goals concurrently (Algorithm 1, line 11). We set the hyperparameter top_p = 0.9 and sampling temperature = 0.9 by

[1] The price per M input/output tokens: DeepSeek-v3.2 ($0.27/$0.40), Qwen3-235B ($0.072/$0.464), Gemini 2.5 flash ($0.30/$2.50), GPT-5 ($1.25/$10.00).

default according to [29]. For cvc5, we run 4 threads in parallel, one thread per option configuration. Other tools use their default thread configurations without additional parallelization settings.

4.1 RQ1. Effectiveness

The results are reported in Table 1, including the number of tasks solved by each tool with the time limits 1200 s and 360 s, as well as the average solving time of the solved tasks. Recall that Racer cannot directly verify StandardDT, AutoProofBM and IndBen (hence is marked by N/A). For reference, a full run of LLM4IND over all benchmarks consumed approximately 11.5M tokens using Qwen, corresponding to an estimated cost of about $4.

Overall, LLM4IND solved significantly more tasks than the baselines, regardless of the time limit. For instance, under 1200 s, LLM4IND solved 232 more tasks than cvc5 and 182 more tasks than Vampire. This confirms the effectiveness of our approach. On StandardDTLIA, LLM4IND also solved 94 more tasks than Racer. On IndBen, Vampire solved more tasks than LLM4IND, but we will see later in Table 2 that LLM4IND surpasses Vampire when more powerful LLMs (Gemini and GPT-5) are used instead of Qwen.

It is not surprising that LLM4IND's solving time is higher than the traditional logic-based approaches, because it recursively queries the LLM and invokes the SMT solver. Arguably, the time remains acceptable for verification tasks (on average about 100 s), especially when producing a correct result is more critical.

Table 1. Comparison with baseline SMT solvers

Benchmark	Total	LLM4IND		cvc5		Vampire		Racer	
		<1200 s	<360 s	<1200 s	<360 s	<1200 s	<360 s	<1200 s	<360 s
StandardDT	241	**212**	**200**	150	150	160	160	N/A	N/A
StandardDTLIA	168	**134**	**127**	76	76	58	56	40	40
AutoProofBM	141	**65**	**61**	34	34	3	3	N/A	N/A
IndBen	156	114	104	34	34	**122**	**118**	N/A	N/A
Total	706	**525**	**492**	293	293	343	337	40	40
Avg time (s)		97.30	58.66	**3.38**	**3.38**	19.39	9.26	6.09	6.09

We note that the recent tools AutoProof [44] and CCLemma [26] have reported promising results on lemma generation for inductive reasoning. However, these tools support neither SMTLIB2 nor CHC format inputs. As an *indirect* comparison using the results reported in [44], AutoProof solved 161 tasks while LLM4IND solved 180 tasks using their benchmarks.

4.2 RQ2. Ablation Study

To study the contribution of our prompt and filtering designs, we compare our prompt strategies with the naive prompt strategy (cf. Listing 1.2) and evaluate LLM4IND with and without the filter.

Table 2. Results of LLM4IND using different LLMs and naive prompt strategy

Benchmark	Qwen		DeepSeek		Gemini		GPT-5	
	LLM4IND	Naive	LLM4IND	Naive	LLM4IND	Naive	LLM4IND	Naive
StandardDT	**212**	160	**206**	153	**206**	155	**212**	180
StandardDTLIA	**134**	88	**131**	90	**119**	91	**144**	106
AutoProofBM	**65**	42	**63**	43	**68**	49	**67**	46
IndBen	**114**	89	**121**	79	**126**	76	**129**	118
Total	**525**	379	**521**	365	**514**	371	**552**	450

Prompt Design. The results are reported in Table 2 using four different LLMs: Qwen, DeepSeek, Gemini and GPT-5. Here, LLM4IND uses the prompt strategies presented in Sect. 3.3, while the naive one (cf. Listing 1.2) provides only a simple task environment description, requests lemma generation, and specifies output format. For fairness, the naive prompt is invoked up to 6 times, matching the others (2 prompts×3 times/prompt). LLM4IND with our prompt strategies significantly outperforms the naive one, across all LLM models and benchmarks.

Filtering Design. The results are reported in Table 3. Because solving outcomes depend heavily on the quality of LLM outputs, we run each setting (with and without the filter) three times (R1, R2 and R3) and report the average results (Avg). For a fair comparison, the average time is computed over all solved tasks. For per-task time comparison across all four benchmarks, we provide scatter plots in the full version of the paper, visualizing the solving time differences for each proof task.

In general, with the filter, LLM4IND solves more tasks (520.7 vs. 513.7 in total over three runs), indicating that the filtering design is effective. In many cases, without the filter (w.o. filter), incorrect or useless conjectures consume substantial solving time and can lead to time out. In contrast, with the filter (w. filter), some iterations are discarded early, saving time and allowing those instances to be solved successfully.

In detail, while the filter can reduce the solving time on the three benchmarks (i.e., StandardDTLIA, AutoProofBM and IndBen), StandardDT remains an exception. It may be because StandardDT contains relatively simple ADT definitions (primarily common types such as `List`, `Nat`, and `Tree`) for which LLMs can generate correct conjectures quickly, so that filtering provides little benefit but incurs additional overhead. In addition, the last row of Table 3 reports the total tokens (in millions) consumed across the six runs. Overall, enabling the filter results in lower token usage compared to the configuration without the filter, indicating improved cost efficiency.

Table 3. Results of LLM4IND with and without filter over three runs

Benchmark	W. filter				W.o. filter				Avg time (s)	
	R1	R2	R3	Avg	R1	R2	R3	Avg	W. filter	W.o. filter
StandardDT	212	211	207	**210**	204	203	209	205.3	83.91	**67.26**
StandardDTLIA	134	137	135	135.3	134	139	134	**135.7**	**97.53**	112.05
AutoProofBM	65	61	64	**63.3**	60	64	64	62.7	**99.10**	103.61
IndBen	114	111	111	**112**	113	106	111	110	**168.92**	178.64
Total	525	520	517	**520.7**	511	512	518	513.7	107.59	**107.37**
Tokens (M)	11.50	12.08	11.57	11.72	13.37	11.75	12.49	12.54	N/A	N/A

4.3 RQ3. Robustness of LLM4IND

To evaluate the robustness of LLM4IND across different LLMs, we consider DeepSeek, Gemini and GPT-5, besides Qwen. The experimental results are presented in Table 2, from which we observe that LLM4IND consistently improves cvc5, Vampire and Racer, regardless of the LLM model used. Moreover, to assess robustness under different samplings, we set the temperature of Qwen to 0.1, 0.5, 0.9 and 1.3, respectively. We ran three independent runs for each temperature to account for the randomness of LLM outputs. Table 4 reports the per-run counts. For each temperature, we also report the range (max−min over the three runs) and the standard deviation (std) of the total number of solved tasks to quantify variability. The range is at most 9 (e.g., 518–527 for $T = 1.3$), and the std remains below 5 in all cases, indicating low variance across runs. We observe that as the temperature increases, the variability tends to increase slightly (e.g., the std grows from 3.2 at $T = 0.1$ to 4.7 at $T = 1.3$, and the range increases from 6 to 9), which aligns with the expectation that higher temperatures introduce greater randomness into LLM sampling. In summary, the number of solved tasks varies only slightly across runs and temperatures. These results suggest that the inherent randomness of LLMs and the sampling temperature have a minor impact on LLM4IND's performance, demonstrating its robustness over different runs and temperatures.

We also replace cvc5 in our workflow with Vampire (denoted as LLM4IND-V). Table 5 reports the results. Overall, LLM4IND-V solved more proof tasks than each of the three baselines. Compared with Vampire, LLM4IND-V shows improvements across all four benchmarks, demonstrating that our approach can enhance Vampire's performance and that LLM4IND remains robust across different backend solvers. However, compared to cvc5, LLM4IND-V underperforms in two cases. In particular, for StandardDTLIA mixed with ADT and

Table 4. Results of LLM4IND with different sampling temperatures

Benchmark	T = 0.1			T = 0.5			T = 0.9			T = 1.3		
	R1	R2	R3	R1	R2	R3	R1	R2	R3	R1	R2	R3
StandardDT	211	206	210	207	209	207	**212**	211	207	208	209	210
StandardDTLIA	135	139	135	136	138	138	134	137	135	**140**	136	135
AutoProof	60	62	62	61	62	61	**65**	61	64	60	60	**65**
IndBen	114	108	114	110	107	113	114	111	111	**119**	113	112
Total	520	515	521	514	516	519	525	520	517	**527**	518	522
Range Std	6 3.2			5 2.6			8 4.2			9 4.7		

Table 5. Comparison with baseline SMT solvers (Vampire backend)

Benchmark	Total	LLM4IND-V		cvc5		Vampire		Racer	
		<1200 s	<360 s	<1200 s	<360 s	<1200 s	<360 s	<1200 s	<360 s
StandardDT	241	**214**	**206**	150	150	160	160	N/A	N/A
StandardDTLIA	168	**80**	73	76	**76**	58	56	40	40
AutoProofBM	141	19	16	**34**	**34**	3	3	N/A	N/A
IndBen	156	**140**	**140**	34	34	122	118	N/A	N/A
Total	706	**453**	**435**	293	293	343	337	40	40
Avg time (s)		59.89	38.74	**3.38**	**3.38**	19.39	9.26	6.09	6.09

LIA theories, Vampire has limited support, so LLM4IND-V solved fewer tasks than cvc5 under the 360 s limit. For AutoProofBM, 119 (out of 141) proof tasks involve ADT symbols defined in the SMTLIB2 standard. They are supported by cvc5, but not by Vampire. As a result, although our approach improves Vampire's performance, LLM4IND-V still solved fewer tasks than cvc5 on this benchmark.

5 Related Work

Inductive reasoning is supported, with different degrees of human intervention, by many theorem provers. Existing works address automated inductive reasoning mainly from two perspectives. (1) Incorporating *induction schemas* into decision procedures that originally do not support induction. For instance, [40] enabled SMT solvers to automatically handle inductive definitions by integrating induction schemas into quantifier elimination, while works [8,13,17,18,20,38] extended the superposition calculus with induction rules. (2) Generating auxiliary lemmas to enhance inductive reasoning in provers that support induction, such as automated theorem provers for functional languages, including IsaPlanner [11], ACL2 [23], Zeno [43] and HipSpec [7]. Below, we briefly review representative approaches on automated lemma generation.

Theory Exploration. The main idea is to construct a pool of candidate lemmas from available symbols, typically by term enumeration or instantiating formula templates. The candidates are then filtered using heuristics. HipSpec [7] is a representative tool in this category, which uses counterexample and congruence closure to efficiently filter out invalid lemmas. CCLemma [26] borrows this pruning idea and proposes e-graph guided lemma discovery to make theory exploration more goal-directed. In addition, cvc5 [40] also adopts a similar idea: it enumerates conjectures in increasing size, and filters them using techniques based on active conjectures, equivalence-class construction, and ground-facts reasoning. ADTInd [53] employs syntax-guided templates to enumerate lemmas and implements a generalization mechanism for refining them.

Generalization. Generalization techniques identify common sub-terms in the proof goal and replace them with fresh variables to find auxiliary lemmas that are simpler and can generalize the target property. The main heuristic is to select the common sub-term for replacement by introducing new variables in positions where induction can potentially be applied. This approach has been adopted by modern theorem provers such as ACL2 [23]. Zeno [43] applies the same common sub-term technique but combines it with counterexample searching to avoid over-generalization. Vampire [16,25] introduces an *induction with generalization* inference rule within the superposition calculus framework. This rule can handle properties with multiple occurrences of the same induction term, and instantiates induction axioms with logically stronger variants of the property being proved. AutoProof [44] differs from generalization-based approaches, but can be regarded as a goal-oriented method. It transforms the goal into *induction-friendly forms* that guarantee effective use of inductive hypotheses. Moreover, it synthesizes lemmas as equations where one side matches a term in the goal, enabling systematic rewriting in proof. This directed lemma synthesis avoids enumerating useless lemmas and ensures progress toward provable sub-goals.

CHC-Based Methods. These approaches treat inductive definitions as transition systems and solve the problem by synthesizing inductive invariants. Several CHC solvers can handle ADTs, but reasoning about RDFs remains challenging due to the undecidability of the underlying logic. For instance, VeriMAP [9,10] transforms CHCs with ADTs to CHCs over basic types, but the transformation is unsound for UNSAT answers. Racer [22] addresses this challenge by approximating RDFs abstractions. It compiles RDFs to CHCs while preserving satisfiability and replaces RDFs with finite unfoldings parameterized by depth k. It implements an IC3-style algorithm, enabling automatic learning of inductive invariants over ADTs and RDFs.

Neuro-Symbolic Methods. These approaches largely follow a generate-then-verify paradigm, in which neural models propose candidate lemmas, and symbolic tools are used to validate or refute them. Several works apply this idea in deductive verification or ITPs, where LLMs assist users by proposing lemmas that are subsequently checked by a verifier or proof assistant (e.g., Dafny [41] and Verus [51], and ITPs such as Lean4 [47] and Isabelle [1]). These approaches primarily aim to enhance user productivity or guide interactive proof search, rather than enable fully automated reasoning.

Recent work [35] studies LLM-assisted invariant generation for hardware model checking, where LLMs propose candidate inductive invariants over hardware transition systems. Similarly, [14] learns induction predicates for inductive and arithmetic problems using traditional neural models. It focuses on conjectures over inductive definitions encoded as SMT formulas in integer theory, and is not designed as a general LLM-based framework. AquaForte [30] applies LLMs to generate candidate instantiations for quantified uninterpreted functions in SMT solving, primarily targeting satisfiability checking. In contrast, our work focuses on automated proof generation by establishing the unsatisfiability of the negation of the target property.

6 Conclusion

In this work, we have proposed a neuro-symbolic approach that leverages LLMs to generate auxiliary lemmas for inductive reasoning. Through a three-stage workflow, we synergistically integrate LLMs and constraint solvers, design prompt strategies to guide LLMs in generating high-quality candidate lemmas, and filter and validate them systematically. The experimental results on diverse benchmarks demonstrate the effectiveness of our approach.

In the future, extending the approach to handle more general proof tasks and investigating other prompting and inference-time techniques, or agent-based methods, would be promising.

Acknowledgments. This work is supported by the Strategic Priority Research Program of the Chinese Academy of Sciences, Grant No. XDA0320101. T. Chen is partially supported by overseas grants from the State Key Laboratory of Novel Software Technology, Nanjing University (KFKT2023A04, KFKT2025A05).

References

1. Alhessi, Y., et al.: Lemmanaid: neuro-symbolic lemma conjecturing (2025). https://arxiv.org/abs/2504.04942
2. Barbosa, H., et al.: cvc5: a versatile and industrial-strength SMT solver. In: Proceedings of the 28th International Conference on Tools and Algorithms for the Construction and Analysis of Systems (TACAS), Part I, vol. 13243, pp. 415–442 (2022)
3. Bertot, Y., Castéran, P.: Interactive Theorem Proving and Program Development: Coq'Art: The Calculus of Inductive Constructions. Springer, Cham (2013)
4. Blanc, R., Kuncak, V., Kneuss, E., Suter, P.: An overview of the Leon verification system: verification by translation to recursive functions. In: Proceedings of the 4th Workshop on Scala, SCALA@ECOOP, pp. 1:1–1:10. ACM (2013). https://doi.org/10.1145/2489837.2489838

5. Cao, W., et al.: Clause2inv: a generate-combine-check framework for loop invariant inference. Proc. ACM Softw. Eng. **2**(ISSTA), 1009–1030 (2025). https://doi.org/10.1145/3728920
6. Chuharski, J., Collins, E.R., Meringolo, M.: Mining math conjectures from LLMs: a pruning approach (2024). https://arxiv.org/abs/2412.16177
7. Claessen, K., Johansson, M., Rosén, D., Smallbone, N.: Automating inductive proofs using theory exploration. In: Bonacina, M.P. (ed.) CADE 2013. LNCS (LNAI), vol. 7898, pp. 392–406. Springer, Heidelberg (2013). https://doi.org/10.1007/978-3-642-38574-2_27
8. Cruanes, S.: Superposition with structural induction. In: Dixon, C., Finger, M. (eds.) FroCoS 2017. LNCS (LNAI), vol. 10483, pp. 172–188. Springer, Cham (2017). https://doi.org/10.1007/978-3-319-66167-4_10
9. De Angelis, E., Fioravanti, F., Pettorossi, A., Proietti, M.: Solving horn clauses on inductive data types without induction. Theory Pract. Log. Program. **18**(3–4), 452–469 (2018). https://doi.org/10.1017/S1471068418000157
10. De Angelis, E., Fioravanti, F., Pettorossi, A., Proietti, M.: Removing algebraic data types from constrained horn clauses using difference predicates. In: Peltier, N., Sofronie-Stokkermans, V. (eds.) IJCAR 2020. LNCS (LNAI), vol. 12166, pp. 83–102. Springer, Cham (2020). https://doi.org/10.1007/978-3-030-51074-9_6
11. Dixon, L., Fleuriot, J.: Higher order rippling in IsaPlanner. In: Slind, K., Bunker, A., Gopalakrishnan, G. (eds.) TPHOLs 2004. LNCS, vol. 3223, pp. 83–98. Springer, Heidelberg (2004). https://doi.org/10.1007/978-3-540-30142-4_7
12. Dutertre, B.: Yices 2.2. In: Biere, A., Bloem, R. (eds.) CAV 2014. LNCS, vol. 8559, pp. 737–744. Springer, Cham (2014). https://doi.org/10.1007/978-3-319-08867-9_49
13. Echenim, M., Peltier, N.: Combining induction and saturation-based theorem proving. J. Autom. Reason. **64**(2), 253–294 (2020). https://doi.org/10.1007/S10817-019-09519-X
14. Gauthier, T., Urban, J.: Learning conjecturing from scratch (2025). https://arxiv.org/abs/2503.01389
15. Guo, D., et al.: Deepseek-coder: when the large language model meets programming–the rise of code intelligence. arXiv preprint arXiv:2401.14196 (2024)
16. Hajdú, M., Hozzová, P., Kovács, L., Schoisswohl, J., Voronkov, A.: Induction with generalization in superposition reasoning. In: Benzmüller, C., Miller, B. (eds.) CICM 2020. LNCS (LNAI), vol. 12236, pp. 123–137. Springer, Cham (2020). https://doi.org/10.1007/978-3-030-53518-6_8
17. Hajdú, M., Hozzová, P., Kovács, L., Voronkov, A.: Induction with recursive definitions in superposition. In: Proceedings of the Conference on Formal Methods in Computer Aided Design (FMCAD), pp. 1–10. IEEE (2021). https://doi.org/10.34727/2021/ISBN.978-3-85448-046-4_34
18. Hajdú, M., Kovács, L., Rawson, M., Voronkov, A.: The vampire approach to induction (short paper). In: Proceedings of the Workshop on Practical Aspects of Automated Reasoning (IJCAR), vol. 3201 (2022)
19. Hamza, J., Voirol, N., Kuncak, V.: System FR: formalized foundations for the stainless verifier. Proc. ACM Program. Lang. **3**(OOPSLA), 166:1–166:30 (2019). https://doi.org/10.1145/3360592
20. Hozzová, P., Kovács, L., Voronkov, A.: Integer induction in saturation. In: Platzer, A., Sutcliffe, G. (eds.) CADE 2021. LNCS (LNAI), vol. 12699, pp. 361–377. Springer, Cham (2021). https://doi.org/10.1007/978-3-030-79876-5_21
21. Ireland, A.: Productive use of failure in inductive proof. J. Autom. Reason. **16**(1–2), 79–111 (1996). https://doi.org/10.1007/BF00244460

22. K., H.G.V., Shoham, S., Gurfinkel, A.: Solving constrained horn clauses modulo algebraic data types and recursive functions. Proc. ACM Program. Lang. **6**(POPL), 1–29 (2022). https://doi.org/10.1145/3498722
23. Kaufmann, M., Manolios, P., Moore, J.S.: Computer-Aided Reasoning: ACL2 Case Studies, vol. 4. Springer, Cham (2013)
24. Komuravelli, A., Gurfinkel, A., Chaki, S.: SMT-based model checking for recursive programs. Formal Methods Syst. Des. **48**(3), 175–205 (2016). https://doi.org/10.1007/s10703-016-0249-4
25. Kovács, L., Voronkov, A.: First-order theorem proving and VAMPIRE. In: Sharygina, N., Veith, H. (eds.) CAV 2013. LNCS, vol. 8044, pp. 1–35. Springer, Heidelberg (2013). https://doi.org/10.1007/978-3-642-39799-8_1
26. Kurashige, C., et al.: CCLemma: E-graph guided lemma discovery for inductive equational proofs. Proc. ACM Program. Lang. **8**(ICFP), 818–844 (2024)
27. Lattuada, A.: Verus: verifying rust programs using linear ghost types. Proc. ACM Program. Lang. **7**(OOPSLA1), 286–315 (2023). https://doi.org/10.1145/3586037
28. Leino, K.R.M.: Dafny: an automatic program verifier for functional correctness. In: Clarke, E.M., Voronkov, A. (eds.) LPAR 2010. LNCS (LNAI), vol. 6355, pp. 348–370. Springer, Heidelberg (2010). https://doi.org/10.1007/978-3-642-17511-4_20
29. Li, L., Sleem, L., Gentile, N., Nichil, G., State, R.: Exploring the impact of temperature on large language models: hot or cold? CoRR **abs/2506.07295** (2025). https://doi.org/10.48550/ARXIV.2506.07295
30. Lv, K., Dong, Y., Han, R., Jia, F., Ma, F., Zhang, J.: LLM-guided quantified SMT solving over uninterpreted functions (2026). https://arxiv.org/abs/2601.04675
31. Ma, L., Liu, S., Li, Y., Xie, X., Bu, L.: Specgen: automated generation of formal program specifications via large language models. In: Proceedings of the IEEE/ACM 47th International Conference on Software Engineering (ICSE), pp. 16–28 (2025). https://doi.org/10.1109/ICSE55347.2025.00129
32. Moura, L., Ullrich, S.: The Lean 4 theorem prover and programming language. In: Platzer, A., Sutcliffe, G. (eds.) CADE 2021. LNCS (LNAI), vol. 12699, pp. 625–635. Springer, Cham (2021). https://doi.org/10.1007/978-3-030-79876-5_37
33. de Moura, L., Bjørner, N.: Z3: an efficient SMT solver. In: Ramakrishnan, C.R., Rehof, J. (eds.) TACAS 2008. LNCS, vol. 4963, pp. 337–340. Springer, Heidelberg (2008). https://doi.org/10.1007/978-3-540-78800-3_24
34. Paulson, L.C.: Isabelle: A Generic Theorem Prover. Springer, Cham (1994)
35. Peled, R., Kroening, D., Tautschnig, M., Vizel, Y.: Large lemma miners: can LLMs do induction proofs for hardware? (2025). https://arxiv.org/abs/2511.02521
36. Pirzada, M.A.A., Reger, G., Bhayat, A., Cordeiro, L.C.: LLM-generated invariants for bounded model checking without loop unrolling. In: Filkov, V., Ray, B., Zhou, M. (eds.) Proceedings of the 39th IEEE/ACM International Conference on Automated Software Engineering, ASE 2024, Sacramento, CA, USA, October 27 - November 1, 2024, pp. 1395–1407. ACM (2024). https://doi.org/10.1145/3691620.3695512
37. Reger, G., Suda, M., Voronkov, A.: Instantiation and pretending to be an SMT solver with vampire. In: Brain, M., Hadarean, L. (eds.) Proceedings of the 15th International Workshop on Satisfiability Modulo Theories affiliated with the International Conference on Computer-Aided Verification (CAV 2017), Heidelberg, Germany, July 22 - 23, 2017. CEUR Workshop Proceedings, vol. 1889, pp. 63–75. CEUR-WS.org (2017). https://ceur-ws.org/Vol-1889/paper6.pdf

38. Reger, G., Voronkov, A.: Induction in saturation-based proof search. In: Fontaine, P. (ed.) CADE 2019. LNCS (LNAI), vol. 11716, pp. 477–494. Springer, Cham (2019). https://doi.org/10.1007/978-3-030-29436-6_28
39. Ren, Z.Z., et al.: Deepseek-prover-v2: advancing formal mathematical reasoning via reinforcement learning for subgoal decomposition (2025). https://arxiv.org/abs/2504.21801
40. Reynolds, A., Kuncak, V.: Induction for SMT solvers. In: D'Souza, D., Lal, A., Larsen, K.G. (eds.) VMCAI 2015. LNCS, vol. 8931, pp. 80–98. Springer, Heidelberg (2015). https://doi.org/10.1007/978-3-662-46081-8_5
41. Silva, Á., Mendes, A., Ferreira, J.F.: Leveraging large language models to boost Dafny's developers productivity. In: Proceedings of the 2024 IEEE/ACM 12th International Conference on Formal Methods in Software Engineering (FormaliSE), pp. 138–142 (2024)
42. Singher, E., Itzhaky, S.: Theory exploration powered by deductive synthesis. In: Silva, A., Leino, K.R.M. (eds.) CAV 2021. LNCS, vol. 12760, pp. 125–148. Springer, Cham (2021). https://doi.org/10.1007/978-3-030-81688-9_6
43. Sonnex, W., Drossopoulou, S., Eisenbach, S.: Zeno: an automated prover for properties of recursive data structures. In: Flanagan, C., König, B. (eds.) TACAS 2012. LNCS, vol. 7214, pp. 407–421. Springer, Heidelberg (2012). https://doi.org/10.1007/978-3-642-28756-5_28
44. Sun, Y., Ji, R., Fang, J., Jiang, X., Chen, M., Xiong, Y.: Proving functional program equivalence via directed lemma synthesis. In: Proceedings of the 26th International Symposium on Formal Methods (FM), Part I, vol. 14933, pp. 538–557. Springer, Cham (2024). https://doi.org/10.1007/978-3-031-71162-6_28
45. Sun, Z., Du, X., Song, F., Wang, S., Li, L.: When neural code completion models size up the situation: attaining cheaper and faster completion through dynamic model inference. In: Proceedings of the IEEE/ACM 46th International Conference on Software Engineering (ICSE), pp. 1–12 (2024)
46. Sun, Z., et al.: Don't complete it! preventing unhelpful code completion for productive and sustainable neural code completion systems. ACM Trans. Softw. Eng. Methodol. **34**(1), 1–22 (2025)
47. Varambally, S., Voice, T., Sun, Y., Chen, Z., Yu, R., Ye, K.: Hilbert: recursively building formal proofs with informal reasoning (2025). https://arxiv.org/abs/2509.22819
48. Wang, H., et al.: Kimina-prover preview: towards large formal reasoning models with reinforcement learning (2025). https://arxiv.org/abs/2504.11354
49. Wen, C., et al.: Enchanting program specification synthesis by large language models using static analysis and program verification. In: Proceedings of the International Conference on Computer Aided Verification (CAV), pp. 302–328. Springer, Cham (2024). https://doi.org/10.1007/978-3-031-65630-9_16
50. Wu, G., Cao, W., Yao, Y., Wei, H., Chen, T., Ma, X.: LLM meets bounded model checking: Neuro-symbolic loop invariant inference. In: Filkov, V., Ray, B., Zhou, M. (eds.) Proceedings of the 39th IEEE/ACM International Conference on Automated Software Engineering, ASE 2024, Sacramento, CA, USA, October 27 - November 1, 2024, pp. 406–417. ACM (2024). https://doi.org/10.1145/3691620.3695014
51. Yang, C., et al.: Autoverus: automated proof generation for rust code. Proc. ACM Program. Lang. **9**(OOPSLA2), 3454–3482 (2025). https://doi.org/10.1145/3763174

52. Yang, G., Zhou, Y., Chen, X., Zhang, X., Zhuo, T.Y., Chen, T.: Chain-of-thought in neural code generation: From and for lightweight language models. IEEE Trans. Softw. Eng. **50**(9), 2437–2457 (2024)
53. Yang, W., Fedyukovich, G., Gupta, A.: Lemma synthesis for automating induction over algebraic data types. In: Schiex, T., de Givry, S. (eds.) CP 2019. LNCS, vol. 11802, pp. 600–617. Springer, Cham (2019). https://doi.org/10.1007/978-3-030-30048-7_35

Fast SMT-Based Fault Tolerance Verification for Wide Area Networks

Ning Kang, Peng Zhang(✉), Hao Li, and Jianyuan Zhang

Xi'an Jiaotong University, Xi'an, China
kangning@stu.xjtu.edu.cn, p-zhang@xjtu.edu.cn, hao.li@xjtu.edu.cn,
jyzhang0281@stu.xjtu.edu.cn

Abstract. Configurations of routing protocols in wide area networks (WANs) are highly sophisticated and prone to bugs, leading to severe network outages and security breaches. SMT-based network verification can assist operators in checking the configurations, but it still faces scalability challenges when reasoning about failures: to check whether a property holds when no more than k links fail, a verifier needs to explore a tremendous space of failure scenarios. To this end, this paper proposes *VeriBoost*, a method that can leverage the topology features of WANs to reduce the space of failure scenarios, thereby improving the scalability of SMT-based verification on WANs. *VeriBoost* achieves the reduction by pruning links that are irrelevant to a property, and compressing multiple links whose failures have an equivalent impact on the property. Experiments on real WAN topologies show that it speeds up SMT-based verification by 2–47×.

Keywords: SMT · Verification · Wide Area Network · Failures

1 Introduction

Configuring a modern network is difficult and prone to bugs, which can lead to severe network outages and security breaches [1,4,7,23,27,33,37]. These bugs are often latent: misconfigurations are hidden in tens of thousands of configuration lines, and a property violation, e.g., blackhole, is only triggered when some links or nodes fail [4]. To prevent this, SMT-based verifiers [3–5,9,24,30,32] are proposed. They can prove or disprove whether a given property holds when no more than k links or nodes fail simultaneously.

However, scalability is still one of the most important challenges faced by SMT-based verifiers, due to their implicit enumerative manner. To check whether a property holds when no more than k links fail, verifiers need to consider all $\sum_{i=0}^{k} C_l^i$ possible link failure combinations, where l is the number of links. Here, a property holds for failure tolerance k if it is satisfied under any combination of up to k link failures. Such a large SMT solving space significantly limits the scalability of verifiers. Our experiments show that to verify 300 properties with $k = 1, 2, 3$ on a 158-node WAN, SMT-based verifiers take 12 h (see Sect. 5.3).

A. Sampaio and M. Stoelinga (Eds.): FM 2026, LNCS 16557, pp. 133–153, 2026.
https://doi.org/10.1007/978-3-032-26220-2_7

Some previous efforts tried to enhance the scalability of SMT-based verifiers. However, we observe they achieve limited improvement on WANs. For example, Bonsai [32] and Origami [16] exploit the symmetric structure of data center networks (DCNs) to compress the network state, whereas *WAN topologies lack such symmetry.* BiNode [26] assumes that the network configurations follow the Gao-Rexford principle [15], and Trailblazer [21] only considers the failures of links on packet forwarding paths. These methods are *limited by the complex routing policies of WANs.* Moreover, NetSMT [12] is ineffective when the property holds, i.e., when the SMT formula has no solution.

To this end, we ask "Is it possible to reduce the failure space without assuming symmetric topology structures or special routing policies?" Our answer is affirmative based on the following two observations.

1. *Failures of some links are irrelevant to the property of interest.* For example, a link that appears only in loop paths can never be used for the best routes, because there always exists a non-loop (i.e., *simple*) path [35] with strictly higher preference. Generally, we prove that links that do not appear on any simple path are *irrelevant* to the property and can be pruned (Theorem 1).
2. *Failures of some links have an equivalent impact on the property of interest.* For example, if the best route initially propagates along $path_1$, and the failure of either of two links A and B causes it to switch to $path_2$, then these two link failures have the same impact on route propagation of the property. Generally, we show that such an equivalence of links is transitive, and a group of *equivalent* links can be compressed (Theorem 3).

When realizing the SMT solving space reduction approach, we are faced with the problem: *how to efficiently identify irrelevant or equivalent links, without loss of accuracy?* First, to identify irrelevant links, a straightforward way is to find all *simple paths*, and exclude all links that appear on any of those *simple paths.* However, finding all *simple paths* on a graph is shown to be NP-hard [35]. For example, on a WAN topology containing 158 routers (189 links), computing *simple paths* for a pair of nodes using depth-first search takes more than 4 h. Second, to identify equivalent links, we have to analyze the impact of each single link failure, and check whether any pair of links have the same impact. This requires extra verification (by solving an SMT formula), clearly an overkill.

In this paper, we introduce *VeriBoost*, a method that can efficiently reduce the failure space for SMT solvers, and thereby improve the scalability of network verification on WANs. First, *VeriBoost* leverages *point biconnected components* [31] to transform the topology into a concise tree representation (termed *block-cut tree* or *block-point tree* [17]), where each node represents a set of nodes forming a *point biconnected component.* We show that by traversing the tree we can collect all links on all *simple paths* in polynomial time. Second, instead of identifying all equivalent links which requires verification, *VeriBoost* tries to find a subset of equivalent links, solely based on graph features. For example, we show that if a node has a degree of 2, then failing the adjacent links will have the same impact on routing propagation. Based on this fact, *VeriBoost* can already identify many links that are equivalent based on our datasets. Specifically, both methods achieve a 78% reduction in the number of links (see Sect. 5.1), and their computation time is within 700 μs for real WAN topology (see Appendix A [18]).

In summary, our contribution is three-fold:

- We propose *VeriBoost*, a new method that improves the scalability of SMT-based verifiers on WANs, by reducing the SMT solving space.
- We formally prove that the SMT solving space reduction preserves verification correctness.
- We apply *VeriBoost* to SMT-based verification method [8] and show that it achieves a speedup of 2–47× on real WAN topologies, while the state-of-the-art method [12] achieves only a 2–3× speedup.

2 Motivation

2.1 Background

Network Misconfigurations. Network configurations are complex and prone to bugs. Worse still, many bugs in network configurations are latent, and can only be triggered when some nodes or links fail. For example, consider a network where a packet is normally forwarded along $path_1$. If a link on $path_1$ fails, the packet is rerouted to $path_2$. However, if a node along $path_2$ filters traffic and does not permit this packet, it will be dropped, creating a blackhole [4]. To prevent these latent bugs, researchers have developed many SMT-based verifiers that can comprehensively reason about the impact of arbitrary node/link failures. Specifically, they check whether a given property always holds when no more than k nodes/links fail. We refer to such verification as "fault tolerance" or "k-failure" verification, and use *failure scenario* to denote which links are down and which are up (see Definition 1).

SMT-Based Fault Tolerance Verification Methods. model the verification problem as a set of SMT formulas $N \wedge \neg P$, where N encodes the behaviors of the network under a given configuration, and P encodes the property that the network should satisfy. [3–5,24,30,32]. To reason about link failures, the SMT formula becomes $N \wedge \neg P \wedge K$, with K encoding the arbitrary $\leq k$ link failures: $\sum_{l \in L} failed_l \leq k$, where $l \in L$ denotes a link, and $failed_l$ is a pseudo-boolean variable representing whether link l fails or not, with $\sum_{l \in L} failed_l$ representing the numerical sum of the variables $failed_l$, and L is the set of all links in the network. Then, the verifiers try to find a solution satisfying the SMT constraints with off-the-shelf solvers like Z3 [9]. If there is such a solution, it indicates that the property is violated under some link failures; otherwise, the property holds when no more than k links fail. Although SMT-based verifiers do not explicitly enumerate the *failure scenarios*, they still involve $O(\sum_{i=0}^{k} C_{|L|}^{i})$ iterations.

2.2 Related Work

SMT-based verifiers offer flexibility for verifying a wide range of properties, but their scalability is limited, motivating many studies to accelerate them [5,12,16, 21,26]. However, these methods are ineffective for WANs, as summarized below.

Reducing Encoded Variables via Topology Symmetry. Data centers use topologies like multi-stage Clos networks [2], which are regular: symmetric with multiple layers (core, spine, and leaf), and multiple nodes have primary-backup relationships. Based on this regularity, some tools compress the large topology into a smaller one (with many fewer nodes and links) for speedup. For example, Origami [16] and Bonsai [5] compress a primary node and its backup nodes into a single one, since the routing behavior of primary and backup nodes are quite similar. However, such regularity does not hold for WANs. For example, the real WAN topologies from Topology Zoo [20] have no clear central devices or hierarchical structures, because the deployment of devices and links is closely related to geographic locations that are highly random.

Adding Constraints via Simplified Routing Policy Assumptions. Complex routing policies are widely used to control route WANs connecting thousands of Autonomous Systems (ASes) operated by different institutions, such as Internet Service Providers, companies, and universities, through the complex policies of the Border Gateway Protocol (BGP) protocol [15]. Some methods tried to improve the scalability, with simplified assumptions on the routing policies, making them ineffective or raising false positives. For example, BiNode [26] assumes the network follows the Gao-Rexford principle (e.g., an AS prefers routes from customers over those from peers and providers) [15], so as to constrain the search space. However, it is hard to ensure Gao-Rexford principles on WANs, without verification. Trailblazer [21] only considers the failures of links appearing on packet forwarding paths. This implicitly assumes that packets and routes for the same prefix traverse the same path, which does not hold on WANs [19].

Guiding the Solving Process Through Variable Ordering. NetSMT [12] guides the order of SMT solving to quickly find a solution, i.e., a counterexample that violates a property. However, when the property holds, i.e., the SMT problem does not have a solution, the SMT solver still has to explore the whole SMT solving space, making NetSMT ineffective.

Some other methods accelerate SMT-based verification by sacrificing the support for fault tolerance, and are thus not applicable to our target problem. For instance, the modular approaches LightYear [30], Timepiece [3], and Kirigami [32] require manual network partitioning and are thus applicable only to a single *failure scenario*. Moreover, ACORN [24] ignores path information in its encoding and cannot verify properties such as waypointing (i.e., ensuring that a path passes through a specific node), also restricting it to a single *failure scenario*.

3 Overview

3.1 Basic Idea

To speedup fault tolerance verification, our basic idea is to reduce the uncertainty of link state prior to SMT solving, thereby shrinking the failure space for SMT solvers. Specifically, for a given property to be verified, suppose that the state of some links are already known, including those that are "down" ($L_{\mathcal{D}}$) and those that are "up" ($L_{\mathcal{U}}$). We then treat the remaining links as symbolic links $L_{\mathcal{S}} = L \setminus (L_{\mathcal{U}} \cup L_{\mathcal{D}})$, whose status is unknown (either up or down). This allows SMT-based verifiers to encode only the links in $L_{\mathcal{S}}$ as boolean variables, rather than all links, significantly reducing the solving space. Recall that SMT-based verifiers need to check $N \wedge \neg P \wedge K$, where $K \triangleq \sum_{l \in L} failed_l \leq k$. *VeriBoost* transforms K into the following constraint, enabling more efficient verification.

$$K \triangleq \bigwedge_{l \in L_{\mathcal{U}}} failed_l = 0 \bigwedge_{l \in L_{\mathcal{D}}} failed_l = 1 \wedge \sum_{l \in L_{\mathcal{S}}} failed_l \leq k \quad (1)$$

To this end, we present two observations that allow us to identify these "down" and "up" links without assuming any regular topology structure or specific routing policy, enabling applicability to WANs.

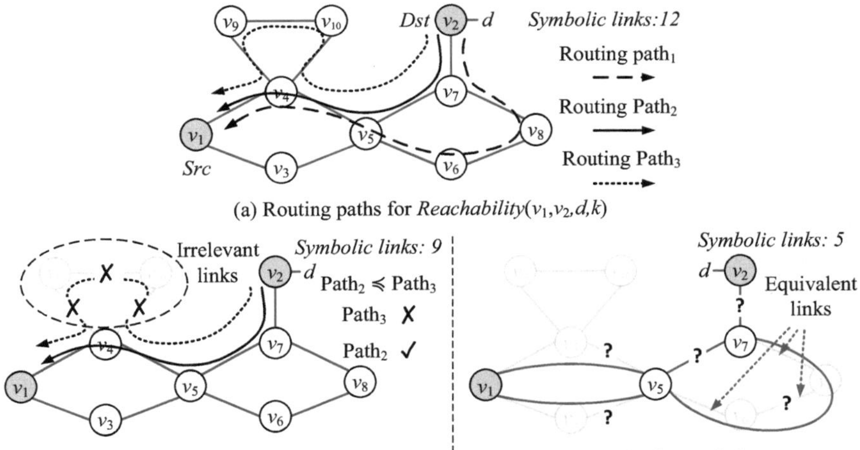

Fig. 1. An example of a BGP network with progressively applied *Link Pruning* and *Link Compression*. Each router has a different AS number and runs BGP to exchange routes. The property to be verified is reachability, that packets from source v_1 can reach destination d on v_2, even if no more than k links fail. Note that the routing path and forwarding path are in opposite directions. Our methods can be applied to any k, hence no specific tolerance is specified here.

Observation 1: For a given property, many failure scenarios are irrelevant. To illustrate this, we consider the example in Fig. 1(a), where each router has a different AS number and runs BGP to select the best routes. According to the best route selection of BGP [13,25], a router will prefer a route with higher local preference and then the AS path length. As specified in [13,25], the local preference is not propagated among different ASes. Now, the best route is determined by the AS path length (with a shorter path being preferred). Then, routes propagated along $path_2$ (the AS path length is 4) are always preferred by node v_1, over the routes propagated along $path_3$ (the AS path length is 7), since $path_2$ is a subsequence of $path_3$. Therefore, the links on $path_3$ (v_4-v_9) will never be used by any best route for the property.

Based on this observation, when verifying $reachability(v_1, v_2, d, k)$, i.e., packets sent from node v_1 can reach prefix d[1] on node v_2 even failing any k links, we can first check whether there are policies that decrease the AS path length (which is quite uncommon), and if not, we can safely prune the links. We summarize that for a given property, a link excluded by all *simple paths*[2] [35] will not propagate the best route. Thus, we compute the set of all links included in *simple paths* and set the complement of these links as "down" status for the verifiers. For example, the three links v_4-v_9, v_4-v_{10}, and v_9-v_{10} are reduced in Fig. 1(b). In this example, the remaining symbolic links amount to $12 - 3 = 9$ links. This effect is more significant in real WAN topology, where the number of links is reduced from 188 to 126 (a $(188 - 126)/188 = 33\%$ reduction) [20], detailed in Sect. 5.1).

Observation 2: For a given property, many failure scenarios have the same equivalent impact. To show this, we consider the example after *Link Pruning* in Fig. 1(b), where routes at node v_8 propagate only in two directions, i.e., to v_6 or v_7. Suppose the best route from v_2 to v_1 is propagated via $path_1$ instead of $path_2$ due to smaller route metric values. When the link v_7-v_8 or v_8-v_6 fails, the best routing path changes from $path_1$ to $path_2$ because the failure of either v_7-v_8 or v_8-v_6 blocks v_5 from selecting the route via v_6. Thus, failing v_7-v_8 and v_8-v_6 individually or simultaneously have the same impact on route propagation for the property, and we refer to them as *equivalent links*.

Based on this observation, when verifying $reachability(v_1, v_2, d, k)$, we can remove one of the variables for those links, v_7-v_8 or v_8-v_6, to create a smaller model. We summarize that, for each set of equivalent links, we randomly select one link as a potentially failed link. The remaining links within this set, equivalent to the potentially failed link, are considered "up" status for the verifiers. For example, four links v_1-v_4, v_1-v_3, v_5-v_6, and v_7-v_8 are reduced in Fig. 1(c). In this example, the remaining symbolic links are $(9 - 4) = 5$ links. This effect is significant in real WAN topology, where the number of links is reduced from 126 to 61 (a $(126 - 61)/188 = 34\%$ reduction), detailed in Sect. 5.1).

[1] The prefix d (i.e., a block of IP addresses, e.g., 112.0.0.1–112.0.0.255) is connected to node v_2. Although the parameter v_2 can be omitted, we retain it for clarity.

[2] Some literature [10,34] uses the terms "walk" and "path" to distinguish between sequences that contain repeated vertices and those that do not, respectively. Other literature [35] uses "path" and "simple path" for the same distinction. For clarity, we adopt the latter terminology, e.g., using "routing path" instead of "routing walk.".

We highlight that the above methods make no assumptions about topology structures or routing behaviors that are hard to guarantee. For example, to ensure that $path_2$ is always preferred over $path_3$, we only need to confirm "the preference of a path does not decrease when prefixed by a link." This, in turn, assumes that "BGP does not propagate local preference across AS," as standardized by [13,25], and that "AS path length never decreases," which can be easily pre-checked. If some routers on the path are configured with routing policies that modify the AS path (which is quite uncommon [4,22,28]), the reduction can be disabled to avoid false positives.

3.2 Challenges

Challenge 1: How to calculate the links that never appear on any *simple path* without enumerating all *simple paths*? From Observation 1, we know that links that never appear on any *simple path* are irrelevant links (Theorem 1). To identify these links, an intuitive approach would be to first find all *simple paths* and then compute the set of links not contained in these paths. However, identifying all *simple paths* is shown to be NP-hard [35], as it requires enumerating all paths between two nodes. Moreover, on the UScarrier dataset, computing all *simple paths* between a single pair of nodes took over 4 h (arbitrarily selecting 10 pairs of far-apart nodes), whereas verifying a property for $k = 3$ required only 200 s (see Sect. 5.3).

Challenge 2: How to judge if two failed links have an equivalent impact on the best routes without calculating the best routes? From Observation 2, we know that two links whose failures have an equivalent impact can be compressed. An intuitive approach to achieve this relies on calculating the best routes. However, implicitly or explicitly computing the best routes would involve performing verification again, creating a chicken-and-egg problem, but we aim to identify equivalent impacts before verification.

4 Details

In this section, we first provide the definitions in Sect. 4.1. We then introduce *Link Pruning* in Sect. 4.2 and *Link Compression* in Sect. 4.3 (with formal proof) to address the challenges described in Sect. 3.2. Finally, we show the optimization.

4.1 Network Model

Property. Let $T = (V, L)$ be a network topology, where V is the set of vertices and L is the set of links. Let $property(v_1, v_2, d, k)$ be the property that packets from source node v_1 to prefix d on node v_2 even if any k links fail, where $v_1 \in V$ is the source (destination) node, and $v_2 \in V$ is the destination (source) node in forwarding (routing). Here, representative properties include [6,19,21,29,38]:

- *Reachability(s,n,d,k)*: packets sent from s can reach d on n.
- *Waypoint(s,n,w,d,k)*: packets sent from s to d on n, passing through w.
- *Loadbalance(s,n,m,d,k)*: packets sent from s to d on n, along m paths.

Failure Scenario. We now clarify the link states. Before verification, links have three states: down, up, and symbolic (unknown). During verification, the verifiers will assign symbolic links a concrete state: either up or failed.

Definition 1. *We define two types of failure scenarios.*

(i) A concrete failure scenario $f = (L_\mathcal{U}, L_\mathcal{D})$ partitions links L into two subsets: $L_\mathcal{U}$ (up) and $L_\mathcal{D}$ (down), i.e., all links have concrete states.
(ii) A symbolic failure scenario $F = (L_\mathcal{U}, L_\mathcal{D}, L_\mathcal{S}, k)$ partitions links L into three subsets: $L_\mathcal{U}$ (up), $L_\mathcal{D}$ (down), and $L_\mathcal{S}$ (symbolic). Here, k is the number of links in $L_\mathcal{S}$ that can fail simultaneously.
(iii) We say a symbolic failure scenario F contains a concrete failure scenario f or $f \in F$, if $F.L_\mathcal{U} \subseteq f.L_\mathcal{U}$, $F.L_\mathcal{D} \subseteq f.L_\mathcal{D}$, and $|f.L_\mathcal{D} \backslash F.L_\mathcal{D}| \leq k$, where $\backslash$ represents the set difference operation.

The third definition states that a *symbolic failure scenario* contains all *concrete failure scenarios* satisfying the links in $F.L_\mathcal{U}$ are up, the links in $F.L_\mathcal{D}$ are down, and no more than k links are failed in $F.L_\mathcal{S}$. Taking Fig. 1(c) as an example, a *symbolic failure scenario* $(L_\mathcal{U}, L_\mathcal{D}, L_\mathcal{S}, 2)$, with $L_\mathcal{S} = \{v_4\text{-}v_5, v_3\text{-}v_5, v_5\text{-}v_7, v_6\text{-}v_8, v_2\text{-}v_7\}$, contains $C_5^2 = 10$ *concrete failure scenarios.*

Based on the notion of *symbolic failure scenario*, most existing network verifiers [4,14] handle a *symbolic failure scenario* of $(\varnothing, \varnothing, L, k)$; while we check $(L_\mathcal{U}, L_\mathcal{D}, L_\mathcal{S}, k)$, and reduce the size of $L_\mathcal{S}$ as much as possible before solving.

Related Routes. Let $\mathcal{R}_f(v_1, v_2, d)$ denote the best routes of the devices among the forwarding and routing paths for the prefix d of *property*(v_1, v_2, d, k) under *failure scenario* f (The rigorous definition is provided in Appendix B [18]).

4.2 Link Pruning

Observation 1 states that for the given property, the links that do not appear on any *simple path* are irrelevant and can be pruned. In the following, we formalize this observation with a theorem and then prove the theorem.

First, we define irrelevant links and *simple paths.*

Definition 2. *Given a network topology $T = (V, L)$, and property(v_1, v_2, d, k) to verify, a link l is said to be "irrelevant", if the best routes related to the prefix d are the same whether l fails or not, on any concrete failure scenario. Formally, this can be expressed as:*

$$\begin{aligned} \text{for } \forall f = (L_\mathcal{U}, L_\mathcal{D}) :\ & \mathcal{R}_{(L_\mathcal{U}\backslash\{l\}, L_\mathcal{D}\cup\{l\})}(v_1, v_2, d) = \\ & \mathcal{R}_{(L_\mathcal{U}\cup\{l\}, L_\mathcal{D}\backslash\{l\})}(v_1, v_2, d) \end{aligned} \tag{2}$$

Note that Definition 2 is protocol-independent, because it is defined solely based on related routes, i.e., the best routes selected by devices among the forwarding and routing paths, without referring to a specific routing protocol. Next, we describe how to identify these irrelevant links.

Definition 3. *For a graph $T = (V, L)$, a path (or walk) from v_0 to v_k (of length k) is a non-empty alternating sequence $v_0, l_1, v_1, l_2, v_2, \ldots, l_k, v_k$ of vertices and links in T such that $l_i = (v_i, v_{i+1})$ for all $i < k$* [10].

Definition 4. *For a graph $T = (V, L)$, a path from $u \in V$ to $v \in V$ is said to be a simple path of T if the vertices in a path are all distinct* [10].

Theorem 1. *For a network topology $T = (V, L)$ and property(v_1, v_2, d, k), if a link does not appear on any simple path of T, then the link is irrelevant.*

To prove this theorem, we make two assumptions: (1) BGP converges to a unique state, which is true in most practical cases [13,29]. (2) The network exhibits monotonicity, meaning that the weight of a path does not decrease when prefixed by a link. This generally holds in common routing protocols, e.g., BGP and open shortest path first (OSPF) [4,22,28]. Here, we introduce the second assumption because Theorem 1 is protocol-specific, as identifying related routes via simple paths relies on the insight that routing protocols prefer loop-free paths, which is specific to the behavior of a specific routing protocol. The formal description of these two assumptions can be found in Appendix B [18].

Proof. (proof sketch) For clarity, we only provide a proof sketch here, and the full proof is in Appendix C [18]. To prove that a link is irrelevant, according to Definition 2, we only need to prove that the best routes among the forwarding path and routing path before and after failing this link remain the same. We prove this by induction, working on the number of routing propagation hops[3].

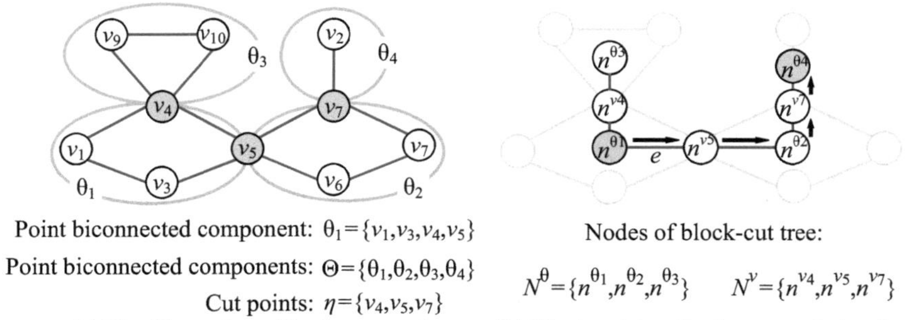

(a) The illustration of basic conception (b) Block-cut tree for the example topology

Fig. 2. The example of the *block-cut tree* for the network in Fig. 1.

[3] We omit the proof for forwarding paths, as it is quite similar.

Then, the problem is how to find all links that do not appear on any *simple paths*. Note that the straightforward way of computing all *simple paths* is NP-Hard [35]. To efficiently identify all irrelevant links, we present a polynomial algorithm. Our algorithm is heavily based on *point biconnected component* (PBC), a concept in graph theory [31]. In the following, we introduce the definition of PBC.

Definition 5. *Given a graph T, and a pair of nodes (u, v), if the deletion of any single node in T (except u and v) does not disconnect u and v, then we say that u and v are biconnected. If all pairs of two nodes in a graph are biconnected, then we say the graph is point biconnected. If a subgraph H of T is maximal point biconnected, meaning it is a point biconnected subgraph of T and no point biconnected subgraph of T contains H as a proper subgraph, then we say that H is a point biconnected component.*

We will use θ_i to represent a PBC, and use Θ to represent the set of all PBCs of the graph. For the example in Fig. 2, one PBC is $\theta_1 = \{v_1, v_3, v_4, v_5\}$, and $\Theta = \{\theta_1, \theta_2, \theta_3, \theta_4\}$.

Definition 6. *A graph is connected if there exists a path between every pair of distinct vertices. A cut point in a connected graph $G = (V, E)$ is a vertex $v \in V$ such that the deletion of v causes the graph to become disconnected.*

We will use η to represent the set of all *cut points* of a graph. For the example in Fig. 2, $\eta = \{v_4, v_5, v_7\}$.

There is a useful property that a cut point is included in two different point biconnected components [31]. Based on it, we define the *block-cut tree* (or *block-point tree*) [17].

Definition 7. *For an undirected graph $T = (V, L)$, a graph $\mathcal{T}$ is the block-cutpoint tree of T if it satisfies: (1) (Block vertices) For each biconnected component θ of T, there is a corresponding vertex n^θ in $\mathcal{T}$. (2) (Cut vertices) For each cut point v of T, there is a corresponding vertex n^v in $\mathcal{T}$. (3) An edge (n^v, n^θ) exists in $\mathcal{T}$ iff v is a vertex in θ.*

According to [17], $\mathcal{T}$ is shown to be a tree. For example, Fig. 2(b) shows the *block-cut tree* of the network topology T in Fig. 2(a).

Algorithm. Alg. 1 summarizes the process to identify all the irrelevant links.

Step1. Computing point biconnected components. First, for a given topology, we obtain *cut points* η and *point biconnected components* (PBCs) Θ using Tarjan's algorithm [31] (Line 2).

Step2. Constructing block-cut tree. Next, we construct the block-cut tree (Line 3). Specifically, given a network topology T with *point biconnected component* set Θ and cut point set η, its block-cut tree $\text{BCT}{\small\text{REE}}(\eta, \Theta)$ is a graph $\mathcal{T} = (\{N^v \cup N^\theta\}, E)$, where:

$$N^\theta = \{n^\theta | \theta \in \Theta\}, N^v = \{n^v | v \in \eta\} \tag{3}$$

Algorithm 1: Links Pruning

input : $T = (V, L)$: the network topology, v_1: the source node of the property, v_2: the destination node of the property.
output : L_{irr}: the set of irrelevant links in L.

1 **Function** IrrelevantLinks(v_1, v_2, T):
2 $\quad \eta, \Theta \leftarrow$ Tarjan(T) ▷ computing PBC
3 $\quad \mathcal{T} \leftarrow$BCTree($\eta, \Theta$) ▷ constructing $\mathcal{T}$
4 $\quad n_1 \leftarrow$ Node2PBC ($v_1, \mathcal{T}$) ▷ src node in $\mathcal{T}$
5 $\quad n_2 \leftarrow$ Node2PBC ($v_2, \mathcal{T}$) ▷ dst node in $\mathcal{T}$
6 $\quad N' \leftarrow$ Traverse($n_1, n_2, \mathcal{T}$) ▷ finding path on $\mathcal{T}$
7 $\quad \Theta' \leftarrow$ Mono($\{\theta | n^\theta \in N'\}$) ▷ checking configs
8 $\quad V' \leftarrow \{v \in \theta | \theta \in \Theta'\}$
9 $\quad L' \leftarrow \{(v_i, v_j) | v_i \in V', v_j \in V', (v_i, v_j) \in L\}$
10 $\quad$ **return** $L \backslash L'$ ▷ returning irrelevant links

$$E = \{e | e = (n^v, n^\theta) \vee (n^\theta, n^v), v \in \theta\} \tag{4}$$

Next, we find nodes n_1 and n_2 in $\mathcal{T}$ that correspond to the PBCs of the source node v_1 and destination node v_2 with Node2PBC (Lines 4–5).

Step3. Identifying irrelevant links. We compute all PBCs on path from n_1 to n_2 on $\mathcal{T}$, with Traverse (Line 6). Note that the source node and destination node may not be within the same PBC. For example, from v_2 to v_1, it may need to pass through three PBCs $\{\theta_1, \theta_2, \theta_4\}$. To confirm monotonicity, we use a function Mono to check that all nodes in a PBC are configured either with pure BGP, where the AS path length never decreases, or with OSPF. If not, we remove all links of the PBC from the down links (Line 7). We then collect all links L' within these PBCs (Lines 8–9). Finally, we return the complement set of L' as irrelevant links (Line 10).

Time Complexity. The complexity of IrrelevantLinks is $O(|V| + |L|)$, because calculating the *point biconnected components* takes $O(|V| + |L|)$ [31], and the remaining part has a lower time complexity. Specifically, we could mark nodes that are configured with pure BGP and without modifications to the AS path during configuration parsing, and then Mono requires $O(|N|)$ in this stage; Since $|E| < |L|$, Traverse traverses each edge in E once, which takes less than $O(|L|)$; Since each link will be added to L' at most once, this requires $O(|L|)$.

Proof of Correctness. We prove that Alg. 1 can find all irrelevant links with the Theorem 2.

Theorem 2. *For* $T = (V, L)$ *and* $property(v_1, v_2, d, k)$, *if* $l \in$IrrelevantLinks (v_1, v_2, T) *in Alg. 1, then* l *does not appear on any simple path from* v_1 *to* v_2.

The full proof is in Appendix D [18].

4.3 Link Compression

Observation 2 states that if the failures of two links have the same impact on the propagation of routes, then these two links are equivalent and can be compressed. We formally define our notion of equivalence as follows.

Definition 8. *Given a network topology $T = (V, L)$, and property(v_1, v_2, d, k) to verify, links l_1 and l_2 are said to be "equivalent" if the best routes related to prefix d are the same either l_1 or l_2 fails, on any concrete failure scenario, formally stated as:*

$$\text{for } \forall f = (L_{\mathcal{U}}, L_{\mathcal{D}}) : \mathcal{R}_{(L_{\mathcal{U}} \setminus \{l_1\}, L_{\mathcal{D}} \cup \{l_1\})}(v_1, v_2, d) = \mathcal{R}_{(L_{\mathcal{U}} \setminus \{l_2\}, L_{\mathcal{D}} \cup \{l_2\})}(v_1, v_2, d) \quad (5)$$

By compressing equivalent links into a single one, we can reduce the SMT solving space. Then, the problem is how to identify the equivalent links? The straightforward way is to fail each of the two links and analyze their impact of these failures, which is equivalent to another round of verification (solving SMT problem or simulating route computation). This clearly is an overkill for this problem. We set out for a less ambitious goal: instead of identifying all equivalent links, we try to find a subset of equivalent links solely based on the graph features. Next, we show the algorithm, and prove its correctness.

Definition 9. *For $T = (V, L)$, a path from $u \in V$ to $v \in V$ is said to be a "trivial path" of T if the degrees of u and v are not 2, and other nodes (except u and v) have degree 2.*

Theorem 3. *For $T = (V, L)$ and property(v_1, v_2, d, k), if two links l_1 and l_2 do not contain either both v_1 or both v_2 and appear on a trivial path of T, then l_1 and l_2 are equivalent.*

Theorem 3, similar to Theorem 1, is protocol-specific, so we also prove that Theorem 3 holds for two widely used protocols, BGP and OSPF.

Proof. (proof sketch) The full proof is in Appendix E [18]. To prove that two links are equivalent, we need to show that the best routes among the forwarding path and routing path after failing these two links remain the same. The proof is similar to the proof of Theorem 1, both using induction. The difference is that the induction is based on the number of endpoints of trivial paths.

Actually, a group of equivalence links has the property that any two links within the group are equivalent, which allows us to compress all the links within the group into a single link. This is due to the transitivity of equality in Definition 8, the equivalence of links also exhibits transitivity: if l_1 and l_2 are equivalent, and l_1 and l_3 are equivalent, then l_2 and l_3 are also equivalent. Due to this transitivity, we know all links in one trivial path are all equivalent.

Algorithm. Due to the simplicity of identifying equivalent links, we provide a narrative description rather than a formal presentation. For a given property, we use a depth-first search to traverse all links and nodes once. This allows us to identify all *trivial paths*. Next, for each *trivial path*, the links on it form a link set S. Then, we randomly select one link as a symbolic link from S, which may potentially fail. The remaining links within S are all treated as equivalent links to this symbolic link, and are treated as up links. We call this method a function $\textsc{EquivalentLinks}(v_1, v_2, T)$.

Time Complexity. The complexity of EQUIVALENTLINKS is $O(|V| + |L|)$, as it traverses all links and nodes twice, i.e., find all trivial paths and symbolic links.

Proof of Correctness. We prove the correctness of EQUIVALENTLINKS following Theorem 3.

4.4 Optimizations

Property Trimming. *VeriBoost* pre-checks the connectivity of the property on the network topology before *Link Pruning* and *Link Compression*. If the tolerance value of a property to be verified is k, there must be more than k disjoint paths in the topology. Here, we use the minimum cut to effectively calculate the number of disjoint paths. If the property fails this check, *VeriBoost* promptly returns it as violated.

5 Evaluation

We evaluate *VeriBoost* on real topologies to address the following questions:

- Is *VeriBoost* effective on different irregular topologies? We show that *VeriBoost* achieves a 40% reduction in symbolic links across all 260 real topologies, and an 80% reduction in 78% of them (Sect. 5.1).
- Is *Veriboost* correct on real datasets? We show that our verification results are identical to the results of existing verifiers for different properties, including reachability, waypointing, and load balancing (Sect. 5.2).
- How does *VeriBoost* speed up SMT-based verifier? We show that it speeds up the SMT-based verifier by 2–47× (Sect. 5.3).
- How do optimizations contribute to *VeriBoost*? We show that *Link Pruning*, *Link Compression* contribute similarly, and *Property Trimming* contributes little; The time for them ranges from microseconds to milliseconds, which is negligible, compared to the seconds-level verification. (For clarity, we present these two experiments in Appendix A [18] and report only the conclusions here.)

Implementation. We implemented *VeriBoost* with 4k lines of Java code, and applied it to the SMT-based verifier.

Approaches for Comparison. We choose Minesweeper [4] as the SMT-based baseline for comparison, because it is one of the most feature-complete SMT-based verifiers and has been widely adopted as a baseline by many works

[5,12,16,21,26]. We also compare with NetSMT [12], the state-of-the-art speedup method for SMT-based verification. We do not compare to other speedup methods for the following reasons. Origami [16] and Bonsai [5] target only data center networks (DCNs) [12]. Trailblazer has false positives due to missing hot edges as analyzed in [19]. As for BiNode, we found it to be slower than Minesweeper on our WAN datasets, which involve solely the BGP protocol or OSPF protocol[4]. This adds overhead without being paid off on our WAN datasets. Other methods, such as ACORN [24], Kirigami [32], Lightyear [30], and Timepiece [3], do not support fault tolerance verification.

Properties. We consider three properties as shown in Sect. 4.1 and select 25 pairs that can possibly hold for each $k = 0, 1, 2, 3$. Thus, we have $3 \times 4 \times 25 = 300$ properties for each network. We verify 300 properties because Minesweeper times out after 12 h on the Uscarrier dataset alone. *For a fair comparison, we select pairs that avoid cases where both nodes fall within a point biconnected component containing only a few nodes to worsen the speedup of VeriBoost.*

Datasets. *VeriBoost* has two stages: (1) reducing the symbolic links and (2) applying the reduction to verifiers. For the first step, we use all the 260 real-world WAN network topologies from the Topology Zoo [20] for the comparison experiments. For the second stage, real-world WAN configurations are typically confidential and not publicly available. Therefore, existing works [6,12] mostly rely on synthesizing configurations from real topologies using the approach in [11]. To avoid arbitrarily selecting unrepresentative topologies, we follow prior work. As summarized in Table 1, Fang et al. [12] considered nine representative topologies for BGP configuration synthesis, while Birkner et al. [6] considered three for OSPF configuration synthesis.

Table 1. The statistics of WAN datasets, where datasets with "-O" are OSPF datasets, and others are BGP datasets. Here, "Lines" denotes the number of lines of network configurations.

Datasets	Ren	Arn	Bic-O	Bic	Esn	Lat	Col-O	Col	Clt	Usc-O	Usc	Cog
Category	Small				Medium				Large			
Nodes	34	35	33	33	69	70	70	70	154	158	158	198
Links	44	47	48	48	80	75	85	85	178	189	189	244
Lines ($\times 10^3$)	4.8	5.0	3.0	6.9	9.2	9.1	6.1	11.5	20.6	13.7	22.0	27.1

Setup. All experiments run on a Linux server with two 16-core Intel Xeon Gold 6226R CPUs @2.9GHz and 256G memory.

[4] To handle the dependency between BGP and OSPF, BiNode replicates the variables associated with each BGP route.

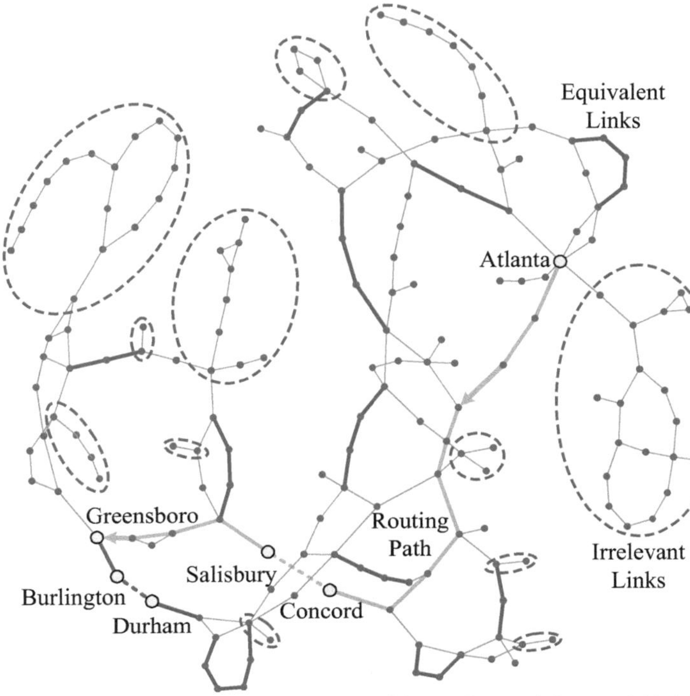

Fig. 3. An example of checking *Reachability(Greensboro, Atlanta, 200.4.219.0/24, 3)* on Uscarrier, a WAN topology from the Topology Zoo. For clarity, The *irrelevant links* and *equivalent links* identified by *VeriBoost* are partially shown.

5.1 Effectiveness of Veriboost

Case Study. Figure 3 shows that *VeriBoost* indeed identifies a large number of irrelevant and equivalent links for *Reachability(Greensboro, Atlanta, 200.4.219.0/24, 3)* on the real large topology of Uscarrier of 158 nodes. For this property, *VeriBoost* identifies the number of 63 irrelevant and 65 equivalent links, and the number of symbolic links is 61.

For $k = 3$, verifiers only need to check $C(61, 3) = O(10^4)$ *failure scenarios*, instead of $C(189, 3) = O(10^6)$. Finally, this property is checked for violation when "Salisbury–Concord" and "Burlington–Durham" fail.

Effectiveness. We evaluate the effectiveness across 260 real WAN topologies. We quantify the effectiveness using the *reduction ratio*, calculated as follows:

$$reduction\ ratio = \frac{\#of\ reduced\ links}{\#of\ original\ links} \tag{6}$$

For each topology, we use *VeriBoost* to compute the worst-case and average number of reduced links across all pairs of properties. Figure 4 shows that for the average case, *VeriBoost* achieves a 40% *reduction ratio* for 100% of the

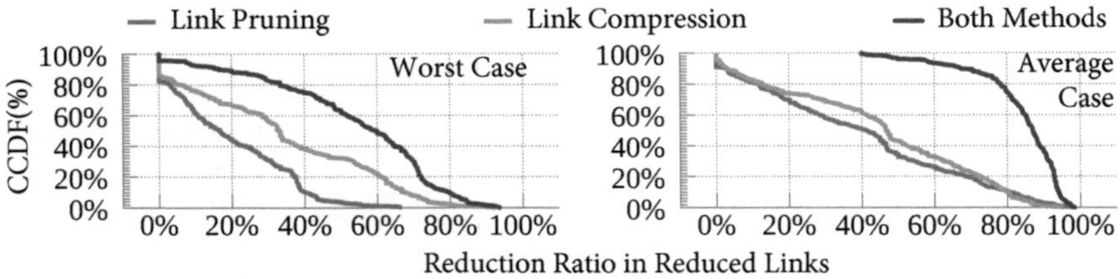

Fig. 4. The complementary cumulative distribution function (CCDF) of the *reduction ratio* in link reduction on 260 real topologies, when applying *Link Compression*, *Link Pruning*, or both methods. Specifically, a point (x, y) on the curve indicates that $y(\%)$ of the topologies have a *reduction ratio* greater than or equal to $x(\%)$.

topologies and a 78% *reduction ratio* for 80% of the topologies. For the worst case, *VeriBoost* achieves a 50% *reduction ratio* for 60% of the topologies. The results indicate that *VeriBoost* is effective in both the average and worst cases.

However, *VeriBoost* achieves only a 20% *reduction ratio* in node reduction for 40% of the topologies, as shown in Appendix A [18], suggesting that it is more effective for link failures than for node failures.

Table 2. The number of satisfied properties of Minesweeper, NetSMT, *VeriBoost*.

Network	BGP									OSPF		
	Ren	Arn	Bic	Esn	Lat	Col	Clt	Usc	Cog	Bic	Col	Usc
$k=1$	94	150	150	112	62	140	100	112	150	151	140	112
$k=2$	44	100	100	62	12	90	50	62	100	100	90	62
$k=3$	12	50	50	12	0	40	0	12	50	50	40	12

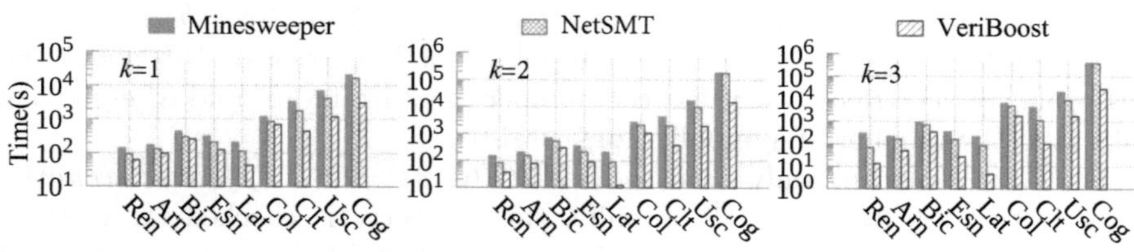

Fig. 5. Comparison of SMT-based verification for BGP datasets: Total time for verifying 300 properties, using Minesweeper, NetSMT, and *VeriBoost*.

5.2 Correctness of Veriboost

We verified 300 properties with $k = 1, 2, 3$. For SMT-based verification, the properties that hold for Minesweeper [4], NetSMT [12] and *VeriBoost* are consistent. The numerical results are summarized in Table 2.

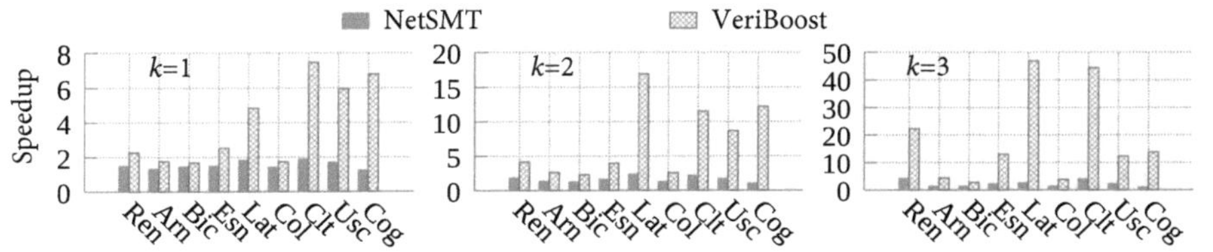

Fig. 6. Comparison of SMT-based verification on BGP datasets: Speedup for verifying 300 properties using NetSMT and *VeriBoost*, compared to Minesweeper.

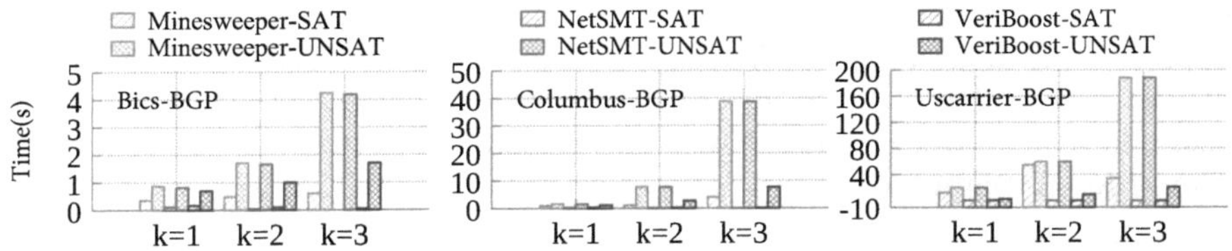

Fig. 7. Comparison of SMT-based verification: Average time for verifying a single property on BGP networks under different tolerance levels using Minesweeper, NetSMT, and *VeriBoost*. The SAT (UNSAT) time corresponds to the average time when the property is violated (holds). Negative values (Y-axis) mainly clarify the differences. The Y-axis starts from negative values mainly to highlight the differences.

5.3 Speedup for SMT-Based Verification

We separately verified 300 properties with $k = 1, 2, 3$ respectively using Minesw-eeper, NetSMT, and *VeriBoost*. Figure 5 (with its corresponding detailed speedup shown in Fig. 6) and Fig. 8 show that *VeriBoost* accelerates Minesweeper by $2\times$ to $47\times$. In contrast, NetSMT only speeds up by 2–$3\times$. Although the speedup of *VeriBoost* varies across different datasets, overall, it increases with the network scale and the tolerance values of the properties being verified, showing that *VeriBoost* benefits more in larger search spaces.

Figure 7 shows the details of the speedup of NetSMT and *VeriBoost* for verifying a single property (We show one dataset for each size and omit the results for OSPF for clarity.). We distinguish between two types of properties: for properties that hold, there are no solutions to the SMT problem (UNSAT); for properties that do not hold, there exist one or more solutions to the SMT problem (SAT). For SAT properties, both NetSMT and *VeriBoost* achieve more than one order of magnitude speedup. For UNSAT properties, NetSMT provides

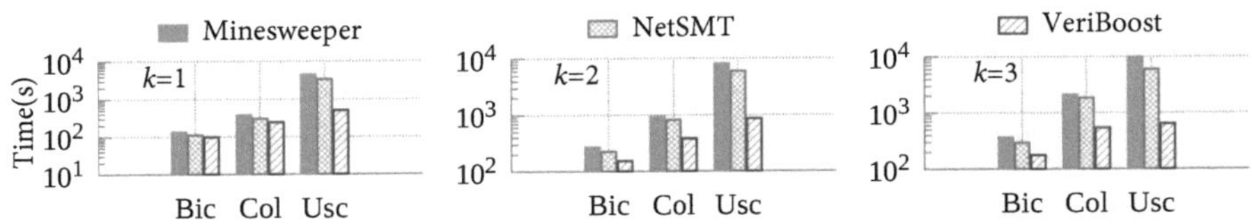

Fig. 8. Comparison of SMT-based verification for OSPF datasets: Total time for verifying 300 properties, using Minesweeper, NetSMT, and *VeriBoost*.

no improvement, whereas *VeriBoost* still delivers roughly an order of magnitude speedup.

6 Conclusion

Summary. We propose *VeriBoost*, a tool that can accelerate SMT-based verifiers in WANs by reducing the SMT solving space before verification. Specifically, *VeriBoost* uses *Link Pruning* to prune irrelevant links and *Link Compression* to compress equivalent links in WANs. We implement *VeriBoost* in SMT-based method, and evaluate *VeriBoost* on real topologies. Results show that *VeriBoost* can scale better to large WANs and is faster than the state-of-the-art tool by an order of magnitude.

Future Work. The basic idea of reducing link status unknown to verifiers, i.e., *Link Pruning* and *Link Compression*, is not peculiar to SMT-based verifiers. Exploring the effectiveness of applying *VeriBoost* to other types of verifiers, including simulation-based [14,37], graph-based [1], and hybrid verifiers [36,38], is left for future work.

Acknowledgement. This work is supported by the National Natural Science Foundation of China (No. 62272382 and No. 62572381). Peng Zhang is the corresponding author of this paper.

Data Availability Statement. The artifact, including the source code and experiment logs, is available at https://github.com/XJTU-NetVerify/veriboost/.

References

1. Abhashkumar, A., Gember-Jacobson, A., Akella, A.: Tiramisu: fast multilayer network verification. In: 17th USENIX Symposium on Networked Systems Design and Implementation (NSDI 2020), pp. 201–219 (2020). https://doi.org/10.23919/ifipnetworking55013.2022.9829765
2. Al-Fares, M., Loukissas, A., Vahdat, A.: A scalable, commodity data center network architecture. ACM SIGCOMM Comput. Commun. Rev. **38**(4), 63–74 (2008). https://doi.org/10.1145/1402958.1402967
3. Alberdingk Thijm, T., Beckett, R., Gupta, A., Walker, D.: Modular control plane verification via temporal invariants. Proc. ACM Program. Lang. **7**(PLDI), 50–75 (2023). https://doi.org/10.1145/3591222
4. Beckett, R., Gupta, A., Mahajan, R., Walker, D.: A general approach to network configuration verification. In: Proceedings of the Conference of the ACM Special Interest Group on Data Communication, pp. 155–168 (2017). https://doi.org/10.1145/3098822.3098834
5. Beckett, R., Gupta, A., Mahajan, R., Walker, D.: Control plane compression. In: Proceedings of the 2018 Conference of the ACM Special Interest Group on Data Communication, pp. 476–489 (2018). https://doi.org/10.1145/3230543.3230583

6. Birkner, R., Drachsler-Cohen, D., Vanbever, L., Vechev, M.: Config2spec: mining network specifications from network configurations. In: 17th USENIX Symposium on Networked Systems Design and Implementation (NSDI 2020), pp. 969–984 (2020)
7. Brown, M., Fogel, A., Halperin, D., Heorhiadi, V., Mahajan, R., Millstein, T.: Lessons from the evolution of the batfish configuration analysis tool. In: Proceedings of the ACM SIGCOMM 2023 Conference, pp. 122–135 (2023). https://doi.org/10.1145/3603269.3604866
8. Brown, M., Fogel, A., Halperin, D., Heorhiadi, V., Mahajan, R., Millstein, T.: The open-source code of batfish (2023). https://github.com/batfish/batfish
9. de Moura, L., Bjørner, N.: Z3: an efficient SMT solver. In: Ramakrishnan, C.R., Rehof, J. (eds.) TACAS 2008. LNCS, vol. 4963, pp. 337–340. Springer, Heidelberg (2008). https://doi.org/10.1007/978-3-540-78800-3_24
10. Diestel, R.: Graph Theory. Springer (print edition); Reinhard Diestel (eBooks) (2024). https://doi.org/10.1007/978-3-662-53622-3
11. El-Hassany, A., Tsankov, P., Vanbever, L., Vechev, M.: Netcomplete: practical {Network-Wide} configuration synthesis with autocompletion. In: 15th USENIX Symposium on Networked Systems Design and Implementation (NSDI 2018), pp. 579–594 (2018)
12. Fang, X., et al.: Network can help check itself: accelerating SMT-based network configuration verification using network domain knowledge. In: IEEE/International Conference on Computer (2023). https://doi.org/10.1109/infocom52122.2024.10621215
13. Feamster, N., Rexford, J.: Network-wide prediction of BGP routes. IEEE/ACM Trans. Networking **15**(2), 253–266 (2007). https://doi.org/10.1109/tnet.2007.892876
14. Fogel, A., et al.: A general approach to network configuration analysis. In: 12th USENIX Symposium on Networked Systems Design and Implementation (NSDI 2015), pp. 469–483 (2015)
15. Gao, L., Rexford, J.: Stable internet routing without global coordination. IEEE/ACM Trans. Networking **9**(6), 681–692 (2001). https://doi.org/10.1145/345063.339426
16. Giannarakis, N., Beckett, R., Mahajan, R., Walker, D.: Efficient verification of network fault tolerance via counterexample-guided refinement. In: Dillig, I., Tasiran, S. (eds.) CAV 2019. LNCS, vol. 11562, pp. 305–323. Springer, Cham (2019). https://doi.org/10.1007/978-3-030-25543-5_18
17. Harary, F.: Graph Theory (on Demand Printing of 02787). CRC Press (2018). https://doi.org/10.1201/9780429493768
18. Kang, N.: The appendix of veriboost (2026). https://xjtu-netverify.github.io/papers/VeriBoost/veriboost_final_version.pdf
19. Kang, N., Zhang, P., Li, H., Wen, S., Ji, C., Yang, Y.: Network specification mining with high fidelity and scalability. In: 2023 IEEE 31st International Conference on Network Protocols (ICNP), pp. 1–11. IEEE (2023). https://doi.org/10.1109/icnp59255.2023.10355598
20. Knight, S., Nguyen, H.X., Falkner, N., Bowden, R., Roughan, M.: The internet topology zoo. IEEE J. Sel. Areas Commun. **29**(9), 1765–1775 (2011). https://doi.org/10.1109/jsac.2011.111002
21. Liu, Y., Subotic, P., Letier, E., Mechtaev, S., Roychoudhury, A.: Efficient SMT-based network fault tolerance verification. In: International Symposium on Formal Methods, pp. 92–100. Springer (2023). https://doi.org/10.1007/978-3-031-27481-7_7

22. Lopes, N.P., Rybalchenko, A.: Fast BGP simulation of large datacenters. In: Enea, C., Piskac, R. (eds.) VMCAI 2019. LNCS, vol. 11388, pp. 386–408. Springer, Cham (2019). https://doi.org/10.1007/978-3-030-11245-5_18
23. Moss, S.: Microsoft azure outage blamed on wan router IP change (2023). https://www.datacenterdynamics.com/en/news/microsoft-azure-outage-blamed-on-wan-router-ip-change/
24. Raghunathan, D., Beckett, R., Gupta, A., Walker, D.: Acorn: network control plane abstraction using route nondeterminism. In: FMCAD, pp. 261–272 (2022). https://doi.org/10.34727/2022/isbn.978-3-85448-053-2_33
25. Rekhter, Y., Li, T., Hares, S.: RFC 4271: a border gateway protocol 4 (BGP-4) (2006). https://doi.org/10.17487/rfc4271
26. Shao, X., Chen, Z., Holcomb, D., Gao, L.: Accelerating BGP configuration verification through reducing cycles in SMT constraints. IEEE/ACM Trans. Networking **30**(6), 2493–2504 (2022). https://doi.org/10.1109/tnet.2022.3176267
27. Sharwood, S.: Facebook rendered spineless by buggy audit code that missed catastrophic network config error (2021). https://www.theregister.com/2021/10/06/facebook_outage_explained_in_detail/
28. Sobrinho, J.L.: An algebraic theory of dynamic network routing. IEEE/ACM Trans. Networking **13**(5), 1160–1173 (2005). https://doi.org/10.1109/tnet.2005.857111
29. Steffen, S., Gehr, T., Tsankov, P., Vanbever, L., Vechev, M.: Probabilistic verification of network configurations. In: Proceedings of the Annual Conference of the ACM Special Interest Group on Data Communication on the Applications, Technologies, Architectures, and Protocols for Computer Communication, pp. 750–764 (2020). https://doi.org/10.1145/3387514.3405900
30. Tang, A., et al.: Lightyear: using modularity to scale BGP control plane verification. In: Proceedings of the ACM SIGCOMM 2023 Conference, pp. 94–107 (2023). https://doi.org/10.1145/3603269.3604842
31. Tarjan, R.: Depth-first search and linear graph algorithms. SIAM J. Comput. **1**(2), 146–160 (1972). https://doi.org/10.1109/swat.1971.10
32. Thijm, T.A., Beckett, R., Gupta, A., Walker, D.: Kirigami, the verifiable art of network cutting. IEEE/ACM Trans. Networking (2024). https://doi.org/10.1109/tnet.2024.3360371
33. Tom Strickx, J.H.: Cloudflare outage on June 21, 2022 (2022). https://blog.cloudflare.com/cloudflare-outage-on-june-21-2022/
34. West, D.B.: Introduction to Graph Theory, vol. 2. Prentice Hall, Upper Saddle River (2001)
35. Yang, R., Gao, R., Zhang, C.: A new algebraic approach to finding all simple paths and cycles in undirected graphs. In: 2015 IEEE International Conference on Information and Automation, pp. 1887–1892. IEEE (2015). https://doi.org/10.1109/icinfa.2015.7279596
36. Ye, F., et al.: Accuracy, scalability, coverage: a practical configuration verifier on a global wan. In: Proceedings of the Annual Conference of the ACM Special Interest Group on Data Communication on the Applications, Technologies, Architectures, and Protocols for Computer Communication, pp. 599–614 (2020). https://doi.org/10.1145/3387514.3406217

37. Zhang, P., Gember-Jacobson, A., Zuo, Y., Huang, Y., Liu, X., Li, H.: Differential network analysis. In: 19th USENIX Symposium on Networked Systems Design and Implementation (NSDI 2022), pp. 601–615 (2022)
38. Zhang, P., Wang, D., Gember-Jacobson, A.: Symbolic router execution. In: Proceedings of the ACM SIGCOMM 2022 Conference, pp. 336–349 (2022). https://doi.org/10.1145/3544216.3544264

Pacing Types for Asynchronous Stream Equations

Florian Kohn, Arthur Correnson(✉), Jan Baumeister, and Bernd Finkbeiner

CISPA Helmholtz Center for Information Security, Saarbrücken, Germany
{florian.kohn,arthur.correnson,jan.baumeister,finkbeiner}@cispa.de

Abstract. Stream-based monitoring is a runtime verification approach where a *monitor* aggregates streams of input data from sensors and other sources to give real-time statistics and assessments of a system's health. One of the central challenges in designing reliable stream-based monitors is to deal with the asynchronous nature of data streams: in concrete applications, the different sensors being monitored produce values at different speeds, and it is the monitor's responsibility to correctly react to the asynchronous arrival of different streams of values. To ease this process, modern frameworks for stream-based monitoring such as RTLola enable users to finely specify data synchronization policies via a system of *pacing annotations*. While this feature simplifies the design of monitors, it can also lead users to write *inconsistent* policies, where synchronization between two streams is explicitly requested via annotations, but cannot always be achieved. To mitigate this issue, this paper presents *pacing types*, a novel type system implemented in RTLola to ensure that monitors for asynchronous streams are free of timing inconsistencies. We give a formal semantics to pacing annotations for a core fragment of RTLola, and present a soundness proof of the pacing type system. For an additional level of guarantees, we machine-checked the soundness proof using the Rocq proof assistant.

1 Introduction

Runtime verification is a safety assurance mechanism in which systems are monitored during their execution. Runtime verification is particularly useful for increasing the reliability of complex cyber-physical systems (CPS) for which obtaining a model for static verification is difficult or impossible. Stream-based monitoring is an approach to runtime verification in which a monitor collects data through input streams, which are then combined and aggregated in output streams to give real-time assessments of a system's health.

While, there exist many formalisms and programming languages to describe stream-based monitors [3,11,13,15,20,23], we focus on the stream-based specification language RTLola. In RTLola, a monitor is specified by listing equations relating input and output streams. Certain boolean output streams, called triggers, indicate whether the system being monitored is behaving as expected.

A. Sampaio and M. Stoelinga (Eds.): FM 2026, LNCS 16557, pp. 154–173, 2026.
https://doi.org/10.1007/978-3-032-26220-2_8

From a system perspective, monitors are passive components. They react to inputs from the system as they arrive. If different system sensors generate inputs at varying rates, the monitor must handle the asynchronous arrival of data. To simplify this process, RTLOLA features a system of *pacing annotations* where each output stream is labeled with additional information indicating when a stream must produce a fresh value in reaction to the asynchronous arrival of new input values. While the pacing annotation system of RTLOLA has proven to be useful in practical applications [3,5,12], it also introduces a critical challenge: ensuring the *consistency* of pacing annotations. Indeed, it is easy to write annotations that are incompatible with one another or that over-constrain a stream by requiring it to produce values at times when its dependencies are not guaranteed to be available. Ensuring consistency of pacing annotations is all the more difficult because the semantics of RTLOLA's pacing annotations have, thus far, never been rigorously formalized. In this paper, we address this issue in two ways. We propose the first formal semantics of RTLOLA's pacing annotations, and we present a type system to check the consistency of pacing annotations.

Contributions. In summary, we make the following contributions:

1. We formalize the essence of RTLOLA's system of pacing annotations. In particular, we give a formal semantics to stream equations decorated with pacing annotations.
2. We present a *type system* to check the consistency of pacing annotations, and formally prove its soundness. Concretely, we prove that well-typed annotated equations have at least one solution.
3. Finally, to obtain an additional level of reliability, we machine-checked the correctness proof of the type system using the Rocq proof assistant.

Structure of the Paper Sect. 2 provides an overview of RTLOLA and its system of pacing annotations. It also illustrates the inconsistency problem. Section 3 formalizes semantics of pacing annotations for a core fragment of RTLOLA, and introduces a type system to check consistency of pacing annotations. Section 5 presents a soundness proof of the type system. Finally, Sect. 6 describes two extensions of the type system.

1.1 Related Work

Runtime verification has been studied since the late 90 s [21]. Specifications are often expressed in temporal logics such as LTL [2] and MTL [1]. Applications include Markov Decision Processes [17,19], cyber-physical systems [3], and software [18]. Stream-based monitoring framework like Lola [11], RTLOLA [4], Tessla [20], and HStriver [15] provide expressive specification languages coupled with efficient monitoring infrastructures.

Synchronous programming languages such as LUSTRE [16], ESTEREL [8], and Zélus [7] share similarities with stream-based languages for monitoring. In particular, both paradigms need to address timing inconsistencies (ensuring that values exists when they are requested). Synchronous languages use notions of *clocks* [10,14,22], and associated static analyses to resolve timing issues. We propose a *type analysis* to address the timing inconsistencies introduced by the expressive pacing annotations of RTLOLA.

We note that type-based approach have already been explored in similar contexts. For example, [6] introduces type systems to detect *non-causal* equations, a property related to the existence of models in stream-based monitoring. Other type systems address additional program properties; for example, MARVeLus employs a refinement type system that reasons about temporal behavior using LTL [24] specifications [9].

2 Asynchronous Streams in RTLola

This section provides a high-level overview of RTLOLA and its pacing annotations using a battery management system as an example. In RTLOLA, monitors are specified by declaring a list of named input and output streams. As an example, consider the following RTLOLA specification which monitors excessive power draw:

```
input battery_lvl
output drain := battery_lvl.prev(or: battery_lvl) - battery_lvl
output warning := drain > 5
```

The input stream `battery_lvl` captures the current battery level measurement. The output stream `drain` computes the charge loss by subtracting the current battery level from the previous one, accessed via `battery_lvl.prev(or:...)`. The argument after `or:` provides a default value for the first evaluation step. This example illustrates the core concepts of input and output streams and the use of stream offsets to express temporal properties.

Handling Asynchronous Inputs. In practice, a monitor often requires access to multiple sensors to produce meaningful verdicts about the system's health. For example, the previous specification can be refined to monitor that the battery does not overheat while it is charging. For this, an additional input stream `temperature` captures the current temperature of the battery being monitored:

```
input battery_lvl: Int
input temperature: Int
output drain := battery_lvl.prev(or: battery_lvl) - battery_lvl
output warning := drain < 0 && temperature > 50
```

Importantly, RTLOLA does not make the assumption that streams are synchronized: different inputs might arrive at different rates. This raises the question of how to interpret an expression like `drain < 0 && temperature > 50` in the previous specification. More precisely, it is not clear what the value of this expression should be if one of the two inputs is not available at the time of evaluation. For example, the monitor could take the last defined value of the stream or wait for all values to be available. However, both solutions would negatively impact the precision of the monitor. Implicitly taking the last defined values would allow to mix measurements performed at very different time points and whose comparison is, therefore, difficult to interpret. Similarly, waiting for all values to arrive might result in dangerous situations being missed. Instead, RTLola permits more fine-grained handling of the asynchronous arrival of data via a combination of *pacing annotations* and a selection of synchronous and asynchronous access operators.

Synchronous and Asynchronous Accesses. To facilitate the handling of multiple inputs arriving at different paces, RTLola distinguishes between *synchronous* and *asynchronous* stream accesses. A synchronous access requires the stream being accessed (i.e., the stream the access refers to) to have a value at the same time as the accessing stream (i.e., the stream that contains the access in its expression). Direct accesses (such as `drain` and `temperature` in the previous example) and `prev` accesses are synchronous by design. This means that in the previous example, the stream `warning` can only produce a value when *both* `drain` and `temperature` are also producing a new value simultaneously. This restriction can be problematic if different sensors generate values at arbitrary, non-overlapping time points. In particular, this could prevent `warning` from ever being evaluated, thus never issuing the intended warning.

To address this issue, RTLola also provides an asynchronous `hold` access. A `hold` access retrieves the last produced value of the accessed stream, regardless of when it was generated. If no such value exists, a default value is returned, similar to `prev`. Unlike `prev` though, the `hold` operator may also reference the current value if the accessed stream produces a value at the same time as the accessing stream. Using a `hold` access the specification in Sect. 2 can be refined as follows:

```
input battery_lvl: Int
input temperature: Int
output drain := battery_lvl.prev(or: battery_lvl) - battery_lvl
output warning := drain < 0 && temperature.hold(or: 0) > 50
```

Replacing the synchronous access to `temperature` in `warning` with a `hold` eliminates the requirement for `temperature` and `battery_lvl` to update simultaneously, allowing `warning` to be evaluated whenever a new battery level is measured. Figures 1a and 1b illustrate the difference between synchronous and asynchronous access on an example trace. Black continuous arrows represent synchronous dependencies, while light gray arrows indicate dependencies due to default values and dashed arrows indicate asynchronous dependencies. On the left, the stream `warning` accesses `temperature` *synchronously*, via `prev`. On the right, `temperature` is accessed *asynchronously* via `hold`.

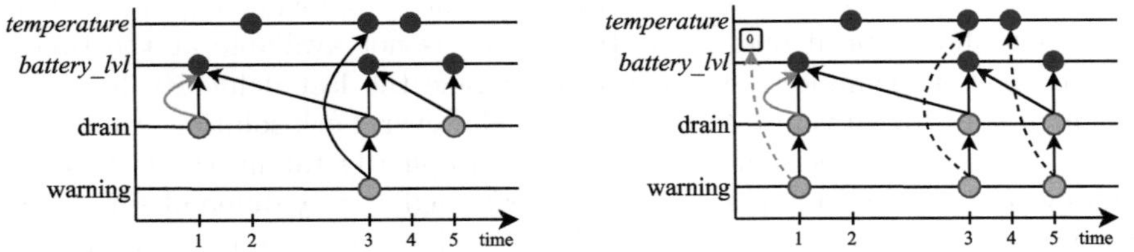

(a) Synchronous access to `temperature` (b) Asynchronous access to `temperature`

Fig. 1. Synchronous VS asynchronous accesses.

We note that with a synchronous access, **`warning`** only produces a value at time three, when both **`temperature`** and **`battery_lvl`** receive new values. Using an asynchronous access instead, **`warning`** also produces values at steps one and five in addition to time step three, resulting in a more precise verdict.

Pacing Annotations. One could argue that, the intent of the **`warning`** stream is still not fully captured. Indeed, even with the help of asynchronous **`hold`** access, **`warning`** only updates when the battery level changes, missing potential safety issues at times two and four in Fig. 1b. As a complement to asynchronous accesses, RTLOLA features *pacing annotations*. The idea is that each output stream will be labeled with an annotation indicating when the stream is expected to evaluate. Concretely, these annotations consists of (positive) boolean formulas over input streams. Our running example can be annotated as follows:

```
input battery_lvl: Int
input temperature: Int
output drain @battery_lvl :=
  battery_lvl.prev(or: battery_lvl) - battery_lvl
output warning @battery_lvl | temperature :=
  drain.hold(or: 0) < 0 && temperature.hold(or: 0) > 50
```

Here, the output stream **`drain`** is annotated with **`@battery_lvl`**, indicating that it is expected to evaluate each time a new **`battery_lvl`** input is received. The case of the **`warning`** stream is more interesting: we annotated it with **`@battery_lvl`** temperature!, indicating that it should be updated whenever either the temperature *or* the battery level changes. We further updated the specification to exclusively employ **`hold`** accesses. As a result of these modifications, the updated specification now produces a *meaningful* temperature warning at *all* time points (see Fig. 2).

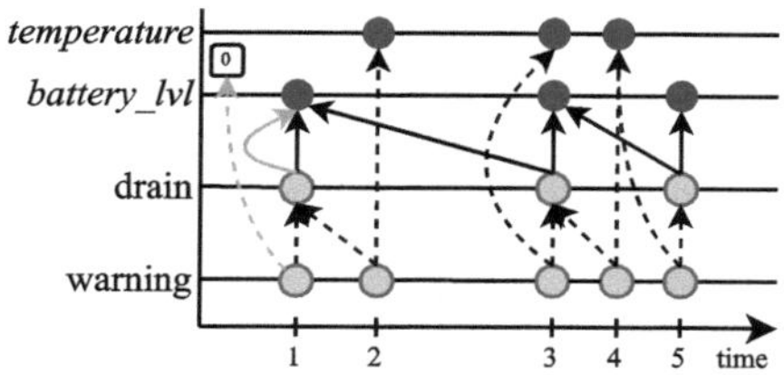

Fig. 2. The effect of disjunctive pacing types.

We note that hold accesses like `temperature.hold(or: 0)` are guaranteed to produce a value, even if the accessed stream `temperature` has not updated in a very long time. As a result, over-using `hold` can degrade the quality of verdicts by producing values that are not connected anymore to the current state of the monitored system. In the previous example, the pacing annotation `battery_lvl` temperature! ensures that the two `hold` accesses produce values that are reasonably up-to-date with the current status of the system: a new warning value is only produced when one of the two monitored sensors updates. Combining pacing annotations with asynchronous accesses in this manner enables balancing the frequency of updates and the interpretability of the calculated value

2.1 The Problem

While pacing annotations offer a convenient way for users to finely specify data-synchronization policies, they also enable users to write *inconsistent* specifications. As an example, consider the specification in Fig. 3a.

```
input a: Int
input b: Int
output x @b := b
output y @a := x
```

(a) An inconsistent specification

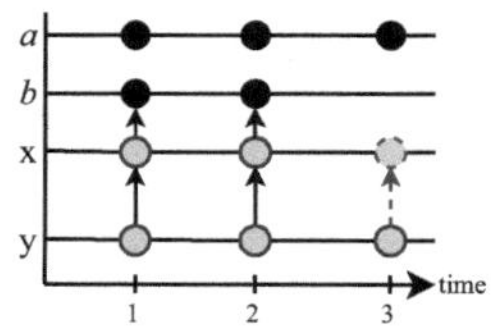

(b) The timing problem visualized

Fig. 3. An example of inconsistent specification.

The pacing annotation `x @b` requires that `x` evaluates whenever `b` receives a new value. Similarly, the annotation `y @a` requires `y` to update whenever `a` receives a new value. As demonstrated in Fig. 3b, these requirements cannot always be met at runtime! At time point three, the stream `y` is expected to have a value (because `a` also has a value). Since `y` contains a direct access to `x`, it cannot evaluate (because `x` does *not* have a value at time point three). The goal of the type system presented in this paper is to reject such invalid specifications, ensuring that a stream value *can* always be computed when it *must* be computed.

We note that one way to "repair" the specification in Fig. 3a would be to replace the synchronous access to `x` in `y` with a `hold` access. However, this change alters the semantics by removing the requirement that `x` must produce a new value whenever `y` produces a new value. With the `hold` operator, `y` keeps producing fresh values (based on old values of `x`) even if `x` stops updating.

2.2 A Type-Driven Solution

To prevent timing inconsistencies, RTLOLA uses a type system. First, we observe that the asynchronous `hold` stream access never causes timing inconsistencies, as it inherently provides a default value when the accessed value is absent. Consequently, the type system focuses on synchronous stream accesses, such as direct or `prev` accesses. More precisely, the type system ensures that a synchronously accessed value is always available when needed at runtime. To enforce this property, it requires that a stream accessed synchronously evaluates at least at the same time points as the stream containing the access.

3 Pacing Types

To formalize pacing annotations and their properties, we focus on a core language modeling the key features of RTLOLA, here dubbed μRTLOLA (for micro RTLOLA). In essence, μRTLOLA programs represent a set of constraints over input and output *stream variables* drawn from (disjoint) sets $\mathbb{V}_{in}$ and $\mathbb{V}_{out}$ respectively. For a fixed set of (named) input streams, these constraints describe which set of (named) output streams are valid reactions to the inputs. Because of this declarative view, we refer to μRTLOLA programs as *specifications*. For clarity, we only formalize constraints over integer-valued streams, but μRTLOLA can be extended to arbitrary data types in a straightforward manner[1].

3.1 Syntax of μRTLOLA

Constraints on stream variables are expressed via a list of equations of the form $x \,@\, \tau := e$. In these equations, $x \in \mathbb{V}_{out}$ is the name of an output stream being constrained, and e is a *stream expression* over input and output variables. Finally, τ is a positive boolean formula over input variables describing when the output stream x must be evaluated depending on the availability of inputs. We call τ a *pacing annotation*.

Stream expressions can be constant values $v \in \mathbb{Z}$, stream variables $x \in \mathbb{V}_{in} \uplus \mathbb{V}_{out}$, arithmetic functions applied to sub-expressions, references to previous values $x.\texttt{prev}(e)$, or *hold* accesses $x.\texttt{hold}(e)$. In $x.\texttt{prev}(e)$ and $x.\texttt{hold}(e)$,

[1] In fact, in our Rocq formalization, μRTLOLA is parameterized by an arbitrary type of values equipped with arbitrary (total) operators.

$x \in \mathbb{V}_{in} \uplus \mathbb{V}_{out}$ is the name of an arbitrary input or output stream, and e is a *default expression* representing what value the expression should take if there is no previous value for x (in the case of `prev`) or if there is no value to hold (in the case of `hold`). The exact syntax of specifications, stream expressions, and pacing annotations is defined below.

Variables	$x \in \mathbb{V}_{\mathtt{in}} \uplus \mathbb{V}_{\mathtt{out}}$
Values	$v \in \mathbb{Z}$
Stream Expressions	$e ::= v \mid x \mid x\mathtt{.prev}(e) \mid x\mathtt{.hold}(e) \mid e_1 + e_2 \mid \ldots$
Pacing Annotations	$\tau ::= x \mid \top \mid \tau_1 \wedge \tau_2 \mid \tau_1 \vee \tau_2$
Equations	$eq ::= x \ @ \ \tau \ := e$
Specifications	$S ::= \epsilon \mid eq \cdot S$

3.2 Semantics of μRTLOLA

Streams. We equip μRTLOLA with a denotational semantics assigning to lists of equations S a relation $[\![S]\!]$ between input and output streams. Importantly, to account for the asynchronous arrival of inputs (and the asynchronous production of outputs), streams are defined as sequences of *optional* values: at each point in time, a stream can either have a value $v \in \mathbb{Z}$, or be undefined denoted by $\bot$. Formally, the set of streams is defined as follows:

$$\mathit{Stream} \triangleq (\mathbb{Z} \uplus \{\bot\})^{\mathbb{N}}$$

When working with streams of optional values, it is often useful to equate streams that are not exactly equal, but only equal at non-$\bot$ positions. Given two streams $w_1, w_2 \in \mathit{Stream}$, we define the relation $w_1 \preccurlyeq w_2$ (read "w_1 *is less defined than* w_2") as

$$w_1 \preccurlyeq w_2 \iff \forall n.\ w_1(n) = \bot \vee w_1(n) = w_2(n)$$

Intuitively, $w_1 \preccurlyeq w_2$ guarantees that w_1 and w_2 are equal at positions where w_1 is defined, but w_2 is unconstrained at positions where w_1 contains $\bot$.

Maps. μRTLOLA equations are interpreted with respect to *stream maps* associating each input and output variable to a stream. We note *InMap* and *OutMap* for input and output stream maps.

$$\mathit{InMap} \triangleq \mathit{Stream}^{\mathbb{V}_{in}} \qquad \mathit{OutMap} \triangleq \mathit{Stream}^{\mathbb{V}_{out}}$$

When proving the soundness of the type system, it will also be useful to consider *partial* output stream maps. We represent them as total maps to optional streams.

$$\mathit{ParMap} \triangleq (\mathit{Stream} \cup \{\ \bot\ \})^{\mathbb{V}_{out}}$$

Given a partial map $\rho \in \mathit{ParMap}$, we note $\mathit{dom}(\rho) \subseteq \mathbb{V}_{out}$ for the domain of ρ (i.e. variables x such that $\rho(x) \neq \bot$), $\emptyset$ for the empty map sending all variables to $\perp$, and $(x \mapsto w)$ the singleton map sending x to the stream w. Finally, given a partial map $\rho_{par} \in \mathit{ParMap}$ and a total map $\rho_{tot} \in \mathit{OutMap}$, we define the *totalization* of ρ_{par} with respect to ρ_{tot} as follows

$$\rho_{par} \cdot \rho_{tot} \triangleq \lambda x. \begin{cases} \rho_{par}(x) & \text{if } x \in \mathit{dom}(\rho_{par}) \\ \rho_{tot}(x) & \text{otherwise} \end{cases}$$

In the rest of the paper, ρ_{in} and ρ_{out} are variables ranging over *InMap* and *OutMap* respectively. We also use the notation ρ_{out} for variables ranging over partial output maps.

Semantics of Stream Expressions. Given a pair of input and output stream maps (ρ_{in}, ρ_{out}), an expression e denotes a stream $[\![e]\!]_{(\rho_{in},\rho_{out})} \in \mathit{Stream}$. To assign a meaning to stream operators, we first introduce the following functions *Last*, *Prev*, and *Hold*. In all definitions below, $w \in \mathit{Stream}$ is a stream of optional values, $n \in \mathbb{N}$ is a time point and $v \in \mathbb{Z} \cup \{\perp\}$ is an optional value.

Definition 1 (Stream Operators).

$$\mathit{Last}(w, n) \triangleq \begin{cases} w(n) & \text{if } w(n) \neq \perp \\ \mathit{Last}(w, n-1) & \text{if } w(n) = \perp \wedge n > 0 \\ \perp & \text{otherwise} \end{cases}$$

$$\mathit{Hold}(w, n, v) \triangleq \begin{cases} \mathit{Last}(w, n) & \text{if } \mathit{Last}(w, n) \neq \perp \\ v & \text{otherwise} \end{cases}$$

$$\mathit{Prev}(w, n, v) \triangleq \begin{cases} \perp & \text{if } w(n) = \perp \\ \mathit{Last}(w, n-1) & \text{if } w(n) \neq \perp \wedge n > 0 \wedge \mathit{Last}(w, n-1) \neq \perp \\ v & \text{otherwise} \end{cases}$$

The intuition behind these functions can be described as follows:

- $\mathit{Last}(w, n)$ is the last defined value of w before the time point n (included), or $\perp$ if w is never defined before n.
- $\mathit{Hold}(w, n, v)$ models `hold` accesses. It returns the last value of w before n (included), or a default value v if w is never defined before n.
- $\mathit{Prev}(w, n, v)$ models `prev` accesses. Since `prev` accesses are synchronous, *Prev* returns $\perp$ if $w(n)$ is undefined. If $w(n)$ is defined, it returns the last defined value before n (excluded), or a default value v if w is never defined before n.

Given a pair of stream maps $\rho = (\rho_{in}, \rho_{out})$ and a stream variable $x \in \mathbb{V}_{in} \uplus \mathbb{V}_{out}$, we use the notation convention $\rho(x)$ to denote the input or output stream associated with x:

$$\rho(x) \triangleq \begin{cases} \rho_{in}(x) & \text{if } x \in \mathbb{V}_{in} \\ \rho_{out}(x) & \text{if } x \in \mathbb{V}_{out} \end{cases}$$

Using these definitions, we can introduce the denotation of stream expressions.

Definition 2 (Denotation of Stream Expressions).

$$\begin{aligned}
[\![c]\!]_\rho(n) &\triangleq c \\
[\![e_1 + e_2]\!]_\rho(n) &\triangleq [\![e_1]\!]_\rho(n) +_\bot [\![e_2]\!]_\rho(n) \\
[\![x]\!]_\rho(n) &\triangleq \rho(x)(n) \\
[\![x\texttt{.prev(}e\texttt{)}]\!]_\rho(n) &\triangleq \mathit{Prev}(\rho(x), n, [\![e]\!]_\rho(n)) \\
[\![x\texttt{.hold(}e\texttt{)}]\!]_\rho(n) &\triangleq \mathit{Hold}(\rho(x), n, [\![e]\!]_\rho(n))
\end{aligned}$$

Where the operator $+_\bot$ is defined as follows:

$$v_1 +_\bot v_2 \triangleq \begin{cases} v_1 + v_2 & \text{if } v_1 \neq \bot \wedge v_2 \neq \bot \\ \bot & \text{otherwise} \end{cases}$$

Semantics of Pacing Annotations. Given an input map ρ_{in}, a pacing annotation τ is interpreted as a set of time points $[\![\tau]\!]_{\rho_{in}} \subseteq \mathbb{N}$. Concretely $[\![\tau]\!]_{\rho_{in}}$ represents all of the time points where the inputs described by the boolean formula τ are defined in ρ_{in}.

Definition 3 (Denotation of Pacing Annotations).

$$\begin{aligned}
[\![x]\!]_{\rho_{in}} &\triangleq \{\, n \mid \rho_{in}(x)(n) \neq \bot \,\} \\
[\![\top]\!]^n_{\rho_{in}} &\triangleq \mathbb{N} \\
[\![\tau_1 \vee \tau_2]\!]_{\rho_{in}} &\triangleq [\![\tau_1]\!]_{\rho_{in}} \cup [\![\tau_2]\!]_{\rho_{in}} \\
[\![\tau_1 \wedge \tau_2]\!]_{\rho_{in}} &\triangleq [\![\tau_1]\!]_{\rho_{in}} \cap [\![\tau_2]\!]_{\rho_{in}}
\end{aligned}$$

Pacing annotation τ can also be understood as describing the set of streams which have defined values exactly at time points $[\![\tau]\!]_{\rho_{in}}$. We write $[\![\tau]\!]^\omega_{\rho_{in}} \subseteq \mathit{Stream}$ for the set of such "τ-paced" streams.

Definition 4 (Well-paced Streams).

$$[\![\tau]\!]^\omega_{\rho_{in}} \triangleq \{\, w \mid \forall n.\ n \in [\![\tau]\!]_{\rho_{in}} \iff w(n) \neq \bot \,\}$$

Semantics of Specifications. Having defined the semantics of expressions and pacing annotations, we can define the semantics of equations and specifications. An equation x @ τ := e denotes a binary relation between input and output maps. A specification (i.e., a list of equations) is interpreted as the intersection of the solutions of its equations.

Definition 5 (Denotation of Equations).

$$[\![x \mathbin{@} \tau \ := e]\!] \triangleq \{\ (\rho_{in}, \rho_{out}) \mid \rho_{out}(x) \in [\![\tau]\!]^{\omega}_{\rho_{in}} \wedge \rho_{out}(x) \preccurlyeq [\![e]\!]_{(\rho_{in},\rho_{out})}\ \}$$

$$[\![eq_1 \cdot \ldots \cdot eq_n]\!] \triangleq [\![eq_1]\!] \cap \ldots \cap [\![eq_n]\!]$$

A pair of stream maps (ρ_{in}, ρ_{out}) satisfies a single equation $x \mathbin{@} \tau \ := e$ under two conditions. First, the stream x should be well-paced w.r.t. τ (i.e.,$\rho_{out}(x) \in [\![\tau]\!]^{\omega}_{\rho_{in}}$). Second, x should be equal to its definition e. We note that the pacing annotation can force x to evaluate to $\bot$ at certain time point, even though e would have had a defined value. Therefore, we only ask x to be less defined than e rather than strictly equal (i.e., $\rho_{out}(x) \preccurlyeq [\![e]\!]_{(\rho_{in},\rho_{out})}$).

Consistency. Equations might have no solutions, or only admit solutions for specific combination of inputs. As discussed in the overview, this issue is worsened by the addition of pacing annotations. Indeed, in the semantics of equations, the additional pacing constraint ($\rho_{out}(x) \in [\![\tau]\!]^{\omega}_{\rho_{in}}$) can turn solvable equations into unsolvable ones. For example, assuming a and b are two input variables, the equations

$$x \mathbin{@} a \ := a$$
$$y \mathbin{@} b \ := x$$

are unsolvable for ρ_{in}'s where a and b are not perfectly synchronized, but would always have a trivial solution if pacing annotations were ignored (just set $\rho_{out}(y) = \rho_{out}(x) = \rho_{in}(a)$). We wish to ensure that annotated equations have solutions for *all* possible inputs. Equations that satisfy this property are called *consistent.*

Definition 6 (Consistency). *A specification S is called* consistent *if, for any input map ρ_{in}, there exists at least one compatible output map ρ_{out}. Formally*

$$Consistent(S) \triangleq \forall \rho_{in}.\ \exists \rho_{out}.\ (\rho_{in}, \rho_{out}) \in [\![S]\!]$$

3.3 Type System for Pacing Annotations

To determine whether a given specification is consistent, we use a type system. Our type system uses typing judgements of the following two forms

$$\Gamma \vdash e : \tau \quad \text{and} \quad \Gamma \vdash S$$

where Γ is a *typing context*, e is a stream expression, τ is a pacing annotation, and S is a list of annotated equations. The first kind of judgement, $\Gamma \vdash e : \tau$, is used to check the pacing of stream expressions e. Judgements $\Gamma \vdash S$ are used to check the consistency of sets of equations. In both judgements, the typing context Γ is a partial map from output variables $x \in \mathbb{V}_{out}$ to annotations τ, representing assumptions made on the pacing of some output streams. In the following, we use the notation convention $\Gamma, x : \tau$ to denote a context where x maps to τ, and we note $(x : \tau) \in \Gamma$ when $\Gamma(x) = \tau \neq \bot$.

Typing Equations. The typing rules for equations are processing the stream equations in a specification S in order of their appearance, extending the typing context after each processed equation. The Empty rule states that an empty list of equations is consistent in any context. The Equation rule processes the first equation of $x \; @ \; \tau \; := e$ of a list. For such an equation to be well-typed, the x should be unbound in Γ (i.e., $x \notin dom(\Gamma)$), making sure that there is at most one stream equation per output variables in the specifications. Further, we require the stream expression e to be well-typed in Γ for the annotated pacing (i.e., $\Gamma \vdash e : \tau$). Finally, the remaining equations should be well-typed in a context where x is assumed to be τ-paced.

Typing Expressions. The more interesting typing rules arise from stream accesses. Each kind of stream access comes associated with two typing rules depending on whether the accessed variable is an input or an output. The rules

Equations

$$\textsc{Empty}\quad \frac{}{\Gamma \vdash \epsilon} \qquad \textsc{Equation}\quad \frac{x \notin dom(\Gamma) \qquad \Gamma \vdash e : \tau \qquad \Gamma, x : \tau \vdash S}{\Gamma \vdash (x \; @ \; \tau \; := e) \cdot S}$$

Expressions

$$\textsc{Const}\quad \frac{v \in \mathbb{Z}}{\Gamma \vdash v : \tau} \qquad \textsc{BinOp}\quad \frac{\Gamma \vdash e_1 : \tau_{must} \qquad \Gamma \vdash e_2 : \tau_{must}}{\Gamma \vdash e_1 + e_2 : \tau_{must}}$$

$$\textsc{DirectOut}\quad \frac{(x : \tau_{can}) \in \Gamma \qquad \tau_{must} \models \tau_{can}}{\Gamma \vdash x : \tau_{must}} \qquad \textsc{DirectIn}\quad \frac{x \in \mathbb{V}_{\texttt{in}} \qquad \tau_{must} \models x}{\Gamma \vdash x : \tau_{must}}$$

$$\textsc{PrevOut}\quad \frac{(x : \tau_{can}) \in \Gamma \qquad \Gamma \vdash e : \tau_{must} \qquad \tau_{must} \models \tau_{can}}{\Gamma \vdash x\texttt{.prev(}e\texttt{)} : \tau_{must}}$$

$$\textsc{PrevIn}\quad \frac{x \in \mathbb{V}_{in} \qquad \Gamma \vdash e : \tau_{must} \qquad \tau_{must} \models x}{\Gamma \vdash x\texttt{.prev(}e\texttt{)} : \tau_{must}}$$

$$\textsc{HoldOut}\quad \frac{x \in dom(\Gamma) \qquad \Gamma \vdash e : \tau_{must}}{\Gamma \vdash x\texttt{.hold(}e\texttt{)} : \tau_{must}} \qquad \textsc{HoldIn}\quad \frac{x \in \mathbb{V}_{in} \qquad \Gamma \vdash e : \tau_{must}}{\Gamma \vdash x\texttt{.hold(}e\texttt{)} : \tau_{must}}$$

Fig. 4. The μRTLola typing rules for pacing types.

for synchronous accesses (i.e., DIRECTOUT, DIRECTIN, PREVOUT, and PREVIN) check that the time points at which the expression *must* be evaluated according to user annotations (τ_{must}) are subsumed by the time points at which the accessed stream *can* be evaluated according to the current typing context (τ_{can}). This is made explicit by side conditions $\tau_{must} \models \tau_{can}$, formally defined as follows:

$$\tau_{must} \models \tau_{can} \triangleq \forall \rho_{in}.\ [\![\tau_{must}]\!]_{\rho_{in}} \subseteq [\![\tau_{can}]\!]_{\rho_{in}}$$

In all rules, τ_{must} corresponds to the pacing type that the target expression is expected to have. If the accessed stream is an output stream, τ_{can} is determined from the typing context. For accesses to input streams, τ_{can} is the stream name itself. An important remark is that checking $\tau_{must} \models \tau_{can}$ can be done using a SAT-solver. Indeed, since pacing annotations are positive boolean formulas, it suffices to check the validity of the boolean implication $\tau_{must} \rightarrow \tau_{can}$.

Since `hold` and `prev` accesses can depend on the value of a default expression e, their associated typing rules require the default expression to be well-paced. For example, for x`.prev`(e) to be τ_{must} paced, e should also be τ_{must} paced. We note that, contrary to synchronous accesses, `hold` accesses do not impose pacing restrictions on their accessed variables: hold accesses x`.hold`(e) are τ-paced for any τ (provided the default expression e is also τ-paced). This coincides with the semantics of `hold`, which is guaranteed to produce a defined value by either returning the last defined value of the accessed stream, or a default value.

4 Examples

We demonstrate how to use our type system by applying it to a simple example. In the following, we suppose that inputs stream variables are $\mathbb{V}_{in} = \{\ a, b\ \}$, and output stream variables are $\mathbb{V}_{out} = \{x, y\}$. In Sect. 3.2, we observed that the following two equations are inconsistent (because y is forced to have a defined value whenever b also does, even though x might be undefined):

$$x \ @ \ a \ := a$$
$$y \ @ \ b \ := x$$

For the purpose of the demonstration, let us try to check $\emptyset \vdash (x \ @ \ a \ := a) \cdot (x \ @ \ b \ := x)$. Using the rule EQUATION two times (one for each equation), we ultimately have to type-check two expressions in two different contexts:

$$\vdash a : a \quad \text{and} \quad x : a \vdash x : b$$

Since a is an input variable and $a \models a$, the first judgement is trivially handled using DIRECTIN. However, the second judgement $x : a \vdash x : b$ cannot be handled. Since x is an output variable, the only rule we could use is DIRECTOUT, but we cannot use it since $a \not\models b$. The (unsuccessful) partial derivation is given below.

This inconsistency can be fixed, for example, by replacing x with x`.hold(`b`)` in the second equation. In this case the stream y would still be guaranteed to have a value whenever a fresh input b is received, but this value will only be an approximation of the stream x. The full typing derivation is given below.

$$\textsc{Equation}\ \dfrac{\textsc{DirectIn}\ \dfrac{a \models a}{\emptyset \vdash a : a} \qquad \textsc{Equation}\ \dfrac{\dots}{x : a \vdash x \ \texttt{@}\ b \ \texttt{:=}\ x\texttt{.hold(}b\texttt{)}}}{\emptyset \vdash (x \ \texttt{@}\ a \ \texttt{:=}\ a) \cdot (x \ \texttt{@}\ b \ \texttt{:=}\ x\texttt{.hold(}b\texttt{)})}$$

$$\dfrac{\textsc{HoldOut}\ \dfrac{x \in \{\, x \,\} \qquad \textsc{DirectIn}\ \dfrac{b \models b}{x : a \vdash b : b}}{x : a \vdash x\texttt{.hold(}b\texttt{)} : b} \qquad \dfrac{}{x : a, y : b \vdash \epsilon}\ \textsc{Empty}}{\dots}$$

We note that this time, we were able to use HoldOut to justify that the second equation $y \ \texttt{@}\ b \ \texttt{:=}\ x\texttt{.hold(}b\texttt{)}$ has a well-paced solution for all inputs. We only needed to check that the default value in the access $x\texttt{.hold(}b\texttt{)}$ (i.e., b) is defined at time points where y must be defined. Since the pacing annotation of the stream y is b, this condition is trivially true ($b \models b$).

5 Soundness Proof

Having defined our type system, we still need to formally prove its soundness. Namely, if a specification is well-typed according to the rules presented in the previous section, is should be *consistent*. We prove soundness by constructing an adequate *semantic model* of our typing judgements.

5.1 Semantic Model

Interpretation of Typing Contexts. To construct a semantic interpretation of typing judgements $\Gamma \vdash e : \tau$ and $\Gamma \vdash S$, the first thing we need is to interpret the context Γ. Given an input map ρ_{in}, a typing context Γ can be understood as describing partial stream maps whose streams are paced as described by the bindings $(x : \tau) \in \Gamma$. We denote this interpretation by $[\![\Gamma]\!]_{\rho_{in}} \subseteq \mathit{Pmap}$.

Definition 7 (Interpretation of Typing Contexts).

$$[\![\Gamma]\!]_{\rho_{in}} \triangleq \{\ \rho_{out} \mid \mathit{dom}(\Gamma) = \mathit{dom}(\rho_{out}) \wedge \forall (x : \tau) \in \Gamma.\ \rho_{out}(x) \in [\![\tau]\!]^{\omega}_{\rho_{in}}\ \}$$

Semantic Typing of Expressions. Intuitively, if $\Gamma \vdash e : \tau$, then for any input map ρ_{in}, and for any output map that is compatible with Γ, e should have a defined value at least at all time points that are described by τ. This semantic interpretation of typing judgements for expressions is denoted by $\Gamma \models e : \tau$.

Definition 8 (Semantic Typing of Expressions).

$$\Gamma \models e : \tau \triangleq \forall \rho_{in}. \forall \rho_{out} \in [\![\Gamma]\!]_{\rho_{in}}. \forall n \in [\![\tau]\!]_{\rho_{in}}.\ \overline{[\![e]\!]}_{(\rho_{in},\rho_{out})}(n) \neq \bot$$

We note that in this definition, output maps ρ_{out} are only *partial maps*, so we cannot use the standard denotation of expressions $[\![e]\!]$ to evaluate e in ρ_{out}. Instead, we use a slightly different *partial semantics* $\overline{[\![e]\!]}$, returning $\bot$ whenever e contains an access to an output variable $x \notin dom(\rho_{out})$. We omit the full formal definition of $\overline{[\![e]\!]}$ here, but we note that this partial semantics *agrees* with the reference semantics in the sense of the following lemma.

Lemma 1 *For any $\rho_{in} \in InMap$, any $\rho^1_{out} \in Pmap$, and any $\rho^2_{out} \in OutMap$,*

$$\overline{[\![e]\!]}_{(\rho_{in},\rho^1_{out})} \preccurlyeq [\![e]\!]_{(\rho_{in},\rho^1_{out}\cdot\rho^2_{out})}$$

In other words, if e has defined values under the partial output map ρ^1_{out}, then it has the same defined values under any totalization $\rho^1_{out} \cdot \rho^2_{out}$.

Semantic Typing of Equations. As for expressions, we construct a semantic interpretation of typing judgements $\Gamma \vdash S$ for equations.

Definition 9 (Semantic Typing of Equations).

$$\Gamma \models S \triangleq \forall \rho_{in}. \forall \rho^1_{out} \in [\![\Gamma]\!]_{\rho_{in}}. \exists \rho^2_{out} \in OutMap.\ [\![S]\!]_{(\rho_{in},\rho^1_{out}\cdot\rho^2_{out})}$$

Intuitively, for any partial output map ρ^1_{out} well-paced according to Γ, the equations in S are expected to have a well-paced solution ρ^2_{out}. As a direct consequence of this definition, if a set of equations S is *semantically* well-paced in an empty context, then it is consistent (adequacy of the semantic model).

Theorem 1 (Adequacy). $\emptyset \models S \implies Consistent(S)$

Proof Suppose $\emptyset \models S$ and let $\rho_{in} \in InMap$. Pick $\rho^1_{out} = (\lambda\_.\ \bot) \in [\![\emptyset]\!]_{\rho_{in}}$, and by definition of $\emptyset \models S$ there must exists a solution ρ^2_{out} to the equations S. □

5.2 Soundness of Typing Rules

Now that we have established the *adequacy* of our semantic model, we can prove that our typing rules are correct with respect to semantic typing.

Theorem 2 (Sound Typing of Expressions). $\Gamma \vdash e : \tau \implies \Gamma \models e : \tau$

Proof. The proof proceeds by a straightforward induction on the derivation $\Gamma \vdash e : \tau$. We refer the curious readers to our Rocq development for details. □

Theorem 3 (Sound Typing of Equations). $\Gamma \vdash S \implies \Gamma \models S$

Proof. The proof proceeds by induction on the derivation $\Gamma \vdash S$. The case $S = \epsilon$ is straightforward. Now suppose $S = x \mathbin{@} \tau := e \cdot S'$, suppose $\Gamma \vdash e : \tau$, and by induction assume $\Gamma, x : \tau \models S'$. For a fixed ρ_{in} and a fixed $\rho^1_{out} \in [\![\Gamma]\!]_{\rho_{in}}$, we have to show the existence of a solution ρ^2_{out} to $x \mathbin{@} \tau := e \cdot S$. By Theorem 2, since $\Gamma \vdash e : \tau$, this implies $\Gamma \models e : \tau$. We can then construct a stream w_e as follows:

$$w_e(n) \triangleq \begin{cases} \overline{[\![e]\!]}_{(\rho_{in}, \rho^1_{out})} & \text{if } n \in [\![\tau]\!]_{\rho_{in}} \\ \bot & \text{otherwise} \end{cases}$$

By definition of $\Gamma \models e : \tau$, we have $w_e \in [\![\tau]\!]_{\rho_{in}}$. Since $\rho^1_{out} \in [\![\Gamma]\!]$, we can deduce that $\rho^1_{out}[x \leftarrow w_e] \in [\![\Gamma, x : \tau]\!]$. Applying $\Gamma, x : \tau \models S'$ to $\rho^1_{out}[x \leftarrow w_e]$, we obtain a solution ρ'_{out} to S'. I.e., $(\rho_{in}, \rho^1_{out}[x \leftarrow w_e] \cdot \rho'_{out}) \in [\![S']\!]$. By definition of w_e and Lemma 1, $\rho^1_{out}[x \leftarrow w_e] \cdot \rho'_{out}$ is also a solution to the equation $x \mathbin{@} \tau := e$. Overall, we have $(\rho_{in}, \rho^1[x \leftarrow w_e] \cdot \rho'_{out}) \in [\![x \mathbin{@} \tau := e \cdot S]\!]$, which concludes the proof. □

As an immediate consequence of Theorem 1 and Theorem 3, we obtain the soundness of our type system for consistency.

Corollary 1 (Type Soundness). $\emptyset \vdash S \implies \mathit{Consistent}(S)$

6 Sound Extensions of the Type System

While the type system presented in Fig. 4 is sound, it is also incomplete: it may reject equations that are consistent. This section describes two simple extensions of the type system to mitigate incompleteness.

Ordering of Equations.
A first source of incompleteness originates from the order-sensitive nature of our type system. As an example, the following two (consistent) equations do not type-check in this order, because y is defined after x and x uses y. However, they are well-typed if we swap x and y.

$$x \mathbin{@} i := y \qquad\qquad \textsc{Reorder}\ \frac{S' \text{ is a permutation of } S \quad \Gamma \vdash S'}{\Gamma \vdash S}$$
$$y \mathbin{@} i := i$$

This issue is easily fixed by introducing the rule Reorder above. This extension is shown to preserve soundness of the type system using the same semantic model as in Sect. 5, and by observing that the semantics of equations is insensitive to re-orderings of the equations.

Self References. Another source of incompleteness stems from the presence of self-references in stream equations. Equations such as $x \mathbin{@} \top := x.\mathtt{prev}(0) + 1$ are consistent, but rejected by the typing rules of Fig. 4. However, such self `prev`-access are very useful to implement counters or other stateful computations. To enable self `prev`-accesses, we extend our type system to make it self-reference-aware. Concretely, we replace judgements $\Gamma \vdash e : \tau$ by judgements $\Gamma \vdash^x e : \tau$, where the additional annotation $x \in \mathbb{V}_{out}$ indicates the name of the stream containing e as a sub-expression. With this change, the rule for equations is updated to track self-references, and an additional rule PrevSelf can be added to support self prev-accesses.

$$\textsc{PrevSelf}\ \frac{x = y \qquad \Gamma \vdash^x e : \tau}{\Gamma \vdash^x y\texttt{.prev(}e\texttt{)} : \tau} \qquad \textsc{Equation}\ \frac{\Gamma \vdash^x e : \tau \qquad \Gamma, x : \tau \vdash S}{\Gamma \vdash x \,\texttt{@}\, \tau \texttt{ := } e \cdot S}$$

All other rules are kept unchanged, and simply ignore the extra annotation. This extension preserves soundness, but the formal proof is significantly more complicated. In particular, the extra annotation x in the typing judgement for expressions needs a semantic interpretation. We omit the formalization here, and refer the curious readers to our Rocq development for a formal proof of soundness of this extension.

We note that beyond self prev-accesses, other types of self-references (like self hold-access, or equations of the form $x = x$) are considered to be anti-patterns even when they have models, and therefore do not need to be covered.

7 Conclusion

In this paper, we formalized the core of RTLola's pacing annotations, a mechanism for data synchronization in stream-based monitoring. We provided a formal semantics interpreting annotated stream equations as relations between input and output streams. We showed that pacing annotations can introduce inconsistencies and addressed this by treating pacing annotations as types. The resulting type system rejects contradictory annotations, and we proved its soundness with respect to the formal semantics.

While the presented fragment of RTLola is already practical, extending it to the whole language remains future work. Another interesting research direction would be to relate the type system to an operational semantics, establishing not only the existence of models but also their computability.

Acknowledgements. This work was supported by the DFG in project 389792660 (TRR 248 – CPEC) and the European Research Council (ERC) Grant HYPER (No. 101055412). Views and opinions expressed are however those of the authors only and do not necessarily reflect those of the European Union or the European Research Council Executive Agency. Neither the European Union nor the granting authority can be held responsible for them. F. Kohn, A. Correnson, and J. Baumeister carried out this work as members of the Saarbrücken Graduate School of Computer Science.

Data Availability Statement. Our Rocq development containing the whole formalization of μRTLola, its type system (and the extensions presented in Sect. 6), and the soundness proof are available online at the following addresses:

https://github.com/acorrenson/pacing-types (*latest updates*)

or

https://doi.org/10.5281/zenodo.18759084 (*artifact*)

References

1. Basin, D., Klaedtke, F., Zălinescu, E.: Runtime verification of temporal properties over out-of-order data streams. In: Majumdar, R., Kunčak, V. (eds.) CAV 2017. LNCS, vol. 10426, pp. 356–376. Springer, Cham (2017). https://doi.org/10.1007/978-3-319-63387-9_18
2. Bauer, A., Leucker, M., Schallhart, C.: Runtime verification for LTL and TLTL. ACM Trans. Softw. Eng. Methodol. (TOSEM) **20**(4), 1–64 (2011)
3. Baumeister, J., et al.: Monitoring unmanned aircraft: specification, integration, and lessons-learned. In: Gurfinkel, A., Ganesh, V., (eds.), Computer Aided Verification - 36th International Conference, CAV 2024, Montreal, QC, Canada, 24-27 July 2024, Proceedings, Part II, volume 14682 of Lecture Notes in Computer Science, pp. 207–218. Springer (2024)
4. Baumeister, J., Finkbeiner, B., Kohn, F., Scheerer, F.: A tutorial on stream-based monitoring. In: Platzer, A., Yvonne Rozier, K., Pradella, M., Rossi, M., (eds.), Formal Methods - 26th International Symposium, FM 2024, Milan, Italy, 9-13 September 2024, Proceedings, Part II, vol. 14934 of Lecture Notes in Computer Science, pp. 624–648. Springer (2024)
5. Baumeister, J., Finkbeiner, B., Schirmer, S., Schwenger, M., Torens, C.: RTLola cleared for take-off: monitoring autonomous aircraft. In: Lahiri, S.K., Wang, C. (eds.) CAV 2020. LNCS, vol. 12225, pp. 28–39. Springer, Cham (2020). https://doi.org/10.1007/978-3-030-53291-8_3
6. Benveniste, A., Bourke, T., Caillaud, B., Pagano, B., Pouzet, M.: A type-based analysis of causality loops in hybrid systems modelers. In: Fränzle, M., Lygeros, J., (eds.), 17th International Conference on Hybrid Systems: Computation and Control (part of CPS Week), HSCC'14, Berlin, Germany, 15-17 April 2014, pp. 71–82. ACM (2014)
7. Benveniste, A., Bourke, T., Caillaud, B., Pouzet, M.: A hybrid synchronous language with hierarchical automata: static typing and translation to synchronous code. In: Chakraborty, S., Jerraya, A., Baruah, S.K., Fischmeister, S., (eds.), Proceedings of the 11th International Conference on Embedded Software, EMSOFT 2011, part of the Seventh Embedded Systems Week, ESWeek 2011, Taipei, Taiwan, 9-14 October 2011, pp. 137–148. ACM (2011)
8. Berry, G., Cosserat, L.: The ESTEREL synchronous programming language and its mathematical semantics. In: Stephen D. Brookes, A. W. Roscoe, and Glynn Winskel, editors, Seminar on Concurrency, Carnegie-Mellon University, Pittsburg, PA, USA, 9-11 July 1984, volume 197 of Lecture Notes in Computer Science, pp. 389–448. Springer (1984)
9. Chen, J., et al.: Synchronous programming with refinement types. Proc. ACM Program. Lang., **8**(ICFP), 938–972 (2024)
10. Colaço, J.-L., Pouzet, M.: Clocks as first class abstract types. In: Alur, R., Lee, I. (eds.) EMSOFT 2003. LNCS, vol. 2855, pp. 134–155. Springer, Heidelberg (2003). https://doi.org/10.1007/978-3-540-45212-6_10
11. d'Angelo, B., et al.: Lola: runtime monitoring of synchronous systems. In: 12th International Symposium on Temporal Representation and Reasoning (TIME'05), pp. 166–174. IEEE (2005)
12. Faymonville, P., Finkbeiner, B., Schirmer, S., Torfah, H.: A stream-based specification language for network monitoring. In: Falcone, Y., Sánchez, C. (eds.) Runtime Verification, pp. 152–168. Springer International Publishing, Cham (2016)

13. Faymonville, P., et al.: StreamLAB: stream-based monitoring of cyber-physical systems. In: Dillig, I., Tasiran, S. (eds.) CAV 2019. LNCS, vol. 11561, pp. 421–431. Springer, Cham (2019). https://doi.org/10.1007/978-3-030-25540-4_24
14. Gautier, T., Le Guernic, P.: SIGNAL: a declarative language for synchronous programming of real-time systems. In: Kahn, G., (eds.), Functional Programming Languages and Computer Architecture, Portland, Oregon, USA, 14-16 September 1987, Proceedings, volume 274 of Lecture Notes in Computer Science, pp. 257–277. Springer (1987)
15. Gorostiaga, F., Sánchez, C.: HStriver: a very functional extensible tool for the runtime verification of real-time event streams. In: Huisman, M., Păsăreanu, C., Zhan, N. (eds.) FM 2021. LNCS, vol. 13047, pp. 563–580. Springer, Cham (2021). https://doi.org/10.1007/978-3-030-90870-6_30
16. Halbwachs, N., Caspi, P., Raymond, P., Pilaud, D.: The synchronous data flow programming language LUSTRE. Proc. IEEE **79**(9), 1305–1320 (1991)
17. Henzinger, T.A., Karimi, M., Kueffner, K., Mallik, K.: Monitoring algorithmic fairness. In: Enea, C., Lal, A., (eds.), Computer Aided Verification - 35th International Conference, CAV 2023, Paris, France, 17-22 July 2023, Proceedings, Part II, volume 13965 of Lecture Notes in Computer Science, pp. 358–382. Springer (2023)
18. Jin, D., O'Neil Meredith, P., Lee, C., Rosu, G.: Javamop: efficient parametric runtime monitoring framework. In: Glinz, M., Murphy, G.C., Pezzè, M., (eds.), 34th International Conference on Software Engineering, ICSE 2012, 2-9 June 2012, Zurich, Switzerland, pp. 1427–1430. IEEE Computer Society (2012)
19. Junges, S., Torfah, H., Seshia, S.A.: Runtime monitors for markov decision processes. In: Silva, A., Leino, K.R.M. (eds.) CAV 2021. LNCS, vol. 12760, pp. 553–576. Springer, Cham (2021). https://doi.org/10.1007/978-3-030-81688-9_26
20. Kallwies, H., Leucker, M., Schmitz, M., Schulz, A., Thoma, D., Weiss, A.: Tessla - an ecosystem for runtime verification. In: Dang, T., Stolz, V., (eds.), Runtime Verification - 22nd International Conference, RV 2022, Tbilisi, Georgia, 28-30 September 2022, Proceedings, volume 13498 of Lecture Notes in Computer Science, pp. 314–324. Springer (2022)
21. Kim, M., Viswanathan, M., Ben-Abdallah, H., Kannan, S., Lee, I., Sokolsky, O.: Formally specified monitoring of temporal properties. In: 11th Euromicro Conference on Real-Time Systems (ECRTS 1999), 9-11 June 1999, York, England, UK, Proceedings, pp. 114–122. IEEE Computer Society (1999)
22. Mandel, L., Plateau, F., Pouzet, M.: Lucy-n: a n-synchronous extension of Lustre. In: Bolduc, C., Desharnais, J., Ktari, B. (eds.) MPC 2010. LNCS, vol. 6120, pp. 288–309. Springer, Heidelberg (2010). https://doi.org/10.1007/978-3-642-13321-3_17
23. Perez, I., Goodloe, A.E., Dedden, F.: Runtime verification in real-time with the copilot language: a tutorial. In: Platzer, A., Yvonne Rozier, K., Pradella, M., Rossi, M., (eds.), Formal Methods - 26th International Symposium, FM 2024, Milan, Italy, 9-13 September 2024, Proceedings, Part II, volume 14934 of Lecture Notes in Computer Science, pp. 469–491. Springer (2024)
24. Pnueli, A.: The temporal logic of programs. In: 18th Annual Symposium on Foundations of Computer Science, Providence, Rhode Island, USA, 31 October - 1 November 1977, pp. 46–57. IEEE Computer Society (1977)

Performance Heuristics for GR(1) Unrealizable Core Computation

Shachaf Cohen and Shahar Maoz(✉)

Tel Aviv University, Tel Aviv, Israel
maoz@cs.tau.ac.il

Abstract. A major challenge of reactive synthesis, an automated process for deriving correct-by-construction reactive systems from temporal specifications, is debugging unrealizable specifications. One way to debug them in the context of GR(1), an LTL fragment that balances efficient synthesis complexity and expressiveness, is the computation of unrealizable cores. Although work has been done to accelerate core computation, it remains a costly operation, despite being commonly used during specification development. In this paper, we first examine a version of the existing `QuickCore` core computation algorithm, where we replace the classic `DDMin` algorithm with `CDD`, a state-of-the-art domain-agnostic minimization algorithm. Then, we present two novel GR(1)-specific heuristics to improve the core computation time by exploiting attributes of the realizability checking algorithm. These heuristics include (1) quickly discarding unneeded justice guarantees, and (2) reusing realizability check results to make subsequent checks faster. We implemented our work on top of the Spectra language and synthesizer and evaluated it over hundreds of specifications. Our evaluation shows that while using CDD in our context was ineffective, the new heuristics improve core computation times for 80% of the specifications, with an average improvement of more than three times compared to the baseline `QuickCore` algorithm.

1 Introduction

Reactive synthesis is an automated process for deriving correct-by-construction reactive systems from temporal specifications [26], rather than manually developing and verifying the implementation against a specification.

We focus on GR(1) [3], a fragment of Linear Temporal Logic (LTL). GR(1) specifications contain initial, safety, and justice assumptions and guarantees. Initial properties define the valid initial states, safety properties specify invariants for all states and transitions, and justice properties describe conditions that must be satisfied infinitely often in any infinite run. GR(1) has gained popularity and was implemented in several tools, e.g., RATSY [2], Slugs [7], and Spectra [19], thanks to its relatively efficient symbolic synthesis algorithm and expressive power, capable of representing almost all of the common specification

A. Sampaio and M. Stoelinga (Eds.): FM 2026, LNCS 16557, pp. 174–192, 2026.
https://doi.org/10.1007/978-3-032-26220-2_9

patterns identified by Dwyer et al. [6,16]. Its balance of expressiveness and computational efficiency makes it suitable for a wide range of applications of reactive synthesis, e.g., in robotics [12,17,30].

A major challenge of reactive synthesis is to debug unrealizable specifications [5,11,13]. One commonly suggested solution is to compute and use an *unrealizable core*, a locally minimal subset of the specification's guarantees that is sufficient for unrealizability. Because realizability checks are expensive—even for GR(1) specifications—core computation is time-consuming, which limits its usability in real-world settings.

In this work, we first examine whether the state-of-the-art domain-agnostic core computation algorithm `CDD` [33] can improve the state-of-the-art GR(1) core computation algorithm `QuickCore` [22]. Then, we present two domain-specific novel heuristics designed to improve the performance of GR(1) core computation time, thus reducing the cost of a common operation performed while developing real-world specifications. For each heuristic, we present its rationale, an overview of its implementation, a discussion of correctness, and an intuition regarding its expected impact. The heuristics are sound, meaning they always return a correct unrealizable core, even though they may return a different core from the one returned by a baseline algorithm (when there are multiple cores).

We have implemented our ideas on top of the Spectra GR(1) synthesizer [19]. We chose Spectra due to its support for GR(1) and unrealizable core computation, its use of existing heuristics [8], and the thousands of specifications it offers for extensive evaluation. In addition, Spectra already has an implementation of `QuickCore` [22], shown to be faster than other, domain-agnostic, core computation algorithms. We evaluated our heuristics on 359 unrealizable specifications from the SYNTECH benchmarks. Our results show performance degradation for the use of CDD, but major performance gains for the use of our new domain-specific heuristics, in particular, an average of core computation at least three times faster than the baseline, with negligible change in core size.

In summary, the contributions of this paper are: (1) An empirical evaluation of CDD in the context of GR(1) core computation; (2) Two novel GR(1)-specific performance heuristics for GR(1) core computation; (3) A comprehensive evaluation on 359 unrealizable specifications.

2 Preliminaries

We provide a short background on GR(1), Spectra, and unrealizable cores, with references to relevant papers for more details and complete formal definitions.

2.1 GR(1)

Since LTL synthesis is computationally expensive (2EXPTIME-complete [26]), authors have suggested LTL fragments with more efficient synthesis algorithms, for example, GR(1) [3,25], whose expressive power covers most of the well-known LTL specification patterns [6,16]. GR(1) specifications include assumptions and guarantees about what needs to hold on all initial states, on all states and transitions (safety), and infinitely often on every run (justice). A GR(1) synthesis problem consists of the following elements [3]:

- $\mathcal{X}$ is a set of input variables controlled by the environment;
- $\mathcal{Y}$ is a set of output variables controlled by the system;
- θ^e is an assertion, i.e., a propositional logic formula, over $\mathcal{X}$ characterizing initial environment states;
- θ^s is an assertion over $\mathcal{V} = \mathcal{X} \cup \mathcal{Y}$ characterizing initial system states;
- ρ^e is an assertion over $\mathcal{V} \cup \mathcal{X}'$, with $\mathcal{X}'$ a primed copy of variables $\mathcal{X}$; given a state, ρ^e restricts the next input;
- ρ^s is an assertion over $\mathcal{V} \cup \mathcal{V}'$, with $\mathcal{V}'$ a primed copy of variables $\mathcal{V}$; given a state and input, ρ^s restricts the next output;
- $J^e_{i \in 1..n}$ is a set of assertions over $\mathcal{V}$ for the environment to satisfy infinitely often (called justice assumptions);
- $J^s_{j \in 1..m}$ is a set of assertions over $\mathcal{V}$ for the system to satisfy infinitely often (called justice guarantees).

GR(1) Realizability. A GR(1) synthesis problem is realizable iff the following LTL formula is realizable:

$$\begin{aligned} \phi_{sr} = &(\theta_e \rightarrow \theta_s) \wedge (\theta_e \rightarrow G((H\rho_e) \rightarrow \rho_s)) \\ &\wedge (\theta_e \wedge G\rho_e \rightarrow (\bigwedge_{i=1}^{n} GFJ^e_i \rightarrow \bigwedge_{j=1}^{m} GFJ^s_j)) \end{aligned} \tag{1}$$

Intuitively, a GR(1) specification is strictly realizable if for every environment that starts with a valid assignment to its variables, the system can answer with a valid assignment to its variables, as long as the environment keeps its safety assumptions the system can keep its safety guarantees, and if the environment fulfills all its safety and justice assumptions, the system can guarantee to fulfill all its justice goals.

GR(1) realizability can be checked using the 3-nested fixed-point algorithm shown in Algorithm 1. Its complexity is measured in number of symbolic operations (implemented, e.g., using BDDs), and is $O(nmN^2)$, where n and m are the number of justice assumptions and guarantees, resp., and N is the state space.

Algorithm 1 GR(1) game algorithm [3] to compute system winning states Z

```
Z = true
while not reached fixed-point of Z do
    for j = 1 to |J^s| do
        Y = false; cy = 0
        while not reached fixed-point of Y do
            start = J^s_j ∧ ◇Z ∨ ◇Y
            Y = false
            for i = 1 to |J^e| do
                X = Z  //better approx. than true, see [3]
                while not reached fixed-point of X do
                    X = start ∨ (¬J^e_i ∧ ◇X)
                end while
                Y = Y ∨ X
                X[j][i][cy] ← X
            end for
            Y[j][cy++] ← Y
        end while
        Z = Y
        Z[j] = Y
    end for
end while
return Z
```

2.2 Spectra

Spectra [19] is a specification language and a synthesis environment that includes a GR(1) synthesizer. It extends GR(1)'s Boolean variables to finite-type variables, e.g., enumerations and bounded integers. Beyond GR(1) with several performance heuristics [8], it includes extensions that are reduced into GR(1), e.g., patterns [16] and triggers [1], and comes with several analyses, e.g., well-separation detection [9,18] and repairs for unrealizable specifications [20]. Since 2015, Spectra has been used by hundreds of final-year CS undergrads in semester-long project classes at Tel Aviv University, specifying and executing robotic and other systems, resulting in the SYNTECH collection of specifications. Spectra is available from https://github.com/spectrasynthesizer.

2.3 Abstract Syntax of a Specification

The notion of a core and some of the heuristics we present depend on the syntax of the specification. We repeat here the abstract syntax definition of a GR(1) specification, from [21].

Definition 1 (Abstract Syntax of a Specification). *A GR(1) specification is a tuple* $Spec = \langle V_e, V_s, D, M_e, M_s \rangle$*, where* V_e *and* V_s *are sets of environment and system variables respectively,* $D : V_e \cup V_s \rightarrow Doms$ *assigns a finite domain to each variable, and* M_e *and* M_s *are the environment and system modules. A module is a triplet* $M = \langle I, T, J \rangle$ *that contains sets of initial assertions* $I =$

$\{I_n\}_{n=1}^{i}$, *safety assertions* $T = \{T_n\}_{n=1}^{t}$, *and justice assertions* $J = \{J_n\}_{n=1}^{j}$ *of the module, where* $i = |I|, t = |T|$ *and* $j = |J|$. *The set of elements of module* $M = \langle I, T, J \rangle$ *is* $B_M = I \cup \{\mathbf{G}T_i\}_{i=1}^{t} \cup \{\mathbf{GF}J_i\}_{i=1}^{j}$.

Relating to the definition in Sect. 2.1, θ^e, θ^s correspond to I of the matching module (M_e, M_s respectively); ρ^e, ρ^s correspond to T of the matching module; and J_i^e, J_i^s correspond to J of the matching module.

2.4 Unrealizable Cores, Delta Debugging, and QuickCore

Given an unrealizable GR(1) specification, an *unrealizable core* is a locally minimal subset of system guarantees for which the specification is unrealizable [11]. Intuitively, an unrealizable specification means that the guarantees are too strong under the assumptions. Therefore, such a subset of guarantees is helpful for debugging unrealizable specifications, as it serves to localize a fault and might help the specification developer pinpoint the conflicting guarantees causing the unrealizability. Formally:

Definition 2 (Unrealizable Core). *Given an unrealizable GR(1) specification Spec* $= \langle V_e, V_s, D, M_e, M_s \rangle$, *an unrealizable core is a subset* $C = B_{M_s'} \subseteq B_{M_s}$ *of guarantees, for which the specification* $\langle V_e, V_s, D, M_e, M_s' \rangle$ *is unrealizable and for every strict subset* $C' = B_{M_s''} \subset C$, *the specification* $\langle V_e, V_s, D, M_e, M_s'' \rangle$ *is realizable.*

We present here three existing algorithms for computing unrealizable cores for GR(1) specifications, `LinearMin`, `DDMin`, and `QuickCore`. All algorithms receive a set to be minimized E as input (in our case, the set of guarantees) and a monotonic criterion `check` (in our case, unrealizability). All find a locally minimal subset of E for the criterion `check`. Monotonicity is required for the correctness of the algorithms, and, indeed, GR(1) realizability is monotonic [22].

LinearMin. The linear minimization algorithm (`LinearMin`) goes over elements of the input set E one by one, and removes an element iff the criterion `check` holds for the set without the element. An example of `LinearMin` can be found in Algorithm 3 lines 33-38.

`LinearMin` has a linear complexity in terms of the number of elements in the set E, $|E|$.

DDMin. The delta debugging algorithm (`DDMin`) was presented in [32] (see Algorithm 2). Its inputs are a set E, the `check` criterion, and the number n of parts of E to check. The initial call to minimize E has inputs E and $n = 2$.

The algorithm starts with $n = 2$ and refines E and n in recursive calls according to different cases (line 3, line 8, and line 14). The computation starts by partitioning E into n subsets and evaluating `check` on each subset *part* ("reduce to subset" - line 1) and its complement ("reduce to complement" - line 6). If `check` holds (line 2 or line 7), the search continues recursively on the

Algorithm 2 Delta debugging algorithm `DDMin` [32]

input: A set E to be minimized
input: A partition size n (initial value $n = 2$)
input: A monotonic criterion `check`
output: A locally minimal subset of E given `check`
1: **for** $part \in partition(E, n)$ **do**
2:     **if check**($part$) **then**
3:         return **ddmin**($part$, 2)
4:     **end if**
5: **end for**
6: **for** $part \in partition(E, n)$ **do**
7:     **if check**($E \setminus part$) **then**
8:         return **ddmin**($E \setminus part$, $max(n-1, 2)$)
9:     **end if**
10: **end for**
11: **if** $n \geq |E|$ **then**
12:     return E
13: **end if**
14: return **ddmin**(E, $min(|E|, 2n)$)

subset *part* (or its complement), until *part* (or its complement) has no subsets that satisfy `check`. If `check` holds neither for any subset *part* nor for the complements, the algorithm increases the granularity of the partitioning to $2n$ (line 14) and restarts or terminates when the granularity is not smaller than the size of E (line 12).

`DDMin` has quadratic worst-case complexity and logarithmic best-case complexity in terms of $|E|$.

Improving the performance of `DDMin` is an active research field, ranging from classic engineering improvements such as parallelization [10] to modified or domain-specific algorithms such as Hierarchical Delta Debugging [24], and more recently Probabilistic Delta Debugging [29] and Weighted Delta Debugging [34]. Most recently, `CDD`, Counter-Based Delta Debugging [33], was suggested. We examine the application of `CDD` to our context in Sect. 3.

QuickCore. Spectra implements a domain-specific algorithm for computing unrealizable cores in GR(1) specifications named `QuickCore` [22] (see Algorithm 3). The `QuickCore` algorithm first finds the core justice guarantees, then finds the safety guarantees, and finally the initial guarantees. The algorithm uses `DDMin` to find the required justice and safety guarantees, and then `LinearMin` without fixed-point computations to find the required initial guarantees. `QuickCore` is based on GR(1)-specific properties to make the core computation faster, mainly that it is possible to compute a core in steps, that realizability checks with fewer justice guarantees are typically much faster, and that it is unnecessary to perform multiple full realizability checks to find the required initial guarantees. Maoz and Shalom [22] evaluated `DDMin` and `QuickCore` on benchmarks from the GR(1) literature and showed that the latter is in most cases significantly faster than the former.

3 Examining Counter-Based Delta Debugging (CDD)

One new, state-of-the-art improvement to the `DDMin` algorithm is `CDD` - Counter-Based Delta Debugging, by Zhang et al. [33]. `CDD` is based on `ProbDD` from [29], and it aims to produce the same efficiency without `ProbDD`'s complex probabilistic model. Zhang et al. analyze the behavior of `ProbDD` and show that probabilities in `ProbDD` serve as monotonically increasing counters, and that probability itself may not be critical for its performance gain. They show that `ProbDD`'s superior performance compared to `DDMin` is due to a reduction in the total number of checks that fail to reduce the size of the set, such as "reduce to subset cases". Based on these observations, Zhang et al. suggest `CDD`, a simplified version of `ProbDD`, shown in Algorithm 4.

The algorithm works in rounds, signifying increasing probability of each element being part of a core. In each round, the algorithm first calculates the size of each subset in line 3. The size is calculated according to the analysis of `ProbDD` [33], which suggested a computation based on a round number and an initial probability p_0 (lines 13-16). The subset size decreases as the number of rounds increases, intuitively meaning that after enough larger subsets were checked, smaller subsets will be required to refine the subset. Lines 4-8 are similar to `DDMin`'s "reduce to complement" case (Algorithm 2 lines 6-10), trying to remove the subset and see if the remaining subset still matches the condition. The subset size keeps decreasing until it reaches 1, meaning each element will be individually removed once.

`CDD` was evaluated in [33] on three different benchmark suites and shown to be approximately 30% faster than `DDMin`, with approximately 52% less queries. This motivates evaluating CDD in the context of GR(1) core computation. Specifically, we propose modifying `QuickCore`'s implementation to use `CDD` instead of `DDMin` (Algorithm 3 lines 22, 29). In order to select the initial probability p_0, we first tried following the suggestions by [29]. However, a preliminary experiment showed that $p_0 = 0.15$ for justice minimization and $p_0 = 3.0/|T_s|$ for safety minimization gave better results, and therefore we use these values in our evaluation.

Note that using `CDD` instead of `DDMin` in `QuickCore` is correct, as `QuickCore` can work with any correct minimization algorithm [22]. In our case `CDD` ensures 1-minimality thanks to the monotonicity of unrealizability checking.

We expect a running time improvement induced by a reduction in the number of redundant realizability checks. However, the running time may also increase because realizability checks with more guarantees, especially justice guarantees, tend to take longer. In addition, it is expected that in some cases different cores will be found because the order of checks is modified.

Algorithm 3 The QuickCore algorithm from [22]. Given an unrealizable specification find an unrealizable core.
The blue code is the implementation of H1, see 4.1

Require: An unrealizable specification $S = \langle V_e, V_s, D, M_e, M_s\rangle$, where $M_e = \langle I_e, T_e, J_e\rangle$ and $M_s = \langle I_s, T_s, J_s\rangle$
Require: j_{last}, the index of the last used justice guarantee in the previous realizability check.
Ensure: An unrealizability core of S

$foundJustsEarly \leftarrow$ **false**
if $j_{last} = 1$ **then**
  $foundJustsEarly \leftarrow$ **true**
  ▷ At most one justice was used to detect unrealizability, do only the cheaper check (no justice at all) to decide
  if $\neg$**Realizable**($\langle V_e, V_s, D, M_e, \langle I_s, T_s, \emptyset\rangle\rangle$) **then**
    $J_s \leftarrow \emptyset$
  else
    $J_s \leftarrow \{J_0^s\}$
  end if
else if $j_{last} > 1$ **then**
  ▷ Guess that $J_{j_{last}}$ is enough for unrealizability
  if $\neg$**Realizable**($\langle V_e, V_s, D, M_e, \langle I_s, T_s, \{J_{jlast}^s\}\rangle\rangle$) **then**
    $foundJustsEarly \leftarrow$ **true**
    if $\neg$**Realizable**($\langle V_e, V_s, D, M_e, \langle I_s, T_s, \emptyset\rangle\rangle$) **then**
      $J_s \leftarrow \emptyset$
    else
      $J_s \leftarrow \{J_{j_{last}}^s\}$
    end if
  else
  ▷ $J_{j_{last}}$ is not enough, just narrow down the search space
    $J_s \leftarrow \{J_1^s, ..., J_{j_last}^s\}$
  end if
end if ▷ Begin by minimizing only justice guarantees
if $\neg foundJustsEarly \wedge$ **Realizable**($\langle V_e, V_s, D, M_e, \langle I_s, T_s, \emptyset\rangle\rangle$) **then**
  $J_c \leftarrow$ **MinWBase**(ddmin, $B_{M_s}, I_s \cup T_s, J_s, \neg$**Realizable**($S$))
else
  if $\neg foundJustsEarly \vee J_s = \emptyset$ **then**
    $J_c \leftarrow \emptyset$
    $M_e \leftarrow \langle I_e, T_e, \emptyset\rangle$
  end if
end if
▷ Continue with minimizing only safety guarantees
$T_c \leftarrow$ **MinWBase**(ddmin, $B_{M_s}, I_s \cup J_c, T_s, \neg$**Realizable**($S$))
▷ End with minimizing only initial guarantees
$w \leftarrow$ **ComputeWinRegion**($\langle V_e, V_s, D, M_e, \langle I_s, T_c, J_c\rangle\rangle$)
$I_c \leftarrow I_s$
$envIni \leftarrow \wedge_{d \in I_e} d$
for $i \in I_c$ **do** ▷ Use linear minimization for initial guarantees
  $I_c \leftarrow I_c \setminus \{i\}$
  if **SysWin**($envIni, \wedge_{d \in I_c} d, w$) **then**
    $I_c \leftarrow I_c \cup \{i\}$
  end if
end for
return $I_c \cup T_c \cup J_c$

Algorithm 4 Counter-Based Delta Debugging algorithm CDD [33]

input: A set E to be minimized
input: An initial probability p_0 that each element is required for unrealizability
input: A monotonic criterion check
output: A locally minimal subset of E given check

$r \leftarrow 0$
do
    $subSize \leftarrow$ ComputeSubsetSize(r, p_0)
    for $part \in partitionBySize(E, subSize)$ **do**
        if check($E \setminus part$) **then**
            $E \leftarrow E \setminus part$
        end if
    end for
    $r \leftarrow r + 1$
while $s > 1$
return E

function ComputeSubsetSize(r, p_0)
    $p_r \leftarrow p_0 \times 1.582^r$
    return $\arg\max_{s \in \mathbb{N}^+}(s \times (1 - p_r)^s)$
end function

4 GR(1)-Specific Performance Heuristics

We now present two new GR(1)-specific performance heuristics for unrealizable core computation. For each, we discuss background motivation, present the heuristic itself, discuss its correctness, and consider its expected impact. We define the heuristics on top of the approach presented in QuickCore [22]; however, they can also be applied in other contexts. Some of them require modifications in the GR(1) realizability checking algorithm (Algorithm 1) and in the QuickCore algorithm (Algorithm 3). The heuristics are independent and each can be applied regardless of the other. Both heuristics are sound and complete, meaning that given an unrealizable specification, they will always output a correct unrealizable core. However, they are not guaranteed to always perform better than QuickCore and, when there are multiple cores, they may return a different core than the one returned by QuickCore.

4.1 H1 - Early Core Justice Guarantees Detection

Motivation. The GR(1) winning states algorithm (Algorithm 1) is a fixed-point algorithm. Each iteration of the Z fixed-point goes over the justice guarantees one by one.

Early detection of unrealizability (eun) is a heuristic suggested in [8], whose rationale is that after the winning states are obtained for each system justice, it is sometimes possible to stop the fixed-point algorithm and declare unrealizability early because some initial states are losing for the system already with regard to the over-approximation of Z computed so far.

Our key new insight, on top of `eun`, is that if the algorithm terminates during the first iteration of the fixed-point, it may be possible to already narrow down the set of justice guarantees required for unrealizability.

Heuristic. Let $J_1, \ldots, J_n$ be the justice guarantees in the specification, and assume that the first iteration of the Z fixed-point finished early due to early detection of unrealizability, immediately after checking J_i $(1 \leq i \leq n)$. From that we can deduce two useful pieces of information:

1. **Immediate elimination of justice guarantees.** Since $J_{i+1}, \ldots, J_n$ were not even used by the algorithm before unrealizability was detected, it is enough to search for the justice guarantees of a core in $J_1, \ldots, J_i$, narrowing down the search space used in `QuickCore`'s justice guarantees search.
2. **Possible detection of a core guarantee.** The addition of J_i makes the algorithm detect that the specification is unrealizable. Although it may be possible that J_i is not actually the cause of the detected unrealizability, preliminary experiments show that in most cases J_i is not only part of a core, but also the only justice required for some core. Therefore, it is possible to speculate that J_i is the only justice required for unrealizability and check it. If the specification with only that justice J_i is indeed unrealizable, it is still required to check that the specification is realizable without J_i before going on to minimize the safety guarantees.

The heuristic is implemented by modifying both the GR(1) algorithm (Algorithm 1) and the `QuickCore` algorithm (Algorithm 3):

- The GR(1) algorithm now also returns a number indicating the last justice used in the computation, or -1 if the Z fixed-point algorithm required more than a single iteration (in which case the heuristic is not applicable).
- A modified `QuickCore` algorithm (colored blue in Algorithm 3) now gets the number of the last used justice computed in the GR(1) algorithm as input. It then removes all the unused justice guarantees from the specification, and checks whether keeping only the last used justice is still unrealizable. If so, it checks whether that justice is required for unrealizability and continues to minimize the safety guarantees. Otherwise, the original justice minimization continues only with the justice guarantees kept after the reduction.

Correctness. First, we will show the correctness of the immediate elimination of justice guarantees. As explained, the removed justice guarantees did not affect the unrealizability detection, and therefore the specification containing only $J_1, ..., J_i$ is unrealizable. Every core of the new specification is also a core of the original specification, and therefore it is indeed enough to compute the core of the new, smaller specification.

Second, the possible detection of a core guarantee speculates that two specific subsets of the guarantees, $\langle I, T, \emptyset\rangle$ and $\langle I, T, \{J_i\}\rangle$, are unrealizable (lines 4, 10, 12). If the first is unrealizable, then no justice guarantee is required for unrealizability. Else, if the second is unrealizable, then the single justice guarantee J_i will be part of the core. Otherwise, if both subsets are realizable, regular core computation will be performed, and will return the correct result.

In any case, the heuristic adds at most two realizability checks, and may, in some cases, not remove any checks. Therefore, the heuristic has no effect on the worst-case time complexity of the algorithm.

Expected Impact. On the one hand, the heuristic will be most effective when there are many justice guarantees, at least one of them is required for a core, and an early unrealizability is detected. On the other hand, the heuristic may degrade performance in presumably rare cases where an early unrealizability is detected for the last justice and that justice is not required for the core. In this case, an extra realizability check where only this justice is used will be performed, without any reduction in the number of guarantees later used for minimization. In theory, there might be other bad cases for the heuristic, i.e., discarding a small core that can be found in the end of the list of justice guarantees in favor of a larger core that is in the beginning. Yet, we have no reason to believe that such cases are frequent. Except for the potential extra realizability check, the heuristic's overheads of running time and memory are expected to be negligible.

The heuristic is only applicable when the specification includes justice guarantees. Otherwise, by definition, it has no effect.

4.2 H2 - Incremental Unrealizability Results

Motivation. During `QuickCore`'s internal `DDMin` executions, many realizability checks are performed over similar sets of guarantees. As also suggested in [8], reusing intermediate results between checks may reduce the time required for later checks.

Heuristic. For each unrealizable check result, we keep the system winning states Z_{prev} and reuse it as the initial value of Y in the GR(1) algorithm (Algorithm 1 line 4) for the subsequent checks (all, by definition, on subsets of the current set). As shown in [8], this yields correct results because Z_{prev} is a safe under-approximation of the correct Y, which is a least-fixed-point.

The incremental approach suggested by Firman et al. [8] was based on several cases, where each case allows a different kind of reuse. This entailed a high overhead in saving previous results beyond just Z_{prev}. Moreover, Firman et al. did not combine the reuse of Z with the `eun` optimization.

In contrast, we would like to keep using `eun` while we apply the reuse of Z_{prev}. However, due to early termination we may get Z_{prev} values that are not subsets of Y. This will lead to false-positive realizability check results - the algorithm may return that some unrealizable specifications are actually realizable - and we will eventually get an over-approximation of a core instead of a core.

To address this and ensure correctness, we add another, final minimization of the resulting unrealizable subset returned by the combination of `eun` and Z_{prev} reuse. Importantly though, for performance, we implement this final minimization using `LinearMin`, as we expect the subset to be a core or almost a core - in which case `DDMin` will perform a quadratic number of checks while `LinearMin` will be linear.

On top of the above, as an additional improvement to make the final `LinearMin` faster, when the resulting core approximation size is ≥ 5, we perform an additional realizability check without `eun`, and then use its Z as a sound initial value for Y for all the remaining checks.

We use this double minimization algorithm to minimize both justice and safety guarantees. For the initial guarantees, we use the over-approximated Z to check whether any initial guarantees are required for unrealizability. If none is required, we avoid the additional realizability check in Algorithm 3 line 30.

Correctness. To prove the correctness of the heuristic, it is enough to show that the result of the first minimization algorithm, initializing Y with the previous (eun) Z, exits with an unrealizable specification which is a subset of the original specification. In that case, the correctness of `LinearMin` is enough to prove that the second minimization yields an unrealizable core.

First, we show that when the algorithm initializes Y with some value Y_0 and returns that a specification is unrealizable, then it is indeed unrealizable. Let Y_0 be the initial value assigned to Y in Algorithm 3 as done in the heuristic. Let Z_f be the winning region computed by the algorithm, initialized with Y_0, and let Z_{win} be the real winning states (obtained when $Y_0 = \emptyset$). Because Y is a least fixed-point, starting from Y_0 yields a greater final value than starting with $\emptyset$. Therefore, we get that $Z_{win} \subseteq Z_f$. If the algorithm returned that a specification S is unrealizable, it means that there is an initial state $i \notin Z_f$, and hence $i \notin Z_{win}$. Therefore, the specification is indeed unrealizable.

Second, because `DDMin` correctly minimizes the specification, the final value returned by the first minimization will be some specification the algorithm found to be unrealizable.

Note that if $Y_0 \subseteq Z_{win}$, because Y is a least fixed-point, we get that $Z_f = Z_{win}$, in which case the approximation is correct. Therefore, starting with $Y_0 = Z_{prev}$, where $Z_{prev} \subseteq Z_{win}$ is the winning states of a larger specification (containing more guarantees), yields correct realizability check results.

The worst-case complexity of the algorithm remains the same, as `DDMin`'s worst-case complexity is $O(n^2)$ realizability checks, and the `LinearMin` adds a linear number of realizability checks.

Expected Impact. The heuristic will be most effective when the core found by the algorithm is much smaller than the original specification, and the Zs computed with `eun` are "close enough" to the correct Z values. When the core over-approximation is "too large", the extra cost of the additional `LinearMin` may be greater than the speedup afforded by the reuse.

5 Evaluation

5.1 Research Question

Our evaluation is guided by the following research question: What is the impact of changing `QuickCore` from using `DDMin` to using `CDD`, and the impact of our two heuristics, H1 and H2, alone and together, on GR(1) unrealizable core computation, in terms of total core computation time, the number of realizability checks, and the resulting core size?

5.2 Specification Corpus, Implementation, and Validation

Specification Corpus. For our evaluations, we used the SYNTECH set of benchmarks, which includes specifications written by CS students during semester-long project classes between 2015 and 2023, modeling robotic and other systems. They include not only final specifications but also many intermediate versions from the specification development process, and so, also many unrealizable specifications. The SYNTECH benchmarks have been used in the evaluation of many previous works, e.g., [1,4,8,9,14,23,31].

SYNTECH includes a total of 1294 unrealizable specifications. To focus on specifications where the need for improvement is significant, we filtered out all specifications where the baseline core computation takes less than 10 s. As a result, we remained with 359 unrealizable specifications.

Implementation. We have implemented our ideas on top of the Spectra GR(1) synthesizer [19] (with the heuristics of [8]), implemented in Java, with a modified version of the CUDD BDD library, implemented in C [28]. The dynamic reordering method used is sifting. Other settings remain unchanged.

Validation. To validate the correctness of our ideas and implementation, after finding the cores, we made additional realizability checks to make sure the result is indeed a core, i.e., independently validating that it is indeed an unrealizable set of guarantees, and that removing any single element from it results in a realizable specification. We checked this for all combinations of heuristics and over all the specifications in the corpus.

5.3 Experiment Setup and Reporting

We ran all experiments on an AWS r5.xlarge instance (representing a PC with up to 3.1GHz CPU and 32GB RAM) with Ubuntu 24.04 64-bit OS, Java OpenJDK 21 64Bit, and CUDD 3 compiled for 64Bit, using one CPU core. Although the core computation algorithm is deterministic, JVM garbage collection and CUDD implementation of dynamic reordering add variance to running times. Therefore, the times we report are the average of 5 runs. A timeout of 2 h was set for the initial realizability check, the core computation, and the core validation.

Table Structure. The results are presented in a table with the following columns:

- Heuristics - The heuristics used for the results in the current row.
- R_t - The geometric mean of the ratios between the total core computation time with the heuristics and the baseline time. The lower the number, the faster the heuristics are compared to the baseline.
- $R_t < 1$ - The percentage of specifications for which the heuristics improved the running time. The higher the number, the faster the heuristics are compared to the baseline.
- R_c - The geometric mean of the ratios between the number of checks with the heuristics and the baseline. A lower value means the heuristics reduce the number of realizability checks performed compared to the baseline.
- $R_c < 1$ - The percentage of specifications for which the heuristics reduced the number of checks.
- R_s - The geometric mean of the ratios between the number of elements in the resulting core with the heuristics and the baseline. A lower number means the heuristics yield a smaller core than the baseline.
- $R_s < 1$ - The percentage of specifications for which the heuristics reduced the core size.
- R_{TO} - The ratio between the number of timeouts with and without the heuristics.
- RT_{TO} - The ratio between the number of timeouts to the total number of specifications in the corpus.

5.4 Results

We aim to assess the impact of `CDD` usage and of each heuristic on the performance of core computation. We use Spectra's original `QuickCore` implementation using `DDMin` as a baseline, and compare the results of `CDD` and for each heuristic and their combinations to it. In all cases, including the baseline, we use Spectra with all the heuristics from [8].

The results of the experiments appear in Table 1. For example, according to Table 1, H1 made unrealizable core computations run 60.5% faster ($R_t = 0.395$), and was faster for 88.5% of the specifications ($R_t < 1 = 0.885$); it reduced 14.7% of the checks ($R_c = 0.853$, reducing the number of checks in 76% of the specifications ($R_c < 1 = 0.760$,); it did not change the core size ($R_s = 1$, and it did not cause additional timeouts compared to the baseline ($R_{TO} = 1$, leaving the total percentage of timeouts at 0.3% ($RT_{TO} = 0.003$).

Table 1. Time, check count, core size, and timeout results for different combinations of CDD and heuristics, over 359 SYNTECH unrealizable specifications. The best values in each column are marked in **bold**.

Heuristic	R_t	$R_t < 1$	R_c	$R_c < 1$	R_s	$R_s < 1$	R_{TO}	RT_{TO}
CDD	1.309	0.378	0.902	0.543	1.015	**0.056**	2.000	0.006
H1	0.395	**0.885**	**0.853**	**0.760**	**1.000**	0.000	**1.000**	**0.003**
H2	0.374	0.765	1.284	0.000	1.016	0.000	**1.000**	**0.003**
H1+H2	0.288	0.796	1.132	0.257	1.016	0.000	**1.000**	**0.003**
CDD+H1+H2	**0.231**	0.809	1.031	0.346	1.022	0.025	3.000	0.008

From the results in Table 1, we make the following key observations:

1. Surprisingly, CDD performs much slower than the baseline for the majority of the specifications ($R_t = 1.309$). While the average number of checks is indeed reduced, the average reduction is only 10% ($R_c = 0.902$), on less than half of the specifications.
2. While both H1 and H2 are effective separately, their combination is even more effective, yielding an average speedup of 71.2% ($R_t = 0.288$), improving 79.6% of the specifications.
3. The combination of CDD with H1 and H2 yields an additional improvement, 76.9% faster ($R_t = 0.231$) than the baseline for 80.9% of the specifications.
4. The heuristics and CDD have a negligible effect, if any, on core size ($1 \leq R_s \leq 1.022$).

To answer our research question:

- Using CDD was found to be harmful to the performance of `QuickCore`, while having a negligible effect on the number of checks and core size.
- Both heuristics and their combination yield a major speedup of up to 71.2% on average, while having a negligible effect on the core size.

6 Related Work

Performance Heuristics for GR(1). Firman et al. [8] suggested performance heuristics for GR(1) synthesis and realizability checking, including incremental heuristics for core computation. However, they found these incremental heuristics ineffective, largely because they were not combined with early unrealizability detection, as in our work. Yatskan et al. [31] suggested additional realizability checking heuristics, as well as heuristics applicable for the detection of non-well-separation. Ma'ayan et al. [15] introduced evolution-aware heuristics, which reuse results from a previous check to accelerate a subsequent realizability check.

However, they did not apply similar techniques to unrealizable core computation, unlike our present work.

GR(1) Unrealizable Core Computation. Cimatti et al. [5] employed `LinearMin` for core computation. Könighofer et al. [11] implemented `DDMin` for core computation in RATSY and showed that it typically outperforms `LinearMin`. Firman et al. [8] incorporated `DDMin` in Spectra, augmenting it with optimizations such as caching. Both `LinearMin` and `DDMin` are domain-agnostic, and do not exploit GR(1)-specific knowledge to further improve computation time.

Maoz and Shalom [22] presented `QuickCore`, a domain-specific algorithm for GR(1) unrealizable core computation that substantially outperforms `DDMin`. However, unlike our approach, `QuickCore` does not reuse results from previous realizability checks to accelerate subsequent ones.

Numerous works aim to improve the performance of domain-agnostic minimization algorithms, particularly `DDMin`. Hodován and Kiss proposed engineering improvements such as parallelization [10]. Misherghi and Su [24] suggested Hierarchical Delta Debugging. Zhou et al. [34] developed Weighted Delta Debugging. Wang et al. [29] presented Probabilistic Delta Debugging, and Zhang et al. [33] used it to build Counter-Based Delta Debugging, `CDD`. All these algorithms are domain-agnostic. Our evaluation showed that compared to `DDMin`, `CDD` does not improve core computation time for GR(1) specifications.

7 Conclusion

We introduced two domain-specific heuristics to enhance the performance of GR(1) unrealizable core computation. The heuristics leverage the observation that previous realizability check results may be used to reduce the number and complexity of future checks. In addition, we examined the use of `CDD`, a domain-agnostic state-of-the-art variant of `DDMin` in the context of GR(1). We implemented and evaluated these algorithms over a large corpus containing hundreds of GR(1) specifications written in the Spectra language. The results showed that compared to the previous implementation of the Spectra synthesizer, based on [8,22], the domain-agnostic algorithm `CDD` does not improve the core computation time, while each of our domain-specific heuristics and their combination are very effective in reducing the total core computation time.

As future work we consider the application of similar heuristics to the computation of cores for kind-unrealizability [9], for non-well-separation [18], for unnecessary assumptions [27], and for the detection of all unrealizable cores.

Acknowledgments. This research was supported by the Israel Science Foundation (grant No. 1954/23, SPECTACKLE) as well as by Len Blavatnik and the Blavatnik Family Foundation. The authors thank Rafi Shalom, Ilia Shevrin, and the anonymous reviewers for helpful comments.

Artifact. An artifact is available in https://doi.org/10.5281/zenodo.18523068. It includes the benchmarks used in our evaluation, raw data, and scripts to run and reproduce the experiments.

References

1. Amram, G., Ma'ayan, D., Maoz, S., Pistiner, O., Ringert, J.O.: Triggers for reactive synthesis specifications. In: 45th IEEE/ACM International Conference on Software Engineering, ICSE 2023, Melbourne, Australia, 14–20 May 2023, pp. 729–741. IEEE (2023). https://doi.org/10.1109/ICSE48619.2023.00070
2. Bloem, R.: RATSY - a new requirements analysis tool with synthesis. In: CAV. LNCS, vol. 6174, pp. 425–429. Springer (2010). https://doi.org/10.1007/978-3-642-14295-637
3. Bloem, R., Jobstmann, B., Piterman, N., Pnueli, A., Sa'ar, Y.: Synthesis of reactive(1) designs. J. Comput. Syst. Sci. **78**(3), 911–938 (2012)
4. Cavezza, D.G., Alrajeh, D., György, A.: Minimal assumptions refinement for realizable specifications. In: Bae, K., Bianculli, D., Gnesi, S., Plat, N. (eds.) FormaliSE@ICSE 2020: 8th International Conference on Formal Methods in Software Engineering, Seoul, Republic of Korea, 13 July 2020, pp. 66–76. ACM (2020). https://doi.org/10.1145/3372020.3391557
5. Cimatti, A., Roveri, M., Schuppan, V., Tchaltsev, A.: Diagnostic information for realizability. In: Logozzo, F., Peled, D.A., Zuck, L.D. (eds.) Verification, Model Checking, and Abstract Interpretation, 9th International Conference, VMCAI 2008, San Francisco, USA, 7–9 January 2008, Proceedings. Lecture Notes in Computer Science, vol. 4905, pp. 52–67. Springer (2008). https://doi.org/10.1007/978-3-540-78163-9_9
6. Dwyer, M.B., Avrunin, G.S., Corbett, J.C.: Patterns in property specifications for finite-state verification. In: ICSE, pp. 411–420. ACM (1999)
7. Ehlers, R., Raman, V.: Slugs: extensible GR(1) synthesis. In: Chaudhuri, S., Farzan, A. (eds.) CAV 2016. LNCS, vol. 9780, pp. 333–339. Springer, Cham (2016). https://doi.org/10.1007/978-3-319-41540-618
8. Firman, E., Maoz, S., Ringert, J.O.: Performance heuristics for GR(1) synthesis and related algorithms. Acta Informatica, 37–79 (2019). https://doi.org/10.1007/s00236-019-00351-9
9. Gorenstein, A., Maoz, S., Ringert, J.O.: Kind controllers and fast heuristics for non-well-separated GR(1) specifications. In: Proceedings of the 46th IEEE/ACM International Conference on Software Engineering, ICSE 2024, Lisbon, Portugal, 14–20 April 2024, pp. 28:1–28:12. ACM (2024). https://doi.org/10.1145/3597503.3608131
10. Hodován, R., Kiss, Á.: Practical improvements to the minimizing delta debugging algorithm. In: Maciaszek, L.A., Cardoso, J., Ludwig, A., van Sinderen, M., Cabello, E. (eds.) Proceedings of the 11th International Joint Conference on Software Technologies (ICSOFT 2016) - Volume 1: ICSOFT-EA, Lisbon, Portugal, 24–26 July 2016, pp. 241–248. SciTePress (2016). https://doi.org/10.5220/0005988602410248
11. Könighofer, R., Hofferek, G., Bloem, R.: Debugging formal specifications: a practical approach using model-based diagnosis and counterstrategies. Int. J. Softw. Tools Technol. Transf. **15**(5–6), 563–583 (2013). https://doi.org/10.1007/S10009-011-0221-Y

12. Kress-Gazit, H., Fainekos, G.E., Pappas, G.J.: Temporal-logic-based reactive mission and motion planning. IEEE Trans. Robotics **25**(6), 1370–1381 (2009)
13. Kuvent, A., Maoz, S., Ringert, J.O.: A symbolic justice violations transition system for unrealizable GR(1) specifications. In: Bodden, E., Schäfer, W., van Deursen, A., Zisman, A. (eds.) Proceedings of the 2017 11th Joint Meeting on Foundations of Software Engineering, ESEC/FSE 2017, Paderborn, Germany, 4–8 September 2017, pp. 362–372. ACM (2017). https://doi.org/10.1145/3106237.3106240
14. Ma'ayan, D., Maoz, S.: Using reactive synthesis: an end-to-end exploratory case study. In: 45th IEEE/ACM International Conference on Software Engineering, pp. 742–754. IEEE (2023). https://doi.org/10.1109/ICSE48619.2023.00071
15. Ma'ayan, D., Maoz, S., Ringert, J.O.: Evolution-aware heuristics for GR(1) realizability checking. In: 40th IEEE/ACM International Conference on Automated Software Engineering, ASE, pp. 266–277. IEEE (2025)
16. Maoz, S., Ringert, J.O.: GR(1) synthesis for LTL specification patterns. In: ESEC/FSE, pp. 96–106. ACM (2015)
17. Maoz, S., Ringert, J.O.: Synthesizing a lego forklift controller in GR(1): a case study. In: Cerný, P., Kuncak, V., Madhusudan, P. (eds.) Proceedings Fourth Workshop on Synthesis, SYNT 2015, San Francisco, CA, USA, 18th July 2015. EPTCS, vol. 202, pp. 58–72 (2015). https://doi.org/10.4204/EPTCS.202.5
18. Maoz, S., Ringert, J.O.: On well-separation of GR (1) specifications. In: Proceedings of the 2016 24th ACM SIGSOFT International Symposium on Foundations of Software Engineering, pp. 362–372 (2016)
19. Maoz, S., Ringert, J.O.: Spectra: a specification language for reactive systems. Softw. Syst. Model. **20**(5), 1553–1586 (2021). https://doi.org/10.1007/s10270-021-00868-z
20. Maoz, S., Ringert, J.O., Shalom, R.: Symbolic repairs for GR(1) specifications. In: Atlee, J.M., Bultan, T., Whittle, J. (eds.) Proceedings of the 41st International Conference on Software Engineering, ICSE 2019, Montreal, QC, Canada, 25–31 May 2019, pp. 1016–1026. IEEE / ACM (2019). https://doi.org/10.1109/ICSE.2019.00106
21. Maoz, S., Shalom, R.: Inherent vacuity for GR(1) specifications. In: ESEC/FSE, pp. 99–110. ACM (2020). https://doi.org/10.1145/3368089.3409669
22. Maoz, S., Shalom, R.: Unrealizable cores for reactive systems specifications. In: 43rd IEEE/ACM International Conference on Software Engineering, ICSE 2021, Madrid, Spain, 22–30 May 2021, pp. 25–36. IEEE (2021). https://doi.org/10.1109/ICSE43902.2021.00016
23. Menghi, C., Tsigkanos, C., Pelliccione, P., Ghezzi, C., Berger, T.: Specification patterns for robotic missions. IEEE Trans. Softw. Eng. **47**(10), 2208–2224 (2021). https://doi.org/10.1109/TSE.2019.2945329
24. Misherghi, G., Su, Z.: HDD: hierarchical delta debugging. In: Osterweil, L.J., Rombach, H.D., Soffa, M.L. (eds.) 28th International Conference on Software Engineering (ICSE 2006), Shanghai, China, 20–28 May 2006, pp. 142–151. ACM (2006). https://doi.org/10.1145/1134285.1134307
25. Piterman, N., Pnueli, A., Sa'ar, Y.: Synthesis of reactive (1) designs. In: International Workshop on Verification, Model Checking, and Abstract Interpretation, pp. 364–380. Springer (2005)
26. Pnueli, A., Rosner, R.: On the synthesis of a reactive module. In: POPL, pp. 179–190. ACM Press (1989)
27. Shalom, R., Maoz, S.: Which of my assumptions are unnecessary for realizability and why should I care? In: 45th IEEE/ACM International Conference on Soft-

ware Engineering, ICSE 2023, Melbourne, Australia, 14–20 May 2023, pp. 221–232. IEEE (2023). https://doi.org/10.1109/ICSE48619.2023.00030
28. Somenzi, F.: CUDD: CU Decision Diagram Package Release 2.3. 0. University of Colorado at Boulder, p. 621 (1998)
29. Wang, G., Shen, R., Chen, J., Xiong, Y., Zhang, L.: Probabilistic delta debugging. In: Spinellis, D., Gousios, G., Chechik, M., Penta, M.D. (eds.) ESEC/FSE '21: 29th ACM Joint European Software Engineering Conference and Symposium on the Foundations of Software Engineering, Athens, Greece, 23–28 August 2021, pp. 881–892. ACM (2021). https://doi.org/10.1145/3468264.3468625
30. Wete, E., Greenyer, J., Wortmann, A., Flegel, O., Klein, M.: Monte Carlo tree search and GR(1) synthesis for robot tasks planning in automotive production lines. In: 24th International Conference on Model Driven Engineering Languages and Systems, MODELS 2021, Fukuoka, Japan, 10–15 October 2021, pp. 320–330. IEEE (2021). https://doi.org/10.1109/MODELS50736.2021.00039
31. Yatskan, R., Shevrin, I., Maoz, S.: Performance heuristics for GR(1) realizability checking and related analyses. In: Gurfinkel, A., Heule, M. (eds.) Tools and Algorithms for the Construction and Analysis of Systems - 31st International Conference, TACAS 2025, Held as Part of the International Joint Conferences on Theory and Practice of Software, ETAPS 2025, Hamilton, ON, Canada, 3–8 May 2025, Proceedings, Part I. Lecture Notes in Computer Science, vol. 15696, pp. 40–59. Springer (2025). https://doi.org/10.1007/978-3-031-90643-5_3
32. Zeller, A., Hildebrandt, R.: Simplifying and isolating failure-inducing input. IEEE Trans. Softw. Eng. **28**(2), 183–200 (2002)
33. Zhang, M., Xu, Z., Tian, Y., Cheng, X., Sun, C.: Toward a better understanding of probabilistic delta debugging. In: 47th IEEE/ACM International Conference on Software Engineering, ICSE 2025, Ottawa, ON, Canada, April 26 - May 6, 2025, pp. 2024–2035. IEEE (2025). https://doi.org/10.1109/ICSE55347.2025.00117
34. Zhou, X., Xu, Z., Zhang, M., Tian, Y., Sun, C.: WDD: weighted delta debugging. In: 47th IEEE/ACM International Conference on Software Engineering, ICSE 2025, Ottawa, ON, Canada, April 26 - May 6, 2025, pp. 1592–1603. IEEE (2025). https://doi.org/10.1109/ICSE55347.2025.00071

A General Framework for Robust Quantitative Semantics of Signal Temporal Logic

Jiawei Chen[1(✉)], José Luiz Vargas de Mendonça[1], Konstantinos Mamouras[2], and Jean-Baptiste Jeannin[1]

[1] University of Michigan, Ann Arbor, MI, USA
chenjw@umich.edu
[2] Rice University, Houston, TX, USA
jeannin@umich.edu

Abstract. Quantitative semantics of Signal Temporal Logic (STL) play an important role in both the falsification and control synthesis for dynamical systems by assigning numerical quantities to truth values. Recently, several different quantitative semantics have been proposed, offering better performance in many cases. Yet a general, systematic understanding of the structure and properties of quantitative semantics is missing. In this paper, we develop a general framework to model quantitative semantics. We focus mainly on soundness, which requires that the quantitative semantics of a statement is positive when the statement is true, and negative when the statement is false. This ensures that counterexamples will not be missed during verification. We derive simple, necessary conditions in our framework for soundness. We show how several recently proposed quantitative semantics fit in our framework, and how others do not, typically because they do not strictly satisfy soundness. We implement various quantitative semantics, including existing semantics from literature, in our framework and compare their effectiveness as objective functions for optimization-based falsification on both novel and existing benchmarks.

Keywords: Signal Temporal Logic · Quantitative Semantics · Cyber-Physical Systems

1 Introduction

Metric temporal logic (MTL) [23] and signal temporal logic (STL) [27] are widely used for specifying correctness properties for cyber-physical systems [6]. The standard semantics for MTL and STL is qualitative, specifically Boolean. The Boolean semantics only tells us whether a property φ holds or not with respect to a system trace. This is, however, not informative enough. In many cases, we would also like to know how *robustly* the system trace satisfies (or falsifies) a desired property. This can be useful, for example, when determining the degree

A. Sampaio and M. Stoelinga (Eds.): FM 2026, LNCS 16557, pp. 193–213, 2026.
https://doi.org/10.1007/978-3-032-26220-2_10

to which a system's safety is resilient against unexpected disturbances. Faella, Legay, and Stoelinga [14] propose a quantitative framework for linear temporal logic (LTL) that can be verified via model checking. Fainekos and Pappas [15] define a real-valued quantitative semantics for MTL and STL that captures the notion of robustness of satisfaction.

In the semantics of Fainekos and Pappas, a truth value is a real number or $\pm\infty$ with the following interpretation: a positive number indicates truth, a negative number indicates falsity, and the truth value 0 is ambiguous. The logical and temporal connectives are interpreted using the operation min for conjunction and max for disjunction. Negation is interpreted using the unary $-$ operation. A key property of this min-max semantics is that it is *sound*, which roughly means that it agrees with the standard qualitative (i.e., Boolean) semantics. If ρ is the min-max semantics of Fainekos and Pappas and the Boolean semantics is given with the satisfaction relation (denoted by $\models$), then the soundness property can be formulated as follows:

1. $\rho(\varphi, \sigma, t) > 0$ implies $\sigma^t \models \varphi$ and
2. $\rho(\varphi, \sigma, t) < 0$ implies $\sigma^t \not\models \varphi$

for every temporal formula φ, trace σ and time instant t, where we write $\sigma^t \models \varphi$ to mean that the trace σ satisfies the formula φ at time instant t.

The real-valued quantitative semantics of MTL/STL has found use in the *falsification* of cyber-physical systems, which is about finding design errors using formal specifications of correct/safe system behavior. Given a system model M and a temporal formula φ, the falsification problem asks whether there exists an input signal σ_i so that the system trace $M(\sigma_i)$ falsifies the desired property φ. Solving this problem amounts to a search in the space of input signals. When a (sound) quantitative semantics is available, then falsification can be reduced to a minimization problem: find an input signal σ_i that minimizes the truth value $\rho(\varphi, M(\sigma_i))$. Several different optimization approaches have been used to solve this problem: Monte-Carlo stochastic optimization [1,32], cross-entropy [34], and other iterative approaches [2]. Breach [11] and S-TaLiRo [4] are two state-of-the-art tools for the falsification of temporal properties of cyber-physical systems.

The robustness (quantitative) semantics of MTL and STL are also useful for solving the controller synthesis problem for cyber-physical systems, where the goal is to find parameter values for a controller so that the system satisfies the specification as robustly as possible [33]. The controller synthesis problem can be cast as a robustness maximization problem.

After the proposed min-max quantitative semantics of Fainekos and Pappas [15], which is often described as "spatial" robustness, several other quantitative semantics have been proposed for MTL and STL. Donzé and Maler [10] consider an extension of spatial robustness that also accounts for temporal displacement. Akazaki and Hasuo [3] proposed an extension of MITL with averaged temporal operators. Another average-based robustness was explored in [26,30]. The original robustness semantics [15] uses max (resp., min) for interpreting disjunction (resp., conjunction), which are not smooth functions. Since smoothness is a valuable property in the context of falsification and synthesis, many authors have considered smooth variants of the robustness semantics [16,17,33].

Similarly, the TeLEx tool [21] uses a smooth version of a *tightness* semantics in order to learn STL formulas from only positive examples. TeLEx uses smooth functions for the temporal operators and atomic predicates, which means that the semantics is differentiable. This smoothness property enables the effective use of gradient-based optimization methods.

Recent developments [35,36] employ Monte Carlo Tree Search to overcome some challenges of a purely optimization-based approach to falsification, in particular the "scale problem" when combining formulas. Nevertheless, these methods still employ quantitative semantics, so the generalization of these semantics, and especially the formalization of soundness conditions, remains relevant.

This proliferation of quantitative semantics for MTL/STL raises the question of whether it is possible to develop a general mathematical framework that (1) can describe a large class of quantitative semantics, and (2) encompasses previously considered semantics. In the context of such a framework, it would be important to develop a methodology for easily checking different properties of a semantics, especially its soundness. Soundness is a fundamental property that ensures agreement with the standard Boolean semantics and is essential for avoiding incorrect results, especially in the context of falsification. A generalization of quantitative semantics has been proposed previously [5,19], based on semirings and an automata-based construction, but it is limited to discrete-time models. Monitoring with algebraic robustness semantics is considered in [8,29] (using lattices) and in [28] (using semirings). In contrast, our proposed generalization and soundness criteria can be extended to continuous-time temporal logics, and operate directly on the original unmodified specification. While many of the semantics we examined may be implemented in the aforementioned discrete-time framework, there exist some which benefit from a more granular approach that allows mixing of different integrators of the same subclass, e.g. using a combination of min and multiplication. In addition, we introduce soundness conditions to check the individual member functions in our generalization. Regardless of the method, a general framework is important for methodical comparisons of semantics, independent of their original implementations.

Contributions. Our main contributions extend previous work-in-progress [20] in the following ways:

- We describe a mathematical framework for robustness quantitative semantics for MTL/STL that allows us to easily check that a semantics is sound.
- We show how this new framework can generalize many generalizations of quantitative semantics that are found in previous work.
- Using our framework, we identify a mistake in previous work [17], showing the effectiveness of our framework at ensuring soundness of quantitative semantics. We also show how to fix that semantics.
- Using Breach, we compare experimentally the effectiveness of several different robustness semantics in the context of CPS falsification.

Paper Outline. Section 2 gives some preliminary definitions for the syntax of MTL/STL, its standard qualitative (Boolean) semantics, and the widely used min-max quantitative semantics of [15]. Section 3 presents our generalization of robustness semantics for MTL/STL and Sect. 4 details sufficient conditions for a semantics to be sound. In Sect. 5, we present an experimental comparison of robustness semantics in the context of CPS falsification.

2 A Generalization of Quantitative Semantics

2.1 Syntax and Qualitative Semantics

We consider properties of a real-valued signal over a discrete or continuous domain $\mathbb{T}$. In the continuous-time case, $\mathbb{T} = \mathbb{R}_+$; in the discrete-time case $\mathbb{T} = \mathbb{N}$. We denote by I a bounded interval over $\mathbb{T}$, and by $\mathbb{I}$ the set of all such bounded intervals. In the continuous case, we assume that the signal is integrable on any bounded interval $I \in \mathbb{I}$. A signal σ over n variables $x_1, ..., x_n$ is a function from $\mathbb{T}$ to $\mathbb{R}^n$. We write $l(\sigma)$ to mean a function of those n variables with real values. We also write $\overline{\mathbb{R}} = \mathbb{R} \cup \{-\infty, +\infty\}$ and similarly $\overline{\mathbb{R}}_+ = \mathbb{R}_+ \cup \{+\infty\}$ and $\overline{\mathbb{R}}_- = \mathbb{R}_- \cup \{-\infty\}$.

Syntax. An STL formula φ is given by:

$$\varphi, \psi ::= \top \mid l(\sigma) \geq 0 \mid l(\sigma) > 0 \mid \neg\varphi \mid \varphi \wedge \psi \mid \psi \mathbf{U}_I \varphi$$

The qualitative semantics of STL, $\sigma^t \models \varphi$, defines the truth value of σ at time t with respect to φ, and is standard [10, 15]:

- It is always true that $\sigma^t \models \top$.
- $\sigma^t \models l(\sigma) \geq 0$ (resp. > 0) if and only if the value of l at time t, $l(\sigma[t]) \geq 0$ (resp. > 0).
- $\sigma^t \models \neg\varphi$ when $\sigma^t \not\models \varphi$.
- $\sigma^t \models \varphi_1 \wedge \varphi_2$ when $\sigma^t \models \varphi_1$ and $\sigma^t \models \varphi_2$.
- $\sigma^t \models \varphi_1 \mathbf{U}_I \varphi_2$ if and only if there exists $t' \in I$ such that $\sigma^{t+t'} \models \varphi_2$ and for all $t'' \in [t, t+t']$, $\sigma^{t''} \models \varphi_1$.

Operators $\bot$, $\vee$, $\mathbf{F}_I$ and $\mathbf{G}_I$ are derived from the above using the standard equivalences.

2.2 Standard Quantitative Semantics

The standard quantitative semantics $\rho_0(\varphi, \sigma, t) \in \overline{\mathbb{R}}$ is defined for a specification φ, a trace σ over $\mathbb{T}$, and a time $t \in \mathbb{T}$ [10, 15]. It is convenient to split ρ_0 into its positive and negative parts $\rho_0^+ \in \overline{\mathbb{R}}_+$ and $\rho_0^- \in \overline{\mathbb{R}}_-$ to facilitate adding robustness values, following the convention of existing literature [3, 17]. Here, $\rho_0(\varphi, \sigma, t) = \rho_0^+(\varphi, \sigma, t) + \rho_0^-(\varphi, \sigma, t)$, where ρ_0^+ and ρ_0^- are:

$$\begin{aligned}
\rho_0^+(\top, \sigma, t) &= +\infty \\
\rho_0^-(\top, \sigma, t) &= 0 \\
\rho_0^+(l(\sigma) \geq 0, \sigma, t) &= \rho_0^+(l(\sigma) > 0, \sigma, t) = \max(0, l(\sigma[t])) \\
\rho_0^-(l(\sigma) \geq 0, \sigma, t) &= \rho_0^-(l(\sigma) > 0, \sigma, t) = \min(0, l(\sigma[t])) \\
\rho_0^+(\neg\varphi, \sigma, t) &= -\rho_0^-(\varphi, \sigma, t) \\
\rho_0^-(\neg\varphi, \sigma, t) &= -\rho_0^+(\varphi, \sigma, t) \\
\rho_0^+(\varphi \wedge \psi, \sigma, t) &= \min(\rho_0^+(\varphi, \sigma, t), \rho_0^+(\psi, \sigma, t)) \\
\rho_0^-(\varphi \wedge \psi, \sigma, t) &= \min(\rho_0^-(\varphi, \sigma, t), \rho_0^-(\psi, \sigma, t)) \\
\rho_0^+(\psi \mathbf{U}_I \varphi, \sigma, t) &= \sup_{t' \in I} \left(\min(\rho_0^+(\varphi, \sigma, t + t'), \inf_{t'' \in [t, t+t']} \rho_0^+(\psi, \sigma, t'')) \right) \\
\rho_0^-(\psi \mathbf{U}_I \varphi, \sigma, t) &= \sup_{t' \in I} \left(\min(\rho_0^-(\varphi, \sigma, t + t'), \inf_{t'' \in [t, t+t']} \rho_0^-(\psi, \sigma, t'')) \right)
\end{aligned}$$

An essential property of quantitative semantics is soundness:

Theorem 1. *if* $\rho_0(\varphi, \sigma, t) > 0$ *then* $\sigma^t \models \varphi$, *and if* $\rho_0(\varphi, \sigma, t) < 0$ *then* $\sigma^t \not\models \varphi$.

Proof. This result is proved by structural induction on the STL formula [10, 15].

2.3 A General Framework

A goal of this work is to establish a generalized form for quantitative semantics, parameterized by unary functions $\nu : \mathbb{R} \to \overline{\mathbb{R}}_+$ and $\mu : \mathbb{R} \to \overline{\mathbb{R}}_-$, binary integrators α, β, ζ, η : $\overline{\mathbb{R}}_+ \times \overline{\mathbb{R}}_+ \to \overline{\mathbb{R}}_+$, as well as time integrators Γ, Δ, Θ, Ξ : $\mathbb{I} \times (\mathbb{T} \to \overline{\mathbb{R}}_+) \to \overline{\mathbb{R}}_+$. Using those operators, we can define a generic $\rho(\varphi, \sigma, t) = \rho^+(\varphi, \sigma, t) + \rho^-(\varphi, \sigma, t)$ with:

$$\begin{aligned}
\rho^+(\top, \sigma, t) &= +\infty \\
\rho^-(\top, \sigma, t) &= 0 \\
\rho^+(l(\sigma) \geq 0, \sigma, t) &= \rho^+(l(\sigma) > 0, \sigma, t) = \nu(l(\sigma[t])) \\
\rho^-(l(\sigma) \geq 0, \sigma, t) &= \rho^-(l(\sigma) > 0, \sigma, t) = \mu(l(\sigma[t])) \\
\rho^+(\neg\varphi, \sigma, t) &= -\rho^-(\varphi, \sigma, t) \\
\rho^-(\neg\varphi, \sigma, t) &= -\rho^+(\varphi, \sigma, t) \\
\rho^+(\varphi \wedge \psi, \sigma, t) &= \alpha(\rho^+(\varphi, \sigma, t), \rho^+(\psi, \sigma, t)) \\
\rho^-(\varphi \wedge \psi, \sigma, t) &= -\beta(-\rho^-(\varphi, \sigma, t), -\rho^-(\psi, \sigma, t)) \\
\rho^+(\psi \mathbf{U}_I \varphi, \sigma, t) &= \mathop{\Gamma}_{t' \in I} \zeta\left(\rho^+(\varphi, \sigma, t + t'), \mathop{\Delta}_{t'' \in [t, t+t']} \rho^+(\psi, \sigma, t'') \right) \\
\rho^-(\psi \mathbf{U}_I \varphi, \sigma, t) &= -\mathop{\Theta}_{t' \in I} \eta\left(-\rho^-(\varphi, \sigma, t + t'), \mathop{\Xi}_{t'' \in [t, t+t']} - \rho^-(\psi, \sigma, t'') \right)
\end{aligned}$$

Note that, as also mentioned in related work, the quantitative semantics of the formulas $f(\sigma) \geq 0$ and $f(\sigma) > 0$ are the same, hence those two formulas cannot be distinguished from the value of their quantitative semantics alone [3].

In order to clarify the presentation, we follow Haghighi et al. [17] in writing $\rho(\varphi, \sigma, t) = \rho^+(\varphi, \sigma, t) + \rho^-(\varphi, \sigma, t)$, and we have the following properties, which we can prove by induction on the definitions of ρ, ρ^+ and ρ^-:

Lemma 1. *For any φ, σ and t, $\rho^+(\varphi, \sigma, t) \geq 0$ and $\rho^-(\varphi, \sigma, t) \leq 0$.*

Proof. By structural induction and using the sign constraints on the codomains of ν, μ, α, β, ζ, η, Γ, Δ, Θ and Ξ. A detailed proof is available in the extended version of the paper.

From those definitions we can derive the definition of $\perp$ and $\vee$ from $\wedge$ and $\neg$; and $\mathbf{F}_I$ and $\mathbf{G}_I$ from $\mathbf{U}_I$, under mild assumptions on ζ, η, Δ and Ξ:

$$\rho^+(\perp, \sigma, t_k) = \rho^+(\neg\top, \sigma, t_k) = 0$$

$$\rho^-(\perp, \sigma, t_k) = \rho^-(\neg\top, \sigma, t_k) = -\infty$$

$$\rho^+(\varphi \vee \psi, \sigma, t_k) = \rho^+(\neg(\neg\varphi \wedge \neg\psi), \sigma, t_k) = \beta(\rho^+(\varphi, \sigma, t_k), \rho^+(\psi, \sigma, t_k))$$

$$\rho^-(\varphi \vee \psi, \sigma, t_k) = \rho^-(\neg(\neg\varphi \wedge \neg\psi), \sigma, t_k) = -\alpha(-\rho^-(\varphi, \sigma, t_k), -\rho^-(\psi, \sigma, t_k))$$

$$\rho^+(\mathbf{F}_I\varphi, \sigma, t_k) = \rho^+(\top\mathbf{U}_I\varphi, \sigma, t_k) = \mathop{\Gamma}_{k' \in I} \zeta\left(\rho^+(\varphi, \sigma, t_{k+k'}), \mathop{\Delta}_{k'' \in [k, k+k']} + \infty\right)$$

$$\rho^-(\mathbf{F}_I\varphi, \sigma, t_k) = \rho^-(\top\mathbf{U}_I\varphi, \sigma, t_k) = -\mathop{\Theta}_{k' \in I} \eta\left(-\rho^-(\varphi, \sigma, t_{k+k'}), \mathop{\Xi}_{k'' \in [k, k+k']} 0\right)$$

$$\rho^+(\mathbf{G}_I\varphi, \sigma, t_k) = \rho^+(\neg\mathbf{F}_I\neg\varphi, \sigma, t_k) = \mathop{\Theta}_{k' \in I} \eta\left(\rho^+(\varphi, \sigma, t_{k+k'}), \mathop{\Xi}_{k'' \in [k, k+k']} 0\right)$$

$$\rho^-(\mathbf{G}_I\varphi, \sigma, t_k) = \rho^-(\neg\mathbf{F}_I\neg\varphi, \sigma, t_k) = -\mathop{\Gamma}_{k' \in I} \zeta\left(-\rho^-(\varphi, \sigma, t_{k+k'}), \mathop{\Delta}_{k'' \in [k, k+k']} + \infty\right)$$

Remark 1. As long as $\mathop{\Delta}_{k'' \in [k, k+k']} + \infty = +\infty$ and $\zeta(x, +\infty) = x$ for any $x \geq 0$, we can simplify the following two definitions:

$$\rho^+(\mathbf{F}_I\varphi, \sigma, t_k) = \mathop{\Gamma}_{k' \in I} \rho^+(\varphi, \sigma, t_{k+k'})$$

$$\rho^-(\mathbf{G}_I\varphi, \sigma, t_k) = -\mathop{\Gamma}_{k' \in I} - \rho^-(\varphi, \sigma, t_{k+k'})$$

Remark 2. As long as $\underset{k''\in[k,k+k']}{\Xi} 0 = 0$ and $\eta(x,0) = x$ for any $x \geq 0$, we can simplify the following two definitions:

$$\rho^-(\mathbf{F}_I\varphi, \sigma, t_k) = -\underset{k'\in I}{\Theta} - \rho^-(\varphi, \sigma, t_{k+k'})$$
$$\rho^+(\mathbf{G}_I\varphi, \sigma, t_k) = \underset{k'\in I}{\Theta} \rho^+(\varphi, \sigma, t_{k+k'})$$

3 Semantics Instantiations

Our generalization can instantiate semantics that replicate or closely match those presented in existing literature, in terms of the aforementioned generic functions (ν, μ, α, β, ζ, η, Γ, Δ, Θ, Ξ). The particular instantiations we chose to represent each semantics is shown in Table 1.

Several semantics were already implemented in Breach, which we adapted to our generalization by splitting the computation of ρ into that of ρ^+ and ρ^-. **Max** [11] emulates the default Breach semantics, which utilizes `min` and `max` for integrators, and `min` and `max` with respect to 0 for the "rectifiers" ν and μ which help split the single robustness scores into positive and negative components.

Const [11] uses the same integrators as Max, but the rectifier functions ν and μ produce constants which depend only on the sign of the predicate's robustness score. This is represented by $A \times \text{sgn}(x)$, where A is a fixed constant ($A = 100$ in the original Breach implementation, which we replicate), and $\text{sgn}(x)$ returns 1 if $x > 0$, 0 if $x = 0$, and -1 if $x < 0$. This simulates Boolean STL semantics, as all scores have the same absolute value regardless of their degree of satisfaction [11].

Add [11] is similar to Max except in certain cases where `max` is replaced by the "Koen &+" operator [9,25]. This integrator is a piecewise function $f(x,y)$ that takes on the behavior of the harmonic mean in certain cases, addition in others, and min everywhere else:

$$f(x,y) = \begin{cases} x+y & x<0 \wedge y<0 \\ \frac{1}{\frac{1}{x}+\frac{1}{y}} & x>0 \wedge y>0 \\ \min(x,y) & \text{otherwise} \end{cases}$$

TeLEx is similar to Max, with the exception of the Peak and Expand transformations defined in the TeLEx project [21]. Since the goal of TeLEx is to learn tight specifications in the absence of negative examples, these transformations cause the semantics to score minimally satisfied specifications higher. The Peak transformation (P) is applied within the rectifier, causing the robustness score to increase sharply for small positive values close to zero. The Expand transformation (E) is a multiplicative coefficient applied to certain integrators used in $\mathcal{U}$ and depends on the length of the time horizon ($t_2 - t_1$). We note that the Contract function (C) from the original paper is unused in our instantiation, as we choose to define all our temporal operators in terms of $\mathcal{U}$. The transformations have adjustable parameters β and γ that can influence the degree to which the scores are transformed. The transformation functions are defined as follows:

$$P(x) = \frac{1}{x + e^{-\beta x}} - e^{-x} \qquad E(\gamma, t_1, t_2) = \frac{2}{1 + e^{-\gamma(t_2 - t_1 + 1)}}$$

We note that TeLEx semantics were designed for specification learning, rather than falsification. For our benchmark evaluation, we arbitrarily chose the values $\beta = 1$ and $\gamma = 0.01$.

Cumulative is based on the smooth cumulative robustness semantics of Haghighi, Mehdipour, Bartocci, and Belta [17], where the max function has been "smoothened" in a way resembling LogSumExp: $\max_\beta(x, y) = \frac{1}{\beta}\ln(e^{\beta x} + e^{\beta y})$. The min function is defined as $\min_\beta(x, y) = -\max_\beta(-x, -y)$. β in this instantiation represents the degree to which these functions approximate max and min, with higher values having a "sharper" response that more closely resembles their non-smooth counterparts. These operators are then generalized to trace integrators which, unique to the semantics we considered from literature, integrate over time rather than simply choosing an extreme value. We note that this semantics is not sound; $\min_\beta$ tends to be an over-estimation, resulting in $\min_\beta(x, 0) > 0$ for some values of $x < 0$. We arbitrarily chose $\beta = 10$ for our evaluations but note that the operator remains unsound regardless of its particular value. This makes falsification more difficult, as true falsifying traces may be incorrectly assigned positive robustness values

The use of summation in this semantics' definition of $\mathcal{U}$ is intended to more strongly reward traces which satisfy a specification for longer periods of time, rather than considering only a maximally-satisfying point. We note that this semantics requires that both $\rho^+(\varphi \mathbf{U} \psi, \sigma, t_k)$ and $\rho^-(\varphi \mathbf{U} \psi, \sigma, t_k)$ employ summation despite the expected multiplication or min in the computation of ρ^-.

We also note that integrator functions can be mixed to form novel semantics, provided they are sound. **Sum-Product, Sum-Min, Max-Product** show the three other possible permutations of mixing max or min and summation or multiplication for the integrators. We note that Sum-Min bears some superficial resemblance to Cumulative semantics [17].

β, η, Γ, and Ξ are thought of as "optimistic" integrators, as they tend to favor robustness scores with larger magnitudes, and are usually instantiated with something resembling max or addition. Making these operators unnecessarily strong, with min or multiplication, preserves soundness but may make the semantics overly conservative, to the point where all scores are zero and thus no optimization progress can be made. **MinOnly** illustrates this, by replacing all non-rectifier operators with min.

All of the semantics instantiations we have introduced so far use either non-smooth rectifiers, or unsound, smooth approximations. We introduce **Smooth-Rect** semantics, which uses the Max semantics but with sound, smooth rectifiers R_+ and R_-:

$$R_+(x) = \begin{cases} 0 & \text{if } x \le 0 \\ xe^{-1/x} & \text{if } x > 0 \end{cases} \qquad R_-(x) = \begin{cases} 0 & \text{if } x \ge 0 \\ xe^{1/x} & \text{if } x < 0 \end{cases}$$

We create **Smooth-1** as a smooth alternative to Sum-Product, using the differentiable, sound rectifiers R_+ and R_-. Differentiability can be useful in conjunction with gradient-based solvers [21,24]. Although this should provide a sound differentiable semantics in theory, the use of products may lead to numerical overflow in practice.

Table 1. Semantics instantiations

	Smooth/ Sound	ν	μ	α	β
Max [11]	N/Y	$\max(\cdot, 0)$	$\min(\cdot, 0)$	min	max
Const [11]	N/Y	$\max(A \times \mathrm{sgn}(\cdot), 0)$	$\min(A \times \mathrm{sgn}(\cdot), 0)$	min	max
Add [11]	N/Y	$\max(\cdot, 0)$	$\min(\cdot, 0)$	Koen &+	Koen &+
TeLEx [21]	N/Y	$\max(P(\cdot), 0)$	$\min(P(\cdot), 0)$	min	max
Cumulative [17]	Y/N	$\max_\beta(\cdot, 0)$	$\min_\beta(\cdot, 0)$	$\min_\beta$	$\max_\beta$
Sum-Product	N/Y	$\max(\cdot, 0)$	$\min(\cdot, 0)$	$\times$	$+$
Sum-Min	N/Y	$\max(\cdot, 0)$	$\min(\cdot, 0)$	min	$+$
Max-Product	N/Y	$\max(\cdot, 0)$	$\min(\cdot, 0)$	$\times$	max
MinOnly	N/Y	$\max(\cdot, 0)$	$\min(\cdot, 0)$	min	min
SmoothRect	N/Y	$R_+(\cdot)$	$R_-(\cdot)$	min	max
Smooth-1	Y/Y	$R_+(\cdot)$	$R_-(\cdot)$	$\times$	$+$

	ζ	η	Γ	Δ	Θ	Ξ
Max [11]	min	max	max	min	min	max
Const [11]	min	max	max	min	min	max
Add [11]	min	max	max	min	min	max
TeLEx [21]	min	max	$E \cdot \max$	min	min	$E \cdot \max$
Cumulative [17]	$\min_\beta$	$\max_\beta$	$\sum$	$\min_\beta$	$\sum$	$\max_\beta$
Sum-Product	$\times$	$+$	$\sum$	$\prod$	$\prod$	$\sum$
Sum-Min	min	$+$	$\sum$	min	min	$\sum$
Max-Product	$\times$	max	max	$\prod$	$\prod$	max
MinOnly	min	min	min	min	min	min
SmoothRect	min	max	max	min	min	max
Smooth-1	$\times$	$+$	$\sum$	$\prod$	$\prod$	$\sum$

4 Soundness of Semantics

Sound quantitative semantics ensures that no falsifying traces for a particular system and specification are incorrectly deemed safe with a positive robustness score. For falsification, soundness ensures that falsifying traces that exist in the search space can indeed be found. On the other hand, soundness in parameter or control synthesis ensures that unsafe controls are not admitted, or that they can at least be readily identified. The consequence of unsoundness in falsification is that unsafe traces may be "masked", making them more difficult or even impossible to find. While the absence of counterexamples in a falsification search cannot and should not imply safety, unsoundness ultimately introduces additional uncertainty that can confound the design of safe systems.

We now introduce novel, sufficient soundness conditions on the aforementioned functions. Recall that functions α, β, ζ, η, Γ, Δ, Θ and Ξ only operate on nonnegative values. To avoid ambiguity, we use $+$ and $\times$ to represent binary addition and multiplication, and $\sum$ and $\prod$ to represent their trace counterparts.

(1) If $x \leq 0$ then $\nu(x) = 0$. Equivalently, if $\nu(x) > 0$ then $x > 0$.
Example of possible ν: $\max(\cdot, 0)$.
(2) If $x \geq 0$ then $\mu(x) = 0$. Equivalently, if $\mu(x) < 0$ then $x < 0$.
Example of possible μ: $\min(\cdot, 0)$.
(3) If $x \geq 0$, $y \geq 0$, if $x = 0$ **or** $y = 0$ then $\alpha(x, y) = 0$. Equivalently, if $x \geq 0$, $y \geq 0$ and $\alpha(x, y) > 0$, then $x > 0$ **and** $y > 0$.
Examples of possible α: min, $\prod$ (discrete product).
(4) If $x \geq 0$, $y \geq 0$, if $x = 0$ **and** $y = 0$ then $\beta(x, y) = 0$. Equivalently, if $x \geq 0$, $y \geq 0$ and $\beta(x, y) > 0$, then $x > 0$ **or** $y > 0$.
Examples of possible β: max, sum $\sum$.
(5) If all $x_k = 0$, then $\left(\underset{k}{\Gamma} x_k\right) = 0$. Equivalently, if all $x_k \geq 0$ and $\left(\underset{k}{\Gamma} x_k\right) > 0$, then there exists k for which $x_k > 0$.
Examples of possible Γ: max, $\sum$ (discrete sum), $\int$ (integral).
(6) If there exists a k for which $x_k = 0$, then $\left(\underset{k}{\Delta} x_k\right) = 0$. Equivalently, if all $x_k \geq 0$ and $\left(\underset{k}{\Delta} x_k\right) > 0$, then all $x_k > 0$.
Examples of possible Δ: min, $\prod$.
(7) ζ follows the requirements of α, and η the ones of β. Θ follows the requirements of Δ, and Ξ the ones of Γ.

Using those conditions, we can now prove a novel generic soundness theorem:

Lemma 2. *Under conditions (1)–(7), if $\rho^+(\varphi, \sigma, t) > 0$ then $\sigma^t \models \varphi$, and if $\rho^-(\varphi, \sigma, t) < 0$ then $\sigma^t \not\models \varphi$.*

Proof. The two statements are proved jointly by structural induction on φ, using the definition of φ given in Sect. 2.3 and applying conditions (1)–(7) as needed. Condition (1) on ν is used for the case $\rho^+(l(\sigma) \geq 0, \sigma, t)$ which is defined with ν; condition (2) on μ is similarly used for the case $\rho^-(l(\sigma) \geq 0, \sigma, t)$ which is defined with μ. Condition (3) on α is used for the $\wedge$ case on ρ^+, and condition (4) on β is used for the $\wedge$ case on ρ^-. Conditions (5)–(7) are used for the $\mathbf{U}_I$ case on ρ^+, and condition (7) is used for the $\mathbf{U}_I$ on ρ^-. A detailed proof is available in the extended version of the paper.

Theorem 2. *Under conditions (1)–(7), if* $\rho(\varphi, \sigma, t) > 0$ *then* $\sigma^t \models \varphi$*, and if* $\rho(\varphi, \sigma, t) < 0$ *then* $\sigma^t \not\models \varphi$*.*

Proof. Since by definition $\rho(\varphi, \sigma, t) = \rho^+(\varphi, \sigma, t) + \rho^-(\varphi, \sigma, t)$ and by Lemma 1 $\rho^-(\varphi, \sigma, t) \leq 0$, if $\rho(\varphi, \sigma, t) > 0$ then $\rho^+(\varphi, \sigma, t) > 0$. Therefore by Lemma 2 we get $\sigma^t \models \varphi$. Symmetrically, since by Lemma 1 $\rho^+(\varphi, \sigma, t) \geq 0$, if $\rho(\varphi, \sigma, t) < 0$ then $\rho^-(\varphi, \sigma, t) < 0$. Therefore by Lemma 2, $\sigma^t \not\models \varphi$.

Corollary 1. *Under conditions (1)–(7), we cannot have both* $\rho^+(\varphi, \sigma, t) > 0$ *and* $\rho^-(\varphi, \sigma, t) < 0$*. That is, if* $\rho^+(\varphi, \sigma, t) > 0$ *then* $\rho^-(\varphi, \sigma, t) = 0$*; and likewise if* $\rho^-(\varphi, \sigma, t) < 0$ *then* $\rho^+(\varphi, \sigma, t) = 0$*. In other words,* $\rho^+(\varphi, \sigma, t)$ *is the positive part of* $\rho(\varphi, \sigma, t)$*, i.e.,* $\rho^+(\varphi, \sigma, t) = \max(0, \rho(\varphi, \sigma, t))$*; and symmetrically* $\rho^-(\varphi, \sigma, t)$ *is the negative part of* $\rho(\varphi, \sigma, t)$*, i.e.,* $\rho^-(\varphi, \sigma, t) = \min(0, \rho(\varphi, \sigma, t))$*.*

Proof. If $\rho^+(\varphi, \sigma, t) > 0$ and $\rho^-(\varphi, \sigma, t) < 0$, then by Lemma 2, both $\sigma^t \models \varphi$ and $\sigma^t \not\models \varphi$, which is not possible.

A Counterexample to the Soundness Claimed in Existing Work

Our framework is particularly effective at ensuring that existing and future quantitative semantics are sound. As an example of this effectiveness, we have found what we believe to be an error in an existing paper describing cumulative quantitative semantics [17]. The error was found while trying to fit the semantics of this paper to our generalized semantics. Below we provide a counterexample to their theorem as well as proposed resolution.

The paper mentions proving the soundness of the semantics by structural induction on φ, proving both $\rho^+(\varphi, \sigma, t_k) > 0 \Rightarrow \sigma^{t_k} \models \varphi$ and $\rho^-(\varphi, \sigma, t_k) < 0 \Rightarrow \sigma^{t_k} \not\models \varphi$ simultaneously. However after verification with that proof strategy, there seems to be one case that does not verify, namely the case $\rho^-(\psi \mathbf{U}_I \varphi, \sigma, t_k)$. We contacted the authors multiple times but were not able to get an answer to our technical questions.

A Counterexample. The example takes a single variable x, with $\sigma^{t_0}(x) = 1$, $\sigma^{t_1}(x) = 3$, $\sigma^{t_2}(x) = -5$, and the formula $\varphi = \neg((x \geq 0)\mathbf{U}_{[0,2]}(x - 2 \geq 0))$. For that example, clearly $\sigma^{t_0} \models (x \geq 0)\mathbf{U}_{[0,2]}(x - 2 \geq 0)$ and thus $\sigma^{t_0} \not\models \varphi$, and yet $\rho^+(\varphi, \sigma, t_0) = 8 > 0$. A complete derivation is available in the extended version of the paper.

A Proposed Fix. We believe, the issue stems from the combination of the definitions of $\rho^+(\neg\varphi, \sigma, t_k)$ and $\rho^-(\psi \mathbf{U}_I \varphi, \sigma, t_k)$. A solution might be to maintain the definition of $\rho^+(\psi \mathbf{U}_I \varphi, \sigma, t_k)$, but change the definition of $\rho^-(\psi \mathbf{U}_I \varphi, \sigma, t_k)$ to:

$$\rho^-(\psi \mathbf{U}_I \varphi, \sigma, t_k) = \max_{k' \in I}(\min\{\rho^-(\varphi, \sigma, t_{k+k'}), \min_{k'' \in [k, k+k']} \rho^-(\psi, \sigma, t_{k''})\})$$

This uses the standard definition of quantitative semantics, but for ρ^-. Although not symmetric or smooth anymore, we believe that the two definitions are still compatible, and are able to prove soundness through our generalized framework.

5 Experimental Evaluation

Our experiments seek to answer three research questions:

RQ1: With all other solver details being equal, does a particular semantics consistently outperform the others in falsification?
RQ2: Does our generalization enable the creation of novel semantics that compare favorably to existing ones?
RQ3: Is the general framework able to replicate the results of existing benchmarking experiments?

We implemented the generalized semantics in MATLAB, by modifying the existing Breach toolkit [11] to calculate the positive (ρ^+) and negative (ρ^-) parts of ρ separately for semantics that integrate via summing [17]. We then implemented our generalized semantics and the instantiations given in Table 1. The source code for our modifications can be found at https://github.com/jchen-cs/breach Apart from the ARCH 2020 replication experiment [12], all evaluations were carried out on two workstations with Intel Xeon E3-1245 v3 CPUs running MATLAB r2024a on Ubuntu 22.04.4. The evaluation of a given benchmark problem and its semantics was run entirely on a single system.

Benchmarks

To answer **RQ1** and **RQ2**, we evaluated our semantics instantiations on falsification benchmarks partially inspired by existing examples in literature. We ensured that each benchmark had a falsifying trace so that all problems were feasible. The evaluations were done using Breach's default Nelder-Mead optimizer. The *autotrans*, *AFC*, *Path*, and *Proj* benchmarks were limited to 500 objective function evaluations, while *Path2* was limited to 1000 to account for its complexity. The Automatic Transmission and Fuel Control benchmarks were tested with the solver's corner search phase disabled, since they could be trivially falsified by extreme values. We did not impose a time limit on the benchmarks. In addition to our main evaluation, we also replicated a subset of an existing paper's experimental evaluation [12].

Automatic Transmission. This benchmark, originally from [18], simulates the drivetrain of a vehicle with an automatic transmission, and has been used in existing works such as S-TaLiRo [1] and others [3,4,31]. We treat the system's throttle input as a piecewise constant function of uniform intervals and parameterize amplitudes. The specification to falsify is that the vehicle speed never exceeds 100 mph. The maximum throttle value was constrained to 70% to minimize "trivial" falsifications in the quasi-random search phase. This increased the likelihood that falsifications occur during the optimization phase, which is reliant on performance differences between semantics.

Abstract Fuel Control. The Abstract Fuel Controller was originally implemented in Simulink [22], and was included as a sample benchmark in Breach. The model simulates an engine controller that regulates the air/fuel ratio based on engine speed and throttle position. For this model, we use the specification defined in the Breach sample that requires the air/fuel ratio to be within a specific range for the duration of the simulation. The engine speed is kept constant throughout the simulation, while the throttle position is controlled by a single pulse generator, parameterized by the maximum pedal angle (amplitude) and the duration for which that position is held (pulse period). As with the previous benchmark, the parameter space was constrained to reduce trivial falsifications.

Projectile Trajectory. The Projectile benchmark simulates a point mass being launched into a ballistic trajectory with a given initial velocity and launch angle. Safety is defined as the projectile avoiding a small area some distance away from the origin. As a result, most combinations of velocity and angle are safe.

Dubins Path. This benchmark simulates an autonomous vehicle with a Dubins steering model (such as a car or airplane) that travels straight and then turns at a constant angular velocity, all while avoiding a stationary obstacle. The adjustable parameters are the vehicle's transverse velocity (u_v), the time at which the turn occurs (t_{turn}), and the angular velocity of the turn (u_ω). The obstacle is small and relatively far away from the vehicle's starting position, so while a counterexample is feasible, it is unlikely to occur and thus makes falsification challenging.

Complex Dubins Path. The Path Planning benchmark was inspired by control synthesis benchmarks in existing literature [17,31], which simulates an autonomous vehicle in a two-dimensional space along with an obstacle that should be avoided (Fig. 1). As before, we use a Dubins vehicle model with controllable transverse and angular velocities. Unlike the previous benchmark, this scenario uses discretized dynamics and treats the inputs as piecewise constant functions with uniform intervals. We set the discretization interval of the control function to 0.1 s, meaning that the 1.1 s simulation provides eleven possible transverse and angular commands within the simulation timespan. The safety specification requires that the vehicle never enter the unsafe region at any point during the simulation. The dynamics are governed by the following equations that are updated every $\Delta t = 0.1$s:

$$
\begin{aligned}
x[k+1] &= x[k] + v[k]\cos(\theta)\Delta t \\
y[k+1] &= y[k] + v[k]\sin(\theta)\Delta t \\
\theta[k+1] &= \theta[k] + \omega[k]\Delta t
\end{aligned}
$$

where k refers to the state for an arbitrary time-step and $k+1$ the next state.

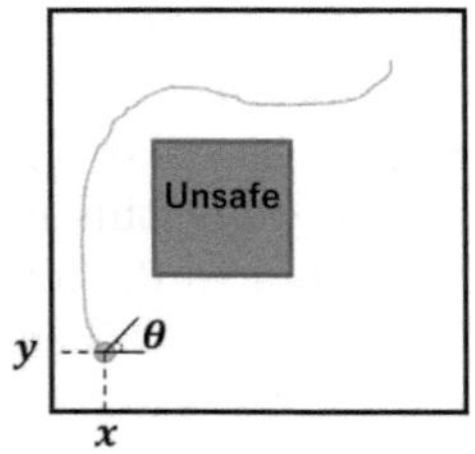

Fig. 1. Path planning model diagram.

ARCH 2020 Replication. To answer **RQ3** and provide additional insights on **RQ2**, we replicated a portion of the experimental evaluation conducted in an existing paper [12], but with our modified version of Breach and, in the interest of time, a subset of the aforementioned semantics. We limited our evaluation to the automatic transmission benchmarks (AT), and used the same propositions and naming convention as existing literature [13]. In addition to the max, additive, and constant semantics in the original evaluation, we also included TeLEx-like, Sum-Product, and Sum-Min. As before, we used Breach with the built-in Nelder-Mead solver. We ran each combination of semantics and benchmark 10 times and set a maximum of 300 objective function calls. These experiments were performed on a 2017 iMac Pro with an 18-core 2.3 GHz Intel Xeon W processor and 128 GB of 2666 MHz DDR4 memory, running MATLAB r2024a on macOS 13.4.1.

6 Results

We performed falsification on each of the benchmarks, and for each semantics recorded the number of objective function evaluations until first falsfication (Table 2). In each benchmark column, the runs with the lowest number of iterations are displayed in **boldface**. Runs that did not find a counterexample within the allotted number of objective evaluations are marked DNF.

Among the benchmarks tested, we did not notice any clear patterns establishing the superiority of any particular semantics. However, the semantics that required fewer iterations to falsify a particular benchmark tended to be efficient in other benchmarks as well (**RQ1**). Constant semantics consistently required a large number of evaluations to falsify, which was consistent with both our own expectations and other published results [12]. Many of the top-performing semantics in each benchmark finished with similar or identical numbers of evaluations. The falsifications were all achieved well within the solver's optimization phase, rather than its random sampling phase, which suggests that differences in these semantics did not significantly affect falsification performance in our evaluation. The Smooth-1 semantics, while able to eventually falsify, required a relatively large number of evaluations; examining its traces revealed numerical overflow was occurring. Since the Nelder-Mead algorithm is gradient-free, we believe the benefits of a smooth semantics may be realized by comparison with gradient-based optimization. Nevertheless, the novel Sum-Min and SmoothRect semantics were comparable to existing semantics (**RQ2**).

The minimum-only semantics (minonly) did not find a falsifying trace in any benchmark. Examining the trace of its search showed that the robustness score returned by the semantics remained zero. Although the min-only semantics is sound, it can be interpreted as overly conservative. The semantics based on [17] also did not find any falsifications. This could possibly be caused by the unsoundness of the semantics, as the reported robustness values do indeed change but remained positive. This would result in some true counterexamples being skipped in the search. Although it is possible to adjust the "softness" of the max and min functions used in these semantics [17], the adjustment would have to be tailored to a specific problem and the scale of its objective function values. The semantics inspired by TeLEx [21] found falsifying traces in all benchmarks, though often requiring more iterations. This is unsurprising as this semantics was optimized for policy learning rather than falsification. Unlike other instantiations of ν and μ, TeLEx features non-monotonic functions, which may complicate the search as robustness scores decrease in magnitude towards both zero and infinity. Product-based semantics (such as sum-product, max-product, and smooth1) tended to on occasion perform poorly. Examining their robustness values showed that these semantics frequently produced either very large or small values, and at times approached or exceeded the system's numerical limits. This may produce similar results to the constant semantics when large values saturate the robustness computation.

The results of the ARCH 2020 replication experiment are shown in Table 3. Each entry shows the number of successful falsifications out of 10, as well as the average number of objective function calls until successful falsification (parenthesized). The run with the most successful falsifications is displayed in **boldface**, with ties broken by the number of evaluations. Most of the AT benchmarks were falsified consistently by all semantics, and required on average a low number of evaluations until a falsifying trace is found. Compared to published results, Max, Const, and Add semantics generally performed similarly in relative performance in our replication experiment with similar trends in the mean evaluations until falsification. Therefore, we believe that, for **RQ3**, the semantics implemented in our generalization compare similarly to those from existing evaluation. In addition, sum-min semantics had the highest falsification rate of all semantics, and had the lowest average number of evaluations per successful falsification in 4 out of the 9 benchmarks. This indicates that, for the benchmarks we evaluated, this semantics is comparable to, and sometimes better than existing semantics from the literature (**RQ2**).

Table 2. Objective Evaluations until Falsification

	autotrans	AFC	Proj	Path	Path2
Max [11]	**65**	**43**	**26**	55	**352**
Const [11]	205	63	127	178	560
Add [11]	**65**	**43**	**26**	55	365
TeLEx [21]	205	**43**	348	358	**352**
Cumulative [17]	DNF	DNF	DNF	DNF	DNF
Sum-Product	397	63	359	**53**	560
Sum-Min	**65**	**43**	**26**	55	365
Max-Product	397	63	359	**53**	560
MinOnly	DNF	DNF	DNF	DNF	DNF
SmoothRect	**65**	**43**	**26**	55	**352**
Smooth-1	397	63	359	173	560

Table 3. ARCH Replication Experiment, Falsifications out of 10 (Mean evaluations per falsification)

	AT1	AT2	AT51	AT52	AT53
Max [11]	8 (151.4)	10 (11)	10 (34.1)	**10 (11.8)**	10 (16.1)
Const [11]	0 (-)	10 (16.8)	**10 (8.1)**	10 (16.8)	10 (8.8)
Add [11]	5 (156.2)	10 (15.2)	10 (36.4)	10 (20.3)	10 (16.3)
TeLEx [21]	0 (-)	10 (13.7)	10 (14.8)	10 (17.9)	10 (12.8)
Sum-Product	0 (-)	**10 (6.6)**	10 (16.2)	10 (16.9)	10 (7.3)
Sum-Min	**9 (173.6)**	10 (13.6)	10 (9.7)	10 (14)	**10 (6.2)**

	AT54	AT6a	AT6b	AT6c
Max [11]	10 (84.9)	10 (56)	9 (49.7)	10 (29.9)
Const [11]	10 (55)	9 (87.7)	**10 (39.9)**	10 (57)
Add [11]	10 (69.5)	**10 (20.8)**	10 (45.2)	10 (33.8)
TeLEx [21]	10 (97.2)	10 (62.8)	10 (70.7)	10 (56.2)
Sum-Product	10 (84.7)	8 (134.9)	7 (50.6)	10 (53.3)
Sum-Min	**10 (49.1)**	10 (55.2)	10 (55.5)	**10 (25.3)**

7 Conclusion and Future Work

The data we collected in our experimental evaluation does not appear to favor any particular semantics for all problems. This is to be expected given the diversity of problems and the relatively small dataset. Nevertheless, our generalization and implementation thereof provides a method to directly compare various semantics

Therefore, concurring with other evaluations [12], we believe that system designers employing falsification should try several quantitative semantics, or experiment to find semantics that perform best for their application. Using our generalization, one may even develop bespoke semantics for a particular problem, with the assurance that its correctness can be easily checked using our soundness conditions. Unlike in existing work, this is facilitated by the general framework as it allows direct comparisons of semantics in the same environment, leaving other implementation details constant.

Our current experimental evaluation only examines the use of quantitative semantics in falsification tasks. Nevertheless, a generalized semantics framework can be employed in other application areas, such as control synthesis [7,17] or policy learning [21]. As with falsification, it is likely that performance will depend on choice of semantics or properties inherent to particular instantiations. It is possible that semantics that perform well in falsification are not necessarily optimal choices for other tasks. Additional experiments using benchmarks representative of these classes of problems will be necessary.

Our falsification experiments can also be expanded somewhat by examining other solvers or optimization techniques. Presently, we use Breach's Nelder-Mead optimizer, a gradient-free method, since many of the evaluated semantics are not differentiable. Other works have proposed using gradient-based solvers, paired with differentiable semantics, to potentially speed up the search process [21,24]. We believe this will allow us to, for instance, evaluate the performance improvement of pairing smooth semantics with a gradient-based solver. In addition to soundness and differentiability, we may consider other "meta-properties" of semantics instantiations and their effects on performance. For example, monotonicity may be suggestive of efficiency when paired with a gradient-following solver, while kernel size may be proportional to falsification difficulty

By developing a generalization of quantitative semantics, we not only enable easier direct comparisons of existing and novel quantitative semantics, but also provide a way to express useful meta-properties about them, such as soundness. We believe that by enabling direct comparisons of quantitative semantics, our framework will enable system designers to make informed decisions about their choice of semantics in optimization-based falsification.

Acknowledgements. We thank Koen Claessen, Pierre-Loïc Garoche, Elias Ghisellini, Umut Kurnaz and Marc Pouzet for valuable discussions on earlier versions of this work. We also thank Johan Lidén Eddeland, Sajed Miremadi, and Knut Åkesson for providing us the source code for their ARCH 2020 evaluation, and the anonymous reviewers for their feedback. This research was supported in part by National Science Foundation grants CCF-2348706, CCF-2426474, and CCF-2319572.

References

1. Abbas, H., Fainekos, G., Sankaranarayanan, S., Ivančić, F., Gupta, A.: Probabilistic temporal logic falsification of cyber-physical systems. ACM Trans. Embed. Comput. Syst. **12**(2s) (2013). https://doi.org/10.1145/2465787.2465797
2. Abbas, H., Winn, A., Fainekos, G., Julius, A.A.: Functional gradient descent method for metric temporal logic specifications. In: 2014 American Control Conference, pp. 2312–2317. IEEE (2014). https://doi.org/10.1109/ACC.2014.6859453
3. Akazaki, T., Hasuo, I.: Time robustness in MTL and expressivity in hybrid system falsification. In: International Conference on Computer Aided Verification, pp. 356–374. Springer (2015)
4. Annpureddy, Y., Liu, C., Fainekos, G., Sankaranarayanan, S.: S-TaLiRo: a tool for temporal logic falsification for hybrid systems. In: Abdulla, P.A., Leino, K.R.M. (eds.) TACAS 2011. LNCS, vol. 6605, pp. 254–257. Springer, Heidelberg (2011). https://doi.org/10.1007/978-3-642-19835-9_21
5. Bartocci, E., Bortolussi, L., Loreti, M., Nenzi, L.: Monitoring mobile and spatially distributed cyber-physical systems. In: Proceedings of the 15th ACM-IEEE International Conference on Formal Methods and Models for System Design, pp. 146–155. ACM, Vienna Austria (2017). https://doi.org/10.1145/3127041.3127050. https://dl.acm.org/doi/10.1145/3127041.3127050
6. Bartocci, E., et al.: Specification-based monitoring of cyber-physical systems: a survey on theory, tools and applications. In: Bartocci, E., Falcone, Y. (eds.) Lectures on Runtime Verification. LNCS, vol. 10457, pp. 135–175. Springer, Cham (2018). https://doi.org/10.1007/978-3-319-75632-5_5
7. Cardona, G.A., Kamale, D., Vasile, C.I.: Mixed integer linear programming approach for control synthesis with weighted signal temporal logic. In: Proceedings of the 26th ACM International Conference on Hybrid Systems: Computation and Control, pp. 1–12. ACM, San Antonio TX USA (2023). https://doi.org/10.1145/3575870.3587120. https://dl.acm.org/doi/10.1145/3575870.3587120
8. Chattopadhyay, A., Mamouras, K.: A verified online monitor for metric temporal logic with quantitative semantics. In: Deshmukh, J., Ničković, D. (eds.) RV 2020. LNCS, vol. 12399, pp. 383–403. Springer, Cham (2020). https://doi.org/10.1007/978-3-030-60508-7_21
9. Claessen, K., Smallbone, N., Eddeland, J., Ramezani, Z., Åkesson, K.: Using valued booleans to find simpler counterexamples in random testing of cyber-physical systems. IFAC-PapersOnLine **51**(7), 408–415 (2018). https://doi.org/10.1016/j.ifacol.2018.06.333. https://linkinghub.elsevier.com/retrieve/pii/S2405896318306633
10. Donzé, A., Maler, O.: Robust satisfaction of temporal logic over real-valued signals. In: Chatterjee, K., Henzinger, T.A. (eds.) FORMATS 2010. LNCS, vol. 6246, pp. 92–106. Springer, Heidelberg (2010). https://doi.org/10.1007/978-3-642-15297-9_9
11. Donzé, A.: Breach, a toolbox for verification and parameter synthesis of hybrid systems. In: Touili, T., Cook, B., Jackson, P. (eds.) CAV 2010. LNCS, vol. 6174, pp. 167–170. Springer, Heidelberg (2010). https://doi.org/10.1007/978-3-642-14295-6_17

12. Eddeland, J.L., Miremadi, S., Åkesson, K.: Evaluating optimization solvers and robust semantics for simulation-based falsification. In: Frehse, G., Althoff, M. (eds.) 7th International Workshop on Applied Verification of Continuous and Hybrid Systems (ARCH20). EPiC Series in Computing, vol. 74, pp. 259–266. EasyChair (2020). https://doi.org/10.29007/f4vs
13. Ernst, G., et al.: ARCH-COMP 2019 category report: falsification. EPiC Ser. Comput. **61**, 129–140 (2019)
14. Faella, M., Legay, A., Stoelinga, M.: Model checking quantitative linear time logic. Electron. Notes Theor. Comput. Sci. **220**(3), 61–77 (2008). https://doi.org/10.1016/j.entcs.2008.11.019. https://linkinghub.elsevier.com/retrieve/pii/S1571066108004568
15. Fainekos, G.E., Pappas, G.J.: Robustness of temporal logic specifications for continuous-time signals. Theoret. Comput. Sci. **410**(42), 4262–4291 (2009)
16. Gilpin, Y., Kurtz, V., Lin, H.: A smooth robustness measure of signal temporal logic for symbolic control. IEEE Control Syst. Lett. **5**(1), 241–246 (2020)
17. Haghighi, I., Mehdipour, N., Bartocci, E., Belta, C.: Control from signal temporal logic specifications with smooth cumulative quantitative semantics. In: 2019 IEEE 58th Conference on Decision and Control (CDC), pp. 4361–4366. IEEE (2019)
18. Hoxha, B., Abbas, H., Fainekos, G.: Benchmarks for temporal logic requirements for automotive systems. In: ARCH14-15. 1st and 2nd International Workshop on Applied veRification for Continuous and Hybrid Systems, pp. 25–18 (2015). https://doi.org/10.29007/xwrs. https://easychair.org/publications/paper/4bfq
19. Jakšić, S., Bartocci, E., Grosu, R., Ničković, D.: An algebraic framework for runtime verification. IEEE Trans. Comput.-Aided Des. Integr. Circ. Syst. **37**(11), 2233–2243 (2018). https://doi.org/10.1109/TCAD.2018.2858460. https://ieeexplore.ieee.org/document/8493500/
20. Jeannin, J.B., Chen, J., de Mendonça, J.L.V., Mamouras, K.: Work-in-progress: towards a theory of robust quantitative semantics for signal temporal logic. In: 2022 International Conference on Embedded Software (EMSOFT), pp. 11–12. IEEE (2022)
21. Jha, S., Tiwari, A., Seshia, S.A., Sahai, T., Shankar, N.: TeLEx: learning signal temporal logic from positive examples using tightness metric. Formal Methods Syst. Des. **54**(3), 364–387 (2019). https://doi.org/10.1007/s10703-019-00332-1. http://link.springer.com/10.1007/s10703-019-00332-1
22. Jin, X., Deshmukh, J.V., Kapinski, J., Ueda, K., Butts, K.: Powertrain control verification benchmark. In: Proceedings of the 17th International Conference on Hybrid Systems: Computation and Control, HSCC 2014, pp. 253–262. Association for Computing Machinery, New York, NY, USA (2014). https://doi.org/10.1145/2562059.2562140
23. Koymans, R.: Specifying real-time properties with metric temporal logic. Real-Time Syst. **2**(4), 255–299 (1990). https://doi.org/10.1007/BF01995674
24. Leung, K., Aréchiga, N., Pavone, M.: Backpropagation through signal temporal logic specifications: infusing logical structure into gradient-based methods. Int. J. Robot. Res. **42**(6), 356–370 (2023). https://doi.org/10.1177/02783649221082115
25. Lidén Eddeland, J., Claessen, K., Smallbone, N., Ramezani, Z., Miremadi, S., Åkesson, K.: Enhancing temporal logic falsification with specification transformation and valued booleans. IEEE Trans. Comput.-Aided Des. Integr. Circ. Syst. **39**(12), 5247–5260 (2020). https://doi.org/10.1109/TCAD.2020.2966480. https://ieeexplore.ieee.org/document/8957695

26. Lindemann, L., Dimarogonas, D.V.: Robust control for signal temporal logic specifications using discrete average space robustness. Automatica **101**, 377–387 (2019). https://doi.org/10.1016/j.automatica.2018.12.022. https://www.sciencedirect.com/science/article/pii/S0005109818306289
27. Maler, O., Nickovic, D.: Monitoring temporal properties of continuous signals. In: Lakhnech, Y., Yovine, S. (eds.) FORMATS/FTRTFT -2004. LNCS, vol. 3253, pp. 152–166. Springer, Heidelberg (2004). https://doi.org/10.1007/978-3-540-30206-3_12
28. Mamouras, K., Chattopadhyay, A., Wang, Z.: Algebraic quantitative semantics for efficient online temporal monitoring. In: TACAS 2021. LNCS, vol. 12651, pp. 330–348. Springer, Cham (2021). https://doi.org/10.1007/978-3-030-72016-2_18
29. Mamouras, K., Chattopadhyay, A., Wang, Z.: A compositional framework for quantitative online monitoring over continuous-time signals. In: Feng, L., Fisman, D. (eds.) RV 2021. LNCS, vol. 12974, pp. 142–163. Springer, Cham (2021). https://doi.org/10.1007/978-3-030-88494-9_8
30. Mehdipour, N., Vasile, C.I., Belta, C.: Arithmetic-geometric mean robustness for control from signal temporal logic specifications. In: 2019 American Control Conference (ACC), pp. 1690–1695. IEEE (2019)
31. Mehdipour, N., Vasile, C.I., Belta, C.: Average-based robustness for continuous-time signal temporal logic (2019)
32. Nghiem, T., Sankaranarayanan, S., Fainekos, G., Ivancić, F., Gupta, A., Pappas, G.J.: Monte-Carlo techniques for falsification of temporal properties of non-linear hybrid systems. In: Proceedings of the 13th ACM International Conference on Hybrid Systems: Computation and Control, HSCC 2010, pp. 211–220. ACM, New York, NY, USA (2010). https://doi.org/10.1145/1755952.1755983
33. Pant, Y.V., Abbas, H., Mangharam, R.: Smooth operator: control using the smooth robustness of temporal logic. In: 2017 IEEE Conference on Control Technology and Applications (CCTA), pp. 1235–1240. IEEE (2017)
34. Sankaranarayanan, S., Fainekos, G.: Falsification of temporal properties of hybrid systems using the cross-entropy method. In: Proceedings of the 15th ACM International Conference on Hybrid Systems: Computation and Control, HSCC 2012, pp. 125–134. ACM, New York, NY, USA (2012). https://doi.org/10.1145/2185632.2185653
35. Sato, S., Waga, M., Hasuo, I.: Constrained optimization for hybrid system falsification and application to conjunctive synthesis. IFAC-PapersOnLine **54**(5), 217–222 (2021). https://doi.org/10.1016/j.ifacol.2021.08.501. https://linkinghub.elsevier.com/retrieve/pii/S2405896321012763
36. Zhang, Z., Lyu, D., Arcaini, P., Ma, L., Hasuo, I., Zhao, J.: Effective hybrid system falsification using Monte Carlo tree search guided by QB-robustness. In: Silva, A., Leino, K.R.M. (eds.) CAV 2021. LNCS, vol. 12759, pp. 595–618. Springer, Cham (2021). https://doi.org/10.1007/978-3-030-81685-8_29

Quantitative Monitoring of Signal First-Order Logic

Marek Chalupa[1], Thomas A. Henzinger[2], N. Ege Saraç[3(✉)], and Emily Yu[4]

[1] Zeroth Research, London, UK
marek.chalupa@zeroth.org
[2] Institute of Science and Technology Austria, Klosterneuburg, Austria
tah@ista.ac.at
[3] CISPA Helmholtz Center for Information Security, Saarbrücken, Germany
ege.sarac@cispa.de
[4] Leiden University, Leiden, The Netherlands
z.yu@liacs.leidenuniv.nl

Abstract. Runtime monitoring checks, during execution, whether a partial signal produced by a hybrid system satisfies its specification. Signal First-Order Logic (SFO) offers expressive real-time specifications over such signals, but currently comes only with Boolean semantics and has no tool support. We provide the first robustness-based quantitative semantics for SFO, enabling the expression and evaluation of rich real-time properties beyond the scope of existing formalisms such as Signal Temporal Logic. To enable online monitoring, we identify a past-time fragment of SFO and give a pastification procedure that transforms bounded-response SFO formulas into equisatisfiable formulas in this fragment. We then develop an efficient runtime monitoring algorithm for this past-time fragment and evaluate its performance on a set of benchmarks, demonstrating the practicality and effectiveness of our approach. To the best of our knowledge, this is the first publicly available prototype for online quantitative monitoring of full SFO.

Keywords: Signal first-order logic · Robustness-based quantitative semantics · Online runtime monitoring

1 Introduction

Runtime verification checks executions of a system against formal specifications by monitoring the system's behavior during runtime. Online monitors are lightweight pieces of software that operate in parallel with the system, checking partial traces on the fly as the execution unfolds. In contrast, offline monitoring analyzes the complete trace only after the system has finished executing.

Previous work [5] introduced Signal First-Order Logic (SFO) as a rich specification language for expressing real-time properties of real-valued signals.

A. Sampaio and M. Stoelinga (Eds.): FM 2026, LNCS 16557, pp. 214–233, 2026.
https://doi.org/10.1007/978-3-032-26220-2_11

SFO offers greater expressiveness than the widely used Signal Temporal Logic (STL) [27], enabling more precise modeling of complex system behaviors. For example, SFO allows the formalization of bounded stabilization properties: following a significant disturbance, the signal must return to its pre-disturbance value within 10 timestamps. This level of expressiveness goes beyond traditional temporal logics and makes SFO well suited for reasoning about the dynamic behavior of hybrid systems. Technically, SFO combines first-order logic with linear arithmetic and uninterpreted unary function symbols to represent real-valued signals over time. Its syntax supports quantification over both time and value variables, allowing for rich and flexible property specification. However, this expressiveness comes at a cost: formulas whose satisfaction depends on unbounded future behavior are in general not efficiently monitorable online, which makes the full logic ill suited for practical runtime monitoring.

So far, the semantics of SFO have been defined only in Boolean terms, i.e., a system trajectory either satisfies or violates a specification [5]. In many real-world applications, especially those involving continuous dynamics and numerical quantities, such binary outcomes are often insufficient. Quantitative semantics, which measure how well or to what extent a specification is satisfied or violated, offer more informative feedback and can significantly improve decision making in practice.

At a high level, we develop the semantic foundations and algorithmic machinery needed for online quantitative monitoring of SFO. In particular, we make the following contributions:

1. *Quantitative semantics for SFO.* We introduce a robust quantitative semantics for SFO, generalizing the standard Boolean semantics and enabling quantitative evaluation of real-valued signals.
2. *Past-time fragment and pastification.* We define a syntactic fragment of SFO, called past-time SFO (pSFO), which is well suited for online monitoring. To enable its use in practice, we provide a pastification procedure that transforms bounded-response SFO formulas into equisatisfiable pSFO formulas.
3. *Online monitoring algorithm.* We propose an efficient algorithm for online monitoring of pSFO formulas over piecewise-linear signals. Our approach computes robustness values symbolically using polyhedral representations.
4. *Implementation and experiments.* We implement our monitoring algorithm in a prototype tool and evaluate its performance on several benchmark scenarios. The results demonstrate the effectiveness of our approach.

Related Work. Signal Temporal Logic (STL) was introduced as a specification language for continuous signals [27], and there is now a large body of work on its quantitative semantics and monitoring algorithms. Robustness-based semantics for STL [14,15], based on signed distance to violation, has since been supported by efficient offline and online monitoring algorithms [11–13,19,25,28]. Our work is complementary to this STL-based line: we lift robustness-based monitoring from STL to full SFO and identify a past-time, bounded-response fragment that still admits efficient online algorithms.

Beyond STL, the work most closely related to ours is Signal First-Order Logic (SFO) [5]: Bakhirkin et al. define a Boolean semantics for SFO and an offline monitoring algorithm, but do not release an implementation or report experimental results. Menghi et al. [30] introduce Restricted Signal First-Order Logic (RFOL), a syntactic fragment of SFO with quantitative semantics. RFOL forbids quantification over value variables, enforces that each subformula contains at most one free time variable, and requires all formulas to be closed. In contrast, we define a generic robustness-based quantitative semantics for *full* SFO and identify a fragment tailored for efficient online monitoring.

The extension of STL with freeze quantifiers (STL*) [10] can record the current signal value for later comparison. While STL* can express relative changes such as local extrema and oscillations, SFO is strictly more expressive: it can capture additional classes of signal-based properties, including punctual derivative constraints that STL* cannot express; see [9] for a detailed taxonomy and an expressiveness comparison of STL, STL*, and SFO.

In a different direction, Bakhirkin and Basset [4] propose a quantitative extension of STL with generalized sliding-window and until operators, tailored to express and monitor bounded-stabilization properties. However, this extension remains quantifier-free: it cannot bind time or value parameters and reuse them later in the formula, so it cannot capture SFO specifications that quantify over dynamically chosen event durations and relate these durations across different signals, such as rise-time properties (e.g., if signal f has a positive edge then signal g subsequently has a positive edge whose rise time is within 10% of that of f; see [5, Ex. 3]). Silvetti et al. [34] extend this framework with integral-based and filtered sliding-window operators, which strengthens its ability to express cumulative and conditional quantitative requirements (e.g., average power consumption restricted to periods where a mode is active). These arithmetic extensions are largely orthogonal: SFO extended with suitable arithmetic operators can encode such sliding-window constructs and, in addition, express duration-parameterized properties that these formalisms cannot capture.

Orthogonal to logic-specific robustness monitoring, recent work develops logic-independent notions of quantitative monitorability, characterizing when (and how well) quantitative trace properties can be monitored (or approximated from finite prefixes) [20,22,24]. Relatedly, quantitative analogs of the safety-liveness dichotomy and corresponding decompositions have been established for quantitative properties and automata [6,7,23]. These general frameworks provide semantic criteria for what is monitorable (possibly approximately), whereas we give a robustness semantics for full SFO and exact online monitoring algorithms.

To our knowledge, this paper is the first to provide a generic robustness semantics for full SFO, together with an online monitoring algorithm for a practically relevant past-time bounded-response fragment.

2 Signal First-Order Logic (SFO)

In this section, we recall Signal First-Order Logic (SFO) and its standard Boolean semantics [5]. A signal is a function $w : \mathbb{T} \to \mathbb{R}$ mapping a temporal domain $\mathbb{T} \subseteq \mathbb{R}$ to real values. Let $\mathcal{F}$ be a set of function symbols and $\mathcal{X} = \mathcal{T} \cup \mathcal{R}$ be a set of variables, partitioned into time variables $\mathcal{T}$ and value variables $\mathcal{R}$. Both time terms τ and value terms ρ evaluate to real numbers. A valuation ν assigns values to variables $x \in \mathcal{X}$, denoted $[\![x]\!]_\nu$. Function symbols remain uninterpreted at the logical level; a signal trace w assigns a real-valued signal $[\![f]\!]_w$ to each function symbol $f \in \mathcal{F}$.

Definition 1 (Syntax). *The syntax of an SFO formula ϕ is defined by the following grammar:*

$$\phi ::= \theta < \theta \mid \neg\phi \mid \phi \vee \phi \mid \exists r \in R.\, \phi \mid \ \exists s \in I.\, \phi \qquad \theta ::= \rho \mid \tau$$
$$\rho ::= r \mid f(\tau) \mid n \mid \rho - \rho \mid \rho + \rho \qquad \tau ::= s \mid n \mid \tau - \tau \mid \tau + \tau$$

where $R \subseteq \mathbb{R}$ and $I \subseteq \mathbb{R}$ are intervals with rational end points, $n \in \mathbb{Q}$ is a rational number, $r \in \mathcal{R}$ is a value variable, $s \in \mathcal{T}$ is a time variable, and ρ and τ denote value and time terms, respectively.

Definition 2 (Boolean Semantics). *Consider an SFO formula ϕ, a trace w, a valuation ν, and a term θ. The satisfaction relation $w, \nu \models \phi$ and the value $[\![\theta]\!]_{w,\nu}$ are defined inductively as follows:*

$$\begin{aligned}
w, \nu \models \theta_1 < \theta_2 &\Leftrightarrow [\![\theta_1]\!]_{w,\nu} < [\![\theta_2]\!]_{w,\nu} \\
w, \nu \models \neg\phi &\Leftrightarrow w, \nu \not\models \phi \\
w, \nu \models \phi_1 \vee \phi_2 &\Leftrightarrow w, \nu \models \phi_1 \textit{ or } w, \nu \models \phi_2 \\
w, \nu \models \exists r \in R.\, \phi &\Leftrightarrow \exists a \in R : w, \nu[r \leftarrow a] \models \phi \\
w, \nu \models \exists s \in I.\, \phi &\Leftrightarrow \exists a \in I : w, \nu[s \leftarrow a] \models \phi
\end{aligned}
\qquad
\begin{aligned}
[\![n]\!]_{w,\nu} &= n \quad (n \in \mathbb{Q}) \\
[\![x]\!]_{w,\nu} &= \nu(x) \quad (x \in \mathcal{X}) \\
[\![\theta_1 \pm \theta_2]\!]_{w,\nu} &= [\![\theta_1]\!]_{w,\nu} \pm [\![\theta_2]\!]_{w,\nu} \\
[\![f(\tau)]\!]_{w,\nu} &= [\![f]\!]_w([\![\tau]\!]_{w,\nu})
\end{aligned}$$

We use standard syntactic abbreviations: $\phi \wedge \psi \equiv \neg(\neg\phi \vee \neg\psi)$, $\phi \rightarrow \psi \equiv \neg\phi \vee \psi$, and $\forall x \in S.\, \phi \equiv \neg\exists x \in S.\, \neg\phi$. Abbreviations for derived comparison operators ($\leq$, $\geq$, $=$) and absolute values ($|\rho|$) are defined as expected.

Remark 1 (Well-definedness). Formulas are interpreted over traces w with a temporal domain $\mathbb{T} \subseteq \mathbb{R}$. We only define the satisfaction relation $w, \nu \models \phi$ for pairs (w, ν) for which all signal accesses in ϕ are well-defined: whenever a term $f(\tau)$ occurs in ϕ, we require that the interpreted time $[\![\tau]\!]_{w,\nu}$ lies in $\mathbb{T}$. If this side condition fails, we leave $w, \nu \models \phi$ undefined. In particular, quantifiers range over those instantiations that yield well-defined signal accesses in the quantified subformula; if no such instantiation exists, the formula is undefined. Whenever we write $w, \nu \models \phi$, we implicitly assume that this side condition holds.

At the semantic level, a trace w interprets each function symbol f as a concrete real-valued function $[\![f]\!]_w$ on its temporal domain $\mathbb{T}$. We do not impose any additional assumptions on how these functions are represented or how values between samples are obtained; the semantics depends only on the resulting pointwise function values at the times where the formula accesses the signals.

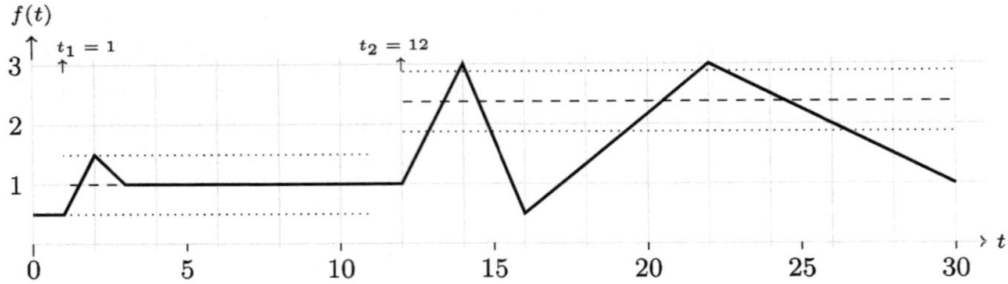

Fig. 1. Illustration of the quantitative semantics for the bounded stabilization property given in Example 1. For convenience, we use piecewise-linear interpolation. Upward arrows indicate rising edges at $t_1 = 1$ and $t_2 = 12$. Following t_1, the signal f stabilizes rapidly within the ± 0.5 tolerance band around $r_1 = 1$. Conversely, following t_2, the signal exhibits wide oscillations that violate the band. The quantitative evaluation considers all 8-unit windows starting within 10 units of t_2, identifying the window with the minimum range to determine the maximal robustness value.

3 Quantitative Semantics of SFO

We introduce a quantitative semantics for SFO, assigning a robustness value to each formula relative to a specific signal and valuation. This score quantifies the degree to which a signal satisfies or violates a formula. We further establish the soundness of this semantics with respect to the Boolean semantics.

Definition 3 (Quantitative Semantics). *Given an SFO formula ϕ, a trace w, and a valuation ν, the* robustness value $\mu(w, \nu, \phi) \in \mathbb{R} \cup \{-\infty, +\infty\}$ *is defined inductively as follows.*

$$\begin{aligned}
\mu(w, \nu, \theta_1 < \theta_2) &= [\![\theta_2]\!]_{w,\nu} - [\![\theta_1]\!]_{w,\nu} \\
\mu(w, \nu, \neg\phi) &= -\mu(w, \nu, \phi) \\
\mu(w, \nu, \phi_1 \vee \phi_2) &= \max\{\mu(w, \nu, \phi_1), \mu(w, \nu, \phi_2)\} \\
\mu(w, \nu, \exists r \in R.\phi) &= \sup_{a \in R} \mu(w, \nu[r \leftarrow a], \phi) \\
\mu(w, \nu, \exists s \in I.\phi) &= \sup_{a \in I} \mu(w, \nu[s \leftarrow a], \phi)
\end{aligned}$$

The quantitative semantics follows the well-definedness convention from Remark 1: $\mu(w, \nu, \phi)$ is defined only when all signal accesses needed to evaluate ϕ are defined. In particular, in the quantified clauses, the supremum ranges only over those instantiations for which the quantified subformula has a defined robustness value; if no such instantiation exists, then $\mu(w, \nu, \phi)$ is undefined.

The definition above is independent of any particular interpolation scheme. When a trace is given as sampled data, an interpolation choice fixes the concrete functions $[\![f]\!]_w$ on $\mathbb{T}$. We illustrate the quantitative semantics of SFO using the bounded stabilization example from the introduction, using piecewise-linear interpolation purely for exposition.

Example 1 (Bounded Stabilization). Let $b : \mathbb{T} \to \{0,1\}$ be a Boolean mode signal and $f : \mathbb{T} \to \mathbb{R}$ be a real-valued signal. A *rising edge* of b at time t is

$$\uparrow b[t] \equiv b(t) = 1 \wedge \exists c \in (0,\infty).\forall c' \in (0,c).b(t-c') = 0.$$

Note that since Boolean signals are piecewise constant and right-continuous, isolated single-point spikes do not arise. The *bounded stabilization* requirement states that whenever b rises at t, the signal f must enter and remain within a tolerance band ± 0.5 around some value r. This stabilization must occur no later than $c \leq 10$ time units after t and persist for at least 8 time units:

$$\phi \equiv \uparrow b[t] \to \exists r \in \mathbb{R}.\exists c \in [0,10].\forall d \in [0,8].|f(t+c+d) - r| \leq 0.5.$$

Quantitative semantics allow for computing the degree to which a signal and valuation satisfy this property. For this example, we consider only time points where a rising edge occurs and focus on the consequent of the implication, denoted by ψ. Given a signal w and a valuation ν, the quantitative semantics $\mu(w,\nu,\psi)$ is given by:

$$\sup_{r\in\mathbb{R}} \sup_{c\in[0,10]} \inf_{d\in[0,8]} (0.5 - |f(t+c+d) - r|).$$

This expression captures the optimal robustness over all candidate stabilization levels $r \in \mathbb{R}$ and settling delays $c \in [0,10]$. Specifically, the inner term represents the worst-case deviation from the ± 0.5 band around r over a window of size 8 starting at $t+c$. Intuitively, for a given t, we seek a target value r and a delay $c \in [0,10]$ that minimize the signal's deviation from r in the subsequent 8-unit window. The robustness of ψ is upper-bounded by 0.5 (indicating perfect stabilization) and is unbounded from below.

Consider the signal in Fig. 1. After the first rising edge at $t_1 = 1$, the signal stabilizes perfectly around $r_1 = 1$ within 2 time units. Thus, the parameters $r = 1$ and $c = 2$ (for any $d \in [0,8]$) maximize the robustness value to 0.5. Conversely, after the second rising edge at $t_2 = 12$, the signal oscillates widely and slowly. To evaluate quantitative satisfaction here, we examine windows of size 8 starting within 10 time units of t_2. For each candidate delay $c \in [0,10]$, we consider the segment $[t_2+c, t_2+c+8]$ and compute the range of f. The robustness value for each segment is $0.5 - \frac{1}{2}(\max f - \min f)$, and the overall robustness is the supremum over all feasible c and r. For the signal in Fig. 1, the smallest range of f over any such 8-unit window occurs in the interval $[19,27]$, which places the peak at $t = 22$ in the window while the minima occur at the endpoints. Over this interval, the minimum and maximum values of f are 1.75 and 3.0, respectively, yielding a range of 1.25. Consequently, the optimal stabilization target is $r_2 = (1.75+3.0)/2 = 2.375$, and the corresponding robustness is $\mu(w,\nu[t \leftarrow 12],\psi) = 0.5 - \frac{1.25}{2} = -0.125$.

This negative value indicates that no matter how parameters r and c are chosen, the signal fails to remain within the ± 0.5 band around any constant level for a full 8-unit interval starting within 10 units of the rising edge. Intuitively, the robustness value quantifies the signal's overshoot in the "best" window: even in the window with minimal range, the signal deviates by 0.125 outside the allowed band at its worst point. This represents the minimal uniform perturbation required to repair the signal to satisfy the property. Shifting the maximum down and the minimum up by 0.125 compresses the range sufficiently to meet the ± 0.5 requirement. Equivalently, the existential quantifiers synthesize parameters: for each fixed c, the maximizing r is the midrange of f on $[t+c, t+c+8]$, and the maximizing pair $(r^\star, c^\star)$ witnesses the best robustness value [2]. □

Remark 2 (Units and Compatibility). Because atomic predicates are interpreted as differences (i.e., $\theta_2 - \theta_1$), each atomic robustness value carries the units of the compared quantity, and we assume that atomic predicates are dimensionally well-formed. When a formula combines predicates over signals with different physical units, the max aggregator in disjunctions compares incompatible quantities, so the resulting robustness value lacks a direct physical interpretation. This limitation is inherited from the standard robustness semantics for STL [14,15] and is not specific to SFO. A practical mitigation is to evaluate all predicates on dimensionless quantities, e.g., by normalizing signals and thresholds using known operating bounds. Alternatively, one can adopt an explicitly normalized predicate-level score [30].

Finally, we relate the signed robustness in our quantitative semantics to the standard Boolean semantics.

Theorem 1 (Soundness of Quantitative Semantics). *Let ϕ be an SFO formula, w a signal trace, and ν a valuation. If $\mu(w, \nu, \phi) > 0$, then $w, \nu \models \phi$. If $\mu(w, \nu, \phi) < 0$, then $w, \nu \not\models \phi$.*

4 Past-Time SFO (pSFO)

Online monitoring is simplest for specifications whose evaluation at time t depends only on the prefix observed up to t. Past-time specifications have this property, so a monitor can emit verdicts (or robustness values) immediately, without buffering future samples. In this section, we isolate a past-time fragment of SFO and provide a procedure to translate SFO formulas with bounded future lookahead to pSFO formulas.

We call an SFO formula ϕ *temporal* if it has exactly one free time variable t, and every signal access is anchored at t: whenever $f(\tau)$ occurs in ϕ, the time term τ contains t with coefficient 1 (equivalently, τ can be rewritten as $t + \delta$ where t does not occur free in δ). This excludes references to fixed absolute times and ensures that each signal access is at a well-defined offset from the reference time t.

Definition 4 (Access time terms and pSFO). *Let ϕ be a temporal SFO formula with unique free time variable t. Let M_ϕ be the set of time terms τ such that some occurrence $f(\tau)$ appears in ϕ. For $\tau \in M_\phi$ and valuation ν, define the* offset *of τ relative to t by $\mathit{off}_\nu(\tau) = [\![\tau]\!]_{\nu[t\leftarrow 0]}$. We call ϕ a* past-time SFO *(pSFO) formula if for every $\tau \in M_\phi$ and every valuation ν of bound time variables consistent with their quantification intervals (evaluated relative to $t = 0$), we have $\mathit{off}_\nu(\tau) \le 0$.*

Intuitively, $\mathit{off}_\nu(\tau)$ is the time shift encoded by τ relative to the reference time t. The pSFO condition says that every signal read happens at time $t+\mathit{off}_\nu(\tau) \le t$, i.e., never strictly in the future.

The *bounded-response* fragment of SFO consists of formulas where every time-quantifier interval I in $\exists s \in I.\, \psi$ is bounded, i.e., $\inf I \neq -\infty$ and $\sup I \neq \infty$. In this fragment, the offsets induced by bound time variables range over bounded sets, so the formula's time dependence can be summarized by finite lookahead and history bounds.

Below, we provide a pastification procedure for temporal bounded-response SFO formulas. Note that this does not preclude unbounded time entirely: the free time variable t ranges over the entire temporal domain, whereas quantified times act as *relative offsets* and are constrained to bounded ranges. This allows the expression of safety (and co-safety) properties whose violations (and satisfactions) are witnessed by input segments of bounded length, as is common in practice (e.g., when enforcing deadlines).

The key quantity for pastification is the *forward horizon*, an upper bound on how far into the future the formula may read relative to t. Dually, the *backward horizon* bounds how far into the past it may read. For bounded-response formulas, both quantities are finite.

Definition 5 (Horizons). *Let ϕ be a temporal bounded-response formula. For each $\tau \in M_\phi$, let $\ell_\tau = \inf_\nu \mathit{off}_\nu(\tau)$ and $u_\tau = \sup_\nu \mathit{off}_\nu(\tau)$, where ν ranges over valuations of the bound time variables consistent with their quantification intervals. We define the* forward *and* backward *horizons, respectively, as $H^+(\phi) = \max_{\tau\in M_\phi} \max(0, u_\tau)$ and $H^-(\phi) = \max_{\tau\in M_\phi} \max(0, -\ell_\tau)$, with the convention that $H^+(\phi) = H^-(\phi) = 0$ when $M_\phi = \emptyset$.*

Thus, $H^+(\phi)$ is the maximal future offset of any signal access in ϕ, and $H^-(\phi)$ is the maximal history length needed to evaluate ϕ at a given reference time. We now define the pastification operator. Without loss of generality, we assume that bound variables are named apart from free variables.

Definition 6 (Pastification). *For $d \ge 0$, the pastification of ϕ is $\Pi_d(\phi) = \phi[t \mapsto t - d]$, obtained by substituting every free occurrence of t by $(t - d)$.*

Pastification turns future signal reads into past reads by shifting the reference time backwards inside the formula. Concretely, if a signal access is at $t + \delta$ in ϕ, then it becomes $(t - d) + \delta = t + (\delta - d)$ in $\Pi_d(\phi)$; choosing d large enough makes all these new offsets non-positive.

Example 2 (Pastification for bounded stabilization). Recall the bounded stabilization property from Example 1: The rising-edge predicate $\uparrow b[t]$ contains an unbounded time quantifier, so ϕ is not bounded-response. However, the consequent $\psi \equiv \exists r \in \mathbb{R}.\ \exists c \in [0,10].\ \forall d \in [0,8].\ |f(t+c+d) - r| \leq 0.5$ is bounded-response and can be pastified. Here $M_\psi = \{t+c+d\}$, and for any valuation ν consistent with $c \in [0,10]$ and $d \in [0,8]$ we have $\mathit{off}_\nu(t+c+d) = [\![t+c+d]\!]_{\nu[t\leftarrow 0]} = \nu(c)+\nu(d)$. Hence $\ell_{t+c+d} = 0$ and $u_{t+c+d} = 18$, so $H^+(\psi) = 18$ (and $H^-(\psi) = 0$). Applying Π_{18} yields

$$\Pi_{18}(\psi) \equiv \exists r \in \mathbb{R}.\ \exists c \in [0,10].\ \forall d \in [0,8].\ |f(t-18+c+d) - r| \leq 0.5.$$

Now every access time has offset $\mathit{off}_\nu(t-18+c+d) = \nu(c)+\nu(d)-18 \leq 0$, so $\Pi_{18}(\psi)$ is a pSFO formula. Evaluating $\Pi_{18}(\psi)$ at time t only reads f at times up to t, whereas evaluating ψ at time t would need f up to $t+18$. Moreover, ψ and $\Pi_{18}(\psi)$ are equisatisfiable via the reference-time shift $t \mapsto t+18$. □

Choosing d to be at least the forward horizon of the input ensures that all time references lie in the past, i.e., a pSFO formula. Finally, we establish that pastification preserves satisfiability up to a constant shift of the reference time.

Theorem 2 (Equisatisfiability under Pastification). *Let ϕ be a temporal bounded-response SFO formula with free time variable t, and let $h \geq H^+(\phi)$. For every trace w and valuation ν, we have $w,\nu \models \phi$ iff $w,\nu' \models \Pi_h(\phi)$, where $\nu'(t) = \nu(t)+h$ and $\nu'(x) = \nu(x)$ for all $x \neq t$.*

5 Online Quantitative Monitoring of pSFO

We present an online quantitative monitoring algorithm for bounded-response pSFO formulas, which capture a broad class of safety and co-safety properties. For any such formula, the monitor operates over a sliding window corresponding to the formula's backward horizon. As execution progresses, robustness values are updated incrementally using exclusively the data within this window. The construction is compositional, enabling the aggregation of monitors for pSFO subformulas while enforcing a bounded memory footprint. Furthermore, the outputs of these monitors can be combined via Boolean operations, extending the monitoring capability beyond safety and co-safety.

Following the well-definedness convention of Remark 1, the algorithm evaluates robustness only over valuations for which all signal accesses fall within the available trace domain; quantifier ranges that partly exceed the domain are effectively trimmed to the well-defined region.

When ϕ is a temporal formula with a unique free time variable t, we write $\mu(w, t_0, \phi)$ for $\mu\big(w, \nu[t \leftarrow t_0], \phi\big)$ to denote the *pointwise robustness* of ϕ at time $t_0 \in \mathbb{T}$, where ν is any valuation of the remaining (non-temporal) free variables of ϕ; if t is the only free variable, the choice of ν is irrelevant. Analogously, we write $w, t_0 \models \phi$ for $w, \nu[t \leftarrow t_0] \models \phi$. Since monitoring evaluates a formula at each time point along a trace, this pointwise view is the one we adopt throughout.

Definition 7 (Online Monitoring Problem). *The* online monitoring problem *for a temporal pSFO formula ϕ over a stream $w_1, w_2, \ldots$ of input segments requires producing an output segment v_i for each w_i. In* Boolean monitoring, *v_i is a Boolean segment indicating whether ϕ holds pointwise. In* quantitative monitoring, *v_i is an extended-real-valued segment providing the pointwise robustness of ϕ, that is, a partial function $t \mapsto \mu(w, t, \phi) \in \mathbb{R} \cup \{-\infty, +\infty\}$ where defined.*

The core idea of the algorithm is to represent signals, terms, and formulas geometrically as collections of convex polyhedra defined by linear constraints. Each signal is encoded as a list of polyhedra over time and value variables; arithmetic operations translate into intersections and projections on these polyhedra; and Boolean connectives become polyhedral case-splits. This geometric perspective, which enables Boolean monitoring [5], forms the foundation for our quantitative extension: by maintaining an auxiliary variable that tracks the robustness of each subformula, we extract quantitative information directly from the polyhedral representation.

Throughout this section, we assume the input signal is sampled at regular time intervals and linearly interpolated between samples. The assumption of regular sampling is a presentational convenience: the polyhedral operations do not require uniform sampling, and extending to irregular time steps would only affect bookkeeping, not the algorithms themselves. Although uniform sampling simplifies certain specifications (for instance, property P2 in the obstacle avoidance experiments in Sect. 6 approximates a derivative using the fixed step size), SFO and our algorithm could be adapted to monitor such properties without this assumption.

We denote by $\sqcup$, $\sqcap$, NEGATE, and ELIMINATE the standard operations of union, intersection, complement, and projection, applied to individual polyhedra or lists thereof. Lists of polyhedra are maintained in increasing order of the upper bound of the free time variable t. The intersection operation denotes intersections between (i) two polyhedra, (ii) a polyhedron and a list (pairwise), and (iii) two ordered lists of polyhedra. These operations are efficiently implemented via the double description method [18,32].

Each function f is interpreted as a signal, represented by a list of convex polyhedra $\mathcal{P}_f$ parameterized by two free variables: t_f (time) and v_f (the value of f). These polyhedra are updated upon the arrival of new signal segments; consequently, we treat them as read-only global variables within helper functions.

Finally, we assume that input pSFO formulas are given in a normalized form: all first-order quantifiers are pushed inward as far as possible, and derived propositional connectives are eliminated; negations are pushed inward through propositional structure as far as possible, without pushing them across quantifiers. This preprocessing preserves the semantics and simplifies the monitoring construction, since quantified variables then appear in local scopes and each quantifier block can be handled independently. The rewriting is linear in formula size, and pushing quantifiers inward only narrows their scope to the subformulas that mention the bound variable, without duplicating any subformula, so normalization does not increase formula size.

Algorithm 1. MONITOR

Require: pSFO formula ϕ with backward horizon h; sampling period Δ; initial signal values w_0; stream of signal segments $w_1, w_2, \ldots$
Ensure: stream of segments $v_1, v_2, \ldots$ encoding the robustness of w for ϕ

$\mathcal{P}_f \leftarrow$ {constraints from w_0} for all $f \in \mathcal{F}$
for $i \geq 1$ **do**
  Receive w_i
  $\mathcal{P}_f \leftarrow \mathcal{P}_f \sqcup$ {constraints from w_i} for all $f \in \mathcal{F}$
  $P_d \leftarrow \{\, t \in [(i-1)\Delta, i\Delta) \,\}$
  $(\mathcal{P}, v) \leftarrow$ FORMULAROBUST(ϕ, P_d)
  Output $v_i \leftarrow (\mathcal{P}, v)$
  Drop from $\mathcal{P}_f$ all signal pieces ending before $i\Delta - h$ for all $f \in \mathcal{F}$

Algorithm 4. ELIMINATEBYSUP

Require: list of polyhedra $\mathcal{P}$; current robustness variable v; quantified variable x; interval $I \subseteq \mathbb{R}$
Ensure: list of polyhedra $\mathcal{P}'$ (without x) and robustness variable v' for the quantified formula on the current segment

$\mathcal{P}' \leftarrow \emptyset$
$v' \leftarrow$ FRESHVAR()
for all $P \in \mathcal{P}$ **do**
  $P_I \leftarrow P \sqcap \{x \in I\}$
  $(P_I', o) \leftarrow$ NORMALIZEOBJ(P_I, v, x)
  $(\mathcal{Q}, v') \leftarrow$ PLPMAXIMIZE(P_I', o, x, v')
  $\mathcal{P}' \leftarrow \mathcal{P}' \sqcup \mathcal{Q}$
$\mathcal{P}' \leftarrow \{(Y, v') \in \mathcal{P}' \mid \neg\exists \hat{v} : (Y, \hat{v}) \in \mathcal{P}' \wedge \hat{v} > v'\}$
return $(\mathcal{P}', v')$

Algorithm 2. FORMULAROBUST

Require: pSFO formula ϕ; domain-constraints P_d for the current segment
Ensure: list of polyhedra $\mathcal{P}$ and robustness variable v for a subformula on the current segment

if $\phi \equiv (\theta_1 < \theta_2)$ **then**
  $(\mathcal{P}', v') \leftarrow$ TERM$(\theta_2 - \theta_1, P_d)$
  return $(\mathcal{P}', v')$
else if $\phi \equiv \neg\phi'$ **then**
  $(\mathcal{P}', v') \leftarrow$ FORMULAROBUST(ϕ', P_d)
  $v'' \leftarrow$ FRESHVAR()
  $\mathcal{P}'' \leftarrow \mathcal{P}' \sqcap \{\, v'' = -v' \,\}$
  return (ELIMINATE$(\{v'\}, \mathcal{P}''), v'')$
else if $\phi \equiv \exists r \in R : \phi'$ **then**
  $(\mathcal{P}', v') \leftarrow$ FORMULAROBUST$(\phi', P_d \sqcap \{r \in R\})$
  $(\mathcal{P}'', v'') \leftarrow$ ELIMINATEBYSUP$(\mathcal{P}', v', r, R)$
  return $(\mathcal{P}'', v'')$
else if $\phi \equiv \exists s \in I : \phi'$ **then**
  $(\mathcal{P}', v') \leftarrow$ FORMULAROBUST$(\phi', P_d \sqcap \{s \in I\})$
  $(\mathcal{P}'', v'') \leftarrow$ ELIMINATEBYSUP$(\mathcal{P}', v', s, I)$
  return $(\mathcal{P}'', v'')$
else if $\phi \equiv \phi_1 \vee \phi_2$ **then**
  $(\mathcal{P}_1, v_1) \leftarrow$ FORMULAROBUST(ϕ_1, P_d)
  $(\mathcal{P}_2, v_2) \leftarrow$ FORMULAROBUST(ϕ_2, P_d)
  $v' \leftarrow$ FRESHVAR()
  $\mathcal{P}^1_{\max} \leftarrow \mathcal{P}_1 \sqcap \mathcal{P}_2 \sqcap \{\, v_1 \geq v_2,\ v' = v_1 \,\}$
  $\mathcal{P}^2_{\max} \leftarrow \mathcal{P}_1 \sqcap \mathcal{P}_2 \sqcap \{\, v_1 < v_2,\ v' = v_2 \,\}$
  $\mathcal{P}' \leftarrow$ ELIMINATE$\big(\{v_1, v_2\},\ \mathcal{P}^1_{\max} \sqcup \mathcal{P}^2_{\max}\big)$
  return $(\mathcal{P}', v')$

Algorithm 3. TERM

Require: a term θ, a polyhedron P_d collecting domain constraints
Ensure: list of polyhedra $\mathcal{P}$ and robustness variable v for the term on the current segment

if $\theta \equiv n$ **then**
  $v \leftarrow$ FRESHVAR()
  return $(\{\, v = n \,\} \sqcap P_d, v)$
else if $\theta \equiv r$ **then**
  $v \leftarrow$ FRESHVAR()
  return $(\{\, v = r \,\} \sqcap P_d, v)$
else if $\theta \equiv s$ **then**
  $v \leftarrow$ FRESHVAR()
  return $(\{\, v = s \,\} \sqcap P_d, v)$
else if $\theta \equiv \theta_1 \pm \theta_2$ **then**
  $(\mathcal{P}_1, v_1) \leftarrow$ TERM(θ_1, P_d)
  $(\mathcal{P}_2, v_2) \leftarrow$ TERM(θ_2, P_d)
  $v \leftarrow$ FRESHVAR()
  $\mathcal{P} \leftarrow (\mathcal{P}_1 \sqcap \mathcal{P}_2) \sqcap \{\, v = v_1 \pm v_2 \,\} \sqcap P_d$
  return (ELIMINATE$(\{v_1, v_2\}, \mathcal{P}), v)$
else if $\theta \equiv f(\tau)$ **then**
  $v \leftarrow$ FRESHVAR()
  return $(\mathcal{P}_f[t_f \mapsto \tau,\ v_f \mapsto v] \sqcap P_d, v)$

Algorithm 5. PLPMAXIMIZE

Require: polyhedron P over Y and x; affine objective o over (Y, x); quantified variable x; robustness v'
Ensure: list of polyhedra $\mathcal{Q}$ over Y with $v'(y) = \sup\{o(y, x) \mid (y, x) \in P\}$

$(\alpha, \beta) \leftarrow$ SPLITCOEFF(o, x)
$(\mathcal{L}, \mathcal{U}, P_0) \leftarrow$ ISOLATEBOUNDS(P, x)
$P_Y \leftarrow$ ELIMINATE$(\{x\}, P)$; $\mathcal{Q} \leftarrow \emptyset$
$G^+ \leftarrow P_0 \sqcap \{\alpha > 0\}$
$G^- \leftarrow P_0 \sqcap \{\alpha < 0\}$
$G^0 \leftarrow P_0 \sqcap \{\alpha = 0\}$
if $G^+ \neq \emptyset$ **then**
  if $\mathcal{U} = \emptyset$ **then**
    $\mathcal{Q} \leftarrow \mathcal{Q} \sqcup (G^+ \sqcap P_Y \sqcap \{v' = +\infty\})$
  else
    for all $u \in \mathcal{U}$ **do**
      $A_u \leftarrow \bigsqcap_{u' \in \mathcal{U}} \{u \leq u'\}$
      $F_u \leftarrow \bigsqcap_{\ell \in \mathcal{L}} \{\ell \leq u\}$
      $\mathcal{Q} \leftarrow \mathcal{Q} \sqcup (G^+ \sqcap A_u \sqcap F_u \sqcap P_Y \sqcap \{v' = \alpha u + \beta\})$
if $G^- \neq \emptyset$ **then**
  if $\mathcal{L} = \emptyset$ **then**
    $\mathcal{Q} \leftarrow \mathcal{Q} \sqcup (G^- \sqcap P_Y \sqcap \{v' = +\infty\})$
  else
    for all $\ell \in \mathcal{L}$ **do**
      $A_\ell \leftarrow \bigsqcap_{\ell' \in \mathcal{L}} \{\ell \geq \ell'\}$
      $F_\ell \leftarrow \bigsqcap_{u \in \mathcal{U}} \{\ell \leq u\}$
      $\mathcal{Q} \leftarrow \mathcal{Q} \sqcup (G^- \sqcap A_\ell \sqcap F_\ell \sqcap P_Y \sqcap \{v' = \alpha\ell + \beta\})$
if $G^0 \neq \emptyset$ **then**
  $\mathcal{Q} \leftarrow \mathcal{Q} \sqcup (G^0 \sqcap P_Y \sqcap \{v' = \beta\})$
return $(\mathcal{Q}, v')$

MONITOR (Algorithm 1) outlines the high-level procedure. The monitor maintains a history of signal polyhedra $\mathcal{P}_f$. Upon receiving segment w_i, it appends constraints to $\mathcal{P}_f$, restricts the reference time to the current segment by constructing P_d, and calls FORMULAROBUST. The result is a list of polyhedra $\mathcal{P}$ together with a distinguished robustness variable v; this pair encodes the piecewise-affine robustness function on the current segment. Finally, using the finite backward horizon $h = H^-(\phi)$, the monitor discards signal pieces whose (right) endpoint is strictly smaller than $i\Delta - h$, which are never needed to evaluate ϕ for any $t \in [(i-1)\Delta, i\Delta)$.

FORMULAROBUST (Algorithm 2) recursively constructs polyhedra for subformulas, mapping logical operators to linear constraints. *Negation* introduces a fresh output variable and enforces $v'' = -v'$, then immediately projects out the old robustness variable; this keeps the representation functional (one designated robustness variable per returned subformula piece). *Disjunction* encodes the max operator by case-splitting on $v_1 \geq v_2$ versus $v_1 < v_2$ over the intersection of the two operand regions. *Existential quantification* computes the supremum of the robustness over the quantified variable's interval via ELIMINATEBYSUP.

TERM (Algorithm 3) transforms arithmetic terms into polyhedra annotated with a value variable. In all base cases (constants and variables), the returned polyhedron is conjoined with P_d so that any currently active domain constraints (e.g., quantifier bounds) are preserved. Sums and differences intersect subterm regions and add the defining equality for the output variable, then project away intermediate variables. For function terms $f(\tau)$, we substitute τ for the signal time variable t_f and constrain the signal value variable v_f to equal the fresh output variable.

ELIMINATEBYSUP (Algorithm 4) eliminates a quantified variable x by taking the supremum of the current robustness variable v over an interval I. It first restricts to the closure $[\inf I, \sup I]$, which does not affect the supremum. Then, it applies a normalization step: NORMALIZEOBJ(P_I, v, x) extracts from P_I the affine expression $o(Y, x)$ that equals v on that piece and substitutes it to eliminate the robustness variable, yielding a reduced polyhedron P_I' over (Y, x) together with the explicit objective o. The reduced polyhedron and explicit objective are then passed to PLPMAXIMIZE. Finally, the last line removes dominated points so that, for each parameter valuation Y, only maximal v' values remain; this implements the supremum over a union of pieces as the pointwise maximum of the per-piece suprema.

PLPMAXIMIZE (Algorithm 5) performs parametric maximization of an affine objective $o = \alpha x + \beta$ over the feasible x-interval induced by a polyhedron $P(Y, x)$. The optimum is achieved at an extremal bound of x determined by the sign of α; if the relevant side is unbounded, the supremum is $+\infty$. If for some y there is no feasible x (i.e., the section P_y is empty), then $y \notin P_Y$ and the procedure produces no output polyhedron at y, leaving the robustness undefined there (cf. Remark 1).

Lines 1–6. SPLITCOEFF(o, x) decomposes the affine objective into $o = \alpha x + \beta$, where α and β are affine in the parameters Y. ISOLATEBOUNDS(P, x) extracts the affine lower bounds $\mathcal{L}$ ($x \geq \ell(Y)$), upper bounds $\mathcal{U}$ ($x \leq u(Y)$), and the x-independent guard P_0. Projecting out x yields $P_Y = \{y \mid \exists x.(y, x) \in P\}$, defining parameter regions where the feasible x-interval is nonempty. The algorithm partitions the parameter space by the sign of α: G^+ (positive slope, optimum at an upper bound), G^- (negative slope, optimum at a lower bound), and G^0 (zero slope).

Lines 7–14 ($\alpha > 0$). For positive slopes, the supremum occurs at the maximal feasible x. If $\mathcal{U} = \emptyset$, the feasible region is unbounded above and the objective diverges to $+\infty$, yielding $v' = +\infty$ on $G^+ \sqcap P_Y$. Otherwise, for each candidate $u \in \mathcal{U}$, we isolate the region where u is the tightest upper bound (A_u) and feasible w.r.t. all lower bounds (F_u). On this region the supremum is $v' = \alpha u + \beta$.

Lines 15–24 ($\alpha < 0$ and $\alpha = 0$). Symmetrically, for negative slopes the supremum occurs at the minimal feasible x. If $\mathcal{L} = \emptyset$, the feasible region is unbounded below and the objective diverges to $+\infty$ as $x \to -\infty$, yielding $v' = +\infty$ on $G^- \sqcap P_Y$. Otherwise, for each $\ell \in \mathcal{L}$, the region $A_\ell \sqcap F_\ell$ isolates the tightest feasible lower bound, and the supremum is $v' = \alpha\ell + \beta$. Finally, for $\alpha = 0$, the objective is independent of x, so $v' = \beta$ on $G^0 \sqcap P_Y$.

Example 3 (Supremum Elimination). Let $\psi \equiv \exists c \in [0, 2] . 0 < f(t - c)$ and consider the samples $f(0) = 0$, $f(1) = -3$, $f(2) = -1$, $f(3) = 1$ with piecewise-linear interpolation: $f(s)$ is given by $-3s$ for $s \in [0, 1)$, and by $2s - 5$ for $s \in [1, 2)$ and $s \in [2, 3)$. We encode f as the list $\mathcal{P}_f$ of convex polyhedra over (t_f, v_f): $P_f^{(1)} = \{0 \leq t_f < 1, v_f = -3t_f\}$, $P_f^{(2)} = \{1 \leq t_f < 2, v_f = 2t_f - 5\}$, and finally $P_f^{(3)} = \{2 \leq t_f < 3, v_f = 2t_f - 5\}$. We monitor on the current segment with bounds $\mathcal{P}_d = \{2 \leq t < 3, 0 \leq c \leq 2\}$. Applying TERM to the atom $0 < f(t - c)$ substitutes $\tau := t - c$ for t_f and introduces a fresh value variable v. For each piece of $\mathcal{P}_f$, this produces constraints in (t, c, v), which we intersect with $\mathcal{P}_d$ and simplify for $t \in [2, 3)$:

$$\begin{aligned}
&(\mathrm{A})\ \ 2 \leq t < 3,\ \ t - 1 \leq c \leq 2,\ \ v = 3c - 3t \\
&(\mathrm{B})\ \ 2 \leq t < 3,\ \ t - 2 \leq c \leq t - 1,\ \ v = -2c + 2t - 5, \\
&(\mathrm{C})\ \ 2 \leq t < 3,\ \ 0 \leq c \leq t - 2,\ \ v = -2c + 2t - 5.
\end{aligned}$$

Hence, FORMULAROBUST returns their union with v as the atomic robustness.

The quantifier $\exists c \in [0, 2]$ is handled by ELIMINATEBYSUP, which first conjoins $c \in [0, 2]$ and then normalizes each piece: NORMALIZEOBJ extracts the affine objective $o(t, c)$ (here, simply $o = v$ as given by the displayed equalities) and projects out v. Then PLPMAXIMIZE writes $o = \alpha c + \beta$ (via SPLITCOEFF) and isolates the affine lower and upper bounds on c.

For (A) we have $t - 1 \leq c \leq 2$, so $\mathcal{L} = \{t - 1\}$, $\mathcal{U} = \{2\}$ and $\alpha = +3 > 0$ (the G^+ branch). Among the candidate upper bounds, the guard $A_u \wedge F_u$ with $u = 2$ is the only nonempty one (feasible for all $t \in [2, 3)$, since $t - 1 \leq 2$), so the supremum is realized at $u = 2$, yielding

$$v'_A(t) = \alpha u + \beta = 3 \cdot 2 + (-3t) = -3t + 6 \quad (t \in [2, 3)).$$

For (B) we have $t-2 \le c \le t-1$, so $\mathcal{L} = \{t-2\}, \mathcal{U} = \{t-1\}$ and $\alpha = -2 < 0$ (the G^- branch). Among the candidate lower bounds, the guard $A_\ell \wedge F_\ell$ with $\ell = t-2$ is the only nonempty one (feasible for all $t \in [2,3)$, since $t-2 \ge 0$ and $t-2 \le t-1$), so the supremum is realized at $\ell = t-2$, yielding

$$v'_B(t) = \alpha\ell + \beta = -2(t-2) + (2t-5) = -1 \quad (t \in [2,3)).$$

For (C) we have $0 \le c \le t-2$, so $\mathcal{L} = \{0\}, \mathcal{U} = \{t-2\}$ and again $\alpha = -2 < 0$ (the G^- branch). Among the candidate lower bounds, the guard $A_\ell \wedge F_\ell$ with $\ell = 0$ is the only nonempty one (feasible for all $t \in [2,3)$, since $0 \le t-2$), so the supremum is realized at $\ell = 0$, yielding

$$v'_C(t) = \alpha\ell + \beta = -2 \cdot 0 + (2t-5) = 2t-5 \quad (t \in [2,3)).$$

Because $t \in [2,3)$ and $c \in [0,2]$, every evaluation time $\tau = t - c$ lies in $[0,3)$, so all three pieces $P_f^{(1)}, P_f^{(2)}, P_f^{(3)}$ are feasible on this segment.

Eliminating c over the union corresponds to taking the pointwise maximum of the three affine outcomes $v'_A(t) = -3t+6$, $v'_B(t) = -1$, and $v'_C(t) = 2t-5$. For all $t \in [2,3)$ we have $v'_B(t) = -1 \le \max\{v'_A(t), v'_C(t)\}$, so v'_B never attains the supremum. Hence the robustness is given by

$$g(t) = \mu(w,t,\psi) = \max\{-3t+6,\ 2t-5\}.$$

The intersection $-3t+6 = 2t-5$ occurs at $t = 11/5 = 2.2$, so we obtain

$$g(t) = \begin{cases} -3t+6, & t \in [2, 11/5), \\ 2t-5, & t \in [11/5, 3). \end{cases}$$

Equivalently, over (t, v'), the monitor outputs

$$\{2 \le t < 11/5,\ v' = -3t+6\} \sqcup \{11/5 \le t < 3,\ v' = 2t-5\}.$$

Note that if we instead quantify $\exists c \in [1,2]$, then for $t \in [0,1)$ we have $\tau = t - c \in [-2,0)$, which lies outside the available trace domain. In this case, the polyhedral construction yields an empty projection onto t for that region, and therefore v' is undefined on $[0,1)$ (cf. Remark 1). □

Finally, we show that our monitoring algorithm is correct by establishing the correctness of each subroutine. We first prove PLPMAXIMIZE correct by characterizing the supremum of an affine function over a polyhedral interval, which is always realized at a boundary determined by affine guards. This result extends directly to ELIMINATEBYSUP, which applies the same mechanism to each piece of a polyhedral union and then takes pointwise maxima via dominated-point removal. We prove afterward that TERM is correct by structural induction on terms, since each case preserves the intended affine relation between the output variable and the term's value. Similarly, we handle FORMULAROBUST by induction on formulas, showing that each logical operator is correctly mapped to its quantitative counterpart; in particular, disjunction intersects the domains of both operands, enforcing the uniform well-definedness convention (Remark 1), while existential quantification ranges only over instantiations for which the robustness of the quantified subformula is defined. The monitor's correctness follows by composing these results: (i) at each iteration the stored signal pieces cover the backward horizon window $[t-h, t]$ for every t in the current segment, so all subroutine preconditions are met, and (ii) garbage collection preserves this invariant for all future iterations while ensuring bounded memory.

Theorem 3 (Correctness of Monitor (Algorithm 1)). *Let ϕ be a pSFO formula with a backward horizon of h, and let w be a piecewise-linear signal sampled with period Δ. For every $i \geq 1$ and every $t \in [(i-1)\Delta, i\Delta)$ such that $\mu(w, t, \phi)$ is defined,* Monitor *outputs v_i satisfying $v_i(t) = \mu(w, t, \phi)$. For every $t \in [(i-1)\Delta, i\Delta)$ such that $\mu(w, t, \phi)$ is undefined, $v_i(t)$ is undefined. Moreover, whenever $v_i(t)$ is defined, it depends only on the restriction of w to $[t-h, t]$.*

6 Experiments

We implemented our algorithm in Python using *SymPy* [31] for symbolic mathematics and *pplpy* [1], a Python wrapper for the *Parma Polyhedra Library* [3]. The implementation[1] provides an SFO parser, polyhedra representations and operations, and the monitoring algorithm.

Our experiments address a basic feasibility question: can the proposed online monitor compute robustness values fast enough to keep up with the sampling rate of representative cyber-physical benchmarks? Concretely, we measure the per-segment runtime of robustness computation and examine how it scales with specification complexity, in particular when moving from horizon-0 (history-free) constraints to longer-horizon formulas, and to formulas with nested quantifiers or Boolean combinations of multiple properties.

All experiments were executed on a MacBook Pro M4 (macOS Sequoia 15.7.1) with 48 GB of memory and 4.5 GHz CPU. Although our algorithm is online, experiments were run offline on pre-recorded CSV files for reproducibility.

6.1 Benchmarks and Properties

(1) *Dynamic Obstacle Avoidance.* The first benchmark [17,33] involves a parcel delivery drone flying through an urban environment populated by 8 other drones that act as obstacles. The ego drone is controlled by a neural-network policy, while the remaining drones follow pre-defined trajectories. The environment's state space is eight-dimensional, defined by the vector $x = [x, y, z, v_x, v_y, v_z, \theta_x, \theta_y]$, which includes the drone's three spatial coordinates, three velocity components, and two angular variables representing roll and pitch. The control inputs are angular accelerations of θ_x and θ_y, as well

[1] https://github.com/ista-vamos/artifact-fm-qsfo

as the vertical thrust T. The control interval is $0.1s$. We check three different properties in this benchmark:

P1: *Safe velocity.* From time t_s onward, the absolute value of velocity in every direction must be less than $v_{max}m/s$. This property corresponds to the formula $t > t_s \implies (|v_x(t)| < v_{max} \wedge |v_y(t)| < v_{max} \wedge |v_z(t)| < v_{max})$.

P2: *Motion smoothness.* The time-derivative of velocity is bounded: $|\dot{v_x}| \leq a_{max}$, $|\dot{v_y}| \leq a_{max}$, $|\dot{v_z}| \leq a_{max}$. Because we have fixed sampling of $0.1s$, we can approximate $\dot{f}$ as $\frac{(f(t)-f(t-0.1)}{0.1}$ and then rewrite the whole inequality to the expression $|f(t)-f(t-0.1)| \leq c$ where $c = 0.1 \cdot a_{max}$. We monitor the conjunction of these inequalities for each derivative above.

P3: *Obstacle separation.* For all obstacles, the distance in every direction must be greater than d_{min} at all times.

(2) *Altitude Control.* This benchmark [21] models a simplified F-16 fighter jet that must maintain a safe altitude above the ground. The state space is represented by the vector $x = [v_a, \alpha, \beta, \phi, pitch, yaw, \dot{\phi}, \dot{pitch}, \dot{yaw}, n, e, alt, p_{lag}]$ which includes the airspeed (v_a), angle of attack (α), sideslip angle (β), roll (ϕ), pitch, yaw, roll rate, pitch rate, yaw rate, northward and eastward displacements, altitude (alt), engine power lag. The sampling time of this environment is fixed at 0.033 s. We monitor the following properties:

P1: *Safe altitude.* The altitude shall satisfy $300m \leq \mathrm{alt}(t) \leq 13000m$ at all times.

P2: *Safe airspeed.* The airspeed shall remain within a safe range $v_{\min} \leq v_a(t) \leq v_{\max}$ for all t.

P3: *Altitude recovery.* Whenever $\mathrm{alt}(t) < 500\,\mathrm{m}$, then within τ_{rec} the altitude shall be at least 700 m and remain so for at least τ_{hold}. We monitor the formula $alt(t) < 500 \implies \exists t_r \in [0, \tau_{rec}].\forall t_h \in [0, \tau_{hold}].\ alt(t + t_r + t_h) \geq 700$.

6.2 Results and Discussion

Table 1. The average time in seconds of computing the robustness value for each new segment.

Avoidance				*Altitude Control*			
P1	*P2*	*P3*	*P1 + P2 + P3*	*P1*	*P2*	*P3*	*P1 + P2 + P3*
0.002	0.005	0.014	0.021	0.002	0.002	0.210	0.532

We generated 100 simulation traces from both scenarios with randomized initial states. For *Dynamic Obstacle Avoidance*, each simulation either stops when the drone collides into nearby drones or reaches a maximum of 1000 simulation steps, whereas the *Altitude Control* simulation has a uniform length of

1000 steps. The results of our experiments are summarized in Table 1. For the *Dynamic Obstacle Avoidance* scenario, the time of analyzing each observed segment is smaller than the control time (0.1 s) by a margin. This means that the monitor is timely enough to be used for making control decisions. One of the reasons for such low computation times is that formulas of properties *P1* and *P3* have horizon 0, which means we do not have to remember and analyze any history. For property *P2* (and combination of all properties *P1* + *P2* + *P3*), the horizon is 0.1 which means remembering one previous segment is sufficient for evaluating the formulas.

For the *Altitude Control* benchmark, first two properties are very similar to the first property of the avoidance benchmark, and therefore the times are similar. Property *P3*, however, is more interesting as the formula contains quantifiers. We used constants $\tau_{rec} = 10$ and $\tau_{hold} = 10$, which means that the formula has horizon 20. Since sampling rate is 0.033 *s*, the monitor has to remember more than 600 last segments. Still, it was able to analyze each new segment in around 0.2 *s*. Monitoring all three properties together (*P1* + *P2* + *P3*) increases the runtime further as the monitor has to compute symbolically minima and maxima over conjunctions and disjunctions. These computation times are not low enough to be sufficient for control, but are sufficient for raising warnings.

7 Conclusion

We introduced the first robustness-based quantitative semantics for Signal First-Order Logic (SFO) and an online monitoring algorithm for a past-time, bounded-response fragment. Our approach combines pastification with symbolic polyhedral computations to obtain robustness values for expressive SFO specifications over piecewise-linear signals. A prototype implementation and experiments on representative benchmarks demonstrate that quantitative SFO monitoring is feasible in practice and scales to non-trivial specifications.

Promising directions for future work include using robustness as an objective for controller synthesis, for example as a reward signal in reinforcement learning for cyber-physical systems [29]. We also aim to develop compositional and decentralized monitoring for multi-agent and distributed settings, where approximate monitoring can improve scalability [8]. Another natural extension is integrating SFO monitoring with runtime enforcement [16] (a.k.a. shielding [26]), where the monitor's low latency relative to the control period could enable a feedback loop that steers the controller away from violations.

Acknowledgments. We thank the anonymous reviewers for their helpful comments. This work was supported by the European Research Council (ERC) Grants VAMOS (No. 101020093) and HYPER (No. 101055412), and by the Advanced Research and Invention Agency under the Safeguarded AI programme (MSAI-PR01-P047).

References

1. PPLPY: Python interface to the parma polyhedra library. https://pypi.org/project/pplpy/. Accessed 05 Dec 2025
2. Asarin, E., Donzé, A., Maler, O., Nickovic, D.: Parametric identification of temporal properties. In: Khurshid, S., Sen, K. (eds.) RV 2011. LNCS, vol. 7186, pp. 147–160. Springer, Heidelberg (2012). https://doi.org/10.1007/978-3-642-29860-8_12
3. Bagnara, R., Hill, P.M., Zaffanella, E., Bagnara, A.: The parma polyhedra library. Sci. Comput. Program. **72**, 1–20 (2006)
4. Bakhirkin, A., Basset, N.: Specification and efficient monitoring beyond STL. In: Vojnar, T., Zhang, L. (eds.) TACAS 2019. LNCS, vol. 11428, pp. 79–97. Springer, Cham (2019). https://doi.org/10.1007/978-3-030-17465-1_5
5. Bakhirkin, A., Ferrère, T., Henzinger, T.A., Nickovic, D.: The first-order logic of signals: keynote. In: Brandenburg, B.B., Sankaranarayanan, S. (eds.) Proceedings of the International Conference on Embedded Software, EMSOFT 2018, Torino, Italy, 30 September–5 October 2018, p. 1. IEEE (2018)
6. Boker, U., Henzinger, T.A., Mazzocchi, N., Saraç, N.E.: Safety and liveness of quantitative automata. In: Pérez, G.A., Raskin, J.F. (eds.) 34th International Conference on Concurrency Theory, CONCUR 2023, Antwerp, Belgium, 18–23 September 2023, vol. 279 of LIPIcs, pp. 17:1–17:18. Schloss Dagstuhl - Leibniz-Zentrum für Informatik (2023)
7. Boker, U., Henzinger, T.A., Mazzocchi, N., Saraç, N.E.: Safety and liveness of quantitative properties and automata. Log. Methods Comput. Sci. **21**(2) (2025)
8. Bonakdarpour, B., Momtaz, A., Nickovic, D., Saraç, N.E.: Approximate distributed monitoring under partial synchrony: balancing speed & accuracy. In: Ábrahám, E., Abbas, H. (eds.) Runtime Verification - 24th International Conference, RV 2024, Istanbul, Turkey, 15–17 October 2024, Proceedings, vol. 15191 of Lecture Notes in Computer Science, pp. 282–301. Springer, Heidelberg (2024). https://doi.org/10.1007/978-3-031-74234-7_18
9. Boufaied, C., Jukss, M., Bianculli, D., Briand, L.C., Parache, Y.I.: Signal-based properties of cyber-physical systems: taxonomy and logic-based characterization. J. Syst. Softw. **174**, 110881 (2021)
10. Brim, L., Dluhos, P., Safránek, D., Vejpustek, T.: Stl*: extending signal temporal logic with signal-value freezing operator. Inf. Comput. **236**, 52–67 (2014)
11. Deshmukh, J.V., Donzé, A., Ghosh, S., Jin, X., Juniwal, G., Seshia, S.A.: Robust online monitoring of signal temporal logic. Formal Methods Syst. Des. **51**(1), 5–30 (2017). https://doi.org/10.1007/s10703-017-0286-7
12. Dokhanchi, A., Hoxha, B., Fainekos, G.: On-line monitoring for temporal logic robustness. In: Bonakdarpour, B., Smolka, S.A. (eds.) RV 2014. LNCS, vol. 8734, pp. 231–246. Springer, Cham (2014). https://doi.org/10.1007/978-3-319-11164-3_19
13. Donzé, A., Ferrère, T., Maler, O.: Efficient robust monitoring for STL. In: Sharygina, N., Veith, H. (eds.) CAV 2013. LNCS, vol. 8044, pp. 264–279. Springer, Heidelberg (2013). https://doi.org/10.1007/978-3-642-39799-8_19
14. Donzé, A., Maler, O.: Robust satisfaction of temporal logic over real-valued signals. In: Chatterjee, K., Henzinger, T.A. (eds.) FORMATS 2010. LNCS, vol. 6246, pp. 92–106. Springer, Heidelberg (2010). https://doi.org/10.1007/978-3-642-15297-9_9

15. Fainekos, G.E., Pappas, G.J.: Robustness of temporal logic specifications for continuous-time signals. Theor. Comput. Sci. **410**(42), 4262–4291 (2009)
16. Falcone, Y., Mounier, L., Fernandez, J.-C., Richier, J.-L.: Runtime enforcement monitors: composition, synthesis, and enforcement abilities. Formal Methods Syst. Des. **38**(3), 223–262 (2011)
17. Fossen, T.I.: A survey on nonlinear ship control: from theory to practice. IFAC Proc. Vol. **33**(21), 1–16 (2000)
18. Fukuda, K., Prodon, A.: Double description method revisited. In: Deza, M., Euler, R., Manoussakis, I. (eds.) CCS 1995. LNCS, vol. 1120, pp. 91–111. Springer, Heidelberg (1996). https://doi.org/10.1007/3-540-61576-8_77
19. Gol, E.A.: Efficient online monitoring and formula synthesis with past STL. In: 5th International Conference on Control, Decision and Information Technologies, CoDIT 2018, Thessaloniki, Greece, 10–13 April 2018, pp. 916–921. IEEE (2018)
20. Gorostiaga, F., Sánchez, C.: General monitorability of totally ordered verdict domains. Innov. Syst. Softw. Eng. **21**(2), 673–686 (2025)
21. Heidlauf, P., Collins, A., Bolender, M., Bak, S.: Verification challenges in f-16 ground collision avoidance and other automated maneuvers. In: ARCH@ ADHS 2018 (2018)
22. Henzinger, T.A., Mazzocchi, N., Saraç, N.E.: Abstract monitors for quantitative specifications. In: Dang, T., Stolz, V. (eds.) Runtime Verification - 22nd International Conference, RV 2022, Tbilisi, Georgia, 28–30 September 2022, Proceedings, vol. 13498 of Lecture Notes in Computer Science, pp. 200–220. Springer, Heidelberg (2022)
23. Henzinger, T.A., Mazzocchi, N., Saraç, N.E.: Quantitative safety and liveness. In: Kupferman, O., Sobocinski, P. (eds.) Foundations of Software Science and Computation Structures - 26th International Conference, FoSSaCS 2023, Held as Part of the European Joint Conferences on Theory and Practice of Software, ETAPS 2023, Paris, France, 22–27 April 2023, Proceedings, vol. 13992 of Lecture Notes in Computer Science, pp. 349–370. Springer, Heidelberg (2023). https://doi.org/10.1007/978-3-031-30829-1_17
24. Henzinger, T.A., Saraç, N.E.: Quantitative and approximate monitoring. In: 36th Annual ACM/IEEE Symposium on Logic in Computer Science, LICS 2021, Rome, Italy, 29 June–2 July 2021, pp. 1–14. IEEE (2021)
25. Jakšić, S., Bartocci, E., Grosu, R., Nguyen, T., Ničković, D.: Quantitative monitoring of STL with edit distance. Formal Methods Syst. Des. **53**(1), 83–112 (2018). https://doi.org/10.1007/s10703-018-0319-x
26. Könighofer, B., et al.: Shield synthesis. Formal Methods Syst. Des. **51**(2), 332–361 (2017)
27. Maler, O., Nickovic, D.: Monitoring temporal properties of continuous signals. In: Lakhnech, Y., Yovine, S. (eds.) FORMATS/FTRTFT -2004. LNCS, vol. 3253, pp. 152–166. Springer, Heidelberg (2004). https://doi.org/10.1007/978-3-540-30206-3_12
28. Maler, O., Nickovic, D.: Monitoring properties of analog and mixed-signal circuits. Int. J. Softw. Tools Technol. Transf. **15**(3), 247–268 (2013)
29. Meng, Y., Fan, C.: Signal temporal logic neural predictive control. IEEE Rob. Autom. Lett. **8**(11), 7719–7726 (2023)
30. Menghi, C., Nejati, S., Gaaloul, K., Briand, L.C.: Generating automated and online test oracles for simulink models with continuous and uncertain behaviors. In: ESEC/SIGSOFT FSE, pp. 27–38. ACM (2019)
31. Meurer, A., et al.: Sympy: symbolic computing in python. PeerJ Comput. Sci. **3**, e103 (2017)

32. Motzkin, T.S., Raiffa, H., Thompson, G.L., Thrall, R.M.: 3. The Double Description Method, pp. 51–74. Princeton University Press, Princeton (1953)
33. Qin, Z., Zhang, K., Chen, Y., Chen, J., Fan, C.: Learning safe multi-agent control with decentralized neural barrier certificates. arXiv preprint arXiv:2101.05436 (2021)
34. Silvetti, S., Loreti, M., Nenzi, L.: Modular and online monitoring of temporal logic specification with integral and filter. In: Könighofer, B., Torfah, H. (eds.) Runtime Verification - 25th International Conference, RV 2025, Graz, Austria, 15–19 September 2025, Proceedings, vol. 16087 of Lecture Notes in Computer Science, pp. 120–139. Springer, Heidelberg (2025). https://doi.org/10.1007/978-3-032-05435-7_8

Complexity of Consistency Testing for the Release-Acquire Semantics

R. Govind[1,2,5], S. Krishna[3], Sanchari Sil[4], and B. Srivathsan[4,5(✉)]

[1] The Institute of Mathematical Sciences, Chennai, India
govind@imsc.res.in
[2] Homi Bhabha National Institute, Anushaktinagar, Mumbai, India
[3] IIT Bombay, Mumbai, India
krishnas@cse.iitb.ac.in
[4] Chennai Mathematical Institute, Chennai, India
sanchari@cmi.ac.in
[5] CNRS, ReLaX, IRL 2000, Chennai, India
sri@cmi.ac.in

Abstract. In a seminal work, Gibbons and Korach [9] studied the complexity of deciding whether an observed sequence of reads and writes of a multi-threaded program admits a sequentially consistent interleaving. They showed the problem to be NP-hard even under strong syntactic restrictions. More recently, Chakraborty *et al.* [6] considered the problem for weak memory models and proved that NP-hardness remains even when the number of threads, the number of memory locations, and the value domain are all bounded.

In this paper we revisit the problem for the release-acquire variants of the C11 memory model. Our main positive result is that consistency testing can be done in polynomial-time when each memory location is written by at most one thread (multiple readers are allowed). Notably, this restriction is already NP-hard for the model of sequential consistency. We complement our upper bound with tight hardness results: we show the problem to be NP-hard when two threads may write to the same location; furthermore, allowing three writers per location rules out $2^{o(k)} \cdot n^{\mathcal{O}(1)}$ algorithms under the Exponential Time Hypothesis, where k denotes the number of threads, and n the number of memory operations.

1 Introduction

Multi-threaded programs, in which threads interact through shared memory operations, are prevalent across software systems. Modern hardware and compilers apply various optimizations (like instruction reordering, store buffering, etc.) to enhance performance. These optimizations can cause threads to observe writes to a shared memory location in an order that differs from the program order. *Memory models* formalize the rules to constrain which values reads may observe, and allow the programmer to be aware of the memory behaviour. These models are implemented via hardware or software protocols. As the correctness

A. Sampaio and M. Stoelinga (Eds.): FM 2026, LNCS 16557, pp. 234–251, 2026.
https://doi.org/10.1007/978-3-032-26220-2_12

of concurrent programs depends on the chosen memory model, a faithful implementation of memory models is essential.

One of the fundamental results in this subject is the work of Gibbons and Korach [9] on *testing shared memories*, which presents a way to test whether an observed execution of a multi-threaded program *is consistent with* the (rules of the) underlying memory model. In their work [9], Gibbons and Korach consider *sequential consistency* (SC) [17], the most common and simple-to-understand memory model. SC uses an interleaving semantics to explain memory operations: any observed execution across the threads is identical to that of some global interleaving of the threads' operations that preserves the program order of each thread, and each read observes the value of the last write (in this interleaving) made to the memory location. Deciding whether an observed execution can be justified by such an interleaving is the classical *consistency testing* problem for SC.

VERIFYING SEQUENTIAL CONSISTENCY (The VSC-problem [9])

Instance: A variable set $\mathbb{X}$, a value set $\mathbb{V}$, a finite collection of nonempty sequences $S_1, S_2, \ldots, S_k$, each consisting of a a finite set of memory operations of the form "$read(x, d)$" or "$write(x, d)$" with $x \in \mathbb{X}$ and $d \in \mathbb{V}$.

Question: Is there a sequence S, an interleaving of $S_1, S_2, \ldots, S_k$ such that for each "$read(x, d)$" there is a preceding "$write(x, d)$" in S with no other "$write(x, d')$" in between.

The set of sequences $S_1, \ldots, S_k$ forms our observed execution. A "Yes" answer to the question above indicates that the observed execution admits an SC behaviour, while a "No" answer shows the memory implementation can violate sequential consistency.

Gibbons and Korach [9] established that the VSC problem is NP-complete, even under a number of natural syntactic restrictions (see Table 1 of [9]). Subsequent work has extended these complexity results along two complementary directions: (i) to weaker memory models that arise in programming languages, hardware, and distributed systems [3,4,7,8,20,22,24]; and (ii) to algorithmic techniques for stateless model-checking of multi-threaded programs under sequential consistency [1]. Consistency testing also appears in the context of comparing memory models [14].

In this work, we revisit this consistency testing problem for the Release-Acquire semantics (RA), which form the core of the memory model adopted by the C11 programming language. We also investigate its variants, the strong release acquire (SRA) [15] and the weak release acquire (WRA) [13]. For all these models, we make use of an axiomatic semantics that help reason about the observed executions as graphs and the rules of the model as constraints over special kinds of edges [15]. Closely related to these memory models are Relaxed [21,23] and causal consistency models [3,5] which employ axioms which differ mildly from the release-acquire variants. From [6] we know that the analogous consistency testing problem is NP-complete for all these models even when the number of threads, the number of memory locations and the domain of values are kept bounded. In this paper, we consider a different dimension to bound.

Contributions. We prove that the consistency testing problem under the release-acquire model and its variants can be solved in $\mathcal{O}(n^3)$ (where n is the total number of memory operations in the instance) for 1-WRITER systems: these are programs where atmost one thread writes to a shared variable, but multiple threads can read from it. To the best of our knowledge, this is the first known polynomial-time algorithm for the consistency testing problem under the release-acquire semantics. Single-writer systems appear in several contexts, for instance, in producer-consumer models or CREW (concurrent-read-exclusive-write) distributed systems [12]. Notably, for sequential consistency, the problem for single-writer systems is known to be NP-hard [9], and from [1] it is known to have a polynomial-time procedure when the number of threads is kept bounded. Here, we show polynomial-time under no restriction, for the release-acquire semantics. As our second contribution, we present new lower bounds for the consistency testing problem under the release-acquire model and its variants:

- For 1-WRITER programs, we prove that there is no $\mathcal{O}(n^{\frac{3}{2}-\epsilon})$-time algorithm, under the Boolean Matrix Multiplication (BMM) hypothesis, for any $\epsilon > 0$.
- When we consider 2-WRITER programs (those where each variable is written into by atmost two threads), the problem continues to be NP-complete.
- Furthermore, when we consider 3-WRITER programs, the problem admits a conditional lower bound under the Exponential Time Hypothesis (ETH): there is no $2^{o(k)} \cdot n^{\mathcal{O}(1)}$ algorithm, where k is the number of threads, and n is the total number of operations.

The results provide a tight complexity picture while bounding the number of writer threads per memory location, with 1-WRITER giving polynomial-time, 2-WRITER being NP-hard and 3-WRITER allowing a stronger conditional lower bound.

Structure of the Paper. In Sect. 2, we present the axiomatic definitions for the release-acquire semantics and its variants. In Sect. 3, we present the polynomial-time procedure for single-writer systems and the hardness result under the BMM hypothesis. In Sect. 4, we present the hardness results for the case of multiple-writers. In this paper, we talk about the memory models RA, SRA and WRA. The corresponding results for the closely related memory models Relaxed and the models for causal consistency are presented in an extended version [10]. Finally, in Sect. 5, we conclude the paper with some directions for future work. Missing proofs can be found in [10].

2 Preliminaries

Unlike SC, where consistency is checked via an interleaving, release-acquire semantics requires reasoning with two special types of relations on the reads and writes: a reads-from relation rf (mapping each read to the write it observes) and a per-location modification order mo (ordering writes to the same variable). The verification question becomes: does there exist rf and mo on the observed events that satisfy the axioms for the release-acquire memory model? The presentation in this section follows [6,15] and introduces these preliminaries.

Events. An *event* is a memory operation: either a read $\mathtt{r}(x, v)$ or write $\mathtt{w}(x, v)$, where $x \in \mathbb{X}$ is a shared memory location and $v \in \mathbb{V}$ is a value that is either read-from or written-to the memory location. These are the analogues of "*read*(x, v)" and "*write*(x, v)" that were used in [9] and recalled in the Introduction. We sometimes use the term *variable* for a memory location. For an event e, we write $e.\mathsf{op}$, $e.\mathsf{var}$ and $e.\mathsf{val}$ respectively for the memory operation ($\mathtt{r}$ or $\mathtt{w}$), the shared memory location and the value associated with the event. We sometimes write $\mathtt{r}(x)$ in short for a read event $\mathtt{r}(x, v)$ when the value v is not important to the context. Similarly, we write $\mathtt{w}(x)$ for a write event $\mathtt{w}(x, v)$. When the variable x is also clear from the context, we simply write $\mathtt{r}, \mathtt{w}$ for $\mathtt{r}(x), \mathtt{w}(x)$.

Events are spread across a set of *threads*. For a set of events E, we write $\mathsf{E.R}$ and $\mathsf{E.W}$ for the set of all read and write events in E respectively, and $\mathsf{E.R}_x$, $\mathsf{E.W}_x$ for the read and write events involving variable x. When the set of events E is clear from the context, we simply write R_x and W_x.

Execution Graph. An *execution graph* [16] (also known as candidate execution [2]) is a tuple $\mathsf{X} = \langle \mathsf{E}, \mathsf{po}, \mathsf{rf}, \mathsf{mo} \rangle$ where E is a set of events and po, rf, mo are binary relations over E. The *program order* po is a partial order that enforces a total order on events of the same thread. The *reads-from relation* $\mathsf{rf} \subseteq (\mathsf{E.W} \times \mathsf{E.R})$ associates write events e_1 to read events e_2. It can be seen as an edge $e_1 \xrightarrow{\mathsf{rf}} e_2$ from (the node corresponding to) e_1 to (the node corresponding to) e_2 in the execution graph X. Indeed, it must be that $e_1.\mathsf{var} = e_2.\mathsf{var}$, $e_1.\mathsf{val} = e_2.\mathsf{val}$ and every read event has a unique writer (rf^{-1} is a function). The *modification order* $\mathsf{mo} \subseteq (\mathsf{E.W} \times \mathsf{E.W})$ is the union of modification orders mo_x for $x \in \mathbb{X}$, where each mo_x is a total order over the set W_x of write events that involve variable x. For a read r, let $\mathsf{rf}^{-1}(r)$ denote the write associated by rf to r, in other words, it is the write from which r reads from according to rf.

A *partial execution graph* $\overline{\mathsf{X}} = \langle \mathsf{E}, \mathsf{po} \rangle$ is an abstraction of execution graphs which does not contain the modification order and the reads-from relation.

Consistency. Consistency of an observed execution graph with respect to a weak memory model can be characterized using certain properties of the combination consisting of the po, rf and mo relations – these properties are called *consistency axioms.* In this work, we focus on the release-acquire memory model (RA) and its close variants strong release-acquire (SRA) and weak release-acquire (WRA). In Sect. 2.1, we formally define the consistency axioms. Before that, we state our problem of interest, the analogue of the VSC problem [9] for release-acquire semantics.

VERIFYING WEAK MEMORY CONSISTENCY (The V$\mathcal{M}$-problem)
Instance: A partial execution graph $\overline{\mathsf{X}} = \langle \mathsf{E}, \mathsf{po} \rangle$ and a memory model $\mathcal{M} \in \{\mathsf{RA}, \mathsf{SRA}, \mathsf{WRA}\}$
Question: Do there exist rf and mo relations on E such that the complete execution graph $\mathsf{X} = \langle \mathsf{E}, \mathsf{po}, \mathsf{rf}, \mathsf{mo} \rangle$ satisfies the consistency axioms of the memory model $\mathcal{M}$?

Notations for Relations. Let S be a binary relation over the set of events E. The reflexive, transitive, reflexive-transitive closures, and inverse relations of S are denoted as $S^?$, S^+, S^*, S^{-1} respectively. The relation S is *acyclic* if S^+ is irreflexive. We write $\mathsf{irr}(S)$ and $\mathsf{acy}(S)$ to denote that relation S is irreflexive and acyclic respectively. The identity relation on a set B is denoted as $[B]$: $[B](x, y) \triangleq x = y \wedge x \in B$. The composition of two relations S_1 and S_2, is denoted by $S_1; S_2$. For a shared variable x, we let $S_x = [\mathsf{E}_x]; S; [\mathsf{E}_x]$ be the restriction of S to all events of E on x, with $\mathsf{E}_x = \{e \in E \mid e.\text{var} = x\}$.

Derived Relations. The *happens-before relation* hb is the transitive closure of po and rf. In our descriptions, we view hb as a path consisting of $\xrightarrow{\mathsf{po}}$ and $\xrightarrow{\mathsf{rf}}$ edges. Happens-before (hb) where all the rf edges are projected to an individual variable x is denoted as hb_x. More precisely: $\mathsf{hb} \triangleq (\mathsf{po} \cup \mathsf{rf})^+$, and $\mathsf{hb}_x \triangleq (\mathsf{po} \cup \mathsf{rf}_x)^+$.

2.1 Consistency Axioms

The axioms are broadly classified as *coherence* and *causality cycles*. We will describe them in detail below. Table 1 summarizes the list of axioms and indicates the axioms for each memory model, and Fig. 1 gives scenarios violating the various axioms.

Coherence. The modification order mo enforces an 'SC-per-variable' constraint in any execution X, that is, memory accesses per shared variable are totally ordered. Coherence axioms essentially describe rules on this modification order mo.

Write-coherence enforces that mo agrees with hb for each variable (denoted as $\mathsf{irr}(\mathsf{mo}_x; \mathsf{hb})$ in Table 1 and a violation of it is shown in Fig. 1(b)): intuitively, if a write $\mathtt{w}(x, v_1)$ happens-before $\mathtt{w}(x, v_2)$, then x is updated to v_1 first, and then v_2. As the order of updates to x is encoded by the modification order mo_x, the axiom requires mo_x to align with hb. A stronger version of it, called *strong-write-coherence*, requires the entire mo (and not just mo_x) to align with hb, in other words, it requires the acyclicity of $\mathsf{po} \cup \mathsf{rf} \cup \mathsf{mo}$– again, notice that this requirement is across variables, as $\mathsf{mo} = \bigcup_{x \in \mathbb{X}} \mathsf{mo}_x$. Execution graph in Fig. 1(e) violates strong-write-coherence, while it does not violate write-coherence.

Read coherence enforces that a read $\mathtt{r}(x)$ cannot read from a write $\mathtt{w}_1(x)$ if there is an *intermediate* write $\mathtt{w}_2(x)$ that happens-before $\mathtt{r}(x)$, i.e. $(\mathtt{w}_2(x), \mathtt{r}(x)) \in \mathsf{hb}$ (denoted as $\mathtt{w}_2(x) \xrightarrow{\mathsf{hb}} \mathtt{r}(x)$), and "intermediate" refers to mo, i.e., $(\mathtt{w}_1(x), \mathtt{w}_2(x)) \in \mathsf{mo}$ (denoted as $\mathtt{w}_1(x) \xrightarrow{\mathsf{mo}} \mathtt{w}_2(x)$). Said differently, if

Table 1. Models WRA, RA, SRA.

WRA	porf-acyclicity weak-read-coherence	$\mathsf{irr}(\mathsf{mo}_x; \mathsf{hb})$ (write-coherence) $\mathsf{irr}(\mathsf{rf}^{-1}; \mathsf{mo}_x; \mathsf{hb})$ (read-coherence) $\mathsf{acy}(\mathsf{hb} \cup \mathsf{mo})$ (strong-write-coherence) $\mathsf{irr}(\mathsf{hb}; [\mathsf{W}_x]; \mathsf{hb}; \mathsf{rf}_x^{-1})$ (weak-read-coherence)
RA	porf-acyclicity write-coherence read-coherence	SRA porf-acyclicity strong-write-coherence read-coherence

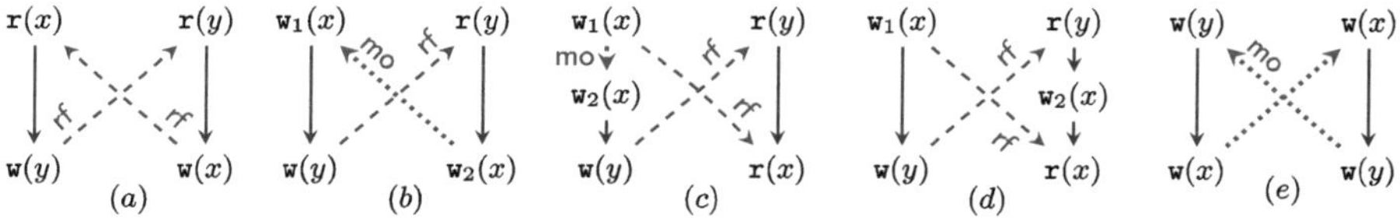

Fig. 1. Violations of coherence and causality. Values in the $\mathtt{r}$ and $\mathtt{w}$ operations suppressed for clarity. Black edge denotes po: (a) porf-acyclicity is violated (b) write-coherence is violated: $\mathtt{w}_1(x)$ hb $\mathtt{w}_2(x)$ while $\mathtt{w}_2(x)$ mo $\mathtt{w}_1(x)$, (c) read-coherence is violated: $\mathtt{w}_1(x)$ rf $\mathtt{r}(x)$ with $\mathtt{w}_1(x)$ mo $\mathtt{w}_2(x)$ and $\mathtt{w}_2(x)$ hb $\mathtt{r}(x)$, (d) weak-read-coherence is violated: $\mathtt{w}_1(x)$ rf $\mathtt{r}(x)$ with $\mathtt{w}_1(x)$ hb $\mathtt{w}_2(x)$ and $\mathtt{w}_2(x)$ hb $\mathtt{r}(x)$, (e) strong-write-coherence is violated: $\mathsf{hb} \cup \mathsf{mo}$ cycle.

$\mathtt{w}_1(x)$ and $\mathtt{w}_2(x)$ happen-before a read $\mathtt{r}(x)$, then $\mathtt{w}_1(x) \xrightarrow{\mathsf{mo}} \mathtt{w}_2(x)$, implies $\mathtt{r}(x)$ cannot read from $\mathtt{w}_1(x)$. The mo relation makes $\mathtt{w}_1$ stale for $\mathtt{r}$. Figure 1(c) shows a violation of read-coherence. In a weaker form of read-coherence called *weak-read-coherence* [15] (Fig. 1(d)), "intermediate" relates to hb (and not mo), i.e., we have $\mathtt{w}_1(x) \xrightarrow{\mathsf{hb}} \mathtt{w}_2(x)$: in other words, if $\mathtt{w}_1(x) \xrightarrow{\mathsf{hb}} \mathtt{w}_2(x) \xrightarrow{\mathsf{hb}} \mathtt{r}(x)$, then $\mathtt{r}$ cannot read from $\mathtt{w}_1$.

Causality Cycles. A causality cycle consists of po and rf orderings. The C11 models explicitly mandate porf-acyclicity [15,19], that is, $\mathsf{acy}(\mathsf{po} \cup \mathsf{rf})$ (Fig. 1(a)).

An execution graph $\mathsf{X} = \langle \mathsf{E}, \mathsf{po}, \mathsf{rf}, \mathsf{mo} \rangle$ is consistent with memory model $\mathcal{M}$ if it satisfies all the axioms corresponding to $\mathcal{M}$ as shown in Table 1. We write $\mathsf{X} \models \mathcal{M}$ if X is consistent with $\mathcal{M}$. For a partial execution graph $\overline{\mathsf{X}} = \langle \mathsf{E}, \mathsf{po} \rangle$, we write $\overline{\mathsf{X}} \models \mathcal{M}$ if there exists an rf and mo such that $\mathsf{X} \models \mathcal{M}$ for the complete execution graph $\mathsf{X} = \langle \mathsf{E}, \mathsf{po}, \mathsf{rf}, \mathsf{mo} \rangle$. Hence, the $\mathsf{V}\mathcal{M}$ problem can be stated as: given a partial execution graph $\overline{\mathsf{X}}$ and a memory model $\mathcal{M}$, does $\overline{\mathsf{X}} \models \mathcal{M}$?

We conclude this section with a useful lemma describing the relationship between the three models RA, SRA and WRA.

Lemma 1. *Let* $\mathsf{X} = \langle \mathsf{E}, \mathsf{po}, \mathsf{rf}, \mathsf{mo} \rangle$ *be an execution graph. Then:* $\mathsf{X} \models \mathsf{SRA}$ *implies* $\mathsf{X} \models \mathsf{RA}$*, and* $\mathsf{X} \models \mathsf{RA}$ *implies* $\mathsf{X} \models \mathsf{WRA}$.

Proof. It is easy to see that if X satisfies strong-write-coherence, then it satisfies write-coherence also. Hence $\mathsf{X} \models \mathsf{SRA}$ implies $\mathsf{X} \models \mathsf{RA}$. For RA to WRA, we show that if weak-read-coherence is violated, then either read-coherence or write-coherence is violated: suppose weak-read-coherence is violated, there is a cycle $\mathtt{r}(x) \xrightarrow{\mathsf{rf}^{-1}} \mathtt{w}_1(x) \xrightarrow{\mathsf{hb}} \mathtt{w}_2(x) \xrightarrow{\mathsf{hb}} \mathtt{r}(x)$. The mo relation can order $\mathtt{w}_1$ and $\mathtt{w}_2$ in either way. If $\mathtt{w}_1(x) \xrightarrow{\mathsf{mo}} \mathtt{w}_2(x)$, then read-coherence is violated, and if $\mathtt{w}_2(x) \xrightarrow{\mathsf{mo}} \mathtt{w}_2(x)$, then write-coherence is violated. □

3 Polynomial-Time Algorithm for *1-WRITER* Systems

We now consider programs where, for every variable, there is a single thread which writes to it. Any number of threads can read from a variable. An execution graph $\mathsf{X} = \langle \mathsf{E}, \mathsf{po}, \mathsf{rf}, \mathsf{mo} \rangle$ is said to be a 1-WRITER execution graph if for each memory location x, the entire set of its write events W_x belong to one thread.

Overview of this Section. The fact that we work with 1-WRITER execution graphs simplifies several axioms of Table 1, allowing for a polynomial-time procedure. Recall however that for sequential consistency, the 1-WRITER assumption offers no specific advantage and still yields a NP-hard complexity [9].

SRA *and* RA *Reduce to* WRA. Firstly, we note that consistency checking for the SRA and RA memory models reduce to checking for WRA in 1-WRITER execution graphs (Lemma 2). Moreover, since all writes of a memory location lie in a single thread, the mo relation on writes gets fixed to the unique order that agrees with po. Therefore, the problem boils down to detecting if there is an rf that satisfies *porf-acyclicity* and *weak-read-coherence*, the two axioms needed for WRA consistency. So we focus on these two axioms for the rest of the section.

A partial order on rf . Secondly, since all writes to a memory location x are totally ordered by po, there is a natural partial order induced on the rf relations. A reads-from relation rf_1 is said to be smaller than rf_2, written as $\mathsf{rf}_1 \sqsubseteq \mathsf{rf}_2$, if for every read r, the write event $w_2 = \mathsf{rf}_2^{-1}(r)$ appears po-later than $w_1 = \mathsf{rf}_1^{-1}(r)$ (we say w_2 is po-later than w_1 if $w_1\ \mathsf{po}^* w_2$). In other words, rf_2 associates each read r to a write that appears later in the program order compared to the write that rf_1 associates it to. We observe two useful properties of this partial order: (1) if an rf violates porf-acyclicity, then every $\mathsf{rf}^\dagger$ larger than it also violates porf-acyclicity, and (2) among all the rf that satisfy weak-read-coherence, there is a unique minimum element $\mathsf{rf}_{\min}$.

Algorithm. Using these two properties, we design an algorithm that starts with the smallest rf and iteratively computes larger reads-from relations until a WRA-consistent rf is found. If this rf additionally satisfies porf-acyclicity, we have found a reads-from relation that witnesses the WRA-consistency. Else, we conclude that no such rf exists.

The two useful properties appear in Sect. 3.1 and the algorithm is described in Sect. 3.2. Section 3.3 presents a lower bound for 1-WRITER systems.

3.1 Structural Properties of *1-Writer* Execution Graphs

Our first observation is that for 1-Writer systems, checking SRA and RA reduces to checking WRA. For WRA the only axioms are porf-acyclicity and weak-read-coherence. For RA one must also ensure write-coherence and read-coherence, in addition to porf-acyclicity. In a 1-Writer setting, write-coherence is enforced by choosing mo to agree with po (so all writes to a location are ordered by the program order of their single writer).

A violating cycle for read-coherence has the form

$$\mathtt{r}(x) \xrightarrow{\mathsf{rf}^{-1}} \mathtt{w}(x) \xrightarrow{\mathsf{mo}_x} \mathtt{w}'(x) \xrightarrow{\mathsf{hb}} \mathtt{r}(x).$$

When mo agrees with po, the mo_x edge is simply po^+, since both $\mathtt{w}_x$ and $\mathtt{w}'_x$ are in the same thread. Now, a violating cycle for weak-read-coherence has the form

$$\mathtt{r}(x) \xrightarrow{\mathsf{rf}^{-1}} \mathtt{w}(x) \xrightarrow{\mathsf{hb}} \mathtt{w}'(x) \xrightarrow{\mathsf{hb}} \mathtt{r}(x),$$

and, since $\mathtt{w}(x)$ and $\mathtt{w}'(x)$ come from the same thread, the hb step between $\mathtt{w}(x)$ and $\mathtt{w}'(x)$ can likewise be instantiated by po^+. Hence read-coherence coincides with weak-read-coherence in the 1-Writer setting. Moreover, having an mo agreeing with po reduces strong-write-coherence to porf-acyclicity, culminating in this lemma.

Lemma 2. *Let* $\mathsf{X} = \langle \mathsf{E}, \mathsf{po}, \mathsf{rf}, \mathsf{mo}\rangle$ *be a* 1-Writer *execution graph. Then, for* $\mathcal{M} \in \{\mathsf{SRA}, \mathsf{RA}\}$*, we have* $\mathsf{X} \models \mathcal{M}$ *iff* mo *agrees with* po *and* $\mathsf{X} \models \mathsf{WRA}$.

As a consequence, the consistency checking of 1-Writer under SRA, RA and WRA boils down to checking porf-acyclicity and weak-read-coherence violations. Secondly, the problem of synthesizing both rf and mo boils down to just synthesizing rf, since mo is fixed to the order that respects po. Nevertheless, there are still exponentially many rf relations possible. The next lemma entails that there is a "smallest" rf among all WRA-consistent rf s in the $\sqsubseteq$ order on rf s defined in the overview of this section. Given $\mathsf{rf}_1, \mathsf{rf}_2$, we write $\min(\mathsf{rf}_1, \mathsf{rf}_2)$ for the reads-from relation which associates every read r to the po-smaller write between $\mathsf{rf}_1^{-1}(r)$ and $\mathsf{rf}_2^{-1}(r)$.

Lemma 3. *Let* rf_1 *and* rf_2 *be two reads-from relations over a partial* 1-Writer *execution graph* $\mathsf{X} = \langle \mathsf{E}, \mathsf{po}, \mathsf{mo}\rangle$ *s.t.* mo *agrees with* po.

- *Suppose* $\mathsf{rf}_1 \sqsubseteq \mathsf{rf}_2$*. Then, if* rf_1 *induces a* $\mathsf{po} \cup \mathsf{rf}_1$ *cycle, then* rf_2 *induces a* $\mathsf{po} \cup \mathsf{rf}_2$ *cycle.*
- *Suppose both* rf_1 *and* rf_2 *satisfy weak-read-coherence. Then,* $\min(\mathsf{rf}_1, \mathsf{rf}_2)$ *also satisfies weak-read-coherence.*

Suppose S is the set of all rf that satisfy weak-read-coherence. The second item of Lemma 3 implies that S contains a unique minimum element, which we denote by $\mathsf{rf}_{\min}$. By the first item of Lemma 3, if $\mathsf{rf}_{\min}$ violates porf-acyclicity then every rf in S also violates it. Therefore, WRA-consistency checking reduces to computing $\mathsf{rf}_{\min}$ and testing it for porf-acyclicity. The algorithm in the next section performs exactly this construction and test.

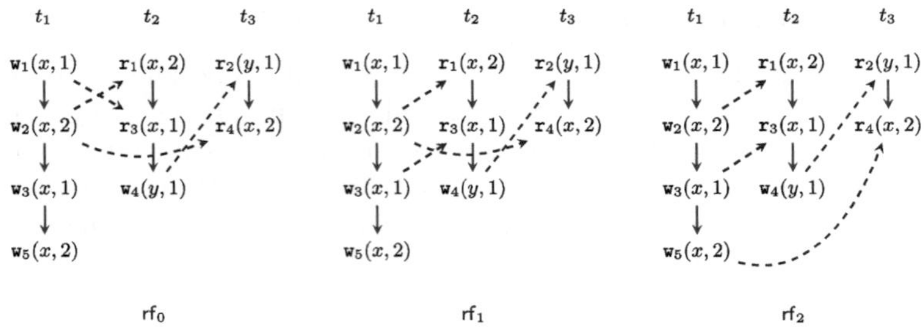

Fig. 2. From left: $\mathsf{rf}_0 \sqsubseteq \mathsf{rf}_1 \sqsubseteq \mathsf{rf}_2$. rf_0 has a violation of weak-read-coherence: $\mathtt{r}_3 \xrightarrow{\mathsf{rf}_0^{-1}} \mathtt{w}_1 \xrightarrow{\mathsf{po}} \mathtt{w}_2 \xrightarrow{\mathsf{hb}_0} \mathtt{r}_3$. rf_1 has a violation of weak-read-coherence: $\mathtt{r}_4 \xrightarrow{\mathsf{rf}_1^{-1}} \mathtt{w}_2 \xrightarrow{\mathsf{po}} \mathtt{w}_3 \xrightarrow{\mathsf{hb}_1} \mathtt{r}_4$. rf_2 does not have any more violations of weak-read-coherence. porf-acyclicity is also satisfied by rf_2. The underlying partial execution graph is WRA-consistent.

3.2 Algorithm

Our algorithm starts from an rf_0 which maps every read r to the po-smallest write in its writer thread, such that the value of the chosen write matches with that of r, and in case the writer thread is the same as the one where r is present, the mapped write lies above r. This relation rf_0 may not satisfy weak-read-coherence. In that case, a violating cycle is identified and rf_0 is modified to rf_1 as explained below. Assume we have iteratively constructed upto rf_i, and rf_i does not satisfy weak-read-coherence.

From rf_i emphto rf_{i+1} . Pick a violating cycle for weak-read-coherence:

$$r_i \xrightarrow{\mathsf{rf}_i^{-1}} w_i \xrightarrow{\mathsf{hb}_i} w_i' \xrightarrow{\mathsf{hb}_i} r_i$$

where all three events r_i, w_i, w_i' are over the same variable, say x_i, and hb_i is $(\mathsf{po} \cup \mathsf{rf}_i)^+$. Recall that since we are in the 1-Writer case, the intermediate hb_i can be replaced with po^+. Modify rf_i to eliminate the violating cycle: choose a write for r_i which appears on or after w_i', that is, pick w_i'' such that w_i' $\mathsf{po}^* w_i''$ and w_i'' is a write on the same variable x_i, writing the same value as the read event r_i. If no such write exists, conclude that there is no rf for $\overline{\mathsf{X}}$ that satisfies weak-read-coherence. Else, let the resulting rf be called rf_{i+1}. If rf_{i+1} does not satisfy weak-read-coherence, we update rf_{i+1} as above. Else, rf_{i+1} satisfies weak-read-coherence. We declare $\mathsf{rf}_{\mathsf{min}} = \mathsf{rf}_{i+1}$.

Finally, we test whether $\mathsf{rf}_{\mathsf{min}}$ satisfies porf-acyclicity. If it does, we have a witness for WRA-consistency. Else, we conclude (due to Lemma 3) that $\overline{\mathsf{X}}$ is not WRA-consistent. We formalize this discussion in Algorithm 1. Here is the key invariant maintained by the algorithm: at each step i where the update function is called on rf_i, we can show that $\mathsf{rf}_i \sqsubseteq \mathsf{rf}^\dagger$ for every $\mathsf{rf}^\dagger$ that satisfies weak-read-coherence. Figures 2 and 3 illustrate the running of the algorithm on two different partial execution graphs.

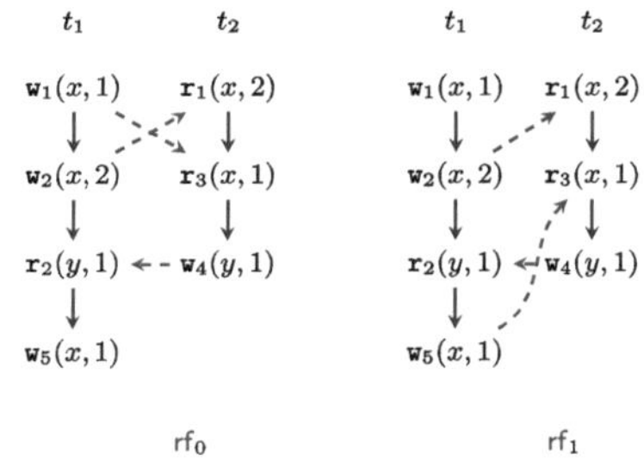

Fig. 3. From left: $\mathsf{rf}_0 \sqsubseteq \mathsf{rf}_1$. rf_0 has a violation of weak-read-coherence: $\mathtt{r}_3 \xrightarrow{\mathsf{rf}_0^{-1}} \mathtt{w}_1 \xrightarrow{\mathsf{po}} \mathtt{w}_2 \xrightarrow{\mathsf{hb}} \mathtt{r}_3$. rf_1 has no weak-read-coherence violation but does not satisfy porf-acyclicity: $\mathtt{w}_5 \xrightarrow{\mathsf{rf}_1} \mathtt{r}_3 \xrightarrow{\mathsf{po}} \mathtt{w}_4 \xrightarrow{\mathsf{rf}} \mathtt{r}_2 \xrightarrow{\mathsf{po}} \mathtt{w}_5$. The partial execution graph is not WRA-consistent.

Theorem 4. *A partial execution graph* $\overline{\mathsf{X}} = \langle \mathsf{E}, \mathsf{po} \rangle$ *of a one-writer system is consistent for* WRA *iff Algorithm 1 returns* $\overline{\mathsf{X}}$ *is* WRA*-consistent. Moreover, the algorithm runs in time* $\mathcal{O}(n^3)$ *where* $n = |E|$ *and returns the smallest* rf *that is* WRA*-consistent, if one exists.*

3.3 Hardness for *1-Writer* Systems

The major contributor to the complexity of Algorithm 1 is verification of whether a given rf satisfies weak-read-coherence. This can be posed as a graph problem: there are n nodes corresponding to the events, and there are three kinds of edges po, rf and rf^{-1}; does there exist a cycle of the form $r \xrightarrow{\mathsf{rf}^{-1}} w \xrightarrow{\mathsf{po}^+} w' \xrightarrow{(\mathsf{po} \cup \mathsf{rf})^*} r$? For each read event r and its corresponding write event w, we need to check if there exist a w' such that (1) there is a path of po edges from w to w', and (2) there is a path from w' to r over a combination of po and rf edges. For each read, this can be achieved in $\mathcal{O}(n)$ time by a graph search. Hence, overall, across all possible reads, this algorithm takes $\mathcal{O}(n^2)$ time. Therefore, each update to an rf costs $\mathcal{O}(n^2)$ and there can be at most n updates (as each update moves the write of a read to a po-later write). This gives a complexity of $\mathcal{O}(n^3)$. As our next observation, we prove a conditional lower bound of $\mathcal{O}(n^{\frac{3}{2}-\epsilon})$ on the time to check for weak-read-coherence in a one-writer program with n events. This also transfers as a lower bound for the WRA-consistency checking. Due to Lemma 2, it also gives a lower bound on the consistency checking for SRA and RA as well. To show the lower bound, we make use of the Boolean Matrix Multiplication hypothesis and a lower bound result on finding triangles in graphs.

Boolean Matrix Multiplication (BMM) Hypothesis. There is no $\mathcal{O}(n^{3-\epsilon})$ algorithm for Boolean Matrix Multiplication, for any $\epsilon > 0$.

Theorem 5 (Theorem 1.3 of [26]). *Under BMM hypothesis, there is no* $\mathcal{O}(n^{3-\epsilon})$ *algorithm for detecting a triangle in an (undirected) graph with* n *vertices, for any* $\epsilon > 0$.

Algorithm 1: WRA-consistency checking in 1-WRITER systems

Function CheckConsistency($\overline{\mathsf{X}}$):
  rf ← Initialize($\overline{\mathsf{X}}$)
  while rf *violates weak-read-coherence* **do**
    rf ← Update($\overline{\mathsf{X}}$, rf)
  /* Now rf = $\mathsf{rf}_{\min}$ */
  if rf *satisfies porf-acyclicity* **then**
    return $\overline{\mathsf{X}}$ is WRA-consistent
  else
    return $\overline{\mathsf{X}}$ is not WRA-consistent
Function Initialize($\overline{\mathsf{X}}$):
  for *each read* $r = \mathtt{r}(x, v)$ **do**
    $t \leftarrow$ thread containing r
    $t_x \leftarrow$ writer thread for x
    if $t \neq t_x$ **then**
      $\mathsf{rf}_0^{-1}(r) \leftarrow$ po-smallest write w of the form $\mathtt{w}(x, v)$ in t_x
      if *there is no such write* **then**
        return $\overline{\mathsf{X}}$ is not WRA-consistent
    if $t = t_x$ **then**
      $\mathsf{rf}_0^{-1}(r) \leftarrow$ po-smallest write w of the form $\mathtt{w}(x, v)$ in t_x s.t. w $\mathsf{po}^+ r$
      if *there is no such write* **then**
        return $\overline{\mathsf{X}}$ is not WRA-consistent
  return rf_0
Function Update($\overline{\mathsf{X}}$, rf_i):
  Pick a cycle violating weak-read-coherence: $r_i \xrightarrow{\mathsf{rf}_i^{-1}} w_i \xrightarrow{\mathsf{hb}_i} w_i' \xrightarrow{\mathsf{hb}_i} r_i$
  Let $r_i = \mathtt{r}(x, v)$, $w_i = \mathtt{w}(x, v)$, $w_i' = \mathtt{w}(x, v')$
  $t \leftarrow$ thread of r_i
  $t_x \leftarrow$ writer thread of variable x
  if $t \neq t_x$ **then**
    $\mathsf{rf}_{i+1}^{-1}(r_i) \leftarrow$ po-smallest write w_i'' of the form $\mathtt{w}(x, v)$ in t_x s.t. w_i' $\mathsf{po} * w_i''$
    if *there is no such write* **then**
      return $\overline{\mathsf{X}}$ is not WRA-consistent
  if $t = t_x$ **then**
    $\mathsf{rf}_{i+1}^{-1}(r_i) \leftarrow$ po-smallest write w_i'' of the form $\mathtt{w}(x, v)$ in t_x s.t. w_i' $\mathsf{po}^* w_i''$ $\mathsf{po}^+ r_i$
    if *there is no such write* **then**
      return $\overline{\mathsf{X}}$ is not WRA-consistent
  $\mathsf{rf}_{i+1}^{-1}(r') \leftarrow \mathsf{rf}_i(r')$ for all $r' \neq r_i$
  return rf_{i+1}

We reduce triangle detection in a graph with n vertices to weak-read-coherence checking in a 1-WRITER execution graph with n^2 events. This results in the following proposition.

Proposition 6. *Under BMM, for any $\epsilon > 0$, there is no $\mathcal{O}(n^{\frac{3}{2}-\epsilon})$ algorithm for deciding if a given* 1-Writer *execution graph* $\mathsf{X} = \langle \mathsf{E}, \mathsf{po}, \mathsf{rf}, \mathsf{mo} \rangle$ *satisfies the weak-read-coherence axiom.*

Given graph $G = (V_G,\ E_G)$ with n vertices, we build an execution graph $\overline{\mathsf{X}}_G = \langle \mathsf{E}, \mathsf{po} \rangle$ of a single-writer program with n^2 events, such that $\overline{\mathsf{X}}$ is consistent w.r.t. $\mathcal{M}$ for iff G has no triangles. Therefore, if the consistency problem can be solved in $\mathcal{O}(n^{\frac{3}{2}} - {}^{\epsilon})$ for some ϵ, we can solve triangle detection on n-node graphs in $\mathcal{O}((n^2)^{\frac{3}{2}} - {}^{\epsilon}) = \mathcal{O}(n^{3-\frac{\epsilon}{2}})$, contradicting the BMM hypothesis.

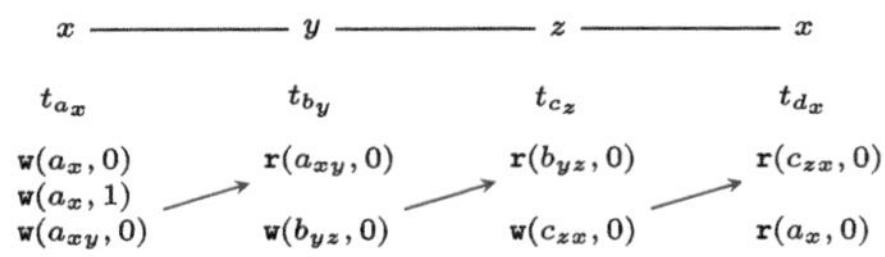

Fig. 4. From triangles to violations of weak-read-coherence.

Our idea for the reduction is as follows. We will construct $\overline{\mathsf{X}}$ in such a way that for every $\mathtt{r}(x, v)$ there is a unique $\mathtt{w}(x, v)$ present in the entire program. Therefore, the rf relation is fixed. If G has a triangle, we will have a shared variable a and a cycle of the form: $\mathtt{w}(a, 0) \xrightarrow{\mathsf{po}} \mathtt{w}(a, 1) \xrightarrow{\mathsf{hb}} \mathtt{r}(a, 0) \xrightarrow{\mathsf{rf}^{-1}} \mathtt{w}(a, 0)$. This cycle violates weak-read-coherence. If $\overline{\mathsf{X}}$ with the fixed rf violates weak-read-coherence, then G has a triangle. The hb path $\mathtt{w}(a, 1) \xrightarrow{\mathsf{hb}} \mathtt{r}(a, 0)$ should reflect the presence of a triangle in G.

The key aspect of the construction is depicted in Fig. 4. Suppose G contains a triangle over vertices x, y, z. We consider a variable a_x for which we extract an hb path as above. To force the hb path to have length 3, we consider four copies $t_{a_x}, t_{b_x}, t_{c_x}, t_{d_x}$ for each vertex x. When there is edge (x, y), thread t_{a_x} writes to t_{b_y}. Similarly, (y, z) makes t_{b_y} write to t_{c_z} and finally the edge (z, x) makes t_{c_z} write to t_{d_x} which contains $\mathtt{r}(a_x, 0)$. The full construction is explained in [10].

4 Hardness in the Presence of Multiple Writers

While having a single writer per location induced a polynomial-time algorithm for the consistency check, we show that having two writers per thread makes the problem NP-complete. The NP upper bound follows by guessing an rf and mo and then checking whether the complete execution graph $\mathsf{X} = \langle \mathsf{E}, \mathsf{po}, \mathsf{rf}, \mathsf{mo} \rangle$ satisfies the consistency axioms. When rf and mo are already given, this test can be done in polynomial-time [25]. Hence the NP upper bound follows. The hardness for our $\mathsf{V}\mathcal{M}$ problem comes from the fact that we need to *synthesize* suitable rf and mo (the analogue in the sequential consistency setting would be the synthesis of an interleaving). For proving hardness, we provide a reduction from 3-CNF-SAT: given a boolean formula φ over k variables and m clauses, with each clause being a disjunction of three literals, is there a satisfying assignment for φ? We provide two reductions.

1. Formula φ to a partial execution graph $\overline{\mathsf{X}}_\varphi$ with $\mathcal{O}(k)$ threads, and containing at most *three threads* writing to a location (3-Writer).
2. Formula φ to a partial execution graph $\overline{\mathsf{Y}}_\varphi$ with $\mathcal{O}(k+m)$ threads, and containing at most *two threads* writing to a location (2-Writer).

In both the reductions, we achieve the property that φ is satisfiable iff $\overline{\mathsf{X}}_\varphi$ (correspondingly $\overline{\mathsf{Y}}_\varphi$) are $\mathcal{M}$-consistent, for $\mathcal{M} \in \{\mathsf{SRA}, \mathsf{RA}, \mathsf{WRA}\}$. This shows that the consistency checking problem is NP-complete with a bounded number of writers per location, in fact, even with two writers. The reason we present the first reduction to 3-Writer is because we get a stronger conditional bound: since the number of threads in $\overline{\mathsf{X}}_\varphi$ is linearly related to the number of variables in φ, we get a parameterized reduction that helps proving a conditional lower bound under the Exponential Time Hypothesis (ETH). In the reduction to 2-Writer systems, the number of threads depends both on k (number of variables in the formula) and m (the number of clauses). Hence this reduction cannot be used for the conditional lower bound. We will first describe the reduction to three writers and its implications, and then follow up with the second reduction.

Notation. Let the variables used by φ be $x_1, \ldots, x_k$. The clauses are $C_1, \ldots, C_m$. Each clause C_j consists of three literals $(\ell_j^1, \ell_j^2, \ell_j^3)$, where each literal is either a variable x_i or its negation $\neg x_i$. We define $S_{x_i} := \{j \mid C_j \text{ contains } x_i\}$ and $S_{\neg x_i} := \{j \mid C_j \text{ contains } \neg x_i\}$ as the set of (indices of) clauses containing the literals x_i and $\neg x_i$ respectively.

4.1 Reduction to 3-*Writer* Systems

We consider execution graphs where for each variable x there are at most three threads writing to x, that is, at most three threads can have statements of the form $\mathtt{w}(x, d)$. We will call them 3-Writer execution graphs. Our objective is to show a hardness result stronger than NP-hardness, for the V$\mathcal{M}$ problem under the assumption stated below.

Exponential Time Hypothesis (ETH). There is no $2^{o(k)} \cdot (k+m)^{\mathcal{O}(1)}$ algorithm for 3-CNF-SAT, where k is the number of variables, m is the number of clauses [11,18].

We prove the following theorem in this section.

Theorem 7. *Under ETH, there is no $2^{o(k)} \cdot n^{\mathcal{O}(1)}$ algorithm for the V$\mathcal{M}$-problem with $\mathcal{M} \in \{\mathsf{SRA}, \mathsf{RA}, \mathsf{WRA}\}$, k being the number of threads, and n the number of events. The result holds even while restricting the input instance to* 3-Writer *execution graphs.*

Proof Strategy. Given a 3-CNF formula φ with k threads and m clauses, we construct a partial execution graph $\overline{\mathsf{X}}_\varphi = \langle \mathsf{E}, \mathsf{po} \rangle$ that satisfies the following properties:

- $\overline{\mathsf{X}}_\varphi$ is 3-Writer,

- the number of threads in $\overline{\mathsf{X}}_\varphi$ is $\mathcal{O}(k)$,
- the number of events in $\overline{\mathsf{X}}_\varphi$ is $\mathcal{O}(k+m)$,
- $\overline{\mathsf{X}}_\varphi \models \mathcal{M}$ iff φ is satisfiable, for $\mathcal{M} \in \{\mathsf{SRA}, \mathsf{RA}, \mathsf{WRA}\}$.

T_i^1	T_i^0
$\mathtt{w}(s_i, 1)$	$\mathtt{w}(s_i, 1)$
$\mathtt{w}(s_i, 2)$	$\mathtt{w}(s_i, 2)$
$\mathtt{w}(v_i, 1)$	$\mathtt{w}(v_i, 0)$

T_{x_i}	$T_{\neg x_i}$
$\mathtt{r}(v_i, 1)$	$\mathtt{r}(v_i, 0)$
$\mathtt{w}(c_j, 1)$ $\forall j \in S_{x_i}$	$\mathtt{w}(c_j, 1)$ $\forall j \in S_{\neg x_i}$

T_f
$\mathtt{r}(c_1, 1)$
$\dots$
$\mathtt{r}(c_m, 1)$
$\mathtt{r}(s_1, 2)$
$\mathtt{r}(s_1, 1)$
$\dots$
$\mathtt{r}(s_k, 2)$
$\mathtt{r}(s_k, 1)$

Fig. 5. Threads in the reduction, for $i \in \{1, 2, \dots, k\}$

Thus, if there was a $2^{o(k)} \cdot n^{\mathcal{O}(1)}$ algorithm for the V$\mathcal{M}$ problem, we could solve 3-CNF-SAT in $2^{o(k)}.(k+m)^{\mathcal{O}(1)}$ time: from φ, construct the V$\mathcal{M}$ instance in $\mathcal{O}(k+m)$ time, and then use the $2^{o(k)} \cdot (k+m)^{\mathcal{O}(1)}$ procedure for V$\mathcal{M}$, thereby contradicting ETH. To show the last item above, we make use of what is known in the literature as a *range reduction* [6]: using Lemma 1, it is enough to show that if φ has a satisfying assignment, then $\overline{\mathsf{X}} \models \mathsf{SRA}$, and if φ does not have a satisfying assignment, then $\overline{\mathsf{X}} \not\models \mathsf{WRA}$.

The Reduction. For every variable x_i, we introduce four threads T_i^1, T_i^0, T_{x_i} and $T_{\neg x_i}$, as shown in Fig. 5. Apart from these, there is a special thread T_f. A concrete example of the reduction is presented in [10].

Here is an intuitive description of the reduction. The statements $\mathtt{w}(v_i, 1)$ and $\mathtt{w}(v_i, 0)$ in T_i^1 and T_i^0 encode variable x_i being set to *true* and *false* respectively. Thread T_{x_i} has a read statement $\mathtt{r}(v_i, 1)$ (reading *true* for x_i) and sets all clauses containing x_i to true, using the statement $\mathtt{w}(c_j, 1)$ for all $j \in S_{x_i}$. Similarly, thread $T_{\neg x_i}$ reads $\mathtt{r}(v_i, 0)$ (*false* for x_i) and sets all clauses containing $\neg x_i$ to true. Finally the thread T_f needs to make sure that all clauses are set to true, hence the read statements $\mathtt{r}(c_1, 1), \dots, \mathtt{r}(c_m, 1)$. The main challenge is to encode a valid assignment, where every variable is assigned a unique truth value: to do this, we need to ensure that there are no two distinct reads $\mathtt{r}(c_j, 1)$ and $\mathtt{r}(c_\ell, 1)$ in T_f which read from some T_{x_i} and its negation $T_{\neg x_i}$. This is where the rest of the statements, using s_i become useful, as we elaborate in the proofs below.

Lemma 8. *If* $\overline{\mathsf{X}}_\varphi \models \mathsf{WRA}$*, then* φ *has a satisfying assignment.*

Proof. If $\overline{\mathsf{X}}_\varphi \models \mathsf{WRA}$ there is an rf function which satisfies porf-acyclicity and weak-read-coherence. We need to prove that this rf encodes a valid satisfying assignment. This can be shown by proving that there are no two distinct reads $\mathtt{r}(c_j, 1)$ and $\mathtt{r}(c_\ell, 1)$ which read from T_{x_i} and $T_{\neg x_i}$ for some variable x_i, since it gives the following assignment: $A(x_i) = true$ if some read event $\mathtt{r}(c_j, 1)$ reads from the write in T_{x_i}; $A(x_i) = false$ otherwise. Moreover, the assignment A is satisfying since every clause is made true by this assignment.

Let us now prove the property. Suppose $\mathtt{r}(c_j, 1)$ and $\mathtt{r}(c_\ell, 1)$ read from T_{x_i} and $T_{\neg x_i}$ respectively. Since $\mathtt{r}(c_j, 1)$ reads from T_{x_i}, there is an hb path from T_i^1 to $\mathtt{r}(c_j, 1)$ due to the rf edges: $\mathtt{w}(v_i, 1) \xrightarrow{\mathsf{rf}} \mathtt{r}(v_i, 1)$ and $T_{x_i} : \mathtt{w}(c_j, 1) \xrightarrow{\mathsf{rf}} \mathtt{r}(c_j, 1)$. Similarly, there is an hb path from T_i^0 to $\mathtt{r}(c_\ell, 1)$. Hence there is an hb path from both T_i^1 and T_i^0 to the event $\mathtt{r}(s_i, 1)$ in T_f. Since the only matching write $\mathtt{w}(s_i, 1)$ appears above $\mathtt{w}(s_i, 2)$ in these threads, any assignment to $\mathtt{r}(s_i, 1)$ violates weak-read-coherence, a contradiction to the hypothesis that rf satisfies weak-read-coherence. □

Lemma 9. *If φ has a satisfying assignment, then $\overline{\mathsf{X}}_\varphi \models \mathsf{SRA}$.*

Proof. Suppose φ has a satisfying assignment A. We need to synthesize rf and mo such that the resulting execution graph satisfies porf-acyclicity, strong-write-coherence and strong-read-coherence. Since A is a satisfying assignment, for every clause C_j there is a literal ℓ_j that is made true. Let $\mathsf{rf}^{-1}(\mathtt{r}(c_j, 1))$ be the write event $\mathtt{w}(c_j, 1)$ in T_{ℓ_j}. Assigning writes this way introduces an hb path from $T_i^{A(x_i)}$ to T_f for all variables that are indeed being utilized to make the clauses true. Importantly, there is no hb path created from $T_i^{\neg A(x_i)}$ to the thread T_f. Therefore, the read event $\mathtt{r}(s_i, 2)$ can read from the write in $T_i^{A(x_i)}$ and $\mathtt{r}(s_i, 1)$ can read from the unused thread $T_i^{\neg A(x_i)}$. This gives an rf which satisfies weak-read-coherence. By design of the execution graph, the rf also satisfies porf-acyclicity. For the lemma, we require to prove strong-write-coherence and read-coherence. For this we need to also synthesize an mo. Consider any mo relation that respects the following orders:

- $T_i^{A(x_i)} : \mathtt{w}(s_i, 1) \xrightarrow{\mathsf{mo}} T_i^{\neg A(x_i)} : \mathtt{w}(s_i, 1) \xrightarrow{\mathsf{mo}} T_i^{\neg A(x_i)} : \mathtt{w}(s_i, 2) \xrightarrow{\mathsf{mo}} T_i^{A(x_i)} : \mathtt{w}(s_i, 2)$,
- For each clause j there are three threads writing $\mathtt{w}(c_j, 1)$. For each clause C_j choose a specific literal ℓ_j that is made true by the satisfying assignment A. The mo relation orders the other write statements $\mathtt{w}(c_j, 1)$ before $T_{\ell_j} : \mathtt{w}(c_j, 1)$.

It can be shown that the constructed rf and mo satisfy strong-write-coherence and read-coherence (see [10] for the full proof). □

4.2 Reduction to *2-Writer* Systems

The reduction in Sect. 4.1 involved 3-Writer systems. Mainly, each clause variable c_j was written by three threads corresponding to its three literals. Here, our objective is to prove hardness even when there are atmost two threads writing to each variable. We will now reduce 3-CNF-SAT to the consistency testing problem for 2-Writer systems. However, in this reduction, a formula φ with k threads and m clauses yields an execution graph with $\mathcal{O}(k + m)$ theads. While this still proves NP-hardness, it does not give the stronger hardness under ETH, parameterized by the number of threads.

Theorem 10. *The* $\mathcal{VM}$ *problem is* NP*-complete for* $\mathcal{M} \in \{\mathsf{SRA}, \mathsf{RA}, \mathsf{WRA}\}$*, even when restricted to* 2-WRITER *execution graphs.*

The key challenge in proving the hardness is to reduce the number of threads writing to each clause variable c_j from three (one for each literal in the clause) to at most two, while still encoding the satisfiability of the 3-CNF formula. This can be achieved by introducing one auxiliary variable d_j per clause. Suppose $C_j = (\ell_j^1 \vee \ell_j^2 \vee \ell_j^3)$. In all the reductions of Sect. 4.1, there are three threads $T_{\ell_j^1}, T_{\ell_j^2}$ and $T_{\ell_j^3}$. We make the following changes to constructions of $\overline{\mathsf{X}}_\varphi$ in Sect. 4.1.

- Replace $\mathtt{w}(c_j, 1)$ in $T_{\ell_j^3}$ with $\mathtt{w}(d_j, 1)$, i.e., in the third literal, the write is on d_j.
- Add a new thread T_j with two events as follows: $\mathtt{r}(c_j, 1)$; $\mathtt{w}(d_j, 1)$
- Finally, in T_f, replace all $\mathtt{r}(c_j, 1)$ with $\mathtt{r}(d_j, 1)$

Notice that due to the introduction of the new threads T_j, the total number of threads also depends on the number of clauses m. Let the resulting partial execution graph be called $\overline{\mathsf{Y}}_\varphi$. In [10] we prove the correctness of the reduction.

5 Conclusion

Testing shared memory implementations is a key problem in concurrency, whose computational foundations have been laid by the work of [9] for sequential consistency. Ever since, the work has been applied to various other memory models and also used in stateless model-checking algorithms for concurrent programs. We have contributed to the study of this problem under the release-acquire semantics. We have presented a trichotomy of the landscape, with 1-WRITER solved in polynomial-time, 2-WRITER being NP-complete, and 3-WRITER having a stronger lower bound under the Exponential-Time-Hypothesis (ETH). Our presentation works for RA and its close variants SRA and WRA.

A milder variant of the problem studies the complexity when the reads-from relation rf is also given. Then the question is to synthesize an mo such that the axioms can be satisfied. This "rf-given" problem can be solved in polynomial-time [25], under no restrictions on the execution graphs. When rf is not provided apriori, while the general problem is known to be NP-hard [6], no subclass or bounded variant with a polynomial-time algorithm is known in the literature (to the best of our knowledge). We have shown that the 1-WRITER subclass lends itself to a polynomial-time algorithm.

Our results can also be extended to the memory model called Relaxed, whose consistency axioms are a relaxed version of write-coherence and read-coherence, where essentially hb is replaced with po. This makes the consistency check easier for the 1-WRITER algorithm. For the hardness, we need to make small changes, but we still obtain the same trichotomy of results. The extended version [10] provides the details. Another set of memory models that are closely related w.r.t the analysis techniques are the models CC, CM and CCv coming from distributed databases. It is known that CC coincides with WRA while CCv coincides with SRA [15]. Some care needs to be taken for CM, but yet again, we get the same trichotomy of results. The extended version [10] has the details.

Adapting these complexity theoretic insights to improve algorithmics for the verification of concurrent programs under the release-acquire semantics, is one direction for future work.

References

1. Agarwal, P., Chatterjee, K., Pathak, S., Pavlogiannis, A., Toman, V.: Stateless model checking under a reads-value-from equivalence. In: Silva, A., Leino, K.R.M. (eds.) CAV 2021. LNCS, vol. 12759, pp. 341–366. Springer, Cham (2021). https://doi.org/10.1007/978-3-030-81685-8_16
2. Batty, M., Owens, S., Sarkar, S., Sewell, P., Weber, T.: Mathematizing C++ concurrency. In: POPL 2011, pp. 55–66. ACM (2011). https://doi.org/10.1145/1926385.1926394
3. Bouajjani, A., Enea, C., Guerraoui, R., Hamza, J.: On verifying causal consistency. In: POPL, pp. 626–638. ACM (2017)
4. Bui, T.L., Chatterjee, K., Gautam, T., Pavlogiannis, A., Toman, V.: The reads-from equivalence for the TSO and PSO memory models. Proc. ACM Program. Lang. **5**(OOPSLA), 1–30 (2021)
5. Burckhardt, S.: Principles of eventual consistency. Found. Trends Program. Lang. **1**(1–2), 1–150 (2014). https://doi.org/10.1561/2500000011
6. Chakraborty, S., Krishna, S.N., Mathur, U., Pavlogiannis, A.: How hard is weak-memory testing? Proc. ACM Program. Lang. **8**(POPL), 1978–2009 (2024)
7. Chen, Y., et al.: Fast complete memory consistency verification. In: HPCA, pp. 381–392. IEEE Computer Society (2009)
8. Chini, P., Saivasan, P.: A framework for consistency algorithms. In: Saxena, N., Simon, S. (eds.) 40th IARCS Annual Conference on Foundations of Software Technology and Theoretical Computer Science, FSTTCS 2020, 14–18 December 2020, BITS Pilani, K K Birla Goa Campus, Goa, India (Virtual Conference). LIPIcs, vol. 182, pp. 42:1–42:17. Schloss Dagstuhl - Leibniz-Zentrum für Informatik (2020). https://doi.org/10.4230/LIPICS.FSTTCS.2020.42
9. Gibbons, P.B., Korach, E.: Testing shared memories. SIAM J. Comput. **26**(4), 1208–1244 (1997)
10. Govind, R., Krishna, S., Sil, S., Srivathsan, B.: Complexity of consistency testing for the release-acquire semantics (2026). https://hal.science/hal-05534072, hAL preprint, hal-05534072
11. Impagliazzo, R., Paturi, R., Zane, F.: Which problems have strongly exponential complexity? J. Comput. Syst. Sci. **63**(4), 512–530 (2001)
12. JáJá, J.: An Introduction to Parallel Algorithms. Addison–Wesley (1992)
13. Kokologiannakis, M., Lahav, O., Sagonas, K., Vafeiadis, V.: Effective stateless model checking for C/C++ concurrency. Proc. ACM Program. Lang. **2**(POPL) (2017). https://doi.org/10.1145/3158105, https://doi.org/10.1145/3158105
14. Kokologiannakis, M., Lahav, O., Vafeiadis, V.: Kater: automating weak memory model metatheory and consistency checking. Proc. ACM Program. Lang. **7**(POPL) (2023). https://doi.org/10.1145/3571212
15. Lahav, O., Boker, U.: What's decidable about causally consistent shared memory? ACM Trans. Program. Lang. Syst. **44**(2) (2022). https://doi.org/10.1145/3505273

16. Lahav, O., Vafeiadis, V., Kang, J., Hur, C.K., Dreyer, D.: Repairing sequential consistency in C/C++11. In: PLDI 2017, pp. 618–632 (2017). https://doi.org/10.1145/3062341.3062352, technical Appendix Available at https://plv.mpi-sws.org/scfix/full.pdf
17. Lamport, L.: Time, clocks, and the ordering of events in a distributed system. Commun. ACM **21**(7), 558–565 (1978)
18. Lokshtanov, D., Marx, D., Saurabh, S.: Lower bounds based on the exponential time hypothesis. Bull. EATCS **105**, 41–72 (2011). http://eatcs.org/beatcs/index.php/beatcs/article/view/92
19. Luo, W., Demsky, B.: C11Tester: A Race Detector for C/C++ Atomics, pp. 630–646. Association for Computing Machinery, New York, NY, USA (2021). https://doi.org/10.1145/3445814.3446711
20. Manovit, C., Hangal, S.: Completely verifying memory consistency of test program executions. In: HPCA, pp. 166–175. IEEE Computer Society (2006)
21. Margalit, R., Lahav, O.: Verifying observational robustness against a c11-style memory model. Proc. ACM Program. Lang. **5**(POPL) (2021). https://doi.org/10.1145/3434285
22. Mathur, U., Pavlogiannis, A., Viswanathan, M.: The complexity of dynamic data race prediction. In: LICS, pp. 713–727. ACM (2020)
23. Norris, B., Demsky, B.: Cdschecker: checking concurrent data structures written with C/C++ atomics. In: Proceedings of the 2013 ACM SIGPLAN International Conference on Object Oriented Programming Systems Languages & Applications, OOPSLA 2013, pp. 131–150. Association for Computing Machinery, New York, NY, USA (2013). https://doi.org/10.1145/2509136.2509514
24. Qadeer, S.: Verifying sequential consistency on shared-memory multiprocessors by model checking. IEEE Trans. Parallel Distrib. Syst. **14**(8), 730–741 (2003)
25. Tunç, H.C., Abdulla, P.A., Chakraborty, S., Krishna, S., Mathur, U., Pavlogiannis, A.: Optimal reads-from consistency checking for c11-style memory models. Proc. ACM Program. Lang. **7**(PLDI) (2023). https://doi.org/10.1145/3591251
26. Vassilevska Williams, V., Williams, R.R.: Subcubic equivalences between path, matrix, and triangle problems. J. ACM **65**(5), 27:1–27:38 (2018)

TAP Track

Towards Formally Verified Smart Contracts Compilation

Elvira Albert, Samir Genaim, and Enrique Martin-Martin(✉)

Complutense University of Madrid, Madrid, Spain
elvira@fdi.ucm.es, {sgenaim,emartinm}@ucm.es

Abstract. Many compilation stages of smart contracts on the Ethereum blockchain have been transitioned to the intermediate language YUL. Tasks such as smart contract optimization and bytecode generation are—or will soon be—performed directly at the YUL level in the compilers for the higher-level languages such as Solidity. In this paper, we develop a formal semantics of YUL programs in Rocq, suitable for verification, which allows formal reasoning at the level of YUL code or generation tools processing YUL programs. Our semantics is *expressive* enough to be the basis for formal verification tools, and *simple* enough to make the development of such tools feasible. In order to prove its adequacy for verification, we develop in Rocq a checker (and associated soundness proofs), based on our semantics, able to verify the results of the liveness analysis stage of the official Solidity compiler `solc`, which opens the door towards formally verified Ethereum's smart contracts compilation. Experiments on more than 1,500 smart contracts show that we are able to automatically verify `solc`'s liveness analysis results in negligible time.

1 Introduction

Smart Contracts. Ethereum [36] is a decentralized blockchain with smart contract functionality. A *smart contract* is a self-executing program that automates the actions required in a blockchain transaction, eliminating the need for a central authority or intermediary. Once transactions are done, they are traceable and cannot be undone. Because smart contracts execute agreements, they can be used for many different purposes: the smart contracts introduced by Ethereum are considered a fundamental building block for decentralized finance (DeFi) and non-fungible token (NFT) applications. On Ethereum, smart contracts are written in Turing-complete programming languages—such as Solidity [8], Vyper [18] or Fe [1]—and compiled into low-level bytecode to be executed by the Ethereum Virtual Machine [36] (EVM). Even though smart contracts are stored on secure blockchains like Ethereum, they are not immune to security

E. Martin-Martin, S. Genaim and E. Albert—Supported partially by Ethereum Foundation (Grant ID AOC-2616), Spanish MCI, AEI, FEDER (EU) projects PID2021-122830OB-C41, PID2024-157044OB-C31, TEC-2024/COM-235 and PR17/24-31926 Funded By Madrid(UCM Action-6th Framework.

A. Sampaio and M. Stoelinga (Eds.): FM 2026, LNCS 16557, pp. 255–274, 2026.
https://doi.org/10.1007/978-3-032-26220-2_13

threats. Attackers can exploit code vulnerabilities to steal funds or manipulate contract execution.

Towards Verified Compilation. Formal methods are playing a fundamental role to guarantee security before deploying smart contracts (see [11,15,17,19,20,22, 31,34] and their references), and thus ensure safe transactions. Our focus is on the development of verified optimization and compilation tools which increase the trustworthiness of the compiled bytecode. Many compilation stages of smart contracts on the Ethereum blockchain have been transitioned to the intermediate language `YUL` [9]. The motivation behind `YUL` is to have a common representation for the analysis and optimization stages that can benefit all target platforms equally. By means of the so-called *dialect* [9], `YUL` can be compiled to bytecode for different backends. Tasks such as smart contract optimization and `EVM` code generation are—or will soon be—performed directly at the `YUL` level in the compilers for the higher-level languages such as Solidity, Vyper or Fe compilers.

Formal Verification. As security requirements are critical today in many contexts, formal verification has taken an essential role in the software engineering process to verify and/or certify formal specifications of the software behavior. In order to formally verify the correctness of an analysis or optimization, one can develop a formally verified tool using machine-assisted mathematical proofs (such as done in the Compert certified compiler [25]), or follow the *translation validation* approach [35] in which each of the analysis or optimization results are formally checked to be correct using a verified checker (as applied e.g. in [12,16,27,28,32,33]). As `EVM` compilers—and their built-in analyses and optimizers—are continuously evolving, translation validation is the most suitable approach until the languages and tools stabilize. The last years have witnessed the development of powerful proof assistants such as Rocq (formerly known as Coq) [14,21], Lean [29], and Isabelle/HOL [30] which provide a formal language to write mathematical definitions, executable algorithms and theorems together with environments for semi-interactive development of machine-checked proofs.

Contributions. The goal of this paper is to develop a formal semantics of `YUL` in Rocq and showcase its adequacy for verification by applying it to verify the soundness of the *liveness* analysis results of the most-widely used Solidity compiler, `solc` [3]. A liveness analysis aims at determining the lifetime of each variable in the program, i.e., the points from which an execution can reach an instruction where each variable is used. It is a fundamental analysis in the development of a compiler, as several compiler analyses need to know the lifetime of variables (e.g., deadcode elimination, register allocation, etc.). Our work is a first major step towards the long-term goal of the Ethereum community of having all optimization and static analysis tools for `YUL` programs verified using a proof assistant. The main contributions of the paper are:

1. *YUL formal semantics.* We develop a formal semantics of `YUL` programs in Rocq that is suitable for verification as it allows formal reasoning at the level of `YUL` code and of the analysis/optimization tools processing `YUL` programs.
2. *Verification of liveness analysis results.* We develop a checker based on our semantics to verify the results of the liveness static analysis phase in the `solc` compiler, which opens the door towards formally verified Ethereum's smart contracts compilation.
3. *Implementation and experimental evaluation.* Our implementation, named `FORYU` [6,10], includes the semantics and checker, it is open-source and publicly available. Experimental evaluation on the large-scale set of 1,550 smart contracts utilized for regression testing by the `solc` development team [7] demonstrates the practicality of our approach: the verification process for the liveness analysis results takes on average $3ms$ per smart contract.

2 An Overview of Verified Smart Contract Compilation

This section provides preliminary background on Ethereum smart contracts [36], their intermediate `YUL` representation [3], and it overviews the general framework for verified smart contracts compilation with the `FORYU` components.

`YUL-IR` as a Common Analysis/Optimization Target. A *smart contract* can be seen as a collection of *functions* whose executions are invoked in the Ethereum transactions. A so-called transaction can execute any public function of the contract, and it may terminate successfully, but also in an `error`, or `revert` status (the latter undoing the contract state changes). Regarding the implementation of the functions, besides the regular types of instructions used by any high-level programming language (e.g., assignments, conditionals, iterative constructs, etc.), there are specific instructions that communicate to the *blockchain* (e.g., `sstore` that modifies the contract storage that is replicated in the blockchain), and instructions to invoke other contracts. There coexist several high-level languages to program Ethereum smart contracts: while Solidity is the most popular, Vyper is also widely used, and Fe is being proposed as the next generation language. In order to unify efforts, `YUL` has been proposed as a common intermediate language to carry out contract optimization and bytecode generation in the compilers for the diverse higher-level languages. Figure 1 provides an overview of the overall verification framework, where we can see on the left (component ①) that smart contracts written in the different languages are first translated to the `YUL-IR`. The core idea proposed by Ethereum is that the *Analysis+Optimization* component ④ (that outputs in ⑤ *Liveness*, *Constancy*, *Dependencies* and any other current or future analysis/optimization) is done at the intermediate language `YUL` so that it can be shared by all high-level compilers and only the first translation stage *Trans* ① needs to be specifically developed, as shown in the figure.

The `YUL-IR` in `CFG-SSA` Form. The `YUL-IR` obtained from the source program must be provided by the compilers in CFG (Control Flow Graph) and SSA

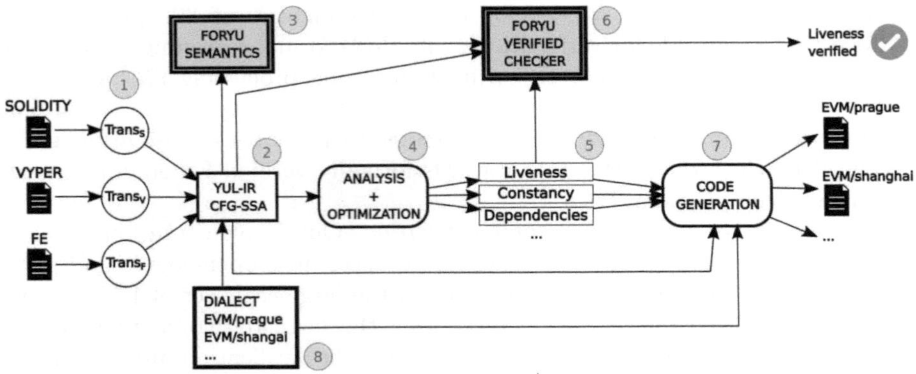

Fig. 1. An overview of verified compilation with the `FORYU` components.

(Single Static Assignment) form ②, named `CFG-SSA` in what follows. Basically, in such `YUL-IR` in `CFG-SSA` form, each function in the smart contract is represented by means of a CFG whose blocks contain a sequence of instructions in SSA and a set of ϕ-functions to handle joint points in the CFG. As usual, there is an entry block and we have conditional and unconditional jumps between blocks, and terminal (or exit) blocks. It is widely known that SSA is a type of intermediate representation (IR) where each variable is assigned exactly once, used in most high-quality optimizing compilers for imperative languages, including LLVM, and many commercial compilers. Converting ordinary code into SSA form is done by replacing the target of each assignment with a new variable, and replacing each use of a variable with the new version of the variable reaching that point. A ϕ-function for a variable is not needed when only one version of the variable is reaching a joint point in the CFG. When there is ambiguity, the ϕ-function is used to combine the values from different control flow paths into a single variable, using the incoming edge to determine which source value to select. The usual notation to declare a ϕ-function is $a_0 = \phi(B_1 : a_1, \dots, B_n : a_n)$ meaning that a_0 takes the value of a_i when the control flows from block B_i with $i = 1, \dots, n$.

Example 1. The left side of Fig. 2 shows a smart contract in the standard `YUL-IR` given by the `solc` compiler. It includes a function `cal`, and the contract's entry code. This contract calculates the interest over a specific period given an initial balance. We can see that the `YUL` code contains the high-level instructions for the control flow (`for` loops and `if` conditional) while it uses the `EVM` opcodes for the operations (e.g., `div`, `add`, etc.) and for storage accesses (e.g., `sstore`). On the right of the figure, we have the `YUL-IR` in `CFG-SSA` form for the function `cal` exactly as generated by `solc`. In this form, the control flow is explicit by means of the CFG blocks (e.g., the inner loop is represented by blocks B5-B8, and the outer loop by block B1-B4) that are connected to their continuations by means of conditional jumps (e.g., the continuation from B1 depends on condition v8 in the outgoing labels) or unconditional (e.g., B8 to B3). Besides, an SSA-transformation has been performed and ϕ-functions introduced at joint blocks when needed (e.g., in B1 and B5). At the CFG level, variables v0–v3 are the input parameters of `cal`, and in block B4 the value of variable v23 is returned.

Our FORYU semantics ③ is defined at the level of the CFGs (example on the right of figure) instead of at the pure abstract syntax tree (AST) representation of YUL programs (on the left), because optimizations and analysis are more efficient when developed at the CFG-SSA level and it is the current trend on the ongoing Ethereum developments. While our semantics could be used for *dynamic* execution by providing a concrete initial state, our focus is on its use within the FORYU verifier ⑥. The verifier takes the YUL-IR ② and the analysis results ⑤ to be verified (our current implementation works on the liveness analysis output, future plans include all analyses) and it returns ✓ if correctness is proven.

```
function cal(bal, rate, basis, years) -> r {
  r := bal
  for { let y := 0 } lt(y, years) { y := add(y, 1) } {
    for { let m := 0 } lt(m, 12) { m := add(m, 1) } {
      let interest := div(mul(r, rate), basis)
      r := add(r, interest)
    }
  }
}
// Contract's entry code
let bal := sload(0)
let rate := calldataload(0)
let basis := calldataload(1)
let years := calldataload(2)
if gt(rate,0x0) { sstore(1,cal(bal,rate,basis,years)) }
```

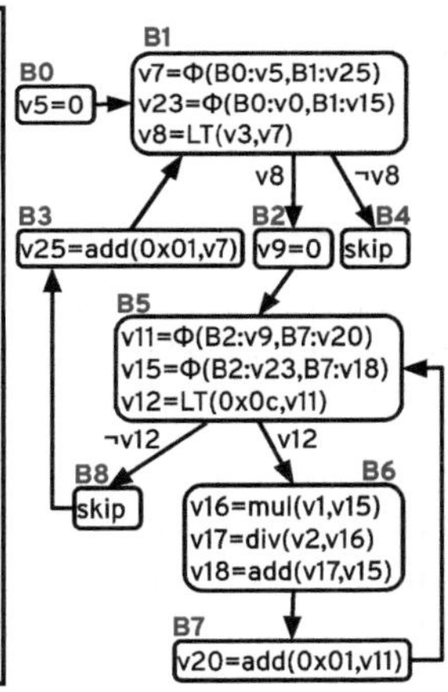

Fig. 2. Left: YUL-IR for smart contract. Right: CFG-SSA form for function `cal`.

Dialects in the YUL-IR. To keep the language simple and flexible, YUL does not have any built-in operation, functions or types in its pure form. These are added together with their semantics when specifying the so-called *dialect* of YUL, which allows specializing YUL to the requirements of different target platforms and feature sets. Currently, there is only one official dialect of YUL, which uses the EVM opcodes as builtin functions and defines only the type u256, which is the native 256-bit type of the EVM (named `EVM/prague` in the figure). There are ongoing projects though aiming at including other dialects (e.g., the `EVM/shanghai` that is a specific version of the EVM to handle the Shanghai hard fork). It can be observed in the figure on the right ⑦ that code generation backends can be developed for specific target code. The YUL-IR of the example in Fig. 2 uses the standard EVM dialect [36], e.g., the EVM opcode `add` is used in the YUL code for addition, `sstore` for a write access in the blockchain storage, etc.

3 The Rocq YUL-Semantics

This section defines a YUL-formal semantics that is expressive enough to be the basis for formal verification tools, and at the same time, simple enough to make the development of such tools feasible. This is not simply a matter of choosing a semantics from among the existing possibilities in the field of programming languages, such as small-step or big-step, but also has to do with the fundamentals of how programs and states are represented and how the data structures are used, as these affect the feasibility of writing proofs for related formal statements. We start by modeling the YUL syntax in Rocq and then we present the essential parts of our semantics based on such YUL models.

3.1 Modeling YUL Syntax

Let us first define the representation of YUL dialects and then model smart contracts. The provided code snippets include only the essential details, with additional explanations integrated via comments. The next Rocq `Module Type` defines the interface for a YUL dialect. It specifies the parameters (data types, functions, and propositions) that any concrete dialect implementation must provide.

```
Module Type DIALECT.
  Parameter value_t : Type. (* Type of basic values *)
  Parameter opcode_t : Type. (* Inductive data type of supported opcodes *)
  Parameter dialect_state : Type. (* The dialect state *)

  Parameter default_value : value_t. (* Value for initializing variables *)
  Parameter is_true_value: value_t → bool. (* Value representing true *)
  Parameter is_false_value: value_t → bool. (* Value representing false *)
  Parameter equal_values: value_t → value_t → bool. (* Equality check *)

  Parameter execute_op_code : (* Execute an opcode *)
    dialect_state → opcode_t → list (value_t) →
    (list value_t * dialect_state * Status.t).
  ...
End DIALECT.
```

All dialect operations are defined by the inductive type `opcode_t` and manipulate data of type `value_t`. The `dialect_state` parameter defines the type for the dialect-specific state, e.g., the different types of storage that the EVM has (EVM uses a heap and the replicated storage). The core functionality is provided by the `execute_op_code` function. It takes the current dialect state, an opcode, and a list of input values. It returns a new dialect state, a list of output values, and an execution status. This status is defined by the following Rocq module:

```
Module Status.
  Inductive t : Type :=
  | Running : t   | Terminated : t   | Reverted : t   | Error : string → t.
End Status.
```

We have implemented the EVM dialect as a module called EVMDialect, and it is based on the semantics defined in [4,5].

A smart contract is modeled using the following Rocq module,[1] which is parameterized over a specific dialect D.

```
Module SmartContract (D: DIALECT).
  Record t : Type := {
    name : string; (* Its name *)
    functions : list Function(D).t; (* List of functions *)
    main: FuncName.t; (* The name of the main function *)
  }.
  ...
  Definition valid_sc: t → Prop := ... (* validity of smart contracts*)
End SmartContract.
```

The core type for a smart contract is SmartContract.t, defined as a Record. This record encapsulates the contract's name, a list of its functions, and the main function name (texttttFuncName.t is defined as a string). The valid_sc parameter enforces structural validity, requiring, among other conditions, that all included functions are valid and that their names are unique. The functions in the contract are modeled using the following Rocq module.

```
Module Function (D: DIALECT).
  Record t : Type := {
    name : FuncName.t; (* Its name *)
    arguments : list YULVariable.t; (* Input parameters *)
    blocks : list Block(D).t; (* List of blocks *)
    entry_block_id : BlockID.t; (* The ID of the entry block. *)
  }.
  ...
  Definition valid_f: t → Prop := ...  (* validity of functions *)
End Function.
```

The core type for a function is Function.t, and it includes its name, a list of input parameters (where YULVariable.t is a unique nat identifier modeling a variable), and the identifier of the initial block (BlockID.t is nat). The valid_f parameter enforces structural validity, requiring, among other conditions, that all blocks have unique identifiers. A block in (the CFG of) a function is modeled using the following Rocq module.

```
Module Block (D: DIALECT).
  Record t : Type := {
    bid : BlockID.t; (* The block's identifier (nat) *)
    instructions : list (Instruction(D).t); (* Its list of instructions *)
    exit_info : ExitInfo(D).t; (* Its kind: branching, return, etc. *)
```

[1] For the sake of conciseness and clarity, we use expressions such as Function(D) to denote the instantiation of the Function module with parameter D, which is not valid Rocq syntax. The code available at [6] contains the actual (less compact) syntax.

```
    phi_function : PhiInfo(D).t; (* Phi function used CFG-SSA *)
  }.
  ...
End Block.
```

The core type for a block is `Block.t`, and it includes a block identifier, a list of instructions, information on how to exit the block, and a ϕ-function that is used to model `CFG-SSA`'s. Instructions are modeled as follows.

```
Module SimpleExpr (D: DIALECT).
  Definition t : Type := YULVariable.t + D.value_t. (* Variable or value *)
End SimpleExpr.

Module Instruction (D: DIALECT).
  Inductive other_inst_t : Type := | ASSIGN.

  Record t : Type := {
    input : list SimpleExpr(D).t; (* Input expressions *)
    output : list YULVariable.t; (* Output variables *)
    op : FuncName.t + D.opcode_t + other_inst_t; (* Operation *)
  }.
End Instruction.
```

The core type for an instruction is `Instruction.t`, and it includes a list of input simple expressions, a list of output variables, and an opcode that can be a function name, a dialect opcode or an assignment. The exit information of a block is modeled using the following Rocq module.

```
Module ExitInfo (D: DIALECT).
  Inductive t : Type :=
  | CondJump (v : YULVariable.t) (bid_t bid_f: BlockID.t) (* Conditional jump *)
  | Jump (bid : BlockID.t) (* Jump to block bid *)
  | ReturnBlock (rs : list SimpleExpr(D).t) (* Return the values of rs *)
  | Terminated. (* Terminate the execution *)
End ExitInfo.
```

The core type for exit information is `ExitInfo.t`, and it indicates how the execution proceeds when reaching the end of a block, in particular, the next block for the case of `CondJump` is decided depending on whether the value of variable `v` is true or false. Finally, a ϕ-function is modeled using the following Rocq module.

```
Module PhiInfo (D: DIALECT).
  Definition t := BlockID.t → list (YULVariable.t * SimpleExpr(D).t).
End PhiInfo.
```

The core type for a ϕ-function is `PhiInfo.t`, and it maps each incoming block identifier to a list of pairs `(v,e)` meaning that when the execution comes from the corresponding block, variable `v` should be assigned the value of `e`.

3.2 The Rocq Small-Step Semantics

We first explain how the states are modeled using the following Rocq module, which is parameterized over a specific dialect D.

```
Module State(D: DIALECT).
  Record t : Type := {
    call_stack : CallStack(D).t; (* The call stack for functions calls *)
    status : Status.t; (* The status of the execution *)
    dialect_state : D.dialect_state; (* The dialect state *)
  }.
  ...
End Module
```

The core type for YUL states is State.t, and it encapsulates a call stack, the status of the execution, and a dialect state that is entirely managed by the dialect D. The variable mapping, stack frame, and call stack are core components for tracking program execution and are modeled in Rocq as follows.

```
Module VarMap (D: DIALECT).
  Definition t := YULVariable.t → D.value_t. (* Maps variables to values *)
  ...
End VarMap

Module StackFrame(D: DIALECT).
  Record t : Type := {
    func_name: FuncName.t; (* The function owning the frame *)
    curr_bid: BlockID.t; (* The currently block executing *)
    pc : nat; (* Position of the next instructions in the current block *)
    vars_map : VarMap(D).t; (* Mapping variables to values *)
  }.
  ...
End StackFrame.

Module CallStack(D: DIALECT).
  Definition t := list StackFrame(D).t. (* Stack of frames *)
  ...
End CallStack.
```

The call stack is modeled as a list of frames, acting as a stack. Each frame contains the following execution context: the name of the function that owns the frame (the function called when the frame was created), the identifier of the block currently executing, the pc (program counter) index of the next instruction to execute, and a map assigning values to local variables.

The small-step semantics is modeled using the following SmallStep module.

```
Module SmallStep (D: DIALECT).
  ...
  (* Performs a single execution step *)
  Definition step (s: State(D).t) (p: SmartContract(D).t) : State(D).t :=
    match s.(State(D).status) with
    | Status.Running ⇒
      match get_next_instruction s p with
      | None ⇒ handle_block_finish s p
      | Some instruction ⇒ execute_instr instruction s p
      end
    | _ ⇒ s
    end.

  (* Performs n execution steps *)
  Fixpoint eval (n: nat) (s: State(D).t) (p: SmartContract(D).t) : State(D).t := ...
End Module
```

Function `step` performs a single execution step as follows. If the state's status is not `Running`, it returns the current state since no transition can be made (we could prohibit this by definition as well, but keeping it this way simplifies the formalization). If the state status is `Running` and there is a next instruction to execute, we execute it using `execute_instr`, and if there is no next instruction it means that the execution is at the end of a block and we use `handle_block_finish` to handle this case. Function `eval` simply executes `step` in a loop `n` times.

4 Verified Liveness Analysis

Notation. A program point (`pp`) is identified by the triple (`fname`,`bid`,`pc`), and it is typically associated with a specific stack frame to indicate the next instruction to be executed within that frame. The `VarSet` module implements a set of `YULVariable.t` elements along with relevant set operations.

4.1 Formal Definition of Dead Variable in Rocq

A variable `v` is considered *dead* at a program point `pp` if its value does not affect any future execution outcome starting from `pp`. This concept is commonly modeled as: If an execution starting from state `st1` at program point `pp` (in the top stack frame) reaches state `st1'` in `n` steps, then starting from any state `st2` that is identical to `st1` except for the value of `v`, an execution of `n` steps will reach state `st2'` such that, (i) `st1'` and `st2'` are at the same program point (relative to the top stack frame); and (ii) the values of all variables that will be accessed in the immediate next step (including instruction inputs, conditional variables, or return variables) are the same in `st1'` and `st2'`. We now formalize this notion of dead variable.[2] This requires defining several propositions about the equivalence of states up to the value of a variable v. The first of such propositions[3] defines the conditions under which two stack frames, sf1 and sf2, are considered equivalent up to the value of v at a given program point (fname,bid,pc). This requires that all variables, except possibly v, hold the same values in both frames, and that the program point of both frames is (fname,bid,pc).

```
Definition same_pp (fname: FuncName.t) (bid: BlockID.t) (pc: nat)
                              (sf1 sf2: StackFrame(D).t) :=
  sf1.func_name = fname ∧ sf2.func_name = fname ∧
  sf1.curr_bid = bid ∧ sf2.curr_bid = bid ∧ sf1.pc = pc ∧ sf2.pc = pc.

Definition equiv_frames_up_to_v (fname: FuncName.t) (bid: BlockID.t) (pc: nat)
  (v: YULVariable.t) (sf1 sf2: StackFrame(D).t) :=
  same_pp fname bid pc sf1 sf2 ∧
  ∀ v', v' ≠ v → (* equality is required for variable different from v *)
    VarMap.get sf1.vars_map v' = VarMap.get sf2.vars_map v'.
```

The next proposition defines when two states, st1 and st2, are considered equivalent up to the value of variable v at a given program point (fname,bid,pc). This equivalence holds when the following conditions are met: the status and underlying dialect states must be equivalent; the top stack frames must be equivalent up to the value of v and at (fname,bid,pc); and all remaining (deeper) frames in the call stack must be identical.

```
Definition equiv_states_up_to_v (fname: FuncName.t) (bid: BlockID.t) (pc: nat)
                                   (v: YULVariable.t) (st1 st2: State(D).t) :=
  (length st1.call_stack > 0) ∧
  st1.status = st2.status ∧ st1.dialect_state = st2.dialect_state ∧
  ∃ sf1 sf2 rsf,
    st1.call_stack = sf1::rsf ∧ st2.call_stack = sf2::rsf ∧
    equiv_frames_up_to_v fname bid pc v sf1 sf2.
```

The next proposition formally defines a set of variables (s) that includes precisely those variables which are immediately accessed in the next execution step, given the block (b) and program counter (pc). This definition distinguishes between two cases: when pc points to an instruction within the block, and when pc points to the end of the block.

```
Definition accessed_vars (b: Block(D).t) (pc: nat) (s: VarSet.t) :=
  ( pc = (length b.instructions) ∧ (* end of block *)
    match b.exit_info with
    | ExitInfo(D).CondJump cv _ _ ⇒
```

[2] This is a simplified version (for the sake of clarity) of a more general property we have formally proven in [6]. In module fm26.v we prove that this version is an instance of the general property.

[3] For clarity, we will use the notation register.field_name to denote access to a field of a record, as in sf1.func_name.

```
      VarSet.Equal s (VarSet.add cv VarSet.empty) (* cv is accessed *)
    | ExitInfo(D).ReturnBlock rvs ⇒
      VarSet.Equal s (list_sexp_to_varset rvs) (* rs are accessed *)
    | _ ⇒ VarSet.Equal s VarSet.empty (* nothing is accessed *)
    end )
  ∨
  ( pc < (length b.instructions) ∧ (* within the block *)
    exists instr,               (* instr.input are accessed *)
      nth_error b.instructions pc = Some instr ∧
      (VarSet.Equal s (list_sexp_to_varset instr.input)) ).
```

The next proposition states that the top stack frames of states `st1` and `st2` agree on the values of all variables that are immediately accessed.

```
Definition equiv_top_frame (p: SmartContract(D).t) (st1 st2: State(D).t) :=
  match st1.call_stack, st2.call_stack with
  | nil,nil ⇒ True (* both call stacks are empty *)
  | sf1::_,sf2::_ ⇒ (* top frames agree on values of accessed variables *)
      let fname := sf1.func_name in
      let bid := sf1.curr_bid in
      let pc := sf1.pc in
        same_pp fname bid pc sf1 sf2 ∧
        ∀ v s b,
          SmartContract(D).get_block p fname bid = Some b →
          accessed_vars b pc s →
          VarSet.In v s →
          VarMap.get sf1.vars_map v = VarMap.get sf2.vars_map v
  | _,_ ⇒ False (* one of the call stacks is empty *)
end.
```

Finally, the following proposition formally defines the conditions under which a variable `v` is considered dead at a given program point (`fname,bid,pc`), it relies on the `eval` function defined in the semantics in Sect. 3.

```
Definition dead_variable (p: SmartContract(D).t) (fname: FuncName.t)
  (bid: BlockID.t) (pc: nat) (v: YULVariable.t) :=
    ∀ (st1 st2 st1': State(D).t) (n: nat),
      equiv_states_up_to_v fname bid pc v st1 st2 →
      SmallStep(D).eval n st1 p = st1' →
      ∃ st2',
        SmallStep(D).eval n st2 p = st2' ∧ equiv_top_frame p st1' st2'.
```

A variable that is not dead is called *live*. We typically compute an over-approximation of live variables and thus any variable not declared live by the analysis is considered dead (i.e., we get an under-approximation for dead variables).

Note that the above definition includes a universal quantification on the length of the trace, meaning that the variable is not accessed independently of that length, and thus also covers non-terminating programs. The same reasoning holds for non-deterministic programs; however, in the context of blockchain, determinism is an essential requirement and hence non-determinism is left out of our formalization.

4.2 Formally Verified Liveness Checker

Our goal is to check that the results of a liveness analysis are sound, i.e., if a program point (`fname,bid,pc`) is associated with a set `s` of live variables, then any variable not in `s` must satisfy the `dead_variable` proposition.

At the level of `YUL-IR` in `CFG-SSA` form, the liveness analysis infers sets of live variables at the entry (`IN`) and exit (`OUT`) of each block (called live-in and live-out sets), and as typical in any liveness analysis is based on solving the following set of equations:

$$\begin{array}{l} \texttt{OUT}(b) = \begin{cases} \texttt{retvars}(b) & b \text{ is a terminal block} \\ \cup_i \phi_{b_i}^{-1}(b, \texttt{IN}(b_i)) & b \text{ has successor } b_1, \ldots, b_n \end{cases} \\ \texttt{IN}(b) = \texttt{BKW}(\texttt{instrs}(b), \texttt{OUT}(b) \cup \texttt{condvars}(b)) \end{array} \qquad (1)$$

The equations define the live-out set `OUT` of a block b based on whether it is a terminal block: (i) If the block is terminal, its live-out set is simply the set of returned variables ($\texttt{retvars}(b)$), if any exist; and (ii) If the block is non-terminal, its live-out set is the union of the live-in `IN` sets of all its successor blocks after appropriate variable renaming using the inverse of the corresponding ϕ-functions. The live-in set of a block b is obtained by propagating backwards (`BKW`) the live-out set of the block, after adding conditional jump variable ($\texttt{condvars}(b)$), if any exists. The propagation is done instruction by instruction in the block ($\texttt{instrs}(b)$), by removing the updated variables (denoted by $\texttt{def}(i)$) and adding the accessed variables (denoted by $\texttt{used}(i)$) as follows:

$$\texttt{BKW}(l, s) = \text{if } l = i :: l' \text{then } \texttt{BKW}(l', (s \setminus \texttt{def}(i)) \cup \texttt{used}(i)) \text{ else } s$$

Note that we could also replace $=$ by $\supseteq$ in (1) to allow correct but less precise over-approximations.

Example 2. The live-in and live-out sets for the blocks in the CFG of Fig. 2 are:

	live-in	live-out		live-in	live-out
B0	{v0,v1,v2,v3}	{v0,v1,v2,v3,v5}	B5	{v1,v2,v3,v7,v11,v15}	{v1,v2,v3,v7,v11,v15}
B1	{v1,v2,v3,v7,v23}	{v1,v2,v3,v7,v23}	B6	{v1,v2,v3,v7,v11,v15}	{v1,v2,v3,v7,v11,v18}
B2	{v1,v2,v3,v7,v23}	{v1,v2,v3,v7,v9,v23}	B7	{v1,v2,v3,v7,v11,v18}	{v1,v2,v3,v7,v18,v20}
B3	{v1,v2,v3,v7,v15}	{v1,v2,v3,v15,v25}	B8	{v1,v2,v3,v7,v15}	{v1,v2,v3,v7,v15}
B4	{v23}	{v23}			

Let us explain some of the blocks: (i) Block B4 is a return block, so its live-out set includes only the return variable v23, and its live-in set is equal to its live-out set since it does not have instructions. (ii) The live-out set of block B2 is obtained from the live-in of B5 after renaming `v11` and `v15` to `v9` and `v23` respectively (according to the ϕ-function). (iii) The live-in set of B6 is obtained by propagating its live-out set backwards, which removes variable `v18` since it is updated and adds v15 since it is accessed. (iii) The live-out set of B5 is the union of the live-in sets of its successors B6 and B8.

The equations in (1) can be modeled in Rocq using the following CoInductive data-type (we do not show all cases).

```
CoInductive live_out (p: SmartContract(D).t) :
  FuncName.t → BlockID.t → VarSet.t → Prop :=
  (* Return block *)
  | lo_ret (fname: FuncName.t) (bid: BlockID.t) ... (sout: VarSet.t):
    SmartContract(D).get_block p fname bid = Some b → (* block exists *)
    Block(D).is_return_block b = Some rs → (* it is a return block *)
    VarSet.Equal sout (list_sexp_to_varset rs) →
    live_out p fname bid sout
  (* A block with a conditional jump *)
  | lout_cjump (fname: FuncName.t) (bid: BlockID.t) ... (sout: VarSet.t):
    SmartContract(D).get_block p fname bid = Some b →
    Block(D).is_cond_jump_block b = Some (cvar, bid_t, bid_f) →
    live_in p fname bid_t s1 →
    live_in p fname bid_f s2 →
    SmartContract(D).get_block p fname bid_t = Some b_t →
    SmartContract(D).get_block p fname bid_f = Some b_f →
    VarSet.Equal sout (VarSet.union (inv_phi b_t.phi_function bid) s1)
                                    (inv_phi b_f.phi_function bid) s2)) →
    live_out p fname bid sout
    ...
  with
  live_in (p: SmartContract(D).t): FuncName.t → BlockID.t → VarSet.t → Prop :=
  | lin (fname: FuncName.t) (bid: BlockID.t) ...  (sout: VarSet.t):
    SmartContract(D).get_block p fname bid = Some b →
    live_out p fname bid s →
    VarSet.Equal sout (prop_bkw b.instructions (add_jump_var b s)) →
    live_in p fname bid sout.
```

The first case of the live_out states that if a block is a return block with rs as its return expressions, then the set of live-out variables is equivalent to (list_sexp_to_varset rs). This set precisely includes all variables in the list rs. The second case of live_out states that if the block is a conditional jump, and s1 and s2 are the live-in sets for the next blocks, then the set of live-out variables is equivalent to the union of s1 and s2 after applying the inverse of the corresponding ϕ-functions. The case of live_in states that if s is the live-out set for the block, then the set of live-in variables is equivalent to propagating s backwards after adding the condition variable if needed (in the case of a block with conditional jump). The use of CoInductive instead of Inductive is because some of the proofs rely on co-induction for the following reason: A direct translation of the equations in (1) into a proof assistant like Rocq might initially lean towards standard Inductive types. However, standard inductive definitions construct strictly well-founded, finite derivation trees. To complete an inductive proof for OUT(b) or IN(b), the derivation chain must physically terminate at a definitive base case, e.g., a terminal block where $\texttt{OUT}(b) = \texttt{retvars}(b)$. Because CFGs naturally contain cycles (loops), this well-foundedness requirement becomes a severe limitation. If block b has a successor b_i, and b_i contains a back-edge leading to b, the evaluation of $\texttt{OUT}(b)$ triggers an infinite dependency chain: $\texttt{OUT}(b) \to \texttt{IN}(b_i) \to \texttt{OUT}(b_i) \to \cdots \to \texttt{OUT}(b)$. To accommodate the cyclic nature of CFGs, we model the equations using `CoInductive` types and use co-indcution in the proofs.

The following lemma guarantees that liveness information as defined above is sound. It states that the complement of `live_in` is a safe under-approximation of the set of dead variables for the corresponding program point. We could state a similar statement for `live_out` as well, but this is not needed since executions that start at the beginning of a block pass through the end of the block (unless there is a call to a non-terminating method or an error in a call to the dialect).

```
Lemma live_in_out_snd:
  ∀ p fname bid b s,
    SmartContract(D).get_block p fname bid = Some b →
    live_in p fname bid s →
    (∀ v, ¬ VarSet.In v s → dead_variable p fname bid 0 v).
```

Now we move to verifying the correctness of a liveness analysis results, in particular that they agree with the above `live_in` and `live_out` propositions. We first define two data types to represent the result of a liveness analysis. The first `fun_live_info_t` is used for the liveness information of a function and it maps blocks to their live-in and live-out sets. The second `sc_live_info_t` is used for a smart contract and it maps functions to their results.

```
Definition fun_live_info_t := BlockID.t → option (VarSet.t * VarSet.t).
Definition sc_live_info_t := FuncName.t → option fun_live_info_t.
```

The following proposition formally establishes when a given result of the liveness analysis is sound. Essentially, the computed live-in and live-out sets must be equal to those that satisfy the `live_in` and `live_out` propositions (although requiring them to be a superset would be valid as well).

```
Definition liveness_info_snd (p: SmartContract(D).t) (r: sc_live_info_t) :=
  ∀ fname bid b f_r in_set out_set,
    SmartContract(D).get_block p fname bid = Some b →
    r fname = Some f_r →
    f_r bid = Some (in_set, out_set) →
    (live_in p fname bid in_set ∧ live_out p fname bid out_set).
```

Our liveness checker is a function `liveness_chk` that takes a smart contract and corresponding liveness analysis results, and returns `true` if the results are a solution to the equation in (1). The lemma `liveness_chk_snd_cmp` establishes that this function is sound and complete, with the proof fully verified in Rocq:

```
Definition liveness_chk (p: SmartContract(D).t) (r: sc_live_info_t) := ...

Lemma liveness_chk_snd_cmp:
∀ (p: SmartContract(D).t) (r: sc_live_info_t),
  SmartContract(D).valid p →
    (liveness_info_snd p r ↔ liveness_chk p r = true).
```

We note that our checker rejects liveness results that are not optimal, i.e., solutions to the equations in (1). As we have mentioned before, we could replace $=$ by $\supseteq$ in (1) to allow correct but less precise over-approximations. This change is straightforward both at the implementation level and in the corresponding proofs.

5 Experimental Evaluation

The different components of the FORYU verification framework have been implemented in the Rocq Prover (Coq v.8.20.1), as well as the complete proofs of all the theoretical results (more than 100 proofs in around 9000 lines of Rocq code). The source code, executables and benchmarks can be found at [6,10]. We have evaluated our implementation using the liveness analysis results produced by the official Solidity compiler `solc`. To this end, we have fully automated the interaction between the compiler and the FORYU liveness verifier by applying the following four-step pipeline: (1) compile the smart contract with the official `solc` compiler to obtain its YUL-IR CFG-SSA representation as a JSON file; (2) translate this JSON file into a Rocq file defining a term of type `SmartContract(EVMDialect)` (see Sect. 3.1); (3) compile the Rocq file together with the `liveness_chk` function and use Rocq's facilities to produce a static binary executable via OCaml (we make sure that the OCaml compiler does not apply any inlining or partial evaluation of the contract); and (4) execute this binary to obtain the output of the liveness checker. Following this pipeline, we used as benchmarks the 1,550 smart contracts contained in the semantic test suite of `solc`,[4] and we successfully validated the liveness analysis results on all of them. The size of these contracts ranges from 2 to 562 blocks (with an average of 44.06 blocks). The total time required to run the compiled liveness checker—step 4 of the pipeline—was 3.576 seconds: 78% of all contracts complete in 2 ms, 20% complete in 3 ms, and the remaining 2% take between 4 and 11 ms. All experiments were conducted on a machine equipped with an Intel i7-4790 CPU at 3.60 GHz and 16 GB of RAM.

In the current setting, we compile the checker into a binary for each contract. We adopted this approach because it easily allows us to measure the time (at OCaml level, not Rocq) required to check each contract once its Rocq representation is generated—which is the key factor in evaluating the checker's efficiency. We plan to streamline the liveness-checking pipeline by generating a specialized

[4] The test suite is available at [7]. We omitted three contracts that rely on experimental features not yet supported by official releases of `solc`.

OCaml implementation of the checker (using Rocq's extraction facilities) that takes as input the Rocq representation of the smart contract and produces the liveness-checking verdict. This specialized checker (combining steps 3 and 4 of the current pipeline) will be compiled once, and the resulting static executable will then be invoked directly, eliminating the dependency on the Rocq Prover for end-users and improving the overall efficiency and usability of the tool.

6 Conclusions, Related and Future Work

There are other ongoing efforts to facilitate formal verification of `YUL` programs, which also require formalization of the `YUL` semantics [2,23]. On one hand, in [23], both a big- and a small-step `YUL` operational semantics are defined together with an interpreter in OCaml. Another important effort is the Clear project [2] by Nethermind Security, developed in Lean, that defines a `YUL` and an `EVM` semantics that share the primitive operators. Using their `YUL` semantics, this work proposes to generate verification conditions that can be automatically verified from knowledge in preconditions and other annotations. In both cases [2,23], the semantics is defined at the level of `YUL` abstract syntax tree, rather than at the level of the `CFG-SSA` as our semantics. As our goal is to verify smart contracts compilation, defining the semantics at the level of the CFG is more adequate as it facilitates reasoning about the compilation stages that are typically developed on this `CFG-SSA` form. As regards the proof assistants, having a formalization in different proof assistants such as Rocq and Lean is not a weakness but a strength, since it extends the community of formal methods in `YUL` to more research groups with experience in the different proof assistants. Moreover, the different proof assistants provide different capabilities, e.g. in Rocq it is easy to export the Rocq functional code to the functional programming language OCaml, thus obtaining more efficient binary programs without dependencies on Rocq, which can be easily integrated into third-party tools.

In summary, we have presented the first steps towards developing a framework for fully verified smart contract compilation in the context of the Ethereum blockchain. Other compilers have already worked towards achieving this goal. An inspiring case is the CompCert project [26], that has investigated the formal verification of realistic compilers usable for critical embedded software. Such verified compilers come with a mathematical, machine-checked proof that the generated executable code behaves exactly as prescribed by the semantics of the source program. By ruling out the possibility of compiler-introduced bugs, verified compilers strengthen the guarantees that can be obtained by applying formal methods to source programs. The main result of the Compcert project is the CompCert C verified compiler, a high-assurance compiler for almost all of the C language (ISO C 2011), generating efficient code for the ARM, PowerPC, RISC-V and x86 processors. This includes the verification of several types of liveness analysis [13,24].

Our future work spawns along several directions. The most immediate plan is to integrate the liveness checker within the `solc` compiler, as described at the end of Sect. 5. We also aim at implementing verified checkers for other stages of the `solc` compilation process with the goal of having the whole compilation fully verified. Finally, let us note that, while our implementation focus has been `solc`, the `FORYU` components are general and could be used to verify other smart contracts compilers, as overviewed in Sect. 2.

References

1. FE: The next generation smart contract language for Ethereum. https://fe-lang.org
2. A formal verification tool for Solidity smart contracts by Nethermind Security. https://github.com/NethermindEth/Clear. Accessed 27 Nov 2025
3. The solc optimizer (2021). https://docs.soliditylang.org/en/v0.8.7/internals/optimizer.html
4. FORVES: Formally Verified EVM Block-Optimizations (2024). https://github.com/costa-group/forves. Accessed 27 Nov 2025
5. coq-of-solidity (2025). https://github.com/formal-land/coq-of-solidity. Accessed 27 Nov 2025
6. FORYU: A Formal Semantics for Yul in Coq (2025). https://github.com/costa-group/foryu. Accessed 27 Nov 2025
7. Semantic tests of solidity (2025). https://github.com/argotorg/solidity/tree/develop/test/libsolidity/semanticTests. Accessed 27 Nov 2025
8. Solidity documentation (2025). https://docs.soliditylang.org/en/latest/index.html
9. Yul (2025). https://docs.soliditylang.org/en/latest/yul.html. Accessed 27 Nov 2025
10. Albert, E., Genaim, S., Martin-Martin, E.: Towards Formally Verified Smart Contracts Compilation (FM'26 Artifact) (2026). https://doi.org/10.5281/zenodo.18612372. Accessed 13 Feb 2026
11. Amani, S., Bégel, M., Bortin, M., Staples, M.: Towards verifying Ethereum smart contract bytecode in Isabelle/HOL. In: CPP, pp. 66–77. ACM (2018)
12. Barrière, A., Blazy, S., Flückiger, O., Pichardie, D., Vitek, J.: Formally verified speculation and deoptimization in a JIT compiler. Proc. ACM Program. Lang. **5**(POPL), 1–26 (2021)
13. Barthe, G., Demange, D., Pichardie, D.: A formally verified SSA-Based Middle-End. In: Seidl, H. (ed.) ESOP 2012. LNCS, vol. 7211, pp. 47–66. Springer, Heidelberg (2012). https://doi.org/10.1007/978-3-642-28869-2_3
14. Bertot, Y., Castéran, P.: Interactive theorem proving and program development - Coq'Art: the calculus of inductive constructions. Texts in Theoretical Computer Science. An EATCS Series. Springer (2004)
15. Cassez, F., Fuller, J., Asgaonkar, A.: Formal verification of the ethereum 2.0 beacon chain. In: Fisman, D., Rosu, G., (eds.) Tools and Algorithms for the Construction and Analysis of Systems - 28th International Conference, TACAS 2022, Held as Part of the European Joint Conferences on Theory and Practice of Software, ETAPS 2022, Munich, Germany, April 2–7, 2022, Proceedings, Part I, vol. 13243 of Lecture Notes in Computer Science, pp. 167–182. Springer (2022)

16. Demange, D., Pichardie, D., Stefanesco, L.: Verifying fast and sparse SSA-based optimizations in Coq. In: Franke, B., (ed.), Compiler Construction - 24th International Conference, CC 2015, Held as Part of the European Joint Conferences on Theory and Practice of Software, ETAPS 2015, London, UK, April 11-18, 2015. Proceedings, vol. 9031 of Lecture Notes in Computer Science, pp. 233–252. Springer (2015)
17. Dill, D.L., et al.: Fast and reliable formal verification of smart contracts with the move prover. In: Fisman, D., Rosu, G., (eds.) Tools and Algorithms for the Construction and Analysis of Systems - 28th International Conference, TACAS 2022, Held as Part of the European Joint Conferences on Theory and Practice of Software, ETAPS 2022, Munich, Germany, April 2-7, 2022, Proceedings, Part I, vol. 13243 of Lecture Notes in Computer Science, pp. 183–200. Springer (2022)
18. Ethereum. Vyper (2024). https://vyper.readthedocs.io. Accessed 27 Nov 2025
19. Grossman, S., et al.: Practical verification of smart contracts using memory splitting. Proc. ACM Program. Lang. **8**(OOPSLA2), 2402–2433 (2024)
20. Hirai, Y.: Defining the Ethereum virtual machine for interactive theorem provers. In: Brenner, M., et al., (eds.) Financial Cryptography and Data Security, pp. 520–535. Springer International Publishing, Cham (2017)
21. INRIA. The Rocq Reference Manual - Version 9.1.0 (2025). https://rocq-prover.org/doc/V9.1.0/refman/index.html
22. ivan71kmayshan27. Coq formalisation of the ethereum virtual machine (wip) (2020). https://github.com/ivan71kmayshan27/coq-evm. Accessed 23 June 2022
23. Koutavas, V., Lin, Y., Tzevelekos, N.: An operational semantics for YUL. In: Madeira, A., Knapp, A., (eds.) Software Engineering and Formal Methods - 22nd International Conference, SEFM 2024, Aveiro, Portugal, November 6-8, 2024, Proceedings, vol. 15280 of Lecture Notes in Computer Science, pp. 328–346. Springer (2024)
24. Léchenet, J.-C., Blazy, S., Pichardie, D.: A fast verified liveness analysis in SSA form. In: Peltier, N., Sofronie-Stokkermans, V. (eds.) IJCAR 2020. LNCS (LNAI), vol. 12167, pp. 324–340. Springer, Cham (2020). https://doi.org/10.1007/978-3-030-51054-1_19
25. Leroy, X.: Formal verification of a realistic compiler. Commun. ACM **52**(7), 107–115 (2009)
26. Leroy, X.: Formal verification of a realistic compiler. Commun. ACM **52**(7), 107–115 (2009)
27. Monniaux, D.: Pragmatics of formally verified yet efficient static analysis, in particular, for formally verified compilers. Int. J. Softw. Tools Technol. Transf. **26**(4), 463–477 (2024)
28. Monniaux, D., Six, C.: Simple, light, yet formally verified, global common subexpression elimination and loop-invariant code motion. In: Henkel, J., Liu, X., (eds.) LCTES '21: 22nd ACM SIGPLAN/SIGBED International Conference on Languages, Compilers, and Tools for Embedded Systems, Virtual Event, Canada, 22 June, 2021, pp. 85–96. ACM (2021)
29. Moura, L., Ullrich, S.: The lean 4 theorem prover and programming language. In: Platzer, A., Sutcliffe, G. (eds.) CADE 2021. LNCS (LNAI), vol. 12699, pp. 625–635. Springer, Cham (2021). https://doi.org/10.1007/978-3-030-79876-5_37
30. Nipkow, T., Wenzel, M., Paulson, L.C. (eds.): Isabelle/HOL. LNCS, vol. 2283. Springer, Heidelberg (2002). https://doi.org/10.1007/3-540-45949-9
31. Schneidewind, C., Grishchenko, I., Scherer, M., Maffei, M.: Ethor: practical and provably sound static analysis of Ethereum smart contracts. In: CCS '20: 2020

ACM SIGSAC Conference on Computer and Communications Security, Virtual Event, USA, 9–13 November 2020, pp. 621–640. ACM (2020)
32. Six, C., Boulmé, S., Monniaux, D.: Certified and efficient instruction scheduling: application to interlocked VLIW processors. Proc. ACM Program. Lang. **4**(OOPSLA), 129:1–129:29 (2020)
33. Six, C., et al.: Formally verified superblock scheduling. In: Popescu, A., Zdancewic, S., (eds.) CPP '22: 11th ACM SIGPLAN International Conference on Certified Programs and Proofs, Philadelphia, PA, USA, January 17 – 18, 2022, pp. 40–54. ACM (2022)
34. Stegeman, L., Solitor: runtime verification of smart contracts on the Ethereum network. Master's thesis, University of Twente (2018)
35. Tristan, J., Leroy, X.: Formal verification of translation validators: a case study on instruction scheduling optimizations. In: Necula, G.C. Wadler, P., (eds.) Proceedings of the 35th ACM SIGPLAN-SIGACT Symposium on Principles of Programming Languages, POPL 2008, San Francisco, California, USA, January 7–12, 2008, pp. 17–27. ACM (2008)
36. Wood, G.: Ethereum: a secure decentralised generalised transaction ledger (Berlin version 8fea825 – 2022-08-22) (2022)

Selective Concolic Testing

Guofeng Zhang[1], Zhenbang Chen[1(✉)], Ziqi Shuai[1], Jun Sun[2], Weijiang Hong[1], Yufeng Zhang[3], Ji Wang[1], and Yang Liu[1]

[1] College of Computer Science and Technology, National University of Defense Technology, Changsha, China
{zhangguofeng16,zbchen,szq,hongweijiang17,wj,liuyang19d}@nudt.edu.cn
[2] Singapore Management University, Singapore, Singapore
junsun@smu.edu.sg
[3] College of Computer Science and Electronic Engineering, Hunan University, Changsha, China
yufengzhang@hnu.edu.cn

Abstract. The principled combination of symbolic execution and random testing lacks a formal foundation, especially in deciding which inputs to symbolize. We propose selective concolic testing, a cost-aware framework that formulates this choice as an optimized policy problem of a MDP (Markov Decision Process). We model program exploration over a finite control-flow graph, where MDP states represent covered statements, actions partition path constraints into symbolic and random fragments, rewards reflect coverage gain, and costs account for SMT solving effort and sampling inefficiency. Our framework yields the first formal characterization of selective symbolization as policy synthesis in a probabilistic system. We prove that exact policy computation is intractable due to the exponential state space and the hardness of solution-density estimation via model counting. Our formulation enables a practical approximation: we partition constraint dependency graphs and use machine learning to predict solver timeouts, guiding per-constraint symbolization decisions. Built on top of KLEE and JFS, our prototype validates the approach on real-world floating-point benchmarks. Results show that selectively symbolizing inputs, guided by predicted solvability and cost, significantly improves coverage efficiency. Our work thus provides both a rigorous theoretical foundation and a practical instantiation for hybrid program analysis.

Keywords: Concolic Testing · Symbolic Execution · MDP · Constraint Solving

1 Introduction

Symbolic execution [8,31] provides a semantically grounded method for precise program analysis, enabling automated test generation [10,12], bug detection [11],

G. Zhang, Z. Chen, Z. Shuai, W. Hong, J. Wang, and Y. Liu—Also with the affiliation: State Key Laboratory of Complex & Critical Software Environment, National University of Defense Technology, Changsha, China

A. Sampaio and M. Stoelinga (Eds.): FM 2026, LNCS 16557, pp. 275–295, 2026.
https://doi.org/10.1007/978-3-032-26220-2_14

and bounded verification [20]. Despite its success in tools such as SAGE [23], PEX [41], and MAYHEM [14], scalability remains limited by path explosion and the intrinsic complexity of constraint solving [13].

Concolic testing [22,38] addresses these limitations by interleaving concrete and symbolic execution: it starts with a concrete input, collects path constraints symbolically, and iteratively generates new inputs via constraint solving to explore new paths. Conceptually, it combines random testing (for initial/-divergent inputs) and constraint-based symbolic testing (for systematic exploration). Random testing is cheap but struggles with low-probability conditions (e.g., magic numbers); symbolic testing is effective for path coverage but incurs high solving overhead, especially for non-linear constraints. An optimal combination would maximize efficiency, yet computing such a strategy is challenging due to its complexity [44]. Prior work optimizes this trade-off for specific goals like statement [44] or branch coverage [9], but assumes full symbolization of all inputs.

We argue that the *policy of input symbolization, i.e.*, which inputs to treat symbolically, is equally critical. Different symbolization choices affect both the feasibility of random testing (via solution probability) and the cost of constraint solving. Thus, selectively symbolizing inputs can outperform approaches that uniformly apply constraint solving to all variables.

To this end, we propose *selective concolic testing*, which formalizes optimal input symbolization as the optimal policy problem of a cost-aware Markov Decision Process (MDP) with respect to statement coverage. For the first time, we cast the problem as a partitioning task based on SMT counting [18] and solving cost. Given the intractability of computing the exact optimal policy, we design an approximate algorithm that combines graph partitioning and machine learning to predict the symbolization decisions of constraints, trading minimal prediction overhead for significant efficiency gains. Implemented on KLEE [10] and JFS [33], our approach is evaluated on two benchmark suites. On GSL, it improves statement coverage by 36.89% over the state-of-the-art [44] and achieves 12.08× speedup for equivalent statement coverage.

Our main contributions are:

- **A formal MDP-based model** for selective concolic testing, grounded in program semantics and cost-aware policy synthesis.
- **An approximation algorithm** that translates the intractable optimal policy problem into a tractable constraint-partitioning task using machine learning.
- **An implementation and evaluation** that the framework aligns with empirical efficiency, validating its practical relevance without compromising theoretical rigor.

2 Preliminaries

This section uses an example program to motivate selective concolic testing. Figure 1 shows a program (denoted by $\mathcal{P}$) that accepts two input variables x and

```
int foo(double x,int y) {
    int z = 0;
    if (x^7 >= 0) z = z+1;
    if (y == 0) z = z+1;
    return z;
}
```

Fig. 1. An example program.

y, where `x^7` means `power(x,7)`. $\mathcal{P}$ has four paths in total. We first show how concolic testing works on $\mathcal{P}$. Then, we show how selective concolic testing works on $\mathcal{P}$.

2.1 Concolic Testing

Starting from input $\{x \mapsto 0, y \mapsto 1\}$ with both variables symbolized, concolic testing executes $\mathcal{P}$ along path condition $PC_1 : x^7 \geq 0 \wedge y \neq 0$. Using DFS, it next targets $PC_2 : x^7 \geq 0 \wedge y = 0$. Solving PC_2 with an SMT solver (e.g., Z3 [37]) may succeed (e.g., yielding $\{x \mapsto 0, y \mapsto 0\}$), but often times out due to the hardness of floating-point (QF_BVFP) constraints, preventing coverage of this path.

If PC_2 is solved, the next target under DFS is $PC_3 : x^7 < 0$, which is similarly expensive. A timeout here halts exploration; otherwise, assuming solution $\{x \mapsto -1\}$, a random value for y (e.g., -1) yields input $\{x \mapsto -1, y \mapsto -1\}$, covering $PC_4 : x^7 < 0 \wedge y \neq 0$. The final target is $PC_5 : x^7 < 0 \wedge y = 0$, which faces the same solvability issues as PC_2.

In theory, a perfect solver would generate three new inputs to cover all paths. In practice, with a 30-second timeout (on a 3.5GHz CPU), KLEE explores only one path; without timeout, full coverage takes 219 s. Pure random testing fares poorly: while $x^7 \geq 0$ holds with probability 0.5, hitting $y = 0$ occurs with probability 2^{-32}. Thus, using random generation for x and constraint solving for y is preferable, which motivates selective concolic testing.

2.2 Selective Concolic Testing

Selective concolic testing optimally combines random generation and constraint solving by partitioning each target path condition PC_n into two parts: $PC_n = PC_n^r \wedge PC_n^c$. It uses random generation for variables in PC_n^r and constraint solving for those in PC_n^c.

For $PC_2 = (x^7 \geq 0) \wedge (y = 0)$, we set $PC_2^r = x^7 \geq 0$ and $PC_2^c = y = 0$. Since $y = 0$ is trivial for SMT solvers and $x^7 \geq 0$ holds with high probability (0.5), this partition enables fast input generation.

Ideally, PC_n^r should have high solution probability and high solving cost, while PC_n^c should have low cost and low probability, which is formally defined as an optimization problem in Sect. 3. Computing exact solution probabilities requires model counting [18], which is intractable. Hence, we propose a lightweight approximation in Sect. 4. For the program in Fig. 1, our implementation completes full path exploration in one second.

3 MDP Theoretical Framework

This section gives a Markov Decision Process (MDP) based theoretical framework for defining the optimized combination of random generation and constraint solving with the selective input variable symbolization.

Definition 1. *A program $\mathcal{P}$ is a tuple (I, N, E), where I represents the program's input variable set, N is the node set of program statements, and $E \subseteq N \times N$ is the edge set representing $\mathcal{P}$'s control flows. Assume the sets of inputs I and non-deterministic choices N are finite.*

Let $\mathsf{pre}(n)$ and $\mathsf{succ}(n)$ denote the predecessor and successor node sets of n. Without loss of generality, assume $\mathcal{P}$ has a single entry node $entry(\mathcal{P})$. A *path* p is a node sequence $\langle n_0, \ldots, n_k \rangle$ with $(n_i, n_{i+1}) \in E$ for $0 \leq i < k$. We write $\mathcal{N}(p)$ for the nodes in p, and $path(n)$ for all paths from $entry(\mathcal{P})$ to n. Note that $path(n)$ may be infinite if $\mathcal{P}$ contains loops. Let $path(\mathcal{P})$ be the set of all program paths, and $PC(p)$ the path condition of p, computed via concolic execution along p. For input i, $\mathcal{P}(i)$ denotes the execution path of $\mathcal{P}$ on i, and by assumption, every node $n \in N$ is reachable: $\forall n \in N$, $\exists i$ s.t. $\mathcal{P}(i) \in path(n)$.

Definition 2 ***(Assignment).*** *Given a variable set V and an SMT theory whose domain is $\mathcal{D}$, an assignment $\mathbb{S} : V \to \mathcal{D}$ is a function that assigns V's each variable a value in $\mathcal{D}$.*

Definition 3 ***(Solution Space).*** *Given an SMT formula φ, the solution space, denoted as $\Omega(\varphi)$, constitutes all assignments satisfying φ.*

If φ is unsatisfiable, $\Omega(\varphi)$ is $\emptyset$. $\Omega(true)$ is the set of any assignments of $V(\varphi)$, where $V(\varphi)$ represents the set of φ's variables. If the underlying SMT theory has an infinite domain, *i.e.*, `QF_LIA` [5], $\Omega(\varphi)$ may be infinite, *e.g.*, $\Omega(x \geq 0)$, where $x \geq 0$ is a linear integer arithmetic formula and x is an integer variable. In symbolic execution, our analysis is restricted to bit-vector theories with bounded width (e.g., 32-bit or 64-bit), ensuring that $\Omega(\varphi)$ does not represent an infinite domain.

Definition 4 ***(Solution Ratio).*** *Given an SMT formula φ, the solution ratio is the rate of all the solutions that satisfy φ in the whole assignment space. We use $Pr(\varphi)$ to denote φ's solution ratio. $Pr(\varphi) = \frac{\#\Omega(\varphi)}{\#\Omega(true)}$, where $\#S$ represents set S's cardinality.*

For example, if $x \geq 0$ is a bit-vector SMT formula and x is a 32-bit signed integer variable, $Pr(x \geq 0)$ is 0.5. In principle, calculating the solution ratio is a problem of model counting [18]. **In this paper, we only consider the programs whose variables are machine variables, which can be precisely represented by bit-vector based SMT theories.**

Let $\mathbb{R}$ be the set of real numbers. Then, we first define Markov Chain.

Definition 5 *(MC). A Markov Chain (MC) is a tuple $(\mathcal{S}, \delta, \mathcal{I})$, where*

- $\mathcal{S}$ *is a finite set of states.*
- $\delta : \mathcal{S} \times \mathcal{S} \rightarrow \mathbb{R}$ *is the transition probability matrix and satisfies the following condition:* $\sum_{u \in \mathcal{S}} \delta(s, u) = 1$ *for each state* s *in* $\mathcal{S}$.
- $\mathcal{I} : \mathcal{S} \rightarrow \mathbb{R}$ *is the initial distribution of* $\mathcal{S}$ *such that* $\sum_{s \in \mathcal{S}} \mathcal{I}(s) = 1$.

Because an MC has a similar structure as a program, we use the similar path notations of programs for Markov Chains. We use $Pr(p, \mathcal{M})$ to represent the probability of the path $p = \langle s_0, .., s_n \rangle$ starting from s_0 in an MC $\mathcal{M} = (\mathcal{S}, \delta, \mathcal{I})$, and $Pr(p, \mathcal{M}) = \Pi_{0 \leq i \leq n-1} \delta(s_i, s_{i+1})$. Besides, we use $\mathsf{Start}(s)$ to represent the set of paths that starts from state s and ends at a sink state s'. For state s, let $\mathsf{succ}(s') = \{s' | \delta(s, s') > 0\}$. We assume $\mathsf{succ}(s) = \emptyset$ for all non-sink states.

Then, we define a program's probabilistic transition graph as an MC. Note that if a prior distribution of the input values is known, our approach can be easily extended to incorporate the information.

Definition 6 *(Probabilistic Transition Graph). Given a program* $\mathcal{P} = (I, N, E)$, $\mathcal{P}$*'s probabilistic transition graph is an MC* $(\mathcal{S}_t, \delta_t, \mathcal{I}_t)$ *such that*

- $\mathcal{S}_t$ *is* $\mathcal{P}$*'s node set* N.
- $\delta_p : \mathcal{S}_t \times \mathcal{S}_t \rightarrow \mathbb{R}$ *such that* $\delta_t(s_1, s_2) = 0$ *if* $(s_1, s_2) \notin E$*; otherwise,* $\delta_t(s_1, s_2)$ *is defined as follows. We consider paths up to length* T*, making* $path(s)$ *finite.*

$$\delta_t(s_1, s_2) = \frac{\sum\limits_{p \in path(s_2) \wedge p = \langle ..., s_1, s_2 \rangle} Pr(PC(p))}{\sum\limits_{p \in path(s_1)} Pr(PC(p))} \tag{1}$$

- $\mathcal{I}_t$ *is the initial distribution such that* $\mathcal{I}_t(entry(\mathcal{P})) = 1$.

Based on a program's probabilistic transition graph, we can formulate the optimized combination problem with respect to statement coverage by a Markov Decision Process (MDP) with cost. First, we define MDP with cost.

Definition 7 *(MDP with cost). A Markov Decision Process (MDP) with cost* $\mathcal{M}$ *is a tuple* $(\mathcal{S}, \mathcal{A}, \beta_0, \delta, \mathcal{S}_f, \mathcal{R}, \mathcal{C})$*, where*

- $\mathcal{S}$ *is a set of states.* $\mathcal{S}_f$ *is the set of final states.* $\beta_0 : \mathcal{S} \rightarrow \mathbb{R}$ *is the initial distribution function such that* $\Sigma_{s \in \mathcal{S}} \beta_0(s) = 1$.
- $\mathcal{A}$ *is a set of actions.*
- $\delta : \mathcal{S} \times \mathcal{A} \rightarrow \mathbb{D}$ *is the transition function for a state* s *given an action* a *that gives the distribution of state transitions, i.e.,* $\delta(s, a) : \mathcal{S} \rightarrow \mathbb{R}$*, where* $\Sigma_{s_1 \in \mathcal{S}} \delta(s, a)(s_1) = 1$.
- $\mathcal{R} : \mathcal{S} \times \mathcal{S} \rightarrow \mathbb{R}$ *is the reward function that gives the reward to a state transition.*

– *$\mathcal{C} : \mathcal{S} \times \mathcal{A} \rightarrow \mathbb{R}$ is the cost function that gives the cost of executing an action at a state.*

MDP provides a general formalism for modeling and verifying probabilistic problems. Given an MDP, a policy represents the choices made for the problem and is defined as follows.

Definition 8 ***(Policy).*** *Given an MDP with cost $\mathcal{M} = (\mathcal{S}, \mathcal{A}, \beta_0, \delta, \mathcal{S}_f, \mathcal{R}, \mathcal{C})$, a policy π is a function in $\mathcal{S} \rightarrow \mathbb{D}$ such that π gives a distribution $\mathcal{D}_s$ of the actions in $\mathcal{A}$ for a state s in $\mathcal{S}$ and $\mathcal{D}_s$ satisfies $\sum_{a \in \mathcal{A}} \mathcal{D}_s(a) = 1$.*

Given an MDP $\mathcal{M}$ and a policy π, the run of the MDP under π is a rollout defined as follows.

Definition 9 ***(Rollout).*** *A rollout $\zeta \sim \pi$ for an MDP with cost $\mathcal{M}$ is the following sequence of the tuples in $\mathcal{S} \times \mathcal{A} \times \mathcal{S} \times \mathbb{R} \times \mathbb{R}$*

$$\langle (s_0, a_0, s_1, r_0, c_0), ..., (s_{n-1}, a_{n-1}, s_n, r_{n-1}, c_{n-1}) \rangle \tag{2}$$

and the sequence is randomly constructed as follows.

– *Use β_0 to sample the initial state s_0.*
– *For each s_i, use $\pi(s_i)$ to sample an action a_i. Then, sample the next state s_{i+1} by $\delta(s_i, a_i)$. Then, the reward r_i is the value $\mathcal{R}(s_i, s_{i+1})$ and the cost c_i is $\mathcal{C}(s_i, a_i)$.*
– *Repeat the second step until s_{i+1} is in the final state set $\mathcal{S}_f$.*

We can calculate the rewards and costs along a rollout. A policy in fact decides a distribution $\mathbb{E}$ of rollouts $\zeta \sim \pi$. Given an MDP with cost, the optimal policy π^* is the one that achieves the maximum reward and minimum cost and is defined as $\pi^* = \arg\max_\pi \mathbb{E}_{\zeta \sim \pi} \left[\sum_{i=0}^{n-1} (r_i - c_i) \right]$.

Then, given an MC of program $\mathcal{P}$, we can define the selective symbolization-based optimized combination with respect to *statement coverage* as an MDP with cost as follows, where $\mathbb{P}(\mathcal{S})$ is the *powerset* of $\mathcal{S}$.

Definition 10 ***(MDP of selective optimized combination).*** *Given a program $\mathcal{P} = (I, N, E)$ whose probabilistic transition graph is $\mathcal{M} = (\mathcal{S}, \delta, \mathcal{I})$, we define the selective concolic testing with respect to statement coverage as an MDP with cost $\mathcal{M}_S = (\mathbb{P}(\mathcal{S}), \mathcal{A}_p, \beta_p, \delta_p, \{S\}, \mathcal{R}_p, \mathcal{C}_p)$, where*

– $\mathcal{A}_p = \{(R_{I_0}, C_{I_1}) \mid p \in path(\mathcal{P}) \land (R_{I_0} \land C_{I_1}) = PC(p) \land I_0 \cup I_1 = I\}$ is the action set, where each action splits $PC(p)$ into two parts: R_{I_0} (constraints on inputs I_0, handled by random generation) and C_{I_1} (constraints on inputs I_1, solved by constraint solving). We denote by $\mathsf{inputs}(a)$ the program inputs generatable by action a.
Let Π_a be the set of terminated paths reachable by $a = (R_{I_0}, C_{I_1})$, with each

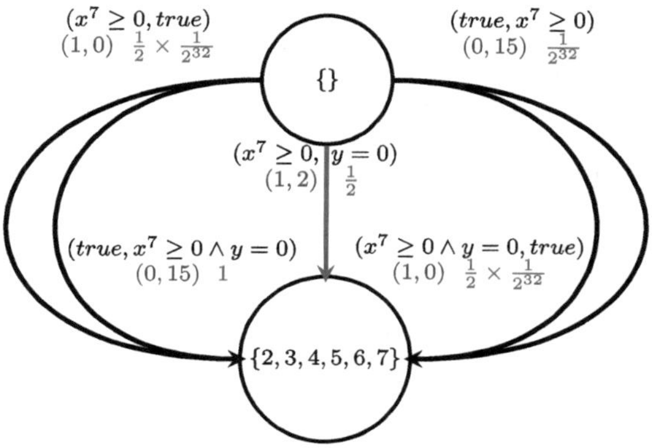

Fig. 2. The *partial* MDP for the selective concolic testing of the example program, where red numbers are probabilities, and blue pairs are the costs.

element $(p, \Pr(p, \mathcal{M}))$ representing a possibly covered terminated path p and its probability. Formally,

$$\Pi_a = \{(p, \Pr(p, \mathcal{M})) \mid \exists i \in \mathsf{inputs}(a) \bullet p = \mathcal{P}(i) \wedge \mathsf{terminated}(p)\},$$

which is derived from $\mathcal{P}$'s probabilistic transition graph.
- β_p is the initial distribution such that $\beta_p(\emptyset) = 1$, *i.e.*, we start from the state where no statement is covered.
- $\delta_p : \mathbb{P}(\mathcal{S}) \times \mathcal{A}_P \to \mathbb{D}$ is the following transition function.

$$\delta_p(N_s, a)(N_t) = \sum_{(p,prob) \in \Pi_a \wedge \mathcal{N}(p) \cup N_s = N_t} prob \tag{3}$$

N_s is the covered statement set of starting state, and N_t is the covered statement set after the path p terminates.
- Only one final state exists, which is $\mathcal{S}$, *i.e.*, all the **reachable** statements are covered.
- $\mathcal{R}_p : \mathbb{P}(\mathcal{S}) \times \mathbb{P}(\mathcal{S}) \to \mathbb{R}$ is defined as $\mathcal{R}_p(N_1, N_2) = \#(N_2 \setminus N_1)$, *i.e.*, the number of the newly covered statements.
- $\mathcal{C}_p : \mathbb{P}(\mathcal{S}) \times \mathcal{A}_p \to \mathbb{R}$ is defined as $\mathcal{C}_p(s, (R_{I_0}, C_{I_1})) = RC(R_{I_0}) + SC(C_{I_1})$, *i.e.*, the cost of random generation for I_0 and the solving cost of the constraint C_{I_1} in the action.

Note that the action set may be infinite. Our formulation generalizes prior work on combining random testing and symbolic execution [9, 44], which restricts actions to those where either I_0 or I_1 is empty.

Assume random generation costs 1, solving $y = 0$ costs 2, and solving $x^7 \geq 0 \wedge y = 0$ (or $x^7 \geq 0$) costs 15. The full MDP $\mathcal{M}_S$ for the program in Fig. 1 has five states and 146 non-zero-probability transitions; Fig. 2 shows only a fragment. For instance, the red transition corresponds to action $(x^7 \geq 0, y = 0)$ with probability 0.5 and cost 3 (*i.e.*, one random generation plus solving $y = 0$). The transition with action $(true, x^7 \geq 0 \wedge y = 0)$ has probability 1 and cost 15. Other transitions—e.g., $(x^7 < 0, y = 0)$, symmetric to the red one—are omitted for brevity.

Under this formalism, selective concolic testing reduces to computing an optimal policy for a cost-aware MDP [21]. A policy π_s is *successful* if it reaches the final state $\mathcal{S}$ (representing full coverage of reachable statements) with positive probability:

$$\Pr_{\zeta \sim \pi_s} \{s_n = \mathcal{S}\} > 0.$$

Let $\Phi_{\mathcal{S}}$ denote all successful policies for $\mathcal{M}_{\mathcal{S}}$. The optimal policy minimizes expected cost (or equivalently maximizes net reward):

$$\pi^* = \arg \max_{\pi \in \Phi_{\mathcal{S}}} \mathbb{E}_{\zeta \sim \pi} \left[\sum_{i=0}^{n-1} (r_i - c_i) \right].$$

For the MDP in Fig. 2, the optimal policy selects $(x^7 \geq 0, y = 0)$ at the initial state (red transition) and $(x^7 \geq 0, true)$ at state $\{2, 3, 5, 6, 7\}$. Its expected value is 2.5, computed as:

$$0.5 \times (6 - 3) + 0.5 \times (5 + 1 - 3 - 1),$$

corresponding to direct termination and a two-step path via $\{2, 3, 5, 6, 7\}$, respectively.

Discussion. The optimized combination problem reduces to computing an optimal policy in a cost-aware MDP. While solvable in time polynomial in the state-space size [44], this becomes infeasible in practice due to: (1) exponential state-space growth with program size and potentially infinite action spaces; (2) solution-ratio estimation requiring model counting, which is harder than standard constraint solving [18]; (3) infinite paths from loops preventing full transition graph construction; (4) high difficulty in accurately predicting SMT solver costs.

To address these, we propose an approximate algorithm yielding a near-optimized strategy:

- **Local greedy optimization:** Since only partial states/actions are observable during concolic execution, we (i) estimate the probabilistic transition graph from observed traces (Eq. 5, Sect. 4.1) using Laplace smoothing [24] for branch probabilities, and bound input depth to avoid non-termination; and (ii) select branches via local greedy reward (Sect. 4.2).
- **Constraint partitioning:** We split path conditions into two parts, one for SMT solving, one for random generation, via graph partitioning and timeout prediction (Sect. 4.3). Algorithm 3 (Sect. 4.4) uses an ML model [17] to predict solver timeouts and guide partitioning.

Algorithm 1: Selective Concolic Testing

SCT($\mathcal{P}, I_0$)
Input: $\mathcal{P} = (I, N, E)$ is a program, I_0 is the initial input.
1: $input \leftarrow I_0$
2: $\mathcal{M}_p \leftarrow (\mathcal{S}_p, \delta_p, \mathcal{I}_p)$
3: $\mathcal{T} \leftarrow \emptyset$
4: **while** true **do**
5: $\quad PC, Path_c \leftarrow$ ConcolicExecute($\mathcal{P}, input$)
6: $\quad \mathcal{T} \leftarrow \mathcal{T} \uplus OTP(PC)$
7: $\quad \mathcal{M}_p \leftarrow$ Refine($\mathcal{M}_p, Path_c$)
8: $\quad PC_n \leftarrow$ ProbSelect($\mathcal{T}, \mathcal{M}_p$)
9: $\quad$ **if** $PC_n = null$ **then**
10: $\quad\quad$ **return**
11: $\quad$ **end if**
12: $\quad input \leftarrow$ SelectiveSolve(PC_n, I)
13: **end while**

4 Approximate Optimized Algorithm

Algorithm 1 shows the high-level approach of selective concolic testing. The inputs are the program $\mathcal{P} = (I, N, E)$ under analysis and the initial input I_0.

We use a map $\mathcal{T}$ to store all the off-the-path branches and their path conditions. In the beginning, $\mathcal{T}$ is empty (Line 3). Then, the algorithm enters the main loop for path exploration. Line 5 executes $\mathcal{P}$ under the current input in a concolic manner [22,38]. The execution returns the current path condition PC and the path $Path_c$ of the current execution. After the execution terminates, the algorithm adds all the off-the-path branches of PC to $\mathcal{T}$ (Line 6). $OTP(PC)$ represents the set of off-the-path branch mappings that are defined as $\{\mathcal{B}(\neg C_j) \mapsto (\bigwedge_{k \in \{1,\ldots,j-1\}} C_k) \wedge \neg C_j \mid 1 \leq j \leq n\}$, where PC is $\bigwedge_{i \in \{1,\ldots,n\}} C_i$ and $\mathcal{B}(\neg C_j)$ denotes the branch (with context information) of the condition $\neg C_j$.

Based on $\mathcal{P}$'s approximate probabilistic transition graph (Sect. 4.1), the algorithm then selects (Sect. 4.2) an off-the-path branch b from $\mathcal{T}$ (Line 8) and returns b's path condition PC_n, *i.e.*, $\mathcal{T}[b]$. In case there exists no unexplored branch, ProbSelect returns $null$, *i.e.*, the path exploration is done, and the algorithm terminates (Line 9–11). Otherwise, the algorithm solves the path condition PC_n by the selective solving algorithm (Line 12) to generate the program input for the next concolic execution. The main loop continues until the computation resource exhausts, *e.g.*, reaching the time limit, which is omitted due to the space limit.

4.1 Approximating the Program's MC

In Algorithm 1, $\mathcal{M}_p$ is the initial probabilistic transition graph of $\mathcal{P}$ at the beginning, where $\mathcal{S}_p$ is N, $\mathcal{I}_p(I) = 1$, and δ_p is as follows.

$$\delta_p(s_1, s_2) = \begin{cases} 0 & (s_1, s_2) \notin E \\ \frac{1}{\#\mathsf{succ}(s_1)} & otherwise \end{cases} \quad (4)$$

Then, after each run of the program, we use the current path $Path_c$ to refine $\mathcal{M}_p$'s transition probabilities (Line 7). δ_p is refined by Laplace estimation [24] and defined as follows, where $\mathsf{EC}(s_1, s_2)$ and $\mathsf{EC}(s_1)$ represents how many times the transition (s_1, s_2) and the statement s_1 have been visited by the executions during concolic testing, respectively.

$$\delta_p(s_1, s_2) = \begin{cases} 0 & (s_1, s_2) \notin E \\ \frac{\mathsf{EC}(s_1,s_2)+1}{\mathsf{EC}(s_1)+\#succ(s_1)} & otherwise \end{cases} \quad (5)$$

4.2 Probability Based Branch Selection

Our algorithm uses transition probabilities to select the off-the-path branch to explore. The key idea is to use the transition probabilities to calculate each branch's reward, which reflects the branch's ability to cover new statements. Here, inspired by the MDP definition of optimized combination (Definition 10) and the value iteration based method [21] for calculating the optimal policy of MDP, we calculate the reward of each statement s (or node) in $\mathcal{M}_p$ as follows.

$$\mathcal{R}(s, \mathcal{M}_p) = \begin{cases} \Sigma_{t\in\mathsf{succ}(s)}(\delta_p(s,t)\times\mathcal{R}(t)) & s\ is\ covered \\ 1+\Sigma_{t\in\mathsf{succ}(s)}(\delta_p(s,t)\times\mathcal{R}(t)) & otherwise \end{cases} \quad (6)$$

Hence, the rewards of uncovered statements are larger than those of covered statements. Then, ProbSelect returns the path condition of the off-the-path branch in $\mathcal{T}$, which is defined as $\arg\max_{(b,PC)\in\mathcal{T}} \mathcal{R}(\mathsf{FST}(b), \mathcal{M}_p)$, where $\mathsf{FST}(b)$ represents the *first statement* of the branch b's basic block.

4.3 Selective Solving Algorithm

We want to optimally combine SMT solving and random generation at the constraint solving level. We formalize the problem as follows. Given a path constraint PC of a program's input I, we need to find the optimized partition of I that is defined as follows, where $\mathcal{P}_I = \{(R, S) \mid R \cup S = I\}$ and $PC \downarrow R$ denotes PC's projection on R.

$$\arg \max_{(R,S)\in\mathcal{P}_I} \frac{Pr(PC \downarrow R)}{SC(PC \downarrow S)} \quad (7)$$

The optimized partition maximizes the probability of random testing and minimizes the solving cost of constraint solving.

Algorithm 2 shows the main procedure of selective solving. The inputs are the path condition PC and the set I of input variables. The output is the solving result, which has the following three cases: `SAT` with a solution, `UNSAT`, and `UNKNOWN`. The algorithm first separates the path condition PC into two parts

Algorithm 2: Selective Solving Algorithm

SelectiveSolve(PC, I)

Input: $\mathcal{PC}$ is a path condition, I is the set of the input variables in PC.

1: $(PC_r, PC_c) \leftarrow$ Partition(PC, I)
2: $(r_1, S_c) \leftarrow$ SMTSolving(PC_c)
3: **if** $r_1 =$ `UNSAT` $\vee\ r_1 =$ `UNKNOWN` **then**
4:     **return** r_1
5: **end if**
6: $(r_2, S_r) \leftarrow$ RandomGeneration(PC, PC_r, S_c)
7: **if** $r_2 =$ `UNKNOWN` **then**
8:     **return** r_2
9: **end if**
10: $S \leftarrow S_c \uplus S_r$
11: **return** S

PC_r and PC_c (Line 1). Algorithm 3 handles the separation. PC_r will be solved by random generation (Line 6), and the algorithm uses the underlying SMT solver to solve PC_c (Line 2). If the SMT solver produces `UNSAT` or `UNKNOWN` (*e.g.*, timeout), the algorithm returns the result directly (Line 4) because the constraint PC_c's `UNSAT` implies the PC's unsatisfiability. If both PC_c and PC_r are `SAT` and can be solved, the algorithm merges their solutions S_c and S_r. $S_c \uplus S_r$ can be defined as follows, where S_c and S_r are the solution maps that give a value to a variable, and $v \hat{\in} S$ represents that v is a key in S.

$$S_c \uplus S_r[v] := \begin{cases} S_r[v] & v \hat{\in} S_r \\ S_c[v] & v \hat{\in} S_c \\ \texttt{Undefined} & otherwise \end{cases} \tag{8}$$

In the algorithm, some input variables may appear in both PC_r and PC_c after partition, which is why we pass the solution of PC_c when using random generation for solving PC_r (Line 6). During the random generation, we will use the solution S_c to assign the values of the PC_r's variables that also appear in PC_c. In our implementation, we employ fuzzing-based solving [33] to solve PC_r. Our selective solving algorithm is sound but incomplete. The input path condition PC may be satisfiable, but our algorithm may produce `UNKNOWN` because we employ random generation (or fuzzing in our implementation) to solve PC_r.

4.4 Partition Algorithm

Algorithm 3 partitions a path condition PC over input variables I into two parts: PC_r (for random generation) and PC_c (for SMT solving), such that $PC = PC_r \wedge PC_c$. The core idea is to (1) model variable dependencies in PC as a weighted graph, (2) partition it using a graph clustering algorithm [29], and (3) assign each sub-constraint to either PC_r or PC_c based on an offline-trained ML model that predicts SMT solver timeout.

Algorithm 3: Constraint Partition

Partition(PC, I)
Input: PC is a path condition set, and I is the set of the variables in PC.
1: $\mathcal{E} \leftarrow \{(v_1, v_2) \mapsto 0 \mid v_1 \neq v_2 \wedge \{v_1, v_2\} \subseteq I\}$
2: $\mathcal{G} \leftarrow (I, \mathcal{E})$
3: **for** condition **do**
4:     $V \leftarrow \mathit{Variables}(C_i)$
5:     **for** each $(v_1, v_2) \in V \times V$ **do**
6:         **if** $v_1 \neq v_2$ **then**
7:             $\mathcal{E}[(v_1, v_2)] \leftarrow \mathcal{E}[(v_1, v_2)] + 1$
8:         **end if**
9:     **end for**
10: **end for**
11: $GS \leftarrow$ GraphPartition$(\mathcal{G})$
12: $PC_r, PC_c \leftarrow \mathit{true}, \mathit{true}$
13: **for** each $g \in GS$ **do**
14:     $PC_g \leftarrow PC \downarrow g$
15:     **if** Predicte(PC_g) is timeout **then**
16:         $PC_r \leftarrow PC_r \wedge PC_g$
17:     **else**
18:         $PC_c \leftarrow PC_c \wedge PC_g$
19:     **end if**
20: **end for**
21: **return** (PC_r, PC_c)

We construct a relation graph $\mathcal{G} = (I, \mathcal{E})$, where each edge (v_1, v_2) is weighted by the number of atomic predicates in PC containing both v_1 and v_2:

$$w(v_1, v_2) = \#\{C_i \mid v_1 \in C_i \wedge v_2 \in C_i\}. \tag{9}$$

Lines 1–10 initialize $\mathcal{E}$ and populate edge weights by iterating over all atomic conditions $C_i \in PC$ and incrementing weights for every distinct variable pair in $\mathit{Variables}(C_i)$. The graph $\mathcal{G}$ is then partitioned into subgraphs GS via GraphPartition (Line 11). For each subgraph $g = (V_g, E_g) \in GS$, we project PC onto g to obtain $PC_g = \bigwedge\{C_i \in PC \mid \mathit{Variables}(C_i) \subseteq V_g\}$ (Line 14). In Lines 13–20, an SVM model [16], trained offline on BVFP formulas, predicts whether solving PC_g will time out. The feature vector is built via Bag-of-Words over SMT operators (e.g., counts of `FAdd`, `FMul`), including the number of BVFP operations. If timeout is predicted, PC_g is added to PC_r; otherwise, to PC_c.

Example: Consider the path condition $x^7 > y \wedge y = z \wedge y = w + 2 \wedge z = w + w$ over input variables $\{x, y, z, w\}$.

We construct a relation graph $\mathcal{G}$ where vertices are variables and edge weights reflect co-occurrence in atomic predicates. Applying graph partitioning yields two subgraphs: $g_1 = \{x, y\}$ and $g_2 = \{y, z, w\}$. The projection of the path condition onto g_1 is $PC_{g_1} = x^7 > y$, which our predictor classifies as likely to

Table 1. Benchmark Introduction

Benchmark	LoC	#Fun	Info
GNU GSL [3]	70725	421	A scientific computing library
Cephes [1]	18307	114	A mathematical routines library
FDLIBM [2]	18499	41	A portable C math library
Total	107531	576 functions	

time out during SMT solving. The projection onto g_2 is $PC_{g_2} = y = z \wedge y = w + 2 \wedge z = w + w$, predicted to be efficiently solvable.

Thus, the path condition is partitioned as:

$$\underbrace{x^7 > y}_{PC_r} \wedge \underbrace{y = z \wedge y = w + 2 \wedge z = w + w}_{PC_c} \tag{10}$$

Solving PC_c with an SMT solver yields a concrete assignment, e.g., $\{y \mapsto 4, z \mapsto 4, w \mapsto 2\}$ (corresponding to S_c in Algorithm 2). To satisfy PC_r, we fix $y = 4$ and use random generation to find x such that $x^7 > 4$, which is straightforward due to the high solution density of this inequality.

5 Evaluation

We implemented our approximate algorithm for C programs in a concolic execution framework [38] built on KLEE [10]. Complex constraints are handled by integrating JFS [33] as the random-testing engine. Path conditions (in KQuery [4]) are partitioned via METIS [29] using their relational graph. To predict solver timeouts, we extract syntactic features with a Bag-of-Words model [48] (counting operator occurrences) and train an SVM using LIBSVM [16]. The model was trained on the SMTLIB2 QF_BVFP benchmark [6] that does not intersect with the experimental evaluation, with an accuracy of 91.1%, avoiding training leakage. We use Z3 [37] as the main solver for its QF_ABVFP support and extended JFS to cover more SMTLIB2 operations.

- **RQ1 (Effectiveness):** How does our method compare to existing optimized combination and heuristic approaches in statement coverage?
- **RQ2 (Efficiency):** How much faster does our method achieve the same statement coverage as prior work?

5.1 Experimental Setup

Benchmarks. Table 1 shows two types of benchmarks for evaluation. **GSL & Cephes** contain complex non-linear floating-point path conditions and intensive control flow, making them challenging for symbolic execution. We evaluate 421

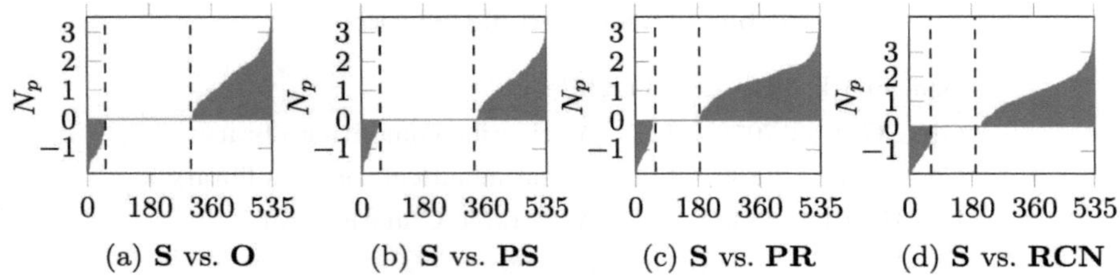

Fig. 3. Experimental results of #**Cov** on GSL and Cephes benchmark

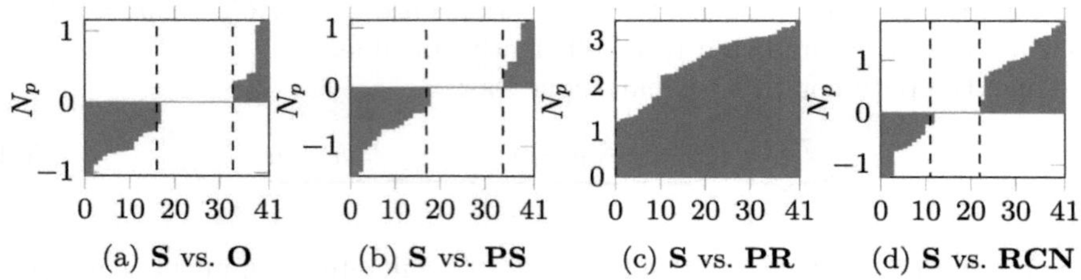

Fig. 4. Experimental results of #**Cov** on FDLIBM benchmark

GSL and 114 Cephes functions (total 535), including both hard cases and small functions that allow full path exploration. **FDLIBM** features linear conditions with low-level bit manipulations typical in embedded systems. To avoid redundancy from type-duplicated logic (e.g., `cos(float)` vs `cos(double)`), we select 41 unique functions.

Baselines. We compare our method (**S**) against four baselines:

- **O**: optimized combination without selective symbolization [44];
- **PS**: probabilistic branch selection with pure Z3 solver;
- **PR**: probabilistic branch selection with pure JFS solver;
- **RCN**: KLEE's default random-cover-new strategy with Z3 solver.

Our evaluation addresses three goals: (1) **Overall performance**: Compare coverage and efficiency of **S** vs. baselines; (2) **Branch selection impact**: Evaluate probabilistic selection via **PS** vs. **RCN**; (3) **Selective solving benefit**: Assess our hybrid solver by comparing **S**, **PS**, and **PR**.

All experiments run for 30 min, repeated 5 times on a 104-core server (Intel Xeon Platinum 8269CY @2.50GHz, 194GB RAM, Ubuntu 18.04).

5.2 Experimental Results

GSL and Cephes. Figure 3 shows statement coverage results across 535 functions, with programs sorted by increasing coverage and relative improvement N_p

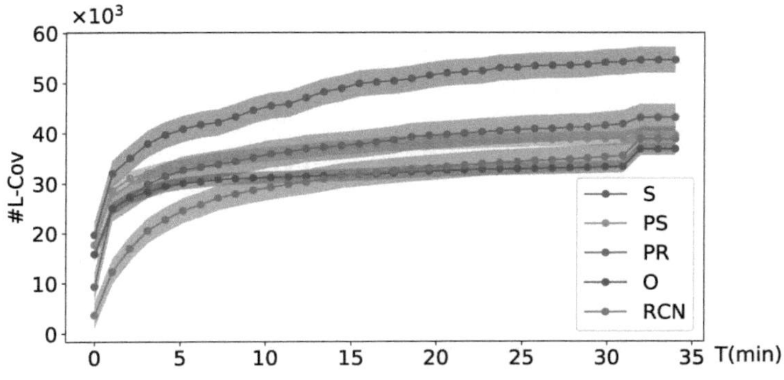

(a) GSL #**Cov** trend

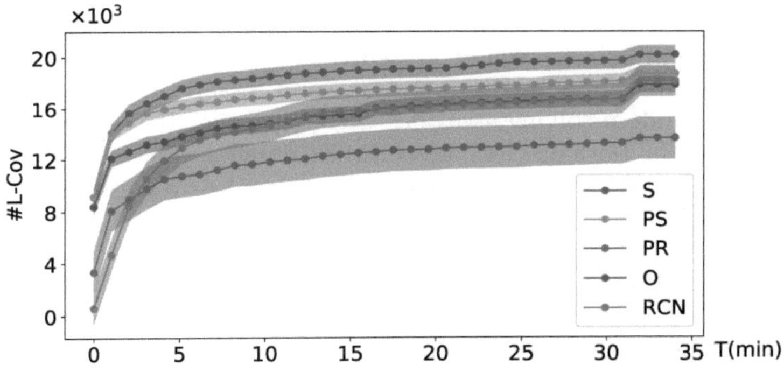

(b) Cephes #**Cov** trend

Fig. 5. Trends of covered statements on **GSL** and **Cephes**. Mean values and 95% confidence intervals over 5 runs are shown in solid lines and shadows, respectively.

(log scale) defined as:

$$N_p = \begin{cases} \log_{10}(R_p \times 100 + 1) & R_p \geq 0 \\ -\log_{10}(-R_p \times 100 + 1) & R_p < 0 \end{cases} \tag{11}$$

where $R_p = (L_p^{\mathbf{S}} - L_p^b)/L_p^b$. $N_p = 0$ indicates equal coverage; $N_p = 1$ and 2 correspond to 10% and 100% gains.

Our method **S** outperforms all baselines on most functions. Compared to **O**, **PS**, **PR**, and **RCN**, **S** achieves higher coverage in 230, 199, 348, and 326 functions, with average improvements of 36.89%, 28.43%, 31.85%, and 31.46%, respectively. Underperformance on some functions stems from inaccurate timeout predictions, causing unnecessary random solving; small functions often terminate early.

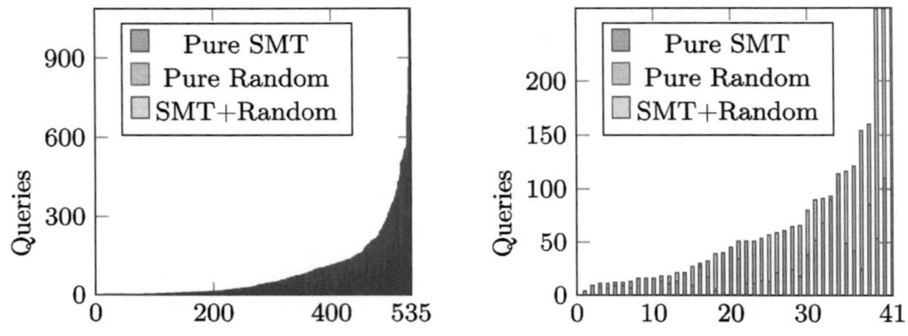

(a) Partition of **GSL and Cephes** (b) Partition of **FDLIBM**

Fig. 6. Partition information of the benchmarks

FDLIBM. Figure 4 shows results on FDLIBM's linear but complex constraints. **S** performs comparably to **PS** (−3.44%) and significantly better than **PR** (157.42%), as random solving struggles with equalities. It matches **O** (−0.05%) and **RCN** (3.44%), both of which favor SMT solving. Despite timeout prediction errors limiting gains, **S** remains competitive, demonstrating robustness across constraint types.

> **Answer to RQ1: S** significantly improves coverage on GSL/Cephes and matches SMT-based methods on FDLIBM, confirming its effectiveness on real-world floating-point programs.

Efficiency. Figure 5 shows cumulative covered statements over time (mean ±95% CI). On GSL, **S** achieves speedups of 12.08×, 10.88×, 3.88×, and 8.76× over **O**, **PS**, **PR**, and **RCN** at peak coverage; similar gains are seen on Cephes. Baselines exhibit delayed coverage spikes after 30 min due to poor state prioritization, whereas **S**, **PS**, and **PR** show smoother, earlier growth thanks to probabilistic branch selection.

> **Answer to RQ2: S** is consistently more efficient at achieving the same coverage.

Constraint Partitioning. Figure 6 shows solving mode distribution. On GSL/-Cephes, 24.03% of constraints used hybrid solving (SMT+random), versus 75.55% pure SMT and 0.42% pure random. On FDLIBM, hybrid use was 50.82%, reflecting its linear nature. Since constraint solving dominates (>90% of runtime), selective symbolization is essential. The low pure-random usage on FDLIBM confirms SMT's suitability, yet prediction errors caused unnecessary hybrid attempts, explaining **S**'s slight underperformance vs. **PS**.

Our method excels on programs with mixed constraint complexity (e.g., GSL and Cephes's hypergeometric functions), where partitioning separates hard floating-point constraints from simple branches. When all constraints are uniformly complex (e.g., FDLIBM's linear cases), gains diminish as random testing becomes less effective.

Limitations. Our approach assumes: (1) program variables have finite domains; (2) the control-flow graph contains no unreachable code. While these hold for mathematical libraries, real-world programs may require preprocessing to satisfy these conditions.

6 Related Work

Our work relates to hybrid approaches combining symbolic execution with fuzzing or random testing [36,40,44,51]. The term **Selective Concolic Testing** was previously used by Lin *et al.* [34] in the context of hardware security for detecting Trojans in SystemC designs. While both approaches aim to improve concolic testing efficiency through selection, they differ fundamentally in objective and mechanism. Lin *et al.* selectively restrict test generation to user-specified code regions to avoid redundant path exploration. In contrast, our work provides a theoretical framework for selective input symbolization, and the practical algorithm partitions based on a cost model derived from machine learning and Markov decision processes.

Concolic testing [22,38] inherently blends both by using concrete values to simplify constraints. Majumdar and Sen [36] interleave random and concolic testing, switching to the latter when coverage stalls; Driller [40] applies a similar idea to binary analysis. Wang *et al.* [44] first formalize the optimized combination for statement coverage as a Markov Decision Process (MDP) [21], while DigFuzz [51] uses probabilistic path prioritization. In contrast, our key insight is that which inputs to symbolize critically affects efficiency. To our knowledge, we are the first to formalize the optimized combination problem under selective symbolization. Unlike Cha *et al.* [15], who mine templates for selective symbolization online, we formally define the optimal policy and use fuzzing for non-symbolized inputs. Marco [25] also models concolic testing as an MDP but uses path divergence probability as cost; we instead model constraint-solving difficulty, making our approach more effective for programs with complex (e.g., floating-point) constraints. Our method complements existing symbolic execution techniques. Search heuristics [7,35,45,50] prioritize branches for coverage or bug finding, while path pruning methods [19,27,42,46] eliminate irrelevant paths, both orthogonal to our constraint-level optimization. We also build on constraint-solving optimizations, such as formula simplification and caching [10], solution reuse [28,43], and domain-specific solvers [17,39,49]. Our algorithm uniquely integrates SMT solving with fuzzing at the constraint level, aligning with hybrid fuzzing [26,30,32,47].

7 Conclusion

Optimally combining symbolic execution and random testing is challenging. We formalize selective concolic testing as an MDP policy optimization problem and propose an approximate algorithm that partitions path constraints: one part is solved by an SMT solver, the other via random generation. Implemented atop KLEE and JFS, our prototype demonstrates significant gains in both effectiveness and efficiency.

Acknowledgments. This research was supported by National Key R&D Program of China (No. 2022YFB4501903) and the NSFC Program (No. 62172429, 62032024, and 62372162), and Jun Sun was supported by the Ministry of Education, Singapore under its Academic Research Fund Tier 2 (Award ID: T2EP20222-0037). Any opinions, findings and conclusions or recommendations expressed in this material are those of the author(s) and do not reflect the views of the Ministry of Education, Singapore.

Data Availability Statement. Our artifact is available at the following URL: https://doi.org/10.5281/zenodo.18603532.

References

1. Cephes mathematical function library (2025). https://www.netlib.org/cephes/. Accessed 12 Nov 2025
2. Freely distributable libm (2025). http://www.netlib.org/fdlibm/. Accessed 12 Nov 2025
3. Gsl - gnu scientific library (2025). https://www.gnu.org/software/gsl/. Accessed 12 Nov 2025
4. Klee's kquery language (2025). https://klee.github.io/docs/kquery/. Accessed 12 Nov 2025
5. Qf_lia logic (2025). https://smtlib.cs.uiowa.edu/logics-all.shtml#QF_LIA. Accessed 12 Nov 2025
6. Smt-lib benchmark (2025). https://smtlib.cs.uiowa.edu/benchmarks.shtml. Accessed 12 Nov 2025
7. Avgerinos, T., Cha, S.K., Hao, B.L.T., Brumley, D.: AEG: automatic exploit generation. In: NDSS 2011, San Diego, California, USA, 6th February–9th February 2011 (2011)
8. Baldoni, R., Coppa, E., D'Elia, D.C., Demetrescu, C., Finocchi, I.: A survey of symbolic execution techniques. ACM Comput. Surv. **51**(3), 50:1–50:39 (2018)
9. Burnim, J., Sen, K.: Heuristics for scalable dynamic test generation. In: 23rd IEEE/ACM ASE 2008), L'Aquila, Italy, 15–19 September 2008, pp. 443–446 (2008)
10. Cadar, C., Dunbar, D., Engler, D.R.: KLEE: unassisted and automatic generation of high-coverage tests for complex systems programs. In: 8th USENIX OSDI 2008, San Diego, California, USA, 8–10 December 2008 Proceedings, pp. 209–224 (2008)
11. Cadar, C., Ganesh, V., Pawlowski, P.M., Dill, D.L., Engler, D.R.: EXE: automatically generating inputs of death. In: Proceedings of the 13th ACM CCS 2006, Alexandria, VA, USA, 30 October–3 November 2006, pp. 322–335 (2006)

12. Cadar, C., et al.: Symbolic execution for software testing in practice: preliminary assessment. In: Proceedings of the 33rd ICSE 2011, Waikiki, Honolulu, HI, USA, 21–28 May 2011, pp. 1066–1071 (2011)
13. Cadar, C., Sen, K.: Symbolic execution for software testing: three decades later. Commun. ACM **56**(2), 82–90 (2013)
14. Cha, S.K., Avgerinos, T., Rebert, A., Brumley, D.: Unleashing mayhem on binary code. In: IEEE SP 2012, San Francisco, California, USA, 21–23 May 2012, pp. 380–394 (2012)
15. Cha, S., Lee, S., Oh, H.: Template-guided concolic testing via online learning. In: Huchard, M., Kästner, C., Fraser, G. (eds.) Proceedings of the 33rd ASE 2018, Montpellier, France, 3–7 September 2018, pp. 408–418. ACM (2018)
16. Chang, C.C., Lin, C.J.: LIBSVM: a library for support vector machines. ACM Trans. Intell. Syst. Technol. **2**, 27:1–27:27 (2011), software available at http://www.csie.ntu.edu.tw/~cjlin/libsvm
17. Chen, Z., et al.: Synthesize solving strategy for symbolic execution. In: ISSTA 2021, Virtual Event, Denmark, 11–17 July 2021, pp. 348–360 (2021)
18. Chistikov, D., Dimitrova, R., Majumdar, R.: Approximate counting in SMT and value estimation for probabilistic programs. In: TACAS 2015, London, UK, 11–18 April 2015. Proceedings, pp. 320–334 (2015)
19. Cui, H., Hu, G., Wu, J., Yang, J.: Verifying systems rules using rule-directed symbolic execution. In: ASPLOS 2013, Houston, TX, USA, 16–20 March 2013, pp. 329–342 (2013)
20. Deng, X., Lee, J., Robby: Bogor/Kiasan: a k-bounded symbolic execution for checking strong heap properties of open systems. In: 21st IEEE/ACM ASE 2006, Tokyo, Japan, 18–22 September 2006, pp. 157–166 (2006)
21. Feinberg, E., Shwartz, A. (eds.): Handbook of Markov Decision Processes - Methods and Applications. Kluwer International Series (2002)
22. Godefroid, P., Klarlund, N., Sen, K.: DART: directed automated random testing. In: Proceedings of the ACM SIGPLAN PLDI 2005, Chicago, IL, USA, 12–15 June 2005, pp. 213–223 (2005)
23. Godefroid, P., Levin, M.Y., Molnar, D.A.: Automated whitebox fuzz testing. In: NDSS 2008, San Diego, California, USA, 10th February–13th February 2008 (2008)
24. Hartley, H.O.: Contributions to Survey Sampling and Applied Statistics. Academic Press, London (1978)
25. Hu, J., Duan, Y., Yin, H.: Marco: a stochastic asynchronous concolic explorer. In: Proceedings of the 46th IEEE/ACM ICSE 2024, Lisbon, Portugal, 14–20 April 2024, pp. 59:1–59:12. ACM (2024)
26. Huang, H., Yao, P., Wu, R., Shi, Q., Zhang, C.: Pangolin: incremental hybrid fuzzing with polyhedral path abstraction. In: 2020 IEEE SP 2020, San Francisco, CA, USA, 18–21 May 2020, pp. 1613–1627 (2020)
27. Jaffar, J., Murali, V., Navas, J.A.: Boosting concolic testing via interpolation. In: The ACM SIGSOFT ESEC/FSE 2013, Saint Petersburg, Russian Federation, 18–26 August 2013, pp. 48–58 (2013)
28. Jia, X., Ghezzi, C., Ying, S.: Enhancing reuse of constraint solutions to improve symbolic execution. In: ISSTA 2015, Baltimore, MD, USA, 12–17 July 2015, pp. 177–187 (2015)
29. Karypis, G., Kumar, V.: A fast and high quality multilevel scheme for partitioning irregular graphs. SIAM J. Sci. Comput. **20**(1), 359–392 (1998)
30. Kim, K., Jeong, D.R., Kim, C.H., Jang, Y., Shin, I., Lee, B.: HFL: hybrid fuzzing on the Linux kernel. In: 27th NDSS 2020, San Diego, California, USA, 23–26 February 2020 (2020)

31. King, J.C.: Symbolic execution and program testing. Commun. ACM **19**(7), 385–394 (1976)
32. Le, H.M.: LLVM-based hybrid fuzzing with libkluzzer (competition contribution). In: FASE 2020, Dublin, Ireland, 25–30 April 2020, Proceedings, pp. 535–539 (2020)
33. Liew, D., Cadar, C., Donaldson, A.F., Stinnett, J.R.: Just fuzz it: solving floating-point constraints using coverage-guided fuzzing. In: Proceedings of the ACM ESEC/SIGSOFT FSE 2019, Tallinn, Estonia, 26–30 August 2019, pp. 521–532 (2019)
34. Lin, B., Chen, J., Xie, F.: Selective concolic testing for hardware trojan detection in behavioral systemc designs. In: 2020 Design, Automation & Test in Europe Conference & Exhibition, DATE 2020, Grenoble, France, 9–13 March 2020, pp. 19–24. IEEE (2020)
35. Ma, K.-K., Yit Phang, K., Foster, J.S., Hicks, M.: Directed symbolic execution. In: Yahav, E. (ed.) SAS 2011. LNCS, vol. 6887, pp. 95–111. Springer, Heidelberg (2011). https://doi.org/10.1007/978-3-642-23702-7_11
36. Majumdar, R., Sen, K.: Hybrid concolic testing. In: 29th ICSE 2007, Minneapolis, MN, USA, 20–26 May 2007, pp. 416–426 (2007)
37. de Moura, L., Bjørner, N.: Z3: an efficient SMT solver. In: Ramakrishnan, C.R., Rehof, J. (eds.) TACAS 2008. LNCS, vol. 4963, pp. 337–340. Springer, Heidelberg (2008). https://doi.org/10.1007/978-3-540-78800-3_24
38. Sen, K., Marinov, D., Agha, G.: CUTE: a concolic unit testing engine for C. In: The 13th ACM ESEC/SIGSOFT FSE 2005, Lisbon, Portugal, 5–9 September 2005, pp. 263–272 (2005)
39. Shuai, Z., Chen, Z., Zhang, Y., Sun, J., Wang, J.: Type and interval aware array constraint solving for symbolic execution. In: 30th ACM SIGSOFT ISSTA 2021, Virtual Event, Denmark, 11–17 July 2021, pp. 361–373 (2021)
40. Stephens, N., et al.: Driller: augmenting fuzzing through selective symbolic execution. In: NDSS 2016, San Diego, California, USA, 21–24 February 2016 (2016)
41. Tillmann, N., de Halleux, J.: Pex-white box test generation for .net. In: Tests and Proofs - 2nd International Conference, TAP 2008, Prato, Italy, 9–11 April 2008. Proceedings, pp. 134–153 (2008)
42. Trabish, D., Mattavelli, A., Rinetzky, N., Cadar, C.: Chopped symbolic execution. In: Proceedings of the 40th ICSE 2018, Gothenburg, Sweden, 27 May–03 June 2018, pp. 350–360 (2018)
43. Visser, W., Geldenhuys, J., Dwyer, M.B.: Green: reducing, reusing and recycling constraints in program analysis. In: 20th ACM SIGSOFT SIGSOFT/FSE 2012, Cary, NC, USA, 11–16 November 2012, p. 58 (2012)
44. Wang, X., Sun, J., Chen, Z., Zhang, P., Wang, J., Lin, Y.: Towards optimal concolic testing. In: ICSE 2018, Gothenburg, Sweden, 27 May–03 June 2018, pp. 291–302 (2018)
45. Xie, T., Tillmann, N., de Halleux, J., Schulte, W.: Fitness-guided path exploration in dynamic symbolic execution. In: IEEE/IFIP DSN 2009, Estoril, Lisbon, Portugal, 29 June–2 July 2009, pp. 359–368 (2009)
46. Yu, H., Chen, Z., Wang, J., Su, Z., Dong, W.: Symbolic verification of regular properties. In: Proceedings of the 40th ICSE 2018, Gothenburg, Sweden, 27 May–03 June 2018, pp. 871–881 (2018)
47. Yun, I., Lee, S., Xu, M., Jang, Y., Kim, T.: QSYM: a practical concolic execution engine tailored for hybrid fuzzing. In: 27th USENIX Security Symposium, USENIX Security 2018, Baltimore, MD, USA, 15–17 August 2018, pp. 745–761 (2018)
48. Zhang, Y., Jin, R., Zhou, Z.: Understanding bag-of-words model: a statistical framework. Int. J. Mach. Learn. Cybern. **1**(1–4), 43–52 (2010)

49. Zhang, Y., Chen, Z., Shuai, Z., Zhang, T., Li, K., Wang, J.: Multiplex symbolic execution: exploring multiple paths by solving once. In: 35th IEEE/ACM ASE 2020, Melbourne, Australia, 21–25 September 2020, pp. 846–857 (2020)
50. Zhang, Y., Chen, Z., Wang, J., Dong, W., Liu, Z.: Regular property guided dynamic symbolic execution. In: ICSE 2015, Florence, Italy, 16–24 May 2015, vol. 1, pp. 643–653 (2015)
51. Zhao, L., Duan, Y., Yin, H., Xuan, J.: Send hardest problems my way: probabilistic path prioritization for hybrid fuzzing. In: NDSS 2019, San Diego, California, USA, 24–27 February 2019 (2019)

EUF-based Solving Dyck-Reachability with Applications to Static Analysis

Yide Du, Zhenbang Chen(✉), Kunlin Liu, Guofeng Zhang, Xudong Wang, Ke Ma, Wei Dong, and Ji Wang

College of Computer Science and Technology, National University of Defense Technology, Changsha, China
{dyd1024,zbchen,klliu18,zhangguofeng17,wangxudong,ke, dongwei,wj}@nudt.edu.cn

Abstract. Static analysis plays a crucial role in program optimization, bug detection, and automated testing. Dyck-reachability provides a foundational formulation for static analysis, as Dyck grammars can model critical properties such as field and context sensitivity, thus offering broad applicability. This paper shows that static analysis problems modeled as Dyck-reachability on bidirected graphs can be encoded into the EUF SMT theory; consequently, all such problems admit efficient formulation and solution via EUF-based SMT solvers. By leveraging the optimized nature of modern SMT solvers, our method achieves efficiency comparable to state-of-the-art graph-based bidirected Dyck-reachability algorithms while eliminating the need for developing complex specialized graph reachability algorithms. Our approach opens new avenues for solving these classical static analysis problems, demonstrating the strong potential of SMT solvers in encoding static analysis solutions.

Keywords: Static Analysis · EUF · SMT · Dyck reachability

1 Introduction

Program analysis is essential for software engineering tasks such as optimization [23], bug detection [20], and automated testing [36]. It enables automated program understanding and is broadly categorized into dynamic and static approaches. Dynamic methods (*e.g.*, fuzzing [9] and runtime monitoring [22]) execute programs to analyze runtime behavior, offering precision but limited coverage and the requirement of specific program execution environments. In contrast, static analysis examines code without execution by employing techniques such as data-flow analysis [3] to over-approximate program behavior through abstract models. These methods benefit from soundness, meaning they guarantee coverage of all possible program behaviors, but face inherent imprecision

Also with the affiliation: State Key Laboratory of Complex & Critical Software Environment, National University of Defense Technology, Changsha, China.

A. Sampaio and M. Stoelinga (Eds.): FM 2026, LNCS 16557, pp. 296–316, 2026.
https://doi.org/10.1007/978-3-032-26220-2_15

due to over-approximation. This imprecision can lead to false positives, and improving precision typically compromises scalability, establishing a fundamental precision-scalability trade-off. Achieving an optimal balance between these competing objectives remains both critical and challenging in practice.

Various static analysis problems including inter-procedural data-flow analysis [35], alias analysis [43], and pointer analysis [39] can be formulated as context-free language (CFL) reachability on graphs [34,35]. In this formulation, programs are modeled as graphs (*e.g.*, control-flow or field-access graphs), and the analysis is framed as a reachability problem under path constraints: it checks for paths where the sequence of edge labels belongs to a specified formal language for instance, a language that requires balanced or matched operations. Constraints such as function call-return consistency and memory load-store matching are thus directly encoded. While providing a natural modeling framework, CFL-reachability's cubic time complexity [25] limits scalability, leading to approximations such as bidirected Dyck-reachability. Defined by the Dyck language grammar [21], this approach operates on bidirected graphs with matched opening/closing labels, forming node equivalence relations that trade precision for efficiency [44]. It has been successfully applied to field sensitivity in alias analysis [44] and call-return matching in data dependence analysis [10,26,27].

The state-of-the-art bidirected Dyck-reachability algorithm achieves a worst-case time complexity of $O(m + n \cdot \alpha(n))$, where n and m denote the number of nodes and edges in the graph, respectively, and $\alpha(n)$ is the inverse Ackermann function [2]. Its average-case complexity is $O(m)$, making whole-program analysis computationally tractable. In contrast, the LLVM compiler[1] employs an alternative algorithm for alias analysis via bidirected Dyck-reachability with complexity $O(n + m \log m)$ [44]. However, implementing static program analysis tools remains challenging. Beyond frontend issues, developers must comprehend intricate data structures and algorithmic details, such as the FDLL (Fast-Doubly-Linked-List) used in graph-based approaches [44], and subsequently abstract and implement the algorithms based on their understanding. This engineering process is labor-intensive and error-prone.

Our key argument in this paper is that well-optimized logical reasoning tools can significantly simplify the design and implementation of static analyses. Inspired by the equivalence relation property of bidirected Dyck-reachability [44], we propose an encoding of such problems using the Equality with Uninterpreted Functions (EUF) SMT theory. This approach provides a straightforward implementation path while enabling direct utilization of highly optimized SMT solvers. Although SAT/SMT techniques are well-established in areas like symbolic execution [5] and path-sensitive analysis [37], to our knowledge, *we are the first to apply SMT solving to static analyses modeled by bidirected Dyck-reachability*, such as pointer and alias analyses. Modern SMT solvers like Z3 [12], CVC5 [6], and Yices [18] incorporate sophisticated algorithms and low-level optimizations, making them ideal for reducing engineering effort while ensuring robustness.

[1] https://github.com/llvm-mirror/llvm/blob/master/lib/Analysis/CFLSteensAliasAnalysis.cpp.

Our evaluation on two representative tasks demonstrates that this EUF-based approach achieves comparable efficiency to an implementation [26] of the state-of-the-art graph algorithm [10] for bidirected Dyck-reachability.

The main contributions of this paper are as follows.

- We propose to utilize SAT/SMT for more static analysis problems. Specifically, we propose a novel approach to solving static analysis problems modeled as bidirected Dyck-reachability problems by EUF SMT theory, which covers a lot of static analysis problems, such as alias and pointer analyses. *Our approach opens a new door to solving these classical static analysis problems and their implementations.*
- We show that any program analysis problem modeled by bidirected Dyck-reachability can be reduced to solving formulas in the EUF SMT theory.
- We evaluated EUF SMT-based, graph-based, and Datalog-based approaches on three static analysis tasks: field-sensitive alias analysis for Java applications and large-scale open-source C programs, and context-sensitive data dependence analysis for Java programs. Experimental results demonstrate that the best-performing SMT solvers, Yices [18], Plat-smt [19] and Egg [42], achieve efficiency comparable to the optimal graph-based algorithm [10]. Notably, some SMT solvers exhibit even better performance in terms of query overhead, highlighting their strong potential for static analysis tasks.

The rest of this paper is organized as follows. Section 2 introduces the necessary background and motivation. Section 3 presents the reduction from bidirected Dyck-reachability to EUF SMT solving. Section 4 provides experimental results. Section 5 provides a discussion. Finally, Sect. 6 concludes the paper.

2 Preliminaries and Motivation

This section provides an overview of EUF SMT theory and bidirected Dyck-reachability. Subsequently, by comparing the solving algorithms for both problems, we present the motivation for this work.

2.1 The Theory of Equality with Uninterpreted Functions

The theory of Equality with Uninterpreted Functions (EUF) is a core component of SMT solvers like Z3 [12] and CVC5 [6], *etc.*. In EUF, uninterpreted functions are defined by their signatures (names and input/output types) while omitting semantic details. This abstraction preserves only the *congruence* property: identical inputs produce identical outputs.

The axioms of EUF logic include the three axioms of equivalence relations and the congruence axiom, which states that functions with the same inputs produce the same outputs.

$$\forall x. \quad x = x \qquad \text{(reflexivity)}$$
$$\forall x, y. \quad x = y \rightarrow y = x \qquad \text{(symmetry)}$$
$$\forall x, y, z. \quad x = y \land y = z \rightarrow x = z \qquad \text{(transitivity)}$$
$$\forall \bar{x}, \bar{y}. \quad \bigwedge_i x_i = y_i \rightarrow f(\bar{x}) = f(\bar{y}) \qquad \text{(congruence)}$$

Abstraction of functions as uninterpreted functions inevitably loses some information. For example, if addition is abstracted as f, the commutative property is not preserved: $f(x, y)$ and $f(y, x)$ are not considered equivalent under EUF congruence, despite $x + y = y + x$ for numbers. However, this paper leverages the congruence property to model parenthesis matching in Dyck language without incurring any precision loss.

We now illustrate the algorithm for solving EUF formulas through an example. Consider the unsatisfiable formula $\varphi : c = f(a) \land d = f(b) \land a = b \land c \neq d$. Its unsatisfiability can be illustrated via the construction of a congruence graph, as shown in Fig. 1. Graph nodes denote terms, with directed edges (labeled by function symbols, *e.g.*, f) representing function applications, and undirected red edges indicating term equivalence. Figure 1a shows the initial graph with all terms and their functional relationships. Subsequently, processing constraint $a = b$ introduces an equivalence edge between a and b (Fig. 1b). By the congruence property (equivalent inputs yield equivalent outputs), terms c and d become equivalent. Consequently, the graph is updated to Fig. 1c. This, however, contradicts the given assertion $c \neq d$, thus proving the unsatisfiability of φ.

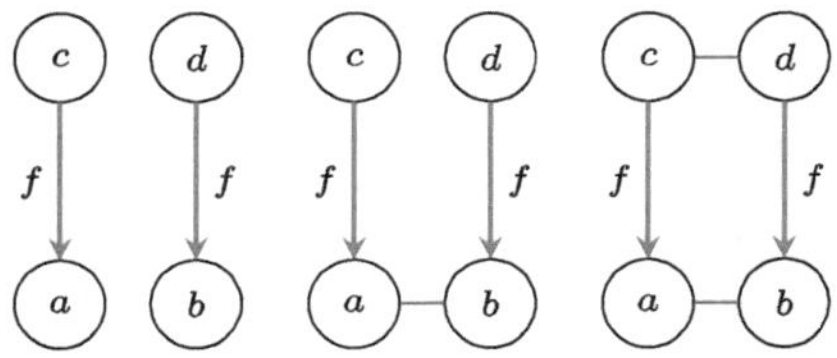

(a) (b) (c)

Fig. 1. Solving EUF SMT Formulas Using Congruence Graphs.

In summary, this algorithm first processes all equivalence relations by iteratively (1) merging equivalent classes and (2) discovering new equivalences via the congruence property, until a fixed point is reached. It then checks each inequality constraint to verify that its operands reside in distinct equivalence classes. The formula is *satisfiable* if no inequality conflicts with any of the computed equivalence classes; otherwise, it is *unsatisfiable*.

2.2 Bidirected Dyck-Reachability and Its Applications in Program Analysis Tasks

The context-free language (CFL) reachability problem on graphs extends the general graph reachability problem. We begin by the following definition:

Definition 1. *(labeled graph) A labeled graph is defined as a tuple $G = (V, E)$, where V is the set of vertices, and $E = V \times \Sigma \times V$ is the set of edges, with Σ being an alphabet.*

For simplicity, we introduce a function $\lambda : V \times V \rightarrow \Sigma$, where $\lambda(u, v) = l$ if and only if $(u, l, v) \in E$. Given a context-free grammar (CFG) $\mathcal{G}$ and a labeled graph G, where $L(\mathcal{G})$ represents the language accepted by the CFG $\mathcal{G}$, we say that two nodes $u, v \in V$ are *CFL-reachable* if and only if the following condition holds:

$$\exists t_1, t_2, \ldots, t_n \in V. \lambda(u, t_1) \cdot \lambda(t_1, t_2) \cdots \lambda(t_n, v) \in L(\mathcal{G}) \tag{1}$$

The Dyck language is a particular type of CFL. Its grammar is as follows, where $\Sigma = \{\epsilon, l_0, r_0, \cdots, l_i, r_i, \cdots\}$ are the terminal symbols, S is the start non-terminal symbol.

$$S \rightarrow l_0 S r_0 \mid \cdots \mid l_i S r_i \mid \cdots \mid SS \mid \epsilon \tag{2}$$

The language defined by this grammar requires balanced matching between terminal symbols, *i.e.*, symbols l_i and r_i appear in pairs, such as in the case of the parentheses matching language. Here, for every i, we refer to l_i as *opening labels* and r_i as *closing labels*, with the notation $\bar{l_i} = r_i$, $\bar{r_i} = l_i$, and $\bar{\epsilon} = \epsilon$.

Definition 2. *(bidirected graph) A bidirectional graph $G = (V, E)$ is a labeled graph where $(u, l, v) \in E$ if and only if $(v, \bar{l}, u) \in E$.*

Consider the Dyck-reachability problem on a general graph G. Relaxing G into a bidirected graph G' by adding reverse edges for each original edge (as Definition 2) yields the bidirected Dyck-reachability problem. This formulation constitutes a relaxation of the general Dyck-reachability problem, which may introduce imprecision. Specifically, if a node pair is Dyck-reachable in graph G, it remains Dyck-reachable in G'. However, the addition of bidirected edges may enable reachability between node pairs that are not connected in the original graph G, thus introducing imprecision. For example, consider the path $a \xrightarrow{l} b \xrightarrow{\bar{l}} c$: a is Dyck-reachable to c, but c is not Dyck-reachable to a in the original graph. However, when relaxed to a bidirected graph (Definition 2), the added edges $a \xleftarrow{\bar{l}} b$ and $b \xleftarrow{l} c$, making c reachable from a.

Theorem 1. *[44] The bidirected Dyck-reachability relation is an equivalence relation.*

This theorem states that the bidirected Dyck-reachability relation is *reflexive* (each node is Dyck-reachable to itself), *symmetric* (if nodes u and v are Dyck-reachable, then so are v and u), and *transitive* (if nodes u and v, and v and k are Dyck-reachable, then so are u and k).

Bidirected Dyck-reachability based field-sensitive alias analysis. Currently, many approaches utilize bidirected Dyck-reachability to model program analysis problems[10,24,26–29,44]. Figure 2 shows an example of alias analysis. Figure 2a shows a Java program, and Fig. 2b illustrates this program's corresponding labeled bidirected graph. In the graph, nodes correspond to program variables. Edges labeled with an *opening labels* represent field write operations, while those with a *closing labels* represent field read operations, with distinct

```
class Tree {
    Tree fst;
    Tree snd; }
void foo(){
    Tree a, b;
    Tree c, d;
    Tree e, f, g;
    ...
    ...
    a.fst = b;
    c.snd = a;
    a.snd = f;
    d = c.snd;
    e = d.fst;
    c = g.fst;
    h = b; }
```

(a) A Java program.

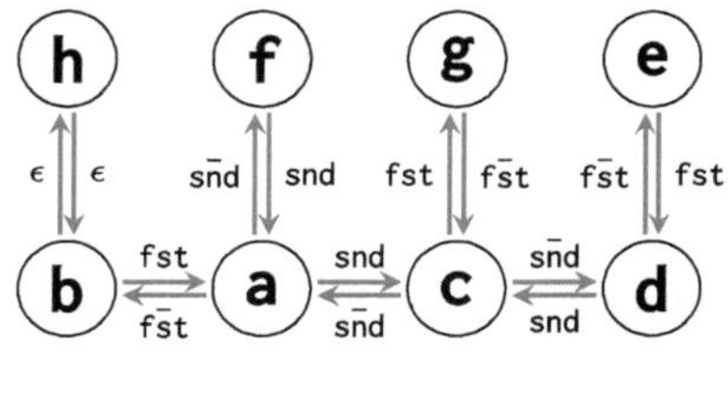

(b) Labeled bidirected graph for alias analysis.

Fig. 2. Bidirected Dyck-reachability based field-sensitive alias analysis.

labels denoting different field accesses. For example, the statement `a. fst=b` is encoded as an edge $\lambda(\mathtt{b}, \mathtt{a}) = \mathtt{fst}$. Consequently, aliases arising from field accesses are represented by paths where opening and closing labels form matched pairs, corresponding to Dyck-reachable paths in the graph. In Fig. 2b, nodes {`a`, `d`} and {`h`, `b`, `e`} are Dyck-reachable, respectively (the equivalence class of each variable with itself is omitted), and the variables in each equivalence class are aliases in the original program.

Relaxing the graph to a bidirected graph introduces some approximation. For example, consider the program statements `a=c` and `a=b`. Under Andersen's pointer analysis [4], `b` and `c` are not considered aliases, but in bidirected Dyck-reachability based approach, `b` and `c` are considered aliases. However, the bidirected Dyck-reachability-based approach is more efficient than Andersen's pointer analysis. To the best of our knowledge, the most efficient algorithm for solving the bidirected Dyck-reachability problem may achieve linear time complexity [10] in many cases, which is significantly more efficient than the cubic lower bound [30] of Andersen's pointer analysis.

In this example, field access operations are modeled via Dyck-reachability. Similar matching properties exist in other program constructs, such as call/return matching in function invocations and reference/dereference matching in C programs. The Dyck-reachability modeling approach thus has broad applications in program analysis problems.

2.3 Motivation

This subsection will first introduce the basic idea of the bidirected Dyck-reachability solving algorithm. Then, we will compare this method with the algorithm for computing the congruence closure of EUF SMT formulas on congruence graphs, motivating our approach.

Figure 3a illustrates a bidirected graph in which the Dyck grammar requires matching parentheses. By observing a Dyck-reachable path between nodes a and c, we deduce that a and c are Dyck-reachable. Figure 3b depicts the subgraph derived from Fig. 3a by retaining only edges labeled with ⦇. In this subgraph, nodes a and c have edges labeled with ⦇ directed toward node b. According to

(a) (b)

Fig. 3. Bidirected graph and its subgraph containing only *opening label* edges.

the definition of a bidirected graph, the existence of such a configuration implies that the complete edge relationship among these three nodes must include the edges shown in Fig. 3a. Moreover, Theorem 1 establishes that the bidirected Dyck-reachability relation is an equivalence relation. Specifically, if nodes a and c are Dyck-reachable, then every node Dyck-reachable from a must also be Dyck-reachable from c, and conversely, every node Dyck-reachable from c must be Dyck-reachable from a. Therefore, nodes a and c can be merged in the graph. Similar to Strongly Connected Component (SCC), a Dyck Strongly Connected Component (DSCC) [10] represents a set of mutually Dyck-reachable nodes, forming a Dyck-reachable equivalence class.

Overall, the state-of-the-art algorithm for solving bidirected Dyck-reachability [10,44] consists of two fundamental steps. The first step identifies configurations where multiple nodes have edges with identical *opening labels* directed toward a common node, as illustrated in Fig. 3b. The second step merges these nodes into the same equivalence class. The process iterates these two steps until no additional node pairs can be merged. As a result, all nodes within a final equivalence class are mutually Dyck-reachable.

By examining this process, we observe an analogy with the congruence closure algorithm used in EUF SMT solving. Specifically, edges labeled with *opening labels* in bidirected graphs correspond to uninterpreted functions in EUF. The rule for merging nodes based on bidirected Dyck-reachability corresponds to the inference of new equivalence classes via the function congruence property in the congruence closure algorithm. Consequently, the resulting DSCCs correspond to the equivalence classes derived in EUF SMT solving.

> **Insight:** *For the bidirected Dyck-reachability problem, if we consider the nodes in the graph as variables and edges with opening label as a function relationship, each edge with an opening label can be encoded as an EUF SMT formula. The equivalence classes obtained in EUF SMT solving correspond to the sets of Dyck-reachable nodes in the bidirected graph.*

3 EUF-based Dyck-Reachability Encoding

In this section, we will present a reduction of the bidirected Dyck-reachability problem to the satisfiability problems of EUF SMT formulas. This demonstrates that all program analysis problems that can be modeled as bidirected Dyck-reachability problem can also be encoded using EUF SMT formulas, allowing them to be solved using SMT solvers.

3.1 Reduction

We present a reduction from the bidirected Dyck-reachability problem to the satisfiability of EUF SMT formulas. We first outline the specifications of both problems, then describe the reduction function that encodes any bidirected Dyck-reachability instance into EUF SMT theory.

Bidirected Dyck-reachability problem.

- **Input:** A Dyck CFG $\mathcal{G}$, a bidirected graph $G = (V, E)$, and a pair of node (u, v) .
- **Output:** u and v are *Dyck-Reachable* or *not*.

Satisfiability of EUF SMT formulas.

- **Input:** An EUF SMT formula.
- **Output:** *SAT* or *UNSAT*

Reduction function R. The reduction function R maps each instance of the Dyck-reachability problem to an EUF SMT formula satisfiability problem. In this mapping, graph nodes correspond to formula variables, each *opening label* in Σ is represented as an uninterpreted function, and edges labeled with *opening labels* or ϵ are encoded as equations. The queried node pairs are represented as inequalities between corresponding variables.

Algorithm 1: *Encoding*$(\mathcal{G}, G)$

Input: A CFG $\mathcal{G}$, a labeled bidirected graph $G(V, E)$
Output: An EUF formula ϕ_{EUF}.
1 $\phi_{\text{EUF}} \leftarrow$ True
2 **foreach** $(s, l, t) \in E$ **do**
3   **if** *l is an opening label in $\mathcal{G}$* **then**
4         $\phi_{\text{EUF}} \leftarrow \phi_{\text{EUF}} \wedge (s = F_l(t))$
5   **if** $l = \epsilon$ **then**
6         $\phi_{\text{EUF}} \leftarrow \phi_{\text{EUF}} \wedge (s = t)$
7 **return** ϕ_{EUF}

As shown in Algorithm 1, the algorithm encodes the input Dyck grammar $\mathcal{G}$ and labeled graph G into an EUF SMT formula. The algorithm initializes ϕ_{EUF} to True (Line 1), then iterates over the edges E of G. For each edge (s, l, t) (Line 2), the encoding depends on the edge label: opening parenthesis l is encoded as $s = F_l(t)$ (Line 4), while ϵ is encoded as $s = t$ (Line 6). After processing all edges, the resulting formula ϕ_{EUF} contains approximately $|E|/2$ conjunctive equality constraints. The reachability query for nodes u and v corresponds to checking the satisfiability of $\phi_{\text{EUF}} \wedge (u \neq v)$, denoted as $R(\mathcal{G}, G, u, v)$. The nodes u and v are Dyck-reachable if and only if $R(\mathcal{G}, G, u, v)$ is *unsatisfiable*, completing the reduction function R.

The reduction algorithm has a time complexity of $O(|E|)$. *However, in practical applications such as field-sensitive alias analysis, a program can be encoded directly as an EUF formula without first being converted to an explicit graph representation.* Furthermore, the EUF SMT-based approach to bidirected Dyck-reachability does not require explicit enumeration of all $|V|^2$ node pairs to determine reachability. SMT solvers utilizing the congruence closure algorithm [16,32] first compute an equivalence closure from ϕ_{EUF}, which *exactly matches* the closure derived by graph-based bidirected Dyck-reachability algorithms. Reachability queries are then handled via *incremental solving*: the base equivalence closure is computed once by solving ϕ_{EUF}, after which constraints like $(u \neq v)$ are pushed to check satisfiability for individual queries and subsequently popped to reset the solver state.

Theorem 2. *Let $\mathcal{G}$ be a Dyck grammar and $G = (V, E)$ a bidirected graph. For any nodes $u, v \in V$, u and v are bidirected Dyck-reachable if and only if the EUF formula $R(\mathcal{G}, G, u, v)$ is unsatisfiable.*

Proof. **Direction1 : When (u, v) are Dyck-reachable in graph G, then $R(\mathcal{G}, G, u, v)$ is unsatisfiable.**
When (u, v) are Dyck-reachable in G, *i.e.*, $\exists t_1, t_2, \ldots, t_{n-1} \in V.\lambda(u, t_1) \cdot \lambda(t_1, t_2) \cdots \lambda(t_{n-1}, v) \in L(\mathcal{G})$, we let p represent this reachable path of length n and *Encoding*$(\mathcal{G}, p)$ denote the formula generated by encoding p. We proceed by induction on the path length n. **Direction 1** holds when the nodes u and v are the same, *i.e.*, when $n = 0$. When $n = 1$, we have $\lambda(u, v) = \epsilon$, and encoding this edge yields $u = v$. This formula establishes the equivalence of u and v, indicating that $R(\mathcal{G}, G, u, v)$ is unsatisfiable.

Assume that **Direction 1** holds for $n < k$. For $n = k$, since p of length n is a reachable path, the labels on p concatenate to form a well-parenthesized string. According to the syntax of the Dyck language (Rule 2), there are two ways to construct the path p:

1. There exists a well-parenthesized path p' of length $n-2$, such that $p = l \cdot p' \cdot \bar{l}$;
2. There exist two well-parenthesized subpaths p_1 and p_2, both satisfying $|p_i| < n$ and $p = p_1 \cdot p_2$.

By the induction hypothesis, *Encoding*$(\mathcal{G}, p')$, *Encoding*$(\mathcal{G}, p_1)$, and *Encoding*$(\mathcal{G}, p_2)$ can all be shown to establish the equivalence of their end-point nodes. Furthermore, by the transitivity (Axiom transitivity) and congruence (Axiom congruence) axioms, it follows that *Encoding*$(\mathcal{G}, p)$ forms a proof sequence for $u = v$, thus $R(\mathcal{G}, G, u, v)$ is unsatisfiable.

Direction2 : If $R(\mathcal{G}, G, u, v)$ is unsatisfiable, then (u, v) are Dyck-reachable in graph G.
When $R(a, b)$ is unsatisfiable, there exists a proof sequence for $u = v$. Based on this proof sequence, we can derive a Dyck-reachable path in the graph G from u to v. The proof process is similar to that of **Direction 1**, and the complete proof can be found in the full version of this paper [17].

3.2 Example

In Sect. 2.2, field-sensitive alias analysis was modeled via Dyck-reachability. In the previous subsection, we present a reduction from bidirected Dyck-reachability problems to EUF SMT solving, demonstrating that EUF-based methods can solve such problems. As an illustration, we now detail the corresponding EUF encoding approaches using the programs in Fig. 2.

The results from encoding the Java program in Fig. 2a using EUF SMT formulas are as follows:

$$\mathtt{h} = \mathtt{b} \wedge \mathtt{b} = F_1(\mathtt{a}) \wedge \mathtt{f} = F_2(\mathtt{a}) \wedge \mathtt{a} = F_2(\mathtt{c}) \wedge \mathtt{c} = F_1(\mathtt{g}) \wedge \mathtt{d} = F_2(\mathtt{c}) \wedge \mathtt{e} = F_1(\mathtt{d})$$

In this case, each program variable maps to a formula variable, and each field is modeled as an uninterpreted function to represent field accesses. Program statements are encoded as follows: assignments like `h=b` as equalities; store operations like `a. fst=b` as $\mathtt{b} = F_1(\mathtt{a})$; and load operations like `d=c. snd` as $\mathtt{d} = F_2(\mathtt{c})$. Solving the formula yields equivalence classes, *e.g.*, $\{\mathtt{a,d}\}$ and $\{\mathtt{h,b,e}\}$ (excluding trivial self-equivalences), matching the Dyck-reachability results in Fig. 2b.

4 Experiment and Evaluation

Solving bidirected Dyck-reachability with EUF SMT solvers only requires designing an appropriate encoding, leveraging their mature implementations and avoiding the need for specialized algorithms. This section evaluates the efficiency of our approach by comparing SMT-based, graph-based, and Datalog-based methods on static analysis tasks. A typical static analysis consists of two phases: program analysis, which computes the *complete analysis results* (*e.g.*, all alias relations), and query resolution, which *answers specific queries* (*e.g.*, determining whether two given variables are aliases). We evaluate all three approaches from both perspectives, addressing the following research questions:

- **RQ1**: Efficiency of analysis, *i.e.*, given a program and a static analysis task, how efficient is the SMT solver-based approach compared to the other two approaches in completing the analysis?
- **RQ2**: Efficiency of querying, *i.e.*, after completing analysis, how efficient is the SMT solver-based approach compared to the other two approaches in answering specific queries?

4.1 Experimental Setup

Baseline. We use graph-based and Datalog-based approaches as baselines. The graph-based baselines include: `FD` by Zhang *et al.* [44] with complexity $O(n+m \cdot \log m)$, and `Optimal` by Chatterjee *et al.* [10] with worst-case complexity $O(m + n \cdot \alpha(n))$. We use the open-source implementation from Krishna *et* *al.* [26], which focuses on dynamic bidirected Dyck-reachability and implements both algorithms. For fairness, we compiled the code with -O3 optimization, whereas the original compilation commands in [26] used no optimization flags.

Datalog tools are also widely used in static analysis, computing relational closures by deriving new *facts* from initial *facts* according to analysis *rules*. For our experiments, we selected two well-known Datalog tools as baselines: Soufflé [15] and Bddbddb [41].

SMT solvers. SMT solvers are widely used in program analysis and verification, with EUF theory playing a central role as it bridges different theories in combined theory solving [31]. To solve EUF formulas, SMT solvers typically employ congruence closure algorithms [13,16,32,33] (core procedures are outlined in Sect. 2.1, though implementations vary). The following SMT solvers were used in our experiments for EUF formula solving:

- **Z3** [12]**:** A high-performance C++ SMT solver from Microsoft Research, is widely adopted in academia and industry. Given its significant impact, we selected Z3 for EUF SMT solving in our experiments.
- **CVC5** [6]**:** A well-known SMT solver developed in C++ by Stanford University and the University of Iowa, it demonstrated outstanding performance in the QF_UF category at SMT-COMP 2024 [38].
- **Yices** [18]**:** Yices, developed in C by SRI International, is an efficient SMT solver that excelled in the SMT-COMP 2024 competition [38], winning both the Single Query Track and the Incremental Track in the QF_UF category.
- **Plat-smt** [19]**:** Plat-smt, a Rust-based QF_UF SMT solver, employs Egg's [42] implementation for congruence closure solving and demonstrated strong performance in that category at SMT-COMP 2024 [38].
- **Egg** [42]**:** Egg is a Rust-based tool for equality saturation that uses e-graphs as its core data structure to support congruence closure for deriving equivalence classes. It innovatively introduces a *rebuilding* operation that maximizes equivalence class merging while delaying the upward search for congruent terms, significantly improving computational efficiency.

Benchmarks. The application of Dyck-reachability in static analysis is well-established [10,24,26–29,44], including the baseline tools used in our experiments. Our initial benchmarks are drawn from these prior works. For further evaluation, we added 10 widely-used open-source C programs. The transformation from program to graph/EUF formulas incurs minimal overhead, as it requires only a single pass over the intermediate representation. Benchmark details are provided below.

Table 1. Field-sensitive Alias Analysis for Java Programs

benchmark	#Locs(K)	#Nodes	#Edges	#Labels
antlr	170	23031	21353	1246
bloat	206	26656	23598	1360
chart	448	51356	44501	3132
eclipse	181	24004	21943	1346
fop	431	46253	39125	2857
hsqldb	156	21646	20271	1160
jython	216	28033	24889	1398
luindex	164	22631	20915	1228
lusearch	175	23344	21569	1275
pmd	193	24586	22522	1322
xalan	155	21574	20186	1152

Field-sensitive alias analysis for Java. Existing work on bidirected Dyck-reachability [10,26,27,44] has selected the DaCapo 2006 [8] benchmark for field-sensitive alias analysis tasks, which have been modeled as bidirected Dyck-reachability problems [44]. Table 1 shows the statistical information for the graphs, including the lines of code (#Locs in thousands), the number of nodes (#Nodes), edges (#Edges), and variable fields (#Labels) in each graph.

Table 2. Field-sensitive Alias Analysis for C Programs

benchmark	#Locs(K)	#Nodes	#Edges	#Labels
git	188	525717	630839	167
libssh2.a	27	73667	81278	346
tmux	49	154992	197807	164
vim	255	489719	612678	512
lighttpd	62	50892	60381	102
sqlite3	268	323952	405705	362
strace	94	140191	164456	92
wrk	601	369384	439726	253
darknet	28	94266	110077	352
libxml2.a	72	159308	196230	199

Field-sensitive alias analysis for C Programs. To further evaluate our approach, we selected 10 real-world C programs, including several large-scale ones, as shown in Table 2. We used lotus's [14] encoding scheme to convert these C programs into graph representations. In this encoding, labels capture not only field accesses but also pointer dereference and reference operations.

Context-sensitive data dependence analysis. Data dependency analysis focuses on identifying Def-Use chains in programs, with applications in various static analysis tasks. Context sensitivity requires distinguishing between different function calls, where a necessary condition for a valid call is the matching of call and return. Existing work models this requirement using bidirected Dyck-reachability [10,26,27]. It builds upon an approach from [40] that represents SPECjvm2008 benchmark programs [1] as bidirected graphs. In this paper, we adopt the same bidirected graph representation following [10]. Table 3 summarizes the statistical information of the graphs derived from these programs, where (#Labels) denotes the number of function calls in each program.

Table 3. Context-sensitive Data Dependence Analysis for Java Programs

benchmark	#Locs(K)	#Nodes	#Edges	#Labels
btree	5	1811	1757	801
check	10	5240	5267	2167
compiler	9	4189	4101	1646
compress	10	4375	4238	1721
crypto	21	6202	6300	2540
derby	21	6116	5948	2358
helloworld	9	4074	3969	1596
mpegaudio	40	9650	9931	3564
mushroom	2	899	809	376
parser	5	1686	1561	690
sample	2	931	834	389
scimark	10	4583	4429	1782
startup	11	5493	5367	2165
sunflow	9	3891	3792	1520
xml	47	23922	24391	9128

Evaluation method. This paper experimentally investigates the efficiency of three approaches: SMT solver-based, graph-based, and Datalog-based approaches. The input for the graph-based method is a graph derived from the compilation of the target program, and the graph models we evaluate are from [27] and [26]. To ensure a unified testing standard, the inputs for both the SMT solver-based approach and the Datalog tools are derived from this graph model.

For the SMT solver-based method, the input is a standard SMT-LIB [7] format file. We convert the existing graph file into an SMT formula using Algorithm 1, with the logic set to QF_UF. The querying functionality is implemented via incremental solving. For instance, to check the equivalence of two variables (u, v)—*i.e.*, whether nodes u and v are Dyck-reachable—the commands (*push* 1), (*assert* (*not* $(= u\ v)$)), (*check-sat*), and (*pop* 1) are appended to the SMT-LIB file. An UNSAT result indicates that u and v are equivalent.

The experiments were conducted on a desktop running Ubuntu 22.04, equipped with an i7-9700 3.00GHz CPU and 16GB of RAM. Each experiment was performed three times, and the reported results are the averages.

4.2 Efficiency of Analysis

This section evaluates SMT-based, graph-based, and Datalog-based methods for computing all equivalence classes, using the programs in Tables 1–3. Results in Tables 4–6 distinguish parsing and solving time (denoted as parsing + solving) for graph-based and SMT tools, with gray values in parentheses showing 10,000-query times. Plat-smt solves each formula immediately upon parsing, interleaving the two phases; the total time is therefore reported. The red-highlighted numbers in the figures indicate the total time (parsing + solving) and query time for the best-performing tool in each category. Since Datalog tools such as Soufflé and Bddbddb explicitly store all relation pairs—requiring $O(n^2)$ storage for n elements in an equivalence class—their analysis and query efficiency is more than two orders of magnitude slower. We therefore omit their results in this section; detailed experimental results can be found in the full version of this paper [17].

Table 4. Results of field-sensitive alias analysis for Java programs, showing total analysis time (Parsing+Solving, ms) and total time for 10k queries (ms, in parentheses). Best results are highlighted in red. TO = Time Out (10 minutes).

benchmark	Graph based(ms)		SMT based(ms)				
	#FD	#Optimal	#Z3	#CVC5	#Yices	#Plat-smt	#Egg
antlr	29.3+10.4(6.0)	16.0+4.8 (6.1)	67.0+431.4 (380.1)	175.8+209.2 (499.3)	25.0+8.6 (28.3)	26.41 (14.8)	20.5+10.9 (4.0)
bloat	32.9+11.2(6.1)	17.7+6.1 (5.9)	75.1+558.7 (419.0)	154.9+245.7 (506.9)	27.5+9.5 (30.4)	29.82 (14.8)	24.4+13.1 (4.2)
chart	65.4+25.2(6.7)	54.9+14.2(6.2)	152.4+294.5 (679.1)	310.1+511.1 (581.8)	53.8+17.9 (30.0)	64.31 (15.0)	50.8+28.8 (4.2)
eclipse	30.7+10.4(6.1)	28.7+5.8 (6.2)	71.5+462.6 (384.6)	143.1+219.9 (505.1)	26.3+9.0 (27.7)	34.77 (14.7)	24.7+11.6 (4.0)
fop	59.4+21.6(6.6)	46.7+11.7(6.1)	138.4+1403.9 (617.7)	280.6+437.9 (539.3)	46.9+16.8 (29.5)	56.37 (15.0)	46.4+25.5 (4.0)
hsqldb	27.7+9.7 (6.0)	15.3+4.4 (6.4)	64.6+408.4 (364.5)	130.0+194.2 (484.5)	27.2+9.2 (28.7)	29.57 (15.7)	19.4+11.1 (4.2)
jython	34.7+12.4(6.1)	18.3+7.0 (6.0)	76.3+642.4 (425.9)	162.7+260.7 (512.1)	28.3+10.1 (28.8)	32.92 (14.6)	26.1+14.7 (4.2)
luindex	37.5+10.0(6.0)	14.9+5.3 (6.1)	66.5+102.4 (372.4)	136.9+205.4 (494.5)	22.6+8.5 (29.1)	28.54 (14.8)	21.4+11.6 (4.1)
lusearch	32.8+10.1(6.1)	15.8+5.4 (6.3)	69.0+107.3 (375.5)	138.3+213.9 (495.5)	26.0+8.5 (28.4)	29.91 (14.7)	21.6+11.7 (4.0)
pmd	38.9+11.2(6.0)	16.8+5.6 (5.9)	69.6+471.8 (403.7)	144.9+225.1 (508.1)	25.1+9.3 (29.2)	30.2 (15.0)	22.4+12.3 (4.3)
xalan	28.4+10.1(6.0)	21.1+4.8 (6.4)	63.7+362.5 (364.9)	129.7+193.0 (483.4)	22.6+8.3 (27.4)	26.21 (14.9)	19.3+11.1 (4.2)
Total Time	417.7+142.3 (67.7)	266.2+75.0(67.6)	914.2+5246.1(4787.4)	1906.9+2916.1 (5610.5)	331.2+115.6 (317.5)	389.0 (164)	297.0+162.3 (45.4)

The data in Tables 4–6 indicate that `Optimal` achieves the best performance in both total and solving time across nearly all benchmarks. The sole exception is in Table 5, where Plat-smt slightly outperforms `Optimal` in total time (3.29 s vs. 3.34 s). Among the SMT solvers, Yices, Plat-smt, and Egg demonstrate performance comparable to `Optimal` across the three benchmarks. Specifically, Plat-smt achieves 88% of `Optimal`'s efficiency in Table 4 (389 ms vs. 341.2 ms), while Yices reaches 65% of `Optimal`'s solving efficiency in the same table (115.6 ms vs. 75 ms) and 90% in Table 5 (1.67 s vs. 1.51 s). Compared to the second-best graph-based algorithm `FD`, Yices and Plat-smt outperform `FD` in both solving and total time in Tables 4 and 5, although they perform slightly worse than

Table 5. Results of field-sensitive alias analysis for C programs, showing total analysis time (Parsing+Solving, s) and total time for 10k queries (s, in parentheses). Best results are highlighted in red. TO = Time Out (10 minutes).

benchmark	Graph-based(s)		SMT-based(s)				
	#FD	#Optimal	#Z3	#CVC5	#Yices	#Plat-smt	#Egg
git	0.77+1.11 (0.007)	0.4+0.32 (0.007)	2.32+85.69 (1.422)	3.85+11.78 (29.286)	0.83+0.43 (0.033)	0.74 (0.016)	0.6+0.5 (0.004)
libssh2.a	0.1+0.12 (0.007)	0.06+0.03 (0.007)	0.26+7.13 (1.203)	0.49+1.15 (0.812)	0.08+0.04 (0.029)	0.09 (0.015)	0.09+0.06 (0.004)
tmux	0.24+0.3 (0.008)	0.12+0.1 (0.008)	0.61+17.51 (2.537)	1.14+3.69 (9.615)	0.21+0.1 (0.033)	0.21 (0.017)	0.23+0.14 (0.004)
vim	0.74+1.16 (0.009)	0.39+0.38 (0.009)	1.76+38.26 (7.219)	3.59+10.93 (32.387)	0.75+0.38 (0.033)	0.72 (0.018)	0.56+0.47 (0.005)
lighttpd	0.07+0.09 (0.007)	0.04+0.03 (0.006)	0.16+2.68 (0.873)	0.34+1.01 (2.204)	0.06+0.03 (0.030)	0.06 (0.015)	0.06+0.04 (0.004)
sqlite3	0.49+0.73 (0.009)	0.26+0.22 (0.008)	1.31+63.69 (4.923)	2.37+7.57 (45.547)	0.48+0.24 (0.038)	0.47 (0.017)	0.48+0.32 (0.004)
strace	0.21+0.26 (0.008)	0.1+0.08 (0.007)	0.55+0.65 (2.13)	0.95+2.86 (2.272)	0.17+0.07 (0.042)	0.17 (0.017)	0.13+0.11 (0.004)
wrk	0.54+0.75 (0.009)	0.29+0.22 (0.008)	1.4+56.88 (5.944)	2.62+8.28 (17.034)	0.54+0.25 (0.035)	0.52 (0.018)	0.5+0.32 (0.004)
darknet	0.14+0.17 (0.007)	0.07+0.05 (0.007)	0.31+3.23 (1.422)	0.6+1.77 (1.346)	0.11+0.04 (0.033)	0.12 (0.016)	0.1+0.06 (0.004)
libxml2.a	0.23+0.31 (0.008)	0.12+0.1 (0.008)	0.61+16.71 (2.467)	1.12+3.52 (8.747)	0.22+0.1 (0.034)	0.2 (0.016)	0.21+0.14 (0.004)
Total Time	3.52+5.0 (0.079)	1.83+1.51 (0.075)	9.29+292.4 (30.14)	17.08+52.57 (149.25)	3.45+1.67 (0.34)	3.29 (0.165)	2.96+2.16 (0.04)

Table 6. Results of context-sensitive data dependence analysis for Java programs, showing total analysis time (Parsing+Solving, ms) and total time for 10k queries (ms, in parentheses). Best results are highlighted in red. TO = Time Out (10 minutes).

benchmark	Graph based(ms)		SMT based(ms)				
	#FD	#Optimal	#Z3	#CVC5	#Yices	#Plat-smt	#Egg
btree	2.9+0.2 (8.6)	2.6+0.3 (5.0)	8.2+7.5 (156.4)	13.2+19.4 (512.3)	2.3+1.2 (29.4)	6.02 (17.4)	1.9+1.3 (4.3)
check	6.8+0.6 (5.9)	6.3+0.7 (5.4)	25.9+25.1 (204.7)	60.8+63.1 (575.0)	7.1+3.2 (33.5)	12.52 (14.2)	7.2+3.4 (4.4)
compiler	5.5+0.5 (6.2)	7.9+0.8 (5.3)	24.1+22.5 (187.6)	29.0+46.9 (533.6)	5.2+2.6 (30.7)	13.73 (14.2)	17.2+4.0 (4.2)
compress	5.9+0.5 (5.3)	7.9+0.8 (5.3)	13.7+19.8 (186.5)	31.6+49.5 (523.3)	5.4+2.6 (30.0)	13.82 (14.3)	7.7+2.9 (4.6)
crypto	9.1+1.0 (5.7)	8.4+1.0 (5.0)	18.2+35.7 (218.3)	45.7+76.7 (591.2)	8.0+4.2 (31.1)	16.38 (14.5)	9.1+3.6 (3.9)
derby	9.0+0.8 (5.4)	7.3+0.9 (5.1)	23.7+27.2 (214.2)	47.8+70.1 (540.6)	7.8+3.9 (31.0)	8.88 (14.7)	15.4+4.2 (4.3)
helloworld	5.7+0.4 (6.4)	4.7+0.6 (5.0)	13.3+20.1 (187.1)	31.6+44.4 (524.7)	5.1+2.5 (30.8)	5.17 (14.4)	5.4+2.3 (4.2)
mpegaudio	12.6+1.3 (6.0)	10.6+1.3 (5.1)	34.3+47.6 (259.8)	69.9+121.5 (581.7)	12.0+5.9 (30.1)	12.46 (14.5)	12.6+5.4 (4.3)
mushroom	2.3+0.2 (4.7)	0.8+0.1 (4.8)	6.3+3.6 (143.7)	12.2+10.7 (487.4)	3.4+1.3 (30.2)	0.97 (14.5)	2.5+1.7 (4.6)
parser	2.8+0.2 (5.1)	1.5+0.2 (4.9)	8.0+6.4 (156.1)	28.3+18.3 (493.3)	7.6+3.4 (34.7)	1.85 (15.4)	5.1+2.4 (6.1)
sample	1.2+0.1 (4.9)	0.8+0.1 (4.8)	6.5+3.5 (150.1)	14.6+11.8 (488.4)	3.4+1.6 (31.3)	1.02 (14.3)	1.8+1.3 (5.0)
scimark	6.2+0.5 (7.2)	4.1+0.5 (5.1)	15.5+22.0 (190.6)	53.4+52.1 (528.9)	17.4+5.9 (31.5)	5.73 (14.3)	10.8+3.2 (4.4)
startup	8.0+0.7 (6.0)	5.2+0.7 (5.2)	35.4+27.3 (208.6)	38.4+67.0 (572.8)	14.5+4.9 (30.4)	6.84 (14.4)	13.9+4.5 (4.1)
sunflow	5.5+0.5 (7.9)	3.5+0.4 (5.0)	12.7+18.2 (185.7)	30.9+42.4 (526.0)	6.4+2.9 (30.5)	5.56 (14.8)	5.4+2.1 (5.3)
xml	37.7+7.1 (6.1)	23.6+5.3 (5.7)	83.6+90.4 (452.4)	184.9+347.1 (580.3)	46.9+13.0 (30.9)	39.1 (14.7)	32.0+16.1 (4.3)
Total Time	121.04+14.43 (91.4)	95.15+13.8(76.7)	329.33+376.9 (3101.8)	692.26+1040.98 (8059.5)	152.31+59.07 (466.1)	150.03 (220.6)	148.22+58.42 (68)

FD in Table 6 yet remain within the same order of magnitude. Notably, Egg is more than twice as fast as FD in Table 5, although it performs slightly worse on other benchmarks. In contrast, Z3 and CVC5 exhibit relatively poor performance across all benchmarks. We now analyze the factors contributing to these performance differences.

Chatterjee *et al.* [10] established that Optimal solves the bidirected Dyck-reachability problem with complexity $O(m + n \cdot \alpha(n))$, proving its optimality. In contrast, known congruence closure algorithms for EUF have complexity $O(n \cdot \log n)$ [16,32,33]. However, since encoding bidirected Dyck-reachability into EUF formulas produces formulas that contain only unary functions with a constant number of function symbols, congruence closure algorithms can achieve lower complexity under this constraint. In Sect. 5.2, we provide a brief comparison between the congruence closure implementation in Egg [42] and the

`Optimal` algorithm. We show that Egg's implementation also attains a complexity of $O(m + n \cdot \alpha(n))$ under the same restriction, and we highlight the notable similarities in their algorithmic designs.

Algorithmic design choices significantly impact performance. First, regarding implementation languages, experiments on union-find sequences show that C and C++ implementations are roughly twice as fast as their Rust counterpart. This directly explains the two-fold performance gap between Egg and `Optimal`, despite their shared asymptotic complexity. Detailed results and analysis are provided in Sect. 5.1. Moreover, various tools employ distinct optimization strategies. For instance, Egg uses a *rebuilding* operation that performs unions first, followed by a single-level upward search for congruent terms, thereby lowering amortized complexity. Yices, on the other hand, flattens EUF formulas into terms of depth at most two [33]. This allows immediate equivalence detection after merging variables: once x and y are merged via $x = y$, terms such as $f(x)$ and $f(y)$ are directly recognized as equal, enabling m and n (from $m = f(x)$ and $n = f(y)$) to be merged without explicit congruence-finding steps.

> **Answer to RQ1:** *On the tasks of field-sensitive alias analysis and context-sensitive data dependence analysis, SMT solvers such as Yices, Plat-smt, and Egg achieve performance comparable to the best graph-based algorithm,* `Optimal`*. These results demonstrate the high solving efficiency of the EUF SMT-based approach, confirming its suitability for practical static analysis.*

4.3 Efficiency of Querying

In this section, we compare the efficiency of SMT-based and graph-based methods for answering a specific type of queries (*i.e.*, the equivalence of two variables). For each program, we generated 10 query sequences with lengths ranging from 1,000 to 10,000 in increments of 1,000, where approximately 50% of the variable pairs are Dyck-reachable. In Tables 4–6, the gray numbers in parentheses report the time per 10,000 queries. Figure 4 plots the trend of query time as query sequence length increases across the three benchmarks; each point shows a tool's average query time for a given length across all programs in that benchmark. Complete results can be found in the full version of this paper [17].

We first compared the query results of the SMT solver-based methods with graph-based methods, and across tens of thousands of queries, both produced identical results, providing experimental support for Theorem 2. The results in Tables 4–6 show that Egg achieves the best performance, with speedups of up to 1.88 × (0.04 s vs. 0.075 s) compared to graph-based methods in Table 5. As shown in Fig. 4, the query time of every tool grows linearly with the query length. While Plat-smt and Yices perform slightly worse than graph-based tools, their average per-query overhead remains within the same order of magnitude. In contrast, Z3 and CVC5 are noticeably slower; nonetheless, their average per-query time remains within an acceptable range.

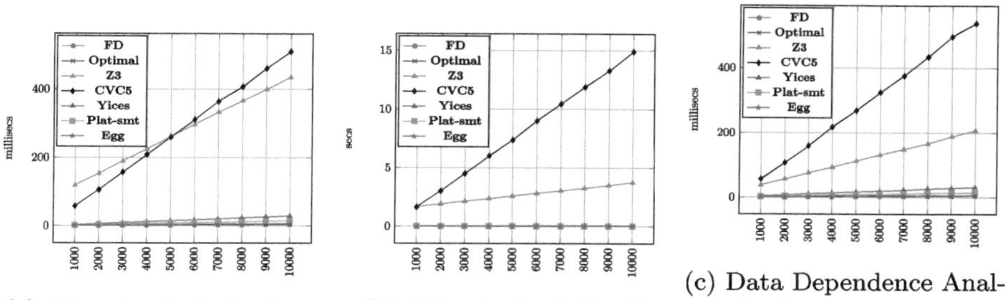

(a) Alias Analysis for Java (b) Alias Analysis for C (c) Data Dependence Analysis for Java

Fig. 4. Average query time for different analysis types

> **Answer to RQ2:** *In terms of query efficiency, Egg outperforms graph-based methods. Although performance varies among solvers, the average per-query overhead remains within acceptable limits. This demonstrates the high efficiency of the EUF SMT-based approach in query processing, making it directly applicable to practical static analysis tasks.*

5 Discussion

Since the union-find operation is fundamental to congruence closure, this section presents the performance of various tools on union-find sequences, along with a concise comparison between the algorithms of the `Optimal` and Egg tools.

5.1 Experimental Results on Union-Find Sequences

This experiment evaluates the performance of various tools on Union-Find sequences, highlighting differences in implementation approaches and programming languages. For a set of 50,000 variables, we randomly generated equality equations ranging from 5,000 to 25,000 in increments of 5,000. These equations involve only variable equalities (representable as ϵ-labeled edges in graph-based tools) and contain no function applications. All results, averaged over three runs and reported in milliseconds, are presented in Table 7.

We first examine the effect of implementation language. Each tool our C++-based `unionfind`, Yices (in C), and Egg (in Rust)implements the classic Union-Find algorithm with path compression and union by rank [11]. The experiments reveal that the C and C++ implementations perform with comparable efficiency, and both outperform the Rust version by more than a factor of two.

Regarding algorithmic implementations, graph-based tools model Union-Find equations as graph edges, use reachability algorithms to identify equivalence classes, and then merge them via standard Union-Find structures, which hinders the overall efficiency. SMT solvers typically employ linked-list–based

Table 7. Results on Union-Find sequences of varying lengths, showing total analysis time (Parsing+Solving, ms). Best results are highlighted in red.

Length	#FD(C++)	#Optimal(C++)	#unionfind(C++)	#Z3(C++)	#CVC5(C++)	#Yices(C)	#Plat-smt(Rust)	#Egg(Rust)
5000	15.16+8.11	6.13+2.95	2.12+1.81	22.53+10.11	33.66+50.97	6.02+1.48	7.05	1.47+3.23
10000	20.65+15.13	11.07+7.56	4.23+3.33	48.0+21.24	64.53+123.3	12.45+2.88	13.64	3.49+7.37
15000	31.02+23.93	16.8+11.65	6.72+5.02	57.48+33.49	120.19+198.31	46.99+5.13	18.02	5.4+10.74
20000	46.79+33.15	20.86+17.59	8.64+5.95	70.39+49.9	126.58+285.8	22.26+6.01	23.88	7.68+13.07
25000	49.71+43.27	25.87+22.2	10.79+7.03	80.58+76.83	151.88+375.4	29.47+6.57	31.76	10.91+16.2
Total Time	163.34+123.59	80.73+61.96	32.49+23.13	278.98+191.57	496.84+1033.78	117.20+22.06	94.37	28.95+50.61

Union-Find algorithms to support backtracking, forgoing path compression and incurring $O(n \log n)$ complexity—leading to degraded performance. In contrast, Yices adopts a hybrid strategy: it applies the optimized Union-Find algorithm for processing pure equalities and switches to a linked-list representation for congruence-derived equivalence classes to enable backtracking, thereby achieving the highest efficiency.

5.2 Comparison of `Optimal` and Egg

This section provides a concise comparison between the congruence-closure algorithms in `Optimal` and Egg. The core procedures of the bidirected Dyck-reachability and congruence-closure algorithms are essentially identical, as both consist of two main steps noted in Sect. 2.3: (1) merging reachable/equivalent nodes, *e.g.*, via $union(a, b)$; and (2) identifying newly reachable/equivalent nodes by examining the predecessor/parent sets of the merged nodes. The algorithm repeats these steps until a fixed point is reached, thereby computing the congruence closure.

The basic operations include $union(a, b)$, merging the predecessor/parent sets of $find(a)$ and $find(b)$, and searching for congruent terms in the merged set. Both `Optimal` and Egg store predecessor/parent sets using maps, where elements are representatives of equivalence classes. Because the number of edge-label/function types is constant, these sets also have constant size—note, however, that if functions with arity greater than one are allowed, the size of parent sets can no longer be bounded by a constant. Consequently, both merging the parent sets and searching for congruent terms require only $O(1)$ time.

`Optimal` uses a worklist that stores pending unions, processes them each iteration, and adds newly discovered equivalent nodes to the worklist for the next round. This corresponds to Egg's *rebuilding* operation, which accumulates multiple unions and then performs a batch upward search for congruent terms via the `rebuild` function only one level upwarditerating to completion. Rebuilding thereby avoids many redundant upward merges, significantly accelerating congruence-closure computation.

In both algorithms, nodes in the worklist to be processed are preprocessed by applying *find* to obtain their equivalence-class representatives, eliminating duplicates and redundant operations. Chatterjee *et al.* [10] proved (Lemma 3.4) that `Optimal` executes $O(m)$ union-find operations. A similar argument shows

Egg also uses $O(m)$ union-find operations. Moreover, because the number of edge-label/function types is constant, a sparse graph with $m = O(n)$ can be constructed in $O(m)$ preprocessing time [10]—a step required by Egg as well. Hence, both algorithms achieve a worst-case time complexity of $O(m+n \cdot \alpha(n))$.

Although designed for different problem domains, the two algorithms exhibit a striking consistency in their underlying principles. This consistency is formally revealed by our proposed encoding of bidirected Dyck-reachability into EUF formulas, as introduced in this paper.

6 Conclusion and Future Work

This paper presents a novel static analysis approach that leverages SAT/SMT solvers. We demonstrate that the EUF theory can encode all problems solvable by bidirected Dyck-reachability, thereby offering a unified framework for various static analysis tasks. Empirically, our SMT-based method achieves efficiency comparable to state-of-the-art graph-based algorithms, with some solvers exhibiting superior query performance. The combined ease of modeling program semantics in EUF and the efficiency of modern solvers underscore the practical viability of this SMT-based paradigm. Further exploration of this paradigm could be a promising direction for future research.

In the future, we plan to further explore the application of SAT/SMT solvers to static analysis. This includes investigating a broader range of problems solvable by EUF SMT theory, examining connections between other SMT theories and static analysis problems, and exploring the potential of SMT solvers as general-purpose static analysis tools.

Data Availability. Our artifact is available at the following URL: https://zenodo.org/records/18630949

Acknowledgments. This research was supported by National Key R&D Program of China (No. 2022YFB4501903) and the NSFC Program (No. 62172429 and 62032024).

References

1. SPECjvm2008 Benchmark Suit (2008). http://www.spec.org/jvm2008/
2. Ackermann, W.: Zum hilbertschen aufbau der reellen zahlen. Math. Ann. **99**(1), 118–133 (1928)
3. Allen, F.E., Cocke, J.: A program data flow analysis procedure. Commun. ACM **19**(3), 137 (1976)
4. Andersen, L.O.: Program analysis and specialization for the c programming language (1994)
5. Baldoni, R., Coppa, E., D'elia, D.C., Demetrescu, C., Finocchi, I.: A survey of symbolic execution techniques. ACM Comput. Surv. (CSUR) **51**(3), 1–39 (2018)
6. Barbosa, H., et al.: cvc5: a versatile and industrial-strength SMT solver. In: International Conference on Tools and Algorithms for the Construction and Analysis of Systems, pp. 415–442. Springer (2022)

7. Barrett, C., et al.: The SMT-lib standard: version 2.0. In: Proceedings of the 8th International Workshop on Satisfiability Modulo Theories (Edinburgh, UK), Vol. 13, p. 14 (2010)
8. Blackburn, S.M., et al.: The DaCapo benchmarks: java benchmarking development and analysis. In: Proceedings of the 21st Annual ACM SIGPLAN Conference on Object-oriented Programming Systems, Languages, And Application, pp. 169–190 (2006)
9. Böhme, M., Pham, V.T., Nguyen, M.D., Roychoudhury, A.: Directed greybox fuzzing. In: Proceedings of the 2017 ACM SIGSAC Conference on Computer and Communications Security, pp. 2329–2344 (2017)
10. Chatterjee, K., Choudhary, B., Pavlogiannis, A.: Optimal Dyck reachability for data-dependence and alias analysis. Proc. ACM Program. Lang. **2**(POPL), 1–30 (2017)
11. Cormen, T.H., Leiserson, C.E., Rivest, R.L., Stein, C.: Introduction to algorithms. MIT press (2022)
12. De Moura, L., Bjørner, N.: Z3: An efficient SMT solver. In: International Conference on Tools and Algorithms for the Construction and Analysis of Systems, pp. 337–340. Springer (2008)
13. Detlefs, D., Nelson, G., Saxe, J.B.: Simplify: a theorem prover for program checking. J. ACM (JACM) **52**(3), 365–473 (2005)
14. Lotus developers: lotus, program analysis and verification framework. https://github.com/ZJU-Automated-Reasoning-Group/lotus
15. Developers, S.: Soufflé, logic defined static analysis. https://souffle-lang.github.io/
16. Downey, P.J., Sethi, R., Tarjan, R.E.: Variations on the common subexpression problem. J. ACM (JACM) **27**(4), 758–771 (1980)
17. Du, Y., et al.: EUF-based solving DYCK-reachability with applications to static analysis (2026). https://doi.org/10.5281/zenodo.18871949
18. Dutertre, B., De Moura, L.: The YICES SMT solver. Tool paper at **2**(2), 1–2 (2006). http://yices.csl.sri.com/tool-paper.pdf
19. Ewert, D.: platsat (2024), toy 'QF_UF' SMT solver written in Rust. https://github.com/dewert99/plat-smt
20. Habib, A., Pradel, M.: How many of all bugs do we find? A study of static bug detectors. In: Proceedings of the 33rd ACM/IEEE International Conference on Automated Software Engineering, pp. 317–328 (2018)
21. Hopcroft, J.E., Ullman, J.D.: Formal languages and their relation to automata. Addison-Wesley Longman Publishing Co., Inc (1969)
22. Jin, D., Meredith, P.O., Lee, C., Roşu, G.: JavaMOP: efficient parametric runtime monitoring framework. In: 2012 34th International Conference on Software Engineering (ICSE), pp. 1427–1430. IEEE (2012)
23. Kildall, G.A.: A unified approach to global program optimization. In: Proceedings of the 1st Annual ACM SIGACT-SIGPLAN Symposium on Principles of Programming Languages, pp. 194–206 (1973)
24. Kjelstrøm, A.H., Pavlogiannis, A.: The decidability and complexity of interleaved bidirected DYCK reachability. Proc. ACM Program. Lang. **6**(POPL), 1–26 (2022)
25. Koutris, P., Deep, S.: The fine-grained complexity of CFL reachability. Proc. ACM Program. Lang. **7**(POPL), 1713–1739 (2023)
26. Krishna, S., Lal, A., Pavlogiannis, A., Tuppe, O.: On-the-fly static analysis via dynamic bidirected DYCK reachability. Proc. ACM Program. Lang. **8**(POPL), 1239–1268 (2024)
27. Li, Y., Satya, K., Zhang, Q.: Efficient algorithms for dynamic bidirected DYCK-reachability. Proc. ACM Program. Lang. **6**(POPL), 1–29 (2022)

28. Li, Y., Zhang, Q., Reps, T.: On the complexity of bidirected interleaved DYCK-reachability. Proc. ACM Program. Lang. **5**(POPL), 1–28 (2021)
29. Li, Y., Zhang, Q., Reps, T.: Fast graph simplification for interleaved-DYCK reachability. ACM Trans. Program. Lang. Syst. (TOPLAS) **44**(2), 1–28 (2022)
30. Mathiasen, A.A., Pavlogiannis, A.: The fine-grained and parallel complexity of andersen's pointer analysis. Proc. ACM Program. Lang. **5**(POPL), 1–29 (2021)
31. Nelson, G., Oppen, D.C.: Simplification by cooperating decision procedures. ACM Trans. Program. Lang. Syst. (TOPLAS) **1**(2), 245–257 (1979)
32. Nelson, G., Oppen, D.C.: Fast decision procedures based on congruence closure. J. ACM (JACM) **27**(2), 356–364 (1980)
33. Nieuwenhuis, R., Oliveras, A.: Fast congruence closure and extensions. Inf. Comput. **205**(4), 557–580 (2007)
34. Reps, T.: Program analysis via graph reachability. Inf. Softw. Technol. **40**(11–12), 701–726 (1998)
35. Reps, T., Horwitz, S., Sagiv, M.: Precise interprocedural dataflow analysis via graph reachability. In: Proceedings of the 22nd ACM SIGPLAN-SIGACT Symposium on Principles of Programming Languages, pp. 49–61 (1995)
36. Sen, K., Marinov, D., Agha, G.: Cute: A concolic unit testing engine for c. ACM SIGSOFT Softw. Eng. Notes **30**(5), 263–272 (2005)
37. Shi, Q., et al.: Pinpoint: fast and precise sparse value flow analysis for million lines of code. In: Proceedings of the 39th ACM SIGPLAN Conference on Programming Language Design and Implementation, pp. 693–706 (2018)
38. SMT-COMP organizers: 2024 SMT competition results. Online Resource (2024). https://smt-comp.github.io/2024/results/
39. Sridharan, M., Bodík, R.: Refinement-based context-sensitive points-to analysis for java. ACM SIGPLAN Notices **41**(6), 387–400 (2006)
40. Tang, H., et al.: Summary-based context-sensitive data-dependence analysis in presence of callbacks. In: Proceedings of the 42Nd Annual ACM SIGPLAN-SIGACT Symposium on Principles of Programming Languages, pp. 83–95 (2015)
41. Whaley, J., Lam, M.S.: Cloning-based context-sensitive pointer alias analysis using binary decision diagrams. In: Proceedings of the ACM SIGPLAN 2004 Conference on Programming Language Design and Implementation, pp. 131–144 (2004)
42. Willsey, M., et al.: Egg: fast and extensible equality saturation. Proc. ACM Program. Lang. **5**(POPL), 1–29 (2021)
43. Yan, D., Xu, G., Rountev, A.: Demand-driven context-sensitive alias analysis for java. In: Proceedings of the 2011 International Symposium on Software Testing and Analysis, pp. 155–165 (2011)
44. Zhang, Q., Lyu, M.R., Yuan, H., Su, Z.: Fast algorithms for DYCK-CFL-reachability with applications to alias analysis. In: Proceedings of the 34th ACM SIGPLAN Conference on Programming Language Design and Implementation, pp. 435–446 (2013)

From Execution to Necessity: Proof-Based Metrics for Code Coverage (Short Paper)

Karl Mattsson and Peter Backeman(✉)

Mälardalen University, Västerås, Sweden
peter.backeman@mdu.se

Abstract. In this short paper we introduce the notion of proof-based coverage. While traditional test coverage metrics measure which parts of the code are executed when a test passes, proof-based coverage aims to measure what parts of the code are required for a test to pass. Through formal verification, a proof of validity of an assertion is created. By mapping back the proof onto the source code, the essential parts can be marked, obtaining a novel coverage metric. We sketch a formalization of this notion, discuss the relationship to previous coverage metrics and show differences on small examples.

Keywords: Software Testing · Test Coverage · SMT · Unsat Core

1 Introduction

Software testing is used for verifying the functionality of a system under test (SUT). For a program, a test suite is created which should exercise and verify the software sufficiently [1], and it is common to use *coverage criteria* to compare test suites [14] and their effectiveness. A coverage criterion describes, by some metric, how much of the SUT is tested by a specific test. Classical coverage criteria have been shown not to strongly correlate with test suite effectiveness [6,7]. This suggests a need to find better coverage criteria. Recent work highlights the importance of assertions [12] and older work suggests that assertions correlate with test suite effectiveness [15].

We propose a new family of coverage metrics, *proof-based coverage*, utilizing output from formal reasoning over a program (e.g., SMT solving). Via analysis of what parts of a program model are required to show the validity of an assertion, we establish a metric which does not focus only on what parts are *present* in a trace (e.g., executed), but what parts are *necessary*. Comparing with more advanced metrics (e.g., slicing-based assertion), it is possible to identify statements with a data- and control-flow dependency which still do not impact the outcome. We sketch a formalization of traditional and proof-based coverage and illustrate the difference with a small example. For brevity, we only present current work regarding line coverage, but we are also considering other metrics (e.g., branch coverage, MC/DC). We show our first steps towards answering the following research questions:

A. Sampaio and M. Stoelinga (Eds.): FM 2026, LNCS 16557, pp. 317–325, 2026.
https://doi.org/10.1007/978-3-032-26220-2_16

- *What is the theoretical relation between traditional coverage metrics and proof-based coverages?*
- *How can proof-based coverage help find untested code?*

Preliminaires. Using model checking [2] as an approach towards verifying C code is a well-established means of program verification (see, e.g., [9]). The C program is translated to a mathematical logic (often via an intermediate representation) which can then be checked for properties. In particular, if the program contains assertions, they can be checked if they are guaranteed to hold.

Satisfiability Modulo Theories (SMT) is the problem of deciding the satisfiability of Boolean formulas considering various theories [3]. A simple example is shown in Fig. 1. It is unsatisfiable as no values for `x` and `y` fulfils all constraints. An SMT solver will try to find values for `x` and `y` fulfilling the constraints, and if a model cannot be found, an *unsatisfiability core* (unsat core) can be provided. An unsat core is a (sometimes minimal) number of constraints which are unsatisfiable. In Fig. 1, `A1` and `A3` form an unsat core, as those by themselves are unsatisfiable. It should be noted that in general an SMT solver does not provide a minimal UNSAT core. For simplicity, we assume that cores provided are minimal, and we have noted (as has others [13]) that we (anecdotally) saw no concrete examples where it was an issue.

```
(assert (! (= (+ x y) 0) :named A1)) ; A1: x + y = 0
(assert (! (= (- x y) 2) :named A2)) ; A2: x - y = 2
(assert (! (> (+ x y) 0) :named A3)) ; A3: x + y > 0
```

Fig. 1. Example SMT formula which is unsatisfiable.

2 Traditional Coverage

Next we sketch a formalization of line coverage (using line and statement interchangeably), restricting ourselves to the following subset of the C language (where `b` and `e` are Boolean and arithmetic expressions over integers)[1]:

p : Program ::= | s ; assert(b) Program
s : Statement ::= | `int v` declaration
| `v = e` assignment
| `if (b){ s } else { s }` conditional
| `s; s` composition

To facilitate our definitions, we use *program graphs* [10]. We study programs at a *statement level*, and utilize program graphs, s.t., each node represents a statement of a program and edges represent control flow. We assume a translation

[1] For simplicity, we use integer arithmetic as an approximation for machine arithmetic.

from C code to a program graph, but leave out details (due to space limitation) and only note that each statement corresponds to one node in the program graph, thus allowing us to use a statement and its node interchangeably. We use $N(P)$ do denote the set of nodes of the program graph of P. When running a program, a sequence statements of the program will be executed. We call this sequence a *trace*, and since we have no loops, each element exists at most once, we can instead treat it as a set of statements.

Definition 1. *The* trace *of P, Trace(P), is the set of statements executed when P is run.*

Consider the fan program in Fig. 2a. Its program graph and trace are shown in Fig. 2b (each node has the line number of its corresponding statement).

```
     | int temp, speed;
Tr/Pr| temp = 28;
Tr/Pr| speed = 3;
     |
Tr/Pr| if (temp > 25) {
Tr/--|   temp = temp + 1;
Tr/Pr|   if (speed > 5) {
--/--|     speed = 5;
--/--|     temp = temp - 2;
     |   } else {
Tr/Pr|     speed = speed + 1;
Tr/--|     temp = temp - 1;
     |   }
Tr/--|   temp = temp - 1;
     | } else {
--/--|   speed = speed - 1;
--/--|   temp = temp + 1;
     | }
Tr/--| temp = temp - 1;
Tr/Pr| assert(speed > 0);
```

(a) FAN CONTROLLER example.

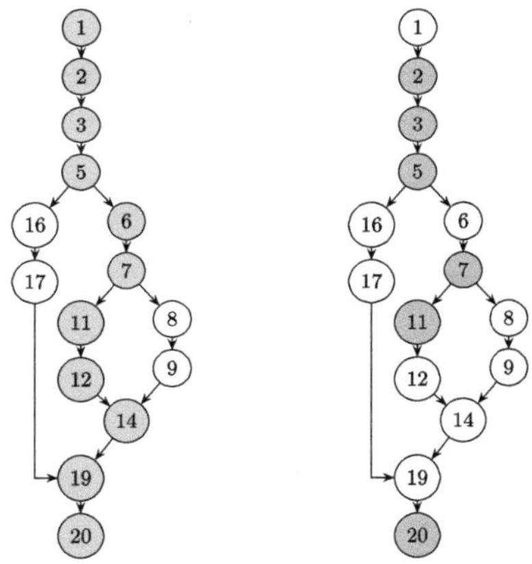

(b) Execution Trace (c) Proof Trace

Fig. 2. C program (a) and program graph with traces (b) and (c). For each line, Tr/Pr means it is in *Trace*(P)/*Proof*(P).

Line coverage measures how many different lines of code have been exercised:

Definition 2. *The* line coverage *of a program P is defined as:*

$$LC(P) = \frac{|\mathit{Trace}(P)|}{|N(P)|}$$

Intuitively, the line coverage is the fraction of statements executed divided by the total number of statements (i.e., the number of nodes in the program graph). For the program in Fig 2a, the line coverage is $LC(P) = \frac{11}{15} \approx 0.69$.

3 Proof-Based Coverage

The idea of *proof coverage* is to treat a program ending with an assert as a verification problem. We convert the program to an SMT formula which is unsatisfiable if the assertion is guaranteed to hold (i.e., model checking). When the formula is found to be unsatisfiable, we study the unsat core to see which constraints were required for unsatisfiability. In a sense, the core can be seen as a proof providing the steps taken to show validity of the program, with the steps being the statements of the source code. Our prototype framework takes as input a C source file, converts the C code to an intermediate single static assignment (SSA) form and finally models it as an SMT formula. Next, it uses a back end SMT solver to find an unsat core (if none is found, the assertion does not hold), and translate it back to the corresponding statements of the program. The set of marked statements correspond to the proof-based coverage of the code. The outline of the approach is shown in Fig. 3. When an SMT model is unsatisfiable, there is possibility for multiple unsat cores, which can mark different statements in the original program. Due to limited space, in this paper we will assume that an SMT solver always returns the same (minimal) unsat core for an SMT model.

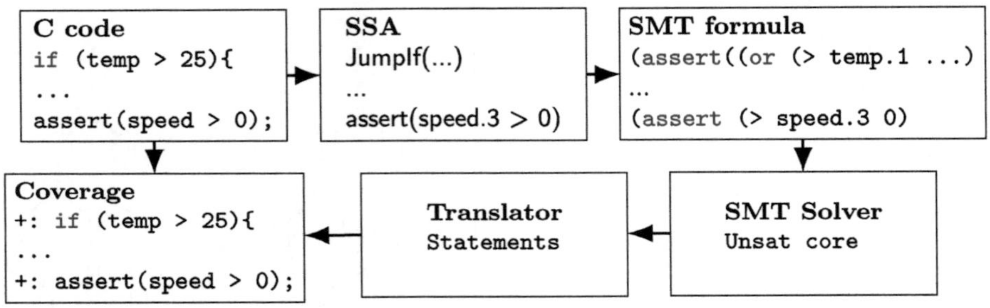

Fig. 3. Overview of the proof-coverage framework workflow.

Definition 3. *The* proof trace *of a program P, denoted Proof(P), is the set of marked statements which are in the minimal unsat core of the translated SMT model.*

A proof trace is analogous to a trace, which leads to a definition of proof line coverage similar to traditional line coverage.

Definition 4. *The* proof line coverage *of a program P is defined as:*

$$PrLC(P,T) = \frac{|Proof(P)|}{|N(P)|}$$

While traditional coverage measures what statements has been executed, proof-based coverage measures which statements has an impact on the truth value of the assertion. This can identify statements which are executed, but they have no effect on the assertion. Consider again the program in Fig. 2a, lines 6, 12, 14 and 19 are covered by line coverage, but not by proof line coverage. For the lines to be covered, the assertion would need to change to also take `temp` into consideration, e.g., `assert(speed > 0 && temp > 0)`.

Non-trace Behaviour. When executing a program the execution trace will be linear, i.e., for every if-statement either the if or the else branch will be executed, but not both. However, proof-based coverage can utilize statements from both branches to prove that an assertion holds, creating *non-trace behaviour.* Consider the program in Fig. 4. The proof line coverage includes lines from both branches of the if-statement. While not inherently problematic, it might be desirable in certain situations to be avoided. Next, we describe next how it can be done.

```
      |int a,b;
Tr/  -|a = 1;
Tr/Pr|if (a == 1)  {
 -/Pr|   b = 1;
     |} else {
 -/Pr|   b = 2;
     |}
Tr/Pr|assert(b > 0);
```

Fig. 4. Example of non-trace behaviour.

Non-trace behaviour can only happen when values in the condition are undefined, i.e., unconstrained, as any concrete assignment of variables will lead to the condition being either true or false. Therefore, to avoid non-trace behaviour we make two additions to the translation: (a) for every variable v in the SMT model we add a meta-variable $v.undef$ which is 1 if v is undefined; (b) when translating if-statements, we allow the constraint to be ignored if $v.undef = 1$ for any v in the condition. Intuitively, this means that if there are undefined variable in the condition, all assignments in the branches also become undefined, meaning those statements can not be in an unsat core, and therefore not be proof-covered.

Lemma 1. *If non-trace behaviour conditions are added, any two statements in the proof of the program share one execution trace.*

We state the following theorem (without proof):

Theorem 1. *With non-trace behaviour conditions,* $\forall(P)\ \mathit{Proof}(P) \subseteq \mathit{Trace}(P)$.

From Theorem 1, Defs. 2 and 4, then clearly follows that proof line coverage is a more conservative metric than traditional line coverage:

Lemma 2. *With non-trace behaviour conditions,* $\forall(P)\ PrLC(P) \leq LC(P)$

We have also outlined a formalization for proof-based branch coverage, following the same shape as for proof-based line coverage introduced above. Due to space constrain it is not described here, but we note that it is non-trivial how to define it. For example, with non-trace behaviour, multiple branches can be covered at the same time. Moreover, if a branch is taken (i.e., the if-condition evaluating to true) but none of the statements included in that branch is covered, should the branch be considered covered?

Short-Circuited Conditions. An if-statement in C is *short-circuited*, i.e., it is evaluated from left-to-right and as soon as the value can be deduced, evaluation is stopped. In our current translation this is not supported as the whole expression is encoded as a Boolean formula. This does not affect soundness (as there are no side-effects) but may affect coverage. To represent short-circuiting, if-statements can be rewritten to semantically equivalent nesting of if-conditions, e.g., `if (A && B){` can be replaced by `if (A){ if (B){`. Other cases can be treated similarly.

4 TCAS Demonstration

We apply our approach[2] to the TCAS use case of the Siemens test suite [5]. The source file is $\tilde{1}75$ lines of C code implementing logic over integers. Provided with the source file is $\tilde{1}500$ test cases which exercises different paths through the program. We adapt to establish the line coverage and proof line coverage of each line. After running all test cases and summarizing coverage, there are two lines which are never marked with proof-based coverage, but multiple times by line coverage (checked using GCOV[3]): for example line 2 shown in Fig. 5. To validate, we modify the program and replace the whole if-statement with `result = 0`. When executing all test-cases on this modified program the output never changed, showing that the line `result = 1` indeed never impacted the outcome of the program. Thus, proof-based coverage can identify a line of code that has no impact on the outcome of any test. This can be interpreted in two ways: (i) the line is non-essential and could be removed from the program, or (ii) there is a need for an additional test that covers the line in a proof-based manner.

```
1 if (Above_Threat && (Vert_Sep>=MINSEP) && (Up_Sep>=ALIM)) {
2     result = 1;
3 } else {
4     result = 0;
5 }
```

Fig. 5. Excerpt from TCAS example (slightly modified to fit our C language).

[2] Source available at https://github.com/ptrbman/proofcov.
[3] https://gcc.gnu.org/onlinedocs/gcc/Gcov.html.

5 Related Work

A very similiary approach is thhe notion of *checked coverage* [11] (also known as assertion coverage [4]). Schuler et al. presents an approach of measuring coverage using program slicing: an approach where executed statements are filtered, and only statements which there are data-flow and control-flow dependencies from target variable(s). In recent work, Koitz-Hristov et al. suggest that while slow, checked coverage can be suitable as coverage metric for smaller pieces of code [8]. Proof-based coverage is very similar in that it only include statements which affects the assertion (which then implies the existence of data- and control-flow dependencies), but the presented approach is different in two important ways. First, a fundamentally different method is used to extract the relevant statements (i.e., unsat core translation vs. program slicing). Secondly, the approach can reason away unimportant dependencies: consider an if-statement where both branches ensures an assertion is fulfilled. A proof-based approach can deduce that the valuation of variables in the if-expression are not relevant (i.e., making assignments to them not covered), while a program slicing approach in general will include them to one of the branches.

Another related work is by Tomb and Joshi who introduce *verification coverage* [13]. The authors propose to use unsatisfiability cores (in the same manner as in this work) to extract lines which, if modified, *could* fail verification. The approach is demonstrated and is shown to be smoothly integrated into verification of Dafny programs, with a reasonable overhead. While Tomb and Joshi focus on deductive verification, in proof-based coverage as presented here is concerned with model-based verification and applied to C code rather than Dafny.

Modified Condition/Decision Coverage (MC/DC) is a coverage criteria also focusing on the *impact* of different conditions. MC/DC requires that for every sub-condition, there must be (at least) one test case where flipping the condition changes the outcome of the test [1]. Proof-based coverage is similar to MC/DC in this manner, but can take a more holistic view of a program and does not focus on a specific condition. Furthermore, proof line coverage does not only reason about if-statements but more general statements as well.

6 Conclusions and Future Work

In this short work in progress-paper we introduce proof-based coverage, where unsatisfiability cores of SMT models are used to mark which statements of a C program are required to ensure an assertion always holds. We show that a proof-based line coverage metric is more conservative than the traditional one. We also demonstrate on the TCAS use case that it can identify lines of code which have no impact on the outcome, even though they are heavily tested.

For future work we plan to present the full formalization of the translation from C to SSA and then to SMT. We are also investigating the implications of short-circuiting, tracking undefined variables and strategies for multiple unsat cores, e.g., find all of them, find the smallest, etc. as well as comparing with other coverage metrics. Moreover, we plan to extend the C subset and test with larger datasets to investigate scaling (which can be problematic for formal analysis).

Acknowledgements. We thank the reviewers for their helpful comments. We acknowledge the support of the Swedish Knowledge Foundation via the Prospekt project RITES - Rich Testing of Software, grant nr. 20230054.

References

1. Ammann, P., Offutt, J.: Introduction to software testing. Cambridge University Press (2008)
2. Baier, C., Katoen, J.P.: Principles of model checking. The MIT Press (2008)
3. Barrett, C., Tinelli, C.: Satisfiability modulo theories. Springer International Publishing, Cham, pp. 305–343 (2018). https://doi.org/10.1007/978-3-319-10575-8_11
4. Chen, J., Bai, Y., Hao, D., Zhang, L., Zhang, L., Xie, B.: How do assertions impact coverage-based test-suite reduction? In: 2017 IEEE International Conference on Software Testing, Verification and Validation (ICST). IEEE Computer Society, Los Alamitos, CA, USA, pp. 418–423 (2017). https://doi.org/10.1109/ICST.2017.45
5. Do, H., Elbaum, S., Rothermel, G.: Supporting controlled experimentation with testing techniques: an infrastructure and its potential impact. Empir. Softw. Eng. **10**(4), 405–435 (2005)
6. Hemmati, H.: How effective are code coverage criteria? In: 2015 IEEE International Conference on Software Quality, Reliability and Security, pp. 151–156 (2015). https://doi.org/10.1109/QRS.2015.30
7. Inozemtseva, L., Holmes, R.: Coverage is not strongly correlated with test suite effectiveness. In: Proceedings of the 36th International Conference on Software Engineering, ICSE 2014, pp. 435–445. Association for Computing Machinery, New York, NY, USA (2014). https://doi.org/10.1145/2568225.2568271
8. Koitz-Hristov, R., Sterner, T., Stracke, L., Wotawa, F.: On the suitability of checked coverage and genetic parameter tuning in test suite reduction. J. Softw. Evol. Process **36**(8), e2656 (2024). https://doi.org/10.1002/smr.2656
9. Kroening, D., Tautschnig, M.: Cbmc–c bounded model checker: (competition contribution). In: Tools and Algorithms for the Construction and Analysis of Systems: 20th International Conference, TACAS 2014, Held as Part of the European Joint Conferences on Theory and Practice of Software, ETAPS 2014, Grenoble, France, April 5–13, 2014. Proceedings 20, pp. 389–391. Springer (2014)
10. Paige, M.R.: Program graphs, an algebra, and their implication for programming. IEEE Trans. Softw. Eng. **1**(1), 286–291 (1975). https://doi.org/10.1109/TSE.1975.6312853
11. Schuler, D., Zeller, A.: Assessing oracle quality with checked coverage. In: 2011 Fourth IEEE International Conference on Software Testing, Verification and Validation, pp. 90–99 (2011). https://doi.org/10.1109/ICST.2011.32
12. Taromirad, M., Runeson, P.: Assertions in software testing: survey, landscape, and trends. Int. J. Softw. Tools Technol. Transfer **27**(1), 117–135 (2025)
13. Tomb, A., Joshi, A.: Static coverage in deductive software verification. In: 2025 Formal Methods in Computer-Aided Design (FMCAD), pp. 251–261 (2025). https://api.semanticscholar.org/CorpusID:280977478

14. Yang, Q., Li, J.J., Weiss, D.: A survey of coverage based testing tools. In: Proceedings of the 2006 International Workshop on Automation of Software Test, pp. 99–103. AST '06, Association for Computing Machinery, New York, NY, USA (2006). https://doi.org/10.1145/1138929.1138949
15. Zhang, Y., Mesbah, A.: Assertions are strongly correlated with test suite effectiveness. In: Proceedings of the 2015 10th Joint Meeting on Foundations of Software Engineering, ESEC/FSE 2015, pp. 214–224, Association for Computing Machinery, New York, NY, USA (2015). https://doi.org/10.1145/2786805.2786858

Active Automata Learning with Noisy Data: From Big to Small Data

Felix Wallner[1(✉)], Bernhard K. Aichernig[1], Benjamin von Berg[2], and Maximilian Rindler[2]

[1] Johannes Kepler University Linz, Linz, Austria
{felix.wallner.fmv,bernhard.aichernig}@jku.at
[2] Graz University of Technology, Graz, Austria
benjamin.vonberg@tugraz.at

Abstract. Active automata learning enables model-based testing and verification of black-box systems by automatically constructing models from observations via interactions with the system. As interactions are usually expensive, active algorithms attempt to perform as few interactions as possible to learn a given system. However, many such algorithms struggle when confronted with noise, such as message loss, when learning otherwise deterministic systems. We investigate and adapt different algorithms to learn deterministic automata in a noisy setting. One of these is a novel active algorithm based on our previous passive Partial Max-SAT algorithm. In our analysis, we demonstrate techniques to lower the required number of interactions and order the evaluated algorithms accordingly. Finally, we show that the necessary interactions can be further reduced when leaving the classical active learning framework.

Keywords: active automata learning · passive automata learning · noise-tolerant · model learning · grammar inference · partial Max-SAT

1 Introduction

Active automata learning is a technique to automatically learn behavioural models in the form of automata, i.e., finite state machines (FSMs), from black-box systems via testing. This process is also known as grammatical inference [17] or model learning [31]. The resulting automata enable the application of advanced testing and verification techniques, such as model checking [14], model-based testing [3], as well as other applications like learning from RNNs [23] or fingerprinting IoT devices [24]. Active automata learning has been applied successfully on a range of systems such as cyber-physical systems [1,19], embedded systems [2,28] and communication protocols [7,15,25]. The two most well established active automata learning libraries are LearnLib [18] and AALpy [22].

In practice, automata learning frequently encounters two important, but often insufficiently emphasized, challenges:

A. Sampaio and M. Stoelinga (Eds.): FM 2026, LNCS 16557, pp. 326–346, 2026.
https://doi.org/10.1007/978-3-032-26220-2_17

(1) *Noise*: Environmental factors, such as message loss, can disrupt otherwise deterministic learning procedures. As a result, we require learning methods that are robust to such disturbances or filter the noise when interacting with otherwise deterministic systems.
(2) *Interaction costs*: In active automata learning, the primary bottleneck is the number of queries issued to the System Under Learning (SUL), as each individual interaction, i.e., step, may be expensive or time-consuming. Consequently, it is essential to minimize the number of system interactions while still guaranteeing correct inference of the target system.

This partially explains why active automata learning is generally preferred over passive automata learning whenever feasible: In the active setting, the learner is capable of interacting directly with the (SUL), thus guiding the acquisition of new data. In contrast, in the passive setting no such interactions are possible and learning depends exclusively on a given set of execution traces. Active learning tends to be more efficient because the learner can query missing data specifically, thereby generally requiring fewer observations [4,31].

We therefore want to actively learn deterministic systems in the presence of noise while minimizing the number of interactions required. Specifically, passive algorithms have already been devised for precisely this setting, i.e., learning deterministic systems in the presence of noise [27,34]. Consequently, we are interested in evaluating whether such passive methods can be adapted to operate (more) efficiently within the active learning framework.

Our contributions are (1) a novel passive-to-active learning algorithm, Active Partial Max-SAT Learning (APMSL) , for which we developed advanced data gathering and processing methods. The algorithm builds on a passive algorithm from previous work [34] and trades computational effort for fewer interactions with the (SUL). (2) To reduce the number of steps even further, we present learning in a setting outside of the classic active automata framework. (3) We demonstrate a number of different active learning methods in the setting with noise and evaluate them based on the number of steps. Our research questions are:

RQ 1: Which learning algorithms are suitable in an active setting with noise?
RQ 2: How do noise-tolerant passive algorithms perform when made active?
RQ 3: Is it possible to skip the equivalence oracle when making a passive algorithm active and stop when reaching a fixpoint in learning?

Structure. After the preliminaries in Sect. 2, we first establish stochastic learning as a baseline in Sect. 3. We demonstrate and evaluate specific noise-tolerant passive algorithms in an active setting in Sect. 4. We then showcase these algorithms outside of the classical automata learning framework in Sect. 5. We end with related work, threats to validity and the conclusion in Sects. 6, 7 and 8 respectively.

2 Preliminaries

2.1 Notation, (FSMs), Noise

Moore Machines. We are interested in learning models for systems that can be described by Moore machines [21]. A Moore machine $\mathcal{M}$ is a type of (FSM) defined as a 6-tuple $\mathcal{M} = (Q, I, O, q_0, \delta, \lambda)$ where Q, I and O are the finite set of states, input and output symbols respectively, $\delta : Q \times I \rightharpoonup Q$ is the partial deterministic transition function and $\lambda : Q \to O$ is the state labeling function. When the transition function δ is total we say that $\mathcal{M}$ is *input-complete.*

Markov Decision Processes. For stochastic learning we consider Markov decision processes (MDPs), which allow modeling noise as part of the automaton. (MDPs) are defined analogous to Moore machines except with δ replaced with the transition function $\Delta : Q \times I \to \mathcal{D}(Q)$ where $\mathcal{D}(Q)$ is the set of probability distributions over Q.

Traces. Let $\mathcal{T}$ be a set of traces where each trace $t \in \mathcal{T}$ is a sequence of alternating outputs and inputs that both start and end with an output: $t \in O \times (I \times O)^*$. The first output of each sequence always corresponds to the output of the initial state, followed by arbitrary many inputs and their corresponding outputs. The function $\textsc{MakeTrace}(i, o) = t$ is used to generate a trace t from $i \in I^n$ and $o \in O^{n+1}$ with the first element of o being the output of the initial state. Given two sequences $s_1, s_2 \in I^*$ we denote $s_1 \cdot s_2 = s'$ as their concatenation. We further say that s_1 and s_2 are a prefix and a suffix of s' respectively.

Random Walks. A random walk is a sequence of independently drawn inputs $i \in I$ with a geometrically distributed length L depending on a stopping probability $s \in [0, 1]$. A random walk has therefore $\mathbb{E}[L] = \frac{1}{s}$ and $Var(L) = \frac{1-s}{s^2}$.

Bisimilarity and Accuracy. Two Moore machines $\mathcal{M}_1$ and $\mathcal{M}_2$ are equivalent iff their initial states are trace equivalent, i.e., permit the same traces. Formally, iff $\forall s \in I^* : \lambda_1^*(q_{0,1}, s) = \lambda_2^*(q_{0,2}, s)$, where $\lambda^*(q, s)$ denotes the output sequence produced when processing the input word s starting from state q. For Moore machines, which are deterministic and have no internal transitions, trace equivalence coincides with bisimilarity, which can be efficiently checked. We will use *bisimilarity* as a measure for correctness, i.e., whether a learned automaton is bisimilar to the ground truth. *Accuracy* is a stochastic measure for the similarity between two languages. Specifically, it is a ratio of how many random walks produce the same output for a given set of n random walks. A similarity score of 1 would mean the automata behave identically under the chosen random walks, while 0 would mean they never produce the same output sequence.

Noise. In this paper, we model noise as taking a wrong transition not defined in the deterministic transition function δ, so called faulty transitions. When a fault occurs an arbitrary successor state is chosen from Q, ignoring δ. The observed trace therefore coincides with the deterministic trace except at those fault steps, where the system may exhibit nondeterministic behavior. This noise model can also capture other noise models. Consider a noise model where the transition

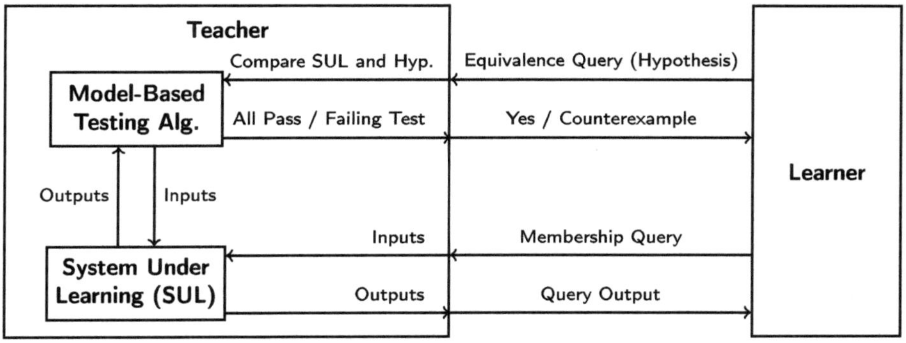

Fig. 1. Interaction between minimally adequate teacher (MAT) and active learner.

function is unaffected but a random output is selected from O instead. Such a fault can be modeled by at most two faulty transitions in our noise model, one entering a different state with that output and one leaving it. Furthermore, we assume noise to appear *sparsely* and *uniformly distributed* across all steps taken on the (SUL). In other words, any step is equally likely to induce a fault as any other, independent of the underlying system. Our goal is to learn a Moore machine from a system that exhibits such noisy behavior with as little interaction with the (SUL) as possible.

2.2 Minimally Adequate Teacher

Active automata learning algorithms usually employ the Minimally Adequate Teacher (MAT) framework, introduced by Angluin [8]. The framework has since been extended to not only cover deterministic finite automata (DFAs) but also Moore machines among others. In this framework, depicted in Fig. 1, the learner can perform two types of queries: The *membership query*, composed of a sequence of inputs, is answered with a sequence of corresponding outputs from the system. Between each membership query the system is also *reset* to its initial state. The *equivalence query* asks whether a given hypothesis corresponds to the (SUL). In case the hypothesis differs from the real system a *counterexample* is provided by the teacher. Importantly, in a black-box setting such an ideal teacher that can answer an equivalence query directly does not exist. Instead the equivalence query is approximated with a model-based conformance testing algorithm, called an *(equivalence) oracle*, as can be seen from Fig. 1. The oracle in turn uses membership queries itself to confirm or refute the equivalence between hypothesis and SUL. Thus, the oracle is also affected by noise.

3 Baseline Stochastic Learning

An intuitive approach for learning in a noisy environment is to employ an active stochastic learning algorithm, such as L^*_{MDP} [30], to learn MDPs instead of

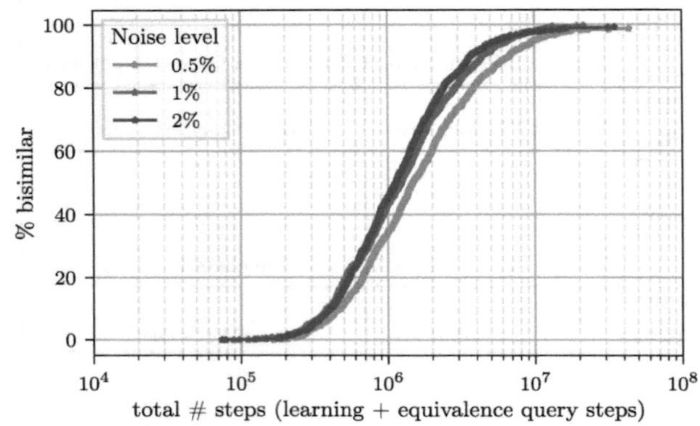

Fig. 2. Number of total steps required by Det-L^*_{MDP} to learn a certain percentage of the benchmarking set correctly, i.e., bisimilar to ground truth, for different noise levels.

deterministic FSMs. Such an algorithm is designed to be *robust* against noise, in the sense that queries are repeated to obtain the exact transition probabilities between states. In this way, noise can be incorporated into the learned automata in the form of low probability transitions, which can then be pruned in a post-processing step to make the model deterministic. In this post-processing step we obtain a Moore machine from the learned MDP by selecting the transition with the highest probability for each q and i to form the deterministic transition function δ from Δ. Formally, $\delta(q,i) := \arg\max_{q' \in Q} \Delta(q,i)(q')$ for each $(q,i) \in Q \times I$, leaving only the most frequent transitions. When applying this to the learned MDP we call this procedure Det-L^*_{MDP}, which provides the deterministic Moore machine we are interested in. Assuming the noise does not dominate the true behavior, this procedure converges to the correct model given enough steps.

Benchmarking. Throughout the paper, we evaluate all algorithms in the AALpy library [22] using a common benchmarking set.[1] This set contains 160 randomly generated Moore machines, five for each combination between five and 15 states, with three inputs and two to four hidden states. Hidden states are states that share outputs with other states, i.e., have non-unique outputs. Noise is modeled as follows: At each step with the SUL a fault occurs with a given probability. In the case of a fault, the transition is redirected to a randomly selected different state instead of the correct successor. We investigated noise levels of 0.5%, 1%, and 2% as they represent low but non-negligible levels of noise, sufficient to disrupt standard deterministic learning algorithms, yet still low enough that the underlying system behavior remains discernible. Each benchmark model is learned a total of five times and we average over the results, for a total of $160 * 3 * 5 = 2400$ *runs* per algorithm.

An important performance metric is the number of steps Det-L^*_{MDP} requires to learn a certain percentage of runs correctly. Using a random walk oracle with

[1] All artifacts and source code is available at [33].

a stopping probability $s = 0.09$, we found that a step budget of 20000 steps per equivalence query is the smallest budget that still allows Det-L^*_{MDP} to learn over 99% of all runs correctly. We show these results in Fig. 2. At 0.5% noise, it takes 600000 steps to learn 20% of all runs correctly, one million to reach 35% and four million steps to reach 80%. Interestingly, Det-L^*_{MDP} requires on average fewer steps to learn with higher levels of noise. The following provides a potential explanation: Under relatively low-noise conditions, the learner might initially miss many noise-induced transitions. This leads to compact hypotheses which are then incrementally extended through equivalence queries. In contrast, with higher level of noise the learner is able to find noise-induced transitions earlier. This results in larger hypotheses early on and faster overall exploration. This gives us a baseline for an existing active algorithm that explicitly handles noise during learning.

However, over one million steps could be considered infeasible for some systems. For example, if each step would take one second in a slow interface, then collecting one million steps would take over eleven days for a single learning run. While we can reduce the number of steps by introducing a step budget, we can already see that this has a negative effect on correctness for Det-L^*_{MDP}.

In essence, we are balancing the number of steps required against the average correctness. While stochastic learning is suitable in the context of noise, intuitively, it should not be *necessary* to use a stochastic learning algorithm to deal with sporadic noise. Basically, stochastic learning was developed for systems that exhibit systemic uncertainty, while we only have a little random noise that should be filtered out. In other words, stochastic learning wastes a lot of effort in learning transition probabilities corresponding to noise that are later pruned. In the next section we will see a different approach compared to stochastic learning that reduces the number of steps required significantly.

4 Noise-Tolerant Learning

The problem of learning automata from data that contains random noise has been studied in the passive setting. Existing approaches [27,34] directly infer deterministic automata from noisy observations, instead of incorporating the noise into the learned automaton like stochastic learners. Since these algorithms are tailored to deterministic systems with noise, they can be more efficient than general stochastic learners. Our goal is to leverage these passive noise-tolerant algorithms in an active setting. By making them active, we also gain control over how (much) data is collected and retain the assurances that an equivalence oracle provides for termination and testing of hypotheses.

4.1 Passive to Active

The idea of using passive algorithms in an active setting has been demonstrated by Grinchtein et al. [16] and by Walkinshaw et al. [32]. This approach has since been applied in several domains, including SMT-based learning [29], stochastic

Algorithm 1. GeneralPassiveToActive

Input:
- params: set containing input alphabet I, extension length el among others
- PassiveAlgorithm: $\mathcal{T} \times$ passiveParams $\rightarrow$ Hyp
- passiveParams: parameter set for PassiveAlgorithm

Output: hyp

```
1: traces ← CreateInitialTraces(I, el)
2: loop
3:     hyp ← PassiveAlgorithm(traces, passiveParams)
4:     if not isInputComplete(hyp, params) then
5:         traces ← traces ∪ GatherInputCompletnessTraces(hyp, params)
6:     else
7:         cex ← equivalenceQuery(hyp, params)
8:         if cex = ∅ then
9:             return hyp
10:        traces ← traces ∪ GatherAdditionalTraces(cex, hyp, params)
11: function CreateInitialTraces(I, el)
12:     traces ← ∅
13:     for all test ∈ I^el do
14:         outputs ← membershipQuery(test)
15:         traces ← traces ∪ makeTrace(test, outputs)
16:     return traces
```

systems [6] and real-time systems [5]. Generally, any passive learning algorithm can be made into an active one. We formalize this in Algorithm 1, where any passive learning algorithm can be given as an input for active learning.

The algorithm takes several parameters, collectively written as *params*. One of them is the extension length (el), which influences the initialization phase, the processing of counterexamples and other data-gathering methods. The initial traces are obtained by issuing membership queries for all input sequences of length el in Lines 1 and 11. With this data, we learn the first hypothesis via the provided passive algorithm with its corresponding *passiveParams*, which may differ between different passive learning algorithms. After receiving the hypothesis, we can sometimes immediately infer that we are missing data in Lines 4, such as when the hypothesis is not input-complete. In this case, we can query the missing traces in Line 5 without performing an expensive equivalence query. Otherwise, we test the hypothesis using an equivalence query in Line 7. If we obtain a counterexample, we process it while gathering additional traces in Line 10, which we discuss in the next section. This procedure repeats until no counterexample is found anymore, terminating the algorithm in Line 9.

4.2 Passive Algorithms

We selected two noise-tolerant passive algorithms designed for specifically learning deterministic automata from noisy data but built on different principles:

RPNI.* The established RPNI* algorithm by Sebban and Janodet [27] infers deterministic automata from noisy traces. We utilize the implementation within the generalized state-merging framework in AALpy [10]. RPNI* works by *merging states* starting from an automaton that exactly represents the input data but does not generalize to new samples. A merge is rejected if it fails a statistical test related to the rate at which steps are misclassified. Instead of greedily accepting merge candidates that pass the test, we apply the merge with the highest confidence. We call the active variant ARPNI*.

Partial Max-SAT Learning. We also evaluate the Partial Max-SAT (PMSAT) learning algorithm from our previous work [34]. A PMSAT solver handles two types of clauses: (1) hard-clauses, which must be satisfied as in regular SAT, and (2) soft-clauses, of which as many as possible must be satisfied as in Max-SAT. The PMSAT learning algorithm uses this framework to fit a deterministic model to noisy traces by satisfying all structural constraints (hard-clauses), while discarding only the minimal amount of conflicting data (soft-clauses). The goal is thus to explain as much of the observed data as possible with a single deterministic automaton. In addition to the deterministic automaton, the algorithm also provides information about the exact faults and the location of the faults in the traces. However, the passive algorithm requires the expected number of states as an additional input for learning. If the solver receives enough computation time, we are guaranteed to find a best-fitting deterministic model with n states. A key contribution of this paper is the development of an active variant of this algorithm, by incorporating both the search for the correct number of states and the additional information about faults. We call this active variant APMSL.

4.3 Active Partial Max-SAT Learning (APMSL)

Due to the requirement that the passive PMSAT learning algorithm must be given a fixed number of states n, and because the true number of states is unknown, we must infer a suitable n during active learning. However, comparing hypotheses learned from both different amount of data and different values of n is difficult. To address this, we adapt the general Algorithm 1 to the specific Algorithm 2 (APMSL) with the following changes: 1) Instead of learning a single hypothesis each round we learn w_{size} hypotheses with different numbers of states, in what we call the *sliding window* W, in Lines 4 and 15, 2) we choose the best model from W according to a heuristic function in Line 9, and 3) we move the sliding window so it is centered on our best hypothesis for the next round in Line 14. Learning multiple hypotheses with the same traces allows for direct comparison between models with different n by adapting the guidelines from the original paper [34] on the best choice of n. While the original guidelines enumerate the desired properties of a learned model, such as low number and frequency of faulty transitions, they do not describe a quantitative metric. To apply these guidelines to our learning procedure directly, we introduce a composite score that aggregates them in Eq. 1. We select the best model based on f_{fault},

Algorithm 2. APMSL

Input:
- params: set containing input alphabet I, extension length el among others
- w_{size}: size of the sliding window to learn

Output: hyp

1: traces ← CreateInitialTraces(I, el)
2: $w_{start} \leftarrow |O|$ ▷ *Window start: smallest number of states in window*
3: **loop**
4: $W \leftarrow$ LearnSlidingWindow(traces, w_{start}, w_{size})
5: **if** $\exists$ hyp $\in W$: **not** isInputComplete(hyp, params) **then**
6: **for all** hyp $\in W$: **do**
7: traces ← traces ∪ GatherInputCompletnessTraces(hyp, params)
8: **else**
9: hyp ← chooseBestModel(W)
10: cex ← equivalenceQuery(hyp, params)
11: **if** cex $= \emptyset$ **then**
12: **return** hyp
13: traces ← traces ∪ GatherAdditionalTraces(cex, hyp, params)
14: $w_{start} \leftarrow |\text{hyp}| - \lfloor \frac{w_{size}}{2} \rfloor$ ▷ *Position |hyp| in the middle of next window*
15: **function** LearnSlidingWindow(traces, w_{start}, w_{size})
16: $W \leftarrow$ new array of size w_{size}
17: **for all** $j \leftarrow 0$ **to** $w_{size} - 1$ **do**
18: $n \leftarrow w_{start} + j$
19: $W\langle j \rangle \leftarrow$ PassivePmsatLearn(traces, n) ▷ *Learn model with n states [34]*
20: **return** W

the multiset of all faulty transition frequencies, and f_{det}, the multiset of all regular deterministic transitions frequencies, where a higher score corresponds to a better model:

$$\text{score} = \begin{cases} \dfrac{1}{\text{mean}(f_{fault})} + \dfrac{1}{\text{sum}(f_{fault})} - \dfrac{1}{\min(f_{det} \setminus \{0\})} & |f_{fault}| > 0 \\ 1 - \dfrac{1}{\min(f_{det} \setminus \{0\})} & |f_{fault}| = 0 \end{cases} \tag{1}$$

The first and second term of the score reward low faulty transition frequencies and less total number of faults respectively. The last term punishes low deterministic transition frequencies. This leads to hypotheses with fewer and less frequent faulty transitions being scored higher, while models containing rarely-observed, i.e., unreliable, deterministic transitions are punished. In chooseBestModel in Line 9, we select the model with the highest score from among the sliding window W. We selected a sliding window size $w_{size} = 5$ for our experiments. Larger windows increase the number of comparisons but also raise computational cost, whereas smaller windows limit the step granularity of the sliding process. Further details of the algorithm are presented in the Master's thesis of Rindler [26], supervised by this paper's other co-authors.

4.4 Robust Equivalence Queries

When each step on the (SUL) may be affected by noise, active learning in the classical MAT setting (see Sect. 2.2) encounters a problem. In particular, the correctness of membership queries can no longer be guaranteed, which the learning algorithm must account for. More critically equivalence queries rely on membership queries themselves and therefore inherit this problem. As a result, the counterexamples they produce cannot be trusted, especially for termination.

A straightforward way to address this problem is to use what we call *robust* membership queries during the equivalence testing. A robust membership query consists of a regular membership query repeated n times, returning a *majority trace*, i.e., the output sequence that occurs in more than 50% of these repetitions. A similar robustness mechanism for equivalence testing was shown to be effective by Pferscher and Aichernig [24]. If no majority trace exists because the noise level is too high, then we cannot return a counterexample and must continue testing to potentially identify a shorter one. Thus, if n is chosen too small relative to the noise level, some existing counterexamples may not be detected.

When performing a robust membership query, it is therefore important to choose a sensible number of repetitions n, which also depends on the length of the query n_t. In particular, the probability of not obtaining a majority trace should be smaller than a given threshold ϵ. Assuming the per step probability for errors due to noise p_e is small, the probability of an erroneous trace becoming the majority trace is negligible. Thus, it is sufficient to consider the probability p_{maj} of the correct trace obtaining a majority. For a single query, the probability of obtaining the correct trace is given as $p_1 = (1-p_e)^{n_t}$ and thus $p_{maj} = B(k >= n/2, p_1)$, where $B(k,p)$ denotes the binomial distributions with k successes and a success probability p. The goal is then to find the smallest n such that $1-p_{maj} < \epsilon$, which can be done via search until a normal approximation of the binomial distribution becomes viable.

Note that during equivalence testing we do not repeat all membership queries robustly. Instead, robust membership queries are only used to validate potential counterexamples, thereby reducing the overall number of executed steps. For our benchmarking set, we selected $n = 20$, which is sufficient to validate all counterexamples for noise levels up to 2%, while keeping the number of repetitions as low as possible.

4.5 Data Gathering

A key aspect of any active automata learning algorithm is the gathering of new data to refine the hypotheses. These methods can be far more sophisticated than simply adding the received counterexample to the traces. We applied and tested different combinations of the following methods:

Input-completeness. One method that can be applied even before invoking an expensive equivalence query with the oracle is input-completeness pre-processing in Lines 4 and 5. Here we verify whether the hypothesis produced by the passive algorithm is input-complete. If it is not, at least one counterexample can immediately be generated without querying the oracle, since we assume the SUL being input-complete. We calculate the shortest prefix p that reaches the state with the missing input i, apply i, and extend the resulting trace with a randomly chosen suffix r from I^{el}. We then add the result of the membership query for $p \cdot i \cdot r$ to our set of traces. This procedure can be applied for every missing input in every state, after which we continue with the next round of passive learning, skipping the oracle in the process. This gives us the chance to learn an input-complete automaton next round and saves us potentially many calls to the equivalence oracle. Because of this, the number of equivalence queries is not necessarily equal to the number of learning rounds as is usually the case in active learning.

Counterexample Processing. Within the (MAT) framework we receive counterexamples from the teacher, which are always added to the traces. Additional traces may be generated around the counterexample in Line 10 if desired. Specifically, we query *each prefix* p of the counterexample appended with either (a) each possible suffix s from I^{el}, or (b) a single, randomly chosen suffix s from I^{el}, thereby specifically collecting data for states and transitions for which our current hypothesis diverges from the (SUL)..

State Prefix Coverage. To leverage the learned hypothesis for targeted exploration of the SUL, we construct additional queries in Line 10 that increase state coverage rather than relying on uninformed random walks. We cycle through all hypothesis states, generating queries by taking each state's prefix, i.e., its shortest access sequence, and appending a random walk. The number of steps executed per round is controlled by a parameter n (written as spc $= n$), while the length of each trace is given by the random walk stopping probability s, which is 0.09 for our benchmarks.

Fault Replay and Reproduction. For APMSL, which provides not only the hypothesis but also detailed information about faults and where they occur, we introduce two new additional data-gathering options: With the *replay* method, we re-execute *each trace* containing a fault at least twice, once as originally observed and once only up to and including the fault-inducing input, followed by a random suffix from I^{el}. With the *reproduce* method, we instead attempt to reproduce each faulty transition in the learned hypothesis. For *each state* in which a fault was reported by APMSL, we access the state via its prefix, apply the fault-inducing input, and append a random suffix from I^{el}.

4.6 Evaluation

The evaluation was performed on the same benchmarking set used for Det-L^*_{MDP} in Sect. 3 and features different combinations of data gathering methods explained in the previous subsection. Again each run on the 160 models was

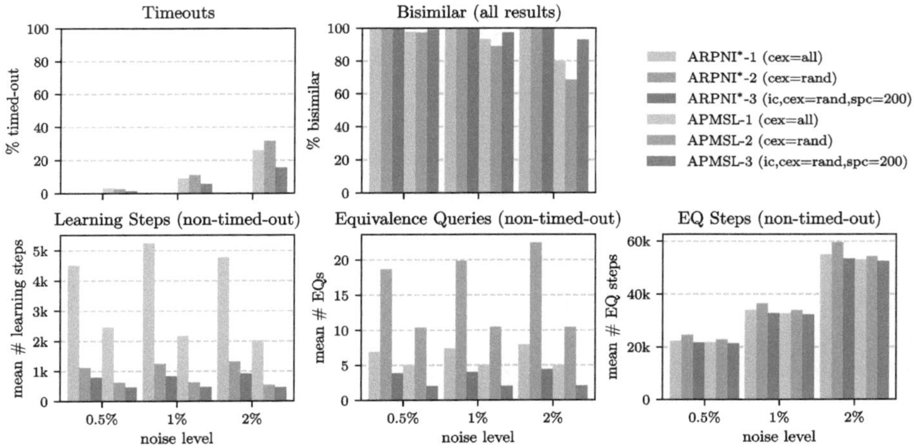

Fig. 3. Benchmarking results with (MAT) for ARPNI* and APMSL with combinations of counterexample processing methods (cex) (see Sect. 4.5), input-completness processing (ic) (see Sect. 4.5) and state prefix coverage (spc) (see Sect. 4.5).

repeated five times. The algorithms also use a robust random walk oracle with a step budget of 10000 steps per equivalence query, excluding counterexample validation steps, which is sufficient to find over 99% of all counterexamples for these benchmarks. Additionally, all individual learning runs were subject to a one hour timeout. This is due to a practical consideration that learning must occur within reasonable time, otherwise learning real systems would be impractical. Our assumption that interactions with the SUL are expensive should not justify taking an arbitrary amount of computational time to compensate. In particular, PMSAT occasionally performs computationally heavy calls that can trigger the timeout. If a timeout is triggered, we terminate the run and evaluate the last produced hypothesis. As a consequence, particularly due to the sliding window requiring multiple PMSAT calls, APMSL can still return the correct model despite timing out, as part of the already computed window may contain the correct hypothesis. In the following figures, we show ARPNI* in blue and APMSL in green. To simplify references to data-gathering configurations, we enumerate them and use the same numbering for both algorithms. Thus, combination-N refers to the same data-gathering method N, applied in both ARPNI*-N and APMSL-N, e.g., combination-4 (ic,spc $= 100$) refers to the combined application of input-completness and state prefix coverage data-gathering.

In Fig. 3, we show data-gathering combinations that include additional counterexample processing (cex) from Sect. 4.5. Combination-1 adds all suffixes while combination-2 adds only fewer random suffixes for each counterexample. We can see that both combinations trade off learning steps against equivalence queries (see the lower left and middle plots). We see that, if more steps are performed during learning, fewer equivalence queries are required. We will discuss the signif-

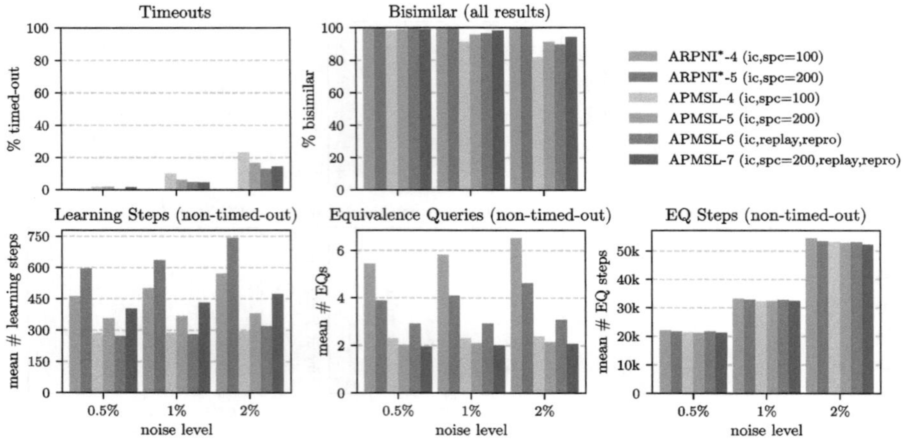

Fig. 4. Benchmarking results with (MAT) for ARPNI* and APMSL with combinations of input-completness processing (ic) (see Sect. 4.5), state prefix coverage (spc) (see Sect. 4.5) and fault replay and reproduction methods (see Sect. 4.5).

icance of this tradeoff in more detail later. Combination-3, by contrast, includes both input-completeness (ic) and state prefix coverage (spc) processing, which substantially lowers both the required amount of learning steps and number of equivalence queries (see the lower left and middle plots). On average, ARPNI*-3 requires only one fifth of the learning steps of ARPNI*-1 and one quarter of the equivalence queries of ARPNI*-2. A similar ratio holds for APMSL. We can thus see that additional gathering of data is beneficial over just additional counterexample processing.

In Fig. 4, we show data-gathering combinations without additional counterexample processing. Here we selected some combinations with different amount of steps in state prefix coverage (spc), with combination-4 only gathering 100 additional steps per round, while combination-5 gathers 200 instead. Specifically for ARPNI*-4 and 5, we can again see that there is a trade-off in the amount of learning steps and equivalence queries (see the lower left and middle plots). We also show off two APMSL variants that include the data-gathering methods from Sect. 4.5, here shortened to replay and repro. APMSL-6 includes just these methods and input-completness processing, while APMSL-7 adds to this additional state prefix coverage data collection. On average, APMSL-6 requires the fewest learning steps and APMSL-7 the fewest equivalence queries across all algorithms and combinations respectively.

Across both figures, we observe that APMSL variants do not learn all runs within the imposed timeframe. This can be seen in the bisimilar and timeout plots respectively (see the upper middle and left plots). However, all APMSL variants learn over 99.8% of all non-timed-out runs correctly and even several that do timeout. For example for 2% noise, APMSL-3 times out for 15.6% of runs yet learns 92.8% of all runs correctly. The corresponding ARPNI* variants require a few hundred to a few thousand additional learning steps, though this difference is minor since the number of learning steps is dwarfed by the number of equivalence query steps. The majority of the equivalence query steps occur during the final *equivalenceQuery* call, which finds no counterexample and therefore executes the full limit of 10000 steps. The remaining equivalence query steps come from validating potential counterexamples, which naturally increases with noise level. It is of note that **the passive-to-active approach requires about two orders of magnitude fewer total steps than the previous** $\boldsymbol{Det - L^*_{MDP}}$ **method.**

The most interesting result is that ARPNI* requires on average more equivalence queries than APMSL. This suggests that APMSL may construct correct hypotheses from less data and with fewer counterexamples than ARPNI*, at the cost of increased computation time. However, any potential advantage gained from the more refined hypotheses produced by APMSL is offset by the cost of the steps required by the equivalence oracle to test the hypotheses. In contrast, ARPNI* constructs hypotheses much more quickly, though potentially less accurately, yet still identifies the correct model with only a small number of additional equivalence queries. Moreover, improvements amounting to a few hundred or even a few thousand learning steps are negligible when compared to the tens of thousands of interactions required by the equivalence oracle. To illustrate, assuming again that each interaction with the (SUL) takes one second, acquiring 20000 steps for the 0.5% case would already require roughly six hours, which only grows with higher noise levels. Additionally, the correct choice of settings for the equivalence oracle is hard. For this benchmarking set, we knew the minimum required step budget for the oracle to learn 99% correctly, which is not usually known. Just doubling our step budget would have in turn more than doubled the required number of equivalence steps, especially for higher levels of noise. In fact, most other oracles, such as the more rigorous W-method oracle [11], would have required orders of magnitude more steps and exacerbated this difference even more. These observations motivated us to consider omitting the oracle calls entirely, an approach explored in the following section.

5 Avoiding Equivalence Queries

We now investigate learning *without* an equivalence oracle, thereby leaving the classical MAT framework. This has two immediate consequences: We no longer obtain counterexamples to guide the gathering of new traces, and we must introduce a new termination criterion. Inspired by machine learning and genetic programming, we adopt the following rule: If the learned hypotheses remain bisimilar for k consecutive learning rounds, the algorithm terminates. Intuitively, once repeated data gathering no longer changes the hypothesis, we have reached a fixpoint and can place more confidence in the resulting hypothesis.

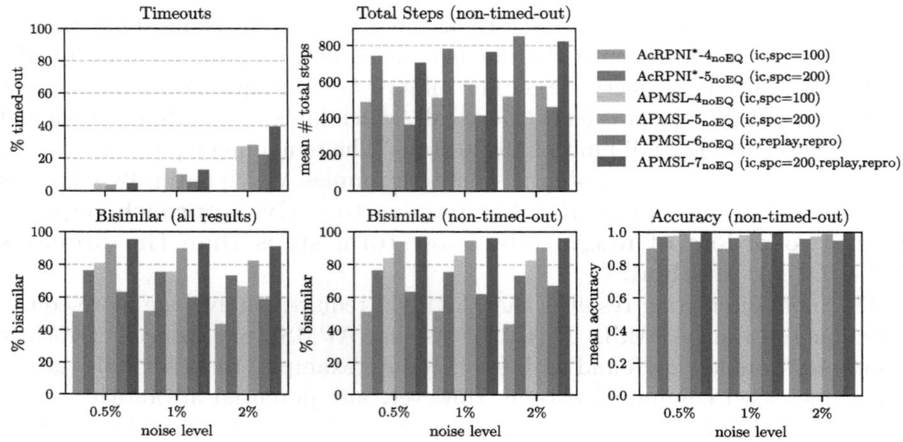

Fig. 5. Benchmarking results *without* (MAT) for ARPNI* and APMSL with combinations of input-completness processing (ic) (see Sect. 4.5), state prefix coverage (spc) (see Sect. 4.5) and fault replay and reproduction methods (see Sect. 4.5).

Naturally, without an equivalence oracle we cannot expect to learn the correct model in every case. However, several practical alternatives exist for validating the learned hypotheses. If a classical equivalence oracle is infeasible, we can instead rely on an alternative teacher, for example, a domain expert who inspects the final model or a model checker that verifies relevant system properties. In this way, a key advantage of automata learning is that it produces formal models, which can be rigorously *verified*, unlike many machine learning techniques, which typically allow only for empirical testing of a trained model. This means that even in the absence of the guarantees usually provided by an equivalence oracle, the resulting models can still be validated using domain knowledge and remain highly useful, especially if they can be learned with significantly fewer steps.

In Fig. 5, we present the results of learning *without* MAT, i.e., equivalence queries, using termination threshold $k = 2$ to minimize the required number of steps. Note that total steps is the same metric as learning steps in the previous section: Without equivalence query steps, the learning effort is all that remains (see upper middle plot). We see that the combinations that relied heavily on counterexamples or collected comparatively little data perform worse, e.g., combination-4 and APMSL-6_{noEQ}. In contrast, APMSL-7_{noEQ}, which was the algorithm that required the fewest equivalence queries in Fig. 4, now achieves the highest correctness, learning up to 95.2% runs correctly for 0.5% noise. Overall, $(\text{APMSL})_{\text{noEQ}}$ performs strongly, likely because it constructs higher-quality hypotheses through additional computation and its lesser reliance on equivalence queries. In fact, we can see a correlation between the correctness with (MAT) and the number of equivalence queries in Fig. 4. The key takeaway is that, *with* an oracle, collecting less additional data is acceptable because the oracle compensates for missing information, whereas *without* an oracle, collecting more data becomes essential for constructing sufficiently accurate hypotheses and avoiding premature convergence. In this setting, investing additional computation time can therefore be justified by the resulting improvement in hypothesis quality. Finally, **the total number of steps drops by roughly two orders of magnitude compared to learning with an equivalence oracle.**

6 Related Work

Passive automata learning in the presence of noise is well studied. Early work by Angluin [9] addresses noisy observations in grammatical inference. State-merging approaches have also been adapted to noisy data, for example most recently within AALpy [10]. This also includes RPNI* [27], which extends classical RPNI to handle noise, as well as (PMSAT) learning [34], both presented in this paper. There are also stochastic passive learning algorithms, such as IO ALERGIA [20]. Noise-tolerant inference has also been explored beyond deterministic finite automata and Moore machines, such as for timed automata [12].

In active learning, noise is often addressed by introducing a filter between the (SUL) and the learner, filtering observations before they reach the deterministic learning algorithm. Such filters typically rely on caching, repeated queries, or resetting and cleaning the learner. Pferscher and Aichernig [24] additionally also apply domain knowledge to filter faults for learning for Bluetooth Low Energy implementations, while Ferreira et al. [13] formalize the mapper in the more general $C\mathcal{E}$AL framework. In essence, these frameworks keep track of the non-determinism due to noise in a cache external to the learning algorithm. Here, the teacher uses the cache to filter noise before the deterministic learning algorithm. By contrast, we focus on dedicated robust algorithms capable of learning in the presence of noise directly.

7 Threats to Validity

A potential threat to validity lies in our choice of *benchmark models*. Our goal was to compare APMSL and ARPNI* within a setting where both algorithms remain practically applicable. The decision to use between five and 15 states, and two to four hidden states were made with this in mind. In our prior work [34], Fig. 3 a shows that the passive (PMSAT) algorithm is particularly sensitive to this parameter, requiring significantly more time as the number of hidden states increases. Our focus here is on the impact of reducing interactions and examining the significance the amount of equivalence queries has on learning. For substantially larger systems or models with many hidden states, ARPNI* is currently the more suitable choice given the capabilities of modern SAT solvers. Nevertheless, we consider the chosen range of two to four hidden states to be representative of many real-world scenarios.

Another potential threat is our choice to report averages mostly over non-timed-out runs. For APMSL our goal was to measure "how many steps are required to terminate", rather than "how many steps are performed in total".

Including timed-out runs would generally make APMSL appear slightly worse at low noise levels and slightly better at high noise levels, largely because frequent timeouts reduce the opportunity to generate steps. This effect is particularly pronounced for equivalence query steps, where including timed-out runs would make APMSL appear significantly better than ARPNI*, due to the majority of steps being taken in the last equivalence query.

8 Conclusion

We presented techniques to reduce the number of interactions, typically the bottleneck, in active automata learning of deterministic automata in noisy environments. To this end, we made two passive algorithms active, including a novel Partial Max-SAT–based learner APMSL extending our previous work, and evaluated several data-gathering strategies. Our results show that augmenting passive-to-active learning with additional data collection methods is generally beneficial, with input-completeness processing and state-prefix coverage performing best. Although APMSL requires fewer learning steps for the given range of our benchmarking set, this advantage is offset by the costly interactions demanded by the equivalence oracle. In practice, ARPNI*-3 offers the most effective trade-off while equivalence queries are available. We further demonstrated that active learning without an equivalence oracle is feasible. By relying solely on data-gathering methods and terminating once a hypothesis stabilizes, inspired by machine learning, we can still obtain useful models. APMSL-7 specifically, which trades computation time for better hypotheses, learned almost the entire benchmarking set correctly. While this setting lacks the guarantees of classical (MAT) learning, the resulting formal models remain amenable to verification techniques such as model checking. This makes the approach attractive in scenarios where interactions with the (SUL) are extremely expensive. Currently, APMSL is successfully being applied in the TASTE project to learn the safety behavior of an autonomous mobile robot. In this setting, steps are expensive and observations of the robot's state sporadically arrive delayed. **In total, we observed a reduction in steps of roughly two orders of magnitude when moving from stochastic learning to passive-to-active with an equivalence oracle, and another two when leaving the MAT framework entirely.**

Acknowledgement. The work in this paper was supported by the Austrian Science Fund (FWF) Cluster of Excellence Bilateral AI under contract number 10.55776/COE12 as well as the Transformative AI-Assisted Testing in Industrial Mobile Robotics (TASTE) project (project number FO999911053) funded by the Austrian Research Promotion Agency (FFG). We would like to thank Martin Tappler for his insights and his help with L^*_{MDP}.

References

1. Aichernig, B.K., et al.: Learning a behavior model of hybrid systems through combining model-based testing and machine learning. In: Gaston, C., Kosmatov, N., Gall, P.L. (eds.) Testing Software and Systems - 31st IFIP WG 6.1 International Conference, ICTSS 2019, Paris, France, October 15–17, 2019, Proceedings. Lecture Notes in Computer Science, vol. 11812, pp. 3–21. Springer (2019). https://doi.org/10.1007/978-3-030-31280-0_1
2. Aichernig, B.K., Burghard, C., Korošec, R.: Learning-based testing of an industrial measurement device. In: Badger, J.M., Rozier, K.Y. (eds.) NFM 2019. LNCS, vol. 11460, pp. 1–18. Springer, Cham (2019). https://doi.org/10.1007/978-3-030-20652-9_1
3. Aichernig, B.K., Mostowski, W., Mousavi, M.R., Tappler, M., Taromirad, M.: Model learning and model-based testing. In: Bennaceur, A., Hähnle, R., Meinke, K. (eds.) Machine Learning for Dynamic Software Analysis: Potentials and Limits - International Dagstuhl Seminar 16172, Dagstuhl Castle, Germany, April 24–27, 2016, Revised Papers. Lecture Notes in Computer Science, vol. 11026, pp. 74–100. Springer (2018). https://doi.org/10.1007/978-3-319-96562-8_3
4. Aichernig, B.K., Muskardin, E., Pferscher, A.: Active vs. passive: a comparison of automata learning paradigms for network protocols. In: Luckcuck, M., Farrell, M. (eds.) Proceedings Fourth International Workshop on Formal Methods for Autonomous Systems (FMAS) and Fourth International Workshop on Automated and verifiable Software sYstem DEvelopment (ASYDE), FMAS/ASYDE@SEFM 2022, and Fourth International Workshop on Automated and verifiable Software sYstem DEvelopment (ASYDE) Berlin, Germany, 26th and 27th of September 2022. EPTCS, vol. 371, pp. 1–19 (2022). https://doi.org/10.4204/EPTCS.371.1
5. Aichernig, B.K., Pferscher, A., Tappler, M.: From passive to active: learning timed automata efficiently. In: Lee, R., Jha, S., Mavridou, A., Giannakopoulou, D. (eds.) NASA Formal Methods. pp. 1–19. Springer International Publishing, Cham (2020). https://doi.org/10.1007/978-3-030-55754-6_1
6. Aichernig, B.K., Tappler, M.: Probabilistic black-box reachability checking (extended version). Formal Meth. Syst. Des. **54**(3), 416–448 (2019). https://doi.org/10.1007/s10703-019-00333-0
7. Aichernig, B.K., Tappler, M., Wallner, F.: Benchmarking combinations of learning and testing algorithms for automata learning. Formal Aspects Comput. **36**(1), 3:1–3:37 (2024). https://doi.org/10.1145/3605360
8. Angluin, D.: Learning regular sets from queries and counterexamples. Inf. Comput. **75**(2), 87–106 (1987). https://doi.org/10.1016/0890-5401(87)90052-6
9. Angluin, D., Laird, P.D.: Learning from noisy examples. Mach. Learn. **2**(4), 343–370 (1987). https://doi.org/10.1007/BF00116829
10. von Berg, B., Aichernig, B.K.: Extending AALpy with passive learning: a generalized state-merging approach. In: Piskac, R., Rakamaric, Z. (eds.) Computer Aided Verification - 37th International Conference, CAV 2025, Zagreb, Croatia, 23–25 July 2025, Proceedings, Part IV. Lecture Notes in Computer Science, vol. 15934, pp. 127–140. Springer (2025). https://doi.org/10.1007/978-3-031-98685-7_6
11. Chow, T.S.: Testing software design modeled by finite-state machines. IEEE Trans. Softw. Eng. **4**(3), 178–187 (1978). https://doi.org/10.1109/TSE.1978.231496
12. Cornanguer, L., Gimenez, P.: TADAM: learning timed automata from noisy observations. In: Papalexakis, V., Riondato, M., Zheleva, E., Weninger, T., Ding, W. (eds.) Proceedings of the 2025 SIAM International Conference on Data Mining,

SDM 2025, Alexandria, VA, USA, 1–3 May 2025, pp. 114–123. SIAM (2025). https://doi.org/10.1137/1.9781611978520.10
13. Ferreira, T., Henry, L., da Silva, R.F., Silva, A.: Conflict-aware active automata learning. In: Achilleos, A., Monica, D.D. (eds.) Proceedings of the Fourteenth International Symposium on Games, Automata, Logics, and Formal Verification, GandALF 2023, Udine, Italy, 18–20th September 2023. EPTCS, vol. 390, pp. 150–167 (2023). https://doi.org/10.4204/EPTCS.390.10
14. Fiterău-Broştean, P., Janssen, R., Vaandrager, F.: Combining model learning and model checking to analyze TCP implementations. In: Chaudhuri, S., Farzan, A. (eds.) CAV 2016. LNCS, vol. 9780, pp. 454–471. Springer, Cham (2016). https://doi.org/10.1007/978-3-319-41540-6_25
15. Fiterău-Broştean, P., Jonsson, B., Merget, R., de Ruiter, J., Sagonas, K., Somorovsky, J.: Analysis of DTLS implementations using protocol state fuzzing. In: Capkun, S., Roesner, F. (eds.) 29th USENIX Security Symposium, USENIX Security 2020, August 12–14, 2020, pp. 2523–2540. USENIX Association (2020). https://www.usenix.org/conference/usenixsecurity20/presentation/fiterau-brostean
16. Grinchtein, O., Leucker, M., Piterman, N.: Inferring network invariants automatically. In: Furbach, U., Shankar, N. (eds.) IJCAR 2006. LNCS (LNAI), vol. 4130, pp. 483–497. Springer, Heidelberg (2006). https://doi.org/10.1007/11814771_40
17. De la Higuera, C.: Grammatical inference: learning automata and grammars. Cambridge University Press (2010)
18. Isberner, M., Howar, F., Steffen, B.: The open-source LearnLib - A framework for active automata learning. In: Kroening, D., Pasareanu, C.S. (eds.) Computer Aided Verification - 27th International Conference, CAV 2015, San Francisco, CA, USA, 18–24 July 2015, Proceedings, Part I. Lecture Notes in Computer Science, vol. 9206, pp. 487–495. Springer (2015). https://doi.org/10.1007/978-3-319-21690-4_32
19. Khosrowjerdi, H., Meinke, K.: Learning-based testing for autonomous systems using spatial and temporal requirements. In: Perrouin, G., Acher, M., Cordy, M., Devroey, X. (eds.) Proceedings of the 1st International Workshop on Machine Learning and Software Engineering in Symbiosis, MASES@ASE 2018, Montpellier, France, pp. 6–15. ACM (2018). https://doi.org/10.1145/3243127.3243129
20. Mao, H., Chen, Y., Jaeger, M., Nielsen, T.D., Larsen, K.G., Nielsen, B.: Learning deterministic probabilistic automata from a model checking perspective. Mach. Learn. **105**(2), 255–299 (2016). https://doi.org/10.1007/s10994-016-5565-9
21. Moore, E.F.: Gedanken-experiments on sequential machines. In: Shannon, C.E., McCarthy, J. (eds.) Automata Studies. (AM-34), Volume 34, pp. 129–154. Princeton University Press (2016). https://doi.org/10.1515/9781400882618-006
22. Muskardin, E., Aichernig, B.K., Pill, I., Pferscher, A., Tappler, M.: AALpy: an active automata learning library. Innov. Syst. Softw. Eng. **18**(3), 417–426 (2022). https://doi.org/10.1007/S11334-022-00449-3
23. Muskardin, E., Aichernig, B.K., Pill, I., Tappler, M.: Learning finite state models from recurrent neural networks. In: ter Beek, M.H., Monahan, R. (eds.) Integrated Formal Methods - 17th International Conference, IFM 2022, Lugano, Switzerland, 7–10 June 2022, Proceedings. Lecture Notes in Computer Science, vol. 13274, pp. 229–248. Springer (2022). https://doi.org/10.1007/978-3-031-07727-2_13
24. Pferscher, A., Aichernig, B.K.: Fingerprinting bluetooth low energy devices via active automata learning. In: Huisman, M., Pasareanu, C.S., Zhan, N. (eds.) Formal Methods - 24th International Symposium, FM 2021, Virtual Event, 20–26

November 2021, Proceedings. Lecture Notes in Computer Science, vol. 13047, pp. 524–542. Springer (2021). https://doi.org/10.1007/978-3-030-90870-6_28
25. Pferscher, A., Wunderling, B., Aichernig, B.K., Muskardin, E.: Mining digital twins of a VPN server. In: Hallerstede, S., Kamburjan, E. (eds.) Proceedings of the Workshop on Applications of Formal Methods and Digital Twins co-located with 25th International Symposium on Formal Methods (FM 2023), Lübeck, Germany, 06 March 2023. CEUR Workshop Proceedings, vol. 3507. CEUR-WS.org (2023). https://ceur-ws.org/Vol-3507/paper6.pdf
26. Rindler, M.: Active automata learning with partial max-SAT from noisy data. Master's thesis, Graz University of Technology, Graz, Austria (2026). https://doi.org/10.3217/t8aqc-x8424
27. Sebban, M., Janodet, J.: On state merging in grammatical inference: A statistical approach for dealing with noisy data. In: Fawcett, T., Mishra, N. (eds.) Machine Learning, Proceedings of the Twentieth International Conference (ICML 2003), 21–24 August 2003, Washington, DC, USA, pp. 688–695. AAAI Press (2003). http://www.aaai.org/Library/ICML/2003/icml03-090.php
28. Smeenk, W., Moerman, J., Vaandrager, F., Jansen, D.N.: Applying automata learning to embedded control software. In: Butler, M., Conchon, S., Zaïdi, F. (eds.) ICFEM 2015. LNCS, vol. 9407, pp. 67–83. Springer, Cham (2015). https://doi.org/10.1007/978-3-319-25423-4_5
29. Smetsers, R., Fiterău-Broştean, P., Vaandrager, F.: Model learning as a satisfiability modulo theories problem. In: Klein, S.T., Martín-Vide, C., Shapira, D. (eds.) LATA 2018. LNCS, vol. 10792, pp. 182–194. Springer, Cham (2018). https://doi.org/10.1007/978-3-319-77313-1_14
30. Tappler, M., Aichernig, B.K., Bacci, G., Eichlseder, M., Larsen, K.G.: L*-based learning of markov decision processes (extended version). Formal Aspects Comput. **33**(4–5), 575–615 (2021). https://doi.org/10.1007/S00165-021-00536-5
31. Vaandrager, F.: Model learning. Commun. ACM **60**(2), 86–95 (2017). https://doi.org/10.1145/2967606
32. Walkinshaw, N., Bogdanov, K., Holcombe, M., Salahuddin, S.: Reverse engineering state machines by interactive grammar inference. In: Proceedings of the 14th Working Conference on Reverse Engineering, pp. 209–218. WCRE '07, IEEE Computer Society, USA (2007). https://doi.org/10.1109/WCRE.2007.45
33. Wallner, F., Aichernig, B., von Berg, B., Rindler, M.: Active automata learning from noisy data - algorithms and artifacts (2026). https://doi.org/10.5281/zenodo.18544018
34. Wallner, F., Aichernig, B.K., Burghard, C.: It's not a feature, it's a bug: fault-tolerant model mining from noisy data. In: Proceedings of the 46th IEEE/ACM International Conference on Software Engineering, ICSE 2024, Lisbon, Portugal, 14–20 April 2024, pp. 29:1–29:13. ACM (2024). https://doi.org/10.1145/3597503.3623346

Correct-by-Construction Dynamic Reachability: A Galois-Connected Approach to Bidirected Dyck Languages

Xiaofei Zhao[1,2](✉)

[1] School of Information Engineering, Lanzhou Petrochemical University of Vocational Technology, No. 1 Shandan Street, Xigu, Lanzhou 730060, Gansu, China
[2] School of Automatization and Electric Engineering, Lanzhou Jiaotong University, No. 88 Anning West Road, Anning, Lanzhou 730070, Gansu, China
xifizhao@gmail.com

Abstract. The integration of precise static analysis into interactive development environments (IDEs) necessitates algorithms that can update analysis results incrementally within milliseconds. However, maintaining *Bidirected Dyck-CFL reachability*—the standard formalism for field-sensitive alias analysis—under dynamic graph mutations remains an open challenge. Standard batch algorithms exhibit prohibiting cubic complexity ($\mathcal{O}(N^3)$), while existing dynamic approaches often lack formal guarantees when handling non-monotonic edge deletions (the "ghost path" problem). In this paper, we present **GC-DBDR**, a novel incremental framework rooted in Abstract Interpretation. We reformulate the dynamic analysis problem not as graph patching, but as computing *Differential Fixpoints* over a lattice. By establishing a rigorous Galois Connection between execution traces and reachability relations, we derive update rules that are *correct-by-construction.* To resolve the asymmetry between monotonic insertions and non-monotonic deletions, we introduce a *Counting-Augmented Abstract Domain* supported by a *Derivation Hypergraph.* This structure operationalizes the *Inverse Abstraction Principle*, ensuring that reachability facts are retracted if and only if their supporting derivation trees are fully invalidated. We provide formal proofs demonstrating that GC-DBDR is sound and complete relative to batch analysis. Empirically, the algorithm achieves an optimal input-output complexity of $\mathcal{O}(\Delta)$, delivering sub-millisecond tail latencies on real-world benchmarks.

Keywords: Abstract Interpretation · Dynamic Graph Algorithms · Galois Connections · Dyck-CFL Reachability · Incremental Static Analysis

1 Introduction

The efficacy of modern software verification hinges on the ability to detect defects early in the development lifecycle, a paradigm increasingly formalized as *continuous reasoning* [22]. At the algorithmic core of precise static analysis—particularly

A. Sampaio and M. Stoelinga (Eds.): FM 2026, LNCS 16557, pp. 347–366, 2026.
https://doi.org/10.1007/978-3-032-26220-2_18

for field-sensitive pointer analysis and interprocedural data-flow tracking—lies the problem of *Context-Free Language (CFL) reachability*. Specifically, modeling the balanced flow of data across function calls and field accesses requires solving reachability over *bidirected Dyck languages* [24,25,32]. This formalism captures the precise matching of open-close parentheses (representing call/return or read/write pairs) while accounting for the bidirectional nature of memory aliasing. However, a fundamental tension exists between the algorithmic complexity of these rigorous analyses and the operational demands of modern software engineering. Standard batch algorithms for Dyck-CFL reachability exhibit a time complexity of $\mathcal{O}(N^3)$ [30,31], or at best combinatorial sub-cubic bounds ($\mathcal{O}(N^3/\log N)$) [3], where N is the size of the program graph. While acceptable for nightly builds, this cubic cost becomes prohibitive in the context of "live static analysis" deployed in industry [7]. In an Integrated Development Environment (IDE), developers expect feedback within milliseconds (typically $< 100\,\text{ms}$) of a code edit [14]. Re-executing a cubic-time algorithm on a codebase spanning millions of lines after every keystroke is computationally infeasible. This performance gap forces existing tools to compromise. They often resort to coarser abstractions (e.g., unification-based approaches [26]), abandon context-sensitivity, or rely on ad-hoc timeout heuristics, thereby sacrificing the formal guarantees required for safety-critical systems. To enable true "analysis-as-you-type" verification without compromising soundness, we require a fundamental shift from stateless re-computation to mathematically sound incremental maintenance [16].

The central algorithmic obstacle in dynamic static analysis is the inherent asymmetry between *incremental* (insertion) and *decremental* (deletion) updates, which we term the *monotonicity gap*. This asymmetry is rooted in the theoretical limitations of maintaining graph properties, as established by Ramalingam and Reps [23]. Edge insertion is semantically constructive. In lattice-theoretic terms, adding an edge e moves the program graph G up the lattice to $G' = G \sqcup \{e\}$. Due to the monotonicity of the abstract semantic function F defined in abstract interpretation [4], the least fixed point grows monotonically: $\mathsf{lfp}(F_G) \sqsubseteq \mathsf{lfp}(F_{G'})$. Algorithms can thus proceed via standard chaotic iteration, propagating new reachability facts from the old fixed point without retracting existing knowledge. Edge deletion, conversely, is non-monotonic. Removing an edge e requires the analysis to transition from a larger fixed point to a smaller one ($\mathsf{lfp}(F_{G'}) \rightarrow \mathsf{lfp}(F_G)$). The fundamental difficulty lies in the *alternative path problem*, often referred to as the "ghost path" problem [16]. When a premise edge e is removed, any abstract fact ϕ (e.g., $u \rightsquigarrow v$) that depended on e becomes suspect. The analyzer must determine—efficiently—whether ϕ is supported by an *alternative derivation tree* (rendering ϕ redundant but still valid) or if e was a bridge in the sole derivation of ϕ (rendering ϕ invalid). Naive approaches to deletion typically incur the cost of re-proving reachability, degrading to $\mathcal{O}(N^3)$ complexity or worse for context-free languages. To achieve $\mathcal{O}(\Delta)$ performance (input-output complexity [23]), the analysis requires a mechanism to distinguish *dependency* from *redundancy* without traversing the graph, a capability absent in standard boolean abstract domains and traditional dynamic transitive closure algorithms [6].

To bridge the monotonicity gap without sacrificing formal guarantees, we propose **GC-DBDR** (Galois-Connected Dynamic Bidirected Dyck Reachability). Our methodology represents a paradigm shift from *algorithmic adaptation*—common in traditional dynamic graph algorithms [23]—to *semantic derivation* [5]. Rather than patching a static graph algorithm with ad-hoc heuristics to handle updates, we reformulate the problem within the framework of abstract interpretation [4]. We treat graph mutations not as structural edits, but as *differential inputs* to the fixpoint equations, building upon principles of incremental fixpoint iteration [12] and differential static analysis [17]. By computationally "differentiating" the abstract semantic function, we derive update rules that are intrinsically sound. It is important to clarify that we use the term *correct-by-construction* to denote that our update algorithms are mathematically derived directly from the differential fixpoint semantics, distinguishing this approach from correctness enforced by type systems or mechanical proof assistants.

This paper makes the following contributions:

1. **Galois-Theoretic Formalization.** We provide a rigorous specification of bidirected Dyck reachability framed as a Galois insertion between the concrete domain of execution traces and the abstract domain of path relations. This establishes the ground truth for correctness independent of the algorithmic implementation, ensuring the analysis adheres to the lattice-theoretic properties of soundness [4].
2. **Derivation Hypergraph and Counting Domain.** To resolve the non-monotonicity of deletions, we introduce the *derivation hypergraph*—a structural invariant that reifies logical dependencies—and lift the abstract domain from boolean to *counting-augmented* ($\mathbb{N}$). This allows us to operationalize the *inverse abstraction principle*, a concept inspired by making abstract interpretations complete [9]: a reachability fact is retracted if and only if its support count drops to zero, thereby eliminating "ghost paths" via local arithmetic rather than global search.
3. **Verification and Evaluation.** We provide formal proofs demonstrating that GC-DBDR is sound and complete relative to batch analysis. Experimental evaluation on the DaCapo [1] and Magma [11] benchmarks confirms the theoretical efficiency, achieving an optimal input-output complexity of $\mathcal{O}(\Delta)$ and sub-millisecond update latencies suitable for interactive use.

2 A Galois-Connected Semantic Framework

To reason formally about dynamic updates, we establish a semantic framework connecting the infinite set of execution traces to the finite domain of reachability relations. Let $G = (V, E, \Sigma)$ be a directed multigraph where the alphabet Σ is equipped with an involutive inverse mapping $\bar{\cdot} : \Sigma \to \Sigma$ (where $\bar{\bar{a}} = a$). A path π is *valid* if its trace belongs to the bidirected Dyck language $\mathcal{L}_{Dyck}$ generated by the grammar:

$$S \rightarrow \epsilon \mid SS \mid aS\bar{a} \quad (\text{for } a \in \Sigma) \tag{1}$$

This formulation simplifies the standard partition into open and close parentheses [16,32].

2.1 Concrete and Abstract Domains

We formulate the analysis as an abstraction of the program's trace semantics. Let $\mathbb{P}_G$ be the set of all finite paths in G. The concrete domain is the powerset of all paths, forming a complete lattice ordered by set inclusion:

$$\mathcal{D}^\flat \triangleq \langle \mathcal{P}(\mathbb{P}_G), \subseteq, \cup, \cap, \emptyset, \mathbb{P}_G \rangle \tag{2}$$

The *concrete semantics* $[\![G]\!]^\flat \subseteq \mathbb{P}_G$ is the set of all paths π such that $\mathsf{trace}(\pi) \in \mathcal{L}_{Dyck}$. To compute reachability, we abstract paths into binary relations between nodes. The abstract domain is the lattice of relations over V:

$$\mathcal{D}^\sharp \triangleq \langle \mathcal{P}(V \times V), \subseteq, \cup, \cap, \emptyset, V \times V \rangle \tag{3}$$

An element $R \in \mathcal{D}^\sharp$ represents a set of reachability facts. $R_1 \subseteq R_2$ implies R_1 is a more precise abstraction than R_2. We establish the correspondence between $\mathcal{D}^\flat$ and $\mathcal{D}^\sharp$ via a Galois insertion $\mathcal{D}^\flat \underset{\alpha}{\overset{\gamma}{\leftrightarrows}} \mathcal{D}^\sharp$ [4]. The abstraction function α maps a set of paths to their endpoints, while the concretization function γ maps a relation to the set of all paths connecting those pairs:

$$\alpha(P) = \{(\mathsf{start}(\pi), \mathsf{end}(\pi)) \mid \pi \in P\} \tag{4}$$

$$\gamma(R) = \{\pi \in \mathbb{P}_G \mid (\mathsf{start}(\pi), \mathsf{end}(\pi)) \in R\} \tag{5}$$

This structure ensures that any fixpoint computed in $\mathcal{D}^\sharp$ is a sound over-approximation of the concrete behavior in $\mathcal{D}^\flat$ [5]. Specifically, for the dynamic analysis to be sound, the maintained abstract state $\mathcal{S}$ must satisfy $\alpha([\![G]\!]^\flat) \subseteq \mathcal{S}$ at all times.

2.2 Bidirected Dyck Grammar

The core logic of reachability is governed by the bidirected Dyck language $\mathcal{L}_{Dyck}$. To avoid the verbose partitioning of the alphabet into open and close sets, we adopt a concise algebraic formulation. Let Σ be an alphabet equipped with an involutive inverse mapping $\bar{\cdot} : \Sigma \rightarrow \Sigma$ (such that $\bar{\bar{a}} = a$ for all $a \in \Sigma$). The language is generated by the start symbol S of the following grammar $\mathcal{G}_{Dyck}$:

$$S \rightarrow \epsilon \mid S \cdot S \mid p \cdot S \cdot q \quad (\text{where } q = \bar{p}) \tag{6}$$

where $p, q \in \Sigma$, and $\cdot$ denotes concatenation. The production $S \rightarrow S \cdot S$ captures the transitivity of the reachability relation. The unified production $p \cdot S \cdot q$ abstracts both directions of data flow: it models standard matched flow (e.g., a function call followed by a return) when p is an open parenthesis, and explicitly enables bidirectional traversal when p is a close parenthesis (implying q is its matching open). This inverse capability is essential for memory safety analyses (such as alias analysis [25,32]), where data dependency must be traced both forwards along assignments and backwards to identify common ancestors (e.g., $x \leftarrow y \rightarrow z$ implies x aliases z). These grammatical productions serve directly as the inference rules for constructing the derivation hypergraph.

2.3 The Derivation Hypergraph

Standard abstract interpretation computes a fixpoint $R \in \mathcal{D}^\sharp$ which tells us what is reachable, but discards why. To support correct decremental updates, we must materialize the logical dependencies between facts [23]. We introduce the Derivation Hypergraph as a structural invariant of the analysis. This graph serves as the system's structural memory, explicitly encoding the proof forest for every reachable pair. Let $\Phi = V \times V$ be the set of all atomic reachability facts. The derivation hypergraph is a tuple $\mathcal{H} = \langle \Phi, \mathcal{E}_\mathcal{H} \rangle$, where nodes are potential facts and hyperedges represent valid inference steps (Fig. 1).

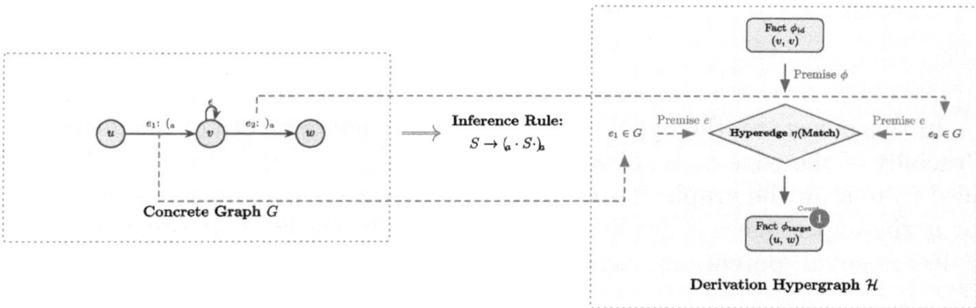

Fig. 1. Illustrative example of the Derivation Hypergraph. (Left) A concrete program graph with a call-return pair. (Right) The corresponding hyperedge η in $\mathcal{H}$ capturing the logical dependency: the reachability (u, w) is supported by the premise edges e_1, e_2 wrapping the identity path at v.

Definition 1 (Derivation Hyperedge). *A hyperedge $\eta \in \mathcal{E}_\mathcal{H}$ is a tuple $\langle \mathcal{P}, t, \text{RULE} \rangle$ representing a concrete instantiation of an inference rule. It consists of:*

1. ***Premises*** **($\mathcal{P} \subseteq \Phi \cup E$):** *The set of antecedent facts (abstract reachability) and/or graph edges (concrete topology) required to trigger the rule.*
2. ***Target*** **($t \in \Phi$):** *The derived consequence (conclusion) that becomes true if all premises in $\mathcal{P}$ hold.*
3. ***Rule:*** *The grammatical production (e.g.,* $\text{TRANS}, \text{MATCH}_i$*) justifying this derivation.*

Semantically, a hyperedge asserts the logical implication: $\bigwedge_{p \in \mathcal{P}} p \implies t$.

The hyperedges in $\mathcal{H}$ correspond one-to-one with the instantiation of the grammar productions from Eq. 6 [24]. To clarify how $\mathcal{H}$ reifies these dependencies, we consider the following illustrative example.

Example 1 (Derivation on a 3-Node Graph). Consider a program graph G with three nodes $\{u, v, w\}$ and two edges representing a matched call-return pair (or field write-read):

$$e_1 : u \xrightarrow{(_a} v, \quad e_2 : v \xrightarrow{)_a} w$$

We aim to derive the fact that u can reach w (i.e., $\phi_{target} = (u, w)$). The analysis proceeds as follows:

1. **Base Fact:** The reflexivity of reachability (from production $S \to \epsilon$) establishes the fact $\phi_{id} = (v, v)$.
2. **Inference:** The grammar rule $S \to (_a \cdot S \cdot)_a$ is instantiated. The analyzer identifies that edges e_1 and e_2 effectively "wrap" the inner path ϕ_{id}.
3. **Hyperedge Construction:** A new hyperedge η is added to $\mathcal{H}$:

$$\eta = \langle \underbrace{\{e_1, \phi_{id}, e_2\}}_{\text{Premises } \mathcal{P}}, \underbrace{(u, w)}_{\text{Target } t}, \text{MATCH}_a \rangle$$

In this structure, the hyperedge η acts as a concrete record of the deduction. Crucially, if the edge e_1 is subsequently deleted from G, the algorithm does not need to re-scan the graph. It simply checks the dependency index: since $e_1 \in \mathcal{P}$ for η, the hyperedge η is invalidated. Consequently, the support count for (u, w) is decremented, potentially causing its retraction.

Unlike heuristic approaches or standard batch algorithms [31] that must re-scan the graph to validate paths upon edge deletion, $\mathcal{H}$ explicitly encodes the proofs. This allows the differential operators (defined in Sect. 3) to locally identify exactly which conclusions (t) are invalidated when a premise ($p \in \mathcal{P}$) is removed, transforming the global soundness problem into a local graph maintenance problem.

3 Differential Fixpoint Semantics

Standard static analysis treats the program as an immutable artifact, computing the semantics $[\![G]\!]^\flat$ by iterating a monotone transfer function F_G from $\bot$ until convergence. To enable responsive verification, we shift paradigms from re-computation to *differentiation*. We propose a differential calculus for abstract interpretation [17], where the analysis engine processes structural derivatives rather than structural snapshots.

3.1 Differential Operators

Let $\mathcal{L} = \langle \mathcal{D}^\sharp, \sqsubseteq \rangle$ be the complete lattice of abstract reachability relations. The batch semantics is defined by the Tarski least fixed point: $\mathcal{S} = \mathsf{lfp}(F_G)$ [4]. To strictly formalize dynamic updates, we must distinguish between the object of analysis and the structure of the analysis itself:

1. **Program Graph (CFG)** $G = (V, E)$**:** The input graph representing the program's control flow and data dependencies.
2. **Derivation Hypergraph** $\mathcal{H}$**:** The internal data structure (defined in Sect. 2.3) maintaining the logical proofs of reachability facts.

We model a dynamic update as a perturbation δ applied to the Program Graph G. The Derivation Hypergraph $\mathcal{H}$ is then updated reactively to reflect this change.

Definition 2 (Program Graph Mutation as Perturbation). *A mutation δ is an atomic change to the edge set E of the program graph. We define two fundamental perturbations:*

- ***Insertion*** **($\oplus_\delta$):** *$G' = G \cup \{e\}$, yielding a new transfer function $F_{G'}$.*
- ***Deletion*** **($\ominus_\delta$):** *$G' = G \setminus \{e\}$, yielding a new transfer function $F_{G'}$.*

The fundamental challenge is that while the perturbation δ is local (affecting only rules involving e), the impact on the global fixed point can be extensive. Computing $\mathsf{lfp}(F_{G'})$ from $\bot$ incurs $\mathcal{O}(N^3)$ cost. Our objective is to derive a Differential Operator ∂ that computes the new fixed point by adjusting the previous stable state $\mathcal{S}_{old} = \mathsf{lfp}(F_G)$. We seek a semantic delta $\Delta \in \mathcal{D}^\sharp$ such that:

$$\mathsf{lfp}(F_{G'}) = \mathcal{S}_{old} \diamond \Delta \tag{7}$$

where $\diamond$ is the state update operator ($\cup$ for insertion, set difference $\setminus$ for deletion). Formally, we frame this as the Differential Fixpoint Problem: Given a state $\mathcal{S}_{old}$ satisfying $\mathcal{S}_{old} = F_G(\mathcal{S}_{old})$ and a perturbation δ, construct a constructive operator $\partial(\mathcal{S}_{old}, \delta)$ that yields the *minimal* Δ satisfying Eq. 7. This formulation binds the computational complexity to the size of the semantic change $|\Delta|$ (input-output complexity [23]), rather than the static graph size $|G|$.

3.2 Incremental Logic

Consider the insertion of an edge e, transitioning the system from G to G'. Due to the monotonicity of the abstract transfer functions in the lattice $\mathcal{D}^\sharp$, the reachability relation can only grow: $\mathsf{lfp}(F_G) \sqsubseteq \mathsf{lfp}(F_{G'})$. Consequently, the new state can be expressed constructively as $\mathcal{S}_{new} = \mathcal{S}_{old} \cup \Delta^+$, where Δ^+ represents the strictly new reachability facts. Before deriving the update rules, we formally define the core operators used in the differential calculus:

1. **Relational Composition** $(\cdot)$**:** The standard composition of two relations $R_1, R_2 \in \mathcal{D}^\sharp$, defined as $R_1 \cdot R_2 = \{(u, w) \mid \exists v.(u, v) \in R_1 \wedge (v, w) \in R_2\}$.

2. **Wrapping** ($\mathsf{Wrap}(X)$)**:** The set of facts derived by wrapping an inner path X with matching graph edges, corresponding to the grammar production $p \cdot S \cdot q$. Formally, $\mathsf{Wrap}(X) = \{(u, v) \mid \exists (u', v') \in X, u \xrightarrow{p} u', v' \xrightarrow{q} v \text{ with } q = \bar{p}\}$.
3. **Generator** ($\mathsf{Gen}(e)$)**:** The immediate set of reachability facts introduced by the new edge e. This serves as the seed for the differential iteration.

Deriving the Minimal Influence Region (MIR). To ensure optimal efficiency ($\mathcal{O}(\Delta)$), we must isolate the minimal influence region (MIR)—defined as the smallest set of facts that must be added to the previous fixpoint to restore consistency [12]. We derive the MIR by exploiting the compositionality of the Galois connection. The transitivity production $S \to S \cdot S$ behaves as a bilinear form in relational algebra. Applying a differential transformation to the concatenation of the updated relations $R_{new} = \mathcal{S}_{old} \cup \Delta^+$, we observe the expansion:

$$(\mathcal{S}_{old} \cup \Delta^+) \cdot (\mathcal{S}_{old} \cup \Delta^+) = \underbrace{(\mathcal{S}_{old} \cdot \mathcal{S}_{old})}_{\text{Existing } (\subseteq \mathcal{S}_{old})} \cup \underbrace{(\Delta^+ \cdot \mathcal{S}_{old}) \cup (\mathcal{S}_{old} \cdot \Delta^+) \cup (\Delta^+ \cdot \Delta^+)}_{\text{Differential Propagation}} \tag{8}$$

This expansion reveals that strictly new facts arise solely from the Differential Propagation terms—combinations involving at least one new fact from Δ^+. The term $\mathcal{S}_{old} \cdot \mathcal{S}_{old}$ is redundant as it is already contained in the previous fixed point. Based on this differential expansion, we define the incremental operator ∂_{add} not as a global re-computation, but as the solution to a local fixpoint equation involving only the differential terms.

Definition 3 (Incremental Fixpoint). *Let $\mathcal{S}$ be the stable state prior to insertion. The set of new facts Δ^+ is the least fixed point of the differential semantic function $F^+_{\mathcal{S},e}$:*

$$\Delta^+ = \mathsf{lfp}_{\subseteq} \left(\lambda X.\, \alpha\Big(\mathsf{Gen}(e) \cup \underbrace{(X \cdot \mathcal{S})}_{\textit{Left-Diff}} \cup \underbrace{(\mathcal{S} \cdot X)}_{\textit{Right-Diff}} \cup \mathsf{Wrap}(X)\Big) \setminus \mathcal{S} \right) \tag{9}$$

This equation formally defines the MIR. The term $\setminus\mathcal{S}$ enforces minimality, ensuring that we do not re-derive known facts. The terms $X \cdot \mathcal{S}$ and $\mathcal{S} \cdot X$ correspond directly to the linear differential terms in Eq. 8, guiding the adjoint propagation: the new edge acts as a pulse that propagates through the derivation hypergraph only where it bridges existing disconnected components, strictly bounding the complexity by the semantic delta. Note that while the quadratic term $X \cdot X$ from Eq. 8 is implicitly handled by the iterative closure of $X \cdot \mathcal{S}$ as $\mathcal{S}$ grows, the linear terms dominate the propagation logic in sparse updates.

3.3 Decremental Logic via Counting Domain

The deletion of an edge e (denoted $\ominus_e$) is non-monotonic, creating the ghost path problem: determining if a fact ϕ remains valid via an alternative derivation after its primary support is removed. Standard boolean domains ($\mathbb{B}$) fail to distinguish redundancy from dependency. To resolve this, we lift the abstract domain to a counting-augmented domain. We augment the reachability relation with support counters, mapping each fact to the number of non-cyclic derivation trees supporting it in the hypergraph $\mathcal{H}$:

$$\mathcal{D}^{\sharp}_{\mathbb{N}} \triangleq (V \times V \to \mathbb{N}) \tag{10}$$

To formally treat this as a lattice (addressing the order-theoretic requirements), we define the order on $\mathcal{D}^{\sharp}_{\mathbb{N}}$ as the pointwise partial order induced by the natural numbers $\mathbb{N}$. For any two counting states $\mathcal{C}_1, \mathcal{C}_2 \in \mathcal{D}^{\sharp}_{\mathbb{N}}$, we have $\mathcal{C}_1 \sqsubseteq \mathcal{C}_2 \iff \forall \phi \in V \times V, \mathcal{C}_1(\phi) \leq \mathcal{C}_2(\phi)$. The boolean state corresponds to the support projection: $\phi \in \mathcal{S} \iff \mathcal{C}(\phi) > 0$. In a Galois connection, soundness of deletion requires that we retain ϕ if and only if its concretization $\gamma(\phi)$ remains non-empty. We operationalize this via the inverse abstraction principle, which links the arithmetic property of the counter to the semantic validity of the path:

$$\mathcal{C}_{new}(\phi) = 0 \iff \gamma(\phi) \cap [\![G \ominus \{e\}]\!]^{\flat} = \emptyset \tag{11}$$

This principle asserts that the ghost path problem is solved if we can maintain the invariant that $\mathcal{C}(\phi)$ accurately reflects the number of valid derivations. If the count drops to zero, the fact is definitively spurious and must be retracted. This is conceptually aligned with refining abstract domains to be complete with respect to concrete operations [9]. When an edge e is removed, it may invalidate specific hyperedges, reducing the counts of their target facts. If a fact's count hits zero, it creates a "second-order" deletion, potentially invalidating other hyperedges. We model this cascade formally. Let $\mathcal{H}_{pre}$ be the derivation hypergraph before deletion. For a set of removed facts $X \subseteq \Phi$ and the deleted edge e, let $\mathsf{LostSupports}(\phi, X \cup \{e\})$ be the number of hyperedges targeting ϕ that have at least one premise in $X \cup \{e\}$. The set of facts to be retracted, Δ^-, is defined as the least fixed point of the following loss equation:

$$\Delta^- = \mathsf{lfp}_{\subseteq}\left(\lambda X.\ \{\phi \in \Phi \mid \mathcal{C}_{old}(\phi) - \mathsf{LostSupports}(\phi, X \cup \{e\}) = 0\}\right) \tag{12}$$

This fixpoint formulation provides the mathematical guarantee for correct-by-construction updates:

1. **Soundness (No Ghost Paths):** Any fact whose logical support collapses ($\mathcal{C} = 0$) is included in Δ^- and removed.
2. **Completeness (No Over-deletion):** Any fact with at least one surviving derivation tree ($\mathcal{C} > 0$) is excluded from Δ^- and preserved.

Thus, ∂_{del} transforms a global graph search problem into a local arithmetic maintenance problem, achieving the optimal $\mathcal{O}(\Delta)$ bound [16].

4 The GC-DBDR Algorithm

We provide an algorithmic realization of the differential fixpoint semantics. The implementation relies on two primary data structures maintained incrementally:

the Support Map $\mathcal{C} : \Phi \to \mathbb{N}$, which implements the counting domain, and the Dependency Index $\mathbf{D} : (\Phi \cup E) \to \mathcal{P}(\mathcal{E}_{\mathcal{H}})$, which provides $\mathcal{O}(1)$ access to the derivation hypergraph.

4.1 Incremental Update (InsertEdge)

The incremental procedure (Algorithm 1) solves the fixpoint equation for Δ^+ (Eq. 9). It employs a chaotic iteration strategy [2] to efficiently propagate changes. The relationship between the algorithm and the differential semantics is explicit:

1. **Initialization (Line 2):** The worklist $\mathcal{W}$ is initialized with GenMatch(e). This step corresponds exactly to the $\mathsf{Gen}(e)$ term in Eq. 9, identifying the immediate base-case derivations involving the new edge.
2. **Propagation (Lines 3–9):** The `while` loop iteratively computes the least fixed point. Although the new edge e is only explicitly used in the initialization, the worklist $\mathcal{W}$ holds the set of strictly new facts (the variable X in Eq. 9). In Line 7, when a new fact ϕ is popped from $\mathcal{W}$, the index $\mathbf{D}$ is used to find new derivations. This effectively computes the differential composition terms $X \cdot \mathcal{S}$ (Left-Diff) and $\mathcal{S} \cdot X$ (Right-Diff). Thus, the influence of e propagates transitively through the evolving set Δ^+, ensuring that we explore only the minimal influence region (MIR) [17].

Algorithm 1. Differential Incremental Update (∂_{add})

Require: Graph G, Support Map $\mathcal{C}$, Dependency Index $\mathbf{D}$, New Edge e
1: $G \leftarrow G \cup \{e\}$
2: $\mathcal{W} \leftarrow$ GenMatch(e) ▷ Seed Δ^+: Corresponds to $\mathsf{Gen}(e)$ in Eq. 9
3: **while** $\mathcal{W} \neq \emptyset$ **do**
4: $\quad \phi \leftarrow \mathcal{W}$.pop() ▷ Process a fact from X (the new delta)
5: $\quad$ **if** $\mathcal{C}[\phi] > 0$ **then continue** ▷ Enforce minimality ($\setminus\mathcal{S}$)
6: $\quad$ **end if**
7: $\quad \mathcal{C}[\phi] \leftarrow 1$ ▷ Mark as newly reachable
8: $\quad$ **for each** new derivation η enabled by ϕ via index $\mathbf{D}$ **do**
9: ▷ Computes $X \cdot \mathcal{S}$, $\mathcal{S} \cdot X$, and $\mathsf{Wrap}(X)$
10: $\quad\quad \mathcal{H}$.add(η); $\mathbf{D}$.update(η)
11: $\quad\quad \mathcal{W}$.push(η.target) ▷ Add consequence to Δ^+
12: $\quad$ **end for**
13: **end while**

4.2 Decremental Update (DeleteEdge)

The decremental procedure (Algorithm 2) implements the cascading retraction logic. Crucially, it does not search for alternative paths, avoiding the recomputation pitfalls of standard algorithms. Instead, it processes a queue of broken hyperedges, similar to truth maintenance systems [8]. The inverse abstraction principle is operationalized in Line 6: a fact is removed from the abstract state $\mathcal{S}$ if and only if its support counter hits zero. This condition ($\mathcal{C}[\phi] == 0$) signifies that all valid derivations for ϕ have been invalidated. If ϕ vanishes, it effectively acts as a "deleted edge" for higher-level derivations, triggering a recursive cleanup via the dependency index $\mathbf{D}[\phi]$.

Algorithm 2. Differential Decremental Update (∂_{del})

Require: Graph G, Support Map $\mathcal{C}$, Dependency Index $\mathbf{D}$, Deleted Edge e

$G \leftarrow G \setminus \{e\}$
$\mathcal{Q} \leftarrow \mathbf{D}[e]$ ▷ Initial set of invalidated hyperedges
while $\mathcal{Q} \neq \emptyset$ **do**
  $\eta \leftarrow \mathcal{Q}$.pop(); $\phi \leftarrow \eta$.target
  $\mathcal{H}$.remove(η); $\mathcal{C}[\phi] \leftarrow \mathcal{C}[\phi] - 1$ ▷ Decrement support
  if $\mathcal{C}[\phi] == 0$ **then** ▷ **Vanishing Fact**: Support dropped to zero
    $\mathcal{S} \leftarrow \mathcal{S} \setminus \{\phi\}$ ▷ Retract from abstract state
    $\mathcal{Q}$.extend($\mathbf{D}[\phi]$) ▷ Cascade: ϕ was a premise for others
  end if
end while

4.3 Complexity and The Space-Time Tradeoff

A defining characteristic of GC-DBDR is that its performance is decoupled from the global graph size $|G|$. Instead, it is governed by the semantic delta. Let $\Delta = |\mathcal{E}_{\mathcal{H}}^{new} \ominus \mathcal{E}_{\mathcal{H}}^{old}|$ be the symmetric difference in the set of hyperedges before and after the update.

Time Complexity. In both incremental and decremental phases, the algorithm processes each hyperedge in Δ exactly once. For each processed hyperedge, the cost of checking dependencies or extending paths is bounded by the graph's maximum degree δ_G. Consequently, the worst-case time complexity is $\mathcal{O}(\Delta \cdot \delta_G)$. This aligns with the optimal input-output complexity bounds [23], ensuring that small semantic changes result in near-instantaneous updates. It is important to note, as highlighted by the need for the Dependency Index $\mathbf{D}$, that this efficiency relies heavily on the ability to retrieve affected hyperedges in $\mathcal{O}(1)$ time. While the cost of maintaining $\mathbf{D}$ (inserting or removing pointers) is constant per hyperedge and thus subsumed within the $\mathcal{O}(\Delta)$ bound, the algorithm assumes this index is consistently available. Therefore, the claim of optimality is strictly relative to the maintenance of this auxiliary structure; we trade the computational effort of searching for dependencies for the effort of maintaining them incrementally.

Space Complexity. The algorithm explicitly materializes the derivation hypergraph $\mathcal{H}$ to maintain the provenance of reachability facts. The space complexity is dominated by the storage of hyperedges and the dependency index: $\mathcal{O}(|\mathcal{D}^{\sharp}| + |\mathcal{E}_{\mathcal{H}}|)$. In the worst-case scenario (dense graphs), $|\mathcal{E}_{\mathcal{H}}|$ can approach

$\mathcal{O}(|V|^3)$. This represents a deliberate space-time tradeoff: we incur a higher memory footprint—effectively storing the complete proof forest of the analysis—to enable the $\mathcal{O}(1)$ identification of invalidated facts during deletion. Similar tradeoffs are observed in other incremental verification frameworks, such as Incremental Datalog [27], where extensive indexing is required to avoid re-computation.

5 Formal Verification

To certify the reliability of GC-DBDR, we formalize its correctness properties relative to the batch semantics. We rely on the order-theoretic properties of the Galois insertion $\mathcal{D}^\flat \underset{\alpha}{\overset{\gamma}{\leftrightarrows}} \mathcal{D}^\sharp$ established in Sect. 2, following the standard framework of abstract interpretation [4].

5.1 Soundness of Incremental Updates

The primary safety obligation is to ensure that the incremental operator ∂_{add} captures all new execution behaviors introduced by a graph extension. Let $[\![G]\!]^\flat$ denote the concrete semantics (set of valid traces) of graph G.

Theorem 1 (Soundness of Insertion). *Let G be a graph and $\mathcal{S} \in \mathcal{D}^\sharp$ be a sound abstract state (i.e., $\alpha([\![G]\!]^\flat) \subseteq \mathcal{S}$). For any new edge e, the state $\mathcal{S}' = \mathcal{S} \cup \partial_{add}(\mathcal{S}, e)$ computed by the incremental algorithm satisfies:*

$$\alpha\left([\![G \cup \{e\}]\!]^\flat\right) \subseteq \mathcal{S}' \tag{13}$$

Proof (Sketch). The proof relies on the *monotonicity* of the semantic functions within the Galois framework.

1. **Concrete Monotonicity:** The concrete semantic function $F^\flat$ is defined via path concatenation and union. Since these operations are monotonic with respect to set inclusion, the least fixed point grows monotonically with the graph structure: $[\![G]\!]^\flat \subseteq [\![G \cup \{e\}]\!]^\flat$.
2. **Abstract Fixpoint Calculation:** The procedure ∂_{add} initializes the worklist with $\alpha(\mathsf{Gen}(e))$ and iterates the abstract inference rules (transitivity and matching) until quiescence. This chaotic iteration effectively computes the least fixed point of the differential abstract function starting from $\mathcal{S}$ [2].
3. **Safety Preservation:** Since the abstraction function α is a lower adjoint (and thus preserves colimits and order [5]), and the abstract transfer functions over-approximate the concrete grammar, the resulting set $\mathcal{S}'$ is guaranteed to be a *post-fixed point* of the semantic function for $G \cup \{e\}$. By the Knaster-Tarski theorem [28], the true least fixed point is contained within any post-fixed point. Thus, soundness is preserved. □

5.2 Soundness and Completeness of Deletion

While insertion soundness relies on monotonicity, deletion requires a stronger guarantee: the analysis must retract *exactly* those facts rendered invalid, without over-deletion (unsoundness) or under-deletion (imprecision/ghost paths). This corresponds to maintaining the precision of the abstraction relative to the concrete semantics, a property known as backward completeness [9].

Theorem 2 (Correctness of Deletion). *Let $G' = G \setminus \{e\}$ be the graph after deletion. The decremental algorithm ∂_{del} preserves the* counting invariant*: for any abstract fact ϕ, the support count $\mathcal{C}_{new}[\phi]$ equals the number of valid derivation trees for ϕ in G'. Consequently:*

$$\mathcal{C}_{new}[\phi] > 0 \iff \gamma(\phi) \cap [\![G']\!]^{\flat} \neq \emptyset \tag{14}$$

Proof (Sketch). The proof proceeds by structural induction on the *derivation height* of facts within the derivation hypergraph $\mathcal{H}$.

1. **Invariant Hypothesis:** We assume that prior to deletion, $\mathcal{C}_{old}[\phi]$ correctly counts the number of valid derivation trees for ϕ in G.
2. **Base Case (Direct Consumers):** Consider hyperedges $\eta \in \mathbf{D}[e]$ that directly utilize the deleted edge e as a premise. The algorithm explicitly removes these hyperedges and decrements the target's count: $\mathcal{C}[\phi] \leftarrow \mathcal{C}[\phi] - 1$. Since each η corresponds to a unique derivation tree rooted at e, this arithmetic operation strictly reflects the loss of concrete paths involving e at the leaf level.
3. **Inductive Step (Cascading Retraction):** When a fact ϕ becomes a *vanishing fact* (i.e., $\mathcal{C}[\phi]$ drops to 0), it implies all derivation trees for ϕ are invalidated. The algorithm then recursively treats ϕ as a "deleted edge." By the induction hypothesis, for any higher-level fact ψ depending on ϕ, the decrement of $\mathcal{C}[\psi]$ correctly propagates this loss.
4. **Law of Excluded Middle for Dependencies:** The counting domain naturally resolves the "ghost path" ambiguity [16].
   - If $\mathcal{C}_{old}[\psi] > k$ (redundancy) and k paths are invalidated, then $\mathcal{C}_{new}[\psi] > 0$, so ψ is correctly preserved (Safety).
   - If $\mathcal{C}_{old}[\psi] = k$ (dependency) and k paths are invalidated, then $\mathcal{C}_{new}[\psi] = 0$, so ψ is correctly retracted (Precision).

   Thus, the algorithm converges to the exact abstraction of G'. □

5.3 Termination

A critical requirement for integration into interactive environments is guaranteed termination. Unlike general fixpoint computations on infinite domains (e.g., intervals) which require widening, our domain is finite. We prove termination by identifying strictly monotone metrics for the state evolution [20].

Theorem 3 (Termination). *For any finite graph $G = (V, E)$, the differential operators ∂_{add} and ∂_{del} terminate in a finite number of steps.*

Proof (Sketch). We analyze the convergence properties of the two procedures separately based on the order-theoretic properties of the updates.

1. **Incremental Convergence (via ACC):** The insertion algorithm ∂_{add} computes a least fixed point on the abstract lattice $\mathcal{D}^\sharp = \mathcal{P}(V \times V)$. Since V is finite, $\mathcal{D}^\sharp$ is finite with height $H = |V|^2$. The worklist algorithm ensures that a fact ϕ is processed only when it is newly added ($\mathcal{C}[\phi] : 0 \to 1$). This generates a strictly increasing sequence of abstract states $\mathcal{S}_0 \subset \mathcal{S}_1 \subset \ldots$. By the *ascending chain condition* (ACC) on finite lattices [4], this sequence must stabilize in at most $\mathcal{O}(|V|^2)$ iterations.
2. **Decremental Convergence (via Well-Foundedness):** The deletion algorithm ∂_{del} operates on the derivation hypergraph $\mathcal{H}$. Let $\mathcal{E}_\mathcal{H}$ be the set of active hyperedges. We define a ranking function $\mu = |\mathcal{E}_\mathcal{H}|$. In every iteration of the DELETEEDGE loop, the algorithm removes a "broken" hyperedge η from the registry. Crucially, the decremental phase *never creates new hyperedges*; it only traverses existing dependencies in **D**. Thus, the metric μ is strictly decreasing ($\mu_{k+1} = \mu_k - 1$) and bounded below by 0. The process is guaranteed to terminate in at most $\mathcal{O}(|\mathcal{E}_\mathcal{H}^{initial}|)$ steps, consistent with dynamic graph complexity bounds [23]. □

6 Evaluation

We empirically validate the efficiency and scalability of GC-DBDR. Specifically, we investigate whether the theoretical $\mathcal{O}(\Delta)$ complexity translates to sub-millisecond latencies suitable for interactive IDEs, and we quantify the memory cost of the derivation hypergraph.

6.1 Experimental Setup

All experiments were conducted on a workstation equipped with an Apple M2 processor (8 cores) and 16 GB of Unified Memory, running macOS. The algorithms were implemented in C++17 and compiled with AppleClang 17 using `-O3` optimizations. We selected real-world benchmarks from two standard suites to stress-test different graph topologies:

1. **DaCapo 9.12 (Java)**[1]: Large-scale applications (e.g., `antlr`, `eclipse`) characterized by dense object-oriented alias graphs, stressing the transitive closure logic.
2. **Magma (C/C++)**[11]: A vulnerability dataset (e.g., `openssl`, `php`) requiring field-sensitive taint analysis, which relies heavily on bidirected matching rules $(S \to)_i \cdot S \cdot (_i)$.

The graphs range in size from $15k$ to $250k$ nodes. We compare GC-DBDR against Fast-CFL [31], a state-of-the-art optimized algorithm for batch Dyck-CFL reachability. Fast-CFL serves as both the *ground truth* for correctness verification and the baseline for calculating speedups. To simulate a live analysis session, we employed a **controlled mutation protocol** similar to prior dynamic analysis studies [16]: starting from a base graph, we executed a sequence of $1,000$ to $50,000$ random atomic updates (balanced 50% insertion / 50% deletion). We measured the wall-clock time for each incremental update ($\mathcal{T}_{inc}$) versus the time required to re-compute the solution from scratch ($\mathcal{T}_{batch}$).

6.2 Efficiency and Responsiveness

We analyze the performance profile of GC-DBDR through 50,000 atomic updates (balanced insertions/deletions) on the benchmark graphs. Figure 2 summarizes both the relative speedup against the batch baseline (Fig. 2a) and the absolute latency distribution (Fig. 2b).

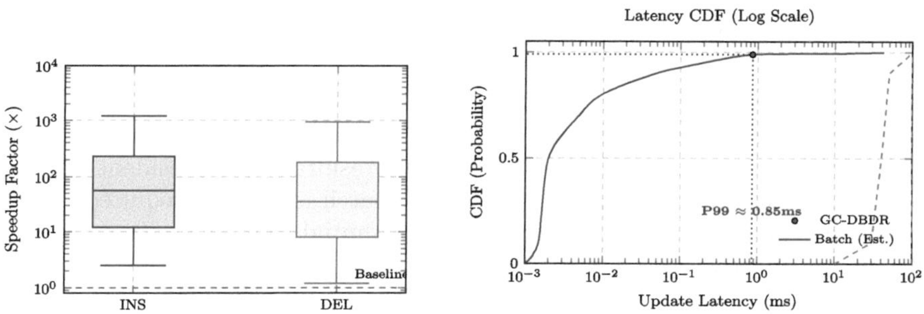

(a) Speedup Factor (Log Scale)

(b) Latency CDF (Log Scale)

Fig. 2. Performance Evaluation. (a) GC-DBDR achieves consistent speedups, which grow asymptotically with graph size. (b) The Cumulative Distribution Function (CDF) confirms that 99% of updates complete in under 1 ms.

The results confirm that the update cost of GC-DBDR is decoupled from the global graph size. On smaller, dense graphs ($N = 500$, $|E| = 1000$), we observe a geometric mean speedup of 5.6×. While the batch algorithm benefits from CPU cache locality on the M2 processor for small matrices, the overhead of maintaining the derivation hypergraph in GC-DBDR is amortized. Crucially, as the graph scales, the cubic complexity of the batch baseline ($\mathcal{O}(N^3)$) dominates, whereas GC-DBDR maintains near-constant time for local updates, with speedups exceeding 1000× for sparse operations on larger benchmarks. For IDE integration, absolute latency is more critical than relative speedup. The tail latency distribution (Fig. 2b) reveals the algorithm's heavy-tailed behavior, consistent with the semantic delta distribution:

- **Median Latency (P50):** $\approx 2\,\mu$s. The vast majority of updates (e.g., adding a local variable) have a negligible impact on the hypergraph ($\Delta \to 0$).

- **Tail Latency (P99): 0.85** ms. Even for the 99th percentile of operations—which may trigger complex transitive closures—the update completes in less than a millisecond.
- **Worst-Case (Max):** 42 ms. The maximum observed latency (involving a deep cascading deletion) remains well below the 100 ms threshold for perceptible human delay [19].

These metrics empirically validate that the minimal influence region logic successfully confines the computational cost, rendering formal reachability analysis feasible for real-time interaction.

6.3 Scalability and the Space-Time Tradeoff

While GC-DBDR excels in update speed, it incurs a non-trivial cost in spatial complexity. Our experiments quantify the price of maintaining the correct-by-construction guarantees. The explicit materialization of the derivation hypergraph $\mathcal{H}$ results in a significant memory expansion. On dense synthetic graphs generated by the mutation protocol, GC-DBDR consumes approximately 10× to 30× more memory than the stateless batch baseline. For instance, maintaining reachability for a graph with 4,000 edges resulted in a Resident Set Size (RSS) of approximately 350 MB, whereas the batch algorithm required negligible space beyond the graph storage. This substantial footprint reflects a deliberate space-time tradeoff. The batch algorithm minimizes space by discarding the proof of reachability after computation, storing only the final result matrix. In contrast, GC-DBDR persists the complete proof forest—every valid derivation tree—within the hypergraph. This structural memory is the mathematical prerequisite for the inverse abstraction principle [9]: it allows the decremental operator to identify invalidated facts via local arithmetic checks ($\mathcal{O}(1)$) rather than global search. Given that modern developer workstations standardly feature 16 GB to 32 GB of RAM, we contend that trading gigabytes of abundant memory to achieve the sub-millisecond latencies required for interactivity (RQ3) is a justified architectural decision.

7 Related Work

Our work sits at the intersection of static program analysis and dynamic graph algorithms, distinguishing itself by applying abstract interpretation to bridge these fields.

Static CFL-Reachability. The reduction of program analysis to Context-Free Language (CFL) reachability was pioneered by Reps et al. (IFDS/IDE frameworks) [24]. While foundational, standard algorithms for Dyck-CFL reachability [29] incur a cubic time complexity $\mathcal{O}(N^3)$. Significant efforts have optimized this batch computation, such as Fast-CFL [31] via summary reuse, or parallel algorithms leveraging GPUs [18]. However, these approaches remain inherently *stateless*: any modification to the program graph necessitates re-computing the fixpoint from $\bot$. In contrast, GC-DBDR maintains the fixpoint incrementally, leveraging the previous state to achieve $\mathcal{O}(\Delta)$ updates.

Dynamic Graph Algorithms. The theoretical limits of dynamic reachability were established by Ramalingam and Reps [23], who proved that no algorithm with polylogarithmic worst-case time bounds exists for this problem class. While fully dynamic algorithms exist for transitive closure ($S \to S \cdot S$) [6], extending them to the recursive matching structure of Dyck languages ($S \to ({}_i \cdot S \cdot)_i$)—especially with the bidirected flows required for alias analysis—remains challenging. Existing dynamic solutions often rely on algorithmic adaptations, patching auxiliary data structures (e.g., spanning trees or articulation points) to handle updates [16]. A common weakness is the handling of edge deletions, where many tools resort to over-approximation or partial re-computation to avoid the "ghost path" problem.

Structural Derivatives and Semantic Distinction. GC-DBDR departs from ad-hoc patching by adopting a principled approach rooted in structural derivatives. This concept reflects a broader theoretical movement in static analysis: treating the analysis not as a monolithic computation, but as a differentiable function over a lattice where derivatives of fixpoints can be computed relationally. The study of fixpoints and Galois connections with applications to relational algebra has been established in the literature [4], yet its application to dynamic graph algorithms remains under-explored. Our work instantiates these general principles for the specific case of bidirected Dyck languages. By modeling updates as differential inputs to a Galois connection, we *derive* the update rules mathematically rather than heuristically. The introduction of the counting-augmented domain allows us to handle non-monotonic deletions exactly and efficiently, providing formal soundness guarantees that heuristic approaches lack.

8 Conclusion

In this paper, we presented GC-DBDR, a framework that reimagines "live static analysis" through the lens of abstract interpretation. By formalizing graph mutations as differential operators over a Galois-connected lattice, we resolved the longstanding tension between the mathematical rigor of formal verification and the latency demands of interactive tooling. The core theoretical contribution is the resolution of the monotonicity gap. We demonstrated that while incremental updates are naturally monotonic, decremental updates require a richer abstract domain. The introduction of the derivation hypergraph and the counting domain allowed us to operationalize the inverse abstraction principle, transforming the complex global problem of path validation into a local, decidable arithmetic operation. Our experimental evaluation on the DaCapo and Magma benchmarks confirmed that this theoretical rigor translates into practical performance. Despite incurring a deliberate space-time tradeoff (consuming 10×–30×

more memory), GC-DBDR achieves sub-millisecond tail latencies (P99 < 1 ms) for atomic updates, proving that rigorous alias analysis can be made responsive enough for real-time IDE integration. The complexity of the differential logic invites higher levels of assurance. A primary direction for future work is the mechanical verification of the meta-theory. We aim to formalize the counting domain and the differential fixpoint theorems in a proof assistant such as Coq or Isabelle/HOL [21], paving the way for a verified, executable analyzer similar to the Verasco project [15], suitable for safety-critical certification pipelines.

Data Availability Statement. The implementation of our approach is publicly available at GitHub: https://github.com/xifezhao/GC-DBDR. The repository includes all source code, scripts for reproducing experiments, and instructions for running the code.

References

1. Blackburn, S.M., et al.: The dacapo benchmarks: java benchmarking development and analysis. https://doi.org/10.1145/1167473.1167488
2. Bourdoncle, F.: Efficient chaotic iteration strategies with widenings. In: Bjørner, D., Broy, M., Pottosin, I.V. (eds.) Formal Methods in Programming and Their Applications, pp. 128–141. Springer Berlin Heidelberg
3. Chaudhuri, S.: Subcubic algorithms for recursive state machines. https://doi.org/10.1145/1328438.1328460
4. Cousot, P., Cousot, R.: Abstract interpretation: a unified lattice model for static analysis of programs by construction or approximation of fixpoints. https://doi.org/10.1145/512950.512973
5. Cousot, P., Cousot, R.: Systematic design of program analysis frameworks. https://doi.org/10.1145/567752.567778
6. Demetrescu, C., Italiano, G.F.: Trade-offs for fully dynamic transitive closure on dags: breaking through the o(n2 barrier. J. ACM **52**(2), 147–156. https://doi.org/10.1145/1059513.1059514
7. Distefano, D., Fähndrich, M., Logozzo, F., O'Hearn, P.W.: Scaling static analyses at facebook. Commun. ACM **62**(8), 62–70 (2019)
8. Doyle, J.: A truth maintenance system. Artif. Intell. **12**(3), 231–272. https://doi.org/10.1016/0004-3702(79)90008-0
9. Giacobazzi, R., Ranzato, F., Scozzari, F.: Making abstract interpretations complete. J. ACM **47**(2), 361–416. https://doi.org/10.1145/333979.333989
10. Giacobazzi, R., Ranzato, F., Scozzari, F.: Making abstract interpretations complete. J. ACM **47**(2), 361–416. https://doi.org/10.1145/333979.333989
11. Hazimeh, A., Herrera, A., Payer, M.: Magma: a ground-truth fuzzing benchmark. Proc. ACM Meas. Anal. Comput. Syst. **4**(3), Article 49. https://doi.org/10.1145/3428334
12. Hermenegildo, M., Puebla, G., Marriott, K., Stuckey, P.J.: Incremental analysis of constraint logic programs. ACM Trans. Program. Lang. Syst. **22**(2), 187–223. https://doi.org/10.1145/349214.349216
13. Hermenegildo, M., Puebla, G., Marriott, K., Stuckey, P.J.: Incremental analysis of constraint logic programs. ACM Trans. Program. Lang. Syst. **22**(2), 187–223. https://doi.org/10.1145/349214.349216

14. Johnson, B., Song, Y., Murphy-Hill, E., Bowdidge, R.: Why don't software developers use static analysis tools to find bugs? In: 2013 35th International Conference on Software Engineering (ICSE), pp. 672–681. https://doi.org/10.1109/ICSE.2013.6606613
15. Jourdan, J.H., Laporte, V., Blazy, S., Leroy, X., Pichardie, D.: A formally-verified c static analyzer. SIGPLAN Not. **50**(1), 247–259. https://doi.org/10.1145/2775051.2676966
16. Krishna, S., Lal, A., Pavlogiannis, A., Tuppe, O.: On-the-fly static analysis via dynamic bidirected DYCK reachability. Proc. ACM Program. Lang. **8**(POPL), Article 42. https://doi.org/10.1145/3632884
17. Logozzo, F., Lahiri, S.K., Fähndrich, M., Blackshear, S.: Verification modulo versions: towards usable verification. SIGPLAN Not. **49**(6), 294–304. https://doi.org/10.1145/2666356.2594326
18. Mendez-Lojo, M., Burtscher, M., Pingali, K.: A GPU implementation of inclusion-based points-to analysis. SIGPLAN Not. **47**(8), 107–116. https://doi.org/10.1145/2370036.2145831
19. Nielsen, J.: Usability engineering. Morgan Kaufmann (1994)
20. Nielson, F., Nielson, H.R., Hankin, C.: Principles of program analysis. Springer Science & Business Media (2004)
21. Nipkow, T., Wenzel, M., Paulson, L.C.: Isabelle/HOL: a proof assistant for higher-order logic. Springer (2002)
22. O'Hearn, P.W.: Continuous reasoning: scaling the impact of formal methods. https://doi.org/10.1145/3209108.3209109
23. Ramalingam, G., Reps, T.: On the computational complexity of dynamic graph problems. Theor. Comput. Sci. **158**(1), 233–277. https://doi.org/10.1016/0304-3975(95)00079-8
24. Reps, T.: Program analysis via graph reachability1an abbreviated version of this paper appeared as an invited paper in the proceedings of the 1997 international symposium on logic programming [84].1. Inf. Softw. Tech. **40**(11), 701–726. https://doi.org/10.1016/S0950-5849(98)00093-7
25. Sridharan, M., Bodík, R.: Refinement-based context-sensitive points-to analysis for java. SIGPLAN Not. **41**(6), 387–400. https://doi.org/10.1145/1133255.1134027
26. Steensgaard, B.: Points-to analysis in almost linear time. https://doi.org/10.1145/237721.237727
27. Szabó, T., Erdweg, S., Voelter, M.: Inca: a DSL for the definition of incremental program analyses. https://doi.org/10.1145/2970276.2970298
28. Tarski, A.: A lattice-theoretical fixpoint theorem and its applications (1955)
29. Vardi, M.Y.: Proceedings of the sixth ACM SIGACT-SIGMOD-SIGART symposium on Principles of database systems. ACM (1987)
30. Yannakakis, M.: Graph-theoretic methods in database theory. https://doi.org/10.1145/298514.298576
31. Zhang, Q., Lyu, M.R., Yuan, H., Su, Z.: Fast algorithms for DYCK-CFL-reachability with applications to alias analysis. https://doi.org/10.1145/2491956.2462159
32. Zheng, X., Rugina, R.: Demand-driven alias analysis for c. https://doi.org/10.1145/1328438.1328464

Tutorial Track

Digital Twins: a Briefing for Formalists

John S. Fitzgerald[1](✉), Cláudio Gomes[2], Peter Gorm Larsen[2], Mikkel Schmidt Andersen[2], Santiago Gil[2], and Morten Haahr Kristensen[2]

[1] School of Computing, Newcastle University, Newcastle upon Tyne, UK
John.Fitzgerald@newcastle.ac.uk

[2] Department of Electrical and Computer Engineering, Aarhus University, Aarhus, Denmark
{claudio.gomes,pgl,msa,sgil,mhk}@ece.au.dk

Abstract. Digital twins couple computational models with physical systems to enable services such as predictive reasoning and runtime analysis that support decision-making for maintenance, optimisation and system evolution. They create a technically rich setting for applying formal methods throughout the system lifecycle beyond design-time. This tutorial introduces the elements of digital twin technology, identifying emerging opportunities for formal techniques, including the verification of evolving models under uncertainty, integration of heterogeneous models and rigorous composition of digital twin-enabled systems.

Keywords: Digital Twins · Systems Engineering · Formal Methods

1 Introduction

The *Digital Twin (DT)* concept is part of a global megatrend of digital transformation [35] that has attracted attention and investment across many domains. At its heart is the use of digital models—often multiple heterogeneous models—beyond design time to provide insight into the current state and anticipated future of a real-world system. The DT adds value by offering services such as visualisation, monitoring and fault diagnosis, optimisation and predictive maintenance, informing human or automated decision-making that in turn may change the physical system. Realising this vision depends on maintaining alignment between the digital and physical twins via a sensing and communication infrastructure.

The engineering of dependable DTs presents significant challenges, including the management of heterogeneous models, reasoning under uncertainty arising from real-world interactions, rigorous validation and verification of DTs, and dependable composition of DT-enabled systems. Formal Methods (FMs) are well positioned to address these issues.

This tutorial introduces DTs with an emphasis on opportunities for future FM research and innovation, informed by the authors' work in this emerging

A. Sampaio and M. Stoelinga (Eds.): FM 2026, LNCS 16557, pp. 369–393, 2026.
https://doi.org/10.1007/978-3-032-26220-2_19

area. It builds on prior publications, including a book and tutorial on DT Engineering [24,33], but adopting FM perspective and extending prior tutorial publications with insights from more recent case studies and roadmaps such as that of the US National Academies of Science, Engineering and Medicine [51].

We outline core DT concepts from the perspective of model-based systems engineering (Sect. 2), give an overview of achievements and challenges in the engineering of DTs (Sect. 3), consider examples formal techniques in DT Engineering to date (Sect. 4) and identify promising directions for future FM contributions (Sect. 5). Section 6 provides a few concluding remarks.

2 Fundamentals

2.1 DT-Enabled Systems

In this tutorial, we adopt a systems perspective. A system is an "arrangement of parts or elements that together exhibit a stated behaviour or meaning that the individual constituents do not" [36]. A *system boundary* separates the system from its *environment*. The system's interactions with that environment determine its observable *behaviour*. Our interest is in DTs of *Cyber-Physical Systems (CPSs)*, whose elements combine computational and physical processes.

A DT is a digital system that replicates a real-world entity termed the *Physical Twin (PT)*[1]. The DT and PT are linked by a *communications infrastructure* through which data flows in both directions, allowing changes in one to update the other. This view aligns with emerging standards [37–39], while explicitly permitting the DT to influence the PT, potentially autonomously. A DT that only mirrors the PT without affecting it is often termed a *Digital Shadow*. We refer to the composition of PT, communications infrastructure and DT as the *DT-enabled system (DTeS)* (Fig. 1).

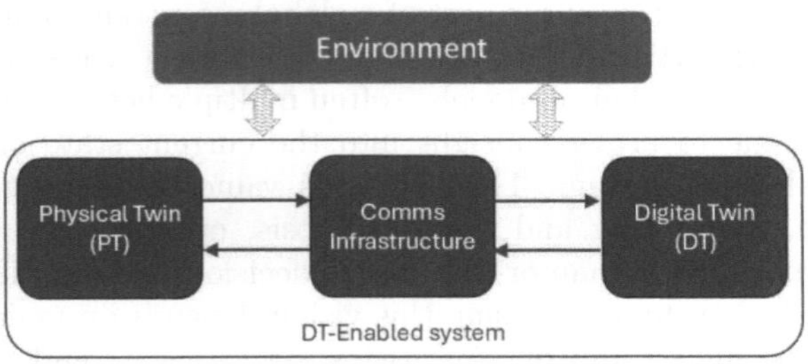

Fig. 1. Simplified overview of a DTeS. External stakeholders omitted.

A DT uses models of the PT alongside incoming data to offer its stakeholders services such as visualisation, monitoring, fault diagnosis, predictive maintenance, what-if analysis and reconfiguration. These services form the building

[1] Despite the name, the *Physical* Twin is often itself a CPS.

blocks of the DT discussed in Sect. 3 below. Note that the DT is different from a model-based controller in that it adds value to that of the PT alone; the PT should operate satisfactorily even if the DT is absent.

2.2 A Basic Example: the Tempeh Incubator

We introduce a small example to illustrate concepts introduced so far. For full details we refer the reader to a previous tutorial [33] and book [24]. Our example is based on an incubator for producing the fermented food tempeh. In tempeh production, cooked soybeans are inoculated with a fungus and fermented at a controlled temperature for 24–48 hrs. The fermentation initially absorbs heat, but later the internal temperature rises as the fungus grows rapidly. Just after the temperature passes a peak, the tempeh is removed for consumption.

The (cyber-)physical incubator (Fig. 2) consists of a styrofoam box with a heatbed, fan and temperature sensors, controlled by a Raspberry Pi. An online video shows the real-world setup[2]. The incubator's control system, implemented in Python, includes a communication server, a low-level driver and a controller that manages the heatbed and fan, with a basic on-off control strategy. A communication server facilitates message exchange between the controller and a low-level driver. The driver manages communications with sensors and actuators, reading current temperature data and checking for commands from the controller via the communication server. The controller is a form of bang-bang controller which keeps temperature between lower and upper limits. A user of this setup can ferment tempeh satisfactorily, but we envisage a DT that extends the incubator's capabilities, interfacing with stakeholders through a connected device such as a smartphone, to offer improved control, higher consistency and better quality of the final product. The services offered by such a DT may include:

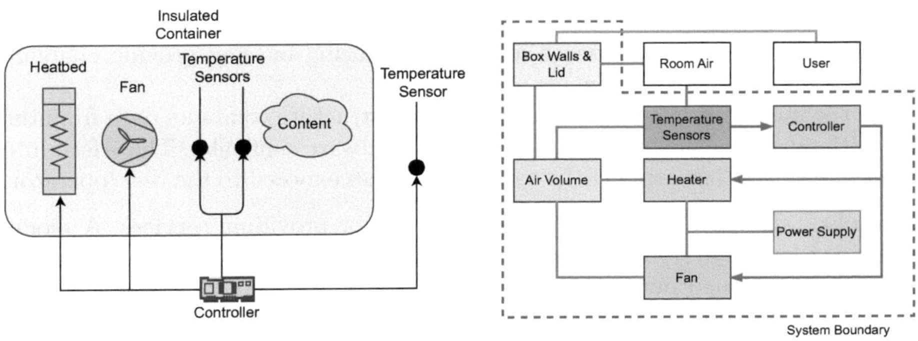

Fig. 2. Informal schematic of tempeh incubator showing cyber-physical setup (left) and abstraction of the CPS's elements and their interactions. From [24].

[2] https://youtu.be/gG4za7iPY0I.

- visualisation of the measured state of the fermentation (e.g., via a dashboard or a 4D graphical view);
- monitoring in order to detect anomalies such as the opening of the incubator lid;
- decision support, for example in selecting from among alternative strategies for replacing incubator components, or trading off parameters (e.g., temperature error and actuator working life);
- reconfiguration, such as switching in alternative control strategies when the incubator is used outside the current assumptions.

Creating a DTeS that can deliver such services requires an understanding of the ecosystem of models that go to make up the DT, each with their own purposes, assumptions, formalisms and levels of abstraction. It must also take into account the connecting infrastructure an other sources of uncertainty and the potential for dynamic adaptation. In all of these areas, there are opportunities for contributions from FMs.

3 The Building Blocks of a DT

The emerging field of *DT Engineering* encompasses the analysis, specification, design, implementation, deployment, maintenance and disposal of DTeSs. It is inherently interdisciplinary in that DTeSs may be developed by teams of specialists in cyber, physical and systems engineering backgrounds; the models and methods used in a DT will reflect this diversity.

DT Services. A DT can be understood as an integration of data, models and the services built on them. Data are acquired from external physical entities and managed within the DT; heterogeneous models capture relevant aspects of those entities and evolve under the DT's control; and services use these artefacts to deliver value, from model calibration and prediction to visualisation and stakeholder interaction. Services may be user-facing or may provide enabling capabilities for other components.

In the incubator example (Fig. 3), a layered structure combines data from the PT with models of the physical plant and the software controller. These feed into *enabler* services that support the top-level services exposed to the user/operator.

Models. DTs use models [43,59] as the basis for providing services. A model is derived from a system of interest, captures a selected subset of its properties and serves a defined purpose in a given context. For DTs of CPSs, such models may describe physical or computational processes in the PT or its environment, including sensors, actuators, networks and computational platforms.

Determining what constitutes a 'good' model for a particular purpose is important but difficult to formalise without committing to specific modelling formalisms (state machines, differential equations, Markov models, etc.). In our work, we emphasise properties such as *relevance*, *substitutability* and *fidelity*. The most critical property is *verifiability*, which is the extent to which one can

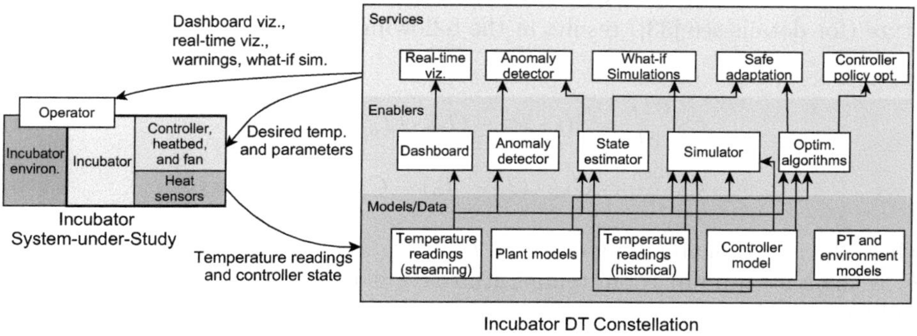

Fig. 3. Informal schematic of incubator DT showing a constellation of services. From [24].

determine whether the model satisfies properties of interest. Further discussion of these qualities is provided in [52].

Models can be classified by their state form (numerical or symbolic), state evolution (continuous or discrete time) and behaviour (deterministic or stochastic). These dimensions guide the choice of modelling formalism for DTs. We regard *simulation* as the process of generating a model's behaviour. *Calibration* is the process of identifying parameter values that align model behaviour with real-world data, while *validation* concerns gaining confidence in a model's qualities by comparing its predictions with observed data. Introductions to modelling can be found in [11,24,61]. A general treatment of simulation is in [12].

An individual DT may require multiple models capturing properties ranging from energy consumption to functional correctness. The resulting model set is necessarily heterogeneous, reflecting contributions from diverse disciplines and formalisms, including:

Physics-based models: Derived from first principles, using algebraic and (ordinary or partial) differential equations to describe physical processes such as heat transfer.

Data-driven models: Predictive models learned from input-output data, employing techniques such as machine learning, deep learning and physics-informed neural networks.

Models for computer-based systems: Using formalisms familiar in FM to describe hardware and software components, typically via discrete or hybrid models.

Model selection for a DT involves balancing factors such as precision, accuracy and the computational cost of the services that rely on model simulation or analysis.

Considering our incubator example, the DT might require a physics-based model of the temperature performance and a discrete-event model of the controller, among other models. A first principles analysis of the incubator temperature (for details see [33]) results in the following ordinary differential equation:

$$C_h \frac{\mathrm{d}T_h}{\mathrm{d}t} = H_h \cdot P_h - G_{hb} \cdot (T_h - T_b) \tag{1a}$$

$$C_b \frac{\mathrm{d}T_b}{\mathrm{d}t} = G_{hb} \cdot (T_h - T_b) - G_{br} \cdot (T_b - T_r) \tag{1b}$$

where T_h is the heater temperature, T_b is the air temperature inside the box, T_r is the known input room temperature; C_h is the heat capacity of the heater and C_b is the heat capacity of the box including contained air. The power to the heatbed P_h is a constant, while H_h is a Boolean variable switching power on/off. Finally, G_{br} is a coefficient of heat transfer between the box and the room while G_{hb} is the coefficient of heat transfer between the heater and the box.

Models of this form are termed *lumped-parameter models* because they assume the model components are concentrated in a single point, in this case assuming that the air temperature is the same throughout the full volume of the incubator. If the variation across the volume is significant for some purposes, it may be appropriate to develop a higher fidelity model based on partial differential equations that account for energy transport in fluids.

It would be natural to use a discrete state-transition model of the controller (Fig. 4). This is based on bang-bang control with a slight variation. It turns on the heatbed until the temperature reaches the setpoint, then waits to prevent overshoot.

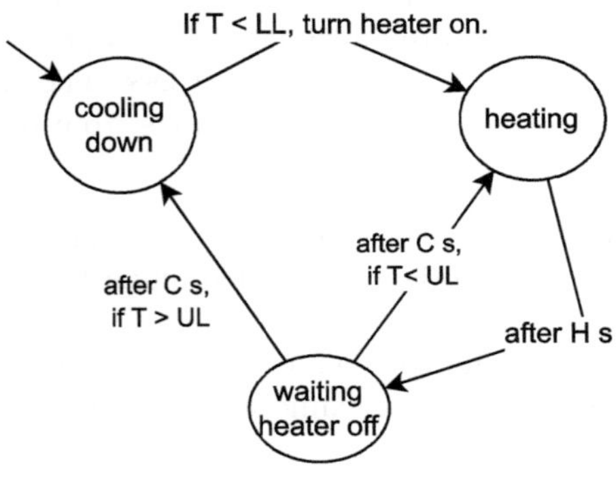

Fig. 4. Controller Statechart. LL and UL are lower and upper temperature limits; H and C are the heating and waiting durations. From [23].

Model-based DT engineering is inherently multi-disciplinary and therefore requires mechanisms for coherently linking diverse models of digital, mechanical, electrical and other processes. Much of our past work has focused on *co-simulation* [34], the coordinated simulation of an ensemble of models coupled via a standardised software interface, as a means of achieving this.

In the incubator example, co-simulation underpins decision-support services. One instance (described in detail in [33]) is a reconfiguration service triggered when the user opens the lid to inspect the incubator, invalidating the system's thermal model and controller. A monitoring service detects the resulting temperature anomaly and the reconfiguration service responds by recalibrating the thermal model and adjusting the controller to the new conditions. It selects a lower temperature setpoint to save energy and avoid the overshoot that may occur when the lid is closed again [64]. Figure 5 illustrates this process, marking the moments when the lid is opened and closed and showing how the controller profile adapts accordingly.

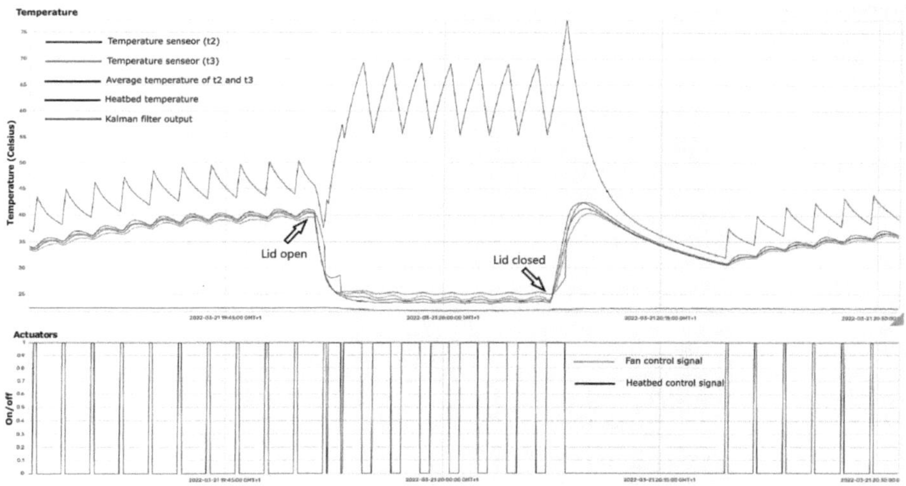

Fig. 5. Experimental result of self-adaptation. Adapted from [22].

This part of the tutorial has introduced core DT concepts and building blocks, illustrated through an incubator study which is available as an open proof-of-concept [33]. We have highlighted the central role of models and the services that exploit them during operational life.

4 Examples of FM Applications in DT Engineering

DT engineering requires methods, tools and processes that can handle the plurality and heterogeneity of models and services that constitute a DT. They must also support maintaining convergence between PT and DT, recalibration and reconfiguration. This section outlines three recent studies at Aarhus University that address these challenges. The topics include dynamic management of updates in a live DT-enabled system, the role of runtime monitoring and verification, and support for DT construction.

4.1 Concrete Batch Plant

This case study concerns **fibo Collect**, a concrete batch plant developed by **Fibo Intercon**. The plant (Fig.6) is a fully automated system responsible for handling, batching and mixing stone, sand, water and cement to produce certified concrete in production environments. The plant operates continuously and customers demand high availability: interruptions or incorrect behaviour can cause production losses and material waste.

As part of its ongoing evolution, fibo Collect is being equipped with a DT to enable closer integration between the physical plant and its digital counterpart. The resulting DT-enabled system is intended to enhance monitoring, support optimisation, facilitate anomaly detection and ultimately increase plant uptime. However, introducing a DT also brings new engineering challenges, most notably the need to update software services safely in a live DT-enabled system.

Fig. 6. fibo Collect concrete batching plant, at Fibo Intercon A/S. Image credit: Fibo Intercon A/S, used with permission.

Because fibo Collect requires a human operator to supervise the system during updates, it is not feasible to perform updates while the plant is offline at night. Instead, new service versions must be deployed while the system remains operational and the transition from old to new services must preserve the plant's availability. These challenges are explored in two complementary works: a probabilistic update scheduler (PUS) presented in [3] and an as-yet unpublished study applying model checking to a swap algorithm for stateful services.

Probabilistic Update Scheduler. These challenges led us to hypothesise that DT-enabled systems evolve through primary states that can be classified as either *safe* or *unsafe*. This perspective gives rise to three key problems: (i) identifying when the DT-enabled system is in a safe state; (ii) determining the duration of each safe state; and (iii) predicting when the system will next enter a safe state.

To tackle these issues, we model each component of the DT-enabled system as a Semi-Markov process (SMP), capturing both its state transitions and associated sojourn times. The SMPs are defined as follows. First, the probability that the chain moves from state i to state j after $n+1$ transitions is given in Eq. (2) and the wait time distribution, adapted from Medhi [49] is given by Eq. (3).

$$Q_{ij}(n) = P(X_{n+1} = j \mid X_n = i) \tag{2}$$

$$W_{ij}(t) = P(T_{n+1} - T_n \leq t \mid X_{n+1} = j, X_n = i) \tag{3}$$

Using the above equations, we can model multiple SMPs corresponding to different parts of the DT-enabled system. This allows us to predict both the current and future behavioural states of the system and to compute the probability that it remains in a safe state over the current time and the next n transitions. To quantify *safety*, a Monte Carlo Simulation approach is used. Starting from an initial state, the system can be simulated for a fixed target duration (e.g., 30 min per simulation) across a large number of independent runs. For each run, it is determined if the system remained in a safe state for the target duration. The probability that the system remains in the set of safe states of $\mathcal{S}$ is estimated as the proportion of simulation runs where all visited states are safe together at the target duration. This probability enables us to assess the confidence in the scheduled update time and determine whether or not a customer should be informed.

Simulating the collection of SMPs yields the probability that the system is currently in a safe state, as well as its likelihood of remaining safe in the future. This information can guide decisions about when to update a DT-enabled system, reducing the risk of performing updates at inappropriate times.

Future work includes the automated verification of such probabilistic models, i.e., checking whether the system satisfies given probabilistic properties. For example, we may wish to verify that the probability of eventually reaching a state in which a service has been successfully swapped meets a required threshold. Using *Probabilistic Computation Tree Logic* (PCTL), this can be expressed as a property such as "eventually a service will be swapped with probability at least 98%," formalised as shown in Eq. (4).

$$P_{\geq 0.98}[\,\Diamond\, \mathit{ServiceSwapped}\,] \tag{4}$$

Swapping Stateful Services. While the PUS determines *when* an update attempt may take place to reduce the risk of system interruptions, it does not address *how* a running DT service can be replaced without disrupting the operation of other DT services or the PT. DT services are often stateful, meaning they cannot be stopped or restarted without losing information, which may lead to errors or system-wide disruptions. To support updates without taking the system offline, we require a method for determining when two instances of the same DT service are behaviourally equivalent. We therefore focused on comparing the input-output behaviour of the existing DT service with that of the candidate replacement. Once the replacement service produced outputs equivalent to those of the original for the same inputs, the system would perform a swap.

To build confidence in this approach, the correctness of the swapping mechanism was specified in PlusCal and translated into TLA+, enabling us to model the services and the swap mechanism. We then used the TLC model checker to verify that the system satisfied the intended swapping properties. These included a range of safety and liveness conditions, among them the liveness property shown in Eq.(5), which asserts that the system should eventually perform the service swap.

$$Liveness \triangleq \Diamond\Box(swapped \Rightarrow swapped') \tag{5}$$

This case study highlights opportunities for integrating FMs into operational DT settings. By combining probabilistic reasoning for update scheduling with a verified mechanism for swapping stateful services, the approach offers a path toward more reliable continuous deployment. Future work includes implementing the formally verified swapping mechanism in the concrete batch plant and deploying it in production, allowing us to further assess its effectiveness in increasing deployment frequency within a DevOps workflow.

4.2 Runtime Verification and an Autonomous Mobile Robot

Digital Twins and Runtime Verification. One advantage of DTs is their ability to support continuous system monitoring, enabling the detection and reporting of anomalies at runtime based on real operational data. Such capabilities provide stakeholders with critical information that can help prevent failures or inform recovery strategies. Monitoring services can also be combined with complementary DT functionalities, such as visualisation, further increasing the value derived from the DT. Research opportunities related to runtime monitoring of DTs, including validity frame monitoring, are discussed in Sect. 5.

Although monitoring services can, in principle, be implemented in any programming language, such ad hoc implementations often lack the precision required for complex or critical applications. A formal approach to monitoring offers clear advantages: it enables the specification of precise, unambiguous system properties with mathematically defined semantics over execution traces. This perspective underpins the field of *Runtime Verification (RV)*, which investigates methods and tools for checking system behaviour at runtime to increase confidence in correctness and reliability.

In the context of DTs, RV allows a designer to write a formal specification that is then used to evaluate the observed behaviour of the DT against the expected properties. The monitor issues verdicts that may be used by the PT, by other DT services, or by a human-in-the-loop, as illustrated in Fig. 7.

The introduction of the Lola language marked the emergence of a new RV paradigm known as *Stream Runtime Verification (SRV)* [16]. In SRV, declaratively specified output streams of verdicts are computed from input streams

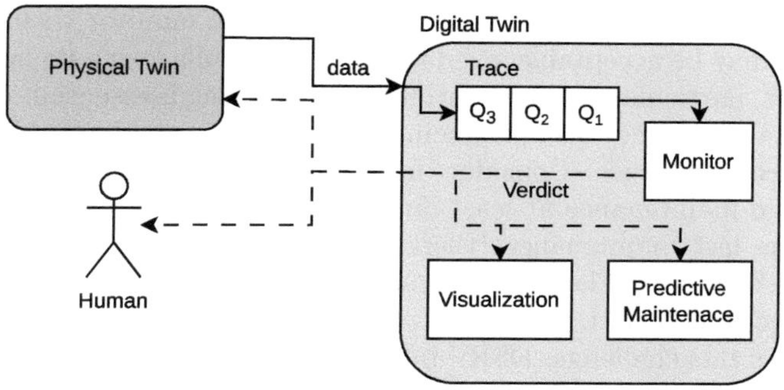

Fig. 7. The basic idea of monitoring within a DT scenario. "Visualization" and "Predictive Maintenance" are examples of other DT services.

via stream equations. Unlike traditional RV languages rooted in temporal logics, SRV formalisms draw inspiration from functional reactive programming, providing a compositional and declarative framework for defining efficient monitors capable of continuously processing and reasoning about real-time data streams.

Lola is a minimal SRV language and most subsequent SRV formalisms implement supersets of its core *stream expressions*. These include constant-valued streams, arithmetic and boolean operators, conditional if–then–else expressions and temporal offset operators. Notable developments in this line of work include RTLola [21], TeSSLa [46] and hLola [13].

In this paper, we adopt the recent language DSRV (pronounced "Deserve"), previously introduced as DynSRV in [42]. Examples of DSRV properties are shown below, where $\texttt{default}(z[N], V)$ returns the Nth previous value of stream z, or V if that value is unavailable. Equation (6) illustrates basic conditional and arithmetic constructs of the language.

Equations (7) and (8) demonstrate counters, where Eq. (7) increments z by 1 indefinitely as is it not bound by any input streams and Eq. (8) increments z by x when a new x-value is received. Equations (9) and (10) demonstrate the implementation of a moving-average filter with $M = 3$, where Eq. (10) defines it recursively.

$$z = \texttt{if}\ x \leq 42\ \texttt{then}\ \text{true}\ \texttt{else}\ y \cdot 2 \leq x \tag{6}$$

$$z = \texttt{default}(z[1], 0) + 1 \tag{7}$$

$$z = \texttt{default}(z[1], 0) + x \tag{8}$$

$$z = \frac{(x + \texttt{default}(x[1], 0) + \texttt{default}(x[2], 0))}{3} \tag{9}$$

$$z = \frac{\texttt{default}(z[1], 0) + (x - \texttt{default}(x[3], 0))}{3} \tag{10}$$

Systems often evolve over time due to changing environments, self-adaptation mechanisms, or other dynamic factors [8] and the RV monitor may need to evolve accordingly. In some situations, restarting the monitor with an updated specification may be acceptable, even though this entails losing its internal state. In other cases, particularly when continuous monitoring is essential, such restarts are unacceptable, as they risk producing incorrect verdicts due to the loss of the complete execution trace. Consider, for example, a property requiring that a robot undergo maintenance at least once every 24 h. If the monitor is restarted 20 h after the last maintenance (thereby losing this timestamp) and the next maintenance occurs 10 h later, the resulting 30-hour gap constitutes a violation that would go undetected.

To address this challenge, DSRV introduces *Dynamically Updated Properties* (DUPs), which allow monitoring properties to evolve at runtime without restarting the monitor. The DUP primitive $\texttt{dynamic}(p)$ evaluates a stream expression p dynamically within the current trace context. As illustrated in Eq. (11), this can be combined with static properties via conjunction: the left-hand side enforces the constraint $T < T_m$ (the time since last maintenance, T, must remain below the fixed threshold T_m), while the right-hand side strengthens this requirement with a dynamic constraint derived from p, defaulting to true when unavailable.

Notably, conjunction permits strengthening, e.g., requiring more frequent maintenance during mission-critical phases, but does not permit relaxation of the window. For a detailed discussion of DUPs and associated design patterns in DSRV, we refer to [42].

$$z = T < T_m \wedge \texttt{default}(T < \texttt{dynamic}(p), \text{true}) \tag{11}$$

Digital Twin of Autonomous Mobile Robot. We demonstrate the use of DSRV in a scenario featuring an Autonomous Mobile Robot (AMR) operating in a warehouse with a DT. The DT provides a visualisation service through which AMR metrics can be inspected and new navigation goals can be set. The AMR requires periodic recharging, which is autonomously managed by a decision-making service within the DT. According to the datasheet, the battery supports at least 2.5 h of operation; the service therefore maximises uptime by initiating recharging as late as possible without risking full depletion.

However, reliably determining when the robot must be scheduled to recharge is non-trivial. Dynamic task priorities, environmental changes (e.g., blocked pathways) and uncertainty in travel-time prediction all contribute to this complexity. To increase confidence in the system's behaviour, the DT continuously monitors the property in Eq. (11). Here, T is an input stream that increments every minute and resets after each recharge, $T_m = 150$ enforces recharging at least every 2.5 h, p provides dynamic updates from DT services or human operators and z is the resulting verdict stream.

The following three scenarios illustrate how this property evolves in tandem with the DT services.

Scenario 1: Trends in the visualisation service reveal that the AMR's battery frequently reaches critical levels during recharge attempts. To mitigate the risk of failing to reach the charging station, operators reconfigure the DT service to schedule recharges every 2 h as a temporary measure. An operator updates the monitor by sending $p = 120$, essentially reducing Eq. (11) to the stricter property $z = T < 120$ until further updates to p are received.

Scenario 2: As a more robust solution, the DT introduces a battery model predicting capacity degradation over time, enabling dynamic T_m adjustment based on projected remaining life. The monitor updates via $p = d$, where d is an input stream from the battery model service carrying expected lifetime values. This strengthens the property to $z = T < T_m \wedge T < d$, ensuring maintenance aligns with actual battery health.

Scenario 3: As trust in the AMR grows, it is assigned increasingly mission-critical tasks (e.g., high-priority orders) within the warehouse, elevating the consequences of battery failure. The DT is updated to incorporate mission-criticality streams c from the task planner and the monitor is updated accordingly via $p = \texttt{if}\ c\ \texttt{then}\ d - 30\ \texttt{else}\ d$, to tighten recharge requirements during critical missions. Alternatively, for greater generality, $p = \texttt{if}\ c\ \texttt{then}\ \texttt{defer}(p[1]) - 30\ \texttt{else}\ \texttt{defer}(p[1])$ updates the dynamic property by reusing the previously assigned value of p via the `defer` primitive. `defer` is a DUP primitive similar to `dynamic` except it only allows dynamically updating a property once.

Our work with DSRV demonstrates how DUPs enable dynamic reconfiguration of monitoring properties in DTs, allowing seamless adaptation without monitor restarts. Several open questions remain:

1. *Practical deployment of DUPs* in real-world settings such as self-adaptive systems and dynamic software updating environments, where specifications must evolve alongside system behaviour.
2. *Formal verification of DUP safety*, in which static meta-properties within a DSRV specification ensure that dynamically introduced properties satisfy required constraints, thereby restricting the space of admissible updates.
3. *Integration with complementary FMs*, using theorem provers or model checkers to verify that dynamically introduced properties preserve overall specification integrity and critical constraints.

This case study illustrates how SRV can support robust and adaptable runtime monitoring in DT-enabled systems. By enabling properties to evolve without interrupting monitoring, DSRV gives a basis for handling dynamic operational contexts. At the same time, the open challenges identified above highlight opportunities for deeper integration between SRV and other FMs.

4.3 Coupling UR and Kuka Robots

It is challenging to engineer DTs for robotic systems, in part because robotics relies on a wide range of modelling principles and methods [58], including kinematics, dynamics, geometry, forces, control and other domain-specific aspects.

The complexity increases further in scenarios involving robotâĂŞrobot cooperation or humanâĂŞrobot collaboration, where multi-system interactions and heterogeneity must be addressed. Compounding this, specialised tooling for DT engineering in robotics remains limited [48], making the process more time- and resource-intensive. Nevertheless, the robotics community already offers several tools that can support DT development, including the Robot Operating System (ROS), which has been widely adopted for this purpose [47, 57], as well as robotic simulators such as Gazebo [47], CoppeliaSim [56] and the Robotics Toolbox [15].

For this example, we consider the Flex-cell (see Fig. 8), a manufacturing cell composed of two robotic arms: a UR5e and a Kuka LBR iiwa 7. Although both arms belong to the same system and appear similar, they differ in their underlying models, control methods, communication and software interfaces, and crucially their simulation environments. As a result, they must be treated as two independent and heterogeneous sub-systems within the Flex-cell case study.

This heterogeneity has made the Flex-cell a useful platform for demonstrating a range of DT engineering methods, including modelling composable DTs for cooperative robotic systems [29], integrating the *robot skills* concept [54] in DTs [32], describing and reporting engineered DTs [31], executing simulations of coupled heterogeneous systems [27] and engineering DTs that combine co-simulation with RoboStar technologies [9, 26].

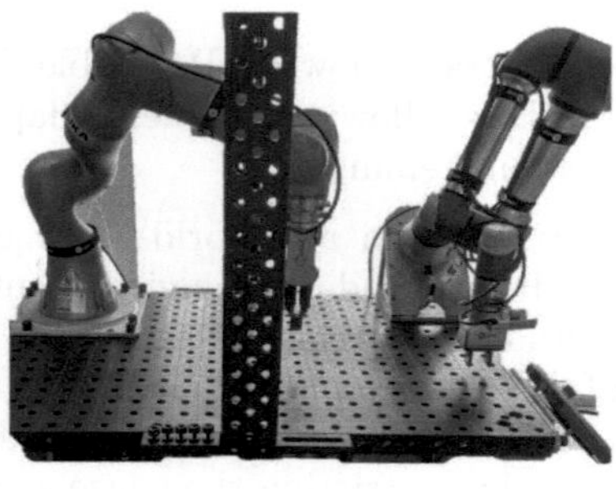

Fig. 8. Flex-cell manufacturing cell. From [27].

When engineering a DT for a composed system with heterogeneous components like the Flex-cell, it is important to consider not only the system-wide aspects, but also the characteristics of each internal subsystem, so the system as a whole behaves as intended. From our experience with this example, several system-level considerations are fundamental to fine-tuning composed DTs, including interoperability and common semantics, co-orchestration of components, common and reusable interfaces and time synchronisation. To address these aspects, we have used ontologies as common vocabularies and structure/knowledge graphs, communication protocols such as MQTT and AMQP that support reusable and scalable patterns, and co-simulation to deliver coordinated execution and time synchronisation.

Intra sub-system aspects are also important to consider when engineering the system as a whole. Among the publications using the Flex-cell as a case study, the work combining co-simulation and RoboStar technologies [26], provides the most detailed treatment of the internal aspects of each of each subsystem. RoboStar offers a unified robotics toolkit that disaggregates the representation of a robotic platform into several formally specified components, including the physical modelling (*p-model*), the control software modelling (*d-model*), the mapping between the control software and the physics (*platform mapping*), and the world assumptions (*RoboWorld*). A key benefit brought by RoboSim [10], one of RoboStar's tools, is support for automatic validation and verification of model properties, and automatic code generation (compliant with the properties already verified).

In [26], co-simulation was used to orchestrate the components at the subsystem level with finer granularity, following a pattern established in earlier Flex-cell studies. More recently, this has been extended with the formalism of *synchronous clocks* to provide better manoeuvrability on both axes of time (continuous-time axis and discrete-event axis) and to improve the observability of state machine computations in the co-simulation of robotic systems [30].

Overall, the Flex-cell case study shows that coupling heterogeneous robotic sub-systems demands attention to both system-level coordination and detailed sub-system modelling, co-simulation and shared semantics. Future work for the Flex-cell include a more detailed design of the cooperative scenarios (covering both the intra sub-system and macro system aspects with a wider set of models with higher fidelity) and extending its capabilities to multi-robot cooperation with other robotic systems and human-robot collaboration.

5 Opportunities for Formal Methods

Our prior work [24] and recent experience (e.g., Sect. 4) suggests that there are many opportunities for FMs research and innovation, both in the construction of dependable DTeSs and within operational DTs. The recent report from the US National Academies of Science, Engineering and Medicine [51] emphasises the need for Verification, Validation and Uncertainty Quantification as a continual process. This gives rise to numerous opportunities for the FMs. We here identify areas where we see the highest level of opportunity for the FM community.

5.1 The Reality Gap, Validity Frames and Uncertainty

There is an unavoidable *reality gap* between a real-world CPS and its representation in a DT. Consequently, when decisions and services depend on DT-based representations, it is essential to characterise and understand the bounds on their accuracy. These bounds are shaped by uncertainties in sensing, latency, modelling and environmental variability. Idealised assumptions—such as noise-free sensors—are often adopted to keep analysis tractable, but recognising such assumptions is crucial for assessing the limitations of DT predictions and their impact on decision-making. The FM community is well positioned to characterise these limits systematically, enabling more principled reasoning about confidence in DT-based analyses.

A further challenge is determining whether DT models can adapt to environmental changes not anticipated when the PT was originally engineered. This relates closely to the notion of *validity frames*, which specify the conditions under which a DT can be expected to provide acceptable predictions [1]. FMs have significant potential to support the definition and analysis of validity frames, offering rigorous guarantees about the conditions under which DT predictions remain dependable.

Techniques for configuring and propagating uncertainty in CPS models have been studied for many years. Continued research is needed at the DT level to address the challenges outlined above. Different application domains employ distinct techniques for representing and propagating uncertainty, including reachability analysis [19], Monte-Carlo simulation [63], stochastic differential equations [5] and sensitivity analysis [14]. Formal techniques can help provide precise representations of uncertainty within DT models, define semantics for how uncertainty evolves across DT layers and offer guarantees about the correctness or robustness of DT-based predictions.

5.2 Assurance at Runtime

As DT-enabled systems increasingly support monitoring, decision-making and autonomous adaptation during operation, confidence in their behaviour can no longer rest solely on offline verification. Instead, assurance must be maintained continuously, in the presence of evolving models, uncertain data and potentially hostile environments. This shifts the focus from static guarantees to runtime assurance mechanisms that combine monitoring, analysis and adaptation. Formal methods provide a natural foundation for such mechanisms, offering precise semantics for observed behaviour, principled ways to detect and respond to deviations, and ways to reason rigorously about safety and correctness as systems evolve in operation.

Towards Verified DTs. DTs are proposed as enablers for self-adaptive CPSs that can modify their behaviour in response to evolving and previously unseen environments. While existing work has explored architectures and adaptation mechanisms, there has been limited progress on providing guarantees for such runtime adaptations. In particular, there is a need for formal foundations, semantics and concepts for RV (as discussed in Sect. 4.2) and monitoring in DT-enabled systems [42,64].

A key challenge in this setting is the development of integrated formal foundations for CPSs that can underpin DT services supporting decision-making, particularly under high uncertainty. This requires unifying frameworks able to accommodate diverse verification tools: provers such as Isabelle [53] and Coq [6] for discrete systems; tools such as KeYmaera X [55] for hybrid models; and model checkers such as Storm [18] for probabilistic behaviour. Such foundations would enable a range of use cases, including probabilistic DTs augmented with Markov and Bayesian models, probabilistic risk models to support operational decisions and the verification of DTs within certification or compliance processes.

Security of DT-enabled Systems. Security is a particular concern for DT verification. The communications infrastructure of a DT-enabled system—including sensors, actuators, networks and the DT itself—creates opportunities for malicious actors to disrupt the PT [2,62]. It is therefore essential to consider how to protect DT-enabled systems from such attacks [45]. While specific countermeasures depend on the deployment context, systematic exploration of attack patterns and mitigation strategies is required [44]. As DT deployments scale to Digital Twin as a Service (DTaaS) platforms hosting many DTs, it becomes critical to ensure strong isolation and protection of each DT's data space.

These challenges present clear opportunities for FMs. Potential contributions include identifying vulnerabilities and exploit paths in DT-enabled systems; developing and specialising FMs and tools for verifying DT security; cataloguing DT-specific attack models and mitigation strategies; and designing low-overhead, formally justified security protocols for bidirectional DTâĂŞPT communication.

5.3 Engineering Automation and Evolution

A defining characteristic of DTs is that, over time, they yield new engineering knowledge about how a PT operates in its actual deployment environment, often in ways that were not anticipated at design time. Exploiting this knowledge to optimise DT services for a specific PT naturally motivates evolution, both of models and of the services built upon them. However, experience shows that such evolution is hindered by limited automation in DT production, a lack of modelling languages with precise semantic foundations and insufficient support for evolving DT-enabled systems safely while they remain operational. These challenges are tightly coupled: automation depends on sound language foundations, while safe evolution presupposes both automated reasoning and explicit models of system structure and behaviour. Together, they define a core set of engineering problems where FMs can play a decisive role.

Automating Digital Twin Production. Creating a dependable DT (and DTeS) from scratch is resource-intensive. Reusable, model-based frameworks are therefore essential for systematic DT engineering. Such frameworks require a well-defined architectural core and a coherent set of language concepts and services, enabling the integration of heterogeneous models that can be adapted and extended for specific DT instances. Precise semantic foundations are needed to ensure that reuse and extension preserve correctness and intended system properties.

Most existing DT frameworks inherit heavily from Internet of Things (IoT) architectures, which focus primarily on data communication between the PT and the DT. Consequently, automation is largely limited to data acquisition and synchronisation [28], with far less support for automated model-based simulation or formal checking within runtime monitors. Providing comparable automation for simulation, analysis and verification would significantly strengthen assurance and enable DT frameworks that move beyond data-centric integration.

Modelling Languages for Digital Twins. Theories and techniques for combining and managing heterogeneous languages are well established in software engineering, yet their systematic application to DT frameworks remains limited. Moyne et al. identify medium- and long-term requirements for DT frameworks, including dedicated languages for DT definition, support for reuse and portability, interoperability and interchangeability of DTs, and integrated validation and verification capabilities [50].

Languages such as the Digital Twin Definition Language (DTDL)[3], an open-source language targeting Azure DTs, offer a pragmatic starting point for automating aspects of DT construction and are being extended with features such as configuration data and services. However, such languages typically lack a formally defined semantics, limiting rigorous analysis and verification. This gap highlights clear opportunities for FMs to provide semantic foundations and enable automated reasoning within DT engineering.

Safe Evolution During Operation. Synchronisation between the PT and DT is fundamental to ensuring that the DT accurately reflects the evolving state of the PT. At the same time, functions such as control, anomaly detection and behavioural optimisation depend on timely information flowing from the DT back to the PT. Achieving scalable, bi-directional communication and synchronisation throughout the lifecycle of DT-enabled systems remains a significant challenge, particularly under fault conditions where increased PT error rates can raise data traffic and latency, potentially undermining the feasibility of interventions. This creates a clear opportunity for the FM community to contribute principled, formally justified communication and synchronisation protocols that minimise information exchange without compromising correctness or responsiveness [60].

A related challenge concerns the automated evolution of software within DT-enabled systems, especially in settings that require the safe deployment of updates while systems remain in use, consistent with DevOps practices. In particular, over-the-air updates should only be applied when the PT is in a state that guarantees safe installation and minimal disruption to its primary function. FMs can play a key role here by enabling rigorous analysis and planning of update strategies, including identifying optimal update points that balance availability, safety and operational efficiency [3].

5.4 The Ecosystem of Models and Digital Twins

A model's abstraction level is determined by its intended use case and the system properties it must reproduce, so engineering a complex CPS typically produces multiple models spanning different viewpoints, assumptions and development stages. DTs create an opportunity to exploit this diversity at runtime, though

[3] https://azure.github.io/opendigitaltwins-dtdl/.

doing so remains challenging. As DT services evolve and new knowledge of the PT emerges, different models—or heterogeneous combinations such as those in Sect. 4.3—may become preferable. Model choice depends not only on original intent but also on whether it can align with live data and operate as a digital shadow. Establishing principled methods to manage and combine models in this dynamic setting is a key challenge for FMs.

Leveraging Multiple Levels of Abstraction and Fidelity. Managing abstraction is a familiar concern in formal methods, where modelling choices are guided by a model's intended purpose and the properties it must reproduce. As a result, multiple models of the same system are often developed, each reflecting a different engineering viewpoint. A DT provides an opportunity to exploit these diverse models during operation as DT services evolve, more is learned about the PT's deployed environment and stakeholder needs change.

This creates the challenge of selecting—potentially autonomously—the most appropriate model, based not only on its purpose but also on its performance, fidelity and our ability to ensure that it remains a faithful representation of the PT [40]. Recent work has explored dynamic model changes [20], including replacing models and altering the DT's structure (i.e., the interactions between models) while maintaining safety guarantees. Some control here may be data-driven, making it essential to ensure safety under data-dependent adaptation [7].

FMs have a clear role to play in analysing how models are characterised and selected and in providing assurance that model switching remains safe over time. A key challenge lies in formalising notions of abstraction and fidelity—concepts that are inherently subtle but whose effects propagate directly to DT services.

Capture and Representation of Causal Relations. The reality gap arises from misalignment between the PT, its environment and their DT representations, motivating the need for formal representations of system operating modes and precise characterisations of the conditions under which transitions between such modes occur. Established formalisms such as Hybrid Automata [25] and Hidden Markov Models [4], together with their underlying theoretical foundations, provide natural starting points for this endeavour.

From a FMs perspective, a key challenge is distinguishing mere correlations from genuine causal relations between system states and observed behaviours. Addressing this may require a principled integration of formal reasoning with the design and analysis of experiments, enabling the formulation, validation and refutation of explicit causal hypotheses.

Enabling Interoperability and Composition. To date, only limited attention has been paid to the composition of DTs or DTeSs. A recent review [41] identifies a spectrum of composition approaches, ranging from loosely coupled DTs that interact through defined interfaces to highly centralised orchestrations. However, when it comes to validating and verifying such compositions, most studies rely on simulation-based or qualitative assessments, with comparatively little focus on scalable formal frameworks.

Full automation of interoperability between DTs presupposes agreed standards for DT interfaces. At present, however, standards, frameworks and regulations specific to DTs remain underdeveloped [17]. In practice, standards tend to emerge from established communities of developers and users, and such communities are only beginning to form. For example, the Digital Twin Consortium[4] operates within the Object Management Group and represents a step toward coordinated efforts. These initiatives must be sustained and extended across different technology stacks so that competition centres on functionality and quality of DT services rather than proprietary choices or alliances. In this regard, the development of SysML Version 2[5] marks significant progress for Model-Based Systems Engineering (MBSE) and offers a promising foundation for future DT-related standardisation.

Other standardisation efforts are underway, but such initiatives are inherently slow, requiring alignment of baseline concepts. As a result, it will take time before new research results can be incorporated into emerging standards. Moreover, interoperability between DTs remains an under-emphasised aspect of current work. Further research is needed at a cognitive layer to understand how collaboration between DTs can be effectively enabled.

6 Conclusions

This tutorial has introduced the fundamentals of DT technology, highlighting opportunities for FMs to advance research and innovation in this rapidly developing field. We have emphasised the central role of models, the added value of services built upon them, and the challenges of maintaining alignment with evolving PTs and environments. Three recent case studies have illustrated the importance of dynamic model updates, the role of runtime verification and the challenges of linking diverse PTs.

We have outlined several research challenges where FMs can make a meaningful contribution. These include understanding the characteristics and applicability limits of the heterogeneous models that constitute a DT, providing assurance at runtime, engineering dependable DTs and exploiting the emerging ecosystem of models and DTs. In many cases, the key contribution lies not in inventing new formalisms, but in developing principled ways to harness existing methods and tools coherently and soundly.

DT technology offers a rare opportunity to place FMs at the heart of digital transformation, enabling the development of systems that are not only more capable, but also more dependable and trustworthy. As the field continues to grow, we hope this tutorial serves as both an entry point and an invitation for the community to engage with the rich and expanding landscape of DT research.

Acknowledgments. We are grateful to many colleagues at Aarhus and Newcastle Universities for supporting our work to date, as well as to numerous agencies funding

[4] https://www.digitaltwinconsortium.org/.

[5] https://github.com/Systems-Modeling/SysML-v2-Release.

our DT research but in particular the Grundfos Foundation for establishing the AU Centre for Digital Twins and the DIGIT BENCH project for funding a part of this research.

Disclosure of Interests. The authors have no competing interests to declare.

References

1. Acker, B.V., Meulenaere, P.D., Vangheluwe, H., Denil, J.: Validity frame–enabled model-based engineering processes. SIMULATION (2023). https://doi.org/10.1177/00375497231205035
2. Alcaraz, C., Lopez, J.: Digital twin: a comprehensive survey of security threats. IEEE Commun. Surv. Tutorials **24**(3), 1475–1503 (2022). https://doi.org/10.1109/COMST.2022.3171465
3. Andersen, M.S., Gomes, C., Thompson, S., Larsen, P.G.: Probabilistic update scheduling for digital twins: a semi-markov approach. In: 2025 ACM/IEEE 28th Intl. Conf. on Model Driven Engineering Languages and Systems Companion (MODELS-C), pp. 250–256. IEEE Computer Society (2025).https://doi.org/10.1109/MODELS-C68889.2025.00043
4. Baum, L.E., Petrie, T.: Statistical inference for probabilistic functions of finite state Markov chains. Ann. Math. Stat. **37**(6), 1554–1563 (1966)
5. Bayram, M., Partal, T., Buyukoz, G.O.: Numerical methods for simulation of stochastic differential equations. Adv. Diff. Eqn. **2018**(17) (2018). https://doi.org/10.1186/s13662-018-1466-5
6. Bertot, Y., Casteran, P.: Interactive Theorem Proving and Program Development. Springer Verlag (2004). https://doi.org/10.1007/978-3-662-07964-5
7. Brunke, L., Greeff, M., Hall, A.W., Yuan, Z., Zhou, S., Panerati, J., Schoellig, A.P.: Safe learning in robotics: from learning-based control to safe reinforcement learning. Annu. Rev. Control Robot. Auton. Syst. **5**, 411–444 (2022). https://doi.org/10.1146/annurev-control-042920-020211
8. Carwehl, M., Vogel, T., Rodrigues, G.N., Grunske, L.: Runtime verification of self-adaptive systems with changing requirements. In: IEEE/ACM Symp Software Engineering for Adaptive and Self-Managing Systems, pp. 104–114 (2023). https://doi.org/10.1109/SEAMS59076.2023.00024
9. Cavalcanti, A., et al.: Robostar technology: a roboticist's toolbox for combined proof, simulation, and testing. In: Cavalcanti, A., Dongol, B., Hierons, R., Timmis, J., Woodcock, J. (eds.) Software Engineering for Robotics, pp. 249–293. Springer (2021). https://doi.org/10.1007/978-3-030-66494-7_9
10. Cavalcanti, A., et al.: Verified simulation for robotics. Sci. Comput. Program. **174**, 1–37 (2019). https://doi.org/10.1016/j.scico.2019.01.004
11. Cellier, F.E.: Continuous system modeling. Springer New York (1991).https://doi.org/10.1007/978-1-4757-3922-0
12. Cellier, F.E., Kofman, E.: Continuous system simulation. Springer New York (2006). https://doi.org/10.1007/0-387-30260-3
13. Ceresa, M., Gorostiaga, F., Sánchez, C.: Declarative stream runtime verification (hLola). In: Oliveira, B.C.d.S. (ed.) Programming Languages and Systems, pp. 25–43. Springer (2020). https://doi.org/10.1007/978-3-030-64437-6_2

14. Chou, Y., Sankaranarayanan, S.: Bayesian parameter estimation for nonlinear dynamics using sensitivity analysis. In: Proc. 28th Intl. Jnt. Conf. on Artificial Intelligence, IJCAI-19, pp. 5708–5714. Intl. Jnt. Confs. on Artificial Intelligence Organization (2019). https://doi.org/10.24963/ijcai.2019/791
15. Corke, P., Haviland, J.: Not your grandmother's toolbox – the robotics toolbox reinvented for python. In: 2021 IEEE Intl. Conf. on Robotics and Automation (ICRA), pp. 11357–11363 (2021). https://doi.org/10.1109/ICRA48506.2021.9561366
16. D'Angelo, B., et al.: LOLA: runtime monitoring of synchronous systems. In: Proc. Intl. Symp. on Temporal Representation and Reasoning, pp. 166–174 (2005). https://doi.org/10.1109/TIME.2005.26
17. David, I., Shao, G., Gomes, C., Tilbury, D., Zarkout, B.: Interoperability of digital twins: challenges, success factors, and future research directions, pp. 27–46. Springer (2024). https://doi.org/10.1007/978-3-031-75390-9_3
18. Dehnert, C., Junges, S., Katoen, J.P., Volk, M.: A storm is coming: a modern probabilistic model checker. In: Majumdar, R., Kunčak, V. (eds.) Computer Aided Verification, pp. 592–600. Springer (2017). https://doi.org/10.1007/978-3-319-63390-9_31
19. Donzé, A.: Breach, a toolbox for verification and parameter synthesis of hybrid systems. In: Touili, T., Cook, B., Jackson, P. (eds.) Computer Aided Verification, pp. 167–170. Springer (2010). https://doi.org/10.1007/978-3-642-14295-6_17
20. Ejersbo, H., Lausdahl, K., Frasheri, M., Esterle, L.: Dynamic runtime integration of new models in digital twins. In: 2023 IEEE/ACM 18th Symp. on Software Engineering for Adaptive and Self-Managing Systems (SEAMS), pp. 44–55 (2023). https://doi.org/10.1109/SEAMS59076.2023.00016
21. Faymonville, P., et al: StreamLAB: stream-based monitoring of cyber-physical systems. In: Dillig, I., Tasiran, S. (eds.) Computer Aided Verification, pp. 421–431. Springer (2019). https://doi.org/10.1007/978-3-030-25540-4_24
22. Feng, H., et al.: Integration Of the mape-K Loop in digital twins. In: 2022 Annual Modeling and Simulation Conference (ANNSIM), pp. 102–113. IEEE, San Diego, CA, USA (2022). https://doi.org/10.23919/ANNSIM55834.2022.9859489
23. Feng, H., Gomes, C., Thule, C., Lausdahl, K., Sandberg, M., Larsen, P.G.: The incubator case study for digital twin engineering (2021). https://doi.org/10.48550/arXiv.2102.10390
24. Fitzgerald, J., Gomes, C., Larsen, P.G. (eds.): The Engineering of Digital Twins. Springer (2024). https://doi.org/10.1007/978-3-031-66719-0
25. Frehse, G.: An introduction to hybrid automata, numerical simulation and reachability analysis. In: Drechsler, R., Kühne, U. (eds.) Formal Modeling and Verification of Cyber-Physical Systems: 1st Intl. Summer School on Methods and Tools for the Design of Digital Systems, Bremen, Germany, September 2015, pp. 50–81. Springer, Wiesbaden (2015). https://doi.org/10.1007/978-3-658-09994-7_3
26. Gil, S., Badyal, A., Miyazawa, A., Larsen, P.G., Cavalcanti, A.: A model-based approach for co-simulation-driven digital twins in robotics. Robot. Auton. Syst. **196**, 105240 (2026). https://doi.org/10.1016/j.robot.2025.105240
27. Gil, S., Kamburjan, E., Talasila, P., Larsen, P.G.: An architecture for coupled digital twins with semantic lifting. Softw. Syst. Model. **24**(5), 1379–1404 (2025). https://doi.org/10.1007/s10270-024-01221-w
28. Gil, S., Mikkelsen, P.H., Gomes, C., Larsen, P.G.: Survey on open-source digital twin frameworks–a case study approach. Softw. Pract. Exper. **54**(6), 929–960 (2024). https://doi.org/10.1002/spe.3305

29. Gil, S., Mikkelsen, P.H., Tola, D., Schou, C., Larsen, P.G.: A modeling approach for composed digital twins in cooperative systems. In: 2023 IEEE 28th Intl. Conf. on Emerging Technologies and Factory Automation (ETFA), pp. 1–8. IEEE (2023). https://doi.org/10.1109/ETFA54631.2023.10275601
30. Gil, S., Miyazawa, A., Cavalcanti, A., Gomes, C.: Application of FMI 3.0 synchronous clocks to co-simulation of robotic systems. In: IEEE/SICE Intl. Conf. on System Integration (SII). IEEE (To Appear 2026)
31. Gil, S., Oakes, B.J., Gomes, C., Frasheri, M., Larsen, P.G.: Toward a systematic reporting framework for digital twins: a cooperative robotics case study. simulation **101**(3), 313–339 (2025). https://doi.org/10.1177/00375497241261406
32. Gil, S., Schou, C., Mikkelsen, P.H., Larsen, P.G.: Integrating Ds. In: IEEE/SICE Intl. Conf. on System Integration (SII). pp. 1124–1131. IEEE (2024). https://doi.org/10.1109/SII58957.2024.10417610
33. Gomes, C., et al.: Digital twin tutorial: the incubator case study. In: Engineering Trustworthy Software Systems: 6th Intl. School, SETSS 2024, Chongqing, China, 14–21 April 2024, Tutorial Lectures, pp. 68–101. Springer (2025). https://doi.org/10.1007/978-981-96-4656-2_3
34. Gomes, C., Thule, C., Broman, D., Larsen, P.G., Vangheluwe, H.: Co-simulation: A Survey. ACM Comput. Surv. **51**(3), 49:1–49:33 (2018). https://doi.org/10.1145/3179993
35. International Council on Systems Engineering: INCOSE Systems Engineering Vision 2035: Engineering Solutions for a Better World (2021)
36. ISO/IEC 15288:2023: Systems and software engineering. system life cycle processes. International Organization for Standardization, Geneva (2023)
37. ISO/IEC 20924:2024: internet of things (IoT) and digital twin. International Organization for Standardization, Geneva (2024)
38. ISO/IEC 23247:2021: automation systems and integration. Digital twin framework for manufacturing. International Organization for Standardization, Geneva (2021)
39. ISO/IEC 30173:2023: digital twin – concepts and terminology. International Organization for Standardization, Geneva (2023)
40. Jones, D., Snider, C., Nassehi, A., Yon, J., Hicks, B.: Characterising the digital twin: a systematic literature review. CIRP J. Manuf. Sci. Technol. **29**, 36–52 (2020). https://doi.org/10.1016/j.cirpj.2020.02.002
41. Khedr, M.T., Fitzgerald, J.S.: The composition of digital twins for systems-of-systems: a systematic literature review (2025).https://doi.org/10.48550/arXiv.2506.20435
42. Kristensen, M.H., Wright, T., Gomes, C., Esterle, L., Larsen, P.G.: DynSRV: dynamically updated properties for stream runtime verification. In: RV 2025 (25th Intl. Conf. on Runtime Verification). Springer (2025).https://doi.org/10.1007/978-3-032-05435-7_7
43. Kühne, T.: What is a Model? In: Language Engineering for Model-Driven Software Development. vol. 04101. Internationales Begegnungs- und Forschungszentrum für Informatik (IBFI) (2005). https://doi.org/10.4230/DagSemProc.04101.15
44. Kulik, T., et al.: A survey of practical formal methods for security. Formal Aspects of Computing (2022). https://doi.org/10.1145/3522582
45. Kulik, T., Gomes, C., Macedo, H.D., Hallerstede, S., Larsen, P.G.: Towards secure digital twins. In: Margaria, T., Steffen, B. (eds.) ISoLA 2022. Lecture Notes in Computer Science, vol. 13704. Springer (2022).https://doi.org/10.1007/978-3-031-19762-8_11

46. Leucker, M., Sánchez, C., Scheffel, T., Schmitz, M., Schramm, A.: TeSSLa: runtime verification of non-synchronized real-time streams. In: Proc. Annual ACM Symp. on Applied Computing, pp. 1925–1933 (2018). https://doi.org/10.1145/3167132.3167338
47. Mattila, J., Ala-Laurinaho, R., Autiosalo, J., Salminen, P., Tammi, K.: Using digital twin documents to control a smart factory: simulation approach with ROS, Gazebo, and twinbase. Machines **10**(4) (2022). https://doi.org/10.3390/machines10040225
48. Mazumder, A., et al.: Towards next generation digital twin in robotics: trends, scopes, challenges, and future. Heliyon **9**(2), e13359 (2023). https://doi.org/10.1016/j.heliyon.2023.e13359
49. Medhi, J.: Stochastic processes. New Academic Science Limited (2012)
50. Moyne, J., et al.: A requirements driven digital twin framework: specification and opportunities. IEEE Access **8**, 107781–107801 (2020). https://doi.org/10.1109/ACCESS.2020.3000437
51. National academies of sciences, engineering, and medicine: foundational research gaps and future directions for digital twins. National Academies Press, Washington, DC (2024). https://doi.org/10.17226/26894
52. Oakes, B., et al.: Examining model qualities and their impact on digital twins. In: Annual Modelling and Simulation Conference, pp. 220–232. Ontario, Canada (2023)
53. Paulson, L.C.: Isabelle: a generic theorem prover. Springer-Verlag LNCS 828 (1994)
54. Pedersen, M.R., Nalpantidis, L., Andersen, R.S., Schou, C., Bøgh, S., Krüger, V., Madsen, O.: Robot skills for manufacturing: From concept to industrial deployment. Robotics and Computer-Integrated Manufacturing **37**, 282–291 (2016https://doi.org/10.1016/j.rcim.2015.04.002
55. Platzer, A.: Logical Analysis of Hybrid Systems - Proving Theorems for Complex Dynamics. Springer (2010)
56. Rohmer, E., Singh, S.P.N., Freese, M.: Coppeliasim (formerly v-rep): a versatile and scalable robot simulation framework. In: Proc. of The Intl. Conf. on Intelligent Robots and Systems (IROS). pp. 1321–1326 (2013).https://doi.org/10.1109/IROS.2013.6696520, www.coppeliarobotics.com
57. Saavedra Sueldo, C., Perez Colo, I., De Paula, M., Villar, S.A., Acosta, G.G.: ROS-based architecture for fast digital twin development of smart manufacturing robotized systems. Ann. Oper. Res. **322**(1), 75–99 (2023). https://doi.org/10.1007/s10479-022-04759-4
58. Siciliano, B., Khatib, O.: Springer Handbook of Robotics. Springer Cham (2016). https://doi.org/10.1007/978-3-319-32552-1
59. Stachowiak, H.: Allgemeine Modelltheorie. Springer-Verlag, Wien and New York (1973)
60. Tang, V., Gomes, C., Lucani, D.E.: Precision on demand: Propositional logic for event-trigger threshold regulation. IEEE Internet Things J. **12**(3), 2674–2689 (2025). https://doi.org/10.1109/JIOT.2024.3476922
61. van Amerongen, J.: Dynamical Systems for Creative Technology. Controllab Products B.V, Enschede (2010)
62. Wang, Y., Su, Z., Guo, S., Dai, M., Luan, T.H., Liu, Y.: A survey on digital twins: Architecture, enabling technologies, security and privacy, and future prospects. IEEE Internet of Things Journal pp. 1–1 (2023).https://doi.org/10.1109/JIOT.2023.3263909

63. Woodcock, J., Gomes, C., Macedo, H.D., Larsen, P.G.: Uncertainty quantification and runtime monitoring using environment-aware digital twins. In: Margaria, T., Steffen, B. (eds.) Leveraging Applications of Formal Methods, Verification and Validation: Tools and Trends. pp. 72–87. Springer (2021).https://doi.org/10.1007/978-3-030-83723-5_6
64. Wright, T., Gomes, C., Woodcock, J.: Formally Verified Self-Adaptation of an Incubator Digital Twin. In: Leveraging Applications of Formal Methods, Verification and Validation. Practice: 11th Intl. Symp., ISoLA 2022, Rhodes, Greece, October 22–30, 2022, Part IV. p. 89–109. Springer (2022)https://doi.org/10.1007/978-3-031-19762-8_7

Reasoning over Relaxed Shared Memory Models: A Tutorial

Brijesh Dongol(✉)

University of Surrey, Guildford, UK
b.dongol@surrey.ac.uk

Abstract. The notion of a *relaxed* (aka *weak*) *memory model* is well known, given that such models are implemented by almost all hardware vendors and embedded within the concurrency semantics of many major programming languages. However, given the sheer volume of work on consistency models, formalisations and tools, the area can be both confusing (and intimidating) for newcomers to get into, with even experts missing new developments. In this paper, we coalesce a recent line of work that has focussed on developing *reasoning principles* for relaxed memory into a single reference, extrapolating their key ideas. This line of work aims to reuse (standard) verification techniques for concurrent programs such as Owicki-Gries, rely-guarantee and refinement that were established in the 1970s and 80 s, which are well known to most formal methods researchers. We aim to explain these ideas in simple terms, explain some of the main developments and discuss open problems and opportunities for further research. Throughout the paper, we will focus on the RC11 memory model, but discuss how these reasoning principles apply to other models. Instead of focussing on relaxed memory litmus tests (which non-experts may not appreciate), we use a novel proof of a non-trivial buffer developed by Lamport as a running example.

1 Introduction

Hardware vendors and language developers have, for many years, forgone Lamport's principle of *sequential consistency (SC)*, where

> *"... the result of any execution is the same as if the operations of all the processors were executed in some sequential order, and the operations of each individual processor appear in this sequence in the order specified by its program."* — *Lamport* [40]

Instead, most architectures implement *relaxed execution models*, which permit greater levels of hardware and compiler optimisation [26]. The (excessive) synchronisation induced by SC architectures was already anticipated by Lamport: "*The requirements needed to guarantee SC rule out some techniques which can be*

The author is supported by VeTSS and EPSRC grants EP/X037142/1, EP/X015149/1 and EP/R025134/2.

A. Sampaio and M. Stoelinga (Eds.): FM 2026, LNCS 16557, pp. 394–419, 2026.
https://doi.org/10.1007/978-3-032-26220-2_20

used to speed up individual sequential processors.", which was later validated by empirical studies (e.g., [26]). As such, relaxed memory hardware has been available since the IBM System/370 Model 158 developed in 1972 [8], while high-level programming languages supporting relaxed behaviours (e.g., causal consistency) have been proposed since the early 1990s [4]. (The latter have inspired the design of concurrency models in modern languages such as Java [42] and C++ [9].)

Interestingly, in the same article [40], Lamport also anticipated that the loss of SC *may be necessary for some applications*, but that these would lead to challenges in reasoning. As he states, without SC:

1. "*designing multiprocess algorithms cannot be relied upon to produce correctly executing programs*" and
2. "*verifying their correctness becomes a monumental task*".

Both of these predictions have turned out to be true.

On design, there is still no clear consensus on the semantics of many high-level languages. For example, current formal models for both C11 (the 2011 C/C++ standard) [38,45] and Java [3,50] are known to have unexpected or even unwanted behaviours, despite the fact that the semantics and formal models of relaxed memory hardware are well established [5,6,48,51]. For this tutorial, we focus on a memory model known as repaired C11 (RC11) [38] which has been designed from the ground up to be well-behaved (e.g., does not allow unintuitive behaviours such as thin-air reads), but is still non-trivial to reason about. The model supports four types of ordering guarantees on memory accesses (i.e., reads and writes), but we only consider the two most interesting ones: relaxed and release-acquire, eschewing non-atomic and SC memory accesses.

On verification, early attempts aimed to bypass the problem altogether by relying on data-race freedom (DRF) guarantees [25]—if a program could be shown to be free from data-races, provided that the memory model of the language supported particular desirable properties (e.g., DRF-SC), one could simply reason about the program as if it were executing on an SC memory model. However, many programs are correct in spite of data races and moreover, reasoning about the absence of data races itself is a non-trivial task.

Our point of departure is the desire to support reasoning about a program that may have data races to avoid restricting a programmer to a specific design paradigm. Here, one line of work is on specialised logics for particular weak memory models, e.g., RC11 [32,56] and Rust [21]. While these are impressive efforts that are supported by (automated) verification tools, their reasoning principles enforce use of particular separation logics.

In this tutorial, we focus on an alternative line of work [10,11,18,20,37,49,61], which have aimed to link verification of relaxed memory programs with classical reasoning frameworks such as Owicki-Gries [44] and rely-guarantee [31], both of which are already well-established within the formal methods community. These frameworks stem from the observation that many modern approaches to formalising relaxed memory use complex state models (e.g., graphs of read/write events, or "views") accompanied by *operational semantics* in which the traces of a program are obtained by *interleaving* the program's threads. This therefore provides an opportunity to use classical reasoning techniques as desired. We note that there are complex models such as the full C11 model, in which compiler optimisations result in intra-thread reorderings [45], requiring consideration of more complex techniques that additionally interleave operations within a thread [60,61], but we do not consider them in this paper.

Our running example is a buffer developed by Lamport [39], which uses reads and writes on two shared variables to synchronise a producer that executes concurrently with a consumer. Interestingly, the algorithm does not require any additional synchronisation primitives such as locks or read-modify-write instructions (e.g., CAS or FAA). As such, our presentation only comprises reads and writes, further simplifying the memory model, helping keep the paper focussed on reasoning techniques. It is straightforward to extend models with reads and writes to accommodate read-modify-write instructions (e.g., see [17,19,20,37]).

Contributions. This paper aims to be instructive, rather than presenting any significant new results. Nevertheless, we do develop some new material. **(1)** We present a model of RC11 (with relaxed and release-acquire accesses) based on an earlier encoding of the release-acquire model [14]. We consider this to be one of the simplest possible encodings, reducing the encoding of the memory model to its essential components. **(2)** We present a new proof of Lamport's buffer in RC11 and show that the properties defined in the original paper can be guaranteed even when using relaxed accesses, which provide very weak ordering guarantees. **(3)** We extend the model and proof of the buffer to model the data written by the producer and read by the consumer, which we show induces release-acquire synchronisation. **(4)** We develop a generalised set of assertions that allow us to reason about "view-transfer" during release-acquire synchronisation, as required in the proof of the full buffer.

Overview. We present background to this work in Sect. 2, where we introduce Lamport's buffer and semi-formally discuss the RC11 memory model and correctness of the buffer under this relaxed memory model. In Sect. 3, we present the operational semantics for the relaxed fragment of RC11 reads and writes as well as the assertions introduced in Sect. 2. We then present the full buffer in Sect. 4, the problems with using relaxed accesses to transfer data between threads and how release-acquire accesses are used to ensure inter-thread memory ordering. Additionally, we introduce the formal model and our new generalised assertions. In Sect. 5, we conclude with some discussions on related work and future challenges.

2 Background

In this section, we motivate this paper and present the informal background to this work. In Sect. 2.1, we present our running example, which is a buffer developed by Lamport in 1977 designed to support message passing from a producer thread to a concurrently executing consumer thread. Interestingly, synchronisation between these threads is achieved without using expensive primitives such as compare-and-swap or locks. Instead, the algorithm uses two shared variables with only increment operations on each variable (see Sect. 2.1 for details). We informally introduce the RC11 memory model in Sect. 2.2, then the behaviour and correctness of the buffer under RC11 in Sect. 2.3.

2.1 Lamport's Buffer Under SC

Lamport's algorithm [39] (given in Fig. 1), assumes a single producer (left thread) and a single consumer (right thread) that communicate through a shared buffer[1]. As in the original paper [39], we only present the *intent* of the buffer and (for now) do not model the reads and writes on the buffer by the producer and consumer threads. In particular, we use counters m_s and m_r (both initially set to 0) to model the number of messages sent and received, respectively. Later (in Sect. 4) we will discuss how the buffer may be implemented and discuss the complications that this induces in the RC11 memory model.

$$\{m_s = m_r = s = r = 0 \wedge 0 < b \wedge \neg\mathtt{js} \wedge \neg\mathtt{jr}\}$$

$$\begin{array}{l||l}
\{n_c \leq s - r \leq b \wedge \neg\mathtt{js} \wedge I\} & \{0 \leq s - r \leq n_p \wedge \neg\mathtt{jr}\} \\
\quad \textbf{while } \textit{true} & \quad \textbf{while } \textit{true} \\
\quad \textbf{inv } n_c \leq s - r \leq b \wedge \neg\mathtt{js} \wedge I & \quad \textbf{inv } 0 \leq s - r \leq n_p \wedge \neg\mathtt{jr} \\
\qquad \{n_c \leq s - r \leq b \wedge \neg\mathtt{js} \wedge I\} & \qquad \{0 \leq s - r \leq n_p \wedge \neg\mathtt{jr}\} \\
\qquad \textbf{wait } s - r < b \textbf{ end}; & \qquad \textbf{wait } s - r > 0 \textbf{ end} \\
\qquad \{n_c \leq s - r < b \wedge \neg\mathtt{js} \wedge I\} & \qquad \{0 < s - r \leq n_p \wedge \neg\mathtt{jr}\} \\
\qquad \langle m_s := m_s + 1; \mathtt{js} := \textit{true}\rangle; & \qquad \langle m_r := m_r + 1; \mathtt{jr} := \textit{true}\rangle; \\
\qquad \{n_c - 1 \leq s - r < b \wedge \mathtt{js} \wedge I\} & \qquad \{0 < s - r \leq n_p + 1 \wedge \mathtt{jr}\} \\
\qquad \langle s := s + 1; \mathtt{js} := \textit{false}\rangle & \qquad \langle r := r + 1; \mathtt{jr} := \textit{false}\rangle \\
\{I \wedge \neg\mathtt{js}\} & \{\neg\mathtt{jr}\}
\end{array}$$

$$\{I\}$$

Fig. 1. Correctness proof of Lamport's buffer, where $n \mathrel{\widehat{=}} m_s - m_r$, $I \mathrel{\widehat{=}} 0 \leq n \leq b$, $n_p \mathrel{\widehat{=}} \textbf{if } \mathtt{js} \textbf{ then } n - 1 \textbf{ else } n$ and $n_c \mathrel{\widehat{=}} \textbf{if } \mathtt{jr} \textbf{ then } n + 1 \textbf{ else } n$. The purpose of the algorithm is to ensure that I is an invariant of the program. The highlighted conjuncts are present in Lamport's proof, but can be omitted.

The threads are synchronised using the shared variables s and r, which are modified by the producer and consumer, respectively. In particular, after sending a message (modelled by the increment to m_s), the producer increments s. After receiving a message (modelled by incrementing m_r), the consumer increments r. Note that s (resp. r) is only modified by the producer (resp. consumer). We assume that at most b messages may be in-flight (i.e., sent but not yet received) at any time, thus the producer only attempts to send a message to the buffer if

[1] To simplify the presentation, we elide a modulus parameter, $k > b$. Lamport's algorithm replaces $s-r$ by $(s-r) \mathsf{mod} k$, and the increments to s and r by $(s+1) \mathsf{mod} k$ and $(r+1) \mathsf{mod}\ k$, respectively. But as Lamport discusses, an instance of this parameter is $k = \infty$, which results in the algorithm in Fig. 1.

$s - r < b$. Similarly, the consumer only attempts to read from the buffer if an item is available, i.e., $s - r > 0$. Variables js and jr are auxiliary variables [44], which are only used in the proof; we discuss these in more detail below. We use $\langle a_1; a_2 \rangle$ a statement that executes a_1 followed by a_2 as a single *atomic* step. The statement **wait** *grd* **end** is used to block the respective thread until the guard *grd* holds.

Memory Model. Like Lamport [39], the program in Fig. 1 assumes a strong SC memory model [40], i.e., each modification to each shared variable, is visible to all other threads immediately after the modification takes place. This means that we can view the execution of the program in Fig. 1 as an *interleaving* of the two threads. Like Lamport, we assume that each line of the program is atomic since the variables modified by each thread are not modified by the other.

Correctness. Lamport also gives a proof of correctness of his algorithm by modelling each thread as a flowchart with nodes representing (atomic) statements and transitions labelled with assertion that holds after each statement has been executed. This is then accompanied by an informal proof discussing why the main invariant, I, holds. I states that the difference between the number of messages sent and the number of messages received is at least 0, and at most b.

Note that both of the increments in the producer (resp. consumer) are accompanied by an update to js (resp. jr) to denote that a message has just been sent (resp. received). We convert Lamport's proof to an Owicki-Gries [44] proof outline, which we then encode and verify in Isabelle/HOL's built-in encoding of the Owicki-Gries framework [43]. The Owicki-Gries framework is well known, so we only describe it informally here. A proof outline is *valid* [44] iff each of the following holds.

1. The precondition of the program implies the initial assertion of each thread. We refer to this proof obligation as *initialisation.*
2. The conjunction of the final assertions of all threads implies the postcondition of the program. We refer to this proof obligation as *finalisation.*
3. For each atomic statement a in each thread, the Hoare triple $\{P\}a\{Q\}$ from the proof outline holds. This proof obligation is known as *local correctness.*
4. Each assertion P of each thread is stable against the execution of the other thread, i.e., for each $\{Q\}a$ from the other thread (where Q is the precondition of the atomic statement a), $\{P \wedge Q\}a\{P\}$ holds. This proof obligation is known as *global correctness.*

Lemma 1. *The proof outline in Fig. 1 is valid.*

This lemma has been verified using Isabelle/HOL using the encoding of Owicki-Gries framework that is part of the standard Isabelle/HOL distribution [43]. This encoding is powerful and is supported by tactics that allow the proof outline to be validated fully automatically in under 1 s. This speed has enabled us to experiment with the assertions, and we have discovered redundant conjuncts in Lamport's original proof [39], which are highlighted in Fig. 1 and have been omitted in our proof.

2.2 The RC11 Memory Model

Although Lamport's algorithm and its proof of correctness (as presented in Sect. 2.1) are straightforward, this simplicity stems from the strong assumption of an SC memory model, which is difficult to achieve in practice. As already discussed, most modern hardware (and programming languages) implement a relaxed memory model whose effects invalidate the proof outline in Fig. 1.

In this paper, we will consider the RC11 memory model in detail, which is a restricted version of the C/C++ 2011 standard [38]. Given that this is a tutorial on relaxed memory *reasoning*, we only discuss the minimal fragments of the model that are needed to address each example. Moreover, we focus on *principles* for building logics to support formal proofs under relaxed memory, whereby frameworks such as Owicki-Gries can be re-used.

Variables in RC11 may either be *non-atomic* or *atomic*. The former type results in undefined behaviour if there is a data race (a read or write on one thread occurring concurrently with a write on another thread). However, we are interested in the behaviour of racy programs such as Lamport's buffer. For this reason, we will assume that all shared variables are atomic. Under RC11, one can use a range of *memory-ordering*[2] restrictions, describing the observation guarantees that different threads can make. We will make use of three of these:

- *relaxed* reads and writes, which does not impose ordering among concurrent accesses,
- *releasing* writes, which induce a happens-before ordering with acquiring reads (see below), and
- *acquiring* reads, which induce a happens-before ordering whenever the read reads from a releasing write. That is, any happens-before information known by the thread executing the releasing write is transferred to the thread that executes the acquiring read.

We will formalise each of these in the upcoming sections, and discuss how they impact Lamport's buffer.

An important feature of the RC11 memory model is that *program order* [5,38] is maintained, i.e., the order of operations within a thread is unchanged. Thus, as in SC, we can think of the execution of an RC11 program as an *interleaving* of the program's atomic statements. The difference is that, due to relaxed memory effects, the modifications in one thread are not guaranteed to be seen by the other threads, unless dictated by the memory-order restrictions.

2.3 Lamport's Buffer Under RC11

We now consider the behaviour of Lamport's buffer under RC11, where (like Lamport [39]) we ignore the reads and writes to the buffer and simply use m_s and m_r to count the number of sends and receives. We will discuss the complications induced by RC11 when modelling the full buffer in the upcoming sections.

[2] See https://en.cppreference.com/w/c/atomic/memory_order.html.

Given the range of memory-ordering options, the first question is: which of these is sufficient to correctly implement the buffer from Fig. 1? It turns out that, because each shared variable is only modified by one of the threads, it is sufficient to use *relaxed* accesses everywhere. This results in the program and correctness proof in Fig. 2. We use highlights to indicate the assertions that change or are introduced due to the execution of each atomic command.

$$\left\{s \overset{p}{=} \mathsf{sv} \wedge r \overset{c}{=} \mathsf{rv} \wedge V_{pc} \wedge b > 0 \wedge \mathsf{sv} - \mathsf{rv} = n \wedge \neg\mathtt{js} \wedge \neg\mathtt{jr} \wedge I\right\}$$

$$\begin{array}{l||l}
P_1 : \left\{\begin{array}{l} s \overset{p}{=} \mathsf{sv} \wedge V_p \wedge \neg\mathtt{js} \wedge I \wedge \\ (\forall \mathsf{rv}.\ r \overset{c}{=} \mathsf{rv} \Rightarrow n_c \leq \mathsf{sv} - \mathsf{rv}) \end{array}\right\} & \left\{\begin{array}{l} r \overset{c}{=} \mathsf{rv} \wedge V_c \wedge \neg\mathtt{jr} \wedge \\ (\forall \mathsf{sv}.\ s \overset{p}{=} \mathsf{sv} \Rightarrow \mathsf{sv} - \mathsf{rv} \leq n_p) \end{array}\right\} \\
\quad \mathbf{s}_p :\overset{\mathtt{X}}{=} s; & \quad \mathbf{r}_c :\overset{\mathtt{X}}{=} r; \\
P_2 : \left\{\begin{array}{l} s \overset{p}{=} \mathbf{s}_p \wedge V_p \wedge \neg\mathtt{js} \wedge I \wedge \\ (\forall \mathsf{rv}.\ r \overset{c}{=} \mathsf{rv} \Rightarrow n_c \leq \mathbf{s}_p - \mathsf{rv}) \end{array}\right\} & \left\{\begin{array}{l} r \overset{c}{=} \mathbf{r}_c \wedge V_c \wedge \neg\mathtt{jr} \wedge \\ (\forall \mathsf{sv}.\ s \overset{p}{=} \mathsf{sv} \Rightarrow \mathsf{sv} - \mathbf{r}_c \leq n_p) \end{array}\right\} \\
\quad \mathbf{do}\ \mathbf{r}_p :\overset{\mathtt{X}}{=} r; & \quad \mathbf{do}\ \mathbf{s}_c :\overset{\mathtt{X}}{=} s \\
P_3 : \left\{\begin{array}{l} s \overset{p}{=} \mathbf{s}_p \wedge V_p \wedge \neg\mathtt{js} \wedge I \wedge \\ (\forall \mathsf{rv}.\ r \overset{c}{=} \mathsf{rv} \Rightarrow n_c \leq \mathbf{s}_p - \mathsf{rv} \wedge \mathbf{r}_p \leq \mathsf{rv}) \end{array}\right\} & \left\{\begin{array}{l} r \overset{c}{=} \mathbf{r}_c \wedge V_c \wedge \neg\mathtt{js} \wedge \\ (\forall \mathsf{sv}.\ s \overset{p}{=} \mathsf{sv} \Rightarrow \mathbf{s}_c \leq \mathsf{sv} \wedge \mathsf{sv} - \mathbf{r}_c \leq n_p) \end{array}\right\} \\
\quad \mathbf{until}\ \mathbf{s}_p - \mathbf{r}_p < b; & \quad \mathbf{until}\ \mathbf{s}_c - \mathbf{r}_c > 0; \\
P_4 : \left\{\begin{array}{l} s \overset{p}{=} \mathbf{s}_p \wedge (\exists \mathsf{rv}.\ r \overset{c}{=} \mathsf{rv}) \wedge \neg\mathtt{js} \wedge I \wedge \\ (\forall \mathsf{rv}.\ r \overset{c}{=} \mathsf{rv} \Rightarrow n_c \leq \mathbf{s}_p - \mathsf{rv} < b) \end{array}\right\} & \left\{\begin{array}{l} r \overset{c}{=} \mathbf{r}_c \wedge \neg\mathtt{jr} \wedge \\ (\forall \mathsf{sv}.\ s \overset{p}{=} \mathsf{sv} \Rightarrow 0 < \mathsf{sv} - \mathbf{r}_c \leq n_p) \end{array}\right\} \\
\quad \langle m_s := m_s + 1; \mathtt{js} := \mathit{true}\rangle; & \quad \langle m_r := m_r + 1; \mathtt{jr} := \mathit{true}\rangle; \\
P_5 : \left\{\begin{array}{l} s \overset{p}{=} \mathbf{s}_p \wedge \mathtt{js} \wedge I \wedge \\ (\forall \mathsf{rv}.\ r \overset{c}{=} \mathsf{rv} \Rightarrow n_c - 1 \leq \mathsf{sv} - \mathsf{rv}) \end{array}\right\} & \left\{\begin{array}{l} r \overset{c}{=} \mathbf{r}_c \wedge \mathtt{jr} \wedge \\ (\forall \mathsf{sv}.\ s \overset{p}{=} \mathsf{sv} \Rightarrow \mathsf{sv} - \mathbf{r}_c \leq n_p + 1) \end{array}\right\} \\
\quad \langle s :\overset{\mathtt{X}}{=} \mathbf{s}_p + 1, \mathtt{js} := \mathit{false}\rangle & \quad \langle r :\overset{\mathtt{X}}{=} \mathbf{r}_c + 1, \mathtt{jr} := \mathit{false}\rangle \\
P_6 : \left\{s \overset{p}{=} \mathbf{s}_p + 1 \wedge \neg\mathtt{js} \wedge I\right\} & \left\{s \overset{p}{=} \mathbf{s}_p + 1 \wedge \neg\mathtt{jr}\right\}
\end{array}$$

$$\{I\}$$

where

$$V_p \mathrel{\widehat{=}} (\forall \mathsf{rv}, \mathsf{v}.\ r \overset{c}{=} \mathsf{rv} \wedge \mathsf{v} \in |p\rangle r \Rightarrow \mathsf{v} \leq \mathsf{rv}) \wedge (\exists \mathsf{rv}.\ r \overset{c}{=} \mathsf{rv})$$
$$V_c \mathrel{\widehat{=}} (\forall \mathsf{sv}, \mathsf{v}.\ s \overset{p}{=} \mathsf{sv} \wedge \mathsf{v} \in |c\rangle s \Rightarrow \mathsf{v} \leq \mathsf{sv})$$
$$V_{pc} \mathrel{\widehat{=}} V_p \wedge V_c$$

Fig. 2. Lamport's buffer under RC11, where n_c, n_p, n and I are defined as in Fig. 1. The assertions are named $P_1 \ldots P_6$ to simplify referencing.

RC11-Based Algorithm. To simplify the discussion, we only consider a single iteration of the loop, but it is straightforward to generalise the program to the looping version since a loop invariant is maintained and the precondition does not over-constrain the initial values of s and r. There are three main differences in the RC11 version of the producer code compared to the SC version (the consumer has similar differences):

1. The read accesses of the shared variables s and r within the guard of the **wait** is replaced by explicit *relaxed reads* ($\mathbf{s}_p :\overset{\mathtt{X}}{=} s$ and $\mathbf{r}_p :\overset{\mathtt{X}}{=} r$) that store the values read into local variables $\mathbf{s}_p$ and $\mathbf{r}_p$, respectively.

2. The **wait** is replaced by an equivalent **do-until** loop that re-reads r whenever the program is due to continue waiting. Note that we do not re-read s since it is never modified by the consumer thread.
3. The update to s is replaced by a *relaxed write* ($s :\stackrel{\mathtt{X}}{=} \mathtt{s}_p + 1$), which increments s. Again, because s is only modified by the producer, we do not need to re-read s, i.e., it is sufficient to use the local copy $\mathtt{s}_p$ read in the first line of the producer thread.

Example 1. For the program in Fig. 2, because all accesses are relaxed, a possible snapshot of the program state is depicted in Fig. 3. The top and bottom chains represents the ordering of values of the shared variables s and r, respectively generated by the writes to each variable. Each arrow represents the *modification order* [5,38], which is an ordering of the values of each variable as the program executes. The modification order is a total order on each variable. The values of s depicted are $\mathsf{sv} - 5$, $\mathsf{sv} - 4$, ..., sv, whereas the values of r are $\mathsf{rv} - 2$, $\mathsf{rv} - 1$ and rv. In Fig. 3, we assume that $\mathsf{rv} = \mathsf{sv} - 4$.

The value(s) that a thread may read are dictated by the thread's current view. Since s is only modified by the producer thread, p, its view of s is the latest (i.e., most up-to-date) value sv. Thus, any read (including a relaxed read) of s performed by p is guaranteed to return sv. However, the consumer thread c's view of s may be *stale*, as is the case in Fig. 3. Here, since c's view is currently $\mathsf{sv} - 3$, a read performed by c may return any of the values in $\{\mathsf{sv} - 3, \mathsf{sv} - 2, \mathsf{sv} - 1, \mathsf{sv}\}$. Similarly, c's view of r is the latest value rv, whereas the producer thread p's view is the stale value $\mathsf{rv} - 2$. Thus, any read of r by c may only return rv, whereas a read of p may return values in $\{\mathsf{rv} - 2, \mathsf{rv} - 1, \mathsf{rv}\}$.

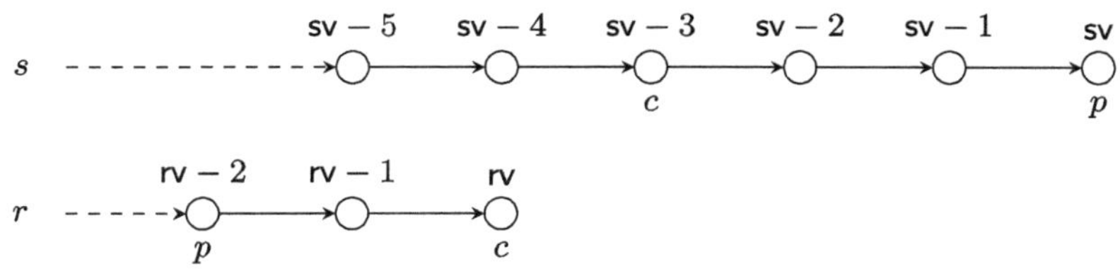

Fig. 3. A possible state of Lamport's buffer under RC11, with each arrow representing the *modification order* [5,38].

The challenge therefore (which was solved in prior works [17,20,37,61]) is to develop a set of *assertions* that capture the phenomena in Fig. 3 at a high level. We discuss these below.

Correctness Under RC11. Validity of the proof outline in Fig. 2 is defined using Owicki-Gries' definition, exactly as in Fig. 2.1, i.e., we check the standard proof obligations: initialisation, finalisation, local correctness and global correctness. The only difference is that the assertions used describe the relaxed memory RC11 state. The proof in Fig. 2 only requires two weak memory assertions:

- The *definite value* assertion, $x \stackrel{\tau}{=} \mathsf{xv}$, which indicates that thread τ's view of x is of the last value, xv.
- The *possible values* function, $\lvert\tau\rangle x$, which returns the set of values that τ may read for x.

Example 2. For the state in Fig. 3 (described in Example 1), we have $s \stackrel{p}{=} \mathsf{sv}$ and $r \stackrel{c}{=} \mathsf{rv}$. Moreover, $\lvert p\rangle r = \{\mathsf{rv}-2, \mathsf{rv}-1, \mathsf{rv}\}$ and $\lvert c\rangle s = \{\mathsf{sv}-3, \mathsf{sv}-2, \mathsf{sv}-1, \mathsf{sv}\}$.

In general, if $x \stackrel{\tau}{=} \mathsf{xv}$, then $\mathsf{xv} \in \lvert\tau\rangle x$ and, in fact, $\lvert\tau\rangle x = \{\mathsf{xv}\}$. Note however that $\lvert\tau\rangle x = \{\mathsf{xv}\}$ does *not* imply $x \stackrel{\tau}{=} \mathsf{xv}$ since the definite-view assertion also requires that τ's view of x is the last-written value.

With this understanding, it is possible to prove the following lemma. We have, in fact, encoded and verified validity in Isabelle/HOL using the encoding by Dalvandi et al. [20].

Lemma 2. *The proof outline in Fig. 1 is valid.*

The main components of the proof outline in Fig. 2 are derived from the SC proof in Fig. 1. The main difference is in the supporting assertions that are used to establish the invariant, I.

Example 3. Consider the assertion P_1. The conjunct $\forall \mathsf{rv}.\ r \stackrel{c}{=} \mathsf{rv} \Rightarrow n_c \leq \mathsf{sv} - \mathsf{rv}$ is used directly to establish I. Note the similarity of this conjunct with the corresponding assertion $n_c \leq s - r$ in Fig. 1. The difference in RC11 is that we cannot directly use the values s and r since these values may be different for different threads.

The value sv is the definite value that p observes of s, which is obtained from the conjunct $s \stackrel{p}{=} \mathsf{sv}$ in P_1 itself. For rv, we cannot fix the value of r using a corresponding conjunct of the form $r \stackrel{c}{=} \mathsf{rv}$ since such an assertion would *not* be stable under the update to r in thread c, i.e., the global correctness proof would fail. Instead, we obtain the value rv using universal quantification, which states that for *any* definite value rv of r as seen by thread c, we have $n_c \leq \mathsf{sv} - \mathsf{rv}$.

As in Fig. 1, the conjunct $\neg \mathtt{js}$ supports the global correctness proof of thread c (establishing a value for n_p. Finally, V_p is used within thread p to establish the conjunct $\mathtt{r}_p \leq \mathsf{rv}$ in P_3 and the conjunct $\exists \mathsf{rv}.\ r \stackrel{c}{=} \mathsf{rv}$ in P_4. The latter is straightforward since it follows directly from V_p, but the former is interesting.

Essentially, we require a condition that allows us to strengthen the for-all condition in P_3, which describes the value read by p in relation to the last value written by c. As depicted in Fig. 3, this value is at most rv (i.e., the value that c can see). To support establishing this conjunct (i.e., $\mathtt{r}_p \leq \mathsf{rv}$) in P_3, we use the first conjunct of V_p, which states that, if the definite value of r is rv (according to c) and p can read a value v, then v must be at most rv. Under this assumption, the read of r by thread p within the **do** loop is guaranteed to establish $\mathtt{r}_p \leq \mathsf{rv}$.

3 Relaxed Reads/Writes

We now discuss the formal background to support the claims in Sect. 2. We start by introducing the fragment of RC11 comprising only *relaxed* read and write accesses to keep the discourse as simple as possible. We introduce the operational semantics in Sect. 3.1, then in Sect. 3.2 formalise the "value" assertions from Fig. 2. Later, in Sect. 4, we consider the more complex release-acquire fragment and the assertions that can be used to reason about message passing under such semantics.

3.1 Operational Semantics

The semantics of the RC11 was originally defined in a graph-based declarative (aka axiomatic [5,38]) style, which provide descriptions of the allowable behaviours of a program in a given memory model. These semantics consider *complete* executions of a program to determine whether or not each execution is allowed by (i.e., possible in) the memory model using axioms over the resulting execution graph.

However, such semantics are not amenable to stepwise verification, as required by frameworks such as Owicki-Gries and rely/guarantee. However, for many axiomatic models (including RC11), it is possible to develop equivalent operational semantics [23,32,33,46,48]. These models can either build directly on the axiomatic graphs [23,34,41] or use an equivalent encoding based on "views" [11,15,32,46,61]. The assertions needed for relaxed memory verification can easily be defined in both operational encodings (e.g., see [60]). In this paper, we opt to present the view-based semantics, which are closer to the assertions that we use, and hence easier to understand.

Language. We assume the extended while language below. We assume Var denotes the set of all variables, which are partitioned into local variables, L, and global variables, G. We further partition the local variables into registers, R_L, and auxiliary variables, A_L. Registers appear in programs, whereas auxiliary variables are only used to support proofs. We also assume a set of values, Val, and a set of thread identifiers, Tid.

Suppose $\mathsf{v} \in \mathit{Val}$, $\mathtt{r} \in R_L$, $\mathsf{a} \in A_L$, $e, e_1, e_2 \in \mathit{Exp}$, $f, f_1, f_2 \in \mathit{Aux}$, $x \in G$, $a \in \mathit{Asgn}$, $c, c_1, c_2 \in \mathit{Com}$ and $b \in \mathit{Exp}$ is boolean valued. Let $\mathtt{M} \in \{\mathtt{X}, \mathtt{R}, \mathtt{A}\}$ define the memory access type. Then the syntax of programs is given by:

$$\mathit{Exp} ::= \mathsf{v} \mid \mathtt{r} \mid \ominus e \mid e_1 \oplus e_2 \qquad\qquad \mathit{Aux} ::= \mathsf{v} \mid \mathtt{r} \mid \mathsf{a} \mid \ominus f \mid f_1 \oplus f_2$$

$$\mathit{Asgn} ::= \mathtt{r} := e \mid \mathtt{r} \overset{\mathtt{M}}{:=} x \mid x \overset{\mathtt{M}}{:=} e$$

$$\mathit{Com} ::= a \mid \langle a; \overline{\mathsf{a}} := \overline{f} \rangle \mid c_1; c_2 \mid \mathbf{if}\ b\ \mathbf{then}\ c_1\ \mathbf{else}\ c_2 \mid \mathbf{do}\ c\ \mathbf{until}\ b$$

$$\mathit{Prog} ::= \lambda \tau \in \mathit{Tid}.\ c$$

Note that all expressions are over local registers only. We have three types of assignments: local assignments ($\mathtt{r} := e$), reads ($\mathtt{r} \overset{\mathtt{M}}{:=} x$) and writes ($x \overset{\mathtt{M}}{:=} e$). The remaining commands are standard.

For simplicity, we assume top-level parallelism though it is possible to support dynamic thread creation [37]. A concurrent program is a mapping from thread identifiers to commands. We assume that the set of registers are further partitioned into disjoint sets of local registers for each thread.

Modularising the Memory Semantics. The state is split into two components: a local component (which has type $\Sigma \mathrel{\widehat{=}} L \to \mathit{Val}$ as in standard SC semantics [43,44]) and a relaxed-memory state (whose type Γ can be instantiated in different ways). We discuss the view-based instantiation of Γ below.

Let $\sigma \in \Sigma$, $\gamma \in \Gamma$ and $\Pi \in \mathit{Prog}$. At the top level, program execution is defined by the following rule:

$$\textsc{Prog}\quad \frac{\langle \sigma, \gamma, \Pi(\tau)\rangle \to_\tau \langle \sigma', \gamma', c'\rangle}{\langle \sigma, \gamma, \Pi\rangle \to \langle \sigma', \gamma', \Pi[\tau \mapsto c']\rangle}$$

Here the program takes a step if some thread τ takes a step. The execution of non-atomic commands is standard. As an example, we present the rule for atomics with auxiliary variables, where $[\![e]\!]_\sigma$ denotes the value of e in the state σ.

$$\textsc{Aux}\quad \frac{\langle \sigma, \gamma, a\rangle \to_\tau \langle \sigma', \gamma', \bot\rangle \overline{\mathsf{a}} = \mathsf{a}_1, \dots, \mathsf{a}_n \overline{f} = f_1, \dots, f_n}{\langle \sigma, \gamma, \langle a, \overline{\mathsf{a}} := \overline{f}\rangle\rangle \to_\tau \langle \sigma'[\mathsf{a}_1 \mapsto [\![f_1]\!]_\sigma, \dots, \mathsf{a}_n \mapsto [\![f_n]\!]_\sigma], \gamma', \bot\rangle}$$

Note that because the auxiliary update must take place in parallel with the execution of a, we evaluate e in the initial local state σ. We use $\bot$ to determine that the command under consideration has terminated.

Finally, we have rules for the three assignment statements. The rule for local assignment $\mathtt{r} := e$ is as standard and updates the local state σ. For the reads and writes, we have the following rules, which are parameterised by the memory ordering restriction $\mathtt{M}$. The premises of both rules for different instances of $\mathtt{M}$ are defined in Figs. 4 and 7 and will be discussed below.

$$\textsc{Rd}\quad \frac{\gamma \xrightarrow{\langle x, \mathsf{read}(v), \mathtt{M}\rangle}_\tau \gamma'}{\langle \sigma, \gamma, \mathtt{r} \overset{\mathtt{M}}{:=} x\rangle \to_\tau \langle \sigma[\mathtt{r} \mapsto v], \gamma', \bot\rangle} \qquad \textsc{Wr}\quad \frac{\gamma \xrightarrow{\langle x, \mathsf{write}([\![e]\!]_\sigma), \mathtt{M}\rangle}_\tau \gamma'}{\langle \sigma, \gamma, x \overset{\mathtt{M}}{:=} e\rangle \to_\tau \langle \sigma, \gamma', \bot\rangle}$$

The Rd read rule presumes that it is possible to transition from γ to γ' using the rule $\gamma \xrightarrow{\langle x, \mathsf{read}(v), \mathtt{M}\rangle}_\tau \gamma'$, where the read returns the value v. Note that the reads may have a side effect of updating the view and hence γ' may not be the same as γ. Similarly, the write rule presumes $\gamma \xrightarrow{\langle x, \mathsf{write}([\![e]\!]_\sigma), \mathtt{M}\rangle}_\tau \gamma'$ and introduces a new write with value $[\![e]\!]_\sigma$ to the weak memory state.

View-Based Relaxed Read/Write Semantics. We provide an instantiation of Γ based on the formalisation in [14]. Each $\gamma \in \Gamma$ is a pair of the form $\langle m, \mathcal{V}\rangle$.

- m is a set of *messages* comprising *all* previously executed writes. Each message is a tuple of the form $\mu = \langle x, \mathsf{v}, t, V, \mathtt{M}\rangle$ where: $x \in \mathit{Var}$ is the variable of the message, $\mathsf{v} \in \mathit{Val}$ is the written value, $t \in \mathbb{Q}$ is the *timestamp*, $V \in \mathsf{View} \mathrel{\widehat{=}} \mathit{Var} \to \mathbb{Q}$ is the message's *view*, and $\mathtt{M}$ is the memory ordering. Timestamps are used to order messages of the same variable, as depicted in Fig. 3. A message's view is used in the release-acquire fragment, and can be ignored for now since it has no use for relaxed reads/writes.
- $\mathcal{V} : \mathit{Tid} \to \mathsf{View}$ is a mapping from each thread to its view.

The operational semantics of relaxed reads/writes on the weak memory state are given by the transitions in Fig. 4. We use the functions $\mathsf{var}(\mu)$, $\mathsf{val}(\mu)$, $\mathsf{ts}(\mu)$, $\mathsf{view}(\mu)$ and $\mathsf{ord}(\mu)$ to retrieve the variable (x), value (v), timestamp (t), view (V) and memory order ($\mathtt{M}$) of a message μ. We lift these projection functions to sets of messages pointwise (e.g., $\mathsf{ts}(S) \mathrel{\widehat{=}} \{\mathsf{ts}(\mu) \mid \mu \in S\}$), and write $m_{|x}$ for the set of all messages $\mu \in m$ with $\mathsf{var}(\mu) = x$.

$$\textsc{Read-Rlx}\quad \frac{e = \langle x, \mathsf{read}(v), \mathtt{X}\rangle \qquad \langle x, v, t, \_, \_\rangle \in m \qquad \mathcal{V}(\tau)(x) \le t \qquad V' = \mathcal{V}(\tau)[x \mapsto t]}{\langle m, \mathcal{V}\rangle \xrightarrow{e}_\tau \langle m, \mathcal{V}[\tau \mapsto V']\rangle}$$

$$\textsc{Write-Rlx-Rel}\quad \frac{e = \langle x, \mathsf{write}(v), \mathtt{M}\rangle \qquad V' = \mathcal{V}(\tau)[x \mapsto t] \qquad \mathcal{V}(\tau)(x) < t \qquad t \notin \mathsf{ts}(m_{|x}) \qquad \mu = \langle x, v, t, V', \mathtt{M}\rangle}{\langle m, \mathcal{V}\rangle \xrightarrow{e}_\tau \langle m \cup \{\mu\}, \mathcal{V}[\tau \mapsto V']\rangle}$$

Fig. 4. Relaxed reads and relaxed/releasing writes in RC11.

A relaxed read by thread τ on shared variable x may read from any message $\langle x, v, t, \_, \_\rangle$ whose timestamp t is at least as large as τ's view of x. The read ignores the view and memory-ordering constraints of the message, but retrieves the value v. As a side effect of this read, the view of τ for x is updated to t.

A relaxed write by thread τ writes to a variable x, adds a corresponding message μ to m with some *fresh* timestamp t that is larger than $\mathcal{V}(\tau)(x)$, the timestamp of the latest message to x that was observed by τ. Then, the thread τ updates its view of x to t. Note that the write rule in Fig. 4 also encodes releasing writes, which we discuss in Sect. 4. Such writes make use of the component V' in each message, which is ignored by relaxed reads/writes.

3.2 Value Assertions and Owicki-Gries

With an operational semantics in place, we now formalise the value assertions used in Fig. 2. The view function and definite-view assertion are now straightforward to define:

$$\begin{aligned}
\mathsf{obs}(\tau) &\mathrel{\widehat{=}} \lambda\alpha.\ \textbf{let}\ \alpha = \langle_, \langle m, \mathcal{V}\rangle\rangle\ \textbf{in}\ \{\mu \in m \mid \mathsf{ts}(\mu) \ge \mathcal{V}(\tau)(\mathsf{var}(\mu))\} \\
|\tau\rangle x &\mathrel{\widehat{=}} \lambda\alpha.\ \mathsf{val}(\mathsf{obs}(\tau)(\alpha)_{|x}) \\
x\langle\!\langle\tau &\mathrel{\widehat{=}} \lambda\alpha.\ \textbf{let}\ \alpha = \langle_, \langle m, \mathcal{V}\rangle\rangle\ \textbf{in}\ \mathcal{V}(\tau)(x) = \max(\mathsf{ts}(m_{|x})) \\
x \stackrel{\tau}{=} e &\mathrel{\widehat{=}} \lambda\alpha.\ \textbf{let}\ \alpha = \langle\sigma, _\rangle\ \textbf{in}\ (x\langle\!\langle\tau)(\alpha) \wedge (|\tau\rangle x)(\alpha) = \{[\![e]\!]_\sigma\}
\end{aligned}$$

Thus $\mathsf{obs}(\tau)$ determines the set of messages that τ can observe in the given state. These are the messages whose timestamp is at least the timestamp of τ's view for $\mathsf{var}(\mu)$. $\lceil\tau\rangle x$ is a function on the state that returns the set of values of x that τ may read. $x\langle\!\langle\tau$ is a predicate that holds iff τ's view of x is the message over x with the highest timestamp. $x \stackrel{\tau}{=} \mathsf{xv}$ holds iff τ is viewing the highest-timestamp message on x and this message has the value xv.

The operational semantics also leads to a natural definition of a Hoare-triple, where $\tau \rhd c$ denotes the command c executed by thread τ.

Definition 1. *Given assertions $P, Q : \Sigma \times \Gamma \to \mathbb{B}$, thread $\tau \in Tid$, command $c \in Com$ and program $\Pi \in Prog$:*

$$\{P\}\tau \rhd c\{Q\} \mathrel{\widehat{=}} \forall \sigma, \gamma.\ P((\sigma, \gamma)) \wedge \langle \sigma, \gamma, c\rangle \to_{\tau}^{*} \langle \sigma', \gamma', \bot\rangle \Rightarrow Q((\sigma', \gamma'))$$

$$\{P\}\Pi\{Q\} \mathrel{\widehat{=}} \forall \sigma, \gamma.\ P((\sigma, \gamma)) \wedge \langle \sigma, \gamma, \Pi\rangle \to^{*} \langle \sigma', \gamma', \lambda\tau.\ \bot\rangle \Rightarrow Q((\sigma', \gamma'))$$

This definition naturally supports decomposition rules from Hoare logic for compound statements, e.g., sequential composition is covered by the following rule.

SEQ
$$\frac{\{P\}\tau \rhd c_1\{Q\} \qquad \{Q\}\tau \rhd c_2\{R\}}{\{P\}\ \tau \rhd c_1; c_2\ \{R\}}$$

The decomposition of proof outlines of concurrent programs is also the same as in Owicki-Gries' framework (see Sect. 2.1), except that each triple must include information about the executing thread since we use thread-specific assertions.

- Local correctness holds if, for each atomic statement a in each thread τ, the Hoare triple $\{P\}\tau \rhd a\{Q\}$ from the proof outline holds.
- Global correctness holds if, for each assertion P of each thread τ, for each $\{Q\}\tau' \rhd a$ in another thread τ', we have $\{P \wedge Q\}\tau' \rhd a\{P\}$.

The main difficulties stem from the proof rules for atomic commands (e.g., relaxed reads and writes). Here, it is necessary to describe how these statements interact with the view-based assertions above [20,61]. We give some examples of such rules below.

RD-POSS
$$\frac{}{\{\lceil\tau\rangle x = S\}\tau \rhd \mathtt{r} \stackrel{\mathtt{M}}{:=} x\{\mathtt{r} \in S\}}$$

RD-DEF
$$\frac{}{\{x \stackrel{\tau}{=} \mathsf{xv}\}\tau \rhd \mathtt{r} \stackrel{\mathtt{M}}{:=} x\{\mathtt{r} = \mathsf{xv}\}}$$

RD-STABLE
$$\frac{}{\{x \stackrel{\tau}{=} \mathsf{xv}\}\tau' \rhd \mathtt{r} \stackrel{\mathtt{M}}{:=} y\{x \stackrel{\tau}{=} \mathsf{xv}\}}$$

WR-LATEST
$$\frac{}{\{x\langle\!\langle\tau\}\tau \rhd x \stackrel{\mathtt{M}}{:=} e\{x \stackrel{\tau}{=} e\}}$$

Rule RD-POSS states that, if the view of τ on variable x is such that it sees the set of values S, then loading the value of x into the local register $\mathtt{r}$ results in a post state where $\mathtt{r} \in S$. By RD-DEF, if the view of τ on x is the last value xv, then the effect of loading x into $\mathtt{r}$ results in a post state where $\mathtt{r}$ has the value xv. Rule RD-STABLE states that a definite-view assertion is stable under the read (by any thread). Note that x and y as well as τ and τ' are not necessarily different. Finally, WR-LATEST states that if τ sees the last value on x, then storing e in x results in a state where, from the perspective of τ, the value of x is definitely e.

These can be extended to cover stability of assertions, e.g., due to assignments on other variables and statements executed by other threads. Soundness of each of these rules is proved by unfolding Definition 1. Such proofs are generally straightforward to encode and prove in theorem provers such as Isabelle/HOL [11,20,61].

4 Modelling and Verifying the Full Buffer

Although Lemma 2 establishes correctness of the main invariant I, there are additional subtleties if the correctness proof includes details about the buffer itself. As Lamport discusses (though does not prove), under SC, the buffer can be modelled and verified if the algorithm includes an array $B[0..b-1]$, where the producer puts each new message into $B[s \bmod b]$, and the consumer takes each message from $B[r \bmod b]$. To simplify the assertions and avoid some tedious modulo calculations, we assume that B is an infinite-sized array. With some additional work, the program and its proof outline can be extended to accommodate a wraparound buffer.

4.1 Synchronising the Buffer in RC11

Under RC11, each $B[i]$ is a shared location, and thus is affected by the relaxed memory semantics. In particular, it may be possible for the consumer to read a stale value of $B[i]$, which is undesirable. To prevent this, we must strengthen the memory-ordering guarantees.

In an implementation, updates to B may involve a large payload, so it is undesirable to introduce extra memory orderings on B directly. Instead, we use the index updates on s to induce *release-acquire* synchronisation, which enables transfer of happens-before information from the producer to the consumer. We require that, upon reading a value k of s, that allows the consumer to read from B (i.e., exit its **do-until** loop), the consumer thread c must see the latest value of $B[k]$. However, the semantics of reads (cf Fig. 4) is that the reader may read from *any* later write. Thus, it may be the case that c jumps ahead by several steps when reading s, i.e., updates its value from k_0 to k where $k - k_0 > 1$. Since it will read from the buffer in order, i.e., $B[k_0]$, then $B[k_0+1]$, up to $B[k-1]$, we require that upon reading k for s, the consumer thread sees the latest values of each of the entries $B[k_0], B[k_0+1], \dots, B[k-1]$.

Release-acquire synchronisation provides precisely this guarantee, which we describe informally using our view model using the example state in Fig. 5.

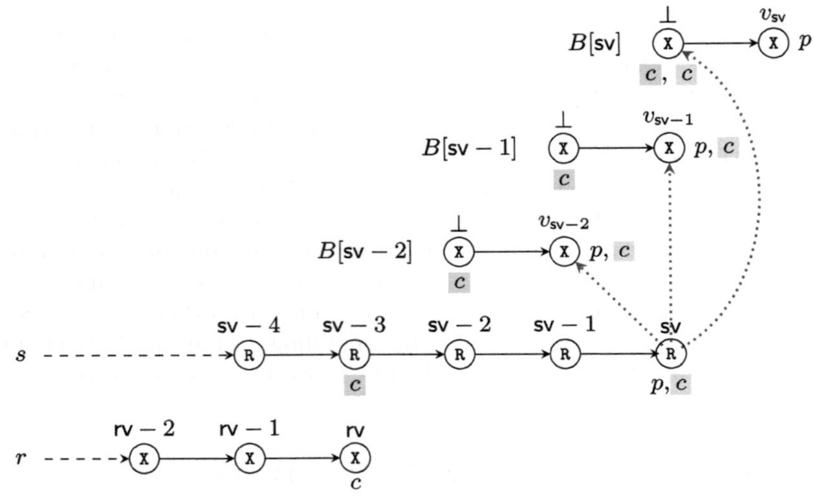

Fig. 5. A possible state of Lamport's buffer under RC11, with the solid arrows representing the *modification order* [5,38] and the dashed arrows representing the message view. In this state p has written the value v_{sv} in $B[\mathsf{sv}]$, but it has not yet updated s.

Example 4. Figure 5 aims to capture the effect of an acquiring read of s (returning the value $\mathsf{sv} - 1$) by thread c. Here, we not only consider the writes on s and r, but also on the buffer indices. We assume that all writes are relaxed (X) except those on s, which are always releasing (R).

The view of c before and after this read are highlighted in red and green, respectively. Notice how the effect of the read is to not only update c's view on s but also its views on the important buffer indices $B[\mathsf{sv}-2]$ and $B[\mathsf{sv}-1]$. Thread c's view of $B[\mathsf{sv}]$ is unchanged. To achieve this in the view model, the key idea is to *store the writing thread's view within each releasing write at the time of writing*. In Fig. 5, when the write of value sv occurs, the view of this write on B is as indicated by the blue dashed arrows. Now, if c reads sv for s using an acquiring read, c not only updates its view of s (cf Fig. 4), it also *inherits* the views of this write.

These observations lead to the program and proof in Fig. 6. To model the send-receive behaviour, we assume an (infinite) array of values, v, that p wishes to send (via B) and that these are read into an array of values w by c. Thus main invariant is thus I_{pc} (see Fig. 6). Given that rv is the value of r definitely seen by c, for all all indices i less than rv, each value received w_i must equal the value sent (v_i).

The proof I_{pc} is supported by new assertions W_p and W_c. Here, W_c describes *view-transfer* during release-acquire synchronisation via its consequent, which is a predicate of the form $\pi \overset{\tau}{\leadsto} \varphi$, stating that, if thread τ performs an *acquiring read* such that π holds, then the state *after* the read is guaranteed to satisfy φ.

$$\{Init \wedge I_{pc} \wedge W_c\}$$

$$
\begin{array}{l|l}
P_1' : \left\{\begin{array}{l} P_1 \wedge (\forall i.\ B[i]\langle\!\langle p\rangle\!\rangle \wedge \\ (\forall i < \mathsf{sv}.\ B[i] \overset{p}{=} v_i) \end{array}\right\} & Q_1' : \{Q_1 \wedge I_{pc} \wedge W_c\} \\
\quad \mathtt{s}_p :\overset{\mathtt{X}}{=} s; & \quad \mathtt{r}_c :\overset{\mathtt{X}}{=} r; \\
P_2' : \{P_2 \wedge W_p\} & Q_2' : \{Q_2 \wedge I_{pc} \wedge W_c\} \\
\quad \textbf{do } \mathtt{r}_p :\overset{\mathtt{X}}{=} r; & \quad \textbf{do } \mathtt{s}_c :\overset{\mathtt{A}}{=} s \\
P_3' : \{P_3 \wedge W_p\} & Q_3' : \left\{Q_3 \wedge I_{pc} \wedge W_c \wedge (\forall k < \mathtt{s}_c.\ B[k] \overset{c}{=} v_k)\right\} \\
\quad \textbf{until } \mathtt{s}_p - \mathtt{r}_p < b; & \quad \textbf{until } \mathtt{s}_c - \mathtt{r}_c > 0; \\
P_4' : \{P_4 \wedge W_p\} & Q_4' : \left\{Q_4 \wedge I_{pc} \wedge W_c \wedge (\forall k < \mathtt{s}_c.\ B[k] \overset{c}{=} v_k)\right\} \\
\quad \langle B[\mathtt{s}_p] :\overset{\mathtt{X}}{=} v_{\mathtt{s}_p}; & \quad \langle w_{\mathtt{r}_c} :\overset{\mathtt{X}}{=} B[\mathtt{r}_c]; \\
\quad\ \mathtt{ms}, \mathtt{js} := \mathtt{ms} + 1, \mathit{true}\rangle; & \quad\ \mathtt{mr}, \mathtt{jr} := \mathtt{mr} + 1, \mathit{true}\rangle; \\
P_5' : \left\{P_5 \wedge W_p \wedge B[\mathtt{s}_p] \overset{c}{=} v_{\mathtt{s}_p}\right\} & Q_5' : \{Q_5 \wedge I_{pc} \wedge W_c \wedge w_{\mathtt{r}_c} = v_{\mathtt{r}_c}\} \\
\quad \langle s :\overset{\mathtt{R}}{=} \mathtt{s}_p + 1; \mathtt{js} := \mathit{false}\rangle & \quad \langle r :\overset{\mathtt{X}}{=} \mathtt{r}_c + 1; \mathtt{jr} := \mathit{false}\rangle \\
P_6' : \left\{\begin{array}{l} P_6 \wedge (\forall i.\ B[i]\langle\!\langle p\rangle\!\rangle \wedge \\ (\forall i < \mathsf{sv} + 1.\ B[i] \overset{p}{=} v_i) \end{array}\right\} & Q_6' : \{Q_6 \wedge I_{pc} \wedge W_c\}
\end{array}
$$

$$\{I \wedge I_{pc}\}$$

where

$$
\begin{aligned}
W_p &\mathrel{\widehat{=}} (\forall i.\ B[i]\langle\!\langle p\rangle\!\rangle \wedge (\forall i < \mathtt{s}_p.\ B[i] \overset{p}{=} v_i) \\
W_c &\mathrel{\widehat{=}} (\forall \mathsf{sv}.\ s \overset{p}{>} \mathsf{sv} \Rightarrow s = \mathsf{sv} \overset{c}{\rightsquigarrow} (\forall k < \mathsf{sv}.\ B[k] \overset{c}{=} v_k)) \\
I_{pc} &\mathrel{\widehat{=}} (\forall \mathsf{rv}.\ s \overset{c}{=} \mathsf{rv} \Rightarrow w_0 w_1 \ldots w_{\mathsf{rv}-1} = v_0 v_1 \ldots v_{\mathsf{rv}-1})
\end{aligned}
$$

Fig. 6. Full model of Lamport's buffer under RC11, where *Init* is the initial precondition in Fig. 2, and $Q_1 \ldots Q_6$ are the counterparts to $P_1 \ldots P_6$ in the consumer thread.

Example 5. Consider the state in Fig. 5 before c performs the read of s, i.e., c's view of B is as indicated in red. Let $(x = \mathsf{xv}) \overset{\tau}{\rightsquigarrow} \varphi$ be a special case of $\pi \overset{\tau}{\rightsquigarrow} \varphi$, where the transition to φ is only guaranteed if τ reads the value xv for x (see (1) for the formal definition). Then in the state before c performs its read, we have:

$$(s = \mathsf{sv}) \overset{c}{\rightsquigarrow} (\forall k < \mathsf{sv}.\ B[k] \overset{c}{=} v_k)$$

i.e., if c were to read the value sv for s using an acquiring read, then after performing this read, its view would be updated so that $(\forall k < \mathsf{sv}.\ B[k] \overset{c}{=} v_k)$ holds, as indicated by the green views.

4.2 Release-Acquire Semantics Proof

We now present the formal definitions that establish the intuitions over view transfer from Sect. 4.1.

Release-Acquire Semantics. Recall that the semantics of releasing writes is already covered by the WRITE-RLX-REL rule in Fig. 4. This time we take note of the fact that the message μ corresponding to the write records the *view* (V') *of the executing thread* (τ). Roughly speaking, this view represents the happens-before information of τ, and by storing it in μ, we allow future releasing reads to learn this information.

$$\text{ACQ-RF-RLX}\quad \frac{\begin{array}{cc} e = \langle x, \mathsf{read}(v), \mathtt{A}\rangle & \langle x, v, t, \_, \mathtt{X}\rangle \in m \\ \mathcal{V}(\tau)(x) \leq t & V' = \mathcal{V}(\tau)[x \mapsto t] \end{array}}{\langle m, \mathcal{V}\rangle \xrightarrow{e}_{\tau} \langle m, \mathcal{V}[\tau \mapsto V']\rangle}$$

$$\text{ACQ-RF-REL}\quad \frac{\begin{array}{cc} e = \langle x, \mathsf{read}(v), \mathtt{A}\rangle & \langle x, v, t, V, \mathtt{R}\rangle \in m \\ \mathcal{V}(\tau)(x) \leq t & V' = \mathcal{V}(\tau) \sqcup V \end{array}}{\langle m, \mathcal{V}\rangle \xrightarrow{e}_{\tau} \langle m, \mathcal{V}[\tau \mapsto V']\rangle}$$

Fig. 7. Transitions of acquiring reads.

The semantics of acquiring reads is given in Fig. 7. There are two rules for acquiring reads, depending on whether the write that it reads from is releasing.

- When the write is not releasing, the transition is covered by the rule ACQ-RF-RLX. The effect of this rule on the memory state is identical to the rule for relaxed reads (rule READ-RLX in Fig. 4).
- When the write is releasing, we use the rule ACQ-RF-REL. The difference here is that the read causes the reading thread's view to synchronise with the view of the write that it reads from, i.e., the new view is determined by $\mathcal{V}(\tau) \sqcup V$, where

$$V_1 \sqcup V_2 \mathrel{\widehat{=}} \lambda x \in \mathit{Var}.\max\{V_1(x), V_2(x)\}.$$

Given $V_1, V_2 \in \mathsf{View}$, $V_1 \sqcup V_2$ defines a new view mapping each variable x to the larger timestamp between $V_1(x)$ and $V_2(x)$.

View-Transfer Assertions. With the release-acquire semantics in place, we formalise the view-transfer assertions. We first define a function filt that filters messages that τ can observe according to a predicate π over messages:

$$\mathsf{filt}(\tau, \pi) \mathrel{\widehat{=}} \lambda\alpha.\ \{\mu \in \mathsf{obs}(\tau)(\alpha) \mid \pi(\mu)\}$$

Thus, for example $\mathsf{filt}(\tau, (\lambda\mu.\ \mathsf{var}(\mu) = x \wedge \mathsf{val}(\mu) = v))$ returns the set of all messages on variable x that thread τ can observe whose value is v. Filters allow one to define assertions such as:

$$x \overset{\tau}{>} \mathsf{xv} \mathrel{\widehat{=}} x\langle\!\langle \tau \wedge \mathsf{filt}(\tau, (\lambda\mu.\ \mathsf{var}(\mu) = x \wedge \mathsf{val}(\mu) > \mathsf{xv})) \neq \emptyset$$

which is used in W_c in Fig. 6. This assertion states that τ's view of x is maximal, and that the value of this last message is greater than xv. Note that if $x\langle\!\langle\tau$, then $|\tau\rangle x$ is a singleton set since τ can read from exactly one message (which is a write on x). If $\mathsf{filt}(\tau, (\lambda\mu.\ \mathsf{var}(\mu) = x \wedge \mathsf{val}(\mu) > \mathsf{xv}))$, then this last message on x has a value greater that xv, precisely when the set returned by filt is non-empty.

The view-transfer assertion is then defined as follows:

$$\begin{aligned} \pi \overset{\tau}{\rightsquigarrow} \varphi \mathrel{\widehat{=}} \lambda\alpha.\ & \textbf{let } \alpha = \langle\sigma, \langle m, \mathcal{V}\rangle\rangle \textbf{ in} \\ & \forall\mu \in \mathsf{filt}(\tau, \pi)(\alpha). \\ & \quad \mathsf{ord}(\mu) = \mathtt{R} \wedge \varphi(\langle\sigma, \langle m, \mathcal{V}[\tau \mapsto \mathcal{V}(\tau) \sqcup \mathsf{view}(\mu)]\rangle\rangle) \end{aligned}$$

First we filter the messages observable to τ satisfying π, and require that all of these are releasing. Moreover, the effect of reading these via an acquiring read results in a state that satisfies φ. In particular, in $\pi \overset{\tau}{\rightsquigarrow} \varphi$, we require that φ holds in the state *after* τ reads a value satisfying π, rather than in the current state. A special case of the view-transfer predicate is the following:

$$(x = \mathsf{xv}) \overset{\tau}{\rightsquigarrow} \varphi \mathrel{\widehat{=}} (\lambda\mu.\ \mathsf{var}(\mu) = x \wedge \mathsf{val}(\mu) = \mathsf{xv}) \overset{\tau}{\rightsquigarrow} \varphi \tag{1}$$

which ensures view transfer whenever τ reads the value xv for x. An instance of this predicate appears in the consequent of W_c in Fig. 6.

Establishing and Using View-Transfer. We make use of view transfer to establish the new assertion φ as a postcondition using the following rule:

ACQ-VIEW-TRANSFER

$$\frac{}{\left\{x = \mathsf{xv} \overset{\tau}{\rightsquigarrow} \varphi\right\}\tau \rhd \mathtt{r} \overset{\mathtt{A}}{:=} x\{\mathtt{r} = \mathsf{xv} \Rightarrow \varphi\}}$$

Given the precondition $x = \mathsf{xv} \overset{\tau}{\rightsquigarrow} \varphi$ and an acquiring read of x into $\mathtt{r}$, if the value of $\mathtt{r}$ in the post state is xv then φ must hold.

Although ACQ-VIEW-TRANSFER is generic, establishing it is non-trivial. We make use of the following rule, which establishes a generalised form of the view-transfer assertion in our earlier work [17,20,23].

REL-DEF-VIEW-SET

$$\frac{\forall i \in S.\ x \neq y_i}{\{\mathsf{xv} \notin |\tau'\rangle x \wedge (\forall i \in S. y_i \overset{\tau}{=} \mathsf{yv}_i)\}\tau \rhd x \overset{\mathtt{R}}{:=} \mathsf{xv}\left\{x = \mathsf{xv} \overset{\tau'}{\rightsquigarrow} (\forall i \in S. y_i \overset{\tau'}{=} \mathsf{yv}_i)\right\}}$$

The rule REL-DEF-VIEW-SET is used to establish $x = \mathsf{xv} \overset{\tau'}{\rightsquigarrow} (\forall i. y_i \overset{\tau'}{=} \mathsf{yv}_i)$ in the post state via the execution of $x \overset{\mathtt{R}}{:=} \mathsf{xv}$ by τ where τ' is an arbitrarily chosen thread (and may include $\tau' = \tau$). Let ψ be the postcondition. If $\tau = \tau'$, then because $(\forall i. y_i \overset{\tau}{=} \mathsf{yv}_i)$ holds in the pre-state and the assignment does not modify any y_i, $(\forall i. y_i \overset{\tau}{=} \mathsf{yv}_i)$ holds in the post-state, which implies ψ. The more interesting case is when $\tau \neq \tau'$. Here, we make use of the assumption that τ' cannot read xv for x. Then, because $(\forall i. y_i \overset{\tau}{=} \mathsf{yv}_i)$ holds, after τ executes a releasing write on x with value xv, if thread τ' then reads xv for x, its state must change so that $(\forall i. y_i \overset{\tau'}{=} \mathsf{yv}_i)$ holds.

Note that the overall transfer of the assertion $(\forall i. y_i \overset{\tau}{=} \mathsf{yv}_i)$ (in thread τ) to the assertion $(\forall i. y_i \overset{\tau'}{=} \mathsf{yv}_i)$ (in thread τ') is via *two* transitions: the first is the releasing write on x (of value xv) and the second is the acquiring read of x by τ' (of the same value xv).

Example 6. We demonstrate the view transfer rules on a simpler message passing litmus test shown in Fig. 8, where the data d written by the writer thread (τ_1) is transferred to the reader thread (τ_2) via a release-acquire synchronisation on the flag (f). This is captured by the postcondition $\mathtt{r} = 1 \Rightarrow \mathtt{s} = 1$, i.e., if the value of f read by the reader is 1, then it also later reads the data d to be 1.

$$\{d\langle\!\langle\tau_1 \wedge 1 \notin |\tau_2\rangle f\}$$

$$\begin{array}{l||l}
\{1 \notin |\tau_2\rangle f \wedge d\langle\!\langle\tau_1\} & \left\{f = 1 \overset{\tau_2}{\leadsto} (d \overset{\tau_2}{=} 1)\right\} \\
\quad d :\overset{\mathtt{X}}{=} 1; & \quad \mathtt{r} :\overset{\mathtt{A}}{=} f; \\
\left\{1 \notin |\tau_2\rangle f \wedge d \overset{\tau_1}{=} 1\right\} & \left\{\mathtt{r} = 1 \Rightarrow d \overset{\tau_2}{=} 1\right\} \\
\quad f :\overset{\mathtt{R}}{=} 1 & \quad \mathtt{s} :\overset{\mathtt{X}}{=} d \\
\{true\} & \{\mathtt{r} = 1 \Rightarrow \mathtt{s} = 1\}
\end{array}$$

$$\{\mathtt{r} = 1 \Rightarrow \mathtt{s} = 1\}$$

Fig. 8. Proof of the message passing litmus test.

The proof in Fig. 8 is almost identical to our prior work [17,20], but makes use of some weaker assertions that we have introduced in this paper. For instance, the initial condition and precondition of $d :\overset{\mathtt{X}}{=} 1$ only require $d\langle\!\langle\tau_1$ instead of the stronger condition $d \overset{\tau_1}{=} 0$ in prior work.

Lemma 3. *The proof outline in Fig. 8 is valid.*

We focus on the local and global correctness of the conditional view assertion:

$$\psi \mathrel{\widehat{=}} f = 1 \overset{\tau_2}{\leadsto} (d \overset{\tau_2}{=} 1)$$

in thread τ_2. For local correctness, if $1 \notin |\tau_2\rangle f$ then ψ is trivially true (since τ_2 could never read the value 1 for f). Hence, the initial condition implies ψ. For the same reason, ψ is stable under $\{1 \notin |\tau_2\rangle f\}\tau_1 \rhd d :\overset{\mathtt{X}}{=} 1$, which establishes the stronger assertion $1 \notin |\tau_2\rangle f$ as a postcondition. Finally, ψ is stable under $\left\{1 \notin |\tau_2\rangle f \wedge d \overset{\tau_1}{=} 1\right\}\tau_1 \rhd f :\overset{\mathtt{R}}{=} 1$ by REL-DEF-VIEW-SET discussed above.

With this formal framework in place, we now state our main result:

Lemma 4. *The proof outline in Fig. 6 is valid.*

Note that this has only been checked manually as it relies on a generalised set of assertions which have not yet been encoded in Isabelle/HOL. We leave this as future work.

5 Related Work and Discussion

There has been immense progress in relaxed memory models in the past few decades, from formal models and compilation correctness results to library abstractions and verification frameworks and tools. This has helped uncover numerous bugs in real software as well as inconsistencies in architecture and language specifications. We present some related works below and discuss some key challenges and opportunities for future work. Given the scale of works on this topic, the review below is almost certainly incomplete. More comprehensive reviews may be found elsewhere, e.g., a recent survey by Su and Colvin [55].

Other Memory Models. This paper has focussed on reasoning techniques for relaxed memory models, whereby the principles for reasoning about inter-thread interference stem from existing frameworks for SC memory (cf [37]). Although this tutorial has focussed on the RC11 memory model, our methods have been applied more generally to other memory models, e.g., total store order (TSO) [22], persistent x86-TSO [11], strong release-acquire [37] and C11 with relaxed dependencies [60,61]. There have also been some attempts to unify reasoning principles across different memory models [22,37].

Yet, there is more work to do as the memory models increase in complexity, e.g., to support future architectures comprising heterogeneous computing [27] or distributed shared memory [6,7]. Here, the types of assertions that are likely to be useful for reasoning are currently unclear. Moreover, when memory models become too relaxed (i.e., they lose program order within a thread), they induce "interference" from within the same thread, which induces many additional complications [60,61]. Exactly how intra-thread interference can be efficiently handled is currently unknown. One possible approach is a generalisation of the technique by Coughlin et al. [16], which uses rely conditions to describe weak memory effects. Another is the development of new separation logics over mixed semantics comprising both axiomatic and operational principles [29].

Refinement and Library Abstraction. Several works have consider the problem of developing abstractions of common data structures and algorithms, over both declarative [7,47,54] and operational [10,19,52] semantics. While the former supports modular reasoning, the latter can be trivially linked to stepwise refinement and standard simulation-based proof techniques.

One difficulty here is that the precise notions of correctness for relaxed libraries (e.g., linearizability [30]) can be unclear since library programs not only affect the happens-before relations within the library, but also the client program's context [24,28,47,54]. Moreover, a total order of operation invocations and responses, as used in the definition of linearizability, has less meaning here since it is possible for an operation to be after another (according to this total order), but still be executed concurrently since the effects of the first are not seen by the second. As such, an alternative set of works have taken (contextual) refinement directly as the correctness condition [13,19,52], *without* a detour through linearizability. Like linearizability, these approaches *use* an atomic sequential specification, but define the happens-before interactions between different libraries and the client context at this abstract level.

More recently, approaches to library construction have placed *compositionality* at the forefront [59], which have discovered that compositionality and linearizability are intrinsically linked. It would be interesting to study these ideas through the lens of relaxed memory models.

Logics. We have focussed on a particular line of work that is based on Owicki-Gries reasoning. Other works have considered generic methods for building rely-guarantee frameworks in causally consistent memory models [37]. Another line of work is based on separation logic, which has been applied to several memory models [21,29,32,56–58]. Although the approaches are different, a common feature between all of these frameworks use bespoke (non-standard) assertions to support reasoning about relaxed memory effects. There are some efforts aimed at unifying reasoning across multiple memory models [22], but these techniques still assume certain restrictions such as causal consistency, which not all memory models satisfy.

Work is therefore needed on verifying more examples to generate libraries of proofs across different memory models. These will, in particular, shed more light into the types of synchronisation patterns that are most common amongst different algorithms, allowing one to develop more efficient strategies for their verification. Such work could also involve development of decidable logics that sacrifice completeness to boost scalability and automation.

Tool Support. The complexity of reasoning about relaxed memory behaviours has meant that developing tool support is critical. However, building tools is challenging since checking safety problems is often undecidable [2,12,36,53]. There are now several model checkers built specifically for program executing over a relaxed memory semantics [28,35,41]. However, even when implementing optimal model checking algorithms based on dynamic partial order reduction [34] or using techniques such as stateless model checking [1], such provers remain bounded in the number of threads, locations and/or values, i.e., are not fully generic. Of course, their push-button (automated verification) capabilities makes them valuable tools in the verification toolchain, so continued research into further optimisations remains critical.

Full correctness proofs have been supported by a wide range of theorem proving tools such as Isabelle/HOL [19,61], Iris [29] and Coq [57], as well as bespoke tools such as Viper [56] that support automatic deductive verification. As with the logics discussed above, further experience with proof tools is required to enable verification to scale, e.g., through more efficient encodings and/or the development or tactics that allow proof obligations to be automatically discharged. Our experiences with Isabelle/HOL [17,20,61] have shown that once the Hoare triples for (atomic) assignment commands have been established as Isabelle lemmas, proofs of Hoare triples from a program's proof outline can typically be automatically discharged using calls to Isabelle's `sledgehammer` tool. Note that `sledgehammer` itself calls a number of internal tactics and SMT solvers to discharge proof obligations. As these tactics and SMT solvers improve, so will our ability to verify relaxed memory programs. However, this experience still pales in comparison to the speed at which Isabelle is able to discharge proofs under SC (see Lemma 1).

Acknowledgements. We thank the anonymous reviewers for suggestions that have helped improve the paper.

References

1. Abdulla, P.A., Aronis, S., Atig, M.F., Jonsson, B., Leonardsson, C., Sagonas, K.: Stateless model checking for TSO and PSO. In: Baier, C., Tinelli, C. (eds.) TACAS 2015. LNCS, vol. 9035, pp. 353–367. Springer, Heidelberg (2015). https://doi.org/10.1007/978-3-662-46681-0_28
2. Abdulla, P.A., Atig, M.F., Bouajjani, A., Ngo, T.P.: A load-buffer semantics for total store ordering. Log. Methods Comput. Sci. **14**(1) (2018). https://doi.org/10.23638/LMCS-14(1:9)2018
3. Adve, S.V., Boehm, H.J.: Memory models: a case for rethinking parallel languages and hardware. Commun. ACM **53**(8), 90–101 (2010). https://doi.org/10.1145/1787234.1787255
4. Ahamad, M., Hutto, P.W., John, R.: Implementing and programming causal distributed shared memory. In: ICDCS, pp. 274–281. IEEE Computer Society (1991). https://doi.org/10.1109/ICDCS.1991.148677
5. Alglave, J., Maranget, L., Tautschnig, M.: Herding cats: modelling, simulation, testing, and data mining for weak memory. ACM Trans. Program. Lang. Syst. **36**(2), 7:1–7:74 (2014). https://doi.org/10.1145/2627752
6. Ambal, G., Dongol, B., Eran, H., Klimis, V., Lahav, O., Raad, A.: Semantics of remote direct memory access: operational and declarative models of RDMA on TSO architectures. Proc. ACM Program. Lang. **8**(OOPSLA2), 1982–2009 (2024). https://doi.org/10.1145/3689781
7. Ambal, G., et al.: A verified high-performance composable object library for remote direct memory access. Proc. ACM Program. Lang. **10**(POPL), 2051–2082 (2026). https://doi.org/10.1145/3776713
8. Batty, M., Memarian, K., Nienhuis, K., Pichon-Pharabod, J., Sewell, P.: The problem of programming language concurrency semantics. In: Vitek, J. (ed.) ESOP 2015. LNCS, vol. 9032, pp. 283–307. Springer, Heidelberg (2015). https://doi.org/10.1007/978-3-662-46669-8_12
9. Batty, M., Owens, S., Sarkar, S., Sewell, P., Weber, T.: Mathematizing C++ concurrency. In: Ball, T., Sagiv, M. (eds.) POPL, pp. 55–66. ACM (2011). https://doi.org/10.1145/1926385.1926394
10. Bila, E.V., Dongol, B.: A verified durable transactional mutex lock for persistent x86-TSO. Formal Methods Syst. Des. **64**(1), 237–282 (2024). https://doi.org/10.1007/S10703-024-00462-1
11. Bila, E.V., Dongol, B., Lahav, O., Raad, A., Wickerson, J.: View-based Owicki–Gries reasoning for persistent x86-TSO. In: ESOP 2022. LNCS, vol. 13240, pp. 234–261. Springer, Cham (2022). https://doi.org/10.1007/978-3-030-99336-8_9
12. Bouajjani, A., Enea, C., Mutluergil, S.O., Tasiran, S.: Reasoning about TSO programs using reduction and abstraction. In: Chockler, H., Weissenbacher, G. (eds.) CAV 2018. LNCS, vol. 10982, pp. 336–353. Springer, Cham (2018). https://doi.org/10.1007/978-3-319-96142-2_21
13. Burckhardt, S., Gotsman, A., Musuvathi, M., Yang, H.: Concurrent library correctness on the TSO memory model. In: Seidl, H. (ed.) ESOP 2012. LNCS, vol. 7211, pp. 87–107. Springer, Heidelberg (2012). https://doi.org/10.1007/978-3-642-28869-2_5

14. Castañeda, A., Chockler, G., Dongol, B., Lahav, O.: What cannot be implemented on weak memory? In: Alistarh, D. (ed.) DISC. Leibniz International Proceedings in Informatics (LIPIcs), vol. 319, pp. 11:1–11:22. Schloss Dagstuhl – Leibniz-Zentrum für Informatik, Dagstuhl, Germany (2024). https://doi.org/10.4230/LIPIcs.DISC.2024.11
15. Cho, K., Lee, S., Raad, A., Kang, J.: Revamping hardware persistency models: view-based and axiomatic persistency models for Intel-x86 and Armv8. In: Freund, S.N., Yahav, E. (eds.) PLDI '21: 42nd ACM SIGPLAN International Conference on Programming Language Design and Implementation, Virtual Event, Canada, 20–25 June 2021, pp. 16–31. ACM (2021). https://doi.org/10.1145/3453483.3454027
16. Coughlin, N., Winter, K., Smith, G.: Compositional reasoning for non-multicopy atomic architectures. Formal Aspects Comput. **35**(2), 8:1–8:30 (2023). https://doi.org/10.1145/3574137
17. Dalvandi, S., Doherty, S., Dongol, B., Wehrheim, H.: Owicki-Gries reasoning for C11 RAR. In: Hirschfeld, R., Pape, T. (eds.) ECOOP. Leibniz International Proceedings in Informatics (LIPIcs), vol. 166, pp. 11:1–11:26. Schloss Dagstuhl – Leibniz-Zentrum für Informatik, Dagstuhl, Germany (2020). https://doi.org/10.4230/LIPIcs.ECOOP.2020.11
18. Dalvandi, S., Doherty, S., Dongol, B., Wehrheim, H.: Owicki-Gries reasoning for C11 RAR (artifact). Dagstuhl Artifacts Ser. **6**(2), 15:1–15:2 (2020). https://doi.org/10.4230/DARTS.6.2.15
19. Dalvandi, S., Dongol, B.: Implementing and verifying release-acquire transactional memory in C11. Proc. ACM Program. Lang. **6**(OOPSLA2), 1817–1844 (2022). https://doi.org/10.1145/3563352
20. Dalvandi, S., Dongol, B., Doherty, S., Wehrheim, H.: Integrating Owicki-Gries for C11-Style memory models into Isabelle/HOL. J. Autom. Reason. **66**(1), 141–171 (2022). https://doi.org/10.1007/S10817-021-09610-2
21. Dang, H., Jourdan, J., Kaiser, J., Dreyer, D.: Rustbelt meets relaxed memory. Proc. ACM Program. Lang. **4**(POPL), 34:1–34:29 (2020). https://doi.org/10.1145/3371102
22. Doherty, S., Dalvandi, S., Dongol, B., Wehrheim, H.: Unifying operational weak memory verification: an axiomatic approach. ACM Trans. Comput. Log. **23**(4), 27:1–27:39 (2022). https://doi.org/10.1145/3545117
23. Doherty, S., Dongol, B., Wehrheim, H., Derrick, J.: Verifying C11 programs operationally. In: Hollingsworth, J.K., Keidar, I. (eds.) PPoPP. pp. 355–365. ACM (2019). https://doi.org/10.1145/3293883.3295702
24. Dongol, B., Jagadeesan, R., Riely, J., Armstrong, A.: On abstraction and compositionality for weak-memory linearisability. In: VMCAI 2018. LNCS, vol. 10747, pp. 183–204. Springer, Cham (2018). https://doi.org/10.1007/978-3-319-73721-8_9
25. Gharachorloo, K., Adve, S.V., Gupta, A., Hennessy, J.L., Hill, M.D.: Programming for different memory consistency models. J. Parallel Distrib. Comput. **15**(4), 399–407 (1992). https://doi.org/10.1016/0743-7315(92)90052-O
26. Gharachorloo, K., Gupta, A., Hennessy, J.L.: Performance evaluation of memory consistency models for shared memory multiprocessors. In: Patterson, D.A., Rau, B. (eds.) ASPLOS, pp. 245–257. ACM Press (1991). https://doi.org/10.1145/106972.106997
27. Goens, A., Chakraborty, S., Sarkar, S., Agarwal, S., Oswald, N., Nagarajan, V.: Compound memory models. Proc. ACM Program. Lang. **7**(PLDI), 1145–1168 (2023). https://doi.org/10.1145/3591267

28. Golovin, P., Kokologiannakis, M., Vafeiadis, V.: RELINCHE: automatically checking linearizability under relaxed memory consistency. Proc. ACM Program. Lang. **9**(POPL), 2090–2117 (2025). https://doi.org/10.1145/3704906
29. Hammond, A., Liu, Z., Pérami, T., Sewell, P., Birkedal, L., Pichon-Pharabod, J.: An axiomatic basis for computer programming on the relaxed arm-a architecture: the axsl logic. Proc. ACM Program. Lang. **8**(POPL), 604–637 (2024). https://doi.org/10.1145/3632863
30. Herlihy, M., Wing, J.M.: Linearizability: a correctness condition for concurrent objects. ACM Trans. Program. Lang. Syst. **12**(3), 463–492 (1990)
31. Jones, C.B.: Tentative steps toward a development method for interfering programs. ACM Trans. Program. Lang. Syst. **5**(4), 596–619 (1983). https://doi.org/10.1145/69575.69577
32. Kaiser, J., Dang, H., Dreyer, D., Lahav, O., Vafeiadis, V.: Strong logic for weak memory: Reasoning about release-acquire consistency in Iris. In: Müller, P. (ed.) ECOOP. LIPIcs, vol. 74, pp. 17:1–17:29. Schloss Dagstuhl - Leibniz-Zentrum für Informatik (2017). 10.4230/LIPICS.ECOOP.2017.17
33. Kang, J., Hur, C., Lahav, O., Vafeiadis, V., Dreyer, D.: A promising semantics for relaxed-memory concurrency. In: Castagna, G., Gordon, A.D. (eds.) POPL, pp. 175–189. ACM (2017). https://doi.org/10.1145/3009837.3009850
34. Kokologiannakis, M., Majumdar, R., Vafeiadis, V.: Enhancing genmc's usability and performance. In: Finkbeiner, B., Kovács, L. (eds.) TACAS. Lecture Notes in Computer Science, vol. 14571, pp. 66–84. Springer, Heidelberg (2024). https://doi.org/10.1007/978-3-031-57249-4_4
35. Kokologiannakis, M., Marmanis, I., Gladstein, V., Vafeiadis, V.: Truly stateless, optimal dynamic partial order reduction. Proc. ACM Program. Lang. **6**(POPL), 1–28 (2022). https://doi.org/10.1145/3498711
36. Lahav, O., Boker, U.: What's decidable about causally consistent shared memory? ACM Trans. Program. Lang. Syst. **44**(2), 8:1–8:55 (2022). https://doi.org/10.1145/3505273
37. Lahav, O., Dongol, B., Wehrheim, H.: Rely-guarantee reasoning for causally consistent shared memory. In: Enea, C., Lal, A. (eds.) CAV. Lecture Notes in Computer Science, vol. 13964, pp. 206–229. Springer, Heidelberg (2023). https://doi.org/10.1007/978-3-031-37706-8_11
38. Lahav, O., Vafeiadis, V., Kang, J., Hur, C.K., Dreyer, D.: Repairing sequential consistency in C/C++11. SIGPLAN Not. **52**(6), 618–632 (2017). https://doi.org/10.1145/3140587.3062352
39. Lamport, L.: Proving the correctness of multiprocess programs. IEEE Trans. Softw. Eng. **3**(2), 125–143 (1977). https://doi.org/10.1109/TSE.1977.229904
40. Lamport, L.: How to make a multiprocessor computer that correctly executes multiprocess programs. IEEE Trans. Comput. **28**(9), 690–691 (1979). https://doi.org/10.1109/TC.1979.1675439
41. de León, H.P., Furbach, F., Heljanko, K., Meyer, R.: BMC with memory models as modules. In: Bjørner, N.S., Gurfinkel, A. (eds.) FMCAD, pp. 1–9. IEEE (2018). https://doi.org/10.23919/FMCAD.2018.8603021
42. Manson, J., Pugh, W.W., Adve, S.V.: The Java memory model. In: Palsberg, J., Abadi, M. (eds.) POPL, pp. 378–391. ACM (2005). https://doi.org/10.1145/1040305.1040336
43. Nipkow, T., Nieto, L.P.: Owicki/Gries in Isabelle/HOL. In: Finance, J.-P. (ed.) FASE 1999. LNCS, vol. 1577, pp. 188–203. Springer, Heidelberg (1999). https://doi.org/10.1007/978-3-540-49020-3_13

44. Owicki, S., Gries, D.: An axiomatic proof technique for parallel programs i. Acta Informatica **6**(4), 319–340 (1976)
45. Paviotti, M., Cooksey, S., Paradis, A., Wright, D., Owens, S., Batty, M.: Modular relaxed dependencies in weak memory concurrency. In: ESOP 2020. LNCS, vol. 12075, pp. 599–625. Springer, Cham (2020). https://doi.org/10.1007/978-3-030-44914-8_22
46. Pulte, C., Pichon-Pharabod, J., Kang, J., Lee, S., Hur, C.: Promising-ARM/RISC-V: a simpler and faster operational concurrency model. In: McKinley, K.S., Fisher, K. (eds.) PLDI, pp. 1–15. ACM (2019). https://doi.org/10.1145/3314221.3314624
47. Raad, A., Doko, M., Rozic, L., Lahav, O., Vafeiadis, V.: On library correctness under weak memory consistency: specifying and verifying concurrent libraries under declarative consistency models. Proc. ACM Program. Lang. **3**(POPL), 68:1–68:31 (2019). https://doi.org/10.1145/3290381
48. Raad, A., Wickerson, J., Neiger, G., Vafeiadis, V.: Persistency semantics of the Intel-x86 architecture. Proc. ACM Program. Lang. **4**(POPL), 11:1–11:31 (2020). https://doi.org/10.1145/3371079
49. Semenyuk, M., Batty, M., Dongol, B.: Verifying read-copy update under RC11. In: Ferreira, C., Willemse, T.A.C. (eds.) SEFM. Lecture Notes in Computer Science, vol. 14323, pp. 301–319. Springer, Heidelberg (2023). https://doi.org/10.1007/978-3-031-47115-5_17
50. Ševčík, J., Aspinall, D.: On validity of program transformations in the java memory model. In: Vitek, J. (ed.) ECOOP 2008. LNCS, vol. 5142, pp. 27–51. Springer, Heidelberg (2008). https://doi.org/10.1007/978-3-540-70592-5_3
51. Sewell, P., Sarkar, S., Owens, S., Nardelli, F.Z., Myreen, M.O.: x86-TSO: a rigorous and usable programmer's model for x86 multiprocessors. Commun. ACM **53**(7), 89–97 (2010). https://doi.org/10.1145/1785414.1785443
52. Singh, A.K., Lahav, O.: An operational approach to library abstraction under relaxed memory concurrency. Proc. ACM Program. Lang. **7**(POPL), 1542–1572 (2023). https://doi.org/10.1145/3571246
53. Singh, A.K., Lahav, O.: Decidable verification under localized release-acquire concurrency. In: Finkbeiner, B., Kovács, L. (eds.) TACAS. Lecture Notes in Computer Science, vol. 14572, pp. 235–254. Springer, Heidelberg (2024). https://doi.org/10.1007/978-3-031-57256-2_12
54. Stefanesco, L., Raad, A., Vafeiadis, V.: Specifying and verifying persistent libraries. In: ESOP (2). Lecture Notes in Computer Science, vol. 14577, pp. 185–211. Springer, Heidelberg (2024). https://doi.org/10.1007/978-3-031-57267-8_8
55. Su, R.C., Colvin, R.J.: Weak memory model formalisms: introduction and survey. Concurr. Comput. Pract. Exp. **38**(2), e70484 (2026). https://doi.org/10.1002/cpe.70484
56. Summers, A.J., Müller, P.: Automating deductive verification for weak-memory programs. In: Beyer, D., Huisman, M. (eds.) TACAS 2018. LNCS, vol. 10805, pp. 190–209. Springer, Cham (2018). https://doi.org/10.1007/978-3-319-89960-2_11
57. Tassarotti, J., Dreyer, D., Vafeiadis, V.: Verifying read-copy-update in a logic for weak memory. In: Grove, D., Blackburn, S.M. (eds.) PLDI, pp. 110–120. ACM (2015). https://doi.org/10.1145/2737924.2737992
58. Turon, A., Vafeiadis, V., Dreyer, D.: GPS: navigating weak memory with ghosts, protocols, and separation. In: Black, A.P., Millstein, T.D. (eds.) OOPSLA, pp. 691–707. ACM (2014). https://doi.org/10.1145/2660193.2660243
59. Vale, A.O., Shao, Z., Chen, Y.: A compositional theory of linearizability. Proc. ACM Program. Lang. **7**(POPL), 1089–1120 (2023). https://doi.org/10.1145/3571231

60. Wright, D., Batty, M., Dongol, B.: Owicki-Gries reasoning for C11 programs with relaxed dependencies. In: Huisman, M., Păsăreanu, C., Zhan, N. (eds.) FM 2021. LNCS, vol. 13047, pp. 237–254. Springer, Cham (2021). https://doi.org/10.1007/978-3-030-90870-6_13
61. Wright, D., Dalvandi, S., Batty, M., Dongol, B.: Mechanised operational reasoning for C11 programs with relaxed dependencies. Formal Aspects Comput. **35**(2), 10:1–10:27 (2023). https://doi.org/10.1145/3580285

Requirements Elicitation, Formalization, and Analysis with FRET: A Tutorial

Anastasia Mavridou[1], Andreas Katis[1], Mari A. Aoki[2], and Marie Farrell[3](✉)

[1] KBR Inc. at NASA Ames Research Center, Moffett Field, CA, USA
[2] NASA Ames Research Center, Moffett Field, CA, USA
[3] University of Manchester, Manchester, UK
marie.farrell@manchester.ac.uk

Abstract. Formal methods enable the verification of system behavior against requirements that are expressed as formal, mathematical properties. However, translating the often ambiguous natural language requirements that are typically produced by developers and engineers into precise mathematical specifications remains a significant bottleneck in the formal verification process. NASA's Formal Requirements Elicitation Tool (FRET) is an open source tool that addresses this challenge by bridging the gap between natural language requirements and formal specifications that are suitable for automated verification. FRET enables practitioners to express requirements in FRETish, a structured natural language, that balances intuitive readability with formal rigor. FRET automatically translates these requirements into formal properties that verification tools can directly process. This tutorial paper introduces FRET and guides readers through expressing requirements in FRETish. We present the tool's key analysis capabilities, including simulation, realizability checking, test-case generation, and automated generation of verification conditions for external formal verification tools. Our goal is to provide a comprehensive guide that helps practitioners, regardless of their formal methods background, to effectively leverage FRET in their verification and validation (V&V) workflows.

This work was partially supported by a Royal Academy of Engineering Research Fellowship.

A. Sampaio and M. Stoelinga (Eds.): FM 2026, LNCS 16557, pp. 420–445, 2026.
https://doi.org/10.1007/978-3-032-26220-2_21

1 Introduction

System development typically begins with a requirements engineering phase in which stakeholders collaborate to elicit and document requirements [6,33,38]. These requirements guide the subsequent development phases from architecture design through implementation, to verification and validation of both hardware and software components. Ultimately, systems are assessed for conformance to these requirements.

In development processes that incorporate formal methods for modeling, specification and verification, requirements serve as the foundation for deriving formal logical models, specifications, and verification conditions [14,37]. However, requirements are typically written in ambiguous natural language which is not amenable to formal analysis. Formal analysis tools require precise formalisms such as temporal logics [2] as input, which are representations that present a significant barrier to practitioners who are accustomed to natural language. This semantic gap between natural language requirement descriptions and formal logical specifications has long been recognized as one of the primary bottlenecks in applying formal methods to real-world systems [35].

NASA's Formal Requirements Elicitation Tool (FRET)[1] addresses this challenge by bridging the gap between ambiguous natural language and formalized logical requirements [17]. FRET provides FRETISH, a structured natural language that enables developers to intuitively express unambiguous requirements. These requirements are automatically translated into temporal logic formulas [18] suitable for a wide range of formal analysis tools [1,5,10,16,20,26,30, 32,36].

Large Language Models (LLMs) are beginning to be integrated into requirements engineering due to their natural language processing capabilities, which offers efficiency gains. However, the inherent uncertain nature of LLMs and their tendency to hallucinate create a distinct tension with the field's need for precision. This is where tools such as FRET remain indispensable, particularly in safety-critical domains where reliance on LLMs is discouraged or strictly regulated. Even in less critical contexts, LLMs demonstrate higher effectiveness when used in conjunction with structured representations such as FRETISH [27].

This tutorial provides a comprehensive guide to FRET and its feature set. We systematically cover:

- **The FRETISH Language:** We present FRETISH's structured syntax, demonstrating how to express requirements ranging from simple safety properties to complex temporal specifications involving timing constraints, conditions and modes.
- **Automated Analysis Capabilities:** We explore FRET's built-in analysis features including:
  - realizability checking to detect conflicting requirements
  - interactive simulation for validating requirement behavior through concrete scenarios

[1] https://github.com/NASA-SW-VnV/fret.

  - test-case generation for creating test suites from formalized requirements
- **Tool Integration:** We demonstrate how FRET integrates with external verification tools, including model checkers and runtime monitors, by automatically generating verification conditions in appropriate formats.
- **Practical Workflows:** We provide concrete examples and case studies illustrating end-to-end workflows from initial requirement elicitation through formalization, analysis, and test-case generation.

We target three primary audiences: (1) practitioners interested in requirement-based V&V methodologies, (2) researchers seeking to incorporate FRET into their work, and (3) students interested in studying requirement formalization and analysis. We target readers who have a basic background in Computer Science, including those who have recently started their studies in this area.

After completing this tutorial, the reader can expect to possess:

- an ability to formalize unambiguous natural language requirements
- a capacity to analyze and verify formalized requirements using the analysis tools supported by FRET as well as external analysis tools
- a curiosity and enthusiasm for specifying verifiable requirements using FRET

2 Background Prerequisites

This section overviews background material on requirements and temporal logic. We begin by defining what constitutes a requirement and examining common requirement categories. We then introduce Linear Temporal Logic (LTL), which is the primary formalism used by FRET for expressing temporal properties.

2.1 What is a Requirement?

A requirement specifies a property that a system must satisfy. Requirements typically follow the pattern: "*given certain conditions, a system **shall** exhibit some behavior*". The modal verb ***shall*** serves as a key indicator that a statement expresses a requirement rather than a preference or recommendation.

Requirements can be classified into several categories [33]:

Functional Requirements specify what a system must do, defining its behavior in response to specific inputs, events, or conditions [23]. E.g.,: "The system shall authenticate users before granting access."

Safety Requirements protect systems from accidental harm by preventing hazardous behaviors, detecting safety violations, and ensuring appropriate responses to unsafe conditions. E.g.,: "An autonomous vehicle shall never exceed the posted speed limit" [24].

Performance Requirements define measurable criteria for system quality attributes including speed, throughput, reliability, accuracy, and resource utilization [7]. E.g.: "The system shall process transactions within 2 s."

These categories represent common requirement types, though many others exist (e.g., security, usability, maintainability requirements).

2.2 Linear Temporal Logic (LTL)

LTL [2] is a logic for reasoning about how things change over time. Instead of just saying "the door is closed," LTL lets us say things like "the door will *eventually* be closed" or "the alarm will *always* be off." This makes it suitable for describing system behaviors that unfold across multiple steps.

LTL formulas are built from simple building blocks:

- **Atomic propositions**: Basic facts about the system state, like `door_closed` or `high_speed`.
- **Boolean operators**: Standard logical connectives like *not* ($\neg$) and *conjunction* ($\wedge$).
- **Temporal operators**: Special operators that let us talk about time.

The key temporal operators are:

- $X\phi$ (neXt): ϕ is true in the very next moment/state
- $F\phi$ (Finally): ϕ will be true at some point in the future
- $G\phi$ (Globally): ϕ is true now and remains true forever
- $\phi_1 \mathcal{U} \phi_2$ (Until): ϕ_1 stays true until ϕ_2 becomes true

For example, we might write $G(\texttt{button\_pressed} \rightarrow F\ \texttt{light\_on})$ to say: "Whenever the button is pressed, the light will eventually turn on."

The until operator can have two different semantics: *strong until* guarantees that ϕ_2 eventually becomes true, while *weak until* allows ϕ_1 to hold forever if ϕ_2 never occurs.

Formally, LTL syntax is defined as follows:

$$\phi := \ true \mid a \mid \neg\phi \mid \phi_1 \wedge \phi_2 \mid X\phi \mid F\phi \mid G\phi \mid \phi_1 \mathcal{U} \phi_2$$

where a represents any atomic proposition.

Beyond properties expressible using the standard LTL operators above, FRET supports constraints over bounded time horizons such as $F_{[0,10]}\phi$ ("ϕ becomes true within 10 time units") and $G_{[0,7]}\phi$ ("ϕ remains true for at least 7 time units"). Finally, LTL semantics over execution traces of finite length are also supported, via the automated translation to finite-trace future-time or past-time LTL [18].

3 Running Example: Finite State Machine

As a running example throughout this tutorial, we use requirements from the Finite State Machine (FSM) component of the Lockheed Martin Cyber-Physical Challenge Problems (LMCPS) [12,13,25]. The FSM represents an abstraction of an advanced autopilot system that interacts with an independent sensor platform to ensure safe autonomous operation in the vicinity of hazards.

Table 1 presents a representative subset of the FSM requirements from LMCPS. These requirements are written in natural language, which is intuitive but often ambiguous and therefore unsuitable for formal analysis. Despite their apparent simplicity, closer examination reveals significant ambiguities in these

Table 1. FSM requirements from LMPCPS

Req ID	Description
[FSM-001]	Exceeding sensor limits shall latch an autopilot pullup when the pilot is not in control (autopilot mode) and the system is supported without failures (not apfail).
[FSM-002]	The autopilot shall change states from TRANSITION to STANDBY when the pilot is in control (standby).
[FSM-003]	The autopilot shall change states from TRANSITION to NOMINAL when the system is supported and sensor data is good.
[FSM-004]	The autopilot shall change states from NOMINAL to MANEUVER when the sensor data is not good.
[FSM-005]	The autopilot shall change states from NOMINAL to STANDBY when the pilot is in control (standby).
[FSM-006]	The autopilot shall change states from MANEUVER to STANDBY when the pilot is in control (standby) and sensor data is good.

$$
\begin{aligned}
&(G(\neg(\neg\mathtt{autopilot}\wedge X\mathtt{autopilot})\vee\\
&\quad X((\mathtt{autopilot}\wedge X(\neg\mathtt{autopilot}))V((\neg(\neg\mathtt{apfail}\wedge\mathtt{exceedingSensorLimits})\\
&\qquad\wedge(X(\neg\mathtt{apfail}\wedge\mathtt{exceedingSensorLimits})\wedge\neg(\mathtt{autopilot}\wedge X(\neg\mathtt{autopilot}))))\\
&\quad\rightarrow(X\mathtt{autopilot\_pullup}\wedge\neg(\mathtt{autopilot}\wedge X(\neg\mathtt{autopilot})))\wedge((\neg\mathtt{apfail}\wedge\mathtt{exceedingSensorLimits})\\
&\qquad\rightarrow\mathtt{autopilot\_pullup}))))\\
&)\wedge(\mathtt{autopilot}\rightarrow((\mathtt{autopilot}\wedge X(\neg\mathtt{autopilot}))V\\
&\quad((\neg(\neg\mathtt{apfail}\wedge\mathtt{exceedingSensorLimits})\wedge(X(\neg\mathtt{apfail}\wedge\mathtt{exceedingSensorLimits})\\
&\qquad\wedge\neg(\mathtt{autopilot}\wedge X(\neg\mathtt{autopilot}))))\rightarrow(X\mathtt{autopilot\_pullup}\wedge\neg(\mathtt{autopilot}\wedge X(\neg\mathtt{autopilot})))\\
&\quad\wedge((\neg\mathtt{apfail}\wedge\mathtt{exceedingSensorLimits})\rightarrow\mathtt{autopilot\_pullup})))))
\end{aligned}
$$

Fig. 1. Formalized version of [FSM-001] requirement.

natural language requirements. Consider requirement **[FSM-001]**. It is unclear whether the FSM must latch an autopilot pullup every time the condition *holds* (sensor limits are exceeded) or only when it *becomes* true, that is, on the transition from false to true. Similarly, requirement **[FSM-002]** does not specify whether the autopilot must transition from "TRANSITION" to "STANDBY" instantly when "standby" holds, or whether some delay is acceptable. Without explicit timing constraints, multiple interpretations are equally valid: the transition could occur immediately, within a bounded interval (e.g., within 2 s), or eventually at some unspecified future time.

This example illustrates the fundamental challenge that FRET addresses. On one hand, we have ***how** developers write requirements*: in natural language that is intuitive but ambiguous. On the other hand, we have ***what** verification tools require*: formalized temporal logic specifications (see Fig. 1) that are rigorous and unambiguous but unintuitive and difficult to write, even for formal methods experts. FRET bridges this gap by providing a structured language that retains the readability of natural language while eliminating ambiguity.

4 Installation Instructions

FRET can be installed natively on MacOS and Linux, while Windows installations require an Ubuntu instance using Windows Subsystem for Linux (WSL). Sections 4.1 and 4.2 provide step-by-step instructions for all three operating systems. Section 4.3 provides instructions on how to install dependencies for optional features.

4.1 MacOS and Linux Installation Instructions

1. **Node.js**: Use any version between v16.16.x and v20.19.x
   - Installation instructions are available at https://nodejs.org/.
2. **Python**: Use any version between v3.10.x and v3.13.x.
   - Installation instructions are available at https://www.python.org/.
3. **Ubuntu users only**: Ensure that the following libraries are installed, based on the Ubuntu version:
   - Ubuntu 24.04 or later: `libgtk-3-0t64 libdrm2 libgbm1 libnss3 libx11-xcb1 libasound2t64`
   - Older than Ubuntu 24.04: `libgtk-3.0 libdrm2 libgbm1 libnss3 libx11-xcb1 libasound2`

Install and Run. Follow these four simple steps to install and run FRET:

1. Clone the FRET repository:

```
git clone \url{https://github.com/NASA-SW-VnV/fret.git}
```

2. Navigate to the fret-electron directory:

```
cd fret/fret-electron
```

3. Install FRET dependencies:

```
npm run fret-install
```

4. Start FRET:

```
npm start (or npm start -- --no-sandbox, on Ubuntu 22.04 or later)
```

4.2 Windows Installation Instructions

Prerequisites

1. Windows 10 or 11 machine.
2. An Ubuntu instance installed using Windows Subsystem for Linux 2 (WSL 2). An internet connection is required to install WSL 2, the Ubuntu instance and the associated dependencies.

Install and Run. To install, follow the steps below:

1. Open a Windows Powershell terminal and run the following command:
   `wsl -install`
   This will install WSL 2, choosing Ubuntu as the default instance. After the command is complete, the Ubuntu instance will be available by searching for either 'WSL' or 'Ubuntu' in the Start Menu. You may be asked to create a username and password when running the instance for the first time. Make sure to store this information as it is important in order to continue with the installation.
2. Open a terminal in the Ubuntu instance and create a `Documents` folder under the home directory by running the following command:
   `mkdir ~/Documents`
3. In the Ubuntu terminal install `git` by running:
   `sudo apt install git`
4. Follow all of the steps of the guide in Sect. 4.1, **including all prerequisites**. Windows currently installs a version of Ubuntu that is 24.04 or later by default, so make sure that the listed 24.04 libraries are installed.

4.3 Optional Dependencies

This section covers the installation of dependencies for the following optional features: 1) The LTLSIM interactive simulator, 2) realizability checking and 3) test case generation. The instructions are the same regardless of the operating system (Windows users must install the dependencies within the WSL instance).

1. **Install NuSMV**:
   - Download from: http://nusmv.fbk.eu/.
   - Add the NuSMV binaries directory to the `PATH` environment variable.
2. **Add LTLSIM to PATH**:
   - Add the following directory to your `PATH` environment variable:
     `$PATH_TO_FRET/fret/tools/LTLSIM/ltlsim-core/simulator`
   - Note: `$PATH_TO_FRET` is the path location where the FRET git repository was cloned (not the path to the FRET binary).
3. **JKind**:
   - Install Java version 8 or later. OpenJDK installation instructions can be found at https://openjdk.org/install/.

   - Download the files `jkind.jar`, `jkind` and `jrealizability` from https://github.com/andreaskatis/jkind-1/releases/latest to a folder of your choosing.
   - Add the folder containing the downloaded files to `PATH`.
4. **Kind 2**:
   - Download version 2.2.0 from: https://github.com/kind2-mc/kind2/releases/tag/v2.2.0 **Note: Version 2.3.0 is NOT supported.**
   - Extract the downloaded package and add extracted folder to `PATH`.
5. **Z3 Theorem Prover**:
   - Download from https://github.com/Z3Prover/z3/releases. **Note: Latest known supported version is 4.15.3.**
   - Extract the downloaded file and add the path to the `bin` folder to `PATH`.

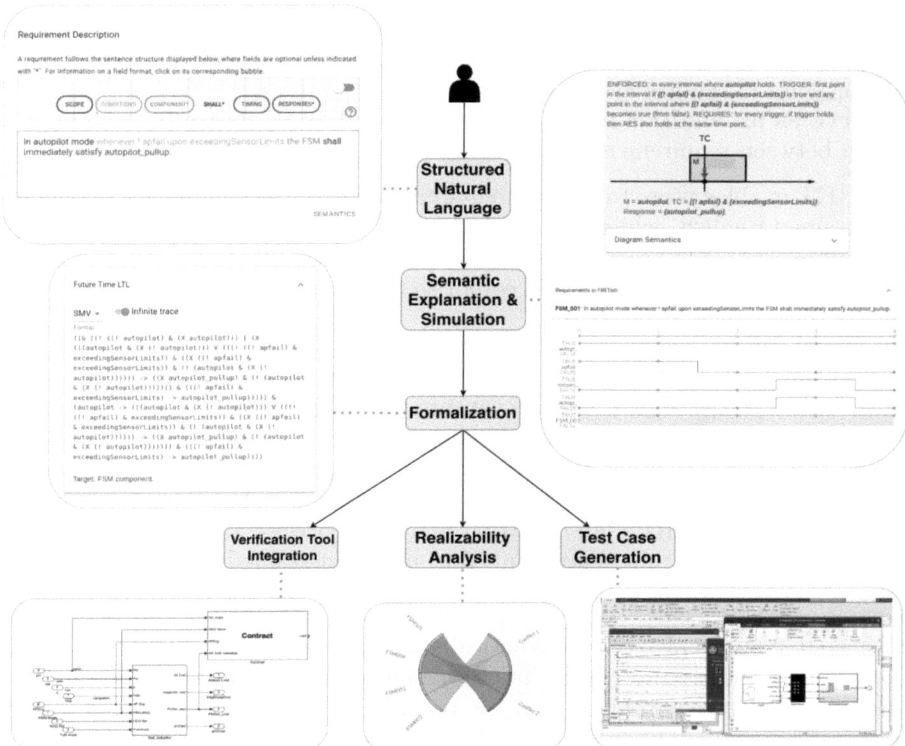

Fig. 2. The FRET workflow incorporates (1) Structured Natural Language, (2) Semantic Explanation and Simulation, (3) Formalization, (4) Verification Tool Integration, (5) Realizability Analysis, and (6) Test Case Generation.

5 Introducing FRET

Modern systems are inherently complex and heterogeneous, integrating diverse components across hardware, software, and communication layers. In such systems, requirement errors that escape early development phases can propagate through design and implementation, leading to costly failures, extensive redesigns, and safety risks. Early detection of specification defects is therefore crucial to ensuring system correctness and reducing development costs.

The foundation for early error detection lies in *precise requirements.* Ambiguous, informal requirements written in natural language, i.e., as English statements, are prone to misinterpretation, incompleteness, and logical inconsistencies. To prevent requirement errors at design, practitioners need a rigorous yet accessible method for capturing system specifications that balances precision with usability. To this end, the mission of FRET is to:

1. **Provide an intuitive platform for capturing precise requirements** that enables domain experts to express specifications without requiring expertise in formal methods.
2. **Enable early verification and validation (V&V) during requirements elicitation and authoring phases**, catching errors before they propagate to implementation.
3. **Serve as an enabler to a variety of formal analysis tools**, bridging the gap between requirements engineering and formal verification techniques.

FRET enables users to express requirements in structured natural language (constrained English) while maintaining mathematical precision, making formal methods accessible to practitioners without sacrificing rigor. The *core capabilities* of FRET are shown in Fig. 2 and comprise the following.

Structured Natural Language. FRET introduces FRETISH [17,18], a structured natural language that carries unambiguous, formally defined semantics. This controlled language allows users to capture requirements that are both human-readable and machine-analyzable. By constraining the expressiveness of natural language within a well-defined grammatical structure, FRETISH ensures that requirements are accessible to domain experts while remaining amenable to formal analysis. Each FRETISH construct maps to a precise formal interpretation, eliminating the ambiguity that is inherent in unconstrained natural language.

Semantic Explanation & Simulation. Once a requirement is expressed in the FRETISH language, FRET automatically generates explanations in various formats to help the user to clarify its precise meaning [17]. These include natural language descriptions that explain the requirement's semantics, as well as diagrammatic representations that visualize the temporal behavior that is specified by the requirement. Additionally, FRET provides interactive simulation capabilities that allow users to explore requirement behavior under different scenarios and input sequences. This combination of textual, visual, and interactive feedback helps users to understand and validate the precise meaning of their specifications, facilitating iterative refinement during the elicitation process.

Compositional Formalization. FRET formalizes requirements that are expressed in FRETish into multiple formal logic representations which ensures compatibility with different analysis tools. The translation is performed in a compositional manner, meaning that complex specifications can be built from simpler atomic components of the FRETish language [18]. This compositionality of the formalization process makes the framework inherently extensible.

Compositional Realizability Analysis. FRET analyzes requirements to detect logical inconsistencies, providing feedback to the users [20]. A requirement set is unrealizable if no implementation can satisfy all of the requirements simultaneously, often due to contradictory or overly constrained specifications. This automated analysis helps to catch specification errors early in the development process, before resources are invested in design and implementation. Furthermore, FRET employs a decomposition algorithm that tackles the high computational complexity of realizability checking by breaking the analysis into smaller, more tractable sub-tasks, while maintaining soundness [26].

Requirements-Based Test Case Generation. FRET automatically generates test cases directly from FRETish requirements, ensuring systematic requirement coverage. To this end, FRET supports a requirement-based coverage metric that measures how thoroughly the test suite exercises the requirements set, providing quantifiable assurance that critical behaviors specified in the requirements have been validated [21]. This automation reduces the manual effort required for test design and helps ensure that testing activities are aligned with requirements.

Integration with Verification Tools. FRET connects with multiple verification tools including:

- Runtime monitoring tools (e.g., Copilot [31], R2U2 [36]) to check requirements during system execution, enabling online detection of requirement violations in deployed systems [30].
- SMT solvers (e.g., Z3 [11]) for satisfiability and realizability checking.
- Model Checkers (e.g., nuXmv [3], Kind 2 [5], Simulink® Design Verifier™, PLCVerif [9]) for verifying models against formal specifications, enabling exhaustive analysis of system behavior in the design phase [1,16].

This integration enables a complete requirements-to-verification workflow, supporting both design-time analysis, runtime and implementation assurance.

6 Specification and Formalization

FRET allows users to write requirements in FRETish, a structured natural language, which is precise yet understandable. FRET then generates explanations in various formats and provides a simulator that helps users validate the meaning. We use the requirement below from our running example (Sect. 3):

Table 2. Breakdown of each FRET field when translating requirements.

Field	Mandatory	Defines	Example
Scope	Optional	Execution portion where the requirement is enforced	`in autopilot mode`
Condition (holding)	Optional	Condition under which the response applies	`whenever !apfail`
Condition (regular)	Optional	Condition triggering the response	`upon exceeding sensor limits`
Component	Mandatory	System part where the requirement is levied	`the FSM`
shall	Mandatory	Expresses required behavior conformance	shall
Timing	Optional	When the response occurs relative to scope/condition	`immediately`
Response	Mandatory	Desired system achievement	satisfy `autopilot_pullup`

> **[FSM-001]** Exceeding sensor limits shall latch an autopilot pullup when the pilot is not in control (autopilot mode) and the system is supported without failures (not apfail).

A FRETISH requirement consists of six sequential fields, though only three are mandatory. The six fields are as follows: 1) `scope` specifies the portion of the execution where the requirement is enforced; 2) `condition` is a Boolean expression that describes what triggers the response to occur at the time the expression's value becomes true, or is true at the beginning of the `scope` interval; 3) `component` (mandatory field) is the part of the system that the requirement is levied upon; 4) shall (mandatory field) is the word used to express that the component's behavior must conform to the requirement; 5) `timing` specifies when the `response` should happen, subject to the constraints defined in the `scope` and `condition` fields; and 6) `response` (mandatory field) is a Boolean expression that describes the behavior that the `component` must satisfy.

FRETISH provides eight options for the scope field: `global`, `after`, `only after`, `in`, `not in`, `only in`, `before`, `only before`. The default `global` scope means *always* and is used if the field is omitted; the others are with respect to when the system is in a particular mode or satisfies a Boolean expression.

There are two types of conditions that can be described by the optional condition field. A regular condition, introduced by the keywords `upon`, `when`, `where`, or `if`, describes a requirement that holds after a condition becomes true (either when the condition becomes true after being false, or the condition is true at the beginning of a scope). A holding condition, introduced by the `whenever` keyword, means that the response holds whenever the condition holds.

FRETISH supports 11 timing options: `immediately` (meaning at the same time point), `at the next timepoint`, `finally` (or synonymously, `at the last timepoint`), `eventually`, `always`, `never`, `within` N time steps, `for` N time steps, `after` N time steps, `until` Boolean stop condition, and `before` Boolean stop condition. When the timing field is omitted, the default value is `eventually`.

Boolean expressions are used in the `while` scope, condition, `until` and `before` timings, and response fields. Boolean operators include the standard logical symbols (! (negation), & (conjunction), | (disjunction), `xor` (exlusive or), -> or => (implication), <-> or <=> (equivalence)), arithmetic operations (+, −, *, /, mod, ^) and arithmetic relations (=, !=, <, >, <=, >=). FRETish also supports predefined temporal predicates: (1) `preBool`, `preInt`, and `preReal` return the value of an expression at the previous time point; (2) `persisted`, `occurred`, `persists`, and `occurs` express how conditions behave over time windows. `persisted(n,p)` means that p has held true for the previous n time points, as well as the current time point, in the current scope interval. `occurred(n,p)` means that the proposition p held at least once in the previous n time points, or at the current time point. `persists` and `occurs` are the future-time equivalents.

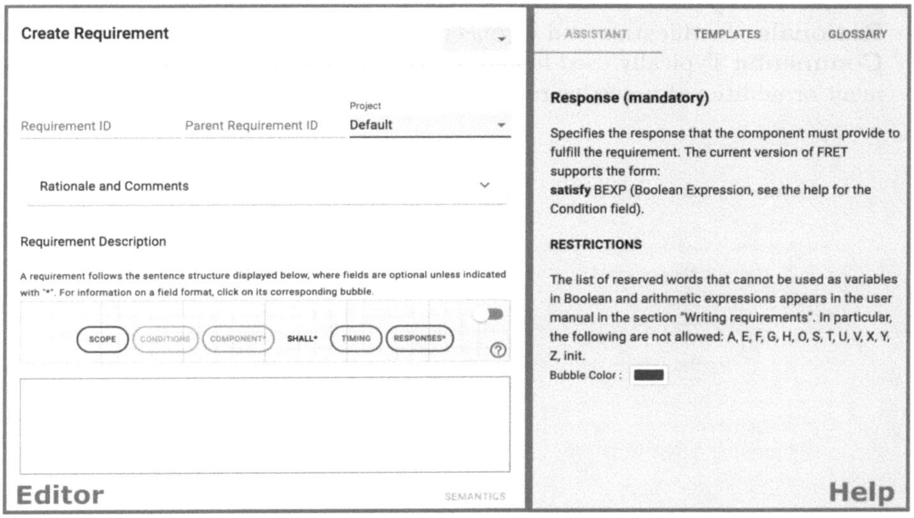

Fig. 3. A view of the FRET User Interface for creating requirements, showing the Editor on the left and the Help panel on the right.

When writing a requirement in FRETish, FRET parses the requirement and automatically color codes the parts that are associated with each field. We take the running example and deconstruct and translate the requirement into FRETish. The specific part of the execution where this requirement comes into play is `in autopilot mode` (the scope) and "the system" that concerns this requirement for this particular example is `the FSM` (the component). The triggering condition here is when the sensor limits are exceeded and the system is not apfail. The condition field can be written as: `whenever ! apfail upon exceedingSensorLimits`. The requirement describes that the `autopilot_pullup` (the response) should happen `immediately` once the condition becomes true in the scope. Now **[FSM-001]** is written in FRETish as:

> `In autopilot mode` `whenever ! apfail upon exceedingSensorLimits` `the FSM` shall `immediately` satisfy `autopilot_pullup`.

We capture the breakdown of this requirement in Table 2.

The FRET UI for the requirements editor is divided into two main parts as shown in Fig. 3. The editor pane on the left, where users compose and edit requirements, and a help panel on the right (assistant tab) that provides guidance, examples, and semantics explanations.

The editor pane has the following fields:

- **Requirement ID:** Unique identifier for the requirement.
- **Parent Requirement ID:** Identifier of the parent requirement.
- **Project:** Drop-down menu to select the requirement's project.
- **Rationale:** Justification and engineering reasoning behind the requirement.
- **Comments:** Typically used to capture the original natural language requirement or additional notes for reference.

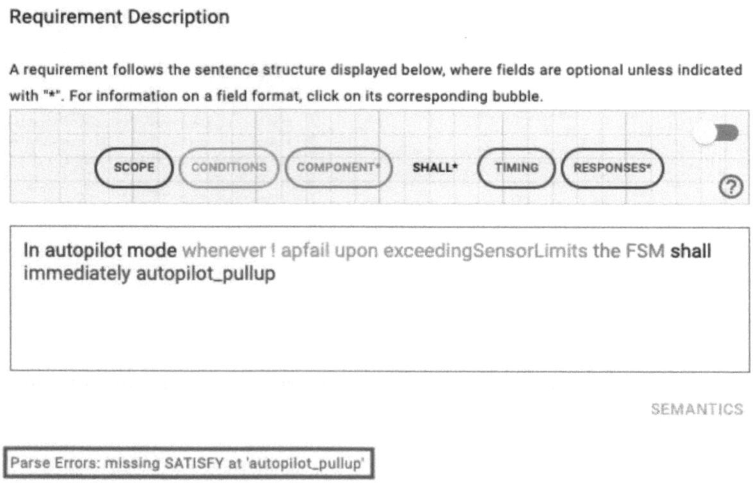

Fig. 4. Example requirement parsing error.

Each FRETISH field has a corresponding field bubble in the editor. Clicking on any bubble displays in the assistant tab: a description of the field and its syntax rules, as well as multiple examples demonstrating proper usage. Figure 3 shows the assistant tab for the `response` field.

As users type in a FRETISH requirement, the FRET editor provides real-time feedback through 1) color coding: each field is dynamically highlighted in a distinct color as it is recognized and 2) parser feedback: Error messages and warnings appear in red at the bottom of the editor. Figure 4 shows an example where a parse error message appears because the `response` field does not meet the required format (i.e. the word satisfy is required before the Boolean expression).

Once a requirement is successfully parsed, clicking `SEMANTICS` reveals:

- *English explanation* of what the requirement means

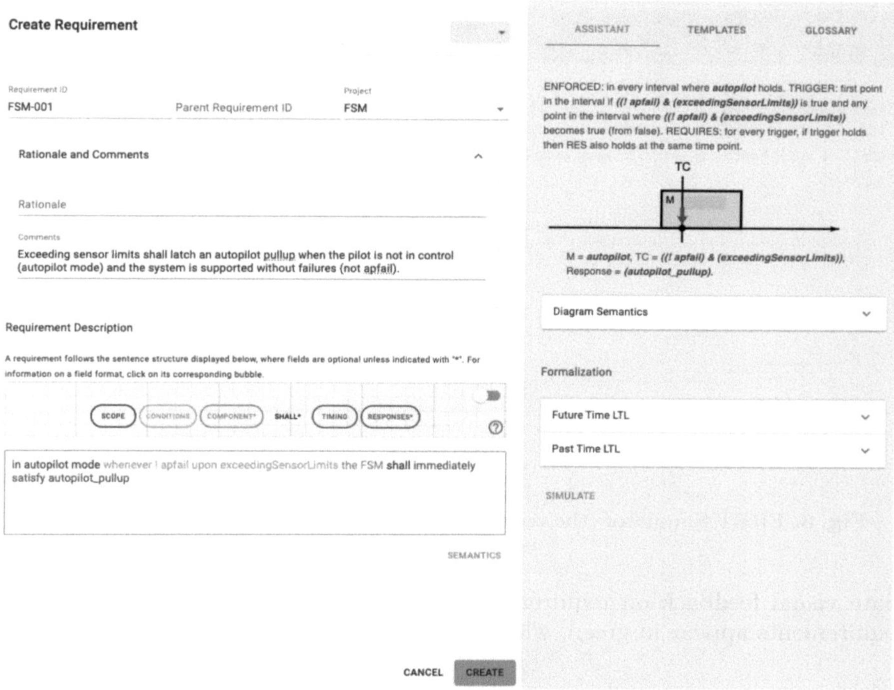

Fig. 5. The diagrammatic explanation is shown in the assistant tab.

- *Temporal Logic Formalization* in Future time LTL and Past time LTL
- *Diagrammatic explanations*

Let's explain the diagrammatic explanation as seen in Fig. 5. The requirement is only relevant within the grayed box, in this case, when the component is in autopilot mode, as described by the scope field. The vertical line shows the triggering condition sensorLimitsExceeded. The green arrow states that the pullup response must hold at the same time point that the condition becomes true from false, as expressed by the timing field immediately. A key of the symbols and colors used in the diagrammatic explanation can be found in the accordion menu under "Diagram Semantics". Temporal Logic Formalizations can be found in the accordion menus under "Future Time LTL" and "Past Time LTL".

FRET includes a requirements visualizer, available by clicking `SIMULATE` at the bottom of the semantics view. As shown in Fig. 6, the visualization uses graphical signal representation, displaying signals as waveforms over discrete time steps along a horizontal timeline. For a FRETish requirement, the visualizer displays temporal traces for each signal (variable) involved as binary (high/low) or multi-valued waveforms, followed by the requirement's valuation shown as a separate trace at each time point. The visualizer provides imme-

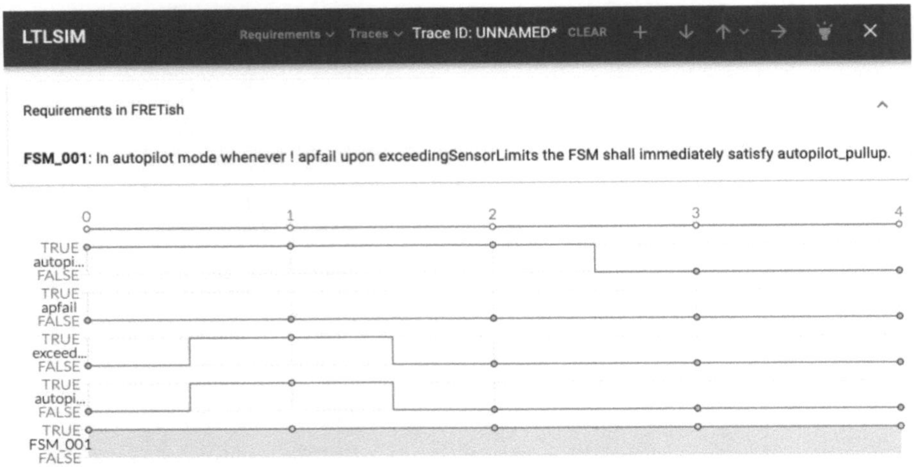

Fig. 6. FRET Simulator: the requirement is satisfied for the chosen valuation.

diate visual feedback on requirement valuation through color coding: satisfied requirements appear in green, while violated ones appear in red in Fig. 7.

7 FRET-Supported Analysis

Formalized requirements offer significant advantages beyond eliminating ambiguity. They enable automated analysis to detect conflicting constraints and support the generation of implementation-agnostic test suites with maximal requirement coverage. Next, we present FRET's Analysis Portal and its two core features: 1) Realizability Checking and 2) Test Case Generation. Both features require users to provide additional variable data through the Variable Mapping UI (Fig. 9).

7.1 Variable Mapping

Both realizability checking and test case generation require knowledge of each variable's type i.e. whether it is an `Input`, `Output` or `Internal` signal, as well as its data type e.g. `boolean`, `integer`, etc. The Variable Mapping interface in the Analysis Portal presents the list of variables in a tabular view, with which users can interact to complete the necessary information.

Figure 9 shows the Variable Mapping UI, with completed variable information for our running example. The tabular view displays variables from requirements **[FSM-001]** to **[FSM-006]** with their types. `Input` and `Output` variables require data types. `Internal` variables are meant to capture complex expressions over input and output variables and as such additionally require an assignment expression to be provided (shown in the Description column).

In addition to the tabular view, the Variable Mapping interface is where users can export requirements to a format that is compatible with external verification

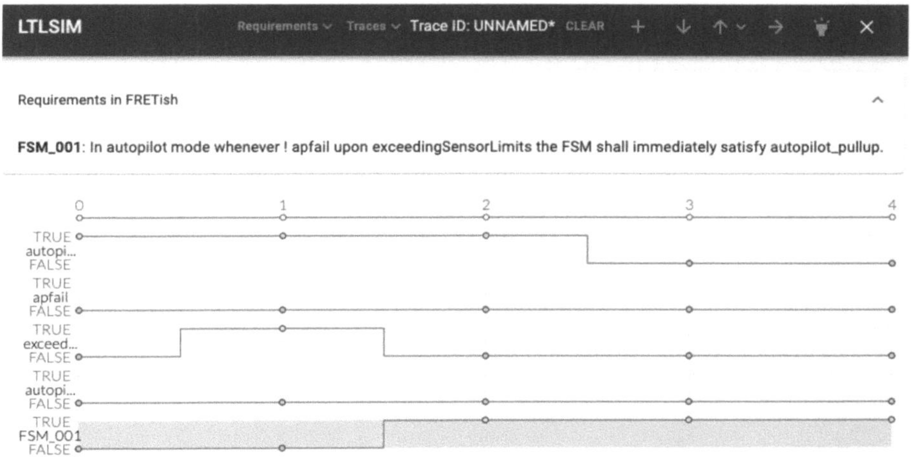

Fig. 7. FRET Simulator: the requirement is not satisfied for the chosen valuation.

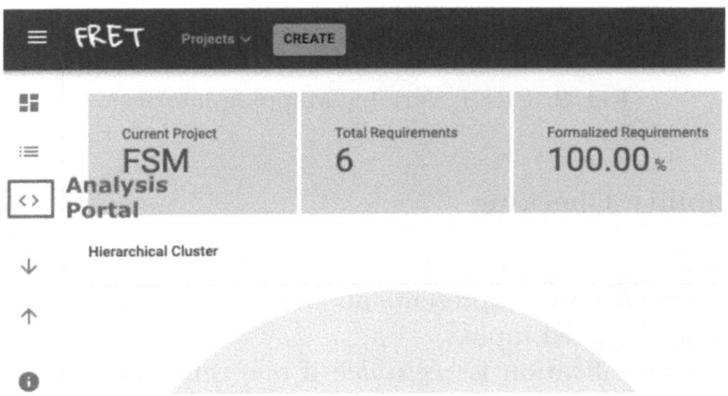

Fig. 8. Button to enter the Analysis Portal.

tools, including CoCoSpec [4] and NASA's CoPilot [31] (see Sect. 8). Additionally, users can export *test obligations* in the CoCoSpec and NuSMV/nuXmv [3,8] formats (for more details, see Sect. 7.3).

Each variable is marked as complete relative to the selected export language (top left of Fig. 9), indicating that: 1) all required information has been provided for subsequent analysis, and 2) the completion status for the specific export format. Once all of the variable mappings are complete, users can export requirements, check realizability, and generate test cases.

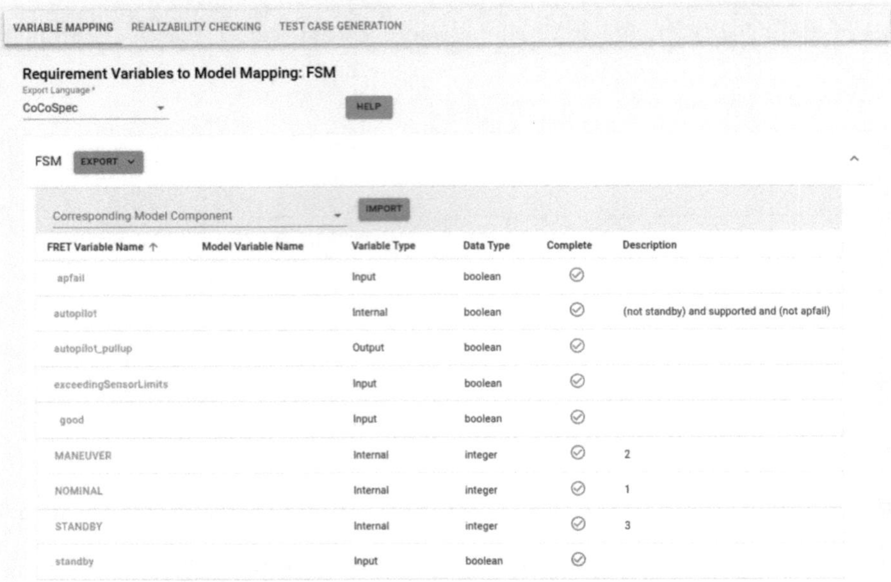

Fig. 9. FRET Variable Mapping interface.

7.2 Realizability Checking

FRET's realizability checking interface analyzes specifications for conflicting behaviors. The specification represents an abstract system that defines required outputs based on observed inputs.

Intuitively, a specification is *realizable* if one can implement a system that always conforms to the specification by providing values to variables of type `Output`, *for any* values assigned to variables of type `Input`. Realizability is strictly stronger than traditional consistency (satisfiability) checking, which only requires the specification be satisfiable for *some* inputs.

Figure 10 provides an overview of FRET's realizability checking interface, where users can check the specification, diagnose sources of unrealizability and export reports. Our running example of requirements **[FSM-001]** to **[FSM-006]** (from Table 1) is displayed in a tabular view that enables users to select which requirements to consider for analysis.

In general, realizability checking is computationally expensive, with the problem being undecidable for specifications that admit arithmetic over unbounded numbers. An important element towards taming this complexity is FRET's specification decomposition feature [26]. Intuitively, this feature is a pre-processing step where the specification is analyzed to identify whether it can be split into disjoint sub-specifications. If successful, each sub-specification can be independently checked for realizability. The decomposition algorithm is sound with respect to realizability, meaning that compositional analysis will always yield

Fig. 10. FRET Realizability Checking interface.

the same result as its monolithic counterpart, where no decomposition is performed.

Figure 10 shows how our running example of six requirements can be decomposed into two disjoint sub-specifications *CC0* and *CC1*, where the latter is a singleton containing only requirement **[FSM-001]**. Compositional analysis is therefore possible at this point and the result is shown in Fig. 11, declaring the specification as unrealizable, due to unrealizability of *CC0* (indicated by the red circled cross mark icon in the top left of Fig. 11). For unrealizable (sub-)specifications, users can further proceed with diagnosing the source by inspecting minimal sets of conflicting requirements, accompanied by an execution trace that demonstrates the conflict [20]. Figure 11 shows such a one-step conflict trace between **[FSM-002]** and **[FSM-003]** which, under common values for inputs `good`, `standby`, `supported` and `state` (which is an input feedback signal monitoring the machine's current state), disagree on the value of output `STATE`. In this trace, a state change from `TRANSITION` to `NOMINAL` is performed (each mapped to integer values 0 and 1), thus satisfying only **[FSM-003]**.

Realizability checking analysis and diagnosis supports requirements with Boolean and linear arithmetic constraints over unbounded integers and reals [19]. Satellite features include exporting and importing of reports and simulation of conflicts and realizable specifications [34].

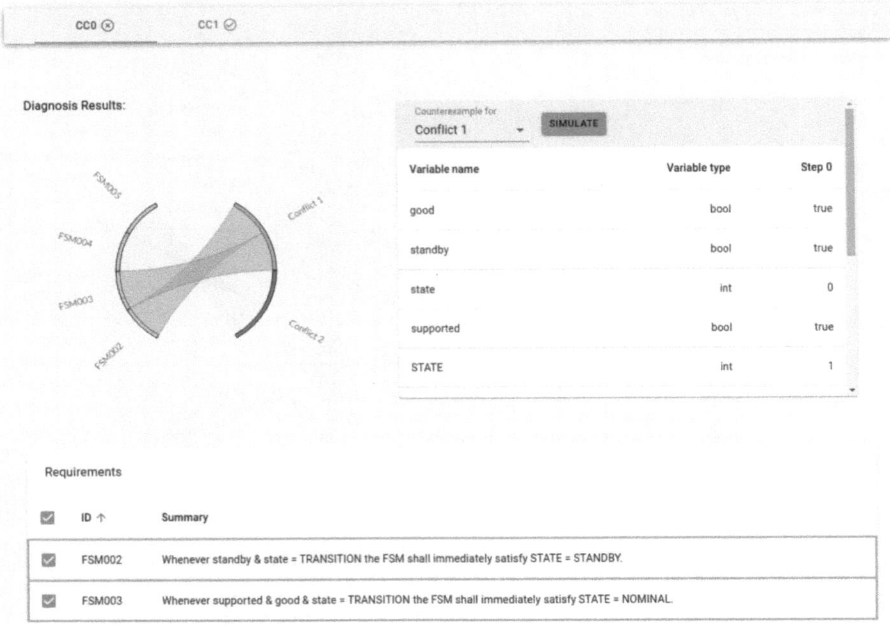

Fig. 11. Diagnosis view for unrealizable specifications.

7.3 Test Case Generation

System testing is an integral element of the development cycle, where typically tests are manually or automatically created in order to establish maximum coverage of the implementation and expose feature regressions. *Requirements-based* testing enforces this principle by incorporating the formalized requirements to be part of the actual test case generation process. FRET's test case generation feature facilitates requirements-based testing by 1) providing means to export *test obligations* that need to be met in order to achieve requirements coverage and 2) usage of these test obligations to automatically generate test cases using only the information that is provided in the specification.

Test obligations can be exported in CoCoSpec and NuSMV/nuXmv format using the Export function in Variable Mapping (Fig. 9). Each requirement is independently analyzed based on its variables, generating one test obligation per variable appearance in the formalization. Each obligation extends the original requirement with constraints that expose a specific variable's impact on requirement satisfaction.

Figure 12 shows an example obligation derived from requirement **[FSM-001]**, expressed in CoCoSpec format. The obligation is a past-time LTL formula, extending the corresponding requirement formula with constraints that assess the impact of variable `autopilot_pullup`, based on the FLIP coverage criterion [21,29]. When used as a property in a model checking task for the target

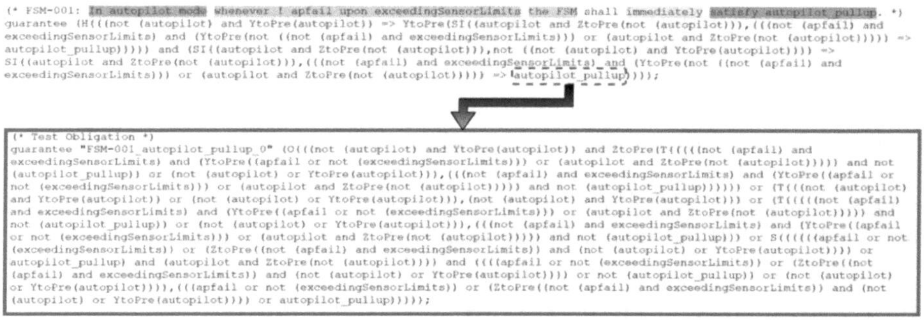

Fig. 12. Test Obligation for **[FSM-001]** in CoCoSpec format.

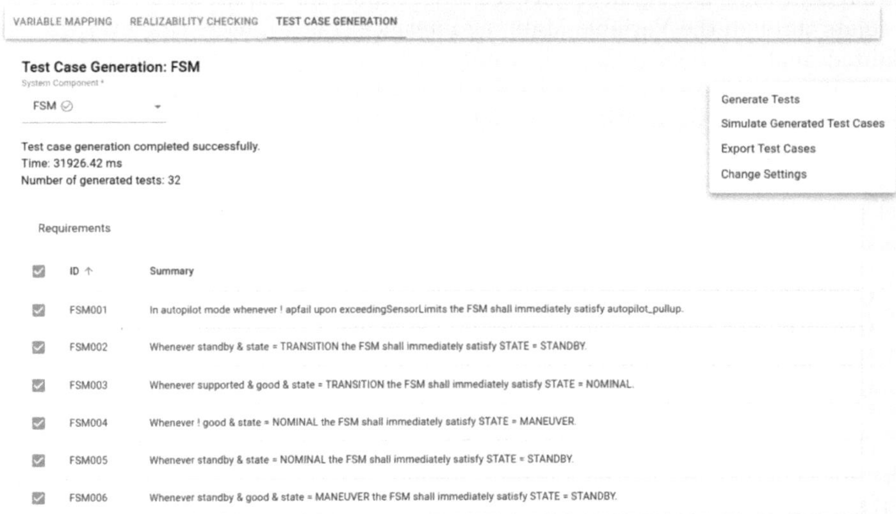

Fig. 13. FRET Test Case Generation interface.

system, the obligation is covered if and only if the task yields a test in the form of counterexample. If no counterexample can be generated, further investigations need to take place as, for example, the requirement could be vacuously true with respect to the implementation, or a bug may exist in the system-under-test.

Finally, Fig. 13 shows the test case generation interface in FRET where users can generate, simulate and export test suites that achieve requirements coverage without relying on the system implementation. This implementation-agnostic design supports the development and refinement of requirements early in the development cycle, providing nuanced information regarding the behavior that is allowed by the specification in the form of executable tests.

In the case of our running example, 32 tests are generated from the corresponding set of test obligations, each targeting a different part of the specification, thus achieving maximal requirement coverage. Furthermore, the derived test suite is minimal in size; if the backend model checking engine returns the same test for more than one obligation, these tests are collapsed into one. With the successful generation of tests, users can proceed with viewing and simulating the tests through the LTLSIM interface, as well as export the tests in JSON format, enabling straightforward integration to standard system testing workflows.

8 Connection with External Analysis Tools

FRET serves as a requirements hub that connects to various verification tools throughout the development lifecycle. By exporting requirements in different formats through the Variable Mapping interface (Fig. 9), users can leverage specialized analysis tools without learning each tool's specification language. This section demonstrates how to use FRET with four representative tools, covering both design-time verification and runtime monitoring.

The first step includes completing the variable mapping as described in Sect. 7.1. Once all variables are marked complete, the user may select their target export format from the dropdown menu in the Variable Mapping interface (see Fig. 14a), i.e.,

- **CoCoSpec:** For Simulink model verification via CoCoSim
- **Copilot:** For runtime monitoring of embedded C systems
- **R2U2:** For runtime monitoring with strict resource constraints
- **SMV:** For model checking with NuSMV, nuXmv and PLCVerif [1]

Then, the user may click the EXPORT button (shown in Fig. 14b) to generate the specification file in their chosen format. FRET automatically translates FRETISH requirements into the target tool's specification language.

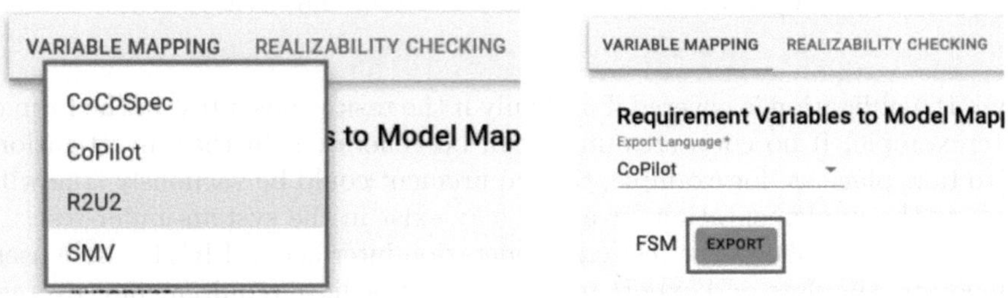

(a) Dropdown to select export language. (b) Export button turns blue when variable mapping is complete.

Fig. 14. Exporting requirements through the Variable Mapping interface.

8.1 Design-Time Verification Tools

Design-time verification enables early detection of requirement violations before system deployment, reducing development costs and safety risks.

- **CoCoSim** [4] automatically verifies Simulink® and Stateflow® models against formal requirements. After exporting FRETISH requirements to CoCoSpec format, CoCoSim translates both the model and requirements into Lustre, then uses the Kind toolset [5] for verification. When verification fails, CoCoSim generates counterexamples showing precisely which input sequences violate the requirements.
- **PLCVerif** [1,9,16] verifies industrial control programs written in IEC 61131-3 languages such as Structured Text and Ladder Diagrams. Users export FRETISH requirements in SMV format, which PLCVerif uses alongside the translated PLC program for formal verification via nuXmv. This integration enables automated verification of safety-critical industrial control systems against their requirements.

8.2 Runtime Monitoring Tools

While model checking verifies designs before deployment, runtime monitors continuously check requirements during actual system execution. This is essential for systems operating in unpredictable environments or where exhaustive pre-deployment verification is infeasible.

- **Copilot** [30,32] generates efficient C code monitors for embedded systems. After exporting requirements in Copilot format, the Copilot compiler produces monitors that integrate directly into the target system. These monitors continuously verify requirements during execution and can trigger alerts or automated recovery actions upon detecting violations. E.g., requirement **[FSM-001]** becomes a runtime monitor verifying that the autopilot correctly latches pullup commands whenever sensor limits are exceeded.
- **R2U2** [36] provides real-time monitoring optimized for small UAVs and spacecraft with limited computational resources. R2U2 supports both LTL and MTL properties, making it ideal for timing-critical requirements. Monitors execute with minimal computational overhead, making R2U2 ideal for systems with severe resource limitations. R2U2 has been demonstrated on NASA UAV missions for monitoring critical flight properties.

8.3 Choosing the Right Tool

Tool selection depends on verification objectives and system constraints.

- **Model verification:** Use CoCoSim or PLCVerif to formally model check Simulink and PLC models, respectively.

– **Runtime assurance:** Use R2U2 or Copilot to monitor deployed systems.

The key advantage of FRET is that requirements are formalized once in FRETISH, then exported to multiple tools as needed. This unified approach ensures consistency across verification activities while leveraging each tool's specialized capabilities, enabling comprehensive V&V workflows from early design through operational deployment.

9 FRET Extensions from the Community

Beyond the core FRET platform that is developed at NASA, several community extensions address specialized verification needs. Here are two examples:

Mu-FRET [15,22] extends FRET with requirements refactoring capabilities analogous to software refactoring. By leveraging FRET's formal analysis tools, Mu-FRET enables users to restructure and improve requirements while automatically verifying that refactoring operations preserve the original semantics. This ensures that requirement sets remain maintainable as specifications evolve throughout the development lifecycle[2].

SLEECinFRET [28] integrates Social, Legal, Ethical, Empathetic, and Cultural (SLEEC) requirements, i.e., a specialized class for autonomous systems, into FRET. This extension expands the FRETISH syntax to support SLEEC's language, automatically translating SLEEC requirements into FRETISH format. Translated requirements can then utilize all of FRET's built-in analysis tools, including realizability checking, enabling formal verification of ethical and societal constraints alongside traditional functional requirements[3].

10 Conclusion

This tutorial introduced FRET, NASA's Formal Requirements Elicitation Tool, which bridges natural language requirements and formal verification. Through our FSM autopilot example, we demonstrated how FRET's structured natural language, FRETISH, eliminates ambiguity while maintaining readability.

We explored FRET's core capabilities: automated formalization into temporal logic, interactive simulation, compositional realizability checking for detecting conflicts, and requirements-based test case generation. FRET's integration with external verification tools enables end-to-end workflows from requirement elicitation through implementation assurance.

After completing this tutorial, readers should be able to: formalize unambiguous requirements in FRETISH, leverage FRET's analysis capabilities to detect requirement defects early, generate verification conditions for external tools, and apply FRET in their V&V workflows. For practitioners new to formal methods, FRET provides an accessible entry point to rigorous requirements engineering.

[2] https://github.com/valu3s-mu/mu-fret.

[3] https://github.com/mirgit/fretForSleec.

References

1. Ádám, Z., et al.: From natural language requirements to the verification of programmable logic controllers: integrating FRET into PLCverif. In: NASA Formal Methods, pp. 353–360. Springer (2023)
2. Baier, C., Katoen, J.P.: Principles of model checking. MIT press (2008)
3. Cavada, R., et al.: The nuXmv symbolic model checker. In: International Conference on Computer Aided Verification, pp. 334–342. Springer (2014)
4. Champion, A., Gurfinkel, A., Kahsai, T., Tinelli, C.: CoCoSpec: a mode-aware contract language for reactive systems. In: De Nicola, R., Kühn, E. (eds.) SEFM 2016. LNCS, vol. 9763, pp. 347–366. Springer, Cham (2016). https://doi.org/10.1007/978-3-319-41591-8_24
5. Champion, A., ebsout, A., Sticksel, C., Tinelli, C.: The Kind 2 model checker. In: International Conference on Computer Aided Verification, pp. 510–517. Springer (2016)
6. Cheng, B.H., Atlee, J.M.: Research directions in requirements engineering. In: Future of Software Engineering, pp. 285–303. IEEE (2007)
7. Chung, L., Nixon, B.A., Yu, E., Mylopoulos, J.: Non-Functional requirements in software engineering, vol. 5. Springer (2012)
8. Cimatti, A., et al.: NuSMV Version 2: an opensource tool for symbolic model checking. In: International Conference on Computer-Aided Verification. Springer (2002)
9. Darvas, D., Vinuela, E.B., Adiego, E.F.: PLCverif: A tool to verify PLC programs based on model checking techniques. In: International Conference on Accelerator and Large Experimental Physics Control Systems. JACoW (2015)
10. De Ferro, C.M., Mavridou, A., Dille, M., Martins, F.: Simplifying requirements formalization for resource-constrained mission-critical software. In: International Conference on Dependable Systems and Networks Workshops, pp. 263–266 (2023)
11. De Moura, L., Bjørner, N.: Z3: an efficient SMT solver. In: International Conference on Tools and Algorithms for the Construction and Analysis of Systems, pp. 337–340. Springer (2008)
12. Elliott, C.: On example models and challenges ahead for the evaluation of complex cyber-physical systems with state of the art formal methods V&V, Lockheed Martin Skunk works. In: Safe & Secure Systems and Software Symposium, pp. 9–11 (2015)
13. Elliott, C.: An example set of cyber-physical V&V challenges for S5, Lockheed Martin Skunk Works. In: Safe & Secure Systems and Software Symposium, pp. 12–14 (2016)
14. Farrell, M., Luckcuck, M., Monahan, R., Reynolds, C., Sheridan, O.: Adventures in FRET and specification. In: International Symposium on Leveraging Applications of Formal Methods, pp. 106–123. Springer (2024)
15. Farrell, M., Luckcuck, M., Sheridan, O., Monahan, R.: Towards refactoring FRETish requirements. In: NASA Formal Methods Symposium, pp. 272–279. Springer (2022)
16. Fink, X., Mavridou, A., Katis, A., Adiego, B.F.: Verifying PLC programs via monitors: extending the Integration of FRET and PLCverif. In: NASA Formal Methods, pp. 427–435. Springer (2024)
17. Giannakopoulou, D., et al.: Formal requirements elicitation with FRET. In: Requirements Engineering: Foundation for Software Quality, p. 1 (2020)

18. Giannakopoulou, D., Pressburger, T., Mavridou, A., Schumann, J.: Automated formalization of structured natural language requirements. Inf. Softw. Technol. **137**, 106590 (2021)
19. Katis, A., et al.: Validity-Guided synthesis of reactive systems from assume-guarantee contracts. In: Tools and Algorithms for the Construction and Analysis of Systems, pp. 176–193. Springer (2018)
20. Katis, A., Mavridou, A., Giannakopoulou, D., Pressburger, T., Schumann, J.: Capture, analyze, diagnose: realizability checking of requirements in FRET. In: International Conference on Computer Aided Verification, pp. 490–504. Springer (2022)
21. Katis, A., Mavridou, A., Pressburger, T.: A streamlined, formal approach to requirements-based testing. In: NASA Formal Methods, pp. 159–179. Springer (2025)
22. Luckcuck, M., Sheridan, O., Farrell, M., Monahan, R.: Mu-FRET: a catalogue and tool for requirement refactoring. Software and Systems Modeling, To Appear (2026)
23. Malan, R., Bredemeyer, D.: Functional requirements and use cases. Bredemeyer Consulting, pp. 335–1653 (2001)
24. Martins, L.E.G., Gorschek, T.: Requirements engineering for safety-critical systems: a systematic literature review. Inf. Softw. Technol. **75**, 71–89 (2016)
25. Mavridou, A., et al.: The Ten Lockheed martin cyber-physical challenges: formalized, analyzed, and explained. In: International Requirements Engineering Conference, pp. 300–310. IEEE (2020)
26. Mavridou, A., Katis, A., Giannakopoulou, D., Kooi, D., Pressburger, T., Whalen, M.W.: From partial to global assume-guarantee contracts: compositional realizability analysis in FRET. In: Huisman, M., Păsăreanu, C., Zhan, N. (eds.) FM 2021. LNCS, vol. 13047, pp. 503–523. Springer, Cham (2021). https://doi.org/10.1007/978-3-030-90870-6_27
27. Mendoza, D., Mavridou, A., Katis, A., Trippel, C.: Automating requirements formalization: using LLMs and low-complexity distinguishing traces for semantic validation. In: IEEE/ACM International Conference on Software Engineering (2026)
28. Mirani, M., Inverardi, P., Pelliccione, P., Raimondi, F., Troquard, N.: Extending FRET with SLEEC Rules: formalization, obligation inference, and monitoring. In: International Conference on Tools and Algorithms for the Construction and Analysis of Systems, To appear (2026)
29. Pecheur, C., Raimondi, F., Brat, G.: A formal analysis of requirements-based testing. In: International Symposium on Software Testing and Analysis, pp. 47–56. ACM (2009)
30. Perez, I., Mavridou, A., Pressburger, T., Goodloe, A., Giannakopoulou, D.: Automated translation of natural language requirements to runtime monitors. In: Tools and Algorithms for the Construction and Analysis of Systems, pp. 387–395. Springer, 2022
31. Perez, I., Mavridou, A., Pressburger, T., Goodloe, A., Giannakopoulou, D.: Automated translation of natural language requirements to runtime monitors. In: International Conference on Tools and Algorithms for the Construction and Analysis of Systems, pp. 387–395. Springer (2022)
32. Perez, I., Mavridou, A., Pressburger, T., Will, A., Martin, P.J.: Monitoring ROS2: from requirements to autonomous robots. In: International Workshop on Formal Methods for Autonomous Systems (2022)
33. Pohl, K.: Requirements Engineering: An Overview. RWTH, Fachgruppe Informatik Aachen (1996)

34. Pressburger, T., Katis, A., Dutle, A., Mavridou, A.: Authoring, analyzing, and monitoring requirements for a Lift-Plus-Cruise aircraft. In: Requirements Engineering: Foundation for Software Quality, pp. 295–308. Springer (2023)
35. Rozier, K.Y.: Specification: the biggest bottleneck in formal methods and autonomy. In: Blazy, S., Chechik, M. (eds.) VSTTE 2016. LNCS, vol. 9971, pp. 8–26. Springer, Cham (2016). https://doi.org/10.1007/978-3-319-48869-1_2
36. Rozier, K.Y., Schumann, J.: R2U2: tool overview. In: International Workshop on Competitions, Usability, Benchmarks, Evaluation, and Standardisation for Runtime Verification Tools, number ARC-E-DAA-TN48849 (2017)
37. Stachtiari, E., Mavridou, A., Katsaros, P., Bliudze, S., Sifakis, J.: Early validation of system requirements and design through correctness-by-construction. J. Syst. Softw. **145**, 52–78 (2018)
38. Van Lamsweerde, A.: Requirements engineering in the year 00: a research Perspective. In: International Conference on Software Engineering, pp. 5–19 (2000)

Mixed Flow-Sensitive Static Analysis: Engineering Modularity

Helmut Seidl[1(✉)], Vesal Vojdani[2], Julian Erhard[3], and Michael Schwarz[4]

[1] Technical University of Munich, Garching, Germany
helmut.seidl@tum.de
[2] University of Tartu, Tartu, Estonia
vesal.vojdani@ut.ee
[3] CISPA Helmholtz Center for Information Security, Garching, Germany
julian.erhard@cispa.de
[4] National University of Singapore, Singapore, Singapore
m.schwarz@nus.edu.sg

Abstract. Flow-sensitive and flow-insensitive analyses of programs occupy opposite ends of a spectrum. Between these extremes lie *mixed flow sensitive* approaches, where some aspects of program behavior are analyzed flow-insensitively and others flow-sensitively. Mixed flow-sensitivity arises, for example, in the analysis of multi-threaded code or code withnon-local control flow. Another instance is global store widening for efficient analysis of functional languages and some forms of pointer analysis. While mixed flow-sensitive analyses are common in the literature, the formulation of the particular analysis problem and the means to solve it are often tightly coupled. *Side-effecting constraint systems* provide a generic mechanism for describing mixed flow-sensitive analyses, thus decoupling the analysis definition from solver algorithm details. The abstract interpreter GOBLINT realizes this decoupling, and allows defining mixed flow-sensitive analyses independently of generic solvers. We indicate how the precision of specified analyses can be improved using *digests* (on the side of the formulation of the analysis problem) and suitable update rules (on the side of the solver). We explain how developers can use GOBLINT to implement their own mixed flow-sensitive analyses.

Keywords: abstract interpretation · static analysis · flow-sensitivity · modularity · mixed flow-sensitivity

1 Introduction

Static analysis by abstract interpretation is an automated technique for establishing invariants about all executions of a program without running it. Static analysis has been applied successfully in settings such as avionics, where program correctness is crucial and the consequences of missing even a single bug can range from tremendous monetary damage to the loss of lives.

A. Sampaio and M. Stoelinga (Eds.): FM 2026, LNCS 16557, pp. 446–470, 2026.
https://doi.org/10.1007/978-3-032-26220-2_22

As the question that static analysis tries to solve, namely "Does this program ever go wrong?", is undecidable, it sidesteps the issue by overapproximating concrete program behavior: If the overapproximation does not go wrong, neither will the program. Designing an appropriate overapproximation is far from trivial: make it too coarse, and your analysis terminates almost instantly but establishes few meaningful invariants. If you hardly overapproximate, your analyzer will be quite precise but may fail to terminate for all but the most simple programs. This trade-off already appears when choosing the kind of invariants. Consider, e.g., *int* variables: Just tracking properties `x==c` for constants `c` (*constant propagation*) may be cheaper than tracking properties `l<=x`, `x<=u` for `l`, `u` $\in \mathbb{Z}$ (*intervals*), while tracking relationships between variables will be even more expensive.

A similar trade-off occurs when choosing what invariants are attached to: in *flow-sensitive* analysis, program properties (usually represented as *abstract values*) are associated with control locations. This yields information such as "Whenever the program is at location u, it holds that `x==42`." At the opposite end of the flow-sensitivity spectrum are *flow-insensitive* analyses: instead of invariants that hold when the program reaches specific program locations, they compute invariants that hold throughout the entire execution. Andersen's points-to analysis [3], e.g., establishes for each pointer expression which addresses it may evaluate to at any time during the execution. *Partial flow-sensitivity* on the other hand, tries to strike a compromise between *flow-insensitive* and *flow-sensitive* analyses by computing flow-sensitive invariants only for some program points, while collecting a single invariant for the remaining ones [52, 53, 57].

In this tutorial, we present a technique that encompasses and goes beyond partial flow-sensitivity, namely *mixed flow-sensitivity* [71]: the choice is not only which program points to treat flow-sensitively, but which *aspects* of program execution to treat flow-sensitively and which flow-insensitively. The tutorial is aimed at readers with some familiarity with abstract interpretation, as might be gained from the first chapters of an introductory textbook [45, 55, 67], though we try to keep it accessible to readers without this background. Should readers, after working through this tutorial, seek a comprehensive overview over most aspects of abstract interpretation, we recommend the textbook by Cousot [11].

While the term *mixed flow-sensitivity* was coined only recently, it covers many approaches from the literature. One instance is *global store widening* [69], which applies flow-sensitive analysis not to complete states but to specific parts of states only, while treating the remaining parts flow-insensitively.

Example 1 (Global store widening). Consider the program in Fig. 1, written in a stylized version of C, that maintains the two global variables `g`,`h`, and has one (non-recursive) procedure `f`, called by the *main* procedure. In principle, the values of `g` and `h` could be analyzed flow-sensitively. For efficiency, we may choose to analyze the values of one or both of them flow-insensitively. If we do this for `g`, we find three contributions to `g`, namely, 1, −17, and 42 by the initializer, by procedure `main`, and by procedure `f`, respectively. This set of values suffices to falsify the error condition at line 11. □

Sets of integer values may grow unboundedly, which in practice requires further abstraction. One natural abstraction is to over-approximate sets of integers with enclosing *intervals*, and apply *widening* [12] when intervals are found to grow during fixpoint iteration.

Example 2. Let us again consider Example 1 where we over-approximate the set of possible values of g with intervals. When processing the initializer for `g` (line 1), the interval recorded for `g` is $[1, 1]$. When encountering the assignment of -17 to `g` (line 9), the analyzer would detect an *insufficient* lower bound. The standard *widening* operator for intervals as proposed in [12], would therefore decrease that lower bound to $-\infty$. Finally, when the assignment of 42 to `g` is encountered (line 5), the standard widening for intervals also abandons the upper bound of the interval, resulting therefore in the interval $[-\infty, \infty]$ for `g`—implying that we fail to deduce that `g<=42` holds in line 10. □

The issue from Example 2 can be resolved by applying widening not to the values accumulated for `g`, but *per origin*, i.e., to the contributions from the same control location separately. Since each origin contributes at most one value to `g` in this example, no widening is needed. Accordingly, the resulting interval for `g` is $[-17, 42]$. Thus, precision loss due to widening as in Example 2 can be mitigated by such solver adaptations, independently of the analysis.

The minimalistic program from Fig. 1 is also instructive because it shows another precision loss caused by flow-insensitivity.

Example 3. Consider a flow-insensitive analysis of the other global variable `h` in the program from Fig. 1. For `h`, contributions are provided in the initializer and in line 4 only. Since a flow-insensitive analysis cannot distinguish whether a statement executes once or multiple times, it infers for `h` the interval $[0, \infty]$ (given that integers are of arbitrary precision). □

The precision loss from Example 3 can be mitigated by means of *digests* [61]. In that approach, the analysis tracks how often global variable `h` has been written. A counter bounded by 2 already suffices to exhibit the final value $[1, 1]$ for `h`.

```
1  int g = 1;
2  int h = 0;
3  void f() {
4    h = h+1;
5    g = 42;
6  }
7  void main() {
8    g = -17;
9    f();
10   if(g > 42) {
11     error();
12   }
13 }
```

Fig. 1. Sequential program with global variables.

Global store widening for shared data has not only been used to improve scalability, but is also helpful for *incremental* static analysis where it *decouples* analysis of the increment from its wider context as much as possible—thus enabling reuse of previous analysis results [20,64]. This decoupling potential of mixed flow-sensitivity is crucial for thread-modular static analysis [27,42–44,60,62,63,65,72,75,79]: Collecting the invariant that approximates thread interactions flow-insensitively effectively modularizes the analysis of individual threads, allowing their local states to be tracked flow-sensitively without considering all threads potentially running concurrently.

Example 4 (Thread-modular analysis). Consider the multi-threaded program from Fig. 2 where two threads manipulate the shared global variable `g`. If `g>3` holds, a function `error()` is called.

A mixed flow-sensitive analysis may track the values of the local *int* variable `y` of each thread flow-sensitively, while accumulating values of the global variable `g` flow-insensitively. Assume again that we use intervals to abstract sets of possible values of program variables. The initial value of `g` is $[0, 0]$ (line 1). Assume the analysis first considers the thread created in line 8. The abstract value corresponding to the program point after executing the assignment `y = 2;` results in the local state $\{\mathtt{y} \mapsto [2, 2]\}$. The subsequent line 4 provides a contribution of $[2, 2]$ to `g`. This contribution is then combined with the initial value $[0, 0]$ to obtain $[0, 2]$—while the local program state of the thread remains unchanged. The increment of `y` in the last line then results in the local state $\{\mathtt{y} \mapsto [3, 3]\}$.

Now consider the *main* thread. When checking the first condition (line 10), the analysis accesses the interval recorded for `g`. Since that interval contains the value 2, the *then* branch may be entered. The assignment `g = y;` is therefore considered reachable (line 11). Since the local state before this assignment is $\{\mathtt{y} \mapsto [1, 1]\}$, the effect of this assignment is a contribution of $[1, 1]$ to `g`, while the local program state again remains the same. Since the interval $[1, 1]$ is included in the interval $[0, 2]$, the interval for `g` will not change. The final result $[0, 2]$ for `g` certifies that the condition in line 13 is necessarily false—implying that the call `error()` is unreachable.

Observe that this differs from a fully flow-insensitive analysis. In that case, the value 3 of `y` would also be contributed to the abstract value for `g`, as the values of variables at different program points would not be distinguished. □

```
1  int g = 0;
2  void thread() {
3    int y = 2;
4    g = y;
5    y++;
6  }
7  void main() {
8    create(thread);
9    int y = 1;
10   if(g == 2) {
11     g = y;
12   }
13   if(g > 2) error();

14 }
```

Fig. 2. Simple multi-threaded program.

Mixed flow-sensitivity is also the method of choice for straightening out the analysis of *irregular* control-flow. This is the case, e.g., for the analysis of `setjmp`/ `longjmp` in C [22], where `setjmp` stores a reference to the current stack frame and program point in a *store buffer*, which allows calls to `longjmp` to return there and take a different branch of execution.

Flow-sensitive analysis may compute not only one invariant per program point, but also distinguish multiple copies using additional entities. One common entity is an abstraction of the *calling-context* of the current procedure [21,68] or, more generally, some abstraction of the execution history recording, e.g., the set of mutexes which have been acquired or (some abstraction of) the sequence of threads that have already been created. Such extra entities have been called *digests* [61]. Interestingly, such refinement is not restricted to flow-sensitive analysis; it can also be used in flow-insensitive analysis to recover some lost precision.

Table 1 gives an overview of representative instances of mixed flow-sensitive analyses both from the previous examples and from the broader literature, classifying them by their flow-sensitive and flow-insensitive components.

These works provide specialized solutions for specific (classes of) analysis problems. Viewing them through the common lens of mixed flow-sensitivity allows us to provide a generic and modular framework to specify, implement, and solve such analysis problems, without having to re-invent the wheel for each new analysis instance. In this tutorial, we present such a framework for mixed flow-sensitive abstract interpretation.

The rest of this tutorial is structured as follows. In Sect. 2, we give a gentle introduction to side-effecting constraint systems, a convenient formalism for mixed flow-sensitive analyses. Section 3 then shows how such constraint systems can be derived from control-flow graphs of programs and discusses procedure calls. Section 4 discusses how such program analyses may be refined using *digests*. Section 5 covers solver algorithms and a recent technique for recovering precision lost at globals due to widening. Section 6 discusses how to implement further mixed flow-sensitive analyses in the static analysis framework GOBLINT. Finally, Sect. 7 concludes the tutorial and extends an invitation to static analysis researchers to view their own analyses through the lens of mixed flow-sensitivity and to build on our framework where appropriate.

2 Framework

Side-effecting constraint systems [4] are a formalism for describing mixed flow-sensitive analyses. They distinguish between constraint variables for which information is collected flow-sensitively (*locals* $\mathcal{L}$) and flow-insensitively (*globals* $\mathcal{G}$). To avoid confusion with program variables, we refer to constraint variables as *unknowns*. The set of all unknowns is denoted by $\mathcal{X} = \mathcal{L} \cup \mathcal{G}$. The *constraints* of the system specify how the globals and locals are semantically related. Each unknown may have its own domain of *abstract values*. For convenience, we assume these domains to be *complete lattices*. It has been observed, however, that many practical domains are not complete lattices [25]. Side-effecting constraint systems and GOBLINT therefore also permit more general partial orders

Table 1. Overview of representative instances of mixed flow-sensitive analyses with their flow-sensitive and flow-insensitive çomponents.

Instance type	**Flow-sensitive**	**Flow-insensitive**
Side-effecting constraint systems (generic) [4,65]	Program points with context	Shared/global invariants
Partial/selective flow sensitivity [53,57]	Facts tracked flow-sensitively for selected CFG regions (strong-update set)	Merged facts for collapsed regions (weak-update set)
Global-store widening & abstract machines [31,32,51,69]	Control component of states (program point, expression, continuations)	Single widened store (joined values), optionally effect summaries
Incremental / interactive fixed points [20,64]	Local invariants (and dependencies) for updated program points recomputed	Shared/global invariants accumulated across edits and re-analysis
Thread-modular (interferences) [27,42–44,72,75]	Per-thread flow-sensitive state (thread-local control/locals) under assumed interference	Flow-insensitive interference summary over shared memory with event (locking) history
Thread-modular (side-effects) [60–63,71,79]	Thread state refined by local history-abstraction (locks, traces/digests)	Shared memory and scheduling/parallelism invariants
History refinements and trace partitioning [19,21,35,54,68]	Locals indexed by a history key (call context, trace partition, etc.)	Concurrent versions refine global component of the base analysis with key
Nonlocal control flow (`setjmp`/`longjmp`) [22]	Program points with context	Arrival points for irregular returns
Strong-update points-to analysis [9,39,73,74]	Flow-sensitive refinement for singleton points-to sets (enabling strong updates).	Sound flow-insensitive fallback for other pointer dereferences.

that do not offer least upper or greatest lower bounds, but just *quasi*-join and *quasi*-meet operations.

Formally, a side-effecting constraint system is a tuple $(\mathcal{L}, \mathcal{G}, \mathcal{D}^{\mathcal{X}}, \mathcal{C})$ with

Locals $\mathcal{L}$. A (potentially infinite) set of *locals* $\mathcal{L}$ at the granularity at which program flow is considered, e.g., the level of program points.

Globals $\mathcal{G}$. A (potentially infinite) set of *globals* $\mathcal{G}$, where contributions arising from the constraints for locals are collected flow-insensitively.

A family of domains $\mathcal{D}^{\mathcal{X}}$. A family $\mathcal{D}^{\mathcal{X}} = (\mathcal{D}_{[x]})_{[x]\in\mathcal{X}}$, where for each $[x] \in \mathcal{X}$, $\mathcal{D}_{[x]}$ is a complete lattice;

Constraints $\mathcal{C}$. A (potentially infinite) set of constraints over the unknowns and their domains.

For convenience, we denote all occurring partial order relations by $\sqsubseteq$. As the sets of locals and globals often coincide with other sets (e.g., the set of program points for locals or the set of global variables in the program in Fig. 2), we enclose them in square brackets whenever we refer to unknowns. We allow different domains for different unknowns, since program points of different procedures may, e.g., record invariants referring to different sets of local variables. Likewise, invariants collected for global unknowns, such as variables g and h in Fig. 2, may require different domains, e.g., due to their different types.

Let $\mathcal{D} := \bigcup_{[x]\in\mathcal{X}} \mathcal{D}_{[x]}$. A mapping $\eta : \mathcal{X} \to \mathcal{D}$ is called a *valuation* if $\eta\,[x] \in \mathcal{D}_{[x]}$ for each $[x] \in \mathcal{X}$. Let $\mathcal{E}$ denote the set of all such valuations. A local unknown $[u]$ may be constrained by one or more constraint functions f of type

$$f : \mathcal{E} \to (\mathcal{E} \times \mathcal{D}_{[u]})$$

Given a valuation η, the value $f\,\eta = (\eta', d)$ consists of

- a valuation η' providing finitely many contributions $\eta'\,[g] \neq \bot$ to globals $[g]$;
- a single contribution d to the local $[u]$.

Here, $\bot$ denotes the least element in each respective domain. By omitting entries with $\bot$ results, the map η' can be represented as the finite map $\{[g] \mapsto \eta'\,[g] \mid [g] \in \mathcal{G}, \eta'\,[g] \neq \bot\}$. The constraint in $\mathcal{C}$ for a local $[u]$ and constraint function f on valuations η then is given by

$$(\eta, \eta\,[u]) \sqsupseteq f\,\eta \tag{1}$$

Crucially, this per-local view exposes the mixed flow-sensitive nature of constraints: the finite set of contributions to globals generated alongside the contribution to the left-hand side local $[u]$ allows to accumulate information at globals flow-insensitively while following the dependencies between locals. These contributions to globals are also called *side-effects* of the constraint.

A valuation in $\mathcal{E}$ is a *solution* of a side-effecting constraint system if it satisfies all constraints of the system.

Example 5. Consider again Example 1. The program from Fig. 1 has unknowns $[i], 3 \leq i \leq 13$, representing the abstract state *before* executing line i, and the global unknowns $[\texttt{g}]$ and $[\texttt{h}]$ corresponding to the global program variables g and h, respectively. Since neither procedure uses local program variables, we may choose $\mathcal{D}_{[i]} = \{\bot \sqsubset \top\}$—just to express whether a program point is reachable ($\top$) or not ($\bot$). For globals $[\texttt{g}]$ and $[\texttt{h}]$, on the other hand, we use the complete lattice $\mathbb{I}_\bot$ of integer *intervals* together with $\emptyset$ as least element. For the unknowns $[4]$ and $[5]$, we have, e.g., the constraints

$$\begin{aligned}(\eta, \eta\,[4]) \;\sqsupseteq\; & \textbf{if}\;(\eta\,[3] = \bot)\;\textbf{then}\;(\emptyset, \bot)\\ & \textbf{else}\;(\{[\texttt{h}] \mapsto \eta\,[\texttt{h}] +^\sharp [1,1]\}, \top)\end{aligned}$$

$$\begin{aligned}(\eta, \eta\,[5]) \;\sqsupseteq\; & \textbf{if}\;(\eta\,[4] = \bot)\;\textbf{then}\;(\emptyset, \bot)\\ & \textbf{else}\;(\{[\texttt{g}] \mapsto [42, 42]\}, \top)\end{aligned}$$

which contribute abstract values to $[\texttt{h}]$ and $[\texttt{g}]$, respectively, whenever the program points 4 and 5 are reachable. □

3 From CFG to Mixed Flow-Sensitive Analysis

We use side-effecting constraint systems to define abstract interpretations with mixed flow-sensitivity. We assume that each procedure or function in the program is represented by a control-flow graph. Example 5 illustrates one way side-effecting constraint systems can define mixed flow-sensitive analyses for programs in C-like languages. There, the set of locals contains all program points that have a predecessor in the control-flow graph. It also contains a dedicated local $[\_\mathsf{main}]$ to provide an initial contribution to the `main` procedure which we will elaborate on shortly. The set of globals is given by aspects of program state meant to be analyzed flow-insensitively. In the case of global store widening, these may, e.g., include an unknown $[g]$ for each selected global variable g.

Let N be the set of program points, Act the set of possible program actions, and $E \subseteq N \times \mathsf{Act} \times N$ the set of program edges. For unknowns corresponding to program points with a predecessor, one constraint is introduced per incoming edge in the control-flow graph. Accordingly, the set $\mathcal{C}$ contains the following constraints:

$$(\eta, \eta\,[v]) \sqsupseteq [\![(u, \mathsf{act}, v)]\!]^\sharp\,(\eta\,[u])\,\eta, \qquad ((u, \mathsf{act}, v) \in E) \tag{2}$$

For each control-flow edge $e \equiv (u, \mathsf{act}, v)$, we thus require a function

$$[\![e]\!]^\sharp : \mathcal{D}_{[u]} \to \mathcal{E} \to (\mathcal{E} \times \mathcal{D}_{[v]})$$

that realizes the abstract semantics of e. It takes the abstract state from predecessor unknown $[u]$, and a valuation for possibly accessed further locals and globals, and returns a set of contributions to globals as well as an abstract state for the local corresponding to the end point v of e. This function is also called the *transfer function* of e.

Example 6. For the multi-threaded program in Fig. 2, the constraint for $[u_{11}]$ is

$$(\eta, \eta\,[u_{11}]) \sqsupseteq [\![(u_{10}, \texttt{(g==2)?}, u_{11})]\!]^\sharp\,(\eta\,[u_{10}])\,\eta$$

where we use an OCaml-like notation for defining

$$\begin{aligned}[\![(u_{10}, \texttt{(g==2)?}, u_{11})]\!]^\sharp\, d\, \eta = \;& \textbf{if } d = \bot \textbf{ then } (\emptyset, \bot)\\ & \textbf{else if } 2 \in \eta\,[\mathsf{g}] \textbf{ then } (\emptyset, d)\\ & \textbf{else } (\emptyset, \bot)\end{aligned}$$

No side-effect is provided, and $[u_{11}]$ receives the value of $[u_{10}]$ if u_{10} is reachable and 2 is among the possible values of g. Otherwise, $[u_{11}]$ receives the value $\bot$. □

Of particular interest are *call* edges $(u, f(args), v)$ for some procedure f and arguments *args*. Assume that $\mathsf{start}_f, \mathsf{ret}_f$ are the start and end points, respectively, of f. Assume further that we are given functions

$$\begin{aligned}\mathsf{enter}^\sharp_{f,args} &: \mathcal{D}_{[u]} \to \mathcal{D}_{\mathsf{start}_f}\\ \mathsf{combine}^\sharp_f &: \mathcal{D}_{[u]} \to \mathcal{D}_{\mathsf{ret}_f} \to \mathcal{D}_{[v]}\end{aligned}$$

where $\mathsf{enter}^\sharp_{f,args}$ maps the abstract state before the call to the start state of the called procedure f, and $\mathsf{combine}^\sharp_f$ takes the abstract states before the call and at the end point of f to determine the abstract state after the call. When the language supports *dynamic* procedure calls, as in C or Java, the procedures possibly called at a call edge are only revealed at analysis time. As a remedy, Apinis et al. [4] suggest realizing contributions to the start points of potentially called procedures via *side-effects*. Following this idea, we obtain for the call edge $e \equiv (u, f(args), v)$, the transfer function

$$
\begin{aligned}
[\![e]\!]^\sharp\, d\, \eta \;=\;& \mathbf{if}\, d = \bot \;\mathbf{then}\; (\emptyset, \bot) \\
& \mathbf{else\ let}\; \sigma = \{[\mathsf{start}_f] \mapsto \mathsf{enter}^\sharp_{f,args}\, d\}\;\mathbf{in} \\
& \quad \mathbf{let}\; d' = \mathsf{combine}^\sharp_f\; d\; (\eta\,[\mathsf{ret}_f])\;\mathbf{in} \\
& \quad (\sigma, d')
\end{aligned} \tag{3}
$$

Note that under this formalization, the abstract values of procedure start points are collected flow-insensitively. In other words, start points are treated as *globals*. To provide an abstract state for the initial call to main, we introduce the dedicated local $[\_\mathsf{main}]$ whose right-hand side function also may provide side-effects for initializing program variables treated as globals:

$$(\eta, \eta\,[\_\mathsf{main}]) \;\sqsupseteq\; (\sigma_0, \eta\,[\mathsf{ret}_{\mathsf{main}}]) \tag{4}$$

where σ_0 maps $[\mathsf{start}_{\mathsf{main}}]$ to some designated initial state. In Example 1, for instance, we would have $\sigma_0 = \{[\mathtt{g}] \mapsto [1,1], [\mathtt{h}] \mapsto [0,0], [7] \mapsto \top\}$.

Flow-insensitive analysis of data-structures shared among multiple threads is a convenient way to decouple flow-sensitive analyses of the individual threads. The downside is that only very imprecise results are to be expected for the global unknowns. Therefore, methods are sought to fine-tune the analysis so that better results for global unknowns can be hoped for. Such methods are presented in the next section.

4 Refining Mixed Flow-Sensitive Analyses

Instead of inferring invariants only for program points and globals, we may infer *refined* invariants that also take into account some extra property a of the execution history. We mention three prominent instances of this idea:

- *Context-sensitive* analysis where multiple instances of procedures are differentiated according to properties of their (abstract) calling-contexts [21,68].
- *Trace-partitioning* according to guard predicates [19,54], an idea closely related to *predicate abstraction* as considered for software model-checking [24].
- *Digests* where properties not only of the sequential trace of the current thread, but also of the full concurrently reaching trace are used for refining analysis results [60,61,63].

We first describe how *context-sensitivity* is realized in our setting, and then elaborate on digests, which generalize context-sensitivity, trace partitioning, and other approaches such as the *view* framework [35].

Context-Sensitivity. To increase precision, it may be desirable to distinguish between multiple instances of a procedure, whether by properties of actual arguments or by call sites. To achieve this, unknowns $[u]$ for program points from some procedure f are conceptually replaced by unknowns $[u, c]$ where c is an (abstract) *calling-context* representing the call stack leading to the current call of f. Accordingly, invariants computed for $[u, c]$ only need to take into account executions reaching u that satisfy the restriction imposed by context c. A classical example is the k-callstring approach [68] where procedure calls are distinguished based on the k-most recent calls on the call stack. Callstrings of length 1, e.g., just keep calls from distinct call sites apart.

Mixed flow-sensitive analysis can naturally be refined with any such form of context-sensitivity. Let $\mathbb{C}$ be a set of *contexts*. Then, each unknown $[u]$, u a program point, is conceptually split into the collection of unknowns $[u, c], c \in \mathbb{C}$. Likewise, every control-flow edge $e \equiv (u, \mathsf{act}, v)$ is split up into the collection of control-flow edges $(e, c) \equiv ([u, c], \mathsf{act}, [v, c])$, each of which gives rise to a constraint

$$(\eta, \eta\,[v, c]) \sqsupseteq [\![(e, c)]\!]^\sharp\,(\eta\,[u, c])\,\eta \tag{5}$$

When the edge e is just an assignment or a guard, the transfer function for (e, c) equals the transfer function for e. For constructing the transfer functions for a *call* edge $e \equiv (u, f(\mathit{args}), v)$, we additionally require a function

$$\mathsf{context}^\sharp_{u,f,\mathit{args}} : \mathcal{D}_u \to \mathbb{C} \to \mathbb{C}$$

which computes from the local state at u and the current context the new context for the start point of f. For each current context c, the transfer function for the *call* edge e from the previous section is refined to

$$\begin{array}{rcl} [\![(e, c)]\!]^\sharp\; d\; \eta & = & \textbf{if}\; d = \bot\; \textbf{then}\; (\emptyset, \bot) \\ & & \textbf{else let}\; c' = \mathsf{context}^\sharp_{u,f,\mathit{args}}\, d\, c\; \textbf{in} \\ & & \quad \textbf{let}\; \sigma = \{[\mathsf{start}_f, c'] \mapsto \mathsf{enter}^\sharp_{f,\mathit{args}}\, d\}\; \textbf{in} \\ & & \quad \textbf{let}\; d' = \mathsf{combine}^\sharp_f\; d\; (\eta\,[\mathsf{ret}_f, c'])\; \textbf{in} \\ & & \quad (\sigma, d') \end{array} \tag{6}$$

Finally, assume we are given some element $c_0 \in \mathbb{C}$ representing the initial context of `main`. Then the constraint (4) for initialization becomes

$$(\eta, \eta\,[\_\mathsf{main}]) \sqsupseteq (\sigma_0, \eta\,[\mathsf{ret}_{\mathsf{main}}, c_0]) \tag{7}$$

where σ_0 now maps $[\mathsf{start}_{\mathsf{main}}, c_0]$ to some initial state d_0.

Example 7. For 1-callstring context-sensitivity, $\mathbb{C}$ is the set of call sites in the program plus the element $\_\mathsf{main}$, $c_0 = \_\mathsf{main}$, and $\mathsf{context}^\sharp_{u,f,\mathit{args}}\,\_\,\_ = u$. □

Example 8. To realize *partial tabulation* of summary functions [68], we use the abstract states at start points of functions to distinguish function calls and set:

$$\mathbb{C} = \bigcup_f \mathcal{D}_{[\mathsf{start}_f]} \qquad c_0 = d_0 \qquad \mathsf{context}^\sharp_{u,f,\mathit{args}}\, d\, \_ = \mathsf{enter}^\sharp_{f,\mathit{args}}\, d$$ □

Indeed, distinguishing function calls by contexts, be it via callstrings or partial tabulation, has been shown to considerably increase analysis precision [40,41]. Realizing context-sensitivity in this way provides modularity as different kinds of context-sensitivity can be used without changing the rest of the analysis specification.

Digests. Contexts can be understood as abstractions of sequences of calls (and preceding states) leading to a program point. *Digests* generalize this idea by considering not just abstractions of the sequence of calls, but the full, possibly concurrent trace reaching a program point. This consists not only of steps executed by a given thread t so far, but also of all *events* from other threads that the current program state of t may depend on. The idea is then to compute, for each such digest A from some set of digests $\mathcal{A}$, a refined invariant describing the program states that can be reached along traces satisfying A. This applies not only to local, but also to global unknowns. Some combinations of digests may correspond to situations that cannot happen in one concrete execution, making them *incompatible*. Such reasoning can be exploited to increase precision, e.g., by excluding certain values of globals from being read.

```
int g = 0;
void thread() {
  g = -12;
}
```

```
void main() {
  g = 42;
  assert(g > 0);
  create(thread);
  g = -5;
}
```

Fig. 3. Example program for illustrating multi-threaded mode digests.

Example 9. Consider the multi-threaded program from Fig. 3. Consider tracking whether any threads besides *main* have already been created. To this end, consider the set of digests $\mathcal{A} = \{\mathsf{ST}, \mathsf{MT}\}$. An analysis in Example 4 refined with this set of digests, may track whether a given *write* to a global happened while the program was in single-threaded mode (ST) or in multi-threaded mode (MT). When *reading* the value of the global g in single-threaded mode, i.e., with digest ST, then only values written with digest ST must be taken into account, as *writes* in multi-threaded mode have definitely not happened yet. □

Refinement with digests can be realized by employing a *reduced cardinal power*-like construction [10,13,26] which, for an unknown $[x]$, is given by the complete lattice $\mathcal{D}_{[x]}^{\mathcal{A}} = \mathcal{A} \to \mathcal{D}_{[x]}$ of mappings from digests to elements of $\mathcal{D}_{[x]}$. Note that in this case, not only locals but also globals are refined depending on digests. Transfer functions can then be defined by a case distinction on digests. For each digest, both a map of digests to abstract states for the successor program point, and the contributions to digests of globals are computed. Further, when accessing the value of a global $[g]$, a *compatibility* relation on digests may be provided that restricts the set of digests that must be taken into account. We elide a formal treatment here and instead illustrate the idea with the example from Example 9.

At the start of `main`, the abstract state maps ST to the initial state, with MT mapped to $\bot$. The side-effect to the global $[\mathsf{g}]$ also consists of a map, which here maps ST to the initial value $[42, 42]$ and MT to $\bot$. Once the thread is created, both at line 10 and throughout the newly created thread, only digest MT maps to a non-$\bot$ value. Thus, both *writes* in multi-threaded mode only affect the MT component of the global $[\mathsf{g}]$. When g is read in line 9, only the ST component of $[\mathsf{g}]$ is considered, as the only non-$\bot$ digest at this point is ST, and MT is not *compatible* with ST. As a consequence, the assertion in line 9 holds.

Other commonly used digests include sets of held mutexes, mutex types, more expressive thread *ids*, or modification counts for globals. For these instances, as well as a more semantic definition of digests, see [60,61].

5 Solvers for Mixed Flow-Sensitive Analyses

So far, we have presented side-effecting constraint systems as a convenient formalism for *expressing* mixed flow-sensitive analysis problems. This is in line with our vision for how a static analyzer for real-world programming languages should be built [79]: An analysis engineer designing a mixed flow-sensitive analysis should not have to think about the actual algorithm for inferring valid analysis results, i.e., how constraint sets are eventually solved. Instead, they should focus on appropriately tailoring abstract domains and specifying the constraints for the individual program entities of the analysis in a declarative manner. Both tasks pose challenges of their own. A clear separation of concerns is therefore necessary to realize high-quality analysis software.

Multiple solvers are conceivable for solving side-effecting constraint systems [2,65,66]. Here, we briefly discuss common techniques before sketching one of the more elaborate solvers, namely, the *top-down* solver **TD**, enhanced with support for side-effects [66].

5.1 Leveraging Existing Solvers

A natural idea is to repurpose an existing solver for context-sensitive analysis. This can be done by maintaining a mapping from globals to their accumulated values while the solver runs. If global values change after a full run, the solver is re-invoked with the new global values. This procedure continues until the values of all globals have stabilized. This idea, which corresponds to nested fixpoint computations, has been advocated for thread-modular analysis [42–44,65,75]. This basic approach can be made more efficient when dependencies between globals and locals are either known ahead of time or tracked during analysis to allow for only re-computing potentially affected locals [72], respectively globals in settings such as the *blackboard architecture* [1,28,34,56] where essentially all unknowns are globals and locals are simulated by tagging some globals, e.g., with program points.

5.2 Widening and Narrowing

Whenever lattices used by an analysis admit infinite (or very long, think, e.g., of the interval abstract domain over 32-bit integers) ascending chains, to obtain termination one has to give up on computing the least fixed-point (in cases where it exists), and instead compute an overapproximation thereof. This is typically achieved by using *widening* operators that enforce that the ascending chains considered by the analysis are finite, followed by another *narrowing* iteration to possibly regain some lost precision [12,14,15]. A classical interval-domain example is to disallow finite bounds from increasing and instead going to one of the infinities as a bound whenever this is encountered. A corresponding narrowing operator for intervals may only get rid of infinities, but refrain from improving finite bounds to ensure that the narrowing iteration terminates as well. Applying widening indiscriminately can lead to a significant loss of precision. For flow-sensitive analyses where the control-flow graph is known ahead of time, a common technique is to pre-compute loopheads and apply widening and subsequently narrowing only there. This basic idea reaches its limits for interprocedural analysis or in the presence of mixed flow-sensitivity.

Alternative approaches for flow-sensitive analysis are based on self-observation [18,78] of the solver during its execution to dynamically detect unknowns involved in cycles and applying widening/narrowing only at these. An extension of such ideas to flow-insensitively analyzed globals is not straightforward. In fact, a variety of approaches to applying widening and, perhaps, also narrowing at globals are conceivable. To decouple this design decision from the remaining solver, the generic concept of *update rules* has been proposed [71].

5.3 Recovering Precision at Globals: Update Rules

An update rule is invoked by the solver when a contribution d to some global g is triggered. The rule then determines the next value to be stored for g. In the most naïve variant, the returned value is obtained by applying the widening operator to the old value of g and the new contribution d. As demonstrated in Example 1, this may result in unacceptably imprecise results. As a first improvement, the update rule maintains a data structure that separates individual contributions to each global g according to their *origins*, i.e., the locals where the contribution to g was provided. Widening then is only applied per origin, with the new value for g obtained as the join over the contributions from all origins. This is sufficient to recover a precise result in Example 1. Stemmler et al. [71] propose further refined strategies that, under certain conditions, are even able to completely revoke previous, outdated, contributions which may be *toxic* in the sense that they cause loss of precision at other unknowns—even including locals.

5.4 A Custom-Built Solver: Top-Down Solver with Side-Effects

The idea of the top-down solver **TD** goes back to early work on interprocedural analysis, originally targeted at logic languages such as PROLOG [37,48,49] and is still used today in the CIAO system [29,30]. Similar solvers have also been employed for the analysis of OCAML programs [38,46]. As observed in [66], the solver **TD** can be extended with widening intertwined with narrowing, as well as with support for side-effects in constraint systems. Therefore, it has become the solver of choice in the analyzer GOBLINT for multi-threaded C.
TD is a *local* solver, in that it starts from some initial unknown $[x]$ and explores the constraint system only as much as is necessary to determine the value of $[x]$. In our formulation from Sect. 2, we choose [_main] as this unknown. For exploration, **TD** recursively queries the values of unknowns encountered in the right-hand side of a constraint. Only when for them the best possible solution has been found, the evaluation of the right-hand side proceeds. To avoid infinite descent into unknowns, it maintains a set *called* of unknowns where querying has been started but has not yet been completed. The current values of these unknowns are returned immediately. During the recursive exploration of unknowns, the algorithm dynamically tracks dependencies between unknowns. Interestingly, the set *called* not only prevents nontermination, but also detects cyclic dependencies between unknowns and thus candidate unknowns where widening / narrowing should be applied. Recently, **TD** with side-effects has been parallelized [36] and its correctness has been formally verified in ISABELLE [77].

6 GOBLINT: Mixed Flow-Sensitivity in Practice

There is a wealth of static analysis frameworks based on abstract interpretation. Commercial tools include ASTRÉE [16], ABSINT AIT [23], POLYSPACE [47], and ADACORE CODEPEER [76], covering industrial use cases from runtime-error proofs to timing analysis and security bug detection. Research tools include WALA [59], FRAMA-C [5], MOPSA [33], CODEX [70], LISA [50], DOOP [8], and many others. While all of these systems have their unique strengths, we consider the GOBLINT system, which is, to our knowledge, the only general framework supporting mixed flow-sensitive analyses.

GOBLINT targets multi-threaded C programs. Its modular design decouples analysis specification from the algorithmic technicalities of the solver, enabling researchers to focus and innovate on each aspect separately. GOBLINT has served as a testbed for ideas from multiple generations of PhD students. Since 2021, it has participated in the Competition on Software Verification (SV-COMP) [6], where it has repeatedly achieved top results in verifying the absence of data races. GOBLINT has never produced a false negative on the SV-COMP benchmark suite, and in 2026 it won the silver medal in the *C.TrueOverall* category [58], making it the second-best correctness prover for C programs. GOBLINT is open source and publicly available.[1] Below we list three ways GOBLINT can be enhanced:

[1] https://goblint.in.tum.de and https://github.com/goblint/analyzer.

- *Enhancements of solvers* for side-effecting constraint systems. Novel update rules, e.g., could be added, as could alternative strategies for fixpoint iteration itself.
- *Extensions of the analysis framework.* This means adding support for further language concepts or introducing general mechanisms that boost expressiveness or efficiency of other analyses, often by generically lifting existing analysis specifications.
- *Novel analyses* for specific program properties.

We focus on the last item: adding new mixed flow-sensitive analyses to GOBLINT. A (potentially updated) version of this section and the code to follow along are available at https://goblint.in.tum.de/tutorial.

Designing Analyses. GOBLINT allows running several analyses jointly, each of which can be specified separately, and each analysis may interact with others through a query system. The specification for a new mixed flow-sensitive analysis in GOBLINT consists of the following parts:

- An abstract domain $\mathcal{D}_{\mathcal{L}}$ that describes the information tracked per flow-sensitive unit (e.g., per context, per thread, per CFG node)
- A set of globals $\bar{\mathcal{G}}$ which are the global variables associated with this specific analysis. Note that this does not include the start points of procedures or threads, which are handled separately by the framework.
- An abstract domain $\mathcal{D}_{\bar{\mathcal{G}}}$ over which the new globals range.
- A set of transfer functions, one for each kind of basic statement in the control-flow graph, such as assignments and guards.
- For function calls, only the functions `enter` and `combine` must be provided for passing values into calls and returning results, respectively. From these, the framework constructs the constraints for call edges (cf. (3) and (6)) and also takes care of inferring sets of possibly called functions. Similarly, thread creation is already handled by the framework.

Example 10. Consider again Example 4. To realize this analysis, we set:

- The local domain $\mathcal{D}_{\mathcal{L}}$ to be maps from the local variables to non-$\bot$ intervals, extended with a $\bot$ element to represent unreachable states, i.e., $\mathcal{D}_{\mathcal{L}} = (\mathsf{Var} \to \mathbb{I})_{\bot}$ where Var is the set of local program variables, i.e., {y} in our example.
- The set of globals $\bar{\mathcal{G}}$ corresponds to the set of *global* program variables, i.e., $\bar{\mathcal{G}} = \{[g], [h]\}$ for our example program.
- The domain $\mathcal{D}_{\bar{\mathcal{G}}}$ for globals then is given by the domain of intervals $\mathbb{I}_{\bot}$. □

Figures. 4 and 5 show the stylized signature of a mixed flow-sensitive analysis in the GOBLINT framework. For implementor convenience, the framework already deconstructs edges and supplies the arguments of transfer functions individually.

The abstract domains, for locals and the one for globals, are usually not built from scratch, but assembled from existing building blocks—such as products of various kinds, lifters, and *map* domains—in a modular way.

```
(** Man(ager) is passed to transfer functions and allows accessing the
    the context, read values from globals, side-effect values to globals,
        and query information from other analyses *)
type ('g,'c,'v) man =
{   ask        : 'a. 'a Queries.t -> 'a Queries.result
                 (* To communicate with other analyses *)
  ; edge       : MyCFG.edge
  ; orig_node  : MyCFG.node
  ; dest_node  : MyCFG.node
  ; context    : 'c
  ; global     : 'v -> 'g
  ; sideg      : 'v -> 'g -> unit
}
```

Fig. 4. Simplified type of the manager passed to transfer functions in GOBLINT.

```
module type SimplifiedSpec = sig
  module V : UnknownSet        (** Set of globals. *)

  module G : Lattice           (** Domain for globals. *)
  module D : Lattice           (** Domain for locals. *)

  module C : Printable         (** Context information. *)

  val name : string            (** Name of the analysis. *)

  val startstate : D.t         (** Initial local state for main. *)
  val startcontext: C.t        (** Initial context for main. *)

  val query : (G.t, C.t, V.t) man -> D.t -> 'a Queries.t -> 'a Queries.result

  (** A transfer function which handles the assignment of a rval to a lval, i.e.,
      it handles program points of the form "lval = rval;" *)
  val assign: (G.t, C.t, V.t) man -> D.t -> lval -> exp -> D.t

  (** A transfer function which handles conditional branching yielding the
      truth value passed as a boolean argument *)
  val branch: (G.t, C.t, V.t) man -> D.t -> exp -> bool -> D.t

  ...

  (** For a function call "lval = f(args)" or "f(args)",
      enter returns the initial state of the callee. *)
  val enter   : (G.t, C.t, V.t) man -> D.t -> lval option -> fundec -> exp list -> D.t

  (** Combines the states before and after the call. *)
  val combine: (G.t, C.t, V.t) man -> D.t -> D.t -> lval option -> fundec -> exp list -> D.t

  (** Compute the context for a function call, given the local state and context at the caller,
      the called function and the local state inside the callee *)
  val context: (G.t, C.t, V.t) man -> (D.t * C.t) -> fundec -> D.t -> C.t

  (** Compute the start state of a new thread starting with the function given by fundec *)
  val threadenter: (G.t, C.t, V.t) man -> D.t -> fundec -> exp list -> D.t
end
```

Fig. 5. Signature for defining a mixed flow-sensitive analysis in GOBLINT.

Example 11. Consider again Example 4 and the definitions given in Example 10. Figure 6 shows how the abstract domains for that analysis can be defined in GOBLINT using lifters and primitive building blocks. Figure 7 then shows what the definition of the assignment transfer function may look like for a very simple expression language. If the left-hand side variable of the assignment is some program variable g whose set of possible values is analyzed *flow-insensitively*, the abstract value d for the right-hand side is *contributed* to the global unknown $[g]$ as a side-effect. This is conveniently achieved by the call `man.sideg` g d. The collection of all pairs for which `man.sideg` was called is delegated to the framework where global unknowns are eventually modified by the solver. □

```
module Analysis: SimplifiedSpec = struct
    module V = GlobalVariables

    module G = LiftBot(Intervals)
    module D = LiftBot(MapLattice(LocalVariables)(Intervals))

    module C = Unit

    let name = "fmTutorialAnalysis2"

    let startstate = D.top
    let startcontext = ()

    let context _ _ _ _ = ()
    let query _ _ _ = Queries.Result.top

    ...

end
```

Fig. 6. OCAML definition for the analysis from Example 4 in GOBLINT.

In the GOBLINT framework, multiple analyses may be activated in parallel. A user may, e.g., activate an interval analysis for numerical properties together with a lockset analysis for inferring sets of protecting mutexes for shared data. In each analysis A, all locals as well as the start points of functions take values from the same lattice $\mathcal{D}_{\mathcal{L},A}$ while it may introduce its own set $\mathcal{G}_A$ of globals taking values from some lattice $\mathcal{D}_{\mathcal{G},A}$. The domains $\mathcal{D}_A$ for locals of all activated analyses are combined into one *product* lattice

$$\mathcal{D}_{\mathcal{L}} = \prod_A \mathcal{D}_{\mathcal{L},A}$$

In contrast, globals introduced by distinct analyses are considered distinct. Their domains therefore are combined into the *sum*

$$\mathcal{D}_{\mathcal{G}} = \sum_A \mathcal{D}_{\mathcal{G},A}$$

```
let rec eval_exp man d exp =
  match exp with
  | Const i -> Intervals.of_int i
  | Lval lval when Var.is_local lval -> D.find d lval
  | Lval lval -> man.global lval
  | BinOp (Plus, e1, e2) ->
      let v1 = eval_exp man d e1 in
      let v2 = eval_exp man d e2 in
      Intervals.add v1 v2
  | _ -> Intervals.top  (** Simplified for illustration *)


let assign man d lval rval =
  if D.is_bot d then d (** Predecessor unreachable, do nothing *)
  else
        (** Evaluate the rval expression to an abstract value *)
      let v = eval_exp man d rval in
      if Var.is_local lval then
          (** Update binding for lval *)
          D.replace d lval v ()
      else
          (** Publish contribution to global variable *)
          (man.sideg lval v; d)
```

Fig. 7. OCAML definition for the assign transfer function of the analysis from Example 4 in GOBLINT.

of their respective domains. In this lattice, the elements of distinct domains $\mathcal{D}_{\mathcal{G},A}$ are considered as incomparable where new elements $\overline{\top}$ and $\underline{\bot}$ are introduced to serve as the top and bottom elements of the sum domain, respectively.

The transfer functions of the combined analysis are then defined by applying the individual transfer functions point-wise per component of the locals.

Context-Sensitivity and Digests. To realize context-sensitivity in GOBLINT, the framework requires the analysis to specify the type of *contexts* to be used (Fig. 5) by supplying an appropriate module, as well as supplying a function `context` that computes the context for the callee from the caller context and the state at the call site and a value `startcontext` which is the start context with which to analyze `main`. The framework then automatically replaces locals $[u]$ by locals $[u, \prod_A c_A]$ for contexts $\prod_A c_A$ computed by the activated analyses A that are context-sensitive.

Example 12. With the definitions from Fig. 8, a variant of the analysis from Example 4 is obtained that is context-sensitive with 1-callstring sensitivity. □

Digests in their more general form, however, are much more flexible. The heavy lifting cannot be delegated to the framework in this case. Instead, the analysis

```
module C = MyCfg.Fundec

let startcontext = "_main"
let context man _ _ = man.orig_node
```

Fig. 8. OCAML definition for making the analysis from Example 4 in GOBLINT context-sensitive (1-callstring).

developer may prefer to define the local domain $\mathcal{D}_{\mathcal{L}}$ as a reduced cardinal power domain from digests to abstract values. For globals, the same construction can be employed, or alternatively (where the set of digests is small and statically known) each global g may explicitly be split into pairs $\langle g, B \rangle$, B a digest. For specific digest instances, namely *ego-lane digests* [60], the framework offers support for automatically maintaining the digests along the control flow graph via its legacy support for path-sensitive analyses, providing a modular way of leveraging *ego-lane digests.*

Interaction Between Multiple Analyses. To return meaningful results, analyses may require information computed by some other analysis. For instance, an analysis tracking relationships between variables may require points-to information to remain sound in the presence of variable accesses through pointers. An optimal way of transferring information within a product lattice is through *reduction* of the product [17]. This means each component in the product is maximally improved given the values of other components, while still describing the same set of concrete states. A limited form of such a reduction is that, if according to one analysis, the current program point is *unreachable*, i.e., if the corresponding component has the value $\bot$, then the program point is already considered as generally unreachable, implying that all components of the abstract local state are set to $\bot$, thus effectively making all analyses aware of unreachable code. This reduction is naturally supported by GOBLINT.

Since realizing full mutual reduction is expensive and introduces tight coupling between the different analyses, several abstract interpreters instead allow analyses to *query* other analyses for information they have already computed [7,33]—both to provide information needed to give transfer functions and to perform an *ad hoc* reduction between analyses.

GOBLINT implements a limited version of this idea via the functions `ask` and `query` (cf. Figs. 4 and 5). Whenever an analysis requires information from another analysis, it calls `man.ask`, providing a query object that describes the needed information. The framework then dispatches the query to all analyses by calling their respective `query` functions, and collects the results. Here, answers form a lattice, and the answers from all analyses are combined using the *meet* operation of the answer lattice. This allows analyses to co-operate by sharing information with each other through a well-specified interface, while still maintaining an overall modular design.

7 Conclusion and Open Questions

In this tutorial, we outlined how mixed flow-sensitivity provides a common lens through which many analyses can be viewed. We also showed how side-effecting constraint systems offer a framework for expressing scalable, possibly incremental, thread-modular analyses in a declarative manner, enabling a separation of concerns between describing what analyses are supposed to compute and the way the analysis problem is solved starting from that declarative description. Furthermore, we presented GOBLINT—an extensible, modular framework that allows researchers and practitioners to specify their own mixed flow-sensitive analyses in OCAML. We also indicated how GOBLINT can serve researchers working on fixpoint engines to test their algorithms on real-world analysis problems.

Our mixed flow-sensitive framework is a jumping-off point for several interesting directions. Beyond designing additional analyses and developing new solvers, one exciting direction is applying mixed flow-sensitive analysis in novel domains, such as real-time embedded systems or for programming paradigms making extensive use of non-local control flow, as is the case for actor-based languages such as ERLANG or languages using (algebraic) effects such as EFF, KOKA, or (recently) OCAML. We believe there is value in viewing existing analyses through the lens of mixed flow-sensitivity, and hope this may spark new ideas. Where appropriate, GOBLINT may serve as an environment for quickly prototyping and evaluating such analyses.

Acknowledgments. We would like to thank everyone who has contributed to GOBLINT over the years, and in particular Ralf Vogler who maintained GOBLINT for many years and Simmo Saan who is the most active current maintainer. This work was supported by the Estonian Research Council grant PRG2764.

References

1. Afonso, V., et al.: Going native: using a large-scale analysis of android apps to create a practical native-code sandboxing policy. In: 23rd Annual Network and Distributed System Security Symposium, NDSS 2016, The Internet Society (2016). https://doi.org/10.14722/ndss.2016.23384
2. Amato, G., Scozzari, F., Seidl, H., Apinis, K., Vojdani, V.: Efficiently intertwining widening and narrowing. Sci. Comput. Program. **120**, 1–24 (2016). https://doi.org/10.1016/J.SCICO.2015.12.005
3. Andersen, L.O.: Program analysis and specialization for the C programming language, University of Copenhagen (1994). Ph.D. thesis
4. Apinis, K., Seidl, H., Vojdani, V.: Side-effecting constraint systems: a swiss army knife for program analysis. In: Programming Languages and Systems, pp. 157–172. Springer, Berlin Heidelberg (2012). https://doi.org/10.1007/978-3-642-35182-2_12
5. Baudin, P., et al.: The dogged pursuit of bug-free c programs: the Frama-c software analysis platform. Commun. ACM **64**(8), 56–68 (2021). https://doi.org/10.1145/3470569

6. Beyer, D., Strejček, J.: Improvements in software verification and witness validation: SV-COMP 2025. In: Proceeding TACAS (3). LNCS, vol. 15698, pp. 151–186. Springer (2025). https://doi.org/10.1007/978-3-031-90660-2_9
7. Blazy, S., Bühler, D., Yakobowski, B.: Structuring abstract interpreters through state and value abstractions. In: Bouajjani, A., Monniaux, D. (eds.) VMCAI 2017. LNCS, vol. 10145, pp. 112–130. Springer, Cham (2017). https://doi.org/10.1007/978-3-319-52234-0_7
8. Bravenboer, M., Smaragdakis, Y.: Strictly declarative specification of sophisticated points-to analyses. In: Proceeding of the 24th Annual ACM SIGPLAN Conf. on Object-Oriented Programming, Systems, Languages, and Applications, OOPSLA 2009, October 25–29, 2009, Orlando, Florida, USA, pp. 243–262, ACM (2009). https://doi.org/10.1145/1640089.1640108
9. Cai, Y., Zhang, C.: A cocktail approach to practical call graph construction. Proc. ACM Program. Lang. **7**(OOPSLA2), 1001–1033 (2023). https://doi.org/10.1145/3622833
10. Cortesi, A., Costantini, G., Ferrara, P.: A survey on product operators in abstract interpretation. In: Semantics, Abstract Interpretation, and Reasoning about Programs: Essays Dedicated to David A. Schmidt on the Occasion of his Sixtieth Birthday, EPTCS, vol. 129, pp. 325–336. (2013). https://doi.org/10.4204/EPTCS.129.19
11. Cousot, P.: Principles of abstract interpretation, MIT Press (2021)
12. Cousot, P., Cousot, R.: Abstract interpretation: A unified lattice model for static analysis of programs by construction or approximation of fixpoints. In: Conference Record of the Fourth ACM Symposium on Principles of Programming Languages, Los Angeles, California, USA, January 1977, pp. 238–252, ACM (1977). https://doi.org/10.1145/512950.512973
13. Cousot, P., Cousot, R.: Systematic design of program analysis frameworks. In: Sixth Annual ACM Symposium on Principles of Programming Languages, pp. 269–282 (1979). https://doi.org/10.1145/567752.567778
14. Cousot, P., Cousot, R.: Abstract interpretation frameworks. J. Log. Comput. **2**(4), 511–547 (1992). https://doi.org/10.1093/LOGCOM/2.4.511
15. Cousot, P., Cousot, R.: Comparing the Galois connection and widening/narrowing approaches to abstract interpretation. In: Bruynooghe, M., Wirsing, M. (eds.) PLILP 1992. LNCS, vol. 631, pp. 269–295. Springer, Heidelberg (1992). https://doi.org/10.1007/3-540-55844-6_142
16. Cousot, P.: The astrée analyzer. In: Programming Languages and Systems (ESOP 2005). LNCS, vol. 3444, pp. 21–30. Springer (2005). https://doi.org/10.1007/978-3-540-31987-0_3
17. Cousot, P., Cousot, R., Mauborgne, L.: The reduced product of abstract domains and the combination of decision procedures. In: Foundations of Software Science and Computational Structures (FOSSACS 2011), pp. 456–472, LNCS 6604. Springer (2011). https://doi.org/10.1007/978-3-642-19805-2_31
18. Cousot, P., Giacobazzi, R., Ranzato, F.: A^2i: abstract^2 interpretation. Proc. ACM Program. Lang. **3**(POPL), 42:1–42:31 (2019). https://doi.org/10.1145/3290355
19. Dhurjati, D., Das, M., Yang, Y.: Path-sensitive dataflow analysis with iterative refinement. In: Yi, K. (ed.) SAS 2006. LNCS, vol. 4134, pp. 425–442. Springer, Heidelberg (2006). https://doi.org/10.1007/11823230_27
20. Erhard, J., et al.: Interactive abstract interpretation: reanalyzing multithreaded C programs for cheap. Int. J. Softw. Tools Technol. Transf. **26**(6), 647–667 (2024). https://doi.org/10.1007/S10009-024-00768-9

21. Erhard, J., Schinabeck, J.F., Schwarz, M., Seidl, H.: Context gas and friends: taming context-sensitivity on the fly. Int. J. Softw. Tools Technol. Transf. **27**(2), 289–307 (2025). https://doi.org/10.1007/S10009-025-00803-3
22. Erhard, J., Schwarz, M., Vojdani, V., Saan, S., Seidl, H.: When long jumps fall short: control-flow tracking and misuse detection for nonlocal jumps in C. Int. J. Softw. Tools Technol. Transf. **26**(5), 589–605 (2024). https://doi.org/10.1007/S10009-024-00764-Z
23. Ferdinand, C., Heckmann, R.: Verifying timing behavior by abstract interpretation of executable code. In: Borrione, D., Paul, W. (eds.) CHARME 2005. LNCS, vol. 3725, pp. 336–339. Springer, Heidelberg (2005). https://doi.org/10.1007/11560548_34
24. Flanagan, C., Qadeer, S.: Predicate abstraction for software verification. In: Conf. Record of POPL 2002: The 29th SIGPLAN-SIGACT Symposium on Principles of Programming Languages, Portland, OR, USA, 16–18 January 2002, pp. 191–202, ACM (2002). https://doi.org/10.1145/503272.503291
25. Gange, G., Navas, J.A., Schachte, P., Søndergaard, H., Stuckey, P.J.: Abstract interpretation over non-lattice abstract domains. In: Logozzo, F., Fähndrich, M. (eds.) SAS 2013. LNCS, vol. 7935, pp. 6–24. Springer, Heidelberg (2013). https://doi.org/10.1007/978-3-642-38856-9_3
26. Giacobazzi, R., Ranzato, F.: The reduced relative power operation on abstract domains. Theor. Comput. Sci. **216**(1–2), 159–211 (1999). https://doi.org/10.1016/S0304-3975(98)00194-7
27. Gotsman, A., Berdine, J., Cook, B., Sagiv, M.: Thread-modular shape analysis. In: PLDI '07, pp. 266–277. ACM (2007). https://doi.org/10.1145/1250734.1250765
28. Helm, D., Kübler, F., Reif, M., Eichberg, M., Mezini, M.: Modular collaborative program analysis in OPAL. In: ESEC/FSE '20, pp. 184–196. ACM (2020). https://doi.org/10.1145/3368089.3409765
29. Hermenegildo, M.V., et al.: An overview of ciao and its design philosophy. Theory Pract. Log. Program. **12**(1–2), 219–252 (2012). https://doi.org/10.1017/S1471068411000457
30. Hermenegildo, M.V., Puebla, G., Bueno, F., López-García, P.: Integrated program debugging, verification, and optimization using abstract interpretation (and the Ciao system preprocessor). Sci. Comput. Program. **58**(1), 115–140 (2005). https://doi.org/10.1016/j.scico.2005.02.006
31. Horn, D.V., Might, M.: Abstracting abstract machines. In: SIGPLAN International Conference on Functional Programming, pp. 51–62. ACM (2010). https://doi.org/10.1145/1863543.1863553
32. Horn, D.V., Might, M.: Systematic abstraction of abstract machines. J. Funct. Program. **22**(4–5), 705–746 (2012). https://doi.org/10.1017/S0956796812000238
33. Journault, M., Miné, A., Monat, R., Ouadjaout, A.: Combinations of reusable abstract domains for a multilingual static analyzer. In: Proceeding of VSTTE 2019. LNCS, vol. 12031, pp. 1–18. Springer (2019). https://doi.org/10.1007/978-3-030-41600-3_1
34. Keidel, S., Helm, D., Roth, T., Mezini, M.: A modular soundness theory for the blackboard analysis architecture. In: Programming Languages and Systems. LNCS, vol. 14577, pp. 361–390. Springer (2024). https://doi.org/10.1007/978-3-031-57267-8_14
35. Kim, S., Rival, X., Ryu, S.: A theoretical foundation of sensitivity in an abstract interpretation framework. ACM Trans. Program. Lang. Syst. **40**(3), 13:1–13:44 (2018). https://doi.org/10.1145/3230624

36. Koçal, A.R., Schwarz, M., Saan, S., Seidl, H.: Same engine, multiple gears: parallelizing fixpoint iteration at different granularities. In: Proceeding TACAS 2026, Lecture Notes in Computer Science (2026). https://doi.org/10.1007/978-3-032-22749-2_9
37. Charlier, L., Van Hentenryck, B., P.: A Universal top-down fixpoint algorithm, Brown University (1992). CS-92-25 Technical Report
38. Lermusiaux, P., Montagu, B.: Detection of uncaught exceptions in functional programs by abstract interpretation. In: European Symposium on Programming. LNCS, vol. 14577, pp. 391–420. Springer (2024). https://doi.org/10.1007/978-3-031-57267-8_15
39. Lhoták, O., Chung, K.A.: Points-to analysis with efficient strong updates. In: Proceeding of the 38th ACM SIGPLAN-SIGACT Symposium on Principles of Programming Languages, POPL 2011, Austin, TX, USA, 26–28 January 2011, pp. 3–16, ACM (2011). https://doi.org/10.1145/1926385.1926389
40. Mangal, R., Naik, M., Yang, H.: A correspondence between two approaches to interprocedural analysis in the presence of join. In: Shao, Z. (ed.) ESOP 2014. LNCS, vol. 8410, pp. 513–533. Springer, Heidelberg (2014). https://doi.org/10.1007/978-3-642-54833-8_27
41. Martin, F.: Experimental Comparison of Call String and Functional Approaches to Interprocedural Analysis. In: Jähnichen, S. (ed.) CC 1999. LNCS, vol. 1575, pp. 63–75. Springer, Heidelberg (1999). https://doi.org/10.1007/978-3-540-49051-7_5
42. Miné, A.: Static analysis of run-time errors in embedded real-time parallel C programs. Log. Methods Comput. Sci 8(1), (2012). https://doi.org/10.2168/LMCS-8(1:26)2012
43. Miné, A.: Relational thread-modular static value analysis by abstract interpretation. In: McMillan, K.L., Rival, X. (eds.) VMCAI 2014. LNCS, vol. 8318, pp. 39–58. Springer, Heidelberg (2014). https://doi.org/10.1007/978-3-642-54013-4_3
44. Miné, A.: Static analysis of embedded real-time concurrent software with dynamic priorities. Electr. Notes Theor. Comput. Sci. **331**, 3–39 (2017). https://doi.org/10.1016/j.entcs.2017.02.002
45. Miné, A.: Tutorial on static inference of numeric invariants by abstract interpretation. Found. Trends Program. Lang. **4**(3–4), 120–372 (2017). https://doi.org/10.1561/2500000034
46. Montagu, B.: The design and implementation of an abstract interpreter for OCAML programs. In: ML Family 2023-Higher-order, Typed, Inferred, Strict: ACM SIGPLAN ML Family Workshop, pp. 1–4. (2023)
47. Munier, P.: Polyspace. In: Industrial Use of Formal Methods, pp. 123–153. Wiley (2012). https://doi.org/10.1002/9781118561829.ch4
48. Muthukumar, K., Hermenegildo, M.: Determination of variable dependence information at compile-time through abstract interpretation. In: North American Conference on Logic Programming, pp. 166–189. MIT Press (1989)
49. Muthukumar, K., Hermenegildo, M.: Deriving a fixpoint computation algorithm for top-down abstract interpretation of logic programs. MCC , (1990). ACT-DC-153-90 Technical Report
50. Negrini, L., Ferrara, P., Arceri, V., Cortesi, A.: Lisa: a generic framework for multi-language static analysis. In: Challenges of Software Verification, Intelligent Systems Reference Library, vol. 238, pp. 19–42. Springer (2023). https://doi.org/10.1007/978-981-19-9601-6_2
51. Nicolay, J., Stiévenart, Q., De Meuter, W., De Roover, C.: Effect-driven flow analysis. In: Enea, C., Piskac, R. (eds.) VMCAI 2019. LNCS, vol. 11388, pp. 247–274. Springer, Cham (2019). https://doi.org/10.1007/978-3-030-11245-5_12

52. Park, J., Lee, H., Ryu, S.: A survey of parametric static analysis. ACM Comput. Surv. **54**(7), 149:1–149:37 (2022). https://doi.org/10.1145/3464457
53. Rinetzky, N., Ramalingam, G., Sagiv, S., Yahav, E.: On the complexity of partially-flow-sensitive alias analysis. ACM Trans. Program. Lang. Syst. **30**(3), 13:1–13:28 (2008). https://doi.org/10.1145/1353445.1353447
54. Rival, X., Mauborgne, L.: The trace partitioning abstract domain. ACM Trans. Program. Lang. Syst. **29**(5), 26 (2007). https://doi.org/10.1145/1275497.1275501
55. Rival, X., Yi, K.: Introduction to static analysis: an abstract interpretation perspective, MIT Press (2020)
56. Roth, T., Helm, D., Reif, M., Mezini, M.: Cifi: versatile analysis of class and field immutability. In: 36th IEEE/ACM International Conference on Automated Software Engineering, ASE 2021, pp. 979–990. IEEE (2021). https://doi.org/10.1109/ASE51524.2021.9678903
57. Roy, S., Srikant, Y.N.: Partial flow sensitivity. In: High Performance Computing - HIPC 2007, 14th International Conference Goa, India, 18–21 December 2007, Proceedings, pp. 245–256, LNCS 4873. Springer (2007). https://doi.org/10.1007/978-3-540-77220-0_25
58. Saan, S., et al.: Goblint: a portfolio for mixed flow-sensitive abstract interpretation (competition contribution). In: Proceeding TACAS 2026, Lecture Notes in Computer Science (2026). https://doi.org/10.1007/978-3-032-22749-2_26
59. Santos, J.C.S., Dolby, J.: Program analysis using WALA (tutorial). In: Proceeding of ESEC/FSE 2022, pp. 1819–1819, ESEC/FSE '22, ACM (2022). https://doi.org/10.1145/3540250.3569449
60. Schwarz, M.: Thread-modular abstract interpretation: the local perspective, Technical University of Munich (2025). Ph.D. Thesis
61. Schwarz, M., Erhard, J.: The digest framework: concurrency-sensitivity for abstract interpretation. Int. J. Softw. Tools Technol. Transf. **26**(6), 727–746 (2024). https://doi.org/10.1007/S10009-024-00773-Y
62. Schwarz, M., Saan, S., Seidl, H., Apinis, K., Erhard, J., Vojdani, V.: Improving thread-modular abstract interpretation. In: Drăgoi, C., Mukherjee, S., Namjoshi, K. (eds.) SAS 2021. LNCS, vol. 12913, pp. 359–383. Springer, Cham (2021). https://doi.org/10.1007/978-3-030-88806-0_18
63. Schwarz, M., Saan, S., Seidl, H., Erhard, J., Vojdani, V.: Clustered relational thread-modular abstract interpretation with local traces. In: European Symposium on Programming. LNCS, vol. 13990, pp. 28–58. Springer (2023). https://doi.org/10.1007/978-3-031-30044-8_2
64. Seidl, H., Erhard, J., Vogler, R.: Incremental abstract interpretation. In: Di Pierro, A., Malacaria, P., Nagarajan, R. (eds.) From Lambda Calculus to Cybersecurity Through Program Analysis. LNCS, vol. 12065, pp. 132–148. Springer, Cham (2020). https://doi.org/10.1007/978-3-030-41103-9_5
65. Seidl, H., Vene, V., Müller-Olm, M.: Global invariants for analysing multi-threaded applications. In: Proceeding-Estonian Academy Of Sciences Physics Mathematics, vol. 52, pp. 413–436. Estonian Academy Publishers (2003)
66. Seidl, H., Vogler, R.: Three improvements to the top-down solver. Math. Struct. Comput. Sci. **31**(9), 1090–1134 (2021). https://doi.org/10.1017/S0960129521000499
67. Seidl, H., Wilhelm, R., Hack, S.: Compiler design - analysis and transformation. Springer (2012). https://doi.org/10.1007/978-3-642-17548-0
68. Sharir, M., Pnueli, A.: Two Approaches to Interprocedural Data Flow Analysis. New York University, Courant Institute of Mathematical Sciences (1978)

69. Shivers, O.G.: Control-flow analysis of higher-order languages or taming lambda, Carnegie Mellon University (1991). Ph.D. Thesis
70. Simonnet, J., Lemerre, M., Sighireanu, M.: A dependent nominal physical type system for static analysis of memory in low level code. Proc. ACM Program. Lang. **8**(OOPSLA2), 30–59 (2024). https://doi.org/10.1145/3689712
71. Stemmler, F., Schwarz, M., Erhard, J., Tilscher, S., Seidl, H.: Taking out the toxic trash: recovering precision in mixed flow-sensitive static analyses. Proc. ACM Program. Lang. **9**(PLDI) (2025). https://doi.org/10.1145/3729297
72. Stiévenart, Q., Nicolay, J., de Meuter, W., de Roover, C.: A general method for rendering static analyses for diverse concurrency models modular. J. Syst. Softw. **147**, 17–45 (2019). https://doi.org/10.1016/J.JSS.2018.10.001
73. Sui, Y., Xue, J.: On-demand strong update analysis via value-flow refinement. In: Proceeding of the 2016 24th ACM SIGSOFT International Symposium on Foundations of Software Engineering, FSE' 2016, pp. 460–473. ACM (2016). https://doi.org/10.1145/2950290.2950296
74. Sui, Y., Xue, J.: Value-flow-based demand-driven pointer analysis for C and C++. IEEE Trans. Software Eng. **46**(8), 812–835 (2020). https://doi.org/10.1109/tse.2018.2869336
75. Suzanne, T., Miné, A.: Relational thread-modular abstract interpretation under relaxed memory models. In: Ryu, S. (ed.) APLAS 2018. LNCS, vol. 11275, pp. 109–128. Springer, Cham (2018). https://doi.org/10.1007/978-3-030-02768-1_6
76. Taft, S.T.: Sound and precise static analysis using a generalization of static single assignment and value numbering. Int. J. Softw. Tools Technol. Transfer **26**, 621–632 (2024). https://doi.org/10.1007/s10009-024-00762-1
77. Tilscher, S., Graß, A., Seidl, H.: Verifying a solver for mixed flow-sensitive analyses. In: NASA Formal Methods Symposium, Lecture Notes in Computer Science (2026)
78. Tilscher, S., Stade, Y., Schwarz, M., Vogler, R., Seidl, H.: The top-down solver - an exercise in a^2i. In: Challenges of Software Verification, Intelligent Systems Reference Library, vol. 238, pp. 157–179. Springer (2023). https://doi.org/10.1007/978-981-19-9601-6_9
79. Vojdani, V., Apinis, K., Rõtov, V., Seidl, H., Vene, V., Vogler, R.: Static race detection for device drivers: the goblint approach. In: 31st IEEE/ACM International Conference on Automated Software Engineering, pp. 391–402. ACM (2016). https://doi.org/10.1145/2970276.2970337

Formally Explaining Neural Network Classification

Tomáš Kolárik[1(✉)], Grigory Fedyukovich[1,3], Faezeh Labbaf[1], Fabrizio Leopardi[1], Natasha Sharygina[1], and Michael Wand[2]

[1] University of Lugano (USI), Lugano, Switzerland
tomas.kolarik@usi.ch
[2] SUPSI, IDSIA, Lugano, Switzerland
[3] Florida State University, Tallahassee, USA

Abstract. Neural networks (NNs) are the core of AI-based technologies. However, the degree of reliability in performing the task is an open problem. The explainability of a central task of NNs, classification, is of immense importance. While at the rise of AI-based reasoning, explainability of the NN classification has mostly been done using statistical methods, nowadays, a more reliable trend of formal logic-based methods is gaining popularity. The advantage of the formal approach is that it gives strict and provable guarantees of the classification. Formal methods is a mature field that has delivered a number of efficient computational solutions already applied in the analysis of software and hardware systems. Formal explainability methods naturally have the ability to reuse existing techniques and tools for a newly emerging field of formal explainability of NN classification.

This paper surveys existing efforts to compute explanations of neural network classification based on logical abductive reasoning. The abduction approach is crucial for generalizing the results, capturing the underlying behavior of the classifier. We present the existing techniques as instances of a general formalization that allows contrasting them against each other. In addition, we discuss the issue of the quality of explanations, focusing on their key metrics and factors. As an illustrative example, the paper also presents a practical framework, SpEXplAIn, which automatically computes Space Explanations, the most general abduction-based explanations for classifying NNs with provable guarantees of the behavior of the network in continuous areas of the input feature space. The tool leverages an SMT solver compatible with a range of flexible Craig interpolation algorithms and unsatisfiable core generation, and is applicable to a wide range of applications.

1 Introduction

Classifying neural networks (NNs) are of immense interest and importance in the research and industry. Nonetheless, despite the enormous complexity, these machine learning models resemble black boxes and the reasons behind the classification remain unclear. Explainable artificial intelligence (XAI) reasons through

A. Sampaio and M. Stoelinga (Eds.): FM 2026, LNCS 16557, pp. 471–496, 2026.
https://doi.org/10.1007/978-3-032-26220-2_23

the classification process and provides human-understandable interpretations: Given a NN and a sample point, that is, an assignment of all features of the NN to concrete values, the problem is to explain the classification outcome computed by the NN. Many approaches to XAI have been developed over the years:

- The analysis at the *unit (neuron) level* [23,76], which, however, fails to capture global properties or dependencies between units.
- *Gradient-based* methods [4,68], which locally describe the network behavior around a particular sample.
- *Inversion* methods [49], which provide global, but only approximate, views.
- *Model-agnostic* methods [5,48], including SHAP [43], LIME [61] and Anchors [62], which, however, often yield logically inconsistent explanations [51].

While useful and widely used these methods *lack rigor*. In this paper, we focus on alternative rigorous XAI approaches that are based on first-order logic. Relying on constraint solving and Satisfiability Modulo Theories [56,57] (SMT), they deliver explanations accompanied by *provable guarantees* regarding their correctness. These XAI tools leverage computational engines like Marabou [41] or SMT solvers [6,10,12,13,16,21]. The explanations are obtained through *abductive reasoning*, which originates from program verification [1,19,58,59]. The key insight is that *changing* the classification outcome for the input sample point changes the *satisfiability* of the logical formula that encodes the XAI-problem. Thus, if a relaxation of the input constraints in the formula does not affect the unsatisfiability of the formula, then this relaxation gives a *logically guaranteed generalization* of the explanation. Early methods in this domain perform rather naïve relaxation of input constraints based on exhaustive enumeration of features and without taking the proof of unsatisfiability into account [32,50,51,65,67,73]. They yield explanations where the behavior outside the space of the underlying dataset or samples generated from the model remains unknown. In contrast, interval-based explanations [35] are represented by hyperrectangular spaces, but they still cannot reflect feature relationships to precisely approximate decision boundaries. Thus, all existing approaches to logic-based XAI are more restrictive than what the original notion of abductive reasoning could in theory offer.

A breakthrough in logic-based XAI has been recently enabled by the concept of *Space Explanations* [44] that exploits the proofs of unsatisfiability of logical formulas as opposed to the structure of these formulas. It yields explanations where subsets of the feature space associated with the formulas are correct by construction and, in practice, much more general than the ones produced by the earlier methods. There is, however, no unifying framework to formulate the outcome expectations of the logic-based approaches, given the diversity of logic-reasoning tools that are used at the background. Consequently, even the notion of abduction is often misused. In our overview, we bridge the gaps in existing formalizations of logic-based methods to obtain explanations and bring them under the same umbrella. This paper uses a unified formalization for logical abduction-based explanations of NNs, allows a comprehensive comparison of the different approaches, provides a unified encoding for the corresponding explainability problems, gives an overview of existing algorithms solving the addressed problems, and provides an extensive discussion on the quality of abduction-based explanations.

Moreover, the paper demonstrates the first tool to automatically generate space explanations called SpEXplAIn[1]. At the heart of SpEXplAIn, *Craig interpolation* [15] extracts a reason why the given input cannot change the classification from the logical proof. The tool is flexible and accepts optional configurations from the user, allowing them to tune various properties of the explanations, such as the logical strength or the size of the formula, and it can generalize existing space explanations. Notably, the tool can provide human-recognizable interpretations by plugging in different visualization methods. SpEXplAIn is applicable to a wide range of applications which the paper illustrates on several real-life case studies and showcases that the explanations can be visualized and provide meaningful feedback to the user.

2 Background

A classification task involves mapping input data into a predefined set of classes. In this context, we are given a set of *features* $\mathcal{F} = \{1, \ldots, m\}$ and a finite set of classes $\mathcal{K} = \{c_1, c_2, \ldots, c_K\}$. Each feature $i \in \mathcal{F}$ takes values from a discrete or continuous *domain* $\mathcal{D}_i$. The set of domains is represented by $\mathcal{D} = \{\mathcal{D}_1, \ldots, \mathcal{D}_m\}$. The *feature space* is defined as $\mathbb{F} = \mathcal{D}_1 \times \mathcal{D}_2 \times \cdots \times \mathcal{D}_m$.

The notation $\mathbf{x} = (x_1, \ldots, x_m)$ is used for an arbitrary point in feature space, where each x_i is a *feature variable* taking values from $\mathcal{D}_i$. Where clear from context, we will abbreviate feature variables as features. The notation $\mathbf{s} = (s_1, ..., s_m)$ represents a specific *sample point* in the feature space (i.e., $\mathbf{s} \in \mathbb{F}$), where each s_i is a constant representing one concrete value from $\mathcal{D}_i$.

A *classifier* $\mathcal{M} = (\mathcal{F}, \mathcal{D}, \mathbb{F}, \mathcal{K}, \kappa)$ is characterized by a *classification function* $\kappa : \mathbb{F} \to \mathcal{K}$ that maps the feature space $\mathbb{F}$ into the set of classes $\mathcal{K}$. An *instance* is denoted as a pair $(\mathbf{s}, c)$, where $\mathbf{s} \in \mathbb{F}$ and $c \in \mathcal{K}$ is a prediction $c = \kappa(\mathbf{s})$. The classification function κ is usually estimated from a set of sample points for which the correct class c is known (the *training data*).

(Classifying) Neural Networks are a specific class of general classifiers. A (feedforward) neural network is represented by a graph $G = (E, V)$ with the set of edges E and the set of vertices V, representing the neurons and their weighted connections, respectively. The neurons are partitioned into layers $V_0, \ldots, V_L$ where each V_l contains neurons $v_i^{(l)}$, $i \in \{1, \ldots, n^{(l)}\}$. Layer V_0 is the *input* layer, V_L is the *output* layer, and layers $V_1, \ldots, V_{L-1}$ are *hidden* layers. The edges correspond to the pairs connecting every neuron of layer V_l to every neuron in the next layer V_{l+1}, for all $0 \leq l < L$, i.e., edge $e_{i,j}^{(l)}$ connects i-th neuron in layer l with j-th neuron in layer $l+1$. Every edge is assigned a numerical weight $w_{i,j}^{(l)}$. Moreover, every neuron $v_i^{(l)}$ in a hidden layer is assigned a

[1] https://github.com/usi-verification-and-security/spexplain.

bias $b_i^{(l)}$. We assume that each input neuron $v_i^{(0)}$ corresponds to feature $i \in \mathcal{F}$, i.e., is represented by feature variable $x_i \in \mathcal{D}_i$, and that each output neuron $v_i^{(L)}$ corresponds to class c_i. We assume that each domain $\mathcal{D}_i$ is either a finite set or a closed interval $\mathcal{D}_i = [L_i, U_i]$ where both the lower and upper bounds are finite. This is always true as long as the (training) data set is finite. The restriction to finite intervals is required since a neural network is always trained on a finite set of training data, thus it will learn to expect inputs in a finite range; passing an input (far) outside this range will inevitably cause unpredictable behavior.

Given an input sample $\mathbf{s} = (s_1, \ldots, s_m)$, the activations $x_i^{(l)}$ for each neuron $v_i^{(l)}$ are computed as follows. In the case of the input layer, $x_i^{(0)} = s_i$; for hidden layers, $y_i^{(l)} = \sum_j x_j^{(l-1)} w_{j,i}^{(l-1)} + b_i^{(l)}$ and $x_i^{(l)} = ReLU(y_i^{(l)})$, where $ReLU(x) = \max\{x, 0\}$ is the rectifier function. In the case of the output layer, no activation function is applied[2] . The class c_i assigned to input sample $\mathbf{s}$ is given by the output neuron with maximum activation, formally, $\kappa(\mathbf{s}) = c_i \Leftrightarrow i = \arg\max_j \{y_j^{(L)}\}$.

Example 1. A simple neural network maps features $\mathbf{x} = (x_1, x_2, x_3) \in [0, 4]^3$ to classes $\{c_1, c_2\}$. It has 3 input neurons, 2 hidden neurons with *ReLU* activation, and 2 output neurons. Using fixed weights and zero biases, they are defined by

$$x_1^{(1)} = ReLU(2x_1 + x_3) \wedge o_{c_1} = y_1^{(2)} = x_1^{(1)} - 4x_2^{(1)}$$
$$x_2^{(1)} = ReLU(-x_1 + x_2 - x_3) \wedge o_{c_2} = y_2^{(2)} = -x_1^{(1)} + 4x_2^{(1)}.$$

The network predicts c_1 if $o_{c_1} \geq o_{c_2}$, otherwise c_2. For instance, with sample $\mathbf{s} = (1, 1, 3)$, we get $o_{c_1} = 5$, $o_{c_2} = -5$, so $\mathbf{s}$ is classified as c_1.

3 Abduction-Based Explanations

The explanations of NNs studied in this paper are obtained through abductive reasoning which, given an observation O, infers an explanation φ that entails O. Formally, for vectors of variables $\mathbf{x}$ and $\mathbf{y}$, and two formulas $T(\mathbf{x}, \mathbf{y})$ and $O(\mathbf{x}, \mathbf{y})$, the *abduction problem* is to find a formula $\varphi(\mathbf{x})$, such that: (1) $\varphi(\mathbf{x}) \wedge T(\mathbf{x}, \mathbf{y}) \nRightarrow \bot$, and (2) $\varphi(\mathbf{x}) \wedge T(\mathbf{x}, \mathbf{y}) \Rightarrow O(\mathbf{x}, \mathbf{y})$. For example, if $T \equiv x > 0$ and $O \equiv x+y > 0$, then one possible solution for the abduction problem is $\varphi_1 \equiv y \geq 0$ (the weakest), and another one is $\varphi_2 \equiv y = 10$ (a stronger). In our context, φ is an explanation, vectors $\mathbf{x}$ and $\mathbf{y}$ are the input and output variables of the network, respectively. T encodes the network and the constraints over the input variables, and O contains the conditions over the output, e.g., the classification. *Weaker* formulas give more information about the classifier. However, since the explanations serve for humans to better understand the behavior of the NN, it is essential that the explanations are also meaningful and interpretable—refer to Sect. 6 for more details. Explanation methods are distinguished mainly in how they solve the abduction problem and the type of solutions they provide.

---

[2] During the training, usually the softmax function is used in the last layer; since this is a monotonic function, it can be ignored here.

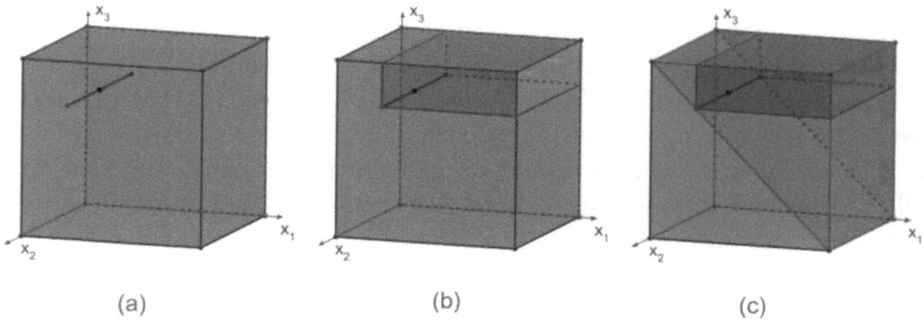

Fig. 1. Impact spaces associated to different explanations in domain $[0,4]^3$

Abduction-based explanations have been studied by several works in different forms [32,34,35,44][3]. To unify all different forms of such explanations, this section starts with a general formalization of explanations introduced in [44] and called *space explanations*, which directly stems from the abduction problem. Then, the section discusses special cases, some of which are widely used in the literature.

Definition 1 (Space Explanation). *Given a classifier $\mathcal{M}$ computing a classification function κ from feature space $\mathbb{F}$, and a class c, a* space explanation *of c is a formula φ s.t. $\forall \mathbf{x} \in \mathbb{F} \,.\, \varphi(\mathbf{x}) \implies \kappa(\mathbf{x}) = c$ holds.*

This formalism can be viewed as a set of *rules*, where a space explanation represents a sufficient condition for the classification: if the condition holds for the given features, then it must be classified to class c. The set representation of a space explanation φ is its *impact space* $\mathbb{F}_\varphi := \{\mathbf{x} \in \mathbb{F} \mid \varphi(\mathbf{x})\}$, and $\mathbb{F}_\varphi \subseteq \{\mathbf{x} \in \mathbb{F} \mid \kappa(\mathbf{x}) = c\}$.

Given the network in Example 1, a space explanation for class c_1 is e.g. $\varphi \equiv x_1 + x_3 \geq 4$. Figure 1 (c) shows the impact space $\mathbb{F}_\varphi$ as a pink prism.

The above formalization is general enough to cover any other abduction-based explanation definition by possibly restricting formula φ. In the following, we introduce them as special cases of space explanations.

Explanations as a Subset of Features. The concept of abductive explanations was first introduced by [32] as a subset of conditions over the input feature variables that guarantee the output conditions, i.e., the classification. This abductive explanation, that is restricted compared to the original notion of abduction, can be formalized as a special case of space explanation as follows:

Definition 2. *An* abductive explanation *$\varphi_\mathcal{X}$ is a space explanation for a classifier $\mathcal{M}$ and a class c such that $\varphi_\mathcal{X} := \bigwedge_{i \in \mathcal{X}} x_i = s_i$ where $\mathcal{X} \subseteq \mathcal{F}$ is a subset of features. $\varphi_\mathcal{X}$ is* subset-minimal *iff $\forall \mathcal{X}' \subsetneq \mathcal{X} \,.\, \varphi_{\mathcal{X}'}$ does not constitute a space explanation anymore.*

[3] Refer to Sect. 5 for other formalisms beyond abductive reasoning.

That is, any input sample that coincides with **s** on features from $\mathcal{X}$ but can have arbitrary values for features from $\mathcal{F} \setminus \mathcal{X}$ is also classified to class c. In the case $\varphi_\mathcal{X}$ is subset-minimal, subset $\mathcal{X}$ is required to be irreducible.

For example, given the NN in Example 1 and instance $((1,1,3), c_1)$, a subset-minimal abductive explanation is $\varphi_\mathcal{X}$ with $\mathcal{X} = \{x_1, x_3\}$. Thus, if $x_1 = 1$ and $x_3 = 3$, then for any value assigned to x_2 from the domain, the classification is still c_1. However, if any feature from $\mathcal{X}$ is removed, then neither $x_1 = 1$ nor $x_3 = 3$ alone guarantees the classification anymore. The corresponding impact space is a line shown in Fig. 1 (a).

Explanations as a Set of Intervals. Another special case of space explanations specifies an interval for selected input features (which is a subset of the corresponding domain interval), offering higher granularity over Definition 2.

Definition 3 (Interval Explanation). *An* interval explanation φ_I *is a space explanation for a classifier* $\mathcal{M}$ *and a class* c *s.t.* $\varphi_I := \bigwedge_{i\in\mathcal{X}} l_i \le x_i \le u_i$ *where* $\mathcal{X} \subseteq \mathcal{F}$ *and* $I = \{[l_i, u_i] \mid i \in \mathcal{X},\ l_i, u_i \in \mathcal{D}_i\}$.

Thus, $[l_i, u_i] \subseteq \mathcal{D}_i$ for all $i \in \mathcal{X}$. The definition does not allow φ_I to represent multiple discontinuous intervals (which could together still be a subset of $\mathcal{D}_i$). The associated impact space is a hyperrectangle expressed as the Cartesian product of intervals over each feature i: $\mathbb{F}_{\varphi_I} = \prod_{i\in\mathcal{F}} [l_i, u_i]$, $[l_i, u_i] = \mathcal{D}_i$, $i \in \mathcal{F} \setminus \mathcal{X}$.

For example, given the NN in Example 1, an interval explanation for class c_1 with $I = \{[1,4], [3,4]\}$ and $\mathcal{X} = \{1,3\}$ is $\varphi \equiv (1 \le x_1 \le 4) \wedge (3 \le x_3 \le 4)$. The impact space of the explanation is shown in Fig. 1 (b).

A notion of maximal interval explanations depends on multicriterial optimization, resulting in ambiguity of what should be maximized—considering both the size of $\mathcal{X}$ and of particular intervals. Also, given a subset $\mathcal{X}$, many combinations of differently sized intervals would exist because it would highly depend on the order and the extent of the particular expansions. A possibility would be to require the resulting volume to be maximal.

There is a similar notion of interval explanations called *inflated explanations* [35]. However, the definition differs in that it uses a general concept of *maximal sets*, meaning that it also admits discrete domains or subsets of continuous space (which are not necessarily continuous, i.e., connected)[4], and that they cannot be expanded further. Moreover, they explicitly restrict the explanations to origin from a subset-minimal abductive explanation. Still, in all cases, the resulting explanations fall into space explanations.

Distance-Restricted Explanations. Literature often considers explanations that focus only on a small vicinity around the sample point rather than the entire feature space [31,34,73]. They are defined as follows.

[4] Despite the general definition, [35] incorrectly considers only intervals in the continuous domain.

Definition 4. *A* distance-restricted abductive explanation $\varphi_{\mathcal{X},\epsilon}$ *is a space explanation for a classifier $\mathcal{M}$ and an instance* $(\mathbf{s}, c)$ *s.t.* $\varphi_{\mathcal{X},\epsilon} := \|\mathbf{x} - \mathbf{s}\|_p \leq \epsilon \wedge \bigwedge_{i \in \mathcal{X}} x_i = s_i$ *where ϵ is a distance value and $\|.\|_p$ is a norm.*

Distance-restricted abductive explanations do not fall into abductive explanations: effectively, such an explanation forms a space in the feature space and no longer depends solely on the features. Nonetheless, if the uniform norm is used ($p = \infty$), they can be expressed as interval explanations, which can naturally restrict the neighborhood around the sample point, resulting in a hyperrectangular space. However, this is not possible with other norms, such as the Euclidean norm ($p = 2$) where the resulting space corresponds to a hypersphere. In all cases, distance-restricted abductive explanations still fall into space explanations.

Summary. Abductive, interval, and inflated explanations are special cases of space explanations, where the formula φ is restricted. This constrains the shape of the corresponding impact space. While a general space explanation can have an arbitrarily shaped impact space, the impact space of both abductive and interval explanations takes the form of an affine rectangle. Abductive explanations also always keep some of the features fixed to a constant. Figure 1 compares impact spaces for the three types of explanations discussed above.

4 Technical Foundations

The computation of abduction-based explanations for NNs follows a structured two-phase methodology. First, the target ML model is encoded in first-order logic, e.g., real arithmetic (LRA) in Satisfiability Modulo Theories (SMT). Second, abductive reasoning is applied to identify a formula that entails the model's prediction for a given data instance. Coming up with such a formula involves one or more entailment queries where tools like SMT solvers or specialized verifiers act as oracles to verify the entailment. Different methods vary in how they approach the abduction problem and in the types of tools used to compute explanations. We start with a possible encoding for various approaches that yield different types of abduction-based explanations. Then, we survey the approach that leverages Craig interpolation to efficiently produce highly general explanations, followed by approaches that generate more structured and restricted forms of explanations using NN verification tools.

4.1 Encoding

While all theoretical foundations hold in both cases, we instantiate all the techniques in the continuous domain and omit the discrete domains. All approaches to NN explainability use the following encoding schema. Section 2 gives the definition of all layers of a NN, where all neurons are variables and the weights and biases are constants. The encoding of this NN is $\psi_{\mathcal{M}}$. The encoding of $ReLU(x)$ depends on the concrete approach: in the most precise scenario, it is an ITE-term; some approaches, typically NN verifiers (e.g. [70]), apply abstractions [69], but at the cost of additional verification calls to check that the correctness has not been affected. The domains $\mathcal{D}$ of the features are encoded as $\psi_{\mathcal{D}}$, which is a conjunction of intervals for possible values of every feature.

Let the formula $\neg\psi_c \equiv \bigvee_{i\in\mathcal{K}} y_i^{(K)} > y_{j_c}^{(K)}$ represent a constraint that the outcome of the classification is *not* class $c \in \mathcal{K}$, then with no additional restrictions, formula $\psi \equiv \psi_{\mathcal{M}} \wedge \psi_{\mathcal{D}} \wedge \neg\psi_c$ is *satisfiable.* Now suppose a sample point $\mathbf{s} \in \mathbb{F}$ classified as class c. Using the encoding $\varphi_{\mathbf{s}} \equiv \mathbf{x} = \mathbf{s}$, formula $\varphi_{\mathbf{s}} \wedge \psi$ is in turn *unsatisfiable.* For explainability, this fact lets to weaken formula $\varphi_{\mathbf{s}}$ while still preserving the unsatisfiability.

4.2 Craig Interpolation for Explanations

Given an unsatisfiable formula $A \wedge B$, a *Craig interpolant* is a formula I such that $A \Rightarrow I$, $I \wedge B$ is unsatisfiable, and I uses only the common variables of A and B. We denote interpolant I computed by an interpolation procedure *Itp* from formulas A and B by $I = \mathit{Itp}(A, B)$. Hence, given encoding ψ and a space explanation φ of c, interpolant $I = \mathit{Itp}(\varphi, \psi)$ is also a space explanation of c satisfying $\varphi \implies I$ (and hence $\mathbb{F}_\varphi \subseteq \mathbb{F}_I$). This offers a universal means of generalization of existing space explanations (e.g. trivially a sample $\varphi_{\mathbf{s}}$), still guaranteeing the classification.

Several interpolation algorithms, e.g. [2,8] for quantifier-free LRA, typically yield explanations in the form of bounds on linear combinations of features such as $\bigwedge_j \sum_{i\in\mathcal{F}} \alpha_{j,i} x_i \leq \beta_j$ ($\alpha_{j,i}, \beta_j \in \mathbb{Q}$). However, these algorithms are "blind" in the context of neural network classification and they do not allow more granular control of the form of the interpolants, such as specific shapes. While post-processing and simplifying the formula is often beneficial, several methods (cf. Section 6) exist to interpret even complex formulas in a user-friendly manner.

Furthermore, interpolation algorithms can benefit from special cases [26] where given $\mathit{Itp}(A, B)$, formula A can be partitioned, leaving a part of A unchanged. For example, if a space explanation can be split into two parts $\varphi_A \wedge \varphi_B$, and given a Craig interpolation procedure *Itp* with $I = \mathit{Itp}(\varphi_A, \varphi_B \wedge \psi)$, then $I \wedge \varphi_B$ is a logically weaker space explanation. Hence, it is possible to, for example, focus the interpolation only on selected features that appear only in φ_A, excluding the rest appearing only in φ_B from the expansion.

Different Craig interpolation procedures offer various properties of the interpolants, in particular, their *strength* [20,63]. Moreover, the *dual* approach Itp' to an existing procedure *Itp* s.t. $\mathit{Itp}'(A, B) = \neg\mathit{Itp}(B, A)$ can be used to produce (usually) weaker explanations. The algorithms usually range over either the propositional part or a specific theory. An example of interpolation rules over the Boolean abstraction of the formula is *McMillan* algorithm [53] (Itp_M) that typically yields conjunctions of constraints, while its weaker dual interpolation scheme (Itp'_M) typically yields disjunctions ($\mathit{Itp}_M(A, B) \implies \mathit{Itp}'_M(A, B)$).

In the case of quantifier-free LRA, given a system of unsatisfiable linear inequalities, a widely used technique is based on Farkas' lemma, producing *Farkas interpolants* (Itp_F), or their logically stronger *decomposed* variants [8]

(Itp_{DF}), where Itp_F yields a single inequality, while Itp_{DF} a conjunction of inequalities. The duals of these algorithms Itp'_F and Itp'_{DF} yield weaker interpolants and Itp'_{DF} yields a disjunction of inequalities. Since these algorithms are orthogonal to McMillan algorithms, we can even produce conjunctions of disjunctions or vice versa. Furthermore, there is an infinite family of interpolants [2] (Itp_f) with the logical strength in between Itp_F and Itp'_F interpolants, parametrized by a rational factor $f \in [0, 1]$.

Hence, a number of flexible interpolation algorithms are available, offering different strengths and structures of the resulting formula. The presented arithmetic algorithms exhibit the following relationship [2,8]:

$$Itp_{DF} \implies Itp_F \implies Itp_f \implies Itp'_F \implies Itp'_{DF}$$

such that each is applied to the same arguments (A, B).

Unsatisfiable Cores. Given formulas A and B such that $A = \bigwedge_{i \in \mathcal{I}} a_i$ and $A \wedge B$ is unsatisfiable, an *unsatisfiable core* of A is a formula $U = \bigwedge_{i \in \mathcal{J}} a_i$, $\mathcal{J} \subseteq \mathcal{I}$ such that $U \wedge B$ is still unsatisfiable. We denote unsatisfiable core U computed by an algorithm *Ucore* from formulas A and B by $U = Ucore(A, B)$. If $Ucore(A, B)$ produces a formula over the common variables to A and B, it is a special case of Craig interpolation.

Most modern SMT solvers [6,13,16,30] implement computation of unsatisfiable cores. However, different techniques offer different trade-offs between the computational effort and size of the core. Some techniques [14] add very little overhead on top of proving unsatisfiability but offer no guarantees about the size of the core. Other techniques compute *irreducible* unsatisfiable cores [27,39], but are typically much more computationally intensive.

Generalizing Existing Explanations. The concept of Craig interpolation allows the *iterative construction* of space explanations using a sample point or existing abductive, interval, or inflated explanation as a starting point. [44] lists several strategies and suggests some combinations in order to control the logical strength of the resulting explanations, as well as their form, including possible simplifications using the unsatisfiable core extraction.

4.3 Verification for Explanations

Given a specification of a neural network f of the form $\{precondition\}$ $\mathbf{r} = f(\mathbf{x})$ $\{postcondition\}$, where *precondition* is a predicate over variables $\mathbf{x}$, and *postcondition* is a predicate over variables $\mathbf{x}$ and $\mathbf{r}$, the verification problem consists of checking if for all the values of variables for which *precondition* holds, *postcondition* also holds. Many state-of-the-art NN explanation methods use verification tools to compute conditions over input features that guarantee the output [7,32,34,73]. The explanation approach is reduced to a sequence of verification problems with different preconditions until a suitable (and in practice, weaker) one is found. In a nutshell, these techniques perform a naive relaxation of input constraints based on exhaustive enumeration of features, heavily depending on selecting a specific order of the features. Each verification problem is defined by *precondition* as a conjunction of an input φ (starting with the sample point $\varphi_{\mathbf{s}}$) and the domain constraints $\psi_{\mathcal{D}}$, and the *postcondition* as the constraint that the classification is not class c, $\neg\psi_c$ (recall Sect. 4.1).

NN verification and abstraction literature is thoroughly reviewed in [40]. It classifies the existing approaches into either "Reduction" class [3], "Branch-and-Bound" class [74], or classes that apply abstractions to the input domains [46,69] or the NN architecture [22]. Reduction approaches solve the verification problem by reducing it to an instance of another problem, such as SMT (most closely related to abduction, to be discussed in the next paragraph), MILP, Convex Optimization, and BDD. Branch-and-bound techniques work by splitting the input or activation space into independent branches. Since each branch is verified separately, this often increases the likelihood of finding a proof or counterexample. One part of abstraction-based approaches use "abstract interpretation" (e.g. [69]) to over-approximate the input domains (e.g., to intervals, zonotopes, or polytopes) and propagate the properties through the layers of the NN. Other approaches instead perform NN's structural abstraction (e.g. [22]) by reducing the size of the NN while maintaining guarantees of the verification, either by exploiting syntactic or semantic information regarding the network's topology.

Among existing tools for NN verification, $\alpha, \beta-$CROWN [70,74], VeriNet [28], Marabou [41,71], MN-BaB [24] are most used. A survey paper [42] gives details about other tools. Many of these approaches use an SMT solver under the hood. For instance, Marabou uses the CVC5 [6] SMT solver to check that every assignment of variables for which the *precondition* holds can only be extended to an assignment for which *postconditon* holds. In order to solve the verification problem, it loops by trying to satisfy the currently unsatisfied constraints at each iteration. The main verification loop exploits the well-known simplex algorithm.

The explanation algorithms discussed in this section leverage a *robustness oracle* to answer the robustness query: whether a constrained adversarial example exists when certain features are fixed to their original values [7,32,34,73]. Formally speaking, given a classifier $\mathcal{M}$ computing a classification function κ from feature space $\mathbb{F}$, and constraint $\varphi(\mathbf{x})$, a *constrained adversarial example* for instance $(\mathbf{s}, c)$ exists if $\exists\, \mathbf{x} \in \mathbb{F} \,.\, \kappa(\mathbf{x}) \neq c \wedge \varphi(\mathbf{x})$. If an adversarial example does not exist, then the classifier is *constrained robust.* Robustness is a special case of the verification problem.

Finding a constrained adversarial example in the vicinity of a sample point is equivalent to solving the negation of the problem in Definition 1, searching for a counterexample $\mathbf{x}$ with $\varphi = \varphi_{\mathcal{X},\epsilon}$ from Definition 4.

Compared to Craig-interpolation-based methods, the methods discussed in this subsection *restrict* the *shape* of the impact space associated with the explanations, disallowing the approximation of arbitrarily complex decision boundaries. Consequently, the former methods can *capture relationships* among features, offering meaningful insights into the *reasons* behind a classification. On the other hand, interpolation methods are in a sense "blind" (i.e., unaware of their application context), and their output still benefits from further post-processing.

Computing Explanation as a Subset of Features. Methods that aim at subsets of features typically use brute force or binary search, with optimizations.

Brute-Force Methods. A straightforward approach to solve the problem is by assuming that all features are initially fixed to their original values. That is, if $\mathcal{X}$ initially contains all features $\mathcal{F}$, then no adversarial example exists. A brute-force algorithm then loops through each feature i in the set of all features $\mathcal{F}$. In each iteration, it tentatively removes the current feature i from the set of fixed features $\mathcal{X}$. It then makes a call to a robustness oracle (like Marabou) to check if a constrained adversarial example exists when only the features remaining in the set $\mathcal{X}$ are fixed. If the oracle returns true (an adversarial example exists), it means feature i was necessary to maintain the original prediction. The algorithm then adds feature i back to the set $\mathcal{X}$. If the oracle returns false (no adversarial example exists when i is unfixed), it means feature i was not necessary to prevent the adversarial example, and it can be safely left unfixed. After iterating through all features, the remaining set $\mathcal{X}$ is a subset-minimal abductive explanation. The algorithm needs $\Theta(|\mathcal{F}|)$ calls to the robustness oracle. To potentially find smaller subset-minimal explanations, heuristic models of feature importance can be employed, checking less important features first. Variants of this method are studied in recent works, for example, VeriX [73] or [29].

Binary Search. Methods from another group do not check each feature sequentially, but instead use binary search on the set of features being considered ($\mathcal{W}$). It maintains a set $\mathcal{X}$ of features already identified as part of the abductive explanation and a set $\mathcal{W}$ of features still under consideration. In an inner loop, it performs a binary search over the features in $\mathcal{W}$ (which are assumed to be ordered). In each step of the binary search, it checks a *chunk* of features. Specifically, it calls the oracle to see if an adversarial example exists when the features in $\mathcal{X}$ plus a subset of features from the beginning of $\mathcal{W}$ (e.g., $\mathcal{W}_{1..t}$) are fixed. Based on whether the oracle returns true (adversarial example exists) or false (no adversarial example), the binary search narrows down the range within $\mathcal{W}$ that contains the relevant feature. Once the inner binary search loop finishes, a relevant feature is found and moved from $\mathcal{W}$ to $\mathcal{X}$. This process repeats in an outer loop until all features initially in $\mathcal{W}$ have been examined. This method is studied in [7,34,72].

This binary search structure reduces the number of oracle calls within the inner loop to $O(\log |\mathcal{W}|)$, where $|\mathcal{W}|$ is the number of features currently being searched. If the largest abductive explanation has size kM, the outer loop runs $O(k_M)$ times. Therefore, the total number of sequential oracle calls is reduced to $O(k_M \log m)$, where m is the total number of features $|\mathcal{F}|$. This can offer better performance than the greedy algorithm if the largest abductive explanation is considerably smaller than the total number of features.

Viewing the algorithm as analyzing *chunks* of features highlights a possibility for parallelization, where checks on different chunks could be run simultaneously. This idea is further developed in the *SwiftXplain* algorithm [34].

Optimizations. The robustness queries used in the mentioned algorithms can be expensive due to the complexity of neural networks, especially when many features are set to free. Various techniques have been proposed to optimize the explanation process.

Ordering features heuristically [7,72,73] The order of processing features influences both explanation size and time. Prioritizing features that are more likely to appear in the final explanation generally improves efficiency and makes the explanations smaller. Analyzing the network to establish a heuristic ordering of features can significantly enhance the explanation process. Scoring methods such as LIME [61] and SHAP [48], and a feature-level sensitivity method [73] can be employed as a preprocessing step.

Using contrastive examples [7] A *contrastive example* (or *contrastive explanation*) is a subset of features $\mathcal{C} \subset \mathcal{F}$ s.t. altering them is sufficient to cause a misclassification of a given instance $(\mathbf{s}, c)$: $\exists \mathbf{x} \in \mathbb{F} \,.\, \kappa(\mathbf{x}) \neq c \wedge \bigwedge_{i \in \mathcal{F} \setminus \mathcal{C}} x_i = s_i$. The technique identifies contrastive singletons—features that, if changed individually, cause a misclassification—and immediately marks them as part of the explanation (i.e., fixes them), thus avoiding unnecessary robustness queries. Contrastive examples of size two are also exploited: if one feature in such a pair is removed from the explanation (i.e., freed), the other should be included.

Local-Singleton Search [7] Freeing features enlarges the search space for the underlying verifier and thus slows down the process. The core observation is that input points near the original input that cause misclassification tend to be clustered. To take advantage of this, instead of letting the values of previously freed features be arbitrary during verification queries, the method fixes them to the values from a previously found counterexample. Then, it searches for local singletons: features whose individual perturbation leads to misclassification, assuming all other features stay fixed at the counterexample values.

Computing Interval and Inflated Explanations. Inflated explanations can be computed using a greedy algorithm that starts from a subset-minimal abductive explanation. In the continuous domain, it systematically expands the associated interval for each feature, strictly following a selected order while maintaining the same classification outcome. The resulting explanations represent maximal intervals. However, the intervals associated with the features appearing at the beginning of the ordering are usually larger than the latter, and the volume of the resulting impact space is typically not maximal. This is due to the greedy expansion of the former intervals in the ordering which is too restrictive with respect to the remaining features, leaving lower flexibility for their expansion.

Moreover, a modified version of Algorithm 3 from [35] can be used for computing interval explanations (Definition 3) if the stopping condition is relaxed to e.g. a predefined number of expansion steps. Also, since we assume that all domains are bounded (Sect. 2), the linear search in the algorithm should be replaced by binary, as suggested by the authors. The resulting explanations can be easier to compute and may yield higher volumes as well, although without guarantees of the maximality of the intervals or the volume. Such an algorithm can also be generalized so that it accepts any interval explanation as input.

5 Related Work

Explanation methods for neural networks range from gradient-based and attribution techniques to formal logic-based approaches. This section expands on methods that are not based on logic or abduction.

Approaches Not Based on Logic. Classical approaches to explain classifying NNs mostly fall into one of the following categories: *Analysis at the unit (neuron) level* [23, 76], which however fails to capture global properties or interdependencies between units, *gradient-based methods* [4,68], which describe the network behavior around a particular sample; and inversion methods [49], which are computationally efficient and can provide global views, but are still only approximating. There is also a range of *Model-agnostic* methods [5,48,61,62], which can be applied not only to NNs, but in principle to any classifier. Bottleneck-based approaches [64,75] address explainability through architectural constraints. Bottleneck architectures constrain information flow through compressed intermediate layers: either by forcing the model to focus on the most prediction-relevant input information, or by using human-defined concept layers. These models improve explainability by making the model's decision more transparent in the intermediate representations.

Other Logic-based Explanations. Beyond abduction-based explanations, which aim to identify sufficient reasons for a classifier's decision, *contrastive* or *counterfactual* explanations [54] have also gained attention. It explains why another decision was not made—also known as a necessary reason. Several works have explored counterfactual reasoning in the context of AI systems [18,36,55, 60]. There are other logic-based approaches that are not abduction-based, for instance, [25] introduces an approach that maps the neural activations to semantic concepts, which are then used to induce logic-based theories. Furthermore, in the context of image classification, [11] introduced *compositional* explanations as the set $\mathcal{E}$ of superpixels (i.e., clusters of adjacent pixels considered as a unique feature) that are the most *relevant* for the classification task. Last, but not least, recent work has addressed the practical computation of probabilistic abductive explanations for various families of classifiers [33, 37].

Other Surveys. A few surveys have been done on formal AI explanation and verification. [45] surveys early works on utilizing formal methods for the explanation, verification, and testing of machine learning systems. [52] provides a technical survey of logic-based XAI, discussing its origins, the current topics of research, and emerging future research areas. [43] surveys NN verification tools from a machine learning perspective, while [40] reviews and classifies the literature on NN verification, which we used in Sect. 4.3.

6 Quality of Space Explanations

Since based on logic, space explanations enable rigorous measurement and comparisons of the quality of explanations, which typically depends on multiple properties of the underlying formula and is application-specific. Such quality metrics are based on comparing formulas that represent explanations, which often have different shapes. For example, interval explanations are easy to interpret, but can sometimes be too restricted given the complexity of the feature space. We identify two main aspects of the quality that are often contradictory: (human) *interpretability*, and *generality* of the explanations. More general explanations may convey richer information about the network [44] but often come at the cost of reduced readability. The challenges of practical human interpretation of complex explanations, particularly in high-dimensional feature spaces, and their comparison, were addressed in [47] by applying classical methods from geometry, demonstrating the usability of the explanations for domain experts without the need for expertise in logic.

6.1 Interpretability

Explanations can be communicated in various ways, mainly by reading the formula, or using a visualization.

Formula Readability. Human readability of the formula is, for example, important in the domains of natural sciences where an expert in the domain verifies whether the obtained rules are compatible with their knowledge. Explanations that constrain individual features (Definitions 2–4) result in short and structurally simple formulas that are easy to interpret. In contrast, non-restricted space explanations can be harder to grasp. The readability depends on the size of the formula, which can be defined as the number of terms, number of features involved, number of characters, etc.

Visualization. Visualization can provide intuitive insights for the classification even for complex formulas. However, it may be difficult to quantify or compare them with other visualizations. A direct way is to visualize the impact space which is usually high-dimensional. There are several ways to overcome this, such as using *slices* or *projections* for selected

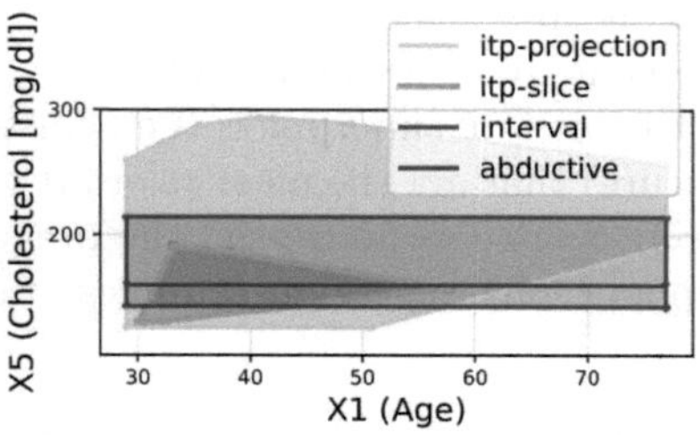

Fig. 2. Projection and slice visualization of three types of explanations

pairs or triplets of features, resulting in a 2D or 3D space. Slices fix all other dimensions, providing a precise view into the selected dimensions, but conclusions on the excluded dimensions are very limited. Projections can be computed by quantifier elimination, which is expensive but can also be approximated using linear programming. They offer an intuitive understanding of the geometrical differences between impact spaces. They still do not capture the complete information contained in higher-dimensional spaces: they may result in too large area because the information from the other dimensions is collapsed. An example of such visual comparisons is shown in Fig. 2: using two visualization methods, projection and slice, it shows three types of explanations from [44] for a heart-attack risk model, focusing on two features of the patients (age and cholesterol). For interval and abductive explanations, both visualization methods result in the same area because the impact space of these explanations is always a hyperrectangle.

There are other means of dimensionality reduction as well [9], such as PCA, that exploits the data distribution within a dataset of sample points. A special case of visualization applies to the domain of image recognition, where particular features represent pixels. There, the explanations can be visualized using some projections on particular pixels (i.e., features), and the user can then decide which ones identified as responsible for the classification make more sense. An example of such visual comparisons appeared in [73] for abductive explanations.

Following [47], further means of visual interpretation include *extraction of the maximum diameter*, identifying a significant linear relation between features, and *extraction of the* (Euclidean) *minimum norm point*, which illustrates the flexibility of particular pixels in explanations for image-classification datasets: the darker pixels compared to the original image, the higher flexibility given by the explanation, reflecting the relevance of the pixels for classification, and also the size of the impact space (see an example in Fig. 7). Both methods compute objects that are easy to visualize but at the same time represent correct explanations that are subsets of the original explanation.

6.2 Generality

The generality is characterized by the size of the impact space, reflecting the ability to cover multiple sample points. The size of the continuous space is estimated based on selected factors such as the volume. Moreover, some applications may favour the importance of certain input features over others. The choice of the used metric is application- or even explanation-specific: for example, volume estimation is robust and comprehensive, but only if the accuracy is meaningful and given that the computation for the explanation is feasible.

Subset Checks. A direct comparison of explanations is to check whether one is a subset of another (e.g., using implication checks). While providing rigorous comparisons, explanations often intersect but are not entirely subsumed by one another. As a result, many explanations remain incomparable [44].

Number of Fixed Features. A rough aspect of the generality is based on the number of features that are fixed to a single value by the formula, not allowing any flexibility. For example, a sample point fixes all features, resulting in the least general explanation possible, and abductive explanations (Definition 2) only constrain features to be fixed or entirely free within the domain.

Volume. Computing a relative volume, that is, a volume after normalizing particular domains, estimates the size of an impact space. Particular domains can also be adjusted according to the importance of some features. While the volume is easily computable for hyperrectangles, it is non-trivial for more general high-dimensional spaces, even if they are convex. Approximations can be obtained based on symbolic reasoning [66] or Monte Carlo methods [47], but may still be expensive. The volume is zero if any feature is fixed in the explanation, but one can compare the number of fixed features and compute the volume only within the rest. Since computing the volume in 2D or 3D spaces is feasible, another possibility is to exhaustively compare volumes of all slices or projections (cf. visualization) for all or selected pairs or triples of features.

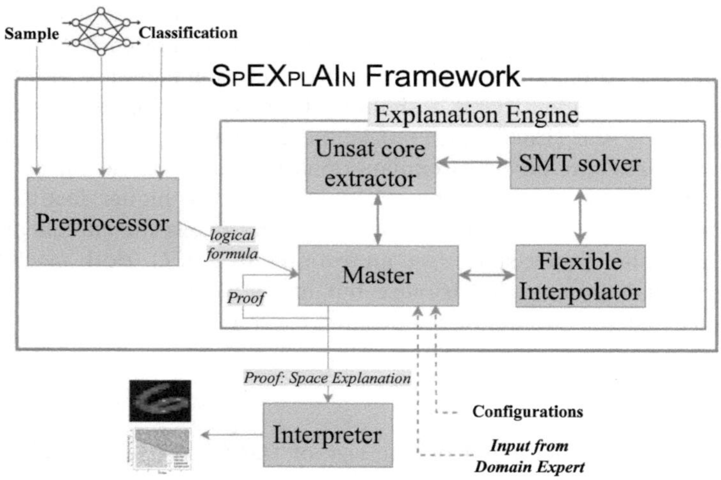

Fig. 3. Space explanations computation scheme

7 Illustration of a Practical Framework *SpEXplAIn*

The SpEXplAIn framework builds on the most general abduction-based concept of *Space Explanations*. The tool is available online[5][6] with detailed instructions

[5] https://github.com/usi-verification-and-security/spexplain.
[6] https://verify.inf.usi.ch/spexplain.

on how to build and run the tool. Based on SMT, it automatically generates explanations of NN classification in the form of logical formulas with provable guarantees about the classification and covering complex-shaped subsets of the feature space, enabling close decision boundary approximation. Starting from a sample point or an existing space explanation, the tool delivers explanations from logical proofs about the classification constructed on the fly. *Craig interpolation* utilizes the proof for expanding the space around the sample point and capturing the reasons behind the classification. Importantly, it preserves the guarantees about the classification, and can iterate for generalizing the explanation further. Figure 3 gives the framework overview with three main parts:

- **Preprocessor** encodes the NN, the sample point, and its classification into a logical formula in SMT, as decribed in Sect. 4.1.
- **Explanation Engine** receives the logical encoding and generalizes the constraints on the input features while preserving guarantees on the classification. Following Sect. 4.2, it outputs a proof that represents a space explanation. Internally, the Master subcomponent instructs the Interpolator and Unsat-core extractor subcomponents for iterative explanation generalization. The engine is optionally configurable to tailor the explanation to the specific needs of the user.
- **Interpreter** automatically transforms the resulting space explanation into a human-recognizable form, for example, using a visualization (cf. Section 6).

The engine contains four components:

- **Master** coordinates the user inputs and configuration with other components. and also handles the actual explanations and proofs (pass and receive them from the other components).
- **SMT solver** is the core constraint solver, available on demand, answering queries from other components. It automatically determines whether logical formulas are satisfiable, and can generate proofs when they are not.
- **Interpolator** takes an unsatisfiable formula of the form $\psi \wedge \varphi$ and uses Craig interpolation to weaken φ. Initially, the interpolator lets φ be $\varphi_{\mathbf{s}}$ and produces a generalization over the sample point. Optionally, it continues to weaken the generalization iteratively.
- **Unsat core extractor** receives an unsatisfiable conjunctive formula represented as a set of constraints and extracts a smaller unsatisfiable subset in the form of an unsatisfiable core. This results in a shorter and weaker formula, meaning that the explanation is generalized and simplified.

The engine is instantiated by the OpenSMT2 solver [10,30] which integrates a range of techniques into a unified framework, combining standard SMT solving with the computation of unsatisfiable cores and Craig interpolants of varying sizes and logical strengths. It also provides a naive implementation of computing feature-based explanations (Definition 2 and 3).

Users may optionally specify configurations and a sequence of strategies based on their domain knowledge. This provides flexibility to tailor the explanation to

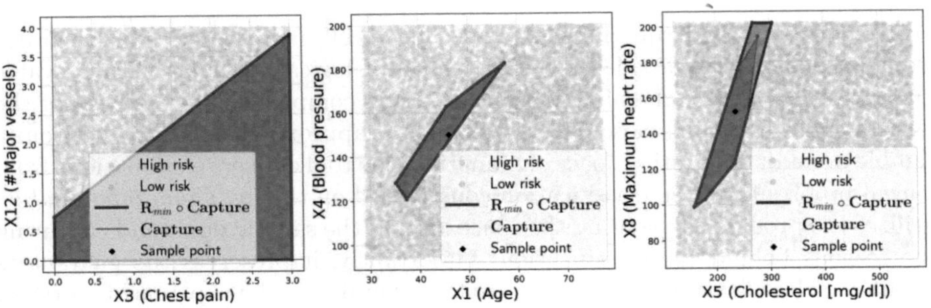

Fig. 4. Approximation of decision boundaries and feature relations using strategies **C** and $\mathbf{R}_{min} \circ \mathbf{C}$ within selected pairs of features.

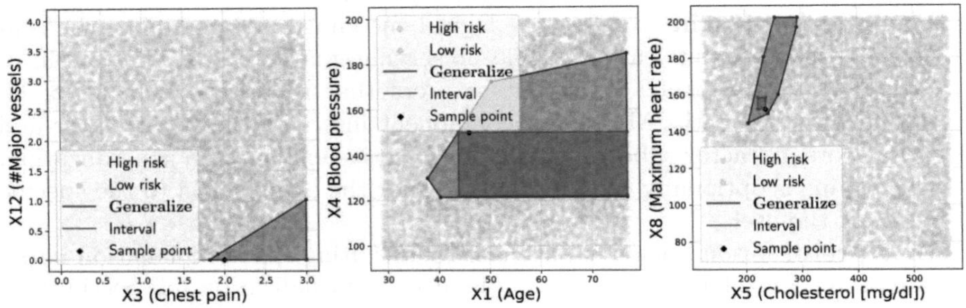

Fig. 5. Generalizing an interval explanation

their specific needs, but it is not required for the framework to function. The available strategies—detailed below—can be applied in sequence when provided.

- **Generalize (G).** Generalizes an arbitrary space explanation (including e.g. $\varphi_{\mathbf{s}}$, abductive or interval-based explanations) by calling the Interpolator. The strategy is parametric in the interpolation method with different logical strengths. We use arithmetic interpolation algorithms from Sect. 4.2: `weak` $\mapsto$ Itp'_F, `mid` $\mapsto$ Itp_f using $f = 0.5$, and `strong` $\mapsto$ Itp_F, all integrated with McMillan's propositional interpolation algorithm Itp_M.
- **Reduce (R).** Calls the Unsat core extractor to simplify an explanation. When applied on top of strategy **G** (denoted as $\mathbf{R} \circ \mathbf{G}$), the resulting explanation is shorter and more readable. Optionally, minimal reduction ($\mathbf{R}_{min}$) arrives at an irreducible explanation, possibly with significant overhead.
- **Capture (C).** Similarly to **G**, calls the Interpolator but only on the part of the initial formula $\varphi_{\mathbf{s}}$ with constraints over selected features, hence capturing their mutual relationships. Moreover, using **Reduce** is more efficient on top of strategy **C** than **G** (i.e., $\mathbf{R} \circ \mathbf{C}$ vs. $\mathbf{R} \circ \mathbf{G}$) since the interpolation is focused.

The interpreter transforms the resulting space explanation into a user-friendly representation that also captures meaningful insights into the classifica-

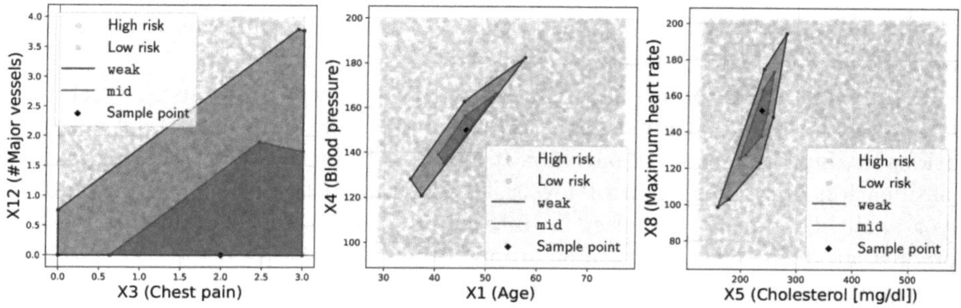

Fig. 6. Explanations produced by **G** with different logical strengths

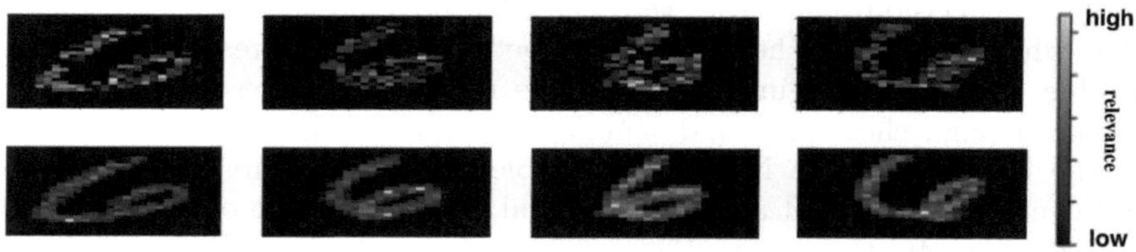

Fig. 7. Visualization of space explanations of MNIST dataset computed by strategy **G** with algorithm `weak` (above) and `strong` (below).

tion. We demonstrate several visualizations of the generated space explanations in the experiments below.

Evaluation. SpEXplAIn has been evaluated [44] and compared to the state-of-the-art abduction-based XAI tool VeriX[7] on test cases of different sizes and different nature, such as medical domain datasets ("heart-attack risk" and "obesity risk") and image classification (MNIST). Thanks to the efficiency of SMT solvers, SpEXplAIn applies in realistic settings. Using the direct encoding of the NN (cf. Sect. 4.1) with no further optimizations, [44] demonstrated the usefulness of the explanations generated by SpEXplAIn but also the scalability challenges, showing that while using this encoding, SpEXplAIn is less sensitive to the number of input features than VeriX, but more sensitive to the number of hidden layers in the NN. In this paper, we present a snapshot of the experiments to illustrate the benefits of SpEXplAIn, focusing on the quality of explanations. More details, such as the runtimes, are available in [44]. Notably, its main concept of space explanations enables the preservation of important domain information. Moreover, the flexibility of tuning the algorithms to deal with logic formulas accommodates the specific needs of the user. The tutorial

[7] VeriX showed it produces (sound) explanations that involve fewer pixels compared to SHAP and Anchors, but using a significantly higher computation time.

accompanying the paper will show further experiments and comparisons[8], using also other models and datasets.

We illustrate the application of SpEXplAIn on *tabular data* using the Heart-attack risk dataset [38]. The task is to predict the risk of a heart attack based on various medical indicators of patients. It contains 13 input features and 2 possible classification outcomes: high or low risk. We trained the NN using one hidden layer with 50 neurons and used a dataset with 303 sample points. Below, we demonstrate that our explanations are meaningful to the user and often confirm the intuition behind the relations of the medical indicators.

SpEXplAIn automatically approximates decision boundaries among all pairs of input features. Figure 4 illustrates the relation between selected pairs (e.g. age and blood pressure) using strategy **Capture** and `weak` interpolation algorithm. Notably, the decision boundaries that can even be non-convex are well approximated while preserving the relationship between the features (cf. yellow and light-blue colors). The figure shows feature relations that resemble real-world phenomena (e.g., chest pain and #major vessels) and at the same time stem from the behavior of the NN with provable guarantees. Moreover, the same experiment illustrates that applying minimal **Reduce** on top of **Capture** (i.e., $\mathbf{R}_{min} \circ \mathbf{C}$) is useful to improve the accuracy of the obtained explanations (red and blue areas in Fig. 4). Another benefit of performing the reduction is that the resulting explanations are more compact and thus easier to communicate.

Furthermore, SpEXplAIn can generalize already existing space explanations, as demonstrated in Fig. 5. In this experiment, an interval explanation (cf. Sect. 3) is first computed based on techniques from related work [35] (see red rectangles) and then widened using strategy **Generalize** (see blue polygons), again with `weak` algorithm. Notably, the generalized explanations approximate the behavior of the NN much more closely than the prior work which is restricted to simplistic feature-based explanations.

Figure 6 illustrates the capabilities of SpEXplAIn to generate explanations of various logical strengths. The use of weaker interpolation algorithms (e.g. `weak`, blue area) results in explanations with larger space compared to stronger ones (e.g. `mid`, red area). Note that the shapes are similar for all sizes.

Overall, these experiments confirm that SpEXplAIn captures feature relations meaningfully, which could be useful for the user.

The other illustration uses the *image classification dataset* MNIST [17], which represents benchmarks with a different kind of sample points, where particular features correspond to pixels arranged into a matrix and are often related locally. The MNIST dataset collects grayscale images of handwritten digits with 784 pixels ranging from 0 to 255. We trained the model using a hidden layer with 200 neurons and experimented on a dataset with 50 sample points.

For this domain, the common task is to identify pixels that are relevant for classification. We used SpEXplAIn to experiment with this task by identifying explanations of different relevance for the visualization of digits. The

[8] See more experiments with optimizations and more layers or neurons in the NNs at:https://verify.inf.usi.ch/content/experiments-spexplain.

SpEXplAIn capability of generating explanations of different strengths is very useful for this task. For example, here is a snapshot of experimentation with digit 6 for which we use strategy **G**. For each explanation, Fig. 7 illustrates visualizations for particular points extracted using the minimum Euclidean norm, that is, the point closest to the original image (cf. Sect. 6.1). The figure enables the comparison of interpolation algorithms `weak` and `strong`. The lighter the color, the higher participation in the explanation it represents for particular pixels of the digits. Notably, while both approaches accurately resemble the actual intended digit, applying algorithm `weak` results in explanations that are more fuzzy upon visualization compared to algorithm `strong`. Such observations suggest that stronger interpolation algorithms suit image classification better because they yield more compact, smoother, and easier-to-interpret explanations.

8 Conclusion

This paper surveys the state of the art in logical abduction-based explanations, a concept of provably correct explanations of NN classification w.r.t. a formal specification of the model. In this area, it establishes a unified theoretical framework and encoding and list the state-of-the-art approaches, allowing their comprehensive comparison. Furthermore, it suggests multiple metrics and factors for estimation of the quality of explanations, allowing their comparison as well. Finally, it illustrates SpEXplAIn, a practical framework for computing most general abduction-based explanations for neural networks. SpEXplAIn preserves provable guarantees of the classification and generates substantially more informative explanations than prior work. Based on SMT solving techniques such as Craig interpolation, it approximates complex decision boundaries and identifies relations between input features that are visualized in a user-friendly way. We illustrate several real-world experiments from different domains with different requirements, highlighting the usefulness of the explanations and the flexibility of their generation.

Acknowledgement. This work was conducted as part of the "Formal Reasoning on Neural Networks" project funded by the Hasler Foundation, Switzerland.

References

1. Albarghouthi, A., Dillig, I., Gurfinkel, A.: Maximal specification synthesis. SIGPLAN Not. **51**(1), 789–801 (2016). https://doi.org/10.1145/2914770.2837628
2. Alt, L., Hyvärinen, A.E.J., Sharygina, N.: LRA interpolants from no man's land. In: Strichman, O., Tzoref-Brill, R. (eds.) HVC, pp. 195–210. Springer International Publishing, Cham (2017)
3. Amir, G., Wu, H., Barrett, C., Katz, G.: An SMT-based approach for verifying binarized neural networks. In: Groote, J.F., Larsen, K.G. (eds.) Tools and Algorithms for the Construction and Analysis of Systems, pp. 203–222. Springer International Publishing, Cham (2021)

4. Bach, S., Binder, A., Montavon, G., Klauschen, F., Müller, K.R., Samek, W.: On pixel-wise explanations for non-linear classifier decisions by layer-wise relevance propagation. PLOS One **10**(7) (2015)
5. Baehrens, D., Schroeter, T., Harmeling, S., Kawanabe, M., Hansen, K., Müller, K.R.: How to explain individual classification decisions. J. Mach. Learn. Res. **11** (2010)
6. Barbosa, H., et al.: cvc5: a versatile and industrial-strength SMT solver. In: Fisman, D., Rosu, G. (eds.) TACAS, pp. 415–442. Springer International Publishing, Cham (2022)
7. Bassan, S., Katz, G.: Towards formal XAI: formally approximate minimal explanations of neural networks. In: TACAS, pp. 187–207 (2023)
8. Blicha, M., Hyvärinen, A.E.J., Kofroň, J., Sharygina, N.: Decomposing Farkas interpolants. In: Vojnar, T., Zhang, L. (eds.) TACAS, pp. 3–20. Springer International Publishing, Cham (2019)
9. Bohn, B., Garcke, J., Griebel, M.: Algorithmic mathematics in machine learning. SIAM (2024)
10. Bruttomesso, R., Pek, E., Sharygina, N., Tsitovich, A.: The OpenSMT solver. In: Esparza, J., Majumdar, R. (eds.) TACAS 2010. LNCS, vol. 6015, pp. 150–153. Springer, Heidelberg (2010). https://doi.org/10.1007/978-3-642-12002-2_12
11. Chockler, H., Kelly, D.A., Kroening, D., Sun, Y.: Causal explanations for image classifiers. arXiv preprint arXiv:2411.08875 (2024)
12. Christ, J., Hoenicke, J., Nutz, A.: SMTInterpol: an interpolating SMT solver. In: Donaldson, A.F., Parker, D. (eds.) SPIN. Lecture Notes in Computer Science, vol. 7385, pp. 248–254. Springer (2012). https://doi.org/10.1007/978-3-642-31759-0_19
13. Cimatti, A., Griggio, A., Schaafsma, B.J., Sebastiani, R.: The MathSAT5 SMT solver. In: Piterman, N., Smolka, S.A. (eds.) TACAS 2013. LNCS, vol. 7795, pp. 93–107. Springer, Heidelberg (2013). https://doi.org/10.1007/978-3-642-36742-7_7
14. Cimatti, A., Griggio, A., Sebastiani, R.: Computing small unsatisfiable cores in satisfiability modulo theories. J. Artif. Intell. Res. **40**, 701–728 (2011)
15. Craig, W.: Three uses of the Herbrand-Gentzen theorem in relating model theory and proof theory. In: Journal of Symbolic Logic, pp. 269–285 (1957)
16. De Moura, L., Bjørner, N.: Z3: an efficient SMT solver. In: TACAS, pp. 337–340. TACAS'08/ETAPS'08, Springer-Verlag, Berlin, Heidelberg (2008). https://doi.org/10.1007/978-3-540-78800-3_24
17. Deng, L.: The MNIST database of handwritten digit images for machine learning research. IEEE Signal Process. Mag. **29**(6), 141–142 (2012). https://doi.org/10.1109/MSP.2012.2211477
18. Dhurandhar, A., et al.: Explanations based on the missing: towards contrastive explanations with pertinent negatives. In: NIPS, pp. 590–601. NIPS'18, Curran Associates Inc., Red Hook, NY, USA (2018)
19. Dillig, I., Dillig, T., Li, B., McMillan, K.L.: Inductive invariant generation via abductive inference. In: Hosking, A.L., Eugster, P.T., Lopes, C.V. (eds.) OOPSLA, pp. 443–456. ACM (2013). https://doi.org/10.1145/2509136.2509511
20. D'Silva, V., Kroening, D., Purandare, M., Weissenbacher, G.: Interpolant strength. In: Barthe, G., Hermenegildo, M. (eds.) VMCAI, pp. 129–145. Springer, Berlin Heidelberg, Berlin, Heidelberg (2010)
21. Dutertre, B.: Yices 2.2. In: Biere, A., Bloem, R. (eds.) CAV, pp. 737–744. Springer International Publishing, Cham (2014)

22. Elboher, Y.Y., Gottschlich, J., Katz, G.: An abstraction-based framework for neural network verification. In: Lahiri, S.K., Wang, C. (eds.) CAV, pp. 43–65. Springer International Publishing, Cham (2020)
23. Erhan, D., Bengio, Y., Courville, A., Vincent, P.: Visualizing Higher-layer Features of a Deep Network. University of Montreal, Tech. rep. (2009)
24. Ferrari, C., Muller, M.N., Jovanovic, N., Vechev, M.: Complete verification via multi-neuron relaxation guided branch-and-bound. arXiv preprint arXiv:2205.00263 (2022)
25. Ferreira, J., de Sousa Ribeiro, M., Gonçalves, R., Leite, J.: Looking inside the black-box: logic-based explanations for neural networks. In: KR, pp. 432–442 (2022). https://doi.org/10.24963/kr.2022/45
26. Gurfinkel, A., Rollini, S.F., Sharygina, N.: Interpolation properties and SAT-based model checking. In: Van Hung, D., Ogawa, M. (eds.) ATVA, pp. 255–271. Springer International Publishing, Cham (2013)
27. Guthmann, O., Strichman, O., Trostanetski, A.: Minimal unsatisfiable core extraction for SMT. In: FMCAD, pp. 57–64 (2016). https://doi.org/10.1109/FMCAD.2016.7886661
28. Henriksen, P., Lomuscio, A.: Efficient neural network verification via adaptive refinement and adversarial search. Ph.D. thesis, Ph. D. Dissertation. Imperial College London (2019)
29. Huang, X., Marques-Silva, J.: From robustness to explainability and back again. arXiv preprint arXiv:2306.03048 (2023)
30. Hyvärinen, A.E.J., Marescotti, M., Alt, L., Sharygina, N.: OpenSMT2: an SMT solver for multi-core and cloud computing. In: Creignou, N., Le Berre, D. (eds.) SAT 2016. LNCS, vol. 9710, pp. 547–553. Springer, Cham (2016). https://doi.org/10.1007/978-3-319-40970-2_35
31. Ignatiev, A., Narodytska, N., Asher, N., Marques-Silva, J.: From contrastive to abductive explanations and back again. In: AIxIA, pp. 335–355. Springer-Verlag, Berlin, Heidelberg (2020). https://doi.org/10.1007/978-3-030-77091-4_21
32. Ignatiev, A., Narodytska, N., Marques-Silva, J.: Abduction-based explanations for machine learning models. In: AAAI. AAAI'19/IAAI'19/EAAI'19, AAAI Press (2019). https://doi.org/10.1609/aaai.v33i01.33011511
33. Izza, Y., Huang, X., Ignatiev, A., Narodytska, N., Cooper, M., Marques-Silva, J.: On computing probabilistic abductive explanations. Int. J. Approximate Reasoning **159**, 108939 (2023). https://www.sciencedirect.com/science/article/pii/S0888613X23000701
34. Izza, Y., Huang, X., Morgado, A., Planes, J., Ignatiev, A., Marques-Silva, J.: Distance-restricted explanations: theoretical underpinnings & efficient implementation. In: KR. KR '24 (2024). https://doi.org/10.24963/kr.2024/45
35. Izza, Y., Ignatiev, A., Stuckey, P.J., Marques-Silva, J.: Delivering inflated explanations. In: AAAI. AAAI'24/IAAI'24/EAAI'24, AAAI Press (2024). https://doi.org/10.1609/aaai.v38i11.29170
36. Izza, Y., Marques-Silva, J.: Efficient contrastive explanations on demand. CoRR abs/2412.18262 (2024). https://doi.org/10.48550/ARXIV.2412.18262
37. Izza, Y., Meel, K.S., Marques-Silva, J.: Locally-minimal probabilistic explanations (2024). https://arxiv.org/abs/2312.11831
38. Janosi, A., Steinbrunn, W., Pfisterer, M., Detrano, R.: Heart Disease. UCI Machine Learning Repository (1989). https://doi.org/10.24432/C52P4X
39. Junker, U.: QUICKXPLAIN: preferred explanations and relaxations for over-constrained problems. In: National Conference on Artifical Intelligence, pp. 167–172. AAAI'04, AAAI Press (2004)

40. Kanav, S., Křetínský, J., Rieder, S.: A Literature Review on Verification and Abstraction of Neural Networks Within the Formal Methods Community. Springer Nature Switzerland, Cham (2025). https://doi.org/10.1007/978-3-031-75778-5_3
41. Katz, G., et al.: The marabou framework for verification and analysis of deep neural networks. In: Dillig, I., Tasiran, S. (eds.) CAV, pp. 443–452. Springer International Publishing, Cham (2019)
42. König, M., Bosman, A.W., Hoos, H.H., van Rijn, J.N.: Critically assessing the state of the art in neural network verification. J. Mach. Learn. Res. **25**(12), 1–53 (2024). http://jmlr.org/papers/v25/23-0119.html
43. König, M., Bosman, A.W., Hoos, H.H., Van Rijn, J.N.: Critically assessing the state of the art in neural network verification. J. Mach. Learn. Res. **25**(1) (2024)
44. Labbaf, F., Kolárik, T., Blicha, M., Fedyukovich, G., Wand, M., Sharygina, N.: Space explanations of neural network classification. In: Piskac, R., Rakamaric, Z. (eds.) CAV. Springer International Publishing (2025)
45. Larsen, K., Legay, A., Nolte, G., Schlüter, M., Stoelinga, M., Steffen, B.: Formal methods meet machine learning (f3ml). In: ISoLA, Part III. p. 393–405. Springer-Verlag, Berlin, Heidelberg (2022). https://doi.org/10.1007/978-3-031-19759-8_24
46. Laurel, J., Yang, R., Singh, G., Misailovic, S.: A dual number abstraction for static analysis of Clarke Jacobians. Proc. ACM Program. Lang. **6**(POPL) (2022). https://doi.org/10.1145/3498718
47. Leopardi, F., Labbaf, F., Kolárik, T., Wand, M., Sharygina, N.: Interpreting logical explanations of classifying neural networks. In: ESANN (2026). to appear
48. Lundberg, S.M., Lee, S.I.: A Unified Approach to Interpreting Model Predictions. In: NIPS (2017)
49. Mahendran, A., Vedaldi, A.: Understanding deep image representations by inverting them. In: CVPR, pp. 5188–5196 (2015)
50. Malfa, E.L., Michelmore, R., Zbrzezny, A.M., Paoletti, N., Kwiatkowska, M.: On guaranteed optimal robust explanations for NLP models. In: IJCAI, pp. 2658 – 2665 (2021)
51. Marques-Silva, J.: Logic-Based Explainability in Machine Learning. Springer Nature Switzerland, Cham (2023). https://doi.org/10.1007/978-3-031-31414-8_2
52. Marques-Silva, J.: Logic-based explainability: past, present and future. In: ISoLA, Part IV, pp. 181–204. Springer-Verlag, Berlin, Heidelberg (2024). https://doi.org/10.1007/978-3-031-75387-9_12
53. McMillan, K.L.: Interpolation and SAT-based model checking. In: Hunt, W.A., Somenzi, F. (eds.) CAV, pp. 1–13. Springer, Berlin Heidelberg, Berlin, Heidelberg (2003)
54. Miller, T.: Contrastive explanation: a structural-model approach. Knowl. Eng. Rev. **36**, e14 (2021). https://doi.org/10.1017/S0269888921000102
55. Mothilal, R.K., Sharma, A., Tan, C.: Explaining machine learning classifiers through diverse counterfactual explanations. In: ACM FAT, pp. 607–617. FAT* '20, Association for Computing Machinery, New York, NY, USA (2020). https://doi.org/10.1145/3351095.3372850
56. de Moura, L., Bjørner, N.: Satisfiability modulo theories: introduction and applications. Commun. ACM **54**(9), 69–77 (2011)
57. Nieuwenhuis, R., Oliveras, A., Tinelli, C.: Solving SAT and SAT modulo theories: from an abstract Davis–Putnam–Logemann-Loveland procedure to DPLL(T). J. ACM (JACM) **53**(6), 937–977 (2006). https://doi.org/10.1145/1217856.1217859
58. Prabhu, S., D'Souza, D., Chakraborty, S., Venkatesh, R., Fedyukovich, G.: Weakest precondition inference for non-deterministic linear array programs. In: Finkbeiner,

B., Kovács, L. (eds.) TACAS, Part II. Lecture Notes in Computer Science, vol. 14571, pp. 175–195. Springer (2024). https://doi.org/10.1007/978-3-031-57249-4_9
59. Prabhu, S., Fedyukovich, G., D'Souza, D.: Maximal quantified precondition synthesis for linear array loops. In: Weirich, S. (ed.) ESOP, Part II. Lecture Notes in Computer Science, vol. 14577, pp. 245–274. Springer (2024). https://doi.org/10.1007/978-3-031-57267-8_10
60. Prabhushankar, M., Kwon, G., Temel, D., AlRegib, G.: Contrastive explanations in neural networks. CoRR abs/2008.00178 (2020). https://arxiv.org/abs/2008.00178
61. Ribeiro, M.T., Singh, S., Guestrin, C.: "why should i trust you?": Explaining the predictions of any classifier. In: KDD, pp. 1135–1144. KDD '16, Association for Computing Machinery, New York, NY, USA (2016). https://doi.org/10.1145/2939672.2939778
62. Ribeiro, M.T., Singh, S., Guestrin, C.: Anchors: high-precision model-agnostic explanations. In: AAAI (2018)
63. Rollini, S.F., Sery, O., Sharygina, N.: Leveraging interpolant strength in model checking. In: Madhusudan, P., Seshia, S.A. (eds.) Computer Aided Verification, pp. 193–209. Springer, Berlin Heidelberg, Berlin, Heidelberg (2012)
64. Saxe, A.M., et al.: On the information bottleneck theory of deep learning. In: ICLR. OpenReview.net (2018). https://openreview.net/forum?id=ry_WPG-A-
65. Seshia, S.A., Sadigh, D., Sastry, S.S.: Towards verified artificial intelligence. Commun. ACM **65**(7), 46–55 (2022)
66. Shaw, A., Sarkar, U., Meel, K.S.: Efficient volume computation for SMT formulas. In: Proceedings of the 22nd International Conference on Principles of Knowledge Representation and Reasoning, pp. 544–554 (2025). https://doi.org/10.24963/kr.2025/53
67. Shih, A., Choi, A., Darwiche, A.: A symbolic approach to explaining bayesian network classifiers. In: IJCAI, pp. 5103 – 5111 (2018)
68. Simonyan, K., Vedaldi, A., Zisserman, A.: Deep inside convolutional networks: visualising image classification models and saliency maps. In: Workshop Proceedings of the ICLR (2014)
69. Singh, G., Gehr, T., Püschel, M., Vechev, M.: An abstract domain for certifying neural networks. Proc. ACM Program. Lang. **3**(POPL) (2019). https://doi.org/10.1145/3290354
70. Wang, S., et al.: Beta-CROWN: efficient bound propagation with per-neuron split constraints for neural network robustness verification. Adv. Neural. Inf. Process. Syst. **34**, 29909–29921 (2021)
71. Wu, H., et al.: Marabou 2.0: a versatile formal analyzer of neural networks. In: CAV, pp. 249–264. Springer (2024)
72. Wu, M., Li, X., Wu, H., Barrett, C.: Better verified explanations with applications to incorrectness and out-of-distribution detection (2024). https://arxiv.org/abs/2409.03060
73. Wu, M., Wu, H., Barrett, C.: VeriX: towards verified explainability of deep neural networks. In: Oh, A., Neumann, T., Globerson, A., Saenko, K., Hardt, M., Levine, S. (eds.) Advances in Neural Information Processing Systems. vol. 36, pp. 22247–22268. Curran Associates, Inc. (2023)
74. Xu, K., et al.: Fast and complete: Enabling complete neural network verification with rapid and massively parallel incomplete verifiers. arXiv preprint arXiv:2011.13824 (2020)
75. Yamaguchi, S., Nishida, K.: Explanation bottleneck models. CoRR abs/2409.17663 (2024). https://doi.org/10.48550/ARXIV.2409.17663

76. Zeiler, M.D., Fergus, R.: Visualizing and understanding convolutional networks. In: ECCV, pp. 818 – 833 (2014)

The SLEEC Framework for Normative Requirements Engineering

Pedro Ribeiro[1(✉)], Radu Calinescu[1], Ana Cavalcanti[1], Marsha Chechik[2], Sinem Getir Yaman[1], Lina Marsso[3], Isobel Standen[4], and Beverley Townsend[5]

[1] Department of Computer Science, University of York, York, UK
{pedro.ribeiro,radu.calinescu,ana.cavalcanti,sinem.getir.yaman}@york.ac.uk
[2] University of Toronto, Toronto, Canada
marsha.chechik@cs.toronto.edu
[3] Polytechnique Montréal, Montréal, Canada
lina.marsso@polymtl.ca
[4] Department of Philosophy, University of York, York, UK
isobel.standen@york.ac.uk
[5] York Law School, University of York, York, UK
bev.townsend@york.ac.uk

Abstract. Autonomous agents are increasingly deployed in sensitive, human-centric domains—such as healthcare, assistive care, and emergency response—where their decision-making must align with complex human norms. These translate into Social, Legal, Ethical, Empathetic, and Cultural (SLEEC) requirements that are often nuanced and context-dependent, challenging traditional software engineering paradigms. Our tutorial paper presents a comprehensive, tool-supported methodology for managing the SLEEC requirements lifecycle, covering elicitation, well-formedness validation, and conformance verification of software design models against SLEEC requirements. We demonstrate the use of our methodology and associated tools through application to a robot-assisted dressing system, providing a guide for researchers and engineers to bridge the gap between abstract human norms and verifiable system designs.

Keywords: Normative requirements · Autonomous agents · Formal verification and validation · SLEEC

1 Introduction

Rapid technological advances, particularly in Artificial Intelligence (AI), enable the development of applications offering significant societal and economic benefits. However, the deployment of the increasingly sophisticated AI-enabled systems underpinning many of these applications—ranging from cyber-physical to agentic systems—comes with significant normative implications of a social, legal, ethical, empathetic, and cultural (SLEEC) nature [4,15].

A. Sampaio and M. Stoelinga (Eds.): FM 2026, LNCS 16557, pp. 497–523, 2026.
https://doi.org/10.1007/978-3-032-26220-2_24

The importance and complexity of these implications are particularly evident in domains where AI-enabled systems interact with vulnerable users, such as health and assistive care [20]. An assistive-dressing robot [26], for instance, must be aware of and navigate complex normative trade-offs, e.g., to ensure both user privacy by closing curtains during dressing, and user autonomy if asked to keep them open by an anxious user. In healthcare, chatbots provide medical advice and machine learning (ML) solutions perform diagnostics, raising concerns regarding accountability and fairness. Beyond care, collaborative robots assist human operators with dangerous manufacturing tasks, and self-driving cars navigate with varying levels of autonomy in public spaces. Each case poses SLEEC challenges; for example, the collaborative robot must balance operational efficiency with the psychological wellbeing of its human partner, while autonomous vehicles must make ethically charged decisions in uncertain environments.

There is wide awareness of the need to address these normative concerns, reflected in numerous international initiatives aimed at ensuring the responsible development and deployment of AI-enabled systems. These efforts span a broad spectrum, from high-level ethical frameworks and principles, such as the UNESCO Recommendation on the Ethics of AI [24], to more concrete standards and legislative instruments like the IEEE 7000 series [14] and the EU AI Act [6]. Nevertheless, a significant gap exists between these useful but general normative documents, which are discussing application-independent issues, and the definition of a concrete set of well-formed, unambiguous and verifiable normative requirements for a concrete AI-enabled system [19,22]. Furthermore, verifying that a system's design actually complies with such requirements remains a challenge for its developers and regulators [1,13,25].

Our tutorial paper presents an integrated suite of tool-supported methods designed to address this gap. It has been developed over almost a decade via an international collaboration [7,9–11,16,21,22,25], and has already sparked the interest of multiple research groups. In presenting this tutorial, we benefit from the experience of having presented tutorial sessions on SLEEC requirements engineering across several major conferences in Software Engineering.

The paper is organised as follows. In Sect. 2, we introduce our definition of normative requirements, and the rule-based language we use to capture them, which has a mathematical semantics [25]. Section 3 describes our collaborative process for eliciting well-formed SLEEC rules, which is based on formal verification tools, yet accessible for use by teams of multi-disciplinary stakeholders. In Sect. 4, we illustrate our notion of conformance used for the formal verification of designs against SLEEC rules. We conclude with a brief summary and provide pointers to additional material on the SLEEC framework in Sect. 5.

2 The SLEEC Paradigm

2.1 Normative Requirements

In this tutorial, we distinguish between high-level normative principles (such as autonomy, non-maleficence, and privacy) and the specific normative requirements (i.e., *SLEEC rules*) that operationalise them. While principles provide abstract evaluative standards covered extensively by existing international guidelines and regulations [6,14,24], SLEEC rules formulate the explicit constraints that an autonomous agent must satisfy to comply with these principles within a given context—and are significantly more difficult to elicit, formalise, and verify.

In our methodology, SLEEC rules are treated as non-functional requirements that augment an agent's functional specifications. Functional requirements define what an agent should achieve, whereas SLEEC rules constrain the admissible state space and action selection process. Specifically, these rules limit an agent's degrees of freedom by filtering out plans that violate normative constraints, ensuring the agent executes a normatively compliant plan from the possible alternatives [22]. In the next section, we describe the language used to specify these rules as formally verifiable properties.

2.2 The SLEEC Language

The SLEEC language, introduced in [25], is a domain-specific language (DSL) for specifying normative requirements for autonomous agents. Co-developed by an interdisciplinary team of computer scientists, engineers, ethicists, lawyers, roboticists, and social scientists, the language supports the creation of SLEEC specifications comprising two parts: a definition block, and a rule block.

Definition Block. This block declares the capabilities of the agent that have normative relevance because they:

1. raise normative concerns, leading to normative requirements;
2. provide information about the status of the world to support decisions;
3. provide facilities for the agent to react in a norm-sensitive way.

For example, in the robot assistive dressing scenario considered in our paper, the robot's capability to start dressing a user raises a modesty concern. The robot also has a capability to measure room temperature to ensure that decisions regarding when to undress the user consider their (thermal) comfort. Finally, the robot has a capability to close the curtains, so that it can act to protect the user's modesty when asked to start dressing. The process presented in the next section describes how we can iteratively elicit rules based on available and suggested capabilities.

Listing 1.1 illustrates these concepts in a SLEEC specification, showing how agent capabilities are represented by either **`event`**s or **`measure`**s. For instance, the **`event`** `UserFallen` corresponds to the agent detecting that its user has fallen, while the **`measure`** `userDistressed` corresponds to the robot's capability to estimate a user's distress level on a discrete **`scale`** (`low`, `medium` or `high`). Together, these two constructs are used to provide the *vocabulary* for the rule block, allowing normative rules to be defined over the agent's capabilities.

Rule block. This block contains the SLEEC rules for the autonomous agent, specified over the **`event`**s and **`measure`**s defined in the previous block. These rules encode the normative requirements of the agent by mapping environmental

```
def_start // Capabilities
  event CurtainOpenRqt // Triggers and responses
  event CurtainsOpened
  event DressingAbandoned
  event DressingComplete
  event DressingStarted
  event HealthChecked
  event RefuseRequest
  event RetryAgreed
  event SupportCalled
  event UserFallen
  measure assentToSupportCalls : boolean // Measures
  measure emergency : boolean
  measure roomTemperature : numeric
  measure userUnderDressed : boolean
  measure userDistressed : scale(low, medium, high)
  constant MAX_RESPONSE_TIME = 60 // Constants
def_end
rule_start
 Rule1 when CurtainOpenRqt then CurtainsOpened within 60 seconds
       unless userUnderDressed then RefuseRequest within 30 seconds
       unless userDistressed > medium then CurtainsOpened within 60 seconds

 Rule2 when DressingStarted and userUnderDressed
            then DressingComplete within 2 minutes
       unless roomTemperature < 19 then DressingComplete within 90 seconds
       unless roomTemperature < 17 then DressingComplete within 60 seconds

 Rule3 when UserFallen then HealthChecked within 30 seconds
       otherwise SupportCalled within MAX_RESPONSE_TIME seconds

 Rule4 when DressingAbandoned then RetryAgreed within 2 minutes
       otherwise { SupportCalled unless not assentToSupportCalls }
rule_end
```

Listing 1.1. Sample SLEEC specification for the robot-assisted dressing system

triggers to required agent responses. A SLEEC rule has the basic form

id when triggerEvent [**and triggerGuard**] **then response** [**within time**]

where:

- **id** is a unique rule identifier;
- **triggerEvent** is an **event** that must be monitored by the agent;
- **triggerGuard** is an optional Boolean expression over the agent's **measure**s;
- **response** is an **event** specifying the action that the agent shall perform when the **triggerEvent** occurs and, if specified, the **triggerGuard** is true;
- if provided, **time** specifies a deadline for the **response** to happen.

For example, `Rule1` from Listing 1.1 applies when the event `CurtainOpenRqt` occurs, and `Rule2` applies when the event `DressingStarted` occurs and, additionally, the Boolean measure `userUnderDressed` is true.

To accommodate situations where a **response** may be infeasible within the required time, the **otherwise** construct can be used to specify an alternative response. In Rule3, the **response** requires the occurrence of the **event** HealthChecked in 30 s, but provides an alternative to have SupportCalled if there is a timeout.

Importantly, a rule can be followed by one or more *defeaters* [5], introduced by the **unless** keyword, and specifying circumstances that preempt the original **response** and may provide an alternative. The general form of a defeater is

unless defeaterGuard [**then defeaterResponse**]

where:

- **defeaterGuard** is a Boolean expression over the agent's **measure**s;
- if provided, **defeaterResponse** represents an **event** specifying an alternative action for the agent.

If a SLEEC rule includes defeaters, and any **defeaterGuard** from its defeaters is true, then—instead of responding as specified by the base rule **response**—the agent shall act as specified by the outermost defeater whose **defeaterGuard** holds, performing its **defeaterResponse** if present, or ignoring the rule altogether otherwise.

The use of defeaters is illustrated by two rules from the SLEEC specification in Listing 1.1. In Rule1, the first **unless** specifies a defeater that preempts the response CurtainsOpened if the **measure** userUnderDressed is true, while a second defeater takes precedence over both the base-rule response *and* the first defeater if the **measure** userDistressed is greater than medium (i.e., high). Additionally, in Rule4, curly brackets are used to specify a defeater that applies to the alternative response of SupportCalled defined by the **otherwise** construct.

The elicitation of SLEEC rules is supported by a family of tools [7,11] that provide validation, first through parsing and type-checking, and, second, through the ability to identify and diagnose conflicts and redundancies between rules by exploiting the formal semantics of SLEEC as described later in Sect. 3. Figure 1 depicts one of these tools, SLEEC-TK, with a subset of rules from Listing 1.1 being edited in a file named tutorial.sleec whose content is validated as the user modifies the SLEEC specification. In this case, an error is shown in the pane at the bottom, indicating that CurtainsOpenRqt is mistyped. Related, syntactic elements can be clicked on to look up their definition, as indicated by the hand cursor and the highlighted capability CurtainsOpened. Instructions on how to install SLEEC-TK are available online[1], with a set of example projects that can be imported in the tool available from its repository[2].

SLEEC Semantics. Rules are given semantics using two distinct, but complementary, formal approaches, to support detection of conflicts, redundancies, and other well-formedness issues, as well as conformance verification of a design against a set of SLEEC rules. In the first approach, the behaviour of SLEEC

[1] https://github.com/UoY-RoboStar/SLEEC-TK.
[2] https://github.com/UoY-RoboStar/SLEEC-TK/tree/main/examples.

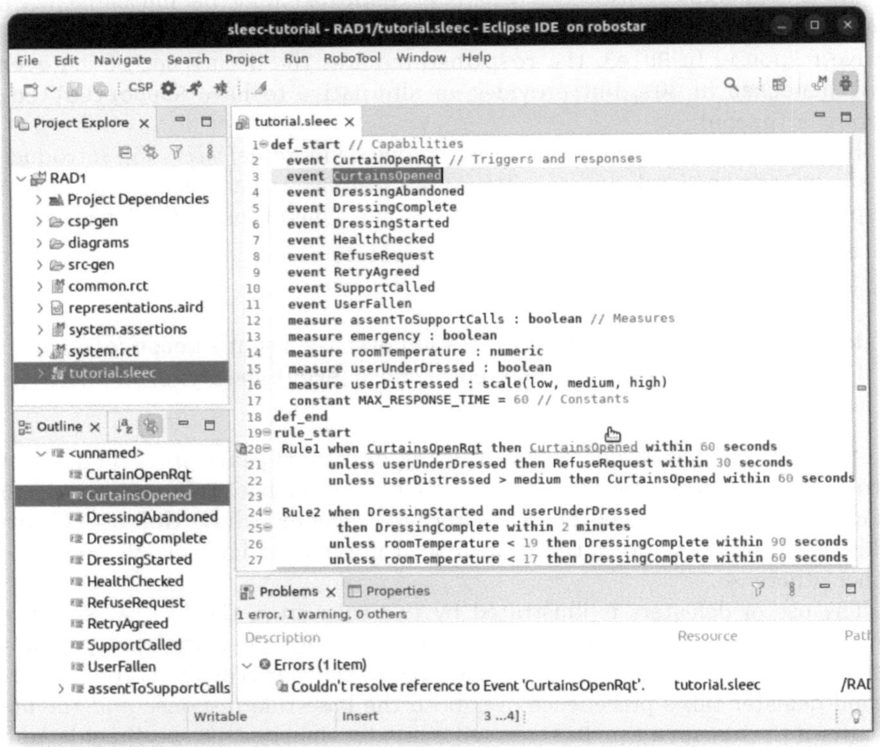

Fig. 1. SLEEC-TK screenshot showing a subset of the SLEEC rules from Listing 1.1 being edited within the file `tutorial.sleec`, which is opened in the top-right pane. The bottom-righ pane summarises syntactic and type errors.

rules is captured using the process algebra *tock*-CSP [3], a discrete-timed variant of CSP [18] supported by the model checker FDR [12] for formal verification. In the second approach, the semantics are enhanced using SAT-based reasoning grounded in first-order logic with relational objects (FOL*) [8], providing more comprehensive checks for requirements validation, including various types of conflicts, situational conflicts, redundancy, restrictiveness, and insufficiency [7].

3 Eliciting and Validating SLEEC Rules

As illustrated in Fig. 2 our SLEEC framework employs a three-stage normative requirements engineering workflow. In stage (i), a team of *SLEEC experts* comprising multi-disciplinary stakeholders, identify normative principles, capabilities and measures relevant for the autonomous agent under development—all captured in a SLEEC document. Through the use of our framework's tools (e.g., see Fig. 1), the SLEEC experts ensure that the rules are correctly formulated, that is, they are syntactically correct and use the available capabilities as specified. The elicitation stage includes consulting a broad range of agent stakeholders, as well as normative documents, such as regulations and standards.

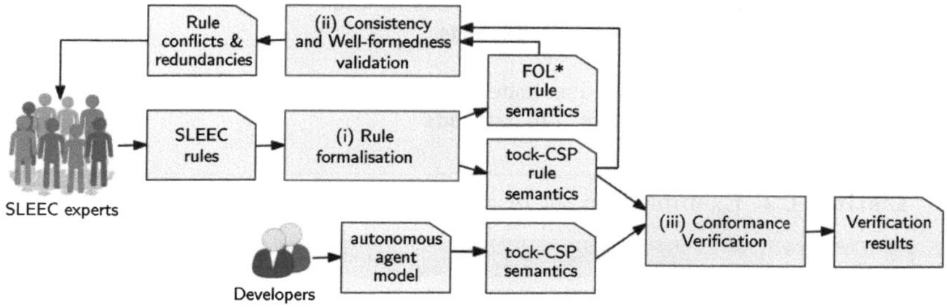

Fig. 2. Elicitation, validation and verification workflow of normative requirements

In stage (ii), rules are validated to identify any conflicts, redundancies and other well-formedness issues using the two complementary approaches mentioned earlier, with feedback provided to SLEEC experts for iterative revision of the proposed rules, until a set of valid rules is obtained. In Sect. 3.1, we describe our framework's tool-supported techniques for pair-wise validation of rules based on their *tock*-CSP semantics [25], while in Sect. 3.2 we present our FOL*-based techniques and tool for detecting other well-formedness issues across a set of rules. Finally, once a valid set of SLEEC rules is obtained, the autonomous agent model provided by its developers is verified for compliance with the SLEEC rules in stage (iii). This verification process is detailed in Sect. 4.

3.1 Pair-Wise Consistency Validation

As mentioned, the second stage in our workflow involves validating SLEEC rules to identify any pair of rules that are conflicting or redundant. To enable this validation, each SLEEC rule's semantics is formally defined as a CSP process [18] that specifies a pattern of interaction over CSP events capturing reading of measures, events (i.e., rule triggers and responses), and the passage of time as marked by *tock*. These patterns can be characterised mathematically as a set of traces, that is, finite sequences of CSP events.

The semantics of a SLEEC rule encompasses two phases. First, there is a monitoring phase before a trigger occurs, where events used in the responses of a rule are unconstrained, followed by a response phase, if any responses are applicable. The response phase is followed again by the monitoring phase. So, following a trigger event, the measures used in the expression of the trigger and defeaters, if any, are read immediately. Then, the trigger's expression, if it exists, is evaluated. For example, the following *CSP trace* (i.e., sequence of interactions including SLEEC events, readings of measures, and the passage of discrete time as captured by `tock` events) can be observed for `Rule1` in Listing 1.1:

```
rule_start
 RuleA when CurtainOpenRqt then CurtainsOpened within 2 minutes
       unless userUnderDressed then RefuseRequest within 3 seconds
       unless userDistressed > medium then CurtainsOpened within 6 seconds

 RuleB when CurtainOpenRqt and userUnderDressed then
       not CurtainsOpened within 60 seconds

 RuleC when CurtainOpenRqt and userUnderDressed then
       not CurtainsOpened within 120 seconds
rule_end
```

Listing 1.2. Example of a set of rules with conflicts and redundancies.

Trace 1 for `Rule1`

```
CurtainOpenRqt, userUnderDressed.false, userDistressed.medium, tock,
CurtainsOpened
```

In this scenario, the trigger event `CurtainOpenRqt` is observed, followed by the reading of the measures `userUnderDressed` and `userDistressed`, with values `false` and `medium`, respectively, indicating that neither *defeater* applies, followed by one `tock` event indicating the passage of one second, and finally the response event `CurtainsOpened` is observed. A complete account of the *tock*-CSP semantics of SLEEC specifications can be found in [25].

Given this semantics, conflicts arise for rules whose restrictions cannot be jointly satisfied, whereas a redundancy indicates that a rule can be discarded as the restrictions it imposes are covered by another rule. We define both notions more precisely in what follows and illustrate how our tools can be used to identify and diagnose them.

Conflicts. Two rules are conflicting if there is a scenario where both apply but impose restrictions that cannot be simultaneously satisfied. For example, both `RuleA` and `RuleB` from Listing 1.2 have the event `CurtainOpenRqt` as a trigger. Consider the scenario where a user asks for the curtains to be opened, is underdressed, and is exhibiting a `high` distress level. In these circumstances, there is a conflict between the requirement to open the curtains **`within`** `6,s` (`RuleA`), as the user is highly distressed, and to not open them for `60,s` (`RuleB`) for modesty reasons. Stakeholders can be guided by our tools to automatically identify such inconsistencies and revise their rules accordingly.

Checking with SLEEC-TK. The first tool at our disposal is SLEEC-TK [25], which implements the calculation of the *tock*-CSP [3] semantics for SLEEC. Checking for conflicts between `RuleA` and `RuleB` of Listing 1.2 in SLEEC-TK invokes model checking with FDR in the background, and results in the following counterexample being generated as a CSP trace.

Trace 2 showing conflict between `RuleA` and `RuleB`

```
CurtainOpenRqt, userUnderDressed.true, userUnderDressed.true,
userDistressed.high, tock, tock, tock, tock, tock, tock
```

This trace specifies a scenario where after the occurrence of the trigger event `CurtainOpenRqt`, the measures `userUnderDressed` and `userDistressed` are read, with values `true` and `high`, respectively, followed by the passage of 6 s, as indicated by the six `tocks`, leading to a conflict. We observe that measures are read more than once with the same value in a conflicting trace, allowing for independent reading of measures by two rules. The trace suggests that resolving the conflict requires relaxing the deadline of `RuleA` or adding defeaters to `RuleB`.

Process-Algebraic Notion of Conflict. In what follows, we explain how conflicts can be diagnosed for a pair of rules based on their traces. For example, the following trace can be observed for `RuleB`:

Trace 3 for RuleB

```
CurtainOpenRqt, userUnderDressed.false, CurtainOpenRqt
```

Here, the measure `userUnderDressed` is read as being `false` after the triggering event `CurtainOpenRqt`, the response does not apply, and so the rule can be triggered again. On the other hand, if the trigger's expression evaluates to true, or there is none, the measures associated with defeaters, if any, are evaluated following the reverse order in which they are written. For example, for `RuleA`, the expression `userDistressed > medium` is evaluated first, and only when it is false is the expression for the next defeater (**`unless`** `userUnderDressed`) evaluated. So, the following trace can be observed for `RuleA`, noting that both measures are read at the same time, though in an order unrelated to that of evaluation:

Trace 4 for RuleA

```
CurtainOpenRqt, userUnderDressed.true, userDistressed.low, RefuseRequest
```

In this scenario, the second defeater (**`unless`** `userDistressed > medium`) does not apply, but the first defeater (**`unless`** `userUnderDressed`) does, and so the response `RefuseRequest` can be observed afterwards, with a maximum of 3 `tocks` in between corresponding to the imposed deadline (**`within`** `3,s`).

Our notion of conflict for a pair of rules `r1` and `r2`, adapted from [25] is recast below in terms of traces, where a joint trace captures the behaviour of two rules, with triggers and responses performed in agreement and the value of measures read being the same.

Definition 1. *Rules* `r1` *and* `r2` *are conflict free, if, and only if, for every joint trace* t_1 *of* `r1` *and* `r2`*, there is a joint trace* t_2 *of* `r1` *and* `r2` *after* t_1 *that contains at least one event and at least one event different from tock.*

In other words, conflict-freedom requires that for every joint trace t_1, there is an extension t_2 of t_1 that can make timed progress for both rules, namely by having at least one event and a non-*tock* event. This ensures that enforcing both rules does never lead to a deadlock (where no further events are possible), or to a scenario in which there is no deadlock, but only passage of time can be observed. The latter is a *timed deadlock*, where time can progress via *tock*, but no other event is possible.

The mechanisation of our notion of conflict [25] relies on composing the *tock*-CSP semantics of the two rules under examination in parallel, synchronising on their common alphabets, that encompass trigger and response events, while ensuring uniform passage of time by also synchronising on *tock*s. Without loss of generality, this composition is defined in a context which ensures that the values of measures do not change over time, and that the values of measures read by both rules are the same, though they can be read independently, i.e., in a different order. A conflict arises if, and only if, the composition has a (timed) deadlock, with a counter-example trace identifying a scenario where no further joint progress is possible. Our tool automatically produces assertions for identifying the presence of such deadlocks using the CSP model checker FDR [12].

Redundancies. A pair of non-conflicting rules that have overlapping alphabets may be redundant. Informally, a rule is redundant if all restrictions it imposes are covered by another rule. For example, `RuleB` and `RuleC` from Listing 1.2 have the same alphabet and are not conflicting; however, `RuleB` imposes a restriction on **`not`** `CurtainsOpened` for `60,s`, which is weaker than the restriction imposed by `RuleC` for a longer period of `120,s`, so `RuleB` is redundant with respect to `RuleC`. Therefore, `RuleB` can be removed from the set of normative requirements, as an implementation that satisfies `RuleC` will also satisfy `RuleB`.

Checking with SLEEC-TK. Checking for redundancies with SLEEC-TK is performed via model checking with FDR. For each pair of SLEEC rules `r1` and `r2` with overlapping alphabets, the tool considers checks for redundancies in both directions, that is, whether `r1` is redundant with respect to `r2` and vice-versa.

Unlike for conflict checking, a counterexample trace is only produced when a rule is not redundant, corresponding to a scenario not constrained by the other rule being checked. For example, in the case of `RuleB` and `RuleC` in Listing 1.2, checking whether `RuleC` is redundant with respect to `RuleB` produces the trace:

Trace 5 showing redundancy of `RuleC` with respec to `RuleB`

userUnderDressed.true, CurtainOpenRqt, ...
(userUnderDressed.false, tock)^{60} ..., CurtainOpenRqt

This trace corresponds to a scenario where the measure `userUnderDressed` is initially read as `true`, the trigger `CurtainOpenRqt` happens, followed by 60 occurrences of reading `userUnderDressed` as `false` and `tock`, with the trigger happening again afterwards. Here, the trace indicates that `RuleB` may apply again `60,s` after being triggered, yet `RuleC` continues to require that the response `CurtainsOpened` should not happen for the full `120,s` from the initial trigger. On the other hand, checking whether `RuleB` is redundant with respect to `RuleC` produces no trace, indicating that the restrictions imposed by `RuleB` are fully covered by `RuleC`. Next, we elaborate on the formal notion of redundancies based on traces.

Process-Algebraic Notion of Redundancies. Our definition of redundancy [25] considers a pair of conflict-free rules and is also based on their traces. It is reproduced below, where $t \upharpoonright E$ denotes the trace obtained from t by removing all events that are not in the set E, and $\alpha_{\mathsf{E}}(\mathsf{r})$ is the alphabet of the rule r, that is, the set of CSP events capturing the rule's trigger and responses.

Definition 2. *For conflict-free rules* r1 *and* r2*, rule* r2 *is redundant with respect to* r1 *if, and only if, for every trace* t_1 *of* r1*, there is a trace* t_2 *of* r2*, such that, (1)* $t_1 \upharpoonright \alpha_{\mathsf{E}}(\mathsf{r}_1) = t_2 \upharpoonright \alpha_{\mathsf{E}}(\mathsf{r}_1)$*, and (2) for every event* e *of* $\alpha_{\mathsf{E}}(\mathsf{r}_1)$ *in a position* i *of these traces, for every measure* m *in* $\alpha_{\mathsf{M}}(\mathsf{r}_1)$*, the value of* m *recorded in* t_1 *and* t_2 *at position* i *are the same.*

In words, we characterise a rule r2 as redundant if every trace of r1, with events restricted to triggers and responses of just r1, via $t_1 \upharpoonright \alpha_{\mathsf{E}}(\mathsf{r}_1)$, can also be performed by r2 when restricted to the same events (1), and the traces agree on the value of measures recorded in relation to a trigger or response event (2). This is so that reading of measures can be performed in different orders, but their values must be consistent when comparing two rules for redundancy. In summary, if every behaviour allowed by r1 is also allowed by r2, then r2 imposes no additional restrictions, and so is redundant.

The mechanisation of redundancy checking is stated in terms of traces refinement [18] between the *tock*-CSP semantics of a pair of rules, r1 and r2, in a context where all measures used by either rule are read, once at the beginning of every trace, and after every *tock*, before any other events, encoding condition (2). Similarly, the context also encodes condition (1) by ensuring that events not in $\alpha_{\mathsf{E}}(\mathsf{r}_2)$ are unrestricted when comparing the refinement of r1's semantics in context against that of r2. This is equivalent to the formulation in Definition 2.

In summary, the traces of the process that characterises a rule identify the behaviours allowed by the rule. So the smaller that set of traces, as captured by the notion of traces refinement, the more restrictive is that rule. This concludes our discussion of pair-wise consistency validation of rules. In what follows we discuss set-wise validation of rules.

3.2 Set-Wise Well-Formedness Validation

Pairwise conflict and redundancy checks provide important guarantees, but they do not capture all *well-formedness issues* (WFIs) that arise from the *combined* effect of multiple rules. To support the identification of these additional WFIs, our SLEEC framework includes a second validation tool, LEGOS-SLEEC, whose functionality augments that of SLEEC-TK.

LEGOS-SLEEC[3] enables *set-wise verification* of SLEEC rulesets [7], allowing stakeholders to reason about the global behaviour induced by a set of normative rules. Specifically, the tool supports the analysis of five classes of WFIs: (i) *vacuous conflicts*, where a rule can never be triggered without violating the ruleset; (ii) *situational conflicts*, where inconsistencies arise only under specific contextual conditions; (iii) *set-wise redundancies*, where a rule does not restrict

[3] The LEGOS-SLEEC tool and instructions for its installation are available online at https://zenodo.org/records/14714867.

behaviour beyond what is already enforced; (iv) *insufficiencies*, where normatively undesirable behaviours remain possible despite the rules; and (v) cases of *over-restrictiveness*, where desirable or intended behaviours are inadvertently ruled out. Together, these checks allow stakeholders to assess whether a SLEEC ruleset is coherent, sufficiently protective, and aligned with its intended purpose.

For each WFI, LEGOS-SLEEC provides: (i) a formal check at the level of the full ruleset, and (ii) an explanation either as a witness trace (when the property is satisfiable), or as a causal unsatisfiability proof (when the property is impossible).

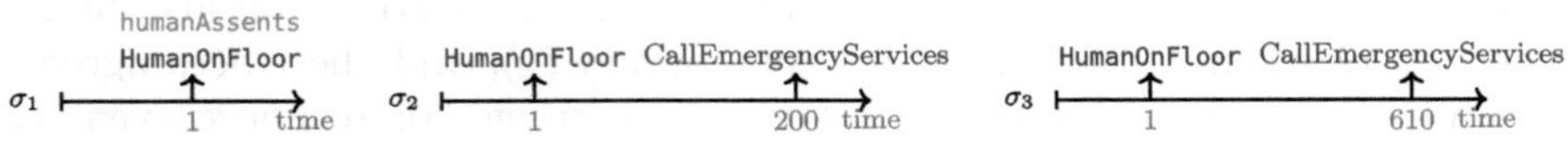

Fig. 3. Examples of traces for the ALMI robot.

Trace Semantics (rule sets). LEGOS-SLEEC adopts a trace-based semantics. Given a SLEEC specification over a set of events E and a set of measures M, a *trace* represents a *finite* sequence of *states*

$$\sigma = (\mathcal{E}_1, \mathbb{M}_1, \delta_1), (\mathcal{E}_2, \mathbb{M}_2, \delta_2), \ldots (\mathcal{E}_n, \mathbb{M}_n, \delta_n),$$

where, for every time point $i \in [1, n]$

1. $\mathcal{E}_i \in E$ is a set of events that occur at time point i;
2. $\mathbb{M}_i : M \to \mathbb{N}$ assigns every measure in M to a concrete value at time point i (with appropriate mapping of Boolean measures to $\{0, 1\}$, scale measures with three values to $\{0, 1, 2\}$, etc.), and
3. $\delta_i : \mathbb{N}$ captures the time value of time point i (e.g., the second time point can have the value 30, for 30 s).

We assume that the time values in the trace are strictly increasing (i.e., $\delta_i < \delta_{i+1}$ for every $i \in [1, n-1]$). Given a measure assignment $\mathbb{M}_i$ and a term t, let $\mathbb{M}_i(t)$ denote the result of substituting every measure symbol m in t with $\mathbb{M}_i(m)$. Since $\mathbb{M}_i(t)$ does not contain free variables, the substitution results in a natural number. Similarly, given a proposition p, we say that $\mathbb{M}_i \models p$ if p is evaluated to $\top$ after substituting every term t with $\mathbb{M}_i(t)$, where $\top$ stands for Boolean true and $\bot$ for false.

Example. Consider the trace σ_1 shown in Fig. 3, corresponding to $(\mathcal{E}_1, \mathbb{M}_1, \delta_1)$, where $\mathcal{E}_1 = \texttt{HumanOnFloor}$, $\mathbb{M}_1(\texttt{humanAssents}) = \top$, and $\delta_1 = 1$. While the trace σ_2 shown in Fig. 3, corresponds to $(\mathcal{E}_1, \mathbb{M}_1, \delta_1), (\mathcal{E}_2, \mathbb{M}_2, \delta_2)$, where $\mathcal{E}_1 = \texttt{HumanOnFloor}$, $\mathbb{M}_1(\texttt{humanAssents}) = \bot$, $\delta_1 = 1$, $\mathcal{E}_2 = \texttt{CallEmergencyServices}$, $\mathbb{M}_2(\texttt{humanAssents}) = \bot$, and $\delta_2 = 200$.

Trace Fulfilment. A trace σ *fulfils* an obligation by evaluating whether the required event occurrence (or non-occurrence) holds within the designated time frame starting from the triggering time point.

- A *positive obligation* ("*e within t*") is fulfilled if the specified event e occurs within the designated time window of length t from the starting time point.
- A *negative obligation* ("*not e within t*") is fulfilled if the specified event e does *not* occur at any time throughout the designated time window.
- A *conditional obligation* ("$p \Rightarrow ob$") is fulfilled if either the precondition p does not hold at the triggering time point, or the subsequent obligation ob is fulfilled.
- An *obligation chain* ("cob_1^+ *otherwise* $cob_2^+ \ldots cob_m$") is fulfilled if the first obligation is fulfilled; otherwise, if it is violated, the subsequent obligation in the chain must be fulfilled from the violation point onward.

A rule r is fulfilled if, *whenever its conditions hold at any time point*, the associated obligation (or obligation chain) is fulfilled from that time point onward.

Example. Consider the negative obligation

```
then not CallEmergencyServices within 600 seconds
```

of rule `R1` in Listing 1.3. We analyze the traces σ_1, σ_2, and σ_3 shown in Fig. 3. In σ_2 and σ_3, the triggering event `HumanOnFloor` occurs at $\delta_1 = 1$ and the measure `humanAssents` is $\perp$, thus activating the obligation. In σ_2, `CallEmergencyServices` occurs at $\delta_2 = 200$. Since $200 - 1 = 199 \leq 600$, the event occurs within the forbidden 600-second window; hence, the negative obligation is *violated.* In σ_3, `CallEmergencyServices` occurs at $\delta_2 = 610$. Since $610 - 1 = 609 > 600$, the event occurs outside the 600-second window; therefore, the negative obligation is *fulfilled.*

Rule Set Fulfilment. A trace σ fulfils a ruleset Rules if it fulfils every rule in the set. A SLEEC ruleset Rules induces a language of accepted behaviours, denoted $L(\mathsf{Rules})$, defined as: $L(\mathsf{Rules}) = \{\, \sigma \mid \forall r \in \mathsf{Rules},\ \sigma \models r \,\}$. That is, $L(\mathsf{Rules})$ is the largest set of traces such that every trace in the set respects every rule in Rules. For example, if $\mathsf{Rules} = \{R_1, R_2, R_3\}$, then: $L(\mathsf{Rules}) = L(R_1) \cap L(R_2) \cap L(R_3)$. This set-based view enables uniform definitions of all well-formedness issues (WFI) in terms of emptiness or non-emptiness of language intersections, and supports diagnostics in the form of concrete witness traces (counter-examples) or unsatisfiability explanations. In the following, we present each class of well-formedness issue.

Vacuous Conflicts. A rule is *vacuously conflicting* when triggering it in any feasible situation necessarily leads to a violation of the ruleset. Intuitively, such a rule is unachievable, that is, its trigger can never be realised without contradicting other rules.

```
rule_start
 R1 when HumanOnFloor and not humanAssents then
    not CallEmergencyServices within 600 seconds

 R2 when OpenCurtainRequest and (not underDressed) then OpenCurtain within 30
     seconds

 R3 when SmokeDetectorAlarm then CallEmergencyServices within 300 seconds

 R4 when DressingStarted and (roomTemperature < MIN_TEMP)
    and userUnderDressed then DressingComplete within 1 minute

 R5 when SmokeDetectorAlarm then CallEmergencyServices within 7 minutes

 R6 when SmokeDetectorAlarm then NotifyCaregiver within 2 minutes

 R7 when NotifyCaregiver then CallEmergencyServices within 5 minutes

 R8 when SmokeDetectorAlarm then not CallEmergencyServices within 10 minutes
rule_end
```

Listing 1.3. Example of a set of rules with well-formedness issues.

Trace-Based Definition. Let Rules be a ruleset and let r be a rule in Rules with trigger $\mathit{Trig}(r)$. The rule r is vacuously conflicting (w.r.t. Rules) if no trace in $L(\mathsf{Rules})$ contains an occurrence of its trigger under satisfied trigger conditions. Equivalently, $L(\mathsf{Rules}) \cap L(\mathit{Trig}(r)) = \emptyset$, where $L(\mathit{Trig}(r))$ denotes traces in which $\mathit{Trig}(r)$ can occur with its associated measures satisfying the trigger predicate.

Example. Let Rules consist of the rules in Listing 1.3. Rules `R6` and `R7` together imply that whenever `SmokeDetectorAlarm` occurs, `CallEmergencyServices` must occur within 7 min (2 + 5 min). This follows *transitively*: `R6` requires that `SmokeDetectorAlarm` triggers `NotifyCaregiver` within 2 min, and the occurrence of `NotifyCaregiver` in turn triggers `R7`, which requires the response `CallEmergencyServices` within 5 additional minutes. Thus, the response of one rule (`R6`) activates the trigger of another (`R7`), propagating the obligation. However, `R8` prohibits the response `CallEmergencyServices` within 10 min after `SmokeDetectorAlarm`. Since 7 min $\leq$ 10 min, any trace that triggers `SmokeDetectorAlarm` necessarily violates either `R7` or `R8`. Therefore, no trace in $L(\mathsf{Rules})$ can contain a feasible occurrence of `SmokeDetectorAlarm` satisfying these rules. Hence, `R6` (and symmetrically `R8`) is vacuously conflicting, as $L(\mathsf{Rules}) \cap L(\mathit{Trig}(\mathsf{R6})) = \emptyset$.

Checking with LEGOS-SLEEC. The tool checks whether $L(\mathsf{Rules}) \cap L(\mathit{Trig}(r))$ is empty and, if so, produces a causal unsatisfiability proof pinpointing why the trigger cannot occur without violating the ruleset, as shown below.

Diagnostic for Vacuous Conflicting SLEEC Rules

Conflicting SLEEC rule:
R8: when SmokeDetectorAlarm then not CallEmergencyServices within 10 min.

Because of the following SLEEC rules:
R6: when SmokeDetectorAlarm then NotifyCaregiver within 2 min.
R7: when NotifyCaregiver then CallEmergencyServices within 5 min.

Situational Conflicts. A *situational conflict* exists when there is *at least one feasible situation* (trace prefix) in which the ruleset can be triggered into an inconsistent state, i.e., a history of events and measures, always leads to a conflict with some other rules in the future. Unlike vacuous conflicts, these are *witnessed* by a feasible trace.

Trace-Based Definition. A ruleset Rules has a situational conflict if there exists a trace σ that is consistent with the ruleset up to some point, but cannot be extended to a full trace in $L(\mathsf{Rules})$ once certain triggers occur under certain measure valuations. Operationally, LEGOS-SLEEC reports a witness trace demonstrating a feasible path to inconsistency.

Example. Consider the rules `R1` and `R3` shown in Listing 1.3. `R1` conflicts with `R3` in a situation where the event `HumanOnFloor` occurs without human consent for calling emergency services (**`not`** `humanAssents`), while `SmokeDetectorAlarm` is also triggered. This results in `R3` blocking the response of `R1`, as illustrated below in trace σ_4.

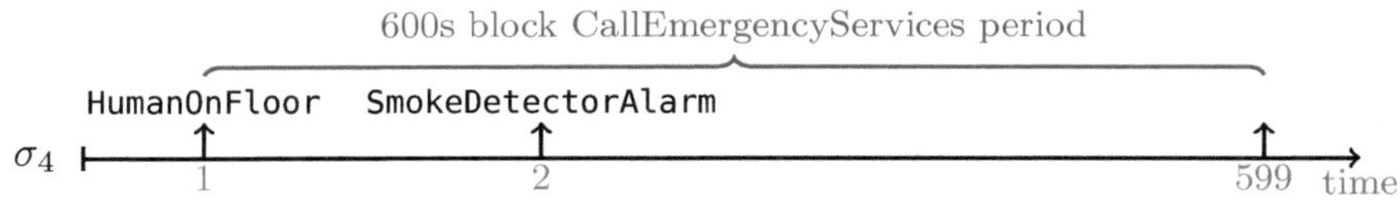

Checking with LEGOS-SLEEC. The tool produces a concrete witness trace (scenario) that exhibits the conflict (shown below), which can be used as guidance during repair (e.g., adding defeaters, introducing priorities, or refining triggers).

Diagnostic for Situational Conflict

Situational conflict under situation:
at time 1: HumanOnFloor
at time 1: Measure (humanAssents = false)
at time 2: SmokeDetectorAlarm
at time 2: blocked_CallEmergencyServices

Conflicting SLEEC rule:
R3: when SmokeDetectorAlarm then CallEmergencyServices within 300 sec.

Because of the following SLEEC rule:
R1: when HumanOnFloor and not humanAssents then
not CallEmergencyServices within 600 sec.

Set-Wise Redundancy. A rule is *set-wise redundant* if removing it does not change what the ruleset allows the system to do; that is, it imposes no additional restriction beyond the remaining rules.

Trace-Based Definition. Given Rules and a rule $r \in \mathsf{Rules}$, the rule r is set-wise redundant if $L(\mathsf{Rules}) = L(\mathsf{Rules} \setminus \{r\})$. Equivalently, $L(\mathsf{Rules}) \subseteq L(\mathsf{Rules} \setminus \{r\})$ always holds (by monotonicity), and redundancy holds when also $L(\mathsf{Rules} \setminus \{r\}) \subseteq L(\mathsf{Rules})$.

Example. Consider rules R5, R6, and R7. Together, R6 and R7 imply that whenever `SmokeDetectorAlarm` occurs, `CallEmergencyServices` must occur within 7 min. Rule R5 explicitly requires that when `SmokeDetectorAlarm` occurs, `CallEmergencyServices` must occur within 7 min. Since this obligation is already enforced by the combination of R6 and R7, adding R5 does not further restrict the accepted behaviour of the ruleset. Hence, R5 is redundant, that is: $L(\{\mathsf{R6},\mathsf{R7},\mathsf{R5}\}) = L(\{\mathsf{R6},\mathsf{R7}\})$.

Checking with LEGOS-SLEEC. LEGOS-SLEEC checks redundancy at the rule-set level and provides an explanation (reproduced below) showing why the removed rule does not exclude any additional traces beyond those already excluded by the remaining rules.

Diagnostic for Set-wise Redundant SLEEC Rules

Redundant SLEEC rule:
R3 when SmokeDetectorAlarm then CallEmergencyServices within 300 seconds

Because of the following SLEEC rules:
R6: when SmokeDetectorAlarm then NotifyCaregiver within 2 minutes
R7: when NotifyCaregiver then CallEmergencyServices within 5 minutes

```
concern_start
 C1 when SmokeDetectorAlarm and ((not userDisablesAlarm) or alarmRestarts) then
     not CallEmergencyServices within 1 minute
concern_end
purpose_start
 P1 when HumanOnFloor and (userUnconscious and (not humanAssents)) then
    CallEmergencyServices within 5 minutes
purpose_end
```

Listing 1.4. Example of SLEEC concerns and purpose.

Insufficiency. A set of rules are insufficient if adhering to those rules can still allow executing an undesirable behavior,5which we call *concern*. Concerns capture undesirable behaviours that must be avoided (e.g., hazards, violations, or harms). Concerns are specified in a SLEEC-like form (such as `C1` shown in Listing 1.4), and are treated as sets of traces $L(c)$ representing bad behaviours.

Trace-Based Definition. A ruleset Rules is *insufficient* for preventing a concern c iff $L(\mathsf{Rules}) \cap L(c) \neq \emptyset$. That is, there exists at least one feasible trace that satisfies all rules yet realises the undesirable behaviour.

Example. Consider the concern `C1` shown in Listing 1.4. `C1` can arise even when all the rules in Listing 1.3 are respected, as illustrated below in σ_5.

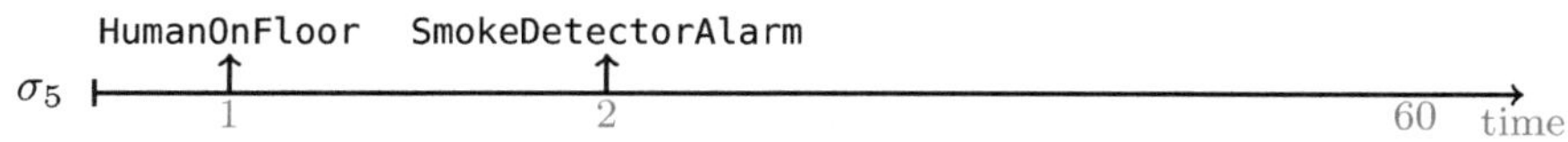

Checking with LEGOS-SLEEC. The tool searches for a witness trace satisfying both the ruleset and the concern. The returned trace is a concrete diagnostic scenario that guides strengthening the ruleset (e.g., adding missing obligations or refining defeaters).

Diagnostic for Insufficiency

Concern raised:
C1 when SmokeDetectorAlarm and ((not userDisablesAlarm) or alarmRestarts) then not CallEmergencyServices within 1 minute
under the situation:
at time 1: HumanOnFloor
at time 1: Measure (humanAssents = false, userDisablesAlarm = false, alarmRestarts = false)
at time 2: SmokeDetectorAlarm
at time 2: Measure (userDisablesAlarm = false, alarmRestarts = false)

Over-Restrictiveness A ruleset is *overly restrictive* if it prevents the realisation of desirable or intended behaviours, even though these behaviours are consistent with the system's purpose. Purposes capture functional goals or intended outcomes of the system and are specified (such as `P1` shown in Listing 1.4), similarly to concerns, as sets of traces $L(p)$ representing desirable behaviours.

Trace-Based Definition. A ruleset Rules is *overly restrictive* with respect to a purpose p iff $L(\mathsf{Rules}) \cap L(p) = \emptyset$. That is, no feasible trace satisfying all rules realises the intended behaviour.

Example. Consider a system purpose `P1` shown in Listing 1.4, stating that when a person is detected on the floor and is unconscious without explicitly assenting, emergency services must be called within 5 min. However, consider the ruleset shown in Listing 1.3, which contains `R1`, a rule prohibiting calling emergency services within 600 s whenever a person is on the floor and has not assented. Since 600 s corresponds to 10 min, this prohibition directly blocks the intended 5-minute call in all feasible traces. As a result, the ruleset becomes overly restrictive with respect to this purpose.

Checking with LEGOS-SLEEC. LEGOS-SLEEC checks whether the intersection $L(\mathsf{Rules}) \cap L(p)$ is empty. If so, the tool produces a causal unsatisfiability explanation identifying the subset of rules responsible for blocking the intended behaviour, thereby supporting rule relaxation or refinement.

Diagnostic for Restrictive SLEEC Rules

Blocked system purpose:
P1 when HumanOnFloor and (userUnconscious and (not humanAssents))
then CallEmergencyServices within 5 minutes

Because of the following SLEEC rules:
R1 when HumanOnFloor and not humanAssents then
not CallEmergencyServices within 600 seconds

4 Conformance Verification of Design Models

The implementation of a SLEEC-aware system must conform to the relevant SLEEC rules. We now show, via an example, what is required of a conforming implementation. In Sect. 4.1, we present a design for an assistive-dressing robot described in a diagrammatic domain-specific language called RoboChart [17]. This is a language for defining control software for robotic applications. Via the example, we illustrate the different roles of events and measures in considering conformance. Section 4.2 describes how we can use our tool to check that a RoboChart design satisfies a set of SLEEC rules. Finally, in Sect. 4.3, we present our mathematical notion of conformance based on CSP refinement.

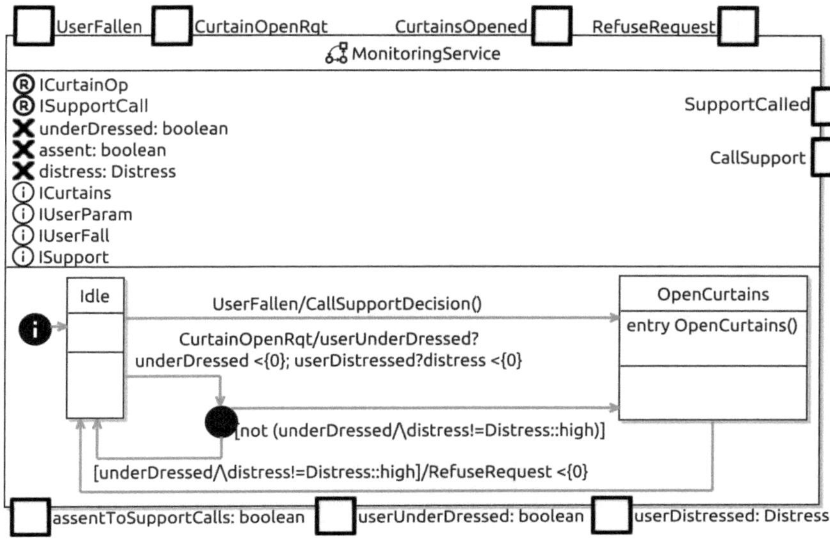

Fig. 4. MonitoringService state machine for the assistive-dressing robot

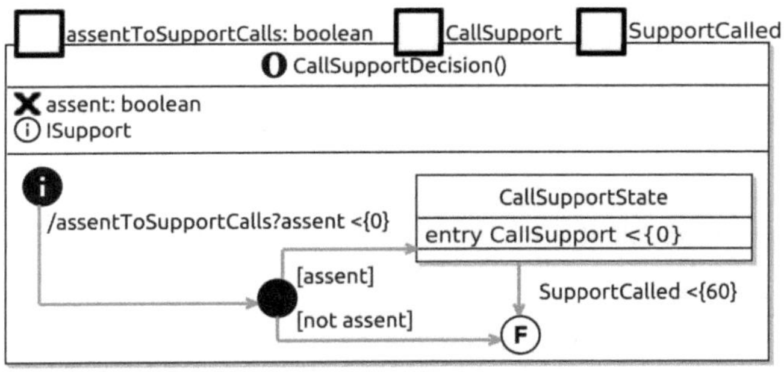

Fig. 5. CallSupportDecision operation

4.1 Design

A RoboChart model defines behaviour based on a collection of (parallel) state machines and neural network components grouped to define a RoboChart component called module. For our example, we present in Figs. 4, 5, and 6 three RoboChart state machines for part of the software of an assistive-dressing robot. The machine in Fig. 4, called MonitoringService, defines a monitor that accepts requests to open the curtains (via a capability CurtainOpenRqt) and UserFallen signals. Around the box of the machine, all capabilities used in this agent are indicated by small squares. At the top half of the box, above the actual machine, a context of declarations defines these capabilities and local variables.

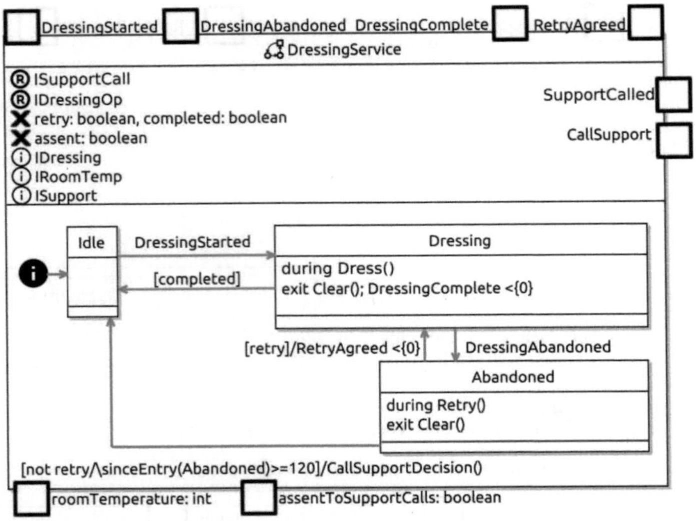

Fig. 6. DressingService state machine for the assistive-dressing robot

At the start, indicated by the black circle with an i inside, MonitoringService is in the state Idle. In that state, two outgoing transitions define behaviour when a CurtainOpenRqt and when UserFallen is signalled. For UserFallen, the outgoing transition calls a software operation CallSupportDecision(). This operation is defined by another state machine presented in Fig. 5.

In CallSupportDecision(), at the start, the measure assentToSupportCalls is read and the result is recorded in the variable assent. With <{0}, we indicate that the reading of the measure is immediate, since all measures are assumed to be always readily available. A junction, represented by a black circle, defines behaviour when there is assent and when there is not (not assent). If there is, CallSupportDecision() moves to the CallSupport state. If there is not, CallSupport-Decision() moves to the final state: white circle with an F.

In CallSupport, the entry action raises the event CallSupport immediately. That capability establishes a communication with the support team, with an input via supportCalled indicating when the contact is successful, before CallSupportDecision() terminates. The deadline of 60 time units (seconds) requires that any actions involved in implementing the CallSupport capabilities, such as establishing a phone connection and dialling, are completed within that time.

Finally, the state machine DressingService in Fig. 6 deals with the interactions between the human and the robot during dressing. In the Idle state, a request to start the dressing via the capability represented by an event DressingStarted is accepted. At that point, DressingService moves to a state Dressing, where the actual dressing is carried out via an operation Dressing A result can be signalled via a variable completed, when DressingService goes back to Idle, of via an event DressingAbandoned, when DressingService transitions to the state Abandoned. Upon exitting of Dressing, a Clear() operation resets the robot and signals that the dressing is completed, via the event DressingComplete.

In Abandoned, the operation Retry() negotiates an attempt to try again. Success is recorded in a variable retry and signalled via RetryAgreed. If that does not happen, and more than 120 time units (corresponding to two minutes) have gone by, CallSupportDecision() is called, before DressingService moves back to Idle.

RoboChart has a tool, called RoboTool[4], which supports modelling via editing and validation facilities. RoboTool also includes support for verification that RoboChart models conforms to a SLEEC rule, with such functionality also available as part of SLEEC-TK. In the next sections, we present this tool and the technique that it implements. The complete RoboChart model for our robot is available in our repository of supplementary material[5].

4.2 Verification

Using our technique, we get confirmation that Rule1 in Listing 1.1, for instance, is satisfied by our design of assistive-dressing robot. Figure 7 shows a screenshot of RoboTool showing the result of the verification.

At the top of Fig. 7, we show the RoboTool editing panel, where the state machine MonitoringService in Fig. 4 is visible. This is the part of our RoboChart model that enforces `Rule1`. Underneath, the editing panel we show a few lines of a file highlighted on the left panel. It is called rad.assertions, and defines a document using a simple controlled natural language to specify verification checks of SLEEC rules. The first check shown is introduced by a clause **`timed sleec assertion`**. This clause gives a name to the check, A1, and then the name of the RoboChart package, RAD, and module, Software, defining a design for the control software of our robot. MonitoringService is part of that module. Finally the assertion states that `RAD::Software` **`conforms to`** `Rule1`. We recall that `Rule1` is in the file tutorial.sleec that is in the Eclipse project.

At the bottom of Fig. 7, we show the report that RoboTool generates after checking this assertion using the model checker FDR [12]. At the rightmost column, we see that the result of the verification is true. As said, `Rule1` is enforced by MonitoringService. As explained in the previous section, in its initial state, MonitoringService accepts a CurtainOpenRqt, and then checks the measures userUnderDressed and userDistressed. As dictated by the defeaters of `Rule1`, the results of those checks lead to a call to a software operation OpenCurtains() which effectively opens the curtains, or to a RefuseRequest event. That refusal event happens immediately. Since we reading of measures is also immediate, and calculation of boolean expressions also do not take time, the deadline for a RefuseRequest given in `Rule1` is satisfied. Satisfaction of the other deadlines is ensured by the definition of the operation OpenCurtains() omitted here.

[4] https://robostar.cs.york.ac.uk/robotool/.

[5] https://github.com/UoY-RoboStar/SLEEC-TK.

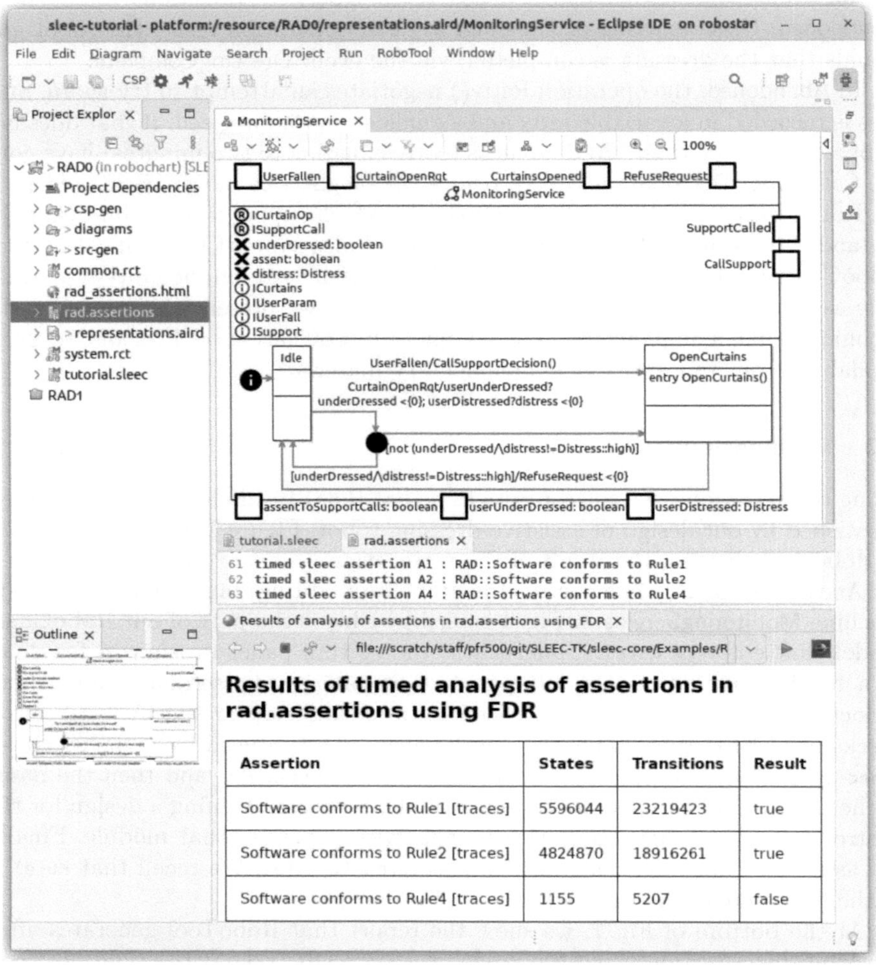

Fig. 7. Verification of `Rule1`, `Rule2` and `Rule4` in Listing 1.1 using RoboTool

The report in Fig. 7 also shows that `Rule4` is not satisfied. As shown in Fig. 4, the design engineer has specified a behaviour in which support is called in no more than 1 min, if the user falls and there is consent. Even though this is a reasonable course of action, it means that `Rule4` is not satisfied at least in the scenario described by the trace below, which our tool indicates as possible.

Trace 6 as counter-example of non-conformance against `Rule4`

```
assentToSupportCalls.true, DressingStarted.in, UserFallen.in,
DressingAbandoned, SupportCalled
```

The trace indicates that there is assentToSupportCalls, the dressing process has started (DressingStarted), the user has fallen (UserFallen), and the dressing has been abandoned (DressingAbandoned). The issue is that Rule4 dictates that when dressing is abandoned, there should be a period of 2 min in which a protocol for agreeing on a new attempt (RetryAgreed) should take place. The defined design, however, allows for an immediate call to support, as indicated by the final event SupportCalled in the trace above, because the user has fallen.

Using our tool, we can get these scenario descriptions, which help the identification of at least one situation in which a rule is violated. This may lead to changes in the design, or to revision of rules and other requirements.

In Fig. 7, we note that Rule2 is also satisfied. This rule has two defeaters (see Listing 1.1), both based on the roomTemperature. The motivation is that the temperature should inform how fast the dressing process needs to proceed. The designer, however, has decided to meet Rule2 by always carrying out the dressing as fast as possible. So, Rule2 requires use of the measure roomTemperature to decide on a response, but a correct design does not need to take that measure.

In this case, taking the measure is not needed because there is a response that is satisfactory in all cases predicted in the rule, and that response in chosen. In other situations, it might be because the value of the measure is already available. For example, if our robot were to control the room temperature, it might take that measure routinely, and not need to take it again when the dressing is started.

In the next section, we explain our definition of conformance, which describes formally the set of designs that are correct.

4.3 Conformance

Conformance of a given System Under Verification (SUV) against a SLEEC rule is considered only when a rule is "relevant", that is, the rule only refers to events representing capabilities of the SUV. This is to ensure that the rule is applicable, that is, it can be triggered and respond using capabilities available in the SUV.

Our definition of conformance for an SUV with respect to a SLEEC rule assumes that the SUV can be given a semantics as a *tock*-CSP process. For the RoboChart machines in Figs. 4, 5, and 6, this can be automatically calculated using RoboTool so that conformance can be established by model checking with FDR. It is based on the notion of traces refinement for *tock*-CSP, that ensures that events of the SUV occur in the order and time specified by the rules, respecting time budgets and deadlines. Our notion of conformance assumes that the value of measures do not change in a time unit.

The formal definition is reproduced [25] below, where $[\![\mathsf{r}]\!]_{\mathsf{R}}$ is the *tock*-CSP semantics for a rule r, $\alpha_{\mathsf{E}}(\mathsf{r})$ is the alphabet of the rule, that is, the set of CSP events capturing the rule's trigger and responses, and $\alpha_{\mathsf{M}}(\mathsf{r})$ is the set of measures used by the rule.

Definition 3. *An SUV conforms to a rule* r*, written* $\mathsf{r} \models SUV$*, where SUV is the tock-CSP semantics of SUV, if, and only if, for every trace* t_1 *of the process* $SUV;Stop$*, there is a trace* t_2 *of* $[\![\mathsf{r}]\!]_{\mathsf{R}}$*, such that: (1)* $t_1 \upharpoonright \alpha_{\mathsf{E}}(\mathsf{r}) = t_2 \upharpoonright \alpha_{\mathsf{E}}(\mathsf{r})$ *and;*

(2) for every event e of $\alpha_E(r)$ *in a position i of these traces, for every measure m in* $\alpha_M(r)$*, the value of m recorded in* t_1 *and* t_2 *at position i are the same.*

In words, it corresponds to traces refinement in *tock*-CSP, where the specification is defined in terms of the process $[\![r]\!]_R$ that defines the semantics of a rule r. Similarly to the check for redundancy, refinement disregards the measures, however, we require that the values of the measures recorded in the specification and the SUV to be the same.

In the definition above, we consider the CSP process $SUV ; Stop$, rather than just SUV, as the process that gives semantics to a rule is non-terminating, whereas the SUV could terminate. Therefore, to avoid flagging a problem with termination, SUV is sequentially composed ($;$) with the *tock*-CSP process $Stop$ that deadlocks, but allows passage of time.

The filtering (1) in our definition of conformance allows an SUV to engage in additional events and read additional measures, as the comparison is based on the alphabet of events $\alpha_E(r)$ and measures $\alpha_M(r)$ of a rule. The mechanisation of conformance adopts a similar pattern to that of redundancy (Definition 2), introducing a context on both CSP processes in the refinement.

5 Conclusion

SLEEC [22], its language [25], its semantics [8,25], and its techniques and tools [7, 11] provide a powerful basis to study, design, and verify SLEEC properties of autonomous agents. This tutorial paper provides a starting point to understand the in-depth published work on elicitation and operationalisation of SLEEC principles, concerns, and requirements (rules).

As explained here, elicitation of normative requirements, however, is not solely a task for computer scientists. It potentially requires the involvement of lawyers, psychologists, sociologists, philosophers, domain experts, and users, or their representatives. To develop the SLEEC framework, we have collaborated within a multi-disciplinary team to define a notation and devise techniques that are accessible across all these disciplines (therefore, catering for non-technical stakeholders) and across domains. SLEEC provides a common language and structure to discussions, with tool-supported validation and verification based on (counter)examples providing valuable input, and feedback on intermediate versions of a SLEEC ruleset and/or autonomous agent design.

The SLEEC tools are freely available for use by the academic community. SLEEC-TK [11] is released open source under the Eclipse Public Licence (EPL-2.0), while LEGOS-SLEEC [7] is licenced under CC-BY-A.

The SLEEC paradigm has attracted the attention of the community worldwide. Besides our contributions to academic and industrial conferences, other researchers have taken up the work. For example, in [23], the authors have studied SLEEC in the context of Prolog implementations, and the project presented in [2] proposes a fuzzy-logic representation for SLEEC rules. In South Korea, there is an effort to use SLEEC to capture the rules of the road embedded in traffic regulations, contributing to the development of SLEEC-aware monitors for autonomous vehicles.

There is much that still needs to be done. On the elicitation side, a richer language that considers prioritisation of rules [21] may facilitate describing their required application in the case of conflicts. On the engineering side, the issue of explainability, based on the interpretation of the low-level decisions and actions of an agent in terms of the rules, can have a significant impact on the acceptability of that agent.

We invite colleagues to contribute to our efforts. Our goal is to enable the cost-effective development of trustworthy autonomous agents that can be assured to act according to human values. For that, we need the contribution of scientists and engineers with a wide range of backgrounds.

Acknowledgements. This work has received funding from the UKRI project EP/V026747/1 'Trustworthy Autonomous Systems Node in Resilience', the RAi UK project 'Disruption Mitigation for Responsible AI', an IVADO Distinction Starting Grant, the Royal Academy of Engineering grant CiET1718/45, the EU Horizon project RoboSAPIENS under agreement number 101133807, and the Bridging Responsible AI Divides (BRAID) programme with funds received from the Arts and Humanities Research Council under grant 873 AH/X007146/1.

References

1. Alechina, N., Dastani, M., Logan, B.: Norm specification and verification in multi-agent systems. J. Appl. Logics **5**(2), 457–489 (2018)
2. Assadi, Z., Inverardi, P.: Fuzzy representation of norms (2026). https://arxiv.org/abs/2601.04249
3. Baxter, J., Ribeiro, P., Cavalcanti, A.L.C.: Sound reasoning in tock-CSP. Acta Informatica **59**, 125–162 (2022). https://doi.org/10.1007/s00236-020-00394-3
4. Boltz, N., Getir Yaman, S., Inverardi, P., de Lemos, R., Landuyt, D.V., Zisman, A.: Human empowerment in self-adaptive socio-technical systems. In: SEAMS'24, pp. 200–206. ACM (2024)
5. Brunero, J.: Reasons and defeasible reasoning. Philos. Q. **72**(1), 41–64 (2021). https://doi.org/10.1093/pq/pqab013
6. European Union: Regulation (EU) 2024/1689 of the European Parliament and of the Council of 13 June 2024 laying down harmonised rules on artificial intelligence (Artificial Intelligence Act). Official Journal of the European Union **L 2024/1689** (2024). http://data.europa.eu/eli/reg/2024/1689/oj
7. Feng, N., et al.: Analyzing and debugging normative requirements via satisfiability checking. In: 46th IEEE/ACM International Conference on Software Engineering, pp. 1–12 (2024)
8. Feng, N., Marsso, L., Sabetzadeh, M., Chechik, M.: Early verification of legal compliance via bounded satisfiability checking. In: Proceedings of the 34th International Conference on Computer Aided Verification (CAV'23). LNCS, Springer (2023)
9. Feng, N., et al.: Normative requirements operationalization with large language models. In: 32nd IEEE International Requirements Engineering Conference, pp. 129–141 (2024). https://doi.org/10.1109/RE59067.2024.00022

10. Feng, N., et al.: Towards a formal framework for normative requirements elicitation. In: 2023 38th IEEE/ACM International Conference on Automated Software Engineering (ASE), pp. 1776–1780. IEEE (2023)
11. Getir Yaman, S., Ribeiro, P., Burholt, C., Jones, M., Cavalcanti, A., Calinescu, R.: Toolkit for specification, validation and verification of social, legal, ethical, empathetic and cultural requirements for autonomous agents. Sci. Comput. Program. **236**, 103118 (2024)
12. Gibson-Robinson, T., Armstrong, P., Boulgakov, A., Roscoe, A.: FDR3 — a modern refinement checker for CSP. In: Ábrahám, E., Havelund, K. (eds.) Tools and Algorithms for the Construction and Analysis of Systems. LNCS, vol. 8413, pp. 187–201 (2014)
13. Hashmi, M., Governatori, G., Wynn, M.T.: Normative requirements for regulatory compliance: an abstract formal framework. Inf. Syst. Front. **18**(3), 429–455 (2016)
14. IEEE Standard 7000/2021: Model Process for Addressing Ethical Concerns during System Design (2021). https://ieeexplore.ieee.org/document/9536679
15. Inverardi, P.: The challenge of human dignity in the era of autonomous systems. In: Werthner, H., Prem, E., Lee, E.A., Ghezzi, C. (eds.) Perspectives on Digital Humanism, pp. 25–29. Springer, Cham (2022). https://doi.org/10.1007/978-3-030-86144-5_4
16. Kleijwegt, A., Getir Yaman, S., Calinescu, R.: Tool for supporting debugging and understanding of normative requirements using LLMs. In: 33rd IEEE International Requirements Engineering Conference, pp. 576–579. IEEE (2025)
17. Miyazawa, A., Ribeiro, P., Li, W., Cavalcanti, A., Timmis, J., Woodcock, J.: RoboChart: modelling and verification of the functional behaviour of robotic applications. Softw. Syst. Model. **18**, 1–53 (2019). https://doi.org/10.1007/s10270-018-00710-z
18. Roscoe, A.W.: Understanding Concurrent Systems. Texts in Computer Science, Springer (2011). https://doi.org/10.1007/978-1-84882-258-0
19. Ryan, M., Stahl, B.C.: Artificial intelligence ethics guidelines for developers and users: clarifying their content and normative implications. J. Inf. Commun. Ethics Soc. **19**(1), 61–86 (2021)
20. Tana, C., et al.: Smart aging: integrating AI into elderly healthcare. BMC Geriatrics (2025). https://doi.org/10.1186/s12877-025-06723-w
21. Townsend, B., Parnell, K.J., Yaman, S.G., Nemirovsky, G., Calinescu, R.: Normative conflict resolution through human-autonomous agent interaction. J. Responsible Technol. **21**, 100114 (2025). https://doi.org/10.1016/j.jrt.2025.100114
22. Townsend, B., et al.: From pluralistic normative principles to autonomous-agent rules. Mind. Mach. **32**(4), 683–715 (2022)
23. Troquard, N., De Sanctis, M., Inverardi, P., Pelliccione, P., Scoccia, G.L.: Social, legal, ethical, empathetic, and cultural rules: compilation and reasoning. Proc. AAAI Conf. Artif. Intell. **38**(20), 22385–22392 (2024). https://doi.org/10.1609/aaai.v38i20.30245
24. UNESCO: Recommendation on the Ethics of Artificial Intelligence (2021). https://unesdoc.unesco.org/ark:/48223/pf0000380455
25. Yaman, S.G., Ribeiro, P., Cavalcanti, A., Calinescu, R., Paterson, C., Townsend, B.A.: Specification, validation and verification of social, legal, ethical, empathetic and cultural requirements for autonomous agents. J. Syst. Softw. **220**, 112229 (2025). https://doi.org/10.1016/J.JSS.2024.112229
26. Zhu, J., Gienger, M., Franzese, G., Kober, J.: Do you need a hand? – a bimanual robotic dressing assistance scheme. IEEE Trans. Rob. **40**, 1906–1919 (2024). https://doi.org/10.1109/TRO.2024.3366008

Probabilistic Model Checking Taken by Storm

A Tutorial on the Probabilistic Model Checker Storm

Matthias Volk[1(✉)], Linus Heck[2], Sebastian Junges[2], Joost-Pieter Katoen[3], and Tim Quatmann[3]

[1] Eindhoven University of Technology, Eindhoven, The Netherlands
m.volk@tue.nl
[2] Radboud University, Nijmegen, The Netherlands
[3] RWTH Aachen University, Aachen, Germany

Abstract. This tutorial paper presents a hands-on perspective on probabilistic model checking with the Storm model checker. Storm is a decade-old model checker that excels in performance and a rich Python-based ecosystem, which makes it easy to integrate in various workflows. This tutorial focuses on Markov decision processes (MDP), which are popular in a variety of fields. It demonstrates the basic workflow, from Python-based modeling, model checking with a variety of properties, to the extraction of policies. Further, it showcases the support for recent topics that focus on different types of uncertainty, such as interval MDP and POMDP, and the ability to quickly implement simple algorithms on top of existing data structures.

1 What is Probabilistic Model Checking?

The verification of safety-critical systems requires faithfully modeling uncertainty about system behavior, whether it is due to noisy sensors, hardware degradation, or randomization as used in distributed protocols. The use of probabilistic models has developed as a promising direction to support verification of the aforementioned systems [25]. In particular, *probabilistic model checking (PMC)*, an extension of classical (non-probabilistic) model checking [6], allows to verify whether a model with probabilistic behavior satisfies the given requirements. Given a Markov model and a formal property of interest, (classical) PMC checks whether the probability of the model satisfying the property is above (or below) a given threshold and returns the yes/no answer. Moreover, probabilistic model checkers can also compute the probability—up to some precision guarantee—and therefore answer queries such as *"what is the maximal probability of collecting all resources before a (randomly determined) deadline"* or *"what is the expected time to collect all resources?"*

Historically, there has been a variety of probabilistic model checkers, including E-(MC)^2 [20], MRMC [24], Modest-mcsta [14], Prism [28], EPMC / iscasMC [12], Storm [18], UPPAAL [29], and PET [32]. Application areas of PMC are surveyed in [26] and the capabilities of PMC tools have been presented for various tools, such as Prism [11], Modest [13] and Storm [19].

A. Sampaio and M. Stoelinga (Eds.): FM 2026, LNCS 16557, pp. 524–549, 2026.
https://doi.org/10.1007/978-3-032-26220-2_25

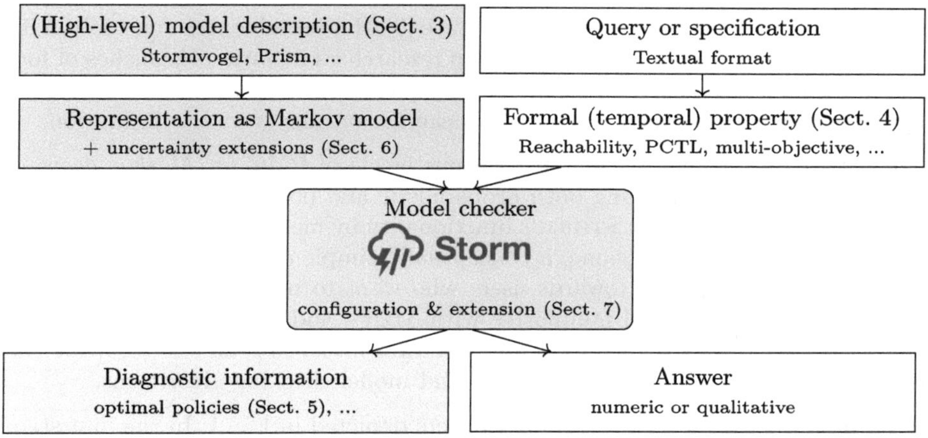

Fig. 1. Probabilistic model checking (extended from [19] with tutorial-specific content in small font)

The classical workflow of probabilistic model checking is depicted in Fig. 1 and consists of the following three main steps in sequence:

1. *Model building* (indicated in blue color in Fig. 1). First, the model is specified precisely. The de-facto standard are high-level model descriptions, using, e.g., the MODEST language [7], the PRISM language [28], or the JANI interchange format [8]. From this high-level description of the model, the underlying explicit Markov model—including transition matrices and state labels—is automatically created. Alternatively, the underlying model can already be specified explicitly, typically originating from external model construction.
2. *Model checking* (orange color). Using preprocessing steps, the model is reduced and put into a standard form. In that form, the model checker must solve a set of, e.g., Bellman equations [33], using highly optimized routines. Solving these equations typically translates into an answer of the original query on the original model.
3. *Diagnostic information* (green color). In addition to the qualitative or numerical answer, further diagnostic information can be produced, although it may induce significant overhead.

Modern usage of probabilistic model checking typically goes beyond such linear workflows. The different steps become more and more intertwined and embedded into reasoning loops in an effort to solve real problems. For example, in addition to the answer—*"what is the winning probability"?*—diagnostic information yields the optimal strategy and for this optimal strategy, additional queries can then be used to refine, e.g., the original model.

Storm. The development of such loops in the PMC workflow is eased by the ability to do rapid prototyping in Python. The open-source model checker

STORM [18] has long excelled in performance [9]. Recent advances and tools in its ecosystem now provide a rich API to help researchers in various branches of formal methods to use probabilistic model checking in their own tools and methods. We refer to STORM's website www.stormchecker.org for general information.

This tutorial. We will present the various facets of PMC for *Markov decision processes (MDPs)* combining both probabilistic and nondeterministic behavior. The tutorial demonstrates STORM's functionality by means of the running example of the *Orchard* board game, a conceptually simple children's game.

The tutorial is aimed towards users who want to make first steps applying PMC. We assume general familiarity with MDPs and recap relevant notation in Sect. 2. We refer to [23] for an introduction into PMC and to [5,6] for the relevant background on Markov models and model checking algorithms.

Outline. The tutorial will cover the aspects depicted in Fig. 1. In the first step, Sect. 3, we demonstrate how to model the board game in a high-level manner. In the second step, Sect. 4, we specify and check a variety of properties, such as reachability probabilities, rewards and multi-objective queries. In the third step, Sect. 5, we use the diagnostic information provided by STORM to obtain e.g., the optimal strategies for winning the Orchard game. In the fourth step, Sect. 6, we show how our (classical) MDP model can be extended by including additional sources of uncertainty. We cover *interval MDP* which allow uncertainty in the transition probabilities, and *partially observable MDP (POMDP)* which allow uncertainty about the exact state by means of restricted observations. Lastly, in Sect. 7, we demonstrate how the STORM model checker can be configured to user needs, and how to extend STORM by writing simple Python algorithms.

Data. We provide the models, code examples and instructions of this tutorial as supplementary material online[1]. These Jupyter notebooks can be executed interactively alongside this tutorial and allow to easily explore the STORM ecosystem. The full code examples are also available in the extended version [36].

2 Preliminaries

We briefly recap Markov decision processes, policies, and properties. An introduction to the topic can be found in [5,6,16]. In the tutorial presentation, we will elaborate these definitions.

MDPs. We model our game as a *Markov decision process (MDP)* which combines probabilistic and nondeterministic behavior. Let $\mathsf{Distr}(X)$ denote distributions over finite sets X.

Definition 1 (Markov decision process (MDP)). *An* MDP *is a tuple* $\mathcal{M} = (S, s_0, Act, \mathbf{P}, AP, L)$ *with a finite set* S *of* states *and an initial state* s_0*, a finite set* Act *of* actions*, a partial transition function* $\mathbf{P}\colon S \times Act \nrightarrow \mathsf{Distr}(S)$ *mapping state-action pairs to distributions over successors, and a finite set* AP *of* atomic propositions *with a state labeling function* $L\colon S \to 2^{AP}$.

[1] https://doi.org/10.5281/zenodo.18955665.

The partial transition function clarifies that not all actions are necessarily enabled in all states. In particular, the set of *enabled actions* in a state s is given by $Act(s) = \{\alpha \in Act \mid \mathbf{P}(s,\alpha) \text{ is defined}\}$. A *choice* is a pair of state and enabled action. As is customary, we write $\mathbf{P}(s,\alpha,s')$ for $\mathbf{P}(s,\alpha)(s')$ if defined and assume $\mathbf{P}(s,\alpha,s') = 0$ otherwise. Thus, an MDP transitions from state $s \in S$ with action $\alpha \in Act$ to state $s' \in S$ with probability $\mathbf{P}(s,\alpha,s')$. A *discrete-time Markov chain (DTMC)* is an MDP with one choice per state.

Rewards. An MDP can be naturally extended with *rewards* (also called costs). We use action-rewards here which associate a non-negative reward $R(s,\alpha) \in \mathbb{R}_{\geq 0}$ with each pair of state s and enabled action $\alpha \in Act(s)$. Visiting an action accumulates the associated rewards. An MDP can have multiple reward structures.

Policies. A *policy* (also called *strategy* or *scheduler*) on an MDP $\mathcal{M}$ is a function $\sigma\colon (S \times Act)^* \times S \to Act$ mapping paths to actions, such that $\sigma(\ldots \cdot s) \in Act(s)$. A special case of such (deterministic) policies are *memoryless* policies where the action choice depends solely on the last state. Policies resolve nondeterminism. A (memoryless) policy σ on an MDP $\mathcal{M}$ *induces* a DTMC $\mathcal{M}^\sigma$ by applying the policy in each state. The probability measure $\Pr_s^{\mathcal{M}^\sigma}$ on an MDP $\mathcal{M}$ is defined via the given policy σ. Similar constructions can be made for general policies.

Specifications. Probabilistic model checkers use (various extensions to) *probabilistic computation tree logic (PCTL)* or *linear temporal logic (LTL)* to express properties of interest [5,6]. The standard queries quantify universally, or existentially, over policies and ask, e.g., whether the probability to satisfy a temporal path property is above a threshold for every policy. The probability to satisfy a path property for a specific policy is defined by a standard probability measure on the DTMC induced by the MDP and the policy.

3 Step One: Modeling

We use a simple board game as a running example throughout this tutorial. Board games typically involve probabilistic behaviour, e.g., by throwing dice, and nondeterministic behavior based on the players' choices. The analysis of a board game typically involves optimizing the players' strategies. Board games are therefore a good model for demonstrating the various modeling and model checking capabilities of STORM.

3.1 The Orchard Game

The Orchard game (also called Obstgarten, Le verger, Boomgaard, El Frutal, Første Frukthage) is a simple children's game. In the following, we model the *First Orchard* game variant based on the official rule book available online.[2]

The game is played cooperatively. There are four types of *fruit*—apples, pears, cherries and plums[3]. For each type of fruit, there is a tree with four

[2] https://cdn.hff.de/image/upload/v1/22/30/28/14/22302814.pdf.

[3] We use the grape symbol as there is no plum symbol.

pieces of this fruit. There is a *basket* for collecting fruit. As antagonist, there is a *raven* which reaches the orchard after five steps. The goal of the game is to collect all fruit before the raven reaches the orchard. In each round, a player throws a six-sided dice and performs an action depending on its outcome:

- If the outcome is a type of fruit, then the player picks a piece of this type from the tree and places it in the basket. If no fruit is left on the tree, the player cannot pick anything.
- If the outcome is the fruit basket, then the player may pick any fruit.
- If the outcome is the raven, the raven moves one step towards the orchard.

The players win iff they collect all fruit before the raven arrives at the orchard.

3.2 The Orchard Model

Before starting the actual modeling, we make some general modeling decisions. First, we decide which parts are relevant for the model. In the Orchard game, we need to keep track of the remaining fruit on the trees and the position of the raven. As the game is cooperative, it is irrelevant how many players are playing and we assume one player. We also do not need to keep track of the basket as only the remaining fruit on the trees are relevant for the winning condition.

In our Orchard model, we divide each round into two distinct phases: first throwing the dice and then acting on the outcome. This allows to separate the probabilistic dice throw from the player's (potentially nondeterministic) action.

Lastly, to facilitate analysis of different variants of the game, we parameterize the model in terms of fruit types, number of fruit and distance of the raven.

3.3 Defining the Orchard MDP in Stormvogel

We use the Stormvogel library [30] for our initial model declaration. Stormvogel[4] is a Python library providing easy-to-use interfaces for probabilistic model checking. A main feature of Stormvogel is the bird API, which allows to specify the model using Python code. This enables quickly declaring models by using the familiar Python language instead of a dedicated modeling language with a steeper learning curve. While the API allows natural modeling, the user must still pay attention in order to obtain small and finite models.

Defining an MDP in Stormvogel. To create an MDP model in Stormvogel, we need to provide all ingredients of the tuple $\mathcal{M} = (S, s_0, Act, \mathbf{P}, AP, L)$. Starting from the initial state s_0 of the MDP, the bird API of Stormvogel automatically generates the successor states by following the transition function $\mathbf{P}$ for each enabled action $Act(s)$. Exploring all newly created states allows to subsequently generate the complete state space S of the MDP and Stormvogel keeps track of all states found so far. If a successor state already exists, it does not need to be explored further. The state space exploration terminates once all

[4] https://github.com/stormchecker/stormvogel.

```
# Main class for the orchard game
class Orchard(stormvogel.bird.State):
    def __init__(self, fruit_types, num_fruits, raven_distance):
        self.trees = {fruit: num_fruits for fruit in fruit_types}
        self.raven = raven_distance
        self.dice = None
```

Listing 1: An extract of the Orchard game class.

states have been explored. The check whether a state already exists is based on the *equality function* of a state. Two Python state objects which are equal w.r.t. the equality function must represent the same (logical) state and the equality function must consider all relevant information to distinguish different states.

It is the responsibility of the user to ensure that the state space exploration terminates, i.e. that the state space is finite. In our Orchard example, the game will eventually end and therefore only finitely many distinct states are explored. In general, there might be infinitely many states that would need exploration—for instance due to unbounded variables like counters, lists, etc.—potentially resulting in non-termination of the BIRD API. STORMVOGEL prevents such issue by using a configurable threshold `max_size` on the explored state space.

State Representation. We start the modelling by defining general data structures for the Orchard game, the different types of fruit, the possible outcomes of the dice and the game end. The corresponding Python code can be found in [36].

We introduce a class `Orchard` which represents the current state of the game, see Listing 1. The full class can be found in [36]. This class inherits from the STORMVOGEL `State` class. The `Orchard` object is initialized with a list of configuration parameters such as the considered types of fruit, the number of fruit per tree, and the distance of the raven. The game initializes the variable `trees` which keeps track of the remaining number of fruit per tree. It also keeps track of the outcome of the `dice`, i.e. 🧺, 🐦 or a fruit. In case of a fruit, the second entry of the `dice` tuple denotes the specific type of fruit which was thrown. Value `None` represents that the dice needs to be thrown first.

The `Orchard` class also defines various functions for ease of modeling. For instance, `game_state` returns whether the game is ongoing or who has won. Function `pick_fruit` picks the given fruit from the tree (if available). Function `next_round` resets the dice for the next round, `move_raven` moves the raven. Modeling these functions separately also yields a modular design which allows to easily modify specific behavior, such as the movement of the raven.

Initial State. The initial state s_0 is given by the initialization of our `Orchard` class. At first, we use a smaller configuration of the game with only two different types of fruit, two pieces of fruit per tree, and a distance of two for the raven.

```
>>> init_small = Orchard([Fruit.APPLE, Fruit.CHERRY],
...                      num_fruits=2, raven_distance=2)
>>> print(init_small)
2🍎, 2🍒, 2←🐦
```

```
89  # Define available actions
90  def available_actions(state):
91      if state.game_state() != GameState.NOT_ENDED:
92          return ["gameEnded"]
93      if state.dice is None:
94          return ["nextRound"]
95      if state.dice[0] == DiceOutcome.FRUIT:
96          return ["pick{}".format(state.dice[1].name)]
97      if state.dice[0] == DiceOutcome.BASKET:
98          available_fruits = []
99          # Choice over available fruits
100         for fruit in state.trees.keys():
101             if not state.is_tree_empty(fruit):
102                 available_fruits.append(fruit)
103         return [
104             "choose{}".format(fruit.name) for fruit in available_fruits
105         ]
106     if state.dice[0] == DiceOutcome.RAVEN:
107         return ["moveRaven"]
```

Listing 2: The available actions.

Printing the initial state of the simplified game shows that there are two apples and two cherries on the trees, and the raven is two steps away from its goal.

Available Actions. The transition function returning successor states requires information which actions are enabled in each state. Listing 2 defines $Act(s)$ for any state s. In the Orchard game, we use the following actions:

- `gameEnded` (ll. 91–92): the only action available once the game has ended. This action represents the self-loop for such final states.
- `nextRound` (ll. 93–94): the dice must be thrown again.
- `pickF` (ll. 95–96): the dice outcome is `F`∈{, , , } and the player picks the corresponding fruit `F` from the tree.
- `chooseF` (ll. 97–105): dice outcome and fruit `F`∈{, , , } is chosen.
- `moveRaven` (ll. 106–107): dice outcome and the raven must be moved next.

In most states, a single action is available. The only exception is dice outcome , for which we list all fruit which are still available for choosing by the player.

Transition Function. Stormvogel defines the transition function **P** via function `delta` which takes as arguments a state and an available action, and returns the distribution over successor states as a sparse list of nonzero transition probabilities for different target states. We define the transition function for the Orchard game in Listing 3. If the game has ended, we add a self-loop to the same state with probability one (ll. 113–115). Otherwise, we perform each round in two phases: first throwing the dice and then acting on the outcome. These are two separate transitions in the model.

In the first phase (ll. 117–139), the dice outcome is `None` and the player must throw the dice. We consider the outcomes for the different fruit types plus two additional outcomes and . Each outcome has the same uniform probability `fair_dice_prob`, which is $1/|fruit\_types|+2$. The rest of the first phase (ll. 124–139)

```
# The transition function
def delta(state, action):
    if state.game_state() != GameState.NOT_ENDED:
        # Game has ended -> self loop
        return [(1, state)]

    if state.dice is None:
        # Player throws dice and considers outcomes
        outcomes = []
        # Probability of fair dice throw over
        # each fruit type + 1 basket + 1 raven
        fair_dice_prob = 1 / (len(state.trees.keys()) + 2)

        # 1. Dice shows fruit
        for fruit in state.trees.keys():
            next_state = deepcopy(state)
            next_state.dice = DiceOutcome.FRUIT, fruit
            outcomes.append((fair_dice_prob, next_state))

        # 2. Dice shows basket
        next_state = deepcopy(state)
        next_state.dice = DiceOutcome.BASKET, None
        outcomes.append((fair_dice_prob, next_state))

        # 3. Dice shows raven
        next_state = deepcopy(state)
        next_state.dice = DiceOutcome.RAVEN, None
        outcomes.append((fair_dice_prob, next_state))
        return outcomes

    elif state.dice[0] == DiceOutcome.FRUIT:
        # Player picks specified fruit
        fruit = state.dice[1]
        next_state = deepcopy(state)
        next_state.pick_fruit(fruit)
        next_state.next_round()
        return [(1, next_state)]

    elif state.dice[0] == DiceOutcome.BASKET:
        assert action.startswith("choose")
        # Player chooses fruit specified by action
        fruit = Fruit[action.removeprefix("choose")]
        next_state = deepcopy(state)
        next_state.pick_fruit(fruit)
        next_state.next_round()
        return [(1, next_state)]

    elif state.dice[0] == DiceOutcome.RAVEN:
        next_state = deepcopy(state)
        next_state.move_raven()
        next_state.next_round()
        return [(1, next_state)]
```

Listing 3: The transition function `delta`.

creates the different successor states depending on the dice outcomes. To this end, we create the `next_state` as a copy of the current state, and set the dice outcome `next_state.dice`. Afterwards, we store the successor state together with the corresponding probability in the list `outcomes` of successor states.

The remainder of the function handles the second phase (ll. 141–162), where the player acts depending on the dice outcome `state.dice`. If the dice outcome is a fruit, the player performs `pick_fruit(fruit)` on the `next_state` and then finishes the turn with `next_round` (ll. 141–147). If the dice outcome is , the player needs to choose a fruit (ll. 149–156). This choice is modeled by the different available actions `chooseF` in this state. Based on the chosen `action` given as argument to the `delta` function, we can extract the chosen fruit. For example, action name `chooseAPPLE` corresponds to . In the successor state, we pick the chosen fruit and then continue with the next round. Lastly, if the dice outcome is , the raven moves (ll. 158–162).

Labeling. We use a labeling L to associate atomic propositions AP to states which satisfy them. These labels help identify states which fulfill a certain property which is relevant to our analysis. The labeling for the Orchard game can be found in [36]. The function `labels()` returns a list of atomic propositions per state. As we are mainly interested in the winner of a game, we introduce two labels `PlayersWon` and `RavenWon` which mark each state with the party who won. We also label each state with the string representation of the game state to help understanding and investigating the resulting model as the internals of a state become visible. For scalable analysis, such kind of internal information is typically omitted in the final version of a model.

Rewards. Lastly, we specify a reward which counts the number of rounds. The reward function can be found in [36]. Each state in which the dice is thrown has an action reward of one. All other actions have reward zero.

Build Model. All ingredients for the model have been specified and we can build the MDP model for the Orchard game.

```
>>> orchard_simple = stormvogel.bird.build_bird(
... modeltype=stormvogel.ModelType.MDP, init=init_small,
... available_actions=available_actions, delta=delta,
... labels=labels, rewards=rewards)
>>> print(orchard_simple.summary())
ModelType.MDP model with 90 states, 7 actions, and 33 distinct labels.
```

The function specifies the type of model—`MDP` here–and the MDP tuple entries.

We can also easily build the full model by configuring the initial state accordingly and calling `build_bird()` as before. We need to increase the `max_size` of the explored state space as the resulting model has around $22,500$ states[5].

[5] Stormvogel has a small default `max_size` to quickly spot issues with toy examples used in teaching.

```
>>> init_game = Orchard([Fruit.APPLE, Fruit.CHERRY, Fruit.PEAR, Fruit.PLUM],
... num_fruits=4, raven_distance=5)
>>> orchard = stormvogel.bird.build_bird(..., max_size=100000)
>>> print(orchard.summary())
ModelType.MDP model with 22469 states, 11 actions, and 3 distinct labels.
```

Some of the analyses in the following require a representation of the Orchard model in STORM data structures. We can simply translate the model:

```
>>> import stormpy
>>> orchard_storm = stormvogel.mapping.stormvogel_to_stormpy(orchard)
```

3.4 Defining the Orchard MDP in the Prism Language

We showed how to create an MDP model via STORMVOGEL. STORM also supports creating models from other input formats such as the PRISM language or the JANI interchange format. In this tutorial, we also provide a PRISM model for our Orchard game. The full specification can be found in [36]. The PRISM specification follows the same structure as the STORMVOGEL model. It can be configured via global constants such as `NUM_FRUIT` and built via STORMPY. The current game state is represented by variables such as `raven` or `apple`.

```
>>> prism_program = stormpy.parse_prism_program('orchard_stormvogel.pm')
>>> constants = "NUM_FRUIT=4, DISTANCE_RAVEN=5"
>>> prism_program = stormpy.preprocess_symbolic_input(prism_program,
... [], constants)[0].as_prism_program()
>>> options = stormpy.BuilderOptions()
>>> options.set_build_state_valuations()
>>> options.set_build_choice_labels()
>>> options.set_build_all_labels()
>>> options.set_build_with_choice_origins()
>>> orchard_prism = stormpy.build_sparse_model_with_options(prism_program,
... options)
>>> print("Model with {} states and {} transitions".format(
... orchard_prism.nr_states, orchard_prism.nr_transitions))
Model with 22469 states and 44954 transitions
```

In the remainder, `orchard_simple` refers to the game with only two fruit types, `orchard` is the STORMVOGEL model of the full game, `orchard_storm` its corresponding representation in STORM, and `orchard_prism` is that model, built from the PRISM specification. The latter three are semantically equivalent.

3.5 Model Inspection

Before applying model checking, we inspect the model to spot modeling issues. We suggest to use a combination of simulations and visualizations. STORMVOGEL also allows to generate simulation runs through the model using random or user-provided policies and it can expose models via a gym-compatible interface.

STORMVOGEL visualizes a model with `stormvogel.show(orchard_simple)`. The interactive visualization is based on JavaScript and is therefore available via a web browser. An extract of the resulting state space is shown in Fig. 2. States are depicted as ellipses, actions by blue boxes and transitions by arrows. Labels are given inside the states and actions, and the rewards are indicated by the

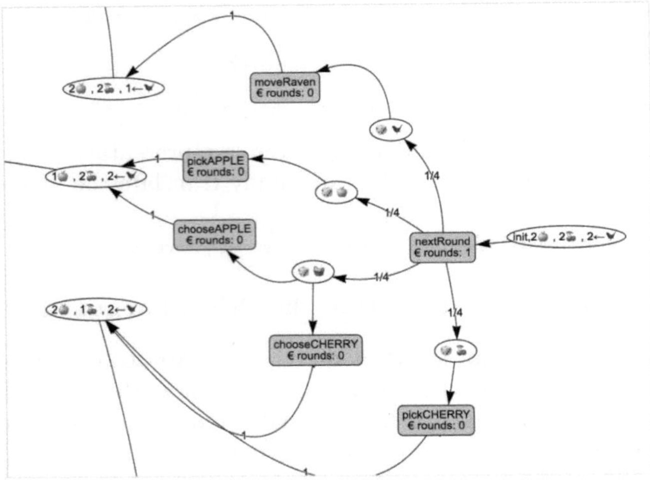

Fig. 2. State space of the simplified MDP.

€ symbol. Starting from the initial state on the right, the only possible action is `nextRound`, which corresponds to throwing the dice. It has reward one. The four outcomes are depicted by the four successor states with symbol 🎲 and are each reached with transition probability 1/4. Afterwards, the corresponding action can be performed. For example, from state 🎲🍎, action `pickAPPLE` is possible and leads to a state 1🍎, 2🍒, 2←🐦 where only one apple is remaining. For larger state space sizes, STORMVOGEL has the option to interactively *explore* the state space by clicking on the successor states or by connecting it to the simulator.

4 Step Two: Model Checking

After creating and inspecting the model, we will now apply model checking to the Orchard MDP. While, strictly speaking, model checking asks whether a property holds on an MDP, we will see that modern probabilistic model checking tools, including STORM, can actually go beyond such queries. The section is structured based on the type of properties that we consider.

4.1 Reachability

One of the simplest properties for MDPs is a reachability query *"what is the maximal probability to reach a specific set of states, described by* φ*?"* and is denoted in probabilistic model checking as $\mathbb{P}_{\max=?}(F\,\varphi)$. The probability operator $\mathbb{P}_{max=?}(\psi)$ denotes the maximal probability, ranging over all policies, that in the induced Markov chain, a path models a path formula ψ. In this case $\psi := F\,\varphi$ (also $\Diamond\varphi$) represents finally reaching a state which satisfies a Boolean formula φ

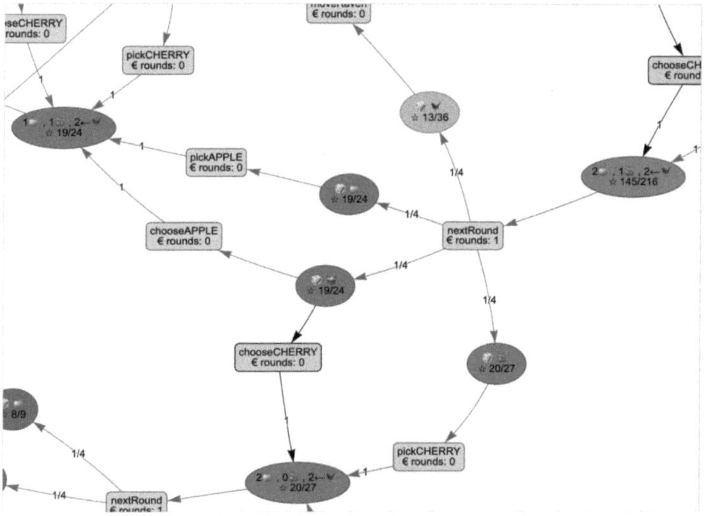

Fig. 3. Visualization of the winning probability on the simplified MDP.

over atomic propositions *AP*. In Storm and in Stormvogel, we specify formulas in the de-facto standard Prism representation, e.g., `'Pmax=? [F"Label"]'`.

In the Orchard game, we are mostly interested in the *"maximal probability of winning the game"*. Using the label `PlayersWon`, we express this property as $\mathbb{P}_{\max=?}(F\,\texttt{PlayersWon})$, i.e. maximizing the probability of reaching a state where the players have won. Notably, for many queries, we indeed get a result for every state and we therefore clarify in the code that we are interested in the probability from the initial state.

```
>>> prob_players_won = 'Pmax=? [F "PlayersWon"]'
>>> result = stormvogel.model_checking(orchard_simple, prob_players_won)
>>> print(result.get_result_of_state(orchard_simple.get_initial_state()))
0.5711805425946498
```

We can also visualize the result of this query on the MDP state space.

```
>>> vis = stormvogel.show(orchard_simple, result=result)
```

An extract is depicted in Fig. 3. In each state, ☆ indicates the computed value from that state onward. The state is also colored with a red gradient depending on the result—the darker a state the higher its winning probability. For instance, from the state 2🍏, 1🍒, 2←🐦 on the right, the players have a winning probability of $^{145}/_{216} = 0.671$, whereas from the 🎲🐦-successor, the probability is only $^{13}/_{36} = 0.3611$, as the raven is then one step closer to the orchard.

The model checking also returns a memoryless policy σ which ensures the maximal winning probability. For each state s, the policy chooses one action $\sigma(s)$. This action is highlighted in red color in the Stormvogel visualization. For example, from state 🎲🧺, the player has the option to either `chooseAPPLE` or

```
def model_check(model, prop):
    formula = stormpy.parse_properties(prop)[0]
    result = stormpy.model_checking(model, formula, only_initial_states=True)
    return result.at(model.initial_states[0])
```

Listing 4: Function for model checking calls.

`chooseCHERRY`. The optimal policy chooses , because the successor state has a winning probability of $19/24 = 0.7916$ which is higher than for , $20/27 = 0.7407$.

For the full model `orchard`, the winning probability is 0.6314.

4.2 Total Rewards

Probabilistic model checkers commonly handle reward-based queries like *"what is the maximal expected total reward until reaching a specific set of states?"* which can be expressed as $\mathbb{R}^{rew}_{\max=?}(F\ \varphi)$, where φ is again describing a set of states and *rew* refers to the name of a reward structure.

In the Orchard game, the *"maximal expected total number of rounds until the game ends"* is described by $\mathbb{R}^{rounds}_{\max=?}(F\ (\texttt{PlayersWon} \lor \texttt{RavenWon}))$. Abbreviating the code to get the result for the initial state as above, we simply write:

```
>>> reward_prop = 'R{"rounds"}max=? [F "PlayersWon" | "RavenWon"]'
>>> print(stormvogel.model_checking(orchard, reward_prop).get_res[..])
22.339089182046724
>>> reward_prop = 'R{"rounds"}min=? [F "PlayersWon" | "RavenWon"]'
>>> print(stormvogel.model_checking(orchard, reward_prop).get_res[..])
20.882789624542582
```

Model checking reveals that the game lasts between 20.88 and 22.34 rounds on average—depending on the chosen strategy of the players.

4.3 Beyond

Storm supports several queries that go significantly beyond the classical queries. While we used the Stormvogel model checking wrapper around Storm, for such properties we must often operate directly on the (somewhat more intricate) stormpy API. We use Listing 4 for simplicity.

Reward-Bounded Reachability Probabilities [15]. $\mathbb{P}_{\max=?}(F^{rew \leq \lambda}\ \varphi)$ ask for the probability to reach a φ-state while having accumulated at most λ reward. In the Orchard game, we can compute the winning probability for varying number of rounds and plot the result in Fig. 4.

```
>>> probabilities = []
>>> for k in range(41):
... win_steps = 'Pmax=? [F{"rounds"}<=' + str(k) + ' "PlayersWon"]'
... probabilities.append(model_check(orchard_storm, win_steps))
```

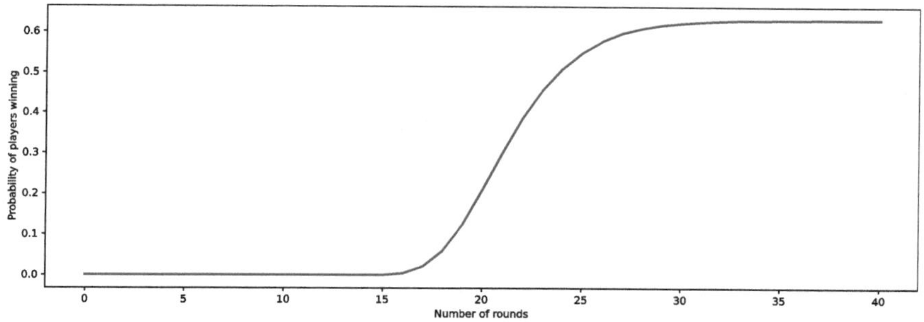

Fig. 4. Winning probability within given number of rounds.

Conditional Reachability Probabilities [10]. $\mathbb{P}_{\max=?}(F\,\varphi \mid F\,\theta)$ express the probability of reaching a φ-state, conditioned under the event that a θ-state is reached. We compute the maximal winning probability under the condition that the raven is eventually only one step away from the orchard. Checking the property `'Pmax=? [F"PlayersWon" || F "RavenOneAway"]'` yields 0.3198 which is significantly lower than the overall winning probability of 0.63.

Multi-objective Queries [34] $\mathit{multi}(\Phi_1, \ldots, \Phi_m)$ allow to compute possible tradeoffs between various (single-objective) properties $\Phi_1, \ldots, \Phi_m$.

For the Orchard game, the multi-objective query asks for the maximal expected number of rounds of the game among all policies that induce a winning probability of at least 60%.

```
>>> query = 'multi(R{"rounds"}max=? [F ("PlayersWon" | "RavenWon")], \
... P>=0.60 [F "PlayersWon"])'
>>> print(model_check(orchard_storm, query))
21.79468486
```

We see that using a nearly optimal winning strategy also reduces the expected number of rounds played—which is 22.3391 when considering all policies.

5 Step Three: Working with Policies

As outlined in the introduction, the use of modern probabilistic model checking increasingly asks for diagnostic information, e.g., the policies that optimize the objectives that are being checked. We consider their extraction, the analysis of such policies with respect to alternative objectives, and the extraction of policies that can be concisely represented.

5.1 Computing Policies

In this tutorial, we are interested in understanding aspects of an optimal policy to win the Orchard game, as computed by Storm. We already saw in Fig. 3 how to infer optimal choices from the visualization. However, this is only feasible for smaller state spaces and we consider a more general approach. First, we inform STORM that we want to compute an optimal policy as a by-product of the model checking call, by adding the argument `extract_schedulers=True`. Computing policies can induce overhead due to the additional bookkeeping—we also note that this bookkeeping is not implemented for all types of properties. The output of the model checking call is a quantitative result, which now also holds a policy.

```
>>> formula = stormpy.parse_properties('Pmax=? [F "PlayersWon"]')[0]
>>> result = stormpy.model_checking(orchard_storm, formula,
... extract_scheduler=True)
>>> print(result.scheduler)
Fully defined memoryless deterministic scheduler:
model state: choice(s)
          0 0
          1 0
          ... ...
```

Policies for maximal reachability probabilities as in `formula` are memoryless and deterministic, i.e. a mapping from states to actions. The STORM representation mildly deviates: It represents for every state s (index) the local index x of the choice. These local indices encode the x'th choice for state s, using some internal ordering. Note that two choices in different states can yield the same action. The policy can be queried to understand for every state which action is selected. We use the high-level state information present in `orchard_prism`, cf. Sect. 3.4, and obtain a more informative representation of the policy:

```
>>> for state in orchard_prism.states:
... choice = result.scheduler.get_choice(state)
... action_index = choice.get_deterministic_choice()
... action = state.actions[action_index]
... print(f"In state {state.valuations} choose action {action.labels}")
...
In state [... & apple=4 & pear=4 & cherry=4 & plum=4 & raven=5]
    choose action {'chooseAPPLE'}
...
In state [... & apple=3 & pear=4 & cherry=4 & plum=4 & raven=5]
    choose action {'choosePEAR'}
...
```

We see that initially—when all trees are full—the player chooses 🍎 for the dice outcome 🧺. After picking 🍎, the optimal strategy chooses 🍐 instead.

Some properties, such as reward-bounded reachability probabilities, require history-dependent schedulers, which are typically represented using a finite-state machine—also in STORM. In this tutorial, we do not explore these policies further. While the policy above is represented on the level of the input, it necessarily is a very large object, as it depends on the number of underlying states. This problem is amplified for policies that require memory. In Sect. 5.3, we therefore circumvent the explicit representation of policies.

5.2 Analyzing the Induced Submodel

After obtaining a policy, a standard question is to analyze the MDP and policy with respect to additional properties. For example, we may want to calculate the probability to collect all before the raven arrives using the current policy and contrast this probability with a policy that optimizes collecting. To support analyzing policies, we can create the induced Markov chain $\mathcal{M}^\sigma$ by applying policy σ on MDP $\mathcal{M}$.

```
>>> # Get induced model
>>> induced = orchard_prism.apply_scheduler(result.scheduler, True)
>>> # Analysis (allCherriesPicked means cherries = 0 and raven > 0)
>>> all_cherries = 'Pmax=? [F "allCherriesPicked"]'
>>> print(f"Prob for fixed: {model_check(induced, all_cherries)}")
>>> print(f"Prob for optimal: {model_check(orchard_prism, all_cherries)}")
Prob for fixed: 0.7726100987523447
Prob for optimal : 0.9120560407842777
```

We see that the policy optimized for collecting yields a higher probability of picking all than the submodel induced by the overall winning strategy.

5.3 Compact Policies

The policy in Sect. 5.1 was represented as an explicit list. There are various ways to compress a given policy: Via decision tree learning (available via the DTCONTROL tool [3]) one can create heuristically small trees for a policy, whereas DTMAP [2] (via the PAYNT tool [1]) allows to find a minimal decision tree for a given policy. Both tools allow interfacing with STORM. By default, we can find a decision tree with 15 layers, which is not particularly helpful. The challenge is that—for this model—there is no small decision tree for the policies that STORM generates. The root cause here combines two challenges:

1. The predicates that the decision tree is allowed to use is not adequate.
2. The policy computed by STORM makes (arbitrary) choices that yield policies that cannot easily be compressed.

Solving either of these challenges is insufficient. In Listing 5, we manually add adequate predicates and run a computationally expensive decision tree synthesis task in PAYNT [2] to find policies that satisfy a property and are representable as a small tree. By enforcing predicates for each fruit type that express that the type has the most remaining fruit in a tree, we find a small decision tree within a minute: One should pick if there are more than other remaining fruit, or otherwise if they are more than others, etc.

6 Step Four: MDPs with More Uncertainty

While MDPs are the standard model to describe processes that are subject to both nondeterminism and probabilistic uncertainty, not every process can adequately be represented as an MDP. We briefly show how STORM can go beyond the verification of MDPs by means of imprecise probabilities and partial observations, while noting that support for such extensions is still actively developed.

```python
import paynt
# Create the predicates (and their valuations) that the DT can use.
svt = stormpy.StateValuationTransformer(orchard_prism.state_valuations)
extra_definitions = _declare_extra_orchard_variables(expression_manager)
for (v, definition) in extra_definitions:
    svt.add_boolean_expression(v, definition)
# Create a new MDP with the valuations above
new_mdp = stormpy.set_state_valuations(orchard_prism, svt.build(False))
# Setup paynt and run synthesis
colored_mdp_factory = paynt.dt.DtColoredMdpFactory(new_mdp)
task = paynt.dt.DtTask([formula], tree_depth)
```

Listing 5: Searching for small decision trees with PAYNT

6.1 Imprecise Probabilities

Transition probabilities in MDPs are often an abstraction of more complicated processes, such as dice rolls in the Orchard game. Instead of precise point estimates for probabilities, STORM supports interval estimates in two flavors.

Interval MDPs. *Interval MDPs (iMDPs)* replace the precise probabilities in MDPs with intervals of possible probabilities [4]. For the sake of this tutorial, one can consider interval MDPs as a set of MDPs, covering all contained point estimates that yield proper successor distributions.[6] We replace the dice roll probabilities from the point estimates $1/6$ to the intervals $[5/36, 7/36]$. This can be done by changing all occurrences of `fair_dice_prob` in Listing 3 with the interval `Interval(fair_dice_prob-(1/36), fair_dice_prob+(1/36))`. No further change is needed, and STORMVOGEL will build an interval MDP.

Uncertainty in interval MDPs can be resolved in two ways. First, in the *cooperative* or *angelic* interpretation, the uncertainty is in favor of the policy. Therefore, if the policy maximizes the probability to win Orchard, the dice probability will also be chosen in order to maximize the probability to win. Second, in the *robust* or *demonic* interpretation, the uncertainty is resolved against the policy. If the policy maximizes the probability to win, the dice probability will be chosen in order to minimize the policy's probability to win.

We show how to check an interval MDP in Listing 6. In the cooperative setting, the player can achieve a maximal winning probability of 0.7961, while in the robust setting, the player can only achieve a maximal winning probability of 0.4315. It turns out that in Orchard, modifying the dice probabilities by $1/36$ has a much larger influence on the winning probability than the player's strategy.

Parametric MDPs. The interval MDP for Orchard bounds the dice probabilities within an interval *at each state*. Different values may be optimal at different states, and thus, the dice probabilities may differ per state. If we want to have uncertain but fixed dice probabilities, we use a *parametric MDP (pMDP)* [22]. We take the simplified two-fruit variant `orchard_simple` and parametrize it as

[6] Interval MDPs have two semantics: *static* and *dynamic uncertainty*. These semantics coincide for indefinite reachability probabilities and expected rewards [21]. The static semantics used here make it easier to explain the difference to parametric MDPs.

```
# Parse properties
properties = stormpy.parse_properties('Pmax=? [F "PlayersWon"]')
task = stormpy.CheckTask(properties[0].raw_formula)
# Set cooperative resolution mode, alternatively: ROBUST
task.set_uncertainty_resolution_mode(
    stormpy.UncertaintyResolutionMode.COOPERATIVE)
# Check model
env = stormpy.Environment()
stormpy_result = stormpy.check_interval_mdp(orchard_stormpy, task, env)
```

Listing 6: Checking an interval MDP.

$$\begin{aligned}&\big(24{\cdot}p^7{+}84{\cdot}p^6q{-}80{\cdot}p^6{+}114{\cdot}p^5q^2\\&\quad{-}268{\cdot}p^5q{+}90{\cdot}p^5{+}75{\cdot}p^4q^3{-}344{\cdot}p^4q^2\\&\quad{+}283{\cdot}p^4q{-}30{\cdot}p^4{+}24{\cdot}p^3q^4{-}211{\cdot}p^3q^3\\&\quad{+}333{\cdot}p^3q^2{-}85{\cdot}p^3q{+}3{\cdot}p^2q^5{-}62{\cdot}p^2q^4\\&\quad{+}181{\cdot}p^2q^3{-}85{\cdot}p^2q^2{-}7{\cdot}pq^5{+}45{\cdot}pq^4\\&\quad{-}35{\cdot}pq^3{+}4{\cdot}q^5{-}5{\cdot}q^4\big)/(p{-}1)^3\end{aligned}$$

Fig. 5. Winning probability in the pMC.

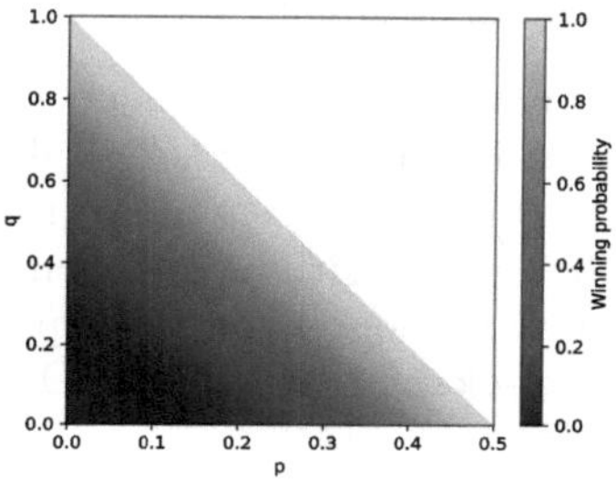

Fig. 6. Plot of winning probability.

follows: the parameter p describes the probability to roll or , the parameter q describes the probability to roll , and the probability to roll is described by $1-2p-q$. Similar to the iMDP, we can again replace the transition probabilities in Listing 3 with these functions over the parameters. We apply the optimal winning policy from the non-parameterized game on the pMDP and obtain the induced *parametric Markov chain (pMC)*. Computing the winning probability on the pMC yields the rational function in Fig. 5. The function is plotted in Fig. 6. We can see that if $1 - 2p - q = 0$, the player will win with probability one, and if $p = q = 0$, the player will win with probability zero.

The rational function describing the winning probability is already very large in this simplified model with only 90 states and generally explodes. However, Storm also exposes methods to analyze the pMC without calculating this function, such as [22]: (1) Finding feasible parameter instantiations satisfying a given property, (2) verifying that all instantiations in a region satisfy a given specification, (3) partitioning the parameter space into satisfying and violating regions, and (4) analyzing the pMC for monotonicity.

6.2 Partially Observable MDPs

In MDP verification, we consider maximizing over all policies. We already saw in Sect. 5.3 that this may not always be adequate. In particular, the policy has perfect information about the current state, which influences the choice. In many (cyberphysical or distributed) systems, the policy should only depend on states that are known by the agent. Similarly, in the Orchard game, the state of the trees may not be visible to the players, however, we do assume that we know which types of fruit can still be picked. To model this setting faithfully, the policy should not depend on the state, but instead on a set of observations—leading to a *partially observable MDP (POMDP)* [35].

Definition 2 (POMDP). *A* POMDP *is a triple* $\langle \mathcal{M}, Z, o \rangle$, *where* M *is an MDP with states* S *as before,* Z *is a finite set of* observations, *and* $o : S \to Z$ *assigns observations to states such that* $Act(s) = Act(s')$ *implies* $o(s) = o(s')$.

In POMDPs, we consider observation-based, history-dependent policies which are of the form $\sigma : (Z \times A)^* \times Z \to A$[7]. Intuitively, as we must quantify over this infinite set, the verification of POMDPs is indeed undecidable [31]; and different algorithms focus on maximizing, e.g., the reachability probability or on proving bounds on these probabilities. The API therefore has no regular `model_checking` method for POMDPs. We can print a POMDP after building it as before:

```
>>> print(pomdp)
Model type: POMDP (sparse)
States: 22469
Transitions: 44954
Choices: 29354
Observations: 546
...
Choice Labels:  11 labels
   * nextRound -> 3749 item(s)
   * chooseCHERRY -> 2500 item(s)
   * pickAPPLE -> 3120 item(s)
   * pickCHERRY -> 3120 item(s)
   * pickPLUM -> 3120 item(s)
...
```

We see that the POMDP now contains information regarding the number of different observations, as well as information regarding the action names for the choices in the model (as they matter semantically). The semantics of a POMDP can be given in terms of a belief MDP, which we can partially explore and verify. We omit relevant details for the different options here, the full code is available in the supplementary material.

```
>>> belmc = BeliefExplorationModelCheckerDouble(pomdp, belexpl_options)
>>> result = belmc.check(properties[0].raw_formula, [])
>>> print(f"Result in: [{result.lower_bound}, {result.upper_bound}]")
Result in [0.63135...,0.63135...]
```

First, note that the result approximates the true optimum from below and above. The result here is tight, as the underlying belief MDP is sufficiently simple. Second, note that the result coincides with the fully observable MDP case! That is correct as a policy can track how much fruit is available from every type.

[7] Alternative representations, e.g., belief-based policies, are also used.

We can consider a modified Orchard game, where another player randomly steals fruit before the game starts. This makes the game easier, but now an observation-based policy will perform worse than the state-based policies. The verification time for the POMDP is now significantly higher than for the MDP.

```
>>> mdp_res = stormpy.model_checking(pomdp_steal,winprop,
... force_fully_observable=True)
>>> print(mdp_res.at(pomdp_stealing.initial_states[0]))
0.66574...
>>> belmc = BeliefExplorationModelCheckerDouble(pomdp_steal, belexpl_options)
>>> result = belmc.check(winprop.raw_formula, [])
>>> print(f"Result in: [{result.lower_bound}, {result.upper_bound}]")
Result in [0.65986...,0.65986...]
```

7 Next Steps

Above, we showcased various analysis approaches for different model types in the previous sections. This section presents typical use cases for the Python API: Using symbolic model representations, using preprocessing algorithms, configuring solvers, and implementing own prototypes on top of existing routines.

Model Representation. So far, we have represented the MDP through an explicit (sparse) transition matrix. Instead, we can build the MDP symbolically and represent it through *decision diagrams (DD)* [27].

```
>>> orchard_symbolic = stormpy.build_symbolic_model(prism_program)
>>> print(orchard_symbolic)
Model type: MDP (symbolic)
States: 22469 (88 nodes)
Transitions: 44954 (946 nodes)
...
```

The symbolic MDP has the same size as the sparse MDP, but instead of explicitly storing nearly 45,000 transitions, the function is represented through a DD with 946 nodes. We can perform model checking on this symbolic representation:

```
>>> formula = stormpy.parse_properties('Pmax=? [F "PlayersWon"]')[0]
>>> symbolic_result = stormpy.model_checking(orchard_symbolic, formula,
... only_initial_states=True)
>>> filter = stormpy.create_filter_initial_states_symbolic(orchard_symbolic)
>>> symbolic_result.filter(filter)
>>> print("Maximal probability: {}".format(symbolic_result))
Maximal probability: 0.631356
```

Bisimulation. Bisimulation minimization can be applied to minimize the state space without affecting the model behavior [6].

```
>>> formula = stormpy.parse_properties('Pmax=? [F "PlayersWon"]')
>>> orchard_bisim = stormpy.perform_bisimulation(orchard_prism, formula,
... stormpy.BisimulationType.STRONG)
>>> print("Model with {} states and {} transitions".format(
... orchard_bisim.nr_states, orchard_bisim.nr_transitions))
Model with 956 states and 2446 transitions
```

For the orchard model, we can reduce the state space size by 95% and thereby speed-up subsequent analysis on the reduced model.

Model Checking Algorithms. The underlying model checking algorithms in STORM are configured through so-called *environments*. For example, we can set the precision requirement from the default value of 10^{-6} to 0.1 and see how that affects the resulting probability.

```
>>> env = stormpy.Environment()
>>> prec = stormpy.Rational(0.1)
>>> env.solver_environment.minmax_solver_environment.precision = prec
>>> result = stormpy.model_checking(orchard_prism, formula[0], environment=env)
>>> print(result.at(orchard_prism.initial_states[0]))
0.5815061686029693
```

We can also change the underlying algorithm to e.g., policy iteration or optimistic value iteration [17]. All methods provide the same result, but their timings differ.

```
>>> methods = [stormpy.MinMaxMethod.value_iteration,
... stormpy.MinMaxMethod.policy_iteration,
... stormpy.MinMaxMethod.optimistic_value_iteration]
>>> for m in methods:
... env = stormpy.Environment()
... env.solver_environment.minmax_solver_environment.method = m
... start = time.time()
... result = stormpy.model_checking(orchard_prism, formula[0],
... environment=env, extract_scheduler=True)
... print(f"Method: {m}")
... print(f"Result: {result.at(orchard_prism.initial_states[0])}")
... print(f"Time: {time.time() - start:.2}s")
Method: MinMaxMethod.value_iteration
Result: 0.6313572986959962
Time: 0.015s
Method: MinMaxMethod.policy_iteration
Result: 0.6313573066006364
Time: 0.24s
Method: MinMaxMethod.optimistic_value_iteration
Result: 0.6313576699315776
Time: 0.018s
```

Demo: Writing an LP-based MDP Model Checker. One further step is extending STORMPY with own algorithms. Extending STORMPY allows to profit from existing data structure and efficient algorithms when integrating own ideas. We exemplify this with a fundamental approach to solve MDPs using a *linear program (LP)* solver, an approach that is popular due to the flexibility to add additional constraints. The LP encoding below computes reachability probabilities. Given a target label $B \in AP$, this LP characterizes the variables $(x_s)_{s \in S}$ as follows [6, p. 856]. Minimize $\sum_{s \in S} x_s$ such that:

(1) If $B \in L(s)$, then $x_s = 1$,
(2) If there is no path from s to any state with label B, then $x_s = 0$,
(3) Otherwise, $0 \leq x_s \leq 1$, and for all actions $\alpha \in Act(s)$:

$$x_s \geq \sum_{s' \in S} \mathbf{P}(s, \alpha, s') \cdot x_{s'}.$$

```
# These are the target states
players_won = stormpy.parse_properties('"PlayersWon"')[0].raw_formula
psi_states = stormpy.model_checking(orchard_prism, players_won).get_truth_values()

# These are the states that can never reach the target states
phi_states = stormpy.BitVector(orchard_prism.nr_states, True)
prob0max_states = stormpy.compute_prob01max_states(
    orchard_prism, phi_states, psi_states
)[0]
                                    ⋮

# Encode LP
for state in range(num_states):
    if psi_states.get(state): # Case 1
        m.addCons(state_vars[state] == 1)
    elif prob0max_states.get(state): # Case 2
        m.addCons(state_vars[state] == 0)
    else: # Case 3
        for row in orchard_prism.transition_matrix.get_rows_for_group(state):
            summed_prob = 0
            for transition in orchard_prism.transition_matrix.get_row(row):
                prob = transition.value()
                next_state = transition.column
                summed_prob += prob * state_vars[next_state]
            m.addCons(state_vars[state] >= summed_prob)
                                    ⋮
```

Listing 7: Encoding an LP from an MDP (extract). The full code is given in [36].

Listing 7 shows how to create a linear program that computes the reachability probability based on this encoding. We first compute all states that satisfy the requirements in (1) and (2) using graph algorithms provided in STORMPY. We then construct and solve the LP by inspecting the model's transition matrix.

8 Conclusion

This tutorial presents usage of the probabilistic model checker STORM, via its Python bindings. It demonstrates how to specify an MDP, analyze it w.r.t. to various properties of interest and extract additional diagnostic information such as winning policies. The tutorial focuses on MDPs and the most used properties, and purposefully omits many functionalities in STORM, in particular with respect to continuous-time models. It focuses on using STORM, rather than extending it, and thus does not discuss internal data structures, nor details on effective modeling in PRISM or JANI.

Acknowledgments. This tutorial builds upon tools in the STORM ecosystem, especially on PAYNT and STORMVOGEL. We thank Christian Hensel, Alexander Bork, and Luko van der Maas as well as all other contributors to the STORM tool (See github.com/moves-rwth/storm?tab=readme-ov-file#authors for a complete list.) We thank Roman Andriushchenko, Filip Macák, and Milan Češka for the development of

PAYNT, especially for the API changes for this tutorial. We thank Pim Leerkes and Ivo Melse for their support in developing STORMVOGEL.

This work has been partially supported by the NWO Open Science Fund and the NWO VENI Grant UNICORN (232.211).

Data availability. The models and Jupyter notebooks of this tutorial are publicly available in the artifact at doi.org/10.5281/zenodo.18955665.

References

1. Andriushchenko, R., Češka, M., Junges, S., Katoen, J.-P., Stupinský, Š: PAYNT: A Tool for Inductive Synthesis of Probabilistic Programs. In: Silva, A., Leino, K.R.M. (eds.) CAV 2021. LNCS, vol. 12759, pp. 856–869. Springer, Cham (2021). https://doi.org/10.1007/978-3-030-81685-8_40
2. Andriushchenko, R., Ceska, M., Junges, S., Macák, F.: Small decision trees for MDPs with deductive synthesis. In: CAV (2). LNCS, vol. 15932, pp. 169–192. Springer, Cham (2025). https://doi.org/10.1007/978-3-031-98679-6_8
3. Ashok, P., Jackermeier, M., Křetínský, J., Weinhuber, C., Weininger, M., Yadav, M.: dtControl 2.0: Explainable Strategy Representation via Decision Tree Learning Steered by Experts. In: CAV 2021. LNCS, vol. 12759, pp. 326–345. Springer, Cham (2021). https://doi.org/10.1007/978-3-030-72013-1_17
4. Badings, T., Simão, T.D., Suilen, M., Jansen, N.: Decision-making under uncertainty: beyond probabilities. Int. J. Softw. Tools Technol. Transf. **25**(3), 375–391 (2023). https://doi.org/10.1007/S10009-023-00704-3
5. Baier, C., de Alfaro, L., Forejt, V., Kwiatkowska, M.: Model Checking Probabilistic Systems. In: CAV 2021. LNCS, vol. 12759, pp. 963–999. Springer, Cham (2018). https://doi.org/10.1007/978-3-319-10575-8_28
6. Baier, C., Katoen, J.P.: Principles of Model Checking. MIT Press (2008)
7. Bohnenkamp, H.C., D'Argenio, P.R., Hermanns, H., Katoen, J.P.: MODEST: a compositional modeling formalism for hard and softly timed systems. IEEE Trans. Software Eng. **32**(10), 812–830 (2006). https://doi.org/10.1109/TSE.2006.104
8. Budde, C.E., Dehnert, C., Hahn, E.M., Hartmanns, A., Junges, S., Turrini, A.: JANI: Quantitative Model and Tool Interaction. In: Legay, A., Margaria, T. (eds.) TACAS 2017. LNCS, vol. 10206, pp. 151–168. Springer, Heidelberg (2017). https://doi.org/10.1007/978-3-662-54580-5_9
9. Budde, C.E., Hartmanns, A., Klauck, M., Křetínský, J., Parker, D., Quatmann, T., Turrini, A., Zhang, Z.: On Correctness, Precision, and Performance in Quantitative Verification. In: Margaria, T., Steffen, B. (eds.) ISoLA 2020. LNCS, vol. 12479, pp. 216–241. Springer, Cham (2021). https://doi.org/10.1007/978-3-030-83723-5_15
10. Ceska, M., Junges, S., van der Maas, L., Macák, F., Quatmann, T.: Fast computation of conditional probabilities in MDPs and Markov chain families (2026). submitted
11. Forejt, V., Kwiatkowska, M., Norman, G., Parker, D.: Automated Verification Techniques for Probabilistic Systems. In: Bernardo, M., Issarny, V. (eds.) SFM 2011. LNCS, vol. 6659, pp. 53–113. Springer, Heidelberg (2011). https://doi.org/10.1007/978-3-642-21455-4_3
12. Hahn, E.M., Li, Y., Schewe, S., Turrini, A., Zhang, L.: ISCASMC: A Web-Based Probabilistic Model Checker. In: Jones, C., Pihlajasaari, P., Sun, J. (eds.) FM 2014. LNCS, vol. 8442, pp. 312–317. Springer, Cham (2014). https://doi.org/10.1007/978-3-319-06410-9_22

13. Hartmanns, A.: An overview of Modest models and tools for real stochastic timed systems. In: MARS. EPTCS **355**, pp. 1–12 (2022). https://doi.org/10.4204/EPTCS.355.1
14. Hartmanns, A., Hermanns, H.: The Modest Toolset: An Integrated Environment for Quantitative Modelling and Verification. In: Ábrahám, E., Havelund, K. (eds.) TACAS 2014. LNCS, vol. 8413, pp. 593–598. Springer, Heidelberg (2014). https://doi.org/10.1007/978-3-642-54862-8_51
15. Hartmanns, A., Junges, S., Katoen, J.P., Quatmann, T.: Multi-cost bounded tradeoff analysis in MDP. J. Autom. Reason. **64**(7), 1483–1522 (2020). https://doi.org/10.1007/S10817-020-09574-9
16. Hartmanns, A., Junges, S., Quatmann, T., Weininger, M.: A practitioner's guide to MDP model checking algorithms. In: TACAS (1). Lecture Notes in Computer Science, vol. 13993, pp. 469–488. Springer (2023). doi: 10.1007/978-3-031-30823-9_24
17. Hartmanns, A., Kaminski, B.L.: Optimistic Value Iteration. In: Lahiri, S.K., Wang, C. (eds.) CAV 2020. LNCS, vol. 12225, pp. 488–511. Springer, Cham (2020). https://doi.org/10.1007/978-3-030-53291-8_26
18. Hensel, C., Junges, S., Katoen, J.P., Quatmann, T., Volk, M.: The probabilistic model checker Storm. Int. J. Softw. Tools Technol. Transf. **24**(4), 589–610 (2022). https://doi.org/10.1007/S10009-021-00633-Z
19. Hensel, C., Junges, S., Quatmann, T., Volk, M.: Riding the Storm in a probabilistic model checking landscape. In: Principles of Verification (2). Lecture Notes in Computer Science, vol. 15261, pp. 98–114. Springer (2024). doi: 10.1007/978-3-031-75775-4_5
20. Hermanns, H., Katoen, J.-P., Meyer-Kayser, J., Siegle, M.: A Markov Chain Model Checker. In: Graf, S., Schwartzbach, M. (eds.) TACAS 2000. LNCS, vol. 1785, pp. 347–362. Springer, Heidelberg (2000). https://doi.org/10.1007/3-540-46419-0_24
21. Iyengar, G.N.: Robust dynamic programming. Math. Oper. Res. **30**(2), 257–280 (2005). https://doi.org/10.1287/MOOR.1040.0129
22. Junges, S., et al.: Parameter synthesis for Markov models: covering the parameter space. Formal Methods Syst. Des. **62**(1), 181–259 (2024). https://doi.org/10.1007/S10703-023-00442-X
23. Katoen, J.P.: The probabilistic model checking landscape. In: LICS, pp. 31–45. ACM (2016). https://doi.org/10.1145/2933575.2934574
24. Katoen, J.P., Zapreev, I.S., Hahn, E.M., Hermanns, H., Jansen, D.N.: The ins and outs of the probabilistic model checker MRMC. Perform. Evaluation **68**(2), 90–104 (2011). https://doi.org/10.1016/J.PEVA.2010.04.001
25. Kwiatkowska, M., Norman, G., Parker, D.: Probabilistic model checking and autonomy. Annu. Rev. Control. Robotics Auton. Syst. **5**, 385–410 (2022)
26. Kwiatkowska, M., Norman, G., Parker, D.: Probabilistic model checking: Applications and trends. In: Principles of Formal Quantitative Analysis. Lecture Notes in Computer Science, vol. 15760, pp. 158–173. Springer (2025). doi: 10.1007/978-3-031-97439-7_7
27. Kwiatkowska, M.Z., Norman, G., Parker, D.: Probabilistic symbolic model checking with PRISM: a hybrid approach. Int. J. Softw. Tools Technol. Transf. **6**(2), 128–142 (2004). https://doi.org/10.1007/S10009-004-0140-2
28. Kwiatkowska, M., Norman, G., Parker, D.: PRISM 4.0: Verification of Probabilistic Real-Time Systems. In: Gopalakrishnan, G., Qadeer, S. (eds.) CAV 2011. LNCS, vol. 6806, pp. 585–591. Springer, Heidelberg (2011). https://doi.org/10.1007/978-3-642-22110-1_47

29. Larsen, K.G., Pettersson, P., Yi, W.: UPPAAL in a nutshell. Int. J. Softw. Tools Technol. Transf. **1**(1–2), 134–152 (1997). https://doi.org/10.1007/S100090050010
30. Leerkes, P., Melse, I., Heck, L., Volk, M., Junges, S.: Stormvogel: probabilistic model checking for almost everyone (2025). https://repository.ubn.ru.nl/handle/2066/325553
31. Madani, O., Hanks, S., Condon, A.: On the undecidability of probabilistic planning and related stochastic optimization problems. Artif. Intell. **147**(1–2), 5–34 (2003). https://doi.org/10.1016/S0004-3702(02)00378-8
32. Meggendorfer, T.: PET - A partial exploration tool for probabilistic verification. ATVA. Lecture Notes in Computer Science **13505**, 320–326 (2022). https://doi.org/10.1007/978-3-031-19992-9_20. Springer
33. Puterman, M.L.: Markov Decision Processes: Discrete Stochastic Dynamic Programming. Wiley Series in Probability and Statistics (1994). Wiley. https://doi.org/10.1002/9780470316887
34. Quatmann, T.: Verification of multi-objective Markov models. Ph.D. thesis, RWTH Aachen University, Germany (2023). https://publications.rwth-aachen.de/record/971553
35. Spaan, M.T.J.: Partially observable Markov decision processes. In: Reinforcement Learning, Adaptation, Learning, and Optimization, vol. 12, pp. 387–414. Springer (2012). https://doi.org/10.1007/978-3-642-27645-3_12
36. Volk, M., Heck, L., Junges, S., Katoen, J.P., Quatmann, T.: Probabilistic model checking taken by storm. CoRR abs/2603.15559 (2026). https://arxiv.org/abs/2603.15559

Distributed Runtime Verification in Proximity-Based Networks: A Tutorial on the Aggregate Programming Approach

Giorgio Audrito[1(✉)], Ferruccio Damiani[1], Giordano Scarso[1], Volker Stolz[2], and Gianluca Torta[1]

[1] Department of Computer Science, University of Turin, Torino, Italy
{giorgio.audrito,ferruccio.damiani,giordano.scarso,gianluca.torta}@unito.it

[2] Department of Computer science, Electrical engineering and Mathematical sciences, Western Norway University of Applied Sciences, Bergen, Norway
Volker.Stolz@hvl.no

Abstract. Distributed runtime verification (DRV) addresses the problem of checking the correctness of distributed systems during execution, coping with partial knowledge, dynamic topologies, and the absence of global time. These challenges are particularly prominent in proximity-based networks, such as those arising in IoT and Far Edge computing scenarios, where large numbers of devices interact through local communication. This tutorial presents an approach to DRV based on Aggregate Programming (AP), a paradigm for designing distributed collective systems via high-level abstractions over computational fields. We show how temporal and spatial properties (expressed in past-CTL and SLCS, respectively) can be systematically compiled into aggregate monitors grounded in the eXchange Calculus and executed using the FCPP C++ framework and simulator for AP. The tutorial combines conceptual foundations with practical guidance: participants learn how to specify spatio-temporal properties, generate corresponding monitors, and execute them in a 3D simulation environment. Examples are drawn from ongoing industrial collaborations and research projects, which we use to illustrate realistic monitoring scenarios and motivate open challenges for AP-based DRV.

Keywords: Aggregate Programming · Distributed Runtime Verification · Self-Stabilisation

1 Introduction

What is Distributed Runtime Verification? *Runtime verification (RV)* is a program analysis paradigm that focuses on observing a system during execution in order to check whether its behaviour conforms to a given specification, using monitors generated from formal specifications [26]. *Distributed runtime verification (DRV)* extends this paradigm to distributed systems, encompassing both

A. Sampaio and M. Stoelinga (Eds.): FM 2026, LNCS 16557, pp. 550–575, 2026.
https://doi.org/10.1007/978-3-032-26220-2_26

the monitoring of distributed systems and the use of distributed infrastructures to perform monitoring. As a verification technique, runtime verification emphasizes the systematic generation of monitors from formal specifications, enabling properties to be stated precisely and providing formal guarantees about the correctness of the monitoring outcomes. Distribution, however, introduces additional challenges for verification, as it requires addressing issues such as synchronization, communication faults, and the absence of a unique global time [23].

What are Proximity-based Networks? *Proximity-based networks* are (possibly dense and large) networks of devices (possibly mobile, ephemeral, capable of computation, communication and interaction with the environment and people) where:

- the topology is dynamic, induced by the proximity (physical or logical) of the devices; and
- each device can receive/send messages to/from devices in the neighbourhood.

These arise naturally in relation to Far Edge Computing, which brings together all locally installed IT systems capable of operating independently of the Cloud and can form the basis for decentralised IT architectures. Far Edge Computing represents an opportunity for the EU's industrial ecosystem, which is strong in professional Internet of Things (IoT) applications [29].

What is AP? *Aggregate Programming (AP)* [15] is an established paradigm for engineering distributed systems that has been actively developed for over a decade [37]. Its execution model (typically implemented within a runtime) assumes that each device repeatedly executes asynchronous sense–compute–interact rounds. The paradigm is formally grounded in AP calculi, such as the *Field Calculus (FC)* [14] and its recent extension *eXchange Calculus (XC)* [5]. These are functional core languages designed to express computations over distributed data structures, called *computational fields* [27], that map sets of devices to values, representing collective inputs and outputs. This formal foundation enables rigorous reasoning about aggregate programs, including the derivation of guarantees and proofs (for example, establishing that a collective computation is self-stabilising, meaning that it eventually converges to the correct result once its inputs have stabilized [36]). Aggregate Programming languages are realized through standalone or embedded domain-specific languages (DSLs) [37], which are often accompanied by libraries of reusable functions capturing common patterns of collective behaviour. Using these languages and libraries, programmers write a single, global program that specifies the overall behaviour of the collective as a composition of functions. This program can then be deployed over a network of devices, either simulated or, through middleware support, made of physical systems [17]. By separating the logical specification of the system from its physical deployment, AP also supports partitioning applications into deployable components, enabling computation to be offloaded across the (Far) Edge–Cloud Continuum [22]. Currently, the main way to analyse AP programs is by simulation. The AP-based DRV techniques discussed here, which can of course be applied in particular to AP systems, can add a valuable tool for (on-line) checking of formal properties.

Practical Support for AP. The main practical support for AP is given by: the Scala internal DSL, library and toolchain *ScaFi* [32] (see also [7,16,18,19]); the C++ internal DSL, library and toolchain *FCPP* [35] (see also [2,11–13]); and the Kotlin internal DSL, compiler plugin, library and toolchain *Collektive* [33] (see also [21]). FCPP allows programmers to embed XC programs in C++ applications, and deploy them on networks of physical devices, as well as a simulator. It is used in a number of industrial case studies currently developed in collaboration with the companies Concept Reply (https://www.reply.com/concept-reply/en), Synesthesia (https://synesthesia.it), Eurix (https://www.eurixgroup.com), ST Microelectronics (https://www.st.com/content/st_com/en.html), and Torino Airport (https://www.aeroportoditorino.it/en)[1] – preliminary results related to these activities have been published [8,28,31].

This Tutorial, its Accompanying Material, and Further Reading. This tutorial caters to all who want to learn about DRV in proximity-based networks using AP: newcomers to the field as well as experienced researchers in formal methods outside of distributed programming and RV. It offers an overview of DRV, of the past-CTL (an extension of past-LTL [24]) temporal logic and the SLCS [20] spatial logic for specifying monitors over proximity-based networks, on AP, on how to generate AP monitors from past-CTL and SLCS formulas, and on how to run them in the FCPP simulator. Participants of the live conference tutorial gain hands-on experience with the FCPP framework. By the end of the tutorial, they will be able to actively use FCPP for specifying (simple) monitors, and running them on the FCPP simulator. Videos, slides, and all examples of the conference tutorial, as well as the FCPP framework itself, are available for download at https://fcpp.github.io/quickstart/tutorial-fm-2026.html. For further reading, two journal papers present the AP approach for past-CTL monitors [9] and SLCS monitors [6], a journal paper presents the formal foundation of XC [5], and the FCPP website (https://fcpp.github.io) provides an up-to-date list of papers related to DRV by AP.

Paper Organization. Sect. 2 presents DRV from a theoretical perspective. Section 3 introduces aggregate programming with the XC language and FCPP. Finally, Sect. 4 describes an automatic synthesis of AP monitors for spatio-temporal formulas, and presents experiments of DRV in FCPP. Section 5 provides a recollection of AP case studies and how DRV could be applied to them.

2 The Theory of DRV in Proximity-Based Networks

Distributed runtime verification is a sub-field of the more general field of runtime verification, where specifications over events as atomic propositions are evaluated on traces or streams [26]. Popular specification languages include variations on the Linear Time Logic LTL, and regular expressions. Events may be generated through state changes or execution flow, such as method calls.

[1] See, e.g., the RoboAPP [34] and the RoboNG [30] cascade funding projects of the Spoke 1 "Sustainable mobility and aerospace" of the PNRR NODES (https://ecs-nodes.eu/en); and the CN AgriTech (https://agritechcenter.it).

In distributed runtime verification [23], agents (representing the part of the system to verify at each device) are generally considered *remote* to each other: as constituents of the whole system, they are assumed to execute independently and occasionally synchronise or communicate with each other via the underlying communication platform. A *local trace* of *events* corresponds to a sequence of sets of values for observables, as defined through the sensors of an agent, or derived values from those. Since agents may appear or disappear over time from the overall system, traces from different agents are not aligned in time in the sense that for a particular index/position in each trace, these events did not necessarily happen at the same time. Accordingly, logic formulas cannot state properties over single traces (except in the degenerate case of specifying a property only on the history of a single agent). Therefore, we will first introduce a suitable computational model through augmented event structures, before we define our logics for the temporal and the spatial fragments which will allow us to draw conclusions from the shared histories of agents, that evolve intermittently only locally on each device in-between synchronizations with peers within communication range. The partially or fully evaluated spatio-temporal properties as seen from a particular agent can hence be seen as a compressed history wrt. to the values of these properties.

2.1 The Execution Model

In earlier work, we have identified Lamport's notion of *event structures* as a suitable view on the overall system [6,25]. This representation has been used successfully already before, to model the partial order relation between events arising from model causality (communication across agents, or agent internal computation steps, where agents are identified by device identifiers) [14,39]. We paraphrase the necessary background from [6]:

Definition 1 (Augmented Event Structure). *An* augmented event structure $\mathbb{E} = \langle E, \rightsquigarrow, < \rangle$ *is a finite or countably infinite set of* events E *together with a neighbouring relation* $\rightsquigarrow \subseteq E \times E$ *and a causality relation* $< \subseteq E \times E$*, such that the transitive closure of* $\rightsquigarrow$ *forms the irreflexive partial order* $<$*, and the set* $X_\epsilon = \{\epsilon' \in E \mid \epsilon' < \epsilon\} \cup \{\epsilon' \in E \mid \epsilon \rightsquigarrow \epsilon'\}$ *is finite for all* ϵ *(i.e.,* $<$ *and* $\rightsquigarrow$ *are locally finite). Thus, we say that* ϵ' *is a neighbour of* ϵ *iff* $\epsilon' \rightsquigarrow \epsilon$*, and that* $\mathcal{N}(\epsilon) = \{\epsilon' \in E \mid \epsilon' \rightsquigarrow \epsilon\}$ *is the set of neighbours of* ϵ*.*

Figure 1 depicts a sample augmented event structure, showing the $\rightsquigarrow$ relation, and how the $<$-relation partitions events into "causal past" (red), "causal future" (cyan), and non-ordered "concurrent" (black) subspaces with respect to any given event (in Fig. 1, colours reflect causality with respect to the doubly-circled event in magenta). In principle, an execution at ϵ can depend on information from any event in its past and its results can influence any event in its future. Causality is uniquely induced by neighbouring (the $\rightsquigarrow$ relation), dictating when an event can *directly* influence (by message-passing) another. Intuitively, every $\rightsquigarrow$ relation correspond to the send and receive of a message: in order for $\epsilon_1 \rightsquigarrow \epsilon_2$

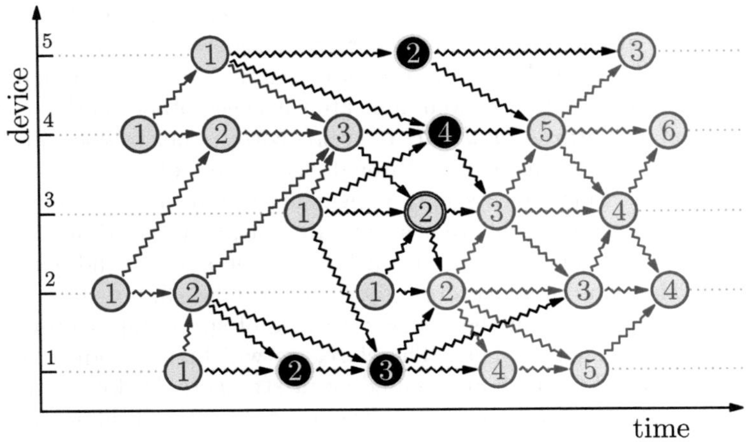

Fig. 1. Example of an augmented event structure, comprising events (circles), neighbour relations (arrows), devices (ordinate axis). Colours indicate causal structure with respect to the doubly-circled event (magenta), splitting events into past (red), future (cyan) and concurrent (non-ordered, in black). The numbers within events represent a space-time value associated with that event, namely a counter value which is reset if the device crashes, see device 2. Note that the set of neighbouring events $\mathcal{N}(\epsilon)$ of the doubly-circled event has three elements: event 1 at the same device (its previous firing), event 3 at device 4, and event 1 at device 2. Figure taken from [3]. (Color figure online)

to hold, execution of event ϵ_1 on a device δ_1 must result in a message which reaches a device δ_2 before its execution of ϵ_2. The name *neighbouring* reflects that message exchanges happen on devices that are close to each other (in some physical or logical sense).

Any sequence of computation events and message exchanges between them can be represented as an augmented event structure, however, not all event structures are physically realisable by a distributed system following the firing model described at the beginning of this section. We refer the reader to [6] for a detailed definition of the subset of *realisable event structures* which additionally satisfy coherence constraints on linearity and uniqueness (a single device only has a single history), as well as impersistence ("reboots" lose history) and computation immediacy (acyclic event relation).

2.2 Temporal Logic: Past-CTL

Past-CTL is a variant of the more standard future-time Computation Tree Logic (CTL). Here, operators look back in time, and we have chosen corresponding names dual to their well-known future counterparts. Note that past-CTL is suitable for monitoring, since it refers only to the past and each formula therefore has a definite truth value at any given time. In contrast, future-time logics such as CTL are more suitable for tasks such as model checking.

Atomic propositions get a (possibly different) truth value for each particular event. Path quantifiers distinguish if properties hold on *all* or *some* paths starting from some initial event of a device and ending in the current event. Unquantified formulas are hence linear properties on the past on a single device.

$\mathrm{Y}\,\phi$	*yesterday*, ϕ held in the previous event on the considered path (on the same device for Y, on every neighbour device for AY, on some device for EY)
$\psi\,\mathrm{S}\,\phi$	*since*, the second argument held in some past event in the considered path, and its first argument has held since then
$\mathrm{P}\,\phi$	*previously*, ϕ held in some past event on the considered path
$\mathrm{H}\,\phi$	*historically*, ϕ held in every past event on the considered path

Operators $\mathrm{Y}, \mathrm{EY}, \mathrm{S}, \mathrm{AS}, \mathrm{ES}$ can be considered as primitive, while the other operators can be derived from them: $\mathrm{AY}\,\phi \triangleq \neg\,\mathrm{EY}\,\neg\phi$; $\mathrm{P}\,\phi \triangleq \top\,\mathrm{S}\,\phi$ (similarly for AP, EP with AS, ES); $\mathrm{H}\,\phi \triangleq \neg\,\mathrm{P}\,\neg\phi$ (similarly for AH, EH with EP, AP).

Some of the properties will hence be monotonic, in the sense that an intermediate verdict fixes all future verdicts, by quantification across paths (A/E) over the past (P/H).

From our work in [9], we present two example properties using the above past-CTL operators. Let us assume a property a_i indicates that a device (e.g. a drone) is in an area which we identify through its index i. Then the following formula will indicate success in the sense that all areas have been handled (assuming that drones come periodically into contact with each other so that knowledge propagates): $\bigwedge_{i\in N} \mathrm{EP}\,a_i$. A slightly more elaborate construction would be $\bigwedge_{i\in N} \mathrm{AH}\,\neg(a_i \wedge \mathrm{EY}(\mathrm{EP}\,a_i))$, which asserts that no device is servicing an area that has already been serviced by someone. In both cases we have the side-effect that every device updates and spreads the list of areas that have been handled with its neighbours in each round, propagating that knowledge through the network.

2.3 Spatial Logic: SLCS

The physical distribution of nodes in space over time (each node can be in proximity of different sets of nodes at any given time during its history) requires another dimension along which we would like to express interesting properties. Here, we have settled on Ciancia et al.'s Spatial Logic of Closure Space (SLCS) [20], which is based on two main modalities: $\Box\,\phi$ (holds in the interior of points where ϕ holds) and $\Diamond\,\phi$ (holds in the closure of points where ϕ holds). We have shown that this logic is able to express properties of discrete networks of devices [6]. A pictorial representation of this concept is given in Fig. 2, where we have a set of nodes satisfying some property P, the *closure* $\Diamond P$ of all nodes adjacent to those satisfying P, and those where some P is reachable, $\mathcal{F}P$. Next, we present the local and global modalities of SLCS, which we will later map to equivalent constructs in AP (either as elementary functions or as macros on top of a functionally complete subset of operators).

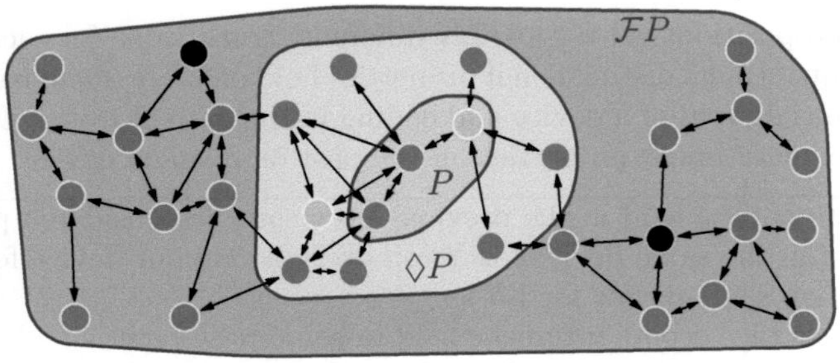

Fig. 2. Pictorial representation of a closure space induced by a graph [6]. Nodes are coloured black/grey/yellow according to which set of atomic propositions holds, edges are the proximity relation. Shaded areas satisfy (different closures of) some example property P.

Local modalities	
$\Diamond\,\phi$	*closure*, true at points with *some* neighbour satisfying ϕ
$\Box\,\phi$	*interior*, true at points with *all* neighbours satisfying ϕ
$\partial\,\phi$	*boundary*, true at points with some (but not all) neighbours satisfying ϕ
$\partial^-\,\phi$	*interior boundary*, true at points satisfying ϕ with some neighbour not satisfying it
$\partial^+\,\phi$	*closure boundary*, true at points not satisfying ϕ with some neighbour satisfying it
Global modalities	
$\phi\,\mathcal{R}\,\psi$	*reaches*, true at the ending points of paths whose starting point satisfies ψ and where ϕ holds
$\phi\,\mathcal{T}\,\psi$	*touches*, true at the ending points of paths whose starting point satisfies ψ and where ϕ holds in the rest of the path
$\phi\,\mathcal{U}\,\psi$	*surrounded by*, true at points in an area A satisfying ϕ, whose neighbours satisfy ψ
$\mathcal{G}\,\phi$	*everywhere*, true at points where ϕ holds in *every* incoming path
$\mathcal{F}\,\phi$	*somewhere*, true at points where ϕ holds in at least *one point* of an incoming path

Note that there is a functionally complete minimal set with $\Diamond$, $\mathcal{R}$ and Boolean operations which allows us to express all other operators as equivalences. Local modalities: $\Box\,\phi \equiv \neg\Diamond\neg\phi$, $\partial\,\phi \equiv (\Diamond\,\phi)\wedge\neg(\Box\,\phi)$, $\partial^-\,\phi \equiv \phi\wedge\neg(\Box\,\phi)$, $\partial^+\,\phi \equiv (\Diamond\,\phi)\wedge\neg\phi$. Global modalities: $\phi\,\mathcal{T}\,\psi \equiv \phi\,\mathcal{T}\,\Diamond\,\psi$, $\phi\,\mathcal{U}\,\psi \equiv \phi\wedge\Box\neg(\neg\psi\,\mathcal{R}\,\neg\phi)$, $\mathcal{F}\,\phi \equiv \top\,\mathcal{R}\,\phi$.

As a sample application of SLCS in a smart home scenario, consider the following property to monitor: *air conditioning and lights are on whenever the room is not empty, off otherwise*. Consider the atomic propositions where *(i)* P is true on points that are sensing the presence of people; *(ii)* D is true on points that are the monitored electrical devices (air conditioning, lights); *(iii)* O is true on electrical devices that are on.

Fig. 3. Screenshot from an FCPP-simulation of the *cluster*-scenario, where node shape (circle/diamond), size (small/large) and colour (green/red) indicate a node's properties for *warning*, *cluster* and truth-value of the final property, respectively. The simulation is overlaid on top of a city-plan.

If we only want to consider the presence of people in the immediate vicinity, the considered property can be written as $\neg D \vee (O \Leftrightarrow \Diamond P)$. When also considering people farther away, the property can be written as $\neg D \vee (O \Leftrightarrow \mathcal{F}\, P)$. In the sample closure space of Fig. 2, a possible evaluation of these properties could be the following, where different colours are used for points where D is false (grey), D is true and O is false (black), D and O are true (yellow): The green area denotes the nodes for which P is true, i.e., the nodes that do perceive at least a person in the room. The cyan area ($\Diamond P$) is given by the nodes that have at least a neighbour within the green area. The blue area ($\mathcal{F}\, P$) is given by the nodes for which there exists a path to a node in P ("somewhere"); this includes all the nodes of the example.

In the practical session, we will use a more dynamic example where we combine AP properties with temporal and spatial properties: let us first assume the following two space-time variables: *warning* which is true if there are more than five neighbours in our direct vicinity, and another that indicates that a device is member of a *cluster* when at least three neighbours have the *warning*-flag set. We will later see that both flags can easily be derived through AP features that just rely on counting direct neighbours. Based on this, we can flag when we become part of a cluster by observing the transition of our corresponding variable through the (degenerate, as it is just per-device) past-CTL formula: $\mathit{cluster\_start} := (\mathrm{Y}\, \neg \mathit{cluster}) \wedge \mathit{cluster}$. Similarly, we define $\mathit{cluster\_end}$. Through an SLCS-formula a device can also observe if all its neighbours "agree" that they are part of a *cluster*: $\mathit{all\_cluster} := \mathcal{G}\, \mathit{cluster}$. And finally we can use this property in turn in a path-quantified past-CTL formula which states that "no new neighbour joined since everyone observed formation of the cluster": $\neg$*cluster_start* AS *all_cluster*. Figure 3 graphically illustrates a given situation in the FCPP-simulator.

3 Aggregate Programming for Proximity-Based Networks

Aggregate programming [15,37] is a computational paradigm for engineering distributed systems composed of large numbers of interacting devices, typically deployed in a physical or logical space and connected through neighbourhood relations. The paradigm has been developed to address the intrinsic complexity of systems such as the Internet of Things (IoT), pervasive computing infrastructures, and spatially situated cyber–physical systems, where devices operate asynchronously, may be mobile, and interact primarily through local communication. At its core, aggregate programming promotes a *global* view of computation: instead of describing individual device behaviours and their message exchanges explicitly, developers specify the intended collective behaviour of the system as a whole. Communication, synchronisation, and coordination among devices are abstracted away and handled implicitly by the computational model and its semantics. This approach enables programs to be expressed compositionally and declaratively, while remaining robust with respect to changes in network size, topology, and device density.

An aggregate program P is executed repeatedly and asynchronously by every device in the network. Each device δ follows its own local firing cycle, with no assumption of global synchronisation or shared clocks. A single execution of P on a device is referred to as a *computation round* (or *firing*), and can be informally described as consisting of the following phases:

1. **Context acquisition:** the device gathers contextual information, including sensor readings, locally stored state from the previous firing, and messages received from neighbouring devices since the last execution.
2. **Local evaluation:** the aggregate program P is evaluated using the acquired context as input, according to the operational semantics of the language.
3. **State update and communication:** the result of the evaluation is stored locally, possibly exposed to actuators, and may induce a broadcast to neighbouring devices for use in their subsequent firings.
4. **Sleep:** the device waits for its next round while listening for new messages.

Through the continuous repetition of these rounds across devices and over time, a coherent *global behaviour* emerges at the network level [36]. This can be formalised through the same notion of *Augmented Event Structure* introduced in Sect. 2.1, with one event for each round, connecting events $\epsilon_1 \rightsquigarrow \epsilon_2$ when the broadcast resulting from round ϵ_1 was considered in the context acquisition phase of round ϵ_2. This behaviour can be understood as the evolution of distributed data over space and time, rather than as the result of explicit message-passing protocols, and formalised accordingly as follows.

Definition 2 (Space–Time Value). *Given an augmented event structure* $\mathbb{E}$ *and a universe of values* $\mathbf{V}$*, a* space–time value *is a function* $f : E \to \mathbf{V}$*, associating a value with each event in* $\mathbb{E}$*. We denote as* $\mathbf{V}(\mathbb{E})$ *the set of such space–time values.*

From this perspective, an aggregate computation denotes a transformation of space–time values. This leads naturally to the notion of space–time functions.

Definition 3 (Space–Time Function). *An* n-ary space–time function *in an augmented event structure* $\mathbb{E}$ *is a partial function* $\mathtt{f} : \mathbf{V}(\mathbb{E})^n \twoheadrightarrow \mathbf{V}(\mathbb{E})$*, mapping tuples of input space–time values to an output space–time value defined over the same event structure.*

Not all space–time functions are implementable by aggregate programs: realizable functions must respect causality and be computable through local interactions [3]. Nevertheless, this abstraction provides a powerful semantic foundation for reasoning about distributed computations, linking the local execution of programs on individual devices with their global, collective meaning.

3.1 The eXchange Calculus

The *exchange calculus* (XC) [4,5] is a compact functional language that provides a foundational model for aggregate programming [15,37]. Its design goal is to express distributed collective computations at a high level, while abstracting away low-level concerns such as message passing, neighbourhood management, device identities, and network size. To this end, XC relies on a single communication primitive capable of expressing all forms of neighbourhood interaction, called *exchange*, and on the central data abstraction of *neighbouring values*.

A neighbouring value (or *nvalue* for short) is used to represent both information collected from nearby devices during a computation round, as well as information to send back to neighbour devices as result of an *exchange* routine. Formally, it is defined as a finite map from device identifiers to *local literals*, together with a distinguished default value, written as $\mathtt{w} = \ell[\delta_1 \mapsto \ell_1, \ldots, \delta_n \mapsto \ell_n]$, where δ_i, ℓ_i are neighbour devices with an explicitly associated value, and ℓ denotes the default local literal, used whenever no explicit value is associated to a neighbour device. Local values (such as simple numbers) are treated uniformly as nvalues with an empty mapping, written $\ell[]$. As a result, every value manipulated in XC can be regarded as a nvalue.

Neighbouring values are crucial in the exchange of information with neighbours. Given a nvalue, we can also retrieve the local value in it for the current device via built-in `self(w)`, which when evaluated on device δ returns the local value $\ell = \mathtt{w}(\delta)$ stored in `w` for δ (which could be either the default or an explicit value). Besides these basic usages, the manipulation of nvalues comes in mostly two main forms. First, the point-wise lifting of local built-in operators to nvalues, applying the built-in independently for each device entry. For instance, $2[\delta_1 \mapsto 1, \delta_2 \mapsto -1] + 4[\delta_1 \mapsto 3, \delta_3 \mapsto 5]$ produces as result $6[\delta_1 \mapsto 4, \delta_2 \mapsto 3, \delta_3 \mapsto 7]$ (the sums for δ_2 and δ_3 use default values 2 and 4, respectively). Second, nvalues can be *folded* into a single local value, via built-in `nfold(f,w)` which combines the values associated with the currently *aligned* neighbour devices using the binary function `f`. For instance, `nfold`$(+, 2[\delta_1 \mapsto 1, \delta_2 \mapsto -1]])$ in a context where the current device is δ_1 and the only aligned neighbour is δ_3 will produce 1 (value for the current device, always considered aligned) +2 (value for δ_3) $= 3[]$.

In XC, when evaluating any sub-expression, we consider as *aligned* only the neighbours that in their last round evaluated that *same sub-expression*, both as position in the AST (Abstract Syntax Tree of the overall program), and as stack frame of function calls. This *alignment* process is not only used in `nfold` operations, but it is also crucial to ensure that multiple *exchanges* of information can be compositionally combined, while automatically matching the information pertinent to each of them, without generating unpredictable interferences.

Such exchange of information is formalised through a single built-in function called *exchange*, with the following form:

$$\texttt{exchange}(\texttt{e}_i\texttt{, (n) => return e}_r\texttt{ send e}_s\texttt{)}$$

where $\texttt{return e}_r\texttt{ send e}_s$ is syntactic sugar for pair construction $\texttt{pair}(\texttt{e}_r, \texttt{e}_s)$, to make explicit the role of the two expressions. Its evaluation proceeds as follows:

- First, expression $\texttt{e}_i$ is evaluated into a local value $\ell[]$, to be used as a starting value for new interactions.
- Then, a nvalue `w` is built, mapping each aligned neighbour δ' to the last value shared by δ' for the current device δ in this exchange expression, using the ℓ just computed as default. Notice that if it is the first execution of `exchange` on the current device δ, the default ℓ also provides the value for the current device δ in `w` (otherwise, the value shared by δ for itself in its previous round is used).
- Expressions $\texttt{e}_r$ and $\texttt{e}_s$ then are evaluated by substituting `w` for `n` in their bodies, resulting in values $\texttt{w}_r$ and $\texttt{w}_s$ respectively.
- Finally, value $\texttt{w}_r$ is returned as the value of the whole `exchange` expression in the program, and value $\texttt{w}_s$ is sent to neighbours, who will use it in their subsequent rounds to construct their nvalue `w`.

This exchange construct is sufficiently general to model any pattern of communication and information sharing between neighbour devices as a defined function on top of it. In this tutorial, we will sometimes use one of these patterns called `nbr`, that can be defined as follows:

```
def nbr(init, value) {
  exchange(init, (n) => return n send value)
}
```

The `nbr` function sends to neighbours the provided argument `value` (which is a nvalue, so it may encode different values for different neighbours). Then, it

Syntax:

$$\mathtt{e} ::= \mathtt{x} \mid \mathtt{fun}\ \mathtt{x}(\overline{\mathtt{x}})\{\mathtt{e}_0\} \mid \mathtt{e}_0(\overline{\mathtt{e}}) \mid \mathtt{val}\ \mathtt{x} = \mathtt{e}_1; \mathtt{e}_2 \mid \ell \mid \mathtt{w} \qquad \text{expression}$$

$$\mathtt{w} ::= \ell[\overline{\delta} \mapsto \overline{\ell}] \qquad \text{nvalue}$$

$$\ell ::= \mathtt{b} \mid \mathtt{fun}\ \mathtt{x}(\overline{\mathtt{x}})\{\mathtt{e}_0\} \mid \mathtt{c}(\overline{\ell}) \qquad \text{local literal}$$

$$\mathtt{b} ::= \mathtt{s} \mid \mathtt{exchange} \mid \mathtt{nfold} \mid \mathtt{self} \mid \mathtt{mux} \mid \ldots \qquad \text{built-in function}$$

Syntactic sugar:

$$(\overline{\mathtt{x}})\ \texttt{=>}\ \mathtt{e} ::= \mathtt{fun}\ \mathtt{y}(\overline{\mathtt{x}})\{\mathtt{e}\}\ \textit{where } \mathtt{y} \textit{ is fresh}$$

$$\mathtt{def}\ \mathtt{x}(\overline{\mathtt{x}})\{\mathtt{e}\} ::= \mathtt{val}\ \mathtt{x} = \mathtt{fun}\ \mathtt{x}(\overline{\mathtt{x}})\{\mathtt{e}\};$$

$$\mathtt{retsend}\ \mathtt{e} ::= \mathtt{val}\ \mathtt{y} = \mathtt{e};\ \mathtt{return}\ \mathtt{y}\ \mathtt{send}\ \mathtt{y}\ \textit{where } \mathtt{y} \textit{ is fresh}$$

$$\mathtt{if}(\mathtt{e})\{\mathtt{e}_\top\}\ \mathtt{else}\ \{\mathtt{e}_\bot\} ::= \mathtt{mux}(\mathtt{e}, ()\ \texttt{=>}\ \mathtt{e}_\top, ()\ \texttt{=>}\ \mathtt{e}_\bot)()$$

Fig. 4. Syntax of the eXchange Calculus (XC) language.

collects from neighbours the values that they sent to the current device, building a neighbouring value `w` (using the value of `init` as default), which is returned by the whole function. Thus, the `nbr` function returns a view of the `value` argument on neighbours.

Figure 4 illustrates the syntax of expressions $\mathtt{e}$ in XC (we use $\overline{\mathtt{x}}$ to denote a sequence of 0 or more elements), which can be either:

- a variable name $\mathtt{x}$;
- a (possibly recursive) function definition, which may contain free variables;
- a function call, where the overline $\overline{\mathtt{e}}$ is used to denote a sequence $\mathtt{e}_1, \ldots, \mathtt{e}_n$;
- a let-like expression, where the value $\mathtt{e}_1$ is substituted in $\mathtt{e}_2$ for the free vriable $\mathtt{x}$;
- a local literal ℓ, that is either the name of a built-in function $\mathtt{b}$, a defined function *without* free variables, or a data constructor $\mathtt{c}$ applied to a number of local literals (possibly none);
- a *nvalue* $\mathtt{w}$, as defined previously.

In order to ensure that XC programs are not dependent of a specific network configurations, nvalues are not allowed to directly appear in source programs, but only arise during expression evaluation. In addition to the basic syntax, we will also exploit some syntactic sugar (Fig. 4, bottom) for expressing anonymous functions and defining named functions, for returning and sending a same expression in an `exchange`, as well as for `if`/ `else` branching based on a condition. The latter exploits built-in $\mathtt{mux}(\mathtt{e}, \mathtt{e}_\top, \mathtt{e}_\bot)$, a multiplexer function returning either the value of $\mathtt{e}_\top$ or $\mathtt{e}_\bot$ depending on whether $\mathtt{e}$ is true or false. Note that expressions on different branches of an `if` *do not* align in the sense explained above (since they are in different parts of the AST).

Example 1 (Distance-To). Consider the following example of a function computing in each device its distance in hops from its closest *source*:

```
def dist(source) {
  exchange(∞, (d) =>
    retsend mux(source, 0, nfold(min, d, ∞)+1)
  )}
```

The `dist` function calculates hop-count distances from the nearest device where `source` is true using a single `exchange` construct. This construct collects distance estimates for neighbouring devices into a neighbouring value `w` (which maps neighbour devices to their distance estimates, defaulting to ∞ when no information is available), and assigns it to the variable `d`. The inner anonymous function body returns zero for source devices. For other devices, as in the traditional Bellman-Ford algorithm, it returns the minimum value from the neighbours' estimates `nfold(min, d, ∞)` increased by one.

This version of `nfold` with an additional argument works similarly as the `nfold` previously explained, except that it uses the value of its third argument (∞ in this case) as value for the current device, thus ignoring the value for the current device in the second argument (`d` in this case).

3.2 The FCPP Framework: a C++ DSL and Toolchain for AP

The XC programming language design has so far been implemented by several practical frameworks (see Sect. 1). In this tutorial, in order to experiment with space-time monitors we will use FCPP,[2] which is a C++-based domain-specific language and execution framework for aggregate programming. This framework provides a practical environment for implementing, simulating, and deploying aggregate programs on real distributed systems, ranging from simulations to networks of embedded devices. The framework can simulate asynchronous execution, message delays, communication failures, device mobility, and all kinds of dynamic changes in network topology and composition, making it suitable for modelling realistic distributed environments. FCPP is designed as a faithful implementation of the aggregate programming model and, in particular, of the semantics of the field calculus and the exchange calculus. Since it includes an implementation of both spatial and temporal logic operators within its standard library of aggregate algorithms, it is a natural choice for experimenting with the theory introduced in this tutorial.

A distinctive strength of FCPP lies in its versatility across different execution environments. The framework supports:

Interactive 3D graphical simulations, where large networks of devices are visualised in space and their aggregate state evolves in real time. This environment is particularly useful for debugging, teaching, and qualitative analysis of spatial behaviours.

[2] Monitors used in this paper are available at: https://github.com/fcpp-experiments/past-ctl-monitoring.

Batch simulations with statistical analysis, enabling repeated runs under varying parameters and initial conditions. These simulations can automatically produce plots and aggregated metrics, supporting quantitative evaluation of algorithmic properties such as convergence time, resilience, and stability.

High-performance processing of large graph-based datasets, where aggregate programs are executed over massive static or dynamic network topologies in HPC environments. In this setting, FCPP can be used to analyse complex relational data through aggregate computations.

Deployment on real distributed systems, including networks of embedded devices, robots, or IoT nodes.

FCPP is designed so that the same aggregate programs can be reused on multiple execution environments, possibly with minimal adaptation, thanks to the separation between program logic and execution platform. These multiple usage modes make FCPP both a research tool for studying aggregate algorithms and a practical platform for engineering distributed collective systems. In this tutorial, we will exclusively employ the 3D graphical simulation environment. This setting provides immediate visual feedback on the evolution of space-time values and allows us to illustrate monitoring and runtime verification techniques in a controlled yet expressive scenario. The other execution modes remain fully compatible with the techniques discussed here, but fall outside the scope of the present exposition.

At the language level, FCPP realises aggregate constructs through templated C++ functions and classes. The library provides macros and constructs to enable expressing programs with a syntax closely resembling that of XC, while allowing the underlying runtime system to manage communication, state persistence, and alignment of sub-computations, as shown in this code snippet.

```
FUN bool monitor(ARGS, int source_id) { CODE
    bool close = dist(CALL, node.uid == source_id) < 10;
    field<bool> neigh_close = nbr(CALL, close);
    return all_hood(CALL, neigh_close);
}
```

Several macros are used in this snippet. `FUN` hides a template argument for the type of the node. `ARGS` hides arguments for the necessary execution context (the `node` object, and a representation of the stack trace). `CODE` hides updating of the stack trace with the current function call. `CALL` passes the execution context along in function calls. The `node` object is not only implicitly used within constructs such as `nbr` or `all_hood` (which in XC would be `nfold` with logical and), but also explicitly in order to directly access to sensors and actuators available in the node (such as `node.uid` in the example above, to access a unique identifier for the node). This code snippet also shows the usage of neighbouring values, which are represented through a dedicated `field<T>` class, with built-in operators providing support for both local value extraction, as well as point-wise and folding operations.

<table>
<tr><td>$\top$</td><td><code>true</code></td><td>q</td><td><code>q()</code></td><td>$\neg\phi$</td><td><code>!</code>ϕ</td><td>$\phi_1 \vee \phi_2$</td><td>ϕ_1 <code>|</code> ϕ_2</td></tr>
<tr><td>$\Diamond\,\phi$</td><td colspan="2"><code>nfold(|, nbr(false, </code>ϕ<code>))</code></td><td>$\phi_1\,\mathcal{R}\,\phi_2$</td><td colspan="4"><code>if (</code>ϕ_1<code>) {dist(</code>ϕ_2<code>)<D} else {false}</code></td></tr>
<tr><td>$\mathrm{Y}\,\phi$</td><td colspan="2"><code>self(nbr(false, </code>ϕ<code>))</code></td><td>$\mathrm{EY}\,\phi$</td><td colspan="4"><code>nfold(|, nbr(false, </code>ϕ<code>))</code></td></tr>
</table>

<table>
<tr><td>$\phi_1 \mathrm{S}\, \phi_2$</td><td><code>exchange(false, (old) => retsend </code>ϕ_2<code> | (</code>ϕ_1<code> & self(old)))</code></td></tr>
<tr><td>$\phi_1 \mathrm{AS}\, \phi_2$</td><td><code>exchange(false, (old) => retsend </code>ϕ_2<code> | (</code>ϕ_1<code> & nfold(&, old)))</code></td></tr>
<tr><td>$\phi_1 \mathrm{ES}\, \phi_2$</td><td><code>exchange(false, (old) => retsend </code>ϕ_2<code> | (</code>ϕ_1<code> & nfold(|, old)))</code></td></tr>
</table>

Fig. 5. Translation of a primitive set of SLCS and past-CTL operators into XC.

4 Monitors for Past-CTL and SLCS Formulas

4.1 Generating XC Monitors

Formulas of SLCS and past-CTL logics can be translated into XC Boolean expressions, and then evaluated during the execution of a program. Figure 5 shows a possible translation, by recursion on sub-formulas. We translate atomic propositions q into function calls `q()`, which produce a result by accessing the readings of sensors and/or performing simple local computations. Logical formulas without temporal/spatial modalities are easily translated by mapping the logical operators to their XC representations. In Fig. 5 (first line), we show the translations of the operators $\neg$ and $\vee$.

For SLCS we consider the translation of primitive modalities $\Diamond$ and $\mathcal{R}$ (second line in Fig. 5). The representation of formula $\Diamond\,\phi$ should be understood as follows:

- the most recent Boolean values of ϕ received by the neighbours (including the current device) are collected into an nvalue with `nbr`
- the folding operator `nfold` applies logical "or" to the nvalue, returning true iff at least one of the elements in the nvalue is true

The representation of formula $\phi_1\,\mathcal{R}\,\phi_2$ requires a function `dist`, and a value `D`, where:

- `dist(`ϕ_2`)` performs an aggregate computation of the distance between the current device and the closest device where ϕ_2 holds (see Sect. 3.1)
- `D` is an upper bound to the network diameter (either fixed at design time or estimated with an aggregate computation)

Notice that, as function `dist` is executed within the `if` branch where ϕ_1 is true, it will only receive messages from neighbours for which ϕ_1 is also true. Thus, that function call computes the shortest distance from a point satisfying ϕ_2, restricted within the region where ϕ_1 is true. The XC expression for formula $\phi_1\,\mathcal{R}\,\phi_2$ evaluates to true if such a distance is within the network (i.e., `< D`), to false otherwise.

Regarding formal guarantees, [36] shows that each monitor self-stabilises to the value of the abstract formula over the limit topology of the device network, and specific bounds on convergence time have also been obtained in [10].

For Past-CTL, we consider the translation of primitive modalities Y, EY, S, ES, and AS. The representation of $\mathrm{Y}\,\phi$ extracts with `self` the value for the current device from the nvalue obtained by exchanging the Boolean value of ϕ with neighbours. In other words, we are taking the value of ϕ at the previous round on the current device. Interestingly, the translation of EY ϕ is the same as that of $\Diamond\,\phi$ discussed above, since *neighbour devices* can only be accessed through *previous events*. The representation of formula $\phi_1\,\mathrm{S}\,\phi_2$ is a bit longer:

- the `old` argument of the lambda function passed to `exchange`, receives the most recent Boolean values of the formula computed by the neighbours (including the current device)
- the value of the formula in the current device is true, either if: ϕ_2 is currently true; or, the S formula was true on the device in the previous round (the value is obtained with `self` from `old`), and ϕ_1 is currently true

The translations of $\phi_1\,\mathrm{ES}\,\phi_2$ and $\phi_1\,\mathrm{AS}\,\phi_2$ are very similar to the one we just describe, except that they require that the formula was true in at least one, (resp. all) neighbours in the previous round.

Interestingly, for past-CTL the monitors correspond exactly to the associated formulas, i.e., they *always* return the correct values.

4.2 FCPP Implementation

Each past-CTL and SLCS operator has been implemented in FCPP as an aggregate function, in the namespace `fcpp::coordination::logic`. For example, the Y past-CTL operator is implemented as `FUN bool Y (ARGS, bool f)`, where `FUN` and `ARGS` are C++ macros explained in Sect. 3.2, and the argument `f` takes the Boolean value of the formula ϕ to which Y is applied, as computed in the current round and device. Clearly, since XC and its FCPP translation are in *functional* style, logic operators with two or more arguments are applied using *prefix* instead of *infix* notation. For example, to evaluate formula $\phi_1\,\mathcal{R}\,\phi 2$, the following calls are issued:

```
f1 = ...; f2 = ...;
f = R(CALL, f1, f2);
```

We point out that, although we have shown how all past-CTL and SLCS operators can be realized in terms of a few operators chosen as *primitive*, FCPP implements each operator from scratch, to improve efficiency.

Among the operators of Past-CTL and SLCS, the ones that offer a wider range of possible implementations are the global SLCS operators $\mathcal{R}$, $\mathcal{T}$, $\mathcal{U}$, $\mathcal{G}$, and $\mathcal{F}$. The reason is that, as seen for the $\mathcal{R}$ primitive operator above, they all depend on a `dist()` function which can be implemented in several ways, trading-off the network convergence time, the resources overhead, and other factors (see [1] for details).

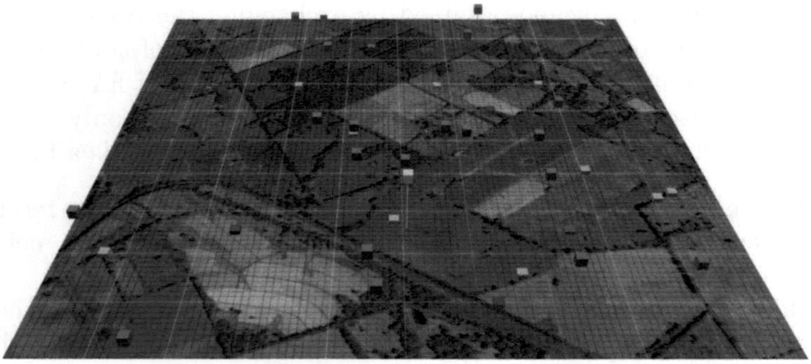

Fig. 6. Screenshot of the FCPP simulation of the drones recognition example.

4.3 Experimenting with the FCPP Framework

In an example scenario implemented in the FCPP simulator[3], we defined an outdoor area subdivided into $N = 4$ square areas, which had to be handled (e.g., observed with a camera) by a given set of 50 drones (Fig. 6). Each sub-area had a communication tower at its middle (brown/red cubes), and drones swept the environment performing random tasks (green cubes are drones heading to a destination, yellow ones are handling their goal), interleaved by recharging (blue) and waiting for tasks (grey) periods. The four towers issued "handling"-requests at random times, instructing the closest waiting drones to reach their area.

Proposition a_i for $i = 1 \ldots N$ is true iff the current drone is now handling area i. The properties below are formulated in terms of those observables, and how drones actually move or communicate are external factors for the purpose of our example. The existential quantifiers also encode that at the same point in time, different drones may have different views of the global system. Consider the properties that have already been discussed in Sect. 4: (i) every area gets eventually handled by a drone – $\bigwedge_{i \in N} \mathrm{EP}\, a_i$; (ii) no drone is handling an area that it knows to be already handled – $\bigwedge_{i \in N} \mathrm{AH}\, \neg(a_i \wedge \mathrm{EY}(\mathrm{EP}\, a_i))$.

The monitoring of the first property in FCPP is based on function:

```
FUN bool logic::area_handled(ARGS, bool handling) { CODE
    return EP(CALL, handling);
}
```

That is, we have an aggregate function `logic::area_handled` which takes as input a Boolean value telling whether *an* area is being *handled now* by the current drone. And returns true iff this drone, or any other drone which has already been able to propagate the information, has handled the area up to now. Interestingly, the calls to `logic::area_handled` are issued within a `LOOP`

[3] Introduced in paper [9], and available with other examples on GitHub, at https://github.com/fcpp-experiments/past-ctl-monitoring.

macro, which guarantees through alignment (Sect. 3.1) that the information for each area is kept separate (i.e., if a drone handles area i this does not affect the value of `logic::area_handled` for area $j \neq i$). Finally, the values of the function call for each area are simply $\wedge$-ed to get the value of the complete property.

The monitoring of the second property in FCPP is based on function:

```
FUN bool no_redundancy(ARGS, bool handling) { CODE
    return AH(CALL, !(handling & EY(CALL, EP(CALL, handling))));}
```

While the property is more complex, the implementation follows the same lines as that of the first property.

Simulations. The concern of (visually appealing) simulation and gathering statistics from them is orthogonal to the challenges of programming with FCPP, with or without monitors. While visualization will always have to be customized for the scenario at hand, the FCPP library offers handy operations to specify scenarios, such as deploying a group of agents of a particular type with a certain distribution in a starting area for that group, and FCPP fragments for group members following a leader on a programmed or random walk.

5 Case Studies in AP and Challenges for AP-Based DRV

In this section, we briefly examine some industrial case studies that we have addressed with the FCPP framework, and use them to identify interesting challenges for AP-based DRV, focusing in particular on the need for more expressive monitor specification logic. Note that we have *not yet* implemented any DRV functionality in the use cases. However, incorporating it is expected to require negligible effort compared to the development of the AP algorithms, and even more so relative to implementing simulation scenarios or the software modules for physical devices. Deployments on physical devices have been limited in scale due to costs and other practical constraints. Simulations using the Gazebo simulator allowed us to increase the scale (e.g., up to eight Jackal robots in the Torino Airport outdoor use case), but they were still constrained by Gazebo's high computational requirements. In contrast, lightweight simulations within FCPP enable scaling to hundreds of thousands of devices on a high-end laptop.

5.1 Some Industrial Case Studies in FCPP

Torino Airport Indoor. In the RoboApp project [34], SAGAT (the company that manages Torino Airport, https://www.aeroportoditorino.it/en) proposed a compelling use case, involving the boarding gates instead of outdoor areas. In particular, it can happen that, due to several reasons, long queues of people form at the gates, overflowing out of the serpentine barriers into the transit areas. Such queues should be promptly managed by human operators, since they get in the way of people passing through, but it is often the case that queues form while no one is (yet) attending the gate. We proposed to exploit a team of Autonomous

Fig. 7. Queue detection with the RoboAPP CV pipeline.

Mobile Robots (AMRs) in charge of monitoring the gates, equipped with sensors and software for navigating and observing the environment with a camera, as well as with an FCPP application for distributed control and coordination. When the schedule (which is currently fixed) requires that a gate is monitored, the team executes a distributed XC algorithm to choose two (or more) robots that will jointly execute the task. The chosen robots navigate from the parking areas (where they gather when idle) to predefined locations close to the gate, where they can start their observations.

Detecting a queue near a gate (see Fig. 7) is a challenging computer-vision task, requiring a non-trivial (Computer Vision) CV pipeline executed on-board Jetson Nano computers. Each robot perceives the scene from its own point of view, and may draw different conclusions about the presence of an overflowing queue. An FCPP application running onboard, implements a distributed XC algorithm to reach a consensus among the robots (e.g., a distributed majority voting). However, since even the consensus among multiple points of view is sometimes unreliable, we have added an additional component to the system, namely an AI Agents [38] application running on a PC on the local network, based on a multi-modal Small Language Model (SLM). The AI application is invoked to make a final decision when the team of robots agrees that there is *likely* a critical situation to notify to the human operators.

Torino Airport Outdoor. The RoboNG project [30] addressed another use case supplied by SAGAT, which proposed a relevant problem faced by airports, namely the timely detection and removal of FODs (Foreign Object Debris). FODs are small objects that can clutter the parking and taxiing areas used by airplanes, with potentially disastrous effects. Continuously inspecting the relevant areas for FOD detection is a repetitive, costly, and error-prone task when performed by humans. The RoboNG project proposed the prototype implementation of a system able to perform FOD detection in a (semi) autonomous way.

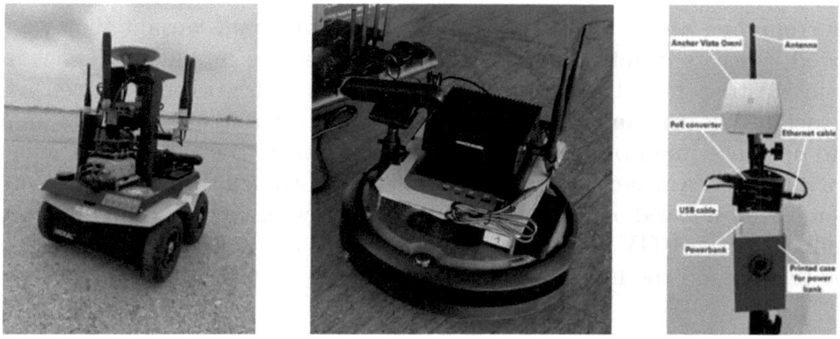

Fig. 8. The Jackal by Clearpath (left), the Create3 by iRobot (center) equipped for RoboNG; a Sewio UWB anchor (right).

In particular, a team of Autonomous Mobile Robots (AMRs) is charged with covering the relevant areas while exploiting on-board sensors to detect potential FODs, and send alerts to human operators. In order to be able to navigate the environment, perceive it, and communicate with the operator each robot is equipped with a high-resolution camera, a CV pipeline for the detection of FODs, and autonomous navigation software (supporting obstacle and collision avoidance), powered by a Lidar and/or an additional camera. The AMRs communicate with each other via a WiFi connection to a local wireless network supplied by a router, and host on-board a FCPP application that enables communication and coordination among robots. As the main *robot brain*, the FCPP application gets information from the CV pipeline, and accesses the local robot to get telemetry and send control commands. Figure 8 shows the two robots employed for the prototype system implementation. The Jackal, produced by Clearpath Robotics, is a powerful small-size rover for outdoor exploration. The Create3, produced by iRobot, is a basic platform for research, usable mainly indoor. The distributed covering algorithm was implemented as an XC program running as an on-board FCPP-based application.

Smart Agriculture. Among the many partners of the Agritech National Research Center for Technology in Agriculture, Spoke 6 (https://agritechcenter.it/), we have cooperated in particular with the Department of Agricultural, Forest and Food Sciences (DISAFA) of the University of Turin. Most relevant here, is the study of the employment of swarms of drones in agriculture. Guided by the domain experts at DISAFA, we have implemented several simulations of swarms of autonomous drones, controlled and coordinated with AP algorithms implemented in XC. We have focused in particular on the following problem: given an autonomous vehicle (either aerial or ground), named *worker*, in charge of following a pre-determined trajectory across a crops area to perform its task (e.g., weeding, or trimming), devise algorithms to coordinate a swarm of drones that must accompany the worker along its trajectory providing surveillance of the surrounding areas, for safety purposes. An important variant that we have explored, consists in introducing multiple workers: the problem becomes much more interesting (from the AP point of view) when each worker of a given set has its associated swarm, and individual drones can move from one swarm to another one to optimize monitoring. In [28], we have reported the results of a cooperation with DISAFA, where we have experimentally studied the effectiveness of the implementation based on a single worker and lightweight simulation in providing surveillance for a UGV in a vineyard, taking into account the detailed 3D map of the terrain and the field-of-view of the cameras on the drones.

The MAcA Living Lab. The MAcA Living Lab was established at the MAcA Environmental Museum (https://www.acomeambiente.org/en/) in Turin, Italy, operating within the framework of the European NODES Project, Spoke 4 "Digital Innovation Toward Sustainable Mountain". The museum encompasses approximately 3,500 square meters of indoor and outdoor space, and is an ideal setting for exploring BIM-centric Digital Twins for sustainable cultural and educational facilities. The museum has already been equipped with a heterogeneous IoT sensor network, including sensors for: weather, humidity, air quality, indoor parameters (e.g., temperature and CO_2), and presence/activity detection. More recently, we have started to deploy an RTLS (Real-Time Localization System) based on Ultra-wideband (UWB) anchors – Fig. 8 (right) – and tags, to be complemented also by UWB-enabled smartphones. We will exploit XP distributed algorithms both to optimize real-time localization through cooperation, and to perform on-device aggregate computation of spatial-dependent measures (e.g., crowd density) and collective behavior analysis.

5.2 Challenges for AP-Based DRV

Past-CTL and SLCS provide a solid logical foundation for aggregate-based DRV, applied to complex collective systems such as those outlined in Sect. 5.1. Crucially, runtime verification becomes increasingly important with the introduction of AI modules in those systems, as it can make them less deterministic and less reliable. From our case studies, several expressive and methodological limitations or the presented logics emerge, that highlight important research challenges and suggest directions for future investigation.

A first limitation concerns the expressive power of *past-CTL*. By construction, past-CTL only allows reasoning over the past evolution of space-time values. This restriction prevents the specification of inherently forward-looking properties such as eventual convergence, stabilization guarantees, or liveness conditions. On the other hand, it is a natural limitation for runtime monitoring, where future behaviour is unknown and only the past is monitorable. Future work may devise a predictive semantics and possibly automatic translations of (certain classes of) forward-looking properties (in logics such as CTL) to past-CTL, closing this gap. In the Torino Airport Outdoor case study, we could e.g., require that "every cell will eventually be visited", which is the main goal of the swarm in that scenario.

A second issue lies in the limited interaction between temporal and spatial dimensions. In past-CTL, spatial reasoning is mediated through aggregate constructs, but the logic does not offer first-class operators for structured spatial quantification (e.g., reasoning over dynamically defined regions, boundaries, or topological features). Conversely, *SLCS* provides a robust spatial logic based on closure and neighbourhood relations, yet it lacks intrinsic temporal expressiveness. When used in combination, the two logics still exhibit a conceptual separation between time and space, by also relying on two incompatible semantic spaces, making it difficult to express genuinely spatio-temporal patterns such as moving fronts, persistent corridors, or region-wide synchronization phenomena. In the Torino Airport Indoor case study, we could e.g., require that when a gate opens, a robot should be available and able to reach it. A more integrated spatio-temporal logic for aggregate systems remains an open research direction.

Another challenge concerns quantitative reasoning. Both past-CTL and SLCS are primarily qualitative: they express whether a property holds, but not to what degree, with what robustness margin, or under which quantitative constraints. However, many aggregate applications require reasoning about numerical thresholds, convergence rates, probabilistic guarantees, or robustness against perturbations. Quantitative semantics could be incorporated in many different ways, including: spatio-temporal operators with metric bounds, counting constraints, or parametric first-order predicates. Each of these extensions would significantly enhance the applicability of AP-based DRV. For example, in the Smart Agriculture case study, we could require that two workers (e.g., UGVs) never stay closer than R meters for more than a period of time of T seconds.

An additional and largely unexplored direction concerns extending spatio-temporal logics to explicitly reason about *graph edges*, rather than only node-based properties. Both past-CTL and SLCS interpret formulas over devices, with spatial operators implicitly relying on adjacency, yet many aggregate behaviours are inherently edge-centric: links may carry weights or reliability annotations, may fail dynamically, and often determine properties such as flow consistency, cut detection, or path optimality. Supporting edge predicates and mixed node–edge formulas would therefore broaden the range of specifiable behaviours. Such an extension would also promote higher-level graph abstractions to first-class logical entities: for instance, *regions* could be characterised through connectivity constraints, and *shortest paths* could be captured directly through edge-aware predicates. In the MAcA Living Lab case study, we could require that each person can follow a path of wireless point-2-point links to an emergency exit, without relying on the wired network being operational. Enriching the logic with such custom graph predicates could be done by leveraging the graph-theoretic structure of aggregate computations, but defining a coherent semantics for this while preserving locality and monitorability remains an open challenge.

Acknowledgments. This publication is part of the project NODES, which has received funding from the MUR – M4C2 1.5 of PNRR funded by the European Union-NextGenerationEU (Grant agreement no. ECS00000036). This study has been supported by the Italian PRIN project "CommonWears" (2020HCWWLP) and was carried out within the Agritech National Research Center and received funding from the European Union Next-GenerationEU (PIANO NAZIONALE DI RIPRESA E RESILIENZA (PNRR) – MISSIONE 4 COMPONENTE 2, INVESTIMENTO 1.4 – D.D. 1032 17/06/2022, CN00000022). Volker Stolz is supported by Norwegian Research Council project CroFlow: Enabling Highly Automated Cross-Organisational Workflow Planning (grant no. 326249) and EU HORIZON MSCA Doctoral Network "Towards an Understanding of Artificial Intelligence (TUAI)" (grant no. 101168344). This manuscript reflects only the authors' views and opinions, neither the European Union nor the European Commission can be considered responsible for them.

Disclosure of Interests. The authors have no competing interests to declare that are relevant to the content of this article.

References

1. Aguzzi, G., Audrito, G., Viroli, M.: Optimising aggregate monitors for spatial logic of closure spaces properties. In: Proceedings of the 7th ACM International Workshop on Verification and Monitoring at Runtime Execution, pp. 25–31. VORTEX 2024. ACM (2024). https://doi.org/10.1145/3679008.3685544
2. Audrito, G.: FCPP: an efficient and extensible field calculus framework. In: Proceedings of the 1st International Conference on Autonomic Computing and Self-Organizing Systems, ACSOS, pp. 153–159. IEEE Computer Society (2020). https://doi.org/10.1109/ACSOS49614.2020.00037
3. Audrito, G., Beal, J., Damiani, F., Viroli, M.: Space-time universality of field calculus. In: Coord. Models and Languages. LNCS, vol. 10852, pp. 1–20. Springer, Cham (2018). https://doi.org/10.1007/978-3-319-92408-3_1
4. Audrito, G., Casadei, R., Damiani, F., Salvaneschi, G., Viroli, M.: Functional programming for distributed systems with XC. In: 36th European Conference on Object-Oriented Programming, ECOOP 2022. LIPIcs, vol. 222, pp. 20:1–20:28. Schloss Dagstuhl (2022). https://doi.org/10.4230/LIPIcs.ECOOP.2022.20
5. Audrito, G., Casadei, R., Damiani, F., Salvaneschi, G., Viroli, M.: The exchange calculus (XC): a functional programming language design for distributed collective systems. J. Syst. Softw. **210** (2024). https://doi.org/10.1016/J.JSS.2024.111976
6. Audrito, G., Casadei, R., Damiani, F., Stolz, V., Viroli, M.: Adaptive distributed monitors of spatial properties for cyber-physical systems. J. Syst. Softw. **175**, 110908 (2021). https://doi.org/10.1016/j.jss.2021.110908
7. Audrito, G., Casadei, R., Damiani, F., Viroli, M.: Computation against a neighbour: addressing large-scale distribution and adaptivity with functional programming and Scala. Log. Methods Comput. Sci. **19**(1) (2023). https://doi.org/10.46298/lmcs-19(1:6)2023
8. Audrito, G., Damiani, F., Rinaldi, S., Tagliabue, L.C., Testa, L., Torta, G.: Aggregate programming for customized building management and users preference implementation, pp. 147–172. Springer, Cham (2023). https://doi.org/10.1007/978-3-031-15160-6_7
9. Audrito, G., Damiani, F., Stolz, V., Torta, G., Viroli, M.: Distributed runtime verification by past-CTL and the field calculus. J. Syst. Softw. **187**, 111251 (2022). https://doi.org/10.1016/j.jss.2022.111251
10. Audrito, G., Damiani, F., Torta, G.: Real-time guarantees for SLCS monitors in XC. In: VORTEX Proceedings of VORTEX 2024. pp. 32–37. ACM (2024). https://doi.org/10.1145/3679008.3685545

11. Audrito, G., Rapetta, L., Torta, G.: Extensible 3d simulation of aggregated systems with FCPP. In: Coordination Models and Languages. LNCS, vol. 13271, pp. 55–71. Springer, Cham (2022). https://doi.org/10.1007/978-3-031-08143-9_4
12. Audrito, G., Terraneo, F., Fornaciari, W.: FCPP+Miosix: scaling aggregate programming to embedded systems. IEEE Trans. Parallel Distributed Syst. **34**(3), 869–880 (2023). https://doi.org/10.1109/TPDS.2022.3232633
13. Audrito, G., Torta, G.: FCPP to aggregate them all. Sci. Comput. Program. **231**, 103026 (2024). https://doi.org/10.1016/J.SCICO.2023.103026
14. Audrito, G., Viroli, M., Damiani, F., Pianini, D., Beal, J.: A higher-order calculus of computational fields. ACM Trans. Comput. Logic **20**(1), 5:1–5:55 (2019). https://doi.org/10.1145/3285956
15. Beal, J., Pianini, D., Viroli, M.: Aggregate programming for the Internet of Things. IEEE Comput. **48**(9) (2015). https://doi.org/10.1109/MC.2015.261
16. Casadei, R., Aguzzi, G., Pianini, D., Viroli, M.: Programming (and learning) self-adaptive & self-organising behaviour with ScaFi: for swarms, edge-cloud ecosystems, and more. In: IEEE International Conference on Autonomic Computing and Self-organizing Systems, ACSOS 2023, Toronto, Canada, September 25-29, 2023. pp. 33–34. IEEE (2023). https://doi.org/10.1109/ACSOS-C58168.2023.00032
17. Casadei, R., Fortino, G., Pianini, D., Placuzzi, A., Savaglio, C., Viroli, M.: A methodology and simulation-based toolchain for estimating deployment performance of smart collective services at the edge. IEEE Internet Things J. **9**(20), 20136–20148 (2022). https://doi.org/10.1109/JIOT.2022.3172470
18. Casadei, R., Viroli, M.: Towards aggregate programming in Scala. In: 1st PMLDC Workshop, New York, NY, USA, pp. 5:1–5:7. ACM (2016). https://doi.org/10.1145/2957319.2957372
19. Casadei, R., Viroli, M., Aguzzi, G., Pianini, D.: SCAFI: a scala DSL and toolkit for aggregate programming. SoftwareX **20**, 101248 (2022). https://doi.org/10.1016/j.softx.2022.101248
20. Ciancia, V., Latella, D., Loreti, M., Massink, M.: Specifying and verifying properties of space. In: Diaz, J., Lanese, I., Sangiorgi, D. (eds.) TCS 2014. LNCS, vol. 8705, pp. 222–235. Springer, Heidelberg (2014). https://doi.org/10.1007/978-3-662-44602-7_18
21. Cortecchia, A.: Multiplatform self-organizing systems through a kotlin-mp implementation of aggregate computing. In: 2024 IEEE International Conference on Autonomic Computing and Self-organizing Systems Companion (ACSOS-C), pp. 155–157 (2024). https://doi.org/10.1109/ACSOS-C63493.2024.00048
22. Dustdar, S., Pujol, V.C., Donta, P.K.: On distributed computing continuum systems. IEEE Trans. Knowl. Data Eng. **35**(4), 4092–4105 (2023). https://doi.org/10.1109/TKDE.2022.3142856
23. Francalanza, A., Pérez, J.A., Sánchez, C.: Runtime verification for decentralised and distributed systems. In: Bartocci, E., Falcone, Y. (eds.) Lectures on Runtime Verification. LNCS, vol. 10457, pp. 176–210. Springer, Cham (2018). https://doi.org/10.1007/978-3-319-75632-5_6
24. Gigante, N., Montanari, A., Reynolds, M.: A one-pass tree-shaped tableau for ltl+past. In: LPAR-21. EPiC Series in Computing, vol. 46, pp. 456–473. EasyChair (2017). https://doi.org/10.29007/3HB9
25. Lamport, L.: Time, clocks, and the ordering of events in a distributed system. Commun. ACM **21**(7), 558–565 (1978). https://doi.org/10.1145/359545.359563
26. Leucker, M., Schallhart, C.: A brief account of runtime verification. J. Logic Algebraic Program. **78**(5), 293–303 (2009). https://doi.org/10.1016/j.jlap.2008.08.004

27. Mamei, M., Zambonelli, F., Leonardi, L.: Co-Fields: towards a unifying approach to the engineering of swarm intelligent systems. In: Petta, P., Tolksdorf, R., Zambonelli, F. (eds.) ESAW 2002. LNCS (LNAI), vol. 2577, pp. 68–81. Springer, Heidelberg (2003). https://doi.org/10.1007/3-540-39173-8_6
28. Nitti, L., et al.: Drone-swarm based surveillance system for autonomous machine safety functionality. In: Precision Agriculture 2025. LNCS, vol. 2577, pp. 492–598. Springer, Cham (2002). https://doi.org/10.1163/9789004725232_077
29. Saint-Martin, L., et al.: Study on the economic potential of far edge computing in the future smart Internet of Things. Final study report, European Commission: Decision Etudes & Conseil, Directorate-General for Communications Networks, Content and Technology (2023). https://doi.org/10.2759/05608
30. Santer REPLY SpA: The RoboNG project (2023). https://ecs-nodes.eu/en/1-aerospace-and-sustainable-mobility/progetti-imprese/robong. Accessed 9 Feb 2026
31. Testa, L., Audrito, G., Damiani, F., Torta, G.: Aggregate processes as distributed adaptive services for the industrial internet of things. Pervasive Mob. Comput. **85**, 101658 (2022). https://doi.org/10.1016/j.pmcj.2022.101658
32. University of Bologna: ScaFi Aggregate Programming Toolkit (2018). https://scafi.github.io/. Accessed 9 Feb 2026
33. University of Bologna: Collektive (2024). https://collektive.github.io. Accessed 9 Feb 2026
34. University of Turin: The RoboAPP project (2024). https://ai4future.unito.it/roboapp/ and https://ecs-nodes.eu/en/1-aerospace-and-sustainable-mobility/progetti-accademici/roboapp. Accessed 9 Feb 2026
35. University of Turin: FCPP (2021). https://fcpp.github.io/. Accessed 9 Feb 2026
36. Viroli, M., Audrito, G., Beal, J., Damiani, F., Pianini, D.: Engineering resilient collective adaptive systems by self-stabilisation. ACM Trans. Model. Comput. Simul. **28**(2), 16:1–16:28 (2018). https://doi.org/10.1145/3177774
37. Viroli, M., Beal, J., Damiani, F., Audrito, G., Casadei, R., Pianini, D.: From distributed coordination to field calculus and aggregate computing. J. Log. Algebraic Methods Program. **109** (2019). https://doi.org/10.1016/j.jlamp.2019.100486
38. Wang, L., et al.: A survey on large language model based autonomous agents. Front. Comp. Sci. **18**(6), 186345 (2024). https://doi.org/10.1007/s11704-024-40231-1
39. Winskel, G.: Event structure semantics for CCS and related languages. In: Nielsen, M., Schmidt, E.M. (eds.) Automata, Languages and Programming, 9th Colloquium, Aarhus, Denmark, July 12-16, 1982, Proceedings. LNCS, vol. 140, pp. 561–576. Springer (1982). https://doi.org/10.1007/BFB0012800

Industry Day Track

Verifiably UX Compliant and User-Intent Layout Generation (Extended Abstract)

Joana Coutinho(✉), Alexandre Lemos, and Pedro Resende

OutSystems, Linda-a-Velha, Portugal
{joana.coutinho,alexandre.lemos,pedro.resende}@outsystems.com

Abstract. Low-code platforms accelerate development of mission-critical applications, enabling users with less technical backgrounds to become proficient developers. AI-powered multi-agent systems further boost this experience. Large language models (LLMs) excel at understanding user intent, but alone cannot guarantee that design rules are followed - formal methods can provide such guarantees. We propose a hybrid approach that leverages both LLMs and a maximum satisfiability solver to generate layouts that comply with UX guidelines (e.g., widget ordering and grouping rules) while respecting user intent. We evaluate the approach in production against an LLM-only baseline.

Keywords: Layout Screen Generation · UX guided generation · SAT

1 Introduction

Low-code platforms enable rapid development of mission-critical software. In OutSystems, developers create applications with user interfaces, business logic, and data models through a visual programming language, aided by AI assistance (e.g., natural language (NL) to SQL [1]).

Within OutSystems, *Mentor* [8] provides a vibe coding experience where users create low-code applications through NL interaction with AI agents. Unlike other tools, such as Replit[1] and Lovable[2], we consistently generate applications that are valid, runnable, and deployable. To guarantee these, we employ a set of validation tools that constrain the behavior of the models.

Mentor is a multi-agent system that exists during the whole development life cycle [5], from an empty app to iterating on an already generated application. One main aspect of *Mentor* is the ability to generate user interfaces for the application. To generate effective screens, several considerations must be taken into account, including data visualization principles (e.g., selecting appropriate widgets), user intent (i.e., the functionality the user expects from the screen), and user experience (UX) guidelines to ensure usability and visual appeal. There

[1] https://replit.com/.
[2] https://lovable.dev/.

A. Sampaio and M. Stoelinga (Eds.): FM 2026, LNCS 16557, pp. 579–584, 2026.
https://doi.org/10.1007/978-3-032-26220-2_27

are different types of screens, and this work focuses on dashboards. However, this work can be extended to other types of screens.

Originally, dashboards were generated in a two-stage approach using only LLMs. The first stage focused on understanding user intent and selecting the appropriate visualizations for the data (e.g., plots, lists), utilizing structured outputs and iterative validation to ensure accuracy. The second stage created the layout based on the visualizations, user intent, and UX guidelines (e.g., counters should be on the top of the screen). This approach did not guarantee that the guidelines were followed, which is very important for our stakeholders.

In this work, we propose a hybrid approach to generate layouts that uses a maximum satisfiability (SAT) solver to generate the layout. The solver creates a placement for all visualization items in a grid, subject to the user intent constraints and the weighted UX guidelines. This approach allows us to ensure the most important guidelines are met and, as a byproduct, to easily add new visualizations without risking changing the whole screen.

2 UX Compliant Layout Generation

The dashboard screen is a grid with columns and rows of flexible size. Prior work on constraint-based layout optimization [2,9,10] formulates widget placement as a combinatorial problem but does not integrate NL-driven intent. More recently, LLMs have been applied to UI generation [4,6], yet these approaches lack formal guarantees on guideline compliance due to their probabilistic nature. We combine both directions: we encode the dashboard layout as a placement problem in conjunctive normal form (CNF), using an LLM only to extract user intent and a MaxSAT solver to enforce UX guidelines. We have only one decision variable $vis_{r,c,v}$ that represents the placement of visualization item v in row r column c.

The hard constraints ensure that (i) every element is placed in the grid, (ii) no two elements share the same cell, and (iii) elements occupy consecutive columns matching their size (e.g., plots can span multiple columns). For example, constraint (ii) is formalized as:

$$\forall r, c: \sum_{v} vis_{r,c,v} \leq 1 \tag{1}$$

We encode UX guidelines as soft constraints provided by our internal UX team. There are three main types of visualization elements: plots (lines, bars, etc.), lists, and counters. Examples of the constraints are: (i) Counters should be above plots and plots should be above tables; (ii) Visualization elements should be of the same type if they are placed on the same row; (iii) Only small lists (less than 5 columns) should share a row with other elements; (iv) Counters should be split evenly per row. For example, (i) is encoded as: for every counter v_c and plot v_p, if the counter is placed in a row below the plot, the clause is violated:

$$\forall v_c \in counters,\ v_p \in plots,\ c,\ r_c > r_p:\ (\neg vis_{r_c,c,v_c} \lor \neg vis_{r_p,c,v_p}) \quad [w] \tag{2}$$

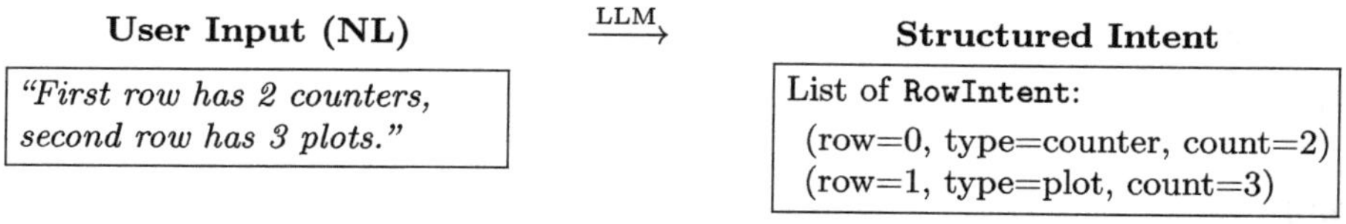

Fig. 1. Example of user intent extraction.

The MaxSAT solver maximizes the total weight of satisfied soft constraints, where each UX guideline is assigned a weight w reflecting its priority. To promote compact screens, we start by forcing the number of rows to 1. If no valid solution is found, we increase the allowed number of rows up to the number of visualization elements.

We use an LLM to understand the user intent. The LLM generates structured JSON objects that serve as an intermediate representation between NL and CNF. These objects can then be converted to SAT clauses deterministically. We analyze the current user requests and define the following representations: maximum number of visualizations per row; requesting a visualization in a specific place in a grid; and groupings per row (e.g., first row has two plots, or plot X and plot Y are on the same row). These user intent requests are encoded as hard constraints. However, if no solutions are found, the problem is solved again with these constraints as soft constraints. Each representation is then converted to a cardinality constraint over the respective visualization variables (e.g., a grouping per row becomes an exactly-k constraint). Figure 1 illustrates the intent extraction step with a concrete example.

A key design decision is how to bridge NL user intent to SAT clauses. We consider three approaches: (i) have the LLM generate a structured intermediate representation that is deterministically converted to CNF (as illustrated in Fig. 1), (ii) have the LLM generate CNF directly, or (iii) expose the PySAT API as LLM tools. Option (ii) is fast but hard to debug in production; option (iii) is the most flexible but too expensive, as it exposes many functions to the agent. We choose (i): while less scalable—new constraint types must be defined a priori—LLMs produce more reliable structured output than raw CNF, and errors are easier to trace.

3 Results

In this section, we compare the LLM-only baseline with the hybrid approach in terms of quality and execution time.

3.1 Experimental Setup

We use PySAT with the RC2 solver and the RC2Stratified algorithm [3]. We use GPT 4.1 [7] for all LLM calls. The dataset is composed of 90 requirement files, where 30 do not have user intent, 30 have user intent, but they are compliant with UX guidelines, and 30 have user intent that goes against the guidelines. All experiments were run 10 times to account for non-determinism.

Table 1. Summary statistics across all 90 instances (10 runs each). Top: end-to-end execution time comparison. Bottom: MaxSAT instance sizes for the hybrid approach.

	mean	std	min	25%	50%	75%	max
Execution Time							
LLM Baseline – Total (s)	58.8	31.8	30	42	51.5	58.75	120
Hybrid – SAT only (s)	1.5	7.1	0.003	0.056	0.084	0.2	47.9
Hybrid – Total (s)	35	7.3	20	31.25	35	38.5	42
MaxSAT Instance Size (hybrid only)							
# variables	418.2	257.5	22	228	391	569	1 346
# hard clauses	2 459.5	2 743.2	22	639	1 434	3 561	12 918
# soft clauses	1 293.6	1 335.9	10	312	1 008	1 665	6 690

3.2 User Intent

We can cover all use cases that our customers are requesting and that our UX team designed. We manually analyze the intent. Only four instances out of 90 had issues in understanding the intent in 2 out of 10 runs. On average, the intent creation takes 35 s (Table 1).

3.3 MaxSAT

The average execution time of the SAT solver is 1.5 s, negligible compared to the LLM call time. Since RC2 is a complete solver, optimality is guaranteed. The instances are small, with at most 12k hard clauses.

3.4 Hybrid vs LLM

The LLM-based approach fails to generate a valid dashboard 33% of the time (broken hard constraints), and the system has to fall back to a default dashboard. Of the remaining valid dashboards, 21% violate at least one UX guideline or do not follow user intent. The hybrid approach always produces a valid dashboard; its only failure mode is when the LLM misinterprets the user intent, which occurs in 6.6% of the runs. The hybrid approach takes on average 35 s to run, while the LLM-based approach takes on average 59 s (including retries).

4 Conclusion and Future Work

We propose a hybrid approach to generate complex layouts for dashboards using LLMs and SAT. The hybrid approach allows us to respect user intent while ensuring UX compliant layouts. We showcase our approach with a dataset of 90 requirement files. The results show that the hybrid approach is twice as fast and reduces generation issues by 48%.

For future work, we plan to expand this approach to other types of screens, leveraging the low-code abstractions to reduce the domain of possible UI elements. Furthermore, we plan to explore the option to generate diverse solutions to provide a different experience to the user.

References

1. Aparicio, S., et al.: Natural language to SQL in low-code platforms (2023). https://arxiv.org/abs/2308.15239
2. Dayama, N.R., Todi, K., Saarelainen, T., Oulasvirta, A.: GRIDS: interactive layout design with integer programming. In: Conference on Human Factors in Computing Systems (CHI), pp. 1–13. ACM (2020)
3. Ignatiev, A., Morgado, A., Marques-Silva, J.: RC2: an efficient maxsat solver. J. Satisf. Boolean Model. Comput. **11**(1), 53–64 (2019)
4. Jiang, Y., Schoop, E., Swearngin, A., Nichols, J.: ILUVUI: instruction-tuned language-vision modeling of UIS from machine conversations. In: Proceedings of the 30th International Conference on Intelligent User Interfaces (IUI) (2025). https://doi.org/10.1145/3708359.3712129
5. Lemos, A., Coutinho, J.: AI mentor system: building a technical debt dashboard for low code. In: International Conference on Software Maintenance and Evolution (ICSME), pp. 679–684. IEEE (2025). https://doi.org/10.1109/ICSME64153.2025.00068
6. Lu, Y., et al.: Designing and developing user interfaces with AI: advancing tools, workflows, and practices. In: Proceedings of the Extended Abstracts of the CHI Conference on Human Factors in Computing Systems. CHI EA (2025). https://doi.org/10.1145/3706599.3706736
7. OpenAI: Gpt-4 technical report (2024). https://arxiv.org/abs/2303.08774
8. OutSystems: Mentor: AI-driven dev for turning ideas into apps fast. https://www.outsystems.com/low-code-platform/mentor-ai-app-generation/. Accessed 25 Nov 2025
9. Swearngin, A., Wang, C., Oleson, A., Fogarty, J., Ko, A.J.: Scout: rapid exploration of interface layout alternatives through high-level design constraints. In: Conference on Human Factors in Computing Systems (CHI), pp. 1–13. ACM (2020)
10. Zeidler, C., Weber, G., Stuerzlinger, W., Lutteroth, C.: Automatic generation of user interface layouts for alternative screen orientations. In: Bernhaupt, R., Dalvi, G., Joshi, A., Balkrishan, D.K., O'Neill, J., Winckler, M. (eds.) INTERACT 2017. LNCS, vol. 10513, pp. 13–35. Springer, Cham (2017). https://doi.org/10.1007/978-3-319-67744-6_2

ConFixer: Robustness Semantics Based Configuration Bug Fixing for Automated Driving Systems

Xiaodong Zhang[1], Songyang Yan[2], and Zijiang Yang[1,3(✉)]

[1] University of Science and Technology of China, Hefei, China
zhangxiaodong@ustc.edu.cn
zhangxiaodong@ustc.edu.cn, yang@synkrotron.ai
[2] Xi'an Jiaotong University, Xi'an, China
tayyin@stu.xjtu.edu.cn
tayyin@stu.xjtu.edu.cn
[3] Synkrotron, Inc., Xi'an, China
yang@synkrotron.ai

Abstract. Automated Driving Systems (ADSs) coordinate multiple modules (e.g., perception, prediction, planning, and control) and expose a large configuration surface that engineers must tune for different platforms and operational conditions. In industrial practice, misconfigured parameters are a common source of unsafe or law-violating behaviors, yet fixing such configuration bugs remains largely overlooked. Debugging is difficult because parameter effects propagate through long, non-linear pipelines (often involving learning-based components), and manual trial-and-error provides little guidance in high-dimensional spaces while risking regressions in previously passing scenarios.

We propose *ConFixer*, an automated approach for repairing ADS configuration bugs against formal correctness specifications, including traffic laws encoded in signal temporal logic (STL). ConFixer uses robustness semantics to derive gradient-like signals that localize bug-relevant parameters and guide fine-tuning toward compliance. To support adoption in scenario-based validation workflows, ConFixer evaluates candidate fixes on large scenario suites and prioritizes regression-safe solutions. Our evaluation on Baidu Apollo with the LGSVL simulator shows that ConFixer fixes 173 configuration bugs without introducing regressions.

Keywords: Automated driving systems · Configuration repair · Signal temporal logic · Robustness semantics · Regression testing

1 Introduction

Autonomous vehicles promise substantial societal benefits, including reduced congestion, improved safety, and enhanced energy efficiency [29]. Delivering these

X. Zhang and S. Yan—These authors contributed equally to this work.

A. Sampaio and M. Stoelinga (Eds.): FM 2026, LNCS 16557, pp. 585–604, 2026.
https://doi.org/10.1007/978-3-032-26220-2_28

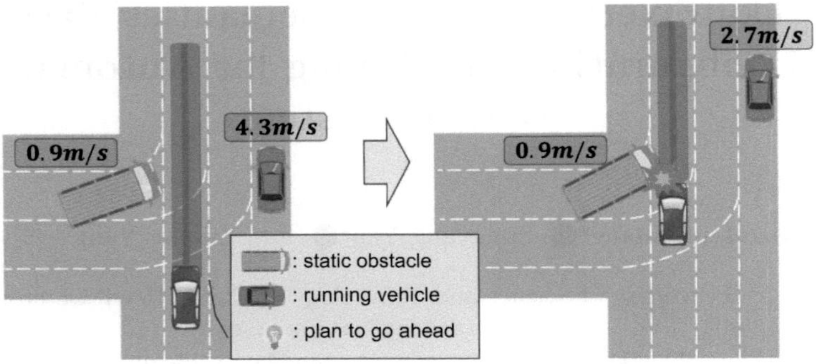

Fig. 1. A collision caused by breaking the fourth clause of Article #51 of Chinese Traffic Law: "vehicles shall pass through in order when the green signal is on."

benefits in practice, however, requires Automated Driving Systems (ADSs) that behave correctly not only in typical situations but also across diverse operational design domains, road structures, and dense traffic interactions. Industrial ADS stacks therefore expose a large configuration surface: engineers routinely tune parameters to adapt perception thresholds, prediction behaviors, and planning constraints to different sensors, vehicle platforms, maps, and driving policies. This configurability is essential for productization, but it also creates a persistent reliability bottleneck: small parameter mistakes can silently alter downstream decisions and only manifest as unsafe or illegal behaviors after a long closed-loop execution in simulation or on-road testing.

Misconfiguration is not a niche edge case. Empirical studies [17] show that a significant portion of ADS failures arises not only from implementation errors or incorrect documentation but also from misconfigured parameters. In fact, more than 27% of reported bugs in two widely used ADS stacks, Baidu Apollo [5] and Autoware [4], relate to incorrect configurations. In industrial settings, such bugs are expensive: they trigger time-consuming debug cycles with repeated scenario replay, and they are notoriously regression-prone—a parameter tuned to address one failure can easily degrade previously passing scenarios, forcing extensive re-validation before release. This is particularly problematic for safety- and compliance-critical requirements, where organizations increasingly rely on scenario-based validation suites and continuous-integration gates to control risk.

Figure 1 illustrates a representative failure mode. Configuration parameters govern the behavior of perception, planning, and control modules in an ADS. For example, setting `still_obstacle_speed_threshold` too high may classify moving objects as stationary, leading to incorrect predictions and downstream planning decisions. In the shown scenario, a slightly misconfigured speed threshold caused the ego vehicle (the ADS-equipped vehicle under test) to ignore a moving school bus and collide, violating the fourth clause of Article #51 of Chinese Traffic Law [19]. This kind of bug is challenging to diagnose in industrial pipelines because the mis-set parameter is far from the emergent failure: its effects are filtered through multiple modules (often including machine learning components) and may only become visible after a long sequence of interactions. While default configurations may suffice in easy scenes, stringent, specification-based checks—such as those derived from traffic laws [38]—often expose these issues by turning near-misses into concrete, reproducible violations. We refer to such issues as *configuration bugs.* Addressing them is important for industrial practice: automated configuration repair reduces costly manual tuning cycles, helps teams maintain larger regression suites without slowing iteration, and produces evidence (via robustness trends and passing-scenario validation) that supports compliance-oriented review and release decisions.

Debugging configuration bugs is challenging for three reasons. First, causality is obscured by long and nonlinear data flows: a parameter change can propagate through many modules and only affect the final driving outcome after seconds of closed-loop evolution. Traditional techniques such as taint analysis [3,11,25,50] are ill-suited because the relevant effect is not a single corrupted value, but a behavioral deviation that emerges over time. Second, trial-and-error tuning is impractical at industrial scale. Apollo 7.0 [5] alone exposes over 1,000 parameters, and engineers typically receive only coarse feedback from a failing run (e.g., collision/no collision, or a violated/not-violated rule), providing little guidance on how to adjust values. Third, any attempted fix must be *regression-safe*: changes must address the failure while preserving correctness across the broad set of previously passing scenarios that form the basis of release and safety validation.

Existing research offers valuable foundations but does not fully solve the repair problem faced in practice. ADS testing research has made rapid progress in generating challenging scenarios—both *dynamic* scenarios by orchestrating surrounding traffic participants and *static* scenarios by manipulating obstacles or perception-related inputs—to reveal failures [7,10,13,21,22,24,35,37,38,44, 47,51]. These approaches are effective for uncovering issues, but they commonly assume default or fixed ADS configurations; consequently, they may miss *configuration-specific* failures where parameter settings interact with scenarios to undermine robustness. Recent work also studies configuration problems directly: Chen et al. [9] introduced CONFVE, which searches for problematic configurations using genetic algorithms, but does not automatically synthesize fixes that satisfy formal requirements while preventing regressions. In parallel, formal methods provide complementary assurance mechanisms, including runtime monitoring and enforcement; for example, REDriver [39] specifies traffic-law properties in a language based on signal temporal logic (STL), monitors planned trajectories with quantitative semantics, and uses gradient-driven repair to steer trajectories away from likely violations. Together, these gaps motivate a method that can *repair* the offline configuration layer against formal specifications while remaining regression-safe across scenario suites.

We present *ConFixer*, an automated method to identify and fix ADS configuration bugs under correctness specifications, including traffic laws. ConFixer is built on a practical observation: to repair configurations efficiently and safely, engineers need not only a failing example but also a *directional signal* that indicates which parameters matter and how they should change. To provide this signal, ConFixer uses a specification language rooted in STL [23,33] together with robustness semantics to quantify *how close* an execution is to satisfying a requirement. This quantitative feedback turns pass/fail validation into an optimization-guiding score and enables ConFixer to approximate a gradient of system behavior with respect to configuration parameters.

ConFixer operationalizes this idea as an end-to-end workflow aligned with industrial validation practices. Given an initial configuration, an STL specification, and a set of failing and passing scenarios, ConFixer (i) localizes bug-relevant configuration points by systematically perturbing parameters and measuring robustness changes; (ii) leverages *unchanged propositions* in violation formulas to skip scenarios that cannot be affected by a parameter change, improving efficiency; (iii) tunes parameter values guided by the approximated gradients; and (iv) validates and ranks candidate fixes using a broad scenario set (e.g., generated via fuzzing), explicitly prioritizing fixes that remove the original violations while avoiding regressions on previously passing cases. This makes ConFixer suitable for integration into iterative, scenario-driven ADS development workflows, where regression safety and repeatable evidence are prerequisites for adopting any configuration update.

We implemented ConFixer on Apollo and the LGSVL simulator, and evaluated it on traffic-law specifications and real misconfigured parameters. Across 173 failing scenarios from the LawBreaker dataset, ConFixer fixes every confirmed misconfiguration without inducing new failures, demonstrating that gradient-guided, specification-driven configuration repair can deliver both effectiveness and regression safety at industrially meaningful scale. We make the implementation artifacts, including source code and case videos, publicly available [49].

Key contributions:

- A specification-driven framework, *ConFixer*, that automatically repairs ADS configuration bugs that lead to safety or traffic-law violations.
- A gradient-inspired, causal localization method that isolates bug-relevant parameters using robustness changes and skips irrelevant scenarios via unchanged propositions.
- A robustness-guided tuning and validation workflow that explicitly targets regression-free fixes using broad scenario sets (including fuzzing-based scenarios).
- An implementation on Apollo and LGSVL that fixes 173 confirmed configuration bugs with no regressions.

2 ConFixer

2.1 Framework Overview

Framework. Figure 2 shows ConFixer's architecture. It takes as input the original configuration parameters, law specification in STL, and both failed and

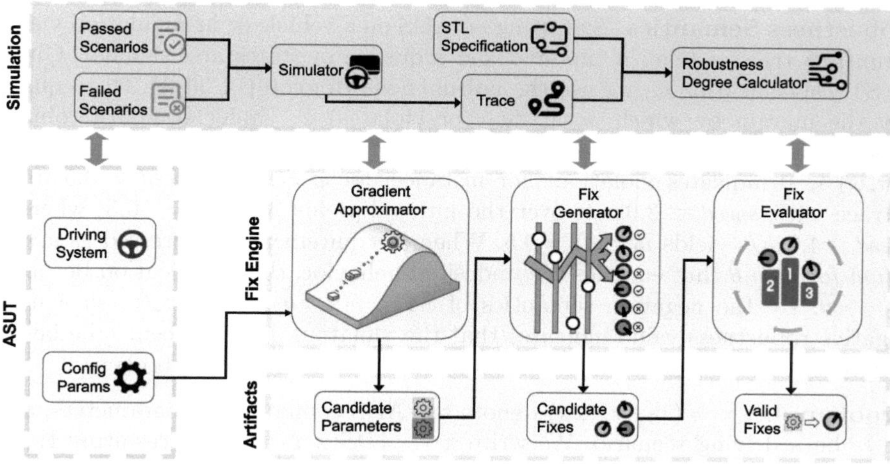

Fig. 2. The Framework of ConFixer.

passed scenarios. It outputs valid fixed values for misconfigured parameters via the following steps:

Step-0: `Robustness Degree Computation` calculates how far the current configuration is from satisfying a given STL formula by using robustness semantics on scenario traces. This measure guides subsequent steps.
Step-1: `Gradient Approximator` determines which parameters may cause the violation. It approximates each parameter's impact on the ADS by tuning parameters individually and checking if the specification becomes satisfied on failed scenarios. To reduce overhead, ConFixer skips scenarios whose relevant propositions remain unchanged in the violation formula.
Step-2: `Fix Generator` synthesizes candidate fixes for parameters flagged by `Gradient Approximator`. A fix is considered a candidate once it passes all previously failed scenarios.
Step-3: `Fix Evaluator` validates each candidate fix under a diverse set of fuzzing-based scenarios, then ranks valid fixes based on their effectiveness.

2.2 Problem Definition

Background. An ADS typically comprises modules for perception, localization, prediction, planning, and control, which jointly determine vehicle behavior [41]. Such modular pipelines expose numerous configuration parameters whose values critically affect safety and correctness. For example, Apollo's planning module comprises predefined scenarios with subtasks [2]; its *lane change decider* uses *change_lane_success_freeze_time* to enforce a cooldown before initiating a subsequent lane change.

Robustness Semantics. Executing an ADS on a vehicle or in simulation yields a runtime trace π (e.g., a time-indexed sequence of states and signals). Given an STL specification φ, we use the **robustness degree** [14,30,31,34] to quantify the margin by which π satisfies (or violates) φ. Under standard robustness semantics [34], $\rho(\pi, \varphi) \in \mathbb{R}$, where $\rho(\pi, \varphi) \geq 0$ indicates satisfaction and $\rho(\pi, \varphi) < 0$ indicates violation. For instance, for $\varphi = F_{[0,6]}(\mathit{speed} > 3.5\,\mathrm{m/s})$, a trace with $\mathit{speed} \equiv 3.0\,\mathrm{m/s}$ over the interval yields $\rho(\pi, \varphi) = -0.5$, whereas $\mathit{speed} \equiv 4.0\,\mathrm{m/s}$ yields $\rho(\pi, \varphi) = 0.5$. When a requirement is expressed as a *violation formula* $\bar{\theta}$ that encodes the undesired behavior, the specification becomes $\varphi = \neg\bar{\theta}$. By the negation semantics of STL, $\rho(\pi, \neg\bar{\theta}) = -\rho(\pi, \bar{\theta})$, so a non-negative robustness value indicates that the violation did *not* occur (the law is obeyed), while a negative value indicates that it did (the law is broken).

Problem. Let $\mathbf{x} = (x_1, \ldots, x_n)$ denote the ADS configuration parameters, and let ts be a driving scenario. We write $\pi = \mathcal{ADS}(\mathbf{x}, ts)$ for the resulting trace, and φ for the target STL specification. Define the scenario-specific robustness function $f_{ts}^{\varphi}(\mathbf{x})$ as the robustness degree of the produced trace:

$$y = \rho\big(\mathcal{ADS}(\mathbf{x}, ts), \varphi\big) = f_{ts}^{\varphi}(\mathbf{x}) \in \mathbb{R}. \tag{1}$$

Given a configuration $\mathbf{x}_0$ that violates φ on ts (i.e., $f_{ts}^{\varphi}(\mathbf{x}_0) < 0$), our goal is to find a revised configuration $\mathbf{x}^\star$ such that $f_{ts}^{\varphi}(\mathbf{x}^\star) \geq 0$, while keeping the change to $\mathbf{x}_0$ minimal. We refer to the partial derivatives $(\frac{\partial y}{\partial x_1}, \ldots, \frac{\partial y}{\partial x_n})$ (when defined) as **tuning directions**: a zero derivative suggests limited local influence on robustness, whereas the sign indicates whether increasing or decreasing the parameter tends to increase y.

2.3 Main Algorithm

Algorithm 1 summarizes ConFixer. It takes as input a default configuration $\mathbb{C}$ (i.e., a set of parameter–value pairs), a set of STL specifications Φ, and a scenario suite $\mathbb{TS}$. For each $\phi \in \Phi$, we partition scenarios into those that *pass* and *fail* under $\mathbb{C}$, denoted $\mathbb{TS}[pass][\phi]$ and $\mathbb{TS}[fail][\phi]$, using the robustness score $f_{ts}^{\phi}(\mathbb{C})$ (Eq. 1). ConFixer then (i) identifies parameters that are likely to influence the violation of ϕ via a lightweight, sign-based gradient approximation (Line 2), (ii) synthesizes candidate fixes by adjusting only those parameters (Line 3), and (iii) validates and ranks the resulting fixes using fuzzed scenarios, favoring fixes with high robustness and low overhead (Line 4).

2.4 Gradient Approximator

This component identifies configuration parameters that are likely to contribute to violating a specification. Since the ADS is a black box, ConFixer uses *finite perturbations* to approximate the *direction* (sign) of the robustness change rather than exact derivatives. Let $\langle x_i, v_i \rangle \in \mathbb{C}$ be a parameter–value pair and let $\mathbb{C}_i^+$ (resp., $\mathbb{C}_i^-$) denote the configuration obtained by updating x_i to $v_i + \Delta^+$ (resp., $v_i - \Delta^-$), while keeping all other parameters unchanged. For a scenario ts

Algorithm 1: ConFixer($\mathbb{C}$, Φ, $\mathbb{TS}$)

Input: $\mathbb{C}$: default ADS configuration parameters;
Φ: set of specifications; $\mathbb{TS}$: testing scenarios.
Output: $\mathbb{C}_{fix}$: well-tuned parameter values.

for $\phi \in \Phi$ **do**
  $grad$ = GradientApproximator($\mathbb{C}$, $\mathbb{TS}$, ϕ);
  $\mathbb{V}^c_{fix}$ = FixGenerator($\mathbb{C}$, $\mathbb{TS}[fail][\phi]$, $grad$);
  $\mathbb{C}_{fix}$ = ValidateAndPrioritize($\mathbb{V}^c_{fix}$);
end

and specification ϕ, we compare $f^{\phi}_{ts}(\mathbb{C})$, $f^{\phi}_{ts}(\mathbb{C}^+_i)$, and $f^{\phi}_{ts}(\mathbb{C}^-_i)$ to infer a sign $grad_i \in \{-1, 0, +1\}$ indicating whether decreasing or increasing x_i is more likely to improve robustness. In doing so, before using $f^{\phi}_{ts}(\mathbb{C}^+_i)$ or $f^{\phi}_{ts}(\mathbb{C}^-_i)$ to determine $grad_i$, ConFixer first checks each candidate perturbed configuration $\mathbb{C}' \in \{\mathbb{C}^+_i, \mathbb{C}^-_i\}$ with Algorithm 2; only perturbations that pass this check are considered in the sign inference.

To avoid regressions, Algorithm 2 filters perturbations by requiring that they (i) preserve satisfaction on previously passed scenarios (Lines 2–9), and (ii) improve at least one previously failing scenario (Lines 11–19). If a perturbation causes any passed scenario to violate ϕ (i.e., $f^{\phi}_{ts}(\mathbb{C}') < 0$), it is rejected. Here, H_{info} is a per-perturbation cache used to avoid redundant simulator runs: it records, for the atomic predicates appearing in ϕ, whether their monitored valuations under $\mathbb{C}'$ have changed (and, when evaluated, the corresponding robustness value), so `DoesRemainUnchanged` can skip scenarios whose relevant predicate valuations appear unchanged.

Evaluating f^{ϕ}_{ts} typically requires running the simulator and monitoring signals, which is costly. To reduce redundant evaluations, `DoesRemainUnchanged` (Algorithm 3) caches information about the propositions of ϕ and skips scenarios whose relevant proposition valuations appear unchanged under the current perturbation.

2.5 Fix Generator

Given the set of tuning directions $grad$, ConFixer synthesizes candidate fixes by incrementally updating only the selected parameters. For a parameter x_i with direction $grad_i$, ConFixer searches along that direction using a step size $\eta > 0$, while respecting user-defined bounds $v_i - \Delta^- \leq x_i \leq v_i + \Delta^+$. At each step, the updated configuration is re-evaluated on $\mathbb{TS}[fail][\phi]$ (and guarded by the harmlessness check on $\mathbb{TS}[pass][\phi]$). The search terminates when all previously failing scenarios satisfy ϕ (i.e., $f^{\phi}_{ts}(\mathbb{C}') \geq 0$ for all $ts \in \mathbb{TS}[fail][\phi]$), or when the bound is reached (Fig. 3).

Algorithm 2: IsHarmlessAndRelevant($\mathbb{C}'$, $\mathbb{TS}$, ϕ)

Output: Boolean indicating whether $\mathbb{C}'$ is harmless on passed scenarios and helpful for failed ones under ϕ.

$H_{info} = \emptyset$;
for $ts \in \mathbb{TS}[pass][\phi]$ **do**
    if *DoesRemainUnchanged*(H_{info}, ϕ) **then**
        continue;
    end
    if $f_{ts}^{\phi}(\mathbb{C}') < 0.0$ **then**
        return false;
    end
    Update H_{info} with ϕ and $f_{ts}^{\phi}(\mathbb{C}')$;
end
$atLeastOne = false$;
for $ts \in \mathbb{TS}[fail][\phi]$ **do**
    if *DoesRemainUnchanged*(H_{info}, ϕ) **then**
        continue;
    end
    if $f_{ts}^{\phi}(\mathbb{C}') \geq 0.0$ **then**
        $atLeastOne = true$;
    end
    Update H_{info} with ϕ and $f_{ts}^{\phi}(\mathbb{C}')$;
end
return $atLeastOne$;

2.6 Fix Evaluator

This module includes fix validation and prioritization. Algorithm 4 validates each candidate fix on a diverse set of *fuzzed* scenarios. We adopt a genetic algorithm (GA) that iteratively mutates and recombines scenarios to generate more challenging cases around the original failures. Concretely, the GA runs for G generations with population size P, initialized from $\mathbb{TS}[fail][\phi]$. In each generation, it selects the bottom 50% of scenarios with respect to robustness (i.e., those with the smallest f_{ts}^{ϕ}) and applies crossover/mutation to produce the next generation (Lines 4–5). For each fix, ConFixer reports a pass rate over the fuzzed population (Line 7); a fix is deemed valid if this rate exceeds a user-defined threshold. Concretely, a fuzzed scenario is counted as a passing test if it satisfies ϕ under the fix (i.e., $f_{ts}^{\phi}(\mathbb{C} \oplus fix) \geq 0$), so $PassRate[\phi]$ is the fraction of passing scenarios among the $G \times P$ fuzzed individuals generated across all generations.

After validation, ConFixer prioritizes fixes by balancing two objectives: **effectiveness** (high robustness on fuzzed scenarios) and **overhead** (small deviation from the original configuration), following the common "minimal-change" princi-

Algorithm 3: DoesRemainUnchanged(H_{info}, ϕ)

Output: Boolean indicating whether the key propositions of ϕ appear unchanged (cached).

P = Extract atomic propositions/predicates in ϕ;
for $p \in P$ **do**
  if *($H_{info}[p]$ == null) or ($H_{info}[p]$ is changed)* **then**
    return false;
  end
end
return true;

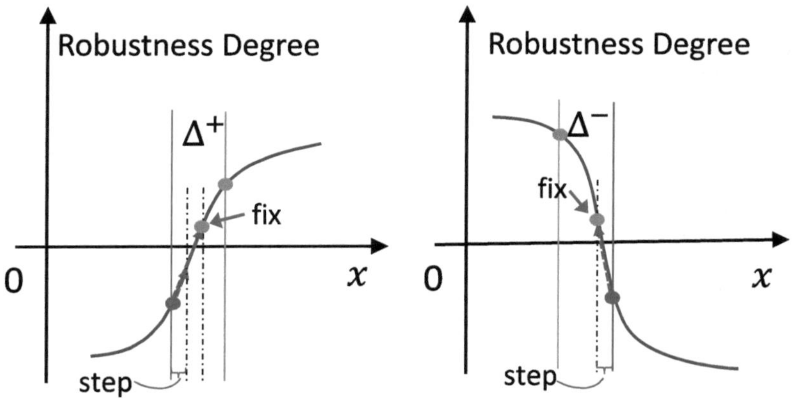

Fig. 3. Incremental tuning based on gradient signs.

ple in runtime enforcement. A higher score $p(fix)$ indicates stronger conformance to ϕ with smaller parameter changes.

$$p(fix) = \frac{\sum_{\langle ts,\phi\rangle \in \mathbb{TS}_{fuzz}} f_{ts}^{\phi}(\mathbb{C} \oplus fix)}{\left|\mathbb{TS}_{fuzz}\right| \times Distance(fix)} \quad (2)$$

Here, $\mathbb{C} \oplus fix$ applies the parameter updates in fix to the original configuration $\mathbb{C}$, $\mathbb{TS}_{fuzz}$ is the fuzzed scenario set, and $\left|\mathbb{TS}_{fuzz}\right|$ is its size. We measure overhead via $Distance(fix)$, e.g., a normalized ℓ_1 distance over the updated parameters: $Distance(fix) = \sum_{x \in dom(fix)} \frac{|fix[x]|}{\max(1, |\mathbb{C}[x]|)}$, so that larger relative changes receive a larger penalty.

Algorithm 4: Validate($\mathbb{V}^c_{fix}$, $\mathbb{TS}[fail][\phi]$)

Output: Testing pass rates for each fix in $\mathbb{V}^c_{fix}$.

```
for fix ∈ V^c_fix do
    GN = Initialize the first generation with P individuals from
      TS[fail][φ];
    for i = 0 to G − 1 do
        Winner = Select bottom 50% from GN based on robustness
          degree;
        GN = Perform mutation and crossover on Winner to obtain P
          individuals;
    end
    PassRate[φ] = Compute (number of passing tests)/(G×P);
end
```

3 Evaluation

3.1 Implementation

We implement ConFixer [49] using the LGSVL simulation framework to evaluate its effectiveness in addressing law-breaking bugs in Baidu Apollo 7.0. We obtain the specific law-breaking scenarios from a prior study [38]. To assess the robustness degree of the STL violation formulae, we use RTAMT [31]. All experiments utilize maps from the LGSVL map store [26], and we employ the LincolnMKZ2017 vehicle model [27].

We conduct all experiments on a dedicated machine with a 16-core Intel i7 3.80 GHz CPU, an NVIDIA RTX 2080 GPU, and 32 GB of system memory.

3.2 Case Study

Case I. Consider the law-breaking scenario depicted in Fig. 1. As analyzed in Sect. 1, the issue arises from a misconfiguration of the `still_obstacle_speed_threshold` (`sost`). Intuitively, `sost` governs when the ADS treats an observed obstacle as still; if `sost` is set too high, the ADS can misclassify a slowly moving vehicle as stationary, leading to incorrect prediction and unsafe planning decisions. This case study demonstrates how ConFixer repairs such a configuration bug in a way that both resolves the violation and remains robust under broader validation.

We start from three inputs: (i) the law-breaking scenario, (ii) the ADS configuration, and (iii) a strict specification of the relevant traffic-law clause. To construct an oracle that is both precise and automatable, we encode the law requirement using STL. As illustrated in Listing 1.1, the violation formula $\bar{\theta}$ captures a breach of the fourth clause of Article #51 in Chinese traffic law (violation formulae are generated following LawBreaker [38]). Specifically, subformula `f1` indicates a green traffic light, signaling the ego vehicle to proceed.

```
f1 = (trafficLightAheadcolor == greenLight);
f2 = (F[0,A](NPCAheadspeed > B m/s));
f3 = (NPCAhead <= C m);
Violation Formula θ̄ = F(f1 and f2 and f3)
```

Listing 1.1. The violation formula of the fourth clause of Article #51 in Chinese traffic law.

Sub-formula `f2` states that the non-player character (NPC) vehicle ahead moves at a minimum speed of B m/s over the next A time steps. Sub-formula `f3` models collision risk by checking whether the ego vehicle becomes too close to the NPC vehicle (distance $\leq C$). For simplicity, we set A, B, and C to 2, 0.5, and 0.5, respectively. The operator `F` denotes "eventually" in temporal logic.

The law-breaking and compliant scenarios are derived from LawBreaker testing results [38]. Let π represent the trace data from this scenario. Because $\bar{\theta}$ encodes the *undesired* behavior, the STL specification used in Eq. 1 is its negation, $\varphi = \neg\bar{\theta}$. The law-complying degree is therefore $y = f_{ts}^{\varphi}(\mathbf{x}) = \rho(\pi, \neg\bar{\theta}) = -\rho(\pi, \bar{\theta})$: it is non-negative when the violation formula is *unsatisfied* (i.e., the law is obeyed) and negative when it is *satisfied* (i.e., the undesired behavior occurred). In this scenario, the law-complying degree is negative, confirming a law-breaking behavior.

We analyze Baidu Apollo 7.0 [5], focusing on configuration parameters within its perception, prediction, planning, and control modules.

(Step-1) Gradient Approximator localizes parameters potentially linked to the violation. ConFixer perturbs each candidate parameter with a coarse change (e.g., by 50%) and replays the same scenario. The resulting change in law-complying degree provides an actionable signal: if perturbing a parameter can move the execution toward compliance, that parameter is likely relevant to the observed law violation.

If the formula is unsatisfied under the modified configuration, the parameter may be relevant to the violation. Otherwise, the next parameter is tested.

ConFixer identifies `sost` as a potentially relevant parameter. Concretely, ConFixer performs a 50% decrease from 0.99 to 0.495 and observes that this perturbation improves compliance, aligning with the intuition that a lower threshold reduces the chance of misclassifying slow-moving vehicles as still obstacles.

(Step-2) Fix Generation then fine-tunes the localized parameter to satisfy the STL specification. Starting from the coarse localization result, ConFixer iteratively adjusts `sost` guided by robustness changes and proposes two candidate fixes: (`sost`, $\downarrow$ 10%) and (`sost`, $\downarrow$ 18%). Both candidates remove the observed violation on the failing scenario, but they differ in how aggressively they alter behavior.

(Step-3) Fix Evaluator assesses each candidate fix across up to 1000 scenarios to guard against regressions and to prioritize stable solutions. The fix (`sost`, $\downarrow$ 10%) achieves a 100% pass rate, whereas (`sost`, $\downarrow$ 18%) achieves 90%. This highlights a common industrial trade-off: a more aggressive parameter change can fix the targeted violation but may reduce robustness elsewhere.

```
f1 = (isAtIntersection);
f2 = (F[0,A](speed < B m/s));
f3 = (inTrafficConflict);
Violation Formula θ̄ = F(f1 and f2 and (not f3))
```

Listing 1.2. The violation formula of the first clause of #22500 of the California Vehicle Code.

ConFixer therefore selects (`sost`, ↓ 10%) as the preferred fix for the violation in Listing 1.1.

Figure 4 illustrates that, post-fix, the ego vehicle no longer perceives the school bus as a stationary obstacle, yielding to it at the T-junction. This fix prevents traffic law violations and collisions without adverse effects on the system.

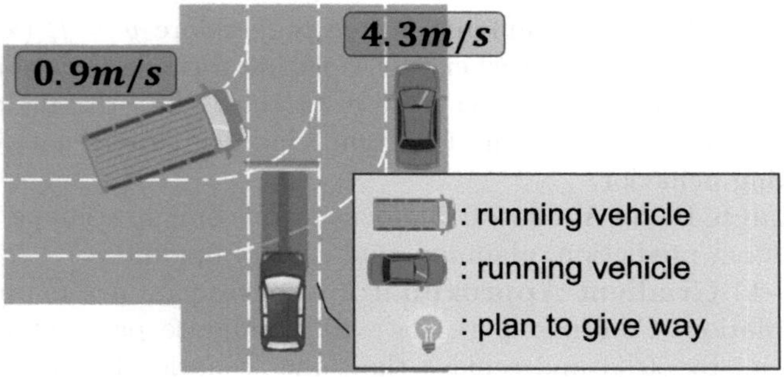

Fig. 4. Post-fix, the ego vehicle yields to the moving school bus at the T-junction.

Case II. This case examines an instance of over-conservative behavior where the ego vehicle stops unnecessarily within an intersection, obstructing a school bus approaching from the left, as depicted in Fig. 5. This behavior violates the first clause of Sect. 22500 of the California Vehicle Code [36], which prohibits stopping, parking, or leaving a vehicle standing within an intersection. In practice, such over-conservatism is problematic: it may appear safe locally, but it can impede traffic flow and create secondary hazards (e.g., blocking cross traffic) that are also regulated by traffic rules.

As illustrated in Listing 1.2, we encode this violation using the formula $\bar{\theta}$. The sub-formula `f1` detects that the ego vehicle is at an intersection, `f2` captures that the ego vehicle becomes (nearly) stationary within a short horizon, and `f3` checks whether the vehicle is actually in traffic conflict. In other words, $\bar{\theta}$ flags the undesired behavior "stopping in an intersection *without* conflict." Here, $A=1$ and $B=0.1$ serve as temporal and speed thresholds, respectively.

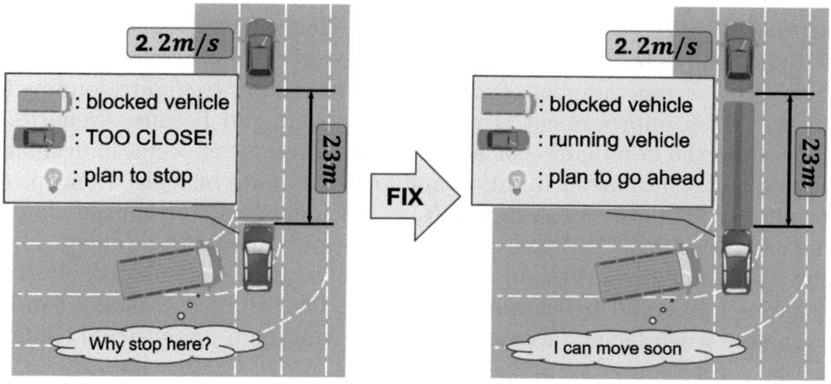

Fig. 5. Top: The ego vehicle stops within the intersection. Bottom: After the fix, the ego vehicle proceeds through the intersection as intended.

Employing the Gradient Approximator, we identify `safe_distance_base` (`sdb`) as the most relevant parameter influencing this behavior. Intuitively, `sdb` governs how conservatively the planner maintains distance to surrounding traffic; if set too high, the planner can overreact and stop even when there is no actual conflict. The fix generation phase proposes two candidate adjustments: (`sdb`, $\downarrow$ 90%) and (`sdb`, $\downarrow$ 50%). Both fixes prevent the ego vehicle from stopping illegally in the failing scenario. However, during the validation and prioritization stage, we find that reducing `sdb` by 90% is overly aggressive, resulting in a 70% scenario failure rate, whereas decreasing `sdb` by 50% achieves a 100% pass rate. ConFixer therefore selects the less aggressive adjustment, which removes the over-conservative stop while preserving safe behavior across the broader scenario set.

As illustrated in the bottom image of Fig. 5, after applying the selected fix, the ego vehicle complies with traffic laws and maintains a safe, continuous flow through intersections.

3.3 Results

We use ConFixer to process 173 failing scenarios derived from the LawBreaker dataset. ConFixer successfully produces a valid fix for each failing scenario, indicating that the proposed gradient-guided search can consistently navigate the configuration space to remove the observed violations. Equally important for industrial practice, these fixes are *regression-safe*: after applying the corrected configurations, the originally failing scenarios are resolved *and* no regressions are introduced in the 1,680 scenarios that were initially passing. This suggests ConFixer can be used as an automated repair loop in scenario-based validation pipelines, where configuration updates must simultaneously eliminate failures and preserve existing safety evidence.

Incorporating H_{info} significantly improves efficiency by reducing the search space. Specifically, H_{info} filters out approximately 70% of irrelevant scenarios, decreasing the average number of scenarios evaluated per configuration bug from 500 to 150. This improvement is practically meaningful because scenario execution is often the dominant cost in ADS validation: fewer scenario evaluations translate directly into faster iteration and lower compute budgets. Conceptually, this result also validates the design of H_{info}: by identifying unchanged clauses in the violation formulas, ConFixer can skip scenarios that cannot be affected by the current parameter perturbation, and focus the expensive simulation effort only on scenarios influenced by parameter changes. Overall, H_{info} makes ConFixer more scalable when facing large scenario suites and high-dimensional configurations.

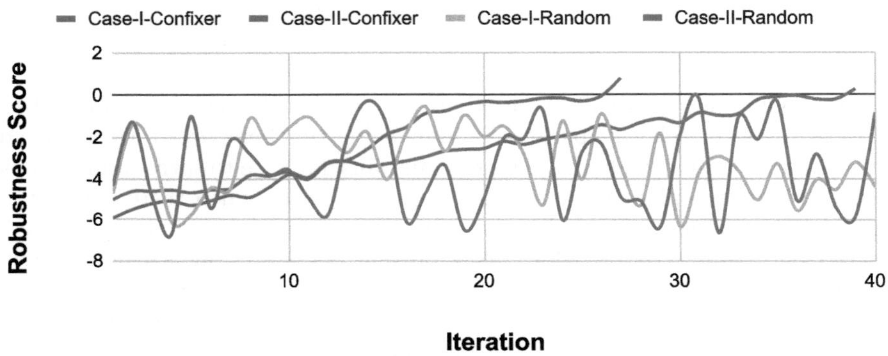

Fig. 6. Robustness Score Trend during Fix Generation.

During the fix generation process, we dynamically record the robustness score y for each iteration. Recall that $y \geq 0$ indicates satisfaction of the STL specification, while $y < 0$ indicates a violation. As shown in Fig. 6, the robustness score trends for both Case I and Case II begin below zero and steadily increase toward the satisfaction boundary, illustrating how ConFixer produces monotonic progress signals (in robustness) even when the final driving outcome is only observable after closed-loop execution. In contrast, the random baseline exhibits large oscillations and remains in the violating region, failing to find a viable solution. These trends support the central claim of ConFixer: robustness-guided gradients provide actionable direction for configuration repair, whereas unguided exploration wastes evaluations and struggles to escape violating configurations.

4 Threats to Validity

ConFixer targets configuration bugs within a single ADS module (e.g., planning) and does not currently address bugs that require coordinated tuning across multiple modules. The main reason is practical scalability and coupling: once parameters from multiple modules are tuned jointly, the effective search space expands substantially, increasing the number of perturbation evaluations needed for sign inference and making it difficult to find a good configuration within a practical time budget. Moreover, cross-module parameters can have coupled or even conflicting effects on specifications (e.g., improving one module's behavior may degrade another's), so a single fixed configuration may not exist that simultaneously satisfies all requirements across scenarios without additional design changes or architectural constraints. ConFixer also does not assume that only one parameter causes a violation: it first selects a *subset* of influential parameters and then performs an incremental, coordinate-style search over that subset. However, this search is greedy and primarily explores parameters one at a time; if a violation is dominated by higher-order interactions among many parameters, more combinatorial strategies (e.g., group perturbations or design-of-experiments style testing) may be needed to efficiently expose and repair such dependencies. Finally, ConFixer's runtime is bounded by the cost of scenario execution; although the harmlessness check and H_{info} caching reduce redundant evaluations, very large parameter sets or slow closed-loop simulations can still make repair expensive. Improving scalability via sensitivity screening, parallel simulation, and surrogate robustness models is a promising direction for future work.

5 Related Work

Existing research on ADS testing predominantly focuses on scenario generation that challenges ADS components under diverse conditions. Many studies generate *dynamic* scenarios by orchestrating surrounding traffic participants to disturb the ADS [10,22,35,38,44,47,51], while others construct *static* scenarios by manipulating obstacles or perception-related inputs to stress the sensing and perception pipeline [7,13,21,24,37]. Complementary to scenario construction, specification-driven testing also emerges as a way to systematically assess component-level performance (e.g., image-recognition behavior) under well-defined criteria [40].

Beyond individual scenes, additional approaches modify road networks to induce lane deviations or challenging route structures [16,52], often leveraging mature ADS stacks such as Baidu Apollo [5] and Autoware [4]. While effective at uncovering failures, these approaches commonly assume default or fixed ADS configurations; consequently, they may miss *configuration-specific* issues where parameter settings interact with scenarios to undermine robustness.

To address configuration problems, Chen et al. [9] introduced CONFVE, which searches for problematic configurations using genetic algorithms but does not automatically produce fixes. Meanwhile, classic bug localization and automated repair techniques predominantly operate on source-code-level semantics and edits [12,15,18,20,28,32,43,45,46,48]. Recent empirical evidence further suggests that ADS bug-fix patterns have distinct characteristics and constraints compared to general-purpose software [8], motivating repair techniques that explicitly account for the operational context and the role of configurations.

In parallel, formal methods provide complementary assurance mechanisms, including online model checking for decision-making safety [42] and scenario modeling languages that help structure scenario development [1]; however, open challenges remain for scaling formal verification to realistic autonomous driving settings [6]. Beyond offline assurance, runtime monitoring and enforcement can further increase confidence after deployment: REDriver [39] specifies rich driving properties (e.g., traffic laws) in an STL-based language, monitors planned trajectories with quantitative semantics, and uses gradient-driven repair to steer the trajectory away from likely violations. ConFixer fills this gap by targeting configuration-induced robustness issues: it not only detects problematic ADS configurations but also automatically synthesizes configuration fixes that improve robustness, complementing scenario generation, configuration search, runtime enforcement, traditional software repair, and formal-methods-based assurance.

6 Conclusion

ConFixer tackles the challenging problem of misconfigured ADS parameters that result in law-breaking or unsafe behaviors. By approximating parameter gradients, it systematically localizes the configuration points most relevant to observed violations and then fine-tunes parameter values to restore compliance with traffic laws. Through extensive simulation-based validation, ConFixer ensures that the applied fixes do not cause regressions on previously passing scenarios.

Case studies on real-world traffic laws and parameters in Apollo 7.0 demonstrate the effectiveness of ConFixer in producing stable, law-compliant fixes. Across 173 failing scenarios from the LawBreaker dataset, ConFixer resolved every misconfiguration without inducing new failures. Its optimization technique, guided by the information on unchanged clauses, substantially reduced the search space. These findings indicate that automated, gradient-based approaches to ADS configuration debugging can serve as a practical and efficient complement to existing software engineering techniques for autonomous vehicles. However, extending ConFixer to address configurations spanning multiple modules remains a key avenue for further study.

Acknowledgments. This work was supported by the National Natural Science Foundation of China (NSFC) under Grants 62232008 and 62032010, and by the NSFC Youth Program under Grant 62402367.

References

1. Aoki, T., Tomita, T., Kawai, T., Kawakami, D., Chida, N.: Modeling language for scenario development of autonomous driving systems. In: 2025 IEEE/ACM 13th International Conference on Formal Methods in Software Engineering (FormaliSE), pp. 11–21. IEEE (2025)
2. Apollo, B.: The configuration files for planning module (2019). https://github.com/ApolloAuto/apollo/tree/master/modules/planning/conf/scenario. Accessed Jan 2024

3. Attariyan, M., Flinn, J.: Automating configuration troubleshooting with dynamic information flow analysis. In: 9th USENIX Symposium on Operating Systems Design and Implementation (OSDI 10) (2010)
4. Autoware: the world's leading open-source software project for autonomous driving (2024). https://github.com/autowarefoundation/autoware. Accessed Nov 2024
5. Baidu: Apollo 7.0 (2021). https://github.com/ApolloAuto/apollo/releases/tag/v7.0.0. Accessed Aug 2024
6. Burgio, P., Ferrando, A., Villani, M.: Open challenges in the formal verification of autonomous driving. arXiv preprint arXiv:2411.14520 (2024)
7. Cao, Y., Bhupathiraju, S.H., Naghavi, P., Sugawara, T., Mao, Z.M., Rampazzi, S.: You can't see me: Physical removal attacks on {lidar-based} autonomous vehicles driving frameworks. In: 32nd USENIX Security Symposium (USENIX Security 23), pp. 2993–3010 (2023)
8. Chen, Y., et al.: A comprehensive study of bug-fix patterns in autonomous driving systems. Proc. ACM Software Eng. **2**(FSE), 380–402 (2025)
9. Chen, Y., Huai, Y., Li, S., Hong, C., Garcia, J.: Misconfiguration software testing for failure emergence in autonomous driving systems. Proc. ACM Software Eng. **1**(FSE), 1913–1936 (2024)
10. Cheng, M., Zhou, Y., Xie, X.: Behavexplor: behavior diversity guided testing for autonomous driving systems. In: Proceedings of the 32nd ACM SIGSOFT International Symposium on Software Testing and Analysis,pp. 488–500 (2023)
11. Clause, J., Li, W., Orso, A.: DYTAN: a generic dynamic taint analysis framework. In: Proceedings of the 2007 International Symposium on Software Testing and Analysis, pp. 196–206 (2007)
12. Cui, Y., Zhang, Z., Ma, Z., Chen, Z., Wang, Y., Shen, Y.: VMD: a visualizable malware detection scheme based on multi-dimensional dynamic behavior information. J. Networking Network Appl. **4**(1), 11–20 (2024)
13. Deng, Y., et al.: A declarative metamorphic testing framework for autonomous driving. IEEE Trans. Software Eng. **49**(4), 1964–1982 (2022)
14. Deshmukh, J.V., Donzé, A., Ghosh, S., Jin, X., Juniwal, G., Seshia, S.A.: Robust online monitoring of signal temporal logic. Formal Methods Syst. Des. **51**(1), 5–30 (2017). https://doi.org/10.1007/s10703-017-0286-7
15. D'Antoni, L., Samanta, R., Singh, R.: Qlose: program repair with quantitative objectives. In: Chaudhuri, S., Farzan, A. (eds.) CAV 2016. LNCS, vol. 9780, pp. 383–401. Springer, Cham (2016). https://doi.org/10.1007/978-3-319-41540-6_21
16. Gambi, A., Mueller, M., Fraser, G.: Automatically testing self-driving cars with search-based procedural content generation. In: Proceedings of the 28th ACM SIGSOFT International Symposium on Software Testing and Analysis, pp. 318–328 (2019)
17. Garcia, J., Feng, Y., Shen, J., Almanee, S., Xia, Y., Chen, Q.A.: A comprehensive study of autonomous vehicle bugs. In: Proceedings of the ACM/IEEE 42nd International Conference on Software Engineering, pp. 385–396 (2020)
18. Goues, C.L., Pradel, M., Roychoudhury, A.: Automated program repair. Commun. ACM **62**(12), 56–65 (2019)
19. Government, C.: Regulations for the implementation of the road traffic safety law of the people's republic of China (2021). http://www.gov.cn/gongbao/content/2004/content_62772.htm. Accessed Mar 2024
20. Gulwani, S., Radiček, I., Zuleger, F.: Automated clustering and program repair for introductory programming assignments. ACM SIGPLAN Notices **53**(4), 465–480 (2018)

21. Guo, A., et al.: CooTest: an automated testing approach for v2x communication systems. In: Proceedings of the 33rd ACM SIGSOFT International Symposium on Software Testing and Analysis, pp. 1453–1465 (2024)
22. Hao, K., et al.: Adversarial safety-critical scenario generation using naturalistic human driving priors. IEEE Trans. Intell. Veh. (2023)
23. Hekmatnejad, M., et al.: Encoding and monitoring responsibility sensitive safety rules for automated vehicles in signal temporal logic. In: Proceedings of the 17th ACM-IEEE International Conference on Formal Methods and Models for System Design. MEMOCODE '19, New York, NY, USA. Association for Computing Machinery (2019). https://doi.org/10.1145/3359986.3361203
24. Jing, P., et al.: Too good to be safe: tricking lane detection in autonomous driving with crafted perturbations. In: 30th USENIX Security Symposium (USENIX Security 21), pp. 3237–3254 (2021)
25. Kemerlis, V.P., Portokalidis, G., Jee, K., Keromytis, A.D.: LIBDFT: practical dynamic data flow tracking for commodity systems. In: Proceedings of the 8th ACM SIGPLAN/SIGOPS conference on Virtual Execution Environments, pp. 121–132 (2012)
26. Lab, L.S.V.: Maps content (2019). https://content.lgsvlsimulator.com/maps/. Accessed Jul 2022
27. Lab, L.S.V.: Maps content (2019). https://content.lgsvlsimulator.com/vehicles/. Accessed Jul 2022
28. Le Goues, C., Nguyen, T., Forrest, S., Weimer, W.: Genprog: a generic method for automatic software repair. IEEE Trans. Software Eng. **38**(1), 54–72 (2011)
29. Litman, T.: Autonomous vehicle implementation predictions: Implications for Transport Planning. Victoria Transport Policy Institute Victoria, BC, Canada (2017)
30. Maler, O., Nickovic, D.: Monitoring temporal properties of continuous signals. In: Lakhnech, Y., Yovine, S. (eds.) FORMATS/FTRTFT -2004. LNCS, vol. 3253, pp. 152–166. Springer, Heidelberg (2004). https://doi.org/10.1007/978-3-540-30206-3_12
31. Ničković, D., Yamaguchi, T.: RTAMT: online robustness monitors from STL. In: Hung, D.V., Sokolsky, O. (eds.) ATVA 2020. LNCS, vol. 12302, pp. 564–571. Springer, Cham (2020). https://doi.org/10.1007/978-3-030-59152-6_34
32. Perry, D.M., Kim, D., Samanta, R., Zhang, X.: Semcluster: clustering of imperative programming assignments based on quantitative semantic features. In: Proceedings of the 40th ACM SIGPLAN Conference on Programming Language Design and Implementation, pp. 860–873 (2019)
33. Raman, V., Donzé, A., Maasoumy, M., Murray, R.M., Sangiovanni-Vincentelli, A., Seshia, S.A.: Model predictive control with signal temporal logic specifications. In: 53rd IEEE Conference on Decision and Control, pp. 81–87 (2014). https://doi.org/10.1109/CDC.2014.7039363
34. Rodionova, A., Alvarez, I., Elli, M.S., Oboril, F., Quast, J., Mangharam, R.: How safe is safe enough? Automatic safety constraints boundary estimation for decision-making in automated vehicles. In: 2020 IEEE Intelligent Vehicles Symposium (IV), pp. 1457–1464. IEEE (2020)
35. Song, R., Ozmen, M.O., Kim, H., Muller, R., Celik, Z.B., Bianchi, A.: Discovering adversarial driving maneuvers against autonomous vehicles. In: 32nd USENIX Security Symposium (USENIX Security 23), pp. 2957–2974 (2023)
36. State of California: California vehicle code section 22500 (2024). https://leginfo.legislature.ca.gov/faces/codes_displaySection.xhtml?lawCode=VEH&sectionNum=22500

37. von Stein, M., Shriver, D., Elbaum, S.: Deepmaneuver: adversarial test generation for trajectory manipulation of autonomous vehicles. IEEE Trans. Software Eng. (2023)
38. Sun, Y., Poskitt, C.M., Sun, J., Chen, Y., Yang, Z.: Lawbreaker: an approach for specifying traffic laws and fuzzing autonomous vehicles. In: 37th IEEE/ACM International Conference on Automated Software Engineering, pp. 1–12 (2022)
39. Sun, Y., Poskitt, C.M., Zhang, X., Sun, J.: Redriver: runtime enforcement for autonomous vehicles. In: Proceedings of the IEEE/ACM 46th International Conference on Software Engineering, pp. 1–12 (2024)
40. Tanaka, K., Aoki, T., Tomita, T., Kawakami, D., Chida, N.: Specification-based testing of the image-recognition performance of automated driving systems. IEEE Access (2024)
41. Taş, Ö.Ş., Kuhnt, F., Zöllner, J.M., Stiller, C.: Functional system architectures towards fully automated driving. In: 2016 IEEE Intelligent vehicles symposium (IV), pp. 304–309. IEEE (2016)
42. Tran, D.D., Hasegawa, A., Riviere, P., Tomita, T., Aoki, T.: Enhancing decision-making safety in autonomous driving through online model checking. In: International Conference on Rigorous State-Based Methods, pp. 231–248. Springer (2025)
43. Wang, K., Singh, R., Su, Z.: Search, align, and repair: data-driven feedback generation for introductory programming exercises. In: Proceedings of the 39th ACM SIGPLAN Conference on Programming Language Design and Implementation, pp. 481–495 (2018)
44. Wang, T., Gu, T., Deng, H., Li, H., Kuang, X., Zhao, G.: Dance of the ads: Orchestrating failures through historically-informed scenario fuzzing. In: Proceedings of the 33rd ACM SIGSOFT International Symposium on Software Testing and Analysis. pp. 1086–1098 (2024)
45. Wang, X., et al.: MTL-transfer: leveraging multi-task learning and transferred knowledge for improving fault localization and program repair. ACM Trans. Software Eng. Methodol. (2024)
46. Xia, C.S., Wei, Y., Zhang, L.: Automated program repair in the era of large pre-trained language models. In: 2023 IEEE/ACM 45th International Conference on Software Engineering (ICSE), pp. 1482–1494. IEEE (2023)
47. Yan, S., et al.: On-demand scenario generation for testing automated driving systems. Proc. ACM Software Eng. **2**(FSE), 86–105 (2025)
48. Zhang, Q., Fang, C., Ma, Y., Sun, W., Chen, Z.: A survey of learning-based automated program repair. ACM Trans. Software Eng. Methodol. **33**(2), 1–69 (2023)
49. Zhang, X., Yan, S., Yang, Z.: ConFixer Data (2026). https://doi.org/10.5281/zenodo.18183868
50. Zhang, X., Yang, Z., Zheng, Q., Hao, Y., Liu, P., Liu, T.: Tell you a definite answer: whether your data is tainted during thread scheduling. IEEE Trans. Software Eng. **46**(9), 916–931 (2018)
51. Zhang, X., et al.: Testing automated driving systems by breaking many laws efficiently. In: Proceedings of the 32nd ACM SIGSOFT International Symposium on Software Testing and Analysis, pp. 942–953 (2023)
52. Zohdinasab, T., Riccio, V., Tonella, P.: Focused test generation for autonomous driving systems. ACM Trans. Software Eng. Methodol. (2024)

Pragmatic Uses of AI in Formal MethodsBased Railway Projects: Early Lessons and Perspectives (Extended Abstract)

Thierry Lecomte(✉), Dalay Almeida, and Norman Maury

CLEARSY, Aix En Provence, France
thierry.lecomte@clearsy.com

Abstract. This extended abstract reports on recent industrial activities exploring pragmatic and controlled uses of AI in railway systems developed with formal methods. First, large language models are used to synthesize and query large-scale railway technical specifications, improving accessibility while requiring systematic human validation. Second, early experiments assess AI-assisted generation of interactive proof scripts for B method proof obligations, showing initial productivity gains but substantial research challenges. Third, AI-based perception is integrated into railway safety functions, where machine learning acts solely as a proposer whose outputs are redundantly checked by a formally verified safety controller. Finally, a proof of concept applies multimodal language models to relay-based railway interlocking systems, automatically transforming relay diagrams into propositional logic specifications applicable to new diagrams. The results confirm that AI-assisted formalization is feasible and accessible, opening promising perspectives for advancing automation and trustworthiness in railway formal verification.

Keywords: formal methods · safety critical · software development · railway

1 Introduction

The integration of machine learning (ML) into safety-critical railway systems raises significant challenges for certification, as current safety standards require transparent, fully specified system designs, whereas ML models are inherently opaque after training. This extended abstract presents several research activities, completed during collaborative projects and in-house investigations. These activities address various safety engineering concerns related to the development of critical railway systems, including the analysis of technical documentation, assistance with interactive formal model verification, the transformation of technical diagrams into formal models, and the use of AI to create a safe image analysis system. Although the activities are diverse and specific, they nevertheless yield a number of common lessons.

A. Sampaio and M. Stoelinga (Eds.): FM 2026, LNCS 16557, pp. 605–609, 2026.
https://doi.org/10.1007/978-3-032-26220-2_29

2 Activities and Results

Four main activities were carried out in 2025 and are summarised below.

2.1 Easing Interaction with Technical Specifications

The work of a software engineer often involves reviewing technical documentation, which is frequently written by other individuals or companies. For example, specifications for embedded railway software can run to several hundred or even thousands of pages. Understanding such content therefore requires a considerable amount of reading time. LLMs were used to facilitate this process by generating summaries and answering technical questions that combine information from multiple sections. The documents analysed in PDF and DOCX format are either proprietary technical documents or public documents (ERTMS/ETCS subsets[1]). To ensure the confidentiality of external documents, the LLMs were operated on the company's internal computing infrastructure.

Overall, the implementation of this feature was perceived as facilitating the understanding of new technical documentation and led to noticeable time savings (up to 30% gain when discovering new technical document). However, limitations were encountered, probably related to the amount of text to be analysed and the presence of numerous images. On several occasions, certain elements were not taken into account in the responses provided by the LLMs. While we were pleasantly surprised by the correct and well-argued response to our first question about the safety of an ETCS feature, the responses provided were often partial, sometimes inaccurate, and occasionally speculative.

The conclusion of this work is that systematic verification of the responses returned by LLMs is necessary to check their accuracy before they can be relied upon for any engineering activity (analysis, development, testing).

2.2 Assistance with Interactive Proof

The research is aimed at demonstrating how AI techniques can be applied to support the B formal method [3] without weakening safety guarantees. A key contribution is the use of machine learning to guide and accelerate formal proof activities, notably through predictive support for automated provers. The work shows that AI can reduce manual proof effort while preserving correctness and traceability. Another important result is the definition of AI architectures suitable for industrial constraints, with fully local inference ensuring data confidentiality. The research also extends AI support to software engineering tasks such as code analysis and project description. Practical demonstrations and experiments illustrate the feasibility of the approach in real engineering workflows.

The results highlight improved efficiency in verification activities. A gain up to 20% was obtained when selecting the best automatic prover to apply instead of

[1] https://www.era.europa.eu/era-folder/archived-set-specifications-3-etcs-b3-r2-gsm-r-b1.

applying static proof tactics. Overall, this work supports the view of trustworthy AI as an enabler of formal methods rather than a replacement. It positions AI assisted formal verification as a realistic path toward scalable, safe industrial systems.

2.3 AI-Based Perception Integrated Into Railway Safety

The research [4] addresses the fundamental challenge of integrating machine learning into safety-critical railway systems governed by strict certification standards. Its main contribution is the proposal and evaluation of a proof of concept architecture that constrains the role of machine learning to perception and candidate generation, while delegating all safety decisions to a formally developed controller. The work recognizes that machine learning models are inherently opaque and therefore incompatible with existing railway safety standards when used directly in vital functions. To overcome this limitation, the research introduces a certified-control-inspired architecture (Fig. 1) in which machine learning outputs are validated by an independent safety controller.

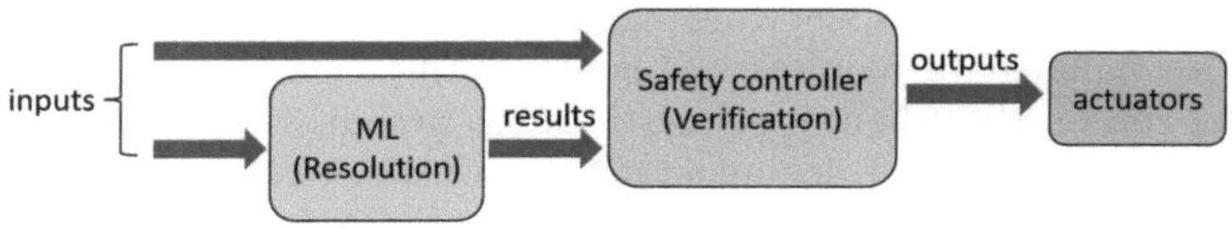

Fig. 1. ML with verifiable results + safety controller.

The safety controller is specified, developed, and verified using the B method, ensuring full traceability from safety requirements to executable code. The approach is illustrated through a free track detection use case, where a convolutional neural network proposes a candidate free track path from camera images. This candidate is then checked by the safety controller against explicit geometric and physical constraints derived from railway domain knowledge. A key safety property guarantees that the computed free track distance is always conservative with respect to real obstacles. The research details how safety properties are systematically refined into verifiable sub properties covering algorithms, code, and execution platform. Experimental results show that unsafe or imperfect machine learning outputs can be safely rejected. Overall, the work demonstrates a credible pathway for the trustworthy use of AI in certified railway systems by combining advanced perception with formally proven safety logic, while clearly emphasizing the necessity of expert validation and further industrial maturation

2.4 Transforming Relay Diagrams Into Formal Specification

In the continuation of research on the formal verification of relay-based diagrams [2], a research effort [1] was aimed at automating the formal analysis of relay based railway interlocking systems, which are still widely deployed in operational networks. The main contribution is a proof of concept showing that a multimodal large language model can directly interpret relay circuit diagrams and generate formal specifications in propositional logic. The approach removes the need for manual reinterpretation or redrawing of diagrams, which are traditionally error prone and time consuming. Using prompt engineering, the model is guided to produce specifications compatible with B method based analyses. Experimental results on academic and industrial SNCF diagrams demonstrate that the AI can correctly capture system behaviour and detect anomalies such as ringbell effects and short circuits. The study highlights both the strengths and limitations of general purpose AI when applied to context specific railway conventions. It shows that modern AI models can act as end to end translators from legacy engineering artefacts to formal models. The work emphasizes that expert validation remains essential for safety critical use.

3 Conclusion and Perspectives

LLMs clearly help engineers throughout the development cycle, even on confidential technical subjects, showing significant recent progress. However, their tendency to hallucinate requires systematic verification, limiting their use to supporting roles rather than direct certification activities. Since late 2025, new applications are emerging—documentation synthesis, test case generation, and technical environment integration—with further breakthroughs expected in the coming months.

Acknowledgements. The work and results described in this article were partly funded by the projects "New Methods for Safety Demonstration" managed by International Union of Railways and H2020-ECSEL "AIDOaRT" focusing on AI-augmented automation.

References

1. Almeida, D., Glemarec, L.: From relay-based railway interlocking circuits to formal specification: an AI-driven approach, pp. 25–34 (2025)
2. Almeida, D., Oliveira, M., Bezerra, P., Bon, P., Collart-Dutilleul, S.: CSP specification and verification of relay-based railway interlocking systems, pp. 97–106 (2022)
3. Butler, M., Körner, P., Krings, S., Lecomte, T., Leuschel, M., Mejia, L.F., Voisin, L.: The first twenty-five years of industrial use of the b-method. In: ter Beek, M.H., Ničković, D. (eds.) Formal Methods for Industrial Critical Systems, pp. 189–209. Springer International Publishing, Cham (2020). https://doi.org/10.1007/978-3-030-58298-2_8
4. Lecomte, T.: Formal development of a safety controller for machine learning outputs in vital railway systems. In: ter Beek, M.H., Teixeira, L. (eds.) Formal Methods: Foundations and Applications, pp. 196–206. Springer Nature Switzerland, Cham (2026). https://doi.org/10.1007/978-3-032-12086-1_11

Formal Verification for Security Certification: From a First Success to Sustainable Industrial Usage

Adel Djoudi[1] and Nikolai Kosmatov[2](✉)

[1] Thales Cybersecurity and Digital Identity, Meudon, France
Adel.Djoudi@thalesgroup.com
[2] Thales Research and Technology, cortAIx Labs, Palaiseau, France
Nikolai.Kosmatov@thalesgroup.com

Abstract. The application of formal methods in an industrial context remains a delicate and time-consuming task, and each successful project contributes to improving productivity. This experience report summarizes five years of applying formal verification for Common Criteria-based security certification of a JavaCard Virtual Machine at Thales. Since the first successful formal verification—directly over the source code with the Frama-C verification platform—was achieved in 2021, five highest-level Common Criteria certifications (at EAL6 and EAL7 levels) have been obtained during the past five years. The strong guarantees of functional correctness and security provided by source code-based verification are essential for security-critical products such as smart cards. This paper traces the evolution of the project from an initial success to the continuous application of formal verification, and discusses the main challenges, applied solutions, and opportunities for further improvement.

Keywords: deductive verification · security properties · JavaCard virtual machine · Frama-C · Common Criteria certification · smart cards

1 Introduction

Many aspects of modern society have become strongly dependent on software-based devices and services. Examples of such critical products include smart cards, which are widely used today for identity documents, banking, and communication. In the presence of increasing cybersecurity risks, it is crucial to guarantee functional correctness and security properties of critical products and services. Formal verification can provide such guarantees.

Although the application of formal methods in industry has become more widespread, industrial users still face difficulties when applying them to large-scale projects while maintaining long-term productivity. Sharing achievements and challenges from these applications within the formal methods community is valuable to both researchers and practitioners. This paper is motivated by this

A. Sampaio and M. Stoelinga (Eds.): FM 2026, LNCS 16557, pp. 610–630, 2026.
https://doi.org/10.1007/978-3-032-26220-2_30

goal and reports on the experience with the formal verification of a critical smart card software at Thales.

In 2021, Thales realized a world-first formal verification [21] of a JavaCard Virtual Machine (JCVM) *directly on the real-life code* using the Frama-C verification platform [30]. It was done for security certification of a smart card product according to the Common Criteria (CC) for Information Technology Security Evaluation [16], which provide an international standard for security certification. It includes seven levels of assurance evalution (EAL1–EAL7). The highest levels (EAL6–EAL7) require a formal Security Policy Model (SPM) along with a rigorous mathematical proof of functional and security properties (including confidentiality and integrity). Overall, since the initial success in 2021 [21], five highest-level certifications of the JavaCard Virtual Machine have been realized by Thales during the past five years [1–5]. For the latest certification [5], Thales incorporated post-quantum cryptography, thus featuring the first quantum-resistant smart card in Europe to receive high-level security certification [41]. The goal is to help government services protect sensitive data—such as those on ID cards, health cards, and driver's licenses—and ensure that citizens' identities remain secure even in the presence of emerging quantum threats.

Contributions. This paper presents an experience report on five years of applying deductive verification for security certification of smart card software at Thales. It traces the evolution of the project from the initial success towards sustained, long-term industrial use. We outline the main difficulties we encountered and the solutions we adopted to improve the overall productivity of formal verification in this project. These solutions encompass both methodological advances and tool enhancements. Our key contributions include the specification of two new Frama-C features—recently implemented by the Frama-C team—namely, the Uncast plugin, which automates code transformations involving pointer casts, and the GenScript solution, which automates proof script generation. In addition, we present a novel open-source Frama-C plugin, Frama-C-lsp[1], which we developed in order to provide a convenient IDE for annotated C code based on VS Code. Finally, we discuss remaining challenges and outline opportunities for further improvements.

Outline. Section 2 presents the context and motivation of this work. Section 3 presents the aspects of the project related to formal specification and organization of source files. A novel VS Code extension for C and ACSL based on the Frama-C-lsp plugin is introduced in Sect. 4. Industrial expectations for productive applications of formal verification are emphasized in Sect. 5. Section 6 provides statistics and outlines challenges related to proof performance. Finally, Sect. 7 presents a conclusion and future work.

[1] Available at https://github.com/ThalesGroup/frama-c-lsp.

2 Context and Motivation

2.1 The Frama-C Verification Platform

Frama-C [30,31] is a verification platform for ISO C99 programs [28] developed by CEA List with contribution of Inria. It offers several analysis tools organized as plugins, working on top of a common kernel. Our project mainly relies on the WP plugin of Frama-C dedicated to modular deductive verification of C code [10].

Frama-C users can specify program annotations in ACSL (ANSI C Specification Language) [8,12], a formal specification language for C programs. Typically, for deductive verification of a C program in Frama-C, each function should be annotated with a *(function) contract* expressed in ACSL. In the contract of a function f, a *precondition* (expressed with an ACSL clause **requires**) defines a property expected to hold before any call to f. It is assumed on entry during the verification of f, but must be proved before any call to f during the verification of a caller. Dually, a *postcondition* (**ensures** clause) of f states a property that must hold after the function call: it must be proved during the verification of f but is assumed after the call to f in a caller. A frame rule (**assigns** clause) in the contract of f specifies the memory locations that f is allowed to modify. In a loop contract, a *loop invariant* (**loop invariant** clause) specifies a property that must hold before the loop and after every loop iteration. Lemmas and local assertions can be used to facilitate the proof by stating a useful intermediate property. A *(local) assertion* (expressed with an ACSL clause **assert**) must hold at the point where it is inserted. Stated at the global level, *lemmas* are proved once and can then be used in different proof contexts. ACSL offers the `integer` type for unbounded mathematical integers in addition to bounded machine types.

Given a C program with a (partial) formal specification expressed by ACSL annotations, the WP plugin [10] tries to prove that the program respects the provided annotations. To do that, WP generates *verification conditions* (or *proof goals*, or *proof obligations*) that are then either proved by WP itself (thanks to its internal formula simplification engine called Qed [18]) or sent via the Why3 tool [26] to external provers (SMT solvers). In our project, we use the Alt-Ergo solver [17]. If all proof goals are proved, the given program is guaranteed to respect the given ACSL specification. When automatic proof fails, the interactive proof editor of WP can be used to manually record a *proof script* by applying predefined proof tactics. Frequently used tactics include: `unfold` to unfold a predicate definition; `bittestrange` and `modmask` to rewrite a bit-level operation, resp., using ranges or modulo operations; `range` to enumerate a finite range of values; `split` and `cut` to split a goal into several subgoals using, resp., either logical operations and conditionals, or an intermediate hypothesis; `overflow` to consider in-range and out-range cases of machine integer conversion; `filter` to erase irrelevant hypotheses in a proof context; `instance` to instantiate a quantified variable to a specific value, etc. A detailed description of tactics can be found in [7]. After the application of a few carefully chosen tactics, SMT solvers can manage to prove the transformed proof goals.

```
1  /*@ requires n ≥ 0 ∧ \valid(t + (0..n-1));
2      assigns nothing;
3      ensures result ≠ 0 ⇔
4          (∀ ℤ j; 0 ≤ j < n ⇒ t[j] == 0);
5  */
6  int all_zeros(int *t, int n) {
7      int k=0;
8      /*@ loop invariant 0 ≤ k ≤ n;
9          loop invariant ∀ ℤ j; 0 ≤ j < k ⇒ t[j] == 0;
10         loop assigns k;
11         loop variant n-k;
12     */
13     while(k < n){
14         if (t[k] ≠ 0)
15             return 0;
16         k++;
17     }
18     return 1;
19 }
```

Fig. 1. Function `all_zeros` with ACSL specification.

To check whether the provided specification contains inconsistencies (e.g. contradictory assumptions), the user can activate their detection by the WP plugin using so-called *smoke tests* [11]. Basically, WP introduces a special check-and-forget assertion of false (that can be expressed in ACSL as **`check`** `\false;`) and tries to prove it with a short timeout. A successful proof indicates an inconsistency. In addition, to ensure that the program under verification cannot provoke *undefined behaviors* (also called *runtime errors*), WP relies on the RTE plugin of Frama-C to generate additional assertions to exclude the risk of undefined behaviors. Their proof is an essential step of formal verification, both to ensure the soundness of verification and to avoid security vulnerabilities due to undefined behaviors.

Finally, the MetAcsl plugin [39,40] of Frama-C can be used to express and verify security properties. It allows the user to specify *global properties* (also called *metaproperties*, or *High-Level Requirements*) and translates them into local assertions (for instance, for each reading or writing operation) that can then be verified by other tools (like WP).

2.2 Example of ACSL Specification

Consider the function `all_zeros`, presented in Fig. 1. It receives an array t and its size n, and returns a nonzero value if all elements of t are zero, and 0 otherwise. We use pretty-printing of some ACSL keywords and symbols, for instance, `integer`, `==>` and `>=` as $\mathbb{Z}$, $\Rightarrow$ and $\geq$.

The contract of the function includes a precondition on line 1, which requires that n is non-negative and the array contains valid memory locations at indices $0..(n-1)$. These constraints must be satisfied by the caller. The assigns clause on line 2 enforces that no non-local variables are modified, ensuring for instance that implementations altering the array would be deemed incorrect. The postcondition on lines 3–4 specifies that the function's return value is nonzero if and only if all array elements are equal to zero.

In the presence of a loop, the loop contract on lines 8–12 is necessary for a successful deductive verification. The loop invariant on line 8 gives an interval of values for k. The loop invariant on line 9 specifies that elements at indices $0..(k-1)$ are zero after each complete iteration (non-interrupted with a return). The loop assigns clause identifies variables that may change within the loop, and the loop variant $(n-k)$ establishes a decreasing positive value providing an upper bound for the number of remaining loop iterations and helping the tool to prove termination of the loop. With this loop contract, Frama-C/WP can verify the correctness of the function, including safe array access on line 14 and the absence of arithmetic overflow on line 16.

2.3 Formal Verification of a JCVM at Thales

JavaCard Virtual Machine (JCVM) [37] is a crucial component present on most current smart cards. Its functional role is to interpret *JavaCard bytecode*, that is, a sequence of *opcodes*, each one specifying one operation of a JavaCard program. Its security role is to ensure mutual *applet isolation* [36]: the applets of mutually non-trusted vendors are not allowed to read or modify each other's data.

In 2021, Thales realized formal verification [21] of a JCVM for Common Criteria certifications of its smart card products. For the first time, the verification was realized directly on the source code, which brings many benefits to the certification process in terms of model understanding and confidence in verified properties [22]. It reduces the gap between the proven properties and the code.

This initial proof relied on several carefully chosen solutions detailed in previous publications [21,23]: *ghost code* to model memory features, to avoid existential quantifiers and to facilitate bit-level reasoning; *code transformations* to address the limitations of Frama-C/WP; *proof scripts* for goals for which automatic proof fails; *preservation predicates* to facilitate reasoning about memory updates; *metaproperties* to express and verify security properties. This initial proof took approximately 3 person-years. We refer the reader to [21] for more detail and technical aspects.

2.4 Motivation: Productivity of Formal Verification in Industry

While an initial proof of a new project is a key milestone—one for which an industrial company may be willing to invest a significant amount of human effort—it does not ensure the sustained, long-term use of formal verification throughout the project. We observe that sustainable application of formal verification requires a higher level of automation across the entire process (including specification writing and comprehension, proof replay, reporting), as well as improved support for maintaining verified code (including the creation and evolution of proof scripts).

Indeed, industrial users must cope with a variety of evolving factors: the evolution of the codebase (e.g., to integrate new features or updates), the evolution of the formal specification (e.g., to express additional properties or to extend the verification scope), the evolution of verification tools (which may break manually written proof scripts or require new ones), and the evolving expectations of evaluators and certification authorities. Each of these changes can trigger substantial proof rework. Rather than performing a one-shot proof of a single project with a fixed set of properties and tools, industrial verification engineers must verify many similar, evolving projects using evolving tools, while carefully tracking all proof-related artifacts (annotations, verification scope, proof scripts, documentation, etc.). Given the size of industrial codebases and the frequency of such evolutions, it is hardly possible to reinvest the same level of effort repeatedly throughout the project lifecycle. Improving the overall productivity of formal verification in industrial contexts is therefore a key requirement for its sustained, long-term application. This concern has been a constant focus of our project over the years.

3 Structured Formal Specification

In this section, we first provide an overview of the main challenges of deductive verification on large-scale industrial projects. We then focus on several challenges and solutions related to the formal specification of the C code and the corresponding properties. Some issues are discussed further in the following sections.

3.1 Deductive Verification Workflow

The theoretical workflow of deductive verification appears straightforward: specify the C code in ACSL and prove it with Frama-C/WP. In practice—especially for large industrial projects involving complex properties—numerous iterations are necessary to reach a meaningful formal specification and address potential proof issues. Such issues include *proof failures* and *warnings* reported by WP, each of which requires careful manual analysis. Figure 2 illustrates the typical workflow of addressing proof issues.

A proof failure may reveal an error in the code or in the ACSL specification (cf. the left and right branches in Fig. 2), both of which must be corrected. *The ACSL specification needs to be complete, precise and free of any inconsistency.* Achieving this is a major challenge in deductive verification. A careful organization and preparation of the annotated code is key to achieve this goal on

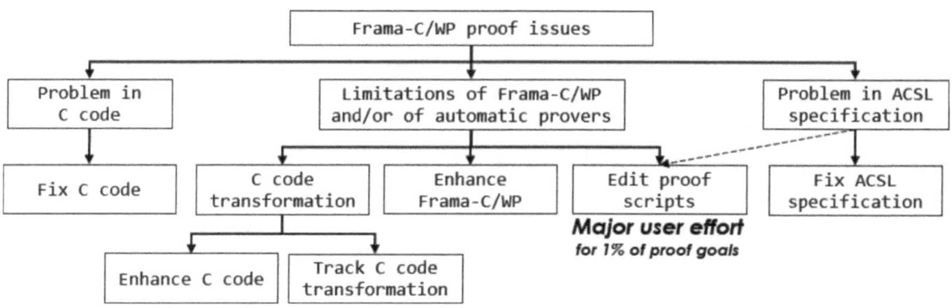

Fig. 2. Handling proof issues during deductive verification with Frama-C/WP.

large projects. A convenient IDE for browsing and understanding the annotated code (further discussed in Sect. 4) is necessary in practice. Some specification issues can be identified by manual review of ACSL annotations (in particular, completeness or precision issues) or by failing smoke tests (in particular, specification inconsistencies, cf. Sect. 2.1).

Due to proof incompleteness [38], even with a correct and complete ACSL specification, *automatic provers may fail to discharge complex proof goals.* This constitutes a second major challenge. In such cases, manually written proof scripts can be used to prove the remaining goals (cf. Sect. 2.1). However, as discussed in Sect. 2.4, in an evolving industrial context, this activity is highly time-consuming and costly in terms of proof maintenance, even when only a small fraction of goals requires proof scripts (1–3% in our project). Proof scripts are futher discussed in Sect. 5.

Last but not least, known limitations of Frama-C or WP in handling certain C and ACSL constructs as well as limitations of the memory model of WP when reasoning about pointer casts, restrict *the set of instructions and properties that can be formally verified.* In such cases, we rewrite the code in an equivalent way to make it suitable for formal verification. These *code transformations* should be introduced in a structural way and carefully tracked until they are either integrated into the production code or rendered obsolete by tool enhancements.

We also regularly provide feedback to the Frama-C developers in order to enhance the tool and extend the set of supported constructs.

3.2 Redundancy of the ACSL Specification: Good and Bad Practices

Repeating the same ACSL clauses in different function contracts is often necessary to specify required properties, which can lead to a certain *redundancy of contracts* and become a source of errors. As our goal is to verify functional and security properties, we express some of them using ACSL predicates. Instances of these predicates must then be included in several function contracts. To avoid redundant clauses in function contracts, we define *parametrized macros* that encapsulate commonly used predicates, and we include these macros in the relevant contracts. The macros are automatically unfolded to the required clauses by the preprocessor.

Another important solution for avoiding redundant clauses and automating the generation of necessary assertions relies on the MetAcsl plugin. We specify many important security properties as globally defined *metaproperties* [39,40], which state confidentiality and integrity properties as constraints on all reading and writing operations. Such metaproperties are then automatically instantiated by the MetAcsl plugin into local checks [39,40] before the corresponding read and write operations.

A common approach to avoid redundant intermediate properties relies on the use of lemmas. Interestingly, *an intensive use of lemmas proved to be inefficient in our project.* We compared a version that makes extensive use of lemmas (61 lemmas) to express intermediate properties with another version that uses only three lemmas and instead relies on 519 local assertions to express such properties where needed. Replacing lemmas with assertions reduced the overall proof time by 1 h 20 min. This result indicates that lemmas can significantly slow down the proof by introducing additional (and often useless) reasoning paths for the solver. Consequently, in our project we favor local assertions (repeated when needed) to simplify proof contexts and improve overall proof efficiency.

3.3 Frama-C/WP Limitations and the Uncast Plugin

In our project, we use the Typed memory model of WP [7] for its ability to provide a good trade-off between expressivity of provable properties and the capacity to discharge proof goals with automatic provers. The major limitation of this memory model is its inability to handle pointer casts and union types. The majority of unsupported operations is related to heterogeneous pointer casts (including byte splits or concatenations) that must be rewritten before the proof.

This observation motivated a new Frama-C plugin, called Uncast[2], that automatically rewrites common patterns of heterogeneous pointer casts into an equivalent code using arithmetic operations supported by the tool. An example of such a rewriting is the following (assuming a big-endian architecture):

```
1 typedef unsigned char u1; typedef unsigned short u2;
2 ...
3 u1 arr[10]; // Read bytes 4,5 of arr as an unsigned short into data:
4 // u2 data = *(u2*)(arr + 4); // before rewriting of casts (original code);
5 u2 data = (u2)( (*(arr + 4))*256 + *(arr + 5) ); // after rewriting of casts.
```

In this example, the real-life code before rewriting (shown on line 4) first constructs a pointer to the byte at offset 4 in array `arr`, than casts it to an unsigned short pointer (`u2*` type) to read the unsigned short value starting at that address, and therefore represented by the two bytes at offsets 4 and 5. The rewriting (line 5) avoids this pointer cast by reading the bytes at offsets 4 and 5 separately in order to compute the required unsigned short value arithmetically, without any pointer casts.

The Uncast plugin was specified by Thales and implemented by the Frama-C team. In addition, the transformations of all patterns were formally verified using Eva [13], the static analysis plugin of Frama-C. The use of Uncast brings significant benefits to our project: *it automates a large number of code transformations and provides strong confidence in their correctness.*

Examples of other limitations of Frama-C/WP include function pointers, unions, setjmp/longjmp instructions, volatile variables, and pointer-to-integer conversion. These constructs still require *manual code transformations* today.

3.4 Analysis Scope vs. Analysis View

In large-scale industrial projects, formal verification is often applied only to a subset of the codebase, excluding less critical components or parts that have not yet been verified. Moreover, the verified code may contain *rewritten parts* or *additional hypotheses* (on the unproven parts or the environment), which should not

2 The Uncast plugin is currently not yet open-source.

pollute the original code. ACSL **admit** clauses can be used to introduce hypotheses as local properties. *Stubs* (i.e. declared functions with ACSL contracts but without function definition) can be used to abstract away low-level or irrelevant features and to introduce hypotheses about them. It is therefore convenient to distinguish and carefully track two aspects of the code: a *scope* and a *view*.

Scope. An *(analysis) scope* consists of the set of function definitions located in a set of input source files that are submitted to formal verification with Frama-C/WP. The scope is selected through a preprocessed macro in order to ignore other functions—considered out of scope—in the input source files. Notice that ACSL contracts of out-of-scope functions called from within the scope functions are needed for the proof, and included as stubs. Once the scope it selected, both its *original* and *analysis* versions (or *views*) should be tracked for traceability.

View. The *(analysis) view* of the scope applies to it all other code transformations required to complete the formal analysis with Frama-C/WP on functions inside the scope. These code transformations are necessary to tackle limitations of Frama-C/WP to handle some C constructs. The analysis view also adds all necessary ACSL annotations and ghost code. We use preprocessed macros to switch the scope from the original view to the analysis view. This enforces the traceability of all code transformations (clearly indicated in the code by the occurrences of the view changing macro) with no impact on the original code.

Source Code Preparation. On request of the development teams, source code preparation must ensure a structured integration of ACSL annotations into the original codebase without any effect on the original code compilation. Frama-C relies on an external C compiler (GCC by default) to preprocess C source files. We use configuration files (called *machdep files*) to set up compilation features for our targeted machine architectures, such as endianness and sizes of integer and pointer types. The code preparation for the analysis requires inclusion of carefully identified source files and analysis-specific headers, and relies on suitable macro definitions.

4 Integrated Development Environment for Frama-C/WP

4.1 Motivation

Frama-C graphical user interface (GUI) provides only a static display of the analyzed C code. Only a limited interactive mode to edit proof scripts and (re)run the proof for goals generated in the current session is provided. This is inconvenient for an intensive usage of Frama-C on large C codebases with ACSL annotations. *Code editing*—typically to add or update ACSL annotations—must be done in a separate text editor with no support for navigation through ACSL constructs. In addition, multiple runs of Frama-C—currently executed in a separate console—are often necessary to complete deductive verification with WP. Running the proof from an IDE would be far more convenient.

Moreover, advanced use of Frama-C requires careful configuration of analysis options, such as command-line options to configure possible interactions between several plugins or options to tune the parsing and processing of targeted C code. The usual practice is to configure these options in a Makefile after getting enough knowledge about option ordering, priorities and code structure. An IDE dedicated to Frama-C, providing a configuration file with patterns of preconfigured options for common analyses, would greatly facilitate configuration setup.

4.2 The IDE Design

The Language Server Protocol (LSP), devised by Microsoft [34], provides a solution to avoid reimplementing the same programming-language features for every code editor and for each language. The LSP specifies a structured communication system between a client (the text editor) and a server that provides language-specific features. This communication is standardized by several message formats built on top of JSON-RPC[3]

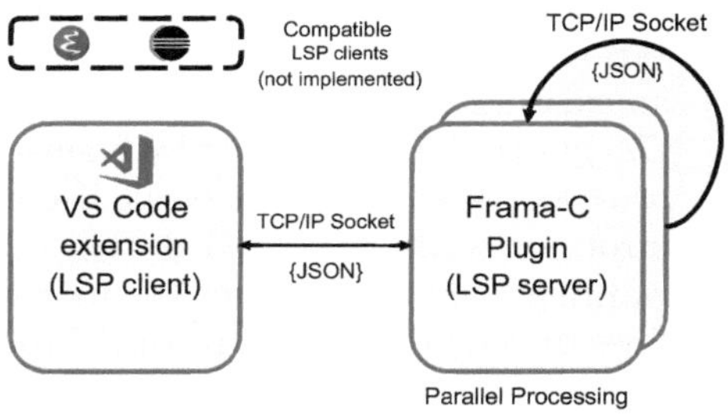

Fig. 3. VS Code extension for C/ACSL.

LSP Server. We implement the C/ACSL LSP server as a new Frama-C plugin, Frama-C-lsp, in order to benefit from the full set of functionalities provided by the Frama-C API for handling C and ACSL. The design of this new LSP plugin constitutes the main contribution of our IDE solution. The plugin implements parallel processing of several Frama-C sessions (cf. right side of Fig. 3). A permanent handler Frama-C session connects to the LSP client through TCP/IP sockets locally on the host machine. It initiates a permanent listening loop on a dedicated connection port to manage incoming and outgoing requests.

Upon receiving a request, the handler session invokes temporary Frama-C sessions with appropriate options to produce the required results. These temporary sessions are executed transparently in the background and send their results back to the permanent handler session, which in turn forwards the response to the client. Communication between the handler session and the temporary Frama-C sessions is also performed via local TCP/IP sockets using the JSON format.

If an error occurs during the execution of a temporary session, the handler catches the exception and returns an appropriate LSP error response. This server architecture avoids any interference with Frama-C's internal code representation throughout IDE usage. Incremental code parsing is handled seamlessly by spawning a new Frama-C session for each request. Our concerns about degraded user

[3] Remote Procedure Call protocol: https://www.jsonrpc.org/.

experience due to significant execution latency were dissipated after several trials, which demonstrated efficient execution performance. Overall, the user experience is largely enhanced compared to running multiple sessions of Frama-C on command line through a separate console.

LSP Client: VS Code Editor. We implement the LSP client as a VS Code extension called `acsl-lsp`[4]. Other editors such as Emacs and Eclipse are also LSP compatible and may implement an equivalent LSP client. In addition to a few useful LSP features, we implement in our VS Code extension other handy features specific to the use of Frama-C:

- *Syntax highlighting*: VS Code allows to apply syntax highlighting based on TextMate (TM) grammar. It allows to define patterns and scopes to highlight various syntax elements such as keywords, strings, and comments. We extend an existing syntax highlighting solution for C and C++ [5].
- *Diagnostics.* A file parsing by Frama-C is performed each time a file is opened or saved. We implement event listeners in our Frama-C plugin to catch any errors or warnings raised during code parsing. All diagnostics (errors and warnings) are displayed to the user in the form of squiggles at corresponding lines of code (and listed in the "Problems" tab).
- *Goto definition/declaration.* The editor moves the focus and navigates to the definition/declaration of symbols within the code. A proper information message is displayed if no definition/declaration is found.
- *Display-CIL.* Display the normalized C program pre-processed by Frama-C.
- *Display-CallGraph.* Display the call-graph computed by the callgraph plugin.
- *Display-Metric.* Display code metrics computed by the metrics plugin.
- *Display-ProofObligation.* Display the proof context(s) computed by WP plugin for a selected ACSL annotation if applicable.
- *Prove*: Trigger proof attempts with WP plugin given a function name, an ACSL annotation label and a timeout in seconds. The proof verdict is displayed in the "Goals" view of the editor.
- *SmokeTests*: Trigger proof attempts with WP plugin for automatically generated smoke tests. Their proof verdict is displayed in the "Goals" view of the editor.

5 Proof Effort and Productivity

In software development, the monitoring of key metrics is useful to optimize code quality, reduce maintenance costs, predict future development efforts and detect deviations from best coding practices. The key metrics for our project during the last five years are summarized in Fig. 4. All metrics are collected on Linux (Ubuntu) machines with 32Gb RAM and 12 cores CPU.

[4] https://marketplace.visualstudio.com/items?itemName=innov-org.acsl-lsp.

[5] https://marketplace.visualstudio.com/items?itemName=ms-vscode.cpptools.

Year	FCv	C	ACSL	Metapr.	Goals	Scripts (auto)	Proof time	GenScript time
2021	25.0	7 175	26 700	54	70 335	2 383 (0)	10 h 07 m	-
2022	25.0	8 211	44 600	48	74 800	936 (0)	09 h 10 m	-
2023	27.1	7 974	27 677	54	82 890	738 (0)	04 h 10 m	-
2024	29.0	9 877	29 638	53	90 477	459 (289)	07 h 11 m	08 h 30 m
2025	30.0	11 223	31 223	58	96 736	913 (647)	06 h 58 m	03 h 55 m

Fig. 4. Main proof metrics for selected milestone versions, indicating: the year; version of Frama-C; size (loc) of C code and ACSL specs; numbers of metaproperties, of all generated proof goals, of goals proved with all scripts (and with automatically generated ones if any); time of complete proof and of GenScript.

5.1 Scope Extension Effort

The initial version in 2021 was proved using Frama-C 25.0. We extended the scope of analyzed C code from 7 175 lines of code up to 11 223 lines in a recent version proved with Frama-C 30.0. The size of ACSL annotations reached a peak of five times the size of analyzed code in 2022. Refactoring of redundant ACSL annotations with global preprocessed macros reduced the size of ACSL annotations to approx. three times the size of analyzed code. Metaproperties remained stable in several consecutive evaluation projects. This confirms that well-designed metaproperties remain relevant without being affected by extensions of the scope of analyzed code. The number of generated proof goals by Frama-C/WP kept a quite linear progression w.r.t. the size of analyzed code. Careful design of ghost code and ACSL annotations [21] along with code refactoring of complex functions allowed to reduce the number of necessary proof scripts (from ∼3% in 2021 to ∼1% in 2025).

5.2 Automation of Proof Script Creation and GenScript

An important concern identified since the first versions of our project was the prohibitive effort to devise and maintain a large number of manual proof scripts (2 383 in 2021). Even if the number of necessary scripts is relatively small compared to the total number of proof goals (and decreased since 2021), it still requires an important effort to maintain them. Proof scripts are fragile and need to be updated even for limited changes in the code and each new version of Frama-C (cf. Sect. 3).

Following our request to the Frama-C developers, a new feature (referred to as GenScript) has been introduced since Frama-C 29.0 (released in 2024) to automatically generate proof scripts. The generation is based on specific user-provided ACSL annotations called *(proof) strategies* [19] that basically indicate a sequence of tactics to try for certain goals. We achieved a very promising automation rate with 289 proof scripts generated automatically out of 459 in 2024 (and 647 out of 913 in 2025). The main advantage of automatically generated scripts is that their generation from strategies is *less fragile* and less dependent on minor code changes or the used Frama-C version. *The effort to design proof strategies is spent only once.*

Our attempts to further exploit automatic script generation are currently limited by some missing features in the current implementation of GenScript that we discuss below in this section. Ideally, there should be no manual proof scripts to maintain in order to be able to fully integrate the formal verification approach with Frama-C/WP in a continuous integration process.

It is interesting to assess and further improve the efficiency of GenScript to generate proof scripts. Currently, the generation time (3 h 55 min spent in our latest milestone) is incomparable with numerous days of human effort that were required to design the scripts manually.

5.3 A Closer Look at Proof Tactics

Figure 5 shows proof tactics usage statistics in our proof scripts for 2023 and 2025 versions, before and after integration of GenScript.

The `unfold` and `split` tactics are the most used in proof scripts. The `instance` tactic—not yet supported by GenScript—is probably the most complicated to apply (the user should correctly instantiate a quantified variable), so a relatively large number of its applications confirms the complexity of proof goals. A large number of applications of other tactics (such as `overflow` or `filter`) is due to a good automation and a clear role.

The ratio column shows that, in average, 1.7 distinct tactics are used in a proof script in both cases, with and without GenScript. However, the total script length becomes relatively high (7.2 tactic applications per script in average) when some scripts are generated automatically with GenScript, while this ratio is lower (4.6 tactic applications per script) for manually edited scripts. This is mainly due to a limited control over multiple applications of tactics with the currently available GenScript implementation. Based on our experimental assessment, we

Tactic	filter	overflow	split	modmask	unfold	shift	bittestrange	cut	range	instance	ratio
All occurr. 2023/fc27.1	179	238	1167	82	1323	68	6	130	2	178	4.6
Scripts 2023/fc27.1	86	128	452	35	315	43	6	63	2	105	1.7
All occurr. 2025/fc30.0	535	998	2043	90	2341	215	36	104	191	210	7.2
Scripts 2025/fc30.0	219	102	531	48	367	58	8	72	63	124	1.7
Auto level 2025/fc30.0	▮▮▮	▮▮▯	▮▯▯	▮▮▯	▮▮▯	▮▮▯	▮▮▯	▮▮▯	▮▮▯	▯▯▯	

Fig. 5. Total number of occurrences of each tactic in proof scripts (line All occurr.) and number of scripts that use at least one occurrence of the tactic (line Scripts), shown for 2023 and 2025 versions. The last column (ratio) shows the average number of tactic applications per script (line All occurr.), and the average number of different tactics per script (line Scripts). The last line indicates the current level of automation with GenScript (estimated by the authors).

provide (in the last line of Fig. 5) a level of satisfaction with the automation currently provided by GenScript for each tactic. An automation of `instance` and a better control over generated scripts for other tactics (in particular for `split`) should allow us to reach *full automation of proof script generation.*

5.4 Getting Proof Verdicts as Fast as Possible

Proof time for all proof goals depends on several parameters. The main parameter remains the complexity of analyzed code. A switch from a little-endian architecture (in 2022) to a big-endian one (in 2023), easier for the solver, reduced the proof time by ~40 min. Code refactoring to reduce *cyclomatic complexity* of a few functions also contributed to improve proof efficiency.

Time of a full proof session of Frama-C/WP over all source files at once is 6 h 58 m for the latest version of the project. This duration is quite high and allows only daily proof runs to get a complete full verdict of proof status. In practice, *we rely on distributed proof sessions on several machines,* each one running Frama-C/WP over only one source file. This distributed (parallel) computation allows us to get a whole proof verdict in less than 30 min, which is better suited for a productive continuous integration and development process.

6 Proof Efficiency and Performance

Frama-C/WP offers a detailed reporting of proof results in json format. For each proof goal, the time spent by Qed and Alt-Ergo to discharge the goal is reported. If a proof script is used to discharge a proof goal, the proof verdict and proof time are reported for each subgoal generated by application of proof tactics. This section highlights some aspects of the project related to proof performance.

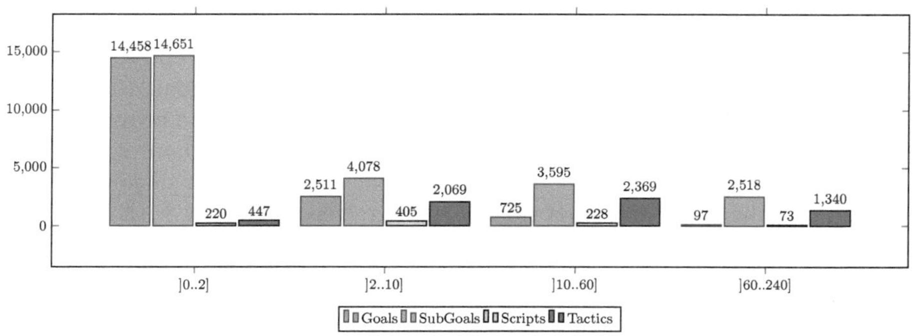

Fig. 6. Number of proof goals and corresponding scripts classified per ranges of proof time.

6.1 Number of Proof Goals

Proof performance depends on the number of goals to be proven. Some design choices at the level of ACSL annotations have a direct impact on the number of generated proof goals and time required to prove them.

Case of Lemmas. Even if lemmas are proved once and used in different proof contexts, we noticed decreased proof performance with lemmas since solvers check the applicability of lemmas in many irrelevant proof contexts (see Sect. 3.2). Hence, we favor assertions to insert intermediate properties only when needed.

Case of Metaproperties. More than half (60%) of the proof goals are generated for metaproperties. The general treatment of a metaproperty with a reading or writing context consists in generating an assertion at each relevant (resp., read or write) instruction in target functions. For complex functions, this treatment can sometimes become inefficient due to a large number of assertions and a complex logic inside the function. In such cases, we rely on a more efficient alternative based on **`assigns...\from...`** clauses. Indeed, an **`assigns...\from...`** clause gives the lists of variables that the function is allowed, resp., to modify and to read. In some cases, it provides sufficient information to justify some metaproperties at the call site, provided that **`assigns...\from...`** clauses are proved. The **`assigns...`** part is directly proved by WP, but the **`\from ...`** part is not yet handled. So, we add (very simple) metaproperties to prove the **`\from ...`** clauses—that is, non-reading of any non-listed variables—in targeted functions. This solution allowed us (again in 2023) to reduce the proof time by 1 h and to avoid costly proof goals inside complex functions.

Case of Scripts. The majority (80%) of proof goals are discharged by Qed alone with negligible reported time (0 s). This is an important achievement of proof automation with Frama-C/WP, but the remaining goals still represent a large number. In this section, we focus only on the (other 20% of) goals that require (in addition to Qed) an SMT prover (in our case, Alt-Ergo) to be proven. We set a global timeout per goal to 240 s to favor automatic proof without scripts.

We consider a partitioning of the set of proof goals into four disjoint subsets according to the necessary proof time to discharge each of them, with four intervals: $]0s..2s]$, $]2s..10s]$, $]10s..60s]$, $]60s..240s]$. Figure 6 shows with blue bars the number of goals in each subset for our latest version. The difference between the number of goals and the number of proof scripts (in brown) gives the number of goals automatically proved with Alt-Ergo without scripts. Only $97 - 73 = 24$ goals are automatically proved by Alt-Ergo in the time range $]60s..240s]$. We still favor automatic proof for them to avoid extra proof scripts to maintain.

220 proof scripts allow to prove corresponding proof goals in the $]0s..2s]$ time range. About half of proof scripts (405) prove corresponding proof goals in the $]2s..10s]$ time range. It is important to monitor the efficiency of proof scripts

because costly scripts are often difficult to maintain and to fix when they are broken.

Proof tactics that split the proof goal into subgoals are often useful but must be used with care to avoid duplication of proofs over generated subgoals instead of reducing the complexity of initial goals. The difference between the number of goals and the number of subgoals (in red) in Fig. 6 illustrated the number of extra subgoals created by tactics.

The ratio between the number of scripts and the number of tactics (black bars) is a relevant indicator on the manual effort required to create proof scripts. While this ratio is reasonable (2 tactics in average) for efficient scripts in the $]0s..2s]$ proof time range, it is dramatically higher for the other scripts (resp., 5, 10, 18 tactics in average per script in $]2s..10s]$, $]10s..60s]$, $]60s..240s]$ time ranges). As a guideline, *the sum of proof time for subgoals of each goal should be less than the considered timeout*, which is the case for our scripts. Otherwise, trying a greater solver timeout could be more efficient (and avoid a script).

6.2 Proof Time

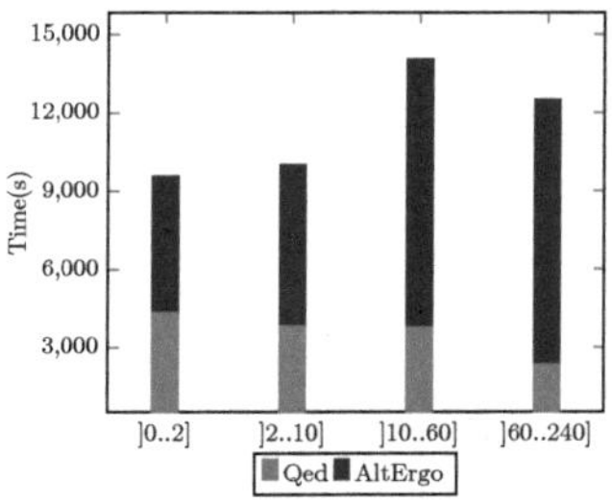

Fig. 7. Proof goals partitioned w.r.t. ranges of proof time.

In our latest version, it takes less than 15 min for Frama-C/WP to generate proof goals and for MetAcsl to instantiate metaproperties. This is negligible compared to 6 h 58 m of total proof time in one Frama-C session. Let us consider the same four subsets of goals as in Sect. 6.2 and focus on the reported CPU time to prove goals (without including time for their generation). The total CPU time to prove all goals (12 h 53 m) is higher than the real time (6 h 58 m) also because we let Frama-C run solvers for some goals in parallel on up to 9 cores. Figure 7 shows the total CPU time spent by Qed and Alt-Ergo to discharge each subset of goals in our latest version.

We observe that the time spent by Qed is not negligible. Unlike for Alt-Ergo, it decreases for costly provable goals but still it is not linearly proportional to the number of goals in each subset. Qed simplifications are very efficient in average for simpler subgoals in the time range $]0s..2s]$ compared to more costly provable goals. We also observed that it is beneficial to disable one Qed simplification (`-no-qed-pruning`) to avoid some costly (and unproductive) simplification attempts. Another observation is that, above 10 s of proof time for each goal, Alt-Ergo takes a long time to discharge the goals. This is correlated to the usage of proof scripts and emphasizes the need to better diagnose the challenging proof contexts for Alt-Ergo. The last but not least observation is that the more the scripts are complex, the more they are costly to prove and penalize the global proof efficiency. Interestingly, it takes negligible CPU time to prove 80% of proof goals discharged by Qed only; and it takes much less CPU time (2 h 40 m) to prove

14 458 goals in the time range $]0s, 2s]$ than the CPU time (10 h 13 m) to prove the remaining 3 333 goals in the other time ranges together.

7 Conclusion

Thanks to the impressive progress made by researchers and tool developers in the past decades, formal verification has become feasible in an increasing number of industrial projects today [27], using various verification tools, for instance, KeY [9,20], VerCors [6,35], Frama-C [24], SPARK [15,25], VCC [32], Dafny [14, 33], VeriFast [29]. However, an initial success—for which investing an important human effort in an industrial verification project can be acceptable—does not guarantee the sustained use of formal verification in long-term projects. This paper presents an experience report on the path towards sustained industrial use of formal verification for the certification of a JavaCard Virtual Machine at Thales. Our achievements have contributed to reducing proof maintenance effort and increasing its performance, and the proposed solutions could benefit to researchers and practitioners in the formal methods community and further enhance proof tools and methodologies.

A finalization of the initiated automatization tasks is mandatory in order to completely endorse the adoption of deductive verification in an industrial continuous development process. In the context of our project, future work includes development and adoption of new versions of GenScript, the automatic script generation mechanism in Frama-C/WP. With a better control over patterns of tactic application and the support of new tactics, the ultimate objective of a 100% automation for proof script generation should be achievable. The proof maintenance effort will be completely under control. The development of more efficient and flexible memory models should reduce the need for code transformations. Another future task would be an advanced study of costly provable goals at the level of SMT solvers in order to suggest possible enhancement of Alt-Ergo performance on our use cases. Beyond the global proof performance improvement, the objective is to examine the ultimate possibility to refine the trade-off between our ACSL design choices and proof performance. Finally, we hope that our open-source IDE solution for Frama-C/WP would benefit to a large number of Frama-C users. It may facilitate the usage of the verification tool by non-expert engineers and students, and we warmly welcome all relevant contributions to further improve it.

Acknowledgement. Part of this work was supported by ANR (grants ANR-22-CE39-0014, ANR-22-CE25-0018, ANR-24-ASM2-0001). The development of the Frama-C-lsp plugin was initiated by Djamila Mohamed during her internship at Thales. The authors thank the Frama-C team members for their support, and the anonymous reviewers for their helpful comments.

References

1. ANSSI: Common criteria evaluation certificate delivered on 2021-10-14 (ref. ANSSI-CC-2021/42, EAL6+). https://messervices.cyber.gouv.fr/visas/ANSSI-CC-2021-42-certificat.pdf
2. ANSSI: Common criteria evaluation certificate delivered on 2022-04-22 (ref. ANSSI-CC-2020/33-S01, EAL7). https://messervices.cyber.gouv.fr/visas/ANSSI-CC-2020-33-S01-certificat.pdf
3. ANSSI: Common criteria evaluation certificate delivered on 2023-08-28 (ref. ANSSI-CC-2023/31, EAL6+). https://messervices.cyber.gouv.fr/visas/ANSSI-CC-2023-31-certificat.pdf
4. ANSSI: Common criteria evaluation certificate delivered on 2023-11-13 (ref. ANSSI-CC-2023/45, EAL7). https://messervices.cyber.gouv.fr/visas/ANSSI-CC-2023-45-certificat.pdf
5. ANSSI: Common criteria evaluation certificate delivered on 2025-09-29 (ref. ANSSI-CC-2025/32, EAL6+). https://messervices.cyber.gouv.fr/visas/ANSSI-CC-2025-32-certificat.pdf
6. Armborst, L., et al.: The VerCors verifier: a progress report. In: Proceedings of the 36th International Conference on Computer Aided Verification (CAV 2024). LNCS, vol. 14682, pp. 3–18. Springer (2024). https://doi.org/10.1007/978-3-031-65630-9_1
7. Baudin, P., Bobot, F., Correnson, L., Dargaye, Z., Blanchard, A.: WP Plug-in Manual (2025). https://frama-c.com/download/frama-c-wp-manual.pdf
8. Baudin, P., et al.: ACSL: ANSI/ISO C specification language (2025). http://frama-c.cea.fr/acsl.html
9. Beckert, B., Sanders, P., Ulbrich, M., Wiesler, J., Witt, S.: Formally verifying an efficient sorter. In: Proceedings of the 30th International Conference on Tools and Algorithms for the Construction and Analysis of Systems (TACAS 2024). LNCS, vol. 14570, pp. 268–287. Springer (2024). https://doi.org/10.1007/978-3-031-57246-3_15
10. Blanchard, A., Bobot, F., Baudin, P., Correnson, L.: Chapter 4: formally verifying that a program does what it should: the WP plug-in. In: Kosmatov, N., Prevosto, V., Signoles, J. (eds.) Guide to Software Verification with Frama-C. Core Components, Usages, and Applications, pp. 187–261. Computer Science Foundations and Applied Logic Book Series, Springer (2024). https://doi.org/10.1007/978-3-031-55608-1_4
11. Blanchard, A., Correnson, L., Djoudi, A., Kosmatov, N.: No smoke without fire: Detecting specification inconsistencies with Frama-C/WP. In: Proc. of the 18th International Conference on Tests and Proofs (TAP 2024), co-located with the 26th International Symposium on Formal Methods (FM 2024). LNCS, vol. 15153, pp. 65–83. Springer (2024). https://doi.org/10.1007/978-3-031-72044-4_4
12. Blanchard, A., Marché, C., Prevosto, V.: Chapter 1: Formally expressing what a program should do: The ACSL language. In: Kosmatov, N., Prevosto, V., Signoles, J. (eds.) Guide to Software Verification with Frama-C. Core Components, Usages, and Applications, pp. 3–80. Computer Science Foundations and Applied Logic Book Series, Springer (2024). https://doi.org/10.1007/978-3-031-55608-1_1
13. Bühler, D., Maroneze, A., Perrelle, V.: Chapter 3: abstract interpretation with the EVA plug-in. In: Kosmatov, N., Prevosto, V., Signoles, J. (eds.) Guide to Software Verification with Frama-C. Core Components, Usages, and Applications, pp. 131–186. Computer Science Foundations and Applied Logic Book Series, Springer (2024). https://doi.org/10.1007/978-3-031-55608-1_3

14. Cassez, F., Fuller, J., Quiles, H.M.A.: Deductive verification of smart contracts with Dafny. In: Proceedings of the 27th International Conference on Formal Methods for Industrial Critical Systems (FMICS 2022). LNCS, vol. 13487, pp. 50–66. Springer (2022). https://doi.org/10.1007/978-3-031-15008-1_5
15. Cluzel, G., Georgiou, K., Moy, Y., Zeller, C.: Layered formal verification of a TCP stack. In: Proceedings of the IEEE Secure Development Conference (SecDev 2021), pp. 86–93. IEEE (2021). https://doi.org/10.1109/SecDev51306.2021.00028
16. Common criteria: common criteria for information technology security evaluation. Part 3: security assurance components. Tech. rep., CCMB-2017-04-003 (2017). https://www.commoncriteriaportal.org/files/ccfiles/CCPART3V3.1R5.pdf
17. Conchon, S., et al.: The Alt-Ergo automated theorem prover. http://alt-ergo.lri.fr
18. Correnson, L.: QED. Computing what remains to be proved. In: NASA Formal Methods (NFM 2014). LNCS, vol. 8430, pp. 215–229. Springer (2014). https://doi.org/10.1007/978-3-319-06200-6_17
19. Correnson, L., Blanchard, A., Djoudi, A., Kosmatov, N.: Automate where automation fails: Proof strategies for Frama-C/WP. In: Proc. of the 30th International Conference on Tools and Algorithms for the Construction and Analysis of Systems (TACAS 2024), Held as Part of the European Joint Conferences on Theory and Practice of Software (ETAPS 2024). LNCS, vol. 14570, pp. 331–339. Springer (2024). https://doi.org/10.1007/978-3-031-57246-3_18
20. de Boer, M., de Gouw, S., Klamroth, J., Jung, C., Ulbrich, M., Weigl, A.: Formal specification and verification of JDK's identity hash map implementation. Formal Aspects Comput. **35**(3), 18:1–18:26 (2023). https://doi.org/10.1145/3594729
21. Djoudi, A., Hána, M., Kosmatov, N.: Formal verification of a JavaCard virtual machine with Frama-C. In: Proceedings of the 24th International Symposium on Formal Methods (FM 2021). LNCS, vol. 13047, pp. 427–444. Springer (2021). https://doi.org/10.1007/978-3-030-90870-6_23, long version available at https://nikolai-kosmatov.eu/publications/djoudi_hk_fm_2021.pdf
22. Djoudi, A., et al.: A bottom-up formal verification approach for common criteria certification: application to JavaCard virtual machine. In: Proceedings of the 11th European Congress on Embedded Real-Time Systems (ERTS 2022) (2022)
23. Djoudi, A., Hána, M., Kosmatov, N.: Chapter 16: Proof of security properties: Application to JavaCard virtual machine. In: Kosmatov, N., Prevosto, V., Signoles, J. (eds.) Guide to Software Verification with Frama-C. Core Components, Usages, and Applications, pp. 659–683. Computer Science Foundations and Applied Logic Book Series, Springer (2024). https://doi.org/10.1007/978-3-031-55608-1_16
24. Dordowsky, F.: An experimental study using ACSL and Frama-C to formulate and verify low-level requirements from a DO-178C compliant avionics project. Electronic Proc. Theor. Comput. Sci. **187**, 28–41 (2015). https://doi.org/10.4204/EPTCS.187.3
25. Dross, C., Moy, Y.: Auto-active proof of red-black trees in SPARK. In: Barrett, C., Davies, M., Kahsai, T. (eds.) NFM 2017. LNCS, vol. 10227, pp. 68–83. Springer, Cham (2017). https://doi.org/10.1007/978-3-319-57288-8_5
26. Filliâtre, J., Paskevich, A.: Why3 – where programs meet provers. In: Proceedings of the 22nd European Symposium on Programming (ESOP 2013). LNCS, vol. 7792, pp. 125–128. Springer (2013). https://doi.org/10.1007/978-3-642-37036-6_8
27. Hähnle, R., Huisman, M.: Deductive software verification: From pen-and-paper proofs to industrial tools. In: Computing and Software Science – State of the Art and Perspectives, LNCS, vol. 10000, pp. 345–373. Springer (2019). https://doi.org/10.1007/978-3-319-91908-9_18

28. ISO/IEC JTC1/SC22/WG14: 9899:TC3: Programming languages—C (2007). http://www.open-std.org/jtc1/sc22/wg14/www/docs/n1256.pdf
29. Jacobs, B., Smans, J., Philippaerts, P., Vogels, F., Penninckx, W., Piessens, F.: VeriFast: a powerful, sound, predictable, fast verifier for C and Java. In: Proceedings of the Third International Symposium on NASA Formal Methods (NFM 2011). LNCS, vol. 6617, pp. 41–55. Springer (2011). https://doi.org/10.1007/978-3-642-20398-5_4
30. Kirchner, F., Kosmatov, N., Prevosto, V., Signoles, J., Yakobowski, B.: Frama-C: a software analysis perspective. Formal Asp. Comput., 1–37 (2015). https://doi.org/10.1007/s00165-014-0326-7
31. Kosmatov, N., Prevosto, V., Signoles, J. (eds.): Guide to software verification with frama-C. Core components, usages, and applications. Comput. Sci. Found. Appl. Logic Book Ser. Springer (2024).https://doi.org/10.1007/978-3-031-55608-1
32. Leinenbach, D., Santen, T.: Verifying the microsoft Hyper-V hypervisor with VCC. In: Proceedings of the Second World Congres on Formal Methods (FM 2009). LNCS, vol. 5850, pp. 806–809. Springer (2009).https://doi.org/10.1007/978-3-642-05089-3_51
33. Leino, K.R.M.: Program Proofs. The MIT Press (2023)
34. Microsoft: Language server protocol. https://microsoft.github.io/language-server-protocol/
35. Oortwijn, W., Huisman, M.: Formal verification of an industrial safety-critical traffic tunnel control system. In: Ahrendt, W., Tapia Tarifa, S.L. (eds.) IFM 2019. LNCS, vol. 11918, pp. 418–436. Springer, Cham (2019). https://doi.org/10.1007/978-3-030-34968-4_23
36. Oracle: Java card platform: runtime environment specification, classic edition, Version 3.1. Tech. rep., Oracle, Oracle (2021). https://docs.oracle.com/javacard/3.1/related-docs/JCCRE/JCCRE.pdf
37. Oracle: Java card platform: virtual machine specification, classic edition, version 3.1. Tech. rep., Oracle, Oracle (2021). https://docs.oracle.com/javacard/3.1/related-docs/JCVMS/JCVMS.pdf
38. Rice, H.G.: Classes of recursively enumerable sets and their decision problems. Trans. Amer. Math. Soc. **74**, 358–366 (1953)
39. Robles, V., Kosmatov, N., Prevosto, V., Rilling, L., Le Gall, P.: MetAcsl: specification and verification of high-level properties. In: Vojnar, T., Zhang, L. (eds.) TACAS 2019. LNCS, vol. 11427, pp. 358–364. Springer, Cham (2019). https://doi.org/10.1007/978-3-030-17462-0_22
40. Robles, V., Kosmatov, N., Prevosto, V., Rilling, L., Le Gall, P.: Methodology for specification and verification of high-level properties with MetAcsl. In: 9th IEEE/ACM International Conference on Formal Methods in Software Engineering (FormaliSE 2021), pp. 54–67. IEEE (2021). https://doi.org/10.1109/FormaliSE52586.2021
41. Thales Press Release: Thales launches Europe's first certified smartcard ready for the quantum age. https://www.thalesgroup.com/en/news-centre/press-releases/thales-launches-europes-first-certified-smartcard-ready-quantum-age

Shift-Left Requirements Verification: Integrating LLMs and Formal Methods for Automotive Systems

Zipong Lim[1,2], Bozhi Wu[3](✉), Yon Shin Teo[2], Shang-wei Lin[4], and Yi Li[1]

[1] Nanyang Technological University, Singapore, Singapore
zipong001@e.ntu.edu.sg, yi_li@ntu.edu.sg
[2] AUMOVIO Singapore Pte. Ltd., Singapore, Singapore
{zi.pong.lim,yon.shin.teo}@aumovio.com
[3] AUMOVIO-NTU Corporate Lab, Singapore, Singapore
bozhi001@e.ntu.edu.sg
[4] Singapore Institute of Technology, Singapore, Singapore
shangwei.lin@singaporetech.edu.sg

Abstract. Requirement defects are a major source of late-stage failures in automotive systems, yet rigorous validation is rarely applied during early development. While formal methods offer strong guarantees, their adoption at the requirements level is limited by high formalization cost and expertise barriers. We present an industry-oriented, shift-left verification approach that integrates Large Language Models (LLMs) with formal methods to enable requirements-level validation. Requirements are classified and decomposed by LLMs, translated into CSP system models and assertions, and refined through a CEGAR-inspired loop using the FDR4 model checker. Validation is decomposed into requirement–assertion pairs supported by natural-language back-translations and confidence scores, preserving expert control without manual formal modeling. Domain knowledge–based validation further leverages historical defect data to identify implicit requirement gaps. We evaluate the approach on three real-world automotive case studies. Results show high automation for small-to-medium systems (80–100% synthesis success for up to $\sim$ 60 requirements), effective expert validation guided by LLM confidence estimates, and 100% detection of known historical defects alongside 22 novel gaps. The workflow completes within 10–35 min per project at negligible cost ($<$ 8 per project), with limited expert effort. Our results demonstrate that LLM+formal method hybridization can provide scalable, rigorous, and industrially viable requirements-level verification, supporting practical shift-left adoption in automotive systems.

Keywords: Requirements Engineering · Requirements Verification · Formal Methods · Large Language Models · Model Checking · CSP · Automotive System

A. Sampaio and M. Stoelinga (Eds.): FM 2026, LNCS 16557, pp. 631–650, 2026.
https://doi.org/10.1007/978-3-032-26220-2_31

1 Introduction

As a Tier-1 automotive supplier, **Aumovio** (formerly Continental Automotive) operates within a complex supply chain in which system requirements are primarily provided by Original Equipment Manufacturers (OEMs). Based on these documents, Aumovio develops safety-critical automotive systems that must meet stringent correctness and compliance requirements. However, at the requirements stage, no formal system model typically exists. Although requirement specifications are extensive and carefully written, they frequently contain ambiguities, inconsistencies, and implicit assumptions that remain undetected until later development phases. Without early-stage validation, such latent defects propagate downstream and often surface only during integration testing or system validation, where their correction incurs substantial cost and risk.

Automotive development traditionally follows the V-model, in which requirements engineering occurs at the earliest phase, while verification and validation activities are deferred to later stages. Decades of industrial experience show that defects discovered during integration or system testing are one to two orders of magnitude more expensive to fix than those detected during requirements engineering [3, 24]. Importantly, many late-stage failures are not caused by implementation errors, but by deficiencies in the requirements themselves, including vague formulations, hidden inconsistencies, missing corner cases, and undocumented assumptions. These observations motivate a shift-left approach that brings rigorous verification techniques into the requirements phase.

Despite its importance, requirements verification faces a fundamental tension between scalability and assurance. Industrial specifications are large, informal, and evolving, making manual formalization prohibitively expensive, while safety-critical systems demand correctness guarantees that cannot be compromised.

On one hand, formal methods offer strong assurance by enabling mathematically precise reasoning about system behavior, and they have been successfully applied at the software and design levels in automotive development [6, 8, 14]. However, their adoption at the requirements stage remains challenging due to the *formalization barrier*: manual translation of informal text into formal specifications requires dual expertise in both domain knowledge and formal methods, which is a combination rarely available in industrial settings. Domain experts with deep system knowledge often lack formal methods training, while formal methods specialists often lack domain expertise. This skills gap makes formal verification approaches difficult to adopt at the requirements phase. On the other hand, recent works [1, 2, 12] leverage Large Language Models (LLMs) to automate requirements analysis, addressing the scalability of the problem. While LLMs excel at processing large volumes of text, they provide no inherent correctness guarantees: their outputs may be fluent and plausible yet semantically incorrect, making them unsuitable as the sole basis for verification in safety-critical contexts. These complementary strengths and weaknesses suggest that neither formal methods nor LLMs alone can adequately address requirements-level verification. Instead, an effective solution must combine the automation and scalability of LLMs with the soundness and rigor of formal methods.

Therefore, we propose a practical hybrid methodology that integrates LLMs with formal methods to enable scalable yet trustworthy requirements verification in industrial automotive settings. Our key idea is to treat LLMs as automation assistants, responsible for accelerating transformation tasks, while delegating all correctness-critical decisions to established formal techniques, specifically Communicating Sequential Processes (CSP) [13] and the FDR4 model checker [10].

The workflow supports *shift-left* verification where free-form natural-language requirements are automatically transformed into formal assertions, checked for correctness and consistency, and used to synthesize a verified system model before architectural or implementation decisions are finalized. Critically, the approach is human-in-the-loop: domain experts retain control over high-level validation while repetitive formal reasoning tasks are automated.

A central technical element is an iterative, counterexample-guided refinement loop inspired by CEGAR. Starting from an initial system model synthesized from validated assertions, all properties are checked in parallel using FDR4. When violations are detected, concrete counterexample traces are fed back to the LLM, which refines the model while preserving previously satisfied properties. To reduce latent requirement gaps, we integrate retrieval-augmented verification that leverages institutional knowledge such as historical defect reports, proactively identifying missing or underspecified requirements.

In summary, this paper makes the following contributions[1]:

- An end-to-end industrial workflow synthesizing formal CSP system models from free-form natural-language requirements through human-LLM-FM collaboration, employing a CEGAR-inspired refinement methodology where LLMs iteratively repair system models based on FDR4 counterexamples.
- A decomposed validation approach transforming the cognitively demanding task of validating complex system models into manageable assertion-requirement pair reviews, combined with retrieval-augmented verification exploiting historical defect knowledge.
- Empirical evaluation on real-world automotive projects demonstrating substantial reductions in expert effort and effective early defect detection.
- Successful trial deployment with an industry partner OEM, where our tool identified a better fix for a previously known issue, demonstrating practical value beyond defect detection to include solution recommendation.

2 Motivation

Automotive systems are typically developed under the V-model, where requirements engineering is expected to provide a correct foundation for all downstream tasks. In practice, however, requirement-related defects remain a major source of late-stage failures. Defects introduced during requirements engineering often

[1] Due to space constraints, additional details and examples can be found in the online appendix: https://sites.google.com/view/shift-left-requirement-supp.

propagate into implementation and are only discovered during integration testing or operation, where remediation costs are substantially higher. Despite informal reviews, requirements are rarely subjected to rigorous validation before being used for development, resulting in a persistent requirements validation gap.

Formal methods offer a principled means to bridge this gap by enabling mathematical validation of system behavior against requirements. However, their industrial adoption in automotive practice remains limited and restricted to later development stages, where verification is applied to artifacts such as models or code. Applying formal methods directly at the requirements level is therefore desirable for shift-left verification, but introduces several fundamental challenges.

Challenge 1: Formalization Barrier. Traditional formal methods rely on manually translating natural-language requirements into formal specifications, requiring expertise in both the application domain and formal modeling languages. In industrial automotive settings, such dual expertise is rarely available: domain experts typically lack formal methods training, while formal methods specialists often lack detailed system knowledge. This mismatch confines formal verification to late development stages and prevents its effective use during requirements engineering.

Challenge 2: Implicit Knowledge Problem. Requirements documents primarily capture *explicit* stakeholder intent and rarely document implicit assumptions, edge cases, or lessons learned from past failures. Consequently, even a formally verified specification may permit undesirable behaviors simply because critical domain constraints were never stated as requirements. In practice, organizations accumulate substantial domain knowledge in the form of historical defect reports, safety standards, and engineering experience, yet this knowledge remains largely tacit and disconnected from requirements-level validation.

Challenge 3: Formal Model Validity. These challenges converge on a fundamental question: how can the *validity* of a formal model be established? As illustrated in Fig. 1, formal specifications must reflect not only stated system requirements R but also relevant domain knowledge D. Traditional formal methods primarily address the $R \rightarrow S$ relationship, ensuring that specifications encode documented requirements, while often neglecting the $D \rightarrow S$ relationship. Many late-stage defects originate from this gap, where a specification satisfies all stated requirements but violates implicit domain constraints.

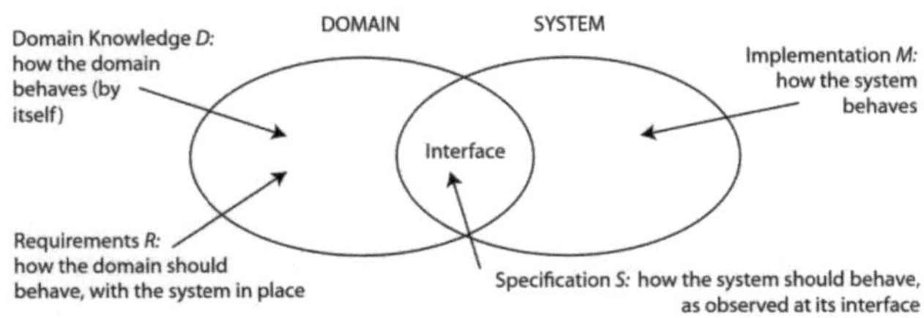

Fig. 1. Relationship between domain knowledge D, requirements R, specification S, and implementation M, adapted from [25].

Our Approach. To address these challenges, we propose an industry-oriented LLM+FM hybrid approach following the principle: LLMs for automation, formal methods for verification, humans for validation. For challenge 1, we decompose validation into assertion-requirement pair reviews with LLM-generated back-translations, shifting human expert effort from low-level formal reasoning to high-level semantic validation. For challenge 2, retrieval-augmented validation transforms tacit organizational knowledge into actionable verification properties. For challenge 3, we address both the $R \rightarrow S$ relationship (through decomposed assertion-requirement reviews) and the $D \rightarrow S$ relationship (through retrieval-augmented validation against historical defect knowledge).

3 Methodology

Our methodology consists of three phases that incrementally construct and validate a formal system specification, with human experts involved at key decision points, as illustrated in Fig. 2.

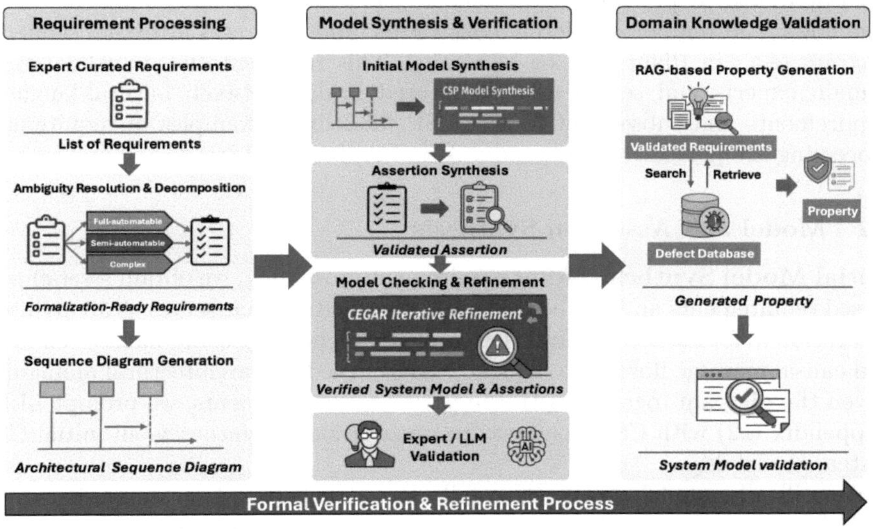

Fig. 2. Overview of the LLM+FM hybrid verification workflow.

3.1 Requirement Processing

OEM requirement documents typically contain heterogeneous content, including tables, flowcharts, diagrams, and free-form text. To extract concrete requirements from such multimodal artifacts, Aumovio employs an internally developed

AI-assisted toolchain based on Large Language Models (LLMs) and Vision–Language Models (VLMs), which converts complex documents into a list of textual requirements. Before formalization, however, these requirements must be further processed to ensure semantic clarity and modeling tractability.

Specifically, we employ an LLM (prompt provided in Appendix L.1) to classify each requirement according to its formalizability into three categories: *Fully Automatable (FA)* requirements that are clear and can be directly formalized; *Semi-Automatable (SA)* requirements that contain ambiguities requiring expert clarification; and *Complex (CX)* requirements that involve substantial domain complexity or multiple interacting concerns. The LLM output is subsequently reviewed and confirmed by domain experts, combining LLM automation efficiency with human validation for reliability. Ambiguities identified in SA requirements are resolved through expert clarification, while CX requirements are decomposed into simpler sub-requirements using a divide-and-conquer strategy. We now have a set of processed requirements that have been classified, disambiguated, and decomposed as needed, suitable for formal modeling.

After ambiguity resolution and decomposition, we construct an intermediate architectural abstraction in the form of a sequence diagram. At this stage, the LLM analyzes the processed requirements to identify system components, message flows, and temporal ordering constraints, and generates a textual sequence diagram (e.g., in PlantUML or Mermaid). This representation is validated by domain experts and serves as a structured bridge between natural-language requirements and subsequent formal CSP modeling. Examples of requirement processing are provided in Appendix A and Appendix B.

3.2 Model and Assertion Synthesis

Initial Model Synthesis. After requirement processing, we obtain a set of processed requirements and a validated sequence diagram that serves as an architectural scaffold for model synthesis. By explicitly identifying components, channels, and causal message flows, the sequence diagram reduces architectural ambiguity. Given the diagram together with the processed requirements, we prompt LLMs (Appendix L.2) with CSP^M syntax documentation to synthesize an initial CSP system model M_0.

Specifically, model construction follows a structured, diagram-guided workflow. First, project-specific datatypes and channels are declared to reflect the vocabulary introduced by the requirements. Next, CSP processes are synthesized to correspond to individual system components identified in the sequence diagram. Inter-component interactions are realized through channel synchronization, and the overall system behavior is defined as a parallel composition of these component processes with appropriate synchronization constraints. This ensures that the synthesized model preserves both the component boundaries and interaction semantics implied by the requirements and architectural context.

For requirements involving timing constraints (e.g., "within 200 ms"), they are abstracted into ordering constraints that are verifiable in untimed CSP. Rather than modeling explicit time, we preserve the essential causal dependencies between events using trace refinement combined with alphabet hiding. For example, in CS1 (Sect. 4), to capture the requirement that `relay.ON` must occur after `ignition.ON`, we define: `AlphaSpec_G3 = {| ignition, relay |}; SPEC_G3 = ignition.ON -> relay.ON -> SPEC_G3; assert SPEC_G3 [T= SYSTEM diff(Events, AlphaSpec_G3)`. Here, alphabet selection is performed by LLM based on the events referenced in each requirement, and alphabet hiding is used to restrict the refinement check to the relevant events, ensuring that the required ordering between `ignition.ON` and `relay.ON` is preserved in all system traces. During expert validation (Sect. 3.2), the back-translation explicitly describes which events are observed and which are hidden, enabling experts to verify that the chosen alphabet correctly captures the requirement's scope. If experts identify that relevant events are incorrectly hidden, the assertion is flagged for revision.

Assertion Synthesis. Before model checking can be performed, the processed requirements must be translated into formal assertions that can be verified against the synthesized system model. To this end, we employ an LLM to translate all processed requirements into CSP assertions in a single batch operation (prompt provided in Appendix L.3). Batch synthesis ensures consistency across assertions and alignment with the datatypes, channels, and process structure introduced during system model synthesis. The generated assertions typically fall into several categories, including trace refinement for ordering constraints, failure refinement for availability properties, deadlock freedom, and determinism.

Following assertion generation, we perform explicit syntax validation of both the initial system model M_0 and the synthesized assertion set using the FDR4 syntax checker. Reported syntax errors, including undefined channels, datatype mismatches, and malformed expressions, are collected and provided to the LLM together with the relevant code fragments and diagnostic messages. The LLM is instructed to correct these errors, and this syntax refinement loop is repeated until all errors are resolved or a maximum of 20 attempts is reached, yielding a syntactically valid system model M_1 and assertion set $\mathcal{A}_1$.

CEGAR-Inspired Iterative Model and Assertion Refinement. After a syntactically valid system model M_1 and assertion set $\mathcal{A}_1$ are obtained, we refine both the model and assertions using a counterexample-guided refinement process inspired by CEGAR. To this end, all assertions $a_i \in \mathcal{A}_1$ are first checked against the current model in parallel, with a timeout of 60 s per assertion. Based on the verification results, the assertions are classified into three groups: $\mathcal{A}_{\text{pass}}$, which contains satisfied assertions; $\mathcal{A}_{\text{fail}}$, which includes violated assertions together with their counterexample traces; and $\mathcal{A}_{\text{timeout}}$, which consists of assertions that exceed the timeout due to state-space explosion.

Assertions in $\mathcal{A}_{\text{pass}}$ require no further action and are treated as invariants that must be preserved throughout subsequent refinement iterations. In contrast, for assertions in $\mathcal{A}_{\text{fail}}$, we apply a CEGAR-inspired global refinement strategy that processes all failures simultaneously in order to avoid local optimization traps, as detailed in Algorithm 1. In each refinement step, the LLMREFINE function constructs a prompt (Appendix L.4) containing the current model, all failed assertions with their counterexamples and corresponding requirements, the set of passing assertions as constraints, and explicit refinement instructions. The LLM proposes updates to the model and the assertions (only when necessary), in a structured diff format. These updates are automatically parsed and applied to produce a refined model M' and assertion set $\mathcal{A}'$. After each update, all assertions are re-verified to detect potential regressions introduced by the refinement. This iterative refinement process continues until one of four termination conditions is met: (1) Success, when all assertions pass; (2) Oscillation, when the same pass–fail pattern recurs; (3) No progress, when the number of passing assertions does not increase for N consecutive iterations; or (4) Maximum iterations, when the limit is reached. Concrete refinement examples are provided in Appendix C.

Assertions in $\mathcal{A}_{\text{timeout}}$ are handled separately through LLM-guided state-space reduction. For each such assertion, we guide the LLM (Appendix L.4) to simplify the model by removing processes not involved in the assertion alphabet, abstracting datatypes to smaller value sets, eliminating internal events not visible to the property, and applying process inlining. The resulting simplified model is then re-verified against the assertion, with up to 10 iterations with increasingly aggressive simplifications. Additional details of the procedure are provided in Appendix D and L.5.

Expert Validation and Golden Standard Establishment. Once the refinement loop yields a system model satisfying all verifiable assertions, we validate assertion correctness through expert review. While refinement guarantees that the model satisfies the assertions ($M \models \mathcal{A}$), it does not ensure that the assertions faithfully capture the intent of the original requirements ($\mathcal{A} \equiv \mathcal{R}$). This validation addresses this gap through human–LLM collaboration.

For each verified assertion $a_i \in \mathcal{A}_{\text{verified}}$, experts review a structured package comprising the original requirement, the corresponding CSP assertion, an LLM-generated natural-language back-translation explaining what the assertion verifies (Appendix L.6), an LLM confidence score (0-1) with rationale (Appendix L.7), and the verification evidence. For example, from CS1 requirement G-2: "System shall provide four grille shutter opening degrees" was formalized as `SPEC_G2 = (actuator.FULL_OPEN -> SPEC_G2) [] (actuator.LEVEL_1 -> SPEC_G2) [] ...` with back-translation "This verifies all four degrees can be commanded" and confidence 0.85. Additional examples are in Appendix E.

Experts assess each assertion–requirement pair for semantic alignment, completeness, abstraction adequacy, and case coverage. Assertions may be *Approved*, *Modified* (triggering re-verification), or *Rejected*. After review, all approved assertions are consolidated into the golden standard specification $\mathcal{A}_{\text{golden}}$, together with a traceability matrix, coverage report, and documented limitations.

By decomposing validation into requirement–assertion pairs rather than validating the system model monolithically, this process enables clear semantic traceability, supports incremental expert review, and provides a robust and scalable validation suitable for industrial practice.

Algorithm 1 Counterexample-Guided System Model and Assertion Refinement

```
Input: Model M, assertions A
Output: Refined M* and A*, or stalemate report
iteration ← 0, no_progress ← 0, history ← ∅
while iteration < MAX_ITER do
    (A_pass, A_fail, A_timeout) ← ParallelCheck(M, A)
    if A_fail = ∅ ∧ A_timeout = ∅ then
        return (M, A)
    end if
    pattern ← ComputePattern(A_pass, A_fail)
    if pattern ∈ history then
        return Stalemate("Oscillation", M, A, A_fail)
    end if
    history ← history ∪ {pattern}
    (M', A') ← LLMRefine(M, A, A_fail, A_pass)
    (A'_pass, A'_fail, _) ← ParallelCheck(M', A')
    if |A'_pass| ≤ |A_pass| then
        no_progress ← no_progress + 1
    else
        no_progress ← 0
    end if
    if no_progress ≥ THRESHOLD then
        return Stalemate("No progress", M', A', A'_fail)
    end if
    M ← M', A ← A', iteration ← iteration + 1
end while
return Stalemate("Max iterations", M, A, A_fail)
```

Final System Model Refinement. With $\mathcal{A}_{\text{golden}}$ established, we perform final model refinement to ensure it satisfies this authoritative set. We re-run parallel verification of M^* against all $\mathcal{A}_{\text{golden}}$. The final refinement follows Algorithm 1 but with: **Golden standard priority**, under which all assertions are treated as inviolable; **No assertion modifications**, meaning that only the model may be refined; and **Stricter stalemate criteria**. When stalemate occurs, diagnostic reports are escalated to experts. Upon successful completion, this phase produces final verified model M_{final} with formal guarantee $\forall a_i \in \mathcal{A}_{\text{golden}}, M_{\text{final}} \models a_i$.

Two-Phase Refinement Design. A key design decision is the separation between *iterative refinement* and *golden standard establishment.* During iterative refinement, both the model and assertions may be modified to resolve verification failures. This flexibility is intentional as early-stage assertions may contain formalization errors or overly strict interpretations that do not reflect requirement intent. However, this flexibility introduces risks of assertion weakening, where the loop converges by relaxing properties rather than correcting the model. To mitigate this, we impose a strict two-phase structure: (1) during iterative refinement, modifications are permitted but all changes are logged and presented to experts; (2) once the golden standard is established through expert validation, assertions become immutable and only model refinement is permitted. This separation ensures that any assertion modifications undergo explicit human review before being accepted as authoritative, preventing silent semantic drift.

3.3 Domain Knowledge-Based Validation

This phase addresses a critical limitation: even when a system model satisfies all explicitly stated requirements, it may still exhibit undesirable behaviors or fail to account for implicitly assumed scenarios. To mitigate this risk, we perform domain knowledge-based validation that leverages institutional knowledge, particularly historical defect reports, to identify requirement gaps.

RAG-Based Approach. We adopt a retrieval-augmented generation (RAG) workflow. Given the current requirements and the verified system model, relevant historical defect databases are retrieved based on semantic similarity. An LLM then generates candidate properties intended to prevent the recurrence of similar defects, formalizes them into CSP assertions, and checks them against the system model. Properties that fail verification are treated as potential requirement gaps.

Defect Database. The defect base is constructed from historical defect reports, lessons learned documents, and test failure reports. Each entry is embedded using pre-trained language models and stored in a vector database. To retrieve relevant defects, we construct a query from current requirements and retrieve top-K most similar defect reports ($K = 10$) using cosine similarity with relevance filtering (threshold > 0.7). Defect database structure is available in Appendix F.

Property Generation and Validation. Retrieved defect reports are fed to an LLM (Prompt in Appendix L.8) with current requirements and golden standard assertions to generate candidate properties $\mathcal{P}_{\text{candidate}}$. Each candidate property includes a natural-language description, an explanation of which historical defect(s) motivated it, an assessment of whether it is already covered by existing requirements, and a severity rating. Domain experts review these properties and produce a validated set $\mathcal{P}_{\text{validated}}$.

Verification and Gap Analysis. Each property $p_i \in \mathcal{P}_{\text{validated}}$ is checked against M_{final} using FDR4. For each gap, we generate a report containing: executive summary, failed property description, historical context, counterexample analysis with FDR4 trace and LLM explanation (Prompt available in Appendix L.9), root cause assessment, and recommended actions. Example gap reports are provided in Appendix G.

Stakeholder Decision-Making. Based on these reports, stakeholders decide whether to *accept and fix* the gap, *accept with alternative mitigation*, *reject* it as not applicable, or *defer* the decision. When gaps are accepted and addressed, new requirements are formalized and added to the golden standard specification, the system model is refined, and domain knowledge-based validation is re-executed. The process terminates when all high-priority gaps are resolved, no new gaps emerge, or practical constraints require freezing the specification.

Overall, this phase does not replace human validation but restructures it into a focused and efficient workflow. By grounding gap detection in historical evidence and validating candidate properties through formal verification, the approach systematically exposes implicit requirement gaps and complements requirements-level verification in industrial settings.

4 Evaluation

4.1 Experimental Setup

We evaluate our methodology through case studies on three real-world automotive projects, assessing its feasibility, correctness, defect discovery capability, and industrial practicality. Specifically, we address the following four research questions: **RQ1:** Can LLMs automatically generate correct CSP system models and assertions from natural language requirements within practical iteration limits? **RQ2:** Do LLM-generated assertions accurately capture requirement intent, and can LLM confidence scores predict formalization quality? **RQ3:** Can domain knowledge-based verification identify real requirement gaps by leveraging historical failure modes? **RQ4:** Is the LLM+FM hybrid approach practical for industrial use in terms of time, cost, and expert effort?

Dataset. Table 1 summarizes three real-world automotive projects with increasing system complexity and diverse functional scope. The cases cover representative subsystem types in V-model development, including local control logic, state-machine–based configuration, and distributed protocol interactions, providing a realistic and varied evaluation setting.

Configuration. All experiments use Claude 4.5 Sonnet as the LLM and FDR4 for model checking, with iteration limits of 20 for both syntax and model-checking refinement loops. Each case study is run five times for consistency. The iteration limit of 20 was determined empirically: beyond this threshold, failed runs consistently exhibited oscillation patterns where the same assertion pass/fail configurations recurred, indicating fundamental conflicts rather than insufficient refinement attempts. Increasing the limit further did not improve success rates for complex systems.

Table 1. Overview of Real-world Automotive Cases. (Details in Appendix H)

Case	#Req	Complexity	Domain	Key Characteristics
CS1	23	Low	Active Grille Shutter Control	Actuator control, sensor integration, power state transitions, diagnostic fault code generation
CS2	56	Medium	Sunroof Auto-Identification	Automatic hardware detection, state-machine-based configuration, CAN communication, non-volatile memory persistence
CS3	83	High	Remote Services Gateway	Multi-ECU distributed communication, remote control protocol compliance, multi-step request–response flows with security handshaking

4.2 RQ1: Synthesis Feasibility

Table 2. Success Rates and Average Iteration Counts (MC = Model Checking).

Case Study	Success Rate	Syntax Iter.	MC Iter.	Total Iter.
CS1 (23 req.)	5/5 (100%)	1.0	5.8	6.8
CS2 (56 req.)	4/5 (80%)	1.6	5.4	7.0
CS3 (83 req.)	2/5 (40%)	4.6	13.4	18.0

Table 2 summarizes the synthesis success rates and average iteration counts across three automotive case studies. CS1 (23 requirements) achieved a 100% success rate in only 6.8 total iterations on average, demonstrating high reliability for small-scale systems. CS2 (56 requirements) reached an 80% success rate; the single failed run exhibited oscillation patterns caused by conflicting assertions rather than synthesis limitations. In contrast, CS3 (83 requirements) showed a reduced success rate of 40% with an average of 18.0 total iterations, reflecting the higher complexity of its distributed multi-ECU architecture. Failed runs in CS3 consistently entered stalemates where violated assertions no longer decreased despite further refinement. More iterations further highlight the increasing difficulty of large-scale synthesis.

Answer to RQ1: LLMs can reliably synthesize correct CSP system models for small-to-medium automotive projects (up to ~60 requirements), achieving success rates between 80% and 100%. For larger and structurally more complex projects (80+ requirements), synthesis success drops to around 40%. This reduced success rate reflects fundamental complexity rather than insufficient iterations: failed runs consistently exhibited oscillation patterns where conflicting assertions could not be simultaneously satisfied. For industrial adoption at larger scales, several mitigation strategies are viable: improved LLM capabilities, enhanced prompting strategies, hierarchical requirement decomposition, and targeted expert intervention for conflicting assertion resolution. The iterative refinement loop is essential, with most successful runs requiring 5-10 total iterations.

4.3 RQ2: Assertion Formalization Accuracy

Table 3 and Table 4 summarize the correctness of LLM-generated assertions and their relationship with confidence scores. Expert validation on CS1 shows that the majority of assertions correctly capture requirement intent: 14 out of 23 assertions (61%) required no changes, accurately modeling event ordering, state transitions, and causal relationships. An additional 22% required only minor adjustments, typically clarifying abstraction choices rather than correcting semantic

Table 3. Distribution of Confidence Scores for Generated Assertions.

Case Study	High (0.80–1.00)	Medium (0.60–0.79)	Low (<0.60)	Avg. Score
CS1 (23)	8 (35%)	11 (48%)	4 (17%)	0.75
CS2 (56)	24 (43%)	30 (54%)	2 (4%)	0.77
CS3 (83)	31 (37%)	50 (60%)	2 (2%)	0.77

errors. Only 17% required major revisions, indicating that substantially incorrect assertions are a minority.

Confidence scores provide supporting evidence of correctness. Assertions accepted without modification exhibit the highest average confidence (0.81), whereas those requiring major revisions have significantly lower confidence (0.59). For example, assertion G-2 was correctly formalized, received a high confidence score (0.85), and was unanimously approved by experts, while assertion G-16 received lower confidence (0.72) alongside an explicit LLM disclaimer that aligned with expert feedback. Detailed expert review data are provided in Appendix J.

Answer to RQ2: Most LLM-generated assertions correctly capture requirement intent, with 61% requiring no expert changes and only a small fraction requiring major revisions. LLM confidence scores correlate strongly with correctness and effectively flag assertions needing closer expert review.

Table 4. Expert Validation Outcomes for CS1 with LLM Confidence Scores.

Category	Count	%	R1	R2	R3	Avg. Conf.
No change required	14	61%	16	18	18	0.81
Minor changes needed	5	22%	3	1	2	0.74
Major revisions required	4	17%	4	4	3	0.59

4.4 RQ3: Defect Detection Capability

Table 5 summarizes the results of domain knowledge–based validation across the three case studies. In total, 29 candidate properties were generated from retrieved historical defect reports. Of these, 23 properties failed FDR4 verification, indicating that the current system model permits behaviors that could lead to defects similar to historical failures. The remaining 6 properties passed verification, meaning the system already satisfies these constraints. The 79% actionable finding rate (23/29) represents properties where formal verification exposed concrete gaps; these were subsequently reviewed by domain experts who

Table 5. Results of Domain Knowledge-based Validation.

Case Study	Retrieved	Failed	Passed	Known Issues	Novel Issues
CS1	11	7	4	0/0	7
CS2	8	8	0	0/0	8
CS3	10	8	2	1/1 (100%)	7
Total	**29**	**23**	**6**	**1/1 (100%)**	**22**

confirmed their validity as genuine requirement deficiencies requiring attention. More importantly, the approach detected 1 out of 1 known historical defect (100%), while uncovering 22 previously undocumented gaps, demonstrating its effectiveness beyond reproducing known issues. Cross-case analysis reveals recurring requirement gaps, dominated by initialization omissions (65%) where safe or well-defined post-reset states are unspecified. Temporal ordering gaps (26%) arise from unconstrained execution order, while obligation-versus-capability gaps (22%) reflect missing mandatory conditions. A smaller fraction involves security and protocol gaps (9%), such as omitted handshakes or validation steps.

Known Defect Detection (CS3). In CS3, a historical integration defect occurred when control requests were issued before completion of a security handshake. While the original OEM fix relied on a fixed delay (500ms), our domain knowledge–based validation retrieved the defect and generated a property enforcing handshake completion before control operations. Verification against M_{final} revealed a violation, confirming that the system could bypass the handshake. Both Aumovio and OEM stakeholders agreed that the handshake-based property provides a superior, timing-independent solution. Details are in Appendix K.

Novel Gap Examples. Among the 22 novel gaps: *CS1 Initialization Gap:* "After power-on, the system SHALL initialize to a known safe state (grille fully open) before accepting control commands" failed, revealing that the model allowed control commands immediately after power-on without guaranteed initialization. *CS2 State Persistence Gap:* "After diagnostic clear of sunroof presence, the system SHALL re-execute detection logic on next ignition cycle" failed, showing the model allowed the system to remain in an indeterminate state.

Answer to RQ3: Domain knowledge–based validation detected 100% of known historical issues and identified 22 novel requirement gaps, achieving a 79% actionable finding rate. The approach is particularly effective at surfacing initialization sequencing and temporal ordering gaps. Beyond defect detection, it enables formally justified improvements to system behavior, demonstrating value that extends beyond reproducing known failures.

Table 6. Objective Metrics for Industrial Practicality Assessment. Expected costs account for synthesis success rates from Table 2.

Case Study	Requirements	Pages	Time	API Cost	Exp. Cost
CS1	23	3	10 min	\$1.80 (\$0.078/req)	\$1.80
CS2	56	11	35 min	\$7.80 (\$0.139/req)	\$9.75
CS3	83	9	20 min	\$3.30 (\$0.040/req)	\$8.25
Avg.	**54**	**7.7**	**22 min**	**\$4.30**	**\$6.60**

4.5 RQ4: Industrial Practicality and Efficiency

Table 6 reports objective metrics assessing the industrial practicality of our method. Across the case studies, the automated pipeline completed within 10–35 min of wall-clock time, even for the largest project with 83 requirements. Notably, processing time scales sub-linearly with requirement document pages, indicating that the approach remains practical as system complexity increases. API costs are consistently low, with all cases remaining well below \$8 per successful run. The "Expected Cost" column accounts for synthesis failure probability (success rates from Table 2), computed as API Cost divided by success rate. Even with potential re-runs, expected costs remain below \$10 per project. In practice, failed synthesis attempts provide diagnostic information (oscillation patterns, conflicting assertions) that can guide targeted expert intervention, potentially reducing subsequent attempt costs.

Comparison with Traditional Workflows. Our approach provides: **Formal guarantees** (mathematical proof that $M_{\text{final}} \models \mathcal{A}_{\text{golden}}$), **Exhaustive checking** (FDR4 explores complete state space within computational limits), **Concrete counterexamples** (precise diagnostic traces guide fixes), and **Traceability** (full requirement-to-assertion-to-model mapping for regulatory compliance). These benefits come at minimal additional cost: $<$ 1 hour expert validation time and $<$ 8 API cost per project.

Answer to RQ4: The LLM+FM hybrid approach demonstrates strong industrial practicality: (1) automated processing completes within 10–35 minutes for projects with 23–83 requirements, (2) API costs remain low ($<$ \$8 per project, $<$ \$0.15 per requirement), (3) and processing time scales sub-linearly with document size. The approach shifts expert effort toward semantic validation rather than formal modeling, enabling practical adoption in industrial settings where domain experts typically lack formal methods training.

5 Related Work

Formal methods have a long history in safety-critical domains, with model checking [4,5,14,20,22], theorem proving [6,8,11,18], and static analysis [7,15,17,

21] increasingly adopted at design and implementation levels in automotive, aerospace, and railways [10,13]. Standards such as ISO 26262 drive formal technique adoption, yet applying these methods at the requirements level remains challenging due to the manual effort for translating requirements into formal specifications.

To reduce this burden, structured specification techniques have been proposed. Controlled natural languages such as Attempto Controlled English restrict grammar and vocabulary [9] to enable deterministic parsing into logic. Requirement boilerplates and patterns, including EARS-style requirements syntax [19] and temporal property specification patterns, provide templates for expressing common behavioral constraints [3,24]. While these techniques improve consistency and analyzability, they impose strict syntactic discipline and require significant process changes and training.

Recent work has explored using LLMs for requirements engineering tasks, including requirements classification [2,12] and ambiguity detection [1]. However, these approaches typically lack formal verification guarantees, relying on LLM outputs without rigorous validation. Our work differs fundamentally by treating LLMs as automation assistants rather than verification authorities, with all outputs validated through formal model checking or expert review.

Retrieval-augmented generation (RAG) has been applied in various software engineering contexts [16,23], but its use for proactive requirement gap detection through historical defect analysis is, to our knowledge, novel. By combining RAG with formal verification, we enable systematic validation against institutional knowledge that would otherwise remain tacit.

6 Discussion and Limitations

Dependence on LLM Quality. The effectiveness of our approach depends on the capabilities of the underlying LLM. LLMs may misinterpret domain-specific terminology or struggle in complex refinement scenarios. To mitigate this risk, LLMs are never treated as trusted verification authorities: all outputs are validated through formal checking or expert review, ensuring that errors are exposed rather than remaining latent. As LLMs improve, we expect corresponding gains in workflow performance without changes to the formal backbone. We informally evaluated several LLMs and found that Claude 4.5 Sonnet provided the best results for CSP synthesis and refinement; however, we did not conduct systematic comparisons across models. Results may vary with different LLMs, and we acknowledge this as a reproducibility consideration.

Timing Abstraction Limitations. Our approach abstracts timing constraints (e.g., "within 200ms") into ordering constraints verifiable in untimed CSP. This abstraction is appropriate for requirements where the essential property is causal ordering (e.g., "A must occur before B"), but may be insufficient for requirements where bounded response time is fundamental. Examples of requirements that may be misrepresented include: watchdog timeouts where the bound itself is safety-critical, jitter constraints requiring bounded variance, and rate-limiting requirements specifying maximum event frequency. In our case studies, timing requirements were predominantly ordering-based, and expert validation did not identify cases where the untimed abstraction materially changed requirement semantics. However, for systems with strict real-time constraints, complementary verification using timed formal methods (e.g., Timed CSP, timed automata) may be necessary. We consider integration with timed verification as future work.

Quality and Coverage of Domain Knowledge. Retrieval-augmented verification is influenced by the quality and coverage of the domain knowledge base. Historical defect records may be incomplete or biased toward certain issue classes, limiting recall. In our evaluation, missed issues were primarily due to insufficient documentation or hardware-dependent failures. Nevertheless, even partial knowledge proved valuable, identifying both known and previously undocumented high-priority gaps, and is expected to improve as organizations accumulate experience.

Human-in-the-Loop Requirement. Although the workflow significantly reduces expert effort, human involvement remains necessary for validating assertions, resolving conflicts, and handling verification stalemates. The approach is designed to support, rather than replace, expert judgment by automating low-level reasoning and providing precise diagnostics.

Scalability and Abstraction. While the method performs well for small-to-medium projects, larger systems may require hierarchical decomposition and compositional verification. The reduced success rate for CS3 (40%) indicates that fully automated synthesis has limitations at scale; however, the diagnostic information from failed attempts (oscillation patterns, conflicting assertions) provides actionable guidance for expert intervention.

Communicating Results to Non-Formal Stakeholders. The current workflow produces formal CSP models and assertions as primary outputs. For stakeholders who rely on natural-language or graphical specifications, we provide several mechanisms: (1) natural-language back-translations accompany each assertion, explaining its meaning in domain terms; (2) FDR4 can generate state transition diagrams from verified models, providing visual representations of system behavior (see Appendix I); and (3) the traceability matrix links formal assertions back to original requirements.

Reproducibility and Artifact Availability. Our evaluation uses proprietary OEM requirements, an internal defect database, and commercial LLM APIs, which limits direct reproducibility. To support replication and adaptation, we provide: complete prompts for all LLM interactions (Appendix L), detailed examples of refinement interactions and gap reports (Appendices C,G), the defect database schema with anonymized examples (Appendix F), and expert review data with inter-reviewer analysis (Appendix J). The methodology is designed to be LLM-agnostic, though performance may vary across models.

7 Conclusion

This paper presented an industry-oriented, shift-left approach to requirements-level verification for automotive systems by integrating large language models with formal methods. The proposed workflow automatically translates natural-language requirements into CSP system models and assertions, refines them through a CEGAR-inspired loop using FDR4 counterexamples, and validates correctness via expert review supported by natural-language back-translations.

Evaluation on real-world automotive case studies shows that the approach is practical and effective for small-to-medium projects, achieving high synthesis success with limited expert effort. Domain knowledge–based validation further extends requirements verification by uncovering implicit gaps using historical defect knowledge. Overall, this work demonstrates that LLM+FM hybridization can enable scalable and trustworthy requirements-level verification without major process changes, providing a concrete step toward practical shift-left adoption in industrial settings.

Acknowledgements. This research/project is supported by A*STAR under the RIE2025 Industry Alignment Fund – Industry Collaboration Projects (IAF-ICP) Funding Initiative (Award: I2501E0045), as well as cash and in-kind contributions from the industry partner(s), and the Singapore Ministry of Education Academic Research Fund Tier 1 (RG12/23).

References

1. Bashir, S., et al.: Requirements ambiguity detection and explanation with LLMs: an industrial study. In: International Conference on Software Maintenance and Evolution (2025)
2. Binkhonain, M., Alfayez, R.: Are prompts all you need? Evaluating prompt-based large language models (LLM)s for software requirements classification. In: Requirements Engineering, pp. 1–21 (2025)
3. Boehm, B., Basili, V.R.: Software defect reduction top 10 list. Found. Empir. Softw. Eng. Legacy Victor R. Basili **426**(37), 426–431 (2005)
4. Clarke, E.M.: Model checking. In: Ramesh, S., Sivakumar, G. (eds.) FSTTCS 1997. LNCS, vol. 1346, pp. 54–56. Springer, Heidelberg (1997). https://doi.org/10.1007/BFb0058022
5. Clarke, E.M., Emerson, E.A., Sistla, A.P.: Automatic verification of finite-state concurrent systems using temporal logic specifications. ACM Trans. Program. Lang. Syst. (TOPLAS) **8**(2), 244–263 (1986)
6. Cook, B., Kroening, D., Sharygina, N.: Accurate theorem proving for program verification. In: International Symposium On Leveraging Applications of Formal Methods, Verification and Validation, pp. 96–114. Springer (2004)
7. Cousot, P., Cousot, R.: Systematic design of program analysis frameworks. In: Proceedings of the 6th ACM SIGACT-SIGPLAN Symposium on Principles of Programming Languages, pp. 269–282 (1979)
8. Darvas, A., Hähnle, R., Sands, D.: A theorem proving approach to analysis of secure information flow. In: International Conference on Security in Pervasive Computing, pp. 193–209. Springer (2005)

9. Fuchs, N.E., Schwitter, R.:. Attempto controlled English (ace). arXiv preprint cmp-lg/9603003 (1996)
10. Gibson-Robinson, T., Armstrong, P., Boulgakov, A., Roscoe, A.W.: FDR3 — a modern refinement checker for CSP. In: Ábrahám, E., Havelund, K., (eds.), Tools and Algorithms for the Construction and Analysis of Systems, vol. 8413, LNCS, pp. 187–201 (2014)
11. Green, C.: Application of theorem proving to problem solving. In: Readings iN Artificial Intelligence, pp. 202–222. Elsevier (1981)
12. Hassani, S., Sabetzadeh, M., Amyot, D.: An empirical study on LLM-based classification of requirements-related provisions in food-safety regulations. Empir. Softw. Eng. **30**(3), 72 (2025)
13. Hoare, C.A.R.: Communicating sequential processes. Commun. ACM **21**(8), 666–677 (1978)
14. Jhala, R., Majumdar, R.: Software model checking. ACM Comput. Surv. (CSUR) **41**(4), 1–54 (2009)
15. Haonan Li, Yu., Hao, Y.Z., Qian, Z.: Enhancing static analysis for practical bug detection: an LLM-integrated approach. Proc. ACM Program. Lang. **8**(OOPSLA1), 474–499 (2024)
16. Li, H., et al.: Retrieval-augmented fine-tuning for improving retrieve-and-edit based assertion generation. IEEE Trans. Softw. Eng. (2025)
17. Lin, S.: LLM-driven adaptive source–CSink identification and false positive mitigation for static analysis. In: Proceedings of the 2025 8th International Conference on Computer Information Science and Artificial Intelligence, pp. 281–285 (2025)
18. Lin, X., et al.: FVEL: interactive formal verification environment with large language models via theorem proving. Adv. Neural. Inf. Process. Syst. **37**, 54932–54946 (2024)
19. Mavin, A., Wilkinson, P., Harwood, A., Novak, M.: Easy approach to requirements syntax (ears). In: 2009 17th IEEE International Requirements Engineering Conference, pp. 317–322. IEEE (2009)
20. McMillan, K.L.: Symbolic model checking. In: Symbolic Model Checking, pp. 25–60. Springer (1993)
21. Nielson, F., Nielson, H.R., Hankin, C.: Principles of program analysis. Springer Science & Business Media (2004)
22. Visser, W., Havelund, K., Brat, G., Park, S.J., Lerda, F.: Model checking programs. Automat. Softw. Eng. **10**(2), 203–232 (2003)
23. Wang, Y., Guo, S., Wei Tan, C.: From code generation to software testing: AI Copilot with context-based RAG. IEEE Softw. (2025)
24. Wiesner, S., Peruzzini, M., Hauge, J.B., Thoben, K.: Requirements engineering. In: Concurrent Engineering in the 21st Century: Foundations, Developments and Challenges, pp. 103–132. Springer (2015)
25. Zave, P., Nelson, T.: Validation of formal models: a case study. In: The Practice of Formal Methods: Essays in Honour of Cliff Jones, Part II, pp. 292–313. Springer (2024)

Formal Verification of Functional Correctness for the OpenHarmony LiteOS-M Kernel

Tianqi Zhao[1], Qinxiang Cao[3], Shenghua Feng[1], Minghui Zhou[1,2](✉), Naijun Zhan[1,2], Yongzhi Cao[1,2], Junfeng Zhao[1,2], Haiyan Zhao[1,2], Hao Wang[4], and Zhenjiang Hu[1,2]

[1] Zhongguancun Laboratory, Beijing, China
zhaotq@zgclab.edu.cn, shenghua.feng@foxmail.com
[2] Peking University, Beijing, China
{zhmh,njzhan,caoyz,zhaojf,zhhy.sei,huzj}@pku.edu.cn
[3] Shanghai Jiao Tong University, Shanghai, China
caoqinxiang@sjtu.edu.cn
[4] Shenzhen Kaihong Digital Industry Development Co., Ltd., Shenzhen, China
wanghao.sei@pku.edu.cn

Abstract. OpenHarmony LiteOS-M, a preemptive operating system (OS) kernel for the Internet of Things (IoT), is widely deployed in safety-critical domains, such as aerospace and transportation. As a rigorous method to assure software safety, formal verification has been applied to OS kernels in industry. However, entirely verified kernels with large codebases are rare, since such verification is typically performed within interactive theorem provers, requiring substantial human effort. In this paper, we present the *functional correctness* verification of LiteOS-M. First, to improve verification efficiency, we design a formal verification platform, Smart Verifier. The platform employs an annotation-based verifier as the front end, while the back end integrates Z3 and Rocq, combining automatic and interactive theorem proving techniques. Second, we tailor two verification methods, expressing program refinement as standard Hoare logic triples and modeling concurrency through state transition systems, to utilize the platform for verifying LiteOS-M. Our verified LiteOS-M kernel consists of 17,000 lines of C. During the code review and verification, we find a total of 17 bugs, all confirmed and fixed by developers.

Keywords: OS kernel · Formal Verification · LiteOS-M · Separation Logic

1 Introduction

In the era of all-scenarios, all-connectivity, and all-intelligence, OpenHarmony LiteOS-M [9]—a preemptive OS kernel originally developed by Huawei for IoT in

A. Sampaio and M. Stoelinga (Eds.): FM 2026, LNCS 16557, pp. 651–669, 2026.
https://doi.org/10.1007/978-3-032-26220-2_32

2020 and presently maintained by the open-source OpenHarmony community—stands out for its small size, low consumption, and high performance. Due to its lightweight nature, the LiteOS-M kernel (referred to as LiteOS-M for short) is widely adopted in microcontroller-based devices with constrained hardware resources. To date, LiteOS-M has been deployed by over 111 vendors spanning industries like aerospace, industrial control, and transportation, and is powering millions of IoT devices globally. Owing to its application in safety-critical domains, even a minor error or vulnerability in LiteOS-M may cause significant failures of the entire software system, potentially resulting in substantial economic losses or posing a threat to human safety. Therefore, ensuring the correctness of LiteOS-M is of paramount importance.

Formal verification, which leverages mathematical proof techniques to establish properties about programs, is a crucial approach to guaranteeing the correctness of system software. Our team is dedicated to advancing research in formal verification. Motivated by the industry's urgent demand and our interest in exploring the practical application of formal verification techniques, we have collaborated with an enterprise focused on the OpenHarmony ecosystem to verify the *functional correctness* of LiteOS-M. Specifically, our goal is to ensure that the kernel's C implementation is free from implementation defects.

It is feasible to formally prove the functional correctness of OS kernels, as shown by notable projects including seL4 [7], Verisoft XT [3,8], CertiKOS [5,6], and μC/OS-II [16]. The seL4 team [7] has been the first to achieve a fully machine-checked functional correctness proof of a high-performance L4-family microkernel in the interactive theorem prover Isabelle/HOL. The kernel is sequential and comprises 8,700 lines of C and 600 lines of assembler. The Verisoft XT project verifies the Hyper-V hypervisor [8] using VCC [3], an automatic verifier for C developed by Microsoft. The hypervisor integrates a small multiprocessor operating system and a virtualization layer, collectively consisting of 100,000 lines of C and 5,000 lines of assembly. However, only 20% of the code is verified [3], leaving uncertainty about the complete verification of the OS kernel. Gu et al. [5] verify the mCertiKOS kernel, which is sequential. By extending the mCertiKOS kernel with various features, the same team presents and verifies the CertiKOS kernel, which supports multi-core concurrency and consists of 6,500 lines of C and assembly. Xu et al. [16] verify key modules of μC/OS-II, which is a commercial preemptive OS kernel designed for single-core environments. The verified modules, including the scheduler, timer interrupt handler, mutexes, and message queues, among others, comprise 3,250 lines of C, constituting roughly 50% of the entire kernel. All the proofs for mCertiKOS, CertiKOS and μC/OS-II are completed in the interactive theorem prover Rocq. Overall, it is evident that completely verified kernels with large codebases are rare. The reasons can be summarized as follows. On the one hand, automated tools seem insufficient to verify entire kernels, since they are typically based on SMT solvers, which are considered black boxes, making them difficult to debug and limiting the expressiveness. On the other hand, interactive theorem provers offer richer logic language and provide a degree of automation through reasoning plugins. However, they remain inefficient and require significant human effort.

With a codebase of 17,000 lines of C, achieving full formal verification for LiteOS-M poses a significant challenge. To improve verification efficiency, we design a formal verification platform named Smart Verifier, which combines automatic and interactive theorem proving techniques. Our platform consists of three main components: an annotation-based verifier, QCP [14] at the front end; an OS verifier OSVAuto [15] at the back end; and a connector module facilitating seamless integration. For the front end, we employ QCP [14], offering a symbolic executor to automatically generate separation logic verification conditions from annotated programs, which are then proved manually by users in Rocq. For the back end, we use OSVAuto [15], a tool tailored for OS verification, which leverages the state-of-the-art SMT solver Z3 [4] for automatic solving of pure propositions. Furthermore, we develop a connector module that translates propositional formulas purified by QCP into a format compatible with OSVAuto, effectively bridging QCP and OSVAuto.

Similar to prior works [6,16], we verify the functional correctness of LiteOS-M through proving *refinement* relationships [1,2], which assert that the low-level concrete implementation strictly follows the high-level abstract specification of the kernel's functionality. We tailor two verification methods to be compatible with our verification platform. First, it is known that refinement relationships are commonly established through relational Hoare logic quadruples. To facilitate the use of annotation-based verifiers, we encode relational Hoare logic quadruples into standard Hoare logic triples by the method proposed in [13]. Second, inspired by the refinement verification of concurrent programs [10,11], we use state transition systems (STSs) to model concurrency at both high and low levels. These verification methods have been implemented and are entirely compatible with our verification platform.

Our verification framework, comprising the platform and the tailored methods, enables us to successfully complete the functional correctness verification of LiteOS-M. The whole verification process takes 12 person-years. During the code review and verification, we discover a total of 17 bugs in LiteOS-M, including concurrency issues, memory errors, numerical overflows, and functional defects. The verification process also contributes to optimize the verification platform. In the scenario of LiteOS-M verification, we propose several new features for QCP, such as support for multi-branch structures and multi-spec derivation, and for OSVAuto, such as support for nested sequences, thereby facilitating the verification process. We also identify and fix numerous bugs in these tools.

This paper is organized as follows. The preliminaries are provided in Sect. 2, including an introduction to the verification tools QCP and OSVAuto. Sect. 3 introduces the LiteOS-M kernel and presents an overview of our verification framework. This framework consists of the verification platform and the verification methods, which are detailed in Sect. 4 and Sect. 5, respectively. Using the framework, we complete the verification of LiteOS-M and summarize the verification efforts and results in Sect. 6. Section 7 concludes this paper.

2 Preliminaries

This section introduces the verification tools QCP [14] and OSVAuto [15], both of which are utilized in our verification platform.

2.1 QCP

```
1  struct list {int head; struct list *tail;};
2  struct list *reverse (struct list *p)
3  /*@ With    (l: list Z),
4       Require sll(p,l)
5       Ensure  sll(__return,rev(l)) */
6  {    struct list *w, *t, *v;
7       w = (void *) 0; v = p;
8       /*@ Inv exists l1 l2,
9           l == app(rev(l1),l2) && sll(w,l1) * sll(v,l2) */
10      while (v) {
11          t = v->tail; v->tail = w; w = v; v = t; }
12      return w;
13 }
```

Fig. 1. An Annotated Program for Verifying Linked-List Reversal in QCP. (&&: logical conjunction; *: separating conjunction; sll(p,l): a separation logic predicate, stating that address p stores a singly linked list whose elements are given by l. We define sll, app and rev in Rocq; QCP only introduces their declarations.)

QCP [14], short for Qualified C Programming Verifier, is an annotation-based verifier for C programs based on separation logic. Figure 6a illustrates the framework of QCP. Given an annotated C program, the symbolic executor in QCP generates verification conditions automatically, expressed as separation logic entailments and thus referred to as *separation logic verification conditions* (SL VCs). Guided by user-defined, soundness-guaranteed *strategies* [12], these SL VCs undergo a purification phase that removes all spatial formulas and reduces them to pure propositions, which we refer to as *propositional verification conditions* (PROP VCs). Theoretically, QCP can purify any SL VC into a PROP VC, provided that users supply enough strategies. Each resulting PROP VC is handled by a lightweight SMT solver, with support for the theories of uninterpreted functions and linear integer arithmetic. When solving a PROP VC, the SMT solver returns "valid", "invalid", or "unknown", and QCP accordingly either generates a proof automatically, reports an error for user correction, or requires users to manually construct a proof for the original SL VC, as purification may remove essential pure facts.

Example 1. Figure 1 presents an implementation of an in-place linked list reversal and its annotations. The annotations on Lines 3–5 describe the specification this function should satisfy: for any list of integers l, if l is stored in a linked list and this link list's head pointer is passed to the function by the program variable p, then the function reverses the linked list and returns the new head pointer __return. The assertion on Lines 8–9 describes a loop invariant, that the program state should satisfy every time the program enters the loop body.

2.2 OSVAuto

```
query sched_status {
    fixes tcb : TaskControlBlock;
    assumes vaild_usedTask_status(tcb.status);
    assumes status_delay_or_pendtime(tcb.status);
    shows tcb.status & (STATUS_DELAY | STATUS_PEND) != 0
    proof { auto }
}
```

Fig. 2. A Query in OSVAuto.

OSVAuto [15] is designed to automatically prove pure propositions without spatial assertions—that commonly arise when verifying OS kernels.

OSVAuto adopts a functional language for describing data structures and expressing verification conditions. On the one hand, it provides native support for defining common data structures in OS kernels, such as structures, enumerations, sequences, and maps. On the other hand, the language allows users to express verification conditions in the form of *queries* [15], each of which consists of a list of type parameters, variables, assumptions, a conclusion, and an automatic proof. An example of such a query is shown in Fig. 2. The query states that for any task control block (TCB), if the TCB status is valid and classified as either delayed or timed pending, it must follow that the status is either delayed or pending. This query can be automatically proven in OSVAuto.

OSVAuto leverages an automation algorithm to prove queries. It employs Z3 [4] as its backend, a state-of-the-art SMT solver that supports a wide range of theories—including arithmetic, arrays, bit-vectors, and uninterpreted functions—and delivers high performance in solving complex constraints.

3 Background and Overview of Our Work

In this section, we introduce the LiteOS-M kernel in Sect. 3.1 and provide an overview of the verification framework in Sect. 3.2.

3.1 LiteOS-M

LiteOS-M is a basic kernel of the OpenHarmony operating system, designed for resource-constrained environments with a focus on small footprint, low power consumption, and high performance. Within OpenHarmony, LiteOS-M plays a central role in maintaining system integrity and managing system resources.

The upper part of Fig. 3 shows the architecture of the LiteOS-M kernel, which includes subsystems such as task management, software timer, scheduler, tick management, inter-process communication, and memory management. LiteOS-M achieves concurrency in a single-core environment by employing preemptive, priority-based scheduling. Within the kernel, execution of a task can be interrupted at any program point (unless interrupts are disabled). Then the control is switched to the interrupt handler. When the handler finishes, it may invoke the scheduler and switch the execution context to the highest priority task, instead of returning to the original interrupted task.

3.2 Overview of the Verification Framework

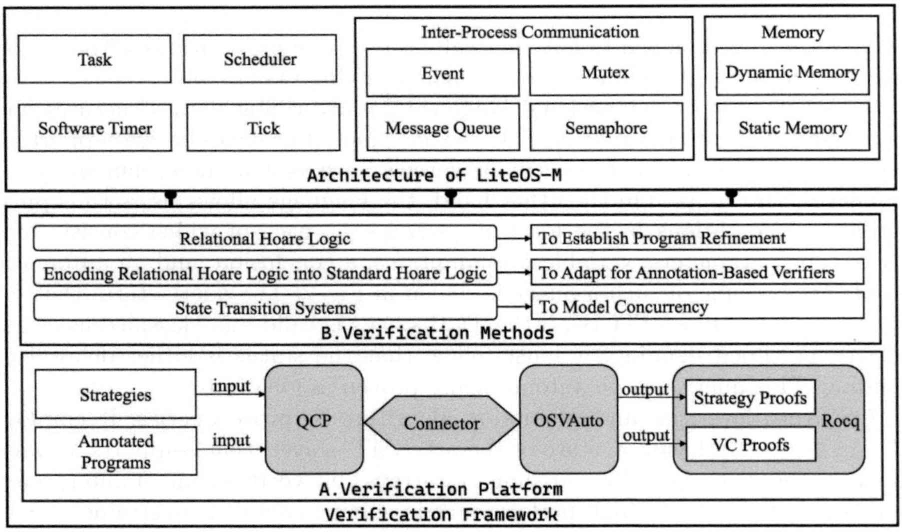

Fig. 3. Verification Framework for LiteOS-M.

OS Correctness. The *functional correctness* of an OS kernel can be described by *refinement* [7,16], which establishes a correspondence between the concrete implementations of application programming interfaces (APIs) and their

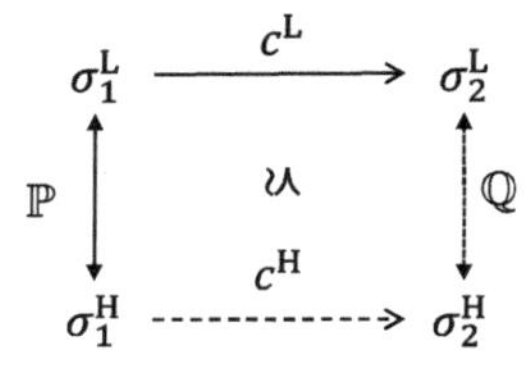

Fig. 4. Program Refinement.

abstract models. Figure 4 presents the refinement relationship between the concrete low-level program c^{L} and the abstract high-level program c^{H}. This refinement can be formalized as a relational Hoare quadruple $\{\mathbb{P}\}\ c^{\mathrm{L}} \precsim c^{\mathrm{H}}\ \{\mathbb{Q}\}$, meaning that given any initial state pair $(\sigma_1^{\mathrm{L}}, \sigma_1^{\mathrm{H}})$ satisfying $\mathbb{P}$, for any final state σ_2^{L} after the execution of c^{L}, there exists a multi-step transition from $(c^{\mathrm{H}}, \sigma_1^{\mathrm{H}})$ to $(\texttt{skip}, \sigma_2^{\mathrm{H}})$ such that $(\sigma_2^{\mathrm{L}}, \sigma_2^{\mathrm{H}})$ satisfies $\mathbb{Q}$. Here, σ_1^{L} and σ_2^{L} are low-level program states, σ_1^{H} and σ_2^{H} are high-level program states, and `skip` represents a no-op high-level program.

The Whole Verification Framework. The lower part of Fig. 3 presents the verification framework for LiteOS-M. Block A presents the verification platform, Smart Verifier (see Sect. 4 for details). It consists of QCP, a connector module, and OSVAuto. Block B shows the verification methods (see Sect. 5 for details). On the one hand, we employ relational Hoare logic to establish program refinement and encode it into standard Hoare logic to facilitate the use of annotation-based verifiers. On the other hand, we utilize STSs to model concurrent behaviors.

A Running Example. Figure 5 presents an API in LiteOS-M, which is designed to retrieve the name of the currently running task, with its annotations. In order to ensure safe access, the task name is read within the critical section, while the value is returned outside the critical section.

We verify this API using Smart Verifier. The annotations on Lines 2–14 describe the specification this function should satisfy. The `With` clause universally quantifies over the model `sg` of C code's global variables, a single high-level program state `st`, and a set of sets of high-level program states `X`. The `Require` clause specifies the following precondition: (1) the global variable `g_taskCBArray` in the C code coincides with its counterpart `sg.g_taskCBArray` in the Rocq model, (2) `sg.g_taskCBArray` is greater than zero, which guarantees that the if-condition on Line 18 holds, (3) from the initial set `LastSeen(st)`, applying the high-level program `abs_curTaskNameGet` results in the final set `X`, and (4) the kernel is outside the critical section before invoking this API. Here, `abs_curTaskNameGet` is the abstraction of the API at the high level, `st` refers to the last-seen high-level state within the critical section during the previous invocation of the API, and `LastSeen(st)` represents a singleton set, where the sole element

```
1  char *LOS_CurTaskNameGet()
2  /*@
3  With (sg: GV) (st: ST) X
4  Require
5      g_taskCBArray == sg.g_taskCBArray &&
6      sg.g_taskCBArray > 0 &&
7      Exec(LastSeen(st),abs_curTaskNameGet,X) &&
8      OutsideCritical(sg,st)
9  Ensure
10     exists st2,
11         __return != 0 &&
12         Exec(LastSeen(st2),return(__return),X) &&
13         OutsideCritical(sg,st2)
14 */
15 {   char *taskName = (char *) 0;
16     /*@ ... */
17     enter_critical();
18     if (g_losTask.runTask != (LosTaskCB *) 0)
19         { taskName = g_losTask.runTask -> taskName; }
20     /*@ ... */
21     exit_critical();
22     return taskName;
23 }
```

Fig. 5. An Annotated Program for Verifying an API in LiteOS-M. (Due to space limitations, the annotations on Line 16 and Line 20 are omitted.)

is the set of reachable high-level states after exiting the critical section, resulting from the nondeterminism in concurrent execution. After the completion of this API call, the postcondition specified in the `Ensure` clause should be satisfied: there exists a high-level state `st2`, such that (1) the return value of the API is nonzero, (2) from the initial set `LastSeen(lvs2)`, applying the high-level program `return(__return)` results in the same final set `X`, and (3) the kernel is outside the critical section after the API invocation. Here, the high-level program `return(__return)` abstracts the code on Line 22, and `st2` refers to the last-seen high-level state within the critical section during this API call.

The formal definitions of `Exec`, `LastSeen` and `OutsideCritical` will be given in Sect. 5, in Definitions 1–3, respectively.

4 Verification Platform

In this section, we provide a detailed description of the verification platform developed and utilized in our framework.

Figure 6b presents our verification platform Smart Verifier. When the lightweight SMT solver in QCP returns "unknown" for some PROP VC, the verification platform submits the PROP VC to OSVAuto for further solving. Since OSVAuto adopts a different language for expressing PROP VCs than QCP, we

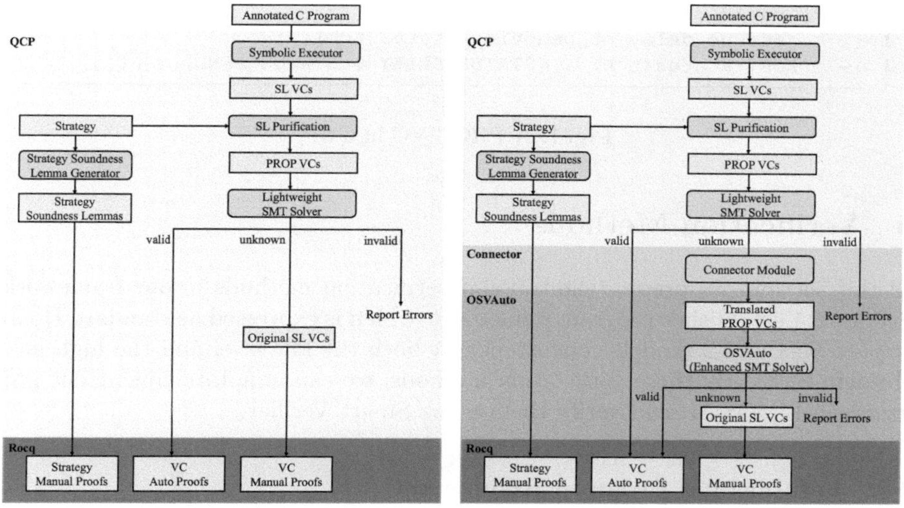

(a) QCP. (b) Smart Verifier.

Fig. 6. Verification Platform. (**Blue rectangles:** files manually completed by users. **Green rectangles:** files automatically generated by tools. **Gray rounded rectangles:** tool components.) (Color figure online)

develop a connector module that translates the PROP VC into a query format compatible with OSVAuto. After appropriate preprocessing, the query is passed to Z3, which offers significantly stronger solving capabilities than the lightweight SMT solver in QCP. The subsequent steps after Z3 solving are identical to those following the lightweight SMT solver in QCP.

The implementation of the connector module is described in detail below. In QCP, a PROP VC is constructed from its abstract syntax tree (AST). The connector module traverses the AST and converts each node into its corresponding expression in the language supported by OSVAuto. This AST-based method performs an entirely syntactic transformation, establishing a one-to-one correspondence in syntax and preserving the semantics of PROP VCs, thereby ensuring the correctness of the transformation.

Example 2. When using QCP alone to verify the scheduler module of LiteOS-M, we encounter a PROP VC (as shown in Fig. 7) that cannot be automatically solved by QCP's built-in SMT solver. This PROP VC asserts that for any TCB, if the status of the TCB is valid and is either delayed or timed pending, then the status must be either delayed or pending. The connector module in Smart Verifier reformulates this PROP VC as a query (as shown in Fig. 2), which is successfully solved by OSVAuto.

```
PROP[ vaild_usedTask_status(status(tcb));
      status_delay_or_pendtime(status(tcb)) ]
|-  PROP[ status(tcb) & (STATUS_DELAY | STATUS_PEND) != 0 ]
```

Fig. 7. A PROP VC in QCP.

5 Verification Methods

In this section, we present the tailored verification methods in our framework. Section 5.1 establishes program refinement, which is expressed as standard Hoare triples. Section 5.2 models concurrency at both the low level and the high level through STSs. By these verification methods, we can annotate concurrent programs in LiteOS-M, and verify them using Smart Verifier.

5.1 Establishing Program Refinement

Let Prog^{L} and Prog^{H} be the set of low-level programs and the set of high-level programs, respectively. Let Σ^{L} and Σ^{H} be the set of low-level program states and the set of high-level program states, respectively. Recalling Sect. 3.2, we use a relational Hoare quadruple $\{\mathbb{P}\}\ c^{\text{L}} \lesssim c^{\text{H}}\ \{\mathbb{Q}\}$ to establish program refinement, where $c^{\text{L}} \in \text{Prog}^{\text{L}}$, $c^{\text{H}} \in \text{Prog}^{\text{H}}$ and $\mathbb{P}, \mathbb{Q} \subseteq \Sigma^{\text{L}} \times \Sigma^{\text{H}}$.

To adapt for annotation-based verifiers, we review how to encode relational Hoare quadruples as standard Hoare triples. In real program verification, a binary assertion $\mathbb{P} \subseteq \Sigma^{\text{L}} \times \Sigma^{\text{H}}$ is usually in a form as

$$\exists \overrightarrow{a}.\ B(\overrightarrow{a}) \wedge [P^{\text{L}}(\overrightarrow{a})] \wedge [P^{\text{H}}(\overrightarrow{a})], \tag{1}$$

where $\overrightarrow{a}$ represents a list of logical variables, B is a pure proposition, $P^{\text{L}} \subseteq \Sigma^{\text{L}}$, and $P^{\text{H}} \subseteq \Sigma^{\text{H}}$. The unary assertions P^{L} and P^{H} can be lifted to the binary assertions $[P^{\text{L}}]$ and $[P^{\text{H}}]$, respectively, in the following manner: for any state pair $(\sigma^{\text{L}}, \sigma^{\text{H}}) \in \Sigma^{\text{L}} \times \Sigma^{\text{H}}$, $(\sigma^{\text{L}}, \sigma^{\text{H}}) \models [P^{\text{L}}]$ holds if and only if (*iff*) $\sigma^{\text{L}} \models P^{\text{L}}$, and $(\sigma^{\text{L}}, \sigma^{\text{H}}) \models [P^{\text{H}}]$ holds *iff* $\sigma^{\text{H}} \models P^{\text{H}}$. Building on this, Theorem 1 presents an encoding method, which relies on the following definition.

Definition 1 (Execution Condition [13, **Definition 18**]**).** *For any* $P^{\text{H}} \subseteq \Sigma^{\text{H}}$, $c^{\text{H}} \in \text{Prog}^{\text{H}}$ *and* $X \subseteq \Sigma^{\text{H}}$, *the* execution condition *is defined as follows:*

$$\texttt{Exec}(P^{\text{H}}, c^{\text{H}}, X) = \exists \sigma^{\text{H}}.\ \left(\sigma^{\text{H}} \models P^{\text{H}}\right) \wedge \left(\forall \sigma.\ ((\sigma^{\text{H}}, \sigma) \in [\![c^{\text{H}}]\!]_{\text{nrm}} \Rightarrow \sigma \models X\right),$$

where $[\![\cdot]\!]_{\text{nrm}}$ *denotes the denotational semantics*[1].

[1] The detailed definition of *denotational semantics* is provided in [13, Definition 6].

Theorem 1 (Encoding Relational Hoare Quadruples [13, Theorem 4]). *In order to prove the following program refinement*[2],

$$\left\{\exists \overrightarrow{a}.\, B_1(\overrightarrow{a}) \wedge [P^{\mathrm{L}}(\overrightarrow{a})] \wedge [P^{\mathrm{H}}(\overrightarrow{a})]\right\} c^{\mathrm{L}} \precsim c^{\mathrm{H}}$$
$$\left\{\exists \overrightarrow{b}.\, B_2(\overrightarrow{b}) \wedge [Q^{\mathrm{L}}(\overrightarrow{b})] \wedge [Q^{\mathrm{H}}(\overrightarrow{b})]\right\} \tag{2}$$

it suffices to prove for any $X \subseteq \Sigma^{\mathrm{H}}$, *the following standard Hoare triple holds.*

$$\left\{\exists \overrightarrow{a}.\, B_1(\overrightarrow{a}) \wedge P^{\mathrm{L}}(\overrightarrow{a}) \wedge \mathtt{Exec}\left(P^{\mathrm{H}}(\overrightarrow{a}), c^{\mathrm{H}}, X\right)\right\} c^{\mathrm{L}}$$
$$\left\{\exists \overrightarrow{b}.\, B_2(\overrightarrow{b}) \wedge Q^{\mathrm{L}}(\overrightarrow{b}) \wedge \mathtt{Exec}\left(Q^{\mathrm{H}}(\overrightarrow{b}), \mathtt{skip}, X\right)\right\} \tag{3}$$

Remark 1. In Definition 1, the execution condition $\mathtt{Exec}(P^{\mathrm{H}}, c^{\mathrm{H}}, X)$ expresses that there exists an initial state satisfying P^{H} such that any final state after executing c^{H} satisfies X.

Given any relational Hoare quadruple $\{\mathbb{P}\}\ c^{\mathrm{L}} \precsim c^{\mathrm{H}}\ \{\mathbb{Q}\}$, by the transformation in Eq. (1), there exist lists of logical variables $\overrightarrow{a}$, $\overrightarrow{b}$ and pure propositions B_1, B_2 such that we can write the relational Hoare quadruple in the form of Eq. (2). Note that Eq. (2) clearly distinguishes between the low-level parts ($[P^{\mathrm{L}}(\overrightarrow{a})], [Q^{\mathrm{L}}(\overrightarrow{b})]$) and the high-level parts ($[P^{\mathrm{H}}(\overrightarrow{a})], [Q^{\mathrm{H}}(\overrightarrow{b})]$). Thus, it is natural to encode the execution of c^{H} into the precondition and postcondition through the execution conditions, as shown in Eq. (3). For any final set X, $\mathtt{Exec}\left(P^{\mathrm{H}}(\overrightarrow{a}), c^{\mathrm{H}}, X\right)$ in the precondition states that from initial set P^{H}, executing c^{H} results in the final set X; while $\mathtt{Exec}\left(Q^{\mathrm{H}}(\overrightarrow{b}), \mathtt{skip}, X\right)$ in the postcondition states that from initial set Q^{H}, performing a skip operation results in the same final set X.

5.2 Modeling Concurrency

In sequential program execution, *determinism* allows the observation of a single program state at any moment. In contrast, concurrent program execution is *nondeterministic*. Although the program resides in a single state at any time, we can only observe a set of reachable program states outside a critical section. Thus, the idea of defining elements in Σ^{H} (or Σ^{L}) as observed sets of high-level (or low-level) states is natural. At the high level, we first construct a set ST of single high-level program states, and then Σ^{H} can be defined as:

$$\Sigma^{\mathrm{H}} = 2^{ST} = \{A \mid A \subseteq ST\}.$$

At the low level, the definition of Σ^{L} is more intricate and will be provided later.

In order to characterize such observed sets of reachable program states, we define a low-level STS and a high-level STS in Fig. 8. Here, the set $GV = GV_1 \times \cdots \times GV_n$, where each GV_i is the set of possible values for global static variable gv_i in the source code.

[2] The relational Hoare quadruple Eq. (2) can be transformed into an equivalent triple form, as used in [13]. Then, Theorem 1 is a straight conclusion of [13, Theorem 4].

Low-Level STS : $(GV \times ST, \rightsquigarrow_{\text{GTrans-L}}, \rightsquigarrow_{\text{RTrans-L}})$
State Space : $GV \times ST$
High-Level STS : $(ST, \rightsquigarrow_{\text{GTrans-H}}, \rightsquigarrow_{\text{RTrans-H}})$
State Space : ST
Transitions : $\rightsquigarrow_{\text{GTrans-L}}, \rightsquigarrow_{\text{RTrans-L}} \subseteq (GV \times ST) \times (GV \times ST)$,
$\rightsquigarrow_{\text{GTrans-H}}, \rightsquigarrow_{\text{RTrans-H}} \subseteq ST \times ST$ such that
(i) $(sg, st) \rightsquigarrow_{\text{GTrans-L}} (sg', st')$ *iff* $sg = sg'$ and $st \rightsquigarrow_{\text{GTrans-H}} st'$;
(ii) $(sg, st) \rightsquigarrow_{\text{RTrans-L}} (sg', st')$ *iff* $sg = sg'$ and $st \rightsquigarrow_{\text{RTrans-H}} st'$;
(iii) $\rightsquigarrow_{\text{RTrans-L}}, \rightsquigarrow_{\text{RTrans-H}}$ satisfy the reflexivity and the transitivity

Fig. 8. The Low-Level STS and the High-Level STS.

Remark 2. Actually, the transitions $\rightsquigarrow_{\text{GTrans-H}}$ and $\rightsquigarrow_{\text{GTrans-L}}$ in Fig. 8 define the state updates allowed for the current thread within a critical section, and transitions $\rightsquigarrow_{\text{RTrans-H}}$ and $\rightsquigarrow_{\text{RTrans-L}}$ define those allowed for other threads outside a critical section. Thus, we require these transitions satisfy conditions (i) and (ii) in Fig. 8. Moreover, we also require that $\rightsquigarrow_{\text{RTrans-L}}$ and $\rightsquigarrow_{\text{RTrans-H}}$ satisfy the reflexivity and the transitivity, corresponding to the case that no other thread updates the state, and the case that the state is updated multiple times by multiple threads, respectively.

Modeling Concurrency at the High Level. Note that the program execution within a critical section is sequential, allowing us to obtain a specific program state. To formalize the set of reachable high-level program states outside a critical section, we define the assertion $\texttt{LastSeen}(st)$, where st refers to the last-seen high-level program state within the critical section.

Definition 2 (Assertion LastSeen). *Given any $st \in ST$, we define the high-level assertion* $\texttt{LastSeen}(st) \subseteq \Sigma^{\text{H}}$ *as follows:*

$$\sigma^{\text{H}} \models \texttt{LastSeen}(st) \textit{ iff } \sigma^{\text{H}} = \{st' \in ST \mid st \rightsquigarrow_{\text{RTrans-H}} st'\}.$$

For an API c^{L}, it is generally the case that the kernel is outside a critical section before and after invoking the API. Thus, when no additional constraints are required, the high-level assertions $P^{\text{H}}, Q^{\text{H}}$ in Eq. (3) can be expressed with the predicate $\texttt{LastSeen}$. Specifically, Theorem 1 becomes that, it suffices to prove for any $X \subseteq \Sigma^{\text{H}}$ and any $st \in ST$, the following standard Hoare triple holds.

$$\left\{\exists \overrightarrow{a}.\ B_1(\overrightarrow{a}) \wedge P^{\text{L}}(\overrightarrow{a}) \wedge \texttt{Exec}\left(\texttt{LastSeen}(st), c^{\text{H}}, X\right)\right\}\ c^{\text{L}}$$
$$\left\{\exists \overrightarrow{b}, st_2.\ B_2(\overrightarrow{b}) \wedge Q^{\text{L}}(\overrightarrow{b}) \wedge \texttt{Exec}\left(\texttt{LastSeen}(st_2), \texttt{skip}, X\right)\right\}$$

Here, $st \in ST$ represents the last-seen high-level program state within the critical section during the previous invocation of c^{L}, and $st_2 \in ST$ represents the one during the current invocation.

Modeling Concurrency at the Low Level. It remains to define the set of low-level program states Σ^{L} and to specify the standard Hoare triples for the statements "`enter_critical()`" and "`exit_critical()`" (Lines 17 and 21 in Fig. 5), which mark the entry to and the exit from a critical section, respectively.

We first refine the state space $GV \times ST$ of the low-level STS into the following set CST, distinguishing the states inside and outside a critical section.

$$CST = \{\mathtt{inside}(sg, st), \mathtt{outside}(sg, st) \mid (sg, st) \in GV \times ST\}$$

We also define a function $\mathcal{T}$ such that for any $(sg, st) \in GV \times ST$,

$$\mathcal{T}(\mathtt{inside}(sg, st)) = \{\mathtt{tt}\}, \ \mathcal{T}(\mathtt{outside}(sg, st)) = \varnothing,$$

which assigns to each element of CST the set of tokens it owns. The set of low-level program states Σ^{L} is defined by

$$\Sigma^{\mathrm{L}} = \{(\mathcal{M}, s_{state}, s_{token}) \mid \mathcal{M} : Addr \to Mem, s_{state} \subseteq CST, s_{token} \subseteq \{\mathtt{tt}\}\},$$

where $\mathcal{M}$ is a memory mapping (assigning each address to a memory state, e.g., no permission, uninitialized, or a concrete value), s_{state} is a set of reachable states, and s_{token} is a set of held tokens.

Definition 3 (Assertions InsideCritical and Outside-Critical). *Given any* $(sg, st) \in GV \times ST$, *we define the low-level assertions* $\mathtt{InsideCritical}(sg, st), \mathtt{OutsideCritical}(sg, st) \subseteq \Sigma^{\mathrm{L}}$ *as follows:*

$$\sigma^{\mathrm{L}} \models \mathtt{InsideCritical}(sg, st) \textit{ iff } \sigma^{\mathrm{L}} = (\mathcal{M}_0, \{\mathtt{inside}(sg, st)\}, \{\mathtt{tt}\}),$$

$$\sigma^{\mathrm{L}} \models \mathtt{OutsideCritical}(sg, st) \textit{ iff}$$

$$\sigma^{\mathrm{L}} = (\mathcal{M}_0, \{\mathtt{outside}(sg', st') \mid (sg, st) \rightsquigarrow_{\text{RTrans-L}} (sg', st')\}, \varnothing),$$

where $\mathcal{M}_0$ maps every address to the "no permission" memory state.

Remark 3. In assertions $\mathtt{InsideCritical}(sg, st)$ and $\mathtt{OutsideCritical}(sg, st)$, st refers to the last-seen high-level program state within a critical section. It is important to note that these two assertions are used to identify low-level program states inside and outside a critical section, without modifying the memory; while any other low-level assertion only cares about the memory, that is, specifying an element in the form of $(\mathcal{M}, CST, \varnothing)$. And, we define the separating conjunction of two low-level assertions as $\{(\mathcal{M}, s_{state}, s_{token})\} \ast\ast \{(\mathcal{M}', s'_{state}, s'_{token})\} = \{(\mathcal{M} \ast \mathcal{M}', s_{state} \cap s'_{state}, s_{token} \cup s'_{token})\}$, where $\ast$ denotes the standard separating conjunction for memory mappings.

Finally, we present standard Hoare triples for "`enter_critical();`" and "`exit_critical();`" in Fig. 9. Every entry to and exit from a critical section requires the accessed shared resources to satisfy a specific invariant.

Remark 4. Inspired by [10,11], we propose an STS-based framework to model concurrency in Sect. 5.2. This basic framework can be tailored by readers to meet specific verification demands. For example, we only present the simplest

$\{$ `OutsideCritical`$(sg, st)\}$
`enter_critical();`
$\{\exists sg_1, st_1.\ (sg, st) \rightsquigarrow_{\text{RTrans-L}} (sg_1, st_1)$ `&&`
`InsideCritical`(sg_1, st_1) `**`
`OsInv`$(sg_1, st_1)\}$

$\{$ `InsideCritical`(sg_1, st_1) `**`
`OsInv`(sg_2, st_2) `&&` $st_2 =$ `Replace`(st_1) `&&`
$(sg_1, st_1) \rightsquigarrow_{\text{GTrans-L}} (sg_2, st_2)\}$
`exit_critical();`
$\{$ `OutsideCritical`$(sg_2, st_2)\}$

Fig. 9. Standard Hoare Triples for the Entry to and the Exit from a Critical Section. (Here, `OsInv`(sg, st) is an invariant that must be satisfied by the shared resources when entering and exiting a critical section; `Replace`(st_1) is a function updating the high-level program state induced by executing the code within the critical section.)

case with a single token `tt`, distinguishing the states inside the critical section from those outside. For more complex scenarios, such as nested critical sections, the token sort can be set as the set of integers: a state outside the critical section has a token value of 0, while a state inside the critical section has a token value equal to its nesting depth.

6 Verification of LiteOS-M

Using the above framework, we successfully complete the verification of the whole LiteOS-M kernel. Section 6.1 constructs the invariant for LiteOS-M. Section 6.3 evaluates the Smart Verifier platform. Finally, Sect. 6.2 presents the verification efforts and identified defects.

6.1 The Invariant for LiteOS-M

As shown in Fig. 10, we construct an invariant `liteos_inv` for the LiteOS-M kernel, which conjuncts the invariants of all kernel modules.

Take the task module invariant `task_inv` as an example (also see Fig. 10). The pure proposition on Line 8 characterizes certain properties that a task maintains when the kernel enters and exits a critical section, such as (1) each task's ID and priority lie within the allowed range; (2) each task must be in one of these five states: running, ready, pending, delayed, or suspended; and (3) each task must appear in one of these three lists: the used task list, the free task list, or the recycled task list. The separation logic assertion on Line 9 states that the used task list whose elements are given by `s.(usedTaskList)` is stored in the corresponding global address variable. The separation logic assertions on Lines 10–11 are similar, describing the storage of the free task list and the recycled task list, respectively.

```
Definition liteos_inv (sg: GV) (st: ST):
    Assertion :=
        task_inv  sg st.(taskst) **
        sched_inv sg st.(schedst) ** ....

Definition task_inv (sg: GV) (s: TaskST):
    Assertion :=
        [| task_vaild_cond_prop sg s |] &&
        store_used_task     sg s.(usedTaskList) **
        store_free_task     sg s.(freeTaskList) **
        store_recycle_task sg s.(recycleTaskList) ** ....
```

Fig. 10. The Invariant for LiteOS-M. (Here, `liteos_inv` is the invariant for LiteOS-M, conjuncting the invariants of all kernel modules, such as `task_inv` for the task module; `ST` defines the type of single high-level program states for LiteOS-M, comprising the types for all kernel modules, such as `TaskST` for the task module.)

6.2 Verification Efforts and Identified Defects

Verification Efforts. The whole LiteOS-M kernel[3] has been verified. It consists of various modules, including task, scheduler, message queue, mutex, semaphore, event, sorted linked list, software timer, tick, static memory and dynamic memory. To enhance access control capabilities, an access control module has been incorporated into LiteOS-M, and we have verified this module as well. These verified modules comprise around 17,000 lines of C code including header files, comments, and blank lines in their original format, which we refer to as *raw code*. After excluding header files, comments, blank lines, and auxiliary functions[4] used for debugging (such as printing functions), the corresponding code amounts to around 5,400 lines, which we refer to as *clean code*.

Table 1 summarizes the size of each verified module, along with a detailed breakdown of the code size for each part in the verification. In addition to the libraries for each module, the verification relies on a *basic library* that includes the global variables, the state transition systems, as well as the invariant and the high-level program state defined for LiteOS-M. The basic library contains 1,046 lines of Rocq code. Table 2 compares the verification of LiteOS-M with two other verified operating systems, namely seL4 [7] and μC/OS-II [16]. Thanks to Smart Verifier and the tailored verification methods, LiteOS-M achieves a ratio of verification-related code to raw code at around $8:1$, and a ratio of verification-related code to clean code at about $27:1$.

Identified Defects. During the code review and verification, we found in total 17 critical bugs and 3 additional defects, all of which have been confirmed and fixed by the developers. The 17 critical bugs include: 10 concurrency bugs, of

[3] https://gitee.com/openharmony/kernel_liteos_m/tree/OpenHarmony-v4.0-Release.

[4] Clean code exclude auxiliary functions for debugging, since these are only used for troubleshooting and do not affect the program's final output.

Table 1. Verified LiteOS-M. (**C code:** lines of source C code; **Raw:** code in the original format; **Clean:** code excluding header files, comments, auxiliary functions, and blank lines. **Library:** lines of Rocq code for the kernel's library, including predicates, functions, and invariants. **Strategy:** lines of strategy code. **Annotation:** lines of annotation code, excluding source C code. **Proof:** lines of Rocq code for manual proofs; **Strategy:** proofs for strategies; **Manual VC:** proofs for manual VCs.)

Verified Module	C Code		Library	Strategy	Annotation	Proof	
	Raw	Clean				Strategy	Manual VC
Task	3722	730	3253	2242	5176	1598	10567
Scheduler	790	509	2042	328	2018	269	7283
Message Queue	2537	747	2137	1948	4473	3671	8254
Mutex	1022	459	2681	643	3463	622	10194
Semaphore	937	338	2735	472	2565	463	6712
Event	782	332	4720	167	1580	188	6342
Sorted Linked List	300	211	1056	237	823	248	3168
Software Timer	1491	620	2416	1232	2708	1062	8998
Tick	1129	244	410	98	485	103	1572
Static Memory	567	174	271	206	326	156	775
Dynamic Memory	3339	827	1430	465	2973	382	10623
Access Control	512	245	1579	696	1062	618	3139
Total	17128	5436	24754	8525	26023	9382	77783

which 8 are race conditions caused by missing critical section protection; 3 numerical overflow errors, all resulting from integer overflows; 3 functional errors, where the corresponding functions fail to implement the intended behavior; and 1 memory error, which is a use-after-free vulnerability. The 3 additional defects are attributed to the setting of an unused lock variable, the passing of unused function parameters, and a flaw in linked list traversal. Furthermore, the two critical bugs we identified are confirmed to exist in the OpenHarmony LiteOS-A kernel as well.

6.3 Evaluation of Smart Verifier

Most current OS verification projects are conducted using interactive theorem provers, such as seL4 [7] in Isabelle/HOL and μC/OS-II [16] in Rocq. These interactive theorem provers cannot automatically generate verification conditions, and provide limited automation through reasoning plugins. We use Smart Verifier to complete the verification of LiteOS-M. On the one hand, given annotated C programs, Smart Verifier can automatically generate verification conditions. On the other hand, Smart Verifier can automatically prove some verification conditions by built-in SMT solvers. Taking the task, scheduler, and dynamic memory modules of LiteOS-M as examples, Smart Verifier automatically proves exceeding 80% of all the generated verification conditions (see Table 3).

Table 2. Comparison of OS Kernel Verification. (**Raw Code:** lines of source code in the original format. **Clean Code:** lines of source code in the clean format. **Verification Code:** lines of verification-related code. **–:** relevant data not found.)

	seL4 [7]	μC/OS-II [16]	LiteOS-M
Concurrency Support	not supported	single-core concurrency	single-core concurrency
Raw Code	9300	3250	17128
Clean Code	–	1300	5436
Verification Code	200000	104808	147513
Verification Code: Raw Code	21: 1	32: 1	8: 1
Verification Code: Clean Code	–	81: 1	27: 1
Verification Efforts (person-years)	20	5.5	12

Table 3. Automation of Smart Verifier. (**# VC:** Number of generated verification conditions (VCs). **# Manual VC:** Number of VCs manually proved in Rocq. **# Auto VC:** Number of VCs automatically proved by SMT solvers.)

	Smart Verifier			
	# VC	# Manual VC	# Auto VC	# Auto VC : # VC
Task	1840	320	1520	83%
Scheduler	836	124	712	85%
Dynamic Memory	1473	240	1233	84%
Total	4149	684	3465	84%

7 Conclusion

We design a formal verification platform and tailor refinement-based verification methods to ensure the functional correctness of LiteOS-M. It is worth noting that although our verification framework is developed to verify LiteOS-M, it is a general verification framework and can be reused to verify other preemptive OS kernels implemented in C.

Our experience verifying LiteOS-M suggests that verifying an existing OS kernel is more challenging than developing one for verification purposes. A considerable amount of time is spent on understanding the source code, identifying defects, and re-verifying after code modifications. Ideally, the development and formal verification of an OS kernel should be conducted simultaneously. Additionally, even with individual module responsibilities, unifying specifications and invariants is time-consuming at the code merging stage. This motivates us to explore more efficient collaborative ways for OS kernel verification tasks.

Acknowledgement. This work is supported by the National Cyber Security–National Science and Technology Major Project (No. 2025ZD1503300) and by Zhongguancun Laboratory. We are deeply grateful to Bohua Zhan for his expert guidance and significant assistance with the verification platform. We would like to acknowledge the contribution of all the team members on this verification project, especially Kai Yang,

Nianyu Li, Qianshan Yu, Wei Xin, Taofei Guo, Wenbo Xu, Mingxuan Zhu, Wenkai Xing, Sen Yan, Zirui Zhou, Zimai Guo, Xinlun Zhang, Weijun Hong, Jiale Xu, and Fanpeng Yang. Thanks to Xiwei Wu for his guidance and support in utilizing the QCP verification tool. Finally, we appreciate the reviewers for their valuable comments and constructive suggestions.

References

1. Abadi, M., Lamport, L.: The existence of refinement mappings. In: Proceedings of the Third Annual Symposium on Logic in Computer Science, LICS '88, Edinburgh, Scotland, UK, 5–8 July 1988, pp. 165–175. IEEE Computer Society (1988)
2. Abrial, J.: Modeling in Event-B - System and Software Engineering. Cambridge University Press (2010)
3. Cohen, E., et al.: VCC: a practical system for verifying concurrent C. In: Berghofer, S., Nipkow, T., Urban, C., Wenzel, M. (eds.) TPHOLs 2009. LNCS, vol. 5674, pp. 23–42. Springer, Heidelberg (2009). https://doi.org/10.1007/978-3-642-03359-9_2
4. de Moura, L., Bjørner, N.: Z3: an efficient SMT solver. In: Ramakrishnan, C.R., Rehof, J. (eds.) TACAS 2008. LNCS, vol. 4963, pp. 337–340. Springer, Heidelberg (2008). https://doi.org/10.1007/978-3-540-78800-3_24
5. Gu, R., et al.: Deep specifications and certified abstraction layers. ACM SIGPLAN Not. **50**(1), 595–608 (2015)
6. Gu, R., et al.: CertiKOS: an extensible architecture for building certified concurrent OS kernels. In: 12th USENIX Symposium on Operating Systems Design and Implementation (OSDI 16), pp. 653–669 (2016)
7. Klein, G., et al.: seL4: formal verification of an operating-system kernel. Commun. ACM **53**(6), 107–115 (2010)
8. Leinenbach, D., Santen, T.: Verifying the Microsoft Hyper-V Hypervisor with VCC. In: Cavalcanti, A., Dams, D.R. (eds.) FM 2009. LNCS, vol. 5850, pp. 806–809. Springer, Heidelberg (2009). https://doi.org/10.1007/978-3-642-05089-3_51
9. OpenAtom Foundation: OpenHarmony Official Website (2020). https://www.openharmony.cn
10. Turon, A., Dreyer, D., Birkedal, L.: Unifying refinement and Hoare-style reasoning in a logic for higher-order concurrency. In: Proceedings of the 18th ACM SIGPLAN International Conference on Functional Programming, pp. 377–390 (2013)
11. Turon, A.J., Thamsborg, J., Ahmed, A., Birkedal, L., Dreyer, D.: Logical relations for fine-grained concurrency. ACM SIGPLAN Not. **48**(1), 343–356 (2013)
12. Wang, Z., et al.: Stellis: a strategy language for purifying separation logic entailments. arXiv preprint arXiv:2512.05159 (2025)
13. Wu, S., Wu, X., Cao, Q.: Encode the ∀ ∃ relational Hoare logic into standard Hoare logic. Proce. ACM Program. Lang. **9**(OOPSLA2), 2454–2481 (2025)
14. Wu, X., et al.: QCP: a practical separation logic-based c program verification tool. arXiv preprint arXiv:2505.12878 (2025)
15. Wu, Y., Xia, B., Xu, J., Zhan, B., Zhao, T.: OSVAuto: automatic proofs about functional specifications in OS verification. arXiv preprint arXiv:2403.13457 (2024)
16. Xu, F., Fu, M., Feng, X., Zhang, X., Zhang, H., Li, Z.: A practical verification framework for preemptive OS kernels. In: Chaudhuri, S., Farzan, A. (eds.) CAV 2016. LNCS, vol. 9780, pp. 59–79. Springer, Cham (2016). https://doi.org/10.1007/978-3-319-41540-6_4

Automated LTL Specification Generation from Industrial Aerospace Requirements

Zhi Ma[1], Xiao Liang[1], Cheng Wen[2](✉), Rui Chen[3,5], Bin Gu[3,5](✉), Shengchao Qin[1,2], Cong Tian[1], and Mengfei Yang[4]

[1] School of Computer Science and Technology, Xidian University, Xi'an, China
mazhi@xidian.edu.cn
[2] Guangzhou Institute of Technology, Xidian University, Guangzhou, China
wencheng@xidian.edu.cn
[3] Beijing Institute of Control Engineering, Beijing, China
[4] China Academy of Space Technology, Beijing, China
[5] Beijing Sunwise Information Technology Ltd., Beijing, China

Abstract. In the development and verification of safety-critical aerospace software, Linear Temporal Logic (LTL) has been widely used to specify complex system properties derived from requirements. However, a significant gap remains in industrial practice: translating natural language (NL) requirements into formal LTL properties is a labor-intensive and error-prone process that requires rare expertise in both aerospace control engineering and formal methods. While recent NL-to-LTL tools (*e.g.*, NL2SPEC, NL2TL, NL2LTL) are capable of automating parts of this process, they often fail on real requirement documents in industrial settings, due to complex domain terminology or implicit temporal and logical structure. To address these challenges, we present AeroReq2LTL, a framework that automates LTL property generation for aerospace requirements using large language models (LLMs), with two key industrial innovations: (i) a data dictionary that normalizes technical jargon into precise atomic propositions; and (ii) a template-based requirement language that makes temporal cues and logical relations explicit before translation. On a real aerospace dataset, AeroReq2LTL achieves 85% precision and 88% recall in LTL generation, and its outputs can be directly consumed by existing verification tools.

1 Introduction

Software verification for safety-critical aerospace control software must provide strong and auditable assurance under tight engineering constraints, including frequent requirement changes, limited verification expertise, and strict expectations on evidence quality. In such settings, formal methods are often adopted because they can offer mathematically grounded guarantees beyond what traditional testing can typically provide. Within this domain, Linear Temporal Logic (LTL) [1] has emerged as the preferred formal specification language for expressing complex system properties. Because aerospace software is inherently

A. Sampaio and M. Stoelinga (Eds.): FM 2026, LNCS 16557, pp. 670–690, 2026.
https://doi.org/10.1007/978-3-032-26220-2_33

reactive—interacting continuously with an environment over time—its correctness depends not only on functional behavior but also on precise temporal properties [2]. LTL provides the mathematical rigor necessary to express critical safety invariants and liveness conditions [3], which are widely supported by existing model checkers and runtime verification tools.

Despite these advantages, the industrial adoption of formal methods is impeded by the formalization bottleneck: the manual translation of natural-language (NL) requirements into precise LTL properties. This process is labor-intensive and error-prone, requiring rare expertise in both aerospace control engineering and formal logic. Furthermore, recent research exploring Large Language Models (LLMs) [4–7] for this task (*e.g.*, NL2SPEC [8], NL2TL [9], NL2LTL [10]) has highlighted a fundamental limitation: while these tools succeed on synthetic benchmarks, they often fail on real industrial documents (as shown in our motivating example and experiments under the industrial setting).

The core challenge lies in the heterogeneous nature of industrial requirements. In practice, aerospace requirement documents contain highly specialized jargon, ambiguous abbreviations [11], and implicit temporal and logical structures that are not explicitly stated in the text. Consequently, existing tools often produce incorrect or incomplete specifications, limiting their applicability in practice.

To address these challenges, we present AeroReq2LTL, a systematic engineering framework designed to automate the generation of LTL specifications from industrial aerospace requirements. AeroReq2LTL leverages the reasoning capabilities of LLMs but enhances them through two key industrial innovations:

- SpaceKG: A domain-specific data dictionary that automatically ingests engineering artifacts (such as interface tables) to resolve technical jargon into precise, code-level atomic propositions.
- SpaceRDL: A structured template language that guides LLMs to uncover implicit temporal keywords and logical relations within the context of aerospace control logic.

Designed for seamless toolchain integration, AeroReq2LTL facilitates a direct connection with the subsequent verification tool to support batch processing and iterative feedback loops. The framework was evaluated using 79 production requirements from real-world spacecraft control systems. Experimental results demonstrate that AeroReq2LTL achieves 85% precision and 88% recall, substantially reducing manual formalization effort and establishing a practical, automated pathway for rigorous software assurance within the aerospace industry.

2 Background and Motivation

2.1 The Complexity of Industrial Requirements

In academic research, the task of translating natural language to Linear Temporal Logic (LTL) is often simplified to mapping short, isolated sentences. Table 1 presents six representative examples adopted in prior work, including

Table 1. Examples of Natural Language and Corresponding LTL Specifications

Natural Language	LTL Specification
[9] Once red, the light cannot become green next.	$G(\text{red} \rightarrow X\neg\text{green})$
[9] Once the light is red, it must remain red until it turns yellow.	$G(\text{red} \rightarrow \text{red}\, U\, \text{yellow})$
[8] If a holds then c is true until b.	$\text{a} \rightarrow (\text{c}\, U\, \text{b})$
[8] If b holds, next c holds until a holds or always c holds.	$\text{b} \rightarrow X((\text{c}\, U\, \text{a}) \vee G\,\text{c})$
[10] Navigate to the green room while avoiding landmark 1.	$(F\,\text{green}) \wedge G(\neg\text{landmark 1})$
[10] Swing by landmark 1 before ending up in the red room.	$F(\text{landmark 1} \wedge (F\,\text{red}))$

NL2LTL [10], NL2SPEC [8], and NL2TL [9]. In these examples, the correspondence between natural language statements and LTL specifications is explicit and largely unambiguous, rendering the translation task relatively straightforward.

However, industrial requirements are fundamentally different, as illustrated by the example simplified from the ACS-LEOS[1] in Fig. 1(a). They are not merely free-form prose but heterogeneous engineering artifacts characterized by three distinct layers of complexity: (1) *Multiple types of context*: A single functional requirement (*e.g.*, `Req13`) cannot be parsed in isolation. Its meaning is bound to the *Interface Description*, which specifies port-level information such as signal names, data types, initial values, and valid ranges, as well as the *Invocation Condition*, which specifies the 128 ms control cycle. (2) Domain-Specific Semantics: Technical terminologies often act as placeholders for complex logic. In `Req13`, the phrase "absolute values of the angular rates on all axes are less than 0.15°/s" is an empirical condition for spacecraft stability. Without the domain knowledge found in the interface port explanations, an automated tool cannot map this intent to the correct variable, `dwCount`. (3) Implicit Temporal Logic: Industrial requirements often omit explicit temporal operators, for example, in `Req13`, the term "switches from...to" implies a state transition that must occur at the next time step (X) once a sustaining condition is met.

2.2 Limitations of General-Purpose LLMs

While LLMs like GPT-4 possess powerful reasoning capabilities, the generated LTL specifications are often unsatisfactory [12,13]. As shown in Fig. 1(b), although the model attempts to incorporate all relevant information in the requirement (`Req13`), the generated LTL specification deviates substantially from the intended semantics. This mismatch stems primarily from three primary reasons. (1) *Contextual Blindness to Engineering Artifacts.* LLMs struggle to connect natural-language requirements with their associated Interface Descriptions.

[1] The Attitude Control Software for Low Earth Orbit Satellites (ACS-LEOS) is responsible for processing sensor data from sunlight and infrared instruments, managing transitions among multiple operating modes, and executing attitude adjustment tasks such as three-axis stabilization, spin stabilization, and sun-pointing control.

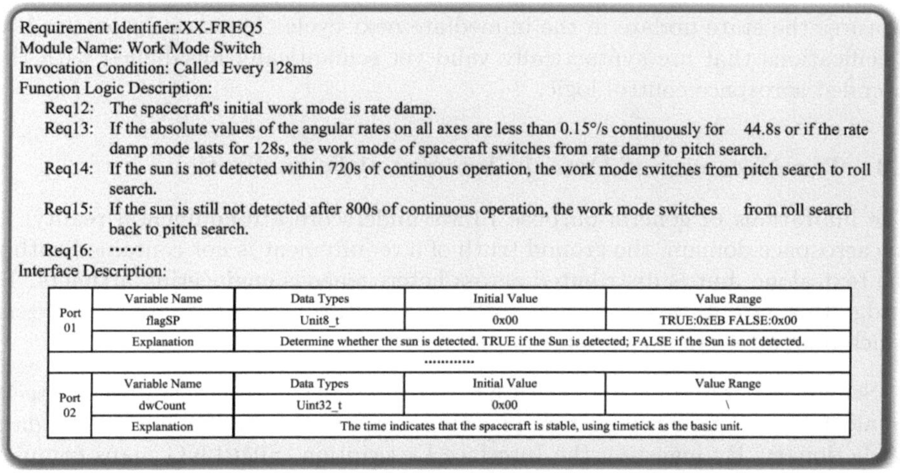

Requirement Identifier:XX-FREQ5
Module Name: Work Mode Switch
Invocation Condition: Called Every 128ms
Function Logic Description:

Req12: The spacecraft's initial work mode is rate damp.

Req13: If the absolute values of the angular rates on all axes are less than 0.15°/s continuously for 44.8s or if the rate damp mode lasts for 128s, the work mode of spacecraft switches from rate damp to pitch search.

Req14: If the sun is not detected within 720s of continuous operation, the work mode switches from pitch search to roll search.

Req15: If the sun is still not detected after 800s of continuous operation, the work mode switches from roll search back to pitch search.

Req16: ······

Interface Description:

Port 01	Variable Name	Data Types	Initial Value	Value Range
	flagSP	Unit8_t	0x00	TRUE:0xEB FALSE:0x00
	Explanation	Determine whether the sun is detected. TRUE if the Sun is detected; FALSE if the Sun is not detected.		
············				
Port 02	Variable Name	Data Types	Initial Value	Value Range
	dwCount	Uint32_t	0x00	\
	Explanation	The time indicates that the spacecraft is stable, using timetick as the basic unit.		

(a) The requirement description for the Mode Switch Function Module of ACS-LEOS.

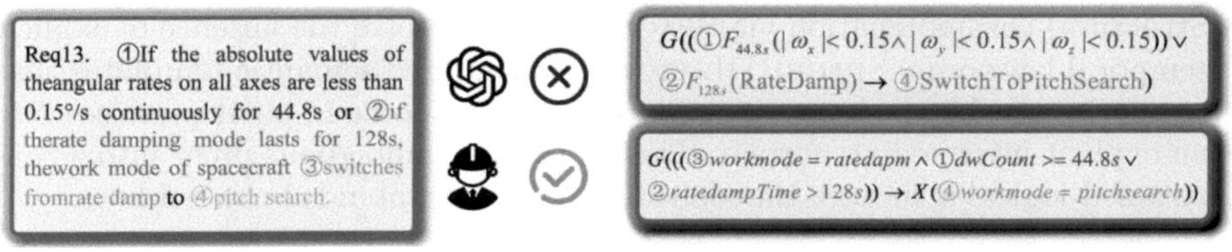

(b) Misalignment of LLM-Generated LTL with Industrial Requirements.

Fig. 1. The complexity and accuracy in NL requirement and its specification.

This results in "grounding errors", where the model cannot link a requirement to the actual system ports or data types defined in the document's tables. (2) *Insufficient Understanding of Domain-specific Terminology.* Industrial requirement statements frequently embed long technical expressions that domain experts interpret as single semantic units. Lacking domain knowledge, LLMs tend to fragment such expressions into multiple isolated noun phrases, yielding atomic propositions that are semantically meaningless and difficult to maintain (*e.g.*, ① in `Req13`, shown in Fig. 1(b)). (3) *Incomplete Modeling of Implicit Temporal Constraints.* Industrial requirements often rely on implicit premises that are not explicitly marked by temporal keywords. In `Req13`, the fragment ③, "*switches from rate damp,*" specifies a prerequisite mode that an LLM may fail to place on the left-hand side of a logical implication. Furthermore, LLMs frequently omit essential temporal operators like X (Next), which are necessary to indicate that a mode transition occurs in the subsequent control cycle. It should be noted that, while the X operator is often considered semantically unpredictable in general asynchronous systems, it is fully safe and deterministic in aerospace supervisory control software with fixed, strictly scheduled execution cycles, where X precisely captures the state update in the immediate next cycle. These omissions lead to specifications that are syntactically valid yet semantically misaligned with the intended aerospace control logic.

2.3 The Necessity of Domain-Integrated Formalization

The limitations of general-purpose LLMs underscore a fundamental reality: in the aerospace domain, the ground truth of a requirement is not contained within the text alone, but is distributed across heterogeneous engineering artifacts. To bridge this gap, a framework must perform multi-source information synthesis, which motivates our two-pronged approach:

- Semantic Grounding via SpaceKG: To prevent the fragmentation of technical terms, the formalization process is grounded in a domain-specific data dictionary. By ingesting the Interface Description, SpaceKG maps complex empirical constraints to precise, code-level atomic propositions. This ensures generated predicates are semantically consistent with actual system signals.
- Structural Disambiguation via SpaceRDL: To resolve the absence of explicit temporal keywords, SpaceRDL provides structured semantic templates that regularize the LLM's reasoning. These templates force implicit premises into an explicit intermediate format. This transforms ambiguous prose into a constrained logical structure that faithfully captures intended control laws.

By integrating SpaceKG and SpaceRDL, the formalization process shifts from a fragile "black-box" translation into a controllable engineering activity.

3 The AeroReq2LTL Framework

3.1 Overall Workflow

As illustrated in Fig. 2, the AeroReq2LTL architecture operates in three consecutive stages, specifically addressing the challenges of *contextual blindness to engineering artifacts*, *insufficient understanding of domain-specific terminology*, *incomplete modeling of implicit temporal constraints* identified in Sect. 2.2.

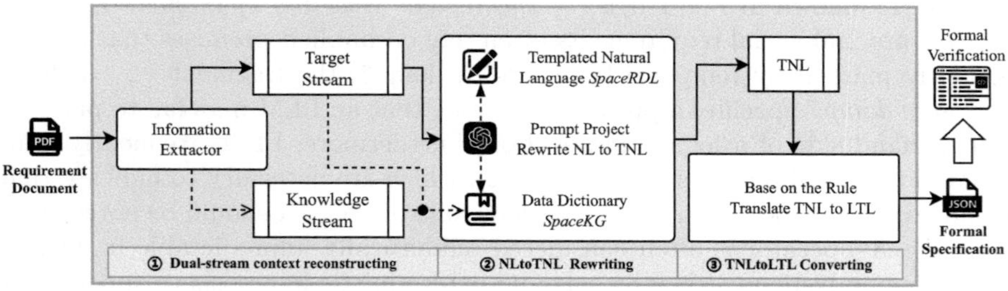

Fig. 2. Workflow of AeroReq2LTL.

Phase ①: Dual-stream Context Reconstructing. To address the issue of contextual blindness—where requirements are misinterpreted when analyzed as isolated sentences—AEROREQ2LTL begins by processing the raw PDF document through a dual-stream extraction strategy. Using a layout-aware information extractor to parse complex industrial artifacts[2], the framework extracts two complementary information streams from the requirement document:

- *Target Stream (Requirement Paragraphs):* Instead of fragmented sentences, the framework extracts requirement statements as complete functional paragraphs. This preservation of paragraph structure allows LLMs to perceive the broader functional logic and explicit invocation conditions (*e.g.*, control cycles). Providing this textual context is essential for LLMs to perform accurate subsequent temporal reasoning.
- *Knowledge Stream (Interface Artifacts):* Simultaneously, AEROREQ2LTL harvests technical metadata from the *Interface Description* tables associated with each module. By extracting signal names, data types, and operational ranges, this stream provides the structured "ground truth" necessary for the subsequent construction of the SPACEKG data dictionary. This ensures that linguistic expressions are grounded in actual engineering constraints.

Phase ②: NL-to-TNL Rewriting. In the second stage, the framework transitions from open-ended natural language to *Templated Natural Language* (TNL). Here, LLMs' functions not as a stochastic generator, but as a semantic parser guided by two domain-specific components:

- *Grounding via SPACEKG*: To prevent terminology fragmentation, the previously extracted target and knowledge stream are ingested into SPACEKG. Through a pipeline of BERT-based classification and expert-defined mapping, complex domain terms are normalized into precise, code-level *Atomic Paradigms.*
- *Structuring via SPACERDL*: To recover implicit temporal constraints, LLMs instantiate semantic structures defined by SPACERDL. By forcing requirements into predefined slots—such as `workmode`, `condition`, `timing`, `action` —the framework makes hidden premises and temporal operators (*e.g.*, X for next-cycle transitions) explicit within the TNL representation.

Phase ③: TNL-to-LTL Converting. The final stage eliminates inference stochasticity by replacing probabilistic LLMs generation with deterministic translation rules. Since the TNL is already structurally consistent and semantically grounded, the framework applies fixed mapping logic to convert TNL constructs into syntactically valid LTL specifications. This two-step process (NL → TNL → LTL) ensures that the final formal properties are syntactically valid and mathematically aligned with the intended aerospace control laws.

[2] We utilize the open-source `PDFplumber` toolkit to robustly handle complex industrial layouts.

3.2 SpaceKG: Domain Grounding for Aerospace Requirements

SpaceKG serves as the foundational semantic grounding layer of AeroReq2LTL, specifically designed to mitigate the terminology fragmentation of industrial documentation. In aerospace requirements, complex engineering predicates—such as "*angular rates* $< 0.15°/s$"—are frequently misinterpreted by text-only LLMs as isolated noun phrases. SpaceKG resolves this by normalizing such expressions into unambiguous *Atomic Paradigms* that link natural language intent directly to system signals.

Knowledge Source and Construction. SpaceKG is constructed from the *Interface Description* sections identified during the context reconstruction phase. Unlike external ontologies, it derives knowledge from the same engineering artifacts that connect requirements to implementation, ensuring that the formalization is grounded in the exact variable definitions used in the software implementation. The construction pipeline proceeds in three steps:

- *Data Retrieval:* Harvesting variable names, data types, and explanatory semantics from interface tables.
- *Intelligent Classification:* Utilizing a BERT-based terminology extractor to categorize entities into domain terms, variables, or concrete values.
- *Expert Mapping:* Collaborating with domain experts to map complex empirical conditions (*e.g.*, "*angular rates* $< 0.15°/s$") into standardized atomic expressions (*e.g.*, `dwCount`).

In practice, variable names are combined with their associated value constraints to form atomic paradigms suitable for formal reasoning. For example, the empirical constraint "*the absolute values of the angular rates on all axes are less than* $0.15°/s$ *continuously for 44.8 s*" is normalized into the atomic expression "$dwCount > 44.8s$", which preserves the engineering intent while eliminating linguistic redundancy. Figure 3 illustrates representative examples of original requirement statements, corresponding SpaceKG entries, and the resulting normalized expressions.

Formal Definition. Formally, SpaceKG is defined as a labeled graph SpaceKG $(\boldsymbol{V}, \boldsymbol{E})$, where $\boldsymbol{V}$ is the set of nodes representing domain entities and $\boldsymbol{E} \subseteq \boldsymbol{V} \times \boldsymbol{V}$ is the set of edges representing semantic relationships among them. Each node $v \in \boldsymbol{V}$ is mapped to a semantic class via a function $\tau : \boldsymbol{V} \rightarrow \boldsymbol{C}$. The entity classes include domain terms (c_{dt}), abbreviations or alternate names (c_{an}), program variable names (c_{vn}), and concrete values (c_{nv}). The values in c_{nv} encompass not only booleans, integers, and floating-point numbers, but also predefined operational modes such as "$RDSM$", indicating the rate-damp work mode.

Edges in SpaceKG are labeled by relationship types $r \in \boldsymbol{R}$ and mapped to pairs of entity classes via a function $\rho : \boldsymbol{R} \rightarrow \boldsymbol{C} \times \boldsymbol{C}$. Two primary relationship types are defined. The first, r_a (assignment/conditional relation), explicitly captures the logical or arithmetic constraints between program variables and their specific concrete values; for example, mapping the informal condition "*sun is not detected*" to the rigorous boolean evaluation "**flagSP** = **FALSE**". The second, r_m (semantic mapping relation), defines the semantic equivalences between formal program variables and various informal domain terms or aliases found in the text; for instance, establishing that both natural language terms "*sun*" and "*sunlight*" semantically refer to the same underlying variable **flagSP**. The second, r_m, captures semantic mappings between program variables and domain terms or their aliases; for example, both "*sun*" and "*sunlight*" map to the variable **flagSP**. Notably, in the aerospace domain, a domain term may correspond to a complete sentence rather than a single noun phrase, reflecting the complexity of engineering conventions in industrial requirement documents.

3.3 *SpaceRDL* and TNL: Explicitizing Implicit Temporal Semantics

The second core contribution of AEROREQ2LTL is SPACERDL, a structured requirement description language designed to make implicit temporal and logical assumptions explicit in aerospace control requirements. As discussed in Sect. 2, industrial aerospace requirements frequently rely on domain conventions—such as control cycles, physical constraints, and work-mode transitions—without explicitly stating temporal operators, which leads to ambiguity in automated formalization.

SPACERDL is derived from an empirical analysis of real aerospace control requirement documents. Although requirement statements exhibit diverse surface forms, their underlying semantics consistently reflect the structured and mode-driven nature of spacecraft control systems. This observation suggests that the apparent linguistic diversity of requirements can be systematically constrained by an explicit semantic structure aligned with the operational logic of aerospace control software. Accordingly, SPACERDL provides a domain-driven semantic scaffold upon which heterogeneous natural-language requirements can be normalized and analyzed.

SPACERDL Schema. Based on the above observation, SPACERDL defines a fixed set of semantic fields that explicitly capture the core elements of aerospace

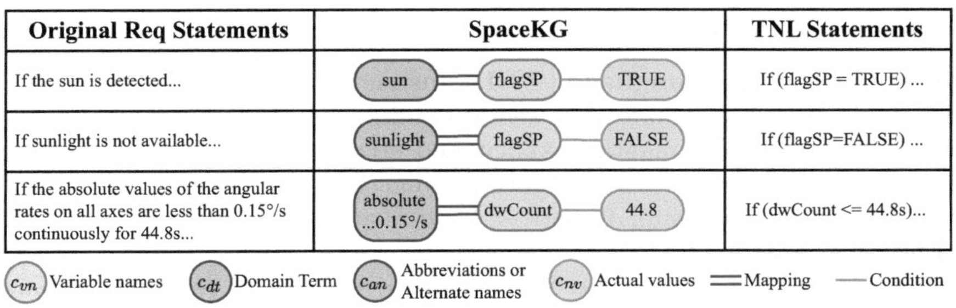

Fig. 3. Examples of requirement statements, corresponding SPACEKG entries, and normalized atomic paradigms used in templated natural language (TNL).

Table 2. Representative SpaceRDL requirement templates and example TNL instantiations abstracted from aerospace industrial requirements.

No.	Type	Requirement Templates and TNL Statements
1	State Change	**Component** shall always satisfy if (**input_state** & **condition**) then next **output_state**.
		'Communicator' shall always satisfy if (**'signal_lost'** & **'mission_phase'**) then next**'backup'**.
2	Bound Check	In **work_mode**, the **component** shall always satisfy **bounds**.
		In **'pitch_search mode'**, the **'gyroscope'** shall always satisfy **'roll_target_velocity = 0'**.
3	Work Mode Change	Upon (**input_work_mode** & **condition**) **component** shall at the next timepoint satisfy **output_work_mode**.
		Upon (**'pitch_search'** & **'sun_not_found'**) **'Spacecraft'** shall at the next timepoint satisfy **'roll_search'**.
4	Command Process	Upon **command** the **component** shall **timing** satisfy **action**.
		Upon **'command_opcode = 29'** the software shall **'immediately'** satisfy **'retract_landing_gear_command'**.
5	State Response	In **work mode** the **component** shall immediately satisfy if **condition** then **response**.
		In **roll_hold mode** the **'Spacecraft'** shall immediately satisfy if **'roll_angle_up'** then **'roll_hold'**.
6	State Maintenance	The **component** shall maintain **state** unless **condition**.
		The **'alarm system'** shall maintain **'active'** unless **'system_disabled'**.
7	....	....

control logic:

$$\text{SPACERDL} = \{\texttt{workmode}, \texttt{condition}, \text{component}^*, \texttt{SHALL}^*, \texttt{timing}^*, \texttt{action}^*\}$$

Fields marked with an asterisk are mandatory. The modal verb `SHALL*` is immutable and enforces declarative requirement expressions. The `component*` field specifies the executor of the control action, while the optional `workmode` field constrains the operational scope under which the requirement applies; if omitted, the requirement is interpreted as globally applicable across all work modes. The `timing*` field explicitly encodes temporal intent using a predefined set of keywords such as *immediately*, *next*, *eventually*, *always*, and *until*, which correspond to standard LTL operators. The `condition` field captures either trigger or sustaining semantics and supports logical composition through conjunctions and disjunctions. Finally, the `action*` field specifies the concrete operational behavior, state transition, or command execution that the target component is mandated to perform once the associated conditions and temporal constraints are satisfied. By enforcing these structural constraints, SPACERDL prevents implicit temporal premises from being silently omitted during formalization.

Templated Natural Language (TNL). Instantiating the fields of SPACERDL yields TNL, an intermediate representation that preserves human readability while making control semantics explicit and structured. Rather than treating requirements as free-form text, TNL constrains each requirement to conform to a predefined semantic template derived from SPACERDL.

Empirical analysis further shows that the majority of aerospace industrial requirements can be categorized into a limited number of recurring semantic patterns. These patterns correspond to fundamental control behaviors—such as state transitions, work-mode switches, boundary enforcement, command processing, and conditional responses—and form the basis of a compact template library. Representative SpaceRDL templates and example TNL instantiations are summarized in Table 2.

By enforcing this template-based structure, TNL serves as a semantic contract between informal natural-language requirements and subsequent formal LTL specifications, substantially reducing ambiguity caused by implicit temporal assumptions and heterogeneous phrasing.

4 Evaluation

The goal of our evaluation is to assess whether AeroReq2LTL can practically and reliably automate LTL specification generation from real industrial aerospace requirements. Accordingly, our evaluation is organized around a real industrial case study, supported by quantitative analysis and targeted ablation experiments.

4.1 Industrial Case and Dataset Construction

Our evaluation is conducted on the *Attitude Control Software for Low Earth Orbit Satellites* (ACS-LEOS), a production aerospace control system developed under strict safety and reliability constraints [14]. ACS-LEOS is responsible for processing sensor data from sunlight and infrared instruments, managing transitions among multiple operating modes, and executing attitude adjustment tasks such as three-axis stabilization, spin stabilization, and sun-pointing control [15,16]. We randomly selected nine control software engineering packages from the ACS-LEOS system as experimental subjects. Each package contains a complete set of requirement documents and the corresponding implementation code. Among them, the most recent package, the *Sun Search Control Software* (SSCS), was chosen as the primary evaluation object. SSCS is a critical autonomous control program that enables a satellite to automatically search for and point toward the Sun when its attitude is unknown or abnormal, thereby restoring solar power supply and ensuring survivability. As a representative high-reliability autonomous control system, SSCS activates different control modes in a staged manner, with explicit temporal dependencies and state-triggering conditions between stages. These characteristics make temporal logic particularly suitable for formally specifying its requirements.

Construction of SpaceKG. The remaining eight historical engineering packages were used to construct the initial SpaceKG. We first applied the *Information Extractor* of AeroReq2LTL to the requirement documents of these packages, resulting in the extraction of 427 requirement statements and 635 domain-specific terms from interface description tables. A BERT-based terminology extractor was then employed to identify and classify domain entities using deep semantic features, including variables, system states, and operating modes. At this stage, the focus was on capturing stable aerospace concepts and commonly used abbreviations (*e.g.*, mapping "rate damping" to `RDSM`).

In addition, aerospace engineers were involved to manually review and refine the automatically generated mappings between domain terms and atomic paradigms, particularly for interface variables with similar names but distinct semantics (*e.g.*, `flagSP` versus `flagSPS`). The validated results were consolidated into the knowledge base for subsequent SSCS requirement analysis. The resulting initial version of SpaceKG contains 407 predefined entity types.

Finally, to mitigate subtle terminology variations across different engineering documents, SpaceKG incorporates an evolution mechanism. When processing new SSCS requirement documents, AeroReq2LTL activates an on-the-fly import module to handle project-specific primitives. In cases where the same domain term corresponds to different variable names, project-local artifacts are prioritized. Through this mechanism, AeroReq2LTL dynamically updates SpaceKG, ensuring that complex empirical conditions (e.g., stability thresholds) are consistently mapped to the actual system ports defined in the current task document.

Establishing Ground Truth. The SSCS requirement document spans 38 pages and contains 79 production requirements across seven functional modules. To establish the ground-truth for evaluation, a panel of aerospace and formal methods experts manually formalized the 79 SSCS requirements. Of these, 76 were successfully translated into LTL, while three were excluded as high-level architectural summaries that do not contain temporal logic and cannot be translated into LTL. Each expert independently performed the translation, followed by a cross-review and discussion phase to finalize the reference LTL specifications used for scoring.

4.2 Evaluation Setup

We implemented AeroReq2LTL as an automation-assisted toolchain that processes raw PDF requirement documents and produces LTL specifications enriched with structured metadata (*e.g.*, module and component scope). The tool provides user interfaces for inspecting and correcting intermediate artifacts (extracted requirements, TNL, and generated LTL), supporting industrial and manual traceability.

Baselines. We compare AeroReq2LTL against several baseline methods across two dimensions: model backend and generation strategy. For model backends, we choose three representative large language models: GPT-3.5, GPT-4o [17], and DeepSeek-V3 [18]. For generation strategies, we consider three methods representing the current landscape of NL-to-LTL translation. (1) SimPro means direct translation using LLMs to assess the baseline"out-of-the-box" reasoning of LLMs; (2) NL2LTL [10]: a Python package developed by IBM Research that uses LLMs to translate NL instructions into LTL formulas. (3) NL2SPEC [8] adopts a template-guided prompting approach that builds LTL expressions in stages, emphasizing interpretability and semantic traceability.

Table 3. Effectiveness of AeroReq2LTL

Methods	Correct	Missing	Wrong	Spurious	Precision	Recall
SimPro + DeepSeek	15	0	61	3	0.19	0.20
SimPro + GPT-3.5	12	0	64	3	0.15	0.16
SimPro + GPT-4o	27	0	49	1	0.35	0.36
NL2LTL + DeepSeek	21	0	55	0	0.28	0.28
NL2LTL + GPT-3.5	18	0	58	0	0.24	0.24
NL2LTL + GPT-4o	37	0	39	0	0.49	0.49
NL2SPEC + DeepSeek	23	0	53	0	0.30	0.30
NL2SPEC + GPT-3.5	23	0	53	0	0.30	0.30
NL2SPEC + GPT-4o	46	0	30	0	0.61	0.61
AeroReq2LTL + DeepSeek	64	0	12	3	0.81	0.84
AeroReq2LTL + GPT-3.5	59	0	17	3	0.75	0.78
AeroReq2LTL + GPT-4o	**67**	**0**	**9**	3	**0.85**	**0.88**

SimPro means direct translation using LLMs with a zero-shot prompt.

Following common practice for fairness, we do not require baselines to adopt SpaceKG naming conventions; if a baseline identifies the correct atomic proposition scope and temporal structure, the result is considered correct.

Evaluation Metrics. We report *precision* and *recall* using statistical correctness criteria. Each requirement is classified into one of four outcomes: *Correct*, *Wrong*, *Missing*, or *Spurious*. A generated LTL property is counted as correct only if it matches the intended logical and temporal semantics; minor operator differences (*e.g.*, X vs. F) are considered *wrong* due to their significant impact on verification outcomes. If a requirement does not describe any checkable temporal behavior and therefore should not yield an LTL specification, then erroneously producing output is considered *spurious*. Conversely, if a requirement is expected to yield an LTL specification but the method fails to produce one, the outcome is classified as *Missing*.

4.3 Evaluation Result

Effectiveness. Table 3 summarizes end-to-end performance on the SSCS case study. Overall, AeroReq2LTL consistently outperforms baseline methods under real industrial document complexity. AeroReq2LTL + GPT-4o performs best with a precision of 85% and a recall of 88%. Across backends, AeroReq2LTL attains 85%, 75%, and 81% precision (GPT-4o, GPT-3.5, DeepSeek-V3 respectively), showing that the framework generalizes beyond a single model while benefiting from stronger LLM reasoning.

Table 4. Ablation Study of SpaceKG and SpaceRDL Using GPT-4o.

Components	Correct	Missing	Wrong	Spurious	Precision	Recall
AeroReq2LTL	67	0	9	3	0.85	0.88
w/o SpaceKG	53	0	23	1	0.69	0.70
w/o SpaceRDL	41	0	35	3	0.52	0.54

In contrast, baseline methods degrade substantially under industrial document complexity. A key reason is that industrial aerospace requirements cannot be interpreted as standalone sentences: correct formalization requires grounding technical jargon and numeric constraints to interface variables, and recovering temporal/logical structure that is often implicit. SimPro and NL2SPEC struggle primarily because they treat requirements as isolated text, failing to resolve the technical jargon and implicit dependencies defined in the document's interface description. While NL2LTL and NL2SPEC avoid *Spurious* outputs by being more conservative, AeroReq2LTL's proactive two-stage rewriting process (NL→TNL→LTL) exposes implicit conditions and timing intent in TNL, and then applies deterministic translation rules. This design improves coverage while preserving precision, with only a small increase in spurious outputs on highly generalized, non-checkable statements.

Ablation Study. We conducted ablation studies to quantify the necessity of SpaceKG and SpaceRDL in the formalization pipeline and to analyze the underlying causes of performance degradation when each component is removed, as shown in Table 4.

As shown in the results, removing SpaceKG causes precision and recall to drop to 69% and 70%, respectively. The primary errors are grounding failures, where domain-specific terms are either fragmented into semantically meaningless propositions or mapped to incorrect yet plausible variables, such as confusing `flagSP` with `flagSPS`. To better understand these failures, we further analyze the extraction accuracy of SpaceKG on the 79 SSCS requirements in Fig. 4. The extracted domain terms, variables, and expressions all demonstrate F1 scores exceeding 0.8, proving SpaceKG's critical role in preventing grounding errors through effective terminology identification and normalization.

Subsequently, removing SpaceRDL leads to a more severe performance degradation, with precision and recall dropping to 52% and 54%, respectively. These errors mostly stem from incorrect temporal structuring, including missing prerequisites on the left-hand side of implications and the incorrect placement or omission of temporal operators such as X. A confusion matrix further analyzes the model's ability to select appropriate semantic templates for requirement statements, as shown in Fig. 5. While AeroReq2LTL correctly classifies 71 statements, several *State-Transition* requirements are misclassified as *Work-Mode-Change*, and non-checkable requirements tend to be forcibly assigned a template. These misclassifications directly account for the temporal-

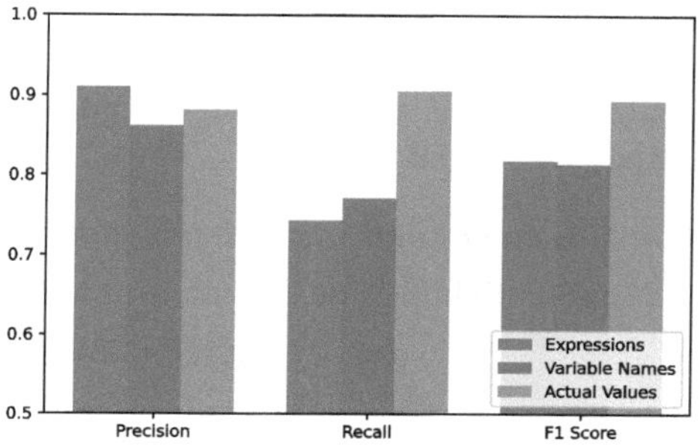

Fig. 4. Accuracy of SpaceKG.

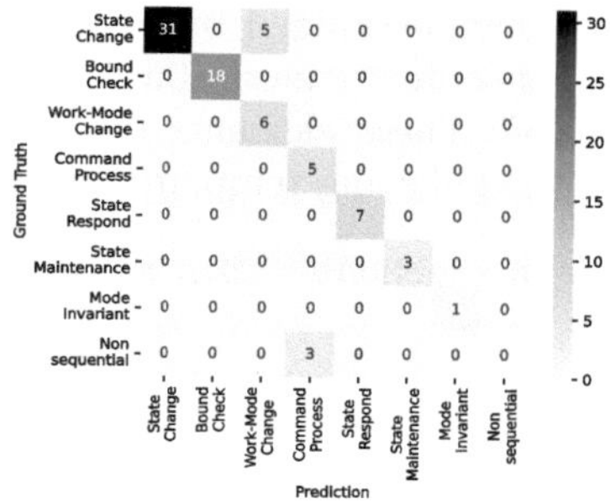

Fig. 5. Accuracy of SpaceRDL.

structure errors observed in the ablation results. These findings confirm that both SpaceKG and SpaceRDL are essential to the accuracy and reliability of AeroReq2LTL.

4.4 End-To-End Case Study and Toolchain Integration

To demonstrate practical usability, we present a detailed walkthrough using `Req39` from the SSCS *Mode Switch* module.

(`Req39`). *In the event that the BeiDou signal is unavailable for a duration exceeding 12.8 s while operating in inertial navigation mode, the navigation computer is required to initiate an immediate transition to the star tracker system.*

Step 1: Context Reconstruction and Knowledge Grounding. The workflow begins by jointly reconstructing textual and engineering context from the SSCS requirement document. The *Information Extractor* identifies the requirement paragraph of `Req39` as the *Target Stream*, while simultaneously extracting the associated *Interface Description* table as the *Knowledge Stream*. Based on this information, AeroReq2LTL determines that the BeiDou signal availability is represented by the boolean variable `flagBD`, the duration timer corresponds to the floating-point variable `deTCount`, and the star tracker system status is denoted by `flagSTS`. Based on this project-local evidence, AeroReq2LTL performs a just-in-time **delta refresh** of SpaceKG, augmenting its domain-level knowledge (*e.g.*, mapping "inertial navigation mode" to `subMode = INM`) with SSCS-specific signal grounding. This step is essential for industrial correctness. Without ingesting interface definitions, a text-only approach cannot associate phrases such as "duration exceeding 12.8 s" with the specific runtime variable `deTCount`, resulting in an underspecified or unverifiable property.

Step 2: Semantic Rewriting (NL → TNL). With the project-adapted SpaceKG in place, AeroReq2LTL invokes the LLM-based semantic rewriter to transform the original natural-language requirement into a structured intermediate representation. During this process, domain phrases are deterministically grounded: "BeiDou signal is unavailable" is mapped to `flagBD = FALSE`, and "duration exceeding 12.8 s" is resolved as the numeric constraint `deTCount >`

12.8. Meanwhile, guided by SpaceRDL templates, the rewriter identifies the operational context ("inertial navigation mode") and the implicit timing keyword "immediate". These elements trigger a *State–Response* template instantiation, yielding the following templated natural language specification:

TNL39: *In (*subMode=INM*), the navigation computer shall immediately satisfy if (*flagBD=FALSE & deTCount > 12.8*) then (*flagSTS=TRUE*).*

Step 3: Deterministic Formal Synthesis (TNL → LTL). In the third stage, the instantiated semantic structure (TNL39) is passed to the deterministic synthesis engine. To ensure rigorous formalization, AeroReq2LTL avoids probabilistic NL-to-LTL generation; instead, it applies formally defined SpaceRDL translation rules that map each TNL field to its corresponding LTL construct , thereby eliminating semantic ambiguity. In particular, the keyword "immediately" is interpreted as a concurrent state implication rather than a future eventuality, and thus no F (Eventually) operator is introduced. The resulting formal specification is:

$$\texttt{LTL39:}\ G((\texttt{subMode} = \texttt{INM} \wedge \texttt{flagBD} = \texttt{FALSE} \wedge \texttt{deTCount} > 12.8) \rightarrow (\texttt{flagSTS} = \texttt{TRUE}))$$

For comparison, a representative text-only baseline such as NL2SPEC abstracts domain predicates into uninterpreted symbols and models the response using a future operator. For instance, the original condition "the BeiDou signal is unavailable for a duration exceeding 12.8 s" is reduced to a symbolic atom (e.g., a), while the response "initiate an immediate transition" is formalized as an eventuality, yielding a specification of the form $G((a \wedge b) \rightarrow F\,c)$. This formulation omits the concrete timing threshold (> 12.8s) and weakens the immediacy constraint by deferring the response to an unspecified future point, rendering the resulting property unsuitable for direct verification in a control-critical setting.

Step 4: Direct Toolchain Integration and Verification. The final step evaluates whether the generated LTL specifications can be directly consumed by an existing industrial verification toolchain. The synthesized property LTL39, together with other correctly generated specifications, is used as drop-in input to the TRACE [19] framework for trace-based runtime verification[3]. TRACE takes LTL formulas as input and performs symbolic execution using KLEE [20,21] to generate execution paths, while GDB [22] is employed to collect concrete execution traces, which are subsequently checked against the temporal properties.

SSCS Verification Results. In the SSCS evaluation, AeroReq2LTL is applied to 76 industrial requirements without manual intervention, of which

[3] TRACE, developed by the Eindhoven University of Technology in the Netherlands, is an automated verification tool that takes LTL formulas as input and utilizes trace-based runtime verification technology to analyze system behavior by checking the execution traces of the system.

Table 5. Verification results of Generated LTL specifications using TRACE.

Sun Search Control System	
Module Name	**LTL properties**
`Remote Control Command Processing`	17/17
`Gyroscope Data Processing`	13/15
`Attitude Determination`	10/12
`Control Calculation`	5/5
`Mode Switch`	6/6
`Fault Handling`	8/8
`Thruster Output`	4/4
Total	63/67

67 are correctly translated into LTL. As summarized in Table 5, the downstream TRACE [19] toolchain successfully verifies 63 of these 67 properties across diverse functional modules. These results demonstrate that AeroReq2LTL produces syntactically valid and semantically precise specifications that are immediately usable by existing verification tools, effectively completing the end-to-end workflow from natural-language requirements to executable software verification.

4.5 Discussion on Scalability and Performance

While the current evaluation demonstrates the effectiveness of AeroReq2LTL on the 79 production requirements of the SSCS module, scaling this approach to the full set of requirements typically found in safety-critical aerospace platforms presents a distinct engineering challenge in and of itself. Complete systems often comprise over 1,000 behavioral requirements, alongside correspondingly massive and complex interface description documents.

In such large-scale industrial scenarios, the performance and scalability of the automated toolchain become critical considerations. The primary computational overheads in AeroReq2LTL stem from the dynamic construction of SpaceKG and the LLM inference latency. To effectively process thousands of requirements, the layout-aware information extraction and SpaceKG entity normalization must be highly optimized. This could be achieved through incremental knowledge base updates—selectively parsing modified interface tables rather than performing full-document re-ingestion. Furthermore, while AeroReq2LTL currently supports batch processing for downstream verification tools like TRACE, the LLM-based NL-to-TNL rewriting phase will necessitate advanced parallel processing and modularized semantic parsing. Addressing these performance bottlenecks and ensuring that generation times remain practical for continuous integration and iterative engineering feedback loops constitute an important direction for scaling AeroReq2LTL to broader, system-level aerospace projects.

5 Related Work

Recent works explore translating natural language requirements into Linear Temporal Logic using Large Language Models [8–10,23–25]. While effective on carefully constructed benchmarks, these approaches often fail in industrial settings such as robotics [24,26], multi-agent systems [27], and software verification [28]. Our analysis of these industrial failures reveals three recurring patterns:

1. Contextual blindness to engineering artifacts. Industrial specifications distribute state-transition logic across long documents. In telecommunications standards such as 3GPP [29], related conditions may appear in distant sections or appendices, making global consistency difficult for LLMs to maintain. As a result, extracted state machines are often fragmented. This challenge underscores the fundamental principle of the "Triptych" approach proposed by Dines Bjørner [30], which asserts that a rigorous domain description is a mandatory prerequisite for any requirements prescription [31]. Without an explicit domain model to provide the necessary context, requirements become isolated and prone to misinterpretation. Wu et al. [32] address this problem through an iterative, business-logic-driven framework that reconstructs implicit requirements by repeatedly aligning partial specifications with system-level objectives.

2. Insufficient understanding of domain-specific terminology [33–35]. In legal texts, LLMs frequently confuse obligation and permission, mistranslating *shall* or ignoring *unless* clauses [34]. Similar issues arise in hardware and protocol specifications, where global state-machine semantics in standards such as Arm are often misinterpreted [29,33]. Such terminological ambiguity often stems from the lack of a formal domain ontology, a gap that Bjørner suggests filling with domain engineering to capture the intrinsic properties of the application environment [36]. Surveys further report frequent errors in temporal semantics [37,38], such as treating *stable* as a static predicate rather than a clocked condition in SystemVerilog Assertions. Huang et al. [39] mitigate these issues via layered semantic analysis that incrementally extracts atomic propositions, mode constraints, and timing information.

3. Incomplete modeling of implicit temporal constraints [40]. Sequence-to-sequence models struggle with nested conditions, recursive triggers, and mode-dependent logic [23]. Flat translations fail to capture hierarchical temporal structure, and text-based metrics such as BLEU are insufficient, as small operator changes (*e.g.*, F vs. G) can drastically alter semantics [23,40]. Trajectory-based benchmarks like VLTL-Bench reveal poor semantic equivalence despite high textual similarity [40]. Consequently, recent work emphasizes structured and verifiable methods, including logic-aware decoding [23], interactive guidance [8,41], conformal guarantees [42], and intermediate semantic representations [32,39].

6 Conclusions

This paper presents AeroReq2LTL, an automated framework for extracting LTL specifications from aerospace requirement documents. By utilizing a large language model along with domain-specific knowledge and standardized templates, the framework improves the accuracy of temporal relationship extraction. AeroReq2LTL integrates seamlessly with existing verification tools like TRACE and CPAchecker and is already being applied in real aerospace software projects. Experimental results and case studies demonstrate its effectiveness, while ablation experiments highlight the challenges in rewriting natural language into templated forms.

Future work will focus on three directions to further improve the framework's generalizability and reliability. First, since SpaceKG is constructed from common engineering artifacts (e.g., interface description tables) rather than domain-specific hardcoding, the toolchain is inherently extensible beyond aerospace. We will apply it to other safety-critical domains with similar documentation standards, such as automotive and medical systems. Second, beyond control-related state transitions and work-mode logic, we will expand AeroReq2LTL to support a wider range of specifications, including functional, safety, and security requirements. Finally, we will further enrich and formalize the SpaceKG model by systematically encoding domain invariants into its core structure, thereby strengthening the rigor and validity of the automated verification process.

Acknowledgements. This work was supported in part by the National Natural Science Foundation of China (Nos. 62192730, 62192734, 62192735, 62302375, 62472339), the China Postdoctoral Science Foundation funded project (No. 2023M723736), the Basic Research Foundation of Shenzhen City (No. JCYJ20250604184202003), and the CCF-Huawei Populus Grove Fund (No. CCF-HuaweiFM202507).

References

1. Rozier, K.Y.: Linear temporal logic symbolic model checking. Comput. Sci. Rev. **5**(2), 163–203 (2011)
2. Merhav, S.: Aerospace sensor systems and applications. Springer Science & Business Media (1998)
3. Bauer, A., Leucker, M., Schallhart, C.: Runtime verification for LTL and TLTL. ACM Trans. Software Eng. Methodol. (TOSEM) **20**(4), 1–64 (2011)
4. Minaee, S., et al.: Large language models: a survey. arXiv preprint arXiv:2402.06196 (2024)
5. Naveed, H., et al.: A comprehensive overview of large language models. ACM Trans. Intell. Syst. Technol. **16**(5), 1–72 (2025)
6. Wen, C., et al.: Automatically inspecting thousands of static bug warnings with large language model: how far are we? ACM Trans. Knowl. Discov. Data **18**(7), 1–34 (2024)

7. Su, J., Deng, L., Wen, C., Qin, S., Tian, C.: CFStra: Enhancing configurable program analysis through LLM-driven strategy selection based on code features. In: Chin, WN., Xu, Z. (eds.) International Symposium on Theoretical Aspects of Software Engineering, pp. 374–391. Springer, Cham (2024). https://doi.org/10.1007/978-3-031-64626-3_22
8. Cosler, M., Hahn, C., Mendoza, D., Schmitt, F., Trippel, C.: nl2spec: Interactively translating unstructured natural language to temporal logics with large language models. In: Enea, C., Lal, A. (eds.) Proceedings of the 35th International Conference on Computer Aided Verification, volume 13965 of Lecture Notes in Computer Science, pp. 383–396, Paris, France, Springer, Cham (2023). https://doi.org/10.1007/978-3-031-37703-7_18
9. Chen, Y., Gandhi, R., Zhang, Y., Fan, C.: NL2TL: transforming natural languages to temporal logics using large language models. In: Proceedings of the Conference on Empirical Methods in Natural Language Processing, pp. 15880–15903. Association for Computational Linguistics, Singapore (2023)
10. Fuggitti, F., Chakraborti, T.: NL2LTL - a python package for converting natural language (NL) instructions to linear temporal logic (LTL) formulas. In: Proceedings of the 37th AAAI Conference on Artificial Intelligence, the 35th Conference on Innovative Applications of Artificial Intelligence, the 13th Symposium on Educational Advances in Artificial Intelligence, pp. 16428–16430, Washington, DC, USA. AAAI Press (2023)
11. Bernstein, J.I.: Design methods in the aerospace industry: looking for evidence of set-based practices. PhD thesis, Massachusetts Institute of Technology (1998)
12. Cao, J., et al.: From informal to formal–incorporating and evaluating LLMs on natural language requirements to verifiable formal proofs. In: Proceedings of the 63rd Annual Meeting of the Association for Computational Linguistics (Volume 1: Long Papers), pp. 26984–27003 (2025)
13. Wen, C., et al.: Enchanting program specification synthesis by large language models using static analysis and program verification. In: Gurfinkel, A., Ganesh, V. (eds.) International Conference on Computer Aided Verification, pp. 302–328. Springer, Cham (2024). https://doi.org/10.1007/978-3-031-65630-9_16
14. Ma, Z., Wen, C., Bin, Yu., Jie, S.: Integrating ensemble learning and large language models for efficient formal verification of IP-based aerospace systems. Inf. Fus. **125**, 103466 (2026)
15. Zhang, S., Liu, H.J., Ma, Z., Liang, X., Wen, C.: Formal verification of aerospace software IP components: a multi-tool case study. In: Proceedings of the 2026 6th International Conference on Computer Network Security and Software Engineering, pp. 1–9 (2026)
16. Ma, Z., et al.: Towards practical requirement analysis and verification: a case study on software IP components in aerospace embedded systems. arXiv preprint arXiv:2404.00795 (2024)
17. OpenAI. GPT-4o technical report. https://openai.com/index/gpt-4o (2024). Accessed May 2025
18. Zhang, S., Zhao, H., et al.: DeepSeek-VL: scaling vision-language models with vision token learner. arXiv preprint arXiv:2405.07927 (2024)
19. Hendriks, M., Geilen, M., Behrouzian, A.R.B., Basten, T., Ara, H.A., Goswami, D.: Checking metric temporal logic with trace. In: 2016 16th International Conference on Application of Concurrency to System Design (ACSD), pp. 19–24 (2016)
20. Cadar, C., Dunbar, D., Engler, D.R., et al.: KLEE: unassisted and automatic generation of high-coverage tests for complex systems programs. In: OSDI, vol. 8, pp. 209–224 (2008)

21. Cadar, C., Nowack, M.: Klee symbolic execution engine in 2019. Int. J. Softw. Tools Technol. Transfer **23**, 867–870 (2021)
22. van Hoey, J.: Program analysis with a debugger: GDB. In: Beginning x64 Assembly Programming: From Novice to AVX Professional, pp. 21–33. Springer, Berkeley, CA (2019). https://doi.org/10.1007/978-1-4842-5076-1_3
23. English, W., Simon, D., Jha, S.K., Ewetz, R.: Grammar-forced translation of natural language to temporal logic using LLMs. In: International Conference on Machine Learning (2025)
24. Liu, J.X., et al.: Lang2LTL: translating natural language commands to temporal specification with large language models. In: Workshop on Language and Robotics at CoRL 2022 (2022)
25. Xu, Y., Feng, J., Miao, W.: Learning from failures: translation of natural language requirements into linear temporal logic with large language models. In: 2024 IEEE 24th International Conference on Software Quality, Reliability and Security (QRS), pp. 204–215. IEEE (2024)
26. Liu, J.X., Shah, A., Konidaris, G.D., Tellex, S., Paulius, D.: Lang2ltl-2: grounding spatiotemporal navigation commands using large language and vision-language models. In: 2024 IEEE/RSJ International Conference on Intelligent Robots and Systems (IROS), pp. 2325–2332 (2024)
27. Shaojun, X., Luo, X., Huang, Y., Leng, L., Liu, R., Liu, C.: Nl2Hltl2Plan: scaling up natural language understanding for multi-robots through hierarchical temporal logic task specifications. IEEE Robot. Autom. Lett. **10**, 10482–10489 (2024)
28. Olausson, T.X., Inala, J.P., Wang, C., Gao, J., Solar-Lezama, A.: Demystifying GPT self-repair for code generation. ArXiv, abs/2306.09896 (2023)
29. Zhang, M., et al.: Automated extraction of protocol state machines from 3gpp specifications with domain-informed prompts and LLM ensembles. ArXiv, abs/2510.14348 (2025)
30. Bjøner, D.: Software Engineering 3: Domains, Requirements, and Software Design. Springer, Berlin, Heidelberg (2006). https://doi.org/10.1007/3-540-33653-2
31. Bjørner, D.: From Domain Descriptions to Requirements Prescriptions (2016)
32. Wu, Z., Chen, X., Jin, Z., Hu, M., Jin, D.: Unlocking the silent needs: business-logic-driven iterative requirements auto-completion. In: 2026 IEEE/ACM 48th International Conference on Software Engineering (ICSE '26), p. 12 (2026)
33. Shih, Y.-A., Lin, A., Gupta, A., Malik, S.: Flag: Formal and LLM-assisted SVA generation for formal specifications of on-chip communication protocols. ArXiv, abs/2504.17226 (2025)
34. Meloche, R., Amyot, D., Mylopoulos, J.: Towards legal contract formalization with controlled natural language templates. In: 2023 IEEE 31st International Requirements Engineering Conference (RE), pp. 317–322. IEEE (2023)
35. Ma, Z., et al.: Bridging natural language and formal specification–automated translation of software requirements to LTL via hierarchical semantics decomposition using LLMs. arXiv preprint arXiv:2512.17334 (2025)
36. Bjørner, D.: Domain engineering: a software engineering discipline in need of research. In: International Conference on Current Trends in Theory and Practice of Computer Science, pp. 1–17. Springer, Berlin, Heidelberg (2000). https://doi.org/10.1007/3-540-44411-4_1
37. Wu, N., Li, Y., Yang, H., Chen, H., Dai, S., Hao, C.: Survey of machine learning for software-assisted hardware design verification: past, present, and prospect. ACM Trans. Design Autom. Electron. Syst. (2024)

38. Abdollahi, M., Yeganli, S.F., Baharloo, M., Baniasadi, A.: A scoping review, challenges, and open issues. Electronics, Hardware design and verification with large language models (2024)
39. Huang, Y., Hu, M., Chen, X., Jin, Z., Xiao, S.: Modeling like peeling an onion: Layerwise analysis-driven automatic behavioral model generation. In: 2026 IEEE/ACM 48th International Conference on Software Engineering (ICSE '26), p. 12 (2026)
40. English, W., Walker, C., Simon, D., Jha, S.K., Ewetz, R.: Verifiable natural language to linear temporal logic translation: a benchmark dataset and evaluation suite. ArXiv, abs/2507.00877 (2025)
41. Mendoza, D., Hahn, C., Trippel, C.: Translating natural language to temporal logics with large language models and model checkers. In: 2024 Formal Methods in Computer-Aided Design (FMCAD), pp. 1–11 (2024)
42. Wang, J., Sundarsingh, D.S., Deshmukh, J.V., Kantaros, Y.: ConformalNL2LTL: translating natural language instructions into temporal logic formulas with conformal correctness guarantees. ArXiv, abs/2504.21022 (2025)

SeaCoral: A Collaborative Test Generation Toolset for Industrial Orchestration of Testing Tools

Nicolas Berthier[1(✉)], Steven de Oliveira[1], Nikolai Kosmatov[2], and Delphine Longuet[2]

[1] OCamlPro, Paris, France
{nicolas.berthier,steven.de-oliveira}@ocamlpro.com
[2] Thales Research and Technology, cortAIx Labs, Palaiseau, France
{nikolai.kosmatov,delphine.longuet}@thalesgroup.com

Abstract. Formal methods have been successfully used to develop advanced test generation techniques. While various test generation tools and test coverage criteria have been proposed, their adoption in industrial practice remains fragmented. This paper presents SeaCoral, a novel open-source toolset for testing C programs, which aims to facilitate the industrial application of diverse testing technologies by integrating them within a single framework. SeaCoral offers a comprehensive set of testing services, ranging from the specification of test objectives for selected test coverage criteria, through test case generation and the detection of uncoverable objectives, to the measurement of the resulting coverage. It integrates a rich set of analyzers: the fuzzer libFuzzer, the model-checker CBMC, the dynamic symbolic execution tool KLEE, and the static analyzer LUncov for detecting uncoverable test objectives. These analyzers enrich each other's results through a shared store of test objectives and a shared corpus of test cases. Particular attention is paid to the careful handling of runtime errors and initialization functions, which are essential for industrial applications. Initial experiments confirm the benefits of the proposed toolset.

Keywords: test generation toolset · fuzzing · model checking · symbolic execution · static analysis · test coverage criteria · KLEE · libFuzzer · CBMC · LUncov · SeaCoral

1 Introduction

Testing [4,31,34,42] remains the most popular software validation technique in industry. Many efficient testing tools have been developed for C programs, based either on formal methods based approaches such as *symbolic execution*, *model checking* and *constraint solving*, or on more heuristic techniques like *fuzzing* (e.g. [1,15–17,19,21,37,40]). Various *test coverage criteria* (a.k.a. *adequacy criteria* or *test criteria*) [2,4,18,27,42] have been proposed to define test objectives and to

A. Sampaio and M. Stoelinga (Eds.): FM 2026, LNCS 16557, pp. 691–710, 2026.
https://doi.org/10.1007/978-3-032-26220-2_34

evaluate the level of code coverage of a given test suite. Achieving a coverage criterion like statement coverage (all statements are executed at least once) or decision coverage (the two branches of all conditional statements are executed at least once) is often used as a confidence metrics for stakeholders. Moreover, some of these criteria are required by existing standards [22,25] for critical software.

Except for fuzzing, the use of these tools in industrial development processes remains limited. Our experience of industrial application suggests that this is due to several reasons: (i) the diversity of the tools and of their approaches makes them difficult to take in hand, each one having its own specifics; (ii) the more user-friendly tools offer light coverage guaranties while an important part of the setup still remains manual; (iii) the tools offering strong guaranties, based on formal methods, need manual writing of harnesses and drivers, as well as fine understanding of the underlying techniques to achieve efficient configuration; (iv) the results come in different tool-dependent formats that are not immediately usable in testing campaigns. The purpose of this work is *to facilitate industrial collaborative applications of existing testing technologies by integrating them within the same framework*. The objective is to propose a simplified access to these different tools through a common interface that minimizes the need for user configuration, and to achieve collaboration between them in such a way as to combine their forces and compensate their weaknesses.

Testing Tool Combination. Platforms simplifying access to several tools through a common interface with minimal configuration exist for fuzzing, for example, OSS-Fuzz [24] proposed by Google, and OneFuzz [33] proposed by Microsoft (which is now dedicated to internal use only). Generalizing this idea to other kinds of testing tools, DeepState [23] offers a unified framework for heterogeneous testing tools, including symbolic execution and fuzzing tools for C/C++. However, the primary objective of these testing platforms is vulnerability detection rather than coverage, and they do not provide mechanisms for enabling tools to collaborate on the same code.

Beyond platforms that allow multiple tools to be used side by side, numerous research efforts focus on combining testing techniques themselves, particularly to enhance fuzzing. It is well known that a major limitation of fuzzing lies in program conditions that have very few satisfying inputs, which are unlikely to be discovered through random mutation. Consequently, combining fuzzing with model checking or symbolic execution is a natural approach. This has led to several tools, such as VeriFuzz [32], FuSeBMC [3] or FDSE [41], which perform remarkably well in the annual testing competition TestComp [38]. Another approach, in which a symbolic execution engine is designed to run in parallel with an existing fuzzer, is exemplified by SymCC [35].

Our goal is not to develop yet another hybrid testing tool, but rather to provide a platform that leverages existing tools and enables them to collaborate on the same code base through shared test objectives.

Labels for Collaborative Test Generation. Recent work [7,8] proposed a generic way to specify test objectives for various coverage criteria with so-called *labels*.

```
//l1: x!=0 && y!=0
//l2: !(x!=0 && y!=0)
if (x!=0 && y!=0)
  ...
```

```
//l1: x!=0 && y!=0
//l2: x!=0 && y==0
//l3: x==0 && y!=0
//l4: x==0 && y==0
if (x!=0 && y!=0)
  ...
```

Fig. 1. Labels for DC (Decision Coverage) on the left, and MCC (Multiple Condition Coverage) on the right.

A *label* ℓ is a pair (l, e) where l is a location in the source code and e a boolean expression. *Covering* a label ℓ corresponds to finding a test that reaches location l while evaluating expression e to true. Satisfiability of many coverage criteria can be reduced to the search of a set of tests covering a given set of labels (e.g. for decision coverage, condition coverage, weak mutations, some variants of the modified condition-decision coverage (MC/DC) [8]). Figure 1 illustrates labels for two usual coverage criteria: decision coverage (DC—on the left) is specified by two test objectives for each conditional statement, where the condition must be evaluated to true and false; multiple condition coverage (MCC—on the right) is specified by 2^n test objectives for each conditional statement composed of n elementary subconditions, since it requires to cover all combinations of truth values of these subconditions.

LTest [5,8] introduced a first testing toolset, where labels were used to express test objectives for various test coverage criteria and to measure the coverage of a test suite, while a dynamic symbolic execution (DSE) tool PathCrawler [39] was used for test generation and static analyzers of the Frama-C platform [10,26] were used to find uncoverable test objectives. It attracted industrial interest [9]. In particular, an experiment on a large industrial code [28] showed the high potential of running several different tools in sequence on a shared version of the code and test objectives, each tool building on the results obtained by the previous one. Tests were generated for 99% of labels for the MC/DC criterion on 82,000 lines of C code, while the remaining labels were proved uncoverable, thus showing the clear interest of such a tool combination for industrial applications.

However, this approach, based on LTest and PathCrawler, has shown important limitations for larger applications. First, PathCrawler remains closed-source and is not supported anymore today. Second, LTest did not integrate other tools, such as model checkers or fuzzers. To circumvent these limitations, a recent work [13] proposed an adaptation of KLEE [15], a popular open-source test generator, to efficiently support test objectives expressed by labels of LTest. It shows that test generation guided by labels reaches a better coverage (often with fewer tests and in limited time), and it demonstrates that other tools can be integrated into the LTest framework.

Contributions. This paper presents SeaCoral [36], a novel open-source toolset for testing C programs. It was designed and developed in a collaboration between Thales and OCamlPro, based on clearly identified industrial needs. The purpose of SeaCoral is to gather the best ideas of previous efforts and to facilitate industrial applications and collaborations of mature testing tools by integrating them into an all-in-one testing platform. SeaCoral offers a complete range of automated testing services, including specification of test objectives for a chosen test coverage criterion, test case generation, detection of *infeasible* (i.e. uncoverable) objectives, and measurement of code coverage of the resulting test suite. It integrates several analyzers based on existing tools: a fuzzer libFuzzer [37], a model checker CBMC [19], a dynamic symbolic execution tool KLEE [15], and a static analyzer for detecting infeasible test objectives LUncov [6,30]. To the best of our knowledge, SeaCoral is *the first toolset offering such a large panel of testing services and efficient open-source tools with a common management of test objectives, test suites and coverage,* and easily extensible thanks to being fully open-source.

SeaCoral generates a suitable harness for each analyzer, and provides a generic mechanism for specifying test objectives and for coverage measurement (based on the notion of *labels* [7]). Moreover, the analyzers can collaborate either by sequential or parallel executions. They enrich each other's results by means of: (i) a shared store of test objectives with their statuses; and (ii) a corpus of tests that enables analyzers to exploit already generated tests. This generated corpus is then translated in a unified readable format (a C file), thus facilitating their reuse outside of the tool. A particular attention is paid to carefully handling runtime errors (RTEs), test initialization functions, and usual preconditions (e.g. variable ranges or array sizes), which are important ingredients for industrial applications. SeaCoral required a significant development effort: 25K lines of OCaml, and 24 person-months. The tool is available as a docker image to ease its use in continuous integration and development chains. The companion artifact [12] includes a docker image with the tool and the illustrative examples of this paper.

Initial experiments confirm the benefits of the toolset. The use of each individual analyzer is much easier through SeaCoral, since most of the manual preparation and configuration required is automated (harness generation, pre-compilation, output formatting). Moreover, the collaboration mechanisms of SeaCoral enable users to execute analyzers sequentially or even in parallel for some of them, so that they enrich each other's results and achieve high coverage.

2 Collaborative Test Generation in Industrial Context

We first present a selection of well-established test-generation tools that covers a large range of analysis techniques. We then advance several factors that impair their adoption in industrial contexts and their interoperability.

2.1 Available Test Generation Tools

CBMC [19] is a bounded model-checker for C and C++ programs. It performs a reachability analysis that can be used to statically check the validity of user-provided assertions. In a first mode, called *assert* mode, the tool tries to prove that every assertion of the program holds and exhibits a counter-example for each falsifiable assertion. In a second mode, called *cover* mode, CBMC automatically generates test cases that satisfy a given code coverage criterion among several usual coverage criteria such as locations, conditions, or decisions, but also user-defined test objectives. In cover mode, CBMC maximizes the coverage of the given criterion by generating several valuations for the inputs of the tested function.

KLEE [15] is a popular test generation tool for C programs based on dynamic symbolic execution. It supports a single coverage criterion, which is the all-path criterion, meaning that its aim is to cover all the feasible execution paths of the (LLVM bitcode of the) program under test. KLEE performs a symbolic execution of the program according to various strategies, and generates a test for each path found to be feasible.

libFuzzer [37] operates by randomly mutating a corpus of test inputs that are stored using arrays of bytes. The fuzzing process consists in a random exploration of the set of test inputs by means of mutations and cross-overs. Each input is associated with a score, that is typically derived from the set of branches and instructions that it executes in the analyzed program. This score influences the likelihood for an input to be subject to a mutation or cross-over. In addition, libFuzzer is able to monitor executed instructions and derive (from comparison instructions for instance) a set of constants that it then uses to perform mutations on the inputs.

LUncov is part of the LTest toolset [5,8]. It efficiently detects uncoverable labels from the results of Frama-C analyzes [6,30], and more precisely of its plugins Eva (based on abstract interpretation) and WP (based on deductive verification).

2.2 Factors that Impair Industrial Adoption

While the above analyzers have reached an industrial maturity level, their use in industrial processes remains limited. Main factors that hinder a wider adoption include the need of expert knowledge to fine-tune the configuration knobs for the analyzers and to write suitable test harnesses, and the lack of available tooling for effectively exploiting their results.

Test generation and code analysis tools rarely work directly on a raw codebase. Instead, they usually require a proper initialization of the inputs of the analyzed program in order to take into account its actual invariants and preconditions. For example, in C, a pointer that is given as an argument of a function can either be a raw pointer or an array of unknown length. It is common in the latter case to have a dedicated argument representing the length of the array.

The definition of the program inputs is done via a *harness*, that is, a specialized program that creates the inputs of interest (some of which may need to be dynamically allocated) and their constraints before calling the analyzed function. While some analyzers like CBMC provide their own harness generation utility , most other tools require tedious manual work for writing harnesses. This phase is even more critical for fuzzers , for which the harness typically needs to convert an array of bytes into a suitable input data structure for the analyzed program. In every case, writing a test harness for programs that manipulate complex data structures with pointers is challenging. Automatic generation of harnesses is still an ongoing research problem [29].

Tuning configuration knobs for an analyzer often requires expert knowledge on the underlying analysis technique. For instance, a bounded model-checker like CBMC is intrinsically parameterized by a limit to the number of loop unfoldings. KLEE's exploration of all execution paths of the program under test sometimes needs a change of strategy to be more efficient. Fuzzers typically offer configurable heuristics that can help them in exploring paths more efficiently.

Finally, none of the aforementioned analyzers returns any actual test in C that can be readily compiled and replayed. Even more, each of them has a quite different output format. In cover mode, CBMC outputs the list of values for the inputs that were marked of interest with a dedicated construct in the harness. By contrast, in assert mode, CBMC constructs the full program inputs via a sequence of assignments in a counter-example that describes corresponding execution path. KLEE outputs a set of `.ktest` files, an internal format that consists in a series of symbolic names (human-readable strings) and binary data. libFuzzer entirely relies on unstructured arrays of bytes. At last, LUncov produces a CSV file that lists locations at which labels are proved uncoverable.

2.3 Factors that Hinder Collaboration Between Analyzers

Collaboration between the analyzers is first impaired by the lack of agreement on the definition of some coverage criteria: even the definition of a simple criterion like branches, understood at the level of LLVM bitcode, does not usually correspond to branches at the source code level (not to mention the case of lazy comparison operators in C).

Next, collaborating by re-using test suites requires common interchange formats for representing the tests, and a set of converters that translate tests to and from the format that each tool accepts or produces. The issue of test conversion is even more striking in the case of fuzzers (and, to a lesser extent, of KLEE), as the format of the tests they produce, or accept as seeds, heavily depends on the test harness.

As is, none of the aforementioned analyzers offer any means to exploit already proven facts about labels. The efficiency of an analyzer may nonetheless be improved if it is able to exploit the fact that some labels have already been covered by a test. Likewise, the uncoverablility of a label may inform analyzers about branches that need not be explored.

At last, analyzers must be cautiously configured to make sure they work under the same hypotheses. Each tool makes assumptions on the analyzed program and adopts default settings that may not be equivalent. For instance, a static analyzer may consider that a branch is unreachable, for example, because it is preceded by an undefined behavior like a buffer overflow (hence does not meet the C standard), while a fuzzer generates a test that actually executes that branch.

SeaCoral facilitates the industrial adoption of these tools by automating the generation of test harnesses and offering a limited set of configuration settings. It is easy to use, and generates tests directly as C files. SeaCoral relies on the notion of labels to define shared tests objectives for every analyzer it integrates. It additionally enables effective collaboration between the analyzers via a shared store of label statuses, and a shared corpus of test cases.

3 SeaCoral on a First Example

To emphasize SeaCoral's ease of use, we illustrate its usage and results with the example of a flasher manager coming from a real-time industrial C program embedded in a car computer. Generated from a Matlab Simulink design [20], it consists of 650 lines of code, with a large loop of 450 lines that is executed at each clock cycle. The function under test checks at runtime that the warning command has priority over other flashing commands. It takes as a parameter a fixed-sized array of 10×6 boolean inputs, representing 6 input signals over a sliding window of 10 cycles (that is, 60 values). Along with 110 global variables, the program has $60 + 110 = 170$ test inputs.

The inputs have rather simple datatypes without complex invariants, so a fuzzer should be able to easily generate a set of tests that cover a fair portion of the program. However, some execution paths are intricate and need constraint solving to be covered. Therefore, we use SeaCoral with libFuzzer first, to quickly get a coverage of the easiest paths. Then we use SeaCoral with a formal tool, for example CBMC, to complete the coverage with tests for rare branches. To be sure that the uncovered test objectives are not coverable, we can even use CBMC in its *assert* mode to prove uncoverability.

SeaCoral works as a command-line tool that can be configured via command-line options or a configuration file. It should be given at least a C program, an entrypoint (the function under test), a coverage criterion, and a set of analyzers to execute on this problem. We launch the following command to execute SeaCoral with libFuzzer on this program:

```
> seacoral --files flashermanager.c --entrypoint prio_warning --criterion MCC
        --tools libfuzzer --libfuzzer-runs 100000
```

The option `--files` accepts a list of files, one of them containing the function under test given with `--entrypoint`. The coverage criterion given by `--criterion` can be any criterion expressible by labels in LTest [5,8].[1] Here we choose the multiple condition coverage criterion (cf. Fig. 1). The option `--tools` accepts a list of analyzers among the available ones (CBMC, KLEE, libFuzzer and LUncov)With the optional parameter `--libfuzzer-runs`, we instruct libFuzzer to perform 100 000 runs.

Without any additional configuration, SeaCoral outputs the following summary of its execution after 3.7 s:

---

[1] See https://git.frama-c.com/pub/ltest/lannotate/-/blob/master/doc/criteria.markdown for the list of all coverage criteria supported by LTest.

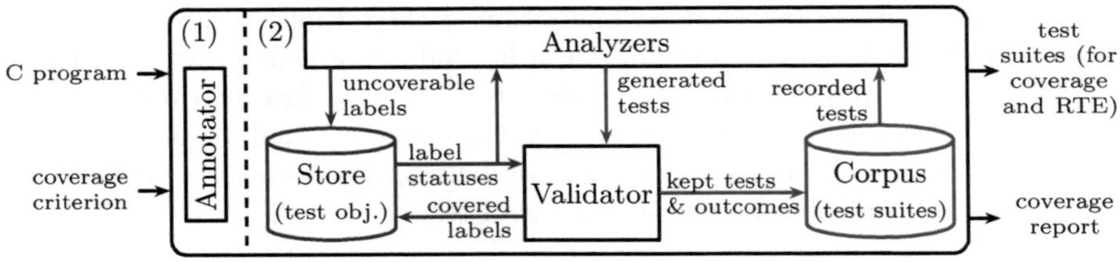

Fig. 2. Overview of SeaCoral with (1) Preparation phase and (2) Analysis phase.

```
Coverage statistics for 'prio_warning':
  cov: 103 (88.8%) uncov: 0 (0.0%) unkwn: 13 (11.2%) with 9 tests
Covered labels: {1, 3, 5, 6, 7, 8, 9, ..., 111, 112, 113, 116}
Uncoverable labels: {}
Crash statistics: rte: none
```

The MCC criterion applied from this entrypoint produces 116 test objectives by adding labels to the whole program. Among these test objectives, 103 were reached by the 9 generated tests (status 'cov'), while 13 test objectives remain unreached (status 'unkwn'). No test objective was proved uncoverable (status 'uncov'), and no runtime error (RTE) was found.

To be able to complete this coverage, we launch SeaCoral a second time on the same file, entrypoint and criterion, this time with CBMC:

```
> seacoral --files flashermanager.c --entrypoint prio_warning --criterion MCC
          --tools cbmc --cbmc-mode assert --cbmc-unwind 61
```

We configure CBMC in *assert* mode, meaning that, at the same time, it tries to generate tests for coverable test objectives and to look for uncoverable ones. We set its *unwind* parameter to 61 ($10 \times 6 + 1$) to make it unfold the loops of the program enough times for its bounded analysis to be complete (see Sect. 5 for details on the use of CBMC). This is one of the very few SeaCoral parameters requiring a bit of user knowledge about the analyzer since it cannot be automatically computed. With this configuration, CBMC resumes the previous execution of SeaCoral with the aim of completing the first coverage with additional tests or uncoverability proofs. In 1.9 s, it covers the 2 last coverable test objectives with 2 tests and proves the remaining 11 to be uncoverable.

In comparison, CBMC in *assert* mode launched alone on this program gives the same results, but in 20 s, instead of less than 6 s when the two analyzers are run in sequence. We explain below how the SeaCoral architecture enables such a smooth collaboration between its analyzers.

4 Design and Implementation of SeaCoral

Overview. SeaCoral operates on a given coverage criterion, a list of C files, and the name f of an *entrypoint* function to be tested. The main processing phases and components of SeaCoral are shown in Fig. 2.

At the *preparation phase* (1), the provided source code with its entrypoint and the coverage criterion are given to an *annotator*. First, it relies on LAnnotate module[2] of LTest [5,8] to produce a version of the source code annotated with labels that encode the coverage criterion (cf. Fig. 1). The labels are then used to initialize a *store* of test objectives that associates each label with its coverage *status*. The preparation phase ends with the *generation of test harnesses* dedicated to *each analyzer*. For a symbolic analyzer, such a harness performs the symbolization of inputs (including global variables). For a fuzzer, the harness translates the array of bytes that encodes the inputs into values for these inputs. Besides, each harness instruments labels to make the analyzer target only the yet-undecided test objectives.

During the *analysis phase* (2), analyzers interact with the store, a *corpus of recorded tests*, and an additional *validator* component. The corpus gathers the set of recorded tests T, and associates each test with an *outcome* that belongs to two categories, thereby partitioning the set T into two disjoint subsets:

T_{cov}**:** the set of *label-covering* tests that terminate successfully;
T_{err}**:** the set of *RTE-triggering* tests inducing undefined behaviors or errors.

To avoid reporting an RTE multiple times, we identify each RTE with an *error summary*, that comprises a broad RTE categorization tag (arithmetic error, memory access error, etc.), and a location in the code where the error occurs.

The validator plays a central role in ensuring the following key invariants :

($\mathfrak{I}_1$) labels reported as covered in the store are actually covered by at least one test in T_{cov};
($\mathfrak{I}_2$) any reported RTE is attached to a statement *in the original source code*;
($\mathfrak{I}_3$) any reported RTE is triggered by exactly one test in T_{err}.

Before making its way into the corpus, any test generated by an analyzer is given to the validator, which is the only component that may add a test into the corpus and record a label as covered in the store. This ensures ($\mathfrak{I}_1$). Analyzers are however able to directly report labels as uncoverable.

The Validator. Analyzers may generate a test simply because it reaches a label, without assessing whether it terminates successfully or identifying an error (if any). Several tests can trigger the *same* RTE, i.e. the same error at the same location in the source code. Industrial applications of testing tools need precise handling of RTEs. The validator provides a unified replay mechanism for (i) *coverage validation*, and (ii) if an RTE occurs, a precise *RTE identification*.

Coverage validation collects the labels that are covered by a given test t using an *instrumented validation program* , that: (i) evaluates the tested function on t; and (ii) performs runtime detection of RTEs using LLVM sanitizers. If it terminates successfully, and if t covers additional labels, then t is inserted into the corpus and the statuses for newly covered labels are submitted into the store. If the program terminates successfully and t covers no new labels, it is discarded.

[2] We extended this module to generate labels transitively in the callees of the entry-point.

An undefined behavior may be induced by the evaluation of a label condition that does not correspond to an RTE in the original source code. Consider for instance a condition `(y != 0 && x/y != 0)` . The annotated code may contain a label with condition `(x/y != 0)` , which triggers a division by 0 when`y == 0` , while it cannot occur in the original code due a lazy evaluation of `&&` .

To filter out such spurious RTEs, any test t for which coverage validation errors out, is subject to a second execution with an *RTE-identification program* where label conditions are not evaluated. If it terminates successfully, then t is discarded. Otherwise, the RTE is legitimate and ($\mathfrak{I}_2$) is satisfied: the validator parses the associated sanitization report to extract a precise error summary. To ensure ($\mathfrak{I}_3$), t is inserted into T_{err} if this summary has not yet been recorded.

Collaboration Mechanisms. Both the store and the corpus are used across (consecutive or parallel) launches of analyzers on the same program, entrypoint, and coverage criterion. In addition, SeaCoral enables any analyzer to reuse recorded tests. It achieves this via dedicated converters that perform the translations of tests to and from each analyzer's input and output formats.

5 Integration of Analyzers

Each SeaCoral analyzer is integrated by means of a dedicated *integration module*. Such a module automates the generation of test harnesses, that define the inputs of interest of the function under test (function arguments and global variables) and their constraints, and that translates the labels into targets for the analyzer. An integration module for an analyzer additionally performs all the necessary preparation phases like compilation or instrumentation of sources, and runs the analyzer itself with the relevant options so that it uses the results of previous runs and contributes only with valid tests for yet uncovered test objectives. At last, such a module handles the analyzer's outputs to consolidate all results in a unified format. We describe in this section some technical details of integration of each of the four analyzers of SeaCoral. The reader is not required to fully understand these details to use the tool and to follow the rest of the paper.

CBMC Integration. SeaCoral is able to interact with CBMC [19] using two different modes, *cover* and *assert* (selected with option `--cbmc-mode` on the command-line). The *cover* mode relies on the test generation capabilities of CBMC while the *assert* mode relies on its reachability analysis.

In *cover* mode, CBMC automatically generates test cases that satisfy a given code coverage criterion . It supports several usual coverage criteria but also user-defined test objectives defined by a built-in `__CPROVER_cover(cond)` statement. Obviously, these objectives have a direct correspondence with labels in SeaCoral. However, to ensure CBMC generates tests that terminate successfully, the harnesses generated by SeaCoral for CBMC in this mode associate each label with a Boolean flag that is only assigned whenever the label is reached and its condition holds. `__CPROVER_cover` calls are then placed at the end of the harness,

Listing 1.1. Label implementation used in harnesses for CBMC.

```
/* One Boolean variable (flag) per label */
static char __sc_cbmc_covered_1 = '\000'          // Flag for label 1
static char __sc_cbmc_covered_2 = '\000'          // Flag for label 2

/* We update the label's flag in the macro */
# define cov_label(expr, id, ...)                  \
  do {                                             \ /* Set the flag for label id */
    if (expr) __sc_cbmc_covered_##id = '\001';     \ /* if expr is satified */
  } while (0)                                        /* at the current location */

/* Function to test, annotated with cov_label(expr,id,...) for each label */
void annotated_entrypoint() {
  ...
}

int main () {
  /* Initialization of variables (not detailed here) */
  ...
  annotated_entrypoint();
  /* Checking if a label is covered */
  __CPROVER_cover(__sc_cbmc_covered_1); // Add assert for label 1 using its flag
  __CPROVER_cover(__sc_cbmc_covered_2); // Add assert for label 2 using its flag
  ...
}
```

guarded with its respective flag. We illustrate the treatment of labels using the `cov_label` macro for CBMC cover mode in Listing 1.1.

In *assert* mode, CBMC performs a reachability analysis that can be used to statically check the validity of user-provided assertions defined by a built-in `__CPROVER_assert(cond)` statement. This mode treats each assertion independently, unlike the cover mode that tries to solve all the test objectives in the same analysis. The assert mode is used in SeaCoral both to generate tests and to detect uncoverable test objectives, by translating each label into an assertion for its negation. If CBMC proves this assertion, then the label is unreachable (under the current assumptions). Otherwise, the counter-example provided by CBMC is exactly a test reaching the label condition. As each assertion is treated separately, this mode makes no effort on minimizing the number of tests generated. It will generate one test for each label. The harness used is similar to the one for the cover mode, except that the calls `__CPROVER_cover(cond)` are replaced with `__CPROVER_assert(!cond)`.

CBMC is a bounded model-checker. Therefore, it makes a bounded analysis of the program by unwinding loops a fixed number of times. In many cases, CBMC is able to determine automatically an upper bound on the number of loop iterations, but if the number of loop iterations highly depends on the data or is unbounded, the bound has to be specified by the user. For this, CBMC provides the option `--unwind n`, where `n` is the maximal number of times any loop condition should be evaluated. The consequence of this bounded analysis is that

Listing 1.2. Label implementation used in harnesses for KLEE.

```
# define cov_label(expr, id, ...)                    \
  do {                                               \
    if (NONDET (id)) {                               \ /* Go to the label or bypass it.
         */
      if (!__sc_is_covered (id))                     \ /* If label id not yet covered,
         */
        klee_assert (!(expr));                       \ /* try to cover it; in any case
         */
      klee_silent_exit (0);                          \ /* exit without going further.
         */
    }                                                \ /* Continue if label was
       bypassed.*/
  } while (0)
```

it is not necessarily complete, if the bound is not sufficient to represent all the initial program paths. This does not affect test generation, but the conclusion of uncoverability for a test objective is conditioned by the completeness of the analysis.

SeaCoral tries to lighten the burden of the user by minimizing the configuration needs, yet since this unwinding parameter cannot be automatically computed, it must be carefully set to obtain meaningful results when using CBMC through SeaCoral. This is one of the very few parameters of SeaCoral that need a bit of understanding of the code under test.

CBMC results are precise and allow to detect uncoverable labels efficiently. Its main drawback comes from its lack of collaboration from and to other analyzers. For instance, it cannot update the shared store during its execution: SeaCoral must wait until CBMC terminates the function's analysis to collect its results, extract test inputs, and submit them to the validator.

KLEE Integration. The integration of KLEE [15] is based on previous work [13]. KLEE symbolically executes the program path by path, and generates a test whenever it reaches a return statement of the main function, an assertion `klee_assert(0)`, or an instruction raising a runtime error like a division by zero or an illegal pointer access.

In the same way as we use `__CPROVER_assert(!cond)` to translate labels to targets for CBMC reachability analysis, we use `klee_assert(!cond)` to transform labels into targets for KLEE's test generation. Several optimizations are necessary to control the number and the length of the paths to explore, and to avoid generating tests for already covered labels (see details in [13]). We end up with the label instrumentation shown in Listing 1.2.

The `NONDET` call separates the paths into two separate branches: one continuing the execution, and one going into a branch in which the label's condition is tested. In both cases, KLEE's exploration is halted in order to cut off the branching introduced by the label: the first calls `klee_assert` , stopping the execution and returning a test; the second calls `klee_silent_exit` , artificially stating that the path is irrelevant and does not need further exploration. Cutting these

Listing 1.3. Label instrumentation for fuzzing.

```
#define cov_label(expr, id, ...)          \
  do {                                    \
    if (expr) __asm__ volatile ("");      \ /* If expr holds, add an empty branch */
  } while (0)
```

branches solves both the path explosion and the growing complexity of the path conditions.

SeaCoral monitors tests generated by KLEE and feeds them to the validator while KLEE is running. Therefore, by updating the store, the execution of tests by the validator plays the role of a *concrete replay* that helps KLEE in avoiding exploration of already covered labels.

KLEE terminates once it finishes the exploration of the whole program, or until a given timeout, whichever happens first. In addition to taking benefit from the store state before and during the analysis, KLEE can be initialized on startup by seeding with tests from the corpus (T_{cov}) from which to start symbolic exploration.

KLEE could be configured to ensure that generated tests do not trigger RTEs by checking, like for CBMC's cover mode, that a label is actually covered *after* the call to the entrypoint function has terminated. Doing this check through the validator actually proved efficient and more reliable, as we found that KLEE's error detection is not as precise as that of LLVM sanitizers (especially regarding dynamic memory).

libFuzzer Integration. libFuzzer [37] is the representative of *purely dynamic analyzers* that we have integrated in SeaCoral. To be efficient, a fuzzer needs to operate on a non-empty fuzzing corpus. Therefore, a first test sampling and retrieval phase constructs a fuzzing corpus that consists in a non-empty set of seeds. This is achieved by:

- re-using every *label-covering* test that is already recorded in SeaCoral's corpus;
- random sampling a given number of additional tests.

Then, a seed triage stage filters out randomly generated tests that trigger RTEs. This triage is necessary to determine whether the fuzzer has any valid seed left, i.e., a test that terminates, to be able to start its execution. This step is to prevent libFuzzer from entering an infinite loop trying to run tests that do not contribute any coverage (i.e., that do not lead to any increment in libFuzzer's internal counters) without any meaningful progress.

libFuzzer runs for a fixed period of time, or a given number of fuzzing iterations. A light instrumentation of labels is used to provide the fuzzer's internal constant monitor with meaningful values for mutations. We show this instrumentation in Listing 1.3, where `__asm__ volatile ("")` is an empty basic block that cannot be optimized out by C compilers: in this way, the branch condition is evaluated by libFuzzer, enabling it to derive meaningful constants.

SeaCoral dynamically retrieves label-covering tests that are recorded into SeaCoral's corpus by other analyzers while libFuzzer is running. It additionally submits tests to the validator as soon as they are persisted by the fuzzer.

LUncov Integration. LUncov is part of the LTest toolset [5,8]. As such, it is already integrated into SeaCoral. It efficiently detects uncoverable labels from the results of Frama-C analyzes [6,30], and more precisely of two of its plugins: Eva [14], an abstract interpretation engine working with various abstract domains, and WP [10], an analysis tool based on the computation of weakest preconditions.

6 Illustrative Examples and Benefits of SeaCoral

This section illustrates the benefits of using SeaCoral through additional examples. Since the analyzers integrated in the toolset are well-known and their capacities have already been demonstrated in previous work (and are preserved in SeaCoral), we focus on the new features enabled by the toolset.

Check UTF-8 encoding. The `checkutf8` function (76 loc) checks whether a buffer of characters is a valid UTF-8 encoding (i.e. it translates to a sequence of valid UTF-8 encoded characters) and has the following signature:

```
int checkutf8 (unsigned char* buffer, int n)
```

We use SeaCoral with KLEE and the condition coverage criterion (CC) to cover each truth value of each subcondition of multiple conditions.

In order to avoid RTEs due to the inconsistency between the size of `buffer` and the value of `n`, SeaCoral offers the option `--array-size-mapping` to constrain the size of an array to equal the value of another parameter.

```
> seacoral --files checkutf8.c --entrypoint checkutf8 --criterion CC --tools klee
        --array-size-mapping buffer:n --max-ptr-array-length 10
```

With this configuration and precondition, SeaCoral launched with KLEE covers 100% of the test objectives required by the CC criterion (i.e., 52 labels) with 21 tests and finds no RTE, as expected. Fig. 3 shows an example of a label-covering test.

```
void test_20 () {
  unsigned char (* buffer) = malloc (sizeof (unsigned char[4]));
  (void) memcpy (buffer, (unsigned char[4]){240, 128, 192, 255},
                 sizeof (unsigned char[4]));
  int n = 4;
  (void) check (buffer, n);
}
```

Fig. 3. Example of a label-covering test generated for function **check**

Array of Queues. We consider a data structure of an array of queues, inspired by industrial code, with its initialization and update functions. A global variable `state` is declared as an array of 5 queues. The `initialize` function sets all pointers of queues to `NULL` and their sizes to 0. The `add` function adds a value at the end of the queue at a given index, while `pop` removes the first value of the queue at the given index if possible and stores the result in its other argument.

```
typedef struct node_t { int data; struct node_t *next; } node;
typedef struct queue_t { node *first; node *last; int length; } queue;

queue state[5];

void initialize();
void add(int val, int ind);
int pop(int ind, int *val);
```

We want to generate tests for the `pop` function with the condition coverage criterion CC.

```
> seacoral --files queue.c --entrypoint pop --criterion CC --tools klee
```

Without any additional configuration, SeaCoral launched with KLEE on function `pop` covers only 5 among the 8 test objectives for CC and finds 1 RTE. Unsurprisingly, the queues of the array `state` do not satisfy the implicit invariants on the length of the queues nor on the consistency of pointers. Furthermore, KLEE does not manage to cover the difficult parts of the function where there is actually a value to pop. The same goes if we try CBMC instead of KLEE.

The problem here is to generate meaningful states that allow to cover specific parts of the code. In such cases, the developer is willing to provide a test initialization function that builds a valid state from which relevant tests can be generated. For this, SeaCoral provides two options: `--ignore-globals state` makes SeaCoral ignore the global variable `state` that is too hard to symbolize due to its invariants; and `--fixtures-init init_valid_state` defines the hand-written function `init_valid_state` as a preamble for test case generation. Here we write this initialization function to create `state` with two queues, of size 1 and 2 respectively. With this configuration, both KLEE and CBMC manage to cover 100% test objectives for CC with 5 tests in less than 3 s. KLEE moreover finds 2 RTEs, when the pointer receiving the value is `NULL` or allocated with no memory.

PKCS#1 Signature Parser. PKCS#1 (Public-Key Cryptography Standard) specifies the usage of RSA algorithm, in particular for creation and verification of signatures. During PKCS#1 v1.5 signature verification, due to a complex structure of the signed message, a dedicated parser must be applied to extract necessary data. The implementation of this parser we target consists of 220 lines of code. Parsers are known to be difficult programs to cover by automatically generated tests, because of the complexity of the input structure and the numerous dependency constraints on the different fields composing it.

The signature of the entrypoint is the following:

```
int pars_PKCS1_lev1(uchar* pad_msg, int pad_msg_sz, uchar* in_msg, int in_msg_sz,
    uint expect_hash_ind)
```

We launch SeaCoral with CBMC with the decision coverage criterion (DC) and the following configuration: we constrain the size of arrays with their size arguments, the size of the largest input to 96 and the size of the largest loop to unfold to 65. These two last values are quite easy to determine by a quick analysis of the specification and the code. Moreover, with `sc_assume` we constrain the values of `expect_hash_ind` between 0 and 3 and the values of `in_msg_sz` and `pas_msg_sz` to be non-negative.

```
> seacoral --files pkcs1.c --entrypoint pars_PKCS1_lev1 --criterion DC
          --tools cbmc --unwind 65 --max-ptr-array-length 96
          --array-size-mapping pad_msg:pad_msg_sz,in_msg:in_msg_sz
```

With this configuration, CBMC runs 17 min and covers all 48 DC labels with 20 tests.

Benefits. These examples highlight the advantages of using SeaCoral. It requires a minimal configuration compared to that required by any of the analyzers used alone. The user does not need to create manually a harness to run any of these analyzers: SeaCoral generates it automatically, with test inputs made symbolic, even when their structure is complex and involves pointers. While all the analyzers generate values for test inputs in a tool-dependent format, the tests generated by SeaCoral are uniformly written (currently, in a C file), and can easily be maintained and replayed for regression checking.

7 Conclusion and Future Work

Industrial use of advanced testing tools based on formal methods still remains limited today. To address this issue, we propose SeaCoral, a new open-source toolset for C program testing, offering a complete range of testing services: annotation of test objectives for a given test criterion, test generation, identification of infeasible objectives and test replay. It provides a convenient framework for applying various tools, and currently integrates a model-checker, a fuzzer, a DSE tool and a static analyzer.

SeaCoral is currently being evaluated with success by Thales on a confidential industrial code of several thousands of lines, containing complex nested data structures and invariants and requiring test initialization functions to test state machine based behaviors.

Ongoing and future work include the maturation of the toolset and its evaluation on other industrial codebases. A visualization of the reached coverage, uncovered and uncoverable test objectives will help the user to understand SeaCoral's results and to find a better configuration. In some cases, user-provided annotations could be used to enrich the chosen criteria and force the generation of a test for a particular statement or branch. A support of hand-written test oracles will complete the coverage report by providing success or failure verdict for the generated tests. An integration of a runtime verification tool (e.g. E-ACSL [11]) could also complete the testing process with formal oracles executed at runtime. Parallel execution of several tools is already possible in SeaCoral and should be further evaluated on codebases of significant size and complexity.

Acknowledgment. Part of this work was funded by the French AID (Defence Innovation Agency) via the RAPID project AutoCouv and by ANR (grants ANR-22-CE39-0014, ANR-24-ASM2-0001). We thank the anonymous referees for their comments and suggestions.

References

1. AFL: American Fuzzy Lop (Accessed Oct 16, 2025). http://lcamtuf.coredump.cx/afl
2. Agrawal, H., DeMillo, R.A., Spafford, E.H.: Debugging with dynamic slicing and backtracking. Softw., Pract. Exper. **23**(6), 589–616 (1993). https://doi.org/10.1002/spe.4380230603
3. Alshmrany, K.M., Aldughaim, M., Bhayat, A., Cordeiro, L.C.: FuSeBMC v4: improving code coverage with smart seeds via BMC, fuzzing and static analysis. Formal Aspects Comput. **36**(2) (2024), https://doi.org/10.1145/3665337
4. Ammann, P., Offutt, J.: Introduction to Software Testing. Cambridge University Press, 1 edn. (2008), ISBN 0521880386, 9780521880381
5. Bardin, S., Chebaro, O., Delahaye, M., Kosmatov, N.: An all-in-one toolkit for automated white-box testing. In: Seidl, M., Tillmann, N. (eds.) TAP 2014. LNCS, vol. 8570, pp. 53–60. Springer, Cham (2014). https://doi.org/10.1007/978-3-319-09099-3_4
6. Bardin, S., et al.: Sound and quasi-complete detection of infeasible test requirements. In: Proceedings of the 8th IEEE International Conference on Software Testing, Verification and Validation (ICST 2015), pp. 1–10, IEEE (2015). https://doi.org/10.1109/ICST.2015.7102607
7. Bardin, S., Kosmatov, N., Cheynier, F.: Efficient leveraging of symbolic execution to advanced coverage criteria. In: Proceedings of the 7th IEEE International Conference on Software Testing, Verification and Validation (ICST 2014), pp. 173–182, IEEE (2014). https://doi.org/10.1109/ICST.2014.30
8. Bardin, S., Kosmatov, N., Marcozzi, M., Delahaye, M.: Specify and measure, cover and reveal: a unified framework for automated test generation. Sci. Comput. Program. **207** (2021). https://doi.org/10.1016/j.scico.2021.102641
9. Bardin, S., Kosmatov, N., Marre, B., Mentré, D., Williams, N.: Test case generation with PathCrawler/LTest: how to automate an industrial testing process. In: Margaria, T., Steffen, B. (eds.) ISoLA 2018. LNCS, vol. 11247, pp. 104–120. Springer, Cham (2018). https://doi.org/10.1007/978-3-030-03427-6_12
10. Baudin, P., Bobot, F., Correnson, L., Dargaye, Z., Blanchard, A.: WP Plug-in Manual (2023). https://frama-c.com/download/frama-c-wp-manual.pdf
11. Benjamin, T., Signoles, J.: Runtime annotation checking with Frama-C: The E-ACSL plug-in. In: Kosmatov, N., Prevosto, V., Signoles, J. (eds.) Guide to Software Verification with Frama-C: Core Components, Usages, and Applications, pp. 263–303, Computer Science Foundations and Applied Logic Book Series, Springer (2024). https://doi.org/10.1007/978-3-031-55608-1_5
12. Berthier, N., de Oliveira, S., Kosmatov, N., Longuet, D.: SeaCoral: a collaborative test generation toolset. Companion Artifact (2025). https://doi.org/10.5281/zenodo.17357288

13. Berthier, N., de Oliveira, S., Kosmatov, N., Longuet, D., Soulat, R.: An efficient black-box support of advanced coverage criteria for Klee. In: Proceedings of the 38th Annual ACM/SIGAPP Symposium on Applied Computing, Software Verification and Testing Track (SAC-SVT 2023), pp. 1706–1715, ACM (2023), https://doi.org/10.1145/3555776.3577713
14. Bühler, D., Maroneze, A., Perrelle, V.: Abstract interpretation with the Eva plug-in. In: Kosmatov, N., Prevosto, V., Signoles, J. (eds.) Guide to Software Verification with Frama-C: Core Components, Usages, and Applications, pp. 131–186, Computer Science Foundations and Applied Logic Book Series, Springer (2024), https://doi.org/10.1007/978-3-031-55608-1_3
15. Cadar, C., Dunbar, D., Engler, D.: KLEE: unassisted and automatic generation of high-coverage tests for complex systems programs. In: Proceedings of the 8th USENIX Symposium on Operating Systems Design and Implementation (OSDI 2008), pp. 209–224, USENIX Association (2008). http://www.usenix.org/events/osdi08/tech/full_papers/cadar/cadar.pdf
16. Cadar, C., et al.: Symbolic execution for software testing in practice: preliminary assessment. In: Proceedings of the 33rd International Conference on Software Engineering (ICSE 2011), pp. 1066–1071, ACM (2011). https://doi.org/10.1145/1985793.1985995
17. Cadar, C., Sen, K.: Symbolic execution for software testing: three decades later. Commun. ACM **56**(2), 82–90 (2013). https://doi.org/10.1145/2408776.2408795
18. Chilenski, J.J., Miller, S.P.: Applicability of modified condition/decision coverage to software testing. Softw. Eng. J. **9**(5), 193–200 (1994). https://doi.org/10.1049/SEJ.1994.0025
19. Clarke, E., Kroening, D., Lerda, F.: A tool for checking ANSI-C programs. In: Jensen, K., Podelski, A. (eds.) TACAS 2004. LNCS, vol. 2988, pp. 168–176. Springer, Heidelberg (2004). https://doi.org/10.1007/978-3-540-24730-2_15
20. Collavizza, H., Vinh, N., Ponsini, O., Rueher, M., Rollet, A.: Constraint-based BMC: a backjumping strategy. Int. J. Softw. Tools Technol. Transfer **16**(1), 103–121 (2014). https://doi.org/10.1007/s10009-012-0258-6
21. Fioraldi, A., Mantovani, A., Maier, D.C., Balzarotti, D.: Dissecting american fuzzy lop: a FuzzBench evaluation. ACM Trans. Softw. Eng. Methodol. **32**(2), 52:1–52:26 (2023). https://doi.org/10.1145/3580596
22. Gigante, G., Pascarella, D.: Formal methods in avionic software certification: the DO-178C perspective. In: Margaria, T., Steffen, B. (eds.) ISoLA 2012. LNCS, vol. 7610, pp. 205–215. Springer, Heidelberg (2012). https://doi.org/10.1007/978-3-642-34032-1_21
23. Goodman, P., Groce, A.: DeepState: a unit test-like interface for fuzzing and symbolic execution (2018). https://github.com/trailofbits/deepstate
24. Google: Documentation for OSS-Fuzz (2025). https://google.github.io/oss-fuzz/
25. ISO Standard: ISO 26262-10:2018 Road vehicles — Functional safety (2018). https://www.iso.org/standard/68392.html
26. Kirchner, F., Kosmatov, N., Prevosto, V., Signoles, J., Yakobowski, B.: Frama-C: a software analysis perspective. Formal Asp. Comput. pp. 1–37 (2015). https://doi.org/10.1007/s00165-014-0326-7
27. Laski, J.W., Korel, B.: A data flow oriented program testing strategy. IEEE Trans. Software Eng. **9**(3), 347–354 (1983). https://doi.org/10.1109/TSE.1983.236871

28. Lavillonnière, E., Mentré, D., Cousineau, D.: Fast, automatic, and nearly complete structural unit-test generation combining genetic algorithms and formal methods. In: Beyer, D., Keller, C. (eds.) TAP 2019. LNCS, vol. 11823, pp. 55–63. Springer, Cham (2019). https://doi.org/10.1007/978-3-030-31157-5_4
29. Liu, Y., et al.: PromeFuzz: a knowledge-driven approach to fuzzing harness generation with large language models. In: Proceedings of the 2025 ACM SIGSAC Conference on Computer and Communications Security, pp. 1559–1573 (2025). https://doi.org/10.1145/3719027.3765222
30. Marcozzi, M., Bardin, S., Kosmatov, N., Papadakis, M., Prevosto, V., Correnson, L.: Time to clean your test objectives. In: Proceedings of the 40th International Conference on Software Engineering (ICSE 2018), pp. 456–467, ACM (2018). https://doi.org/10.1145/3180155.3180191
31. Mathur, A.P.: Foundations of Software Testing. Addison-Wesley Professional (2008)
32. Metta, R., Medicherla, R.K., Chakraborty, S.: BMC+fuzz: efficient and effective test generation. In: Proceedings of the 2022 Conference & Exhibition on Design, Automation & Test in Europe (DATE 2022), pp. 1419–1424 (2022). https://doi.org/10.23919/DATE54114.2022.9774672
33. Microsoft: OneFuzz: a self-hosted fuzzing-as-a-service platform (2023). https://github.com/microsoft/onefuzz
34. Myers, G.J., Sandler, C., Badgett, T.: The Art of Software Testing. Wiley, 3 edn. (2011)
35. . Poeplau, S., Francillon, A.: Symbolic execution with SymCC: don't interpret, compile! In: Proceedings of the 29th USENIX Security Symposium, pp. 181–198, IEEE (2020). https://doi.org/10.5555/3489212.3489223
36. SeaCoral: The SeaCoral Testing Toolset (Accessed 16 Oct, 2025). https://github.com/ocamlpro/seacoral
37. Serebryany, K.: Continuous fuzzing with libFuzzer and AddressSanitizer. In: Proceedings of the IEEE International Conference on Cybersecurity Development, (SecDev 2016), p. 157, IEEE (2016). https://doi.org/10.1109/SECDEV.2016.043
38. TestComp: International competition on software testing (2026). https://test-comp.sosy-lab.org/
39. Williams, N., Marre, B., Mouy, P., Roger, M.: PathCrawler: automatic generation of path tests by combining static and dynamic analysis. In: Dal Cin, M., Kaâniche, M., Pataricza, A. (eds.) EDCC 2005. LNCS, vol. 3463, pp. 281–292. Springer, Heidelberg (2005). https://doi.org/10.1007/11408901_21
40. Zeller, A., Gopinath, R., Böhme, M., Fraser, G., Holler, C.: The Fuzzing Book. CISPA Helmholtz Center for Information Security (2024). https://www.fuzzingbook.org/
41. Zhang, G., Shuai, Z., Ma, K., Liu, K., Chen, Z., Wang, J.: FDSE: enhance symbolic execution by fuzzing-based pre-analysis (competition contribution). In: Beyer, D., Cavalcanti, A. (eds) Fundamental Approaches to Software Engineering. FASE 2024. Lecture Notes in Computer Science, vol 14573, pp. 304–308, Springer, Cham (2024). https://doi.org/10.1007/978-3-031-57259-3_16
42. Zhu, H., Hall, P.A.V., May, J.H.R.: Software unit test coverage and adequacy. ACM Comput. Surv. **29**(4), 366–427 (1997). https://doi.org/10.1145/267580.267590

Efficient Multi-level Mine Dewatering Using Uppaal Stratego

Muhammad Naeem[1(✉)], Cristina Seceleanu[1(✉)], Alf J. Isaksson[2(✉)], and Tiberiu J. Seceleanu[1(✉)]

[1] Mälardalen University, Västerås, Sweden
{muhammad.naeem,cristina.seceleanu,tiberiu.seceleanu}@mdu.se
[2] ABB AB, Västerås, Sweden
alf.isaksson@se.abb.com

Abstract. Effective water management in underground mining requires maintaining safe reservoir levels while minimizing the high energy costs of continuous pumping. Although flexible electricity pricing enables cost-aware operation, traditional threshold-based controllers cannot exploit this flexibility efficiently. This paper presents an industrial case study on efficient mine dewatering using reinforcement-learning-based control synthesized with the Uppaal Stratego framework. A baseline threshold controller is first implemented, followed by a reinforcement-learning controller trained on forecast inflows and day-ahead electricity prices to minimize pumping costs while limiting pump switching. To ensure safety during learning without distorting the optimization objective, we introduce a pre-shield that blocks unsafe transitions. We formally show that this pre-shield is maximally permissive with respect to a monotonicity safety objective. Simulation results demonstrate that the learning-based strategy reduces total energy consumption by up to 40% compared to threshold-based control, while maintaining safe operation in all scenarios.

Keywords: Mine Dewatering · Energy Optimization · Uppaal Stratego · Q-learning · Pump Control

1 Introduction

Dewatering is essential in underground mining to maintain safe working conditions and operational continuity. In deep mines, water is removed through multi-level pumping systems that are highly energy-intensive and can constitute a significant share of total electricity consumption. The increasing availability of flexible electricity pricing motivates the development of energy-aware control strategies for mine dewatering.

This work is part of the *D-RODS* project *Dynamic and Robust Distributed Systems*[1] and focuses on optimizing mine dewatering through formal modeling and strategy synthesis using Uppaal Stratego [8]. In contrast to conventional

[1] https://www.es.mdu.se/projects/615-D_RODS.

A. Sampaio and M. Stoelinga (Eds.): FM 2026, LNCS 16557, pp. 711–729, 2026.
https://doi.org/10.1007/978-3-032-26220-2_35

threshold-based controllers, which ignore electricity prices and inflow variability, our approach integrates online Reinforcement Learning (RL) with external forecasts to enable predictive, price-aware pump scheduling.

The objective is to synthesize a control strategy that achieves energy-efficient and operationally stable dewatering by jointly minimizing pumping *energy consumption* and pump *switching frequency*. These objectives are coupled, as frequent switching reduces short-term energy costs but increases mechanical wear, while fewer switching actions improve reliability at the expense of higher energy usage. Our proposed approach balances this trade-off by learning policies that maintain safe reservoir levels, exploit low-cost electricity periods, and avoid unnecessary switching under varying inflow and pricing conditions.

Existing mine dewatering approaches under time-varying electricity tariffs either rely on fixed rule-based or Model Predictive Control (MPC) controllers [17,19,21], or apply data-driven analysis without formal safety guarantees [16], and therefore do not address how to safely learn cost-efficient pumping strategies under uncertain inflows and dynamic pricing, a gap that this work explicitly targets. We seek a strategy that is *energy-aware*, capable of using inflow forecasts, and ensures operational safety during both learning and execution, through formal guarantees.

In this paper, we make the following contributions:

- **Formal dewatering model:** We develop a timed game model of a multi-level mine dewatering system in UPPAAL STRATEGO. Each level is modeled as a separate process with its own reservoir, inflow, pumps, and outflow, incorporating hourly day-ahead electricity prices and inter-level water transfer.
- **Baseline control strategies:** We implement baselines, a Bang-Bang rule-based controller that turns all pumps on or off based on reservoir limits, and a Price-Aware controller that incorporates electricity price information into the control logic.
- **Q-learning-based strategy synthesis:** We train a Q-learning agent within UPPAAL STRATEGO over a 24-h horizon. The agent determines the optimal number of active pumps based on reservoir levels, inflow rates, and electricity prices, using both random and forecast-driven inflow patterns.
- **Pre-shield for safe learning:** We develop a pre-shield mechanism integrated with the dewatering system to restrict unsafe control actions that could lead to safety violations. This approach avoids the need for large penalty terms in the reward function, which can obscure the goal of minimizing energy consumption. By blocking unsafe actions directly, the pre-shield facilitates safe and efficient learning while maintaining a reward function that accurately reflects energy costs.
- **Evaluation on an industrial case study:** We evaluate the proposed methods on a multi-level mine dewatering use case, part of the *D-RODS*project. Comparative simulations demonstrate that the shielded learning-based strategy maintains reservoir safety while reducing total pumping energy consumption by up to 40% compared to traditional threshold-based control.

Overall, this paper demonstrates how formal modeling, RL, electricity price data, and safety shielding can be integrated to achieve energy-efficient and reliable control for underground mine dewatering.

The remainder of this paper is organized as follows. Section 2 reviews related studies on energy-efficient dewatering and model-based control. Section 3 summarizes the theoretical background on RL, priced timed games, and the UPPAAL STRATEGO framework. Section 4 introduces the mine dewatering case study used for evaluation. Section 5 presents the proposed methodology for strategy synthesis, including learning setup and pre-shield integration, as well as the theorems and lemmas of the pre-shield's permissiveness and generalized physical safety for all future time. Section 6 describes the implementation of the dewatering system in UPPAAL Stratego. Section 7 reports and discusses the simulation results, comparing the proposed controller with baseline strategies. Finally, Sect. 8 concludes the paper and outlines directions for future research.

2 Related Work

Our approach to synthesizing an efficient strategy for mine dewatering combines strategy synthesis for timed systems with the idea of restricting actions to enforce invariants. Much of the closest prior work comes from the literature on shield synthesis and runtime enforcement for reactive and real-time systems.

In contrast to electricity tariff-based scheduling and MPC-based dewatering control [17,19,21], our method integrates cost-aware learning with formal timed-game modeling and an action filter. In contrast to generic (timed) shield synthesis frameworks [5,6] and shielded RL [3], we provide a case-study-driven, deployable encoding in UPPAAL STRATEGO for a realistic dewatering plant and analyze the pre-shield's permissiveness properties in the critical operating bands.

Shielding has also been applied to safe RL, where an automatically synthesized shield restricts the agent's action choices to those that maintain safety during training and execution. This line of work highlights how shielding enforces safety in systems with learned controllers but typically assumes a discrete or abstract model of the environment [13].

Further developments in the shield synthesis literature introduce notions such as k-stabilizing and admissible shields, which bound the duration of deviation after a violation or adapt to cooperative system behavior when strict recovery bounds are unattainable. These notions refine the enforcement mechanism to balance safety against degree of interference, but remain in the domain of discrete reactive system outputs [12].

Unlike classical shield synthesis that operates over discrete output streams of reactive systems, our work focuses on strategy synthesis in a continuous timed domain (dewatering volumes) where actions correspond to ranges of real-valued controller settings rather than Boolean output corrections. By encoding the safety specification as a timed game and the shield as a set of allowable controls at each state, we ensure that safety invariants (e.g., volume bounds) are never violated while preserving as much control flexibility as possible. In contrast to timed shields that enforce invariants on outputs, our shield restricts controller actions to ranges that guarantee safety under physical dynamics, and we formally prove maximal permissiveness relative to these dynamics.

Moreover, while shield synthesis for RL restricts agent action spaces for safety, these approaches generally assume a discrete action model and do not directly address the continuous and hybrid nature of physical process control as in mine dewatering, nor provide formal guarantees on range preservation in strategy synthesis.

More recently, synthesis frameworks such as Uppaal Coshy target automatic generation of compact safety strategies (shields) for continuous and hybrid models by partitioning the state space and solving associated two-player safety games. This work also focuses on efficient and compact representation of the resulting shields to enable deployment in complex systems [7].

3 Preliminaries

This section outlines the key theoretical concepts underlying this work, including reinforcement learning, priced timed games, and the UPPAAL STRATEGO framework used for strategy synthesis.

3.1 Reinforcement Learning (RL)

RL [18] is a computational approach in which an agent learns to make sequential decisions through interaction with an environment. At each time step, the agent observes the current state, takes an action, and receives a reward signal that evaluates the consequence of that action. By repeatedly exploring and updating its policy, the agent aims to maximize its cumulative expected reward over time.

A classical RL algorithm is *Q-learning* [20], which estimates the optimal action-value function for each state–action pair. The Bellman optimality equation expresses this relationship as:

$$q^*(s, a) = \mathbb{E}[R(s, a) + \gamma \max_{a'} q^*(s', a')], \tag{1}$$

where $R(s, a)$ denotes the immediate reward for executing action a in state s, and s' represents the next state. The discount factor $\gamma \in [0, 1]$ controls the weighting between immediate and future rewards, and $\mathbb{E}$ denotes the expectation over possible transitions. Through iterative updates, the algorithm converges to the optimal action-value function $q^*(s, a)$, enabling the agent to select cost-optimal actions in a dynamic environment.

3.2 Priced Timed Games and UPPAAL STRATEGO

UPPAAL STRATEGO allows for the modeling of *Priced Timed Games* (PTG) and facilitates the synthesis of nearly optimal strategies using RL [8]. PTG are two-player games that are played on priced timed automata [4]. We recall the formal definition of PTG as follows.

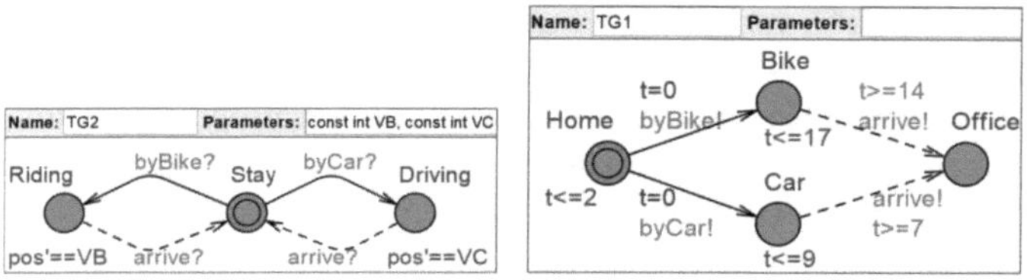

(a) (b)

Fig. 1. An example of NPTG templates in UPPAAL STRATEGO [15]

Definition 1. A Priced Timed Game $\mathcal{G} =< L, l_0, C, \Sigma, E, Inv >$ is a tuple, where L is a finite set of locations, l_0 is the initial location, C is a finite set of non-negative real-valued variables called *clocks*, $\Sigma = \Sigma_c \cup \Sigma_u$ is a finite set of actions (Σ_c and Σ_u denote a set of controllable and uncontrollable actions, respectively), $E \subseteq L \times \mathcal{B}(C) \times \Sigma \times 2^C \times L$ is a finite set of edges, where $\mathcal{B}(C)$ is the set of *guards* over C, that is, conjunctive formulas of constraints of the form $c_1 \bowtie n$ or $c_1 - c_2 \bowtie n$, where $c_1, c_2 \in C$, $n \in \mathbb{N}$, $\bowtie \in \{<, \leq, =, \geq, >\}$, 2^C is a set of clocks in C that are reset on the edge, and $Inv : L \rightarrow \mathcal{B}(C)$ is a partial function assigning invariants to locations.

Intuitively, *controllable* actions are performed by the system, e.g., *pumping* in this paper, and *uncontrollable actions* are performed by the environment.

UPPAAL STRATEGO [8] is a powerful tool for modeling, strategy synthesis, and (statistical) model checking of PTG, in which PTG can be composed in parallel as a *network* of PTG (NPTG) synchronized via *channels*. Figure 1 depicts PTG templates that can be instantiated into NPTG models by assigning concrete values to the parameters of the template, such as `VB` in TG1. The blue circles represent *locations* connected by directional *edges* Double-circled locations are *initial* locations. In UPPAAL STRATEGO, there are *urgent* (encircled "u") and *committed* (encircled "c") locations, where time does not elapse. Committed locations enforce that the next edge must start from them. Edges have assignments for resetting clocks (e.g., `t=0`), updating variables, guarding conditions (e.g., `t>=14`), and synchronization (e.g., `arrive!` and `arrive?`). For instance, in location `Bike` in Fig. 1b, the invariant `t<=17` ensures that clock `t` stays within limits. Derivatives of clocks can be defined by ODEs, such as `pos'==VB` in Fig. 1a.

An NPTG's solution is a *winning strategy* that guides the "system" player on controllable actions at each location to achieve the desired goal, like a driver using different transportation means to reach the office. We can synthesize a winning strategy using the following query:

```
strategy s = minE(cost){ExpList1}->{ExpList2}:<> goal
```
(2)

The strategy `s` aims to minimize the cost function `minE(cost)` within the machine-learning domain. `ExpList1` and `ExpList2` consist of expressions used in a Q-learning algorithm [20]. Previous studies [1,2,11,14] suggest that `ExpList1` should contain discrete variables, while `ExpList2` should have continuous variables (clocks). During execution, UPPAAL STRATEGO randomly simulates the model, sampling execution traces to pass to the learning algorithm. The query specifies that only traces reaching a state where `goal` is satisfied are selected. The algorithm only accesses controllable actions, demonstrating *partial observation* of the state space [9]. Additionally, the latest UPPAAL STRATEGO version supports external C–libraries during learning, simulation, and verification, a crucial feature used in this work (Table 1).

Table 1. Overview of pumping station parameters in the modeled system.

Pumping Station	No. of Pumps	Flow (m^3/min)	Head (m)	Power (kW)	Reservoir (m^3)
Level 1	6	11.5	450	825	1500
Level 2	4	15	350	975	1500
Level 3	4	14.5	280	900	1500
Level 4	6	10.0	250	650	1500

4 Case Study

This case study focuses on the mine dewatering system of a deep underground mine. In such mining environments, it is essential to continuously pump out groundwater to prevent flooding and ensure safe operations. The system includes multiple pumping stations installed at various depths, with each station responsible for lifting water to the next level until it reaches the surface. Figure 2 provides a conceptual overview of this multi-level layout. The system consists of four main pumping levels. Each level includes a reservoir that stores inflowing water before it is pumped upward.

The pumps operate in parallel, with flow capacities at the stations ranging from 10 to 15 m^3/min per pump, and power ratings between 650 and 975 kW. The total nominal flow through the system is approximately 25 m^3/min. Reservoir volumes are around 1,500 m^3 per level, providing limited but valuable buffering capacity that can be exploited for load shifting and energy optimization. In the existing industrial configuration, the pumps are controlled using a simple threshold logic. When the water level in a reservoir exceeds a maximum limit, an additional pump is activated. When the level drops below the minimum threshold, the pumps are turned off in reverse order. This approach ensures safe operation, but overlooks variations in electricity prices or inflow trends, often resulting in unnecessary energy consumption during high-price periods.

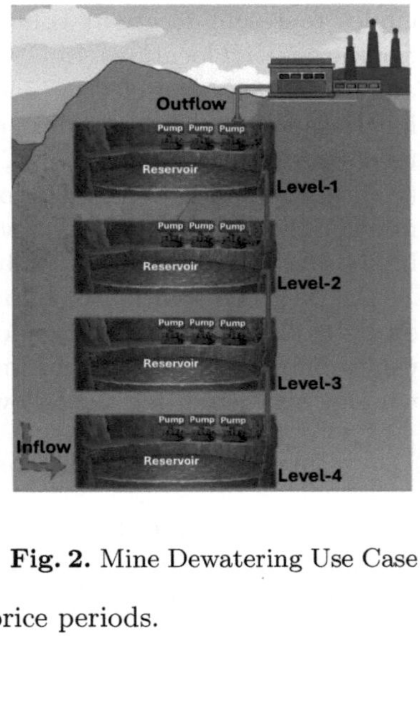

Fig. 2. Mine Dewatering Use Case

4.1 Pump Curve and Energy Models

To effectively simulate the hydraulic performance and electricity demand of the mine dewatering system, this section presents mathematical models for both individual and collective pump behavior in relation to flow rate, pump head, and power consumption [21].

Pump Performance and Operating Point. Centrifugal pump performance is characterized by a curve that relates the head (H) produced by the pump to the volumetric flow rate (Q). The head-flow relationship is approximated using a second-order polynomial:

$$H_p(Q) = aQ^2 + bQ + c, \tag{3}$$

where a, b, and c are empirical coefficients from the pump's data.

The hydraulic system in which the pump operates imposes its own relationship between head and flow, known as the *system curve*. This curve accounts for both the static lift (h_0) and the head losses due to friction, expressed as:

$$H_{sys}(Q) = h_0 + kQ^2, \tag{4}$$

where k is a loss coefficient based on pipe diameter, length, and flow regime.

The steady-state operating point of the system occurs where the pump and system curves intersect, i.e., when $H_p(Q) = H_{sys}(Q)$. Solving for Q yields a quadratic form:

$$(a - k)Q^2 + bQ + (c - h_0) = 0 \tag{5}$$

This equation defines the flow rate at which the pump operates under given system conditions. The analytical formulation enables efficient simulation of flow variations without requiring full hydraulic modeling.

Parallel Pump Configuration. In the mine dewatering network, the demand often exceeds the capacity of a single pump, requiring several identical units to operate in parallel. In this configuration, each pump experiences the same system head while sharing the total flow demand. If n pumps are active, each pump delivers $Q_i = Q_{total}/n$, and the combined head-flow relationship is represented as:

$$H_{parallel}(Q_{total}) = a\left(\frac{Q_{total}}{n}\right)^2 + b\left(\frac{Q_{total}}{n}\right) + c \tag{6}$$

This formulation assumes equal load distribution among all operating units, which is reasonable for pumps of the same type and control mode.

Energy Consumption Model. The total instantaneous power requirement of the system is calculated by adding the electrical power consumption of all active pumps across the four pumping levels. Given that all pumps within a level i have the same rated power P_{Li}, the total power at time t is expressed as:

$$P_{total}(t) = \sum_{i=1}^{4} n_i(t) \cdot P_{Li}, \tag{7}$$

where $n_i(t)$ denotes the number of pumps active at level i at time t. This aggregate formulation enables straightforward integration of energy cost evaluation, particularly under time-varying electricity price conditions.

5 Methodology

This section outlines our methodology for modeling, training, and optimization of the multi-level mine dewatering system using UPPAAL STRATEGO. The proposed workflow is depicted in Fig. 3. The mine dewatering process is modeled as a hierarchical network of pumping levels, each consisting of a reservoir and a set of parallel centrifugal pumps. The reservoirs are hydraulically connected such that the outflow of one level becomes the inflow of the next. Each reservoir has a capacity of 1500 m^3, with a safe operational range between 100 m^3 and 1400 m^3.

The model represents inflow variation, reservoir storage, and outflow dynamics at each level, capturing the fundamental physical behavior of the system. Control actions are represented as the number of pumps activated at each time step, while system observations include inflow rate, reservoir level, and the prevailing electricity price. The learning agent interacts with this environment by selecting the number of active pumps to minimize the cumulative energy cost while maintaining all reservoirs within the safe operating range.

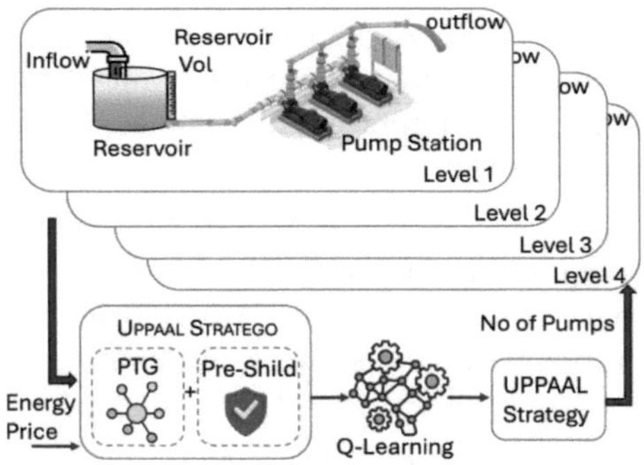

Fig. 3. Overview of the modeling and strategy synthesis workflow.

5.1 Strategy Synthesis

Unlike traditional offline synthesis methods that produce a single static control policy, this work adopts an *online synthesis* approach inspired by recent developments in hybrid stochastic control [10]. In online synthesis, the controller periodically re-computes and updates the control strategy based on the latest operational data and forecast information. The online synthesis framework is illustrated in Fig. 4. At fixed time intervals (15 min), the system collects measurements such as reservoir levels, inflow rates, and day-ahead electricity prices,

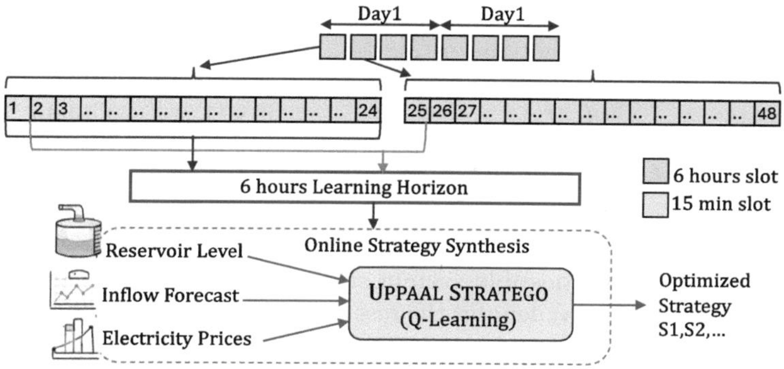

Fig. 4. Illustration of the online strategy synthesis framework

which are then passed to the UPPAAL STRATEGO engine. Using these observations, the tool performs a limited-horizon (6 h) learning phase to synthesize a new control strategy that minimizes the expected energy cost over the next operational horizon.

This strategy synthesis process follows a rolling horizon pattern, where each synthesized strategy remains valid for a defined duration before being replaced by an updated one. The approach ensures that the controller adapts continuously to changing inflow conditions and fluctuating energy prices. Through this *online synthesis* mechanism, the dewatering control system achieves dynamic adaptability and operational robustness.

5.2 Pre-shield Integration for Safe Learning

Safety during the learning process is critical to prevent the controller from exploring unsafe operating regions that could lead to overflow or dry-run conditions. In the proposed framework, a *pre-shield* mechanism is integrated into the model to ensure that the RL agent always selects control actions within safe limits. The pre-shield manages the number of active pumps based on the current reservoir level. It uses a safety evaluation function to determine a safe range for pump activation, returning minimum and maximum values $(n_{\min}, n_{\max})$. This evaluation considers the reservoir volume and inflow rate to maintain safe water levels. When in the safe range, the agent can explore freely, but if the system enters a critical zone, the pre-shield activates to prevent faults.

The reservoir's operating range consists of three regions. The *safe region* includes volumes between $100\,\mathrm{m}^3$ and $1400\,\mathrm{m}^3$, where normal operations occur and the pre-shield remains off. The *critical region* covers $(0–100]\,\mathrm{m}^3$ and $[1400–1500)\,\mathrm{m}^3$, activating the pre-shield to limit pumps and prevent unsafe transitions. If the level drops below $0\,\mathrm{m}^3$ or exceeds $1500\,\mathrm{m}^3$, the system enters a *faulty state*, indicating a violation of safety constraints. The shield prevents actions that could lead to this faulty state.

Formally, the pre-shield $\mathcal{S}$ is defined as a function that maps each system state s to a set of allowed actions $A'(s)$:

$$\mathcal{S} : S \to 2^A, \quad A'(s) = \begin{cases} A(s), & \text{if } V(s) \in [100, 1400], \\ \{a \in A(s) \mid \text{safe}(s, a)\}, & \text{if } V(s) \in (0, 100] \cup [1400, 1500), \\ \emptyset, & \text{otherwise.} \end{cases}$$

Here, S is the set of all possible system states, $A(s)$ is the set of all available actions at state s, and $V(s)$ denotes the reservoir volume in state s. The predicate safe(s, a) evaluates whether taking action a in state s will keep the next state within the physical limits of 0–1500 m^3. In the safe region, the shield is inactive and all actions are available. When the system enters a critical region, the shield filters out actions that could lead beyond the limits.

In the following, we show that the pre-shield is *maximally permissive* with respect to a *monotonicity safety objective*, that is, in the high critical band, enabled actions ensure the reservoir is non-increasing, and in the low critical band, enabled actions ensure the reservoir is non-decreasing.

Correctness and Permissiveness of the Pre-shield. We analyze the pre-shield encoded in the UPPAAL Stratego model and characterize precisely which actions it enables in the critical region.

Theorem 1 (Code-level characterization of enabled actions). *Let $x \in \{0, \ldots, N\}$ be a pump-count action, let q be the per-pump flow rate, and let I denote the inflow used by the model at the decision point. Then the pre-shield enables action x in state s if and only if:*

$$safe_{model}(s, x) \iff \begin{cases} qx \geq I, & \text{if } V(s) \geq 1400, \\ qx \leq I, & \text{if } V(s) \leq 100. \end{cases}$$

Equivalently, the enabled actions are exactly those satisfying $V'(s, x) \leq 0$ in the high critical band and $V'(s, x) \geq 0$ in the low critical band, where $V' = I - qx$ models the reservoir dynamics.

Proof. By construction, the pre-shield computes bounds `min`, `max` such that x is enabled iff $\texttt{min} \leq x \leq \texttt{max}$. In the high band, `min` is the smallest index satisfying $qx \geq I$, and in the low band, `max` is the largest index satisfying $qx \leq I$. Since the reservoir dynamics in the subsequent active phase satisfy $V' = I - qx$, the result follows immediately.

Relation to One-Minute Physical Safety. In the critical region, the model enforces a sampling period of one minute. Assume inflow is bounded by $I \in [20, 30]$ and the reservoir evolves as $V(t) = V_0 + (I - qx)t$ over the interval $t \in [0, 1]$.

Lemma 1 (Robust one-minute physical safety). *An action x keeps the reservoir within physical bounds $[0, 1500]$ for the entire next minute, for all $I \in [20, 30]$, if and only if*

$$\left\lceil \frac{V_0 + 30 - 1500}{q} \right\rceil \leq x \leq \left\lfloor \frac{V_0 + 20}{q} \right\rfloor.$$

Proof. Since $V(t)$ is linear, it suffices to check the endpoint $t = 1$. Worst-case overflow occurs at $I = 30$ and worst-case dry-run at $I = 20$, yielding the stated bounds.

Theorem 2 (Maximal permissiveness for monotonicity safety). *In the critical region $V(s) \in (0, 100] \cup [1400, 1500)$, the pre-shield is maximally permissive with respect to the monotonicity safety objective, as follows:*

- *when $V(s) \geq 1400$, it enables exactly all actions that do not increase the reservoir;*
- *when $V(s) \leq 100$, it enables exactly all actions that do not decrease the reservoir.*

Proof. By Lemma 1, the pre-shield enables precisely those actions for which the reservoir derivative points inward (or is zero). Any action violating this condition would strictly move the reservoir further towards overflow or dry-run, and is therefore unsafe under the monotonicity objective.

Remark 1 (Conservativeness under one-minute semantics). The current pre-shield enforces $qx \geq I$ (resp. $qx \leq I$) in the high (resp. low) critical band, which guarantees monotonic inward drift. However, Lemma 1 shows that one-minute physical safety allows a strictly larger set of actions, including some with outward-pointing derivatives that do not reach the physical bounds within one minute. Consequently, the pre-shield is:

- *non-conservative* with respect to the monotonicity safety objective;
- *conservative* with respect to one-minute physical safety(staying within $[0, 1500]$ for the next minute).

To achieve maximal permissiveness under one-minute physical semantics, the bounds in Lemma 1 must be used instead of derivative-sign constraints.

5.3 General Physical Safety and Forward Invariance

In this subsection, we *generalize* the results shown above for N identical pumps, with respect to the physical safety property, and we also prove the *forward invariance* of the physical safety, that is, the reservoir never overflows and never dries out.

Setup. Let $x \in \{0, 1, \ldots, N\}$ be the number of active (identical) pumps, each with flow rate $q > 0$. The inflow is uncertain but bounded, $I(t) \in [I_{\min}, I_{\max}], \forall t \geq 0$. Over each sampling interval of length 1 min, the control input x is held constant. The reservoir dynamics are $V'(t) = I(t) - qx$. We use $V_{\min} = 0$ and $V_{\max} = 1500$.

Definition 2 (Robust one-step safe action set). *For a given volume $V \in \mathbb{R}$, we define the robust one-step safe action set as follows:*

$$\Gamma(V) \triangleq \left\{ x \in \{0, \ldots, N\} \;\middle|\; \left\lceil \frac{V + I_{\max} - V_{\max}}{q} \right\rceil \leq x \leq \left\lfloor \frac{V + I_{\min} - V_{\min}}{q} \right\rfloor \right\}.$$

Lemma 2 (Robust one-step physical safety for N pumps). *Let the current reservoir volume be V_0 at the beginning of a sampling interval, and let x be held constant on that interval. Then the following are equivalent:*

1. $\forall t \in [0, 1] : V(t) \in [V_{\min}, V_{\max}]$ *for all inflow signals satisfying* $I(t) \in [I_{\min}, I_{\max}]$ *for all* $t \in [0, 1]$;
2. $x \in \Gamma(V_0)$.

Proof. Fix V_0 and a constant action x on $[0, 1]$. Since $I(t) \in [I_{\min}, I_{\max}]$, we obtain the differential inequalities

$$I_{\min} - qx \;\leq\; V'(t) \;\leq\; I_{\max} - qx.$$

Integrating from 0 to $t \in [0, 1]$ yields the enclosure

$$V_0 + (I_{\min} - qx)t \;\leq\; V(t) \;\leq\; V_0 + (I_{\max} - qx)t.$$

Therefore, a sufficient and necessary condition for $V(t) \in [V_{\min}, V_{\max}]$ for all $t \in [0, 1]$ and all admissible inflows is

$$V_0 + (I_{\min} - qx)t \geq V_{\min} \quad \text{and} \quad V_0 + (I_{\max} - qx)t \leq V_{\max} \qquad \forall t \in [0, 1].$$

Because the left-hand sides are affine in t, it suffices to check $t = 1$:

$$V_0 + I_{\min} - qx \geq V_{\min} \quad \text{and} \quad V_0 + I_{\max} - qx \leq V_{\max}.$$

Rearranging gives

$$x \leq \frac{V_0 + I_{\min} - V_{\min}}{q} \quad \text{and} \quad x \geq \frac{V_0 + I_{\max} - V_{\max}}{q}.$$

Since x is integer-valued and must lie in $\{0, \ldots, N\}$, this is equivalent to $x \in \Gamma(V_0)$ by Definition 2.

Definition 3 (Robust physical pre-shield for N pumps). *At the beginning of each sampling interval with current volume $V_k = V(k)$, the pre-shield restricts the controller to actions in $\Gamma(V_k)$. Equivalently, an action x is enabled iff $x \in \Gamma(V_k)$.*

Theorem 3 (Forward invariance of the physical bounds (all time)). *Consider any closed-loop execution where at each integer time $k \in \mathbb{N}$ the chosen action x_k is held constant on $[k, k+1)$ and satisfies*

$$x_k \in \Gamma(V_k), \qquad \text{where } V_k \triangleq V(k).$$

If $V_0 \in [V_{\min}, V_{\max}]$, then for all inflow signals $I(t) \in [I_{\min}, I_{\max}]$ and all times $t \geq 0$,

$$V(t) \in [V_{\min}, V_{\max}],$$

i.e., $[V_{\min}, V_{\max}]$ is forward invariant.

Proof. We prove by induction over sampling steps that $V_k \in [V_{\min}, V_{\max}]$ for all $k \in \mathbb{N}$, and simultaneously that $V(t) \in [V_{\min}, V_{\max}]$ for all $t \in [k, k+1]$.

Base Case. By assumption $V_0 \in [V_{\min}, V_{\max}]$.

Inductive step. Assume $V_k \in [V_{\min}, V_{\max}]$ for some k. By the pre-shield restriction, $x_k \in \Gamma(V_k)$. By Lemma 2 (applied on the interval $[k, k+1]$ with initial value V_k), we obtain

$$\forall t \in [k, k+1]: \quad V(t) \in [V_{\min}, V_{\max}]$$

for all admissible inflow signals $I(t) \in [I_{\min}, I_{\max}]$. In particular, at the endpoint $t = k+1$ this implies

$$V_{k+1} = V(k+1) \in [V_{\min}, V_{\max}].$$

This completes the induction.

Finally, for any $t \geq 0$ there exists $k = \lfloor t \rfloor$ with $t \in [k, k+1]$, and the property above gives $V(t) \in [V_{\min}, V_{\max}]$.

Remark 2 (Feasibility condition). To ensure the shield never deadlocks, it suffices that $\Gamma(V) \neq \emptyset$ for all $V \in [V_{\min}, V_{\max}]$. A simple sufficient condition is

$$qN \geq I_{\max},$$

which guarantees that near $V_{\max}$ the lower bound on x can be met. (At the lower bound $V_{\min} = 0$, feasibility is typically ensured by $x = 0$ whenever $I_{\min} \geq 0$.)

Correctness of the Pre-shield. The correctness of the pre-shield can also be evaluated statistically in UPPAAL STRATEGO rather than through exhaustive verification. Instead of verifying a strict invariant, we estimate the probability of reaching a faulty state using a reachability query of the form `Pr[<=T](<> Fault)`, where `Fault` represents any condition in which the reservoir volume violates physical limits (`Reservoir.Vol < 0` or `Reservoir.Vol > 1500`). If the measured probability is close to zero across repeated simulations, the pre-shield is considered effective and correct. This probabilistic validation gives formal assurance that the learning agent cannot trigger unsafe transitions, even while exploring. The pre-shield allows the RL to concentrate on optimizing energy efficiency, with safety built into the design. By activating only in critical operating regions, the shield enhances learning efficiency, shortens convergence time, and ensures reliable real-time performance under varying inflow and price conditions.

5.4 Control Strategies

We evaluated three control strategies to compare their performance in managing energy consumption and reservoir stability within the dewatering system.

Bang-Bang Control: This baseline approach activates or deactivates all pumps based on reservoir water levels. When the level exceeds the upper threshold, all pumps are turned on until the lower limit is reached. This simple rule-based control ensures operational safety but does not account for energy prices.

Price-Aware Control: The price-aware controller incorporates electricity cost information into the control logic. It first calculates the mean of the day-ahead electricity prices and adjusts the reservoir operation range accordingly. When the current price is below the mean, the controller maintains the water level within a lower range (approximately 100–200 m^3) to take advantage of cheaper energy. Conversely, during high-price periods, the target range shifts upward (around 1300–1400 m^3) to delay pumping until prices decrease. This approach provides moderate energy savings compared to the Bang-Bang controller but often results in frequent pump switching.

Learning-Based Control: The learning-based controller is synthesized using RL. It optimizes pump operation by learning from simulations that include reservoir dynamics, inflow variations, and day-ahead price data. Through iterative exploration, the strategy learns to minimize the total energy cost while maintaining safe reservoir levels and reducing unnecessary pump switching.

6 Modeling in UPPAAL

The complete mine dewatering system is modeled in UPPAAL STRATEGO as NPTG templates that collectively represent the hydraulic, operational, and economic interactions within the mine (https://github.com/muhammednaeem/Efficient-Multi-Level-Mine-Dewatering-Using-UPPAAL-Stratego.git). Figure 5 illustrates the overall model architecture, which integrates five key components: *Data Reader*, *Electricity Prices*, *Water Inflow*, *Objective Function*, and the *Pumping Station with Reservoir*. These components operate concurrently and exchange information through shared variables and synchronization channels, forming a closed-loop system that links inflow dynamics, electricity price variations, and pump control actions.

The *Data Reader* component manages the import of external input data, including day-ahead electricity prices and inflow rates. The model advances in discrete time intervals, controlled by the clock variable *chk_ T*, which synchronizes all processes. Two indexed price variables are maintained: *EP* for the current electricity price and *EP_ 1* for the previous value. This allows the system to capture smooth transitions between time periods and reference recent price history when determining control actions.

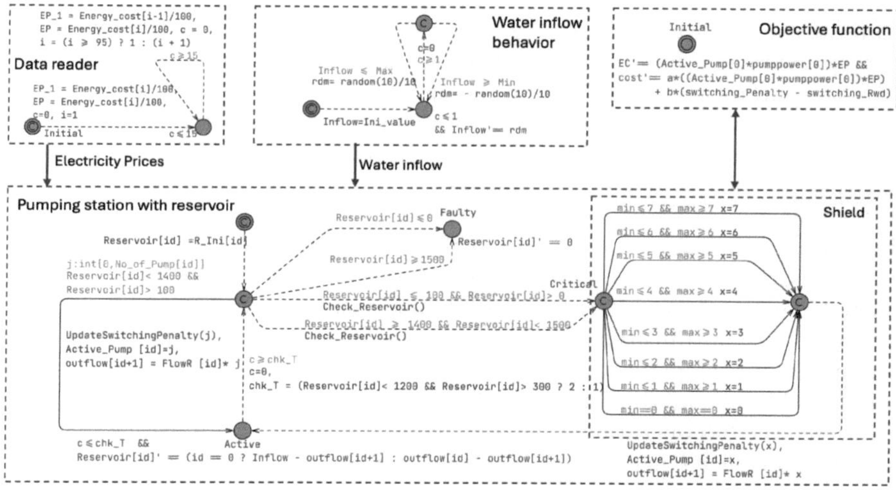

Fig. 5. Overview of the modeled components in UPPAAL STRATEGO.

The *Water Inflow* component represents groundwater inflow to the reservoir. The inflow rate (*Ini_value*) is varied randomly within a defined range to simulate uncertain environmental conditions while keeping behavior physically realistic. Guard conditions ensure that inflow updates occur only at specific intervals, preventing abrupt or non-physical changes.

Energy usage and switching behavior are jointly evaluated through the *Objective Function* component. At each simulation step, the instantaneous energy cost is calculated as:

$$EC' = (\text{Active\_Pump}[0] \times \text{PumpPower}[0]) \times EP,$$

where *Active_Pump* represents the number of running pumps, *PumpPower* is the power rating of each pump, and *EP* is the current electricity price. The total cost function further accounts for switching actions using:

$$\text{cost}' = \alpha\Big[(\text{Active\_Pump}[0] \times \text{PumpPower}[0]) \times EP + \beta \times (\text{switching\_Penalty} - \text{switching\_Rwd})\Big], \quad (8)$$

where α and β are weighting coefficients that balance the objectives of minimizing energy use and limiting switching frequency. The penalty increases the cost when pumps frequently toggle, while the reward (*switching_Rwd*) provides positive reinforcement for maintaining stable operation. This formulation encourages strategies that achieve both energy efficiency and operational smoothness.

The *Pumping Station with Reservoir* component models the physical dynamics of the dewatering system, capturing water storage behavior, pumping operations, and safety constraints. The learning agent determines the number of active pumps at each decision step and re-evaluates its action at every time unit. The reservoir volume evolves continuously according to inflow and outflow dynamics, where the outflow is proportional to the number of active pumps. Each reservoir has a total capacity of $1500\,\mathrm{m}^3$, with a defined safe operating range between $100\,\mathrm{m}^3$ and $1400\,\mathrm{m}^3$. Exceeding these bounds triggers a *Faulty* state, representing either overflow or dry-run conditions.

The function *Check_Reservoir()* enforces safe behavior by dynamically adjusting the allowable range of pump activations (*min* and *max*) based on the current reservoir level and inflow. When the system approaches critical regions (below $100\,\mathrm{m}^3$ or above $1400\,\mathrm{m}^3$), this function restricts the controller's decisions to prevent unsafe transitions. For instance, near overflow, it limits the number of active pumps to prevent excessive inflow accumulation, whereas near depletion, it prohibits additional pumping to avoid dry-run operation. Under normal conditions, the full operational range remains available. This dynamic bounding forms the *pre-shield* mechanism, ensuring safe learning and operation without introducing artificial penalties that could distort the optimization objective.

Complementing this, the *UpdateSwitchingPenalty()* function regulates switching by assigning penalties and rewards based on pump activity. It penalizes frequent pump switching to promote stable operation but suppresses these penalties during sudden electricity price fluctuations or when the reservoir is in critical ranges, allowing the system to adapt appropriately under varying conditions.

7 Simulation Results and Discussion

The developed control strategies were evaluated using the multi-level mine dewatering model introduced in Sect. 6. Simulations were performed over several operational periods to assess the controllers' performance in terms of energy consumption, responsiveness to electricity price variation, and the frequency of pump switching. The results presented here compare three controllers: the *Bang–Bang Controller*, the *Price-Aware Controller*, and the UPPAAL *Controller* synthesized through RL in UPPAAL STRATEGO.

Figure 6 illustrates the performance of the controllers for a short simulation interval, focusing on Level 1 of the pumping station. Subfig. (6a) shows the instantaneous energy consumption in relation to the 15-min electricity price profile, while Subfig. (6b) depicts the corresponding pump activation behavior. The UPPAAL *Controller* clearly outperforms the other approaches by adapting its pumping schedule to price fluctuations, reducing energy use during high-cost periods, and avoiding unnecessary operation when prices are low. The *Price-Aware Controller* also achieves some energy savings by considering electricity price information, but it exhibits frequent switching actions that can lead to mechanical wear and operational inefficiency. In contrast, the *Bang–Bang Controller*, which reacts solely to reservoir thresholds, disregards energy cost trends and results in higher energy usage overall.

Figure 7 presents the cumulative energy consumption over an extended operational period covering a full day. As shown, the UPPAAL *Controller* consistently

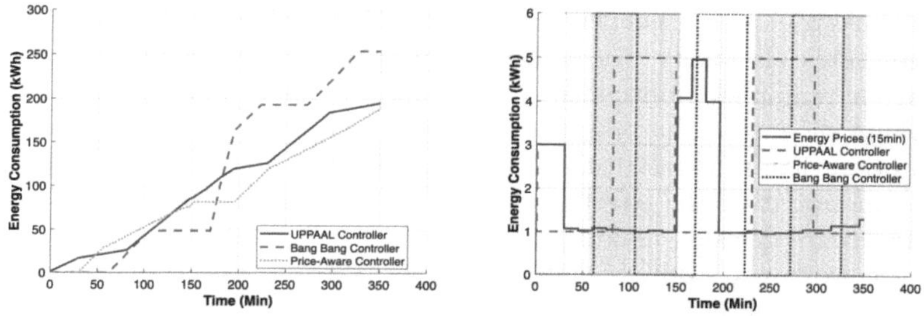

(a) Energy consumption. (b) Pumping operation.

Fig. 6. Short-term performance comparison of controllers at Level 1.

maintains lower total energy usage compared to both the *Bang–Bang* and *Price-Aware* controllers. Although daily performance may vary depending on the inflow dynamics and the hourly electricity price profile, the synthesized learning-based controller achieves an average reduction of up to 40% in total energy consumption relative to the baseline threshold-based control. The results confirm that RL combined with model-based safety shielding enables not only cost-effective but also operationally stable dewatering control.

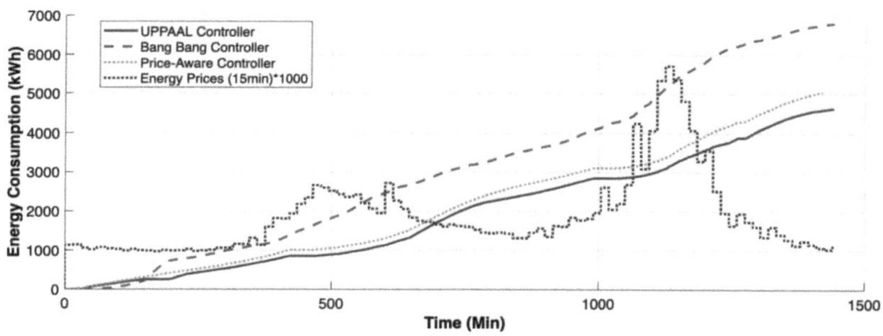

Fig. 7. Cumulative energy consumption over one full day for all controllers.

Overall, the simulation results demonstrate that the integration of electricity price data, formal modeling, and safe RL allows the controller to achieve a balance between energy efficiency and operational stability. By minimizing both the total energy cost and the number of pump switching events, the UPPAAL *Controller* provides a more practical and sustainable control solution for real-world mine dewatering systems.

8 Conclusion and Future Work

This paper presents a formal modeling and reinforcement-learning-based approach for energy-efficient underground mine dewatering, comparing Bang–Bang and Price-Aware controllers with a UPPAAL STRATEGO-synthesized controller that integrates electricity prices, inflow dynamics, and a pre-shield to ensure safety while minimizing both energy consumption and pump switching. Simulation results show that the proposed approach achieves up to 40% lower energy consumption than rule-based strategies while maintaining safe and smooth operation, demonstrating the effectiveness of combining formal modeling, learning, and price-aware decision-making.

Future work will address multi-level coordination, uncertainty-aware optimization, and integration with real-time SCADA systems.

Acknowledgments. We gratefully acknowledge Vinnova's Advanced Digitalization Program via the D-RODS project (nr.: 2023-00244), as well as Mälardalen University via the TSS-INSID project, which supported this research.

References

1. Ali, N., Naeem, M., Ardila, J.P.C., Punnekkat, S.: Safety-aware strategy synthesis for autonomous system of systems with UPPAAL. In: Törngren, M., Gallina, B., Schoitsch, E., Troubitsyna, E., Bitsch, F. (eds.) International Conference on Computer Safety, Reliability, and Security, pp. 73–87. Springer, Cham (2025). https://doi.org/10.1007/978-3-032-02018-5_6
2. Ali, N., Naeem, M., Castellanos-Ardila, J.P., Punnekkat, S.: Formal modeling and strategy synthesis for resource optimization in system of systems. In: 2025 20th Annual System of Systems Engineering Conference (SoSE), pp. 1–6. IEEE (2025)
3. Alshiekh, M., Bloem, R., Könighofer, R., Könighofer, B., Wang, C.: Safe reinforcement learning via shielding. In: Proceedings of the AAAI Conference on Artificial Intelligence (2018). https://arxiv.org/abs/1708.08611
4. Behrmann, G., et al.: Minimum-cost reachability for priced time automata. In: Di Benedetto, M.D., Sangiovanni-Vincentelli, A. (eds.) HSCC 2001. LNCS, vol. 2034, pp. 147–161. Springer, Heidelberg (2001). https://doi.org/10.1007/3-540-45351-2_15
5. Bloem, R., Jensen, P.G., Könighofer, B., Larsen, K.G., Lorber, F., Palmisano, A.: It's time to play safe: shield synthesis for timed systems. In: Formal Methods in System Design (2021). https://arxiv.org/abs/2006.16688
6. Bloem, R., Könighofer, B., Könighofer, R., Wang, C.: Shield synthesis: runtime enforcement for reactive systems. In: Baier, C., Tinelli, C. (eds.) TACAS 2015. LNCS, vol. 9035, pp. 533–548. Springer, Heidelberg (2015). https://doi.org/10.1007/978-3-662-46681-0_51
7. Brorholt, A.H., et al.: UPPAAL COSHY: automatic synthesis of compact shields for hybrid systems. In: Ganty, P., Mansutti, A. (eds.) Reachability Problems, RP 2025. LNCS, vol. 16230. Springer, Cham (2026). https://doi.org/10.1007/978-3-032-09524-4_7
8. David, A., Jensen, P.G., Larsen, K.G., Mikučionis, M., Taankvist, J.H.: Uppaal Stratego. In: Baier, C., Tinelli, C. (eds.) TACAS 2015. LNCS, vol. 9035, pp. 206–211. Springer, Heidelberg (2015). https://doi.org/10.1007/978-3-662-46681-0_16
9. Gu, R., Jensen, P.G., Seceleanu, C., Enoiu, E., Lundqvist, K.: Correctness-guaranteed strategy synthesis and compression for multi-agent autonomous systems. Sci. Comput. Program. **224**, 102894 (2022)

10. Hasrat, I.R., Jensen, P.G., Larsen, K.G., Srba, J.: End-to-end heat-pump control using continuous time stochastic modelling and UPPAAL STRATEGO. In: Aït-Ameur, Y., Crăciun, F. (eds.) Theoretical Aspects of Software Engineering, TASE 2022. LNCS, vol. 13299. Springer, Cham (2022). https://doi.org/10.1007/978-3-031-10363-6_24
11. Jensen, P.G., Larsen, K.G., Mikučionis, M.: Playing wordle with Uppaal Stratego. In: Jansen, N., Stoelinga, M., van den Bos, P. (eds.) A Journey from Process Algebra via Timed Automata to Model Learning. LNCS, vol. 13560. Springer, Cham (2022). https://doi.org/10.1007/978-3-031-15629-8_15
12. Könighofer, B., et al.: Shield synthesis. Form. Meth. Syst. Des. **51**(2), 332–361 (2017). https://doi.org/10.1007/s10703-017-0276-9
13. Könighofer, B., Rudolf, J., Palmisano, A.: Online shielding for reinforcement learning. Innov. Syst. Softw. Eng. **19**, 379–394 (2023). https://doi.org/10.1007/s11334-022-00480-4
14. Naeem, M.: Energy-efficient cyber-physical systems: a model-based approach (2024)
15. Naeem, M., Gu, R., Seceleanu, C., Guldstrand Larsen, K., Nielsen, B., Albano, M.: Energy-efficient motion planning for autonomous vehicles using Uppaal Stratego. In: Chin, W.N., Xu, Z. (eds.) Theoretical Aspects of Software Engineering, TASE 2024. LNCS, vol. 14777. Springer, Cham (2024). https://doi.org/10.1007/978-3-031-64626-3_21
16. Olifant, T., Ramotsoela, T.D., et al.: Parameter influence analysis for deep-level mine dewatering load shifting. Mining **4**(2), 265–283 (2024). https://doi.org/10.3390/mining4020012
17. Romero, D., Guzmán, J.L., Berenguel, M.: Control of mine dewatering systems using centralized and distributed MPC. Appl. Therm. Eng. **80**, 387–399 (2015). https://doi.org/10.1016/j.applthermaleng.2015.01.074
18. Sutton, R.S., Barto, A.G.: Reinforcement Learning: An Introduction. MIT Press (2018)
19. Vosloo, A., Grobler, A.J., Pretorius, J.C.C.: Energy management of water reticulation systems in deep-level mines. Appl. Energy **92**, 291–301 (2012). https://doi.org/10.1016/j.apenergy.2011.11.004
20. Watkins, C.J.C.H.: Learning from delayed rewards. King's College, Cambridge United Kingdom (1989)
21. Šakić, F.: Digital twins for the optimization of resource allocation and scheduling of cyber-physical systems. Master's thesis, Mälardalen University, Sweden (2025)

Correction to: Asynchronous Team Automata

Davide Basile, Maurice H. ter Beek, and José Proença

Correction to:
Chapter 2 in: A. Sampaio and M. Stoelinga (Eds.): *Formal Methods*, LNCS 16557, https://doi.org/10.1007/978-3-032-26220-2_2

In the original version of the book contained a reference error regarding chapter 2. Some references were displayed correctly. This has been corrected.

The updated version of this chapter can be found at
https://doi.org/10.1007/978-3-032-26220-2_2

A. Sampaio and M. Stoelinga (Eds.): FM 2026, LNCS 16557, p. C1, 2026.
https://doi.org/10.1007/978-3-032-26220-2_36

Author Index

A
Aichernig, Bernhard K. 326
Albert, Elvira 255
Almeida, Dalay 605
An, Jie 71
Andersen, Mikkel Schmidt 369
Aoki, Mari A. 420
Archer, Samantha 3
Audrito, Giorgio 550

B
Backeman, Peter 317
Barrett, Clark 3
Basile, Davide 27
Baumeister, Jan 154
Berg, Benjamin von 326
Berthier, Nicolas 691

C
Calinescu, Radu 497
Cao, Qinxiang 651
Cao, Yongzhi 651
Cavalcanti, Ana 497
Chalupa, Marek 214
Chechik, Marsha 497
Chen, Jiawei 193
Chen, Rui 670
Chen, Taolue 111
Chen, Zhenbang 275, 296
Chien, Po-Chun 3
Cohen, Shachaf 174
Correnson, Arthur 154
Coutinho, Joana 579

D
Damiani, Ferruccio 550
de Mendonça, José Luiz Vargas 193
de Oliveira, Steven 691
Dietl, Werner 90
Djoudi, Adel 610
Dong, Wei 296
Dongol, Brijesh 394
Du, Yide 296

E
Erhard, Julian 446

F
Farrell, Marie 420
Fedyukovich, Grigory 471
Feng, Shenghua 71, 651
Feng, Weizhi 111
Finkbeiner, Bernd 154
Fitzgerald, John S. 369

G
Genaim, Samir 255
Gil, Santiago 369
Gomes, Cláudio 369
Govind, R. 234
Gu, Bin 670

H
Heck, Linus 524
Henzinger, Thomas A. 214
Hong, Weijiang 275
Hu, Zhenjiang 651

I
Irfan, Ahmed 3
Isaksson, Alf J. 711

J
Jeannin, Jean-Baptiste 193
Junges, Sebastian 524

K
Kang, Ning 133
Katis, Andreas 420
Katoen, Joost-Pieter 524
Kohn, Florian 154
Kolárik, Tomáš 471
Kosmatov, Nikolai 610, 691
Krishna, S. 234

A. Sampaio and M. Stoelinga (Eds.): FM 2026, LNCS 16557, pp. 731–733, 2026.
https://doi.org/10.1007/978-3-032-26220-2

Kristensen, Morten Haahr 369

L
Labbaf, Faezeh 471
Larsen, Peter Gorm 369
Lecomte, Thierry 605
Lemos, Alexandre 579
Leopardi, Fabrizio 471
Li, Hao 133
Li, Yi 631
Liang, Xiao 670
Lim, Zipong 631
Lin, Shang-wei 631
Liu, Jiaxiang 111
Liu, Kunlin 296
Liu, Yang 275
Longuet, Delphine 691
Lonsing, Florian 3

M
Ma, Ke 296
Ma, Zhi 670
Mamouras, Konstantinos 193
Maoz, Shahar 174
Marsso, Lina 497
Martin-Martin, Enrique 255
Mattsson, Karl 317
Maury, Norman 605
Mavridou, Anastasia 420

N
Naeem, Muhammad 711

P
Perez-Lopez, Áron Ricardo 3
Pfeifer, Wolfram 90
Proença, José 27

Q
Qin, Shengchao 670
Quatmann, Tim 524

R
Resende, Pedro 579
Ribeiro, Pedro 497
Rindler, Maximilian 326
Rodríguez Herrero, Andoni 50

S
Sánchez, César 50
Saraç, N. Ege 214
Scarso, Giordano 550
Schwarz, Michael 446
Seceleanu, Cristina 711
Seceleanu, Tiberiu J. 711
Seidl, Helmut 446
Sharygina, Natasha 471
Shen, Shidong 111
Shuai, Ziqi 275
Sil, Sanchari 234
Song, Fu 111
Srivathsan, B. 234
Standen, Isobel 497
Stolz, Volker 550
Sun, Jun 275

T
Teo, Yon Shin 631
ter Beek, Maurice H. 27
Tian, Cong 670
Torta, Gianluca 550
Townsend, Beverley 497

U
Ulbrich, Mattias 90

V
Vojdani, Vesal 446
Volk, Matthias 524

W
Wallner, Felix 326
Wand, Michael 471
Wang, Hao 651
Wang, Ji 275, 296
Wang, Xudong 296
Wen, Cheng 670

Wu, Bozhi 631
Wu, Zhilin 111

X
Xu, Fanjiang 71

Y
Yaman, Sinem Getir 497
Yan, Songyang 585
Yang, Mengfei 670
Yang, Zijiang 585
Yu, Emily 214

Z
Zhan, Naijun 71, 651
Zhang, Guofeng 275, 296
Zhang, Jianyuan 133
Zhang, Peng 133
Zhang, Xiaodong 585
Zhang, Yufeng 275
Zhao, Haiyan 651
Zhao, Junfeng 651
Zhao, Tianqi 651
Zhao, Xiaofei 347
Zhou, Minghui 651

MIX
Papier aus verantwortungsvollen Quellen
Paper from responsible sources
FSC® C105338

If you have any concerns about our products,
you can contact us on
ProductSafety@springernature.com

In case Publisher is established outside the EU,
the EU authorized representative is:
Springer Nature Customer Service Center GmbH
Europaplatz 3, 69115 Heidelberg, Germany

Printed by Libri Plureos GmbH
in Hamburg, Germany